Lecture Notes in Computer Science 2206
Edited by G. Goos, J. Hartmanis, and J. van Leeuwen

Springer
Berlin
Heidelberg
New York
Barcelona
Hong Kong
London
Milan
Paris
Tokyo

Bernd Reusch (Ed.)

Computational Intelligence

Theory and Applications

International Conference, 7th Fuzzy Days
Dortmund, Germany, October 1-3, 2001
Proceedings

Springer

Series Editors

Gerhard Goos, Karlsruhe University, Germany
Juris Hartmanis, Cornell University, NY, USA
Jan van Leeuwen, Utrecht University, The Netherlands

Volume Editor

Bernd Reusch
University of Dortmund, Computer Science I
44221 Dortmund, Germany
E-mail: reusch@ls1.informatics.uni-dortmund.de

Cataloging-in-Publication Data applied for

Die Deutsche Bibliothek - CIP-Einheitsaufnahme

Computational intelligence : theory and applications ; international conference ; proceedings / 7th Fuzzy Days, Dortmund, Germany, October 1 - 3, 2001. Bernd Reusch (ed.). - Berlin ; Heidelberg ; New York ; Barcelona ; Hong Kong ; London ; Milan ; Paris ; Tokyo : Springer, 2001
(Lecture notes in computer science ; Vol. 2206)
ISBN 3-540-42732-5

CR Subject Classification (1998): I.2, F.1, F.4.3

ISSN 0302-9743
ISBN 3-540-42732-5 Springer-Verlag Berlin Heidelberg New York

Springer-Verlag Berlin Heidelberg New York
a member of BertelsmannSpringer Science+Business Media GmbH

http://www.springer.de

Typesetting: Camera-ready by author, data conversion by Olgun Computergrafik, Heidelberg
Printed on acid-free paper SPIN 10840753 06/3142 5 4 3 2 1 0

Preface

Ten years of „Fuzzy Days“ in Dortmund!

What started as a relatively small workshop in 1991 has now become one of the best known smaller conferences on Computational Intelligence in the world. It fact, it was (to my best knowledge) the first conference to use this term, in 1994, although I confess that another, larger conference was announced first and the trade mark "Computational Intelligence was not coined in Dortmund.

I believe, that the success of this conference is grounded on the quality of its reviewed and invited papers as well as its good organization. From the beginning, we have sent every paper anonymously to five referees, and we have always accepted only around 50% of the papers sent in. This year it was a little less than that.

I would like to thank everybody who helped us by considering Dortmund's Fuzzy Days as the conference at which to appear. I know that among the abstracts not accepted there were some quite good ones, but we were restricted to a fixed number. I also know that referees do a good job but cannot always judge wisely from abstracts. Hence my apologies to those who did not make it this year. Please try again!

I would like to point out that our conference also has a good regional reputation. I am grateful to the City of Dortmund, its Lord Mayor Dr. Langemeyer, the Dortmund project, the DFG – Deutsche Forschungsgemeinschaft, the KVR – Kommunalverband Ruhrgebiet, the Martin-Schmeißer-Stiftung, and the Comline AG/Quantum GmbH for their valuable support.

August 2001 Bernd Reusch

Organization

General Chairman	B. Reusch, Germany
Honorary Chairman	L. A. Zadeh, USA
Program Chairmen	B. Reusch
	R. Rojas
	H.-P. Schwefel

Program Committee

Aizenberg, I., Israel
Albrecht, R., Germany
Berenji, H.; USA
Boquete, L.; Spain
Borisov, A.N.; Latvia
Bosc, P., France
Bouchon-Meunier, B.; France
Brauer, W.; Germany
Butkiewicz, B.; Poland
Cotta Porras, C.; Spain
Damiani, E.; Italy
De Baets, B.; Belgium
Deng, H.; Australia
Di Nola, A.; Italy
Drechsler, R.; Germany
Dubois, D.; France
Esteva, F.; Spain
Fathi, M.; USA
Fernández; F.; Spain
Frank, P.M.; Germany
Friedrich, C. M.; Germany
Glesner, M.; Germany
Gómez-Skarmeta, A.F.; Spain
Gottwald, S.; Germany
Grauel, A.; Germany
Gutiérrez Ríos, J.; Spain
Hajek, P.; Czech Republic
Halgamuge, S.; Australia
Hirota, K.; Japan
Hüllermeier, E.; Germany
Jamshidi, M.; USA
Kacprzyk, J.; Poland
Kerre, E.; Belgium
Klement, P.; Austria
Kóczy, L.; Hungary
Kruse, R.; Germany
Larsen, H.L.; Denmark
Martin, T.; UK
Martyna, J.; Poland
Mesiar, R.; Slovak Republic
Moraga, C.; Germany
Negoita, M.Gh.; New Zealand
Orlowska, E.; Poland
Palade, V.; Romania
Pedrycz, W.; Canada
Peláez Sánchez, J.I.; Spain
Reuter, M.; Germany
Ruspini, E.; USA
Schmitter, E.D.; Germany
Schuster, A.; UK
Sincak, P.; Slovak Republic
Spaanenburg, L.; The Netherlands
Temme, K.-H.; Germany
Thiele, H.; Germany
Trillas, E.; Spain
Valverde, L.; Spain
Wijesoma, W.S.; Singapore

Supporting Organizations

Arbeitsgemeinschaft Fuzzy Logik und Softcomputing Norddeutschland (**AFN**)
Berkeley Initiative in Softcomputing (**BISC**)
City of Dortmund
Deutscher Verband für Schweißen und verwandte Verfahren e. V. (**DVS**)
Dortmund Chamber of Industry and Commerce (**IHK** zu Dortmund)
European Society for Fuzzy Logic and Technologies (**EUSFLAT**)
International Fuzzy Systems Association (**IFSA**)
Information Technology Society within **VDE** (**ITG**)
VDE/VDI-Society of Microelectronics,
Micro- and Precision Engineering (**GMM**)

Organizing Committee

K.-H. Temme	(University of Dortmund)
U. Lippe	(University of Dortmund)
W. Hunscher	(University of Dortmund)

Conference Secretariat

U. Lippe (University of Dortmund)

Table of Contents

Neuro-fuzzy

Soft Computing Systems II

Fuzzy Logic Medical Applications

Neural Networks Theory

Soft Computing Systems III

Invited Lecture

Fuzzy Logic Theory I

Neural Networks Image Processing I

Evolutionary Algorithms Theory I

Fuzzy Logic Theory II

Neural Networks Image Processing II

Evolutionary Algorithms Theory II

Fuzzy Logic Theory III

Neural Networks Applications

Evolutionary Algorithms Applications

Invited Lecture

Fuzzy Logic Systems II

Machine Learning

Soft Computing Systems IV

Posters

Interactive Evolutionary Computation as Humanized Computational Intelligence Technology

Hideyuki Takagi

Kyushu Institute of Design
Shiobaru, Minami-ku, Fukuoka 815-8540 Japan
Tel & Fax +81 92-553-4555
takagi@kyushu-id.ac.jp
http://www.kyushu-id.ac.jp/~takagi/

We first overview the history of computational intelligence in past decades. 1980s was a decade that neural network, fuzzy systems, and evolutionary computation whose seeds were sown in 1960s came into flowers. These computational intelligence technologies were combined, and several fusionor cooperative technologies were proposed since at the end of 1980s. These cooperative models of NN/FS/EC have been widely applied into consumer products and industry systems. We introduce these cooperative models and their real world application in detail.

The second topic is "what comes next?" Although there are several possibility of computational intelligence research, we emphasize that combination of human factor with computational intelligence would be one of the possible research directions. Common AI approach has been to model human functions and installs the model into a computer, but it is difficult to perfectly model, for example, a personal preference model. Here, we would like to stress the different approach that is to directly embed a human as a black box into a target system.

Interactive Evolutionary Computation (IEC) is one of such new directions of computational intelligence research. The IEC is an optimization method that adopts EC among system optimization based on subjective human evaluation. We introduce several variety of IEC applications. Some of them are: graphic arts and animation, 3-D CG lighting, music, editorial design, industrial design, facial image generation, speech processing and synthesis, hearing aid fitting, virtual reality, media database retrieval, data mining, image processing, control and robotics, food industry, geophysics, education, entertainment, social system, and so on.

B. Reusch (Ed.): Fuzzy Days 2001, LNCS 2206, p. 1, 2001.

Microcontroller-Based Fuzzy System to Optimize the Anaerobic Digestion in Biogas Reactors

Steffen Patzwahl[1], Thomas Nacke[1], Dieter Frense[1], Dieter Beckmann[1], Klaus-Dietrich Kramer[2], Tobias Tautz[2], and Gerd-Rainer Vollmer[3]

[1] Institute for Bioprocessing and Analytical Measurement Techniques e.V., Rosenhof, D-37308 Heiligenstadt
Tel.: ++49(0)3606 / 671-163, Fax: ++49(0)3606 / 671-200
info@iba-heiligenstadt
http://www.iba-heiligenstadt.de

[2] Hochschule Harz - University of Applied Sciences, Department of Automation and Computer Science
Friedrichstraße 57-59, D-38855 Wernigerode
Tel.: ++49(0)3943 / 659-317, Fax: ++49(0)3943 / 659-107
kkramer@hs-harz.de, http://www.hs-harz.de

[3] BTN Biotechnologie Nordhausen GmbH, Kommunikationsweg 11, D-99734 Nordhausen
Tel.: ++49(0)3631 / 656961, Fax: ++49(0)3631 / 656997
btn-gmbh@t-online.de

Abstract. This paper describes a fuzzy-system for the optimization of the yield of biogas at biogas plants. The algorithm introduced allows the transformation of a PC supported developed fuzzy-model to a microcontroller-system. This system can be utilised as a robust and efficient unit for system controlling in industry. The transformation of the PC-based software-model to the microcontroller is designed user friendly due to a new fuzzy-shell.

1 Introduction

1.1 Production of Biogas

The production of biogas from organic waste materials is complex - especially at the anaerobic co-fermentation and occurs over a variety of intermediate products. The composition of the substrate is not constant, it depends on material and seasonal fluctuations. To be able to use biogas plants economically for power generation it is necessary to have a continuous gas yield together with an optimum carbon reduction. To reach this objective, the amount of the supplied substrate has to be regulated in dependance of the desired gas amount. During the feeding with substrate a sudden acidification of the reactor should be avoided, which would cause a break off or a total succumbing of the methane production.

1.2 Problems of Control Technology

Conventional control technology for such anaerobic digestion processes reaches its technical limits. They are result of the non-linear correlation of substrate supply and

B. Reusch (Ed.): Fuzzy Days 2001, LNCS 2206, pp. 2–10, 2001.

production of biogas, which are connected to large time constants. These problems are amplified through the order conditional co-fermentation of different organic waste.

The process of anaerobic fermentation is relatively complex and because of non-linearities in the supply of substrate, the amount of biomass and the yield of gas is very difficult to describe in mathematic terms - only with models.

1.3 Technical Conditions of the System Control

Modern processes of fermentation for anaerobic biogas production are supervised by a huge number of measurings and accordingly controlled. Important parameters for the evaluation of this process are the time course measurings of the pH-value, the redox-potential, the concentration of important substrates in the supply and removal of the biogas reactor, metabolites and products of the metabolism or their compounds. Especially the concentrations are not or only very difficult to catch as on-line measurements, which means, that the parameters can not take influence in the control of the system.

Since several years is taken account on various mathematical models to develop a suitable process strategy for the optimization of such kind of fermentation. The models are based on an exact physical and chemical analysis of the initial materials. These models - connected with the applied reactor technique - were used for the development of kinetic models with coeffencies of increase and output. The industrial application although turned out to be difficult, because these systems could react to variations of the substrate composition only in a limited way [Steyer99]. In praxis this led to an organic overload - often connected to the failure of the whole system.

2 Fuzzy Modelling

2.1 Objective

For the optimization of the biogas production and the increase of process security it is necessary to control the substrate supply by on-line measurements. Because knowledge of the process was available (through an expert), a fuzzy-system was favoured for the development of a model.

2.2 Data Acquisition

As basis for developing such a model a database was generated for describing the process. For that a lab-biogas reactor (volume of fermentation: 6 l) was installed with a extensive measurement equipment for on-line controlled data collection. After a first data evaluation in connection with an analysis of correlations the following parameters were classified as process and model relevant:

- the volume of biogas (vob) [dm^3]
- concentration of methane in the biogas (CH4) [vol%]

- concentration of carbon dioxide in the biogas (CO2) [vol%]
- concentration of oxygen in the biogas (O2) [vol%]
- pH-value in the reactor (pH) [-]

Figure 1 shows the time course of the process parameters during a period of 30 days. The supply and removal of substrate took place once a day. This punctual dosage causes the typical curve.

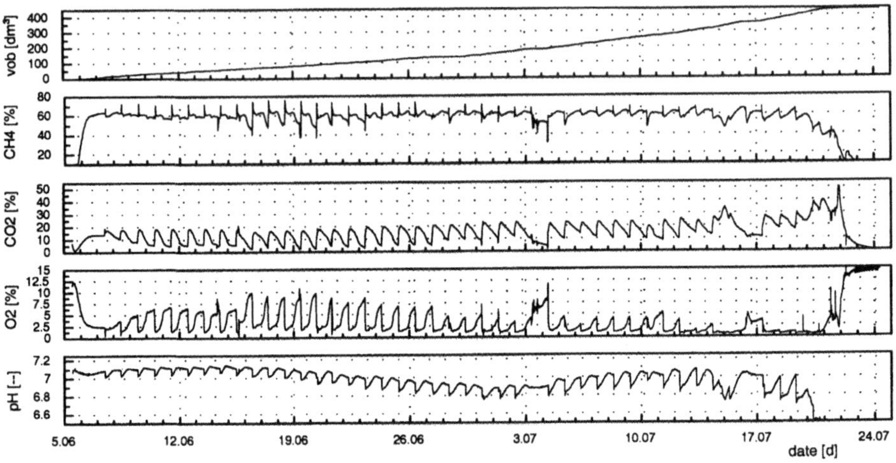

Fig. 1. Time course of relevant parameters

At the time of the supply of substrate, which has because of organic acids a smaller pH-value as the content of the reactor, the pH-value decreased. After approx. 16 hours the pH-value reaches its initial value, which means, that the organic acids are to a great extent reduced. The other parameters showing similar significant courses.

The experiment was so designed, that at the end of it, the biogas production was almost finished. After 4 weeks the system was overloaded in order to reach a bad process situation and decrease the biogas production.

2.3 Data-Mining

Since the model in its final configuration shall find the optimum of the daily supply rates, the time courses have been split in daily curves, describing the curse of parameters between two dosages (about 24 h). Figure 2 shows the time sequel of 20 day curves for the parameter biogas volume. The different daily amounts of developed biogas are shows in figure 2.

On the basis of this daily curves of all five parameters different items were generated for description. Following items were used for the development of the model:

- gas amount per day (ga_d) [dm^3/d)
- mean methane concentration per day (CH4_d) [%]
- mean carbon dioxide concentration per day (CO2_d) [%]
- mean oxygen concentration per day (O2_d) [%]

- ratio of CO_4 and CO_2 concentration per day (CH4/CO2_d) [-]
- change of gas amount between actual and previous day (dga_d) [dm^3/d]
- arithmetic mean of the pH-value (pH_m) [-]

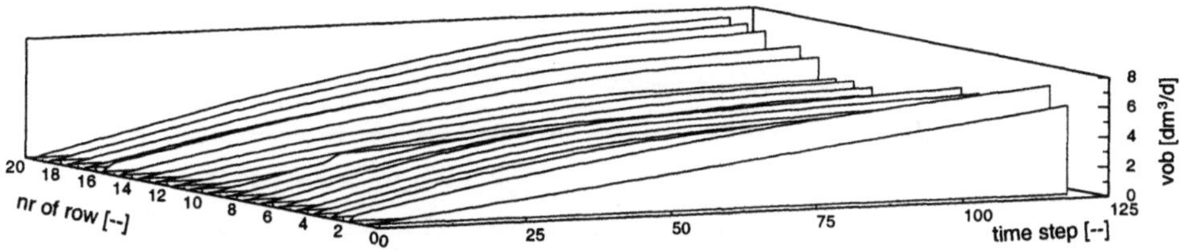

Fig. 2. Courses of the daily curves for the volume of biogas

2.4 Fuzzy-Model

Figure 3 shows the constructed fuzzy-model. The model works with 7 input parameters, which have been generated through the data-mining. The optimum supply rate is designated as output parameter and will be used as manipulated variable in the process control. The control basis (consisting of 22 fuzzy rules) as well as the parameters of the membership function of the linguistic terms have been defined by experts (fuzzy-experts-system).

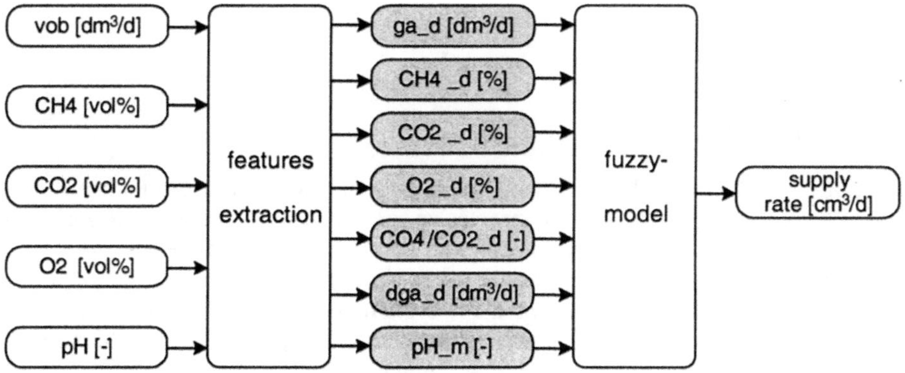

Fig. 3. Fuzzy-model with formation of the variables (structure)

The model relates to the substrates fat and hydrolysis as input materials into the process of fermentation. The fuzzy-model guarantees process security, a process failure will be avoided. The model also optimizes the yield of gas and CH_4.

To verify the model it was installed on a PC, realized with MATLAB™-SIMULINK™ with help of Ilmenauer Fuzzy Toolbox™ [Kuhn94], [Koch96]. Figure 4 shows the results of the simulation with real process data. It traces the time course of 4 of 7 items (inputs into the fuzzy-system) as well as the course of the optimum influent flow (output of the fuzzy-system). After 20 days the hydraulic retention time was shortened and the organic load increased. This resulted in an endanger of the methane developing process (signalized by a strongly decreasing pH-value). Accordingly the system indicated a smaller influent flow as output parameter, to stabilize the process.

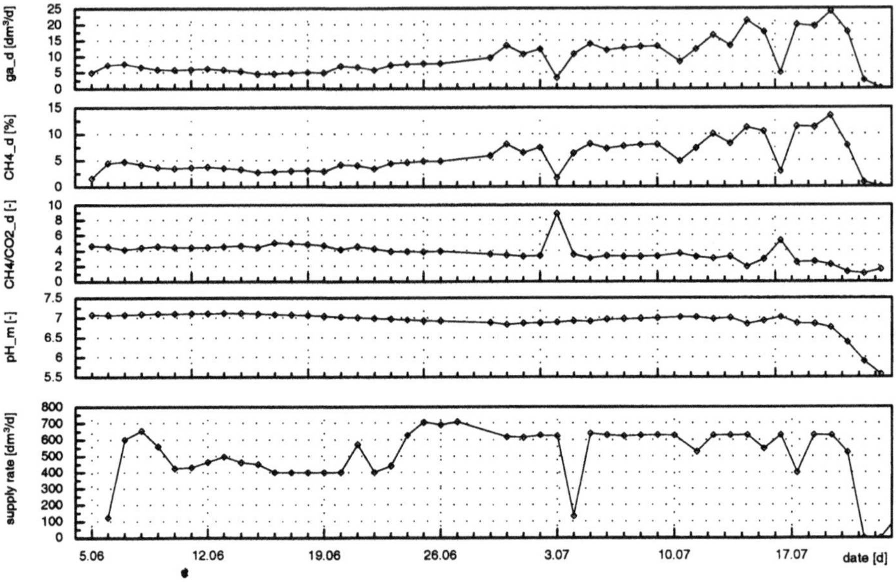

Fig. 4. Result of the simulation

3 Implementation of the Model

3.1 Objective

Operators of biogas plants usually wish robust and practice-orientated control systems. Such systems have to be insensitive to rough environment conditions. Those requirements are better meet by a microcontroller(MC)-system than by a conventional personal computer (PC)-system. At the same time the purchase costs of such a MC-system are lower than the cost for a PC-system. Accordingly the objective was to design such a robust MC-system for biogas plants. The system shall be able to process data in form of sensor signals (analogue voltages) directly. Also the fuzzy-model has to be implemented on the MC, and the manipulated output signal has to be compatible to the system - e.g. as digital control signal.

3.2 Target-Hardware-System

A system with microcontroller 80C167 was favoured as hardware. This system has a instruction cycle time of 100 ns in 16 bit and an efficient and powerful on-chip peripheral equipment (10 bit ADU for 16 channels, 111 digital ports, 2 serial interfaces). We used MiniModul-167 of phytec-Messtechnik GmbH Mainz. Besides the 80C167-microcontroller it has also 1 MByte RAM and 1 MByte Flash-ROM.

Especially for the biogas plant an LCD-display was integrated in the whole system for the display of all important process parameters and the optimum supply rate as well as a key as user interface.

3.3 Fuzzy-Shell

To meet the objective a special program was designed and developed: FuzzySysOnConApp - fuzzy systems on controller applications. It allows to create, determine and test a fuzzy- control(FC)-system very comfortably. The program can also transform the FC-system into a Run-Time(RT)-C-code, which is optimized for the microcontroller 80C167. This code supports the on-chip peripheral equipment of the microcontroller. FuzzySysOnConApp offers following output features:

- data import from a data file, which can be created with usual standard software, for instance Microsoft Excel™
- data export into a data file, which can be read with usual standard software, e.g. Microsoft Excel™
- development and structuring of the fuzzy-system (input, output, terms, rule basis)
- testing of the developed system through manual input of datas or through an imported data file
- generation of a RT-C-source-code of the developed fuzzy-control-system for the MC 80C167 with data collecting over the on-chip analogue-digital converter (ADU).

With the help of a suitable C-compiler for the MC 80C167 (for instance 80C16X - C-Compiler of Tasking's Company) the RT-C-code can be compiled and applied. FuzzySysOnConApp uses for the course of input and output terms the trapezium function. For inferences the procedure Min-MAX Inference is used and for defuzzification the procedure "center of area" (COA). Figure 5 and figure 6 show screen views of the developed software tool FuzzySysOnConApp [Tautz01].

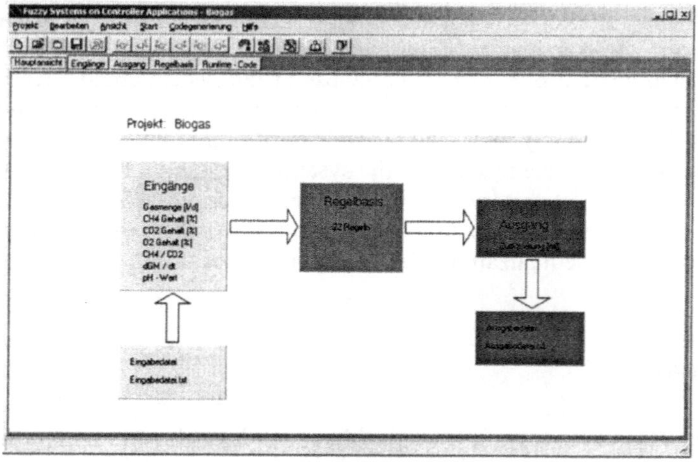

Fig. 5. Main view of FuzzySysOnConApp

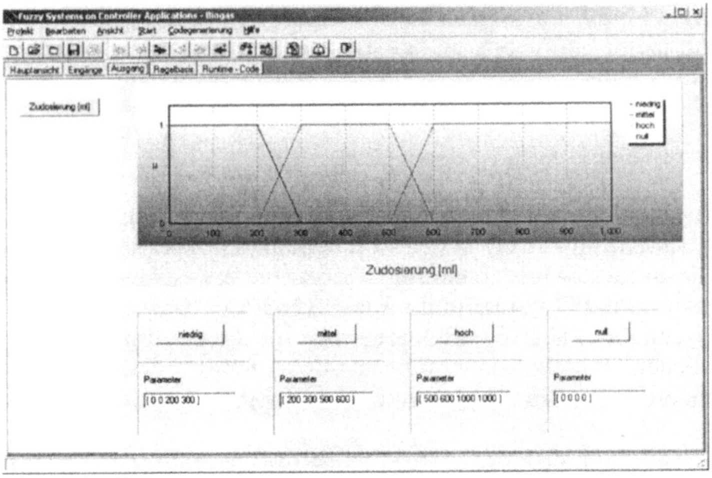

Fig. 6. Screen view - membership function parameter setting

4 Results

As a result comparing inspections of the real and calculated supply rates shall make clear the correct functioning of the fuzzy-control-system. For that both supply rates are faced in figure 7. The real supply rate represents the amount of substrate, which was supplied to the process.

At the beginning of the experiment the calculated influent flow is bigger as the real one. At this point the FC-system tries to maximize the outcome of biogas and calculates the necessary supply rate. However, at the end of the trials the calculated rate is smaller as the real one: here the fuzzy-control-system identifies the bad situation of the process (it tends to collapse, indicated through the decreasing of the pH-value and the outcome of biogas) and throttles the supply of substrate. By using the calculated amount of substrate a collapse of the system could have been avoided. At the interpretation of figure 7 it has to be noticed, that the course of the process was a result of the real supply rates - the instable condition of the process was intended for the purpose of testing the reaction of the system.

First trials in the laboratory showed, that the introduced FC-system allows to get an optimum outcome of biogas. At the moment the verification of the system is going on by using it for the methanization of different substrates.

5 Summary

This paper introduces a complex algorithm, which is able to create a fuzzy-control-system, transform it on a microcontroller-system and apply it to system control. During the development of the model relevant data of the fermentation process will be transformed by sensors into electrical signals and transmissioned to a PC (left side of figure 8). With the help of data mining algorithms data will be processed for further use. The manipulated item (supply of substrates) will be controlled by experts and registrated by the measurement system. The generated data basis can further be processed directly with the developed fuzzy-shell for Mircosoft-Windows™ operating systems. It represents a complex fuzzy-tool, which one the on hand contains the process relevant expert-knowledge in form of a knowledge basis as well as the linguistic terms of the process values and allows on the other hand the readaption of the fuzzy-system.

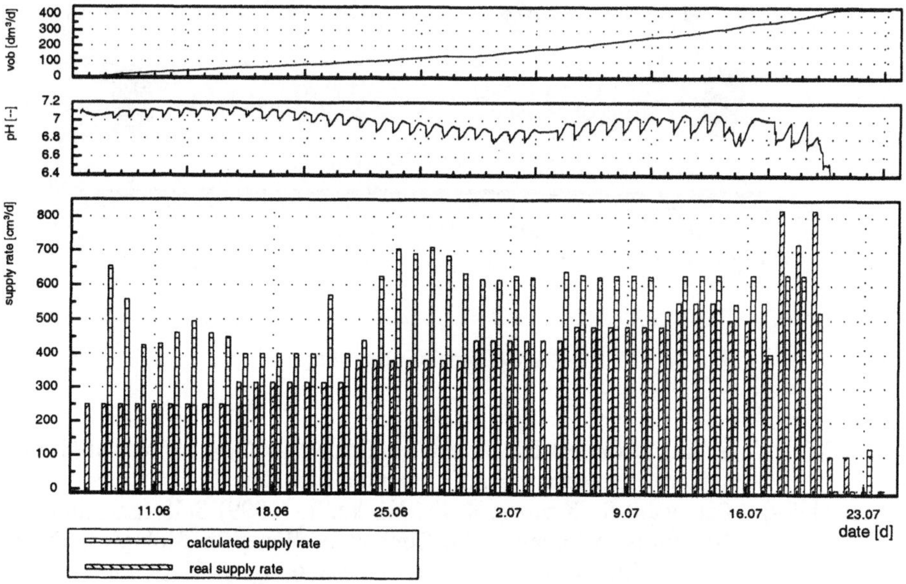

Fig. 7. Comparison of real and calculated supply rates.

The described fuzzy-shell is designed universal, its application is that for not only limited to biogas plants, but can also be used in other processes to control them and to improve process security. This at the PC developed fuzzy-model can also be transformed in a Run-Time-C-code (lower part of figure 8), which then can be used for process control on a microcontroller-based hardware-system - independently of a PC. The fuzzy-system calculates and displays the manipulated variable (supply rate) automatically on the basis of the on-line collected process data. Through those signals the substrate supply will be directly influenced by a pump (right part of figure 8).

The hardware-system on basis of a microcontroller-system meets many demands of plant-operators, who want to apply efficient and robust systems at their facilities.

References

[Koch96] Koch, M., Kuhn, T., Wernstedt, J. (1996) Fuzzy Control. Oldenbourg, München

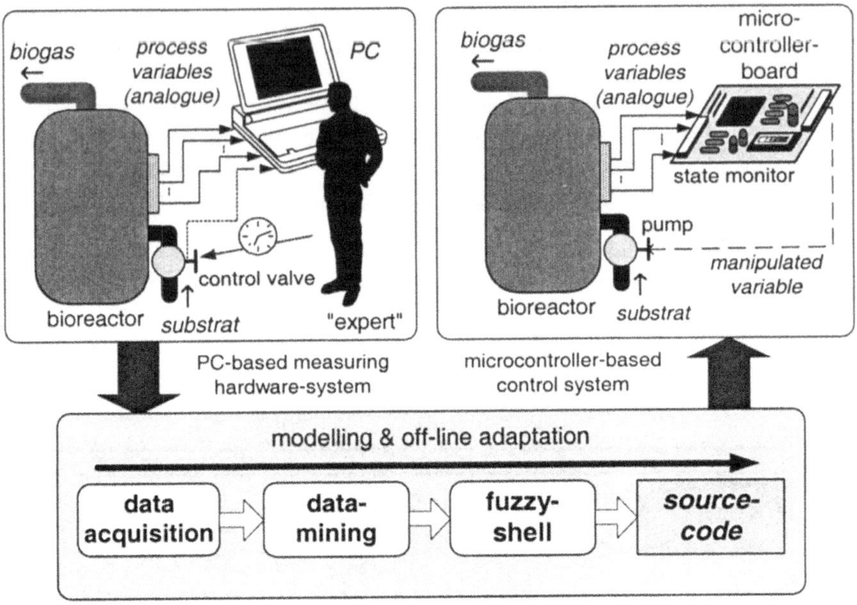

Fig. 8. Complex algorithm

[Kuhn94] Kuhn, T. (1994) Die Ilmenauer Fuzzy Toolbox für Matlab/Simulink. system engineering GmbH ilmenau, Ilmenau

[Patzwahl99] Patzwahl, S., Kramer, K. D., Nacke, T. (1999) Selforganizing Neural Networks to Supervisethe Content of Water in Biological Filtration Plants. 7th European Congress on Intelligent Techniques and Soft Computing, Proceedings of EUFIT '99 Aachen, CD-ROM, 1. Edition Aachen: Verlag Mainz, Wissenschaftsverlag, Aachen

[Steyer97] Steyer, J. P., Rolland, D., Bouvier, J. C., Moletta, R.(1997) Hybrid fuzzy neural network for diagnosis - application to the anaerobic treatment of wine distillery wastewater in a fluidized bed reactor. Water Science & Technology 1997: 36, 6-7, 209-217

[Steyer99] Steyer, J. P., Buffiere, P., Rolland, D., Moletta, R. (1999) Advanced control of anaerobic digestion processes through disturbances monitoring. Water Research 1999: 33, 9, 2059-2068

[Tautz01] Tautz, T. (2001) Entwicklung und praktische Realisierung eines mikrocontroller-basierten CI-Regelungssystems für Biogasanlagen. Diplomarbeit Hochschule Harz, Wernigerode

[Trippe98] Trippe, H., Mauch, W. (1998) Entwicklung von Regelstrategien zur thermophilen zweistufigen Methanisierung fester, ligno-cellulosehaltiger Reststoffe. Korrespondenz Abwasser 1998: 45, 1, 71-78

The Decision of Service Centres Location Problem in Fuzzy Conditions

Tatiana A. Dziouba and Igor N. Rozenberg

Taganrog State University of Radioengineering
Scientific and Research Department of Software and Computing Systems
Scientific and Technical Centre "INTECH"
Oktyabrskaya ploschad, 4, 347922 Taganrog, Russia
tanya@pbox.ttn.ru

Abstract. In this paper a service centres location problem with fuzzy criteria concerning to road network is considered. A review of fuzzy location problems is given in part 1. A short description of problem statement, the basic necessary definitions from fuzzy graph area, the definitions of fuzzy transmission numbers and p-median of fuzzy graph are adduced in part 2. An algorithm of search of p-median of fuzzy graph on the bases of a tree is suggested. A branching and bounding processes are described in part 3. In conclusion the prospects of development of the given problem are described.

1 Introduction

In some cases it is necessary to solve the various optimisation problems on the locality described by geographical map. One of them is a problem of location of several service centres. It needs to locate a given number of service centres so that a sum of the shortest distances or sum of the shortest driving times from the centres to all others objects, which have to be served be minimal. Such a criterion is the most suitable for a problem of location of warehouses in the road network, telephone stations in the telephone network or post sorting departments.

By the decision of such problems it is possible to convert an initial map to a graph, in which a set of vertices represents the places of location of served objects and a set of arcs represents a road network. Then the optimal places of service centre locations will be called p-median of graph. As a result we must show where the service centres have to be located.

Since a driving time from one object to another can be various because of presence of traffic-lights, traffic jam, road slopes and other factors then the driving times between objects can be presented as fuzzy numbers. Besides the served objects at the map can have various importance, cost of construction of service centres and other parameters, which are described fuzzy also.

Before a problem solving we have to define a graph type, which will describe a map. The graph type depends on presence of corresponding fuzzy criteria.

There are three types of fuzzy criteria, which need to optimise: the fuzzy criteria concerning to road network, the fuzzy criteria concerning to polygons describing the served objects and the fuzzy criteria concerning to road network and polygons.

B. Reusch (Ed.): Fuzzy Days 2001, LNCS 2206, pp. 11–17, 2001.

Some statements of location problem were discussed in [1].

In this paper we consider a p-median location problem with fuzzy criteria concerning to road network and we suggest an algorithm, which allows to reduce a number of the calculations in some cases.

2 Conception of p-Median of Fuzzy Graph

Let us consider a map, which has n various locations of objects and we have to locate p service centres, which will serve these objects so that the sum of the shortest driving time from service centres to all other objects be minimal. The driving times between objects i and j $(i,j=1,\dots,n)$ are given as fuzzy triangular numbers $\tilde{c}_{ij}=(c^l_{ij}, c^m_{ij}, c^r_{ij})$, where $c^l_{ij}, c^m_{ij}, c^r_{ij}$ are minimum possible, most possible and maximum possible driving time from object i to object j accordingly. As has been proved in [2], the places of location of service centres must be situated at the p places of location of served objects.

Let us consider now a fuzzy graph corresponding to the map. The vertices of the graph correspond to locations of objects and the arcs correspond to the roads connecting these objects. Let all roads between objects be with two-way traffic. Then a fuzzy graph will be undirected and all arcs in further we shall call the edges of the graph.

First of all we shall give some necessary definitions.

A *fuzzy graph* is a graph $\tilde{G}=(X,\tilde{F})$, where $X=\{x_i\}, i\in I=\{1,2,\dots,n\}$ - graph vertex set, $\tilde{F}=\{<\tilde{c}_{ij}/(x_i,x_j)>\}$ - fuzzy set of edges of the graph, $\tilde{c}_{ij}/(x_i,x_j)$ - a fuzzy value of attribute of considering criterion for edge (x_i,x_j).

A *path* in a fuzzy graph $\tilde{G}=(X,\tilde{F})$ is a sequence of vertices $x_1, x_2, \dots, x_n$ such that for all edges (x_i,x_{i+1}) a value $\tilde{c}\,(x_i,x_{i+1})>0$.

The *fuzzy length* of a path P is a fuzzy number $\tilde{l}\,(P)$, which is equal to sum of all edges including in P, that is $\tilde{l}\,(P)=\sum_{(x_i,x_j)\in P}\tilde{c}_{ij}$, where $\tilde{c}_{ij}$ is fuzzy triangular number, representing an edge length between x_i and x_j.

The *fuzzy distance* $\tilde{d}(x_i,x_j)$ or the shortest fuzzy path between two vertices x_i, x_j is the smallest fuzzy length from all paths from x_i to x_j, $x_i, x_j\in X$.

Let X_p be a subset of vertex set X of fuzzy graph $\tilde{G}=(X,\tilde{F})$ and X_p consist of p vertices [2]. We make some notations now.

$$\tilde{d}\left(X_p,x_j\right)=\min_{x_i\in X_p}[\tilde{d}\,(x_i,x_j)], \tag{1}$$

where $\tilde{d}(x_i, x_j)$ is a length of the shortest fuzzy path between vertices x_i and x_j in graph $\tilde{G}$.

A fuzzy transmission number of the vertex subset X_p is defined as follows:

$$\tilde{\sigma}(X_p) = \sum_{x_j \in X} \tilde{d}(X_p, x_j). \tag{2}$$

A subset $\overline{X}_p$ for which

$$\tilde{\sigma}(\overline{X}_p) = \min_{X_p \subseteq X} [\tilde{\sigma}(X_p)] \tag{3}$$

is called the p -median of graph $\tilde{G}$ [2].

3 Search of p-Median of Fuzzy Graph

Let the matrix of fuzzy distances (of shortest fuzzy driving times) $D = \left\| \tilde{d}(x_i, x_j) \right\|_{n \times n}$ between all vertices of graph $\tilde{G}$ has been generated. Fuzzy triangular number comparison operation is described, for example, in [3]. Then the formulae (1) – (3) can be directly used by the search of p-median of graph $\tilde{G}$. In this case it requires to make C_n^p calculations of value $\tilde{\sigma}(X_p)$. But it is possible only for graph with small number of vertices.

For the graphs with large *n* we suggest to use a branch and bound method [4], [5]. A branching and bounding processes will be described follow.

The method uses a search tree, which is shown on the figure 1. This tree consists of *p* levels. Each node x_i *(i = 1, 2, ..., (n-p+1))* of the first tree level corresponds to subset $X_p^{(1)}{}_{\{i\}}$, for which only one median vertex x_i has been defined and others *(p-1)* vertices are not known yet. Each node x_r of the level *k (k=2,...,(p-1))* corresponds to subset $X_p^{(k)}{}_{\{i,...,q,r\}}$, for which *k* median vertices with the numbers *i, ..., q, r (r = (q+1), ..., (n-p+k))* has been defined and others *(p-k)* vertices are not known yet. The last level *p* consists of the nodes x_t corresponding to subsets $X_p^{(p)}{}_{\{i,...,r,...,t\}}$. They represent the completely generated subsets X_p containing vertices with the numbers *i, ..., r, ..., t.* In other words each a branch of tree from level *p* to level *1* containing the nodes x_i, ..., x_r, ..., x_t (bold lines on the figure 1) corresponds to each complete median subset X_p, e.g. $X_p = \{x_1, x_2, x_3, ..., x_{n-p+p}\}$,

$X_p = \{x_1, x_{n-p+2}, x_{n-p+3}, ..., x_{n-p+p}\}$, $X_p = \{x_2, x_3, x_{n-p+3}, ..., x_{n-p+p}\}$, and so on (see figure 1).

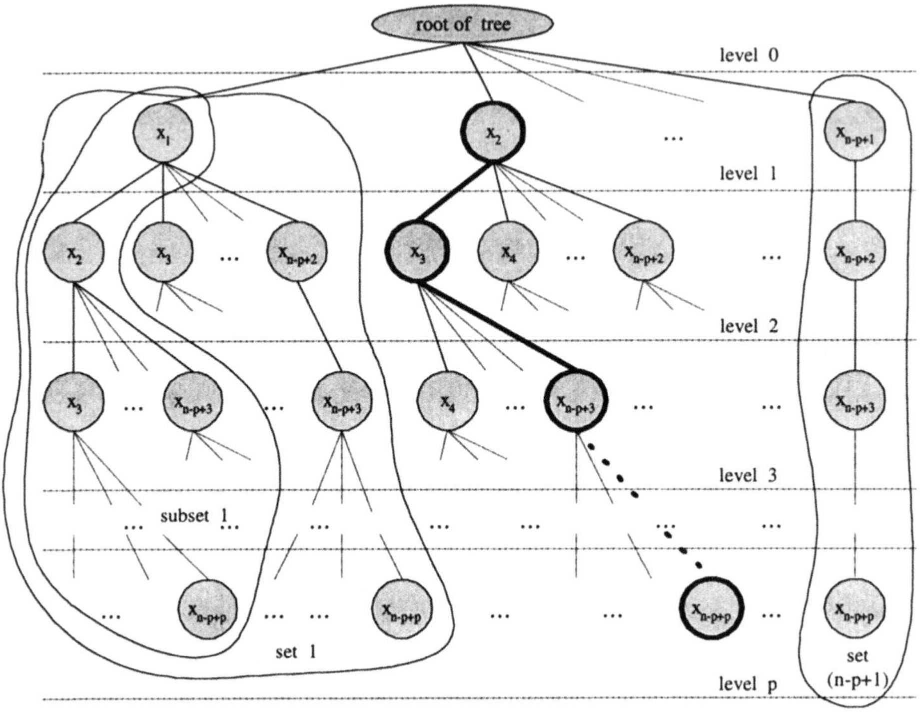

Fig. 1. A search tree for fuzzy p-median problem

There are two basic conceptions, which have to be considered by development of the algorithm based on a search tree: branching rule and method of calculation of lower bound of value $\tilde{\sigma}(X_p)$.

3.1 Lower Bound Calculation

3.1.1 Estimation for Nodes of the First Level of Search Tree

We make some notations by analogy with formulae (1) and (2).

$$\tilde{d}(X_{p\,\{l\}}^{(1)}, x_j) = \min_{i \geq l} \tilde{d}(x_i, x_j), \tag{4}$$

$l = 1, ..., (n-p+1)$, $i \neq j$, $j \neq l$, $i = \overline{1, n}$.

Then an estimation value $\tilde{W}\left(\sigma(X_{p\ \{l\}}^{(1)})\right)$ for set l in consideration that the first vertex in the set will be x_l, is equal to

$$\tilde{W}\left(\sigma(X_{p\ \{l\}}^{(1)})\right)=\sum_{x_j\in U^{(1)}} \tilde{d}\,(X_{p\ \{l\}}^{(1)}, x_j)\,, \tag{5}$$

where $U^{(1)}$ - a subset of vertices x_j of set X, j=1,...,(n-p), for which $\tilde{d}(x_l, x_j) > 0$ and they with the vertex $x_l^{(1)} \in X_p^{(1)}$ give (n-p) minimal lengths of shortest paths.

Let $x_l^{(1)}$ be a vertex from median set $\mathrm{X_p}$. Then in accordance with (1) a column of matrix of shortest distances with the number l doesn't considered. The other columns of this matrix are considered. It is reflected by inequality $j \neq l$ (see formula (4)). That is the estimation is defined not for the specific vertex $x_l^{(1)}$, but for a set of vertices in which the first median vertex is $x_l^{(1)}$, then the rows of shortest distances matrix with the numbers $i < l$ are not considered also.

3.1.2 Estimation for Nodes of the Second Level of Search Tree

Using the cited above reasoning we define the values of estimation for the nodes of the second level of search tree. Make following notations:

$$\tilde{d}(X_{p\ \{l,r\}}^{(2)}, x_j) = \min_{i=l,\ i\geq r} \tilde{d}(x_i, x_j) \tag{6}$$

$r = l+1, \ldots, (n-p+2)$, $j \neq l,\ j \neq r, i \neq j$.

Then an estimation value for the nodes of the second level will be defined as follows:

$$\tilde{W}\left(\sigma(X_{p\ \{l,r\}}^{(2)})\right)=\sum_{x_j\in U^{(2)}} \tilde{d}\,(X_{p\ \{l,r\}}^{(2)}, x_j)\,. \tag{7}$$

Generalize all these formulae for any level k, *k=1,...,(p-1).*

3.1.3 Estimation for Nodes of Level k, k=1,...,(p-1)

We make a notation:

$$\tilde{d}(X_{p\ \{l^{(1)},\ldots,q^{(k-1)},r^{(k)}\}}^{(k)}, x_j) = \min_{x_i\in X_p^{(k)},\ i>r} \tilde{d}(x_i, x_j), \tag{8}$$

Here $x_j \notin X_p^{(k)}$, $r = (q+1), \ldots, (n-p+k)$.

Then estimation value for nodes of level k will be defined as follows:

$$\tilde{W}\left(\tilde{\sigma}(x_l^{(1)},\ldots,x_r^{(k)})\right)=\sum_{x_j\in U^{(k)}}\tilde{d}\,(X_p^{(k)}{}_{\{l^{(1)},\ldots,r^{(k)}\}},x_j) \tag{9}$$

where $U^{(k)}$ - a subset of vertices $x_j \in X$, *j=1,...,(n-p)*, for which $\tilde{d}(x_i, x_j)>0$, $x_i \in X_p^{(k)}$ and which has *(n-p)* minimal values of fuzzy distances.

As you can see from formulas (8) and (9), values of estimation for each branch in the tree will not decrease with increase of tree level number. This fact can be explained as follows. Number of the members of the sum in formula (9) will be constant and will be equal to *(n-p)*. A part of members of the sum corresponds to a part of real value of fuzzy transmission number for already defined vertices $x_i \in X_p^{(k)}$. The other members of the sum are not known yet. As such members we take a minimal elements $\tilde{d}(x_i, x_j)$ for which *i>r* (see formula (8)). By the increase of tree level number a real value of fuzzy transmission number can be equal to previous or increase and the others minimal elements can not be smaller then previous.

3.1.4 Improved Estimations

Obtaining estimations can be improved. We'll use formulas (8) and (9) also taking into account the following reasoning. We have taken into consideration a fact that in accordance with (1) the elements x_{ij} and x_{ji} of shortest paths matrix haven't been included into the value of sum of formula (2) simultaneously. Therefore in such a case if a minimal element of column *j* is x_{ij} then for the column *i* it is necessary to select the next after x_{ji} minimal element.

3.1.5 Estimation for the Nodes of Level p

The last level p consists of the nodes, which represent the completely generated subsets X_p containing *p* vertices. Therefore the estimation for the nodes of this level will be equal to real value of fuzzy transmission number $\tilde{\sigma}(X_p)$ for corresponding subset $X_p^{(p)}$ of vertex set *X* and can be calculated by formulae (1) and (2).

3.2 Branching Process

A branching process will be fulfilled as follows. After the first branching of node of level 1 for each a node of level *k, k=2,...,(p-1)*, the estimations are calculated. Then it is necessary to find a minimal value from all values of estimation of non-branched nodes related to level *k* and other previous levels. Such a node with minimal estimation will be branched as the next.

The optimal solution or p-median of graph $\tilde{G}$ is such a subset $\overline{X}_p \in X$, which is accorded to the node $X_p^{(p)}$ of tree level p with minimal estimation of fuzzy transmission number.

3.3 Calculation Results

The calculation results have shown that in many cases this algorithm and improved estimations (see 3.1.4) allows to obtain the optimal result of p-median problem with fuzzy data using twice less calculations of values of estimations and fuzzy transmission numbers than C_n^p calculations of value $\tilde{\sigma}(X_p)$. A number of considering median subsets for obtaining the best solution of this problem is 5–6 times less in comparison with the consideration of all variants of possible median subsets.

4 Conclusion

In conclusion we would notice that in some cases the vertices of initial graph could have the fuzzy weights also. There can be several criteria concerning to the road network and to object locations. It leads to search of solutions of multicriteria service centres location problems and other problems from this area.

References

1. Rozenberg, I.N.: Solving of Allocation Problem With the Help of Fuzzy Graphs. Proceedings of 3rd Conference "New information technologies. Development and application". TSURE, Taganrog (2000)
2. Christofides, N.: Graph Theory. An Algorithmic Approach. Academic press, London (1975)
3. Bershtein, L.S., Dzuba, T.A.: Construction of a Spanning Subgraph in the Fuzzy Bipartite Graph. Proceedings of EUFIT'98, Aachen (1998) 47-51
4. Goodman, S.E., Hedetniemi, S.T.: Introduction to the Design and Analysis of Algorithms. McGraw-Hill Book Company, New York (1977)
5. Reingold, E.M., Nievergelt, J., Deo, N.: Combinatorial Algorithms. Theory and Practice. Prentice-Hall Inc., New Jersey (1977)

Application of the Biologically Inspired Network for Electroencephalogram Analysis

Minija Tamošiūnaitė[1] and Erinija Prackevičienė[2]

[1] Vytautas Magnus University, Department of Informatics,
Vileikos 8, LT-3035 Kaunas, Lithuania
iimita@vdu.lt
[2] Kaunas University of Technology, Department of Practical Informatics,
Studentų 50, LT–3031 Kaunas, Lithuania
ernagu@if.ktu.lt

Abstract. Architecture of a neural network combining automatic feature extraction with the minimized amount of network training acquired by means of employing of a multistage training procedure is investigated. The network selects prototypical signals and calculates features based on the similarity of a signal to prototypes. The similarity is measured by the prognosis error of the linear regression model. The network is applied for the meaningful paroxysmal activity vs. background classification task and provides better accuracy than the methods using manually selected features. Performance of several modifications of the new architecture is being evaluated.

1 Introduction

The neural network is a black box model, which is supposed to solve the task with as little insight into the subject field as possible. It is widely believed that using neural networks one can achieve reasonable solutions in reasonable time (that is, faster than solving the same task with the help of careful analysis of the subject field, followed by some modeling of the "white-box" type). Nevertheless, a straightforward feeding of a signal into the complex multi-layer network rarely brings success. This is because the network is unable to learn the task due to insufficient data, or due to insufficient time to solve the overall task of global optimization with required precision. For better results one should either carry out a careful feature selection prior to training the network, which is a drawback towards the white-box modeling, or use some type of network specifically suited for the class of problems including the task in consideration.

Looking for more effective and more flexible classification systems (i.e. capable to deal with several problems), research in the field of combination of classifiers and cooperation of neural networks has gained momentum [1, 2, 3]. It was confirmed that using the collective of simple classifiers or simple neural networks, generalization accuracy could be better than that of a single network [2, 3, 4]. While training a simpler network less data is needed to accomplish the training properly, without overtraining. It is possible to choose a collective of simple networks which is less

B. Reusch (Ed.): Fuzzy Days 2001, LNCS 2206, pp. 18–27, 2001.

demanding on the amount of the required data than a corresponding multilayer network. Capability of a network to manage data with small amount of training samples is especially important in solving medical and biological problems, though small sample-size technical problems are sometimes considered as well. Additionally to the economy of data, training of several simple networks is usually less time consuming than training of the overall multi-layer network.

The current paper continues the investigation of the biologically inspired architecture of neural networks, proposed in [5]. The architecture has been designed for signal classification tasks and is combining automatic feature extraction with the minimized amount of network training. The architecture can be applied for a wide class of signals without manual feature selection. The architecture has got its name because of similarity to some information processing circuits of living systems, but it can also be considered from the viewpoint of network cooperation.

The performance of the proposed architecture of a neural network is demonstrated by an application for the signal classification task, arising in automatic electroencephalogram (EEG) analysis. An important task in EEG analysis is to distinguish meaningful paroxysmal activity regions (events) from the background activity in the run of the EEG signal. In the paper classification of event vs. background accuracy using the network with the biologically inspired architecture is analyzed. Several modifications of the algorithm are investigated. The results obtained by the new algorithm are compared with the ones acquired using traditional EEG analysis techniques.

2 Problems of automatic EEG analysis

Electroencephalogram is a signal of electrical brain activity that is normally recorded from the surface of the head. Automatic EEG analysis is applied for solution of several distinct tasks: assistance for a physician in the process of EEG analysis, detection of epileptic seizures in the systems of ambulatory monitoring, EEG-based classification of mental and motor tasks, etc. Usually EEG is recorded as a multichannel signal, consisting of several one-dimensional signals (e.g. 8 or 16). The one-dimensional signals are composed of several types of waveforms that are usually divided into four frequency bands: delta (1-3 Hz), theta (4-7 Hz), alpha (8-13 Hz.), and beta (14-20 Hz). Traditionally, linear methods employing spectral characteristics or autoregression coefficients [6, 7, 8] are used for automatic EEG analysis. Nonlinear methods, especially the chaos theory-based ones, have become popular during the last decade. The chaos-based methods employ parameters like Lyapunov exponents or Kolmogorov entropy [9]. The linear methods have secured the status of the well suited for the EEG analysis, while utility of the chaos-based characteristics remains the object of discussions [9, 10].

Neural network-based methods compose the subclass of nonlinear methods, and are widely employed in EEG analysis. Neural networks have been used for classification of mental tasks [11], detection and prediction of epileptic EEG activity [12, 13, 14, 15, 16, 17, 18], the brain-computer interface studies [19], etc. The types of the employed networks are rather diverse: multilayer perceptrons combined with various feature extractions procedures [15, 18], recurrent neural networks combined with wavelet decomposition [13], self-organizing feature maps [17], recurrent self-organizing maps [11], to mention the few.

In the current study the epileptic EEG of the between-seizure period is analyzed. Specifically, the problem of recognition of the regions containing meaningful paroxysmal activity (events) is considered. The recognition can be performed in two different ways: either the moments of change of the EEG signal can be detected with some parameters exceeding a threshold (e.g. in [20]), or the EEG can be divided into segments, and classification of the segments into several classes performed (e.g. in [18]). In the current study the second approach is utilized. Classification of event segments vs. background segments is carried out. However, the study is not directly dedicated to the analysis of EEG, but rather towards investigation of the specific network architecture, while the EEG analysis is taken as an example.

3 Algorithm

The basic ideas of the proposed architecture have been derived from the coarse estimates of the principles peculiar to organization of multilayer biological networks. The main idea postulates dividing of the multilayer network into separate layers (i.e. single layer subnetworks), and training each layer separately, or defining the weights of some layers *a priori*. For example, some layers could average input signals, the averaging being often employed in biological pattern recognition systems. Other layers could employ the voting procedure, which may be considered as a sort of "non-linear averaging". In addition to the multi-modularity, the network should be insensitive to the noise corrupting the signal. The robustness to noise in the current design is achieved by employing specific similarity to prototype-based features. The mentioned ideas share common aspects with the basics of cooperative neural networks, and the proposed architecture can be considered as a simplified partial case of cooperative networks.

On the basis of these ideas a network for signal classification tasks is designed. A large group of biological, medical and technical signals may be properly characterized by their spectral contents, which otherwise can be expressed by means of linear regression coefficients. The coefficients of the selected prototypical signals in the current design are used for calculation of the prognosis of other signals (i.e. all the signals belonging to the training set in the training phase of the network, and the ones belonging to the test set in the testing phase). The obtained mean prognosis error terms are used as the similarity features. Employing these features decisions of a neural classifier are derived. The average prognosis error is considered to be a robust similarity estimate of the autoregression-based time series.

Data of biomedical and technical signals are usually collected from several parallel channels (sensors). Separate decision-making units are used for separate channels, and the decisions are joined by means of voting in the selected architecture. If a record is single-channeled, information obtained in several time instances may be used as the material for voting (for example when analyzing electrocardiograms one can use data of several consecutive heart cycles). Fig. 1 illustrates the overall network employing similarity feature extraction, classification of signals obtained from different channels, and the voting procedure.

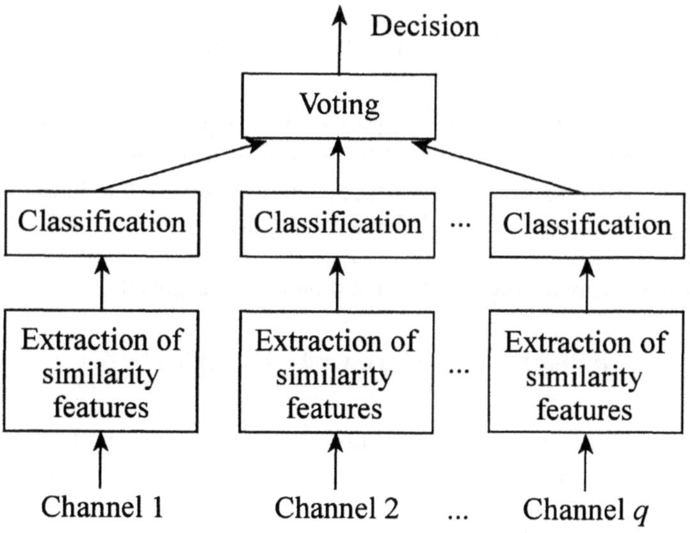

Fig. 1. Scheme of the network designed to process multiple-channel data

The examples of biological time series which can be properly described by autoregressive models are numerous: electroencephalograms, the ST segment of an electrocardiogram, onto which late potentials [21] may appear, rhythmogram of the heart, including time intervals between the two consecutive heart contractions, etc. The algorithm used in the current investigation is adapted to the 16-channel electroencephalogram signal. Let us denote the two sets of signals representing the classes of the backgrounds (ω_1), and the events (ω_2) by $X^{(1)}$ and $X^{(2)}$. Let address the individual segments comprising the classes by subscripts: $x^{(1)}{}_i$ and $x^{(2)}{}_i$, i=1, 2, ..., N, where N denotes the number of segments in a single class. In the current investigation number of samples in both classes is equal. The individual signals comprising the segments are denoted by $x^{(\omega)}{}_i(k)$, where k=1, 2, ..., q, the parameter q denoting the number of channels, and ω representing the class label. In the current investigation the 16-channel EEG is explored, consequently q=16.

Each segment $x^{(\omega)}{}_i$ and each signal $x^{(\omega)}{}_i(k)$ is composed of n points, n in the current investigation being equal 250. The j-th point of the signal is addressed by $x^{(\omega)}{}_i(k,j)$. Training of the network is described by the following seven-step procedure:

1. Several (say p) segments from both classes are selected as prototypical, let us denote them $S^{(1)}$, $S^{(2)}$, ..., $S^{(p)}$. Each prototypical segment is composed of 16 signals. Each signal included in the segment will be referred to as $S^{(l)}(k)$, k=1,2,...,16, where l denotes the number of the segment. Random selection of prototypical segments is employed in the current study, though more elaborate methods for segment selection can be applied as well.

2. The coefficients of the two-sided (non-causal) regressive model are calculated for each signal $S^{(l)}(k)$ of all prototypical segments $S^{(l)}$, l=1, 2, ..., p. The regression model is specified by the following expression:

$$S_{progn}{}^{(l)}(k,j) = \sum_{r=j-z, r\neq j}^{j+z} a^{(l)}(k, r-j) S^{(l)}(k,r), \; j = 1,2,...,n, k = 1,2,...,16\,, \tag{1}$$

where $S_{progn}{}^{(l)}(k,j)$ denotes the prognosis of the j-th point of the prototypical signal $S^{(l)}(k)$ and z determines the order of the regression. The coefficients $a^{(l)}(k,\tau)$, $\tau=\pm1$, ±2, ..., $\pm z$ are obtained in the process of training of the linear single layer perceptron (SLP) for each signal. Thus we obtain $p\times16$ sets of coefficients $\{a^{(l)}(k,\tau)\}$.
3. The coefficients are averaged over each prototypical segment:

$$a^{(l)}(\cdot,\tau) = \sum_{k=1}^{16} a^{(l)}(k,\tau), \; \tau = \pm1, \pm2, ..., \pm z\,, \tag{2}$$

and p sets of prototypical coefficients $a^{(l)}(\cdot,\tau)$ are obtained. Prototypical coefficients are used to make prognosis along each signal $x^{(\omega)}{}_i(k)$, where ω stand for the class label, i for the segment number, and k for the signal number, included in the training set. All the channels in the segment are processed separately.
4. The p mean prognosis errors $e^{(l)(\omega)}{}_i(k)$, l=1, 2, ..., p are calculated for each signal $x^{(\omega)}{}_i(k)$, according to all the p selected prototypes:

$$e^{(l)(\omega)}{}_i(k) = \frac{1}{n} \sum_{j=1}^{n} \left| x^{(\omega)}{}_i(k,j) - x_{progn}{}^{(l)(\omega)}{}_i(k,j) \right|, \tag{3}$$

where

$$x_{progn}{}^{(l)(\omega)}{}_i(k,j) = \sum_{r=j-z, r\neq j}^{j+z} a^{(l)}(\cdot, r-j) x^{(\omega)}{}_i(k,j)\,. \tag{4}$$

The average errors (3) are taken as new features.
5. Together with the obtained mean errors $e^{(l)(\omega)}{}_i(k)$, l=1, 2,..., p, quadratic features of the form e^2 are introduced. In this manner, the set of $2p$ features is formed for each signal.
6. The single-layered perceptron (SLP) is trained to distinguish the two signal classes according to the obtained features. Gradient descent algorithm with regularization term of the form $\lambda(w^Tw-c)^2$, where w – weight vector of the perceptron, λ=0.03, c=10 – parameters, is employed [22]. Prior to training, data transformations are performed in order to obtain statistically uncorrelated features with identical variances.
7. Finally, parameter of the voting procedure (i.e. voting threshold) is defined. The voting threshold *a priori* is put at eight: If at least eight leads out of the 16 comprising the segment vote for the "event", then the whole segment is assigned to be an event.

   The scheme in Fig. 2 illustrates the steps of the neural network training.

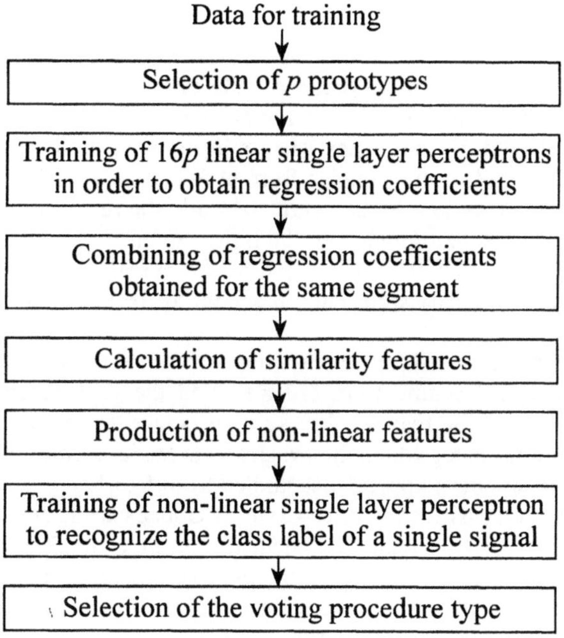

Fig. 2. Scheme of the training algorithm of the neural network.

The steps 1-5 can be interpreted twofold: as training of the first layers of the overall multilayer network, or, otherwise, as the feature extraction steps of the classification algorithm. The training in each step is done separately. In the first step search for the efficient methodology for the selection of prototypes is required. In the second step traditional training of linear perceptrons is employed. In the 4-th step the reliable methodology for evaluating the distance from signals to the selected prototypes must be chosen. In the fifth step one has to choose what kind of nonlinear features should be used. The sixth step is dedicated to the traditional training of the main decision making unit – nonlinear perceptron. In the seventh step the voting is carried out, and the required training in this step is limited to the selection of the voting threshold. Dividing of training of a network into several stages is the main novelty of the architecture. Though the network obtained in the specified way is not strictly optimal because of the imprecise definitions of the training goals of the first, third to 5-th and the 7-th steps, you can form an effective network in this way. It should be taken into account that the corresponding multilayer network trained by back-propagation algorithm most likely would not be globally optimal either.

As the averaging could possibly reduce the specificity of prototypes, the other method of formation of prototypical coefficients $\{a\}$ was explored. Instead of the averaging (1), coefficients of an individual signal from each segment were selected to represent prototypes. Specifically, the signal with the number k=6 was used in the current investigation.

4 Application of the algorithm

The algorithm was applied for solving the EEG event vs. background classification task. 54 event segments and the same amount of background segments of 1 second duration were selected by an expert neurophysiologist in the 16 channel EEG, sampled at 250 Hz. Half of the samples from both classes (i.e. 27+27 samples) were used for the training procedure of the neural network, the rest were used for testing. In order to avoid fitting to the training set, the training and testing of each network was performed 100 times with different division of the overall data into the training and test sets. Prototypical signals were chosen randomly from the training sets of both signal classes.

The new method was tested for several orders of the regression model and different numbers of prototypical segments in order to find the best parameters of the new algorithm. The order of the two-sided regression ranged from the second (1 point in both sides of the element for which the prognosis is made, further denoted as 1+1), to the 8-th (4+4). The number of prototypes ranged from 5 to 13.

The mean error of classification of separate EEG leads at the end of the 6-th step of the algorithm, i.e. prior to voting, has been in the range of 19-20%. The error was averaged over the series of 100 trials with standard deviation up to 2%. Only slight differences of the mean classification error were noticeable with the changing number of prototypes and the changing order of the regression.

After application of the voting procedure the mean classification error dropped to 8-9%. The results obtained for different orders of the regression models and various numbers of prototypes are presented in Table 1. The error tends to increase with the growing order of the regression model, except the lowest (1+1) order. The increase of the error with the number of prototypes is especially visible in the cases of greater order of the AR model (3+3 and 4+4). Standard deviation of the series of 100 tests is around 2-3%, and it makes the differences presented in the table less certain. The significant deviations in different tests may be produced by small sample size and by random selection of prototypes.

Table 1. Mean value of classification errors of EEG segments, averaged over 100 trials.

Order of the model vs. Number of prototypes		5	8	10	13
1+1	mean	*8.2*	*8.2*	*8.6*	*8.4*
	std, %	2.5	2.7	2.8	3.1
2+2	mean	*8.2*	*8.6*	*8.1*	*8.4*
	std, %	2.5	3.1	2.3	2.7
3+3	mean	*8.4*	*8.4*	*9.0*	*8.8*
	std, %	2.9	3.0	3.2	3.0
4+4	mean	*8.9*	*8.8*	*9.4*	*9.6*
	std, %	3.3	3.2	3.4	3.3

The results obtained in the first stage of experiments were improved using the different method of selection of prototypes. While in the first stage of the experiment prototypical regression coefficients of a single segment were obtained by averaging of the coefficients obtained for separate signals included in the segment, in the second stage of the experiment coefficients of a single signal (channel) were selected to represent the whole segment. The results obtained without averaging over the channels are presented in Table 2.

Table 2. Mean value of classification error of EEG segments using the single signal as a prototype, averaged over 100 trials.

Order of the model vs. Number of prototypes		5	8	10	13
1+1	mean	*8.5*	*8.2*	*8.2*	*8.4*
	std, %	2.6	2.8	2.6	2.9
2+2	mean	*7.8*	*7.6*	*7.4*	*8.0*
	std, %	2.6	3.2	2.7	2.8
3+3	mean	*8.3*	*8.0*	*8.2*	*8.3*
	std, %	2.8	2.6	3.0	2.6

Removal of averaging has slightly improved the classification accuracy (by approximately 0.5%). Though the improvement is not big in comparison with the standard deviation, extension of the improvement throughout the different regression orders allows detect the fact of improvement. The improvement caused by the removal of averaging is not universal. Similar test with the ST segment of the ECG has provided opposite results. Averaging for prototypical signals was preferable in the case of ST segment classification. The results may have differed because of the different signal to noise ratio in the two types of signals, the ratio being worse in electrocardiograms. The other possible reason was usage of different types of "channels" in both experiments. The ECG data were obtained from single physical channel, but different heart cycles were used to compose different logical channels. The EEG data investigated in detail in this study were recorded from different physical channels with certain inter-channel differences. It could be stated in conclusion, that one should employ averaging of prototypical signals in the proposed algorithm if the data are obtained from single physical channel and the signal to noise ratio is poor. One should avoid averaging if multiple-channel data with a relatively low noise are provided.

The results obtained by the biologically inspired algorithm were compared with those obtained using the features earlier acknowledged to be important for automatic EEG analysis. These are the variance of the signals, energies of the four basic frequency bands (alpha, beta, delta and theta), coefficients of the one-sided and two-sided autoregressive models, eigenvalues of the singular value decomposition of the EEG segments. Classification based on the standard features was made by an SLP. Some algorithms additionally included voting. The errors obtained using the mentioned methods after careful selection of parameters for each algorithm ranged from 10% to 12.5% for the same set of data [23]. The overall ability of the new method to achieve up to 8% classification accuracy is a significant improvement in comparison to the 10% of errors, which was found to be the limit of classification accuracy for traditional methods.

5 Conclusions

The biologically inspired architecture of neural networks based on the division of the network training procedure into several stages and employing the similarity-to-prototype features proved to be useful in the EEG segment classification task. The network with the proposed architecture has provided better classification accuracy than the neural networks with manually selected features.

Usage of single-channel signals of the EEG for construction of prototypes was found to be more appropriate than averaging of prototypical coefficients over the whole channels. It was hypothesized that prototypes of the EEG may have lost the required specificity in the process of averaging.

The investigated architecture can be employed for analysis of a wide class of signals: these having several channels and an autoregressive structure in each channel. The good examples are medical signals: electroencephalograms and electrocardiograms (for example the ST segment used in late potential analysis). The algorithm may also be applied for non-medical tasks, like image classification, or remote sensing. The algorithm may be improved by using more elaborate selection of the prototypical signals, as compared to the random selection performed in the current study.

References

1. Kittler, J.: Combining Classifiers: A Theoretical Framework. Pattern Analysis & Applic. **1** (1998) 18-27
2. Kittler J., Hatef, M., Duin, R.P.W., Matas, J.: On Combining Classifiers. IEEE Trans. on Pattern Analysis and Machine Intelligence **20** (1998) 226-239
3. Xu, L., Krzyzak, A., Suen, C.: Methods of Combining Multiple Classifiers and Their Application to Handwriting Recognition. IEEE Trans. on Systems Man and Cybernetics **22** (1992) 418-435
4. Saranhli, A., Demirekler, M. A.: Statistical Unified Framework for Rank-Based Multiple Classifier Decision Combination. Pattern Recognition **34** (2001) 865-884
5. Raudys, S., Tamosiunaite, M.: Biologically Inspired Architecture of Feedforward Networks for Signal Classification. In: Ferri, F.J., Inesta, J.M., Amin, A., Pudil, P. (eds.): Advances in Pattern Recognition. Lecture Notes in Computer Science, Vol. 1876. Springer, Berlin Heidelberg New York (2000) 727-736
6. Gotman, J., Flanagan, D., Zhang, J., Rosenblatt, B.: Automatic Seizure Detection in the Newborn: Methods and Initial Evaluation. Electroenceph. clin. Neurophysiol. **103** (1997) 356-362
7. Roessgen, M., Zoubir, A.M. and Boashash, B.: Seizure Detection of Newborn EEG Using a Model-Based Approach. IEEE Trans. Biomed. Eng. **45** (1998) 673-685
8. Kaipio, J.P, Karjalainen, P.A.: Simulation of Nonstationary EEG. Biol. Cybernetics **76** (1997) 349-356
9. Popivanov, D., Mineva, A., Dushanova, J.: Tracking EEG Signal Dynamics During Mental Tasks. IEEE Eng. Med. Biol. **17** (1998) 89-95
10. Stam, C.J., Pijn, J.P.M., Suffczynski, P., Lopes de Silva, F.H.: Dynamics of the Human Alpha Rhythm: Evidence for Nonlinearity? Clin. Neurophysiol. **110** (1999) 1801-1813

11. Anderson, C., Stolz, E., Shamsunder, S.: Multivariate Autoregressive Models for Classification of Spontaneous Electroencephalogram During Mental Tasks. IEEE Trans. Biomed. Eng. **45** (1998) 277-286
12. Pradhan, N., Sadasivan, P.K., Arunodaya, G.R.: Detection of Seizure Activity in EEG by an Artificial Neural Network: A Preliminary Study. Computers and Biomed. Research **29** (1996) 303-313
13. Petrosian, A., Prokhorov, D., Homanet., R. et.al.: Recurrent Neural Network Based Prediction of Epileptic Sezures in Intra and Extracranial EEG. Neurocomputing **30** (2000) 201-218
14. Koskela, T., Varsta, M., Heikkonen, J., Kaski, K.: Temporal Sequence Processing with Recurrent SOM. Proc. of 2nd Int. Conf. on Knowledge-Based Intelligent Engineering Systems KES'98, Adelaide, Australia Vol. 1 (1998) 290-297
15. Varsta, M., Heikonen, J., Del, J., Millan, R.: Epileptic Activity Detection in EEG with Neural Networks. Proc of 3rd Int. Conference on Engineering Applications on Neural Networks EANN (1997) 179-186
16. Weng, W., Khorasani, K.: An Adaptive Structure Neural Networks with Application to EEG Automatic Seizure Detection. Neural Networks **9** (1996) 1223-1240
17. James, C.J., Jones, R.D., Bones, P.J., Carroll, G.J.: Detection of Epileptic Discharges in the EEG by a Hybrid System Comprising Mememtic, Self-Organized Artificial Neural Network and Fuzzy Logic Stages. Clin. Neurophysiol. **110** (1999) 2049-2063
18. Galicky, M., Witte, H., Eiselt, M., Griessbach, G.: Common Optimization of Adaptive Preprocessing Units and a Neural Network During the Learning Period: Application in EEG Pattern Recognition. Neural Networks **10** (1997) 1153-1163
19. Milan, J., Mourino, J., Babiloni F. et.al.: Local Neural Classifier for EEG-Based Recognition of Mental Tasks. In Proc. IEEE-INNS-ENNS International Joint Conference on Neural Networks, Como, Italy, 24-27 July (2000)
20. Thomasson, N., Hoeppner, J., Webber, C.L., Zbilut, J.P.: Reccurence Quantification in Epileptic EEGs. Phys. Letters A **279** (2001) 94-101
21. Simson, M.: Use of Signals in the Terminal QRS Complex to Identify Patients with Ventricular Tachycardia After Myocardial Infarction. Circulation **64** (1981) 235-242
22. Raudys, S.: Evolution and Generalization of a Single Neuron. SLP as Seven Statistical Classifiers. Neural Networks **11** (1998) 283-296
23. E.Pranckevičienė, M.Tamošiūnaitė: Improvement of Classification of EEG Segments by Voting-Based Procedure. Proc. of the Int. Conf. on Biomedical Engineering, October 19-20, Kaunas University of Technology (2000) 91-95

Discrimination between Two Mental States (Rest and Motor Image) Using Wavelet Transform and Neural Networks

Jose L. Martin[1], Luciano Boquete[1], Carmen Diaz[1], Manuel Garcia de Leon[2], and Manuel Mazo[1]

[1] Electronics Department, Universidad de Alcala
Alcala de Henares, Madrid, Spain
{jlmartin, boquete, mazo}@depeca.alcala.es
[2] Neurophysiology Department, Hospital Gomez Ulla
Madrid, Spain

Abstract. This paper presents a method for the processing and classification of electroencephalographic (EEG) signals linked to mental states (rest and motor image) using the wavelet transform of these signals as input information of an LVQ neural network. This system obtained a 70% correct qualification rate in the first recording session, a 50% rate in the second and an 80% rate in the third, with a 75% classification success rate for the whole set of data. These results fall within the average range obtained by other systems which require more information.

1 Introduction

A motor image is part of a wide-ranging phenomenon (motor representation) bound up with the attempt to move and the preparation therefor. It has been suggested that there is a closed functional relationship between an imagined movement and the executed movement. They have a similar duration and similar consequences on such somatic parameters as the breathing rate and heartbeat [1], [2]. Many authors have therefore been prompted to affirm that the simulation of a movement or motor image is the same as the actual realization of that movement, except that the action is blocked.

Distinguishing EEG patterns related to different mental states or even different motor images could allow a paralyzed person to control devices such as wheelchairs or computers.

In this paper, the results of the classification of two different mental states, motor-image and rest, are presented. The proposed system consists of two parts: firstly, EEG signals are processed by means of wavelet transform and then, the resulting information is classified by an ANN. Wavelet transform is a preprocessing method used to obtain relevant pattern from EEG signal. In this case, first level detail coefficients have been employed. Classification has been made by an LVQ Kohonen's ANN.

B. Reusch (Ed.): Fuzzy Days 2001, LNCS 2206, pp. 28–32, 2001.

2 Methods

2.1 Data Collection

EEG signals were collected for the purpose of developing and testing the system mentioned above. The subject of the experiments was a healthy, thirty-five-year-old, right-handed woman. The data were collected using the following procedure:

The subject was asked to relax and try to think of nothing in particular during the first part of the experiment. The beginning of the experiments was indicated with the command "we are starting". Six seconds later, the subject was told to start thinking about a motor action (closing the right hand) with the indication "now". Six more seconds of recording were taken, after which the subject was told to relax once more, this then serving as the start phase for the next attempt.

This process is shown in figure 1.

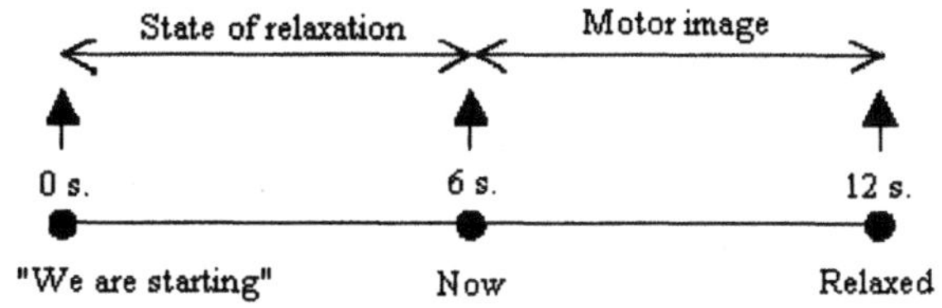

Fig. 1. Experiment protocol

A 19-channel Biologic brain map was used for the EEG recordings. Signals were captured with Ag/AgCl contact electrodes, monopolar in form, i.e., related to one reference electrode: in our case we used the two ears united. The international 10-20 system for electrode placement was used, with information being read from the 19 channels (Fp1, Fp2, F7, F3, Fz, F4, F8, T3, C3, Cz, C4, T4, T5, P3, Pz, P4, T6, 01, 02), with the montage Φ, group A, programme 12 of Biologic.

2.2 Data Choosing

Three recording sessions were done involving an average of 40 attempts per session. 170 records were taken, 80 of which were considered for processing purposes, those with fewest artifacts and wherein the synchronic alpha reading was visible in the first half of the records. In these experiments, the reference elimination method used was the common mean reference, whereby the mean of all the scalp electrodes is subtracted from the potential measured in the electrode in question.

The valid records of each session were divided into two groups: one group used to train the neural network and the other used for simulations. The first session produced 36 valid pairs of signals (rest and action signal), from which 26 were taken to train the network and 10 for simulation. Session 2 produced 20 valid pairs; the network was trained with 14 of them, the remaining pairs being used for simulation. The analysis of session 3 produced 24 valid pairs, from which 17 were used for training and the rest for simulation. Finally, the whole set of signals obtained in the experiments were also separated, 70% of the valid files were used for training and the rest for simulation.

The data processing segment of the rest and motor-image signal is 2 seconds. Two seconds are taken from the six-second segment corresponding to the rest interval and two seconds from the motor action interval immediately after the second six, corresponding to the moment when the subject is told to start thinking about the motor action.

2.3 Data Processing and Classification

Preprocessing: Wavelet transform. Wavelet transform (see equation 1) is specially interesting for EEG analysis since it can detect transient waveforms, that are almost impossible to detect with other analysis. In this case, Discrete Wavelet Transform (DWT) has been employed for preprocessing EEG records. With this transform, filters of different cutoff frequencies are used to analyze the signals at different scales. The signal is passed through a series of low pass and high pass filters in order to separate its low frequency and high frequency components. High frequency components (called details or detail coefficient) are obtained with good time resolution and low frequencies (called approximations or approximation coefficients) with good frequency resolution.

$$WTx(\tau, a) = \frac{1}{\sqrt{a}} \int_{-\infty}^{\infty} x(t)\, h * \left(\frac{t-\tau}{a}\right) dt \quad (1)$$

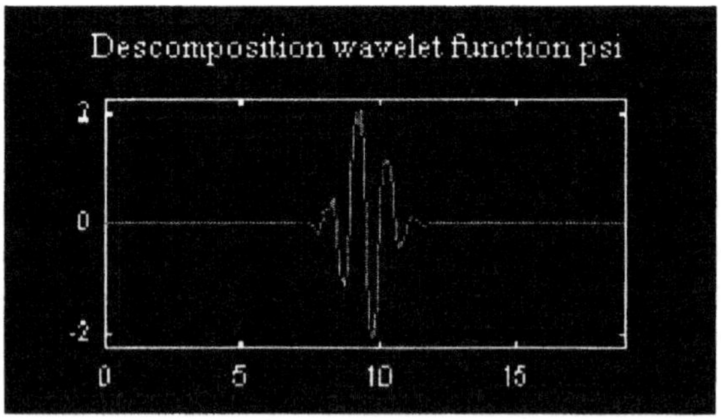

Fig. 2. Representation of the wavelet biorthogonal 3.9

There are different wavelet families (i.e., function h in equation 1) that can be used for this analysis. Wavelet family choosing is a critical task. Results are better when the signal and the wavelet show remarkable similarity.

In this study, the collected data have been processed with the Biorthogonal Wavelet 3.9 [3] (see figure 2) choosing the first detail level wavelet coefficients (cd1) of the channels corresponding to Fp1, Fp2, F7, F3, Fz, F4, F8, C3, Cz, C4 as input information for the classification method (figure 3).

Classification. This classification was performed by Kohonen's LVQ method [4], [5], [6], which is a simple neural classifier based on Kohonen's work with self-organising maps. The network was trained to differentiate between the two mental states (rest and motor action).

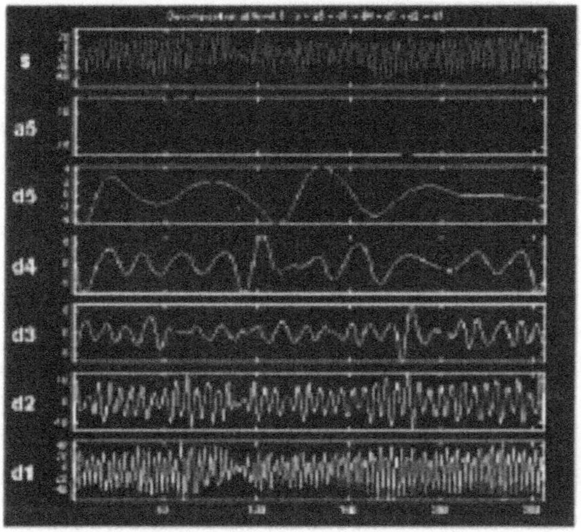

Fig. 3. EEG breakdown of frontal electrode F3 (action file) with biorthogonal wavelet 3.9

3 Results

As it is shown in table 1, the first session gave a correct classification rate of 70% . This result may have been influenced by the fact that the subject had had little training and also by the limited amount of training information given to the network, this being common to all three sessions. Session 2 shows a drop in the success rate, this time being due to the greater number of non-valid files (interference from muscular noise in the frontal electrodes). The subject also admitted to having a lower concentration level in this session. Session 3 shows the best success rate, 80% , with a number of inputs similar to session 1 with artifact-free records. Lastly, a classification is made of the whole set of sessions, giving a 75% classification success rate. In this case the poor results of session 2 affect the overall results.

Table 1. Classification of mental states

Experimental Sessions	Training set pairs/files	Simulation set pairs/files	Correct classification percentage
Session 1	26/52	10/20	70%
Session 2	14/28	6/12	46%
Session 3	17/34	6/12	80%
Total Sessions	57/114	22/44	75%

4 Conclusion and Future Works

The results obtained in most sessions fall within the best classification averages obtained by other authors [7], but in this case with a shorter training time for the experiment subject, and also less input information given to the network.

The use of the Wavelet transform for processing the 2-second signal segments (256 EEG data points) means that network inputs can be effected with half of this information (wavelet coefficients of the first detail level), thus optimizing the latter, i.e., reducing the amount of input information without forfeiting discrimination information.

The results of the second session confirm the need to make a careful choice of records for network training, while the results of the third session indicate the importance of training the subject in the task to be carried out, thus improving classification results.

5 References

1. Decety, J., Jeannerod, M., y Prablanc, C. "The timing of mentally represented actions". Behav. Brain Res. 1989, 34, 35-42.

2. Decety, J. Lindgren, M. "Sensation of effort and duration of mentaly executed actions". Scand. J. Psychol. 1991, 32, 97-104.

3. Mathworks Wavelet Toolbox for Matlab.

4. Kohonen, T.(1988). "Self organization and associative memory" (Springer Verlag. Berlin) 3ra edn. [5] Kohonen, T. (1990). "Self organizing map". Proc. IEEE.78, (9), pp 1464 – 1480.

5. Flotzinger, D.; Kalcher, J.; and Pfurtscheller, G.(1992). "EEG classification by learning vector quantization". Biomediziniche Technik, 37, pp303-309.

6. Neuronal Network. Toolbox. The Mathworks Inc, Howard Demuth, Mark Beale.

7. Jack Culpepper. Discriminating Mental States using EEG Represented by Power Spectral Density.1999.

Decision Making in Diagnosis of Oral Mucosa Lesions

Tatiana Kiseliova[1,3] and Hajo Peters[2]

[1] University of Dortmund, Department of Computer Science I,
D-44221 Dortmund, Germany
phone: +49 231 755 6327
fax:+49 231 755 6555
kiseliova@ls1.informatik.uni-dortmund.de
[2] University of Witten/Herdecke Faculty of Dental Medicine
D-58448 Witten, Germany
phone: +49 2302 926 660
fax:+49 2302 926 661
hajo.peters@hpdent.de
[3] Ivane Javakhishvili Tbilisi State University
Chavchavadze, av. 1., 380028 Tbilisi, Republic of Georgia

Abstract. The problem of decision making in diagnoses of oral mucosa lesions is considered. It is shown that computer-assisted diagnosis systems can be very helpful for clinicians in this branch. In the case of the presented problem for oral mucosa lesions an expert knowledge is used in formalizing the initial information, as well as in the estimation of indicants of the patient. The presented method is used as an alternative to avoid many problems associated with probabilistic and statistical approaches. It handles the expert information in a natural way. The diagnosis is made considering the weight of each symptom. It is shown how the proposed method can be explained from the theory of fuzzy relations point of view.

1 Introduction

The central aim of decision support systems in medicine is to improve the quality of patient care by assisting the clinician to make better decisions.

The advantages of successful decision-support systems is bringing to the individual clinician the experience of many by pooling this clinical experience within databases of the decision-support systems [2]. Further, the use of decision-support systems may increase a clinician's performance simply on the basis of being thorough in the process of data collection [3].

The most common means of data analysis used in the decision-support systems to link patient information to the store of medical knowledge are algorithmic systems, statistical systems, rule-based systems, neural networks and some others [11].

By use of logical classification methods algorithmic systems lead the user to the desired endpoint within a decision trees or flowcharts.

B. Reusch (Ed.): Fuzzy Days 2001, LNCS 2206, pp. 33–39, 2001.

Statistical methods [6, 9] can be used to compare patient's signs and symptoms with a database and to calculate a diagnosis. Successful one has been the probabilistic method, Bayesian classification, that uses the probabilities associated with the occurrence of signs, symptoms or laboratory values to compute a differential diagnosis.

The most common type of decision-support systems are rule-based, moving from premise to conclusion in one or more "*if ... then*" statements. The knowledge in these systems is stored as production rules or heuristics also called rules of thumb [9].

Development of decision-support systems in dentistry has been active for over two decades [1, 8].

These systems have employed all major types of decision-support systems designs.

They covered dental topics like emergencies, orofacial pain, restorative dentistry, oral radiology and oral medicine etc.

2 Medical Problem

In the oral cavity diseases exist which originate from hard tissue such as teeth and bone, or from oral mucosa and soft tissue [7, 10]. The clinical manifestation of common dental diseases like caries and periodontitis can easily be recognized by their characteristic appearance. On the contrary, oral soft tissue alterations can be triggered by different factors which all produce same or similar clinical manifestations. These causing factors include: local traumatic injuries of physical or chemical nature as well as infections of the digestive or respiratory tract and systemic diseases with oral symptoms. Most importantly oropharygeal tumours and HIV-induced oral lesions play an important role in early diagnostic differentiation of oral mucosa lesions.

Oral mucosa lesions are primarily characterized by unphysiological colour changes, traumatized or destroyed surface epithelium or tissue enlargement. Among secondary clinical factors are specific locations, lesion shape and tissue contours, size, consistency and texture, as well as pain or altered function. Contributing factors which must be taken into consideration when a lesion has developed include: age, race and gender of the diagnosed patient as well as oral habits, recent injuries, infections or operations.

The clinical examination of oral mucosa lesions, according to the criteria listed above provides a diagnostic pattern of clinical parameters. These parameters are weighed differently. Primary disease manifestations are more important than contributing factors and therefore have greater influence on the diagnosis.

The condition considered the most likely cause of the lesion is referred to as the working or preliminary diagnosis. The working diagnosis and the less likely diagnostic possibilities provide the basis for additional diagnostic procedures, such as biopsy or clinical laboratory tests, and for the initial clinical management of the condition.

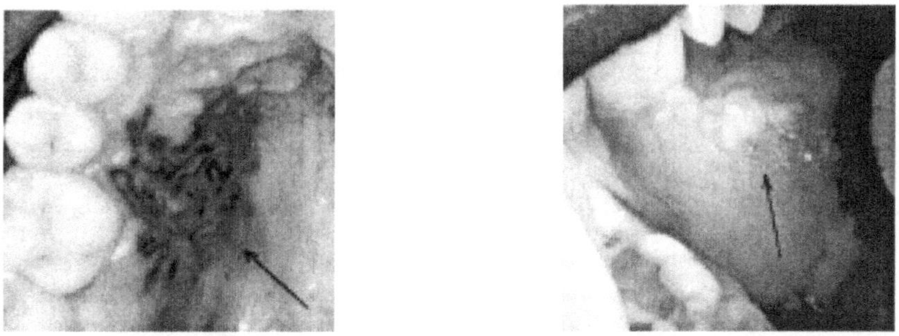

Fig. 1. Examples of red and white oral mucosa lesions

The problem with clinical diagnosis of oral mucosa lesions is to recognize and consider all possible causes. Only after all clinical parameters have been regarded, comparable results of diagnostic certainty can be achieved.

Nevertheless, the final diagnosis of oral mucosa lesions can only be given after histological examination. The quality of the clinical diagnosis can however be significantly improved by a standardized examination protocol and computerized diagnostic assistance, in other words, a decision support system. A diagnostic computer program can aid less experienced clinicians within the broad field of oral mucosa lesions.

3 Representation of the Initial Data

The main goal of this decision support system is a diagnostic task shortly formulated "from symptoms to diagnoses". The specific symptoms of different oral mucosa lesions are collected from the dental literature [7, 10] and presented as a disease-characterizing-table (Table 1).

Each symptom is weighed according to its importance (scaled from 1 to 5) for different diseases. For example colour plays an important role in the differentiation of oral mucosa lesions. Table 1 describes the symptoms for a particular disease called Pseudomembraneous Candidiasis (PC). The sum of all these elements shows the "weight" of a disease. For PC from Table 1 it is 33. About 15 diseases are presented in the investigation, each of them characterized by occurrence or absence of 26 symptoms. It is suggested that the effects of each symptom are isolated from those of the others and overall effect on the outcome could be expressed as an algebraic sum [4]. Therefore at first the examining clinician considers if a symptom is present (1) for a particular patient or not (0). This is done for all 26 symptoms. Each disease-characterizing-table (with differently weighed symptoms) is calculated against this patient examination (see Example 1). This way every disease gets a "label" (a number) that shows belief to this disease for the patient. Assuming that the maximum of these labels corresponds to the most believable disease for the patient, the practical clinician gets an information of the preliminary diagnosis.

Table 1. Importance of symptoms for PC

	Symptoms	Pseudomembranous Candidiasis (PC)
COLOUR	white	5
	red	0
	yellowish	0
	pigmented	0
LOCALISATION	lip	0
	labial angle	0
	buccal plane	5
	vestibule	5
	alveolar ridge / gingiva	0
	tongue	4
	floor of the mouth	3
	palate	4
	palatal arch	0
EXTENSION	pinpoint	0
	wide	4
PAIN	painless	3
	tender	0
	painful	0
...	...	...

4 Mathematical Model

4.1 "Black and White" Symptoms

Let $\Delta = \{\mathbf{D}_1, \ldots, \mathbf{D}_n\}$ be a set of diseases, $\Sigma = \{\mathbf{S}_1, \ldots, \mathbf{S}_m\}$ be a set of symptoms, $\Pi = \{P_1, \ldots, P_p\}$ be a set of patients, w_{ij}, $i = 1, \ldots, n, j = 1, \ldots, m$ take the values from $\{0, \ldots, 4, 5\}$; $w_{ij} = 1, \ldots, 5$ means that symptom $\mathbf{S}_i$ is present for disease $\mathbf{D}_j$ and its weight takes values from 1 to 5; $w_{ij} = 0$ means that symptom S_i is not present; the information about a patient to whom a diagnose is established is present as $P = (p_1, \ldots, p_m)$, where $p_i = 1$ if the patient has symptom S_i and 0 if not; $\delta = (\delta_1, \ldots, \delta_n)$ is a distribution over the diseases, where δ_i, $i = 1, \ldots, n$ can be interpreted as a degree of belief to disease $\mathbf{D}_i$ for this patient [5].

The scheme of the proposed method is:

$$\begin{array}{cccc} & \mathbf{D}_1 & \mathbf{D}_2 \ldots & \mathbf{D}_n \\ \mathbf{S}_1 & w_{11} & w_{12} \ldots & w_{1n} \\ \vdots & & w_{ij} & \\ \mathbf{S}_m & w_{m1} & w_{m2} \ldots & w_{mn} \end{array} \xrightarrow{P=(p_1,\ldots,p_m)} \delta = (\delta_1, \ldots, \delta_n) \tag{1}$$

This can be interpreted as follows: the initial table w_{ij} is compared with the patient set of symptoms (multiplication of the vector of the patient on the matrix) and each element δ_j of the getting vector δ shows the degree of belief to the disease D_j.

Example 1.

$$\begin{array}{cccc} & \mathbf{D}_1 & \mathbf{D}_2 & \mathbf{D}_3 \\ \mathbf{S}_1 & 5 & 0 & 4 \\ \mathbf{S}_2 & 2 & 3 & 5 \\ \mathbf{S}_3 & 0 & 2 & 5 \end{array} \rightarrow (1\ 0\ 1) \times \begin{pmatrix} 5\ 0\ 4 \\ 2\ 3\ 5 \\ 0\ 2\ 5 \end{pmatrix} = (5\ 2\ 9) \tag{2}$$

The clinician can consider $\mathbf{D}_3$ as a more believable disease. (See explanations of the previous section).

4.2 Patient's Information

The "yes/no" symptom‘s representation of the diagnosed patient seems rather insufficient. It is the matter-of-course, that some other possibilities should be assumed, for example, *"I have forgotten to fulfil this place of the table"*, or *"this symptom does not matter at all"*, or *"the symptom is only partly present"*.

Another scale for the investigated patient is introduced, $p_i \in \{-1, 0, 0.3, 0.6, 1\}$: a symptom is *not present* (−1), a symptom is *not defined* (0) (inclusive *"I have forgotten to fulfil this place of the table"* or *"this symptom does not matter at all*). For the symptom‘s appearance three possibilities 0.3, 0.6, 1 are considered: a symptom is *slightly expressed, more expressed, well expressed* correspondingly.

The positive elements in the patient‘s vector enhance the strength of the proposal disease, "0" remains it unchangeable and the negative ones reduces this strength.

Example 2.

$$(0.6 \; -1 \; 1) \times \begin{pmatrix} 5 & 0 & 4 \\ 2 & 3 & 5 \\ 0 & 2 & 5 \end{pmatrix} = (1 \; -1 \; 2.4) \qquad (3)$$

4.3 Fuzzy Relations

This diagnostic process can be described in particular, with fuzzy relations [12, 13].

Assume that the initial matrix is a fuzzy relation $R_{SD} : \Sigma \times \Delta \to [0,\ 1]$.

Note. Due to the definition of fuzzy relations, the scale $0,\ 1, \ldots, 5$ should be mapped to the interval $[0,\ 1]$ as well as scale $-1, 0, 0.3, 0.6, 1$

The information of the diagnosed patient can be presented as fuzzy relation $R_{PS} : \Pi \times \Sigma \to [0,\ 1]$.

The process of establishing of the diagnoses is a composition of fuzzy relations (in particular, a composition of a fuzzy set and a fuzzy relation):

$$R_{PD} = R_{PS} \circ R_{SD} \qquad (4)$$

Operation $\circ$ can be interpreted differently, accordingly to the practical demands. R_{PD} is a fuzzy relation, $R_{PD} : \Pi \times \Delta \to [0,\ 1]$.

In the Example 1 "max"-defuzzification method for the final result is used.

5 Conclusions

In the presented paper the problem of decision making in diagnoses of oral mucosa lesions is considered. It is shown that computer-assisted diagnosis system can be very helpful for clinicians in this branch.

When someone tackles the problem of establishing a medical diagnosis or, more general, constructing the decision support medical system with numerical tabular knowledge base, the first question is if there are enough data to use standard methods, namely, based on Bayesian probability inference, other statistical ones.

In the case of the presented problem for oral mucosa lesions an expert knowledge is used in formalizing the initial information, as well as in the estimation of indicants of the patient. The conditional probability table is absent, independent and dependent symptoms are not established and there are not enough statistical data to use above mentioned methods.

The presented method was used as an alternative to avoid many problems associated with probabilistic and statistical approaches. It handles the expert information in a natural way.

The diagnosis is made considering the weight of each symptom.

In the paper it was shown how the proposed method can be explained from the theory of fuzzy relations point of view.

Because of great interest and enthusiasm of the practical clinicians it was possible to collect almost all necessary information about the subject, to formalize the problem and to use one of the simple mathematical models, that was understood and successfully tested by clinicians.

The method is under practical realization now.

Acknowledgement

The authors would like to thank the anonymous referees for their helpful comments and B. Burghardt for his help in preparing this paper.

References

1. M. F. Abbod, D. G. von Keyserlingk, D. A. Linkens, and M. Mahfouf. Survey of utilisation of fuzzy technology in Medicine and Healthcare. *Fuzzy Sets and Systems*, 120:331–349, 2001.
2. G. Fenyö. Computer-aided diagnosis and decision making in acute abdominal pain. *Dig Dis*, 8:125–37, 1990.
3. IR. Gough. Computer-assisted diagnosis of the acute abdominal pain. *Aust N Z J Surg*, 63:699–702, 1993.
4. G. H. Hall and A. P. Round. Logistic regression - explanation and use. *Journal of the Royal College of Physicians of London*, 28(3), May-June 1994.
5. T. Kiseliova and B. Reusch. Fuzzy decision support method for medical diagnoses. In R. J. Patton and J. Chen, editors, *IFAC Symposium on Fault Detection, Supervision and Safety for Technical Processes, SAFEPROCESS'97*, volume 2, pages 938–943. The University of Hull, August 26–28, 1997.
6. C. Naylon. How to build an inferencing engine. In Richard Forsyth, editor, *Expert Systems, Principles and Case Studies*, chapter 5, pages 85–105. Champman and Hall, 1989.
7. J. J. Pindborg. *Farbatlas der Mundschleimhaut-Erkrankungen.* Deutscher Ärzteverlag, 1993.
8. MA. Siegel, FJ. Firrio, and MW. Finkelstein. Computer applications in oral diagnosis. *Dent Clin North Am*, 37:113–31, 1993.

9. SE. Stheeman, P. van der Stelt, and PA. Mileman. Expert systems in dentistry. past performance -future prospects. *J. Dent*, 20:68–73, 1992.
10. M. Strassburg and G. Knolle. *Farbatlas und Lehrbuch der Mundschleimhauterkrankungen.* Quintessenz, 1991.
11. C. Stuart and D. D. S. White. Decision Support Systems in Dentistry. *Journal of Dental Education*, 60(1):47–63, 1996.
12. K. Tanaka. *An Introduction to Fuzzy Logic for Practical Applications.* ISBN 0-387-94807-4. Springer, 1996.
13. H.-J. Zimmermann. *Fuzzy Set Theory and Its Applications.* ISNB 0-7923-9075-X. Kluwer Academic Publishers, 3300 AH Dordrecht, The Netherlands, 1991.

Fuzzy Logic and the Internet: Lingustic Summarization of Distributed Sets of Data

Janusz Kacprzyk

Systems Research Institute, Polish Academy of Sciences
ul. Newelska 6, 01-447 Warsaw, Poland
kacprzyk@ibspan.waw.pl

As hardware and software limitations become less and less rigid, users tend to maintain and use larger and larger databases. Unfortunately, sooner or later they run into a problem that databases become too large to grasp by a human user, hence useless in practice.

Conventional tools and techniques provide adequate means for the acquisition, processing and reporting of data in the traditional sense. However, for the human being the only natural communication means is natural language which, in spite of its inherent imprecision, does provide useful information. Moreover, natural language can best reflect the very essence (in an extremely condensed form) of information.

Fuzzy logic has been advocated for this purpose for some time, and Yager's (1982) [cf. Kacprzyk and Yager (2000)] approach may be viewed as one of most relevant proposals; we will also employ its idea here. Basically, Yager derives linguistic database summaries as linguistically quantified propositions, e.g., "most of the employees are young and well paid", with a degree of validity (truth, ...), in case of a personnel database.

This may be viewed as a manifestation of the computing with words paradigm [cf. Zadeh and Kacprzyk (1999a, b)], and we present a new approach to such a linguistic summarization explicitly based on that paradigm.

First, since the number of attributes is usually very high, a fully automatic derivation of summaries is not realistic because the system does not know by itself what (class of) summaries the user may be interested in. Instead, we advocate some human assistance [cf. Kacprzyk and Zadrozny (2000)]. First, we assume that the user specifies the subject of the summary and the attributes of interest. Moreover, the user can also specify more sophisticated structures of summaries as, e.g., involving fuzzy relations. Basically, these are equivalent to the formulation of a database query with a fuzzy linguistic quantifier - cf. Kacprzyk and Zadrony's (1997, 2000) FQUERY for Access, a fuzzy logic based add on to Microsoft Access. A recent extension of FQUERY for Access that incorporates elements of Bosc and Pivert's (1992, 1995) SQLf extension of SQL is mentioned and shown that it may increase the efficiency of the summarization process [cf. Kacprzyk and Zadrony (2001)].

Having defined the interesting class of a linguistic summary, we define for each possible instantiation of the summary from that class its degree of quality [cf. Kacprzyk and Yager (2000)], and find the best summary.

B. Reusch (Ed.): Fuzzy Days 2001, LNCS 2206, pp. 40–42, 2001.

We present some results of implementation of the proposed method for the derivation of linguistic summaries of a database concerning sales in a medium size computer retailer in Southern Poland [cf. Kacprżyk and Strykowski (1999)]. The summaries obtained have been very positively accepted by the owner, and have given much valuable insight into diverse aspects of planning and day to day activities.

Notice that the above data summaries are derived on the basis of database contents. However, there also exist some "external" data available from distributed databases and mostly accessible through the Internet. These contain, for instance, data from climatic and weather databases, financial databases, etc. These pieces of data may provide very relevant tools for deriving qualitatively different, more general and relevant types of linguistic summaries, called here linguistic metasummaries that basically summarize own data within a broader context of outside data.

We present a formal problem statement for the derivation of such linguistic metasummaries, and then consider basic issues related to their implementation. In particular, we discuss the implementation through the technology of intelligent agents. We show how both free and commercial external sets of data can be used. We also mention how to deal with problems with data sets that contain, e.g., contradictory information.

We show the derivation of such metasummaries for a computer retailer when, additionally, temperature and precipitation information (available from the Internet, using intelligent agents) is employed.

Clearly, the approach proposed may find many uses in all kinds of human-centered computer systems, notably in e-commerce, decision support, etc.

Bibliography

Bosc P. and O. Pivert (1992) Fuzzy querying in conventional databases, in L.A. Zadeh and J. Kacprzyk (Eds.) *Fuzzy Logic for the Management of Uncertainty*, Wiley, New York, pp. 645–671.

Bosc P. and O. Pivert (1995) SQLf: a relational database language for fuzzy querying, *IEEE Trans. On Fuzzy Syst.* Vol. 3, pp. 1-17.

Bosc P. and O. Pivert (2000) SQLf query functionality on top of a regular relational database management system, in O. Pons, M.A. Vila and J. Kacprzyk (Eds.): *Knowledge Management in Fuzzy Databases*, Physica-Verlag (Springer-Verlag), Heidelberg and New York, pp. 171–190.

Kacprzyk J. and Strykowski P. (1999) Lingusitic summaries of sales data at a computer retailer: a case study. *Proceedings of IFSA'99* (Taipei, Taiwan R.O.C), Vol. 1 pp. 29-33.

Kacprzyk J. and R.R. Yager (2000) Linguistic summaries of data using fuzzy logic. International Journal of General Systems (in press) linguistic summaries of databases, International Journal of Applied Mathematics and Computer Science, Vol. 10, pp. 813–834.

Kacprzyk J. and S. Zadrony (1997a) Fuzzy queries in Microsoft Access v. 2. In: D. Dubois, H. Prade and R.R. Yager (Eds.): Fuzzy Information Engineering - A Guided Tour of Applications, Wiley, New York, pp. 223 - 232. implementation for

Microsoft Access", in T. Andreasen, H. Christiansen i H.L. Larsen (Eds.): Flexible Query Answering Systems, Kluwer, Boston, pp. 247–275. intelligent database querying, in: L.A. Zadeh, J. Kacprzyk (Eds.), Computing with Words in Information/intelligent Systems. Part 2. Applications. Physica-Verlag, Heidelberg, pp. 383-398. the Internet: An implementation", in Proceedings of the Fourth International Conference on Knowledge-Based Intelligent Engineering Systems & Allied Technologies (KES'2000), Brighton, UK, pp. 704-707. querying and mining of association rules, in H.L. Larsen, J. Kacprzyk, S. *Flexible Query Answering Systems. Recent Advances*, Physica-Verlag (Springer-Verlag), Heidelberg New York, pp. 475-484.

Kacprzyk J. and S. *Proceedings of IFSA World Congress*, Vancouver, Canada (in press).

Kacprzyk J. and R.R. Yager (2000) Linguistic Summaries of Data Using Fuzzy Logic, *International Journal of General Systems* (in press)

Yager R.R. (1982) A new approach to the summarization of data. *Information Sciences*, 28, 69–86.

Zadeh L.A. and J. Kacprzyk (eds.) (1999) *Computing with Words in Information/ Intelligent Systems 1. Foundations. 2. Applications.* Physica-Verlag (Springer-Verlag), Heidelberg and New York, 1999.

Internet/WWW: java applets, activeX controls and cookies, in: H.L. Larsen et al. (eds.) *Flexible Query Answering Systems, FQAS'98.* Springer-Verlag, Berlin Heidelberg 1998, pp. 382-392.

Adaptive Color- and Edge-Based Image Segmentation Using Fuzzy Techniques

Christian Leubner

University of Dortmund
Department for Computer Sciences, Computer Graphics Group
Postfach 500500
D-44221 Dortmund, Germany
leubner@ls7.cs.uni-dortmund.de
http://ls7-www.cs.uni-dortmund.de

Abstract. In this contribution a new segmentation technique is presented that is able to segment moving objects in front of a static, but arbitrary background. The system acquires information about the background automatically and is also able to adapt to slight changes within the background, for example caused of changing illumination conditions, automatically. Our technique uses a combination of color- and edge-based segmentation and stores the respective fuzzy information pixelwise. The available knowledge is employed to yield a stable and reliable result using fuzzy techniques.

1 Introduction

Human-machine interaction is an important part of nearly all todays technical devices. To be accepted by the user they should be easy to use and the interaction should be comfortable. Computer vision is one of the key techniques to realize user-friendly interfaces because it enables a natural and wireless interaction for example by observing gestures of the user. The user does not need to learn the handling of specific input-devices like mouse or keyboard and is free to move. He can simply use his arms or his hands to interact with the machine. Thus the ability to find and determine the presence and position of the human body or parts of the human body within a camera image is a central and important task in computer vision. Moreover it is required that the recognition can be performed multiple times per second, ideally in realtime (25 frames per second).

A central task of the image processing pipeline executed in this setting is image segmentation. Segmentation means to separate in a given image an object of interest from the background. In order to achieve a stable segmentation most systems have a couple of restrictions: the user for example must not wear clothes having a specific color, or the background in front of which the segmentation is performed is not allowed to consist of arbitrary colors. Furthermore many systems need specific expert knowledge at least at the start of the system to work properly.

B. Reusch (Ed.): Fuzzy Days 2001, LNCS 2206, pp. 43–54, 2001.

We present a fuzzy image segmentation method which is able to segment reliably a moving object (for example a human arm) in front of a more or less static, but arbitrary background. Our method does not need any manually given information at starting time. The system is able to adapt to slight changes within the background automatically, for example to changing illumination conditions. Our approach uses a combination of both color and edge segmentation to achieve meaningful and reliable results. Fuzzy techniques are applied in order to cope with the fact that digital image data contains a lot of noise and image filtering techniques do not yield perfect results.

Color and edge information is stored for each pixel separately requiring a notable amount of main memory. The segmentation itself is performed on the basis of this knowledge, superseding the need for global segmentation criteria as for example knowledge about the specific color of the human skin. As a consequence our approach does not need any prior information, neither on the background nor on the foreground objects, and can cope with an arbitrary static background situation. Our method yields two different kinds of foreground object information which are combined using fuzzy techniques. The segmentation result is employed to realize an adaptive improvement of the available knowledge by continuously learning the identified background regions. Thus it can be expected that the segmentation results will improve, the longer the algorithm is in use. Nevertheless the background should not contain too many edges because in that case the edge segmentation might not be able to find foreground edges in front of background edges. Furthermore the color segmentation may not work properly if background and foreground object color are almost identical at a large amount of pixels. Generally color and edge segmentation are intended to complement one another, but in extreme cases the segmentation might partly fail.

Segmentation itself is as old as image processing is. A good survey of early used techniques is given in [6] from 1985. More recent surveys are [4], [8] and [11]. The usage of fuzzy techniques for image processing purposes is described in [12].

Because of the variety of known segmentation methods and in order to compare the introduced algorithm to known techniques, it makes sense to concentrate on those techniques which show a similar approach:

- static background
- automatic learning of background information
- segmentation of moving object in front of known background
- adaptive to changes in background
- suitable for interaction purposes
- using both color and edge information.

In [2] the segmentation task is identical. A statistical approach on object and background probability is used to perform the segmentation. In [3] good results were achieved by combining edge detection and a split-and-merge paradigm with motion segmentation in order to find driving cars. In contrast to our approach, the color distribution is calculated for the whole background [2] or at least for squares of the image [3], whereas our technique collects color information for each pixel separately. A closely related work to [2] can be found in [1] and [13]. The described *PFinder* system models the scene as a set of distinct classes. The person is represented as a set of connected blobs, having a spatial and color Gaussian distribution as well as a support map that indicates for each pixel to which class it belongs. Adaptivity is realized by recursively updating the background statistics. In contrast to our approach, this technique also collects information on the color distribution of the foreground objects. The classification of pixels is done by measuring distances in the color space accomplished by spatial priors and connectivity constraints of the blobs. In [10] a new method for color image segmentation and edge linking is proposed. The basic idea is to find contiguous regions by color segmentation and to split and merge regions so that referring to the edge map a consistent and reasonable result is obtained. This method is not intended to segment moving objects and the segmentation time for one image is specified "a few minutes".

Section 2 introduces some formal aspects, section 3 the fuzzy color segmentation and section 4 the fuzzy edge-based segmentation. The combination of both segmentation results is presented in section 5, the derivation of a contour approximation in section 6. A summary of our approach is given in section 7. In section 8 some examples of application are presented. Finally an overall summary and conclusion is in section 9.

2 Denotations

In the following sections the words *foreground* and *background* are often used. Semantically "foreground" denotes the object that has to be found, whereas "background" terms the static scene if there is no foreground object situated in front of the camera.

The application of our technique is divided into an initial *learning phase* and an *application phase*. During the learning phase the system does not have any information, neither about the background nor about any foreground objects. All recorded images are supposed to be background and are "learned" as it is described in the following sections. During the application phase the system has sufficient information and knowledge to perform the segmentation task. Furthermore, background learning is continued on regions not covered by the foreground object.

3 Color-Based Segmentation

In order to store color information the usage of a fuzzy color knowledge base is proposed.

Definition 1 (Fuzzy color knowledge base) *:*

A color space *is a set Q of N_K-tuples with fixed N_K. A discrete image I is defined as a function:*

$$q : (x, y) \longrightarrow Q \tag{1}$$

with

$$(x, y) \in \mathbb{N}_0 \times \mathbb{N}_0, \quad 0 \leq x < w, \quad 0 \leq y < h$$

where $w \in \mathbb{N}$ *denotes the width and* $h \in \mathbb{N}$ *the height of the image in pixels. A color value* $q(\boldsymbol{p}) = (q_1(\boldsymbol{p}), ..., q_{N_K}(\boldsymbol{p})) \in Q$ *of a pixel* $\boldsymbol{p} = (x, y)$ *consists of* N_K *color channels which may have* N_W *different discrete values.*

A fuzzy color knowledge base *for* I *and* Q *is defined by* N_K *arrays* $C_{\boldsymbol{p},i}$ *containing* N_W *entries for numbers* $\mu \in [0,1] \subset \mathbb{R}$ *(with* $i = 1, ..., N_K$*).* $C_{\boldsymbol{p},i}(j)$ *refers to the* j*-th entry of color channel* i *for pixel* $\boldsymbol{p}$ *(*$j = 0, ..., N_W - 1$*) and is considered to be a fuzzy membership value concerning the semantic property "is background color".*

As an example for Q, the *RGB color space* Q_{RGB} is given by a set of 3-tuples: $Q_{RGB} = \{(r, g, b) \mid r, g, b \in \{0, ..., 255\}\}$.
A color value $q(\boldsymbol{p}) = (q_1(\boldsymbol{p}), ..., q_{N_K}(\boldsymbol{p})) \in Q$ is learned by the rule

$$C_{\boldsymbol{p},i}(q_i(\boldsymbol{p})) = \min(C_{\boldsymbol{p},i}(q_i(\boldsymbol{p})) + \sigma, \ 1) \tag{2}$$

where $\sigma \in [0,1] \subset \mathbb{R}$ and $i = 1, ..., N_K$. The parameter σ affects the impact of a single occurrence of a color on the knowledge base.

During the learning phase a sufficient amount of iterations of learning according to this scheme has to be performed.

A previously trained color knowledge base can be employed to perform a pixelwise fuzzy classification of a current video image I_c. For this purpose the membership values of the semantic property "is background color" which are provided by the fuzzy color knowledge base are used to calculate the membership values of the semantic property "is foreground color" by employing a fuzzy "not" operator. The membership $\mu_c(\boldsymbol{p})$ of a color $q(\boldsymbol{p}) = (q_1(\boldsymbol{p}), ..., q_{N_K}(\boldsymbol{p})) \in Q$ for a specific pixel $\boldsymbol{p}$ to the property "is foreground color" can be gained by looking up the corresponding array entries:

$$\mu_c(\boldsymbol{p}) = 1 - \min_{i=1,...,N_K} C_{\boldsymbol{p},i}(q_i(\boldsymbol{p})) \tag{3}$$

Applying $\mu_c(\boldsymbol{p})$ on every pixel $\boldsymbol{p} \in I_c$ yields a fuzzy segmentation of I_c. In order to remove noise in the fuzzy results we propose the usage of an *averaging threshold* filter that averages the membership values of a pixel $\boldsymbol{p}$ and of the pixels $\boldsymbol{p}_i$ of the 8-neighborhood of $\boldsymbol{p}$. The calculated average membership value is compared to a threshold so that the membership value $\mu_{color}(\boldsymbol{p})$ finally is determined as

$$\mu_{color}(\boldsymbol{p}) = \begin{cases} 0 & : \ \frac{1}{9}\mu_c(\boldsymbol{p}) + \frac{1}{9}\sum_{i=1}^{8}\mu_c(\boldsymbol{p}_i) \ \leq \ v_c \\ \mu_c & : \ \frac{1}{9}\mu_c(\boldsymbol{p}) + \frac{1}{9}\sum_{i=1}^{8}\mu_c(\boldsymbol{p}_i) \ > \ v_c \end{cases} \tag{4}$$

where $v_c \in [0,1] \subset \mathbb{R}$ is used as a threshold parameter.

For illustration purposes the result of the fuzzy classification of an image I_c can be transformed into a grey-value image. For example, an 8-bit grey-level image can be obtained by multiplying the fuzzy membership values $\mu_{color}(\boldsymbol{p})$ by 255. In order to obtain reliable segmentation results the fuzzy color classification results are combined with the later explained edge segmentation.

It has to be taken into account that color values that have been learned once may lose their correctness over time. Noise and other influences may accumulate small membership values in the array entries which may grow to large and interfering membership values over a long time. Thus after a certain amount of iterations of learning the following formula is applied:

$$\forall \boldsymbol{p} = (x,y) \text{ with } 0 \le x < w,\ \ 0 \le y < h;\ \ i = 1,...,N_K;\ \ j = 0,...,N_W - 1:$$

$$C_{\boldsymbol{p},i}(j) = \max(0,\ C_{\boldsymbol{p},i}(j) - \rho) \tag{5}$$

where $\rho \in [0,1] \subset \mathbb{R}$. Similar to the parameter σ in formula (2) the value of ρ regulates the speed of aging and dropping of the acquired knowledge.

4 Edge-Based Segmentation

There are a lot of well-known methods for edge detection in the field of image processing (see e.g. [5], p. 416ff). Good results have been achieved by using gradient-based filter functions like the *Sobel* or the *Laplacian* operator. Such an edge operator is applied to each of the N_K color channels of an image separately, so that for every pixel $\boldsymbol{p} = (x,y)$ N_K edge values $e(\boldsymbol{p}) = (e_1(\boldsymbol{p}),...,e_{N_K}(\boldsymbol{p}))$ are obtained. In order to abstract from any specific edge filtering method it is assumed that for the edge values $e_i(\boldsymbol{p}) \in [0,1] \subset \mathbb{R}$ holds (with $i = 1,...,N_K$). $e_i(\boldsymbol{p})$ is considered to be a fuzzy membership value of the semantic property "is edge". Such a general representation can for example be achieved by scaling. If the edge operator for example yields the discrete numbers $\tilde{e}_i(\boldsymbol{p}) \in \{0,...,M\} \subset \mathbb{N}$ the formula

$$e_i(\boldsymbol{p}) = \frac{1}{M}\tilde{e}_i(\boldsymbol{p}) \tag{6}$$

performs the transformation into the required interval $[0,1]$.

Background edge information is collected in a fuzzy edge knowledge base:

Definition 2 (Fuzzy edge knowledge base) *:*

Let I be an image of width w and height h measured in pixels, using a corresponding color space Q with N_K color channels according to definition 1. Then a fuzzy edge knowledge base *is defined as a set of functions*

$$\bar{e}_i:\ (x,y) \longrightarrow [0,1] \tag{7}$$

where $\bar{e}_i(\boldsymbol{p})$ is the average edge value of color channel i at position $\boldsymbol{p} = (x,y)$ (with $i = 1,...,N_K$ and $0 \le x < w,\ \ 0 \le y < h$).

The acquisition of background edge knowledge during the initial learning phase is realized by averaging the edge values of the N_S sample images I_j for every pixel $\boldsymbol{p}$ and color channel i separately (with $i = 1, ..., N_K$ and $j = 1, ..., N_S$):

$$\bar{e}_i(\boldsymbol{p}) = \frac{1}{N_S} \sum_{j=1}^{N_S} e_{i,j}(\boldsymbol{p}) \tag{8}$$

where $e_{i,j}(\boldsymbol{p})$ is the edge value of color channel i in the sample image I_j at the pixel position $\boldsymbol{p}$.

The classification of current video images during the application phase can be performed by comparing the edge knowledge base to the current edge values $e(\boldsymbol{p}) = (e_1(\boldsymbol{p}), ..., e_{N_K}(\boldsymbol{p}))$. Given a current image I_c the following fuzzy membership function to the semantic property "is foreground edge" can be used to classify the image I_c:

$$\mu_e(\boldsymbol{p}) = 1 - \frac{1}{1 + \alpha \cdot \max(0,\ |\bar{e}_i(\boldsymbol{p}) - e_i(\boldsymbol{p})| - \beta)^2} \tag{9}$$

with $\alpha, \beta \in \mathbb{R}$, $\beta \in [0, 1]$. α is used as a regulating factor. β is a threshold parameter so that differences in the edge values are neglected if they are smaller than β. In order to remove noise an averaging threshold filter as explained in section 3 is applied to the fuzzy classified edge membership values $\mu_e(\boldsymbol{p})$:

$$\mu_{edge}(\boldsymbol{p}) = \begin{cases} 0 & : \ \frac{1}{9}\mu_e(\boldsymbol{p}) + \frac{1}{9}\sum_{i=1}^{8} \mu_e(\boldsymbol{p}_i) \ \le \ v_e \\ \mu_e & : \ \frac{1}{9}\mu_e(\boldsymbol{p}) + \frac{1}{9}\sum_{i=1}^{8} \mu_e(\boldsymbol{p}_i) \ > \ v_e \end{cases} \tag{10}$$

where $v_e \in [0, 1] \subset \mathbb{R}$ is used as a threshold parameter. Alternatively for this purpose a *median filter* may be utilized.

The membership function μ_{edge} yields a second fuzzy membership value for every pixel $\boldsymbol{p} \in I_c$. Both membership functions, μ_{color} (see section 3) and μ_{edge}, semantically describe foreground object properties.

5 Reasonable Combination of Fuzzy Knowledge

In the previous sections 3 and 4 two independent fuzzy knowledge bases have been introduced which provide for each pixel $\boldsymbol{p}$ of a current image I_c the membership functions μ_{color} and μ_{edge}. Both membership functions semantically describe foreground object properties as "is foreground color" and "is foreground edge". In order to combine these different results a fuzzy "and" operator as for example the min-function is employed. Thus for every pixel $\boldsymbol{p}$ a fuzzy result to the semantic property "is foreground" can be gained by calculating

$$\mu_{foreground}(\boldsymbol{p}) = \min(\mu_{color}(\boldsymbol{p}),\ \mu_{edge}(\boldsymbol{p})). \tag{11}$$

For some applications it may be necessary to have a two-valued segmentation result. This is achieved by a kind of defuzzification of the fuzzy results. For this purpose the averaging threshold filter can be modified to:

$$\boldsymbol{p} \longrightarrow \begin{cases} 0 & : \ \frac{1}{9}\mu_{foreground}(\boldsymbol{p}) + \frac{1}{9}\sum_{i=1}^{8}\mu_{foreground}(\boldsymbol{p}_i) \ \leq \ v_a \\ 1 & : \ \frac{1}{9}\mu_{foreground}(\boldsymbol{p}) + \frac{1}{9}\sum_{i=1}^{8}\mu_{foreground}(\boldsymbol{p}_i) \ > \ v_a \end{cases} \tag{12}$$

where $\boldsymbol{p}_i$ are the respective pixels which fulfill the 8-neighborhood property to $\boldsymbol{p}$. $v_a \in [0,1] \subset \mathbb{R}$ is used as a threshold parameter.

6 Building of a Contour

After the combination of both segmentation results we aim at deriving a contour approximation in order to separate the foreground object from the background. For this purpose the resulting set of pixels from the combination is used.

Nevertheless it has to be kept in mind that this set of pixels also implies uncertainty that should not be neglected. As soon as one of the segmentation techniques does not work properly, for example because of the mentioned reasons in section 1, some object pixels may miss in the resulting set. Furthermore it cannot be assumed that the contour is closed in the sense of 8-neighborhood of pixels. In order to get a first approximation of the contour that is able to handle even big gaps, the *convex hull* of the edge pixels is calculated. The pixels on the convex hull are connected to a closed contour. Additionally the convex hull is enlarged with respect to the centroid of the foreground object by a fixed factor in order to ensure that small regions of the foreground object which have not been detected, are not clipped by the hull lines. Algorithms for the calculation of the convex hull can for example be found in [9].

The approximation of a contour is a classical problem. A survey of basic techniques can be found in [6]. We do not discuss the approximation of a contour further in this paper. For details we refer to [7].

As it was already explained in section 1 the adaptivity of our system to slight changes within the background is an important aspect of our technique. In order to achieve an adaptive improvement learning of the fuzzy color knowledge base is continued on the background region outside the calculated hull.

7 Summary of the Approach

The segmentation technique presented in this contribution can be summarized as follows:

Learning phase

1. *Color segmentation*
   - Learning of color values for *all* pixels as described in section 3.
2. *Edge segmentation*
   - Learning of edge values for *all* pixels as described in section 4.

Application phase

1. *Color segmentation*
   – Fuzzy classification of current image I_c.
   – Application of an averaging threshold filter and calculation of the membership value μ_{color} to the semantic property "is foreground color" for all pixels.
2. *Edge segmentation*
   – Determination of foreground edges according to section 4.
   – Application of an averaging threshold or median filter and determination of μ_{edge}.
3. *Contour calculation*
   – Calculation of the intersection S of the foreground object pixels resulting from the color and edge segmentation.
   – Application of an averaging threshold filter.
   – Computation of the convex hull of S and connection of the hull pixels to a closed contour.
   – Learning of the color information for the identified background region.

The averaging threshold filters as explained in equations (4), (10) and (12) and the classification membership function in equation (9) are controlled by the parameters v_c, v_e, v_a, α and β. Indeed it is not possible to specify universally valid optimal values for these parameters. The quality of any choice can only be evaluated with respect to the classification quality in specific applications. Because of the freedom of choice there are a lot of adaption possibilities for respective scenarios.

8 Examples of Application and Performance Analysis

Because of the variety of known segmentation techniques it is rather difficult to compare them with our introduced approach. In principle the color-based segmentation is similar to the classical approach to have a static background image that is subtracted from current images in order to perform the segmentation. Our technique supplements this approach by enabling the storage of a large amount of color values which are weighted with respect to the quantity of their occurrence. The edge segmentation is based on measuring the difference between the initially learned average edge values and the edge values in the current image which is also a classical technique. Thus we aim at analyzing the impact of the key benefits of our technique: the combination of color and edge segmentation and the adaptivity.

As a first example of application in figure 1 a typical situation and the sequence of segmentation is shown.

For our example the RGB color space was chosen ($N_K = 3$, $N_W = 256$). The video images have a size of 192×144 pixels. The knowledge bases were trained during the initial learning phase with $N_S = 100$ sample images. After this training some interaction was performed in front of the camera. During

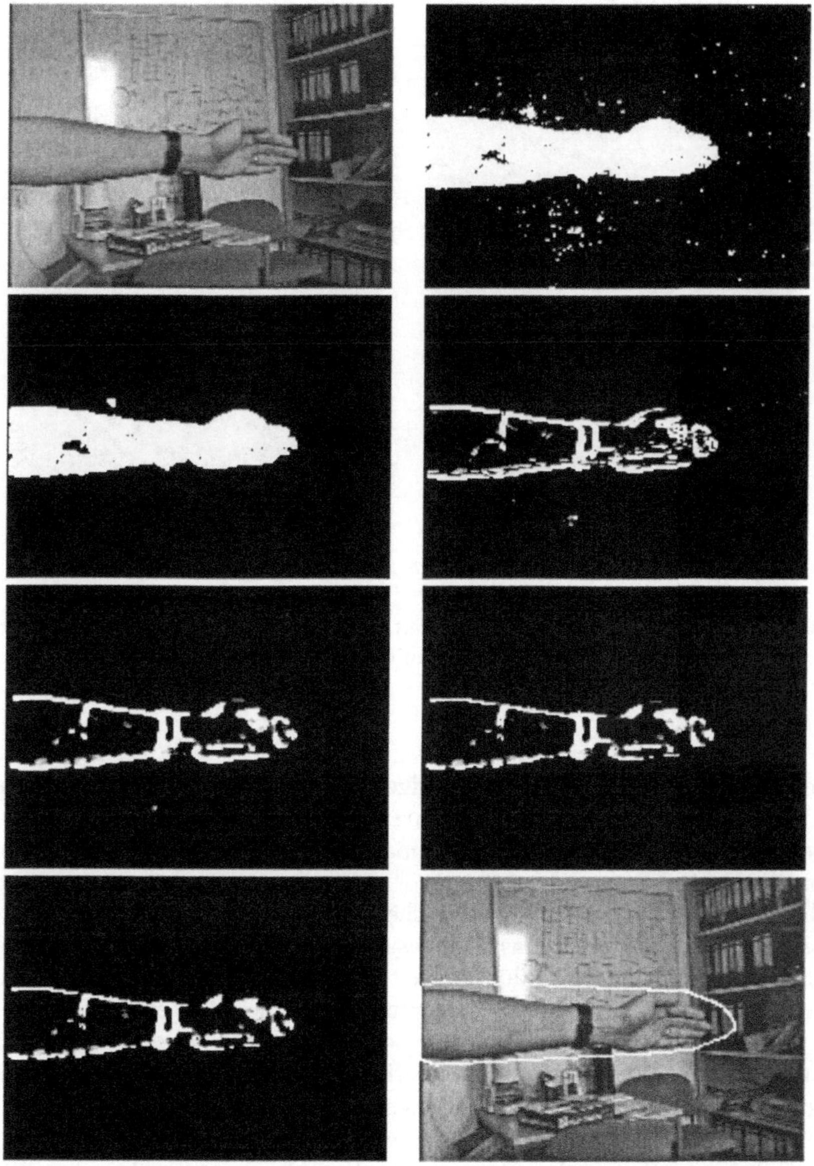

Fig. 1. First row: Original input image (left) and fuzzy color classified image (right). Second row: Noise reduced result (left), fuzzy edge segmented image (right). Third row: Median filtered edge image (left) and fuzzy combination of color and edge segmentation (right). Fourth row: Found object pixels after application of averaging threshold filter (left), enlarged convex hull as contour approximation (superimposed in white).

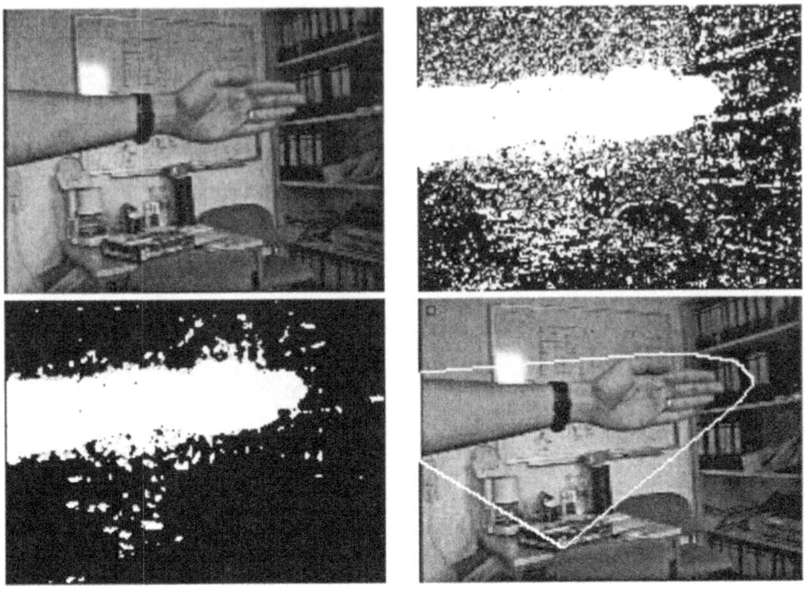

Fig. 2. First row: Original input image (left) and fuzzy color classified image (right). Second row: Noise reduced result (left), faulty segmentation result (right).

this interaction the fuzzy color knowledge base was continuously trained on the identified background regions. The parameter σ, introduced in equation (2), was set to $\sigma = 0.18$. The exemplary image shown in figure 1 was taken after approximately 600 images.

The upper row of figure 1 shows on the left the original color video image and on the right the fuzzy color classified results. For this purpose the fuzzy results were transformed into grey-value images as explained in section 3. Dark colors indicate low, and bright colors large membership values. The averaging threshold filter was applied with $\upsilon_c = 0.66$ (see equation (4)) in order to remove noise (second row, left column). The identification of foreground edges is shown in the second row on the right. The parameters of μ_e in equation (9) were set to $\alpha = 80$ and $\beta = 0.25$. The median filtered edge image is shown on the left in the third row. On the right the result of the combination of color and edge segmentation is shown. This intermediate result is again filtered using the averaging threshold with $\upsilon_a = 0.10$ as parameter of equation (12) (left column, fourth row). The final segmentation result is shown as a white polygon superimposed on the original image.

In order to measure the benefit of the combination of both segmentation methods against their exclusive usage the number of wrongly classified pixels is considered. This analysis shows that the color segmentation (after the noise reduction) still produces four (with an average membership value of 0.86) and the edge segmentation even 15 wrongly classified pixels (with an average mem-

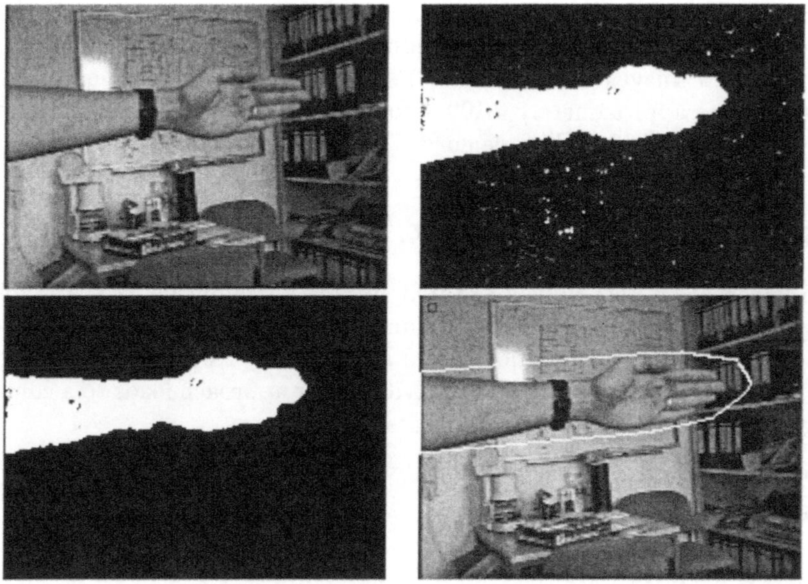

Fig. 3. First row: Same input image as in figure 2 (left) and fuzzy color classified image after approximately further 1000 frames (right). Second row: Noise reduced result (left), correct segmentation result (right).

bership value of 0.14). The fuzzy combination of both segmentation approaches removes these wrongly classified pixels, leaving actual foreground object pixels. Their consideration enables the derivation of the convex hull as an approximated contour which can be used to train the color knowledge base on the background region.

The second key benefit of our approach, the adaptivity to slight changes in the background colors, is illustrated in a second example (figures 2 and 3). As first experiments indicated there are already changes in the background colors when a foreground object enters the scene. In figure 2 the original video image is shown in the upper row on the left. The fuzzy color classified result can be seen on the right and the averaging threshold filtered image in the lower row on the left. The fuzzy color classification was done approximately 80 frames after finishing the initial learning phase with 100 images. As it can be seen there are a lot of wrongly classified pixels which cannot be completely removed by the filtering operation. Due to the combination with the fuzzy edge segmentation whose results are not shown within this example most of these wrong classification results are identified and neglected. Nevertheless some faulty pixels remain, leading to a nonsatisfying contour approximation.

The same input image was again fuzzy color classified approximately 1000 frames later. During this time a continuous movement of the arm in front of the camera was performed. The improvement of the segmentation results is shown in figure 3. The fuzzy color classified image contains only few wrongly classified pixels. As a consequence the contour approximation produces a sufficient result.

Applied to images with a size of 192 × 144 pixels the segmentation of an image requires approximately 0.109 seconds on a Pentium-III 700 MHz system and requires about 22 MB of memory.

9 Summary and Conclusion

The segmentation technique presented in this contribution works stable. Although both knowledge bases provide uncertain information that might contain a lot of errors their combination yields a reasonable result by employing fuzzy logic methods. Furthermore the adaptivity of our approach leads to a continuous improvement of the segmentation results.

References

1. Azarbayejani, A., Wren, C., Pentland, A.: Real-Time 3-D Tracking of the Human Body. M.I.T. Media Laboratory Perceptional Computing Section Technical Report No. 374. (1996)
2. Bichsel, M.: Segmenting simply connected moving objects in a static scene. in: IEEE Pattern Analysis and Machine Intelligence, Vol. 16. (1994)
3. Dubuisson, M.-P. , Jain, A. K.: Contour extraction of moving objects in complex outdoor scenes. In: International Journal of Computer Vision, Vol. 14 (1995) 83–105
4. Girod, B., Greiner, G., Niemann, H. (Edts.): Principles of 3d image analysis and synthesis. Kluwer Academic Publishers (2000)
5. Gonzalez, R.C., Woods, R.E.: Digital Image Processing. Addison-Wesley Publishing (1992)
6. Haralick, R.M., Shapiro, L.G.: Survey: Image segmentation techniques. In: Computer Vision, Graphics, and Image Processing, Vol. 29 (1985) 100–132
7. Leubner, C.: Contour approximation by linking locally calculated hull polygons. Research Report, No. 748, University of Dortmund (2001) `http://ls7-www.cs.uni-dortmund.de/publications/abstracts_2001`
8. Pal, N.R., Pal, S.K.: A review on image segmentation techniques. In: Pattern Recognition, Vol. 26, No. 9 (1993) 1277–1294.
9. Preparata, F.P.: Computational Geometry. Springer-Verlag (1985)
10. Saber, E., Tekalp, A.M., Bozdagi, G.: Fusion of color and edge information for improved segmentation and edge linking. In: Image and Vision Computing, Vol. 15 (1997) 769–780
11. Skarbek, W., Koschan, A.: Colour Image Segmentation - A Survey. Technischer Bericht 94-32, TU Berlin (1994)
12. Tizhoosh, H.R.: Fuzzy-Bildverarbeitung: Einführung in Theorie und Praxis. Springer-Verlag, Berlin Heidelberg New York (1998)
13. Wren, C. , Azarbayejani, A., Darrell, T., Pentland, A.: Pfinder: Real-time tracking of the human body. In: Photonics East. SPIE Proceedings, Vol. 2615 (1995)

Computer Aided Diagnosis System of Meniscal Tears with T1 and T2 Weighted MR Images Based on Fuzzy Inference

Yutaka Hata[1], Syoji Kobashi[1], Yasuhiro Tokimoto[2], Makoto Ishikawa[2], and Hitoshi Ishikawa[3]

[1] Department of Computer Engineering, Himeji Institute of Technology, 2167, Shosha, Himeji, 671-2201, JAPAN
{hata,kobashi}@comp.eng.himeji-tech.ac.jp
http://wwwj3.comp.eng.himeji-tech.ac.jp
[2] Ishikawa Hospital, 784, Bessho, Bessho-cho, Himeji, 671-0221, JAPAN
[3] Kobe University School of Medicine, 7 Tomogaoka, Suma, Kobe, 654-0142, JAPAN

Abstract This paper proposes a computer aided diagnosis system of meniscal tears from 3D human knee MR image with T1-weighted and T2-weighted MR images. The first step of our method is a 3D image registration between both images on the computer display by manual. The second step determines the candidate region of the menisci from T2-weighted MR image aided by fuzzy if-then rules with respect to the location and the intensity. The final step determines the voxels in the menisci from the candidate region by using T1-weighted and T2-weighted MR images. We applied this method to several subjects. All voxels in the menisci of each subject were successfully identified and their 3D surfaces were displayed. Thus, our developed system would improve to diagnose the meniscal tears.

1 Introduction

Three-dimensional (3D) image segmentation is an important issue in bio-medical imaging. The basic problem involves identification of voxels in region of interest (ROI). Magnetic resonance (MR) imaging provides a highly effective means for observing human body. The MR scanner can produce volume (voxel) images with 3D thickness and this MR imaging is accurate in evaluating meniscal tears. Several works described clinical studies to diagnose the meniscal tears by MR imaging. Ref.[1] reviewed some segmentation techniques applicable to menisci. Ref.[2] described the clinical value of volumetric rendering in surgical planning. We also described semi-automated segmentation system of menisci generated with T1-weighted MR images[3]. However, to examine meniscal tears by this system, it is not enough to practical use because of the segmentation accuracy.

In this work, to provide more precise information of meniscal tears, we describe a computer aided diagnosis system of meniscal tears from 3D human knee MR image with T1-weighted and T2-weighted MR images. To do this, we employed fuzzy expert system methodologies such as fuzzy membership functions and fuzzy if-then rules[4][5][6], to obtain information on anatomical locations and intensity information. A knee MR volumetric images composed of many slice images consists of several parts: bone marrow, meniscus, periarticular liquor, cartilage and others.

B. Reusch (Ed.): Fuzzy Days 2001, LNCS 2206, pp. 55–58, 2001.

Generally, in T1-weighted MR image, meniscus and pericular liquor have lower intensity than bone marrow and cartilage, while, in T2-weighted image, meniscus and bone marrow have lower intensity than particular and cartilage. We employ both intensity levels of the T1-weighted and T2-weighted MR images to identify the voxels in menisci with high accuracy. The first step of our work is a 3D image registration between both images on the computer display by manual. We translate T2-weighted image to adjust with T1-weighted image by operating the sliced image on the viewer by manual. The second step determines the candidate region of the menisci from T2-weighted MR image. The candidate region of the menisci is found aided by fuzzy if-then rules on fuzzy membership functions to express information of the location and the intensity. The final step determines the voxels in the menisci from the candidate region using T1-weighted and T2-weighted MR images. To do this, w employ fuzzy if-then rules to express that the menisci exist near the cartilage and the intensity level of the menisci. We applied this method to several subjects. All voxels in the menisci of each subject were successfully identified and their 3D surfaces were displayed. Thus, our developed system would improve to diagnose the meniscal tears.

2 Outline of Our Computer Aided Diagnosis System

We obtained the MR volumetric images from Genesis Signa 1.5 Tesla MRI scanner (General Electric Co.) using surface coil. We employ T1-weighted and T2-weighted MR images. The T1-weighted image acquisition method was T1-weighted 3D SPGR with TR = 100msec, TE = 15msec. The T2-weighted image acquisition method was FL02/T2*COR with TR = 100msec, TE = 20msec. In both images, field of view was 160mm. The matrix was 256 by 256. Each of the volumes was made up of 60 separate slices whose thickness was 1.5mm. We construct the MR volumes of the human knee, which consisted of 256 by 256 by 60 voxels. The intensity value of all intracranial structures ranged between 0 and 255.

First step of our method is the registration between T1-weighted MR image and T2-weighted MR image of a subject. This registration is to find the correspondence between two images in a same 3D coordinate system. We translate T2-weighted image to adjust with T1-weighted image by manual operation on the computer display. In our experiment, since we sequentially obtained the T1-weighted and T2-weighted MR images at the same time, we do not have to do this registration.

Second step of our method is to point out the at least one voxel in each meniscus in T2-weighted MR image manually.

Third step of our method can automatically determine the menisci by the following two steps. The menisci are located between the cartilage of the thigh bone and the cartilage of the shin bone. These cartilages can be easily extracted by intensity thresholding on the T2-weighted MR image. We can set up the volume including all the cartilages. We consider the straight voxels along y-axis, which can be referred by (x, z) in the volume. The set of the straight voxels is called by a unit. First, we attempted to classify every unit into candidate region of the menisci or not so. Fuzzy knowledge representation can be used to express information on the location of the menisci as well as the intensity variance of a unit in T2-weighted MR images. We then produce fuzzy if-then rules for the classification. Every unit can thus be classified by evaluating inference results of fuzzy if-then rules. In the units in the candidate region, we attempted to extract the menisci. Fuzzy knowledge representation can be used to express information on intensity and locations in T1-weighted and T2-weighted MR images. We then produce fuzzy if-then rules for menisci. The value obtained by weighted sum of these inference results was found to be an excellent indicator for identifying the menisci. This value can be used with a region growing procedure to exactly obtain the menisci.

3 Experimental Results

We applied our method to three MR data sets. For each image, we successfully segmented the menisci. In case of segmentation from T1-weighted image only, it required about 20 seconds on each knee MR data sets, running on an SGI OCTANE (R10000, 175MHz). In case of T1 and T2-weighted images, it required about 30 seconds. Quantitative evaluation by a physician shows that this method can successfully segment the menisci. Figures 1 and 2 show the 3-D visualization of the menisci segmented from T1-weighted MR images only as well as the menisci segmented from T1-weighted and T2-weighted MR images. We can understand that the periarticular liquors are eliminated, compared the upper case with lower cases in each figures. Physicians can visually diagnose the meniscal tears. Thus, our system can successfully segment the menisci and then provide the 3D surface for diagnosing meniscal tears with high accuracy.

4 Conclusions

The anatomical structure of menisci is quite complex, so except for our works, there is no works to semi/full-automatically segment the menisci from MR images. In this paper, we described a method to segment the menisci from T1 and T2 weighted MR images. We applied this method to computer aided diagnosis system. Quantitative evaluation by a physicians shows that our system can successfully provide the 3D surface of menisci regions for diagnosing the meniscal tears noninvasively. The future work is to apply our method to the knee CT images.

Acknowledgment

The authors would like to thank Mr. Takashi Sasaki (Sharp, Co.) for coding this software system in part. This research was supported in part by a grant from BISC Program of UC Berkeley.

References

1. C. Fellner, A. Geissler, P. Held, M. Strotzer, W. Treibel, and F. Fellner, "Signal, contrast, and resolution in optimized PD- and T2-weighted turbo SE images of the knee," Journal of Computer Assisted Tomography, vol. 19, no.1, pp. 95-105, 1995.
2. J. H. Mink, M. A. Reicher, J. V. CruesIII, and A. L. Deutsch, *MRI of the knee*, Raven Press, 1992.

3. T. Sasaki, Y. Hata, Y. Ando, and M. Ishikawa, "Fuzzy rule-based approach to segment the menisci regions from MR images," in Proceedings of SPIE Medical Imaging 1999, vol. 3661, pp. 258-263, Feb. 1999.
4. L. A. Zadeh, *Fuzzy Sets and Applications*, John Wiley and Sons, Inc., 1987.
5. A. Kandel, *Fuzzy Expert Systems*, CRC Press, Inc. 1992.
6. Y. Hata, S. Kobashi, S. Hirano, H. Kitagaki, and E. Mori, "Automated segmentation of human brain MR images aided by fuzzy information granulation and fuzzy inference," IEEE Transactions Systems, Man, and Cybernetics, Part C, vol. 30, no. 3, pp. 381-395, 2000.

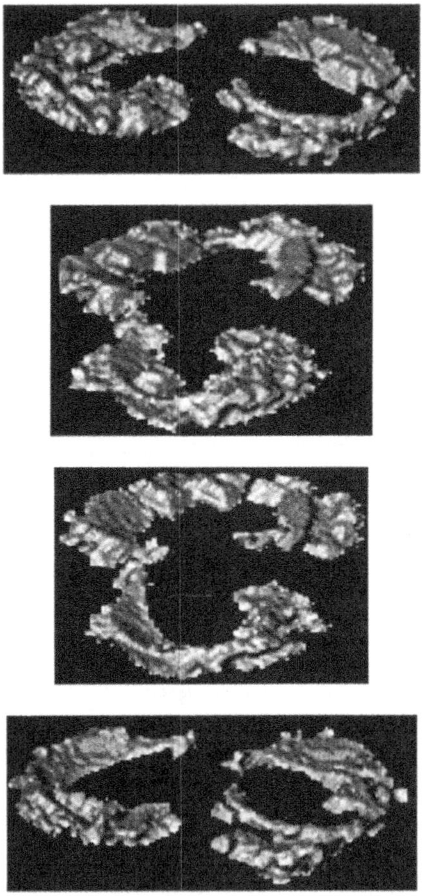

Fig. 1. 3D visualizations of meniscal tears (subject 1); upper case: from T1-weighted image; lower case: from T1 and T2-weighted images.

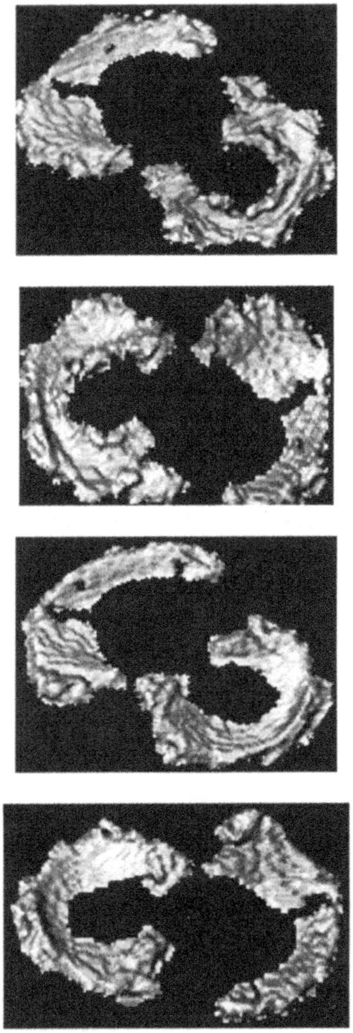

Fig. 2. 3D visualizations of meniscal tears (subject2); upper case: from T1-weighted image; lower case: from T1 and T2-weighted images.

A New Paradigm for Fuzzy Aggregation in Multisensorial Image Processing

Aureli Soria-Frisch

Fraunhofer IPK, Dept. Pattern Recognition, Pascalstr 8-9, 10587 Berlin, Germany
aureli.soria_frisch@ipk.fhg.de

Abstract. This paper presents a new paradigm for image processing in multisensorial computer vision systems based on a new interpretation of the fuzzy integral as fusion operator. Fuzzy integrals offer great chances for the implementation of the fusion stage in multisensorial systems. By exploring these possibilities a new paradigm for image processing in the framework of information fusion in multisensorial systems can be established. This new paradigm, which can be designated as Intelligent Localized Fusion (ILF), is related to Soft-Computing methodologies and the object of this paper. The performance of intelligent localized fusion operators (ILFOs), which are developed under the new introduced paradigm, are exemplary shown in the case of color edge detection in outdoor scenes. Its usage allows in this case the avoidance of false edges due to the presence of shadows.

1 Introduction

Multisensorial computer vision systems are characterized by the inclusion of more than one image information source and the processing of the image signals delivered from them. Some examples of multisensorial computer vision systems can be found in color, multiband remote-sensing or range-visual analysis systems.

The fuzzy integral was introduced by Sugeno [14] to simulate the subjective fusion of multicriteria in human beings. In spite of the extended use of the fuzzy integral in other fields of research, e.g. decision making, the usage of the fuzzy integral in computer vision is almost restricted to the use in the fusion of classificators or the segmentation of color images. Some properties of the fuzzy integral can be exploited in order to cope with difficulties of increasing complexity present in modern image processing application fields.

A new paradigm for the processing of multisensorial images is presented in this paper. The paradigm is based on a novel usage of the fuzzy integral as fusion operator in image processing systems. The paradigm, which receives the name Intelligent Localized Fusion (ILF), uses a local parameterization of the fuzzy integral. This fact leads to the establishment of different fusion strategies while using a unique fusion operator.

The paradigm is exemplary used to accomplish edge detection on outdoor scenes. The operators derived within the paradigm, which can be called Intelligent Localized Fusion Operators (ILFOs), successfully avoid the detection of false edges due to shadows.

B. Reusch (Ed.): Fuzzy Days 2001, LNCS 2206, pp. 59–67, 2001.

The paper is structured as follows: In Section 2 a brief description of fuzzy aggregation operators is given and the new paradigm is presented. In Section 3, the application of the paradigm for outdoor scenes edge detection is presented together with preliminary results. Finally, the conclusions and the projective work can be found in Section 4.

2 A New Fusion Paradigm in Multisensorial Image Processing Systems

In multisensorial systems the reduction of the n-dimensionality introduced in the system by the usage of n sensors, being n bigger than one, is attained by the fusion operator. Thus the fusion operator combines the data coming from the different sources of information into one representational form [8]. But over this basic function the utilization of such an operator will be really useful if it is capable of conferring a quantitative and qualitative gain on information to the system. The quantitative gain on information can be derived from the attainment of a more reliable result or of a smaller error rate. Qualitatively the fusion operation allows the inference of information, which was not available using individually the data from the different sensors [1]. Furthermore it is desired that the fusion operation provides a better understanding of the interaction between the sensors in the system, in order to facilitate the evaluation of the contribution of each information source.

2.1 Fuzzy aggregation in multisensorial processing systems

Fusion paradigms used in the context of computer vision [2] [7] [9][11] are centered on the numerical optimization of the fusion procedure. Fusion operators in this context are limited to the consideration of a very narrow group of them, disregarding recent developments in operator research related specially to fuzzy logic. The semantic interpretation of the images processed is ignored in the fusion procedure. Moreover the characterization of information sources is made in the mentioned systems by a module separated from the fusion operator and mostly based on gaussian and independence suppositions on them [7].

In order to overcome the shortages of classical fusion operators seems unavoidable the introduction of fuzzy aggregation strategies in order to implement the fusion stage. Beyond the fact that most of aggregation operators usually employed in computer vision can be generalized in the framework of fuzzy systems, fuzzy aggregation operators present some special features:

- Adaptability is here referred to the fusion operators capability of reacting to changes in the incoming data. As a consequence the fusion operation is data-driven, what allows the flexible performance of the task in order to cope with non-linear, non-Gaussian, and non-stationary information sources as the most present in practical applications.

- Coincident results of a majority of information sources will be delivered as result of the fusion operation by means of the reinforcement capability of the fusion operator [18].
- The inclusion of global *a priori* knowledge is also an important characteristic to be accomplished by fusion operators. This factor, also called inclusion of metaknowledge by some authors [18], encloses for instance compatibility or importance grade of the information sources. The inclusion of such knowledge increased the intelligence of the fusion operation.
- The information from different sensors inspecting the same object can interact in a redundant, independent or complementary form in multisensorial systems. The characterization of this interaction is an important point to be attained by an intelligent fusion procedure in order to obtain a more complex and complete information. It is worth of note that the characterization of such interaction is allowed only by a part of aggregation operators included in the framework of fuzzy systems.
- The fuzzification of the image data, which is derived from the application of Fuzzy Sets Theory to image processing [16], allows providing it with directly interpretable semantic information.This semantic interpretability can improve the results of image processing operations in real applications [16]. Fuzzy aggregation operators cope successfully with the fusion of fuzzy information.

Particularly in the context of the multisensorial systems all these features can contribute to the attainment of the quantitative and the qualitative gain expected to result from a fusion operation.

The information fusion can be undertaken in different stages of a multisensorial image processing system. The position of the fusion operator in the processing chain leads to different types of fusion, e.g. signal, pixel, feature or segment fusion [12]. In the here presented paradigm the operator acts in the local level, i.e. the fusion undertaken is a pixel fusion. This fact was taken as start point for the development of the intelligent localized fusion (ILF) paradigm.

The key point of the paradigm is the consideration of a non-uniform fusion over the considered image channels. In this way the fusion operators can solve the fusion with different characteristics depending on the area of the image where the fusion is undertaken. Such operators will be referred as Intelligent Localized Fusion Operators (ILFOs). For the sake of understanding a deeper explanation of its implementation is postponed. Previously the theoretical framework of Fuzzy Measure Theory and Fuzzy Integrals [17], which constitutes the most suitable one for the development of ILFOs, will be presented.

2.2 Intelligent Localized Fusion in the theoretical framework of Fuzzy Integrals

The Theory of Fuzzy Measures was built on the conclusions made in Sugenos pioneering work [14]. In an analogous way as fuzzy logic extended classical logic by adding subjectivity to the reason process, fuzzy measures consider the extension of classical measure theory with the introduction of subjectivity. In order to satisfy this attribute, classical measures (i.e. probability measures) are extended by relaxing their additivity axiom. As a consequence fuzzy measures include probability, possibility, necessity, belief and fuzzy-λ measures [17]. One of the differences between the different types of fuzzy measures lies on the way they succeed in characterizing the interaction between individual sources of information. The establishment of a new kind of measures led to the establishment of a new integral, the fuzzy integral.

The fuzzy integral was introduced in order to simulate the process of multicriteria fusion undertaken in the mind [14]. The basic idea of this approach is that when fusing information coming from different sources, the human being undertakes a subjective weighting of the different criteria or factors, and aggregates them in a non-linear fashion. Moreover the fuzzy integral considers the aggregation of fuzzified data and constitutes a theoretical framework where other aggregation operators as average, OWA, weighted median, weighted sum or weighted maximum are hold [4]. There are basically two types of fuzzy integral known as Sugeno's and Choquet's Fuzzy Integral. The difference between them can be found in the used T- and S-norms [4]. While the Sugeno's Fuzzy Integral uses norms of the type maximum and minimum, the Choquet's Fuzzy Integral was introduced as an extension in the framework of Fuzzy Measures Theory of the classical Lagrange's Integral [17].

Sugeno's Fuzzy Integral has successfully been used in information fusion for Computer Vision [15]. For the prototypical development of the paradigm this fusion operator was chosen. The expression of Sugenos Fuzzy Integral is:

$$\mathcal{S}_{\mu}(x_1,\ldots,x_n) = \bigvee_{i=1}^{n} [h_{(i)}(x_i) \wedge \mu(A_{(i)})] \tag{1}$$

where x_i are the information sources to be fused, n the number of information sources, h_i the fuzzifying functions, μ the fuzzy measure coefficients, A_i the fuzzy sets formed by the information sources, $\vee$ the maximum operator, and $\wedge$ the minimum operator. The enclosed subindexes indicate the result of a sorting operation previous to the fusion itself. For a deeper description on algorithmic aspects of Fuzzy Integrals the reader is referred to [4].

One property of fuzzy integrals important for the development of the here presented paradigm is related to their limits. The value of the fuzzy integral has its lower and upper bound in the minimum ($\wedge$) and maximum ($\vee$) operator, respectively:

$$\bigwedge[x_1,\ldots,x_n] \leq \mathcal{S}_{\mu}(x_1,\ldots,x_n) \leq \bigvee[x_1,\ldots,x_n] \tag{2}$$

In this way the possible interval of results delivered by the Sugenos Fuzzy Integral is the largest allowed to a fusion operation. This possible interval of output values for the fusion operation is induced by the value of the fuzzy measures coefficients; e.g. the same result as that of a maximum operator will be achieved when the fuzzy measures coefficients are all the maximum value allowed to the pixel values, while the minimum will be obtained when only $\mu(A_n)$ has this value. This property is exploited in the here presented paradigm in order to achieve a flexible fusion operation.

2.3 "Localized" fuzzy measures: the new aspect of the ILF paradigm

The fuzzy measure coefficients are used in the fuzzy integral for multisensorial fusion to characterize the importance of the information sources and are determined based on *a priori* knowledge of their relevance. This characterization is related to the individual information sources, but also to their possible coalitions or combinations.

The determination of the fuzzy measures is a problem of great complexity. There is no standard procedure for that purpose and in most of the cases this determination is realized empirically. In the here presented framework that is not different. Up to the observation of the different incoming and independently processed images, and after taking into consideration *a priori* knowledge concerning the application on hand the different coefficients of the fuzzy measure are established. Nevertheless a novelty on this aspect is introduced in the here presented paradigm. Normally the coefficients of the fuzzy measures are determined globally in the image, i.e. the fuzzy measure coefficients have a value characterizing the importance of each image channel, and that of their possible coalition, taking the image as a unit.

In the ILF paradigm the used fuzzy measure is not unique. Depending on the region where the fusion is undertaken the fuzzy measure is different, i.e. its coefficients are different (see figure 1). It can be talked then from "localized" fuzzy measures. Taking into consideration the lower and upper bounds of the fuzzy integral's output and its dependence on the value of the fuzzy measures coefficients, the utilization of these "localized" fuzzy measures makes possible operating in the same image with different fusion strategies. The "localized" fuzzy measures exploit the flexibility of the fuzzy integral formerly mentioned and constitutes the main new feature of the ILF paradigm. Other characteristics are the parallel processing and posterior fuzzification of information channels prior to the fusion (see figure 1).

3 Color Edge Detection on Outdoor Color Images

As mentioned in the introduction edge detection is an important previous stage in several image processing algorithms. In the case of edge detection on color images different mathematical approaches have been chosen in order to cope with the vectorial nature of color information [5] [6] [10].

A system with an ILFO for color edge detection following the paradigm formerly detailed and based on the Sugeno's Fuzzy Integral was implemented. The system was applied for detecting edges on color images. The implemented scheme

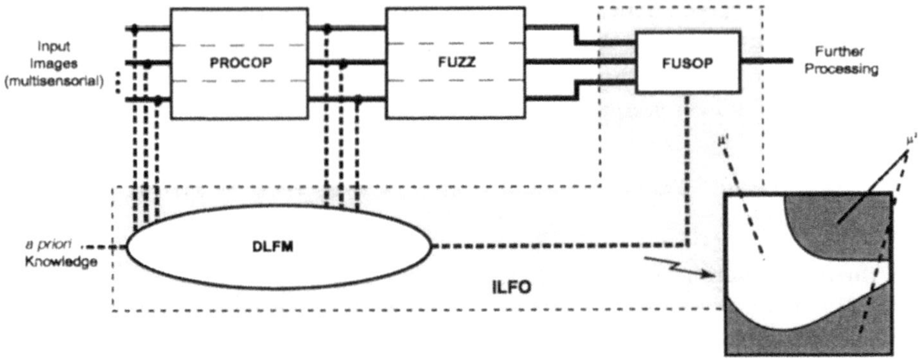

Fig. 1. ILF Paradigm: Multisensorial image information is processed and fuzzified separately and then fused through an ILFO. PROCOP: Image processing operation; FUZZ: Fuzzification; ILFO: Intelligent Localized Fusion Operator, formed by a fusion operator (FUSOP) and the determination of the "Localized" Fuzzy Measures (DLFM). "Localized" Fuzzy Measures: More than one fuzzy measure (μ^1, μ^2, ...), each one for a determined region, are used in a fusion operation.

considers the parallel edge detection on the color channels (RGB) and its separated fuzzification leading to its fusion through an ILFO. Standard algorithms for edge detection on grayvalue images can be used within the here presented framework, simplifying the mathematical complexity of the problem. The result of the edge detection is then fuzzified up to the consideration of its grade of membership to the foreground of the image, what allows filtering the image. More implementation details of the framework for color edge detection can be found in [13].

One important problem when evaluating edge maps of images taken under natural illumination conditions is the appearance of false edges due to the presence of shadows. Such artifacts can for instance bring a camera guided autonomous vehicle to detect a non-existing terrain change. Such a problem is a current term of research in image processing [3]. The here presented paradigm is used to avoid such a problem. "Localized" fuzzy measures are defined for the outdoor scenes, where the regions on shadowed areas present a coefficient configuration different from that of the other areas. Such a configuration allows the fuzzy integral to behave different depending on the image region.

3.1 Preliminary results

The framework for color edge detection was tested on outdoor color images. The results can be found in figures 2 and 3. Two different aspects of the results are worth mentioning.

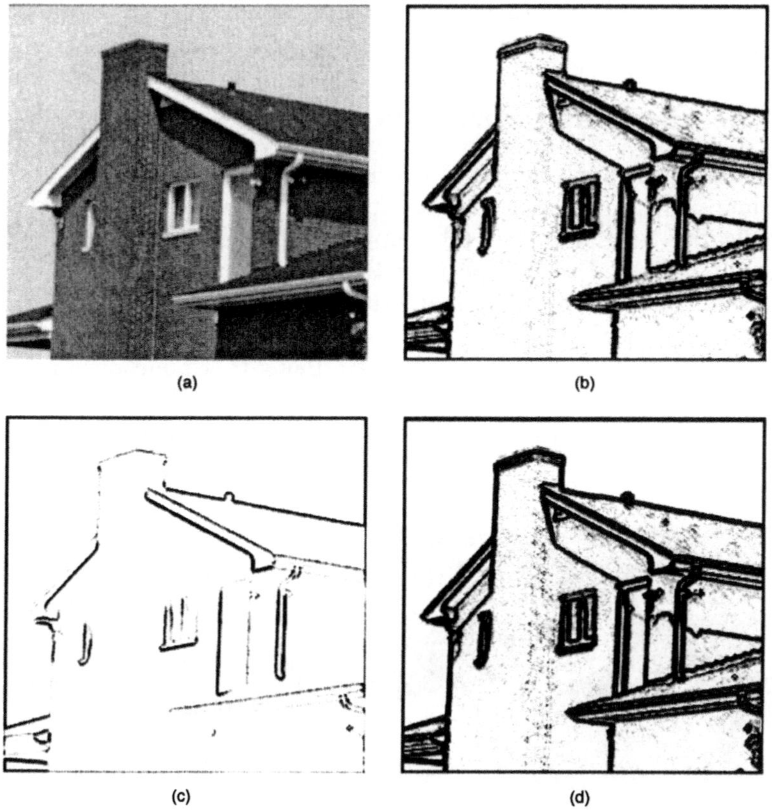

Fig. 2. Fusion results for different values of the fuzzy measures coefficients and Sugeno's fuzzy integral. (a) Input color image. (b) Color edge map for values $\mu_{Red} = 0.6\mu_{Green} = 0.25\mu_{Blue} = 0.8(fuzzy - \lambda measures)$. (c) Fusion with fuzzy integral as minimum. (d) Fusion with fuzzy integral as maximum.

First the utilization of the fuzzy integral allows the fusion output to take all values between those of the minimum and the maximum operators (see figures 2c and 2d). The selection of the coefficients of the fuzzy measures between these two bounds leads to a result which seems less artificial than the mere employment of a maximum operator (see figure 2b).

Finally, the avoidance of false edges due to shadows on images taken under natural conditions of illumination through the the usage of "localized" fuzzy measures is demonstrated (see figure 3). The mask with the region distribution for the different fuzzy measures, which can be observed in figure 3a, was obtained through the treatment of the original image with arithmetical and morphological operations. Thus a mask for two different fuzzy measures was established. The fuzzy measure coefficients for the shadowed regions were set in order for the result of the fuzzy integral to approximate that of a minimum operator. As a consequence the edge due to the roof shadow (see figure 2b) could be avoided (see figure 3b). This result is a direct consequence of the localized parameterization of the fusion operators through the fuzzy measures.

4 Conclusions and Future Work

First results on the application of the here presented paradigm for the detection of edge maps on outdoor scenes are encouraging. The analysis will be extended in this application field. Moreover other application fields will be developed under the same paradigm. For that purpose the treatment of multifocal microscopy images seem a good candidate. Finally the automatic establishment of the fuzzy measure coefficients seem a very important point for the improvement of the adaptability of the here presented approach. Being that point one of the biggest impediments for spreading the usage of the fuzzy integral as fusion operator in image processing applications, will be attached independently of the formerly mentioned research.

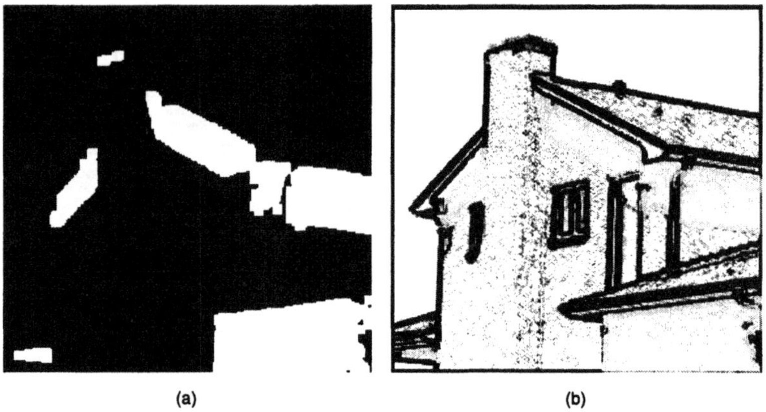

Fig. 3. Fusion with "localized" fuzzy measures with avoidance of shadow false edges. (a) Mask for the fuzzy measures. (b) Color edge maps after fusion.

References

1. M.A. Abidi, R.C. Gonzalez, eds. (1992). *Data Fusion in Robotics and Machine Intelligence*. San Diego: Academic Press.

2. F. Alkoot and J.Kittler (2000). Improving the performance of the Product Fusion Strategy. *Proc. 15th International Conference on Pattern Recognition,*. ICPR 2000, Barcelona, Catalonia.
3. E.D. Dickmanns (1997). Improvements in Visual Autonomous Road Vehicle Guidance 1987-94,in *Visual Navigation: From Biological Systems to Unmanned Ground Vehicles*, Y. Aloimonos ed., Mahwah (New Jersey): Lawrence Erlbaum Associates, Pubs.
4. M. Grabisch, H.T. Nguyen and E.A. Walker (1995). *Fundamentals of Uncertainty Calculi with Applications to Fuzzy Inference*, Kluwer Ac. Pub.
5. M. Koeppen, C. Nowack and G. Rsel (1999). Pareto-Morphology for Color Image Processing. *Proc. of the 11th Scandinavian Conference in Image Analysis*, Greenland, Denmark.
6. R. Krishnamoorti and P. Bhattacharya (1998). Color Edge Extraction Using Orthogonal Polynomials Based Zero Crossings Scheme. *Information Sciences*, 112, 51-65.
7. H. Li, B.S. Manjunath and S.K. Mitra (1995). Multisensor Image Fusion Using the Wavelet Transform. *Graphical Models and Image Processing*, 57 (3): 235-245.
8. R.C. Luo and M.G. Kay eds. (1995). *Multisensor Integration and Fusion for Intelligent machines and systems.* Norwood, NJ: Ablex Publishing Corporation.
9. N. Nandhakumar (1994). Robust physics-based analysis of thermal and visual imagery. *Journal of the Opt. Soc. Am. A*, 1994: 2981-2989.
10. S.-C. Pei and C.-M. Cheng (1999). Color Image Processing by Using Binary Quaternion-Moment-Preserving Thresholding Technique. *IEEE Trans. On Image Processing*, 8 (5) 614-629.
11. R.A. Salinas, C. Richardson, M.A. Abidi and R.C. Gonzalez (1996). Data Fusion: Color Edge Detection and Surface Reconstruction Through Regularization. *IEEE Trans. on Industrial Electronics*, 43(3): 355-363.
12. A. Soria-Frisch and J. Ruiz-del-Solar (1999). Towards a Biological-based Fusion of Color and Infrared Textural image Information. *Proc. of the IEEE W. on Intelligent Signal Processing99*, Budapest, Hungary.
13. A. Soria-Frisch (2000). Intelligent Localized Fusion Operators for Color Edge Detection. *To be presented at the 12th Scandinavian Conference on Image Analysis, SCIA 2001*, Bergen, Norway.
14. M. Sugeno (1974). *Theory of Fuzzy Integral and its applications.* Ph.D. thesis.
15. H. Tahani and J. Keller (1990). Information Fusion in Computer Vision Using the Fuzzy Integral. *IEEE Trans. Systems, Man and Cybernetics*, 20 (3): 733-741.
16. H.R. Tizhoosh (1998). *Fuzzy Bildverarbeitung.* Heidelberg: Springer-Verlag (in german).
17. Z. Wang and G.J. Klir (1992). *Fuzzy Measure Theory*, Plenum Press.
18. R.R. Yager and A. Kelman (1996). Fusion of Fuzzy Information With Considerations for Compatibility, Partial Aggregation, and Reinforcement. *Int. J. of Approximate Reasoning*, 15:93-122.

An Adaptive Learning Algorithm for a Neuro-fuzzy Network

Yevgeniy Bodyanskiy[1], Vitaliy Kolodyazhniy[1], and Andreas Stephan[2]

[1] Artificial Intelligence Department,
Kharkiv State Technical University of Radioelectronics,
14, Lenin Av., Kharkiv, 61166, Ukraine
{Bodyanskiy,Kolodyazhniy}@ieee.org

[2] Dr. Stephan & Partner, System- und Softwarehaus, Grenzhammer, 8,
D-98693, Ilmenau, Germany
Stephan@dr-stephan-und-partner.de

Abstract. The paper addresses the problem of online adaptive learning in a neuro-fuzzy network based on Sugeno-type fuzzy inference. A new learning algorithm for tuning of both antecedent and consequent parts of fuzzy rules is proposed. The algorithm is derived from the well-known Marquardt procedure and uses approximation of the Hessian matrix. A characteristic feature of the proposed algorithm is that it does not require time-consuming matrix operations. Simulation results illustrate application to adaptive identification of a nonlinear plant and nonlinear time series prediction.

1 Introduction

Neuro-fuzzy systems have been an increasingly popular technique of soft computing during recent years. They combine linguistic interpretability of fuzzy systems with learning and universal approximation capabilities of artificial neural networks. The most widely used method of learning in such systems is the error back-propagation [1], based on the gradient descent. Along with its simplicity and satisfactory performance in solving many problems, it has some essential drawbacks, such as slow convergence, sensitivity to noise, and dependence of performance on the heuristically selected learning rate.

The goal of the present work is to develop an efficient learning algorithm with higher rate of convergence without significant increase of computational load.

2 Neuro-fuzzy Network

Let us consider a neuro-fuzzy network, based on a Sugeno-type fuzzy inference system [2] with N_I inputs, one output, and N_R rules:

B. Reusch (Ed.): Fuzzy Days 2001, LNCS 2206, pp. 68–75, 2001.

$$\text{IF } x_1 \text{ IS } A_{1,i}^{(R)} \text{ AND...AND } x_{N_I} \text{ IS } A_{1,N_I}^{(R)} \text{ THEN } y = f_i, \quad i = 1,\ldots,N_R,\ j = 1,\ldots,N_I, \tag{1}$$

where x_j is the j-th input, $A_{j,i}^{(R)}$ is the linguistic label (fuzzy set) on the j-th input in the antecedent of the i-th rule, f_i is the linear model in the consequent of the i-th rule, R is the matrix that determines the structure of the rule base:

$$R = \left(\underbrace{\begin{matrix} R_{1,1} & \cdots & R_{1,N_I} \\ \vdots & \ddots & \vdots \\ R_{N_R,1} & \cdots & R_{N_R,N_I} \end{matrix}}_{\text{antecedent}} \left| \underbrace{\begin{matrix} R_{1,N_I+1} \\ \vdots \\ R_{N_R,N_I+1} \end{matrix}}_{\text{consequent}} \right. \right). \tag{2}$$

The linguistic label $A_{j,i}^{(R)}$ is defined by the membership function

$$\mu_{A_{j,i}^{(R)}}(x_j) = \mu_{i,j}^{(R)} = \begin{cases} \mu_{j,m}(x_j, \omega_{j,m}), & m = R_{i,j}, \quad R_{i,j} > 0, \\ 1, & R_{i,j} = 0, \end{cases} \tag{3}$$

where $\mu_{j,m}$ is the m-th membership function on the input x_j, and $\omega_{j,m}$ is the vector of parameters of the membership function $\mu_{j,m}$.

In case of the widely used Gaussian membership functions

$$\mu_{j,m}(x_j, \omega_{j,m}) = \exp\left(-\frac{(x_j - c_{j,m})^2}{2\sigma_{j,m}^2} \right), \quad \omega_{j,m} = (c_{j,m}, \sigma_{j,m})^T, \tag{4}$$

where $c_{j,m}$ and $\sigma_{j,m}$ are the center and width of the membership function, respectively.

The membership function $\mu_{j,m}$ can be involved in evaluation of more than one rule and can thus have multiple "aliases" $\mu_{i,j}^{(R)}$, which are introduced in order to simplify derivation of the learning algorithm and to preserve generality regardless of the partitioning used.

The system output is calculated according to the equation

$$y = \frac{\sum_{i=1}^{N_R} w_i f_i}{\sum_{i=1}^{N_R} w_i} = \sum_{i=1}^{N_R} \overline{w}_i \Theta_i^T X, \tag{5}$$

$$w_i = \prod_{j=1}^{N_I} \mu_{j,i}^{(R)}, \quad \overline{w}_i = \frac{w_i}{w_\Sigma}, \quad w_\Sigma = \sum_{i=1}^{N_R} w_i, \quad f_i = \Theta_i^T X,$$

where w_i is the firing strength of the i-th rule, $\overline{w}_i$ is the normalized firing strength, Θ_i is the vector of the consequent parameters of the i-th rule, and X is the input vector:

$$\Theta_i = \left(\theta_{0,i}^{(R)}, \theta_{1,i}^{(R)}, \ldots, \theta_{N_I,i}^{(R)}\right)^T, \quad X = \left(1, x_1, \ldots, x_{N_I}\right)^T, \tag{6}$$
$$\theta_{j,i}^{(R)} = \theta_{j,R_{i,N_I+1}}, \quad j = 0, \ldots, N_I .$$

Introducing vectors $\Theta = \left(\Theta_1, \ldots, \Theta_{N_R}\right)^T$, $\Phi = \left(\overline{w}_1 X, \ldots, \overline{w}_{N_R} X\right)^T$, we can transform equation (5) into a compact form

$$y = \Theta^T \Phi . \tag{7}$$

3 Learning Algorithm

The task is to choose at every iteration k such parameters $(\Omega(k), \Theta(k))$, where $\Omega(k)$ is the vector of the antecedent parameters, and $\Theta(k)$ is the vector of the consequent parameters, that would minimize the error function

$$E(k) = \frac{1}{2} e^2(k), \quad e(k) = y^*(k) - y(k), \tag{8}$$

where $y^*(k)$ is the target output, $y(k)$ is the actual system output at time instant k.

3.1 Tuning Antecedents

Tuning of the antecedent parameters is a nonlinear optimization problem, since the system output depends nonlinearly on the components of $\Omega(k)$. So the tuning can be performed using any nonlinear optimization procedure, such as the well-known gradient descent. But this procedure converges relatively slowly. Much higher rate of convergence can be achieved when using second-order procedures, such as the Marquardt algorithm [3]. Let us write the Marquardt algorithm for tuning of the antecedent parameters:

$$\Omega(k) = \Omega(k-1) + \left(J(k)\, J^T(k) + \eta(k)\, I\right)^{-1} e(k)\, J(k), \tag{9}$$
$$\Omega(k) = \left(\Omega_1(k), \ldots, \Omega_{N_R}(k)\right)^T, \Omega_i(k) = \left(\omega_{1,i}^{(R)}(k), \ldots, \omega_{N_I,i}^{(R)}(k)\right)^T,$$
$$\omega_{j,i}^{(R)}(k) = \omega_{j,R_{i,j}}(k), \quad J(k) = \nabla_{\Omega} y(k),$$

where $J(k)$ is the Jacobian matrix (vector in case of a single-output system), which contains first derivatives of the system output with respect to the tuned parameters at time instant k, $\eta(k)$ is a scalar parameter, I is the unity matrix.

The vector $J(k)$ consists of the gradients of the system output with respect to the antecedent parameters of all rules:

$$J(k) = \left(\nabla^T_{\Omega_1} y(k), \ldots, \nabla^T_{\Omega_{N_R}} y(k)\right)^T . \tag{10}$$

In turn, each gradient $\nabla_{\Omega_i} y$ consists of the gradients of the system output with respect to the parameters of the membership functions, involved in evaluation of the i-th rule (the discrete time index k is left out for simplicity):

$$\nabla_{\Omega_i} y = \left(\nabla^T_{\omega^{(R)}_{1,i}} y, \ldots, \nabla^T_{\omega^{(R)}_{N_I,i}} y\right)^T = \left(\frac{\partial y}{\partial c^{(R)}_{1,i}}, \frac{\partial y}{\partial \sigma^{(R)}_{1,i}}, \ldots, \frac{\partial y}{\partial c^{(R)}_{N_I,i}}, \frac{\partial y}{\partial \sigma^{(R)}_{N_I,i}}\right)^T , \tag{11}$$

where $c^{(R)}_{j,i} = c_{j,R_{i,j}}$, $\sigma^{(R)}_{j,i} = \sigma_{j,R_{i,j}}$, $i = 1, \ldots, N_R$. Partial derivatives, constituting the gradients $\nabla_{\Omega_i} y$, can be calculated using the chain rule of differentiation, as it is usually done with the back-propagation procedure.

To reduce computational complexity of equation (9), let us use the Sherman-Morrison formula of matrix inversion:

$$(J\, J^T + \eta\, I)^{-1} J = \left(\eta^{-1} I - \frac{\eta^{-1} I\, J\, J^T I\, \eta^{-1}}{1 + J^T \eta^{-1} I\, J}\right) J . \tag{12}$$

After simple transformations on (12) we obtain:

$$\left(\eta^{-1} I - \frac{\eta^{-1} I\, J\, J^T I\, \eta^{-1}}{1 + J^T \eta^{-1} I\, J}\right) J = \frac{J}{\eta + \|J\|^2} . \tag{13}$$

Substitution of (13) into (9) gives:

$$\Omega(k) = \Omega(k-1) + e(k)\, J(k) \left(\eta(k) + \|J(k)\|^2\right)^{-1} . \tag{14}$$

When $\eta(k) = 0$, this formula can be regarded as a nonlinear modification of the Widrow-Hoff algorithm [4]. To introduce smoothing properties, let us re-write the denominator of equation (14) as follows:

$$q(k) = \eta(k) + \|J(k)\|^2, \quad \eta(k) = \alpha\, q(k-1), \tag{15}$$

where α is a forgetting factor.

Substituting (15) into (14), we obtain:

$$\begin{cases} \Omega(k) = \Omega(k-1) + \Delta\Omega(k), & \Delta\Omega(k) = q^{-1}(k)\, e(k)\, J(k), \\ q(k) = \alpha\, q(k-1) + \|J(k)\|^2, & 0 \le \alpha \le 1, \end{cases} \tag{16}$$

where $\Delta\Omega(k)$ is the update vector for the antecedent parameters of all rules:

$$\Delta\Omega(k) = (\Delta c_{1,1}^{(R)}(k), \Delta\sigma_{1,1}^{(R)}(k), \ldots, \Delta c_{N_I,1}^{(R)}(k), \Delta\sigma_{N_I,1}^{(R)}(k), \ldots$$
$$\ldots, \Delta c_{1,N_R}^{(R)}(k), \Delta\sigma_{1,N_R}^{(R)}(k), \ldots, \Delta c_{N_I,N_R}^{(R)}(k), \Delta\sigma_{N_I,N_R}^{(R)}(k))^T. \tag{17}$$

Since a membership function can be involved in evaluation of more than one rule, the overall update for the parameters of the m-th membership function on the j-th input will be calculated as follows:

$$\Delta c_{j,m} = \sum_{\substack{n=1 \\ \forall n(R_{n,j}=m)}}^{N_R} \Delta c_{j,n}^{(R)}, \quad \Delta\sigma_{j,m} = \sum_{\substack{n=1 \\ \forall n(R_{n,j}=m)}}^{N_R} \Delta\sigma_{j,n}^{(R)}, \quad j = 1, \ldots, N_I \,. \tag{18}$$

3.2 Tuning Consequents

It can be readily seen from (7) that $\Phi = \nabla_\Theta y$. So we can write the learning algorithm for the consequent parameters, similar to (16)-(18):

$$\begin{cases} \Theta(k) = \Theta(k-1) + \Delta\Theta(k), \quad \Delta\Theta(k) = r^{-1}(k)\, e(k)\, \Phi(k), \\ r(k) = \beta\, r(k-1) + \|\Phi(k)\|^2, \quad 0 \le \beta \le 1, \end{cases} \tag{19}$$

$$\Delta\Theta(k) = (\Delta\theta_{0,1}^{(R)}(k), \Delta\theta_{1,1}^{(R)}(k), \ldots, \Delta\theta_{N_I,1}^{(R)}(k), \ldots$$
$$\ldots, \Delta\theta_{0,N_R}^{(R)}(k), \Delta\theta_{1,N_R}^{(R)}(k), \ldots, \Delta\theta_{N_I,N_R}^{(R)}(k))^T, \tag{20}$$

$$\Delta\theta_{j,m} = \sum_{\substack{n=1 \\ \forall n\,(R_{n,N_I+1}=m)}}^{N_R} \Delta\theta_{j,n}^{(R)}, \quad j = 0, \ldots, N_I \,. \tag{21}$$

4 Simulaton Results

To demonstrate the applicability of the proposed algorithm, two experiments are presented. In the first one, a nonlinear dynamic plant is identified online. In the second one, the algorithm is used in batch mode to train the neuro-fuzzy network to predict a chaotic time series.

4.1 Online Learning

The nonlinear plant [5] is governed by the following difference equation:

$$y(k+1) = \frac{y(k)y(k-1)y(k-2)u(k-1)(y(k-2)-1)+u(k)}{1+y^2(k-1)+y^2(k-2)} \,. \tag{22}$$

The plant (22) was identified using the neuro-fuzzy network (1)-(7) with 32 rules, forming a grid partition. The network contained 212 parameters, adapted for 250 time steps using the algorithm (16)-(21) with $\alpha = \beta = 0.95$, $q(0) = r(0) = 1$.

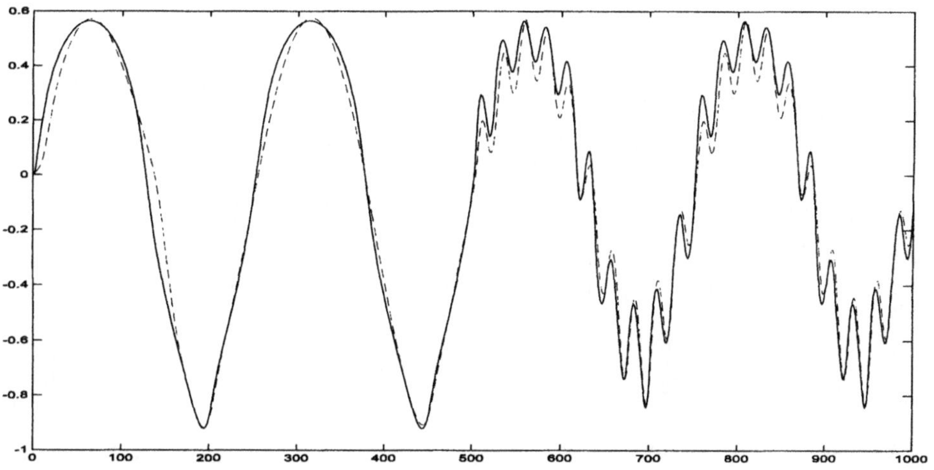

Fig. 1. Outputs of the plant (*solid line*) and neuro-fuzzy network (*dashed line*). The network is not adapted after 250 time steps

The input to the plant and network was $u(k) = \sin(2\pi k/250)$ for $k \leq 500$ and $u(k) = 0.8\sin(2\pi k/250) + 0.2\sin(2\pi k/25)$ for $k > 500$.

The results are shown in Fig. 1. The network output follows the output of the plant immediately, and the error is small even after 500 time steps, when the input signal is changed. In [5], the authors used a neural network with two hidden layers, trained for 100000 time steps using a random input before it could follow the output of the plant (22) satisfactorily. The networks contained 300 adjustable parameters.

In [6], a hybrid fuzzy-neural system with 225 adjustable parameters was trained for 30000 time steps using a random input to identify the same plant.

Thus, the neuro-fuzzy network (1)-(7) with less number of adjustable parameters, adapted using the proposed algorithm (16)-(21), showed superior performance.

4.2 Batch Learning

The time series, used in the second experiment, was generated by the chaotic Mackey-Glass time-delay differential equation [7]:

$$\dot{x}(t) = \frac{0.2\, x(t-\tau)}{1 + x^{10}(t-\tau)} - 0.1\, x(t) \ . \tag{23}$$

The values of the time series (23) at each integer point were obtained by means of the fourth-order Runge-Kutta method. The time step used in the method was 0.1, initial condition $x(0) = 1.2$, delay $\tau = 17$, and $x(t)$ was derived for $t = 0,\ldots,1200$. The values $x(t\text{-}18)$, $x(t\text{-}12)$, $x(t\text{-}6)$, and $x(t)$ were used to predict $x(t+6)$.

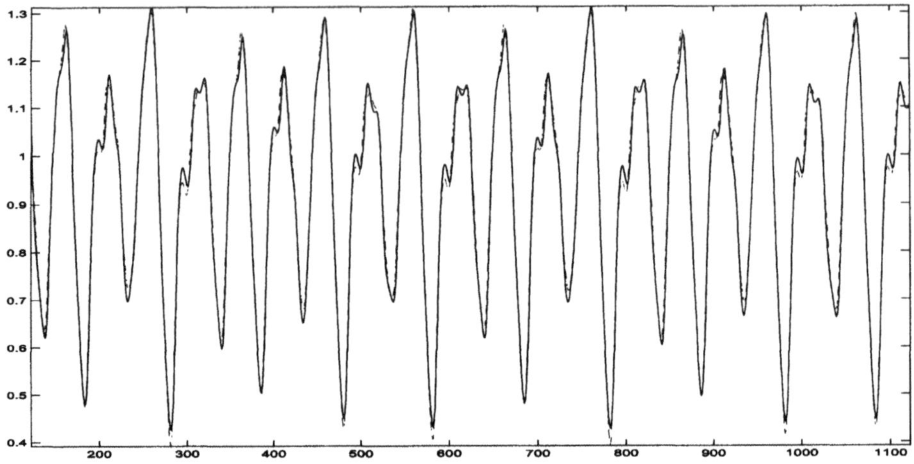

Fig. 2. Mackey-Glass time series (*solid line*) and the output of the forecasting neuro-fuzzy network (*dashed line*), trained in batch mode with the proposed algorithm

From the generated data, 500 values for $t = 118,...,617$ were used as the training data set, and the next 500 for $t = 618,...,1117$ were used as the checking data set.

A 4 input-1 output neuro-fuzzy network (1)-(7) with 2 Gaussian membership functions per input and 16 rules, forming a grid partition, was created. The network was trained for 100 epochs. The parameters of the algorithm (16)-(21) were $\alpha = \beta = 0.98$, $q(0) = 10000$, $r(0) = 1$. The resulting plots are shown in Fig. 2. The two curves, representing the actual and predicted values, are almost indistinguishable.

To estimate the quality of the received results, we used the root-mean-squared errors on the training data set ($\mathrm{RMSE}_{\mathrm{TRN}}$) and checking data set ($\mathrm{RMSE}_{\mathrm{CHK}}$). The same experiment was carried out using the ANFIS learning rules [8]. The results were also compared with those obtained using the Neuro-Fuzzy Function Approximator (NEFPROX) in [9]. All the results are summarized in Table 1.

The ANFIS hybrid learning rule provides the best accuracy among the other methods. It employs the recursive least squares estimate (RLSE) for tuning of antecedent parameters, and is more complex computationally. This complexity increases quadratically as the number of tuned parameters grows, in contrast to the proposed algorithm (16)-(21), in which the computational complexity increases linearly. Thus, for the systems with large number of rules the proposed algorithm will be preferable.

The other ANFIS learning rule, based on the back-propagation only, was unable to provide satisfactory approximation after 100 training epochs.

The NEFPROX provided the results, which are worse than those provided by the proposed algorithm, but it is based on a different architecture (fuzzy perceptron), and is capable of extracting fuzzy rules itself.

Table 1. Comparison of different learning methods

Method	# Epochs	# Rules	$RMSE_{TRN}$	$RMSE_{CHK}$
Proposed	100	16	0.021617	0.021622
Hybrid ANFIS	100	16	0.0083246	0.0083428
Back-propagation ANFIS	100	16	0.30724	0.30717
NEFPROX	216	129	0.0315	0.0332

5 Conclusion

Simulation results proved effectiveness of the proposed algorithm. It can be used for identification of nonlinear plants, and for nonlinear time series prediction. The algorithm is characterized by linear dependence of computational complexity on the number of tuned parameters, as the algorithms based on the conventional gradient descent, providing at the same time higher rate of convergence. The proposed algorithm is simple, and well suited for low-cost hardware implementation.

All the considerations above were based on the assumption that the initial partitioning of the input space is known, and the number of rules does not change while learning of parameters takes place. Further research will be directed to augment the proposed algorithm with the capability of automatic rule extraction from data.

References

1. Rumelhart, D.E., Hinton, G.R., Williams, R.J.: Learning Internal Representation by Error Propagation. In: Rumelhart, D.E., McClelland, J.L. (eds.): Parallel Distributed Processing, Vol. 1. MIT Press, Cambridge, MA (1986) 318–364
2. Takagi, T., Sugeno, M.: Derivation of fuzzy control rules from human operator's control actions. Proc. of the IFAC Symp. on Fuzzy Information, Knowledge Representation and Decision Analysis (1983) 55-60
3. Marquardt, D.W.: An Algorithm for Least Squares Estimation of Non-Linear Parameters. SIAM J. Appl. Math. 11 (1963) 431-441
4. Rojas, R.: Neural Networks. A Systematic Introduction. Springer-Verlag, Berlin Heidelberg New York (1996)
5. Narendra, K.S., Parthasarathy, K.: Identification and Control of Dynamical Systems Using Neural Networks. IEEE Trans. on Neural Networks 1 (1990) 4-27
6. Lin, C.-J.: SISO Nonlinear System Identification Using a Fuzzy-Neural Hybrid System. Int. J. Neural Systems, Vol. 8 (1997) 325-337
7. Mackey, M. C., Glass, L.: Oscillation and chaos in physiological control systems. Science, 197 (1977) 287-289.
8. Jang, J.-S. R.: ANFIS: Adaptive-network-based fuzzy inference systems. IEEE Trans. on Systems, Man, and Cybernetics, Vol. 23 (1993) 665–685
9. Nauck, D., Kruse, R.: A Neuro-Fuzzy Approach to Obtain Interpretable Fuzzy Systems for Function Approximation. Proc. of the IEEE Int. Conf. on Fuzzy Systems (1998) 1106-1111

Fuzzy Probabilistic Neural Networks: A Practical Approach to the Implementation of Baysian Classifier

Farshid Delgosha[1] and Mohammad B. Menhaj[2,3]

[1] P.O. Box 14335-874, Tehran, Iran

Delgosha@mail.dci.co.ir

[2] Dept. of Electrical Eng., Amir Kabir University, Tehran, Iran

menhaj@cic.aku.ac.ir

[3] School of Electrical and Computer Engineering, Oklahoma State University, OK, USA

menhaj@okstate.edu

Abstract. A classifier with the optimum decision, Baysian classifier could be implemented with Probabilistic Neural Networks (PNNs). The authors presented a new competitive learning algorithm for training such a network when all classes are completely separated. This paper generalizes our previous work to the case of overlapping categories. In our new perspective, the network is, in fact, made blind with respect to the overlapping training samples using fuzzy concepts, so the new training algorithm is called Fuzzy PNN (or FPNN). The usefulness of FPNN has been proved by some classification problems. The simulation results highlight the merit of the proposed method.

1 Introduction

Minimization of the decision risk is the criteria used in Baysian classifier. Decision risk is actually measured by an optimum decision variable mathematically derived from the definition of the decision risk. A decision variable, in fact, maps the input-signal space into the decision space with the most distant classes. However, knowledge of conditional probability density functions (PDFs) of all classes is required in its derivation procedure. Some PDF estimators have been proposed in the literature that can be categorized as parametric and non-parametric. Parametrical estimation of PDF is based on the pre-assumption that the process under investigation might be well approximated by one of the known probabilistic distributions. Parameters of the suitably fitted distribution can be calculated with the maximum likelihood method or some proper statistics. However, many actual processes don't lend themselves to this type of modeling. Therefore, non-parametric PDF estimators have been proposed. Parzen introduced a family of non-parametric PDF estimators in [1] that are consistent and asymptotically normal. Parzen estimators are briefly explained in the next Section of this paper. Non-parametric estimators show more robustness in comparison to parametric ones for practical stochastic processes. A thorough discussion of non-parametric estimation techniques can be found in [2] and [3].

B. Reusch (Ed.): Fuzzy Days 2001, LNCS 2206, pp. 76–85, 2001.

The PDF estimator of Parzen can be realized by a neural network type structure called Probabilistic Neural Network (PNN) which was first introduced by Specht in [4] and [5]. In PNN, the number of neurons is directly proportional to the number of sample vectors, so an enormous amount of storage memory is required. Specht introduced the Polynomial Adaline (Padaline) in [5] as a solution to this problem. Data clustering is another solution in which neighboring sample vectors are grouped into one cluster. Each cluster, represented by a prototype vector, is assumed to be governed by a known distribution, mostly Gaussian. The PDF of the whole process is a linear combination of clusters' PDFs. This technique is, in fact, a semi-parametric approach to the PDF estimation. Streit [6] and Traven [7] designed a PNN with this method, and estimated the PDF parameters with the maximum likelihood method. None of the above approaches is based on the learning capabilities of neural networks for parameter adjustment. A competitive-type learning algorithm was proposed in [8] for adjusting the PNN parameters. The merit of that new training algorithm was proved by some classification examples.

In the case of non-overlapping categories, classification results of the PNN are reliable. When there is overlap between classes, the learning algorithm presented in [8] will not converge. For partially overlapped classes common portions must be left out and non-overlapping sections maybe applied to the PNN classifier. Decision-based neural networks, presented in [9], are designed for these situations. These networks make a soft decision in the training phase instead of the hard decision made by traditional classifiers. In the soft decision, the effect of overlapping portions is lessened, or, in other words, the classifier is intentionally made blind with respect to overlapping sections. This paper addresses the fuzzy decision version of the learning algorithm presented in [8]. PNN with the modified learning algorithm is called Fuzzy PNN or FPNN.

This paper is organized as follows: Section 2 is an overview of Parzen PDF estimator, PNN structure and the competitive learning algorithm of [8], which are necessary backgrounds for understanding FPNN. The modified fuzzy learning algorithm is introduced in Section 3. Some signal classification examples in Section 4 show the applicability of FPNN for practical cases. Finally, Section 5 concludes the paper.

2 Parzen PDF Estimator and PNN Structure

Suppose that the PDF of a d-dimensional random process must be estimated. The Parzen PDF estimator is as follows [1]:

$$\hat{f}(\mathbf{x}) = \frac{1}{Nh^d}\sum_{i=1}^{N} K\left(\frac{\mathbf{x}-\mathbf{x}_i}{h}\right) \tag{1}$$

where $\mathbf{x}_1, \ldots, \mathbf{x}_N$ are observation sample vectors, h is a suitably chosen positive number and K is a real valued kernel function satisfying the condition:

$$\int_{\Re^d} K(\mathbf{x})\, d\mathbf{x} = 1\,. \tag{2}$$

Usually a symmetric unimodal PDF, e.g. the multivariate Gaussian density function, is selected for the kernel function K, i.e.

$$K(\mathbf{x}) = \frac{1}{\sqrt{(2\pi)^d |\mathbf{C}|}} \exp\left(-\frac{\mathbf{x}^T \mathbf{C}\mathbf{x}}{2}\right). \tag{3}$$

In the above equation, $\mathbf{C}$ is the covariance matrix and the superscript T denotes matrix transposition. Parzen has proved in [1] that if h in equation (1) is chosen as a function of n such that

$$\lim_{n\to\infty} h(n) = 0 \tag{4}$$

and if $K(\mathbf{x})$ satisfies some certain conditions given in [1], the estimator of (1) will be asymptotically unbiased. Increasing the accuracy of estimation, we must have a large number of sample vectors which as a result requires a large amount of storage memory. The data clustering technique described in [8] can solve this problem. In this method, adjacent sample vectors are grouped as a cluster that is represented by a prototype vector (PV) and the PDF is estimated as [6], [7]:

$$\hat{f}(\mathbf{x}) = \frac{1}{\sqrt{(2\pi)^d |\hat{\mathbf{C}}|}} \sum_{j=1}^{Q} \gamma_j \exp\left\{-\frac{1}{2}(\mathbf{x}-\boldsymbol{\eta}_j)^T \hat{\mathbf{C}}^{-1}(\mathbf{x}-\boldsymbol{\eta}_j)\right\} \tag{5}$$

where Q is the total number of clusters, $\hat{\mathbf{C}}$ is the estimated covariance matrix, γ_j and $\boldsymbol{\eta}_j$ are the proportion and PV of the jth cluster in the estimation, respectively. The γ_j's must obviously sum to unity [8]. The PDF estimator of (5) is a semi-parametric one.

Now suppose that we have M disjoint classes with priori probabilities $p_1, \ldots, p_M$. The pdf of each class may be estimated using the estimator of (5). If ℓ_{jk} is the loss of classifying $\mathbf{x}$ into class j when the correct decision should have been class k, the estimated risk of classifying $\mathbf{x}$ into class j is:

$$\hat{r}_j(\mathbf{x}) = \sum_{\substack{k=1\\k\neq j}}^{M} \ell_{jk}\, p_k\, \hat{f}_k(\mathbf{x}), \qquad j = 1, \ldots, M. \tag{6}$$

Baysian classifier chooses the class with minimum decision risk; i.e. $\mathbf{x}$ is classified into the j^*th class if and only if

$$j^* = \arg\min_{j=1,\ldots,M} \left\{\hat{r}_j(\mathbf{x})\right\}. \tag{7}$$

All the above procedure and formulas can be realized with a neural network type structure depicted in Fig. 1 [6], [7], [8]. In this figure, $\hat{\mathbf{L}}$, an upper triangular matrix, is the square root of the estimated covariance matrix, i.e. $\hat{\mathbf{C}} = \hat{\mathbf{L}}\hat{\mathbf{L}}^{-1}$, which is valid because of positive definiteness of $\hat{\mathbf{C}}$. The inverse of $\hat{\mathbf{L}}$ plays the role of whitening filter in the block diagram of Fig. 1 since

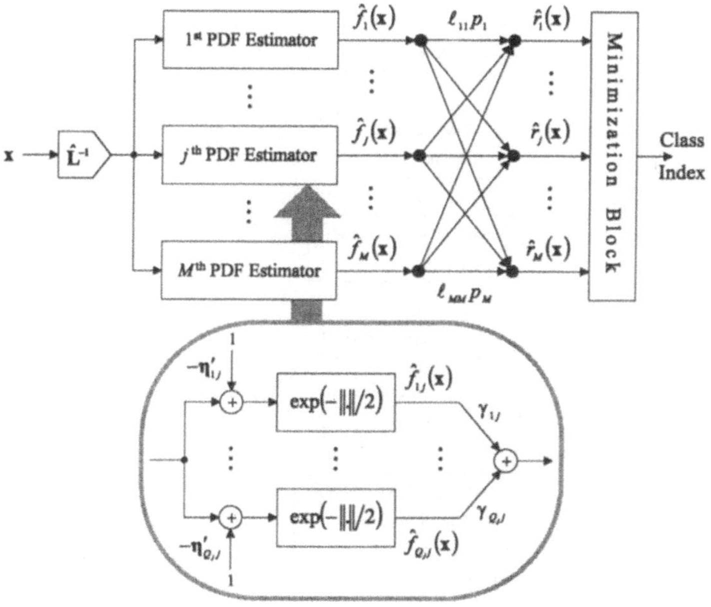

Fig. 1. Block diagram of the PNN

all cross covariances vanish out in the transformed space. Each PDF estimator block consists of some neurons with the exponential activation function. The biases $\boldsymbol{\eta}'_{ij}$ are the whitened versions of prototype vectors:

$$\boldsymbol{\eta}'_{ij} = \hat{\mathbf{L}}^{-1}\boldsymbol{\eta}_{ij}, \qquad i = 1, \ldots, Q_j, \quad j = 1, \ldots, M \tag{8}$$

where $\boldsymbol{\eta}_{ij}$ is the ith prototype vector of the jth class and Q_j is the number of clusters in the jth class. The γ_{ij}'s are the cluster proportions with the restriction $\sum_{i=1}^{Q_j}\gamma_{ij} = 1$ ($j = 1, \ldots, M$). Applying the softmax transformation to the γ_{ij}'s give new variables v_{ij} that are relaxed from the summing to unity restriction:

$$\gamma_{ij} = \frac{e^{v_{ij}}}{\sum_{k=1}^{Q_j} e^{v_{kj}}}, \qquad j = 1, \ldots, M, \quad i = 1, \ldots, Q_j. \tag{9}$$

Flowchart of the competitive learning algorithm proposed in [8] is provided in Fig. 2.

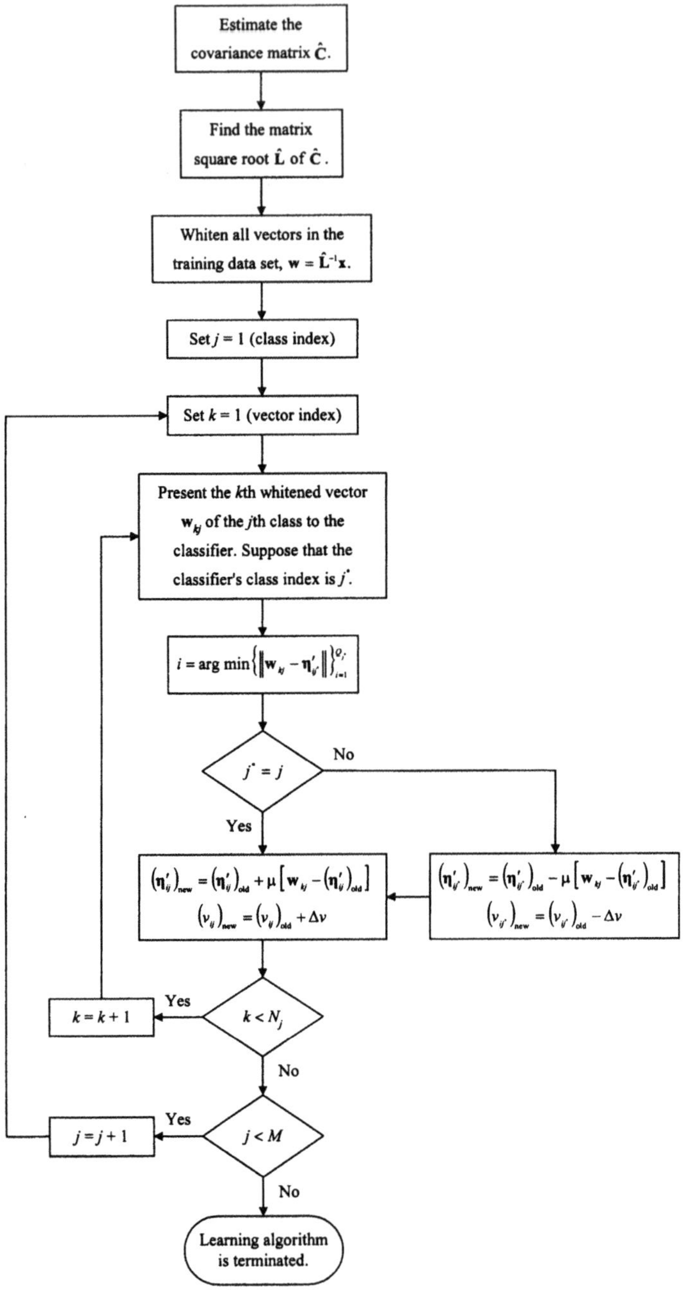

Fig. 2. Flowchart of the competitive learning algorithm used in PNN

3 Fuzzy Probabilistic Neural Networks

To make the PNN insensitive with respect to the overlapping samples, the method presented in [9] is used. Suppose that the sample vector **x**, belonging to the *j*th class, is presented to the PNN and the risk of choosing this class is calculated as $\hat{r}_j(\mathbf{x})$. Let the j'th class has the minimum risk of $\hat{r}_{j'}(\mathbf{x})$ among all classes excluding the correct class *j*, i.e.:

$$j' = \arg\max_{i \neq j}\{\hat{r}_i(\mathbf{x})\} \tag{14}$$

A measure of misclassification can be introduced as:

$$m(\mathbf{x}) = \hat{r}_{j'}(\mathbf{x}) - \hat{r}_j(\mathbf{x}) \tag{15}$$

Two scenarios are of interest: 1) If m is negative, the classification is correctly performed, 2) If m is positive, the sample vector **x** is misclassified. In this case, a small magnitude for m alarms a sample vector near the decision boundary that may have overlap with the region of other classes, and thus can degrade the learning algorithm. To resolve this degradation problem, the effect of misclassification measure must be included in the training process through introduction of a penalty function. The PNN disregards this problem and make use of the following penalty function, which is, in fact, a hard decision:

$$\rho(m) = \begin{cases} 0, & m < 0 \\ 1, & m \geq 0 \end{cases} \tag{16}$$

In other words, the PNN shows equal sensitivity to all sample vectors regardless of their relative distance to decision boundary. A non-sharp (smooth) penalty function (logsigmoidal type), which is defined below, is used by FPNN, i.e. it performs a soft decision.

$$\rho(m; b, s) = \begin{cases} 0, & m < 0 \\ 1/\left(1 + e^{-(m-b)/s}\right), & m \geq 0 \end{cases} \tag{17}$$

In the above equation, b is the bias introduced for more robustness and s is the softness parameter. The curves of the two above penalty functions are depicted in Fig. 3.

The training steps of FPNN are completely similar to those of PNN except the last step: PNN counts the total number of misclassified sample vectors and repeats the training procedure until all the sample vectors are correctly classified. However, FPNN computes the percentage of error and progresses until it falls below some error goal. The misclassification percentage, MP, is set to zero at the beginning of the training procedure and for a misclassified sample vector is incremented by the following update term

$$\Delta\text{MP} = 100 \times \rho / N \tag{18}$$

where N is the total number of sample vectors used in training phase. After presenting all sample vectors to the classifier, MP is compared with preset error goal. If it falls below the error goal, the training process is terminated; otherwise, it is repeated. The merit of the proposed training algorithm is proved in the next Section by some computer-simulated examples.

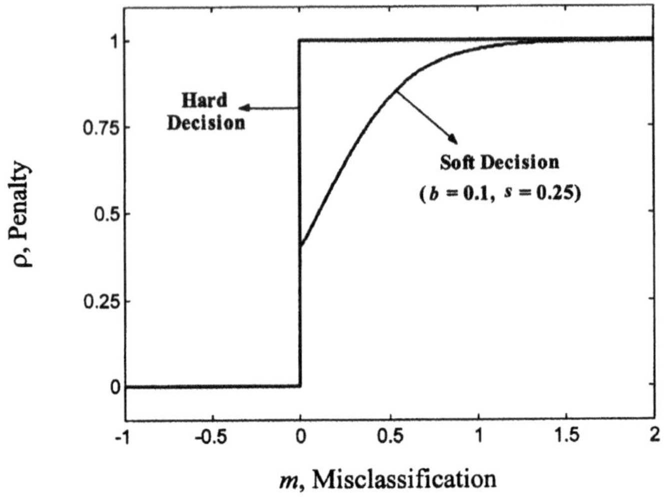

Fig. 3. Hard and soft decision penalty functions

4 Signal Classification Examples

The proposed learning algorithm is tested by two classification problems with overlapping categories.

4.1 Two Simple Overlapping Classes

The first example is the scenario of classifying two overlapping classes with elements normally distributed in the 2-dimentional plane as follows:

$$\text{Class 1:} \quad \mathbf{x} \sim \mathrm{N}\ (0,\ +1,\ 0.75,\ 0.75;\ 0) \tag{19a}$$

$$\text{Class 2:} \quad \mathbf{x} \sim \mathrm{N}\ (0,\ -1,\ 0.75,\ 0.75;\ 0) \tag{19b}$$

where $\mathrm{N}\ (\eta_1, \eta_2, \sigma_1, \sigma_2; r)$ is a two dimensional normal distribution with means η_1 and η_2, variances σ_1 and σ_2, and correlation coefficient r. These two classes along with the assigned PVs of the FPNN and the decision boundary are illustrated in Fig. 4a. Error goal of the employed FPNN has been 10%. The bias and softness parameters of the penalty function have been both 0.1. For comparison purposes, the decision boundary of a 2-layered feedforward neural network with 5 hidden neurons is depicted in Fig. 4b. Logsigmoidal activation functions were used in that network and it was trained with Levenberg-Marquardt learning algorithm [10]. A simple comparison between the two parts of Fig. 4 shows the applicability of FPNN.

4.2 XOR Problem for Overlapping Classes

We have considered the XOR problem as our second simulation example. Fig. 5a shows the 2-dimentinal configuration of this problem that consists of two overlapping classes. The elements of the two classes are normally distributed as follows:

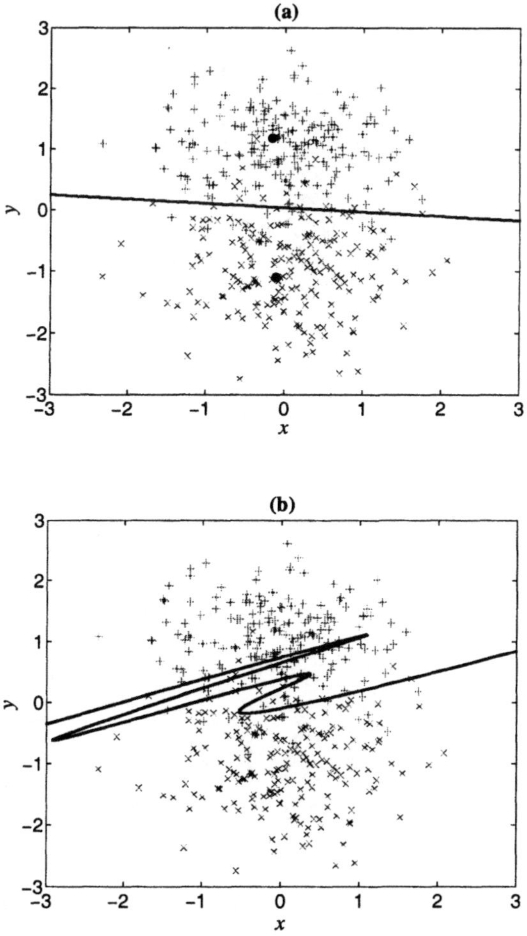

Fig. 4. 2-dimensional representation of the first classification example along with the decision boundary for (a) FPNN, (b) 2-layered feedforward neural network

$$\text{Class 1:} \quad \mathbf{x} \sim \mathrm{N}\left(\mp 1,\ \pm 1,\ 0.5,\ 0.5;\ 0\right) \tag{20a}$$

$$\text{Class 2:} \quad \mathbf{x} \sim \mathrm{N}\left(\mp 1,\ \mp 1,\ 0.5,\ 0.5;\ 0\right) \tag{20b}$$

We have employed an FPNN with 5% error goal for this classification problem. Both bias and softness parameters were considered 0.1. The decision boundaried of FPNN and a 2-layered feedforward neural network (with 5 hidden neurons, logsigmoidal activation functions and trained with Levenberg-Marquardt learning algorithm) are shown in the two parts of Fig. 5. Obviously FPNN has a better classification result than feedforward neural network.

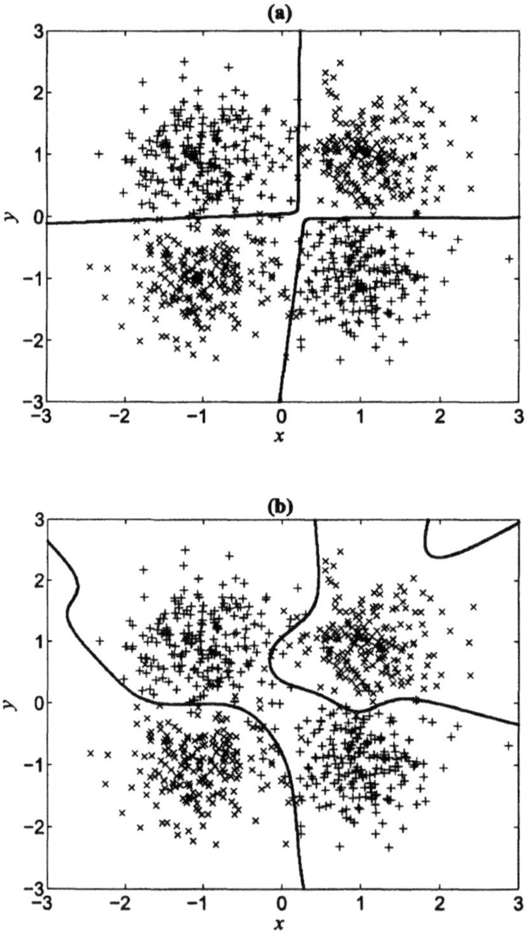

Fig. 5. 2-dimensional representation of the second classification example along with the decision boundary for (a) FPNN, (b) 2-layered feedforward neural network

5 Concluding Remarks

A new fuzzy learning algorithm is presented for probabilistic neural networks for the case of overlapping classes. It is, in fact, an extension to the author's previous work in probabilistic neural network. The new algorithm recognizes overlapping sections and intentionally makes the network blind with respect to those parts. Two computer simulation examples are used to prove the applicability of the new fuzzy algorithm.

References

1. Parzen, E.: On Estimation of Probability Density Function and Mode, Annals of Mathematical Statistics, Vol. 33, (1962) 1065–1076
2. Silverman, B.W.: Density Estimation for Statistics and Data Analysis, Chapman & Hall (1986)
3. Wand, M.P., Jones, M.C.: Kernel Smoothing, Chapman & Hall (1995)
4. Specht, D.F.: Probabilistic Neural Networks for Classification, Mapping, or Associative Memory, Proceeding of the IEEE International Conference on Neural Networks, Vol. 1, San Diego/CA USA (1988) 525–532
5. Specht, D.F.: Probabilistic Neural Networks and Polynomial Adaline as Complementary Techniques for Classification, IEEE Transactions on Neural Networks, Vol. 1, No. 1, (1990) 111–121
6. Streit, R.L., Luginbuhl, T.E: Maximum Likelihood Training of Probabilistic Neural Networks, IEEE Transactions on Neural Networks, Vol. 5, No. 5, (1994) 764–783
7. Traven, H.G.C.: A Neural Network Approach to Statistical Pattern Classification by Semiparametric Estimation of Probability Density Functions, IEEE Transactions on Neural Networks, Vol. 2, No. 3, (1991) 366–377
8. Menhaj, M.B., Delgosha, F.: Probabilistic Neural Networks with Competitive Learning Algorithm, European Congress on Intelligent Techniques and Soft Computing (EUFIT), Germany (1999)
9. Kung, S.Y., Taur, J.S.: Decision-based Neural Networks with Signal/Image Classification Applications, IEEE Transactions on Neural Networks, Vol. 6, No. 1, (1995) 170–181
10. Hagan, M.T., Menhaj, M.B.: Training Feedforward Networks with the Marquardt Algorithm, IEEE Transactions on Neural Networks, Vol. 5, No. 6, (1994) 989–993

Fuzzy-η for Back Propagation Networks

Miroslaw Bedzak

Institute of Control Engineering, Technical University of Szczecin
Sikorskiego 37, 70-313 Szczecin, Poland
bedzak@we.tuniv.szczecin.pl

Abstract. The presented algorithm of fuzzy back propagation used in the teaching phase of a neural fuzzy controller combines the best features of two selected soft computing elements: fuzzy logic and the selected structure of artificial neural network. This approach is the expansion of the classical algorithm of back propagation by use of not only binary, but also any values from the range [0...1] in the teaching sequences and the selection of the value of the teaching factor η using the theory of fuzzy sets.

1 Introduction

Taking pattern by such processes occurring in nervous systems of living organisms as information acquisition, processing and sending, the first models of artificial neural networks (ANN) were worked out at the beginning of the forties. In the neurone model (Fig. 1) proposed by McCulloch and Pitts in 1943 the function f - called the function of activation - was the function of elementary jump.

After the network structure (the number of layers, the number of neurones in each layer) has been selected, the further design and programming of the ANN of multilayer perceptron type proceeds at three stages: the stage of teaching, the stage of testing and the stage of working. Cybenko proved in his paper [1] that the use of two covered layers at least enables to obtain any accuracy if the number of neurones in the layer is sufficiently large and demonstrated that one covered layer is sufficient to approximate any continuous function. The aim of teaching the network in the phase of teaching is to determine weight values $w^{h}_{n^{h+1}n^{h}}$ for neurones of all layers of the network in such a way that at a given input vector $x^{h+1}_{n^{h+1}}$ such values of output signals $y^{h}_{n^{h}}$, are obtained at the network output, which are equal with the determined accuracy to the set values $y^{0}_{n^{0}}$.

One of the algorithms of teaching multilayer networks, that is most often used in practical implementations, is the algorithm of back propagation which determines the strategy of the selection of weight values in the network using gradient methods of

B. Reusch (Ed.): Fuzzy Days 2001, LNCS 2206, pp. 86–91, 2001.

optimization [4],[5],[8]. This algorithm has been elaborated by Werbos (1974), Parker (1982) and Rumelhart (1986) independently.

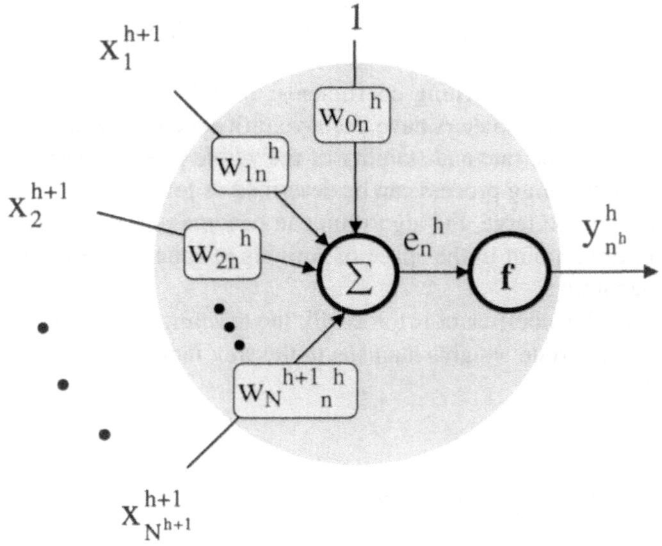

Fig. 1. The model of artificial neurone.

2 Back-Propagation of Algorithm

The name of this algorithm reflects the way of modifying the weights of neurones which consists in calculating the error occurring at the given step of teaching process in the output neurone, and next in sending it back to all neurones whose signals were the input signal for this neurone.The change of the weight of neurones at each step of teaching can be written as follows [4],[6],[7]:

$$w(k+1) = w(k) + \Delta w(k) \tag{1}$$

while the value of the change of the neurone weight is determined as:

$$\Delta w(k) = -\eta \frac{\partial E}{\partial w} + \alpha \Delta w(k-1) \tag{2}$$

The algorithm of teaching ANN of multilayer perceptron type with the method of back propagation is based on the minimization of the quadratic mean error of the following quality coefficient [2],[3]:

$$E = \frac{1}{PN^0}\sum_{p=1}^{P} E^p = \frac{1}{PN^0}\sum_{p=1}^{P}\sum_{n^0=1}^{N^0} E^p_{n^0} = \frac{1}{PN^0}\sum_{p=1}^{P}\sum_{n^0=1}^{N^0}\left(d^p_{n^0} - y^{0p}_{n^0}\right)^2 \tag{3}$$

Finally, the formula for updating the weight coefficient of a neurone in the h-th internal layer of the ANN has the form:

$$\Delta w^{h}_{n^h n^{h-1}}(k) = \eta \delta^{hp}_{n^h} y^{(h-1)p}_{n^{h-1}} + \alpha \Delta w^{h}_{n^h n^{h-1}}(k-1) \tag{4}$$

The proper selection of teaching coefficients: η and α is very important in the teaching process. These parameters have decisive influence on the teaching quality of networks expressed by the rate and stability of the whole process. The influence of η on the course of the teaching process can be described as follows:

- If the value η is too large, the algorithm can become unstable and is not able to attain the optimal point in the space of weights in spite of the big rate of teaching at the beginning,
- If the value of this coefficient is too small, the teaching is sure very slow but the change of the neurone weights happens in the way increasing the stability of the whole process.

3 Fuzzy Back-Propagation of Algorithm

Taking into account the trial of joining the best features of the ANN and fuzzy control to the control strategy of Neuro-Fuzzy type realized in this work, the value of the coefficient η in the phase of teaching can be determined with the method of fuzzy sets theory by means of the following fuzzy conditional statements [9],[10]:

$$\textbf{IF} \ (\text{E is } E_i) \ \textbf{AND} \ (\text{SZW is } SZWj) \ \textbf{THEN} \ (\eta \text{ is } \eta_k) \tag{5}$$

- E_i, SZW_j, η_k - linguistic variables adequately to the error of teaching, the measure of teaching activity and the teaching coefficient of a network,
- E_i, SZW_j, η_k - the values of these linguistic variables represented by respective fuzzy sets.

The value of the teaching error of a network E is calculated according to the formula, while the value of the measure of activity of the teaching process SZW is the sum of the changes of the neurone weights in the respective layer of the ANN. In case of the output layer it is determined by the formula:

$$SZW^0 = \frac{1}{PN^0N^1} \sum_{p=1}^{P} \sum_{n^0=1}^{N^0} \sum_{n^1=1}^{N^1} \left| \Delta w^{0}_{n^1 n^0} \right| \tag{6}$$

The decreasing value of SZW shows that this factor approaches its local minimum. The scheme of the fuzzy back propagation algorithm control is as follows:

Stage I:

- determining the values of the linquistic variables Ei, SZWj and ηk, and the shape of the membership functions of fuzzy sets which represent these variables,
- the set of control rules in the form of fuzzy conditional statements of the type;

Stage II:

- updating the η according to the defined set of rules, making use of fuzzy control, in the final phase of each iteration, i.e. after presenting the entire teaching sequence of the back propagation algorithm.

For general structure of fuzzy conditional statements, presented in (5), adequate linguistic variables:

- E - error of teaching the neural fuzzy controller, expressed by the formula (3),
- SZW - measure of teaching activity, described with the formula (6),
- η - teaching rate factor

are presented in Fig. 2 by means of the membership functions of the following fuzzy sets:

- E: {+1, +2, +3, +4, +5},
- SZW: {+1, +2, +3}
- η: {S1, S2, S3, S4, S5}

defined in Tab. 1 according to denotations from Fig 2.

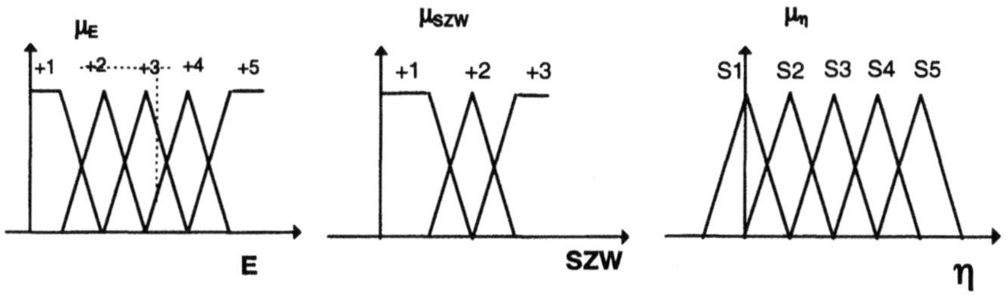

Fig. 2. Membership function of fuzzy sets of E, SZW and η.

Table 1. Rule base table.

		SZW +1	SZW +2	SZW +3
E	+1	S1	S1	S1
E	+2	S3	S2	S1
E	+3	S4	S3	S3
E	+4	S5	S4	S3
E	+5	S5	S5	S5

Fig. 3 shows the result of teaching the ANN accomplishing the function XOR with the method of back propagation characterized by the constant coefficient η and another coefficient determined according to the above mentioned scheme of fuzzy control BP.

The FBP parameters can be formulated in form of decision table and in form of fuzzy conditional statements enabling the realization of the following selection strategy of the value η:

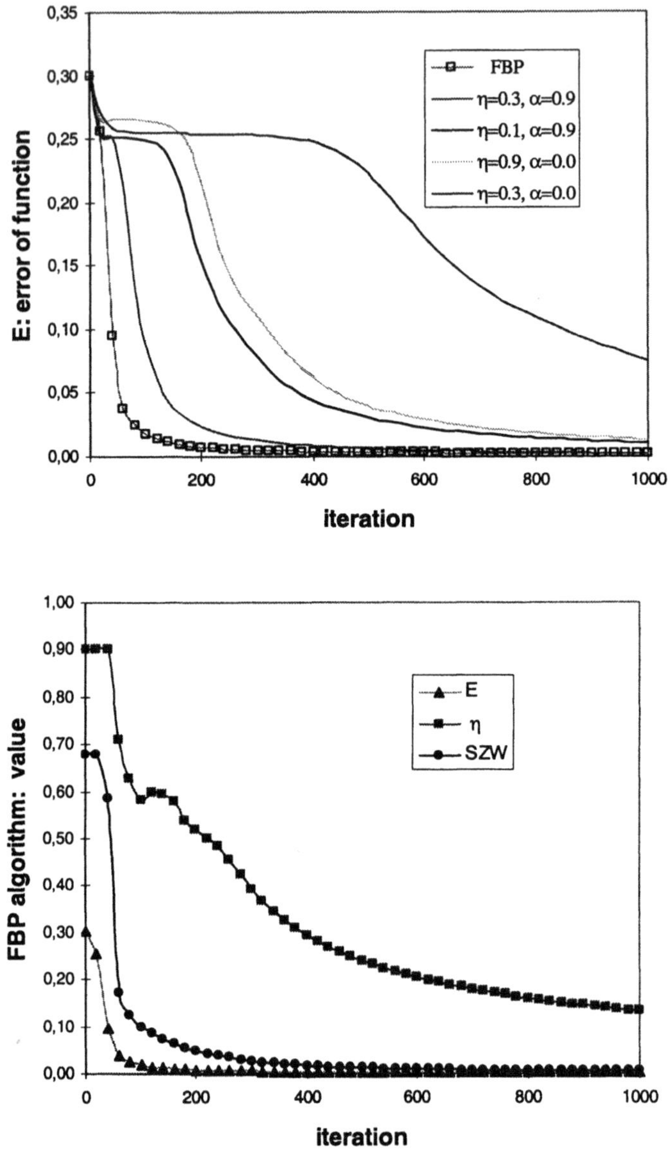

Fig. 3. The result of teaching the ANN.

- If the error of teaching is big and the measure of activity is small, then the teaching coefficient is increased to accelerate teaching process.
- If the error of teaching is small and the measure of activity is big, then the teaching coefficient is decreased to stabilize teaching process.

4 Conclusions

The presented algorithm of fuzzy back propagation (FBP) makes an expansion of classical back propagation algorithm as its teaching sequences employ not only binary but also any values from the interval (0...1), and the values of the teaching factor η are selected with the use of the fuzzy set theory. For the FBP algorithm not to stall, as the classical back propagation algorithm, in the local minimum of the function one should:

- choose the standards in random sequence from the training set
- add the noise by introducing small random changes in the neuron weights, add the noise to the training set data.

The described algorithm of fuzzy back propagation can be enriched by improved self-adjusting mechanisms with the use of not only adaptive change of the teaching coefficient η but also by adaptive selection of the teaching coefficient α (momentum) using fuzzy sets theory including the error of teaching E and the measure of activity SZW.

References

1. Cybenko G.: Approximation by Superpositions of a Sigmoidal Function. Mathematics of Control, Signals and Systems 2 (1989) 303-314
2. Figueiredo M, Gomide F.: Design of Fuzzy Systems Using Neurofuzzy Networks. IEEE Transactions on Fuzzy Systems 7 (1999) 815-828
3. Oh S.H., Lee S.Y.: A new Error function at Hidden Layers for Fast Training of Multilayer perceptrons IEEE Transactions on Neural Networks 7 (1999) 960-964
4. Psaltis D., Sideris A., Yamamura A.: A multilayered neural network controller. IEEE Control System Magazin 4 (1988) 44-48
5. Rumelhart D., Hinton G., Williams R.: Learning representations by backpropagating errors. Nature 323 (1986) 533-536
6. Simpson P.: Artificial neural systems. Pergamon Pres, USA (1990)
7. Song Q, Xiao J., Soh Y.C.: Robust Bacpropagation Training Algorithm for Multilayered Neural Tracking Controller. IEEE Transactions on Fuzzy Systems 8 (1999) 1133-1141
8. Werbos P.: Neurocontrol and fuzzy logic: connections and designs. Inter. j. of Approximation Reasoning 6 (1992) 185-219
9. Zadeh L. A.: Fuzzy sets. Information and Control 8 (1965) 338-353
10. Zimmermann H.: Fuzzy Sets Theory and its Applications. Kluwer, Boston, 2nd edition, (1991)

eaLib – A Java Framework for Implementation of Evolutionary Algorithms

Andreas Rummler and Gerd Scarbata

Technical University of Ilmenau, Institute for Electronic Circuits and Systems, Ilmenau, Germany
arummler@acm.org, gerd.scarbata@inf-technik.tu-ilmenau.de
http://www.inf-technik.tu-ilmenau.de

Abstract. This article gives an overview over *eaLib*, a framework for the implementation of evolutionary algorithms written in Java. After an introduction the kind of genetic representation used in the toolkit is discussed and provided genetic operators are introduced. Thereafter the concept of breaking up an evolutionary algorithm into components and the definition of interfaces for these components is discussed. On that basis a controller model for flexible and fast creation of algorithms is presented. The paper concludes with a section dealing with issues of parallelization of evolutionary algorithms and gives a short outlook on future work.

1 Introduction

Over the last 20 years many kinds of genetic algorithms have been developed, which differ from each other in certain operators or parameters used. A classification of algorithms based on simulated evolution becomes harder each year, because algorithms mixing various different techniques come up. But despite of all the variety that is present now in the field of evolutionary computation all optimization algorithms share several common components.

When we think of those shared items in algorithms a question raises. Is it possible to define some kind of a unified model for algorithms and components that makes it easier to create concrete implementations? Can we create components that share a common interface and that can be easily connected together to form an algorithm?

There are several approaches to create toolkits to support users in creation of evolutionary algorithms. Unfortunately most existing libraries lack either in functionality and/or systematics in some way. Nevertheless some should be mentioned here. MUTANTS is a strongly object-oriented toolkit written by Matthew Caryl in Ada 95, which regrettably has not become a popular programming language these days. The work on MUTANTS has been stopped now since several years and the toolkit is not maintained any longer. The *EA Visualizer* [Bos01] by Peter Bosman is written in Java and is both a general framework and a development environment for interactive visualizations. Results of algorithms can easily illustrated, but the toolkit lacks support of parallel computing. *ECJ*

B. Reusch (Ed.): Fuzzy Days 2001, LNCS 2206, pp. 92–102, 2001.

[Luk01] by Sean Luke is also written in Java, but is a framework focused on Genetic Programming. *EO* [EO001] and *EC++* [ECl01] are frameworks written in C++ which are still growing, but also don't have support for parallel computing (yet?).

There are also a few commercial tools available. ***Discipulus*** [Dis00] by AIM-Learning is a Genetic Programming oriented toolkit which is used for data mining purposes. GepSoft [Gep01] develops AI software based on Gene Expression Programming, where as the *NeuroGenetic Optimizer* [Neu01] by BioComp Systems is focused on optimizing the inputs and the structure of neural networks using genetic algorithms.

The framework *eaLib* discussed in this paper is written in Java. The language Java was chosen for several reasons. Java is widely spread and has become one of the major programming languages. The development kit including compiler and debugger is freely available on a number of different computer platforms. It is object-oriented and relativly easy to learn. The core libraries contain a lot of functionality which can be used directly and need not be retrofitted via external libraries unlike for instance in C++. Another important point is the builtin support for multithreading and distributed computing.

The speed at which a new algorithm can be assembled and the flexibility with that new operators can be defined are of major importance in toolkit design. Normally there is a tradeoff between easy handling and extensibility. By choosing Java as programming language we have to stick to object-oriented design. By using the advantages of object-orientation we have tried to keep the tradeoff as close as possible.

2 Genetic Representation and Evaluation of Individuals

Before we can start the development of a generic toolkit and the implementation of genetic operators we have to spend some thoughts about the requirements and the design of such a framework. Good work about design aspects has been done by Matthew Caryl in [Car97].

First of all we have to think about the genetic representation that should be used. Surely we need the support of the traditional representation forms binary string and real numbered. Second, suitable recombination and mutation operators for these data types must be provided and finally a model for performing selecting and replacing individuals for the creation of an evolution cycle must be found. The creation of an evolution cycle is discussed in the next section, so we attend to the genetic representation first.

Many publications make different usage of the terms *individual*, *chromosome* and *gene*. To avoid confusion with these terms it is necessary to give a definition for the meaning of the words *individual* and *chromosome* which are used throughout this paper. Individuals represent a potential solution to an optimization problem. Individuals contain a number of chromosomes holding information about the details of a problem solution. The set of chromosomes is called genetic representation.

As already mentioned algorithms often use the traditional forms of genetic problem representation like binary string or real-valued. This can be appropriate for some cases and is necessary to implement already existing algorithms, but in general this manner is quite unflexible and not very descriptive for humans. Furthermore almost all binary crossover and mutation operators described show the problem of generating illegal bit strings in case that not the whole range of possible combinations can be used. So a more flexible genetic representation not limited to binary values must be found. But because of lots of operators relying on traditional forms of genetic representation these forms should or better must be provided too.

The structure for the class *Individual* leans closely against the structure of real creatures we know from nature. So an object of the class *Individual* contains an object of the class *ChromosomeSet*. The *ChromosomeSet* itself is a collection of *Chromosomes*. Each chromosome holds a chromosome object of a special type. This can be elementary data types like integer or float numbers or more complex data types like vectors or arrays. Even trees are possible. To create a chromosome with the classical binary representation the class *BinaryString* is provided. Figure 1 shows the UML model of the genetic representation. In this example four types of chromosomes are included.

This type of representation is flexible enough to be used for both simple and complex solution representations. For instance it is no problem to implement the representation form used in messy genetic algorithm as demonstrated in [DG91]. To create an individual holding the representation of the solution of a mathematical function in three dimensions an example of code is given below. The primitive datatype *float* is used.

```
// create a chromosome set with 3 slots
ChromosomeSet mathSet = new ChromosomeSet( 3 );
// add the chromosomes with initial random float values
mathSet.add( new FloatChromosome( (float) 10.0 );
mathSet.add( new FloatChromosome( (float) -20.0 );
mathSet.add( new FloatChromosome( (float) 5.0 );
// create the individual
Individual mathIndividual = new Individual( mathSet );
```

Listing 1.1. creation of an individual

Almost every problem solution we can think of can be expressed in a similiar way - in opposition to the quality appraisal of a solution. This is unique to every optimization task and must therefore be implemented by the user of the toolkit himself. When evaluating an individual we differentiate between the target objective value (called score throughout this paper) and the fitness. The score can be directly derived from the current content of the chromosome set while the fitness is a relative quality appraisal [Poh00].

For that reason the two abstract classes *Score* and *Fitness* have been introduced. Each individual contains one instance of both classes. There are plenty of ways to express the score of an individual. Various different data types can be used. Although normally numbers are utilized, other data types – like for

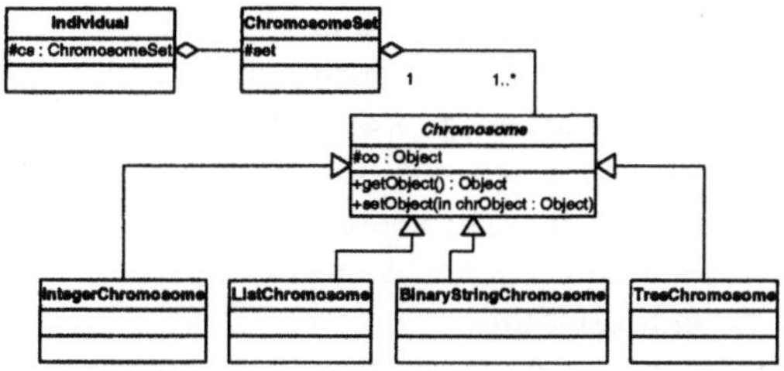

Fig. 1. UML model of the genetic representation used in eaLib

instance a matrix – are conceivable. In addition the way of expression can be different. Given individuals with a score of type integer, we have to define what a good score is. We could say the higher the integer value the higher the score, but in some optimization problems a lower value is a better score. Even the case that a value should be as close as possible to, let's say 100, is conceivable. For that reason there are three different types of scores. The class *Score* has several subclasses, each for a particular datatype. Currently the four data types *int*, *long*, *float* and *double* are supported. For instance for the data type *int* the class *IntegerScore* exists, with the rule the lower the value the better the score. The class *IntegerScore* is subclassed by the classes *ReverseIntegerScore* and *ConvergentIntegerScore*, which represent the circumstances that the higher the value the better the score respectively the closer to a given value the better the score. This distinction had to be done to be able to compare individuals to each other. Figure 2 shows the UML model of an individual including its score and fitness.

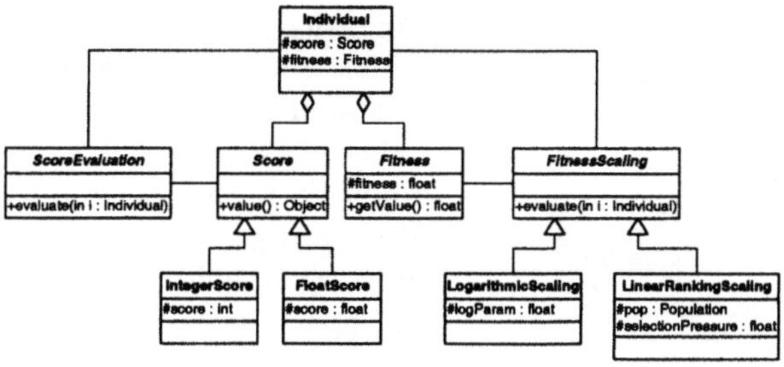

Fig. 2. UML model of score and fitness model and evaluation of both

Other than the various possible definitions of score, a fitness value is always a number. So we have defined the fitness as a positive float number. A bigger value represents a better fitness. This definition should be useful in most of the application cases, otherwise a new class derived from *Fitness* could be introduced by the user. The appropriate methods for comparing various *Fitness* objects must be overloaded.

Also included in figure 2 are two base classes for the evaluation of the score and the fitness. First we discuss the score evaluation. The calculation of the quality of an individual is completely problem specific. For that reason the user has to implement the evaluation himself. The class *ScoreEvaluation* is an abstract base class which defines the method *evaluate(Individual)*. The user is forced to derive a class from that base class and to implement this method which returns a valid score for a given individual.

The fitness evaluation is a quite similiar procedure with the difference that there are several approaches for definining a fitness value. Some of these approaches are already provided by *eaLib*, while the user is able to introduce his own functions. Again there is an abstract base class called *FitnessScaling* which defines the method *evaluate(Individual)*. The method returns an object of type *Fitness* belonging to the evaluated individual.

Several variants of defining a fitness value are already implemented in the toolkit and can be used directly. At the time this paper was written the user had the following choices: the fitness value can be derived directly from the score value via linear, reciprocal, exponential and logarithmic scaling [Poh00]. Furthermore the variants linear [Bak85] and nonlinear ranking scaling [BH91] are provided.

3 Genetic Operators

As already mentioned the *eaLib* toolkit is strongly object-oriented. Object-orientation is very well suited for the desired extensibility. The basic structure of all operators is defined by abstract base classes, while concrete operators are implemented in derived classes. These derived classes can be provided by the toolkit itself or may be completly recreated by a user to fit his own needs.

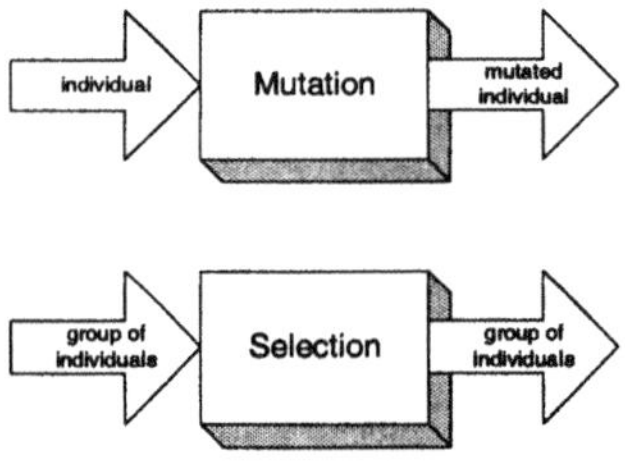

Fig. 3. Processing of individuals with genetic operators

In order to define abstract classes we have to think about what genetic operators have in common. Operators for selection, recombination, mutation and so on seem to be quite different from each other but there is one important thing they have in common: they are processing individuals in some unique way. Some operators are working on a group of individuals (e.g. selection), others process them individually. Figure 3 illustrates that. For instance the mutation operator changes the chromosome set of individuals completely independent from each other, so this operator works in a sequential manner. In contrast to that selection does only make sense when working on a group of individuals. As a result a group of (selected) individuals is produced. As (a more abstract) consequence we can say that every genetic operator takes a *stream of individuals* as an input, processes these individuals in some particular way and produces a *stream of individuals* as a result.

With that knowledge we can define an abstract base class called *GeneticOperator* from which all other genetic operators in the toolkit are derived. To indicate whether an operator is able to process only single individuals or requires a group of individuals the two interfaces *SingleProcessor* and *MultipleProcessor* have been defined. Each operator has to implement at least one of these interfaces. The UML diagram in figure 4 shows this issue with the examples of the operators for mutation and selection.[1]

For a better illustration we will take a closer look at the selection mechanism. One of the implemented selection methods is the well known roulette wheel selection [Bak87]. Let's imagine our operator is to select three pairs for mating out of a stream of ten individuals. The appropriate method *process(Individual[] ia)* of the class *RouletteWheelSelection* takes an array of 10 individuals as an argument. It produces an array of 6 individuals which have been selected by spinning the roulette wheel. The first two individuals represent the first pair, the second two the second and so on. There's no difference in the resulting data structure between for instance 3 selected pairs or 2 selected triples.

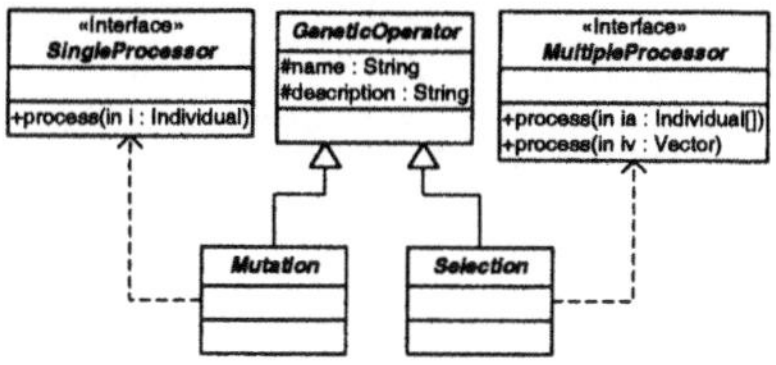

Fig. 4. UML diagram defining the interfaces of genetic operators

All genetic operators provided by *eaLib* are working in the same manner. However, this paper is not the right place to enumerate all of them, for an overview of the implemented operators refer to [Rum01]. The only operator we will pay attention to in the rest of this section is the recombination to explain the functioning in consideration of the genetic representation used in *eaLib*. Due to the fact that individuals can contain a number of chromosomes with different types, we cannot use common crossover procedures directly on an object of the class *Individual*. So the abstract base class *Recombination* is subclassed by two other classes: *IndividualRecombination*, which is responsible for mating two individuals and *ChromosomeRecombination*, which is an abstract base class for recombination operators for different datatypes. An object of the class *IndividualRecombination* must be parameterized to fit the chosen data structure of individuals. It contains a recombination operator for every chromosome in an individual, which can handle the appropriate data type of the chromosome. Derived from the base class *ChromosomeRecombination* are recombination operators for various data types and there we find well known operators there for creating real valued, binary or vector based chromosomes, for instance EX [WSF89], PX [Sys91] or EAX [NK97] operators. Because this condition is also valid for the mutation operator, it is designed in the same way.

4 Architecture of an Algorithm

After having determined how genetic operators can be specified with a common interface, we can do a step further and create an algorithm. One of the easier

[1] At the time this paper was written the data structures array and vector were used to hold the stream of individuals. These will be replaced by several more abstract *IndividualStream* classes, which hide the internal data structures from the user.

(and more common) cases is a genetic algorithm with only one population. Let's take a more abstract look at the structure of such an algorithm. We have already defined streams of individuals, but what path do these streams take? A stream of individuals is created by an initialization operator, enters an evolution cycle and is caught after some termination condition is fulfilled. Reconsidering this with a more abstract view, the stream of individuals is created by a source, flows through and evolves within a gate and is caught by a sink afterwards. Based on the gate we can insert several operators that together create an evolution cycle. The structure for that is given in figure 5.

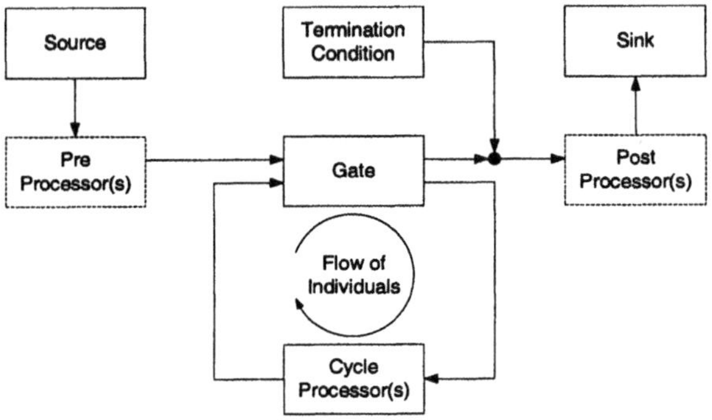

Fig. 5. Controller architecture for a single loop algorithm

A source of individuals is defined by the interface *IndividualSource*. An operator implementing this interface is the initialization operator. The abstract base class for that operator is *Initialization*, which defines the method *createSet*, which returns a chromosome set for a valid individual. For creating an initialization operator, a class must be derived and the abstract method must be implemented. Functions for generating random data of various types are provided by *eaLib*, but special problem related knowledge can also be entered at this point.

After the creation of valid individuals by the source, these individuals can optionally be preprocessed by a number of operators. After the preprocessing stage the stream of individuals enters the gate and evolves based on the processors contained in the evolution cycle. The so-called cycle processors are genetic operators that are needed for the creation of a valid algorithm. The running algorithm is observed by an operator holding information about the termination conditions. When one of the termination conditions is fulfilled, the stream of individuals is (optionally) postprocessed and ends up in a sink, which extracts the result of the algorithm. All the described operators are being hold inside queues in an algorithm controller that controls the execution of the algorithm. All necessary operators for the algorithm can simply be plugged into the controller. It must be assured by the user that the sequence in which the operators are plugged in is reasonable. For a better illustration a piece of Java code is given below (example for solving the TSP). This creates an algorithm with 100 individuals, linear ranking fitness scaling and roulette-wheel selection, which stops after 1000 generations. In this example pure reinsertion is used.

```
protected void setup() {
  // create source (100 individuals , 29 cities )
  setSource( new SalesmanInititialization ( 100, 29 ) );
  // create gate (population, individuals are sorted)
  setGate( new SortedPopulation() );
  // create sink
  setSink( new GenericSink() );
  // create termination condition (stop after 1000 generations)
  addTermination( new MaxGenTermination( 1000, this.gate.getCounter() ) );
  // add cycle operators
  addCycleProcessor( new SalesmanCalculation() );
  addCycleProcessor( new LinearRankingScaling( this.gate ) );
  addCycleProcessor( new RouletteWheelSelection( 100 ) );
  addCycleProcessor( new IndividualRecombination() );
  addCycleProcessor( new PureReinsertion( this.gate ) );
}
```

Listing 1.2. creation of an algorithm

Algorithms with a single evolution cycle can be created very easily that way. But it is possible that an algorithm with a single population and two parallel working evolution cycles should be created (for instance one half of the population should evolve in another way than the other half). In such a case the whole algorithm controller must be rewritten. This is not very flexible, so an other solution had to be found.

As a solution we use the builtin multithreading capabilities of the Java language and additionally created a multithreaded controller. Every genetic operator runs in its own thread and contains a reference to the following operator. After completing the processing of a stream of individuals the stream is passed to the next operator and its method *process()* is invoked within a newly created thread. The current operator quits its work immediately after starting the next one to prevent the construction of an infinite cascade. For creation of parallel evolution cycles special operators have been introduced, which are able to split up a stream, respectively merge several streams into one. The implementation of synchronization capabilities has been neccessary at the stage of merging. With the extension of the multithreaded concept to gates (populations), it is also quite easy to create algorithms based on the island model. For that case a special migration operator is provided.

5 Parallelization of Algorithms

In Java mainly two techniques are provided for distributed computing. The first is RMI (Remote Method Invocation). It is a scheme for distributed objects, that is part of the Java Core Libraries and is Java specific. The other one is CORBA (Common Object Request Broker Adapter). It is a distributed object standard developed by the Object Management Group (OMG). CORBA is a platform- and language-independent generic framework for building of systems relying on distributed objects.

CORBA's big advantage is its language-indepency. CORBA relies on an object interface description that is completly independent from the implementation language. Tasks that are computationally hard are best implemented in languages that can be compiled into native machine code. So it's easy to use C or C++ for that part. The only way to integrate native code into the RMI system is to use the Java Native Interface, whose handling can be a quite tricky. On the other hand, RMI is much easier to master than CORBA. Because of that and because that there are currently no plans to use objects written in other languages than Java we have chosen RMI for algorithm parallelization.

The part of a genetic algorithm that benefits most from a parallelization is the score evaluation. There are cases in which 90 to 95% of the computation time of an algorithm are consumed by the calculation of the target objective value. The implementation of this model of parallelization is often refered to as Master-Slave-Model [Bäc94]. This kind of parallelization is quite easy to implement because the score evaluation of individuals is performed completly independent from each other.

The concept we have chosen is shown in figure 6. The stream of individuals is processed by a genetic operator called *DistributedScoreEvaluation.* All individuals are passed to an *DistributionController* that steers the process of distributing the individuals to compute servers. On these compute servers instances of the already explained class *ScoreEvaluation* are running. These instances evaluate the delivered individuals and return them to the controller. The controller itself collects all individuals and returns them back to the initial distribution operator after all processing by the compute servers is finished.

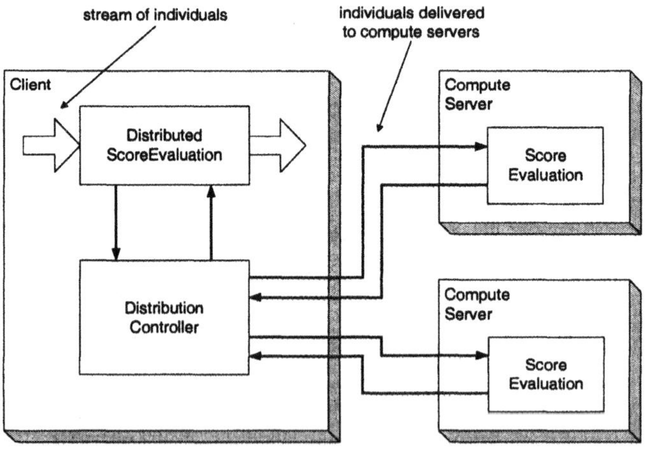

Fig. 6. Distributing the score evaluation in a parallel architecture

The distribution controller runs on the client and controls the distribution as already mentioned. It contains a list of all currently available compute servers. The servers must register themselfs at the distribution controller in order to be used for calculation. Because of the fact that the controller runs independently from the servers, additional compute servers can be added while the algorithm is running.

The implementation model we have chosen is a synchronous model. For that reason, the client (master) has to wait for slow compute servers to finish their calculation. This could be a drawback in some cases. But with similiar servers the speedup of the execution time of the algorithm is nearly proportional to the number of compute servers available. Moreover it should be quite easy to extend this distribution model to other serial working genetic operators like for instance mutation.

6 Conclusion and Future Work

In this paper we have introduced the evolutionary computation framework *eaLib*, which intends to be easy to use and flexible enough for extensions of various kinds. We have shown a general way of representing potential solutions of optimization tasks that is used in the toolkit. A concept for a common interface for genetic operators has been introduced and a short example for the usage has been given. Furthermore the parallelization of genetic algorithms based on a master-slave-model has been shown.

Our aim was to provide a framework to support experimentation with genetic operators and fast creation of algorithms. Now we are hoping for creative feedback from scientists. Based on that feedback the framework could be extended in various directions. From our viewpoint especially the capabilities of parallelizing algorithms are not universal enough. So further development will concentrate on this issue.

References

[Bak85] J. E. Baker. Adaptive Selection Methods for Genetic Algorithms. In *Proceedings of the International Conference on Genetic Algoritms and their Application*, pages 101–111, 1985.

[Bak87] J. E. Baker. Reducing Bias and Inefficiency in the Selection Algorithm. In *Proceedings of the Second International Conference on Genetic Algoritms and their Application*, pages 14–21, 1987.

[Bäc94] Thomas Bäck. Parallel Optimization of Evolutionary Algorithms. In *Parallel Problem Solving from Nature III - International Conference on Evolutionary Computation*, pages 418–427, 1994.

[BH91] T. Bäck and F. Hoffmeister. Extended Selection Mechanisms in Genetic Algorithms. In *Proceedings of the Fourth International Conference on Genetic Algorithms*, pages 92–99, 1991.

[Bos01] Peter Bosman. EA Visualizer. `http://www.cs.uu.nl/people/peterb/computer/eavisualizer.html`, 2001.

[Car97] Matthew Caryl. MUTANTS - A Generic Genetic Algorithm Toolkit for Ada 95. `http://www.cultofcelebrity.com/matthew/projects/mutants/index.html`, 1997.

[DG91] K. Deb and D. E. Goldberg. A Messy Genetic Algorithm in C. Technical Report 91008, University of Illinois at Urbana-Champaign, 1991.

[Dis00] AIMLearning Discipulus. `http://www.aimlearning.com`, 2000.

[ECl01] EClab EC++. `http://www.cs.gmu.edu/~eclab/eclib.html`, 2001.

[EO001] EO Evolutionary Computation Framework. `http://geneura.ugr.es/~jmerelo/EO.html`, 2001.

[Gep01] GepSoft Automatic Problem Solver and Symbolic Regression Toolkit. `http://www.gene-expression-programming.com`, 2001.

[Luk01] Sean Luke. ECJ - A Java-based Evolutionary Computation and Genetic Programming Research System. `http://www.cs.umd.edu/projects/plus/ec/ecj`, 2001.

[Neu01] NeuroGenetic Optimizer. `http://www.biocompsystems.com/pages/neuralnetworkoptimizer.htm`, 2001.

[NK97] Y. Nagata and S. Kobayashi. Edge Assembly Crossover: A High-Power Genetic Algorithm for the Traveling Salesman Problem. In *Proceedings of the Seventh International Conference on Genetic Algorithms and their Application*, pages 450–457. Van Nostrand Reinold, 1997.

[Poh00] Hartmut Pohlheim. *Evolutionäre Algorithmen*. Springer-Verlag Berlin, Heidelberg, New York, 2000.

[Rum01] Andreas Rummler. eaLib API Documentation. `http://www.inf-technik.tu-ilmenau.de/~rummler/ealib.html`, 2001.

[Sys91] G. Syswerda. Schedule Optimization Using Genetic Algorithms. In *Handbook of Genetic Algorithms*, pages 133–140. Van Nostrand Reinold, 1991.

[WSF89] D. Whitley, T. Starkweather, and D. Fuquay. Scheduling Problems and the Travelling Salesman: The Genetic Edge Recombination Operator. In *Proceedings of the Third International Conference on Genetic Algorithms and their Application*, pages 133–140, 1989.

Experimental Evaluation of Fuzzy Controllers for an Outdoor AGV

W.S. Wijesoma and K.R.S. Kodagoda

School of Electrical and Electronic Engineering,
Nanyang Technological University,
Singapore 639798
eswwijesoma@ntu.edu.sg

Abstract. Fuzzy logic control is a relatively new technology and hence it needs rigorous comparative experimental analyses with other well-established conventional control schemes. Further, fuzzy controller stability analysis is a major hindrance for its popularity among control engineers. In this paper, stable fuzzy variable structure systems or Sliding Mode Controllers (SMC) and fuzzy PD/PID controllers for steer and drive control of an Autonomous Guided Vehicle (AGV) are developed and implemented on an actual AGV. The synthesized controllers are further assessed against conventional SMC, PD/PID through actual implementation on a moderately sized outdoor AGV.

1 Introduction

Although, AGVs possess nonlinear dynamics, the application of linear control methods has not been uncommon as evident from the reported literature [1],[2],[3],[4]. Here, the controller design is based on the assumption that the model dynamics are linear or a linearized model of the AGV about a specified operating point. PI, PD and PID control techniques are the most commonly used linear control methods.

Nonlinear methods of AGV control, where the nonlinearities are catered for indirectly has also been reported in the literature. Of these, sliding mode or variable structure systems and fuzzy control methods are the most representative. A multiple sliding surface approach for combined brake and throttle control of an AGV is used by Hedrick et al [5],[6]. It is claimed through simulations that the multiple sliding surface controller is capable of controlling the longitudinal speed of the vehicle effectively. Lee et al [7] proposed a fuzzy-like gain scheduling PD controller for lateral control of an AGV. The control law is realized similar to Sugeno's fuzzy model, but with non-overlapping membership functions. Fuzzy sliding mode controller is a realization of an SMC control law using fuzzy logic. A development of such an FSMC for velocity and acceleration control of an AGV can be found in [7]. One of the advantages of using fuzzy logic in vehicle control is the possibility of effectively incorporating the driver's knowledge in designing the rule base as well as in controller tuning. This is not straightforward in a fuzzy SMC. As the switching surface, s , and its derivative, $\dot{s}$, in [7] cannot be easily interpreted physically, it may

B. Reusch (Ed.): Fuzzy Days 2001, LNCS 2206, pp. 103–117, 2001.

be difficult to design the rule base in terms of the sliding surface and its derivative. Further, tuning of the sliding surface parameters is not intuitive, when compared to tuning of a state space fuzzy logic controller.

Fuzzy logic has a lot of appeal in control. However, fuzzy logic control has a relatively short history. In addition, most of the fuzzy controllers proposed in the literature lack stability analysis and experimental results on real plants. It is also important to note that there are also not much experimental results and analysis reported comparing established and conventional control schemes with fuzzy schemes. In this paper we attempt to address the above issues through the systematic design, analysis and experimentation of fuzzy controllers. In the rest of the paper we describe the particular AGV and its dynamics used as the test bed. A stable SMC is designed for the dynamic system characterized by the AGV dynamics from Lyapunov theory. A stable fuzzy SMC is then derived through fuzzification of the sliding mode control law. A fuzzy PD state space controller is then synthesized by resolving the fuzzy SMC control law. This is followed by experimental results and conclusions.

2 AGV and It's Dynamics

The experimental system shown in Fig. 1 was built using a commercially available *CARRYALL 1* golf-car. In the conversion no attempt was made to alter the drive actuation system, which comprises of a DC series motor (3.1 hp) and a gear assembly (ratio 12.28:1). This is because DC series motors possess speed-torque characteristics suitable for electric vehicle actuation systems [8]. However, the down side of using the existing DC series motor is that it does not provide for negative torques by the application of a negative supply voltage. Accelerations and constant speeds are achieved through application of positive voltages to the DC series motor and active decelerations are achieved through a separate stepper motor and braking mechanism. Outputs of the four encoders mounted on the four wheels of the AGV are used in determining the actual speed of the vehicle. It is assumed that slipping does not exist. As a safety precaution, the vehicle is equipped with a separate parking brake mechanism, which is activated when the vehicle is parked. An emergency switch is used to stop the vehicle in case of any emergency.

The steer system is driven by a permanent magnet DC motor (48V, 650W), which is coupled to the steer shaft by means of a gear assembly (ratio 5:1). The reading of the encoder mounted on the motor shaft is used to calculate the actual steer angle. Limit switches prevent possible over-steer.

An industrial PC provides the required computing power for all three systems. Two servo amplifiers and a stepper motor driver are used to match the motor operating voltage (-48V to +48V) from the digital-to-analogue converter (DAC) output voltage (-10V to +10V).

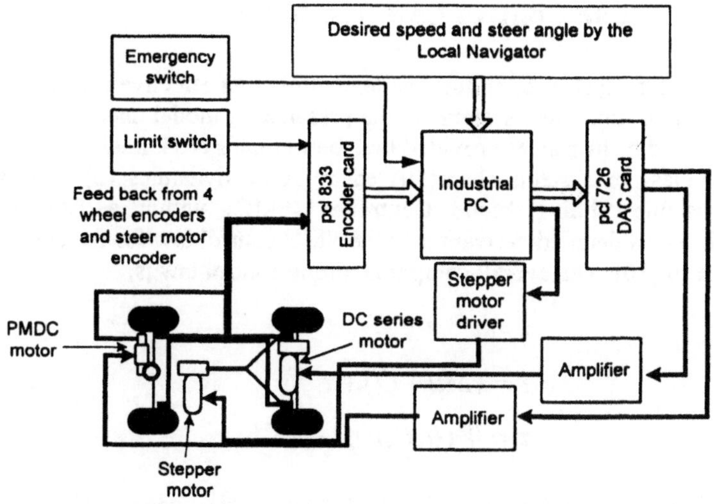

Fig. 1. Hardware structure of the experimental system

The vehicle dynamics of the front-wheel steerable and rear-wheel driven AGV can be expressed as follows [9].

$$\begin{bmatrix} \tau_d - \tau_b \\ \tau_s \end{bmatrix} = \begin{bmatrix} M_v r & Nr \\ N & M_\omega \end{bmatrix} \begin{bmatrix} \dot{v} \\ \dot{\omega} \end{bmatrix} + \begin{bmatrix} q_1 \\ q_2 \end{bmatrix} + \begin{bmatrix} d_1 \\ d_2 \end{bmatrix} \tag{1}$$

where, τ_d, τ_b and τ_s are drive, brake and steer torques respectively.

$\begin{bmatrix} q_1 \\ q_2 \end{bmatrix} = \begin{bmatrix} \left(C_{d1}\omega^2 + C_{d2}v\omega + k_v v + F_{df}\right)r \\ C_{s1}\omega^2 + C_{s2}v\omega + \tau_{sf} \end{bmatrix}$, d_1 and d_2 represents the disturbances.

The remaining parameters are defined in the Appendix.
Further, (1) can be arranged in the form,

$$\tau = M\ddot{\theta} + Q + d \tag{2}$$

where, $\tau = \begin{bmatrix} \tau_d - \tau_b \\ \tau_s \end{bmatrix}$, $M = \begin{bmatrix} M_v r & Nr \\ N & M_\omega \end{bmatrix}$, $Q = \begin{bmatrix} q_1 \\ q_2 \end{bmatrix}$, $\ddot{\theta} = \begin{bmatrix} \dot{v} \\ \dot{\omega} \end{bmatrix}$, M is a positive definite inertia matrix, Q is the vector of Corriolis, centrifugal and frictional forces, $\ddot{\theta}$ is the vector of linear and angular acceleration and d is the disturbance vector.

3 Sliding Mode Control Law Synthesis

Variable structure system or sliding mode control is an effective control methodology for controlling of nonlinear systems in the presence of model uncertainties, parameter fluctuations and disturbances provided that their bounds are known.

For the class of systems characterized by AGV dynamics (2), it is possible to design a stable Sliding Mode Controller (SMC) within a Computed torque framework. For a detailed derivation of the SMC control law for the above dynamics, please refer to [10]. The overall computed torque control law is,

$$\begin{aligned} \tau &= \tau^{ctt} + \tau^{smc} \\ \tau &= (\hat{M}u + \hat{Q}) + \tau^{smc} \\ \tau &= W(\theta, \dot{\theta}, u)\hat{\Omega} + \tau^{smc} \end{aligned} \tag{3}$$

A suitable choice of u is, $u = \ddot{\theta}_d + \Lambda(\dot{\theta}_d - \dot{\theta})$, where, $\ddot{\theta}_d$ represents the desired value of $\ddot{\theta}$ and $\Lambda \in R^{n\times n}$ is a diagonal matrix with elements $\lambda_i (i = 1,\dots,n)$. The choice $(\hat{M}u + \hat{Q})$ represents the computed torque component, τ^{ctt}, where $\hat{M}$ and $\hat{Q}$ are estimates of M and Q respectively. The term τ^{smc} is used to remove the effects of inexact de-coupling as a result of model mismatch and bounded disturbances using sliding mode control theory. $W \in R^{n\times r}$ is a matrix of functions and $\hat{\Omega}$ is the estimate of a suitably chosen parameter vector, $\Omega = [m_c \quad I_{wr} \quad I_c \quad (I_{w\phi} + m_w r^2)]^T$. As detailed in [11] the closed loop equation can now be written as,

$$(\ddot{\theta} - \ddot{\theta}_d) + \Lambda(\dot{\theta} - \dot{\theta}_d) = M^{-1}\left[\tau^{smc} - (W\psi + d)\right] \tag{4}$$

where, $\ddot{\theta}$ and $\ddot{\theta}_d$ are actual linear and angular accelerations and desired linear and angular accelerations vectors, respectively. $\Psi = (\Omega - \hat{\Omega}), \Psi \in R^{r\times 1}$ is the parameter mismatch vector representing any mismatch between the estimated ($\hat{\Omega}$), and actual parameter vector (Ω). Let us define the ith switching plane as,

$$s_i = \left(\dot{\theta}_i - \dot{\theta}_{di}\right) + \lambda_i(\theta_i - \theta_{di}) \tag{5}$$

A condition for the intersection of switching planes, $s_i = 0$, to be attractive can be derived by defining a quasi Lyapunov function $V(t)$ as,

$$V(t) = \frac{1}{2} s^T M s \tag{6}$$

where M is the positive definite inertia matrix of the vehicle. Differentiating (6) with respect to time, we have,

$$\dot{V}(t) = s^T M \dot{s} + \frac{1}{2} s^T \dot{M} s \tag{7}$$

Now the ith component of τ^{smc}, i.e. τ_i^{smc} can be chosen as given below in equation (8) to ensure that $\dot{V}(t) \leq 0$ so as to guarantee the asymptotic stability of s and hence the convergence of tracking errors to zero [10].

$$\tau_i^{smc} = -\operatorname{sgn}(s_i)\left[\sum_{j=1}^{r} \overline{W}_{ij}\overline{\psi}_j + \overline{d}_i + \delta_i\right] - s_i \sum_{j=1}^{n} \frac{\overline{\dot{M}}_{ij}}{2} \tag{8}$$

where,

$$|\psi_j| < \overline{\psi}_j \quad j = 1,\ldots,r \qquad |d_i| < \overline{d}_i \quad i = 1,2 \qquad |\dot{M}_{ij}| < \overline{\dot{M}}_{ij} \quad i = j = 1,2$$

$$|W_{ij}| < \overline{W}_{ij} \quad i = j = 1,2 \qquad \delta_i > 0 \quad i = 1,2$$

Equation (8) can be re-arranged as,

$$\tau_i^{smc} = -\operatorname{sgn}(s_i)\left[K_{i1} + |s_i| K_{i2}\right] \tag{9}$$

where, $K_{i1} = \sum_{j=1}^{r}\left[\overline{W}_{ij}\psi_j + \overline{d}_i + \delta_i\right]$, $K_{i2} = \sum_{j=1}^{n} \frac{\overline{\dot{M}}_{ij}}{2}$

4 Fuzzy Sliding Mode Control Law Synthesis

Fuzzy control offers simple but robust control solutions for nonlinear systems. Further fuzzy logic provides a convenient framework to incorporate the heuristic knowledge about the plant. Also, unlike SMC, fuzzy control approach permits greater flexibility in fine-tuning the controller to achieve arbitrary complex control surfaces. In this section it is shown how an approximate fuzzy SMC can be derived from the SMC control law (9).

It may be noted that K_{i1} and K_{i2} can be chosen as positive constants whose magnitudes depend on the extent of model mismatch and bounds of disturbance d. To eliminate the need for the computed torque component (τ^{ctt}) of the control law (3), we choose K_{i1} and K_{i2} to be sufficiently large. From equation (9) it can be seen that farther the system state is from the sliding surface, larger the magnitude of the torque needed to drive the system on to it. However, the sign of the control torque is opposite to that of s. This property can be exploited to obtain a fuzzy SMC controller from the stable SMC control law (9). This is accomplished through a set of fuzzy rules as follows. For the ith fuzzy SMC torque,

$$\begin{aligned}
&\text{IF } \bar{s}_i \text{ is } PVL \text{ THEN } \tau_i^{fsmc} \text{ is } PVL \\
&\text{IF } \bar{s}_i \text{ is } PL \text{ THEN } \tau_i^{fsmc} \text{ is } PL \\
&\text{IF } \bar{s}_i \text{ is } PM \text{ THEN } \tau_i^{fsmc} \text{ is } PM \\
&\text{IF } \bar{s}_i \text{ is } PS \text{ THEN } \tau_i^{fsmc} \text{ is } PS \\
&\text{IF } \bar{s}_i \text{ is } Z \text{ THEN } \tau_i^{fsmc} \text{ is } Z \\
&\text{IF } \bar{s}_i \text{ is } NS \text{ THEN } \tau_i^{fsmc} \text{ is } NS \\
&\text{IF } \bar{s}_i \text{ is } NM \text{ THEN } \tau_i^{fsmc} \text{ is } NM \\
&\text{IF } \bar{s}_i \text{ is } NL \text{ THEN } \tau_i^{fsmc} \text{ is } NL \\
&\text{IF } \bar{s}_i \text{ is } NVL \text{ THEN } \tau_i^{fsmc} \text{ is } NVL
\end{aligned} \tag{10}$$

where, $\bar{s}_i = -s_i = \left(\dot{\theta}_{di} - \dot{\theta}\right) + \lambda_i \left(\theta_{di} - \theta_i\right)$

Linguistic terms of the input $\bar{s}_i \in \bar{S}$ and output $\tau_i^{fsmc} \in \Gamma$ defined over the universe of discourses are, *PVL* (Positive Very Large), *PL* (Positive Large), *PM* (Positive Medium), *PS* (Positive Small), *Z* (Zero), *NS* (Negative Small), *NM* (Negative Medium), *NL* (Negative Large) and *NVL* (Negative Very Large) respectively. As shown in Fig. 2, the triangular membership functions are chosen for input and output linguistic variables.

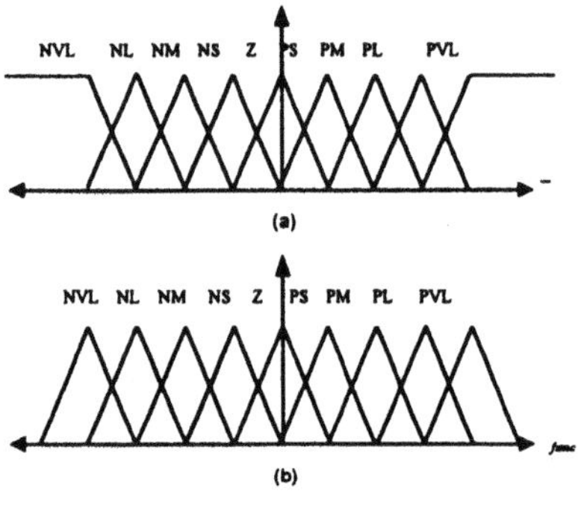

Fig. 2. Triangular membership functions of $\bar{s}_i$ and τ_i^{fsmc}

In general the rule base (10) can be written as,

$$\text{IF } \bar{s}_i \text{ is } A_i^k \text{ THEN } \tau_i^{fsmc} \text{ is } B_i^k \tag{11}$$

where, A_i^k and B_i^k, k=1,..m are the linguistic values of the linguistic variables $\bar{s}_i = -s_i$ and τ_i^{fsmc}.

We choose triangular membership functions, singleton fuzzification, Mamdani inferencing system and center of gravity (COG) defuzzification. Hence the inferred crisp torque output of the fuzzy SMC is,

$$\tau_i^{fsmc\_crisp} = \frac{\sum\limits_{j=1}^{r} b_j^i \int\limits_{\Gamma} \mu_{B_i^j}(\tau_i^{fsmc}) d\tau_i^{fsmc}}{\sum\limits_{j=1}^{r} \int\limits_{\Gamma} \mu_{B_i^j}(\tau_i^{fsmc}) d\tau_i^{fsmc}} \tag{12}$$

b_j^i is the center of the implied fuzzy set corresponding to the rule j, r is the total number of rules and $\int\limits_{\Gamma} \mu_{B_i^j}(\tau_i^{fsmc}) d\tau_i^{fsmc}$ denotes the area under $\mu_{B_i^j}(\tau_i^{fsmc})$.

Fuzzy SMC (12) approximation of the smoothed SMC controller (9) can be arbitrarily controlled by the number of linguistic terms and hence the rules. However, what is really important to ensure stability and convergence of the fuzzy SMC is to guarantee the negative semi-definiteness of the Lyapunov function derivative, i.e. $\dot{V}(t) \leq 0$. This translates to the condition that the crisp fuzzy SMC output is larger or equal in magnitude to the SMC torque for all $\bar{s}$, i.e.

$$\left|\tau_i^{smc}(\bar{s}_i)\right| \leq \left|\tau_i^{fsmc\_crisp}(\bar{s}_i)\right| \quad \forall\, \bar{s}_i \tag{13}$$

5 Fuzzy PD Control Law Synthesis

The heuristic control knowledge is usually available in terms of state variables and their changes. To facilitate inclusion of such knowledge and to add more degrees of freedom for tuning, it is useful to design state based fuzzy controllers. One of such controllers is a fuzzy PD controller. The following illustrates how such a law may be derived from the fuzzy sliding mode controller.

The switching plain, $\bar{s}_i$ can be normalized as,

$$\hat{s}_i = \hat{\dot{e}}_i + \hat{e}_i \tag{14}$$

where, $e_i = (\theta_{di} - \theta_i)$, $\hat{s}_i = \dfrac{\bar{s}_i}{N_{si}}$, $\hat{\dot{e}}_i = \dfrac{\dot{e}_i}{N_{si}}$, $\hat{e}_i = \dfrac{\lambda_i e_i}{N_{si}}$, N_{si} is a positive scalar.

Taking note of the fact that $\hat{s}_i$ is a linear combination of the state variables, $\hat{\dot{e}}_i$ and $\hat{e}_i$, we may re-write the rule base of the fuzzy sliding mode controller (11) in terms of linguistic state variables [11]. Final rule base in state space is as shown in Fig. 3. This diagonal rule base is usually referred to as a standard rule table in the literature. It can be shown that the fuzzy PD controllers designed using the diagonal rule table for the AGV dynamics are stable if,

$$\left|\tau_i^{fpd}(\hat{s}_i)\right| \geq \left|\tau_i^{smc}(\hat{s}_i)\right| \quad \forall \hat{s}_i \tag{15}$$

where, $\left|\tau_i^{fpd}(\hat{s}_i)\right|$ and $\left|\tau_i^{smc}(\hat{s}_i)\right|$ are the magnitudes of the fuzzy PD (or PI with the switching planes defined appropriately) controller's output and output of the SMC respectively. Whilst maintaining this stability condition (equation (15)), parameters of the fuzzy PD/PI controller can now be tuned as well as heuristic rules can be added as required, in order to satisfy the performance criteria.

$\hat{\tau}_i^{fpd}$		$\hat{\dot{e}}_i$								
		NVL	NL	NM	NS	Z	PS	PM	PL	PVL
$\hat{e}_i$	NVL	NVL	NVL	NVL	NVL	NVL	NL	NM	NS	Z
	NL	NVL	NVL	NVL	NVL	NL	NM	NS	Z	PS
	NM	NVL	NVL	NVL	NL	NM	NS	Z	PS	PM
	NS	NVL	NVL	NL	NM	NS	Z	PS	PM	PL
	Z	NVL	NL	NM	NS	Z	PS	PM	PL	PVL
	PS	NL	NM	NS	Z	PS	PM	PL	PVL	PVL
	PM	NM	NS	Z	PS	PM	PL	PVL	PVL	PVL
	PL	NS	Z	PS	PM	PL	PVL	PVL	PVL	PVL
	PVL	Z	PS	PM	PL	PVL	PVL	PVL	PVL	PVL

Fig. 3. Equivalent state space rule base

6 Experimental Results

Experiments were carried out using the test bed designed and developed by Nanyang Technological University (NTU), Singapore (see Fig. 4) to assess the performance of SMC, fuzzy SMC and fuzzy PD controllers. Each controller was given a trapezoidal steer angle profile starting from a zero angle. Fig. 5 shows the SMC performance with a boundary layer. Width of the boundary layer was chosen to reduce the control chatter. Fig. 6 shows the tracking performance of the fuzzy SMC, which was synthesized using the SMC controller. Fuzzy SMC tracking performance is very similar to that of SMC (Fig. 5) in terms of tracking accuracy and control output. Further, it was observed that tighter spacing of membership functions around the origin yields better tracking accuracy, however, with increased control chatter. This is very much similar to the behavior of the equivalent SMC controller with changes in the boundary layer. The tracking performance of the synthesized fuzzy PD controller based on the fuzzy SMC is shown in Fig. 7. It is seen that the performance of the fuzzy PD controller is very similar to the fuzzy SMC and also to the SMC. Again, it was observed that a higher density of membership functions around the origins yields better accuracy and more control chatter. Nature of membership functions around the origin determines the shape and thickness of the effective boundary layer of the equivalent SMC and hence the observed behavior.

Fig. 4. The experimental test-bed designed and developed by NTU

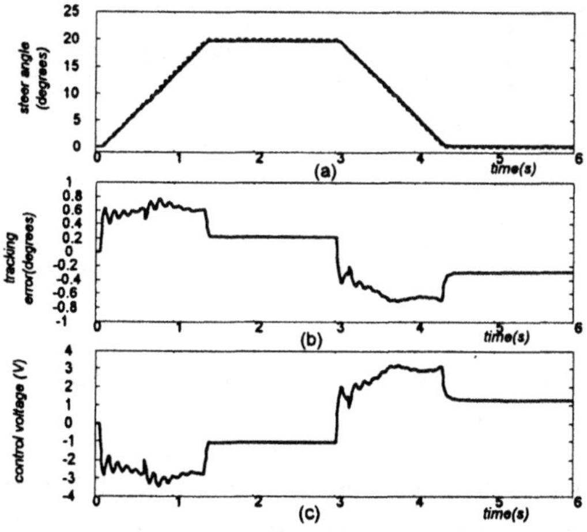

Fig. 5. Steering tracking performance and control voltage-Sliding Mode Controller

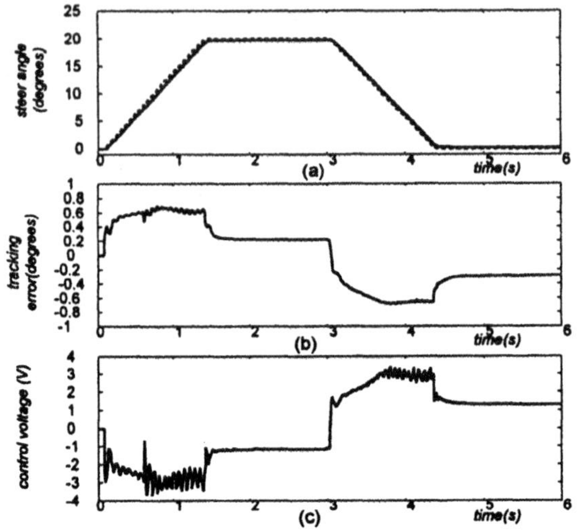

Fig. 6. Steering tracking performance and control voltage-Fuzzy Sliding Mode Controller

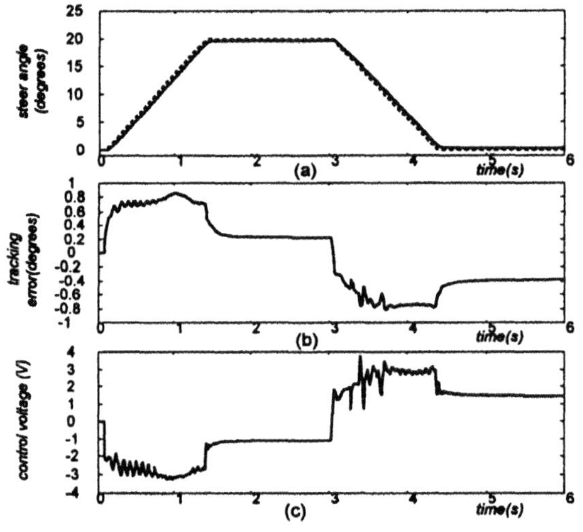

Fig. 7. Steering tracking performance and control voltage-Fuzzy PD Controller

In the designed fuzzy PD control scheme, the steady state error is mainly dependant on the input normalization gain and the distribution of membership functions of the linguistic variable, *angle error,* e_i. Selecting higher normalization gains as well as selecting densely packed membership functions near the origin can minimize the steady state error. However, both methods will increase the control chatter. One of the ways to overcome that problem is to introduce an integral part. Without increasing the number of inputs and hence the size of the rule base, we can incorporate integral action by redefining the normalized switching surface and following the same design procedure reported in Sections 3,4 and 5. Fig. 8 shows the improvement in tracking performance with the new fuzzy P(I+D) controller

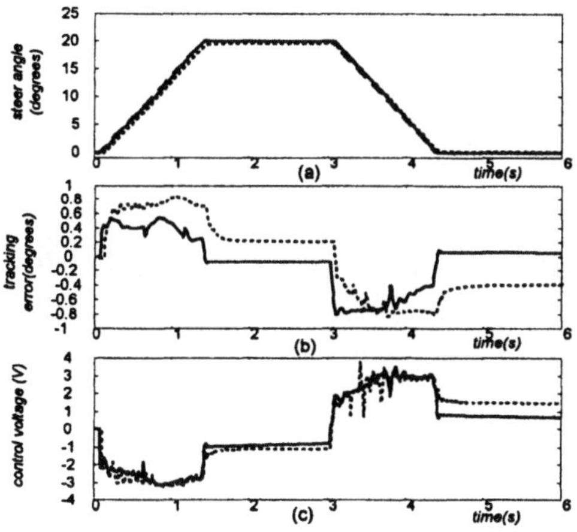

Fig. 8. Steer angle tracking performance of fuzzy PD controller (*dotted curve*) and fuzzy P(I+D) controller (*solid curve*) for the same input trajectory.

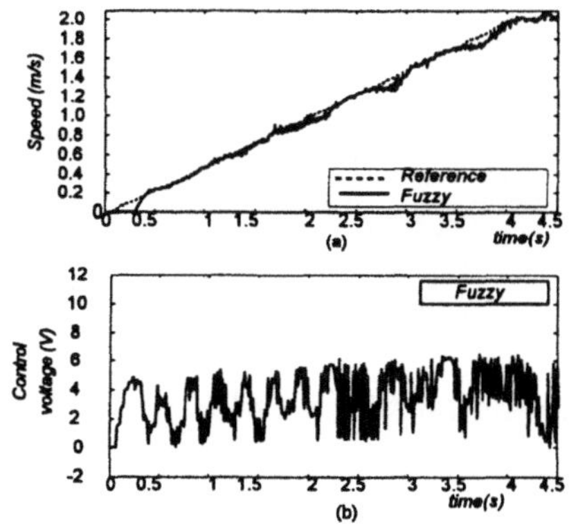

Fig. 9. Speed tracking performance of the FDC for an acceleration of $0.5ms^{-2}$.

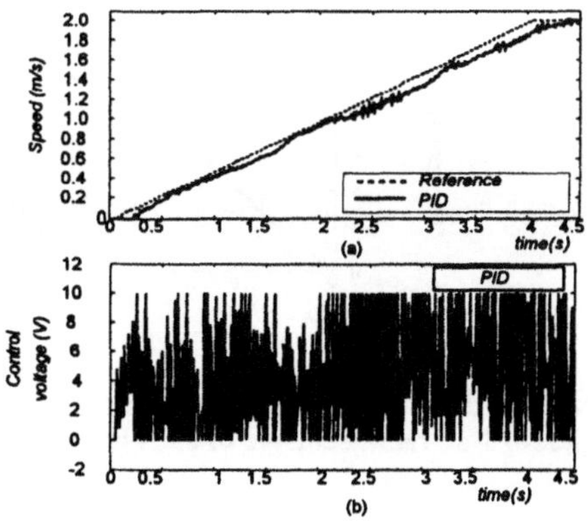

Fig. 10. Speed tracking performance of the PID controller for an acceleration of $0.5ms^{-2}$.

Further experiments were carried out to compare the performance of fuzzy controllers with conventional PID controllers. Fig. 9,10 and 11 shows that the Fuzzy Drive Controller (FDC) is capable of tracking different speed profiles effectively, however PID does not. Fig. 12(a) and (b) shows the high quality tracking performance of steer or Lateral Fuzzy Controller (LAFC) even with change of the operating point.

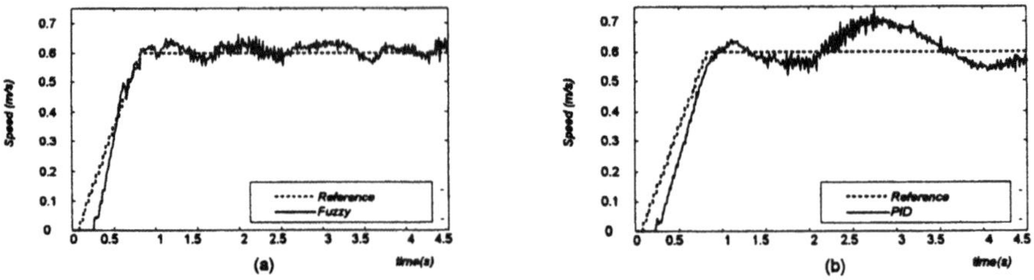

Fig. 11. Speed tracking performance at slower speeds, (a) FDC, (b) PID

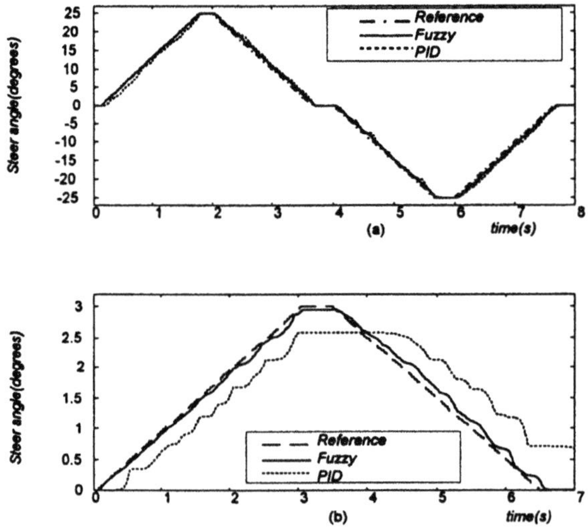

Fig. 12. LAFC and PID controller tracking performance with (a) $15°s^{-1}$ rate of change of steer angle, (b) $1°s^{-1}$ rate of change of steer angle

7 Conclusion

For a class of nonlinear systems characterized by AGV dynamics, stable fuzzy SMC, fuzzy PD control laws can be synthesized using sliding mode control theory. The resultant fuzzy controllers as anticipated yield trajectory tracking performance similar to the equivalent sliding mode controllers from which the fuzzy controllers were synthesized. Another important consequence of the fuzzy control law development given is that it shows how additional state variables, such as a variables derivative and integrals, can be incorporated into the controller by suitably modifying the switching surface. This permits the treatment of linear combinations of state variables (eg. I+D), as inputs to the fuzzy controller thus simplifying the rule base, computational complexity, and memory requirements, especially when the state inputs exceed two. All three controllers, i.e. SMC, fuzzy SMC and fuzzy PDC showed equally good performance.

References

1. Kamga, A., Rachid, A.,: Speed, Steering Angle and Path Tracking Controls for a Tricycle Robot. Procs. of the IEEE Int. Symposium on Computer-aided Control System Design, Dearbom, MI, (1996) 56-61
2. Zalila, Z., Bonnay, F., Coffin, F.: Lateral Guidance of an Autonomous Vehicle by a Fuzzy Logic Controller. IEEE Int. Conf. on Systems, Man and Cybernetics, vol.2, (1998) 1996-2001
3. Seki, Y., Ohya, J., Miyoshi, M.: Collision Avoidance System for Vehicles Applying Model Predictive Control Theory. IEEE/IEE/JSAI Conf. on Intelligent Transportation Systems, ITSC'99, Japan (1999) 453-458
4. Hunt, K.J., Kalkkuhl, J., Fritz, H., Johansen, T.A., Gottsche, T.: Experimental Comparison of Nonlinear Control Strategies for Vehicle Speed Control. Procs. of the IEEE Int. Conf. on Control Applications, Italy (1998) 1006-1010
5. Hedrick, J.K., McMahon, D., Narendran, V., Swaroop, D.: Longitudinal Vehicle Controller Design for IVHS Systems. Vol 3. Procs. of the American Control Conf. , (1991) 3107-3112
6. Hedrick, J.K.: Nonlinear Controller Design for Automated Vehicle Applications, Vol 1. UKACC Int. Conf. on CONTROL'98 (Conf. Publication No. 455), (1998) 23-32
7. Lee, C.D., Kim, S.W., Yim, Y.U., Jung, J.H., Oh, S.H., Kim, B.S.: Longitudinal and Lateral Control System Development for a Platoon of Vehicles. IEEE/IEE/JSAI Conf. on Intelligent Transportation Systems, ITSC'99, Japan (1999) 605-610
8. Senol, H., Gorgun, Aydeniz, M.G., Comparison and Determination the Electrical Motors which are used in Electrical Transportation Systems, Procs. of 9th IEEE Mediterranean Elctrotechnical Conf., MELECON 98, Israel (1998)
9. Xu G.: Dynamic Modeling and Control of Non-holonomic Car-like Vehicle. Technical report, Nanyang Technological University, Singapore (1998)
10. Wijesoma, W.S., Richards, R.J.: Robust Trajectory Following of Robots using Computed Torque Structure with VSS, Vol 52, No. 4. Int. J. Control, 1990
11. Wijesoma, W.S., Kodagoda, K.R.S.: Synthesis of Stable Fuzzy PD/PID Control Laws for Robotic Manipulators from a Variable Structure Systems Standpoint, Lecture Notes in Computer Science, Vol. 1625, Springer-Verlag, Berlin Heidelberg New York (1999) 495-511

Appendix

Fig. 13 shows the schematic diagram of the experimental test-bed.

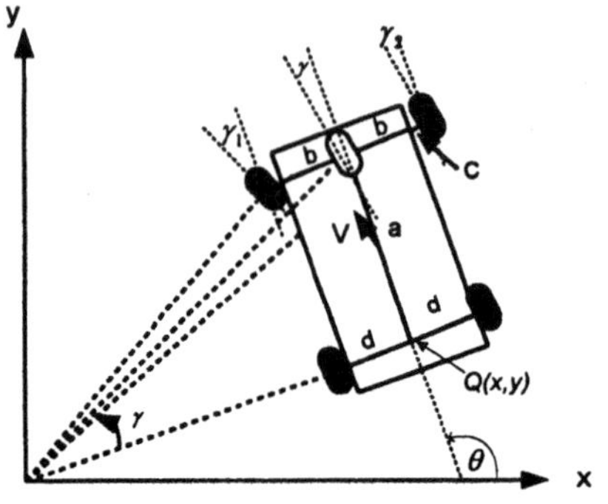

Fig. 13. Schematic diagram of the test-bed

Parameter definitions of the vehicle model

γ - virtual steer angle

v - speed of the vehicle

θ - heading angle

τ_d - driving torque

τ_b - braking torque

τ_s - steer torque

k_v - coefficient of air drag force

k_t - coefficient of turning friction

k_b – coefficient of braking

k_ϕ - coefficient of rolling friction

m_w - mass of a wheel

m_c - mass of the chassis of the vehicle

m - total mass of the vehicle

w - width of a tire

r - radius of a tire

I_c - Inertia of the chassis around the z-axis

$I_{w\phi}$ - Inertia of the tire along the axel

I_{wr} -Inertia of the tire perpendicular to the axel

F_{df} and τ_{sf} terms are related to friction

M_v, M_ω and N are related to inertias of the vehicle.

a, b, c and d are as defined in Fig. 13

$$F_{df} = \mathrm{sgn}(v)\frac{k_\phi m}{2r}(1+\sec\gamma)+\mathrm{sgn}(\dot{\theta})\frac{k_t w(a-y_{cm})mg}{2a^2}\tan\gamma$$

$$\tau_{sf} = \mathrm{sgn}(\dot{\theta}+\dot{\gamma})\frac{k_t w y_{cm} mg}{2a}+\mathrm{sgn}(v)\frac{k_\phi cm}{4r}(\gamma_1'-\gamma_2')$$

$$M_v = \left[\left(1+\frac{x_{cm}}{a}\tan\gamma\right)^2+\left(\frac{y_{cm}}{a}\tan\gamma\right)^2\right]m_c$$

$$+\frac{\tan^2\gamma}{a}(I_c+4I_{\omega r})+\left(a_5^2+a_6^2+a_7^2+a_8^2\right)\left(I_{\omega\phi}+m_w r^2\right)$$

$$M_\omega = \left(\gamma_1'^2+\gamma_2'^2\right)\left(I_{wr}+\frac{c^2}{r^2}I_{w\phi}+m_w r^2\right)$$

$$N = \frac{1}{2}\tan\gamma(\gamma_1'+\gamma_2')I_{wr}-\frac{c}{r}(\gamma_1'a_5+\gamma_2'a_6)\left(I_{w\phi}+m_w r^2\right)$$

$$C_{s1} = \left(\gamma_1'\gamma_1''+\gamma_2'\gamma_2''\right)\left(I_{wr}+\frac{c^2}{r^2}I_{w\phi}+m_w r^2\right),$$

$$C_{d1} = \frac{c}{r}\left(a_5\gamma_1''-a_6\gamma_2''\right)\left(I_{w\phi}+m_w r^2\right)+\frac{1}{a}\tan\gamma\left(\gamma_1''+\gamma_2''\right)I_{wr}$$

$$C_{d2} = \left(x_{cm}+\frac{x_{cm}^2+y_{cm}^2}{a}\tan\gamma\right)\frac{m_c}{a\cos^2\gamma}+\frac{\sin\gamma}{a^2\cos^3\gamma}(I_c+4I_{wr})$$

$$+\left(\gamma_1'a_5a_5'+\gamma_2'a_6a_6'+a_7a_7'+a_8a_8'\right)\left(I_{w\phi}+m_w r^2\right)$$

$$C_{s2} = \left(\gamma_1'+\gamma_2'\right)\frac{I_{wr}}{a\cos^2\gamma}-\frac{c}{r}\left(\gamma_1'^2a_5'+\gamma_2'^2a_6'\right)\left(I_{w\phi}+m_w r^2\right)$$

$$I_c = \frac{m_c}{12}\left(a^2+4d^2\right), I_{wr} = \frac{m_w}{12}\left(w^2+3r^2\right),$$

$$I_{w\phi} = \frac{1}{2}m_w r^2$$

$$\gamma_1 = \arctan\left(\frac{a\sin\gamma}{a\cos\gamma+b\sin\gamma}\right),$$

$$\gamma_2 = \arctan\left(\frac{a\sin\gamma}{a\cos\gamma-b\sin\gamma}\right)$$

$$a_1 = \cos\theta,\quad a_2 = \sin\theta,$$

$$a_3 = \frac{1}{a}\tan\gamma,\quad a_4 = 0,$$

$$a_5 = -\frac{a+c\sin\gamma_1}{r\left(a\cos\gamma_1-b\sin\gamma_1\right)},$$

$$a_6 = \frac{a-c\sin\gamma_2}{r\left(a\cos\gamma_2+b\sin\gamma_2\right)},$$

$$a_7 = -\frac{a+d\tan\gamma}{ar}, a_8 = \frac{a-d\tan\gamma}{ar}$$

Further, a_i' and γ_i' are the first partial derivatives of a_i and γ_i with respect to γ. Similarly, γ_i'' is the second derivative of γ_i with respect to γ.

Time-Series Segmentation and Symbolic Representation, from Process-Monitoring to Data-Mining

Bernard Hugueney[1,2] and Bernadette Bouchon-Meunier[1]

[1] Laboratoire d'Informatique de Paris 6, LOFTI
[2] EDF, Stratégie et Développement-IMA-TIEM-SOAD

Abstract. Data-analysis has undergone an important change from statistical descriptive analysis to data-mining. Information networks and huge data-storage equipments brought data-retrieval to new dimensions. Time-series are especially easy to accumulate as digital sensors can be used to fill databases without any intervention. This is both a boon and a problem as the very amount of data available prevents the user from being able to understand them. One has to build high-level representations of the time-series to be able to extract some information. Segmentation is often used in process-monitoring for similar reasons.
In this paper, we describe step by step difficulties and solutions that we studied when adapting automated time-series segmentation to a real-world example of electric consumption analysis. The data that we want to analyze consist of yearly reports of electric power consumption in 10 minute ticks. We study industrial consumers that have simple processes (ovens, motors) switched either on or off for the duration of the process. Hence we could use this prior knowledge to model the time-series with piecewise constant changing mean models. We then extend the segmentation to a symbolic representation to enable interpretation of the overwhelming number of generated segments.

1 Introduction

Time-series are a special kind of data because using a graphical representation (a chart), we are able to effectively use the great processing power of our visual perception system. The automatic features-extraction that takes places when a chartist looks at time-series makes the graphical plot of the data much more expressive than the raw list of numbers. Nevertheless automated time-series analysis tools have been developed, especially for process monitoring, when the time-series to be considered reveal the state of a system that must be analyzed around the clock with real-time constraints.

The data-mining approach came in with three specific aspects:

1. Data were accumulated, not collected for a specific use of information extraction.

B. Reusch (Ed.): Fuzzy Days 2001, LNCS 2206, pp. 118–123, 2001.

2. Huge amounts of data are available, much more than what can be handled by the user.
3. The user's expectations are not well defined ("useful information").

The first point is important because it means that, at best, the data are not pre-processed and can be very noisy, at worst, a bad pre-processing can be harmful. The second point tells us that one must pay great attention to computational feasibility. Useful algorithms should not have a complexity much greater than linear in order to be applied on the whole data-set. The third point leads us to provide a framework to the user, more than just a tool. This means that different approaches must be provided, as each analysis of the time series gives the user insights leading to better understanding of the data. Of course, relevance criteria can be computed automatically, but given the computing cost of data-mining, we want to be able to use any prior knowledge available (even if we require none).

2 Existing Algorithms for Process-Monitoring

Time-series segmentation is a classical multiple hypotheses testing problem. From a statistical point of view, it is well known that it is possible to estimate the model parameters (segment models and their boundaries) using the *maximum likelihood estimation.* Usually, likelihood L is optimized by minimizing the criterion :

$$-\ln L = \sum_{i=1}^{k} m_i \ln \sigma_i^2 \tag{1}$$

where m_i, σ_i^2 are the number of points and variance of the ith segment from the computed model.

It should be noted that often the goal of the segmentation is not to find the underlying ideal model, but to minimize a cost function. In that case, such cost function should be used, instead of likelihood estimation.

The most computationally efficient approach to segmentation is a hierarchical one, where segments are added by cutting episodes of the previous step of the segmentation. However, the real problem is to estimate the number of segments. A trade-off has to be made between model accuracy, and a *parsimonious principle* because segmenting time-series with one-point segments is very easy, very accurate, but not very useful. A common solution is to use a penalized maximum likelihood criterion. The idea is to add a penalty term to generalized likelihood term in Eq.1. A general form of penalty term is $n(d+1)\gamma(N)$ which is a function of the number of parameters (change-times included) used to describe the segmented signal. The most famous penalty terms are Akaike Information Criterion (AIC [1]): $2n(d+1)$ designed to maximize predictive accuracy of the model, and Bayesian Information Criterion (BIC [4]) or Minimum Description Length (MDL) : $n(d+1)\ln N$ designed to maximize the posterior probability of the chosen model. In [3], only the likelihood is computed, and segments are added until the likelihood criterion becomes stable, that is when adding segments does not *significantly* improves the accuracy. The problem is in the definition of *"significantly"* that takes the form of a threshold that must be set.

3 Segmentation for Data-Mining

The fact the the data had not been collected for our data-analysis prevented from using a BIC criterion, as the sampling period (and the number N of time ticks) was not related to the information that we wanted to extract. AIC criterion and AICC did not yield very good results presumably because our data did not fit the models very well and were very noisy. We decided to use the approach presented in [3] but given the diversity in our time-series, no general threshold could be selected.

We chose a data-driven approach: automatic clustering of the likelihood variations would allow us to select a relevant number of episodes. It is important to note that we do not aim to find *the correct number of segments.* Because unfortunately, no such number exists as shown in Fig. 1. There are numerous potentially relevant scales to consider when segmenting this kind of time-series. Each such scale can be located as a change point on the cost versus number of segments function.

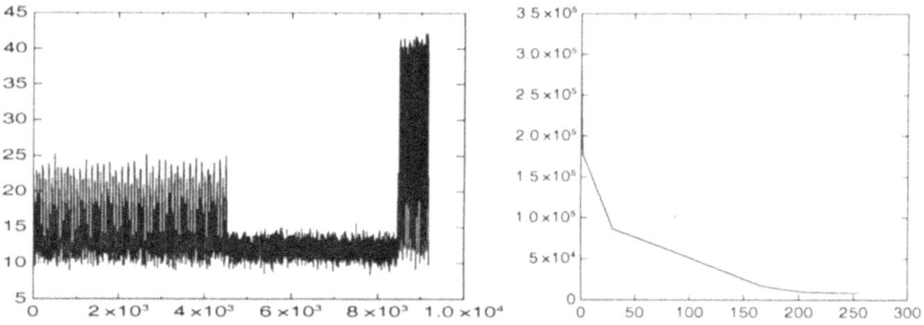

Fig. 1. Sample power consumption and error vs. number of segments.

Instead of trying to find an absolute criterion, we just select a meaningful answer. A relevance feedback from the user or from an automated further step of the data-analysis could lead to a deeper hierarchical clustering selecting other meaningful numbers of segments. We could make effective use of prior knowledge concerning the relevant order of magnitude for the number of segments or duration of episodes. For exemple, a number of segments close to a multiple of the number of weeks or days in the time-series can be relevant if a night/day or week-day/week-end cycle is investigated.

The time-series that we wish to analyze can consist of hundreds of episodes, and another problem arises from the size of the data. Most segmentation algorithms are used (or at least benckmarked) for a limited number of segments (usually half a dozen). Each segment is then very important for the correct fitting of the data and the likelihood criterion is efficient. But as the number of segments increases, the individual contribution of each segment can be diluted in the noise, especially if the time-series are very noisy. The huge amount of data can lead to numerical errors when the built-in precision of the floating-point representation is not enough to prevent round-off errors. The problem is that for efficiency reasons, we must avoid recomputing partial results as much as possible, and keep the built-in floating point representation. If the number of segments is too large, it is difficult to get the correct number of segments because of the noise and the accumulated numerical errors.

On the other hand, if the number of segments is too small, the automatic clustering can become inaccurate and the relative error can be very large. Let $[n_{min}, n_{max}]$ be the number of segments for which we consider the trade-off to be optimal. Thoses values depend on the numerical accuracy of the processor, the noise and relevance of the fitted model, but for our experiments, results were not sensitive to the exact values.

To avoid this problem, we decided to go for a multi-pass segmentation. First we use our change-point detection algorithm on the whole dataset. We can use the resulting episodes as an approximation of the solution. We then split the data set in time-series of k episodes, $k \in [n_{min}, n_{max}]$ is selected so that the change point between the sub-series (that will not be computed again) corresponds to the most certain change point. We just need to store the detected cut-points on a stack to remember the order in which they were found. Cut-points are detected in decreasing likelihood order, so among the $(n_{min} - n_{max})$ candidates for sub-serie boundary selection, we choose the oldest (most likely) one. Each sub-serie is then segmented again. The whole segmentation is constructed by concatenation of the segmented sub-series. This can be iterated until the number of segments is stable.

4 From Segmentation to Symbolic Representation

Episode extraction from time-series is a typical data-analysis problem. We build higher-level abstractions (models) from raw data to gain synthetic and relevant information. But this approach reaches it's limits when the data-sets becomes really huge. As the number of episodes increases, the data reduction from data-points to model-parameters is not enough. Too many episodes cannot be handled by the end-user. We want to go one step further in abstraction. From the many episodes discovered, it is often possible to build representative clusters. In the spirit of computing with words paradigm introduced by L. ZADEH [6], we will go from numbers (model parameters) to for labels of fuzzy sets (episodes clusters). Previous approaches such as in [2] use prior knowledge to create an alphabet of symbols ("spike", "rise", "fall", etc...), but we decide to have a "data-driven" approach and to create the symbolic alphabet from the available data.

First we perform automated clustering of the model parameters. We used a modeling cost function (euclidien distance between the model and the original episod) to evaluate the clustering cost. It is important to use a model-based distance between episods to cluster them, as otherwise, a simple euclidian distance would be fooled by the noise and/or lack of fit. But such noise can be a valuable information in itself. To keep this information, we use fuzzy prototypes. In Fig.2, from an explanation point-of-view, the first two episods are closer to each other than to the third, while a simple euclidian distance would not be able to discriminate them properly.

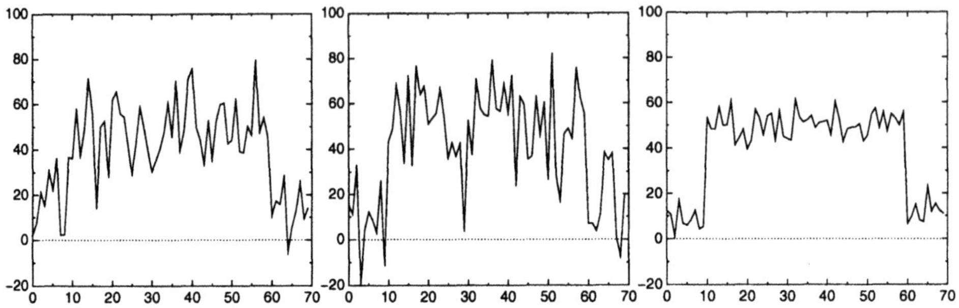

Fig. 2. Three episodes were model-based distance is more appropriate then euclidian one.

It should be taken into account that the cut-points need not be optimal for the new restricted models. The episods should always have to fit to the undelying time-series as closely as possible and as the parameters of the time series get changed when restricted to those of their cluster prototype, their boundaries should be tuned. Unfortunatly, when computing the automatic clustering this is not feasible for efficiency reasons. This adjustment is done after each restriction. Cut-points are moved to minimize a modeling cost function after the episode levels are restricted.

We then compute sub-clusters among the previous clusters, according to episode length. The cost function that we try to minimize is the sum of overlaps and gaps between episods.

Each sub-cluster of episodes can then be represented by a symbol, and each episode can be associated to a symbol(except outliers, but they can be labelled as such). Of course it is important to determine an accuracy level and to be able to label atypical episodes as such. The process is illustrated in Fig.3. A symbol can be considered as the label of the associated cluster, and represents a fuzzy prototype of the cluster.

5 Conclusion and Future Works

There are many uses of the symbolic representation. It can help to visualize long time series. It may also to possible to edit the symbolic representation for predictive analysis. If a symbol corresponds to the consumption of a known

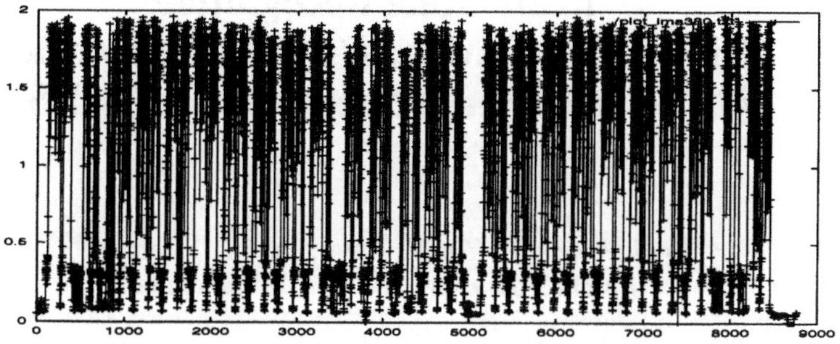

Segments:
{(01/01/1998/02h00-06/01/1998/07h00, mean=0.25487, length=125),
(06/01/199807h00-17/01/1998/03h00, mean=1.46451, length=260),
(17/01/1998/03h00-22/01/1998/22h00, mean=0.26543, length=139),$\cdots$,
(19/12/1998/02h00-01/01/1999/01h00, mean=0.18613, length=312)}
Symbols:
$\{S_1, S_{11}, S_1, S_6, S_2, S_4, S_0, S_7, S_1, S_{12}, S_1, S_{12}, S_1, S_{11}, S_1, S_{11}, S_1, \cdots, S_2\}$

Fig. 3. Original load curve: 8760 hourly ticks, 71 episodes, 12 symbols

device, and one intends to change the consumption of this device, it is possible to edit the prototype, and create predictive time-series. The most interesting part is yet to come, as we aim to further synthesize the information. The next step will be to mine interesting patterns as in [5] from the symbolic representation.

References

1. H. Akaike. Information theory and an extension of the maximum likelihood principle. In B. N. Petrov and F. Cáski, editors, *Second International Symposium on Information Theory*, pages 267–281, Budapest, 1973. Akademiai Kaidó. Reprinted in *Breakthroughs in Statistics*, eds Kotz, S. & Johnson, N. L. (1992), volume I, pp. 599–624. New York: Springer.
2. Michael A. Eisenberg. The kineticist's workbench: Combining symbolic and numerical methods in simulation of chemical reaction mechanisms. Technical report, Massachusetts Institute of Technology, 1991.
3. Valery Guralnik and Jaideep Srivastava. Event detection from time series data. In Surajit Chaudhuri and David Madigan, editors, *Proceedings of the Fifth ACM SIGKDD International Conference on Knowledge Discovery and Data Mining*, pages 33–42, N.Y., August 15–18 1999. ACM Press.
4. G. Schwarz. Estimating the dimension of a model. *The Annals of Statistics*, 6:461–464, 1978.
5. R. Srikant and R. Agrawal. Mining sequential patterns : Generalization and performance improvements. In *EDBT*, March 1996.
6. L. A. Zadeh. Fuzzy logic = computing with words. *IEEE Transactions on Fuzzy Systems*, 4(2):103–111, 1996.

Brain State Recognition Using Fuzzy C-Means (FCM) Clustering with Near Infrared Spectroscopy (NIRS)

Syoji Kobashi[1], Yutaka Hata[1], Yuri T. Kitamura[2,3], Toshiaki Hayakata[4], and Toshio Yanagida[2,3]

[1] Department of Computer Engineering, Himeji Institute of Technology, 2167, Shosha, Himeji, Hyogo, 671-2201, JAPAN
{kobashi, hata}@comp.eng.himeji-tech.ac.jp
{kobashi, hata}@ieee.org

[2] BISC Group, Computer Science Division, University of California at Berkeley, USA

[3] Department of Physiology & Biosignaling, Osaka University Graduate School of Medicine, A-4, 2-2, Yamadaoka, Suita, Osaka, 565-0871, JAPAN
{tsuda, yanagida}@phys1.med.osaka-u.ac.jp

[4] Yanagida Project, Commutations Research Laboratory, 588-2, Iwaoka, Nishi-ku, Kobe, Hyogo, 651-2492, JAPAN

[5] Department of Acute Critical Medicine, Osaka University Graduate School of Medicine, 2-2, Yamadaoka, Suita, Osaka, 565-0871, JAPAN
hayakata@hp-emerg.med.osaka-u.ac.jp

Abstract. Near-infrared spectroscopy (NIRS) is a recently developed method, which can investigate the human brain function with noninvasive, high time resolution, and high portability. However, there are few discussions on post-processing of time series data taken by the NIRS because of the difficulty of understanding the obtained data and the complexity of the human higher-order brain function. This paper discusses on an analysis of such a time series. The analysis method is based on fuzzy c-means (FCM) clustering and wavelet transform, and it divides the time series of a measurement point into some clusters with respect to wavelet coefficients. To evaluate the performance of the method, it has been applied to four healthy volunteers, and three brain-dead patients. The results showed that the proposed method could segment the NIRS time series into some clusters that may represent brain states, and could estimate the number of clusters.

B. Reusch (Ed.): Fuzzy Days 2001, LNCS 2206, pp. 124–136, 2001.

1 Introduction

Functional near-infrared spectroscopy (NIRS) is a recently developed method, which can investigate the human brain function with noninvasive, high time resolution, and high portability [1]. It has been proved that the NIRS measures the cerebral blood volume dynamics closely correlated to the neuron activity of the brain [2], and applied to investigate cortical activation related to motor activity [3], mental tasks [4] and auditory stimulation [5], and so on. However, there are few discussions on post-processing of time series data taken by means of the NIRS because of the difficulty of understanding the obtained data and the complexity of the human higher-order brain function. Only, Gratton *et al.* shown a post-processing method for removing periodical artifacts caused from heart beats [6].

During activation of brain function, there are some stages in human brain. Clare *et al.* studied the brain activation in short-term memory paradigm (Sternberg short-term memory task [7] was used) on functional magnetic resonance imaging (MRI) study [8]. The paradigm consisted of four stages; reception, retention, recall, and match, and rest. The study concluded that homodynamic responses corresponding to those stages could be found in the brain cortex by applying analysis of valiance (ANOVA) to functional MRI (fMRI) data. Our work is to investigate such brain states using the NIRS, which has high time resolution (about 100 msec) in comparison with the fMRI (about 3 sec). In comparison with the previous studies using fMRI, our work aims to decompose a time series into some stages, and estimate the number of stages. They might be fundamental studies in understanding human higher-order brain function.

Fuzzy c-means (FCM) clustering introduced by Bezdek [9] has been used in many research fields. Especially, it has been producing notable results in the field of medical imaging. For example, Hall *et al.* applied FCM clustering to tissue segmentation of human brain MR images [10][11], and also clustering fMRI time series aided by FCM is an interesting application [12]. Although their methods can segment an image into some clusters, it cannot be applied to decomposing time series of a measurement point into some clusters, which is our interest.

The aim of this study is to present a new method for analyzing time series data taken by means of the NIRS. The method is based on fuzzy c-means (FCM) clustering [9], in which the features of each data are extracted by means of wavelet transforms. The method divides the time series of a measurement point into some clusters the may represent the brain states, and estimates the number of clusters. This method has been applied to time course NIRS signals measured for four healthy volunteers, and three brain-dead patients. To evaluate the performance of the proposed method, clustering results and tables of estimated number of clusters are shown. The results showed that our method might obtain the preferred clusters that correspond to brain states, and the number of clusters of brain-dead patients is less than one of healthy volunteers.

2 Materials

The time series data were acquired from an NIRS (Optical topography system, ETG-100; Hitachi Medical Corp., Tokyo, JAPAN), which can measure the temporal changes of the concentration of oxygenated hemoglobin (oxy-Hb), deoxygenated hemoglobin (deoxy-Hb), and total hemoglobin (total-Hb) by using two different near-infrared lasers. The data acquisition was performed with sampling time of 100 msec, wavelength of 780 nm and 830 nm, and two sets of 12 channels mounted on a head shell. The concentrations of oxy-Hb and deoxy-Hb were computed by using the modified Lambert-Beer law [13], and the concentration of total-Hb was computed as sum of oxy-Hb and deoxy-Hb. In the Lambert-Beer law the molar absorption coefficients of oxy-Hb and deoxy-Hb show in Ref. [14] were used. Fig. 1 shows a temporal change of concentration of oxy-HB at a measurement point. In this figure, the subject performed a pre-defined paradigm during the period of shaded area. To measure the concentration of oxy-Hb, deoxy-Hb, and total Hb in the left fronto-central region, we set five incident and four detection positions and defined twelve

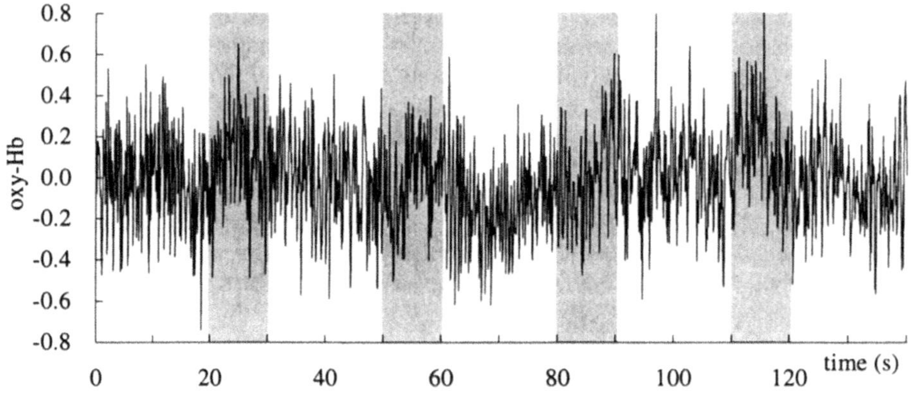

Fig. 1. Temporal change of concentration of oxy-HB.

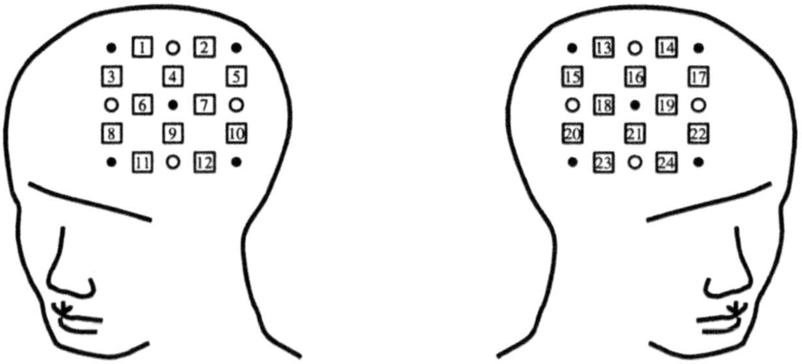

(a) The left temporal region. **(b) The right temporal region.**

Fig. 2. Measurement positions.

measurement positions on the head as shown in Fig. 2(a). In this figure, black circles, white circles, and white squares show the incident positions, the detection position, and the measurement positions, respectively. In a similar way, incident and detection positions were set on the right temporal region shown in Fig. 2(b), the frontal region of head, and the occipital region of head according to region of interests in that study.

3 FCM Clustering for Time Series Analysis

Our method is based on wavelet transform and FCM clustering. The FCM clustering can automatically classify whole data into some clusters aided by fuzzy logic such that each cluster has similar attribute values [9]. The wavelet transform provides an important tool in signal analysis and feature extraction [15][16]. It produces a good local representation of the signal in both the time domain and frequency domain. Unlike the Fourier transform, which is global and provides a description of the overall regularity of signals, the wavelet transforms can decompose a signal such that frequency characteristics and the location of particular feature in a time series may be highlighted simultaneously. Briefly, our method extracts features of a time series data by using the wavelet transforms, and obtains the number of clusters by using unsupervised FCM clustering. The details of the method are the followings.

3.1 Continuous Wavelet Transform Using Mexican hat

Let $x(t)$ be the concentration at time point t. Features of time point p are computed by using continuous wavelet transform defined as

$$c(s,p) = \int_{-\infty}^{\infty} x(t)\psi(s,p,t)dt\,, \tag{1}$$

where s is a scale number, t is a time point, and $\psi(s,p,t)$ is an analyzing wavelet. The analyzing wavelet used is Mexican hat defined as

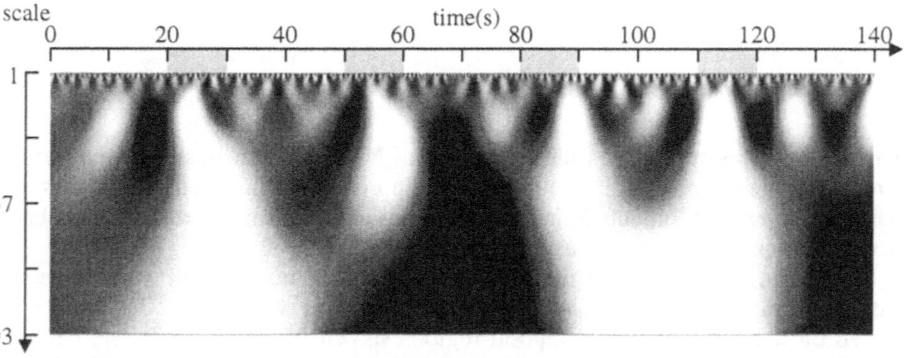

Fig. 3. Wavelet coefficients of the time series data shown in Fig. 1.

$$\psi(s,p,t)=\frac{2}{\sqrt{3}}\pi^{-\frac{1}{4}}(1-\alpha^2)e^{-\frac{1}{2}\alpha^2}\text{, where }\alpha=\frac{t-p}{s}. \tag{2}$$

Fig. 3 shows the wavelet coefficients of the time series data shown in Fig. 1. In this figure, lower coefficients appear as dark, and higher ones appear as bright.

3.2 Fuzzy Clustering with Estimating the Optimum Number of Clusters

In our experiment, scale parameter was set from S_{min}=32 to S_{max}=95 (found by preliminary experiments) to avoid wavelet coefficients corresponding to higher (for artifacts caused from heart beats) or lower frequency (artifacts caused from drift). Our method applied standard FCM clustering [7] to the wavelet coefficients, i.e., number of data=number of samples, number of attributes=64 (=S_{max} – S_{min}+1), and number of clusters=N_c. To estimate the optimal number of clusters we define similarity measure z_{ij} between cluster i and cluster j as

$$z_{ij}=\frac{1}{N_c}\sqrt{\sum_{s=S_{min}}^{S_{max}}\frac{(c_i(s)-c_j(s))^2}{\sigma(s)^2}}, \tag{3}$$

where N_c is the number of clusters, $c_i(s)$ is s-th wavelet coefficient of center vector of i-th cluster, $\sigma(s)^2$ is variance of s-th wavelet coefficient over the all sampling points. The procedure of our method is the followings.

Step 0) Set an initial number of clusters on Nc. In our experiments, initial number was set to 6.
Step 1) Do FCM clustering with the number of clusters of N_c.
Step 2) For all combination of clusters, calculate the similarity measure defined by Eq. (3) and find a pair that has minimum similarity measure.
Step 3) If the similarity measure is smaller than a threshold Th_z (2.4 was used in our experiments), the two clusters are merged, decrease the number of clusters, and go to step 1.
Step 4) Finish the clustering procedure.

4 Experimental Results and Discussion

This method was applied to time series data of NIRS for four healthy volunteers, and three brain-dead patients. This study was approved by the local ethics committee (Osaka University Graduate School of Medicine) and informed consent was obtained from all volunteers and relatives of patients.

To evaluate the clustering performance, we applied our method to the time series data of simple motor task for two right-handed healthy volunteers. The probes were placed on the left and right temporal regions shown in Fig. 2. Subjects performed sequential hand gripping of the left hand for 10 seconds, with rest periods of 20 seconds. The gripping frequency was varied from 1-3Hz. Four epochs were measured. Consequently, one study was composed of 120 seconds of 1200 sample points. Fig. 4 shows a temporal change of oxy-HB at measurement point of #22. The wavelet coefficients are shown in Fig. 5. After applying standard FCM clustering method to the data, center vectors and fuzzy degrees of each cluster were obtained as shown in Fig. 6 and Fig. 7, respectively. Table 1 shows the similarity measures between two clusters. In this case, similarity measure between cluster 2 and 3 was 0.51851, and the clusters were merged. Fig. 8 shows the final clustering result where number of clusters was 2. In this figure, cluster #1 may correspond to task period, and cluster #2 may correspond to rest periods. The results for other measurement points, and for other subject are also shown in Fig. 9, and Fig. 10. These results shows that the clustering procedure seems to obtained preferred results.

To evaluate the estimated number of clusters, we applied the proposed method to NIRS data of three brain-dead patients, and two volunteers. The performed tasks were tabulated in Table 2. The estimated number of clusters for brain-dead patients and for healthy volunteers are tabulated in Table 3, and Table 4, respectively. Table 5 shows the mean values of the estimated number of clusters. As shown in these tables and the mean values, the estimated number of clusters for brain-dead patients is less than one for health volunteers. This suggests that the proposed method may obtain the suitable number of clusters corresponding some brain states.

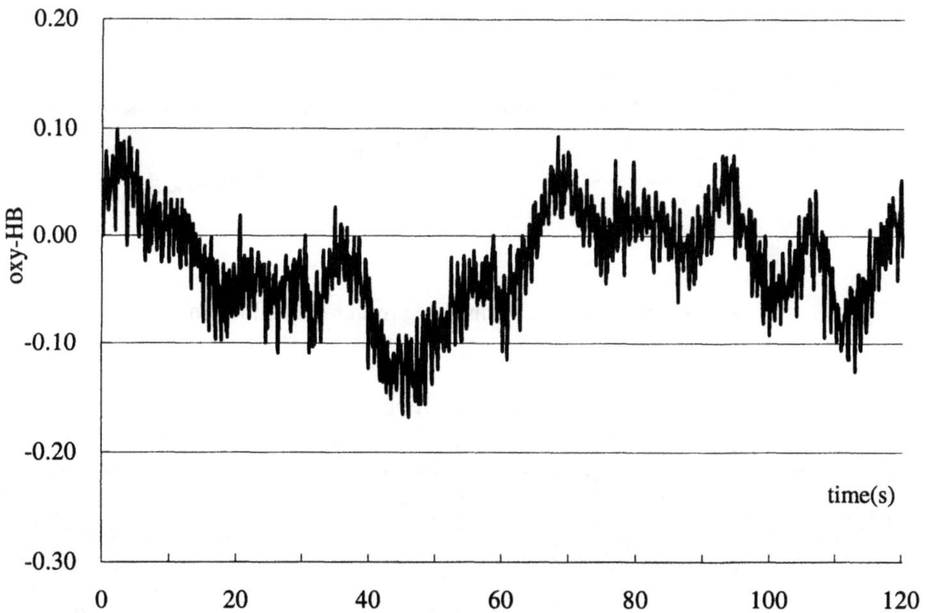

Fig. 4. Temporal change of oxy-HB at measurement point #22.

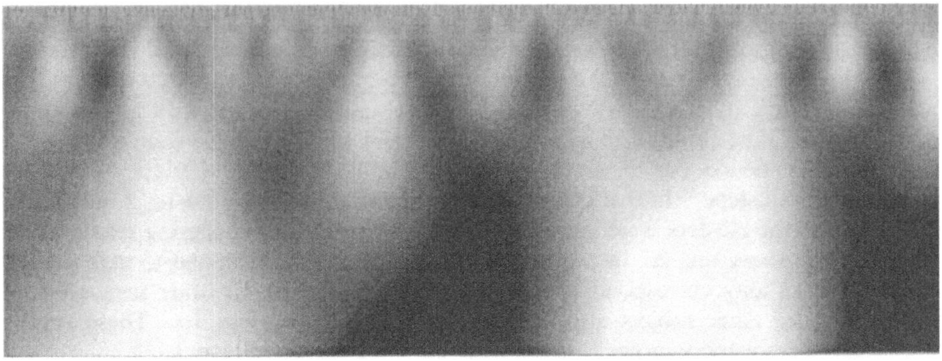

Fig. 5. Wavelet coefficient of time series data shown in Fig. 4. Vertical axis: time(s), Horizontal axis; scale of analyzing wavelet (upper is small scale). Low coefficients appear\ed as black, and high ones appeared as white.

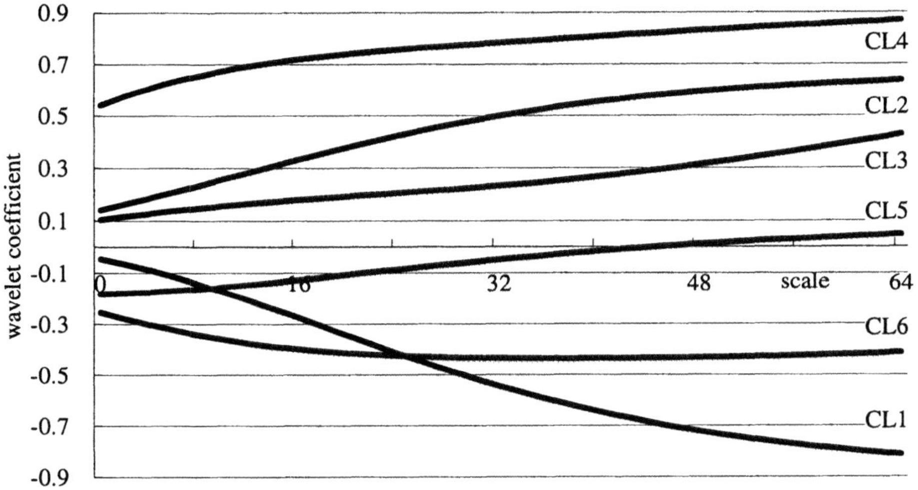

Fig. 6. Center vectors of clustering results before merging.

Table 1. Similarity measures between clusters.

	CL1	CL2	CL3	CL4	CL5	CL6
CL1		2.33823	1.82792	3.41247	1.13586	0.72310
CL2	2.33823		0.51851	1.33160	1.50806	2.36078
CL3	1.82792	0.51851		1.74265	1.05134	1.87152
CL4	3.41247	1.33160	1.74265		2.79125	3.59471
CL5	1.13586	1.50806	1.05134	2.79125		0.87850
CL6	0.72310	2.36078	1.87152	3.59471	0.87850	

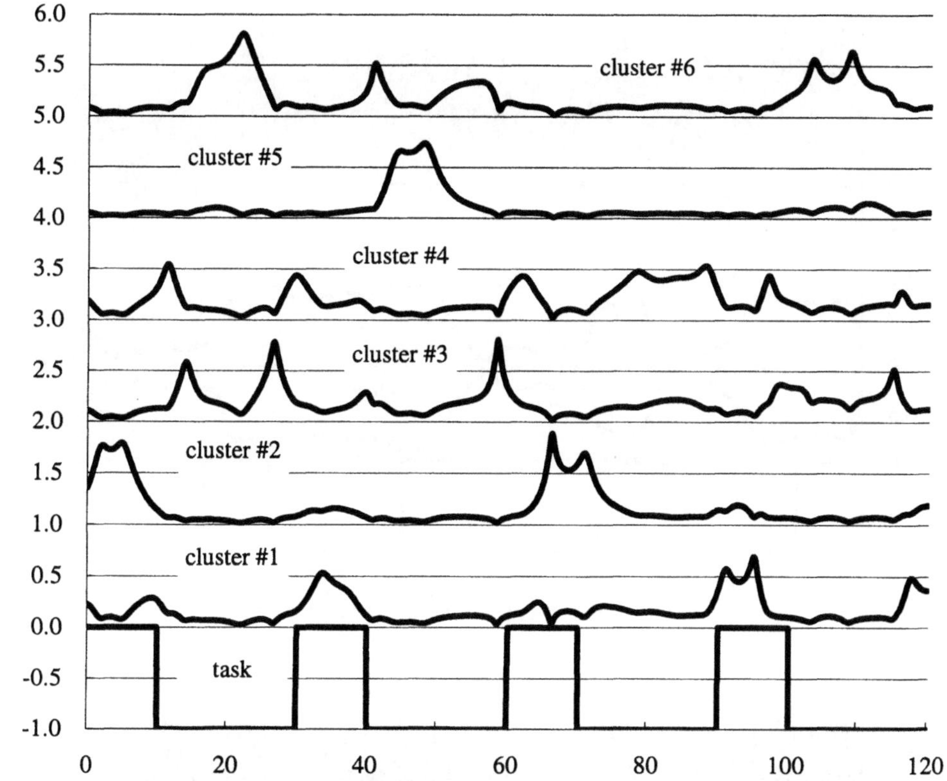

Fig. 7. Clustering results before merging. Vertical axis: time(s); horizontal axis: fuzzy degree; Lowest: performed task (0;task, -1;rest); others: clusters from #1 to #6 (degrees were biased).

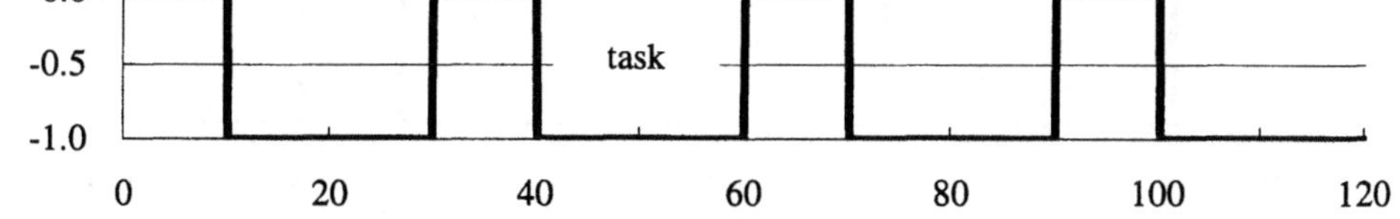

Fig. 8. Final clustering result. Vertical axis: time(s), Horizontal axis: fuzzy degree; Lower: performed task (0;task, -1;rest), Middle: cluster #1, and Upper: cluster #2 (degrees were biased by 1.0).

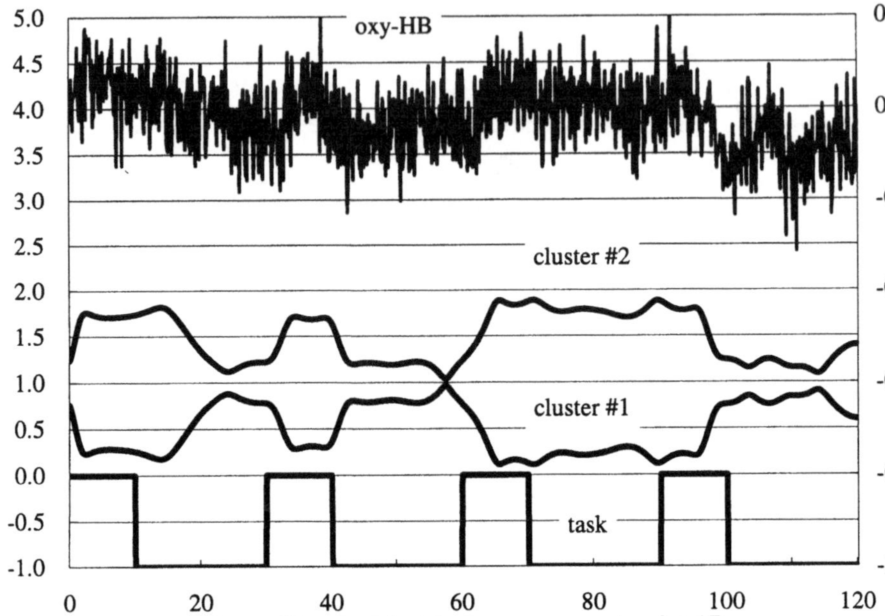

(a) oxy-Hb of subject 1 at measurement point 17.

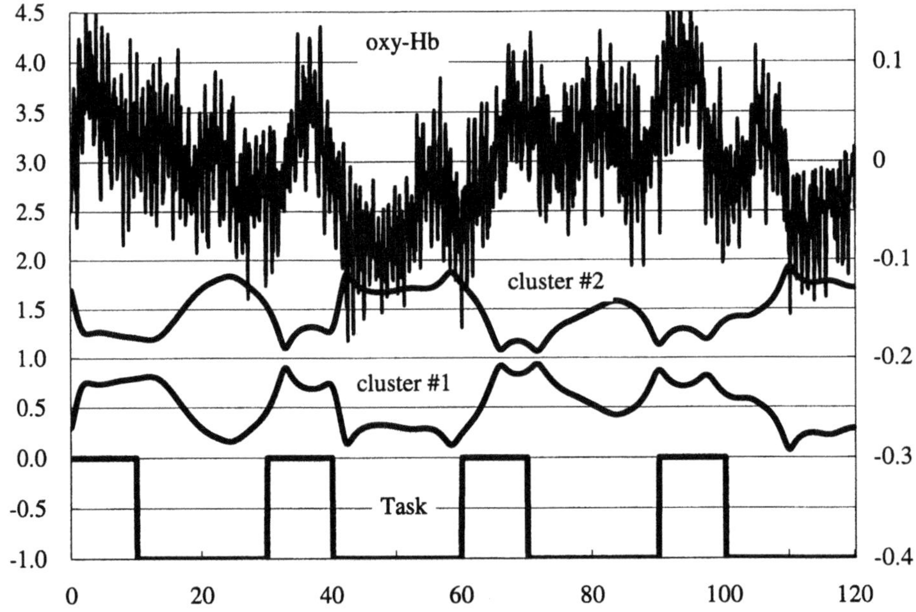

(b) oxy-Hb of subject 1 at measurement point 4.

Fig. 9. Clustering results for subject #1. Vertical axis: time(s), Horizontal axis; left: fuzzy degree; right: concentration, Top; Hb, Bottom: performed task (0;task, -1;rest), Others: clusters (degrees were biased).

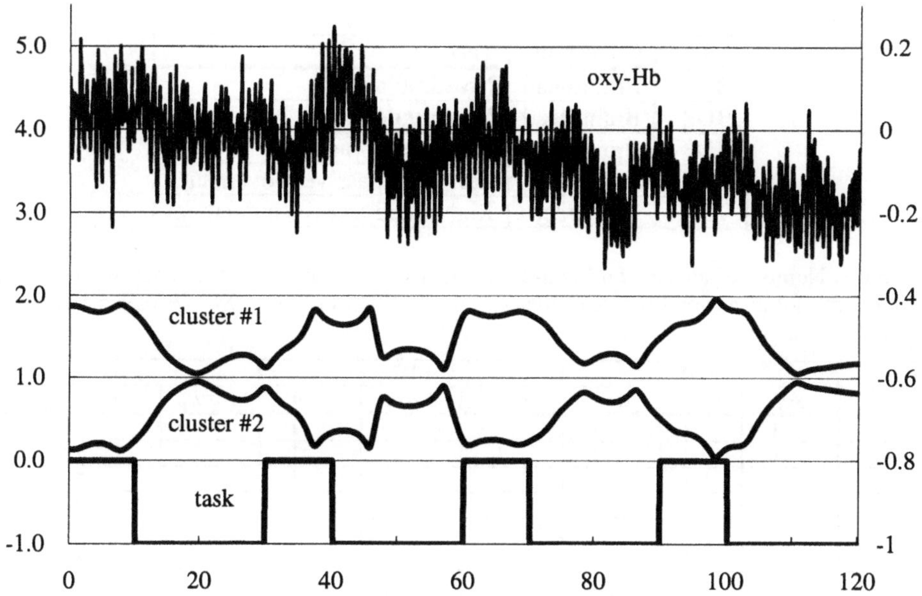

(a) oxy-Hb at measurement point 7.

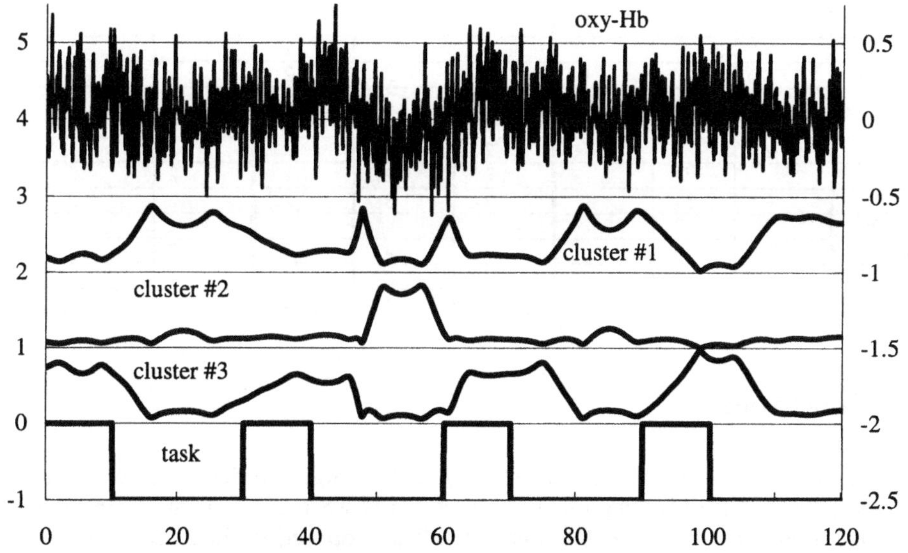

(b) oxy-Hb at measurement point 13.

Fig. 10. Clustering result for subject #2. Vertical axis: time(s), Horizontal axis; left: fuzzy degree; right: concentration, Top; concentration of oxy-Hb, Bottom: performed task (0;task, -1;rest), Others: degrees for each clusters (degrees were biased).

Table 2. Tasks.

Name	Condition	Task
BD1	Brain-dead	No stimulation
BD2	Brain-dead	No stimulation
BD3	Brain-dead	No stimulation
V1	Volunteer	Left hand passive movement
V2	Volunteer	Right hand passive movement

Table 3. Number of clusters for brain-dead patients. '-' denotes that the concentrations of the point were not measured because of the error of the NIRS machine. MP denotes measurement point.

	BD1			BD2			BD3		
MP	oxy	deoxy	total	oxy	deoxy	total	oxy	deoxy	total
1	1	1	1	1	1	1	1	1	1
2	1	1	1	2	2	1	1	1	1
3	1	1	1	1	1	1	1	1	1
4	1	1	1	2	2	1	2	2	1
5	1	1	1	2	2	1	1	2	1
6	1	1	1	1	1	2	2	1	2
7	2	1	1	1	1	2	1	1	1
8	1	1	1	2	2	1	1	1	1
9	1	1	1	1	1	2	1	1	1
10	2	1	1	1	2	2	1	1	1
11	-	-	-	2	1	2	1	1	1
12	-	-	-	1	2	3	1	1	1
13	-	-	-	1	1	2	1	1	1
14	-	-	-	2	2	2	1	1	1
15	-	-	-	2	2	2	2	2	1
16	-	-	-	2	2	2	3	3	2
17	2	2	1	2	2	2	1	2	1
18	-	-	-	2	2	1	2	2	1
19	-	-	-	2	3	2	2	2	1
20	2	2	1	-	-	-	-	-	-
21	6	6	6	4	4	2	3	3	2
22	-	-	-	2	2	2	2	2	1
23	3	3	2	3	3	2	3	4	2

5 Conclusion

There should be various brain states in the human higher-order brain function. This study aims to describing the brain states on the NIRS using a method combining the wavelet transform and the FCM clustering. The proposed method combining wavelet transform and FCM clustering segments a time series data of the NIRS into some clusters, and also estimates the optimum number of clusters by evaluating the similarity measure between clusters. The results suggest that this method obtains the clusters that may represent brain states. In the future, we will apply this method to complex paradigms such as Sternberk short-term memory task [2] for investigating brain states in human higher-order brain function.

Table 4. Number of clusters for healthy volunteers. '-' denotes that the concentrations of the point were not measured because of the error of the NIRS machine. MP denotes measurement point.

	V1			V2		
MP	oxy	deoxy	total	oxy	deoxy	total
1	2	1	2	3	2	2
2	1	1	2	4	3	3
3	2	2	2	3	2	3
4	3	3	2	3	2	3
5	2	1	1	2	1	2
6	3	3	2	2	2	3
7	3	3	2	2	1	2
8	2	2	2	2	2	3
9	3	3	2	2	1	2
10	3	3	2	2	2	3
11	2	2	1	2	2	2
12	3	3	2	3	3	2
13	3	4	1	2	2	3
14	2	2	4	3	2	3
15	3	3	2	3	2	3
16	2	2	2	2	1	3
17	2	2	2	2	2	3
18	1	2	2	-	-	-
19	2	2	1	2	2	2
20	2	3	1	3	2	3
21	2	2	1	3	1	3
22	2	2	1	2	2	2
23	2	3	2	2	2	3

Table 5. Mean values of estimated number of clusters.

	BD1	BD2	BD3	V1	V2
oxy-Hb	1.80	1.87	1.61	2.25	2.48
deoxy-Hb	1.67	1.87	1.70	2.33	1.87
total-Hb	1.47	1.70	1.22	1.79	2.61

Acknowledgements

The authors are thanksfull to the all member of Yanagida Project, Communications Research Laboratory (JAPAN), for giving helpful discussion. This work was supported in part by Ishikawa Hospital Grant, Tateishi Science and Technology Foundation, the Ministry of Education, Culture, Sports, Science and Technology, Grant-in-Aid for Encouragement of Young Scientists, 13780252, 2001, and 12780494, 2000, Grant-in-Aid for Scientific Research from the Ministry of Health, Labour and Welfare, 2000, and the BISC Program of UC Berkeley.

References

[1] F. F. Jobsis, "Noninvasive infrared monitoring of cerebral, and myocardial oxygen sufficiency and circulatory parameters," *Science*, no. 198, pp. 1264-1267, 1977.

[2] A. Villinger, J. Planck, C. Hock, L. Schleinkofer, and U. Dirnagl, "Near infrared spectroscopy (NIRS): a new tool to study hemodynamic changes during activation of brain function in human adults," *Neuroscience Letters*, vol. 153, pp. 101-104, 1993.

[3] V. Toronov, A. Webb A, J. H. Choi, M. Wolf, A. Michalos, E. Gratton, and D. Hueber, "Investigation of human brain hemodynamics by simultaneous near-infrared spectroscopy and functional magnetic resonance imaging," Medical Physics, vol. 28, no. 4, pp. 521-527, 2001.

[4] Y. Hoshi, and M. Tamura, "Near-infrared optical detection of sequential brain activation in the prefrontal cortex during mental tasks," Neuroimage, vol. 5, no. 4, pp. 292-297, 1997.

[5] P. Zaramella, F. Freato, A. Amigoni, S. Salvadori, P. Marangoni, A. Suppjei, B. Schiavo, and L. Chiandetti, "Brain auditory activation measured by near-infrared spectroscopy (NIRS) in neonates," Pediatric Research, vol. 49, no. 2, pp. 213-219, 2001.

[6] G. Gratton, and P. M. Corballis, "Removing the heart from the brain: Compensation for the pulse artifact in the photon migration signal," Psychophysiology, vol. 32, pp. 292-299, 1995.

[7] S. Sternberk, "High speed scanning in human memory," *Science*, vol. 153, pp. 652-654, 1966.

[8] S. Clare, M. Humberstone, J. Hykin, L. D. Blumhardt, R. Bowtell, and P. Morris, "Detecting activations in event-related fMRI using analysis of variance," Magnetic Resonance in Medicine, vol. 42, no. 6, pp. 1117-1122, 1999.

[9] J. C. Bezdek, *Pattern Recognition with Fuzzy Objective Function Algorithms*, Plenum, New York.

[10] M. C. Clark, L. O. Hall, D. B. Goldgof, R. Velthuizen, F. R. Murtagh, and M. S. Silbiger, "Automatic tumor segmentation using knowledge-based techniques," IEEE Transactions on Medical Imaging, vol. 17, no. 2, pp. 187-201, 1998.

[11] L. M. Fletcher-Heath, L. O. Hall, D. B. Goldgof, and F. R. Murtagh, "Automatic segmentation of non-enhancing brain tumors in magnetic resonance images," Artificial Intelligence in Medicine, vol. 21, pp. 43-63, 2001.

[12] C. Goutte, P. Toft, E. Rostrup, F. A. Nielse, and L. K. Hansen, "On clustering fMRI time series," NeuroImage, vol. 9, pp. 297-310, 1999.

[13] G. Gratton, and M. Fabiani, "Dynamic brain imaging: Event-related optical signal (EROS) measures of the time course and localization of cognitive-related activity," Psychomic Bulletin & Review, vol. 5, no. 4, pp. 535-563, 1998.

[14] S. Wray, M. Cope, D. T. Delpy, J. S. Wyatt, and E. O. R. Reynolds, "Characterization of the near infrared absorption spectra of cytochrome aa_3 and hemoglobin for the non-invasive monitoring of cerebral oxygenation," International Journal of Biochemistry, Biophysics and Molecular Biology, vol. 933, pp. 184-192, 1988.

[15] M. Sun, and R. J. Sclabassi, "Wavelet feature extraction from neurophysioological signals," in contribution to Time-Frequency and Wavelet Transforms in Biomedicine, M. Akay, Ed. NJ, IEEE Press.

[16] T. Kalayci, and O. Ozdamar, "Wavelet preprocessing for automated neural network detection of EEG spikes," IEEE Engineering in Medicine and Biology, vol. 14, no. 3, pp. 160-166, 1995.

Microcontroller Based Fuzzy Logic Sevofluorane Anesthesia Control System

A. Yardimci[1], A. Ferikoglu[2], and N. Hadimioglu[3]

[1] Department of Instrumentation& Process Control Technology, TBMYO,
Akdeniz University, Antalya, Turkey
yardimci@teknik.akdeniz.edu.tr

[2] Department of Electrical Engineering, Faculty of Engineering,
Sakarya University, Adapazari, Turkey
af@esentepe.sau.edu.tr

[3] Department of Anesthesia and Reanimation, Faculty of Medicine,
Akdeniz University, Antalya, Turkey

Abstract. In this study, sevofluorane depth of anesthesia was examined through a microcontroller-based fuzzy logic control system according to the blood pressure and heart rate taken from the patient. The system was designed for anesthetic agent, sevofluorane, which is among the first choices of anesthesiologist for inhalation anesthesia. The potential benefits of the systems are as follows:

1) To increase patients' safety and comfort,
2) To direct anesthesiologist' attention to other physiological variables which they have to keep under control by abating their tasks,
3) To make the optimum in the area of anesthetic agent,
4) To help protect the environment by using optimum anesthetic agent,
5) To economize by lessening the costs of an operation.

This study will serve as a guide in developing new anesthesia control systems for patients who are in different agent and different risk groups. All the details concerning the system design were given in paper.

1 Introduction

Intelligent systems have appeared in many technical areas, such as consumer electronics, robotics and industrial control systems. Many of these intelligent systems are based on fuzzy control strategies which desribe complex systems mathematical model in terms of linguistic rules. Since the 1980s new techniques have appeared from which fuzzy logic has been applied extensively in medical systems. The last couple of decades have witnessed a significant developments in control systems theory. In the meantime, developments in electronics and computers have resulted in many application areas of control theory. Although medicine is a science which isn't related to control engineering, it is being affected to such an extent that it is now

B. Reusch (Ed.): Fuzzy Days 2001, LNCS 2206, pp. 137–147, 2001.

possible to use available control techniques for on-line devices, especially during surgical operations and in intensive care units [1].

In medicine, the principle of "measuring everything measurable and trying to make measurable that which hasn't been measurable so far" (Galileo) is still practised. Nevertheless, some fundemantal limitations have already being recognised. Real world knowledge is characterised by incompleteness, inaccuracy, and inconsistency. Fuzzy set theory, which was developed by Zadeh (1965), makes it possible to define inexact medical entitites as fuzzy sets. It provides an excellent approach for approximating medical text [2]. Furthermore, fuzzy logic provides reasoning methods for approximate inference.

Patients have anesthesia so they can undergo a painful procedure (like an operation or test) without feeling any pain. There are several types of anesthetics. With general anesthesia, patients are completely unaware of what is going on. People often speak of this as "being asleep." Anesthesia is different from regular sleep, though, in that the surgeon can perform an operation without waking up the patient. At the end of the operation, the anesthesiologist will make sure that the patient wakes up properly. Regional anesthesia refers to simply being numb in part of the body so the patient doesn't feel any pain. Regional anesthesia can either be used alone or combined with general anesthesia. Sedation, or monitored anesthesia care, provides pain relief and some drowsiness without making the patient completely unconscious.

General anesthesia is the most common type of anesthetic performed at hospitals. There are several different ways to start the anesthetic, a process referred to as "induction of anesthesia." One of the most common types of anesthetic induction used at hospitals is an inhalation induction. This simply means that the patient breathes himself to sleep. The patient, will accompany the anesthesiologist to the operating room. There, after attaching monitors, the anesthesiologist will help hold a clear plastic mask over the patient's nose and mouth. The mask is attached to the anesthesia machine which delivers oxygen and anesthetic gases. The advantage of an inhalation induction is that there are no needles or anything else that is painful until after the patient is completely anesthetized. There are disadvantages, though. Some patients dislike the smell of the anesthetic gases despite the flavorings we use. An inhalation induction is also dangerous in certain medical conditions, like esophageal reflux, for example. An intravenous (IV) induction is another way we induce anesthesia. This is done by giving medicines through a catheter into a vein. IV induction has the advantage of being odorless and is a much faster way to induce anesthesia. The biggest disadvantage is needing to place the IV catheter, which requires a needle. The needle pain can often be prevented or decreased by using a special cream. This is an ointment that numbs the skin if it is left on for an hour to an hour and a half. Less common ways to induce anesthesia are with intramuscular injections or suppositories [3,4]. Monitors help the anesthesiologist watch over the patient closely during the anesthetic. They help keep track of things like how fast the patient's heart is beating, his blood pressure, or how much oxygen is in his blood [5].

In this study a fuzzy logic controller was used to control depth of sevofluorane anethesia, which was taken as a measure of the systolic blood pressure and hearth rate. The main reason for automating the control of depth anesthesia is to release the anesthesiologist so that he or she can devote attention to other tasks as well (controlling fluid balance, ventilation, and drug application) that can't yet be adequately automated and thus to increase the patient's safety.

A primary advantage of an embedded fuzzy logic system is that even complicated functions and adaptive control loops can be implemented with limited resources by using low-cost 8-bit microcontrollers. PIC16C71 microcontroller of Microchip firm was used in the system owing to its cheap price, functionality and its easy availability [6]. The responses of the system to the real values were compared by some anesthesia specialists and they were satisfactory.

2 Patients and Application

The study was approved by the Hospital Ethics Committee. In Akdeniz University, in the operating room of the Emergency and Trauma Hospital, for three and a half months, data were collected from the ASA I-II patients, who underwent an arthroscopy operation, and who were administered sevofluorane. 40 operations were recorded and 25 of them (operation time under 2 hours) were used for the system. 25 patients were studied, 15 females and 10 males, mean age 43.0 (range 16-69) yr, mean weight 68.6 (44-94) kg, mean height 156.2 (148-183) cm. During the operation every five minutes the blood pressure, the heart rate, and the rate of anesthetic agent were recorded. The data base was constructed of the records from a total of 25 patients who were asked permission in advance of the operation and were informed about the study. The membership functions and the rule base of the fuzzy logic system were determined under the inspection of specialists by abiding by the data base information.

The system was designed for anesthetic agent, sevofluorane, which is among the first choices of anesthesiologist for inhalation anesthesia. Sevofluorane is a halogenated volatile anaesthetic with a low blood gas solubility [7]. This facilitate smooth and rapid induction of anesthesia, allow easear titration of anesthetic dose to the desired effect during the maintenance period and permit rapid emergency and recovery at the end of the anesthesia [8,9].

Because a biological process like anesthesia has a nonlinear, time-varying structure and time-varying parameters, modeling it suggests the use of rule-based controllers like fuzzy controllers. The control rules made use of the systolic blood pressure and the heart rate[10,11].

Fuzzy rule-based systems include many aspects of fuzzified values, such as the rules antecedents and consequence. The rules structure are usually of the form of *if.. then.* In its basic form this type of the control is equivalent linguistically to a PI controller, and depending on the output, whether it is incremental or absolute, the controller is known as PI or PD respectively. An example of such a rule is *if the blood pressure is above the target and decreasing slowly, then reduce the drug infusion.* A more sophisticated structure is a PID, where the input, its derivative, and integral are considered as three inputs.The rules are composed either from the expert (anesthesiologist) or crafted by hand depending on the experience of the programmer

[12]. This includes tuning the membership functions in terms of the shape, width and position. This type of controller is widely used and is the most applicable control type in anesthesia [13,14,15,16,17].

Blood pressure is most conveniently measured using a cuff, which should be of the right size for the patient. Systolic pressure can be detected by palpation of the barchial or radial artery or by auscultation. In anesthetic practice, the systolic pressure has greater significance than the diastolic pressure, which isn't frequently recorded, particularly if access to the arm is difficult No "normal" blood pressure can be specified for the anaesthetized patient; in general the systolic pressure should be stable in the range 90-140 mmHg (12.0-18.7 kPa). Measuring the blood pressure of small children may be difficult unless an infant-size cuff is available; a simple tecnique is to squeeze the forearm firmly, inflate the cuff above systolic pressure, and then note the pressure during deflation at which the forearm flushes with blood[9,18,19]. Blood pressure value always contrast to depth of anesthesia. If the depth of anesthesia in deep blood pressure goes down. In other case, it goes up[10,11]. Depth of anaestesia is controlled by using a mixture of drugs that are injected intravenously or inhaled gases. Sevofluorane is widely used, most often in a mixture of 0 to 4 percent by volume of sevofluorane in oxygen and/or nitrous oxide. Figure.1 shows the system block diagram. System has two inputs and one output.

Fig. 1. System block diagram

We used fuzzyTECH 3.16 MP Explorer and MPLAB 3.01 programs for this application. MPLAB is a windows based application which contains a full featured editor, editor, emulator and simulator operating modes and a project manager. MPLAB allows us to edit our source files and one touch assemble and download to PIC16/17. But designer have to write main program byself. Collected data were used to produce membership functions for blood pressure, hearth rate and anesthetics gas out. Figure 2.,3., and 4. shows these membership functions. Linguistic variable ranges on the functions were determined by anesthesia specialist from the collected data.

Rule base includes 17 rules. 8 rules are eliminated by specialist. Eliminated rules are related to conditions which never happened. Input variable ranges imply 8-bit for 256 steps and 16 bits for 65536 steps. We decided to use 8-bit range. Because this range is enough for adequate appraisal.

Most fuzzy logic based application solutions use production rules to represent the relationship between the linguistic variables and to derive actions from sensor inputs. Production rules consist of a precondition (IF-part) and a consequnce (THEN-part). The IF-part can consist of more than one condition linked by linguistic conjunctions like AND and OR. The computation of fuzzy rules is called fuzzy rule inference. The software which we used for this application fuzzyTECH, calculates the inference in two steps: aggregation and composition. Aggregation uses fuzzy logic operators to calculate the result of the IF part of a production rule when the rule consists of more than one input conditions. One of the linguistic conjunctions, AND or OR, links multiple input conditions. Composition uses the fuzzy logic operator, PROD, to link the input condition to the output condition.

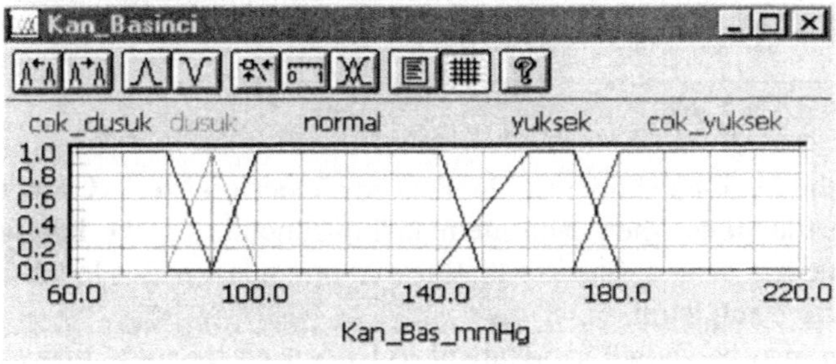

Fig. 2. Blood pressure membership function

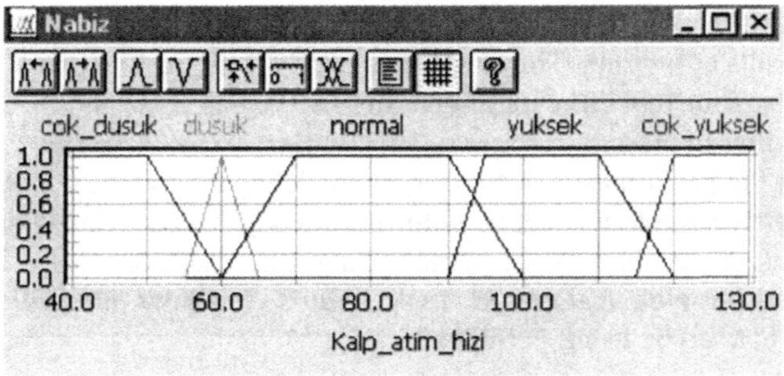

Fig. 3. Hearth rate membership function

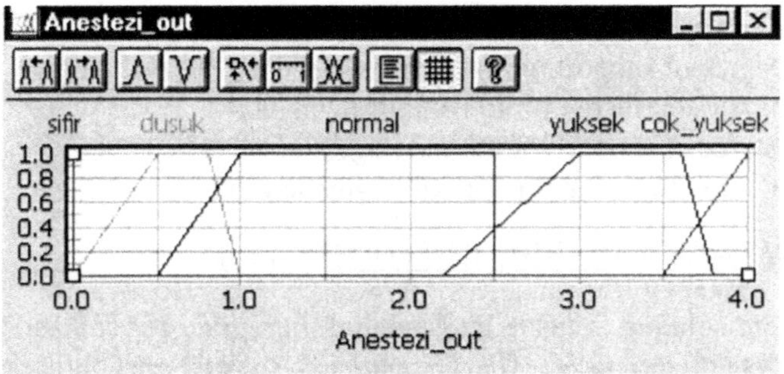

Fig. 4. Anesthetics gas out %

Composition links the validity of the entire condition with the Degree of Support (DoS). Thus, composition, the second calculation step of each production rule, uses the validty of the condition to determine the validity of the consequence. In standart MAX-MIN or MAX-PROD inference methods, the consequnce of a rule is considered equally as true as the condition.

The Center-of-Maximum (CoM) defuzzification method is an approximation of the more computationally-intensive Center-of- Area method. Let R be the linguistic variable to be defuzzified, let μ_{Ri} be the membership functions of all linguistic terms i defined for the base variable internal X ($x \in X$), and let μ_{Ii} be the inference result for every term i. The crisp output value $r \in R$ is computed by the following equation:

$$r = \frac{\sum_i [\mu_{Ii} \bullet \max_x(\mu_{Ri}) \bullet \arg(\max_x(\mu_{Ri}))]}{\sum_i \mu_{Ii}} \tag{1}$$

This method can be shown as identical to Center-of-Area (CoA) using singleton membership functions when each membership function for the linguistic variable to be defuzzified has been defined so that its maximum is 1 and is located at the position of the respective singleton.

CoM can also be shown as identical to CoA used by some fuzzy processors and development systems. In these systems, the membership function overlap in the result is ignored. Thus the individual membership function areas can be calculated during compilation time and therefore these areas only need to be multiplied with the inference results at runtime. This is identical to CoM defuzzification if the maximums of the membership functions represent the individual areas and are located at the individual center of areas.

To check the performance of fuzzy system, debug interactive window is used. The debug interactive window is active with most debug modes. This window shows the input and output variables of the system. Time plot analyser is used in debug mode and also transfer plot analyser is used. Figure. 5. shows interactive test results. Solution is obtained by using Formula 2.

$$\%u = \frac{(D.DoS \times D.An\_out) + (N.DoS \times N.An\_out)}{D.DoS + N.DoS} \tag{2}$$

Where:
%u: Crisp out value (produced by CoM defuzzification),
D.DoS: Degree of support for low linguistic value,
D.An_out: Anesthesia out for low linguistic value,
N.DoS: Degree of support for normal linguistic value,
N.An_out: Anesthesia out for normal linguistic value,

$$\%u = \frac{(0.435 \times 0.64) + (0.315 \times 1.74)}{0.435 + 0.315} = \frac{0.2784 + 0.5481}{0.75} = 1.102$$

The system solution is made by Matlab 4.0 program for comparision Fuzzytech interactive system responses. The comparision showed each program gives same response on same input conditions. Figure. 6. shows 3D surface analysis is obtained from Matlab 4.0 program.

We prefered to use for this system 8-bit PIC16C71 microcontroller. The PIC16C71 is a member of low-cost, high performance, CMOS, fully-static, 8-bit microcontrollers with integrated analog-to-digital (A/D) convertors, in the PIC16C7X mid-range family. The PIC16C71 devices have 36 bytes of RAM and 13 I/O pins. In addition a timer/counter is available. Also a 4-channel high-speed 8-bit A/D is provided. The 8-bit resolution is ideally suited for applications requiring low-cost analog interface, e.g. pressure sensing, thermostat control, etc. Its A/D convertors were used to convert analog blood pressure and hearth rate signals to digital 8-bit data. These digital data were Figure 7., 8., shows the compare between the system response and manuel response during the operations used as crisp real world data by fuzzy system.

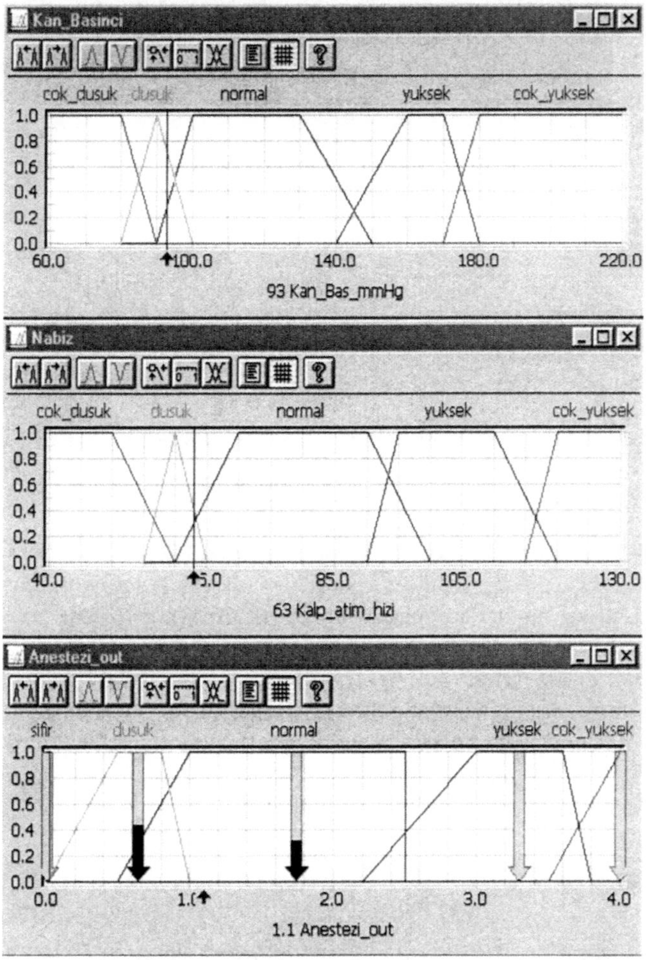

Fig. 5. Interactive system test results for 93mmHg blood pressure and 63 p/m hearth rate values. System anesthetics gas out % 1.1

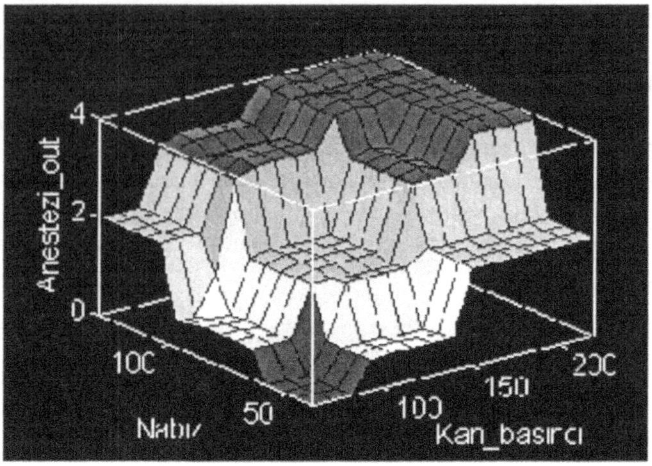

Fig. 6. 3D surface analysis

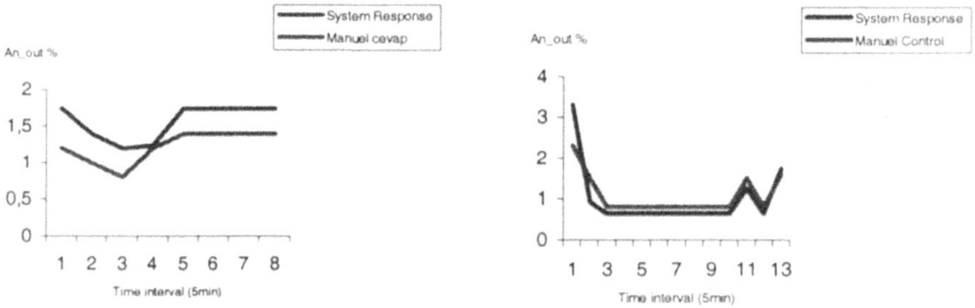

Fig. 7. Anesthesia_out for the youngest (on the left side) and the oldest patient operations

Table 1. shows crisp data manuel records and manuel response during the longest operation. You can also compare system response and manuel response in Table.1.

3 Conclusion

Fuzzy logic simplifies the design of a control strategy by providing an easy to understand and intuitive approach to solve control problems. Fuzzy logic reduces development time, and allows designers to produce control prototypes in less time than would be possible by using conventional methods. The potential benefits which are aimed at the beginning of the study were achieved. The sevofluorane anesthesia fuzzy logic control system can be used as an equipment which controls the depth of anesthesia. If it doesn't seem to ensure the patient's safety as an euipment which works independently without the anesthesiologist, it can easily be used as a monitor which helps keep track of depth of anesthesia. The system is to relase the anesthesiologist so that he or she can devote attention to other tasks that can't yet be adequately automated.

Table 1. Data records during the longest operation

FuzzyTECH Batch Mode, Output for Project ANESTEZI
Patient
Name : Aysen Yilmaz
Sex : F
Age : 22

	BP	HR	SR	Flags	MR
An_1	116	81	1.74	0	1.4
An_2	117	80	1.74	0	1.5
An_3	116	80	1.74	0	1.5
An_4	116	71	1.74	0	1.5
An_5	106	66	1.74	0	1.5
An_6	115	60	1.74	0	1.5
An_7	127	82	1.74	0	1.7
An_8	118	67	1.74	0	1.8
An_9	127	62	1.74	0	1.5
An_10	125	61	1.74	0	1.5
An_11	123	60	1.74	0	1.5
An_12	132	61	1.74	0	1.5
An_13	124	62	1.74	0	1.5
An_14	132	62	1.74	0	1.5
An_15	129	63	1.74	0	1.5
An_16	129	61	1.74	0	1.5
An_17	134	62	1.74	0	1.5
An_18	127	65	1.74	0	2.0
An_19	116	58	1.46	0	1.6
An_20	136	70	1.74	0	2.1
An_21	127	60	1.74	0	1.8
An_22	124	59	1.64	0	2.0
An_23	123	58	1.46	0	2.0
An_24	123	60	1.74	0	1.8
An_25	123	60	1.74	0	1.6
An_26	133	53	0.64	0	0.8
An_27	135	51	0.64	0	0.6
An_28	139	72	1.74	0	1.5

BP:Blood Pressure, HR:Hearth Rate, SR:System Response, MR:Manuel Response

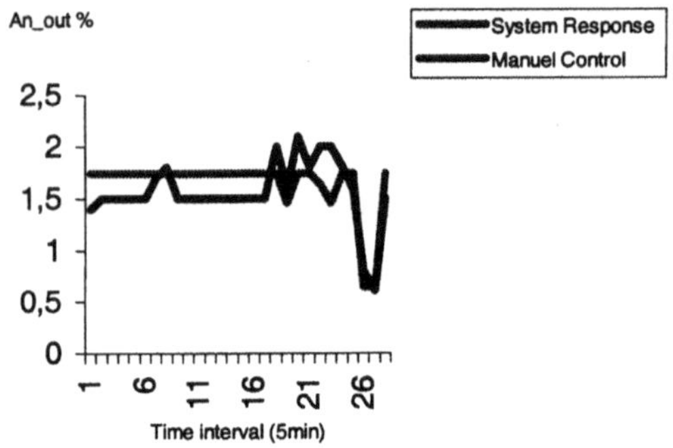

Fig. 8. Anesthesia_out during the longest operation

References

[1].LINKENS, D.A., ABBOD, M.F., MAHFOUF, M., "An initial survey of fuzzy logic monitoring and control utilisation in medicine", *Journal of Biomedical Engineering , Applications, Basis Communications, Special Issues on Control methods in medicine*, Taiwan, August 1999

[2].VIRANT-KLUN, I., VIRANT, J., "Fuzzy Logic Alternative for Analysis in the Biomedical Sciences", *Computers and Biomedical Research* 32, 305-321, 1999.

[3].COLLINS, V.J., "General Anesthesia Fundamental Considerations", 3th Edition, Philadelphia, Lea&Febiger, pp:314-359, 1993

[4].VICKERS, M.D., MORGAN, M., SPENCER, P.S.S., "General Anaesthetics", 7th edition, Butterwurth-Heineman Ltd., Oxford, pp:118-159, 1991

[5].GILBERT, H.C., VENDER, J.S., "Monitoring the Anesthetized Patient", *Clinical Anesthesia*, pp: 742-743, 1992 [6].BENGSTON, J.P., SONADER, H., STENQVIST, O., "Comparision of costs of different anaesthetic techniques" *Acta Anaesthesiologica Scandinavica*, 32: 33-35, 1998

[7].BROWN, B.R., FRINK, E.J., "Whatever happened to sevoflurane?", *Can. Anaesth.*, 39: 207-209, 1992

[8].KATOH, T., IKEDA, K., "Minumum alveolar concentration of sevoflurane in children", *Br. J. Anesth.*, 68:139-141, 1992

[9].KATOH, T., IKEDA, K., "The minumum alveolar concentration (MAC) of sevoflurane in humans", *Anesthesiology*, 66:301-303, 1987

[10].YARDIMCI, A., ONURAL A.,"Fuzzy Logic Control of Child Blood Pressure During Anaesthesia", *Computational Intelligence Theory and Applications, Lecture Notes in Computer Science 1625*,6.Fuzzy Days, 25-27 May 1999:701-704, Springer, Berlin Germany

[11]YARDIMCI, A., FERIKOGLU, A., "Analysis of Biological Process with Microcontroller based Fuzzy Logic Systems", *Workshop on Biomedical Information Engineering Proceedings,* 25-27 June, 2000:261-269 Isik University Istanbul, Turkey

[12].GREENHOW, S. G., LINKENS, D.A., ASBURY, A., "Pilot Study of an Expert System Adviser for Controlling General Anaesthesia", *British Journal of Anaesthesia*, 71: 359-365, 1993

[13].ZBINDEN, A.M., FEIGENWINTER, P.,HUTMACHER, M.,"Fresh gas utilization of eight circle systems", *British Journal of Anaesthesia*, 67: 492-499, 1991

[14].KRAFT, H.H., LEES, D.E., "Closing the loop: How near is automated anesthesia?", *Southern Med. J.*, vol. 77, pp. 7-12, 1984

[15].VISHNOI, R., ROY, R.J., "Adaptive control of closed-circuit anesthesia", *IEEE Trans. Biomed. Eng.*, vol. 38, pp. 39-47, 1991

[16].MILLARD, R.K., HUTTON, P., PEREIRA, E., PRYS-ROBERTS, C., "On using a self-tuning contreller for blood pressure regulation during surgery in man", *Computer Biol. Med.*, vol. 17, no. 1, pp. 1-18, 1987

[17].ISAKA, S., "Control Strategies for Arterial Blood Pressure Regulation", *IEEE Trans. Biomed. Eng.*, vol. 40, pp. 353-363, 1993

[18].BICKFORD, R.G., " Automatic control of general anesthesia", *Electroencephalog. Clin. Neurophysiol.*, Vol.2, pp.93-96, 1950

[19].BICKFORD, R.G., "The use of feedback systems for the control of anesthesia", *Elect. Eng.*, vol.70, pp. 852-855, 1951

Intuitionistic Fuzzy Sets in Some Medical Applications

Eulalia Szmidt and Janusz Kacprzyk

Systems Research Institute Polish Academy of Sciences
ul.. Newelska 6, 01-447 Warsaw, Poland
{szmidt,kacprzyk}@ibspan.waw.pl

1. Introduction

For many real world problems, imperfect, imprecise information is, by nature, part of the problem itself and continuing reasoning without proper modelling tools may led to generating inaccurate inferences.

Traditional methods of analysis are oriented toward the use of numerical techniques. By contrast, much of human reasoning involves the use of variables whose values are fuzzy sets. This observation is the basis for the concept of a linguistic variable, that is, a variable whose values are words rather than numbers.

The use of linguistic variables represents a significant paradigm shift in system analysis. More specifically, in the linguistic approach the focus of attention in the representation of dependencies shifts from difference and differential equations to fuzzy IF-THEN rules in the form *IF X is A THEN Y is B*, where *X* and *Y* are linguistic variables and *A* and *B* are their linguistic values, e.g. *IF Pressure is high THEN Volume is low*.

Description of system behaviour in the language of fuzzy rules lowers the need for precision in data gathering and data manipulation, and in effect may be viewed as a form of data compression.

But there are situations when description by a (fuzzy) linguistic variable given in terms of a membership function only, seems too rough. For example, in decision making problems, particularly in a case of medial diagnosis, sales analysis, new product marketing, financial services, etc. there is a fair chance of the existence of a non-null hesitation part at each moment of evaluation of any unknown object.

Intuitionistic fuzzy sets (Atanassov 1986, 1999) can be viewed in the context as a proper tool for representing hesitancy concerning both membership and non-membership of an element to a set. To be more precise, a basic assumption of fuzzy set theory that if we specify the degree of membership of an element in a fuzzy set as a real number from [0, 1], say *a*, then the degree of its non-membership is automatically determined as 1-*a*, need not hold for intuitionistic fuzzy sets. In intuitionistic fuzzy set theory it is assumed that non-membership should not be more than 1-*a*. The difference let us express lack of knowledge (hesitancy concerning both membership and non-membership of an element to a set). This way we can better model imperfect information – for example, we can express the fact that temperature of a patient changes, and other symptoms arc not quite clear.

In this article we will present intuitionistic fuzzy sets as a tool for reasoning in the presence of imperfect facts and imprecise knowledge. An example of medical diagno-

B. Reusch (Ed.): Fuzzy Days 2001, LNCS 2206, pp. 148–151, 2001.

sis will be presented – having a database, i.e. description of a set of symptoms *S*, and a set of diagnoses *D*, we will describe a state of a patient when knowing results of his/her medical tests. Description of the problem uses the notion of intuitionistic fuzzy set theory. The proposed method of diagnosis involves intuitionistic fuzzy distances as introduced in (Szmidt and Kacprzyk 1997, 2000). Advantages of such an approach are pointed out in comparison with the method presented in (De, Biswas and Roy 2001) where max-min-max composition rule was applied.

2. Medical Diagnosis via Intuitionistic Fuzzy Set Theory

To make a proper diagnosis *D* for a patient with given values of tested symptoms *S*, a medical knowledge base is necessary. In our case a knowledge base is formulated in terms of intuitionistic fuzzy sets. An intuitionistic fuzzy set (Atanassov 1986, 1999), *A* in *X*, is given by:

$$A = \{< x, \mu_A(x), \nu_A(x) > | x \in X\} \quad (1)$$

where: $\mu_A : X \to [0,1]$, and $\nu_A : X \to [0,1]$, with the condition $0 \le \mu_A(x) + \nu_A(x) \le 1 \quad \forall x \in X$ and the numbers $\mu_A(x)$, $\nu_A(x) \in [0,1]$ denote the degree of membership and non-membership of *x* to *A*, respectively. For each intuitionistic fuzzy set in *X*, we will call $\pi_A(x) = 1 - \mu_A(x) - \nu_A(x)$ the *fuzzy intuitionistic index* of *x* in A and it is a hesitation degree of whether *x* belongs to A.

To compare the approach proposed in this article with the method presented in (De, Biswas and Roy 2001), we consider the same data. Let the set of Diagnosis be *D={Viral, Fever, Malaria, Typhoid, Stomach problem, Chest problem}*. The considered set of Symptoms *S={temperature, headache, stomach pain, cough, chest-pain}*. The data are given in Table 1 – each symptom S_i is described by three numbers: membership μ, non-membership ν, intuitionistic fuzzy index π (i.e. hesitation margin). For example, for malaria: the temperature is high (μ = 0.7, ν = 0, π = 0.3), whereas for stomach problem: temperature is low (μ = 0, ν = 0.9, π = 0.1) etc.

The considered set of patients *P={Al., Bob, Joe, Ted}*. The symptoms characteristic for the patients are given in Table 2.

Our task is to make a proper diagnosis for each patient p_i, *i*=1,...,4. The simplest method is to calculate for each patient p_i a distance of his symptoms (Table 2) from a set of symptoms s_j, j=1,...,5 characteristic for each diagnosis d_k, *k*=1,...,5 (Table 1).

Table 1.

	Viral fever	Malaria	Typhoid	Stomach problem	Chest problem
Temperature	(0.4, 0.0, 0.6)	(0.7, 0.0, 0.3)	(0.3, 0.3, 0.4)	(0.1, 0.7, 0.2)	(0.1, 0.8, 0.1)
Headache	(0.3, 0.5, 0.2)	(0.2, 0.6, 0.2)	(0.6, 0.1, 0.3)	(0.2, 0.4, 0.4)	(0.0, 0.8, 0.2)
Stomach pain	(0.1, 0.7, 0.2)	(0.0, 0.9, 0.1)	(0.2, 0.7, 0.1)	(0.8, 0.0, 0.2)	(0.2, 0.8, 0.0)
Cough	(0.4, 0.3, 0.3)	(0.7, 0.0, 0.3)	(0.2, 0.6, 0.2)	(0.2, 0.7, 0.1)	(0.2, 0.8, 0.0)
Chest pain	(0.1, 0.7, 0.2)	(0.1, 0.8, 0.1)	(0.1, 0.9, 0.0)	(0.2, 0.7, 0.1)	(0.8, 0.1, 0.1)

Table 2.

	Temperature	Headache	Stomach pain	Cough	Chest-pain
Al.	(0.8, 0.1, 0.1)	(0.6, 0.1, 0.3)	(0.2, 0.8, 0.0)	(0.6, 0.1, 0.3)	(0.1, 0.6, 0.3)
Bob	(0.0, 0.8, 0.2)	(0.4, 0.4, 0.2)	(0.6, 0.1, 0.3)	(0.1, 0.7, 0.2)	(0.1, 0.8, 0.1)
Joe	(0.8, 0.1, 0.1)	(0.8, 0.1, 0.1)	(0.0, 0.6, 0.4)	(0.2, 0.7, 0.1)	(0.0, 0.5, 0.5)
Ted	(0.6, 0.1, 0.3)	(0.5, 0.4, 0.1)	(0.3, 0.4, 0.3)	(0.7, 0.2, 0.1)	(0.3, 0.4, 0.3)

Table 3.

	Viral fever	Malaria	Typhoid	Stomach problem	Chest problem
Al.	0.28	**0.24**	0.28	0.54	0.56
Bob	0.40	0.50	0.31	**0.14**	0.42
Joe	0.38	0.44	**0.32**	0.50	0.55
Ted	**0.28**	0.30	0.38	0.44	0.54

In (Szmidt and Kacprzyk 1997, 2000) it is proved that the only proper way of calculating distances for intuitionistic fuzzy sets is to take into account all three parameters: membership, non-membership, and hesitation margin. To be more precise, the normalised Hamming distance for all the symptoms of the i-th patient from the k-th diagnosis is equal to

$$l\ (s(p_i),d_k)=\frac{1}{10}\sum_{j=1}^{5}\left(\left|\mu_j(p_i)-\mu_j(d_k)\right|+\left|\nu_j(p_i)-\nu_j(d_k)\right|+\left|\pi_j(p_i)-\pi_j(d_k)\right|\right) \tag{2}$$

The distances (2) for each patient from the considered set of possible diagnoses are given in Table 3 – the lowest distance points out a proper diagnosis – Al suffers from malaria, Bob from stomach problem, Joe from typhoid, whereas Ted from fever.

We obtained the same results, i.e. the same quality diagnosis for each patient when looking for the solution while applying the normalized Euclidean distance [cf. Szmidt and Kacprzyk 1997, 2000].

3. Conclusions

Employing intuitionistic fuzzy sets in databases let us express hesitancy concerning examined objects. The method proposed in this article - making a diagnosis on the basis of calculating distances from a considered case to all considered illnesses, takes into account values of all symptoms. What is more, our approach let us introduce weights for all symptoms (for some illnesses some symptoms can be more important). Such an approach is impossible by the method described in (De, Biswas and Roy, 2001) – the max-min-max rule "neglects" in fact most values basing on extreme ones.

Literature

[1] Atanassov K. (1986) Intuitionistic fuzzy sets, *Fuzzy Sets and Systems*, 20 (1986) 87-96.

[2] Atanassov K. (1999) *Intuitionistic Fuzzy Sets: Theory and Applications.* Physica-Verlag.

[3] De S.K., Biswas R. and Roy A.R. (2001) An application of intuitionistic fuzzy sets in medical diagnosis. *Fuzzy Sets and Systems*, Vol. 117, No.2, pp. 209 – 213.

[4] Szmidt E. and Kacprzyk J. (1997) On measuring distances between intuitionistic fuzzy sets, *Notes on IFS*, Vol.3, No.4, pp.1 - 13.

[5] Szmidt E. and J. Kacprzyk J. (2000) Distances between intuitionistic fuzzy sets, *Fuzzy Sets and Systems*, Vol.114, No.3, pp.505 - 518.

[6] Zadeh L.A. (1965) Fuzzy sets. *Information and Control* 8, pp. 338 - 353.

Interpretation of Trained Neural Networks by Rule Extraction

Vasile Palade[1], Daniel-Ciprian Neagu[2], and Ron J. Patton[1]

[1] The University of Hull
Control and Intelligent Systems Engineering
Cottingham Road, HU6 7RX, Hull, United Kingdom
{V.Palade,R.J.Patton}@eng.hull.ac.uk
[2] University "Dunarea de Jos" of Galati
Department of Computer Science and Engineering
Domneasca Str. 111, Galati 6200, Romania
Dan.Neagu@ugal.ro

Abstract. The paper focuses on the problem of rule extraction from neural networks, with the aim of transforming the knowledge captured in a trained neural network into a familiar form for human user. The ultimate purpose for us is to develop human friendly shells for neural network based systems. In the first part of the paper it is presented an approach on extracting traditional crisp rules out of the neural networks, while the last part of the paper presents how to transform the neural network into a set of fuzzy rules using an interactive fuzzy operator. The rules are extracted from ordinary neural networks, which have not a structure that facilitate the rule extraction. The neural network trained with the well known Iris data set was considered as benchmark problem.

1 Introduction

Artificial neural networks represent an excellent tool that have been used to develop a wide range of real-world applications, especially in case when traditional solving methods fail. They exhibit advantages such as ideal learning ability from data, classification capabilities and generalization for situations not contained in training data set, computationally fastness once trained due to parallel processing, noise tolerance. There were these advantages that made neural networks to be successfully applied to various real-word problems, including: speech recognition, medical diagnosis, image computing, process control and modeling [11], [8], fault diagnosis (a recent survey on using different neural network based techniques in fault diagnosis can be found in [9]). The major shortcoming of neural networks is represented by their low degree of human comprehensibility [13]. Many authors have focused on solving this shortcoming of neural networks, by compiling the knowledge captured in the topology and weight matrix of a neural network, into a symbolic form; some of them into sets of ordinary if - then rules [7], [13], [14], [15], others into formulas from propositional logic or from non-monotonic logics [10], or into sets of fuzzy rules [1], [2], [3], [5], [6]. More transparency is offered by fuzzy neural networks [4], [5], [12], which represent a paradigm that combines the comprehensibility and capabilities of

B. Reusch (Ed.): Fuzzy Days 2001, LNCS 2206, pp. 152–161, 2001.

fuzzy reasoning to handle uncertainty and the capabilities of neural networks to learn from examples.

The paper has the following organization. Section 2 briefly presents an inverting of a neural network, more exactly given the output of the network, how to calculate the input of the network which produces the given output. This input calculation for a given network output is required in the rule extraction method presented in section 3. The inversion method presented in section 2 is a computationally fast inversion method, that optimize the searching for the network inputs which produce the desired network output. Section 3 proposes a method of traditional if-then rule extraction from neural networks, which have not a special structure that facilitates the rule extraction. Many other rule extraction methods reported in the literature rely on and need some special architectures for neural networks in order to be efficiently applied. Our method is based on interval propagation across the network, in a similar way as VIA method proceeds [13]. The rule extraction method is applied on the neural network trained with Iris data set. The main problem encountered when apply this method is the big number of rules required to satisfactory describe the network functioning. That's why we tried to express the network behavior in a more concise way. Section 4 uses the same neural network as in section 3, and a fuzzy rule set is extracted, by introducing an interactive fuzzy operator [1], [5]. Conclusions of the paper are summarized in the last section.

2 Iterative Relation for Neural Network Inversion

An inversion method, which calculates the input of a neural network for a given output, was presented in [11]. An improved and a more computationally efficient version of this method was presented in detail in our previous papers [7], [8].

Given a three layered neural network, with q the number of the network inputs, h the number of hidden nodes, and r the number of network outputs. Within the q inputs of the network, f inputs are considered to be fixed inputs (past inputs and outputs of the process in case when neural network is used for process modeling), and we have to determine, by a searching procedure, the remaining p inputs (p+f=q). Given the output vector y, we have to find the input vector u which produces the output y. By u it is denoted the vector of the p unfixed inputs of the network, and by u_f the vector of the f fixed inputs of the network.

We have the following relations:

$$\begin{aligned} y' &= f^{-1}(y) - \theta^{y} \\ W^{yx} x &= y' \end{aligned} \tag{1}$$

where f is the nonlinear activation function of the network nodes, x the output vector of the hidden nodes, W^{yx} is the weight matrix between hidden layer and output layer, θ^{y} is the bias vector of the output layer. If W^{yx} is a squared matrix, then it is possible to calculate the input u as follows:

$$\begin{aligned} x &= (W^{yx})^{-1} y' \\ x' &= f^{-1}(x) - \theta^{x} \end{aligned}$$

$$W^{xu} u = x' - W_f^{xu} u_f \qquad (2)$$

where W^{xu} is the weight matrix between unfixed inputs and hidden layer, W_f^{xu} is the weight matrix between fixed inputs and hidden layer, θ^x is the bias vector of the hidden layer. When the hidden and the output layers have exactly the same number of nodes (h=r), the inversion of the network consists of solving two linear equation systems ((1) and (2)). In the following, consider d the vector $d = W_f^{xu} u_f$.

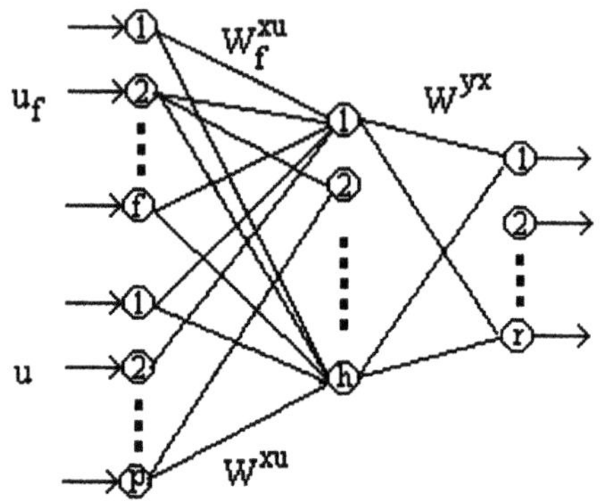

Fig. 1. The three-layered neural network

The general case of most neural models of real-world processes is represented by neural networks with h > r. In this case, it is possible to write the matrix W^{yx} and the linear system (1), by Jordan-Gauss elimination, in the following form:

$$\begin{array}{c} \quad x_1 \; x_2 \quad x_r \; x_{r+1} \; x_h \quad y' \\ \left[\begin{array}{ccc|c|c} 1 & 0 \;\ldots\ldots & 0 & & \\ 0 & 1 \;\ldots\ldots & 0 & & \\ \ldots & \ldots & \ldots & C & Y* \\ \ldots & \ldots & \ldots & & \\ 0 & 0 \;\ldots\ldots & 1 & & \end{array}\right] \end{array} \qquad (3)$$

Partitioning the weight matrix W^{xu} and the vectors x, θ^x and d, as given below:

$$x = \begin{bmatrix} x' \\ x'' \end{bmatrix} \text{ where}$$

$$x' = [x_1, x_2, \ldots\ldots, x_r]^t \text{ and}$$

$$x'' = [x_{r+1}, \ldots., x_h]^t$$

$$W^{xu} = \begin{bmatrix} W^{x'u} \\ W^{x''u} \end{bmatrix} \quad \theta^{x} = \begin{bmatrix} \theta^{x'} \\ \theta^{x''} \end{bmatrix} \quad d = \begin{bmatrix} d^{x'} \\ d^{x''} \end{bmatrix} \text{ where}$$

$W^{x'u}$ is a r x p matrix, and

$W^{x''u}$ is a (h - r) x p matrix

we have the relations:

$$x' = f(W^{x'u}u + d^{x'} + \theta^{x'})$$
$$x'' = f(W^{x''u}u + d^{x''} + \theta^{x''})$$

The input u, which produces the desired output y, is the solution of the equation:

$$x' = y^* - Cx''$$

equivalent with:

$$y^* - Cf(W^{x''u}u + d^{x''} + \theta^{x''}) = f(W^{x'u}u + d^{x'} + \theta^{x'})$$

Expanding [8] the nonlinear function f, through a Taylor series around the point u_k, and solving, the following iterative relations is obtained:

$$\begin{aligned} u = u_k &+ [f'(W^{x'u}u_k + d^{x'} + \theta^{x'}) o W^{x'u} + \\ &+ C(f'(W^{x''u}u_k + d^{x''} + \theta^{x''}) o W^{x''u})]^{-1} \cdot \\ &\cdot [y^* - Cf(W^{x''u}u_k + d^{x''} + \theta^{x''}) - f(W^{x'u}u_k + d^{x'} + \theta^{x'})] \end{aligned} \quad (4)$$

In the previous relation, $f(\cdot)$ and $f'(\cdot)$ are vectors of corresponding orders, where the function f and its derivative f' is applied on each component of the argument vector. The operator o multiplies each element of a matrix row with the corresponding element of the vector.

The matrix which must be inverted in relation (4) is a r x p matrix. If the number of unfixed network inputs is different than the number of network outputs, the inverse from the equation (4) must be replaced with the suitable pseudo-inverse.

3 Rule Extraction by Interval Propagation

In this section, we present a method for rule extraction from neural networks with continuous inputs and outputs. We named this method, presented also briefly in one of our previous paper [7], the *method of interval propagation*. The rules extracted by this method are crisp if – then rules, in the following form:

$$\textit{if} \ \ (a_1 \le x_1 \le b_1) \text{ and } (a_2 \le x_2 \le b_2) \ \ldots\ \text{ and } (a_m \le x_m \le b_m)$$
$$\textit{then} \quad c_j \le y_j \le d_j$$

where $x_1, x_2, \dots, x_m$ are the inputs of the network and y_j is the j output of the network, $j = 1, 2, \dots, n$.

A similar method which tries to extract rules in the same form, out of a trained neural network, is the VIA method, developed by Thrun in [13]. VIA method refines the intervals of all units in the network, layer by layer, by techniques of linear programming, such as Simplex algorithm, propagating the constraints forward and backward through the network. The problem is that, VIA method may fail sometimes to decide if a rule is compatible or not with the network. Also the intervals obtained by VIA method are not always optimal. Our method continues the background ideas of VIA method and eliminates the drawbacks of this method.

Given P a layer in the network, and S the next layer. Every node in layer S calculates the value $x_i = f(\sum_{k \in P} w_{ik} x_k + \theta_i)$, where x_k is the output (activation value) of node k in layer P, x_i the output of node i in the layer S, w_{ik} the weight of the link between node k in layer P and node i in layer S, θ_i the bias of node i, and f the transfer function of the units in the network. The following relations can be written:

$$(\forall)\, k \in P \qquad x_k \in [a_k; b_k]$$

$$(\forall)\, i \in S \qquad x_i' = \sum_{k \in P} w_{ik} x_k + \theta_i$$

$$x_i = f(x_i')$$

For every node $i \in S$, we note with w_{il}^+, $l \in P_i^+$, the positive weights, and with w_{il}^-, $l \in P_i^-$, the negative weights. ($P_i^+ \cup P_i^- = P$). The interval of variation for x_i', $[a_i'; b_i']$, $(\forall) i \in S$, is determined in the following way:

$$a_i' = \sum_{l \in P_i^+} w_{il}^+ a_l + \sum_{r \in P_i^-} w_{ir}^- b_r + \theta_i$$

$$b_i' = \sum_{l \in P_i^+} w_{il}^+ b_l + \sum_{r \in P_i^-} w_{ir}^- a_r + \theta_i$$

and the variation interval for the activation value x_i of node i in layer S is $[a_i; b_i]$, where: $a_i = f(a_i')$ and $b_i = f(b_i')$. In this way, the intervals are propagated, layer by layer, from the input layer to the output layer. So, given the variation intervals for inputs, the intervals of variation for outputs are determined. This is the forward phase. Some of the inputs may be unconstrained, and in this case the intervals are propagated forward across the network layers, assigning the interval of maximum variation ([0; 1]) for unconstrained inputs.

The backward phase appears when it is given the interval of variation for output and eventually for some inputs, and it must be determined the interval for unconstrained inputs. Suppose $x_1, x_2, \dots, x_k$ are the constrained inputs after renumbering, and $x_{k+1}, \dots, x_m$ the unconstrained inputs, and we want to determine rules when $(a_1 \le x_1 \le b_1)$ and ... $(a_k \le x_k \le b_k)$ and $(c_j \le y_j \le d_j)$.

First, it is checked the compatibility of the following rule:

$$\textit{if} \quad (a_1 \le x_1 \le b_1) \text{ and } (a_2 \le x_2 \le b_2) \text{ .... and } (a_k \le x_k \le b_k) \quad \textit{then} \quad c_j \le y_j \le d_j$$

with the network, assigning the maximum interval ([0; 1]) for unconstrained inputs. By forward propagation, the variation interval $[c_j'; d_j']$ for output is determined. If $[c_j';d_j'] \subset [c_j;d_j]$, then the rule given above is a general rule, and it does not have sense to look for the variation intervals of remained inputs. If the intersection $[c_j';d_j'] \cap [c_j;d_j]$ is empty set, then the rule is incompatible with the network. Otherwise, be $y_j^* \in [c_j';d_j'] \cap [c_j;d_j]$.

By inverting the neural network as given in section 2, it is determined the input $x^*=(x_1^*, x_2^*, \ldots x_m^*)$ of the network which produce the output y_j^*. The idea is to find the maximal intervals around the values x_l, l=k+1, ..., m, so that the corresponding rule to be compatible with the network. For example, beginning with input x_l, the right margin b_l of the variation interval is set up to:

$$b_1 = x_1^* + \frac{1 - x_1^*}{2}$$

If the rule with $x_1 \in [x_1^*; b_1]$ is compatible with the network, then the interval is enlarged, otherwise is shrinking, with a technique of dividing intervals into two halves, until the right margin b_l and a_l are determined with a given error. The procedure continues until all the variation intervals for all unconstrained inputs are determined. The hyper-cubs determined at the input depend on the start position – x*, and on the order of the determination of the variation intervals for unconstrained inputs.

Using the method of inverting a neural network described in section 2, the backward phase in VIA method can be reduced, with a very simple calculus, to a forward propagation of the input intervals.

3.1 Case Study

The goal of well-known iris problem is to recognise the type of an iris plant to which a given instance belongs. The data set is composed of 150 records, equally distributed between three classes: setosa, versicolor, and virginica. Two classes are not linearly separable from each other, while the third is linearly separable from the others. The data set is characterised by four attributes: petal length, petal width, sepal length, and sepal width, hence the neural network has four input neurons.

The three possible classes are coded as: 0.1, 0.5 and, respectively, 0.9, such that the application requires a single output neuron. The activation function for network neurons was sigmoid atansig function - $f_{atansig}$.

We trained a neural network with three hidden neurons, shown in figure 2. The input/hidden weights matrix after supervised learning is:

$$W^T = [w_{ij}]^T = \begin{bmatrix} 1.8465 & 5.5406 & 2.7217 & 2.8102 \\ 2.3933 & 2.5549 & -1.6371 & 2.1335 \\ 2.1572 & -6.0267 & 3.2005 & -1.1336 \end{bmatrix}.$$

The hidden/output weights matrix is: $B^T = [\beta_{jk}]^T = [-2.6372 \quad -0.7893 \quad -4.7428]$, while the biases for hidden neurons are: $T^T = [\tau_{jk}]^T = [-3.6773 \quad 0.0046 \quad 3.4464]$.

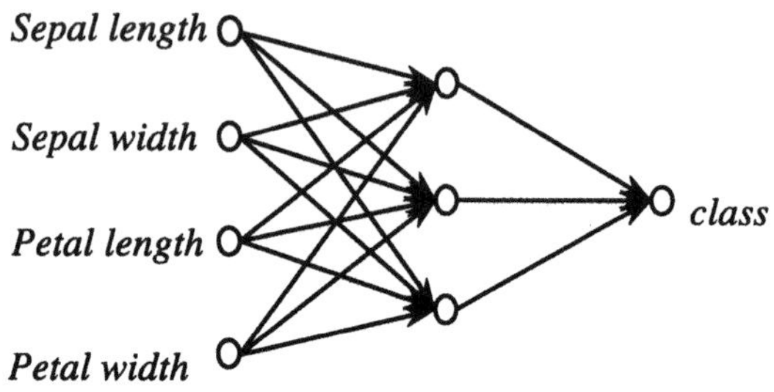

Fig. 2. The three layered neural network for Iris problem

A general form of a rule extracted from trained neural network, with the method presented in the previous section is:

If *sepal lenght* $\in [sl_1;sl_2]$ and *sepal width* $\in [sw_1;sw_2]$ and *petal lenght* $\in [pl_1;pl_2]$ and *petal width* $\in [pw_1;pw_2]$
then $y \in [0;0.2]$ (for setosa)

For versicolor and verginica, the extracted rules have in the consequence the interval [0.4; 0.6], and respectively [0.8; 1.0].

We made studies on the balance between the number of extracted rules (comprehensibility) and the percentage of network functioning covering. First, we tried to extract a number of rules comparable with the total number of training instances. After extracting the more general 150 rules, 94% from the network functioning were covered. On the other hand, with just 25 general rules, it is possible to describe 42% of network functioning.

The extracted rule set proved also good generalization ability, comparable with that of the original network. But, generally, the generalization ability depends on the network training. If the network is properly trained, so that to provide good generalization ability, then the extracted set of rules will have also good generalization ability.

For a better approximation (more than 99%), and covering of almost entire part of the network functioning, the obtained number of extracted rules was very much increased. A big number of rules had to be wasted at the decision boundary between classes, especially at the non-linear border between the two not linearly separable classes.

In fact, this is one of the main drawbacks of approximation techniques by traditional crisp rules, the big number of required rules even for approximating the principal and the most general part of the network functioning (except the neighboring regions to the decision boundaries). In order to obtain a more compact way to describe the network, in the following section we extracted a set of fuzzy rules, which compiles in a more concise way the knowledge embedded in the neural network weights during training.

4 Fuzzy Rule Extraction from Neural Networks

The main disadvantage of most approximation techniques of neural networks by fuzzy rules is the exponential increase of required number of rules for a good approximation. In order to obtain a precise approximation, it must be increased the number of linguistic terms for each input/output variable and consequently the number of fuzzy rules [6]. This causes the lose of the significance of the linguistic terms, as well as of the extracted fuzzy rule set. Of course the number of extracted usual fuzzy rules is not as big as when we extract traditional crisp rules from the neural network, but it still remains high.

In the following part of this section, we introduced a fuzzy interactive operator, in order to express in a very few rules what the neural network learned during training. The meaning of the term "interactive" is given by the inputs correlation embedded in the behavior of the fuzzy operator. Based on the theoretical results presented in [1], it is possible to build a fuzzy rule based system which calculates the same function as a neural network. In this manner, the concept of *f*-duality [1], applied on a three layered feed-forward neural network trained to represent a set of data values, can be used to develop a new class of fuzzy connectives [5].

Let us consider the operation + in $\mathcal{R}$ and the sigmoid function atansig, (used commonly as activation function for neural network nodes):

$f_{atansig}(x) = \frac{1}{\pi}\text{atan}(x) + \frac{1}{2}$, continuous (and bijective) application from $\mathcal{R}$ to (0;1).

The $f_{atansig}$-dual operator of + is o, defined on (0;1) as follows:

$$a\text{o}b = \frac{1}{\pi}\left(\frac{\pi}{2} + \text{atan}\frac{\sin(\pi(a+b-1))}{\cos(\pi(a-0.5))\cos(\pi(b-0.5))}\right)$$

Indeed, it can be proved easily that $f_{atansig}(x_1+x_2) = f_{atansig}(x_1) \text{ o } f_{atansig}(x_2)$. The operator previously defined will be called the *interactive*$_{atan}$-OR operator (for short i_{atan}-OR).

Since f-duality is a general concept, then it could produce other interactive operators (e.g. *i*-OR [1], i_{tanh}-OR [5]), which can be used in knowledge acquisition as well as for motivating the neural inferences.

Based on the properties of i_{atan}-OR [5] and the equivalence theorem proved in [1], it is possible to write the following set of fuzzy rules equivalent with a feed-forward neural network:

$$R_{jk}:\quad \text{IF} \sum_{i=1}^{n} x_i w_{ij} + \tau_j \text{ is } A_{jk} \quad \text{THEN} \quad z_k = \beta_{jk}$$

The fuzzy expressing "x_i is A^i_{jk}" must be interpreted as follows:

"x_i is greater than approximately $r/w_{ij}-\tau_j$" (if $w_{ij}>0$), or

"x_i is lower than approximately $-(r/w_{ij}-\tau_j)$" (if $w_{ij}<0$),

where r is a positive real number obtained from an α-cut (for example 0.9).

The extracted fuzzy rule set equivalent to the neural network trained with Iris data in section 3, obtained by the mechanism given above, is:

R1. IF *sepal-length* is greater than approximately 2.160
i_{atan}-OR
sepal-width is greater than approximately 0.720
i_{atan}-OR
petal-length is greater than approximately 1.465
i_{atan}-OR
petal-width is greater than approximately 1.419
THEN y = -3.6773.

R2. IF *sepal-length* is greater than approximately 1.282
i_{atan}-OR
sepal-width is grater than approximately 1.201
i_{atan}-OR
petal-length is not greater than approximately 1.874
i_{atan}-OR
petal-width is greater than approximately 1.438
THEN y = 0.0046.

R3. IF *sepal-length* is greater than approximately 1.023
i_{atan}-OR
sepal-width is not greater than approximately 0.366
i_{atan}-OR
petal-length is greater than approximately 0.690
i_{atan}-OR
petal-width is not greater than approximately 1.948
THEN y = 3.4464.

The process of classification for a given input is determined by an aggregation computation using the interactive operator. The instance is matched against the rule premises, each rule being fired to a certain degree v_j. The global output is the weighted sum of these degrees: $y = -3.6773v_1 + 0.00462v_2 + 3.4464v_3$.

The class chosen for a given instance is that with the closest numerical value to y.

The three fuzzy rules given above present in a more compact way what the neural network learned by training. Because of the nature of the introduced interactive operator, this set of fuzzy rules is less comprehensible than the rules extracted in section 3, but it expresses in just 3 rules the knowledge captured within the neural network by training. The interactive operator enables us to reformulate fuzzy rules into a more compact way, but also still comprehensible.

5 Conclusions

The problem addressed in this paper is how to interpret the knowledge embedded in a trained neural network into a comprehensible as well as compact (concise) way for human user, using traditional and fuzzy rules. The methods presented in the paper are useful in neural based expert systems, and helps to explain the decisions of the neural network in a more familiar form for human expert. The result is an increased confidence of human user in the actions performed by a neural network. The methods can be used to develop human friendly shells for neural network based applications.

References

1. Benitez, J.M., Blanco, A., Delgado, M., Requena, I.: Neural methods for obtaining fuzzy rules. Mathware Soft Computing, vol.3 (1996) 371-382
2. Ishigami, H., Fukuda, T., Shibata, T., Arai, F.: Structure Optimization of Fuzzy Neural Networks by Genetic Algorithm. Fuzzy Sets and System 71 (1995) 257-265
3. Kosko, B.: Neural Networks and Fuzzy Systems. Prentice - Hall International Inc. (1992)
4. Lin, C.T., George Lee, C.S.: Neural - Network Based Fuzzy Logic Control and Decision System. IEEE Transactions on Computers, vol. 40 No. 12 (1991) 1320-1336
5. Neagu, C.D., Palade, V.: An interactive fuzzy operator used in rule extraction from neural networks. Neural Network World 4, Prague (2000) 675-684
6. Palade, V., Neagu, C.D., Negoita, M.: Genetic algorithm optimization of knowledge extraction from neural networks. Proceedings of the 6th International Conference on Neural Information Processing, Perth - Australia, vol. 2 (1999) 765-770
7. Palade, V., Neagu, C.D., Puscasu, G.: Rule extraction from neural networks by interval propagation. Proceedings of the 4th IEEE International Conference on Knowledge-Based Intelligent Engineering Systems & Allied Technologies-KES2000, Brighton - UK (2000) 217-220
8. Palade, V., Puscasu, G., Neagu, C.D.: Neural network-based control by inverting neural models. Control Engineering and Applied Informatics, no. 1, vol. 1, Bucharest (1999) 25-31
9. Patton, R.J., Uppal, F.J., Lopez-Toribio, C.J.: Soft computing aproaches to fault diagnosis for dynamic systems: a survey. Proceedings of the 4th IFAC Symposium on Fault Detection Supervision and Safety for Technical Processes, Budapest, vol. 1 (2000) 298-311
10. Pinkas, G.: Logical Inference in Symmetric Connectionist Networks. Ph.D. thesis, Washington University (1992)
11. Scott, G.M.: Knowledge - Based Artificial Neural Networks for Process Modeling and Control. Ph.D. thesis, University of Wisconsin (1993)
12. Shann, J.J., Fu, H.C.: A fuzzy neural network for rule acquiring on fuzzy control systems. Fuzzy Sets and Systems 71 (1995) 345 – 357
13. Thrun, S.B.: Extracting Symbolic Knowledge from Artificial Neural Networks. Revised Version of Technical Research Report TR-IAI-93-5, Institut für Informatik III - Universität Bonn (1994)
14. Towell, G., Shavlik, J.W.: The Extraction of Refined Rules from Knowledge - Based Neural Networks. Machine-Learning, vol.13 (1993)
15. Yoo, J.H.: Symbolic Rule Extraction from Artificial Neural Networks. Ph.D. thesis, Wayne State University (1993)

Neural Model Identification Using Local Robustness Analysis*

Héctor Allende[1], Claudio Moraga[2], and Rodrigo Salas[1]

[1] Universidad Técnica Federico Santa María; Departamento de Informática;
Casilla 110-V; Valparaíso-Chile
hallende@inf.utfsm.cl

[2] University of Dortmund; Department of Computer Science;
D-44221 Dortmund; Germany
moraga@cs.uni-dortmund.de

Abstract. A local robustness approach for the selection of the architecture in multilayered feedforward artificial neural networks (MLFANN) is studied in terms of probability density function (PDF) in this work. The method is used in a non-linear autoregressive (NAR) model with innovative outliers. The procedure is proposed for the selection of the locally most robust (around a particular sample) MLFANN architecture candidate for exact learning of a finite set of the real sample. The proposed selection method is based on the output PDF of the MLFANN. As each MLFANN architecture leads to a specific output PDF when its input is a distribution with heavy tails, a distance between probability densities is used as a measure of local robustness. A Monte Carlo study is presented to illustrate the selection method.

1 Introduction

Artificial Neural Networks (ANN) have, by now, a long history and are well established in the applied field as flexible and powerful systems for solving prediction and pattern recognition problems.

That statisticians should be interested in ANN is hardly surprising since the prediction problem plays a central role in statistics. ANN have been used with considerable success in modelling for time series forecasting. However, a mayor weakness of neural network modelling is the lack of established methods for performing test of misspecified models, and test of statistical significance for the various parameters that have been estimated.

Most of the methods used in time series analysis and theoretic approaches rest on two fundamental assumptions; namely, (i) linearity (ii) Gaussianity. Both these assumptions are mathematical idealizations which, in some cases, may be valid, but only as approximations to the real situations.

* Work leading to this paper has been partially supported by the Ministry of Education and Research, Germany, under grant BMBF-CH-99/023 and the Technical University Federico Santa María, Chile, under grant 240022-DGIP.

B. Reusch (Ed.): Fuzzy Days 2001, LNCS 2206, pp. 162–173, 2001.

In the last two decades, non-linear time series have attracted research interest. One of the useful classes of non-linear time series models which has received a great deal of attention is the bilinear model [Sub 81], and [Gab 98]. Another important class of non-linear models is represented by feedforward ANN (FANN) and recurrent networks (ANN with feedback) models. FANN models for non-linear time series analysis have been proposed (see e.g. [COM 94]) and extensively studied (see e.g. [AlM 00]).

In practice, time series data can be contaminated by the presence of a heavy-tailed distribution (non-Gaussianity). In such a case the occurrence of outliers should be respected and robust statistical procedures should be used.

It is well know that when there are outliers in the observations, the method of least squares and maximum likelihood may seriously affect and produce adverse effects on the estimators and on the model identification. The desired statistical procedures used to overcome these situations are the robust estimations instead of the conventional least square procedure. These methods are less sensitive than least square, to typical departures from the ideal assumption of Gaussianity. Robust procedures for estimating the parameters, in presence of outliers have been given by [AlH 92] for linear time series models as well as [CON 96] and [GAB 98] for non-linear time series models.

There have been many papers investigating the accuracy of FANN for forecasting time series (see e.g. [LaF 95] and [FaCH 98]). These authors, concentrated their study mainly on comparing the time series forecasting abilities of traditional methods (linear or bilinear time series models) and FANN, obtaining mixed results. Some favour traditional forecasting methods while others favour FANN. It is difficult to determine the cause of the mixed results since the research methods used are inconsistent. For example, some studies utilized filtered data, "clean data" for FANN, while others allow FANN to use unfiltered data.

In the time series prediction problem, there is a strong culture for testing not only the predictive power of a model on the sensitivity of the dependent variable to changes in the inputs but also the statistical significance of the finding at a specified level of confidence. This is more important in the case of applications where the generating process is dominantly stochastic and only partially deterministic.

The neural networks approach in forecasting problems finds its justification in a central theorem which claims that under certain "reasonable" conditions, any function can be approximated by a FANN with one hidden layer [Whi 92]. FANN are inherently non-linear and approximate well non-linear functions. This "universality" of the FANN approach is given a tremendous applied potential by the existence of a general learning paradigm, the back-propagation algorithm. Unfortunately the basis approximation theorem, in fact a corollary of the Stone – Weierstrass theorem, is non-constructive, that is, it provides no practical guidelines to the construction of a neural network which efficiently approximates a function within pre-established limits: it merely states that such a network exists. Also, the back-propagation algorithm, though extremely useful in practice, is not guaranteed to provide optimal fitting networks; indeed it is a local optimization algorithm i.e., both its speed and the goodness of the resulting fit depend on the initial conditions. Evolutionary design of FANN [HeM 96], [Yao 99] has been shown to alleviate the above mentioned problems. It is however a very time consuming approach.

The above shortcoming may be quite serious in view of the well-known "black box" character of a FANN. No matter how well or how poorly a neural network performs, it is extremely difficult to understand how it works, (see, however, [BCR 97], [Cat 98], [MoT 00], [MoS 01]). In contrast, a traditional statistical model contains in itself an explanation of its successes and failures.

Feedforward Artificial Neural networks have been widely used in nonlinear autoregressive moving average models for time series [CoM 94], [FaCh 98] and [AlM 00]. The ANN architecture that we considered in the next section includes Multilayer FANN with multi-inputs, one hidden layer and single-output. Any continuous function can be approximated with arbitrary accuracy by a network whose structure is sufficiently rich (i.e. it has a sufficient number of hidden neurons).

For neural-based identification, two main issues stand out: one is the choice of the number of neurons in the hidden layer (the FANN architecture) to be adopted for system identification and the other is the choice of the learning algorithm. The learning algorithm is used to control the adjustment of the weights during the training phase, in order to find the best performance in consideration of the training data set. There are generally several one hidden layer FANN architectures candidate for fitting a data set. Then the choice of a specific FANN architecture (number of neurons in the hidden layer) depends on an additional characteristic of the architecture. In this work, the robustness around a particular sample is investigated as such an additional characteristic of the FANN model performance.

In this study, the aim is to find a suitable model which fits the data and which is also locally robust around a given sample. For that purpose the question of how to define local robustness of the FANN architecture is addressed. We assume in the following that the network has already been trained to provide a desired accuracy with the training data set.

The paper is organized as follows: In section 2, the local robustness problem of the FANN is formulated. Section 3 reviews the theory of the three-layered feedforward ANN applied to the modelling of time series forecasting. In Section 4 nonlinear autoregressive models (NAR) of time series are examined. In Section 5, a procedure based on the comparison of the output PDF of each candidate FANN architecture to a given reference PDF is presented for the quantification of the local robustness of an NAR. Last, in section 6, a remark on a possible extension, and conclusions are given.

2 Local Robustness Problem

FANN constitute a very flexible class of overparameterized statistical model. Statistical Inference in MLP networks was developed by [Whi 89]. He showed that if the parameters of ANN are identified they can be consistently estimated by maximum likelihood methods. Moreover, the estimation of the FANN parameters follows an asymptotic normal distribution. This knowledge in principle allows the application of standard asymptotic hypotheses test, such as Wald-test. However, as FANN in general do not encompass the unknown function but only approximate it; they are inherently misspecified models. In [Whi 94] it was proven that the application of standard asymptotic tests is valid even if the models are misspecified. Unfortunately, we have the problem that the parameters of a FANN are not always identified, due to mutual dependencies between them. In such case the parameters are no longer normally distributed and inference is cumbersome. In order to specify a FANN architecture, on-line and off-line techniques have been developed to choose the number of hidden neurons by incremental or decremental procedures. But reaching the right size of the hidden layer by these techniques is not guaranteed and termination criteria clearly lack statistical meaning. In this work, in order to specify different types of FANN architecture candidates, we choose an additional property, the robustness around a particular sample.

The local robustness is investigated as an additional property of the FANN model. An important task is to see how outliers in the data affect the above screening process. Namely the set of selected models should be robust in the sense that they are indifferent to radical change of a small portion of the data or a small change in all of the data.

3 Neural Network Structure

Mathematically, a FANN consists of elementary processing elements (neurons), organized in layers. The layers between the input and the output layers are called "hidden". The number of input units m is determined by the application. The architecture or topology A_λ of a network refers to the topological arrangement of the network connections. We define a class of neural models according to [Vap 95]. A class of neural models is specified by

$$S_\lambda = \{g_\lambda(\underline{x}, \underline{w}), \underline{x} \in \Re^w, \underline{w} \in W\}, \qquad W \subseteq \Re^p \tag{3.1}$$

where $g_\lambda(\underline{x}, \underline{w})$ is a non-linear function of $\underline{x}$ with $\underline{w}$ being its parameter vector, and p is the number of free parameters determined by A_λ, i.e. $p = \rho(A_\lambda)$

A class (or family) of neural models is a set of FANN models which share the same architecture and whose individual members are continuously parameterized by the vector $\underline{w} = (w_1, w_2, ...w_p)^T$. The elements of this vector are usually referred to as weights. For a single-hidden-layer architecture, the number of hidden units λ indexes the different classes of ANN models (S_λ) since it is an unambiguous descriptor of the dimensionality p of the parameter vector $(p = (m+2)\lambda + 1)$.

Given the sample of observations, the task of neural learning is to construct an estimator $g(\underline{x}, \underline{w})$ of the unknown function $\varphi(\underline{x})$

$$g(\underline{x}, \underline{w}) = \gamma_2\left(\sum_{j=1}^{\lambda} w_j^{[2]} \gamma_1\left(\sum_{i=1}^{m} w_{ij}^{[1]} x_i + w_{m+1,j}^{[1]}\right) + w_{\lambda+1}^{[2]}\right) \tag{3.2}$$

where $\underline{w} = (w_1, w_2, ...w_p)^T$ is a parameter vector to be estimated, γ'_s are linearity or non-linearity and λ is a control parameter (number of hidden units). An important factor in the specification of neural models is the choice of base function γ. Otherwise known as 'activation' or 'squashing' functions, these can be any non-linearity as long as they are continuous, bounded and differentiable. Typically γ_1 is a sigmoidal or the hyperbolic tangent. All these functions belong to the family $\Gamma \equiv \{\gamma_1 = \gamma_1(z,k,T,c): z,k \in \Re, T,c \in \Re - \{0\}\}$, where γ_1 is defined as follows:

$$\gamma_1(z,k,T,c) = k + c[1+\exp\{Tz\}]^{-1} \tag{3.3}$$

when $c = 1, k = 0$ and $T = -1$ in equation (3.3) the classical asymmetric sigmoidal activation function is obtained, which is the most commonly used.

The estimated parameter $\underline{\hat{w}}_n$ is obtained by minimizing iteratively a cost functional $L_n(\underline{w})$ i.e.

$$\underline{\hat{w}} = \arg\min\{L_n(\underline{w}) : \underline{w} \in W\}, \quad W \subseteq \Re^p \tag{3.4}$$

where $L_n(\underline{w})$ is for example the ordinary least squares function i.e.

$$L_n(\underline{w}) = \frac{1}{2n}\sum_{i=1}^{n}\left(y_i - g(\underline{x}_i, \underline{w})\right)^2 \tag{3.5}$$

The loss function in equation (3.5) gives us a measure of accuracy with which an estimator A_λ, fits the observed data but it does not account for the estimator's (model) complexity. Given a sufficient large number of free parameter, $p = \rho(A_\lambda)$, a neural estimator A_λ, can fit the data with arbitrary accuracy. Thus, from the perspective of selecting between candidates, model expression (3.5) is an inadequate measure. The usual approach to the selection is the so-called discrimination approach, where the models are evaluated using a fitness criterion, which usually penalizes the in-sample performance of the model, as the complexity of the functional form increases and the degrees of freedom for error become less. Such criteria, commonly used in the context of regression analysis are: the R-Squared adjusted for degrees of freedom, Mallow's C_p criterion, Akaike's AIC criterion, etc.

4 FANN of Non-linear Autoregressive Models (NAR)

A non-linear time series model (NAR) type can be written as

$$x_t = h(x_{t-1}, x_{t-2}, \ldots\ldots, x_{t-p}) + a_t \tag{4.1}$$

where h is an unknown smooth function and a_t denotes the innovation process with $E[a_t / x_{t-1}, x_{t-2} \ldots\ldots] = 0$ and that a_t has finite variance σ^2.

Assume that a FANN provides a non-linear approximation to h given by g_{FANN} ($\underline{x}$, $\underline{w}$), where $\underline{w}$ is the vector of parameters of the FANN. In practice the data that we have to our disposal to represent (4.1) by an FANN can be contaminated by two types of outliers. Innovative outliers are present if a_t is from a distribution F_a with heavier tails than the normal distribution. Additive outliers are present if { x_t } does not satisfy (4.1). Suppose that the NAR process cannot be perfectly observed because a small fraction ε (in practice we usually have $\varepsilon \leq 0.1$) of observations are disturbed by an outlier-generating process $\{U_t V_t\}$ where U_t is a zero-one process with $P(U_t = 1) = \varepsilon$, $P(U_t = 0) = 1 - \varepsilon$, and the random variables V_t independent of U_t, have arbitrary distribution function G_t. Thus the observational model is

$$Z_t = X_t + U_t V_t \tag{4.2}$$

Therefore with probability (1-ε) the NAR process X_t, itself is observed, and with probability ε the observation is the NAR process plus an error with distribution G_t. Contaminated NAR models are denoted by (CNAR).

In this work, the aim is to find a suitable model which fits the data and which is also locally robust around a given sample. For that, it uses a multilayer feedforward network with three layers (several inputs, one hidden layer and a single output) and we propose to characterize the local robustness of $\hat{X}_t^j$ representing the prediction of X_t, with j hidden neurons.

There is a finite number of competing FANN architecture candidates for exactly learning a set of real samples. Each $\Delta(f(\hat{X}_t^j), f(\hat{X}_t))$ FANN architecture gives a particular output probability density function (PDF) when the NAR is non-Gaussian i.e. a CNAR model. The difference between the NAR output PDF and the CNAR output PDF may be used to build a quantitative measure of local robustness. Let $f(\hat{X}_t^j)$ be the FANN prediction PDF and $f(\hat{X}_t)$ be the prediction PDF (normal situation, without outliers). To choose a model, according to the local robustness goal, is equivalent to choose a model for which the Hellinger metric distance is minimal. An approximation of the PDF of the FANN prediction, can be estimated by the method described in [TKK89].

5 Local Robustness of NAR

In this section we present two examples in order to obtain some information concerning of the local robustness behavior of the NAR(1) models. As an example to illustrate such a theory, a set of twenty noisy data points $(x_i, x_{i+1})_{i=1..20}$ was created (as the experimental set) using an arbitrary nonlinear function f_s

$$\begin{cases} x_{i+1} = f_s(x_i) + e_i & i = 1,\dots,20 \\ f_s(x_i) = 10(1 - \exp(-0.5 x_i)) \end{cases} \tag{5.1}$$

where e_i are sampled from a normal distribution $N(0,\sigma^2 I_{10})$ with $\sigma^2 = 1/3$. The function is uniformly sampled on the compact set [0,10]. A single-input single-output (SISO) FANN model is used to fit this data set. For this a training set of five data points selected using a pre-processing procedure, such as optimum experimental design (OED) [HiM 99], is constructed.

According to MSE-criteria, there are three candidate architectures g_{NN2}, g_{NN3}, and g_{NN4} for exact learning of the training set, with respectively N_{h1}, N_{h2}, and N_{h3}, hidden neurons

$$\hat{x}_{i+1} = g_{NN_i}(x, w_i) = \sum_{j=1}^{N_{hi}} w_j^{[2]} \gamma_1 (x w_{1j}^{[1]} + w_{2j}^{[1]}) + w_{N_{hi}+1}^{[2]} \quad where \quad N_{hi} = 2, 3, 4 \tag{5.2}$$

with w_i the vector of the NN and g_{NNi} the architecture. The standard back-propagation training algorithm is used to update the estimates of the weights during the training phase for the different FANN architectures. As can be seen on table 1, these FANN architectures are very similar in the least squares sense, with a small mean square error (MSE) in the observed design region. For the discrimination of the different FANN architecture a quality factor defined by the local robustness around $x_i = 4$ is added. In this case some architectures may appear to be better than others. The FANN output PDF for each candidate architecture is computed using the method described by [TKK89].

A closer view of the local behavior of the FANNs mappings, the selected data points and the actual function are shown on Fig 1.

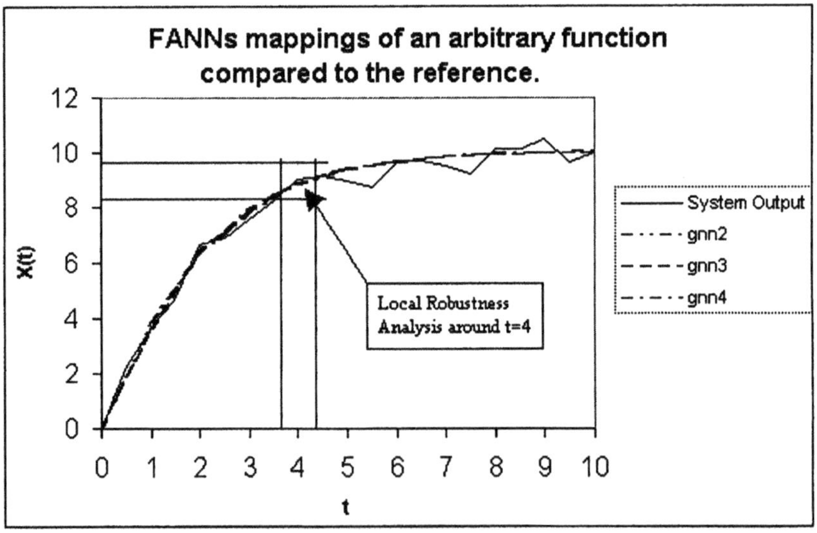

Fig. 1. FANNs mappings of an arbitrary function and selected from architecture compared to the reference output noisy interlapolate points system PDF

Figure 2 shows the output PDFs of the different FANN architectures. In order to choose an architecture, the criterion introduced by the Hellinger metric as a measure of local robustness is computed for each candidate architecture. Results are given in table 1.

Table 1. MSE: mean square error between the FANN and the noisy system outputs on the training set and Δ : distance between FANN output and the reference PDFs.

ANN	g_{NN2}	g_{NN3}	g_{NN4}
MSE	0.0794	0.078	0.0784
Δ	0.193	0.209	0.2942

It appears that the model g_{NN2} is the best one in the sense of this quality factor.

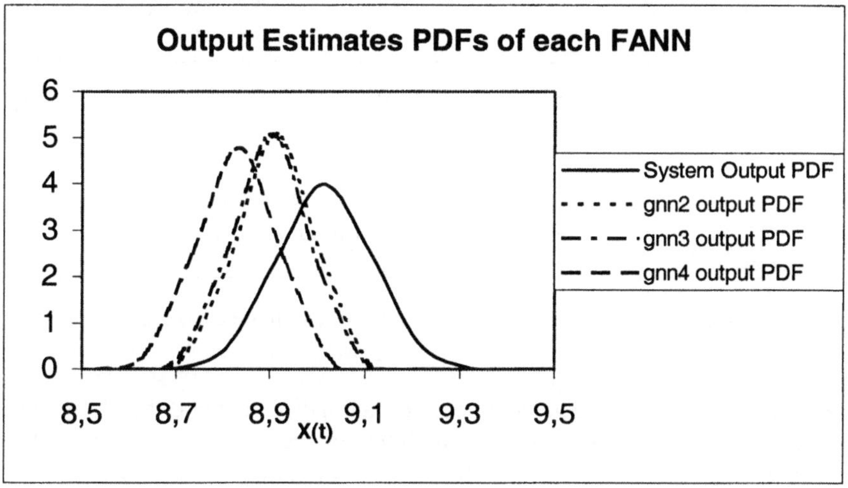

Fig. 2. Output estimates PDFs of each FANN

As second example a NAR(1) is considered using a nonlinear process with innovative outliers

$$x_t = g(x_{t-1}) + a_t \tag{5.3}$$

where $g(x_{t-1}) = 1.5\,X_{t-1}\,\exp[-X_{t-1}^2/4]$ with a_t i.i.d. is normal contaminated with distribution

$$F_a = (1-\alpha)N(0;0.5) + \alpha N(0;5)\text{, and } \alpha = 0\text{; and } \alpha = 0.01 \tag{5.4}$$

The experiment was performed for a fixed sample size of n=200. To avoid initialization effects the first 150 observations were discarded. A single-input single-output FANN model is used to fit this data. For this a training set of sixteen data points selected using a D-optimality criterion, (see [ChC 99]), was constructed.

Figure 3 shows the NAR(1) time series of length n=200 with and without innovative outlier. The initial value on t=0 was $x_0 = 0$, generated by the model (5.3). Figure 3 shows the NAR(1) time series of length n=200 with and without innovative outlier. The initial value on t=0 was $x_0 = 0$, generated by the model (5.3).

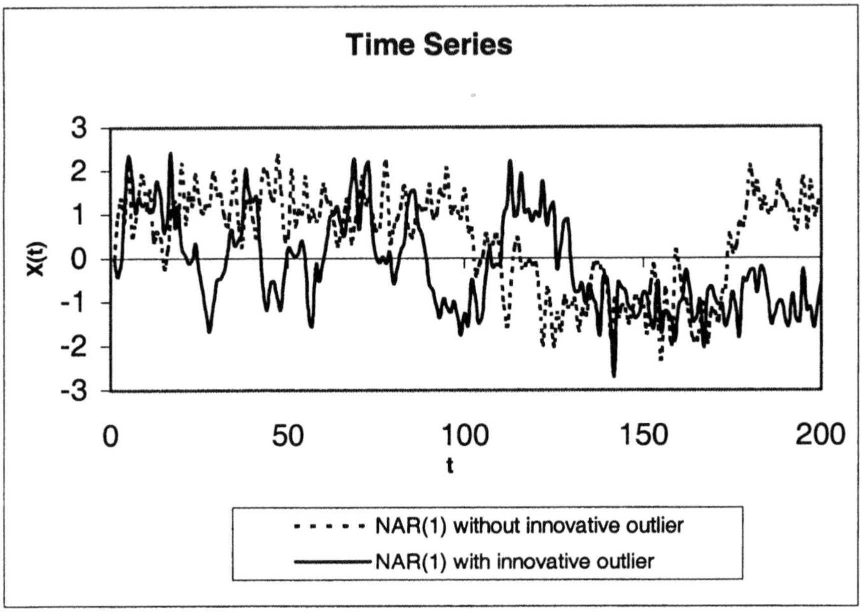

Fig. 3. NAR(1) Time series with and without innovative outliers.

To different FANN candidate architectures a quality factor defined by the local robustness is added. In this case some architectures are discovered to be better than others with respect to the local robustness measure studied as shown in Table 2 and Table 3 for the case of the time series without and with innovative outliers respectively.

Table 2. MSE: mean square error between the FANN and the noisy system outputs on the training set without innovative outliers and Δ: distance between FANN output and the reference PDFs.

ANN	g_{NN8}	g_{NN9}	g_{NN10}	g_{NN11}	g_{NN12}	g_{NN13}	g_{NN14}	g_{NN15}	g_{NN16}
MSE	0.1566	0.1502	0.1522	0.1605	0.1506	0.1594	0.1531	0.1501	0.1501
Δ	0.1418	0.2274	0.1644	0.1581	0.2011	0.1634	0.202	0.3248	0.2666

Table 3. MSE: mean square error between the FANN and the noisy system outputs on the training set with innovative outliers and Δ : distance between FANN output and the reference PDFs.

ANN	g_{NN8}	g_{NN9}	g_{NN10}	g_{NN11}	g_{NN12}	g_{NN13}	g_{NN14}	g_{NN15}	g_{NN16}
MSE	0.0954	0.0928	0.086	0.0783	0.0795	0.061	0.0744	0.0563	0.0587
Δ	0.2239	0.118	0.2246	0.2324	0.2322	0.2283	0.2173	0.221	0.2944

It appears that the models g_{NN8} and g_{NN9} are the best FANN models for the NAR(1) without and with innovative outliers respectively in the sense of this quality factor.

Figure 4 and Figure 5 shows the outputs PDFs of the different FANN architecture

6 Conclusions

A measure of local robustness has been applied to the problem of FANN architecture selection. The method is based in the comparison of the FANN output probability density function. For the output PDF comparison the Hellinger metric distance is used. However, it could be interesting to use a different distance measure. Only FANN architectures with one hidden layer have been considered. It has been shown that all FANN models are not necessarily equivalent in the sense of local robustness, around a particular sample. At present, there is little work on the analysis of the global or local robustness of FANN models with outliers in the input data space. For this reason further research is needed to study robustness properties in FANN with different architectures (e.g., NAR(p); NARMA(p,q)) and other types of outliers (additive outliers and patchy outliers).

Other future work in robust techniques and FANN will center around making neural networks robust to change in the variance noise. Nonlinear models are particularly sensitive to change in the variance. One would expect to see substantial improvements in FANN time series prediction if they can be robustified to the variance.

References

[AlM 00] Allende H., Moraga C.: Time Series Forecasting with Neural Networks. Forschungsbericht N° 727/2000. Universität Dortmund Fachbereich Informatik. (2000)

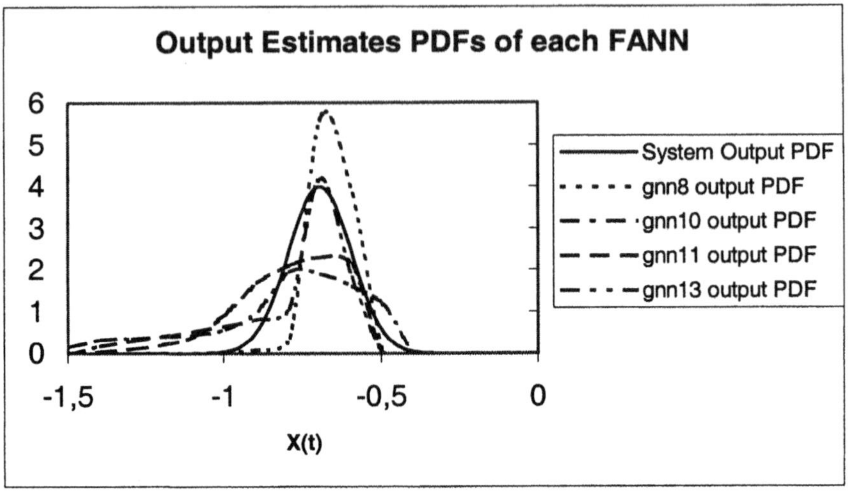

Fig. 4. Output estimates PDFs of the four most robust FANN for the NAR(1) time series without innovative outliers.

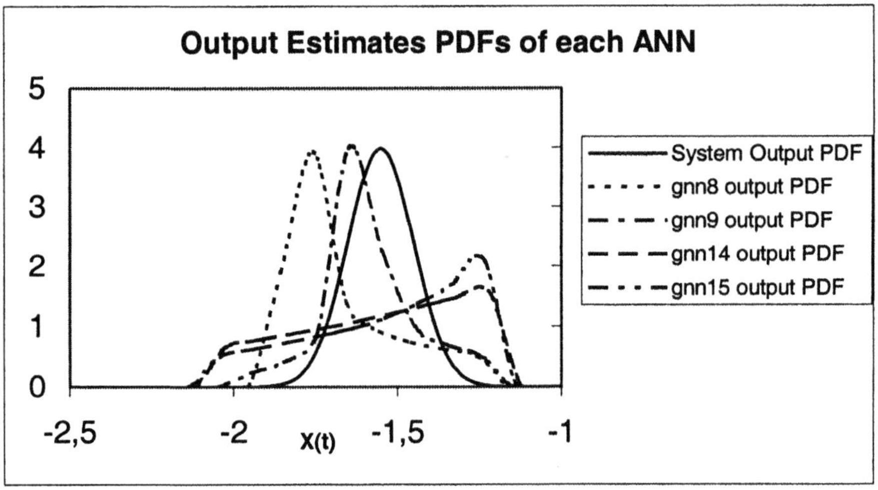

Fig. 5. Output estimates PDFs of the four most robust FANN for the NAR(1) time series with innovative outliers.

[AlH 92] Allende H., Heiler S.: Recursive Generalized M-Estimates for Autoregressive Moving Average Models. Journal of Time Series Analysis. 13, 1-18 (1992)
[CaT 98] Castro J.L., Trillas E.: The logic of neural networks. Mathware and Softcomputing V (1), 23-27, (1998)

[ChC 99] Choueiki M.H. and Mount-Campbell C.A.: Training data development with the optimality criterion. IEEE Transactions on Neural Networks, 10, 56-63 (1999)
[CoM 94] Connor J. T., Martin R. D.: Recurrent Neural Networks and Robust Time Series Prediction. IEEE Transactions of Neural Networks 5 (2), 240-253, (1994)
[CoM 96] Connor J. T.: A robust Neural Networks Filter for Electricity Demand Prediction. J. of Forecasting 15, 437 – 458 (1996)
[FaCh 98] Faraway J., Chatfield C.: Time series with Neural Networks. A comparative study using the airline data. Appl. Statist. 47 (2), 231 – 250 (1998)
[Gab 98] Gabr M.M.: Robust estimation of bilinear time series models. Comm. Statist. Theory and Meth 27 (1), 41 – 53 (1998)
[HaN 97] Hansen J., Nelson R.: Neural Networks and traditional time series methods. IEEE transaction on Neural Networks 8 (4), 863-873. (1997)
[HeM 96] Heider R., Moraga C.; Evolutionary synthesis of neural networks based on graph-grammars. Proc. Int. Conf. On Soft and Intelligent Computing. 119-128, Budapest. ISBN 9634205100, (1996)
[HiM 99] Hisham M. and Mount-Campell C.A. : Training data development with the D-optimality criterion. IEEE Transaction on Neural Networks Vol 10, 56-63 (1999)
[LaF 95] Lachtermacher G. and Fuller J.D.: Backpropagation in Time Series. J. of Forecasting 14, 381-393 (1995)
[MoS 01] Moraga C., Salinas L.: Interpreting neural networks in the frame of the logic of Lukasiewicz. Proceedings Int'l Workconference on artificial and natural neural networks IWANN'2001. Granada, Spain. Springer, (2001)
[MoT 00] Moraga C., Temme K.-H.: S-neural networks are fuzzy models. Proceedings Int'l Conference on Information Processing and Management of Uncertainty in Knowledge based Systems, 1518-1523. Technical University of Madrid, Spain, (2000)
[Sub 81] Subba Rao, T.: On the theory of bilinear time series models. Comm. Statist. Soc. B 43 (2), 244-255 (1981)
[TKK 89] Tierney L., Kass R. and Kadane JB.: Approximative marginal densities of non-linear functions. Biometrika 76, 425-433. Correction in Biometrika 78, 233-234.
[Whi 89] White H.: Learning in Neural Networks: A statistical perspective. Neural Computation 1 425-464 (1989)
[Whi 92] White H.: Artificial Neural Networks: Approximation and Learning Theory, Basil Blackwell, Oxford (1992)
[Whi 94] White H.: Estimate Inference and Specification Analysis, New York: Cambridge University Press, (1994)
[ReZ 99] Referes A. P. N., Zapranis A. D.: Neural Model Identification, Variable Selection and Model adequacy. J. Forecasting 18, 299-322, (1999).
[Yao 99] Yao X.; Evolution of Artificial Neural Networks, Proc. IEEE 87 (9), 47-49. (1999)

Computing with Activity Patterns

Matthias Reuter

University of Dortmund, Chair of Computer Science I
Otto-Hahn-Str. 16, D-44221 Dortmund
Tel: 0231/755 6341, Fax: 0231/755 6555
reuter@ls1.cs.uni-dortmund.de

Abstract. The paper presents a method which uses the activity patterns of simulated neural networks for classification/identification and for combining classification concepts by a chunking process. Furthermore it will be shown that if a neighborhood function similar to the Sinc function is used, the activity of neurons or neuron clusters can be used to compute - based on physical principles like superposition and interference of waves - with activity patterns.

1 Introduction

Even if a large number of neural nets is known none of these nets can solve some basic tasks of organizing classification behavior which biological neural nets can do. Above all none of the simulated nets can join similar classification or identification concepts to new 'meta-concepts (chunks)'. This process, called 'chunking', is well known from cognitive science and even very primitive clusters of biological neurons forming a central nerve system can do chunking. It is a favourite principle of these chunks that on the one hand the different parts of these combined concepts can be activated individually by a special input-pattern but on the other hand the whole chunk can be activated also if only single parts of the input pattern are activating chunked classification concepts. Another problem of the common neural nets is that classification results are solely represented by the activity of the neurons of a special output-layer, why wett even the power of classification of the whole net is restricted by the activity of 'grandmother'-neurons of the output layer. It seems that we are working with a bottle neck which strips the classification behavior of the neural nets to a very simple and localized representative. May be for that reason nature does not use such kind of neurons in brain but activity patterns of an association of neurons which are active at the same time to encode classification behaviour and classification results. By this way it is possible:

- to integrate data from different senses in one global activity pattern,
- to integrate more and more information (aspects) in one activity pattern by modifying its structure step by step,
- to chunk diverse classification concepts by combining so far different activity patterns.

Furthermore it follows from the given number of output neurons that the power of classification is bounded, a fact which leads to the disability of a trained simulated

B. Reusch (Ed.): Fuzzy Days 2001, LNCS 2206, pp. 174–184, 2001.

neural net to discriminate between a noisy input pattern and a really new situation which has to be estimated, whereas biological neural nets first indicate a new situation occurred and try to show similarities to classification concepts afterwards. As far as we understand today all these weak points result from a lack of an organizing method for stored classification concepts, whereat a classification concept is understood as the reaction of the net on a given input pattern.

In this paper a method will be discussed which gives a first hint how such an organization can be done by using memory structures of the net and organizing them by a special learning algorithm.

2 Basic Assumptions

The basic assumption for organizing classification structures/concepts is that we have to characterize the classification behavior of the net by activity patterns which are induced by input patterns and use their structure for computing.

Using these activities of a neural net for analyzing its classification behavior means to be aware of the fact that different kinds of activity patterns occur in a net. At first either the activities of the inter-neural weights, the activities of the neurons or both can be supervised. For selecting the best supervising strategy it may help to take a view on how information is coded and activated in biological nets.

Biological nets (wet nets) encode information in two ways [1]:

- The topology of activity of a more or less complex neural network at juncture k encodes the classification result/classification concept of the momentary constellation of the senses data.
- The firing rate of the neurons encode the intensity of the senses data.

Furthermore we have to be aware of the fact that biological neurons (wet net neurons) fire their signals with a constant amplitude. Therefore the topology of the activity will solely be given by the configuration of the inter-neural weights and the number of input-connections at a juncture k. In periods of about 30 msec brain updates its activity configuration and points out the given activity pattern to stored classification structures/concepts, whereat a classification concept is either a pre-learned, situation relevant interpretation rule induced by the sense data or a pure reflexive action following from a reflex-relevant situation. Another important fact is that wet net neurons quantize the time as they have a fixed time of repolarization. During this time these neurons are 'blind' and can not act in any way. On account of these facts a wet net signal can be regarded as a quantized signal as shown in Fig. 1.

As in addition biological neurons are amenable to a bias-functionality we will define that

- the form of the activity $A_l(k)$ of a simulated neuron l (dry net neuron) at a time interval k follows an indicator function described by the formula

$$A_l(k) = \begin{cases} 1 & \text{if } a > b \\ 0 & \text{else} \end{cases}$$

where a describes the sum of the activities from the neurons i exciting the neuron l and b describes the bias of the neuron l.

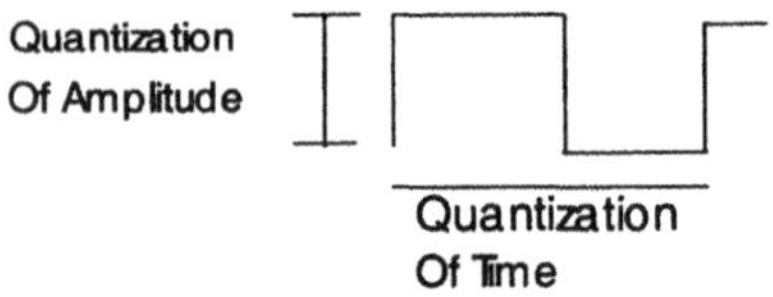

Fig. 1. Principle form of an activity signal of a neuron

As in addition biological neurons are amenable to a bias-functionality we will define that

- the form of the activity $A_l(k)$ of a simulated neuron l (dry net neuron) at a time interval k follows an indicator function described by the formula

$$A_l(k) = \begin{cases} 1 & \text{if } a > b \\ 0 & \text{else} \end{cases}$$

  where a describes the sum of the activities from the neurons i exciting the neuron l and b describes the bias of the neuron l.
- if the activation exciting the neuron l is much higher than b the activity $A_l(k)$ should survive over a longer period in a way that $A_l(k)$ fires with a constant rate over several calculation/updating steps, like shown in Fig. 2.

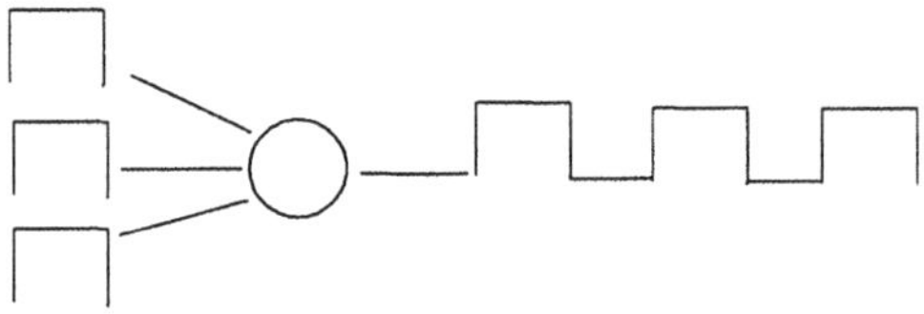

Fig. 2. Principle form of a transferred high intensity signal for a wet net neuron

To encode the 'bias readiness' of the neurons in an adequate form, we will assume, that a non-active neuron lies on a potential equal to zero, a neuron with a negative bias lies on a negative potential value and a neuron with a positive bias lies on a positive potential value as shown in Fig. 3.

3 About the Structure of the Simulated Neural Nets

The idea of 'calculating with activities patterns of neural nets' implies to use special structures of neural nets. These structures may be built up by Kohonen maps, cubs of Kohonen maps or modules of Backpropagation nets. In any case the net has to form a grid of neurons or modules of neurons like shown in Fig. 5, whereat every element of the grid can have an active or a non active status at a given juncture k. The active parts (neurons/modules) of the grid at a juncture k encode the classification result of the net, which is induced by a momentary active input pattern. The 'meaning' of the classification structure - mostly defined be the net designer - represents the classification concept [3]. It is important that the net is structured in a way that more than one activity 'centre' (single neuron, neurons, cluster of neurons) is active to form a significant classification pattern. For better interpretation of these activity patterns, the values of the activation should be bound by [-1,1]. In front of the grid an input layer has to be placed exciting every neuron, every neuron cluster of the grid with input data.

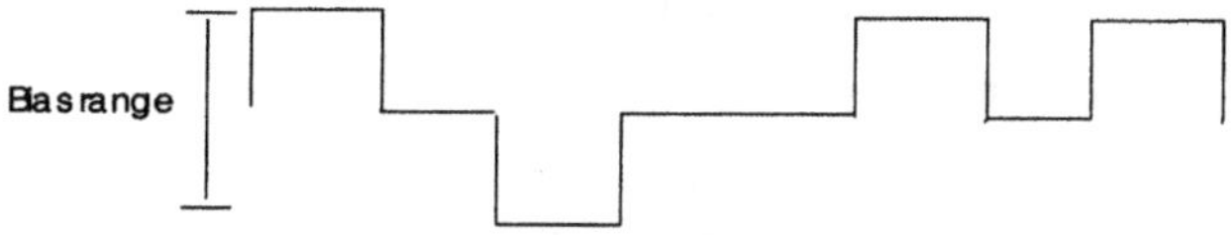

Fig. 3. Shape of neuron potentials with various biases

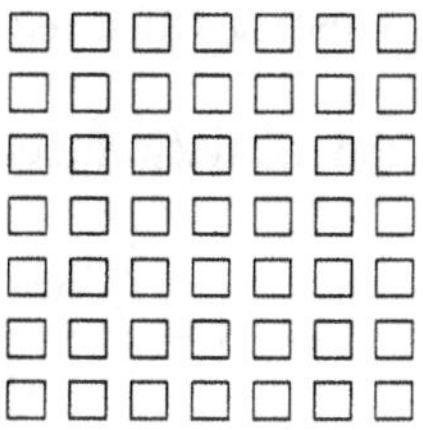

Fig. 5. Non active neuron grid

4 Structure and Task of the Used Neighborhood Function of the Activity Centres

To meet the requirements of computing with activities the neighborhood function of every neuron or neuron cluster should have the form of a Sinc function with rotational symmetry. The Sinc function, also known as mexican had function, corresponds to the Fourier Transform of the 'Tophat' function, which is in turn really just the 'Pupil' function. A multiplication in Fourier space by a Pupil function corresponds to band-limiting an input activation through an aperture, truncating the corners; in direct space, the convolution of the input pattern against the Fourier Transform of the kernel does the same. In short, the Sinc function is

$$Sin(r)/r \ [=1 \ at \ r=0]$$

as shown in Fig. 6 and the convolution kernel can be approximated in matrix form by {{1,-2,1},{-2,4,-2},{1,-2,1}}, whereby it equals the Fourier Transform of the Heaviside Function [1 for r<=1, 0 for r>1].

Especially the existence of several excitatory and inhibitory regions of the Sinc function enables the calculation with activity patterns as now diverse effects of influences between different activity patterns can be taken into account. The number of these regions will be problem-oriented and has to be set before building the net. In the designing phase it also has to be decided, in which dimensions the neighborhood functions act, resp. if the net is regarded as a homogeneous grid on which the neighborhood function acts in a rotational symmetrical way or if special layer symmetries have to be considered. Furthermore the dimension of the Sinc function must be chosen in a way that the maximums and minimums of the function meet neurons on the grid.

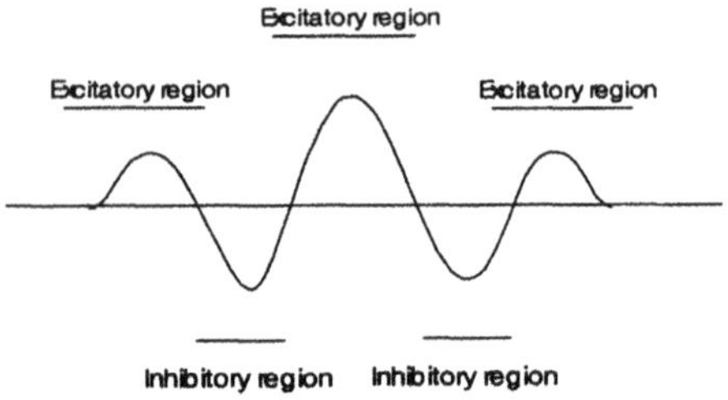

Fig. 6. Shape of the used Sinc neighborhood function

Because of these reasons, the activity function of a neuron should have the following form

$$\psi(\vec{q},\vec{E},k)=\frac{\vec{E}}{\vec{I}}\frac{\sin(d(\vec{q},k))}{d(\vec{q},k)}$$

Here $\psi(\vec{q},\vec{E},k)$ is a function of the position $\vec{q}$ of the active neuron on the grid, a function of the intensity (vector) $\vec{E}$ of the information to be classified, a function of the actual juncture k and a function of the distance d describing the influence of the Sinc function to the adjacent parts of the net. The vector $\vec{I}$ signifies a vector of standardization to meet our activity boundary of [-1,1]. The negative parts of the Sinc function also guarantee that different activity patterns cannot be stored too close to each other. By this way the feature to separate independent activities will not be lost as two activity centres cannot be neighbored directly. Otherwise the positive parts of the Sinc function guarantee that similar activity patterns are close enough to enable chunking effects by providing the positioning of so far not-chunked activity patterns.

For the shape of the envelope of $\psi(\vec{q},\vec{E},k)$ we assume the following formula

$$Env(\psi(\vec{q},\vec{E},k))=e^{-\frac{d(\vec{q},k)^2}{f}}$$

where f denotes a standardization parameter, which defines the focus of influence of an active neuron/cluster of neurons.

If the given neighborhood function is utilized on a neuron with a positive or a negative potential the centre of the Sinc function will be placed in these potentials as shown in Fig. 7.

If this principle is utilized to the shape of Fig. 8, the following activity structure results.

As the global activity $\Psi(\vec{q},\vec{E},k)$ of a grid can be regarded as a superposition of the single activities $\psi_l(\vec{q},\vec{E},k)$, $\Psi(\vec{q},\vec{E},k)$ is given by the formula

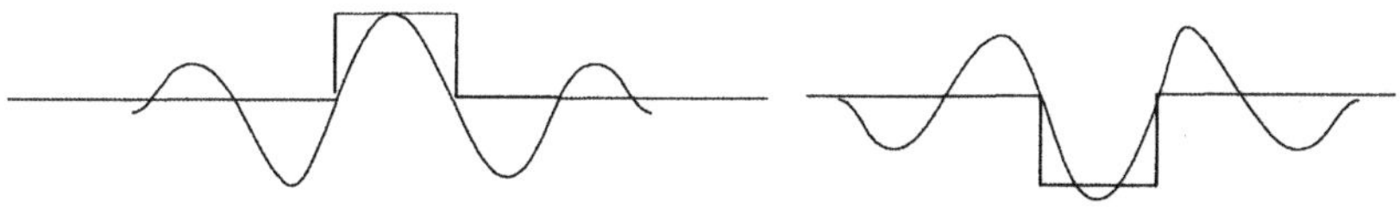

Fig. 7. Form of the neighborhood function correlated with the neuron potentials

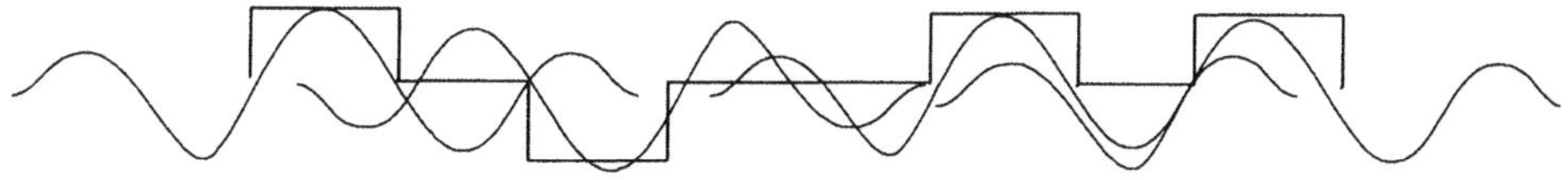

Fig. 8. Form of the neighborhood function correlated with the neuron potentials from Figure 7

$$\Psi(\vec{q},\vec{E},k)=\sum_{l=1}^{L}\psi_l(\vec{q},\vec{E},k)=\sum_{l=1}^{L}\frac{\vec{E}_l}{\vec{I}}\frac{\sin(d(\vec{q}_l,k))}{d(\vec{q}_l,k)}e^{-\frac{d^2(\vec{q},k)}{f}}$$

Therefore the following course for $\Psi(\vec{q},\vec{E},k)$ results from interference effects for our example shown in Fig. 8.

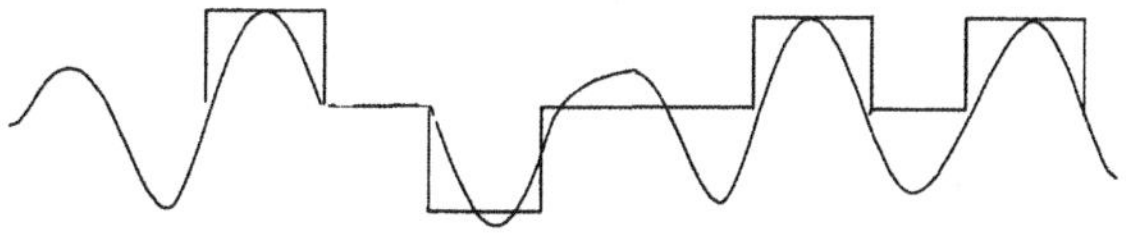

Fig. 9. Form of the activity function $\Psi(\vec{q},\vec{E},k)$ if superposition effects are regarded

In Fig. 10 and Fig. 11 a neural net with different activity patterns is shown, whereat every black point in the grid's matrix represents a centre of activity of a Sinc function. Furthermore every pattern represents a symbol, or a classification result of a symbol encoded through the special constellation of - for example - the three active neurons. The Sinc neighborhood functions of the active neurons show the range of influence of these active symbol representations. Please note that the activation of the neurons forced by the Sinc neighborhood functions is not shown in Fig. 10 and 11 and that the influence of the neighborhood functions is on layer's plane only .

In case that all three patterns are active the net will show the activation structure shown in Fig. 11. It can be pointed out that no superposition effects occur, wherefore no interference effects can be observed.

If neurons are activated as shown in Fig. 12, interference effects will occur, as now the superposition of neighborhood functions leads to a revocation of two activity centres of the net structure. The resulting activity pattern is shown on the right hand side of Fig. 12.

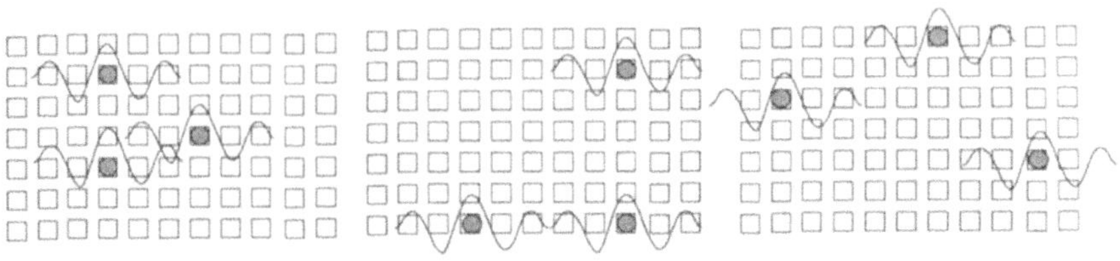

Fig. 10. Different activation states of a neural net grid

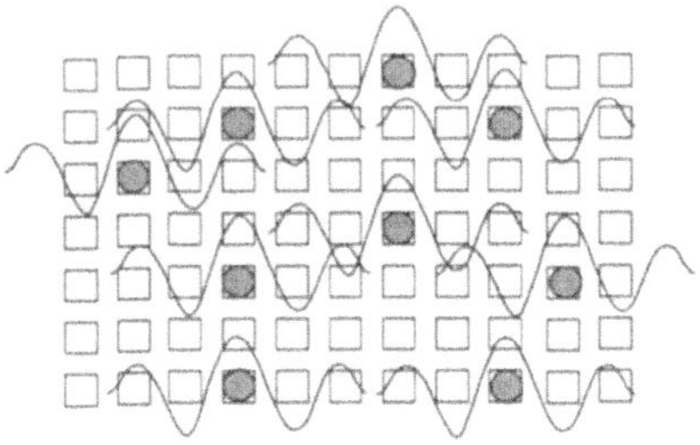

Fig. 11. Activation of the given neural net structure when all stored patterns are active

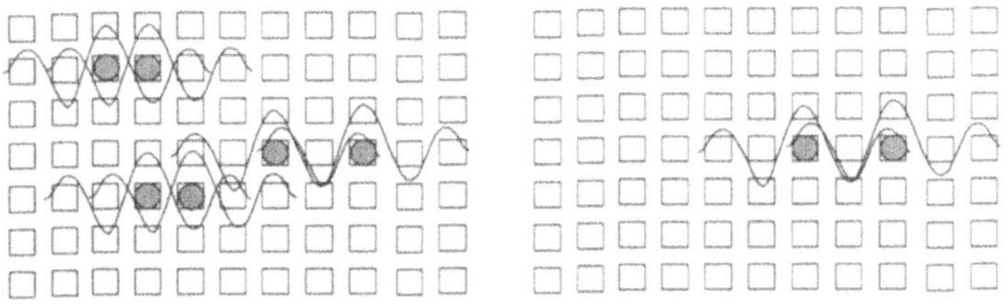

Fig. 42. Superposition effects on activity patterns forced by the neighborhood functions of the active neurons

If neurons are activated as shown in Fig. 12, interference effects will occur, as now the superposition of neighborhood functions leads to a revocation of two activity centres of the net structure. The resulting activity pattern is shown on the right hand side of Fig. 12.

5 Chunking and Activation of Similar Components of an Already Stored Pattern

For the process of chunking and the topological combination of similar features it makes sense to construct two kinds of memory:

- A short time memory
  The short time memory describes the totality of all active neurons on a grid. Besides the active neurons representing the 'pure' momentary input pattern, the short time memory has to contain active neurons which represents components

of already stored patterns which have been activated by foothills of the Sinc function of these neurons which are representing the input pattern also. By this way similarities between stored and activated activity patterns can be taken considered for changing of the net structure during a learning period. Furthermore the 'contents' of the short time memory will be used to produce interference effects by superposing stored activity patterns with activity patterns which are induced by a momentary input pattern. For these reasons, one can say that the short time memory is the 'working platform' of the net.

- A long time memory
  The long time memory describes the structure of the grid when all classification structures are stored after a learning period. Besides the activity patterns $\Psi_n(\vec{q},\vec{E},k)$ which represent the evaluated classification concepts, the according input patterns have to be stored.

We have to be aware of the fact, that these memory structures do not only represent two forms of information memories but two kinds of working states of a net. While the short time memory designates an active net the long time memory designates the 'timeless' power of the classification behavior of the net.

5.1 Example of Topological Neighborhood of Similar Patterns

As similarity of two patterns means that both patterns have some common components, the foothills of the Sinc function can be used to create a topological neighborhood of similar patterns.

The easiest way to use these foothills is given as follows: If a new pattern is presented to the net and this pattern meets a foothill of the Sinc function of a stored pattern, the neuron representing this foothill is selected by the learning algorithm to represent the new activity pattern.

Example: Lets say a net with 11 neurons and a Sinc neighborhood function for each neuron is given and the activity of the neurons is stored in a string of 11 numbers. In addition the following centres of the activity function (representing a classification structure learned before) are stored in the long time memory of the net

$$\Psi_0(\vec{q},\vec{E},k=0)=1,0,0.5,0,0.3,0,0,0.1,0,0,0$$

If an input pattern inducing a net activity function of the form

$$\Psi_1(\vec{q},\vec{E},k=1)=0,0,0,0,0,0,0,0.1,0,0,1$$

we will find that one component of the momentary active pattern Ψ_1 are also included in the store activity function Ψ_0. Therefore the short time memory consists of the stored activity function Ψ_0 and the convolution of Ψ_1 and the Sinc function, whereat the centre of the Sinc function is the conjointly neuron of the functions Ψ_0 and Ψ_1. Next Ψ_0 and Ψ_1 are superposed and the net structure (long time memory) is changed in that way that a modified input pattern

$$\Psi'_1(\vec{q},\vec{E},k=1)=0,0,0,0,0,0,0,1,0,0,0.5$$

is stored. As similarities of activated patterns lead to an active change of the long time memory of the net, a topological structuring on the grid will be submitted, considering common pattern elements.

5.2 Example of a Chunking Process

To initiate a chunking process let us assume the following activity functions $\Psi_n(\vec{q},\vec{E},k)$ represent the long time memory of a net

$$\Psi_1(\vec{q},\vec{E},k=0)=1,0,0,0,0,0,0,0,0,0,0$$
$$\Psi_2(\vec{q},\vec{E},k=3)=0,0,0,0,0,0,1,0,0,0,0$$
$$\Psi_3(\vec{q},\vec{E},k=7)=0,0,0,0,0,0,0,0,0,0,1$$

In this example new patterns have been stored at the junctures *0, 3, 7* and all of them represent totally different classification structures as no common components of them are present.

At juncture 11 the following pattern should activate the net

$$\Psi_5(\vec{q},\vec{E},k=11)=1,1,0,0,0,0,0,0,0,0,0$$

As this pattern matches components of the stored long time memory activity function $\Psi_1(\vec{q},\vec{E},k=0)$, $\Psi_1(\vec{q},\vec{E},k=0)$ will also be activated by $\Psi_5(\vec{q},\vec{E},k=11)$. So $\Psi_1(\vec{q},\vec{E},k=0)$ and $\Psi_5(\vec{q},\vec{E},k=11)$ will build the short time memory .

By superposition of $\Psi_1(\vec{q},\vec{E},k=0)$ and $\Psi_5(\vec{q},\vec{E},k=11)$ a new pattern

$$\Psi'_1(\vec{q},\vec{E},k=11)=1,0,1,0,0,0,0,0,0,0,0,0$$

will be created which contains features from $\Psi_1(\vec{q},\vec{E},k=0)$ and $\Psi_5(\vec{q},\vec{E},k=11)$. After storing this new pattern the long time memory of the net will have a structure as follows

$$\Psi_1(\vec{q},\vec{E},k=0)=1,0,0,0,0,0,0,0,0,0,0$$
$$\Psi'_1(\vec{q},\vec{E},k=11)=1,0,1,0,0,0,0,0,0,0,0,0$$
$$\Psi_2(\vec{q},\vec{E},k=3)=0,0,0,0,0,0,1,0,0,0,0$$
$$\Psi_3(\vec{q},\vec{E},k=7)=0,0,0,0,0,0,0,0,0,0,1$$
$$\Psi_5(\vec{q},\vec{E},k=11)=1,1,0,0,0,0,0,0,0,0,0$$

Now besides $\Psi_1(\vec{q},\vec{E},k=0)$ and $\Psi_5(\vec{q},\vec{E},k=11)$ a 'chunked activity pattern' $\Psi'_1(\vec{q},\vec{E},k=11)$ is part of the long time memory.

A weak point of this method is that step by step all neurons of the net which are able to store classification structures could be combined by chunking. In this case it may be possible that at the end of the learning procedure all stored activity functions are activated if only one feature of an input pattern activates only one part of a classification structure. To suppress such a global activation, razing interference effects should also be implemented. One way to simulate such a negative interference effect is to implement an activation ladder, which counts how often a pattern is activated and how often other patterns are not. If a pattern is not activated over a given period of time the chunk is disbanded and the associated activation function is extinguished out of the long time memory.

It is important to mention again that the effects of creating topological neighborhoods of similar patterns and chunking require a totally new step in the learning phase: The act of superposing the activated contents of the long time memory of the net. The chosen form of this superposition will decide on the form of the net structure.

6 Learning Rules

As we have shown, 'computing with activity patterns' means to select special learning rules for the configuration of the active centres, as now the classification result will be not represented by single output-neurons anymore but by the activity of several activity centres. In this context three questions have to be answered first:

- How many neurons should represent a classification-concept?
- Which value b - marking a relevant activity - should be chosen?
- Which neighborhood function should be chosen?

If these questions are answered, the learning phase launches as follows:

Step a)

- Present a learning pattern to be stored.
- Change the weight vectors $\vec{w}_{ij}$ until only n activity centres, representing a classification concept, have an activity $\geq$ b. (i is a neuron from the input layer, j a neuron on the grid.)
- Refer the neighborhood function to the n neurons.
- Calculate the neuron activity patterns $\psi_{l=1,\ldots,n}(\vec{q},\vec{E},k)$ and the global activity pattern $\Psi_0(\vec{q},\vec{E},k)$.
- Store these activity patterns in the long time memory structure.

Step b)

- Present a new input pattern.
- Look for similarities between the actual input pattern and the input patterns belonging to the already stored activity patterns.
- Activate the stored activity pattern(s) which show similarities.
- Refer the neighborhood function to the n neurons of the stored patterns.
- If applicable force the new activity patterns to place close to the similar already stored activity patterns.
- Inverse the shape of the absolute value of the neighborhood functions of the activity centres representing the stored patterns and 'freeze' the weight vectors of the neurons belonging to the activity pattern. In that way a bias-function is

defined which is strong enough to prevent that the stored activities are destroyed by relearning effects.

- Change the weight vectors $\vec{w}_{ij}$ until n new activity centres have an activity $\geq$ b factoring in similarities. If no significant variances between the two patterns to be combined exist, stop the learning procedure.
- Refer the neighborhood function to the n neurons of the new pattern.
- Analyze the topological neighborhood with the already stored activity patterns and if applicable chunk.
- Calculate the final activity patterns $\Psi_0(\vec{q},\vec{E},k)$, $\psi_{l=1,\ldots,n}(\vec{q},\vec{E},k)$.
- Store these activity patterns in the long time memory structure.

Step c)

- Proceed like step b) until no more patterns should be stored or all relevant neurons have been 'occupied'.

7 Conclusions

To use the all around activity distribution of a neural neuron cluster to identify classification concepts may lead to a substantially new method to deal with neural nets. First time it seems to be possible, to combine so far isolated stored classification concepts to high order classification structures by the method 'computing with activities'. As soon as the first classificators - dealing with activities only - are built and tested we will know, if the principle of the presents of output-neurons can be dropped or if these bottle necks are an restrict part of the dry neural nets.

References

1. Benesch, H.: Der Ursprung des Geistes, dtv, 1972
2. Aizenberg, I. N.: Neural Networks Based on Multi-valued and Universal Binary Neurons: Theory, Application to Image Processing and Recognition, Computational Intelligence, in Lecture Notes in Computer Science, 1625 pp,. 306-317, Springer, Dortmund, 1999
3. Reuter, M.: About the Quantisation of the Neural Nets, in: Lecture Notes in Computer Science, pp. 530-542, Springer, Dortmund, 1999

A GA-Based Algorithm with a Very Fast Rate of Convergence

Navid Seifipour [1] and Mohammad B. Menhaj [1,2]

[1] Dept. of Electrical Eng., Amir Kabir University of Tech., Tehran, Iran
menhaj@cic.aku.ac.ir

[2] Dept. of Electrical & Computer Eng., Oklahoma State University, Oklahoma, USA
menhaj@okstate.edu

Abstract. In this paper we introduce a new model-free optimization method, which is called as On-line Genetic-Based Algorithm (OGA). In order to compare the performance of the OGA with that of the Conventional Genetic Algorithm (CGA), a constraint optimization problem has been considered. The simulation results show that the OGA remarkably outperform the CGA..

1 Introduction

The Genetic Algorithm has been first formulated by Holland [1-2], in 1975. He introduced a novel optimization algorithm that is in-depth different from the two major classes of classical calculus-based and enumerative techniques [3-4].It has been shown that the GA functions as a totally different optimization algorithm than the simulated annealing [5], evolutionary strategies [6] and evolutionary programming [7]. It provides more flexibility and robustness. Because of its ease of implementation procedure, the GA has been successfully used as an optimization tool for a variety of optimization problems such as machine learning [8], delay estimation of sampled signals [9], robotic trajectory planning [10], fuzzy controllers [11], neural networks [12], linear transportable process [13].

The volume of literature published on the applications of the GA in complex, multivariable, and real-world problems is remarkable. However the conventional form of the GA implemented so far, as the authors know, could not be guaranteed to yield an on-line optimum solution, and in fact oftentimes performed rather poorly. Therefore, the applications of the GA for real-time problems are few in numbers. The modifications to the conventional GA proposed in this paper make the GA become (more) suitable for real-time problems.

Since population plays a crucial role in natural evolution, the Conventional Genetic Algorithms (CGA) inspired by natural biologics relies on population. This factor, however, produces some problems, whose optimal cases have not been specified yet. Some of these problems are as follows:

B. Reusch (Ed.): Fuzzy Days 2001, LNCS 2206, pp. 185–193, 2001.

1. Optimum Size of population,
2. Efficient selection method of individuals in population,
3. Efficient crossover operator,
4. Quality of using individual's age in the genetic process.

There are many researchers who have employed different ways to solve the above-mentioned problems by proposing some corrections on the procedure of the simple GA. The interested reader may refer to [14] for optimization of control Parameters, to [15-17] for population size, to [18] for fitness function, to [19] for aging of individuals, to [20-21] for mutation operator, to [22-24] for crossover operator, to [25-26] for migration operator, and to [27-28] for multi-population of the GA.

The paper introduces a new model free based optimization method, which is well suited for real time signal processing task. This is an On-line Genetic-Based Algorithm (OGA) originated from the GA. The remainder of the paper is organized as follows. Section 2 presents a brief review of the CGA. Section 3 fully describes the OGA. To show the effectiveness of the proposed genetic algorithm, in section 4, through an illustrative example the performance of the OGA is investigated and its capabilities is listed. Finally, section 5 concludes the paper.

2 A brief review of CGA

Mimicking the process of biological evolution, GA represents a type of structured random search. In solving a given optimization task, the GA starts with a collection of solutions (i.e. parameter estimates) called by chromosomes. Each individual (chromosome) is evaluated for its fitness. In each iteration of the GA, the most fit chromosomes (parents) are allowed to mate and bear offspring (produce new individuals). These individuals (children) or new parameter estimates provide the basis for the next generation. The conventional GA may be completely described by the following steps [29]:

1- Select a coding scheme to represent adjustable parameters.
2- Initialize and evaluate a population of chromosomes.
3- Repeat the following steps till stopping criteria are met.

 3-1- Reproduction
 3-2- Crossover
 3-3- Mutation
 3-4- Population evaluation

To distinguish our proposed GA-based method that will be fully discussed in section 3 with all the other GA-based techniques developed so far, we denote all the existing GA methods as CGA.

3 The OGA

The CGA uses a crossover operator for recombination of two chromosomes. For this reason two chromosomes are randomly selected from population, (note that this selection is influenced by fitness of chromosomes in the population [30]). This procedure is similar to most natural recombination.

3.1 Main Motivation

Our proposed genetic algorithm (the OGA) is inspired by recombination accomplished by some insects such as bees. To make it more clear we proceed as follows. Consider a hypothetical CGA in which the recombination is done in a way that one chromosome is recombined with the best chromosomes that exists in the present population. In other words, all selected chromosomes are recombined with the best chromosome of the current population called as the queen chromosome. The recombination procedure is illustrated in Figure 1 for a better understanding.

With a deep look at this figure, it becomes clear that the produced chromosomes are indeed the queen with some changes. In other words, the new chromosomes may represent new queens who inherit some parent queen characteristics. Therefore, instead of employing the procedure given in Figure 1, we may use the following procedure for recombination as shown in Figure 2; this represents the core of OGA formulated below.

As observed in Figure 1, to generate a new chromosome some portions of the queen are randomly mutated; this results in new queen candidates as illustrated in Figure 2. Each queen candidate becomes a queen and replaces the original queen if it has a better performance index than that of the previous queen. But, if the performance of the queen candidate gets worse than the current queen's performance, the new chromosome (queen candidate) is disregarded and a part of the original queen is again selected and mutated randomly. Below the procedure of OGA is listed.

1- Select a coding scheme to represent adjustable parameters
2- Take randomly a chromosome as a queen and evaluate its performance
3- Repeat the following steps till stop criteria
    3-1- Mutation
    3-2- Evaluation

By employing this procedure, we can obviously convert the off-line CGA into an on-line algorithm with less computation time, which is well suited for real-time operation. The advantage of the proposed algorithm is that it does not require 1) any determination of the kind of selection operator,2) crossover operator, 3) functions which convert minimization problems to maximization ones, 4) selection of a good population size, and 5) a large memory to store the population. The authors are currently investigating the convergence analysis of the OGA along with its on-line property, which makes it more suitable for real-time control and filtering applications.

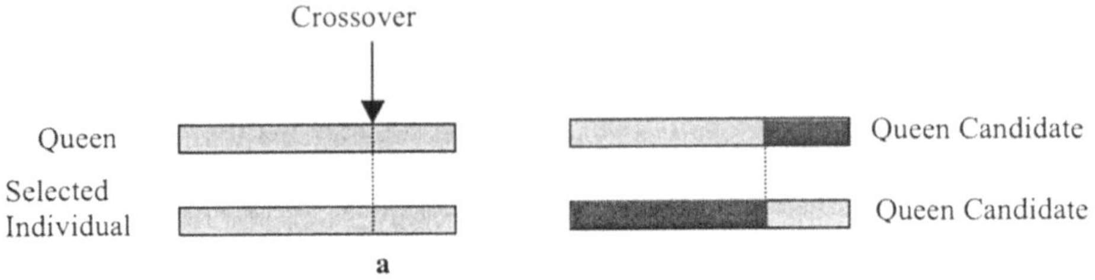

(a) Crossover point selected at the point **a**

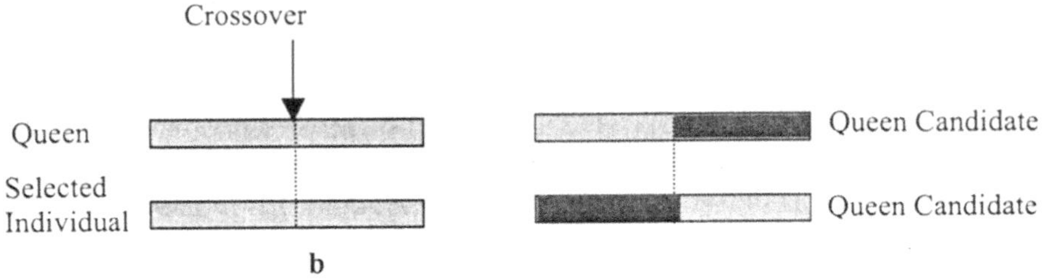

(b) Crossover point selected at the point **b**

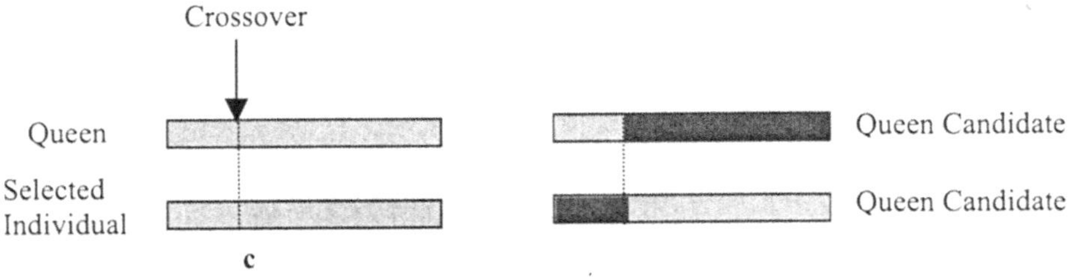

© Crossover point occurs at the point **c**

Fig. 1. Illustration of the OGA

Fig. 2. New Queen Generation

4 An Illustrative Example

Consider the cantilevered beam shown in Fig. 3 [31]. This beam is to be designed for minimum material volume. The set of adjustable design variables includes the width *b* and the height *h* at each of *N* segments, with *N*=5. The objective is to design the beam subject to limits on stress, which are calculated at the left end of each segment, deflection under the load, and the geometric requirement that the height of any segment does not exceed twenty times the width.

The following recursion formula calculate the deflection y_i at the right end of segment *i*:

$$y_o = y'_o = 0, \quad (1)$$

$$y'_i = \frac{P \cdot l_i}{E \cdot I_i} \cdot \left(L + \frac{l_i}{2} - \sum_{j=1}^{i} l_j \right) + y'_{i-1}, \quad (2)$$

$$y_i = \frac{P \cdot l_i^2}{2 \cdot E \cdot I_i} \cdot \left(L + \frac{2 \cdot l_i}{3} - \sum_{j=1}^{i} l_j \right) + y'_{i-1} \cdot l_i + y_{i-1}. \quad (3)$$

In the above, l_i is the length of segment *i*, the deflection y is defined as positive downward, y' represents the slope, *E* denotes Young's module which is the same for all segments, the moment of inertia for segment *i*, I_i, and bending moment at the left end of segment *i*, M_i, are defined by

$$I_i = \frac{b_i . h_i^3}{12} \quad (4)$$

$$M_i = P\left(L + l_i - \sum_{j=1}^{i} l_j \right). \quad (5)$$

And finally the corresponding maximum bending stress becomes

$$\sigma_i = \frac{M_i \cdot h_i}{2 \cdot I_i} \quad (6)$$

Now, we may state the design problem as:

Minimize:

$$V = \sum_{i=1}^{N} b_i \cdot h_i \cdot l_i \tag{7}$$

Subject to:

$$\frac{\sigma_i}{\overline{\sigma}} - 1 \leq 0 \qquad i = 1, \cdots, N, \tag{8}$$

$$\frac{y_N}{\overline{y}} - 1 \geq 0, \tag{9}$$

$$h_i - 20 \cdot b_i \leq 0 \qquad i = 1, \cdots, N, \tag{10}$$

$$b_i \geq 0.1 \qquad i = 1, \cdots, N, \tag{11}$$

$$h_i \geq 0.5 \qquad i = 1, \cdots, N, \tag{12}$$

In the above, $\overline{\sigma}$ and $\overline{y}$ are the allowable bending stress and displacement, respectively.

This is an optimization, indeed a constraint, problem in $n = 2N$ parameters with $N + 1$ nonlinear constraints defined by Eqs. (8) and (9), N linear constraints defined by Eq. (10), and $2N$ side constraints, which are treated here as general inequality constraints, on the design variables defined by Eqs. (11) and (12). It is further assumed that lower bounds of 0.1 are imposed explicitly on b_i and h_i, $i = 1, \ldots, N$ within the optimization procedure to guarantee that the design remains physically meaningful.

The CGA and the OGA solved this problem with five segments (10 design variables). Knowing that because of the above several constraints, the ordinary CGA could poorly solve the problem, we propose a modified CGA in which any individual is not acceptable and only those that rest in the constraints (8)-(12) are valid.

The proposed modified CGA is accomplished in such a way that at first all individuals of the initial population are randomly selected from the allowable solution space. Then, like the CGA discussed in section 2, all permissible individuals are evaluated and the process continues until the crossover operator is done. When the crossover operator combines two permissible individuals, they may produce an offspring outside the allowable space. In this case the offspring is omitted and substituted by a random permissible individual from the allowable space. Then, by this injection of random individuals in new population, it is not necessary to use the mutation operator. If all produced offspring by crossover recombination are valid, the mutation operator operates on population as explained in section 2.

By this procedure we solved the Cantilevered beam problem run on Pentium 133 MHz. The results are listed in Table 1. The (sub)-optimum performance of V*=47811 has been obtained after 3000 revisions on population (3000 epochs) for the optimum design variables as shown in Table 1.

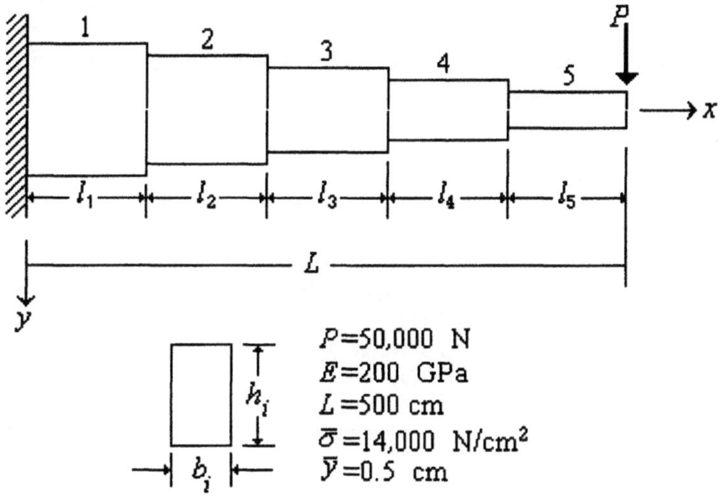

Fig. 3. The Cantilevered Beam

Table 1. Comparision of the CAG With the OGA

Varia-bles	b_1	b_2	b_3	b_4	b_5	h_1	h_2	h_3	h_4	h_5	*Computaion Time (Hours)*
CGA	1.125	3.0	1.125	2.250	5.125	59.000	53.500	37.125	44.125	21.500	~14
OGA	1.75	2.75	1.00	2.00	1.625	31.000	18.250	41.875	17.625	36.75	~1

The cantilevered beam problem was then solved by OGA. The final results of OGA are also listed in Table 1. The (sub)-optimum performance of V^*=24128 is obtained after 3000 revisions on the queen. By comparing the two genetic algorithms, it is obviously seen that the OGA converges to a better solution with fewer amounts of computation time and memory locations.

5 Conclusions

In this paper, a new technique OGA was proposed for function optimization. Its performance has been fully discussed and compared with those of conventional genetic algorithms through on illustrative example. It has been observed that the OGA converged to the global point with less amount of storage memory and time of computations. In contrast, the CGA fails to reach the optimum point. The OGA convergence analysis and properties are currently under investigation and will be reported soon.

References

1. Holland, J. H.: Genetic Algorithms and the Optimal Allocations of Trails, SIAM Journal of Computing, 2(2), (1973) 88-105,.
2. Holland, J. H.: Adaptation In Natural And Artificial Systems, Ann Arbor, MI: The University of Michigan Press, (1975)
3. Scales, E.: Introduction to Nonlinear Optimization, Springer-verlag, N. Y. (1985)
4. Bellman, R. E.: Dynamic Programming, Princetion Univ. Press, princetion, N. J., USA (1957)
5. Kirkpatrick, S., Gellat Jo, C.D., and Vecchi, M. P.: Optimization by Simulated Annealing, Science, Vol. 220, No. 4598, (1983) 671-680
6. Back, T., Hoffmeister, F., and Schwefel, H. P.: A Survey of Evolution Strategies, in R. K. Belew, and L. B. Booker, Eds. Proc. 4th International Conference on Genetic Algorithms, San Mateo, CA: Morgan Kaufmann, (1991) 2-9
7. Fogel, L. J., Owens, A. J. and Walsh, M. J.: Artificial Intelligence Through Simulated Evolution, New York: John Wiley (1966)
8. Fogel, D. B.: System Identification Through Simulated Evolution: A Machine Learning Approach to Modeling, Needham Heights, MA: Ginn Press (1991)
9. Y. T. Chan, J.M. Riley and J.B. Plant, "Modeling of Time Delay and it's Application to Estimation of Nonstationary Delays," *IEEE Trans. Acoust. Speech, Signal Processing,* Vol.: ASSP-29, pp. 577-581, June 1981.
10. Davidor, Y.: Genetic Algorithms & Robotics, A Heuristic Strategy for Optimization, World Scientific, Singapore (1991)
11. Karr, C.L.: Genetic Algorithms for Fuzzy Controllers, A.I. Expert, Vol. 6, No. 2, (1991). 26-33
12. Montana, D. J. Davis, L.: Training Feed-forward Neural Networks Using Genetic Algorithms, Proc. of Int. Joint Conf. on Artificial Intelligence (Detroid), (1989) 762-767
13. Vignaun, G. A., Michalewicz, and Z: A Genetic Algorithms For The Linear Transportation Problem," IEEE Trans. on Systems Man and Cybernetics, Vol. 21, (1991). 445-452
14. Grefenstette, J. J.: Optimization of Control Parameters for Genetic Algorithms, IEEE Trans. on Sys., Man, and Cyb., Vol. Smc-16, No. 1, (1986)
15. Odetayo, M. O.: Optimal Population Size for Genetic Algorithms: An Investigation, IEEE, Colloquium on Genetic Algorithms for Control Systems Engineering, (1993) 2/1-2/4
16. Alander, J. T.: On Optimal Population Size of Genetic Algorithms, CompEuro '92. Computer Systems and Software Engineering, Proc. (1992) 65-70
17. Arabas, J., Michalewicz, Z. and Mulawka, J.: GAVaPS - a Genetic Algorithms with Varying Population Size, Evolutionary Computation, IEEE World Congress on Computational Intelligence, Proceedings of the IEEE Conf., Vol. 1, (1994) 73-78

18. Lima, J., Gracias, A. N., Pereira, H. and Rosa, A.: Fitness Function Design for Genetic Algorithms in Cost Evaluation Based Problems, Proc. of IEEE, International Conf. on Evolutionary Computation, (1996) 207-212
19. Ghosh, A., Tsutsui, S. and Tanaka, H.: Individual Aging in Genetic Algorithms, Conf. on Intelligence Information System, Australian and New Zealand, (1996) 276-279
20. Hesser, J., Manner, R.: Towards an Optimal Mutation Probability for Genetic Algorithms, in Proc. First Int. Workshop on Parallel Problem Solving from Nature, Dortmuntd, (1990) paper A-XII.
21. Oi, X., Palmieri, F.: Theoretical Analysis of Evolutionary Algorithms With an Infinite Population Size in Continuous Space Part I: Basic Properties of Selection and Mutation, IEEE Trans. on Neural Networks, Vol. 5, No. 1, (1994)
22. Oi, X., Palmieri, F.: Theoretical Analysis of Evolutionary Algorithms With an Infinite Population Size in Continuous Space Part II: Analysis of the Diversification Role of Crossover, IEEE Trans. on Neural Networks, vol. 5, No. 1, (1994)
23. Shang, Y. Li, G. J.: New Crossover in Genetic Algorithms, Proc. of IEEE, Third International Conf. on Tools for Artificial Intelligence, TAI '91, (1991) 150-153
24. Coli, M., Gennuso, and Palazzari, G. P.: A New Crossover Operator for Genetic Algorithms, Proc. of IEEE, International Conf. on Evolutionary Computation, (1996) 201-206
25. Potts, J. C., Giddens, T. D. and Yadav, S. B.: The Development and Evaluation of an Improved Genetic Algorithms Based on Migration and Artificial Selection, IEEE Trans. on Syst., Man, and Cyber., vol. 24, Nno. 1, (1994)
26. Moed, M. C. Stewart, C. V. and Kelly, R. B.: Reducing The Search Time of A Steady State Genetic Algorithms Using the Immigration Operator, Proc. of IEEE, Third International Conf. on Tools for Artificial Intelligence, TAI '91, (1991) 500-501
27. Tsutsui, S. Fujimoto, Y.: Phenotypic Forking Genetic Algorithms (p-fGA), Proc. of IEEE, International Conf. on Evolutionary Computation, Vol. 2, (1995) 566-572
28. Tsutsui, S., Fujimoto, Y. and Hayashi, I.: Extended Forking Genetic Algorithms for Order Representation (o-fGA), Proc. of the First IEEE, Conf. on IEEE World Congress on Computational Intelligence, vol. 2, (1994) 566-572
29. Davis, L.: Handbook of Genetic Algorithms, Van Nostrand Reinhold (1991)
30. Koza J.: Genetic Programming: On the Programming of Computers by Means of Natural Selection, Cambridge, MA: MIT Press (1992)
31. Miura, H, Vanderplaats, G. N., Kodiyalam S.: Experiences in Large Scale Structural Design Optimization, Applications of Supper Computer in Engineering: Fluid Flow and Stress Analysis Applications, Elsevier, Amsterdam (1989)

Representing and Learning Conditional Information in Possibility Theory

Gabriele Kern-Isberner

FernUniversität Hagen, Dept. of Computer Science,
P.O. Box 940, D-58084 Hagen, Germany
gabriele.kern-isberner@fernuni-hagen.de

Abstract. Conditionals (*if-then-rules*, *default rules*) are most important objects in knowledge representation and commonsense reasoning. Due to their non-classical nature, however, they are not easily dealt with. In this paper, we present a new approach to represent conditionals inductively in a possibilistic framework. The algebraic theory which underlies this approach proves to guarantee a most appropriate handling of conditional information. Moreover, this novel conditional theory is a very fundamental one, in that it can also be applied to guide possibilistic belief revision and gives rise to a new methodology to learn rules from data.

1 Introduction

Conditionals (*if-then-rules*, *default rules*) are most important objects in knowledge representation and commonsense reasoning. They are closely related to epistemic notions such as plausibility, possibility, probability and the like, in that a conditional "*If A then B*" is generally considered to be accepted if its confirmation, $A \wedge B$, is judged to be more plausible, possible or probable, than its refutation, $A \wedge \neg B$. Conditionals can even be quantified by degrees of belief or by (statistical or subjective) probabilities, expressing to what extent they are taken to be established.

Due to their non-classical nature, however, conditionals are not easily dealt with. On the one hand, to validate conditionals, richer epistemic structures than plain propositional interpretations are needed, as e.g. *probability distributions* (see, e.g., [Par94]) and *possibility distributions* as fuzzy representations of epistemic attitudes (cf. [DP91c,DLP94,BDP99]). On the other hand, inference from conditional knowledge bases is much more difficult than in classically logical environments, since it is not a strictly deductive one. Rather it is based on nonmonotonic reasoning [DP91a,LM92,DP91b,BDP97] and involves belief revision theory [Rot91,FH94,KS91]. A rigid structure seems to be lacking which guides reasoning with conditionals and, the other way around, helps us to identify relevant conditional knowledge (*rules*) from statistical databases, making *knowledge discovery* possible.

The aim of this paper is to present an algebraic theory of conditional structures on possible worlds and to show how this theory can be applied in possibility

B. Reusch (Ed.): Fuzzy Days 2001, LNCS 2206, pp. 194–217, 2001.

theory to obtain adequate inductive representations of conditional knowledge bases by possibility distributions. These possibility distributions can then be used for inferences. The basic idea is to develop a new methodology of making interactions of conditionals transparent, being aware that any attempt to force conditionals into a classical framework will cut down their unique dynamic expressiveness. Rather, conditionals are considered here as agents having effects on worlds so as to establish conditional beliefs. With the notion of *conditional structures*, we can then define what it means that a possibility distribution is *indifferent with respect to a set of conditionals*, thereby respecting the structures imposed by the conditionals under consideration. Note that we will be dealing with (finite) *sets of conditionals*, not with single conditionals, at a time, taking the joint effects of several conditionals into account. Moreover, the same techniques can be applied to handle revisions of possibility distributions by incorporating new conditional beliefs. Last not least, the methodology provides an approach to discovering conditional knowledge in statistical databases (rule extraction) by considering the induced possibility distribution (see, e.g., [GK97]).

As we already said, we will largely avoid classically logical techniques. That does not mean, however, that this new approach to conditionals lacks a solid foundation. Quite to the contrary – the underlying conditional theory is a group theoretical one, e.g., conditional structures are defined via homomorphisms of groups. This embedding into a group theoretical framework, as awkward and strange as it appears at first sight, is the key to handle complex conditional interactions adequately, as it is necessary for inductive representations of conditionals, by representing these interactions as products while allowing cancellations. In search of appropriate structures, homomorphisms are excellent tools, since they are designated to transfer structures. In particular, the kernels of these homomorphisms (as usual) play a prominent role in identifying structures. So, knowledge discovery comes down to elaborating kernels and factorizing groups, inducing operations on conditionals so as to obtain concise ("short") rules compatible with the numerical structure.

This theory is not only devised for possibility theory but actually provides a powerful approach to conditionals which may also be applied in a probabilistic framework or when using *ordinal conditional functions* [Spo88], which are quite similar to possibility distributions, and it even yields essential insights in purely qualitative environments. It is presented in full in [KI99,KI01a], some results for ordinal conditional functions were published in [KI00a], and probabilistic rule extraction is sketched in [KI00b]. Now this paper will present new results for a possibilistic framework.

The paper is organized as follows: In the next section, we will explain the conditional possibilistic framework which we base our considerations upon. Section 3 introduces the central notion of *conditional structures* which will be used in Section 4 to define *conditional indifference.* In Section 5, we will show how this conditional theory may be applied to represent conditional knowledge by possibility distributions in a most appropriate way. Section 6 then brings a precise formalization of the crucial *principle of conditional preservation* for possibilistic belief revision. This principle is also based upon conditional indifference, emphasizing the close connection between inductive knowledge representation and belief revision. In Section 7, the group theoretical framework introduced in Section 3 turns out to be also particularly useful when searching for relevant conditional relationships in data.

2 Conditionals in Possibility Theory

In the sequel, conditionals "*If A then B*" are formally represented as $(B|A)$, expressing a relationship between two (propositional) formulas A, the *antecedent* or *premise*, and B, the *consequent*, of a (finite) propositional language $\mathcal{L}$. *Single-elementary conditionals* are conditionals whose antecedents are conjunctions of literals, and whose consequents consist of one single literal.

Let $(\mathcal{L} \mid \mathcal{L})$ denote the set of all conditionals $(B|A)$ with $A, B \in \mathcal{L}$. To simplify notation, we will write $\overline{A}$ instead of $\neg A$, and AB instead of $A \wedge B$, for formulas $A, B \in \mathcal{L}$. Let Ω be the complete set of interpretations of $\mathcal{L}$, where each $\omega \in \Omega$ is taken to be a *possible world* for $\mathcal{L}$. We will only consider finite sets $\mathcal{R} \subseteq (\mathcal{L} \mid \mathcal{L})$ of conditionals.

A *possibility distribution* is a mapping $\pi : \Omega \rightarrow [0,1]$, where each $\pi(\omega)$ reflects the degree of possibility an intelligent agent assigns to ω. π induces a *possibility measure*

$$\Pi : \mathcal{L} \rightarrow [0,1], \Pi(A) = \max\{\pi(\omega) \mid \omega \models A\},$$

and a dual *necessity measure*

$$N = N_\pi : \mathcal{L} \rightarrow [0,1], N(A) = 1 - \Pi(\overline{A})$$

(cf., e.g., [DLP94]). We will not distinguish between a possibility distribution, π, on Ω, and its induced possibility measure, Π, on $\mathcal{L}$, and will use the same symbol π for both. Furthermore, a possibility degree can also be assigned to a conditional $(B|A)$ by setting

$$\pi(B|A) = \frac{\pi(AB)}{\pi(A)},$$

in full analogy to Bayesian conditioning in probability theory. The corresponding possibility distribution is

$$\pi(\omega|A) = \begin{cases} \dfrac{\pi(\omega)}{\pi(A)}, & \text{if } \omega \models A \\ 0, & \text{otherwise} \end{cases}$$

Note that we also make use of the product operation in $[0,1]$. That means, that our approach is not only based upon comparing numbers, but also takes relations between numbers into account. These numerical relationships encode important information about the (relative) strength of conditionals which proves to be particularly crucial for representation and revision tasks. This amounts to carrying over Spohn's argumentation in [Spo88] to a possibilistic framework (see also [KI99] and [BDP97]).

A conditional $(B|A)$ *is accepted in* π, $\pi \models (B|A)$, iff $\pi(AB) > \pi(A\overline{B})$ (which is equivalent to $\pi(\overline{B}|A) < 1$ and $N_\pi(B|A) = 1 - \pi(\overline{B}|A) > 0$) (cf. [DP94]). So, in accordance with intuition, a conditional $(B|A)$ is accepted in the epistemic state modeled by a possibility distribution, if its confirmation (AB) is considered to be more possible (or plausible) than its refutation $(A\overline{B})$.

We assume all possibility distributions to be normalized, i.e. $\pi(\top) = 1$, and, only for conciseness of presentation, to be positive, i.e. $\pi(\omega) > 0$ for all $\omega \in \Omega$.

3 Conditional Structures

A world $\omega \in \Omega$ is said to *verify* a conditional $(B|A)$, if $\omega \models AB$; it *falsifies* $(B|A)$, if $\omega \models A\overline{B}$; if $\omega \models \overline{A}$, then $(B|A)$ is not applicable to ω. By observing this behavior of worlds with respect to it, each conditional $(B|A)$ can be considered as a generalized (namely three-valued) indicator function on worlds:

$$(B|A)(\omega) = \begin{cases} 1 & : \quad \omega \models AB \\ 0 & : \quad \omega \models A\overline{B} \\ u & : \quad \omega \models \overline{A} \end{cases} \tag{1}$$

where u stands for *unknown* or *indeterminate* [Cal91,DeF74]. Intuitively, representing or incorporating a conditional as a plausible conclusion in an epistemic state means to make – at least some – worlds verifying the conditional more plausible than the worlds falsifying it. In this sense, conditionals to be learned have effects on possible worlds, (more exactly, on their degrees of plausibility or possibility), shifting them appropriately to establish the intended relationship. (1) then provides a classification of worlds for achieving this: On verifying worlds $\omega \models AB$, i.e. $(B|A)(\omega) = 1$, $(B|A)$ possibly has a positive effect, while on falsifying worlds $\omega \models A\overline{B}$, $(B|A)$ possibly has a negative effect; the effects on worlds ω with $(B|A)(\omega) = u$ is unclear. Which worlds will actually be shifted depends on the chosen inductive representation procedure – for the conditional $(B|A)$, all worlds in either of the partitioning sets $AB, A\overline{B}$ and $\overline{A}$ are indistinguishable.

When we consider (finite) sets of conditionals $\mathcal{R} = \{(B_1|A_1), \ldots, (B_n|A_n)\} \subseteq (\mathcal{L} \mid \mathcal{L})$, we have to modify this representation appropriately to identify the effect of each conditional in $\mathcal{R}$ on worlds in Ω. To this end, we replace the numbers 0 and 1 by abstract symbols, $\mathbf{a}_i^+, \mathbf{a}_i^-$, that we associate to each conditional $(B_i|A_i)$ in $\mathcal{R}$. Moreover, we will make use of a group structure to represent the joint impact of conditionals on worlds.

So let

$$\mathcal{F}_\mathcal{R} = \langle \mathbf{a}_1^+, \mathbf{a}_1^-, \ldots, \mathbf{a}_n^+, \mathbf{a}_n^- \rangle$$

be the free abelian group with generators $\mathbf{a}_1^+, \mathbf{a}_1^-, \ldots, \mathbf{a}_n^+, \mathbf{a}_n^-$, i.e. $\mathcal{F}_\mathcal{R}$ consists of all elements of the form $(\mathbf{a}_1^+)^{r_1}(\mathbf{a}_1^-)^{s_1} \ldots (\mathbf{a}_n^+)^{r_n}(\mathbf{a}_n^-)^{s_n}$ with integers $r_i, s_i \in \mathbb{Z}$ (the ring of integers). Each element of $\mathcal{F}_\mathcal{R}$ can be identified by its exponents, so that $\mathcal{F}_\mathcal{R}$ is isomorphic to $\mathbb{Z}^{2n}$ (cf. [LS77]). For each $i, 1 \leqslant i \leqslant n$, we further define a function $\sigma_i = \sigma_{(B_i|A_i)} : \Omega \to \mathcal{F}_\mathcal{R}$ by setting

$$\sigma_i(\omega) = \sigma_{(B_i|A_i)}(\omega) := \begin{cases} \mathbf{a}_i^+ & \text{if} \quad (B_i|A_i)(\omega) = 1 \\ \mathbf{a}_i^- & \text{if} \quad (B_i|A_i)(\omega) = 0 \\ 1 & \text{if} \quad (B_i|A_i)(\omega) = u \end{cases}$$

$\sigma_i(\omega)$ represents the manner in which the conditional $(B_i|A_i)$ applies to the possible world ω. The neutral element 1 of $\mathcal{F}_\mathcal{R}$ corresponds to the non-applicability of $(B_i|A_i)$ in case that the antecedent A_i is not satisfied. The function $\sigma_\mathcal{R} : \Omega \to \mathcal{F}_\mathcal{R}$,

$$\sigma_\mathcal{R}(\omega) = \prod_{1 \leqslant i \leqslant n} \sigma_i(\omega) = \prod_{\substack{1 \leqslant i \leqslant n \\ \omega \models A_i B_i}} \mathbf{a}_i^+ \prod_{\substack{1 \leqslant i \leqslant n \\ \omega \models A_i \overline{B}_i}} \mathbf{a}_i^- \tag{2}$$

describes the all-over effect of $\mathcal{R}$ on ω. $\sigma_\mathcal{R}(\omega)$ is called the *conditional structure of ω with respect to $\mathcal{R}$*. Due to $\mathcal{F}_\mathcal{R}$ being commutative, we will be able to take the simultaneous impact of all conditionals in $\mathcal{R}$ into account, without assuming any order of applying the conditionals.

$\sigma_\mathcal{R}$ is defined on the set Ω, assigning group elements of $\mathcal{F}_\mathcal{R}$ to the possible worlds. Now, to comply with the group theoretical structure in $\mathcal{F}_\mathcal{R}$, we also take the worlds $\omega \in \Omega$ as formal generators of a free abelian group

$$\widehat{\Omega} := \langle \omega \mid \omega \in \Omega \rangle$$

$\widehat{\Omega}$ consists of all products $\widehat{\omega} = \omega_1{}^{r_1} \ldots \omega_m{}^{r_m}$, with $\omega_1, \ldots, \omega_m \in \Omega$, and integers $r_1, \ldots r_m$. Now $\sigma_\mathcal{R}$ may be extended to $\widehat{\Omega}$ in a straightforward manner by setting

$$\sigma_\mathcal{R}(\omega_1{}^{r_1} \ldots \omega_m{}^{r_m}) = \sigma_\mathcal{R}(\omega_1)^{r_1} \ldots \sigma_\mathcal{R}(\omega_m)^{r_m},$$

yielding a *homomorphism of groups* $\sigma_\mathcal{R} : \widehat{\Omega} \to \mathcal{F}_\mathcal{R}$.

Having the same conditional structure defines an equivalence relation $\equiv_\mathcal{R}$ on $\widehat{\Omega}$:

$$\widehat{\omega}_1 \equiv_\mathcal{R} \widehat{\omega}_2 \text{ iff } \sigma_\mathcal{R}(\widehat{\omega}_1) = \sigma_\mathcal{R}(\widehat{\omega}_2) \tag{3}$$

Those elements of $\widehat{\Omega}$ that are balanced with respect to the effects of conditionals in $\mathcal{R}$ are contained in the *kernel of* $\sigma_\mathcal{R}$,

$$\mathit{ker}\, \sigma_\mathcal{R} = \{\widehat{\omega} \in \widehat{\Omega} \mid \sigma_\mathcal{R}(\widehat{\omega}) = 1\}$$

To take the normalizing constraint $\pi(\top) = 1$ into acccount, we also focus on equivalence with respect to $\sigma_{(\top|\top)}$. Since $\sigma_{(\top|\top)}$ simply counts the generators occurring in $\widehat{\omega}$, two elements $\widehat{\omega}_1 = \omega_1^{r_1} \ldots \omega_m^{r_m}$, $\widehat{\omega}_2 = \nu_1^{s_1} \ldots \nu_p^{s_p} \in \widehat{\Omega}$ are $\sigma_{(\top|\top)}$-equivalent, $\widehat{\omega}_1 \equiv_\top \widehat{\omega}_2$, iff $\sum_{1 \leqslant j \leqslant m} r_j = \sum_{1 \leqslant k \leqslant p} s_k$. This means, $\widehat{\omega}_1 \equiv_\top \widehat{\omega}_2$ iff they both are a (cancelled) product of the same number of generators, each generator being counted with its corresponding exponent. In particular, $\omega_1 \equiv_\top \omega_2$ for any possible worlds $\omega_1, \omega_2 \in \Omega$. Therefore, $\mathit{ker}\, \sigma_{(\top|\top)} = \{\widehat{\omega} = \omega_1{}^{r_1} \cdot \ldots \cdot \omega_m{}^{r_m} \in \widehat{\Omega} \mid \sum_{j=1}^m r_j = 0\}$. Set $\mathit{ker}_0\, \sigma_\mathcal{R} := \mathit{ker}\, \sigma_\mathcal{R} \cap \mathit{ker}\, \sigma_{(\top|\top)}$, that is,

$$\mathit{ker}_0\, \sigma_\mathcal{R} = \{\widehat{\omega} = \omega_1{}^{r_1} \cdot \ldots \cdot \omega_m{}^{r_m} \in \widehat{\Omega} \mid \sum_{j=1}^m r_j = 0 \text{ and } \sigma_\mathcal{R}(\widehat{\omega}) = 1\}$$

We will illustrate the benefits of these technical definitions in the next example.

Example 1. Consider the set $\mathcal{R} = \{(d|a),(d|b),(d|c)\}$ of conditionals using the atoms a,b,c,d. Let $\mathbf{a}_1^{\pm}, \mathbf{a}_2^{\pm}, \mathbf{a}_3^{\pm}$ be the group generators associated with $(d|a)$, $(d|b), (d|c)$, respectively, so $\mathcal{F}_{\mathcal{R}} = \langle \mathbf{a}_1^+, \mathbf{a}_1^-, \mathbf{a}_2^+, \mathbf{a}_2^-, \mathbf{a}_3^+, \mathbf{a}_3^- \rangle$. The possible world $\omega = abcd$, for instance, verifies all three conditionals $(d|a),(d|b),(d|c)$, so its conditional structure is $\sigma_{\mathcal{R}}(abcd) = \mathbf{a}_1^+\mathbf{a}_2^+\mathbf{a}_3^+$. For $\omega = ab\overline{c}d$, we calculate $\sigma_{\mathcal{R}}(ab\overline{c}d) = \mathbf{a}_1^+\mathbf{a}_2^+$, since $ab\overline{c}d$ verifies the first two conditionals, and the third conditional is not applicable to it.

Moreover, by conditional structures, it is possible to compare worlds, or sets of worlds, as to their behavior with respect to the conditionals in $\mathcal{R}$. For instance, we have

$$\begin{aligned}\sigma_{\mathcal{R}}(ab\overline{c}d)\sigma_{\mathcal{R}}(a\overline{b}cd)\sigma_{\mathcal{R}}(\overline{a}bcd) &= (\mathbf{a}_1^+\mathbf{a}_2^+)(\mathbf{a}_1^+\mathbf{a}_3^+)(\mathbf{a}_2^+\mathbf{a}_3^+)\\ &= (\mathbf{a}_1^+)^2(\mathbf{a}_2^+)^2(\mathbf{a}_3^+)^2\\ &= (\mathbf{a}_1^+\mathbf{a}_2^+\mathbf{a}_3^+)^2\\ &= \sigma_{\mathcal{R}}(abcd)^2.\end{aligned}$$

That is to say, that the elements of the set $\{ab\overline{c}d, a\overline{b}cd, \overline{a}bcd\}$ show collectively the same conditional behavior as two copies of $abcd$. In the group theoretical framework using $\widehat{\Omega}$, this can be expressed as

$$\sigma_{\mathcal{R}}\left(ab\overline{c}d \cdot a\overline{b}cd \cdot \overline{a}bcd\right) = \sigma_{\mathcal{R}}\left(abcd^2\right)$$

or, respectively, as

$$\sigma_{\mathcal{R}}\left(\frac{ab\overline{c}d \cdot a\overline{b}cd \cdot \overline{a}bcd}{abcd^2}\right) = 1.$$

So $\dfrac{ab\overline{c}d \cdot a\overline{b}cd \cdot \overline{a}bcd}{abcd^2} \in ker\,\sigma_{\mathcal{R}}$, but $\dfrac{ab\overline{c}d \cdot a\overline{b}cd \cdot \overline{a}bcd}{abcd^2} \notin ker_0\,\sigma_{\mathcal{R}}$, since $ab\overline{c}d \cdot a\overline{b}cd \cdot \overline{a}bcd \not\equiv_{\mathsf{T}} abcd^2$ (three generators versus two generators).

4 Conditional Indifference

Conditional structures, on the one hand, reflect clearly the (isolated) impact of each conditional in a set $\mathcal{R}$, and, on the other hand, make the interactions between different conditionals in $\mathcal{R}$ transparent and computable. In this section, we will link the algebraic notion of conditional structures to the numerical properties of possibility distributions $\pi : \Omega \to [0,1]$. To this end, we now focus on the behavior of π with respect to the elements in $\widehat{\Omega}$.

To fit into the group theoretical framework, we extend each such distribution to a homomorphism, $\pi : \widehat{\Omega} \to (\mathbb{R}^+,\cdot)$[1], by setting

$$\pi(\omega_1{}^{r_1} \cdot \ldots \cdot \omega_m{}^{r_m}) = \pi(\omega_1)^{r_1} \cdot \ldots \cdot \pi(\omega_m)^{r_m}$$

This allows us to analyze numerical relationships holding between different $\pi(\omega)$, and to elaborate the conditionals whose structures π follows:

[1] This denotes the group of non-negative real numbers with respect to multiplication

Definition 1. *Let $\mathcal{R} \subseteq (\mathcal{L} \mid \mathcal{L})$ be a set of conditonals, let π be a (positive) possibility distribution. π is said to be* indifferent with respect to $\mathcal{R}$ *iff*

$$\pi(\widehat{\omega}_1) = \pi(\widehat{\omega}_2) \tag{4}$$

whenever $\sigma_{\mathcal{R}}(\widehat{\omega}_1) = \sigma_{\mathcal{R}}(\widehat{\omega}_2)$ for $\widehat{\omega}_1 \equiv_{\top} \widehat{\omega}_2 \in \widehat{\Omega}$.

π being indifferent with respect to $\mathcal{R}$ means that it does not distinguish between different elements $\widehat{\omega}_1, \widehat{\omega}_2$ with the same conditional structure with respect to $\mathcal{R}$. Conversely, if π is indifferent with respect to $\mathcal{R}$, then any inequality $\pi(\widehat{\omega}) \neq 1$ can be explained by an unbalanced conditional structure $\sigma_{\mathcal{R}}(\widehat{\omega}) \neq 1$. Normalizing constraints are taken into account by observing $\equiv_{\top}$-equivalence. Note that we presupposed π to be positive only for technical reasons, in order to make this definition a very concise one. For the general case, cf. [KI01a].

The following proposition provides an obvious characterization of conditional indifference in terms of the kernels of π and $\sigma_{\mathcal{R}}$:

Proposition 1. *Let $\mathcal{R} \subseteq (\mathcal{L} \mid \mathcal{L})$ be a set of conditonals, let π be a (positive) possibility distribution. π is indifferent with respect to $\mathcal{R}$ iff $\ker_0 \sigma_{\mathcal{R}} \subseteq \ker_0 \pi$, where $\ker_0 \pi = \ker \pi \cap \ker \sigma_{(\top \mid \top)}$.*

The notion of conditional indifference captures interactions of conditionals of arbitrary depth by making use of the group homomorphism induced by π. It also handles interactions on the superficial level of the worlds $\omega \in \Omega$:

Proposition 2. *Suppose $\mathcal{R} \subseteq (\mathcal{L} \mid \mathcal{L})$ to be a set of conditonals, let π be a (positive) possibility distribution which is indifferent with respect to $\mathcal{R}$. Let $\omega_1, \omega_2 \in \Omega$ be worlds with identical conditional structures with respect to $\mathcal{R}$, i.e. $\sigma_{\mathcal{R}}(\omega_1) = \sigma_{\mathcal{R}}(\omega_2)$. Then $\pi(\omega_1) = \pi(\omega_2)$.*

This proposition is clear by observing that any two possible worlds are $\equiv_{\top}$-equivalent.

The next theorem gives a simple criterion to check conditional indifference:

Theorem 1. *Let $\mathcal{R} = \{(B_1|A_1), \ldots, (B_n|A_n)\}$ be a set of conditionals.*

A (positive) possibility distribution π is indifferent with respect to $\mathcal{R}$ iff there are positive real numbers $\alpha_0, \alpha_1^+, \alpha_1^-, \ldots, \alpha_n^+, \alpha_n^- \in \mathbb{R}^+$ such that for all $\omega \in \Omega$,

$$\pi(\omega) = \alpha_0 \prod_{\substack{1 \leqslant i \leqslant n \\ \omega \models A_i B_i}} \alpha_i^+ \prod_{\substack{1 \leqslant i \leqslant n \\ \omega \models A_i \overline{B_i}}} \alpha_i^- \tag{5}$$

Proof. According to Proposition 2, the equivalence relation (3) provides a rough classification of the worlds in Ω with respect to the conditionals in $\mathcal{R}$. Obtaining a representation of the form (5) then amounts to checking the solvability of a linear equational system. The proof of this theorem is very similar to the proof of the analogous theorem for probabilistic representation of knowledge given in [KI98].

5 Representing Conditionals by Possibility Distributions

Conditional indifference only focusses on well-behavedness of a possibility distribution with respect to a set of conditionals. It does not include, however, that these conditionals are actually represented. This has to be taken into account in a second step.

Definition 2. *Possibility distributions π which represent a set $\mathcal{R}$ of conditionals, $\pi \models \mathcal{R}$, and are indifferent with respect to it, are called* c-representations *of $\mathcal{R}$.*

Theorem 1 provides an intelligible schema to construct conditionally indifferent distributions. To obtain a c-representation π of $\mathcal{R}$, we simply set up π according to (5) and choose the $\alpha_1^+, \alpha_1^-, \ldots, \alpha_n^+, \alpha_n^-$ appropriately to ensure that π is a possibility distribution that represents $\mathcal{R}$:

Theorem 2. *A (positive) possibility distribution π is a c-representation of a set $\mathcal{R}$ of conditionals, iff there are positive real numbers $\alpha_0, \alpha_1^+, \alpha_1^-, \ldots, \alpha_n^+, \alpha_n^- \in \mathbb{R}^+$ such that for all $\omega \in \Omega$,*

$$\pi(\omega) = \alpha_0 \prod_{\substack{1 \leqslant i \leqslant n \\ \omega \models A_i B_i}} \alpha_i^+ \prod_{\substack{1 \leqslant i \leqslant n \\ \omega \models A_i \overline{B_i}}} \alpha_i^-$$

and for all $i, 1 \leqslant i \leqslant n$,

$$\alpha_i^+ \left(\max_{\omega \models A_j B_j} \prod_{\substack{j \neq i \\ \omega \models A_j B_j}} \alpha_j^+ \prod_{\substack{j \neq i \\ \omega \models A_j \overline{B}_j}} \alpha_j^- \right) > \alpha_i^- \left(\max_{\omega \models A_i \overline{B}_i} \prod_{\substack{j \neq i \\ \omega \models A_j B_j}} \alpha_j^+ \prod_{\substack{j \neq i \\ \omega \models A_j \overline{B}_j}} \alpha_j^- \right) \tag{6}$$

Proof. Let π be a (positive) possibility distributions, and let $\mathcal{R} \subset (\mathcal{L} \mid \mathcal{L})$ be a set of conditionals. π is a c-representation of $\mathcal{R}$ iff π is indifferent with respect to $\mathcal{R}$ and $\pi \models \mathcal{R}$, i.e. iff there are positive real numbers $\alpha_0, \alpha_1^+, \alpha_1^-, \ldots, \alpha_n^+, \alpha_n^- \in \mathbb{R}^+$ such that

$$\pi(\omega) = \alpha_0 \prod_{\substack{1 \leqslant i \leqslant n \\ \omega \models A_i B_i}} \alpha_i^+ \prod_{\substack{1 \leqslant i \leqslant n \\ \omega \models A_i \overline{B_i}}} \alpha_i^-$$

and for all $i, 1 \leqslant i \leqslant n$, $\pi(A_i B_i) > \pi(A_i \overline{B}_i)$, which is equivalent to

$$\max_{\omega \models A_i B_i} \pi(\omega) > \max_{\omega \models A_i \overline{B}_i} \pi(\omega)$$

$$\iff \max_{\omega \models A_i B_i} \alpha_0 \prod_{\substack{1 \leqslant j \leqslant n \\ \omega \models A_j B_j}} \alpha_j^+ \prod_{\substack{1 \leqslant j \leqslant n \\ \omega \models A_j \overline{B}_j}} \alpha_j^- > \max_{\omega \models A_j \overline{B}_j} \alpha_0 \prod_{\substack{1 \leqslant j \leqslant n \\ \omega \models A_j B_j}} \alpha_j^+ \prod_{\substack{1 \leqslant j \leqslant n \\ \omega \models A_j \overline{B}_j}} \alpha_j^-$$

$$\iff \alpha_0 \max_{\omega \models A_i B_i} \alpha_i^+ \prod_{\substack{j \neq i \\ \omega \models A_j B_j}} \alpha_j^+ \prod_{\substack{j \neq i \\ \omega \models A_j \overline{B}_j}} \alpha_j^- > \alpha_0 \max_{\omega \models A_i \overline{B}_i} \alpha_i^- \prod_{\substack{j \neq i \\ \omega \models A_j B_j}} \alpha_j^+ \prod_{\substack{j \neq i \\ \omega \models A_j \overline{B}_j}} \alpha_j^-$$

$$\iff \alpha_i^+ \cdot \max_{\omega \models A_i B_i} \prod_{\substack{j \neq i \\ \omega \models A_j B_j}} \alpha_j^+ \prod_{\substack{j \neq i \\ \omega \models A_j \overline{B}_j}} \alpha_j^- > \alpha_i^- \cdot \max_{\omega \models A_i \overline{B}_i} \prod_{\substack{j \neq i \\ \omega \models A_j B_j}} \alpha_j^+ \prod_{\substack{j \neq i \\ \omega \models A_j \overline{B}_j}} \alpha_j^-$$

Note that the normalizing constant α_0 only has to be chosen appropriately to ensure that actually a normalized possibility distribution is obtained.

Therefore, c-representations of a set $\mathcal{R} = \{(B_1|A_1), \ldots, (B_n|A_n)\}$ of conditionals can be obtained by replacing the formal symbols $\mathbf{a}_i^+, \mathbf{a}_i^-$, associated to each conditional in R and used to reflect the conditional structures of worlds (cf. (2)), by appropriate numerical values. We will call a mapping $I : \{\mathbf{a}_1^+, \mathbf{a}_1^-, \ldots, \mathbf{a}_n^+, \mathbf{a}_n^-\} \to \mathbb{R}^+$, $\mathbf{a}_i^+ \mapsto \alpha_i^+, \mathbf{a}_i^- \mapsto \alpha_i^-, 1 \leqslant i \leqslant n$, such that (6) holds, a *c-interpretation of* $\mathcal{R}$. From c-interpretations, c-representations arise in a straightforward way via the formal tool of conditional structures.

The structure imposed on possible worlds by a c-representation is as finely grained as can be justified by differences in conditional structures. c-representations observe conditional structures and thereby handle interactions of conditionals in a most adequate way. Therefore, they are most appropriate to represent conditional knowledge. We will illustrate this by examples in the following. But first, as a general observation, we will show that all c-representations satisfy the property of *irrelevance*.

Lemma 1. *Suppose $\mathcal{R}$ is a set of conditionals, and let d be an atomic proposition not appearing in any of the conditionals in $\mathcal{R}$. Then, for any c-representation π_c of $\mathcal{R}$, and for any conditional $(B|A) \in (\mathcal{L} \mid \mathcal{L})$, $\pi_c \models (B|A)$ iff $\pi_c \models (B|Ad)$.*

This is easily proved by observing that for any propositional interpretation ω' over $\mathcal{V} - \{d\}$, we have $\sigma_{\mathcal{R}}(\omega' d) = \sigma_{\mathcal{R}}(\omega' \overline{d})$, and therefore, $\pi_c(A\dot{B}) = \pi_c(A\dot{B}d)$ for $\dot{B} \in \{B, \overline{B}\}$.

Let $\mid\!\sim^c$ be the (nonmonotonic) inference relation based on considering all c-representations:

$$\mathcal{R} \mid\!\sim^c (B|A) \quad \text{iff} \quad \pi_c \models (B|A) \text{ for all c-representations } \pi_c \text{ of } \mathcal{R}, \tag{7}$$

where $\mathcal{R} \subseteq (\mathcal{L} \mid \mathcal{L}), (B|A) \in (\mathcal{L} \mid \mathcal{L})$. Then Lemma 1 yields

Corollary 1. *Let $\mathcal{R} \subseteq (\mathcal{L} \mid \mathcal{L}), (B|A) \in (\mathcal{L} \mid \mathcal{L})$, let d be an atomic proposition not appearing in any of the conditionals in $\mathcal{R}$. Then $\mathcal{R} \mid\!\sim^c (B|A)$ iff $\mathcal{R} \mid\!\sim^c (B|Ad)$.*

It is easily shown that $\mid\!\sim^c$ satisfies all basic rationality postulates of system P (cf. [BDP97]). Furthermore, the next example will show that $\mid\!\sim^c$ does not suffer from the so-called *drowning problem* which is a problem for general possibilistic inference (cf. [BDP97]).

Example 2. Let $\mathcal{R}$ consist of the following conditionals:

$r_1 : (w\|p)$	*Penguins have wings.*
$r_2 : (f\|w)$	*Winged animals fly.*
$r_3 : (\overline{f}\|p)$	*Penguins do not fly.*
$r_4 : (b\|p)$	*Penguins are birds.*
$r_5 : (e\|b)$	*Birds lay eggs.*

In Table 1, we list the conditional structures of all possible worlds with respect to $\mathcal{R}$.

Table 1. Conditional structures for Example 2

ω	$\sigma_{\mathcal{R}}(\omega)$	ω	$\sigma_{\mathcal{R}}(\omega)$	ω	$\sigma_{\mathcal{R}}(\omega)$	ω	$\sigma_{\mathcal{R}}(\omega)$
$pwfbe$	$\mathbf{a}_1^+\mathbf{a}_2^+\mathbf{a}_3^-\mathbf{a}_4^+\mathbf{a}_5^+$	$p\overline{w}fbe$	$\mathbf{a}_1^-\mathbf{a}_3^-\mathbf{a}_4^+\mathbf{a}_5^+$	$\overline{p}wfbe$	$\mathbf{a}_2^+\mathbf{a}_5^+$	$\overline{p}\,\overline{w}fbe$	$\mathbf{a}_5^+$
$pwfb\overline{e}$	$\mathbf{a}_1^+\mathbf{a}_2^+\mathbf{a}_3^-\mathbf{a}_4^+\mathbf{a}_5^-$	$p\overline{w}fb\overline{e}$	$\mathbf{a}_1^-\mathbf{a}_3^-\mathbf{a}_4^+\mathbf{a}_5^-$	$\overline{p}wfb\overline{e}$	$\mathbf{a}_2^+\mathbf{a}_5^-$	$\overline{p}\,\overline{w}fb\overline{e}$	$\mathbf{a}_5^-$
$pwf\overline{b}e$	$\mathbf{a}_1^+\mathbf{a}_2^+\mathbf{a}_3^-\mathbf{a}_4^-$	$p\overline{w}f\overline{b}e$	$\mathbf{a}_1^-\mathbf{a}_3^-\mathbf{a}_4^-$	$\overline{p}wf\overline{b}e$	$\mathbf{a}_2^+$	$\overline{p}\,\overline{w}f\overline{b}e$	1
$pwf\overline{b}\overline{e}$	$\mathbf{a}_1^+\mathbf{a}_2^+\mathbf{a}_3^-\mathbf{a}_4^-$	$p\overline{w}f\overline{b}\overline{e}$	$\mathbf{a}_1^-\mathbf{a}_3^-\mathbf{a}_4^-$	$\overline{p}wf\overline{b}\overline{e}$	$\mathbf{a}_2^+$	$\overline{p}\,\overline{w}f\overline{b}\overline{e}$	1
$pw\overline{f}be$	$\mathbf{a}_1^+\mathbf{a}_2^-\mathbf{a}_3^+\mathbf{a}_4^+\mathbf{a}_5^+$	$p\overline{w}\overline{f}be$	$\mathbf{a}_1^-\mathbf{a}_3^+\mathbf{a}_4^+\mathbf{a}_5^+$	$\overline{p}w\overline{f}be$	$\mathbf{a}_2^-\mathbf{a}_5^+$	$\overline{p}\,\overline{w}\overline{f}be$	$\mathbf{a}_5^+$
$pw\overline{f}b\overline{e}$	$\mathbf{a}_1^+\mathbf{a}_2^-\mathbf{a}_3^+\mathbf{a}_4^+\mathbf{a}_5^-$	$p\overline{w}\overline{f}b\overline{e}$	$\mathbf{a}_1^-\mathbf{a}_3^+\mathbf{a}_4^+\mathbf{a}_5^-$	$\overline{p}w\overline{f}b\overline{e}$	$\mathbf{a}_2^-\mathbf{a}_5^-$	$\overline{p}\,\overline{w}\overline{f}b\overline{e}$	$\mathbf{a}_5^-$
$pw\overline{f}\,\overline{b}e$	$\mathbf{a}_1^+\mathbf{a}_2^-\mathbf{a}_3^+\mathbf{a}_4^-$	$p\overline{w}\overline{f}\,\overline{b}e$	$\mathbf{a}_1^-\mathbf{a}_3^+\mathbf{a}_4^-$	$\overline{p}w\overline{f}\,\overline{b}e$	$\mathbf{a}_2^-$	$\overline{p}\,\overline{w}\overline{f}\,\overline{b}e$	1
$pw\overline{f}\,\overline{b}\overline{e}$	$\mathbf{a}_1^+\mathbf{a}_2^-\mathbf{a}_3^+\mathbf{a}_4^-$	$p\overline{w}\overline{f}\,\overline{b}\overline{e}$	$\mathbf{a}_1^-\mathbf{a}_3^+\mathbf{a}_4^-$	$\overline{p}w\overline{f}\,\overline{b}\overline{e}$	$\mathbf{a}_2^-$	$\overline{p}\,\overline{w}\overline{f}\,\overline{b}\overline{e}$	1

Possibility values can be calculated directly from that table. For instance, $\pi(pw\overline{f}be) = \alpha_0\alpha_1^+\alpha_2^-\alpha_3^+\alpha_4^+\alpha_5^+$, where the α's are suitably chosen so as to fulfill (6) for that special set of conditionals. For α_4^+, α_4^- and α_5^+, α_5^-, e.g., we obtain (α_0 has already been cancelled)

$$\begin{aligned}
\alpha_4^+ \cdot \max\{&\alpha_1^+\alpha_2^+\alpha_3^-\alpha_5^+, \alpha_1^+\alpha_2^+\alpha_3^-\alpha_5^-, \alpha_1^+\alpha_2^-\alpha_3^+\alpha_5^+, \alpha_1^+\alpha_2^-\alpha_3^+\alpha_5^-,\\
&\alpha_1^-\alpha_3^-\alpha_5^+, \alpha_1^-\alpha_3^-\alpha_5^-, \alpha_1^-\alpha_3^+\alpha_5^+, \alpha_1^-\alpha_3^+\alpha_5^-\}\\
> \alpha_4^- \cdot \max\{&\alpha_1^+\alpha_2^+\alpha_3^-, \alpha_1^+\alpha_2^-\alpha_3^+, \alpha_1^-\alpha_3^-, \alpha_1^-\alpha_3^+\}
\end{aligned}$$

$$\begin{aligned}
\alpha_5^+ \cdot \max\{&\alpha_1^+\alpha_2^+\alpha_3^-\alpha_4^+, \alpha_1^+\alpha_2^-\alpha_3^+\alpha_4^+, \alpha_1^-\alpha_3^-\alpha_4^+, \alpha_1^-\alpha_3^+\alpha_4^+, \alpha_2^+, \alpha_2^-, 1\}\\
> \alpha_5^- \cdot \max\{&\alpha_1^+\alpha_2^+\alpha_3^-\alpha_4^+, \alpha_1^+\alpha_2^-\alpha_3^+\alpha_4^+, \alpha_1^-\alpha_3^-\alpha_4^+, \alpha_1^-\alpha_3^+\alpha_4^+, \alpha_2^+, \alpha_2^-, 1\}
\end{aligned}$$

So, $\alpha_5^+ > \alpha_5^-$, and therefore,

$$\begin{aligned}
\alpha_4^+\alpha_5^+ \cdot \max\{&\alpha_1^+\alpha_2^+\alpha_3^-, \alpha_1^+\alpha_2^-\alpha_3^+, \alpha_1^-\alpha_3^-, \alpha_1^-\alpha_3^+\}\\
> \alpha_4^- \cdot \max\{&\alpha_1^+\alpha_2^+\alpha_3^-, \alpha_1^+\alpha_2^-\alpha_3^+, \alpha_1^-\alpha_3^-, \alpha_1^-\alpha_3^+\}
\end{aligned}$$

which implies $\alpha_4^+\alpha_5^+ > \alpha_4^-$.

Although penguins are exceptional birds in that they are not able to fly, they should still inhibit the property of laying eggs from their superclass *birds*. This is not guaranteed in general possibilistic inference – the conditional $(e|b)$ is "drowned" by the conditionals $(w|p), (\overline{f}|p)$ which obtain a higher degree of possibility (see [BDP97, p. 273]). By observing conditional structures, however, the impact of $(e|b)$ can be preserved to establish the validity of $(e|p)$: For any c-representation π_c of $\mathcal{R}$, we compute

$$\begin{aligned}
\pi_c(pe) = \alpha_0 \max\{&\alpha_1^+\alpha_2^+\alpha_3^-\alpha_4^+\alpha_5^+, \alpha_1^+\alpha_2^+\alpha_3^-\alpha_4^-, \alpha_1^+\alpha_2^-\alpha_3^+\alpha_4^+\alpha_5^+, \alpha_1^+\alpha_2^-\alpha_3^+\alpha_4^-,\\
&\alpha_1^-\alpha_3^-\alpha_4^+\alpha_5^+, \alpha_1^-\alpha_3^-\alpha_4^-, \alpha_1^-\alpha_3^+\alpha_4^+\alpha_5^+, \alpha_1^-\alpha_3^+\alpha_4^-\}
\end{aligned}$$

$$\begin{aligned}
\pi_c(p\overline{e}) = \alpha_0 \max\{&\alpha_1^+\alpha_2^+\alpha_3^-\alpha_4^+\alpha_5^-, \alpha_1^+\alpha_2^+\alpha_3^-\alpha_4^-, \alpha_1^+\alpha_2^-\alpha_3^+\alpha_4^+\alpha_5^-, \alpha_1^+\alpha_2^-\alpha_3^+\alpha_4^-,\\
&\alpha_1^-\alpha_3^-\alpha_4^+\alpha_5^-, \alpha_1^-\alpha_3^-\alpha_4^-, \alpha_1^-\alpha_3^+\alpha_4^+\alpha_5^-, \alpha_1^-\alpha_3^+\alpha_4^-\}
\end{aligned}$$

By comparing the sets which contribute to the maxima element by element, we see that $\pi_c(pe) > \pi_c(p\overline{e})$, due to $\alpha_5^+ > \alpha_5^-$ and $\alpha_4^+\alpha_5^+ > \alpha_4^-$ (e.g., $\alpha_1^+\alpha_2^+\alpha_3^-\alpha_4^+\alpha_5^+ > \alpha_1^+\alpha_2^+\alpha_3^-\alpha_4^-$ and $\alpha_1^+\alpha_2^+\alpha_3^-\alpha_4^+\alpha_5^+ > \alpha_1^+\alpha_2^+\alpha_3^-\alpha_4^+\alpha_5^-$). This means, that $\pi_c \models (e|p)$, for all c-representations π_c of $\mathcal{R}$. Hence, $\mathcal{R} \mid\!\sim^c (e|p)$.

C-representations handle conditional interactions in a numerical way, but also allow us to use structural arguments when considering conditional entailment.

Our general approach based on conditional structures and c-representations may be improved when combined with the *principle of least commitment* and with the *principle of minimum specificity*, that is, *choose the α's in a least committed and least specific way*. The consequences of these principles are twofold:

1. To allow worlds verifying conditionals to be as normal and least specific as possible, set $\alpha_i^+ = 1$ for $1 \leqslant i \leqslant n$. Since $\mathcal{R}$ is consistent iff there are worlds not falsifying any of the conditionals in $\mathcal{R}$, this implies $\alpha_0 = 1$. Furthermore, to comply with intuition (falsifying conditionals should decrease the possibility of worlds), we choose $\alpha_i^- < 1$ for $1 \leqslant i \leqslant n$.
2. In order not to introduce any relations which are not explicitly sanctioned by the representation postulate $\pi \models \mathcal{R}$, choose $\alpha_1^-, \ldots, \alpha_n^-$ according to (6), but in a least committed way. This amounts to partition $\mathcal{R}$ into sets $\mathcal{R}_1, \ldots, \mathcal{R}_m$ of conditionals, where conditionals in subsets with lower indices can be established more easily than those in subsets with higher indices. We may then choose α_i^- such that for all conditionals in one set $\mathcal{R}_j$, all α_i^- are the same, and the α_i^- are then chosen following some discrete scale, e.g., $\frac{1}{2}, \frac{1}{2^2}, \frac{1}{2^3}, \ldots$, i.e. $\alpha_i^- = \frac{1}{2^j}$ for all conditionals $(B_i|A_i) \in \mathcal{R}_j$.

An *lcc-representation* of $\mathcal{R}$ is a c-representation

$$\pi_{lcc}(\omega) = \prod_{\substack{1 \leqslant i \leqslant n \\ \omega \models A_i\overline{B_i}}} \alpha_i^-$$

of $\mathcal{R}$ whose factors α_i^- satisfy both requirements 1. and 2. above. In particular, $\alpha_i^- < 1$, and

$$\left(\max_{\omega \models A_iB_i} \prod_{\substack{j \neq i \\ \omega \models A_j\overline{B}_j}} \alpha_j^- \right) > \alpha_i^- \left(\max_{\omega \models A_i\overline{B}_i} \prod_{\substack{j \neq i \\ \omega \models A_j\overline{B}_j}} \alpha_j^- \right), \qquad (8)$$

$1 \leqslant i \leqslant n$, for lcc-representations π_{lcc}. Let $\mid\!\sim^{lcc}$ be the inference relation based on considering all lcc-representations:

$$\mathcal{R} \mid\!\sim^{lcc} (B|A) \quad \text{iff} \quad \pi_{lcc} \models (B|A) \text{ for all lcc-representations } \pi_{lcc} \text{ of } \mathcal{R}.$$

The approach we thereby obtain is very similar to the *lcd-approach* of Benferhat, Saffiotti and Smets [BSS00]. Instead of *proto-infinitesimals* and *infinitesimals* which are used in that paper, we use formal symbols $\mathbf{a}_1^-, \ldots, \mathbf{a}_n^-$ and positive real numbers $\alpha_1^-, \ldots, \alpha_n^-$. So our c-interpretations are nearly the same as their interpretations of proto-infinitesimals. Their *auto-deduction principle* is the same as our (quite implicit) representation postulate $\pi \models \mathcal{R}$. They base their considerations upon belief functions, we use possibility distributions (which correspond to plausibility functions). The basic idea to construct infinitesimal lcd-belief functions is to interpret proto-infinitesimals according to a *least committed ϵ-stratification*, where each ϵ-stratification corresponds to a partition of $\mathcal{R}$, and least committedness comes down to minimizing the number of classes, and to assign as many proto-infinitesimals (i.e. α_i^- in our framework) as possible to classes with low index. This amounts exactly to choose $\alpha_1^-, \ldots, \alpha_n^-$ according to (6) in a least committed way, as described above in 2. So our $\mid\!\sim^{lcc}$ and $\mid\!\sim_{lcd}$ of [BSS00] coincide.

The crucial difference between our work and that of Benferhat, Saffiotti and Smets [BSS00] is how the special product form of possibility distributions (see (5)), or belief functions, respectively, is motivated: Benferhat, Saffiotti and Smets derive the product form from Dempster's combination rule, whereas in our framework, (5) arises from an underlying conditional theory, which is much more general and applies not only in a possibilistic environment, but also in other quantitative (probabilistic) and semi-quantitative (ordinal conditional functions) environments and has also important consequences for handling conditionals adequately in a purely qualitative manner (cf. [KI01a]). Actually, in the framework of ordinal conditional functions, the application of our conditional approach turns out to be a generalization of system-Z^* (cf. [GMP93]; see [KI01b]). Note that the ideas of considering conditional structures and using conditional indifference were born when we investigated maximum entropy inference (cf. [KI98]; conditional indifference was termed *conditional preservation* there, as we actually considered probabilistic revision). So the same crucial principle, namely that of conditional indifference, underlies our possibilistic representation approach as well as the probabilistic maximum entropy approach, the generalized system-Z^* and the lcd-approach of Benferhat, Saffiotti and Smets. It is this principle that guarantees a most appropriate handling of conditionals in uncertain reasoning.

In Example 2, we showed that our approach solves the drowning problem; in particular, it also solves the *blocking of property inheritance problem* (cf. [BDP97]). Lemma 1 proves it to observe irrelevance. In the next example, we will show that even the inference relation $\mid\!\sim^{lcc}$ based on lcc-representations does not introduce conditional dependencies where there are none – it respects *ambiguity* (also cf. [BSS00]):

Example 3. Consider the following set of conditionals $\mathcal{R}$:

$(f\|b)$	*Birds fly.*
$(b\|p)$	*Penguins are birds.*
$(\overline{f}\|p)$	*Penguins do not fly.*
$(f\|m)$	*Objects with metal wings fly.*

The factors $\alpha_1^-, \ldots, \alpha_4^-$ of any lcc-representation π_{lcc} of $\mathcal{R}$ must be chosen in a least committed way so as to satisfy (8), which means $\alpha_1^- < 1, \alpha_2^- <$

$\max\{\alpha_1^-, \alpha_3^-\}, \alpha_3^- < \max\{\alpha_1^-, \alpha_2^-\}, \alpha_4^- < 1$. Therefore, lcc-representations are obtained by choosing $\alpha_2^- = \alpha_3^- < \alpha_1^- = \alpha_4^- < 1$.

If we now see Super-Tweety, the penguin with metal wings, waddling around, can we decide whether he flies or not? In our rule base $\mathcal{R}$, there is evidence in favor of and against Super-Tweety's flying, but we are not able to make up our mind. And indeed, this ambiguity is reflected by lcc-representations π_{lcc}: We obtain $\pi_{lcc}(pmf) = \max\{\alpha_3^-, \alpha_2^-\alpha_3^-\} = \alpha_3^-$ and $\pi_{lcc}(pm\overline{f}) = \max\{\alpha_1^-\alpha_4^-, \alpha_2^-\alpha_4^-\} = \alpha_4^- \cdot \max\{\alpha_1^-, \alpha_2^-\} = \alpha_1^-\alpha_4^-$, and α_3^- and $\alpha_1^-\alpha_4^-$ are incomparable. So neither $\mathcal{R} \mathrel{|\!\sim}^{lcc} (f|pm)$ nor $\mathcal{R} \mathrel{|\!\sim}^{lcc} (\overline{f}|pm)$. The same holds if we also take into account that Super-Tweety is a bird – neither $(f|pbm)$ nor $(\overline{f}|pbm)$ can be $\mathrel{|\!\sim}^{lcc}$-entailed from $\mathcal{R}$.

6 Revising Possibility Distributions by Conditional Beliefs

The theory of conditional structures also provides a solid foundation to deal with conditionals appropriately when revising epistemic states by (propositional and conditional) beliefs. Minimality of change is a crucial paradigm for belief revision, and a "principle of conditional preservation" is to realize this idea of minimality when conditionals are involved in change. Minimizing absolutely the changes in conditional beliefs, as in [BG93], is an important proposal to this aim, but it does not always lead to intuitive results (see [DP97]). The idea we will develop here rather aims at *preserving the conditional structure of knowledge* within an epistemic state which we assume to be represented by a possibility distribution π.

We just explained what it means for a possibility distribution, π, to follow the structure imposed by a set of conditionals $\mathcal{R}$ on the possible worlds by introducing the notion of conditional indifference. Pursuing this approach further in the framework of belief revision, a revision of π by simultaneously incorporating the conditionals in $\mathcal{R}$, $\pi^* = \pi * \mathcal{R}$, can be said to preserve the conditional structure of π with respect to $\mathcal{R}$ if the *relative change function* $\frac{\pi^*}{\pi}$ is indifferent with respect to $\mathcal{R}$.

Definition 3. *Let π be a (positive) possibility distribution, and let $\mathcal{R}$ be a (finite) set of conditionals. Let $\pi^* = \pi * \mathcal{R}$ denote the result of revising π by $\mathcal{R}$. π^* is* indifferent with respect to $\mathcal{R}$ and π *iff $\frac{\pi^*}{\pi}(\widehat{\omega}_1) = \frac{\pi^*}{\pi}(\widehat{\omega}_2)$ whenever $\sigma_{\mathcal{R}}(\widehat{\omega}_1) = \sigma_{\mathcal{R}}(\widehat{\omega}_2)$ and $\widehat{\omega}_1 \equiv_{\top} \widehat{\omega}_2$, for $\widehat{\omega}_1, \widehat{\omega}_2 \in \widehat{\Omega}$.*

The principle of conditional preservation is now realized as an indifference property:

Definition 4. *Let π be a (positive) possibility distribution, and let $\mathcal{R}$ be a finite set of conditionals. A revision $\pi^* = \pi * \mathcal{R}$ satisfies the* principle of conditional preservation *iff π^* is indifferent with respect to $\mathcal{R}$ and π.*

So $\pi * \mathcal{R}$ satisfies the principle of conditional preservation if any change in plausibility is clearly and unambigously induced by $\mathcal{R}$. The next theorem characterizes revisions of possibility distributions that satisfy the principle of conditional preservation. The theorem is obvious by observing Theorem 1.

Theorem 3. *Let π, π^* be (positive) possibility distributions, and let $\mathcal{R} = \{(B_1|A_1), \ldots, (B_n|A_n)\}$ be a (finite) set of conditionals in $(\mathcal{L} \mid \mathcal{L})$. A revision $\pi^* = \pi * \mathcal{R}$ satisfies the principle of conditional preservation iff there are positive real numbers $\alpha_0, \alpha_1^+, \alpha_1^-, \ldots, \alpha_n^+, \alpha_n^- \in \mathbb{R}^+$ such that for all $\omega \in \Omega$,*

$$\pi^*(\omega) = \alpha_0 \pi(\omega) \prod_{\substack{1 \leqslant i \leqslant n \\ \omega \models A_i B_i}} \alpha_i^+ \prod_{\substack{1 \leqslant i \leqslant n \\ \omega \models A_i \overline{B_i}}} \alpha_i^- \tag{9}$$

Comparing Theorems 1 and 3 with one another, we see that a possibility distribution π is indifferent with respect to a finite set of conditionals $\mathcal{R}$ iff it can be taken as a revision $\pi_0 * \mathcal{R}$ satisfying the principle of conditional preservation, where $\pi_0(\omega) = 1$ for all $\omega \in \Omega$ is the *uniform (least specific) possibility distribution.*

So knowledge representation and belief revision can be dealt with in a common framework, based on conditionals as crucial pieces of knowledge. At least for this reason, the topic of revising epistemic states by conditional beliefs should not be looked upon as a superfluous artifact, but should be appreciated as providing a most general and enhanced framework for representing and revising knowledge. Due to the fact, that propositions, A, can be considered as degenerated conditionals, $(A|T)$, the theory may also be applied to propositional belief revision. In fact, the principle of conditional preservation, as defined here, can be shown to cover the postulates for iterative belief revision of [DP97] (cf. [KI01a]).

7 Extracting Rules from Possibilistic Data

The concept of conditional structures and the principle of conditional indifference provide a powerful framework for inductive knowledge representation and belief revision in possibility theory. Embedding the techniques into a group theoretical framework makes calculations more convenient and allows an elegant presentation of the basic ideas. In this section, we will show how this link between group theory and conditionals may be used to extract conditional relationships from data.

Let a possibility distribution π be given which has been obtained from statistical data (see, e.g., [GK97]). The aim is to find a most concise set $\mathcal{R}$ of conditionals of which π is a c-representation. This amounts to find $\mathcal{R} \subset (\mathcal{L} \mid \mathcal{L})$ such that $\pi \models \mathcal{R}$ and $ker_0\, \sigma_{\mathcal{R}} \subseteq ker_0\, \pi$ (where π is the induced homomorphism, cf. Proposition 1). So numerical relationships within π – represented by elements of $ker_0\, \pi$ – are taken as manifestations of conditional interactions. These numerical relationships will be translated into equations between group elements. By solving these equations, complex conditional interactions will be disentangled to yield an appropriate set of conditionals.

Our method is based on the following prerequisites:

- We will concentrate on the discovery of single-elementary conditionals, that is, we assume $\mathcal{R}$ to be a set of single-elementary conditionals. This restriction should not be considered as a heavy drawback, bearing in mind the expressibility of single-elementary conditionals.
- To exploit numerical relationships successfully, we must presuppose, that no equation $\pi(\widehat{\omega}) = 1$ holds accidentally, but that any of these equations is clearly induced by $\mathcal{R}$. Therefore, we assume

$$ker_0\ \pi = ker_0\ \sigma_{\mathcal{R}} \tag{10}$$

This last prerequisite (10) is a bit problematic in a possibilistic framework, because possibility values are often leveled for the sake of minimum specificity, and this leveling will usually blur conditional structures. In particular, lcc-representations which minimize conditional dependencies to the debit of conditional clearness, are bad candidates for conditional knowledge discovery. Nevertheless, in spite of these objections, our method is interesting in that it provides a completely new and theoretically profound approach to discover rules from data.

So assume $\mathcal{R} = \{(b_1|A_1), \ldots, (b_n|A_n)\}$ is an existing, but hidden set of single-elementary conditionals, such that (10) holds. Let us further suppose that $ker_0\ \pi$ (or parts of it) is known from exploiting numerical relationships. Let $\sigma_{\mathcal{R}} : \widehat{\Omega} \to \mathcal{F}_{\mathcal{R}} = \langle \mathbf{a}_1^+, \mathbf{a}_1^-, \ldots, \mathbf{a}_n^+, \mathbf{a}_n^- \rangle$ denote a conditional structure homomorphism with respect to $\mathcal{R}$.

Our method is a *bottom-up approach* generalizing conditionals in accordance with the conditional structure revealed by $ker_0\ \pi$. We start with considering basic single-elementary conditionals, which are single-elementary conditionals with antecedents of maximal length. For each atom $v \in \mathcal{L}$, choose an arbitrary, but fixed numbering of the remaining atoms $w \neq v$, $(w_0, w_1, \ldots, w_{\#(atoms)-1})$. Then *basic single-elementary conditionals* are conditionals of the form

$$\psi_{v,l} = (v \mid \bigwedge_j w_j^{\epsilon_j}) \tag{11}$$

with $\epsilon_j \in \{0,1\}, w_j^1 := w_j, w_j^0 := \overline{w_j}, 0 \leqslant j \leqslant \#(atoms) - 1$ and $l = \sum_j \epsilon_j 2^j$. We will abbreviate the antecedent of $\psi_{v,l}$ by $C_{v,l}$. Let

$$\mathcal{B} = \{\psi_{v,l} \mid v \text{ atom in } \mathcal{L}, 0 \leqslant l \leqslant 2^{\#(atoms)-1} - 1\}$$

denote the set of all basic single-elementary conditionals in $(\mathcal{L} \mid \mathcal{L})$, and let $\mathcal{F}_{\mathcal{B}} = \langle \mathbf{b}_{v,l}^+, \mathbf{b}_{v,l}^- \mid v \text{ atom in } \mathcal{L}, 0 \leqslant l \leqslant 2^{\#(atoms)-1} - 1 \rangle$ be the free abelian group corresponding to $\mathcal{B}$ with conditional structure homomorphism $\sigma_{\mathcal{B}} : \widehat{\Omega} \to \mathcal{F}_{\mathcal{B}}$,

$$\sigma_{\mathcal{B}} = \prod_{v,l} \sigma_{v,l}, \quad \sigma_{v,l}(\omega) = \begin{cases} \mathbf{b}_{v,l}^+, & \text{if } \omega = C_{v,l}v \\ \mathbf{b}_{v,l}^-, & \text{if } \omega = C_{v,l}\overline{v} \\ 1, & \text{else} \end{cases}$$

Note that $\sigma_{\mathcal{B}}$ and $\mathcal{F}_{\mathcal{B}}$ are known, whereas $\sigma_{\mathcal{R}}$ and $\mathcal{F}_{\mathcal{R}}$ are not. We only know the kernel, $ker_0\ \sigma_{\mathcal{R}}$, of $\sigma_{\mathcal{R}}$, which is, by assuming (10), the same as the kernel, $ker_0\ \pi$, of π. Now, to establish a connection between what is obvious ($\mathcal{B}$) and what is searched for ($\mathcal{R}$), we define a homomorphism $g : \mathcal{F}_{\mathcal{B}} \to \mathcal{F}_{\mathcal{R}}$ via

$$g(\mathbf{b}_{v,l}^{\pm}) = \prod_{\substack{1 \leqslant i \leqslant n \\ b_i = v,\, C_{v,l} \models A_i}} \mathbf{a}_i^{\pm}, \tag{12}$$

Actually, g is a "phantom" which is not explicitly given, but only known to exist. Its crucial meaning for the knowledge discovery task is revealed by the following theorem and its corollary:

Theorem 4. *Let $g : \mathcal{F}_{\mathcal{B}} \to \mathcal{F}_{\mathcal{R}}$ be as in (12). Then $\sigma_{\mathcal{R}} = g \circ \sigma_{\mathcal{B}}$.*

Corollary 2. *$\widehat{\omega} \in ker_0\ \sigma_{\mathcal{R}} = ker_0\ \pi$ iff $\widehat{\omega} \in \widehat{\Omega}_0$ and $\sigma_{\mathcal{B}}(\widehat{\omega}) \in ker\, g$.*

This means, that numerical relationships observed in π (and represented by elements of $ker_0\ \pi$) translate into group theoretical equations modulo the kernel of g:

Proposition 3. *Let $\widehat{\omega} = \omega_1^{r_1} \cdot \ldots \cdot \omega_m^{r_m} \in \widehat{\Omega}_0$. Then $\sigma_{\mathcal{B}}(\omega_1^{r_1} \cdot \ldots \cdot \omega_m^{r_m}) \in ker\ g$ iff for all atoms v in $\mathcal{L}$,*

$$\prod_{\substack{1 \leqslant k \leqslant m \\ \omega_k = C_{v,l} v}} (\mathbf{b}_{v,l}^{+})^{r_k}, \quad \prod_{\substack{1 \leqslant k \leqslant m \\ \omega_k = C_{v,l} \overline{v}}} (\mathbf{b}_{v,l}^{-})^{r_k} \in ker\, g. \tag{13}$$

So each (generating) element of $ker_0\ \sigma_{\mathcal{R}}$ gives rise to an equation modulo $ker\, g$ for the generators $\mathbf{b}_{v,l}^{+}, \mathbf{b}_{v,l}^{-}$ of $\mathcal{F}_{\mathcal{B}}$. Moreover, Proposition 3 allows us to split up equations modulo $ker_0\ g$ to handle each atom separately as a consequent of conditionals.

The idea of our data mining procedure is to exploit the relations mod $ker\, g$ holding between the group elements $\mathbf{b}_{v,l} \in \mathcal{F}_{\mathcal{B}}$ with the aim to define a finite sequence of sets $\mathcal{S}^{(0)}, \mathcal{S}^{(1)}, \ldots$ of conditionals approximating $\mathcal{R}$:

$$ker_0\ \sigma_{\mathcal{S}^{(0)}} \subseteq ker_0\ \sigma_{\mathcal{S}^{(1)}} \subseteq \ldots \subseteq ker_0\ \sigma_{\mathcal{R}} \tag{14}$$

The set $\mathcal{B}$ of basic single elementary conditionals proves to be an ideal starting point:

Lemma 2. *$\sigma_{\mathcal{B}}$ is injective, i.e. $ker_0\ \sigma_{\mathcal{B}} = \{1\}$.*

So $\sigma_{\mathcal{B}}$ provides the most finely grained conditional structure on $\widehat{\Omega}$: No different elements $\widehat{\omega}_1 \neq \widehat{\omega}_2$ are equivalent with respect to $\mathcal{B}$. So set $\mathcal{S}^{(0)} := \mathcal{B}$ and $g^{(0)} := g$.

By solving the equations modulo $ker\ g^{(t)}$, $t \geqslant 0$, we obtain a finite sequence of sets $\mathcal{S}^{(t)}$ of single-elementary conditionals $\varphi_{v,j}^{(t)} = (v|D_{v,j}^{(t)})$ with corresponding group generators $\mathbf{s}_{v,j}^{(t)\,+}, \mathbf{s}_{v,j}^{(t)\,-}$ such that $ker_0\ \sigma_{\mathcal{S}^{(0)}} \subseteq \ldots ker_0\ \sigma_{\mathcal{S}^{(t)}} \subseteq ker_0\ \sigma_{\mathcal{S}^{(t+1)}} \subseteq \ldots \subseteq ker_0\ \sigma_{\mathcal{R}}$, as desired, together with homomorphisms $g^{(t)}$

describing their relationship to $\mathcal{R}$. Each $\mathcal{S}^{(t+1)}$ arises from $\mathcal{S}^{(t)}$ by eliminating conditionals, or by aggregating conditionals with identical consequents in conjoining their antecedents. These operations are induced by group equations modulo $ker_0\ g^{(t)}$ (equivalence modulo $ker_0\ g^{(t)}$ is denoted by $\equiv_{g^{(t)}}$ in the following, and the $+, -$-signs may be omitted): For instance, if $\mathbf{s}_{v,j}^{(t)} \equiv_{g^{(t)}} 1$, then simply eliminate $\varphi_{v,j}^{(t)}$, and if

$$\mathbf{s}_{v,j_0}^{(t)} \equiv_{g^{(t)}} \mathbf{s}_{v,j_1}^{(t)} \dots \mathbf{s}_{v,j_m}^{(t)}, \tag{15}$$

then

1. eliminate $\varphi_{v,j_0}^{(t)}$ from $\mathcal{S}^{(t)}$;
2. replace each $\varphi_{v,j_k}^{(t)}$ by $\varphi_{v,j_k}^{(t+1)} = \varphi_{v,j_0}^{(t)} \sqcup \varphi_{v,j_k}^{(t)} = (v \mid D_{v,j_0}^{(t)} \vee D_{v,j_k}^{(t)})$, for $1 \leqslant k \leqslant m$; set $D_{v,j_k}^{(t+1)} = D_{v,j_0}^{(t)} \vee D_{v,j_k}^{(t)}$, $1 \leqslant k \leqslant m$;
3. retain all other $\varphi_{w,l}^{(t)}$.

For the formal details and theoretical justifications, cf. [KI00b]. We will briefly illustrate the method by an example.

Example 4. We consider formulas involving the three atomic propositions a - *being a student*, b - *being young*, and c - *being single (i.e. unmarried)*. The distribution π over a, b, c is given as follows:

ω	$\pi(\omega)$	ω	$\pi(\omega)$	ω	$\pi(\omega)$	ω	$\pi(\omega)$
abc	0.720	$ab\overline{c}$	0.900	$a\overline{b}c$	0.12	$a\overline{b}\overline{c}$	0.300
$\overline{a}bc$	0.800	$\overline{a}b\overline{c}$	1.000	$\overline{a}\overline{b}c$	0.400	$\overline{a}\overline{b}\overline{c}$	1.000

Important relationships between possibility values are

$$\pi(\overline{a}b\overline{c}) = \pi(\overline{a}\overline{b}\overline{c}),$$
$$\frac{\pi(abc)}{\pi(ab\overline{c})} = \frac{\pi(\overline{a}bc)}{\pi(\overline{a}b\overline{c})},$$
$$\frac{\pi(a\overline{b}c)}{\pi(a\overline{b}\overline{c})} = \frac{\pi(\overline{a}\overline{b}c)}{\pi(\overline{a}\overline{b}\overline{c})}.$$

We list the twelve basic single-elementary conditionals of $\mathcal{B}$ which we start with:

$$\begin{array}{lll}
\psi_{a,0}=(a \mid \overline{b}\overline{c}) & \psi_{b,0}=(b \mid \overline{a}\,\overline{c}) & \psi_{c,0}=(c \mid \overline{a}\overline{b}) \\
\psi_{a,1}=(a \mid \overline{b}c) & \psi_{b,1}=(b \mid \overline{a}c) & \psi_{c,1}=(c \mid \overline{a}b) \\
\psi_{a,2}=(a \mid b\overline{c}) & \psi_{b,2}=(b \mid a\overline{c}) & \psi_{c,2}=(c \mid a\overline{b}) \\
\psi_{a,3}=(a \mid bc) & \psi_{b,3}=(b \mid ac) & \psi_{c,3}=(c \mid ab)
\end{array}$$

with corresponding generators $\mathbf{b}_{v,l}^{+}, \mathbf{b}_{v,l}^{-}$ of $\mathcal{F}_{\mathcal{B}}$. The numerical relationships yield the following group equations :

$$\sigma_{\mathcal{B}}(\overline{a}b\overline{c}) \equiv_g \sigma_{\mathcal{B}}(\overline{a}\overline{b}\overline{c})$$
$$\sigma_{\mathcal{B}}(abc)\sigma_{\mathcal{B}}(\overline{a}b\overline{c}) \equiv_g \sigma_{\mathcal{B}}(ab\overline{c})\sigma_{\mathcal{B}}(\overline{a}bc)$$
$$\sigma_{\mathcal{B}}(a\overline{b}c)\sigma_{\mathcal{B}}(\overline{a}\overline{b}\overline{c}) \equiv_g \sigma_{\mathcal{B}}(a\overline{b}\overline{c})\sigma_{\mathcal{B}}(\overline{a}\overline{b}c)$$

which imply

$$\mathbf{b}^-_{a,2}\mathbf{b}^+_{b,0}\mathbf{b}^-_{c,1} \equiv_g \mathbf{b}^-_{a,0}\mathbf{b}^-_{b,0}\mathbf{b}^-_{c,0}$$
$$\mathbf{b}^+_{a,3}\mathbf{b}^+_{b,3}\mathbf{b}^+_{c,3} \cdot \mathbf{b}^-_{a,2}\mathbf{b}^+_{b,0}\mathbf{b}^-_{c,1} \equiv_g \mathbf{b}^+_{a,2}\mathbf{b}^+_{b,2}\mathbf{b}^-_{c,3} \cdot \mathbf{b}^-_{a,3}\mathbf{b}^+_{b,1}\mathbf{b}^+_{c,1}$$
$$\mathbf{b}^+_{a,1}\mathbf{b}^-_{b,3}\mathbf{b}^+_{c,2} \cdot \mathbf{b}^-_{a,0}\mathbf{b}^-_{b,0}\mathbf{b}^-_{c,0} \equiv_g \mathbf{b}^+_{a,0}\mathbf{b}^-_{b,2}\mathbf{b}^-_{c,2} \cdot \mathbf{b}^-_{a,1}\mathbf{b}^-_{b,1}\mathbf{b}^+_{c,0}$$

Considering these equations for each atom a, b, c separately and omitting the $\{+,-\}$-signs, we obtain the following equivalences:

$$\begin{array}{l} \mathbf{b}_{a,0} \equiv_g \mathbf{b}_{a,2}, \mathbf{b}_{c,0} \equiv_g \mathbf{b}_{c,1}, \mathbf{b}_{b,0} \equiv_g 1, \\ \mathbf{b}_{a,2} \equiv_g \mathbf{b}_{a,3}, \mathbf{b}_{c,1} \equiv_g \mathbf{b}_{c,3}, \mathbf{b}_{b,0}\mathbf{b}_{b,3} \equiv_g \mathbf{b}_{b,1}\mathbf{b}_{b,2}, \\ \mathbf{b}_{a,0} \equiv_g \mathbf{b}_{a,1}, \mathbf{b}_{c,0} \equiv_g \mathbf{b}_{c,2}. \end{array}$$

This yields

$$\begin{array}{l} \mathbf{b}_{a,0} \equiv_g \mathbf{b}_{a,1} \equiv_g \mathbf{b}_{a,2} \equiv_g \mathbf{b}_{a,3}, \\ \mathbf{b}_{c,0} \equiv_g \mathbf{b}_{c,1} \equiv_g \mathbf{b}_{c,2} \equiv_g \mathbf{b}_{c,3}, \\ \mathbf{b}_{b,0} \equiv_g 1,\ \mathbf{b}_{b,3} \equiv_g \mathbf{b}_{b,1}\mathbf{b}_{b,2}. \end{array}$$

Eliminating $\psi_{b,0}$ and joining conditionals according to these equations results in the set $\mathcal{S} = \{(\overline{a}|\top), (\overline{c}|\top), (b|a), (b|c)\}$ of conditionals of which the above possibility distribution π is a c-representation.

Note that π in Example 4 is *not* an lcc-representation. Actually, the possibility degrees are quite arbitrary. This example should only illustrate the method: Ratios (of first order or of higher orders) are investigated to reveal structures of conditional knowledge.

8 Conditional Structures and Conditional Events

We announced our approach to conditionals as a substantially new one, capturing the nature of conditionals peculiarly well. The results and examples presented above illustrate this, but what is actually new and different, compared with other, established conditional theories?

First formal work on conditionals dates back to the days of Boole. *Conditional events* were investigated particularly intensively in the nineties of the past century, when their crucial meaning for knowledge representation in AI had become apparent (for a historical review, cf. [NW94]; for a survey of conditional logics, cf. [Nut80]). Lewis [Lew76] made clear that conditionals cannot be handled as boolean entities, so new ways of representing and dealing with conditionals had to be explored. A very popular approach, e.g., is the following one (for an overview of important approaches to conditionals, see [DGC94]):

It is generally agreed that two conditionals $(B|A), (D|C)$ are equivalent, $(B|A) \equiv (D|C)$, iff $A \equiv C$ and $AB \equiv CD$. So it seems adequate to define a conditional $(B|A)$ by a set of boolean propositions X all fulfilling $AX = AB$: $(B|A) := \{X \in \mathcal{L} \mid AX = AB\} = \{X \in \mathcal{L} \mid AB \leqslant X \leqslant \overline{A} \vee B\}$, where $\leqslant$ is the usual entailment relation between boolean propositions (see [GN88]). That is, a conditional is identified with an interval of boolean propositions. This gives rise to a partial ordering $\leqslant$ between conditionals which is also frequently understood as an entailment relation between conditionals (cf. [DP91b,BDP97]):

$$(D|C) \leqslant (B|A) \quad \text{iff} \quad CD \leqslant AB \text{ and } A\overline{B} \leqslant C\overline{D} \tag{16}$$

Walker lists in [Wal94] several equivalent representations of conditional events and investigates different ways of defining logical connectives between conditionals. The representation that seems most adequate to compare our approach with others is the one that identifies a conditional $(B|A)$ with an ordered pair of its crucial components $AB, A\overline{B}$:

$$(B|A) \sim (AB, A\overline{B}), \tag{17}$$

so that $(\mathcal{L} \mid \mathcal{L}) \sim \{(X, Y) \mid X, Y \in \mathcal{L}, XY \equiv \bot\}$. Moreover, the appropriateness of the three-valued semantics of conditionals (1) which we base our theory upon is accepted by most other researchers in the field of conditionals. So the essential ingredients to our theory are established and nearly undebatable. The difference lies in the way we make use of them. Our focusing not on logical properties but on the effects conditionals exert on possible worlds suggests another way of relating conditionals to one another. First of all, both positive and negative effects are considered to be equally important, no priority is given to the confirming component AB (as is done in most approaches). So, a conditional $(D|C)$ is called a *subconditional* of $(B|A)$, iff its effect on worlds is in line with $(B|A)$, but possibly $(D|C)$ applies to fewer worlds:

$$(D|C) \sqsubseteq (B|A) \quad \text{iff for all } \omega \in \Omega, \begin{cases} \sigma_{(D|C)}(\omega) = 1 \Rightarrow \sigma_{(B|A)}(\omega) = 1 \\ \sigma_{(D|C)}(\omega) = 0 \Rightarrow \sigma_{(B|A)}(\omega) = 0 \end{cases} \tag{18}$$

which amounts to saying that $(D|C) \sqsubseteq (B|A)$ iff $CD \leqslant AB$ and $C\overline{D} \leqslant A\overline{B}$. Note the essential difference with (16), but we still have $(B|A) \equiv (D|C)$ iff $(B|A) \sqsubseteq (D|C)$ and $(D|C) \sqsubseteq (B|A)$, due to $(B|A) \equiv (D|C)$ iff $AB \equiv CD$ and $A\overline{B} \equiv C\overline{D}$.

Further, we can define connectives $\sqcap, \sqcup$ by setting

$$(B|A) \sqcap (D|C) := \inf\{(B|A), (D|C)\}$$

and

$$(B|A) \sqcup (D|C) := \sup\{(B|A), (D|C)\},$$

where the supremum is defined only if $ABC\overline{D} \equiv A\overline{B}CD \equiv \bot$. It is easy to check that

$$(B|A) \sqcap (D|C) = (BD \mid AC(BD \vee \overline{B}\,\overline{D})) \tag{19}$$

$$(B|A) \sqcup (D|C) = (AB \vee CD \mid A \vee C) \tag{20}$$

Although $(B|A) \sqcup (D|C)$ in its original sense as a supremum is defined only for $ABC\overline{D} \equiv A\overline{B}CD \equiv \bot$, (20) may be used for defining a connective for arbitrary $(B|A), (D|C)$. If the condition $ABC\overline{D} \equiv A\overline{B}CD \equiv \perp$ is not fulfilled, however, then $(B|A) \sqcup (D|C)$ is not the supremum of $(B|A)$ and $(D|C)$ (as can easily be seen by considering $(B|A)$ and $(\overline{B}|A)$). $\sqcap$ and $\sqcup$ are associative and distributive operations on $(\mathcal{L} \mid \mathcal{L})$, if all required suprema exist (cf. [KI01a]). Their connections to the boolean operations $\wedge$ and $\vee$ can be seen more directly by using the representation (17)

$$\begin{aligned}(X,Y) \sqcap (W,Z) &= (X \wedge W, Y \wedge Z),\\ (X,Y) \sqcup (W,Z) &= (X \vee W, Y\overline{W} \vee Z\overline{X}),\\ &= (X \vee W, Y \vee Z) \quad \text{if } XZ \equiv YW \equiv \perp,\end{aligned}$$

that is, the $\wedge$- and $\vee$-operator, respectively, are applied to the confirming and the refuting components separately, respectively. Both $\sqcap$ and $\sqcup$ are truth-functional, with $((B|A) \sqcap (D|C))(\omega) = 1$ iff $(B|A)(\omega) = 1$ and $(D|C)(\omega) = 1$, and $((B|A) \sqcap (D|C))(\omega) = 0$ iff $(B|A)(\omega) = 0$ and $(D|C)(\omega) = 0$, whereas $((B|A) \sqcup (D|C))(\omega) = 1$ iff $(B|A)(\omega) = 1$ or $(D|C)(\omega) = 1$, and $((B|A) \sqcup (D|C))(\omega) = 0$ iff $(B|A)(\omega) = 0$ and $(D|C)(\omega) \neq 1$, or $(D|C)(\omega) = 0$ and $(B|A)(\omega) \neq 1$. Therefore, we obtain the following truth tables for combining conditionals by $\sqcap$ and $\sqcup$, respectively (note that for the supremum, $0 \sqcup 1$ and $1 \sqcup 0$ do not exist):

$\sqcap$	0	1	u
0	0	u	u
1	u	1	u
u	u	u	u

$\sqcup$	0	1	u
0	0	1/*undef.*	0
1	1/*undef.*	1	1
u	0	1	u

These truth-tables reveal some incompatibilities with boolean logic: We have $0 \sqcap 1 = 1 \sqcap 0 = u$, and $0 \sqcup 1$ and $1 \sqcup 0$ are not defined, if $\sqcup$ is taken as a supremum. If one conditional is confirmed and the other refuted, nothing is said how they combine. So, no boolean rules are forced upon the interactions of conditionals. Therefore, the difference between our theory and most other approaches is that we do not expect the connectives $\sqcup$ and $\sqcap$ to observe boolean laws; in particular, though $\sqcup$ and $\sqcap$ are based upon $\vee$ and $\wedge$, respectively, they cannot be taken as extensions of those boolean operators (note, however, that the generalized version of $\sqcup$ does extend $\vee$ and is identical to Calabrese's conditional $\vee$-operator; see [Cal91]). Moreover, $\sqcap$ and $\sqcup$ are not linked by DeMorgan's laws; rather, the negation $(B|A)' := (\overline{B}|A)$ acts as a homomorphism with respect to both connectives: $((B|A) \sqcap (D|C))' = (B|A)' \sqcap (D|C)'$ and $((B|A) \sqcup (D|C))' = (B|A)' \sqcup (D|C)'$.

This breaking with boolean conditional traditions also has consequences for considering facts (i.e., boolean propositions) as conditionals: Usually, a proposition A is identified with the degenerate (two-valued) conditional $(A|\top)$. This appears to be justified by considering (1). Note, however, that actually, stating A means "*A is true*", whereas $(A|\top)$ means "*A is plausible*". This distinction is respected in our framework, in that $A \wedge B$ does not correspond to $(A|\top) \sqcap (B|\top)$ $(= (AB|AB \vee \overline{AB}) \neq (AB|\top)$; the conjunction of two plausible propositions need not be plausible), nor does $(A|\top) \sqsubseteq (B|\top)$ mean $A \subseteq B$; rather, we have $(A|\top) \sqsubseteq (B|\top)$ if and only if $A \equiv B$.

In summary, it can be said that, although the roots of our approach are firmly established in the foundations of conditional theories, the thorough application of the new concept of conditional structures brings forth quite a new theory of conditionals which is non-boolean to a very large extent. We think, however, that the loss of boolean conveniences is more than compensated by gaining a rich methodological framework for handling the dynamic power of conditionals more adequately, in particular in the areas of inductive knowledge representation and belief revision.

To exemplify what it means that c-representations impose conditional structures upon worlds, we state the following proposition which relates $\sqsubseteq$ to $\mid\!\sim^c$-inference (cf. (7)) and extends Corollary 1:

Proposition 4. *$(B|A) \mid\!\sim^c (D|C)$ iff $(D|C) \sqsubseteq (B|A)$, that is, if $(D|C)$ is a subconditional of $(B|A)$.*

Proof. Any possibilistic c-representation of $(B|A)$ has the form

$$\pi(\omega) = \begin{cases} \alpha_0 & \text{if } \omega \models \overline{A}, \\ \alpha_0\alpha^+ & \text{if } \omega \models AB, \\ \alpha_0\alpha^- & \text{if } \omega \models A\overline{B} \end{cases}$$

with positive real numbers $\alpha_0, \alpha^+, \alpha^-$ such that $\alpha^+ > \alpha^-$ (cf. (6)). If $(D|C) \sqsubseteq (B|A)$, then also $\pi \models (D|C)$, and conversely, if $(D|C)$ is not a subconditional of $(B|A)$, then $\alpha_0, \alpha^+, \alpha^-$ can be chosen appropriately so that $\pi \not\models (D|C)$.

So, in the simple case where only one conditional is considered, inferring by observing conditional structures means accepting all subconditionals, too. If the inference is to be based on a set of conditionals, interactions between these conditionals have to be taken into account, and c-entailment should not be expected to be this simple. A thorough investigation of $\mid\!\sim^c$ is a topic of our current research.

9 Conclusion and Outlook

In this paper, we presented an approach to handle conditional knowledge most appropriately in a possibilistic environment. The framework we developed was based on the theory of conditional structures which are apt to make conditional interactions transparent and manageable. Observing conditional structures gives rise to the *principle of conditional indifference* which proves to be a most powerful guideline for inductive knowledge representation as well as for belief revision, where it underlies the *principle of conditional preservation.* Conditional preservation, on the one hand, means to respect prior conditional knowledge as far as possible, and, on the other hand, imposes the structures of the new conditionals onto the revised belief state. It allows us to deal with revision in a most general framework, namely, to revise epistemic states by sets of conditionals.

The link that has been established between conditionals and group theory via conditional structures may also be used to extract rules from possibilistic data, considering knowledge discovery as a process inverse to inductive knowledge presentation. Numerical relationships are taken as reflecting structural dependencies between conditionals, which can be elaborated by group theoretical means. So our approach is numerically based, but not guided by focusing on high possibilities. It does not concentrate on single, isolated rules, but is apt to take highly complex interactions of conditionals into account.

All techniques presented in this paper can be generalized to handle quantified conditionals in a straightforward way. Actually, the application of our theory to inductive knowledge representation, belief revision and knowledge discovery in a richer framework, where the acceptance of conditionals may be specified by values reflecting their degrees of necessity, appears to be even more rewarding. The elaboration of our ideas in that framework is a topic of our present research.

References

[BDP97] S. Benferhat, D. Dubois, and H. Prade. Nonmonotonic reasoning, conditional objects and possibility theory. *Artificial Intelligence*, 92:259–276, 1997.

[BDP99] S. Benferhat, D. Dubois, and H. Prade. Possibilistic and standard probabilistic semantics of conditional knowledge bases. *Journal of Logic and Computation*, 9(6):873–895, 1999.

[BG93] C. Boutilier and M. Goldszmidt. Revision by conditional beliefs. In *Proceedings 11th National Conference on Artificial Intelligence (AAAI'93)*, pages 649–654, Washington, DC., 1993.

[BSS00] S. Benferhat, A. Saffiotti, and P. Smets. Belief functions and default reasoning. *Artificial Intelligence*, 122:1–69, 2000.

[Cal91] P.G. Calabrese. Deduction and inference using conditional logic and probability. In I.R. Goodman, M.M. Gupta, H.T. Nguyen, and G.S. Rogers, editors, *Conditional Logic in Expert Systems*, pages 71–100. Elsevier, North Holland, 1991.

[DeF74] B. DeFinetti. *Theory of Probability*, volume 1,2. John Wiley and Sons, New York, 1974.

[DGC94] D. Dubois, I.R. Goodman, and P.G. Calabrese. Special issue on the conditional event algebra. *IEEE Transactions on Systems, Man, and Cybernetics*, 24(12), 1994.

[DLP94] D. Dubois, J. Lang, and H. Prade. Possibilistic logic. In D.M. Gabbay, C.H. Hogger, and J.A. Robinson, editors, *Handbook of Logic in Artificial Intelligence and Logic Programming*, volume 3, pages 439–513. Oxford University Press, 1994.

[DP91a] D. Dubois and H. Prade. Conditional objects and non-monotonic reasoning. In *Proceedings 2nd Int. Conference on Principles of Knowledge Representation and Reasoning (KR'91)*, pages 175–185. Morgan Kaufmann, 1991.

[DP91b] D. Dubois and H. Prade. Conditioning, non-monotonic logic and non-standard uncertainty models. In I.R. Goodman, M.M. Gupta, H.T. Nguyen, and G.S. Rogers, editors, *Conditional Logic in Expert Systems*, pages 115–158. Elsevier, North Holland, 1991.

[DP91c] D. Dubois and H. Prade. Epistemic entrenchment and possibilistic logic. *Artificial Intelligence*, 50:223–239, 1991.

[DP94] D. Dubois and H. Prade. A survey of belief revision and updating rules in various uncertainty models. *Intern. Journal of Intelligent Systems*, 9:61–100, 1994.

[DP97] A. Darwiche and J. Pearl. On the logic of iterated belief revision. *Artificial Intelligence*, 89:1–29, 1997.

[FH94] N. Friedman and J.Y. Halpern. Conditional logics of belief change. In *Proceedings 12th National Conference on Artificial Intelligence, AAAI-94*, 1994.

[GK97] J. Gebhardt and R. Kruse. Background and perspectives of possibilistic graphical models. In *Proceedings First International Joint Conference on Qualitative and Quantitative Practical Reasoning, ECSQARU-FAPR'97,*, pages 108–121. Springer, 1997.

[GMP93] M. Goldszmidt, P. Morris, and J. Pearl. A maximum entropy approach to nonmonotonic reasoning. *IEEE Transactions on Pattern Analysis and Machine Intelligence*, 15(3):220–232, 1993.

[GN88] I.R. Goodman and H.T. Nguyen. Conditional objects and the modeling of uncertainties. In M.M. Gupta and T. Yamakawa, editors, *Fuzzy Computing - Theory, Hardware and Applications*, pages 119–138. North-Holland, 1988.

[KI98] G. Kern-Isberner. Characterizing the principle of minimum cross-entropy within a conditional-logical framework. *Artificial Intelligence*, 98:169–208, 1998.

[KI99] G. Kern-Isberner. *A unifying framework for symbolic and numerical approaches to nonmonotonic reasoning and belief revision.* Department of Computer Science, FernUniversität Hagen, 1999. Habilitation thesis.

[KI00a] G. Kern-Isberner. Conditional preservation and conditional indifference (preliminary version). In *Proceedings 8th International Workshop on Nonmonotonic Reasoning (NMR'2000)*, Breckenridge, Colorado, 2000.

[KI00b] G. Kern-Isberner. Solving the inverse representation problem. In *Proceedings 14th European Conference on Artificial Intelligence, ECAI'2000*, pages 581–585, Berlin, 2000. IOS Press.

[KI01a] G. Kern-Isberner. *Conditionals in nonmonotonic reasoning and belief revision.* Springer, Lecture Notes in Artificial Intelligence, 2001. (to appear).

[KI01b] G. Kern-Isberner. Handling conditionals adequately in uncertain reasoning. In *Proceedings ECSQARU'01*, 2001. (to appear).

[KS91] H. Katsuno and K. Satoh. A unified view of consequence relation, belief revision and conditional logic. In *Proceedings Twelfth International Joint Conference on Artificial Intelligence, IJCAI-91*, pages 406–412, 1991.

[Lew76] D. Lewis. Probabilities of conditionals and conditional probabilities. *The Philosophical Review*, 85:297–315, 1976.

[LM92] D. Lehmann and M. Magidor. What does a conditional knowledge base entail? *Artificial Intelligence*, 55:1–60, 1992.

[LS77] R.C. Lyndon and P.E. Schupp. *Combinatorial group theory.* Springer, Berlin Heidelberg New York, 1977.

[Nut80] D. Nute. *Topics in Conditional Logic.* D. Reidel Publishing Company, Dordrecht, Holland, 1980.

[NW94] H.T. Nguyen and E.A. Walker. A history and introduction to the algebra of conditional events and probability logic. *IEEE Transactions on Systems, Man, and Cybernetics*, 24(12):1671–1675, 1994.

[Par94] J.B. Paris. *The uncertain reasoner's companion – A mathematical perspective.* Cambridge University Press, 1994.

[Rot91] H. Rott. A nonmonotonic conditional logic for belief revision, part I: Semantics and logic of simple conditionals. In A. Fuhrmann and M. Morreau, editors, *The logic of theory change*, pages 135–181. Springer, Berlin, Heidelberg, New York, 1991.

[Spo88] W. Spohn. Ordinal conditional functions: a dynamic theory of epistemic states. In W.L. Harper and B. Skyrms, editors, *Causation in Decision, Belief Change, and Statistics, II*, pages 105–134. Kluwer Academic Publishers, 1988.

[Wal94] E.A. Walker. Stone algebras, conditional events, and three valued logic. *IEEE Transactions on Systems, Man, and Cybernetics (Special Issue on The Conditional Event Algebra)*, 24(12):1699–1707, 1994.

Extracting Trapezoidal Membership Functions of a Fuzzy Rule System by Bacterial Algorithm

J. Botzheim, B. Hámori, and L.T. Kóczy

Department of Telecommunication and Telematics,
Budapest University of Technology and Economics,
H-1117 Budapest, Pázmány P. sétány 1/d, Hungary
koczy@ttt.bme.hu

Abstract. This paper presents a method of using the so-called "bacterial algorithm" [4, 5] for extracting the rules of a fuzzy system. The class of membership functions is restricted to trapezoidal, as it is general enough and widely used. The algorithm contains the bacterial mutation step allowing both the change of more than one membership function at one time, and the fine-tuning. Further operators are applied for optimising the number of rules in the base, eliminating ineffective rules and contract similar rules into a single one. ...

1 Introduction

In the applications of fuzzy systems [1] one of the most important tasks is to find the optimal rule base. This might be given by a human expert or might be given a priori by the linguistic description of the modelled system. If however neither an expert, nor linguistic descriptions are available, the system has to be designed by other methods based on numerical data [5]. Nature inspired some evolutionary optimisation algorithms suitable for global optimisation of even non-linear, high-dimensional, multimodal, and discontinuous problems. The original genetic algorithm (GA) was developed by Holland [2] and was based on the process of evolution of biological organisms. GA-s use three well-known operators: reproduction, crossover, and mutation. Closely related evolutionary programming (EP) was proposed by Fogel [3]. EP uses selection and mutation operators. A more recent approach is the bacterial algorithm. This includes a modified mutation operator called bacterial mutation, emulating a biological phenomenon of microbial evolution. This gives an alternative by other algorithms, because it is simpler, and it is possible to reach lower error within a short time. In Section 2 the basics of fuzzy systems and evolutionary algorithms are introduced. In Section 3 the bacterial algorithm for trapezoidal fuzzy systems is described. Simulation results can be found in Section 4. Section 5 concludes the paper.

B. Reusch (Ed.): Fuzzy Days 2001, LNCS 2206, pp. 218–227, 2001.

2 Fuzzy Systems and Evolutionary Algorithms

The theory of fuzzy logic was developed by Zadeh in the late 1960s [1]. This was the rediscovering of the multivalued logic created by Lukasiewicz. The main idea was that most of the phenomena of the real world could not be described by two values. The function assigning fuzzy truth degrees between zero and one to elements of a universal set:

$$\mu_A : X \rightarrow [1,0], \tag{1}$$

In the above, *A* is a fuzzy set. A fuzzy rule base contains fuzzy rules R_i:

$$R_i\text{: } IF\ (x_1\ is\ A_{i1})\ AND\ (x_2\ is\ A_{i2})\ AND\ ...\ AND\ (x_n\ is\ A_{in})\ THEN\ (y\ is\ B_i), \tag{2}$$

where A_{ij} and B_i are fuzzy sets, x_j and y are fuzzy inputs and output. The structure of a rule is the following:

$$IF\ Premise\ THEN\ Conclusion, \tag{3}$$

where the premise consists of antecedents linked by fuzzy operator *AND*. There are many alternative possibilities to define the t-norm operator. The most frequently used one is the minimum operation. In the *MAX-MIN* inference method, the activation degree of the premise is:

$$w_i = min\ \{w_{i1}, w_{i2}, ..., w_{in}\}, \tag{4}$$

where w_{ij} is the intersection between x_j input and X_{ij} fuzzy set in the ith rule. In a fuzzy inference system the output is calculated in the following way: the activation degree w_i is computed for each rule in the rule base. Then the ith output fuzzy set is cut-off by w_i in each rule and the union of these cut-off fuzzy sets is composed, where the union means generally the maximum operation. Then a crisp value from the resultant output fuzzy set is made. This process is called defuzzification. There are many ways of defuzzification. The Centre of Gravity (COG) method is used here because it is general and easy to compute. This method calculates the crisp output by the sums of the centre of gravities of the conclusions. Thus, a fuzzy inference system can compute output y of an input vector **x**. The main purpose is to make the best solution possible for each input vector, therefore the optimum rule base need to be found. There are several optimisation algorithms, which were inspired by the processes in nature. The advantage of these algorithms is their possibility to solve and quasi-optimise problems with non-linear, high-dimensional, multimodal, and discontinuous character. The original genetic algorithm (GA) was developed by Holland [2] and based on the process of evolution of biological organisms. The GA use three operators: reproduction, crossover, and mutation. First the population must be initialised. Then the genetic cycle follows: individuals from the population are selected. After this, the selected individuals go through crossover: two parents create two offsprings with the random exchange of bits between the two parents. Finally, the mutation operator alters randomly some bits of the new offsprings. So, a new generation develops and the genetic cycle continues until a good solution to the problem is taken. Evolutionary programming (EP) was proposed by Fogel in the 1960's [3]. EP uses the selection and mutation operators. A newer approach is the bacterial algorithm. This includes a modified mutation operator called bacterial mutation based on a biological phenomenon of microbial evolution.

3 Fuzzy Rule Base Identification by Bacterial Algorithm

The algorithm is similar to that used in [4]. The main difference is that trapezoidal membership functions are used. They are described by four parameters with the four breakpoints of the trapezium. Moreover the membership functions are identified by the two indices *i* and *j*. So, the membership function $A_{ij}(a_{ij}, b_{ij}, c_{ij}, d_{ij})$ belongs to the ith rule and the jth input variable. $B_i(a_i, b_i, c_i, d_i)$ is the output membership function of the ith rule. The relative importance of the jth fuzzy variable in the ith rule:

$$A_{ij}(x_j)=\begin{cases}\dfrac{x_j-a_{ij}}{b_{ij}-a_{ij}}, & if\ a_{ij}<x_j<b_{ij}\\ 1, & if\ b_{ij}\le x_j<c_{ij}\\ \dfrac{d_{ij}-x_j}{d_{ij}-c_{ij}}, & if\ c_{ij}\le x_j<d_{ij}\\ 0, & \text{otherwise}\end{cases}, \tag{5}$$

where $a_{ij} \le b_{ij} \le c_{ij} \le d_{ij}$ must hold.

So the encoding method of a fuzzy system with two inputs and one output, see in Fig. 1.

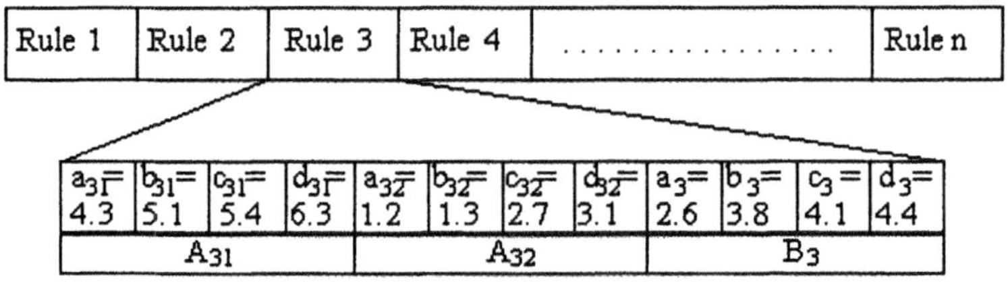

Fig. 1. Fuzzy rules encoded in a chromosome

For example, Rule 3 in Fig. 1 means:

$$\textit{if } x_1 \textit{ is } A_{31}(4.3,5.1,5.4,6.3) \textit{ and } x_2 \textit{ is } A_{32}(1.2,1.3,2.7,3.1) \quad \textit{then } y \textit{ is } B_3(2.6,3.8,4.1,4.4) \tag{6}$$

where A_{ij} and B_i mean the trapezoidal membership function with the four breakpoints. The algorithm is described as follows [4, 5]

3.1 Generating the Initial Bacterium

First the initial (random) bacterium is created. This means that all membership functions in the chromosome must be randomly initialised. The initial number of rules is N_{max}. So, in this step $(k+1)N_{max}$ membership functions are created (k is the number of input variables in the given problem, and each membership function has four parameters).

3.2 Bacterial Mutation

After the initialisation process the algorithm goes through several cyclic steps. First, m -1 copies (clones) of the rule base are generated [4]. Then a certain part of the chromosome [5] is randomly selected and the parameters of this selected part are randomly changed in each clone (mutation). Next all the clones and the original bacterium are evaluated by the error criterion. The best individual transfers the mutated part into the other individuals. This cycle is repeated for the remaining parts, until all parts of the chromosome have been mutated and tested. At the end the best rule base is kept and the remaining m -1 are discharged. The above procedure is repeated once more, however, with different parameters [4]. It is an important question how long is one part suffering mutation and what is the degree of the mutation (expressed as the relative size in terms of the interval). This approach allows both selecting more than one membership function and fine-tuning. The number of mutated membership functions and the mutation degree are external parameters of the bacterial mutation. If selecting more than one membership function is allowed then the local minima in the optimisation process can be avoided. Application of changing two or more membership functions at a time with fine-tuning in the first bacterial mutation and in the second step changing one membership function in the whole interval of the given variable was proposed in [4].

3.3 Fuzzy Operators

Bacterial mutation optimises the membership functions in the rule base. Besides, it is important to optimise the structure of the rule base. Ineffective rules must be eliminated and similar rules must be contracted. The following operators help to optimise the number of rules:

Annihilation. When a membership function becomes too narrow the rule using it must be deleted. The evaluation criterion (Fig. 2) is:

$$l_i \mu_i \left(\frac{a_j + b_j + c_j + d_j}{4} \right) \geq \beta l_j, \tag{7}$$

where l_i and l_j are the lengths of the medians of the membership functions of the given input or output variables in the ith and j[th] rule and β is the annihilation parameter. The larger is the value of β, the more severe is the annihilation criterion.

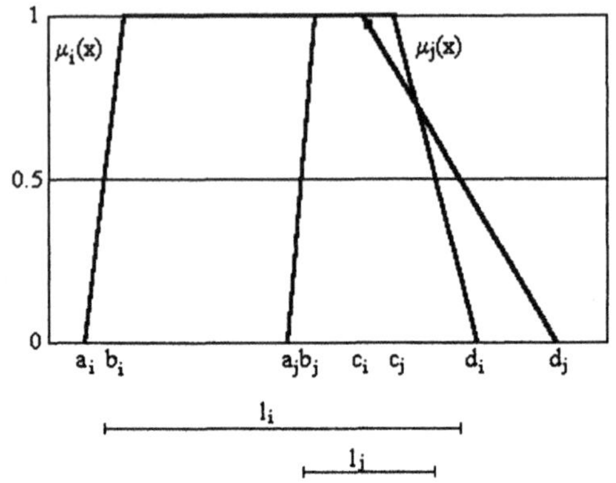

Fig. 2. Annihilation

Fusion. If two membership functions, belonging to the same variable are near to each other, and the difference between the lengths of their median is small enough, then they are fused in a single membership function (Fig. 3).

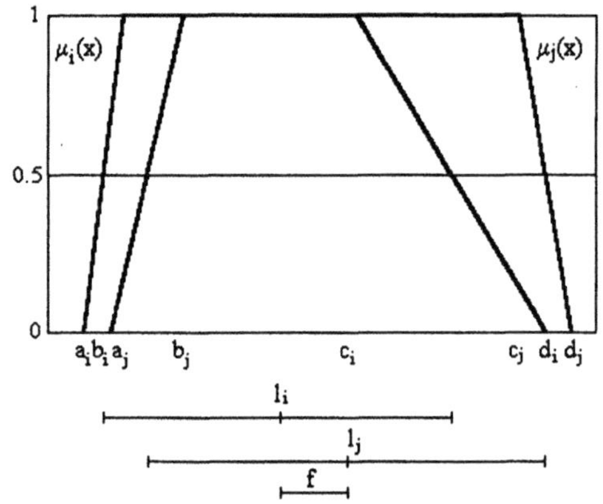

Fig. 3. Fusion

There are two criteria of fusion, accordingly:

$$\left|\frac{l_i}{l_j}-1\right|<\gamma \quad \text{and} \quad |f|<\gamma, \tag{8}$$

where l_i and l_j are the lengths of the median of the membership functions of the given variable in the ith and j[th] rule, and f is the distance between the centres of l_i and l_j. To execute fusion both criteria must be satisfied, but only one parameter, γ is used. The smaller is the value of γ, the more severe is the criterion of fusion. The parameters of the fused membership function will be:

$$z_{fus} = \frac{z_i l_i + z_j l_j}{l_i + l_j}, \qquad (9)$$

where z stands for the four breakpoints (a, b, c and d) of a trapezoidal membership function.

Semantic analysis. If two rules have the same antecedents but a different consequent, the membership functions of the output variable are fused in one output membership function by using the fusion described above [4].

Rule removal. As COG defuzzification is used, removal of a rule [4] generates large error values so it decreases the performance of the fuzzy system. A new method is proposed here where the importance of the remaining rule is increased so it represents the two identical rules in one. For example if the ith and the jth rules are identical then the jth rule is deleted and the parameters of the new output membership function of the ith rule will be calculated as the following:

$$\begin{aligned} m_i &= 2d_i - 2a_i + c_i - b_i \\ n_i &= c_i^2 + 2c_i d_i + 2d_i^2 - 2a_i^2 - 2a_i b_i - b_i^2 \\ a_i' &= \frac{n_i - m_i^2 - c_i m_i}{2m_i + c_i - b_i} \\ d_i' &= a_i' + m_i \end{aligned} \qquad (10)$$

where a_i, b_i, c_i and d_i are the breakpoints of the trapezoidal membership function of the output variable and a_i' and d_i' are the new breakpoints for a_i and d_i. Parameters b_i and c_i remain unchanged.

3.3 Stop Condition

If the bacterium satisfies a stop condition or the maximum generation number is reached then the algorithm ends, otherwise it returns to the bacterial step.

4 Results

4.1 Definition of the Error Function

The widely used error functions in earlier publications defining the error value are as follows [4]:

$$e_1 = \frac{1}{n} \sum_{patterns} |y - \hat{y}| \text{ or } e_2 = \frac{1}{n} \sum_{patterns} (y - \hat{y})^2 , \quad (11)$$

where n is the number of the evaluated patterns, $\hat{y}$ is the desired output of the modelled system for an input pattern and y is the output of the fuzzy system for the same input. The main problem with this formula is that it is not normalised, so the error results cannot be compared with each other. Another formula used [5] is the following:

$$e_{rel} = \frac{1}{n} \sum_{patterns} \left| \frac{y - \hat{y}}{\hat{y}} \right| \quad (12)$$

The problem with this expression is that the error depends on the actual value of the output in a non-normalised way. If the expected output value is smaller, the error seems to be larger with the same difference. In our approach we introduced a different error function in order to reach better approximation of the model:

$$e_{rel} = \frac{1}{n} \sum_{patterns} \frac{|y - \hat{y}|}{I_{\max} - I_{\min}} , \quad (13)$$

where I_{max} is the upper and I_{min} is the lower bound of the interval of the output variable, so the error is normalised by the length of the output interval rather than the actual value of the output. This approach helps with keeping the accuracy of the approximation of the patterns always constant, not being sensitive to the absolute value of the desired output. At the same time it has the advantage of being a normalised value keeping the error within [0,1].

4.2 Simulation Results

Target function. To test the algorithm, the following widely used target function [4, 5] was chosen:

$$y = x_1 + x_2^{0.5} + x_3 x_4 + 2e^{2(x_5 - x_6)} , \quad (14)$$

where

$$x_1 \in [1,5], x_2 \in [1,5], x_3 \in [0,4], x_4 \in [0,0.6], x_5 \in [0,1], x_6 \in [0,1.2] . \quad (15)$$

The training set included 200, the validation set 500 patterns. The algorithm was running for 40 generations with 2 different bacterial operations in each and using 20 clones. Next we summarize some of the interesting points.

Number of changing membership functions. It is possible that the optimisation procedure is trapped in a local optimum and further optimisation becomes impossible when only one membership function is modified in each generation in a rule. This way more than one membership functions have to be changed, as it was experienced through the previous approximation tests. In this method the number of membership functions to change is fixed. Every membership function is allowed to change in one generation. During our tests with the six input variables function two or three changes turned out to be sufficient and larger numbers of changes showed no visible effect on the result as described in Figs. 4 and 5, where $\beta = 5$, $\gamma = 5\%$ and the initial number of rules was 10. The values in the table are averages of 4 results.

MF change	Rule No.	Train. err. [%]	Test err. [%]
1	5,75	4,30425	8,38225
2	4,75	5,66925	9,02325
3	6	4,8755	8,6555
4	4	6,6995	8,78875
5	3,75	6,13475	9,2765
6	4,75	5,646	8,366
7	5,75	5,948	9,28875

Fig. 4. Effect of how many MFs are changed (1).

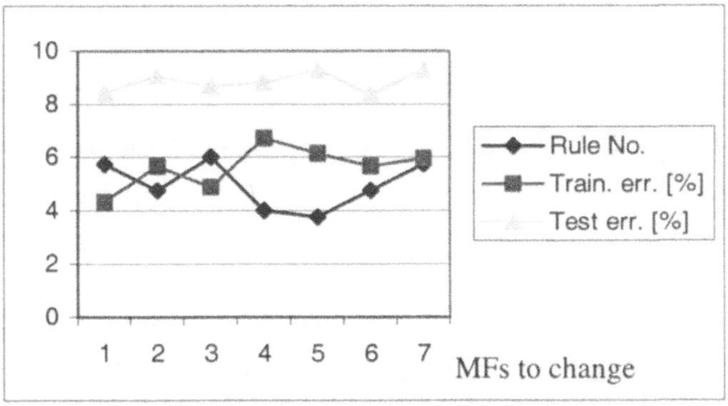

Fig. 5. Effect of how many MFs are changed (2).

We found that if the number of membership functions to be changed was increased, the approximation accuracy of the above mentioned benchmark function was not affected significantly, presumably because of the high complexity of the problem.

Fine-tuning. In our tests fine-tuning the membership functions was included and we found that if a membership function remained in the rule base it meant generally that it was important for the fuzzy system even if it might not be optimal. It was possible to improve the accuracy of a fuzzy system by changing the breakpoints of the trapeziums by small random values.

Reducing complexity of a fuzzy system. The use of the operations fusion, annihilation, semantic analysis, and rule removal proved to be an effective tool to optimise the size of fuzzy rule bases by reducing redundancy. Several tests were taken to identify the optimal parameters to avoid generating "too simple" rule bases and improving performance. Test results are shown in Figs. 6 and 7. γ= 5% and the initial number of rules was 10. The values in this table are averages of 4 results.

β	Rule No.	Train. err. [%]	Test err. [%]
5	4,75	5,66925	9,02325
6	5,75	5,24725	8,6675
7	7,5	4,78775	9,44525
8	9	4,455	9,311
9	8,75	4,55125	8,9235
10	8	4,261	9,01575

Fig. 6. Effect of the annihilation parameter β (1).

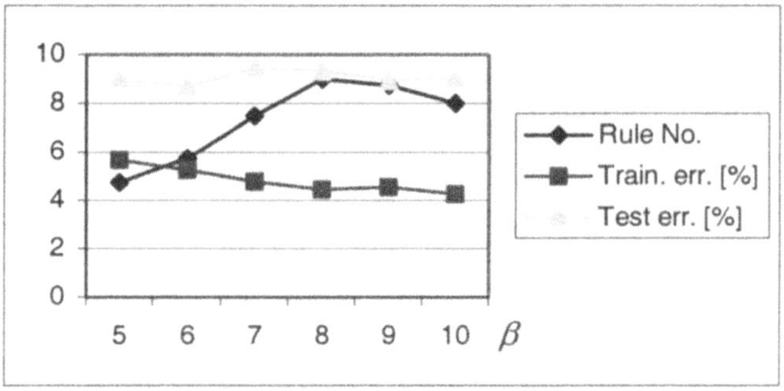

Fig. 7. Effect of the annihilation parameter β (2).

With the increase of the value of β the remaining rule number increased too, because the criteria of annihilation got more severe. As the rule number in the resulting base increased, the error measured during the training phase decreased however the error measured with the validation set did not. We interpreted it that the resulting large rule base was too specific for the training set and had no advantage when applied for an independent pattern set. We concluded that it was not necessary to give β high values as it just generated more complex rule bases with the same capability.

5 Conclusions

In this paper a new algorithm for the optimisation of the membership functions was shown. We intended not only to optimise the rules but also to reduce the size of the rule base. With trapezoidal membership functions the problem could be solved with fewer rules than with triangular ones. By applying fuzzy operators, optimal rule base size could be achieved.

References

1. L.A.Zadeh: Outline of a new approach to the analysis of complex systems and decision processes, IEEE Tr. Systems, Man and Cybernetics **3** (1973), pp. 28-44.
2. J.H.Holland: Adaptation in Nature and Artificial Systems: An Introductory Analysis with Applications to Biology, Control, and Artificial Intelligence, MIT Press, Cambridge, 1992.
3. L.J. Fogel, A.J.Owens, and M.J.Walsh: Artificial Intelligence through Simulated Evolution, Wiley, New York, 1966.
4. M.Salmeri, M.Re, E. Petrongari, and G.C.Cardarilli: A Novel Bacterial Algorithm to Extract the Rule Base from a Training Set, Dept. of Electronic Engineering, University of Rome, 1999.
5. N.E.Nawa, and T.Furuhashi: Fuzzy System Parameters Discovery by Bacterial Evolutionary Algorithm, IEEE Tr. Fuzzy Systems **7** (1999), pp. 608-616.

Possibility Theory, Probability Theory and Multiple-Valued Logics: A Clarification

Prof. Dr. Didier Dubois

University of Toulouse, France

There has been a long-lasting misunderstanding in the literature of artificial intelligence and uncertainty modeling, regarding the role of fuzzy set theory and many-valued logics. The recurring question is that of the mathematical and pragmatic meaningfulness of a compositional calculus and the validity of the exluded middle law. This confusion pervades the early developments of probabilistic logic, despite early warnings of some philosophers of probability. This talk tries to clarify this situation. It emphasizes three main points. First, it suggests that the root of the controversies lies in the unfortunate confusion between degrees of belief and what logicians call "degrees of truth". The latter are usually compositional, while the former cannot be so. This claim is first illustrated by laying bare the non-compositional belief representation embedded in the standard propositional calculus. It turns out to be an all-or-nothing version of possibility theory. This framework is then extended to discuss the case of fuzzy logic versus graded possibility theory. Next, it is demonstrated that any belief representation where compositionality is taken for granted is bound to at worst to collapse to a Boolean truth assignment and at best to a poorly expressive tool. Lastly, some claims pertaining to an alleged compositionality of possibility theory are refuted, thus clarifying a pervasive confusion between possibility theory axioms and fuzzy set basic connectives. Lastly, some connections between numerical possibility and probability will be outlined based on the fact that possibility distributions may encode families of probabilities. This talk is partially based on a paper with Henry Prade, to appear in the Annals of Mathematics for Artificial Intelligence.

B. Reusch (Ed.): Fuzzy Days 2001, LNCS 2206, p. 228, 2001.

Parameterized Petri Net Reformulation of the Fuzzy Production System

Zoran M. Gacovski[1] and Georgi M. Dimirovski[2]

[1]Information Technology Dept. at Ministry of Defence, Skopje, Rep. of Macedonia
[2] Institute of Automation & Systems Engineering (ASE) at Faculty of Electrical Engineering
St.Cyril and St.Methodius University, Karpos 2 BB, PO box 574,
1000 Skopje, R. of Macedonia
Fax: ++ 389-2-234-568,
e-mail: zorang@morm.gov.mk

Abstract. This work presents an original transformation of an organization level of complex systems and its model based on fuzzified Petri net, with rule-based decision-making and appropriate knowledge base. This modeling is an extension of all previous forms of fuzzy-Petri reasoning algorithm, because fuzzified parameterized Petri-nets proposed here, can be used to generate a high level abstract plan, even in the absence of complete information, which may be mapped into a specific plan at a lower level of the hierarchy. Our parameterized fuzzy-Petri net reformulation is following the original conception of G. N. Saridis and is fuzzy-addition to his Petri net co-ordinative controller. So it has general generic feature.

Keywords: Fuzzy reasoning, parameterized fuzzy-Petri-nets, decision making, knowledge base, organization level, complex systems.

1 Introduction

Intelligent Control, according to G. N. Saridis [3], is a multi-level structure obeying the Principle of Increasing Precision with Decreasing Intelligence. Probabilistic models are expressing the uncertainty of reasoning, planning decision making at the organization level, doing the assignment of tasks at the coordination level, and the control activities at the execution level. Entropies are used as measures of the execution of various commands by the Intelligent Machine and they serve for the optimal decision making. We have researched possibilities for two-level Control-system architecture with high-level organization and low-level coordinative layer in the supremal level and for reformulation of the organizing controller as a fuzzzy-parametrized-Petri-net production system.

Parameterized Petri net (PPN) formalism when used in hybrid systems gives us a lot of advantages because of its great power for representation and modeling of parallel and concurent processes, and ability to generate a high level, abstract plan, even in situation of incomplete information. The formalism of parameterized-Petri nets can be used for modeling fuzzy-rule based systems directly, by simply identifying some of the elements (marking function) and features (basically

B. Reusch (Ed.): Fuzzy Days 2001, LNCS 2206, pp. 229–238, 2001.

parameterized places and transitions) of PPNs formalism with the basic elements of a fuzzy knowledge base (KB) (propositions, degree of truth and implication relationships). The formalism is achieved with more specific aspects such as the association of propositions of the KB and places in fuzzy-PPN (FPPN), through the definition of a bijective function, and the association of transitions and the degrees of truth, which make up the KB. Furthermore, a formal separation between the representational scheme (the FPPN itself) and the associated dynamic process (data driven evaluation algorithm) is established using elements defined in the formalism but not included as a part of the FPPN. Our work explores further this line of research and contributes in terms of control-based fuzzy-parameterized-Petri-net computing mechanism which implements a two-layer supervisory controller with organizing-co-ordinating functions and employs fuzzy-rule production system (FPS).

2 Parameterized Petri Nets

Parameterized Petri Nets (PPNs), which were originally proposed by Gracanin, Srinivasan and Valavanis [6], are analytical methodology for representation, modeling, analysis and synthesis of real-time complex hierarchial systems. They incorporate information related to the ordering of a sequence of actions (planning) in a natural way through the introduction of parameterized places, pattern transitions and descriptors. PPNs are also capable of inferring plans at every level of the system hierarchy and allow for the addition of pertinent details/information during subsequent steps (lower levels of hierarchy).

The formal definition of PPNs is as follows:

PPN=(C, D, P_p, T, I, O), where:

$C=\{CV_1, CV_2, ..., CV_{|C|}\}$ - set of all parameter values, where CV_i is a set of values for the i-th parameter and $|C|$ is the number of system parameters;

$D=CV_1 \times CV_2 \times ... \times CV_{|C|}$ - set of all patterns $\underline{x}=(x_1, x_2, ..., x_{|C|})$, where $x_1 \in CV_1,...,x_{|C|} \in CV_{|C|}$;

$P_P=\{PP_1,PP_2,...,PP_{|Pp|}\}$ - a parameterization descriptor, which is set of parameterized places, such that: $\cup_{i=1,|Pp|} PP_i=D$, $PP_i \cap PP_j=\emptyset$, $i \neq j$.

$T=\{pt_1, pt_2, ..., pt_{|T|}\}$ - the set of all pattern transitions, that represents all actions that may take place in the system.

$t=\{pt_{i1}, ..., pt_{i|t|}\} \subseteq T$ - a parameterized transition, which is set of pattern transitions, that satisfy: $|I(pt_i) \cap PP_k| = |I(pt_j) \cap PP_k|$, $|O(pt_i) \cap PP_k|=|O(pt_j) \cap PP_k|$, $\forall i,j,k$: $pt_i \in t$, $pt_j \in t$, $Pp_k \in P_p$.

The number of input and output patterns belonging to a particular parameterized place must be the same for every pattern transition and every parameterized place in the parameterization descriptor. The set of all parameterized transitions is uniquely determined by the parameterization descriptor P_p and the set of all pattern transitions T, i.e. there is a mapping $PT=m_2(P_p, T) \subseteq 2^T$. The parameterized input and output functions are:

$$PI(t)=\{PP_{i1},...,PP_{i|PI(t)|}\} \subseteq P_p,\ PO(t)=\{PP_{j1},...,PP_{i|PO(t)|}\} \subseteq P_p \qquad (1)$$

The number of places in the PPN is not fixed and it depends on the parameterization descriptor which describes the dynamic properties of a PPN. Given the parameterization descriptor, one can accordingly group the pattern transitions to correspond to transitions in a standard PN. Parameterized places correspond to places in the standard PN. Transitions in the standard PN correspond to parameterized transitions. A token corresponds to a pattern of information.

The interaction with the world can be represented as a PPN shown in Fig.1a. Using parameterization, patterns are grouped in several sets and each set is represented with a new place. Pattern transitions are accordingly grouped and each group is represented with a parameterized transition. In order to represent a particular system, a detailed parameterization is necessary. Patterns are grouped in parameterized places according to some criteria. Sets of pattern transitions which consume or produce information from those places are represented as parameterized transitions. A number of different groupings is possible, which produces different PPN representations; however, the system's behavior is the same. This partition of the set of all patterns is a parameterization descriptor. Fig.1b shows such a parameterization.

Def.: A parameterization is a mapping from one PPN to a set of PPNs all representing the same system, in such a way that only the parameterization descriptor P_p is changed:

$$(C, D, P_p, T, I, O) \rightarrow \{(C, D, P_p^{'}, T, I, O)\} \qquad (2)$$

In other words, one parameterization descriptor from the set is chosen in order to produce a new PPN representation of the system.

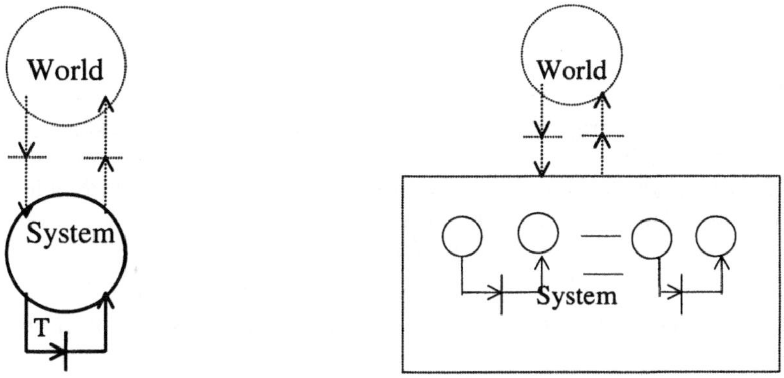

Fig. 1. A general and parameterized representation of a system.

In the most general case the PPN is most detailed with all connected places (very similar to regular PN). The most limited case is those with at most one parameterized place. Somewhere in between is a "useful" parameterization which reduces the number of choices while preserving flexibility. The criteria for the parameterization are based on some desired invariance of PPN properties and the semantics of the application. Parameterization may be restricted to occur only in top-down and bottom-up ways, depending on the necessary level of abstraction/detail required to analyze and simulate a system. If a top-down approach is used, parameterization may be interpreted as an "abstract to specific information" transformation, or *specialization*. When the bottom-up approach is used, a parameterization may be interpreted as a "specific to abstract information" transformation, or *generalization*. The parameterized graph in this case is a tree. In the top-down approach one parameterized place is splitting into two or more new ones, thus creating a new parameterization descriptor, and in bottom-up approach there is a grouping of two or more parameterized places into the same place.

At the highest level of abstraction, the set of all pattern transitions T is represented as a single parameterized transition in the PPN. When all patterns in a pattern transition are present, the pattern transition is enabled and may be activated (fired). If there is more than one enabled pattern transition, only one will be selected and activated. A marking M in the PPN is the same as in the corresponding PN. A single pattern transition is considered as a transition rule with a single pattern transition. So, the parameterized transition may be expressed as a union of rules, or parts of rules. Rules may be viewed as a convenient way to transform knowledge about the system to PPN terms, i.e. derivation of a PPN representation for the system knowledge. All properties of the regular PNs: safeness, boundedness, conservation, reachability and coverability are directly corresponding for the appropriate PPNs.

3 Parameterized Petri Net Model of FPS

We will consider the most generic representation of the rules which make up the KB in a FPS:

R^r: IF $\mathbf{X_1^r}$ **IS** $\mathbf{A_1^r}$ AND ...AND $\mathbf{X_{Mr}^r}$ **IS** $\mathbf{A_{Mr}^r}$ THEN $\mathbf{X_{Mr+1}^r}$ **IS** $\mathbf{B_1^r}$ AND ... AND $\mathbf{X_{Mr+Nr}^r}$ **IS** $\mathbf{B_{Nr}^r}$, (t^r),

where the bold parts are fuzzy propositions and t^r are linguistic values of the truth-variable which qualify the rules. We are going to approach the projection of a KB onto a PPN and we start by identifying the parameter sets of the PPN with linguistic variables and linguistic values from the above rule:

$$C=\{CV_1, CV_2\} \tag{3}$$

$$CV_1=\{X_1^r, \ldots, X_{Mr+Nr}^r\}, \quad CV_2=\{A_1^r, \ldots, B_{Nr}^r\},$$

Patterns of the PPN consist the propositions of the KB by means of bijective function:

$$a: P \to PR, \; p_k \to a(p_k)=pr_k, \quad k=1,\ldots,K, \tag{4}$$

where $PR=\{pr_k\}$ is the set of propositions in the KB and K is the number of propositions in the KB. In the case where a proposition is found several times in different rules of the KB, a different pattern will be assigned to it for each of these appearances in the KB. This is necessary because the rules are qualified by a linguistic value of the truth-variable; the propositions, therefore, cannot be considered identical from the execution point of view. This representation is simplified when the same proposition is found in the consequent part of different rules and the linguistic values that qualify the rules are identical. In this case a single place will be assigned to such propositions.

The description of the meaning of the pattern transitions is more complex, because of the linking rules. In our representation $T=T^R \cup T^C = \{t^1,\ldots,t^R,t^{R+1},\ldots,t^{R+C}\}$. Subset T^R includes the pattern transitions associated with each one of the rules which make up the KB, whereas subset T^C includes the pattern transitions that are associated with the existence of links between propositions. We define the input and the output functions over set T

$$I : T \dashrightarrow f(P) \quad (5)$$

$$O: T \dashrightarrow f(P) \quad (6)$$

which associate to each pattern transition the set of patterns which constitute its input and output, respectively. These functions have a different interpretation depending on the subset of T in which they are considered.

$$\text{If } t^j \in T^R,\ \forall p_i \in P,\ p_i \in I(t^j) <=> a(p_i) \in \text{Antecedent part of } R^j \quad (7)$$

$$\text{If } t^j \in T^R,\ \forall p_i \in P,\ p_i \in O(t^j) <=> a(p_i) \in \text{Consequent part of } R^j \quad (8)$$

$$\text{If } t^j \in T^C,\ p_i \in I(t^j),\ p_k \in O(t^j) <=> a(p_i) \text{ is linked with } a(p_k) \quad (9)$$

Therefore a single pattern transition $t^j \in T^C$ will exist for each of the indermediate variables X_j of the KB.

The graphic representation of the PPN is defined with:

$$A = \bigcup_{t^j \in T} \{ t^j \times O(t^j) \} \cup \{ I(t^j) \times t^j \} \quad (10)$$

We will define a truth function f that assigns to each $t^j \in T^R$ the linguistic truth value associated with rule R^j :

$$f : T^R \dashrightarrow V,\ t^j \dashrightarrow f(t^j) = t^j \quad (11)$$

where V represents the set of linguistic values of the linguistic truth variable. We say that pattern p_l is immediately reachable from pattern p_k if:

$$\exists t^j \in T / p_k \in I(t^j) \text{ and } p_l \in O(t^j) \quad (12)$$

The set of immediately reachable patterns from a given pattern p_k make up the immediate rechability set $IRS(p_k)$. The pattern p_l is reachable from p_k, if p_l is immediately reachable from p_k or p_l is reachable from $p_i \in IRS(p_k)$. The set of all patterns that are reachable from pattern p_k is called the reachability set $RS(p_k)$. Two patterns are adjacent, if:

$$\exists t^j \in T / p_k, p_l \in I(t^j) \quad (13)$$

The concept of adjacency can be extended to pattern transitions in a similar way. The adjacent pattern transition set with respect to pattern $p_i \in P$ is the set:

$$ATS(p_i) = \{ t^j / p_i \in O(t^j) \} \quad (14)$$

The adjacent pattern transitions associated with chains will represent the multiple link situation: several rules establish inferences over the same variable and one or more later rules make use of this variable in its (their) antecedent part. In this case, a pattern transition will be associated with each chain.

4 Fuzzy- PPN Reasoning Components

The fundamental notion for execution of a KB is marked PPN. Marking indicates that the degre of fulfillment (DOF) of the associated proposition is known, so this proposition can be used in the process of obtaining new references. It will be necessary for the DOF's of the different propositions to be available all the time. So we define the fulfillment function:

$$g: P \dashrightarrow [0,1] \quad (15)$$

which assigns to each pattern a real value:

$$g(p) = DOF(a(p)) \quad (16)$$

In our representation tokens are transferred from some patterns to others by means of the activation of pattern transitions, following a basic rule: A pattern transition $t^j \in T$ is active (and will fire) if every $p_i \in I(P)$ has a token. When during the process of firing a pattern transition the token of the input patterns is removed, the information obtained about the DOF of that propositions is preserved in the fulfillment function. The firing of an active pattern transition $t^j \in T^R$ is equivalent to the application of a rule in the process of evaluating the KB. The activation of $t^j \in T^C$ is equivalent to knowing, whether it be through previously performed inferences or through observation, the DOF of propositions $a(p_i)$, $\forall\, p_i \in I(t^j)$. In this case, the DOF for propositions $a(p_k)$, $\forall\, p_k \in O(tj)$ is determined not by the application of rules of the KB, but by essentially the same method as the one used to determine the DOF of a proposition with observed input distribution values. Most of the operations participating in this calculation can be carried out a priori, leading to a significant simplification of the execution process. When all DOF's of the antecedent part of a rule are known and it is executed, the marking function will have placed tokens in all of the input patterns of the corresponding pattern transition, activating it and causing it to fire, which will produce a new marking function.

The definition of the initial marking function M in the PN representation of the KB of a FPS can be established as :

$$M : p \dashrightarrow \{ 0,1 \},\ p_i \dashrightarrow M(p_i) = \begin{cases} 0, & \text{if } g(p_i) \text{ is unknown} \\ 1, & \text{otherwise} \end{cases} \quad (17)$$

The marking function thus makes explicit the requirement that the DOF of a set of propositions must be known before an evaluation of the KB can be carried out. From a given marking function M, the firing of a pattern transition t^j will produce a new marking function M'. The evolution of the marking functions of a PN is described by the transition function tf:

$$tf: M \times T \dashrightarrow M,\ (M,t^j) \dashrightarrow M' \quad (18)$$

where:

$$M'(p_i) = \begin{cases} 0, & \text{if } p_i \in I(t^j) \\ 1, & \text{if } p_i \in O(t^j) \\ M(p_i), & \text{otherwise} \end{cases} \quad (19)$$

and M represents the set of all possible marking functions of the PPN. In the data driven evaluation strategy, the possibility distributions associated with input variables of the KB are initially known through observation. These known distributions will allow certain rules to fire and, as a result, new possibility distributions to be assigned to other (intermediate or output) variables. Repeating the process as many times as necessary, the complet evaluation of the KB is carried out until a possibility distribution is associated to each of the output variables.

The process of executing a KB can be understood as the "propagation" of possibility distributions through the KB, via implication operations (which permit "propagating" distributions from the antecedent part of a rule to the consequent part of the same rule) and via links (which "connect" the consequent part of one or several rules to the antecedent part of other(s)). These evaluation process is carried out following a certain order, which determines at any moment in time the rule(s) that may be applied. The process finishes with the operation of aggregating all the possibility distributions inferred for each output variable into a single final possibility distribution.

Without loss of generality, we analyze a KB which consist of only two chained rules:

R^S: *IF* $X1^S$ *IS* $A1^S$ *AND ... AND* $X_{Ms}{}^S$ *IS* $A_{Ms}{}^S$ *THEN* $X_{Ms+1}{}^S$ *IS* $B_1{}^S$ *AND ...AND* $X_{Ms+Ns}{}^S$ *IS* $B_{Ns}{}^S$ (t^S)

R^T: *IF* $X1^T$ *IS* $A1^T$ *AND...AND* $X_{MT}{}^T$*IS* $A_{MT}{}^T$*THEN* $X_{MT+1}{}^T$ *IS* $B_1{}^T$*AND ...AND* $X_{MT+NT}{}^T$*IS* $B_{NT}{}^T$ (t^T) *(20)*

which are linked by :

$$X_{Ms+1}{}^S = X_1{}^T \quad (21)$$

In order to represent the rule pair in the formalism we have described, we must first define the bijective function α, which relates patterns and propositions to this end, we define the following set of patterns:

$$P= \{ p^r{}_{mr} \mid m_r = 1, \ldots ,Mr+Nr, r=S,T \} \quad (22)$$

and the set of propositions:

$$PR=pr^r{}_{mr} = \{ \text{"}X^r{}_{mr} \text{ IS } A^r{}_{mr}\text{"}, mr<= Mr; \text{"}X^r{}_{mr} \text{ IS } B^r{}_{mr-Mr}\text{"}, mr>Mr\} \quad (23)$$

and, for short we choose:

$$p^r{}_{mr} \text{ ---> } a\,(p^r{}_{mr}) = pr^r{}_{mr} \quad (24)$$

Furthermore, given the simplicity of our KB, the parameterized transition for the rules and links are:

$$T^R= \{ t^S, t^T\} \quad (25)$$

$$T^C = \{ t^3 \} \quad (26)$$

The graphic representation of PPN for the considered rules is shown on Fig. 2:

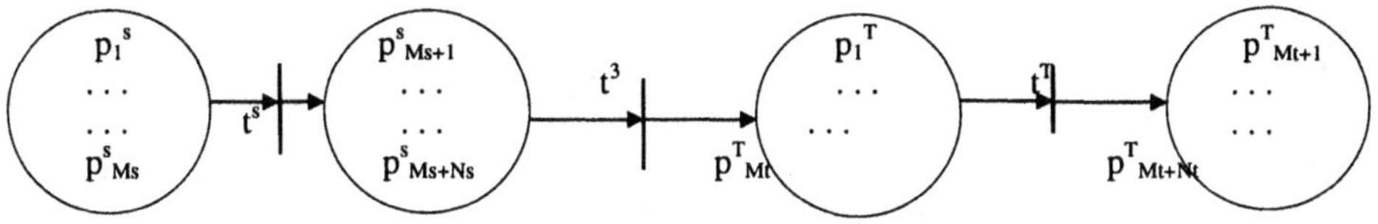

Fig. 2. Representation of the chaining of two rules in the PPN formalism .

More details about the process of obtaining of DOF's can be found in [13] and [14].

5 Reasoning Algorithm

The algorithm will basically consist of two stages: definition of the marking function and production of the DOF's of the corresponding propositions and firing of the active pattern transitions. These stages are sequentually repeated until there are no more active pattern transitions, at in which time the inference process will have ended. Finally, we perform aggregation - assignment of a single possibility distribution to each output variable. Let IP and OP be the sets which group input and output patterns respectively.

Step 1
Initially we assume we know only the DOF's of the propositions which operate on input variables, that is, those associated with input patterns. Therefore, the intial marking function will be:

$$M(p_i) = \{\, 0, \text{ if } p_i \notin IP;\ 1, \text{ if } p_i \in IP \,\} \tag{27}$$

Step 2
We fire the active pattern transitions. Let t^j be any active pattern transition; that is,

$$t^j \in T \mid \forall p_k \in I(t^j),\ M(p_k)=1 \tag{28}$$

The transition function *tf*, which defines the successive marking functions will be as defined with *(14)* and *(15)*. Also, the corresponding DOF's are obtained as follows:

$$\text{If } t^j \in T^R,\ g(p_i) = \bigwedge_{p_k \in I(t^j)} g(p_k),\ \forall p_i \in O(t^j) \tag{29}$$

$$\text{If } t^j \in T^C,\ g(p_i) = \bigvee_{p_k \in I(t^j)} [t^{rk}(\, g(p_k))\ o^{rk}\ m_{pk,pi}],\ \forall p_i \in O(t^j) \tag{30}$$

where:

$$t^{rk} = f(t^{rk}),\ p_k \in O(t^{rk}) \tag{31}$$

Step 3:
Go back to step 2, while:

$$\exists t^j \in T \mid M(p_i) = 1,\ \forall p_i \in I(t^j) \tag{32}$$

Step 4:
For each output variable *X*, its associated possibility distribution $\underline{B}=\{\underline{b}_i\}$, $i=1,\dots,I$, will be:

$$\underline{b}_i = \bigvee_{p_n{}^r \in Px} t^r(g(p_n{}^r))\ o^r\ t^r(b^r{}_{n,i}) \tag{33}$$

where the set P_x of places associated with propositions in which inferences over X are carried out is defined by:

$$Px = \{\, p_n{}^r \in P \mid a(p_n{}^r) = \text{"X IS } Bn^r\text{"} \} \tag{34}$$

6 Conclusions

In this work we present a transformation of an organizing supervisory controller and its model based on fuzzified Petri net, with rule-based decision-making and appropriate knowledge base. This modeling is an extension of all previous forms of fuzzy-Petri reasoning algorithm, because fuzzified parameterized Petri-nets proposed here, can be used to generate a high level abstract plan, even in the absence of complete information, which may be mapped into a specific plan at a lower level of the hierarchy.

It is showed that by using of parameterized-Petri net formalism we can develop a model of FPS as a model for inferencing and conclusion chainings, which is compatibile with well-structured algorithms for data-driven execution of a fuzzified KB. This process, based on a KB execution approach through the compositional rule of inference, transfers most of the computational load to the design, and not execution stage. This allows thr complexity of the execution algorithms to remain independent of the discretization of the discourse universes over which the linguistic variables to be manipulated in the fuzzy production systems are defined. Our parameterized fuzzy-Petri net reformulation is following the original conception of G. N. Saridis and is fuzzy-addition to his Petri net co-ordinative controller. So it has general generic feature.

References

[1] Ashby, W.R. (1952), *Design for a Brain*. Chapmann & Hall, London.

[2] Zadeh, L.A. (1973), " Outline of a new approach to the analysis of complex systems and decision processes", *IEEE Trans. Syst., Man. Cyber.*, vol.3, no. 1, pp.28-44.

[3] Saridis, G. N. (1989). "Analytic formulation of the principle of increasing precision with decreasing intelligence for intelligent machines". *Automatica*, vol. 25, no. 3, pp.461-467.

[4] Stellakis H.M. and K.P. Valavanis (1991). "Fuzzy-logic based formulation of the organizer of intelligent robotic systems", *J. of Intell. & Robot. Syst.*, vol.4, no.1, pp. 1-24.

[5] Dimirovski, G.M., O.L.Iliev, M.K.Vukobratovic, N.E.Gough and R.M.Henry (1993), "Modelling and scheduling control of FMS based on stochastic Petri-nets", In R.Evans (ed.) *Prepr. 12th IFAC World Congress*, The IFAC and Intsn. Engrs. Australia, Sydney, Barton Act, vol. 2, pp. 117-120.

[6] Gracanin D., P.Srinivasan and K.P. Valavanis (1994), "Parametrizied Petri Nets and Their Application to Planning and Coordination in Intelligent Systems", *IEEE Trans. on Systems, Man, and Cybernetics*, vol. 24, No.10, pp. 1483-1497.

[7] Sheng-Ke Yu (1994), "Knowledge representation and reasoning using fuzzy Pr/T net-systems", *Fuzzy Sets and Systems*, No. 75, pp. 33-45.

[8] Bugarin A.J., Barro S. (1994), "Fuzzy reasoning supported by Petri nets", *IEEE Trans.Fuzzy Systems*, vol. 2, pp. 135 - 150.

[9] Zurawski R., M.C.Zhou, (1994). "Petri nets and industrial application: A tutorial", *IEEE Trans. Industr. Elect.*, vol. 41, pp. 567-583.

[10] Dimirovski, G.M., O.Kaynak, A.T. Dinibutin, R. Hanus, (1995), "Contribution to two-level intelligent non-linear control systems", in O. Kaynak, M. Ozkan, N. Bekiroglu, I. Tunay (eds.) *Recent advances in Mechatronics*, Bogazici University, Istanbul, II, pp. 874-881.
[11] Scarpelli H., F. Gomide and R.R.Yager (1996), "A reasoning Algorithm for High-Level Fuzzy Petri Nets", IEEE Trans. on Fuzzy Systems, vol. 4, No.3, pp.282-294.
[12] Dimirovski, G.M., R.Hanus and R.M.Henry (1996), "A two-level system for intelligent automation using fuzzified Petri-nets and non-linear systems", In M.Jamshidi, J.Yuh and P.Dauchez (eds.) *Intelligent Automation and Control: Recent Trends in Development and Applications, TSI Series Volume 4,* TSI Press, Albuquerque NM, pp. 153-160.
[13] Dimirovski, G.M. and Z.M.Gacovski (1996), "An application of fuzzy-Petri-nets to organising supervisory controller", In E.Garcia & P.Albertos (eds.) *Valencia COSY Workshop*, Europan Science Foundation and DISCA-UPV, Valencia, Theme 3, Paper 7.(1-14).
[14] Gacovski, Z.M., G.M. Dimirovski, O. Kaynak and R.M. Henry (1997), "Expert Decision and Control: A Fuzzy-petri-Net Reasoning Simulator". In T. Cifcibasi, M. Karaman & V. Atalay (eds.) *New Trends in Artificial Intelligence and Neural Networks*, The EMO & Baset University, Ankara (TR), pp. 177-184.

Fuzzy Topologies Induced by Fuzzy Relation Based Modifiers

Martine De Cock and E.E. Kerre

Fuzziness and Uncertainty Modelling Research Unit
Department of Applied Mathematics and Computer Science
Ghent University, Krijgslaan 281 (S9), B-9000 Gent, Belgium
{Martine.DeCock,Etienne.Kerre}@rug.ac.be
http://allserv.rug.ac.be/~ekerre

Abstract. In this paper we highlight the specific meaning of images of fuzzy sets under fuzzy relations in the context of fuzzy topology. More precisely we show that fuzzy modifiers taking direct and superdirect images of fuzzy sets under fuzzy pre-orderings are respectively closure and interior operators, inducing fuzzy topologies. Furthermore we investigate under which conditions the same applies to the recently introduced general closure and opening operators based on arbitrary fuzzy relations.

1 Introduction

Images of fuzzy sets under fuzzy relations prove to be very powerful tools in a wide range of applications, varying from fuzzy databases, over fuzzy morphology, fuzzy rough set theory, and the representation of linguistic modifiers, to approximate reasoning schemes [5]. In this paper we will show that they have a specific meaning in the context of fuzzy topology as well. More specifically in Section 2 we will show that fuzzy modifiers taking direct and superdirect images of fuzzy sets under fuzzy pre-orderings are respectively closure and interior operators inducing Chang fuzzy topologies. In Section 3 we will investigate under which conditions the same holds for the general closure and opening operators introduced recently by Bodenhofer [2].

Throughout the paper T will denote a triangular norm with left-continuous partial mappings $T(x,.)$ for all x in $[0,1]$. Furthermore $\vec{T}$ will denote its residual implication defined by

$$\vec{T}(x,y) = \sup\{\lambda \in [0,1] | T(\lambda, x) \leq y\}$$

It can be verified then that $\vec{T}$ is non-increasing and left-continuous in the first argument while non-decreasing and right-continuous in the second. Furthermore for all x, y, and z in $[0,1]$, and $\{y_i \mid i \in I\}$ a family in $[0,1]$:

(1) $T(0,x) = 0$
(2) $T(x, \sup_{i\in I} y_i) = \sup_{i\in I} T(x, y_i)$
(3) $\vec{T}(x,y) = 1$ iff $x \leq y$

B. Reusch (Ed.): Fuzzy Days 2001, LNCS 2206, pp. 239–248, 2001.

(4) $\vec{T}(1, x) = x$
(5) $\vec{T}(x, \inf_{i \in I} y_i) = \inf_{i \in I} \vec{T}(x, y_i)$
(6) $\vec{T}(T(x, y), z) \leq \vec{T}(x, \vec{T}(y, z))$

Let X be a non-empty set. The class of all fuzzy sets on X will be denoted $\mathcal{F}(X)$. As usual union, intersection and complement of fuzzy sets are defined by, for x in X,

$$\bigcup_{i \in I} A_i(x) = \sup_{i \in I} A_i(x)$$

$$\bigcap_{i \in I} A_i(x) = \inf_{i \in I} A_i(x)$$

$$(co\ A)(x) = 1 - A(x)$$

in which $\{A_i | i \in I\}$ is a family of fuzzy sets on X, and A in $\mathcal{F}(X)$. Inclusion for two fuzzy sets A and B on X is defined as:

$$A \subseteq B \text{ iff } A(x) \leq B(x) \text{ for all } x \text{ in } X$$

Definition 1 (Fuzzy modifier). *[16] A fuzzy modifier on X is an $\mathcal{F}(X)$-$\mathcal{F}(X)$ mapping.*

In [4] the following class of fuzzy modifiers based on fuzzy relations is defined:

Definition 2 (Fuzzy relation based modifiers). *Let R be a fuzzy relation on X, i.e. $R \in \mathcal{F}(X \times X)$. The fuzzy modifiers $R{\uparrow}$ and $R{\downarrow}$ on X are defined by, for A in $\mathcal{F}(X)$ and y in X:*

$$R{\uparrow}A(y) = \sup_{x \in X} T(A(x), R(x, y))$$

$$R{\downarrow}A(y) = \inf_{x \in X} \vec{T}(R(y, x), A(x))$$

$R{\uparrow}A$ is also called the direct image of A under R, while $R{\downarrow}A$ is the superdirect image of A under R^{-1} (i.e. the inverse relation of R) [12]. In fuzzy set theoretical settings, fuzzy modifiers are usually associated with the representation of linguistic hedges such as very, more or less, rather,... (see e.g. [7], [13], [16]). The class of fuzzy modifiers based on fuzzy relations as defined above however can be applied to a wide range of other purposes as well. In [4], [10] it is shown that for a suitable fuzzy relation R they correspond to the dilation and erosion operators of fuzzy morphology (used for image processing), while in [6], [10] it is illustrated that they can be used as fuzzy-rough approximators (for dealing with incomplete information). In this paper we will show that for R a fuzzy pre-ordering the R-based fuzzy modifiers defined in Definition 2 also are closure and interior operators inducing fuzzy topologies (in the sense of Chang [3]).

Definition 3 (Fuzzy T-preordering). *A fuzzy relation R on X is called a fuzzy T-preordering w.r.t. a t-norm T iff for all x, y and z in X:*

(FP.1) $R(x, x) = 1$ *(reflexivity)*
(FP.2) $R(x, z) \leq T(R(x, y), R(y, z))$ *(T-transitivity)*

Definition 4 (Fuzzy topology). *A subset τ of $\mathcal{F}(X)$ is called a fuzzy topology on X iff*

(FT.1) $\emptyset \in \tau \wedge X \in \tau$
(FT.2) $O_1 \in \tau \wedge O_2 \in \tau$ *implies* $O_1 \cap O_2 \in \tau$
(FT.3) $(\forall i \in I)(O_i \in \tau)$ *implies* $\bigcup_{i \in I} O_i \in \tau$

Furthermore the class τ' is defined by

$$F \in \tau' \text{ iff } co(F) \in \tau$$

Every element of τ is called an open fuzzy set; *every element of τ' is called a* closed fuzzy set.

Definition 5 (Interior, closure). *Let τ denote a fuzzy topology on X. For A a fuzzy set on X the interior and the closure of A w.r.t. τ are the fuzzy sets $int(A)$ and $cl(A)$ on X defined by*

$$int(A) = \cup\{O | O \in \tau \wedge O \subseteq A\}$$

$$cl(A) = \cap\{F | F \in \tau' \wedge A \subseteq F\}$$

I.e. $int(A)$ is the largest open fuzzy set contained in A, while $cl(A)$ is the smallest closed fuzzy set containing A. int and cl are called the interior operator *and the* closure operator *respectively.*

It can be verified that they satisfy the following properties, for all A and B in $\mathcal{F}(X)$ [11]:

(1) $int(X) = X$
(2) $int(A) \subseteq A$
(3) $int(A \cap B) = int(A) \cap int(B)$
(4) $int(int(A)) = int(A)$

(1') $cl(\emptyset) = \emptyset$
(2') $A \subseteq cl(A)$
(3') $cl(A \cup B) = cl(A) \cup cl(B)$
(4') $cl(cl(A)) = cl(A)$

2 Induced Fuzzy Topologies

To show that the fuzzy relation based modifiers of Definition 2 are closure and interior operators inducing fuzzy topologies on X, we rely on the following theorem [11]:

Theorem 1. *Let f be a fuzzy modifier on X.*

Part I *If for all A and B in $\mathcal{F}(X)$*

(C1) $f(X) = X$
(C2) $f(A) \subseteq A$
(C3) $f(A \cap B) = f(A) \cap f(B)$

then $\tau = \{O | O \in \mathcal{F}(X) \wedge f(O) = O\}$ *is a fuzzy topology on* X. *Furthermore if also*

(C4) $f(f(A)) = f(A)$

for all A *in* $\mathcal{F}(X)$, *then* f *is the interior operator corresponding to* τ.

Part II *If for all* A *and* B *in* $\mathcal{F}(X)$

(C1') $f(\emptyset) = \emptyset$
(C2') $A \subseteq f(A)$
(C3') $f(A \cup B) = f(A) \cup f(B)$

then $\tau = \{O | O \in \mathcal{F}(X) \wedge f(co\ O) = co\ O\}$ *is a fuzzy topology on* X. *Furthermore if also*

(C4') $f(f(A)) = f(A)$

for all A *in* $\mathcal{F}(X)$, *then* f *is the closure operator corresponding to* τ.

Theorem 2. *Let* R *be a reflexive fuzzy relation on* X, *then*

$$\tau = \{O | O \in \mathcal{F}(X) \wedge R{\downarrow}O = O\}$$

is a fuzzy topology on X. *Moreover if* R *is a fuzzy* T*-preordering on* X *then* $R{\downarrow}$ *is the interior operator corresponding to* τ.

Proof. According to Theorem 1 the following conditions have to be fulfilled in order for τ to be a fuzzy topology on X: for all A and B in $\mathcal{F}(X)$:

(C1) $R{\downarrow}X = X$
(C2) $R{\downarrow}A \subseteq A$
(C3) $R{\downarrow}(A \cap B) = R{\downarrow}A \cap R{\downarrow}B$

Furthermore if also (C4) $R{\downarrow}(R{\downarrow}A) = R{\downarrow}A$ for all A in $\mathcal{F}(X)$ then $int = R{\downarrow}$. We will now prove that (C1), (C2) and (C3) hold for a reflexive fuzzy relation R, while (C4) holds for a fuzzy T-preordering. For all y in X:

$$R{\downarrow}X(y) = \inf_{x \in X} \vec{T}(R(y,x), X(x)) = \inf_{x \in X} \vec{T}(R(y,x), 1) = \inf_{x \in X} 1 = 1 = X(y)$$

hence (C1) is valid. Furthermore since R is reflexive

$$R{\downarrow}A(y) = \inf_{x \in X} \vec{T}(R(y,x), A(x)) \leq \vec{T}(R(y,y), A(y)) = \vec{T}(1, A(y)) = A(y)$$

which proves (C2). (C3) can be verified as follows:

$$\begin{aligned} R{\downarrow}(A \cap B)(y) &= \inf_{x \in X} \vec{T}(R(y,x), \min(A(x), B(x))) \\ &= \inf_{x \in X} \min(\vec{T}(R(y,x), A(x)), \vec{T}(R(y,x), B(x))) \\ &= \min(\inf_{x \in X} \vec{T}(R(y,x), A(x)), \inf_{x \in X} \vec{T}(R(y,x), B(x))) \\ &= \min(R{\downarrow}A(y), R{\downarrow}B(y)) \\ &= (R{\downarrow}A \cap R{\downarrow}B)(y) \end{aligned}$$

From (C2) we know that $R{\downarrow}(R{\downarrow}A) \subseteq R{\downarrow}A$ for a reflexive R. To complete the proof for (C4) we show that the other inclusion holds as well for T-transitive R ($\vec{T}$ being the residual implication of T):

$$\begin{aligned} R{\downarrow}(R{\downarrow}A)(y) &= \inf_{x\in X} \vec{T}(R(y,x), \inf_{z\in X} \vec{T}(R(x,z), A(z))) \\ &= \inf_{x\in X} \inf_{z\in X} \vec{T}(R(y,x), \vec{T}(R(x,z), A(z))) \\ &\geq \inf_{x\in X} \inf_{z\in X} \vec{T}(T(R(y,x), R(x,z)), A(z))) \\ &\geq \inf_{x\in X} \inf_{z\in X} \vec{T}(R(y,z), A(z)) \\ &= R{\downarrow}A(y) \end{aligned}$$

□

Theorem 3. *Let R be a reflexive fuzzy relation on X, then*

$$\tau = \{O | O \in \mathcal{F}(X) \wedge R{\uparrow}(co\ O) = co\ O\}$$

is a fuzzy topology on X. Moreover if R is a fuzzy T-preordering on X then $R{\uparrow}$ is the closure operator corresponding to τ.

Proof. According to Theorem 1 the following conditions have to be fulfilled in order for τ to be a fuzzy topology on X: for all A and B in $\mathcal{F}(X)$:

(C1') $R{\uparrow}\emptyset = \emptyset$
(C2') $A \subseteq R{\uparrow}A$
(C3') $R{\uparrow}(A \cup B) = R{\uparrow}A \cup R{\uparrow}B$

Furthermore if also (C4') $R{\uparrow}(R{\uparrow}A) = R{\uparrow}A$ for all A in $\mathcal{F}(X)$ then $cl = R{\uparrow}$. We will now prove that (C1'), (C2') and (C3') hold for a reflexive fuzzy relation R, while (C4') holds for a fuzzy T-preordering. For all y in X:

$$R{\uparrow}\emptyset(y) = \sup_{x\in X} T(\emptyset(x), R(x,y)) = \sup_{x\in X} T(0, R(x,y)) = \sup_{x\in X} 0 = 0 = \emptyset(y)$$

hence (C1') is valid. Furthermore since R is reflexive

$$R{\uparrow}A(y) = \sup_{x\in X} T(A(x), R(x,y)) \geq T(A(y), R(y,y)) = T(A(y), 1) = A(y)$$

which proves (C2'). (C3') can be verified as follows:

$$\begin{aligned} R{\uparrow}(A \cup B)(y) &= \sup_{x\in X} T(\max(A(x), B(x)), R(x,y)) \\ &= \sup_{x\in X} \max(T(A(x), R(x,y)), T(B(x), R(x,y))) \\ &= \max(\sup_{x\in X} T(A(x), R(x,y)), \sup_{x\in X} T(B(x), R(x,y))) \\ &= \max(R{\uparrow}A(y), R{\uparrow}B(y)) \\ &= (R{\uparrow}A \cup R{\uparrow}B)(y) \end{aligned}$$

From (C2') we know that $R\uparrow A \subseteq R\uparrow(R\uparrow A)$ for a reflexive R. To complete the proof for (C4') we show that the other inclusion holds as well for a T-transitive R (relying on the associativity of T):

$$\begin{aligned} R\uparrow(R\uparrow A)(y) &= \sup_{x\in X} T(\sup_{z\in X} T(R(z,x), A(z)), R(x,y)) \\ &= \sup_{x\in X} \sup_{z\in X} T(T(A(z), R(z,x)), R(x,y)) \\ &= \sup_{x\in X} \sup_{z\in X} T(A(z), T(R(x,y), R(z,x))) \\ &\leq \sup_{x\in X} \sup_{z\in X} T(A(z), R(z,y)) \\ &= R\uparrow A(y) \end{aligned}$$

□

From now on we refer to the fuzzy topology defined in Theorem 2 as the fuzzy topology induced by $R\downarrow$ while we say that the fuzzy topology of Theorem 3 is induced by $R\uparrow$.

Remark 1. If R is not reflexive then (C2) and (C2') do not necessarily hold. For instance it can be verified that if R is defined as $R(x,y) = 0$ for all x and y in X, then $R\downarrow A = X$ and $R\uparrow A = \emptyset$, for all A in $\mathcal{F}(X)$. Therefore for $A \neq \emptyset$ and $A \neq X$ neither $R\downarrow A \subseteq A$ (C2) nor $A \subseteq R\uparrow A$ (C2') hold.

Remark 2. From the example given in Remark 1 it is clear that (C4) and (C4') do not hold in general for a non-reflexive fuzzy relation R. Indeed for R the empty relation on X, $R\downarrow(R\downarrow A) = X$ and $R\uparrow(R\uparrow A) = \emptyset$ hold for all A in $\mathcal{F}(X)$, implying that in this case (C4) only holds for $A = X$ and (C4') only holds for $A = \emptyset$. However even the reflexivity of R is not sufficient in order for (C4) and (C4') to hold for arbitrary A as shown in the following example: let $X = [0,1]$ and let A be the fuzzy set on X defined by $A(x) = x$, for all x in X. Furthermore let R be the reflexive fuzzy relation on X defined by

$$R(x,y) = \begin{cases} 1 \; if \; |x-y| < 0.1 \\ 0 \; otherwise \end{cases}$$

for all x and y in X. Then it can be verified that for y in X:

$$\begin{aligned} R\downarrow A(y) &= \inf_{z\in X} \vec{T}(R(y,z), A(z)) \\ &= \inf\{z \mid z \in X \wedge z \in]y-0.1, y+0.1[\} \\ &= \max(0, y-0.1) \end{aligned}$$

Hence $R\downarrow A(1) = 0.9$. Furthermore

$$\begin{aligned} R\downarrow(R\downarrow A)(1) &= \inf_{z\in X} \vec{T}(R(1,z), max(0, z-0.1)) \\ &= \inf\{max(0, z-0.1) \mid z \in]0.9, 1]\} = 0.8 \end{aligned}$$

Likewise

$$\begin{aligned} R\uparrow A(y) &= \sup_{z\in X} T(A(z), R(z,y)) \\ &= \max\{z \mid z \in X \wedge z \in]y-0.1, y+0.1[\} \\ &= \min(1, y+0.1) \end{aligned}$$

Hence $R\uparrow A(0) = 0.1$. Furthermore

$$\begin{aligned} R\uparrow(R\uparrow A)(0) &= \sup_{z\in X} T(min(1, z+0.1), R(z,0)) \\ &= \sup\{min(1, z+0.1) \mid z \in [0, 0.1[\} = 0.2 \end{aligned}$$

Note that R is not T-transitive.

In rough set theory [14] the universe X of objects is equipped with a crisp equivalence relation R expressing the indiscernibility between objects that can be derived from the available (incomplete) information about them. The lower approximation $\underline{R}A$ of a crisp subset of X is defined as the union of the equivalence classes of R included in A. It is the set of objects that necessarily belong to A considering the available information. The upper approximation $\overline{R}A$ of A is the union of all equivalence classes of X that have a non-empty intersection with A. It is the set of all objects that possibly belong to A.

Indiscernibility between objects might be *vague*: e.g. Alberik (28 years old) and Els (29) are indiscernible in age to a higher degree than Alberik (28) and Mike (25). Furthermore one might also want to make statements concerning the degree to which an object necessarily/possibly belongs to a *fuzzy* set A on X. Hence there is a need for *fuzzy* rough set theory ([8], [9], [15]) in which the universe is equipped with a fuzzy T-equivalence relation R, i.e. asymmetrical fuzzy T-preordering. In this case an object y belongs to the lower approximation of a fuzzy set A on X to the degree to which the fuzzy equivalence class of R containing y is included in A, i.e. $\underline{R}A = R\downarrow A$. Likewise y belongs to the upper approximation of A to the degree to which the fuzzy equivalence class of R containing y overlaps with A, in other words $\overline{R}A = R\uparrow A$. The lower and upper approximators $\underline{R}$ and $\overline{R}$ can therefore be seen as the interior and closure operator respectively of a fuzzy topology on X.

3 Generalized Opening and Closure Operators

Note that in the theorems of the previous section the reflexivity of R was sufficient to prove (C2) and (C2') while the reflexivity and the T-transitivity were sufficient for (C4) and (C4'). (C1), (C1'), (C3) and (C3') on the other hand hold for arbitrary R. In [2] the following generalized opening and closure operators are introduced:

Definition 6 (Generalized opening and closure operator). *Let R be a fuzzy relation on X. The generalized opening operator R° and the general closure operator $R^\bullet$ on X are defined by, for A in $\mathcal{F}(X)$:*

$$R^\circ A = R\uparrow(R\downarrow A)$$

$$R^\bullet A = R\downarrow(R\uparrow A)$$

Both of these operators are of course fuzzy modifiers on X. It is proven in [2] that R° satisfies (C2) and (C4) while $R^\bullet$ fulfills (C2') and (C4'). The beauty of these general operators is that they do not require reflexivity or T-transitivity of R for these properties. Ironically enough the reflexivity of R is a suitable requirement to be able to prove (C1) and (C1'), while we can prove (C3) and (C3') assuming T-transitivity. In fact for reflexive R:

$$\begin{aligned} R^\circ X(y) &= R{\uparrow}(R{\downarrow}X)(y) \\ &= R{\uparrow}X(y) \\ &= \sup_{x\in X} T(X(x), R(x,y)) \\ &\geq T(1, R(y,y)) \\ &= 1 = X(y) \end{aligned} \qquad \begin{aligned} R^\bullet\emptyset(y) &= R{\downarrow}(R{\uparrow}\emptyset)(y) \\ &= R{\downarrow}\emptyset(y) \\ &= \inf_{x\in X} \vec{T}(R(y,x), \emptyset(x)) \\ &\leq \vec{T}(R(y,y), 0) \\ &= 0 = \emptyset(y) \end{aligned}$$

Note that these properties also hold for an arbitrary fuzzy relation R if the kernels of all of its foresets Ry are not empty[1]. In other words: if all of its foresets are normalized.

Since $A\cap B \subseteq A$, $A\cap B \subseteq B$, $A \subseteq A\cup B$, $B \subseteq A\cup B$, and the fuzzy modifiers $R{\uparrow}$ and $R{\downarrow}$ are monotonic (cfr. Theorem 2 and 3), it holds that $R^\circ(A\cap B) \subseteq R^\circ A \cap R^\circ B$ and $R^\bullet A \cup R^\bullet B \subseteq R^\bullet(A\cup B)$. The following example shows however that the corresponding equalities do not hold for arbitrary fuzzy relations R.

Example 1. Let $X = [0,1]$. Let A and B be fuzzy sets on X and R a fuzzy relation on X defined by $A(x) = x$, $B(x) = 1-x$, and

$$R(x,y) = \begin{cases} 1 \; if \; |x-y| < 0.1 \\ 0 \; otherwise \end{cases}$$

for all x and y in X. Then it can be verified that for y in X: $R{\downarrow}A(y) = \max(0, y-0.1)$ (cfr. Remark 2) and likewise $R{\downarrow}B(y) = \max(0, 0.9-y)$. Hence

$$\begin{aligned} R^\circ A(0.5) &= \sup_{y\in X} T(R{\downarrow}A(y), R(y,0.5)) \\ &= \sup_{y\in]0.4,0.6[} \; y - 0.1 = 0.5 \end{aligned}$$

and likewise $R^\circ B(0.5) = 0.5$. Therefore $(R^\circ A \cap R^\circ B)(0.5) = 0.5$. On the other hand

$$\begin{aligned} R{\downarrow}(A\cap B)(y) &= \inf_{z\in X} \vec{T}(R(y,z), \min(A(z), B(z))) \\ &= \inf\{\min(z, 1-z) \mid z \in X \wedge z \in]y-0.1, y+0.1[\} \end{aligned}$$

Hence

$$\begin{aligned} R^\circ(A\cap B)(0.5) &= \sup_{y\in X} T(R{\downarrow}(A\cap B)(y), R(y,0.5)) \\ &= \sup_{y\in]0.4,0.6[} \; R{\downarrow}(A\cap B)(y) = 0.4 \end{aligned}$$

Quite analogously it can be verified that

$$(R^\bullet A \cup R^\bullet B)(0.5) = 0.5$$

[1] For y in X the R-foreset of y is the fuzzy set Ry on X defined by $(Ry)(x) = R(x,y)$, for all x in X [1].

while

$$R^{\bullet}(A \cup B)(0.5) = 0.6$$

Note that the relation R in the example above is reflexive and symmetrical, but that the equalities of (C.3) and (C.3') do not hold. They would hold if R was also T-transitive, but as proven in [2], [15], in this case the generalized opening and closure operator coincide with the fuzzy relation based modifiers $R\downarrow$ and $R\uparrow$, i.e. $R^{\circ}(A) = R\downarrow(A)$ and $R^{\bullet}(A) = R\uparrow(A)$, for all A in $\mathcal{F}(X)$. As we have shown in the previous section in this case the operators induce a fuzzy topology on X. The generalized opening and closure operators introduced in [2] however are not in general interior and closure operators inducing a fuzzy topology.

4 Conclusion

We have shown that the fuzzy modifiers taking direct and superdirect images of fuzzy sets under reflexive fuzzy relations induce fuzzy topologies of X. Furthermore if the fuzzy relation is also T-transitive, i.e. it is a fuzzy T-preordering, the fuzzy modifiers correspond to the closure and interior operators. If the fuzzy T-preordering relation is also symmetrical, i.e. it is a fuzzy T-equivalence relation, these operators correspond to the lower and upper approximators of fuzzy rough set theory.

Although the generalized opening and closure operators fulfill some properties for arbitrary fuzzy relations that in the case of the fuzzy modifiers mentioned above only hold for reflexive and T-transitive relations, they are not in general interior and closure operators inducing fuzzy topologies. For T-equivalence relations they coincide with the fuzzy modifiers taking images and therefore induce fuzzy topologies.

Acknowledgement

M. De Cock would like to thank the Fund For Scientific Flanders (FWO) for funding the research reported on in this paper.

References

1. Bandler, W., Kohout, L. J.: Fuzzy Relational Products as a Tool for Analysis and Synthesis of the Behaviour of Complex Natural and Artificial Systems. In: Wang, S. K., Chang, P.P. (eds.): Fuzzy Sets: Theory and Application to Policy Analysis and Information Systems. Plenum Press, New York and London (1980) 341–367.
2. Bodenhofer, U.: Generalized Opening and Closure Operators of Fuzzy Relations. Submitted to FLA'2001 (private communication) (2000)
3. Chang, C. L.: Fuzzy Topological Spaces. Journal of Mathematical Analysis and Applications 24 (1968) 182–190
4. De Cock, M., Kerre, E. E.: A New Class of Fuzzy Modifiers. In: Proceedings of ISMVL2000, IEEE Computer Society (2000) 121–126

5. De Cock, M., Nachtegael, M., Kerre, E. E.: Images under Fuzzy relations: A Master-Key to Fuzzy Applications. In: Ruan, D., Abderrahim, H. A., D'hondt, P., Kerre, E. E. (eds.): Intelligent Techniques and Soft Computing in Nuclear Science and Engineering, Proceedings of FLINS 2000. World Scientific (2000) 47–54
6. De Cock, M., Radzikowska, A., Kerre, E. E.: Modelling Linguistic Modifiers Using Fuzzy-Rough Structures. In: Proceedings of IPMU2000 (2000) 1735–1742
7. De Cock, M., Bodenhofer, U., Kerre, E. E.: Modelling Linguistic Expressions Using Fuzzy Relations. In: Proceedings 6th Int. Conf. on Soft Computing (IIZUKA2000), (CD-ROM) (2000) 353–360
8. Dubois,D.,Prade, H.: Rough Fuzzy Sets and Fuzzy Rough Sets. Int. J. of General Systems, 17(2-3) (1990) 191–209
9. Dubois,D.,Prade, H.: Putting fuzzy sets and rough sets together. In: Roman Słowiński (ed.): Intelligent Decision Support. Kluwer Academic (1992) 203–232
10. Nachtegael, M., Radzikowska, A., Kerre, E. E.: On Links between Fuzzy Morphology and Fuzzy Rough Sets. In: Proceedings of IPMU2000 (2000) 1381–1388
11. Kerre, E. E.: Fuzzy Topologizing with Preassigned Operations, International Congress of Mathematics, Helsinki (1978)
12. Kerre, E. E.: Introduction to the Basic Principles of Fuzzy Set Theory and Some of its Applications. Communication and Cognition, Gent (1993)
13. Kerre, E. E., De Cock, M.: Linguistic Modifiers: an overview. In: Chen, G., Ying, M., Cai, K.-Y. (eds.): Fuzzy Logic and Soft Computing. Kluwer Academic Publishers (1999) 69–85
14. Pawlak, Z.: Rough sets. Int. J. of Computer and Information Science, 11(5) (1982) 341–356
15. Radzikowska, A. M., Kerre, E. E.: A Comparative Study of Fuzzy Rough Sets. To appear in Fuzzy Sets and Systems.
16. Thiele, H.: Interpreting Linguistic Hedges by Concepts of Functional Analysis and Mathematical Logic. In: Proceedings of EUFIT'98, volume I (1998) 114–119
17. Ying-Ming, L., Mao-Kang, L.: Fuzzy Topology. Advances in Fuzzy Systems - Applications and Theory Vol. 9. World Scientific, Singapore, New Jersey, London, Hong Kong (1997)

Advanced Inference Filter Defuzzification

H. Kiendl, P. Krause

Chair of Electrical Control Engineering
University of Dortmund
Germany
{kiendl,krause}@esr.e-technik.uni-dortmund.de

Abstract. In the field of modeling, fuzzy models are one efficient approach for representing technical systems or human control strategies. Fuzzy models have the advantage of supplying a transparent and interpretable model. Conventional fuzzy models are based on fuzzification, inference and defuzzification. The fuzzification and inference operations are theoretically well–established in the framework of fuzzy logic. In contrast, conventional defuzzification methods are essentially empirically motivated. First, we recapitulate the *inference filter* concept, which supplies a new understanding of the defuzzification process and a theoretical framework. Second, we extend this approach to the *advanced inference filter* concept, which leads to a defuzzification method that is better suited to imitate the behavior of a human expert.

1 Introduction

Among many methods for modeling control strategies of human experts or complex technical systems, fuzzy modeling has proven to be a very efficient approach. Fuzzy systems show good modeling quality and they are transparent for users. However, difficulties arise if the obtained fuzzy system exhibits contradictions or ambiguities. In these cases, the fuzzy system may produce an output membership function that recommends quite different ranges of values for the output variable. Clearly, the well–known defuzzification methods [1], such as the center of gravity (COG) defuzzification or the mean of maximum (MOM) defuzzification, aim at generating a well–supported output value considering the output membership function and the purpose of the fuzzy system. However, there are situations where these methods supply an output value that differs from a human expert's choice considering the given application.

To obtain a more appropriate defuzzification KIENDL introduced the inference filter defuzzification method[2, 3]. It is based on the idea of abandoning the conventional interpretation that the output membership function $\mu(y)$ is an attractiveness function, where the function values μ indicate the degree of motivation for taking the corresponding value y as output value. Instead, the inference filter introduces the idea that the attractiveness of each possible output value y_i depends not only on the functional value $\mu(y_i)$, but also on all other values $\mu(y)$, $y \neq y_i$. This idea of a distance effect leads to a filtering process

B. Reusch (Ed.): Fuzzy Days 2001, LNCS 2206, pp. 249–260, 2001.

that transforms $\mu(y)$ into a function $\hat{\mu}(y)$, which can be considered as the real attractiveness function. Consequently, the position of the maximum of the attractiveness function is the most attractive value and is therefore chosen as the output value of the fuzzy system. In Section 2 the main idea and a first approach to realizing the inference filter concept are presented in detail.

The described approach has the advantage of being very flexible, because it allows us to reproduce conventional defuzzification methods or to make arbitrary compromises between them. However, it does not in all situations produce an output value that corresponds to what a human expert would choose as an appropriate output value. In Section 3 we present an advanced filtering process within the framework of the inference filter concept that overcomes these difficulties.

In Section 4 we demonstrate with some examples that the modeling quality of the obtained fuzzy models may be improved by this advanced inference filter defuzzification method.

2 The Inference Filter Concept

2.1 Motivation

We consider a conventional fuzzy system with one or more input variables $x_1 \ldots x_n$ and one output variable y. The purpose of the fuzzy model is to supply the most attractive output value y_D for every given input value. The membership function $\mu(y)$ resulting from inference is usually interpreted as an attractiveness function: its values $\mu(y)$ indicate the degree of recommendation for every possible output value. Following this interpretation, it would be logical to choose that value y (or one of those values y) as the output value of the fuzzy model where $\mu(y)$ becomes maximal (maximum height defuzzification method). However, other defuzzification methods, such as the center of gravity (COG) defuzzification method or the mean of maximum (MOM) defuzzification method, are very effective and are therefore often used [1, 4]. These methods usually supply an output value

$$y_D = \text{defuzz}(\mu(y)) \tag{1}$$

where $\mu(y_D)$ is not generally the maximal value of $\mu(y)$. Therefore, the interpretation of $\mu(y)$ as an attractiveness function is generally not stringent.

This observation leads to the idea of finding a function $\hat{\mu}(y)$ depending on $\mu(y)$ that can stringently be considered as an attractiveness function. We demand that

$$\hat{\mu}(y_D) = \max(\hat{\mu}(y)) \tag{2}$$

holds and that, as the values of function $\hat{\mu}(y)$ increase, y should be rated as more attractive, considering $\mu(y)$ and the applied defuzzification method. We show that the desired function $\hat{\mu}(y)$ can be obtained for well–known defuzzification methods by applying a transformation

$$\hat{\mu}(y) = F_D\{\mu(y)\} \tag{3}$$

to $\mu(y)$, where F_D depends on the defuzzification method. As F_D refines the membership function $\mu(y)$ produced by the inference in the sense that the resulting function $\hat{\mu}(y)$ provides the actual attractiveness, we call F_D an inference filter [3].

The concept of the inference filter supplies a theoretical framework for existing defuzzification methods. Furthermore, it is a basis for a new fuzzy controller structure, in which an inference filter (Eq. (3)) and a simple maximum selector (Eq. (2)) are used instead of conventional defuzzification (Fig. 1).

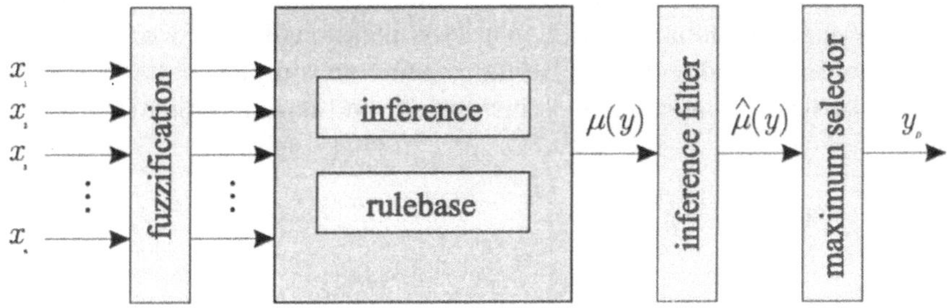

Fig. 1. Fuzzy system structure with an inference filter.

2.2 Convolution Realization of the Inference Filter

The central idea of inference filter defuzzification is that each functional value $\mu(y_i) > 0$ of the output membership function $\mu(y)$ supports not only the value y_i as a possible output value, but also values $y \neq y_i$ in the neighborhood of y_i. Previous realizations of this neighborhood concept were based on a distance function $h(y-y')$, which is convolved with the output membership function $\mu(y)$ to form the attractiveness function

$$\hat{\mu}(y) = \int_{-\infty}^{\infty} \mu(y')h(y-y')dy' \quad . \tag{4}$$

Thus, the inference filter F_D from Eq. (3) is given here by a convolution integral with a selectable distance function $h(y-y_i)$. The selection of this function adjusts the character of the distance effect to the requirements of the application being considered. In [10] the following form of distance function is proposed

$$h_{\gamma,\rho}(y-y') = \begin{cases} \frac{\gamma+1}{2\gamma\rho}\left(1-\left|\frac{y-y'}{\rho}\right|\right)^{1/\gamma} , & \text{if } \; 0<\gamma\leq 1,\; |y-y'|<\rho \\ \frac{\gamma+1}{2\gamma\rho}\left(1-\left|\frac{y-y'}{\rho}\right|^{\gamma}\right) , & \text{if } \; 1\leq\gamma<\infty,\; |y-y'|<\rho \\ 0 , & \text{else} \end{cases} \tag{5}$$

It is characterized by a transparent parameterization, as may be gathered from the following (Fig. 2): the parameter $\rho > 0$ predetermines the "range" of the distance function. The standard approach for setting ρ is $\rho = d \cdot (y_{max} - y_{min})$ where $d = 1$ is the standard value for d, and y_{min} and y_{max} are the minimal and maximal values of the output, respectively. Two cases should be distinguished: $d < 1$ ("limited range") and $d \geq 1$ ("full range"). The parameter $\gamma > 0$ predetermines the shape of the distance function. In the full range case, filtration in respect to $\gamma = 2$, $\gamma = 1$ or $\gamma \to 0$ with the distance function (5) corresponds to the defuzzification methods center of gravity, mean of area, and mean of maximum, respectively. This means that by using the freedom of choice of the parameters of the distance function, the resulting performance is as effective as with these classical defuzzification methods mentioned. However, in many examples, a suitable tuning of the filtering parameters supplies better performance than the classical methods (see for instance an industrial application in [12]).

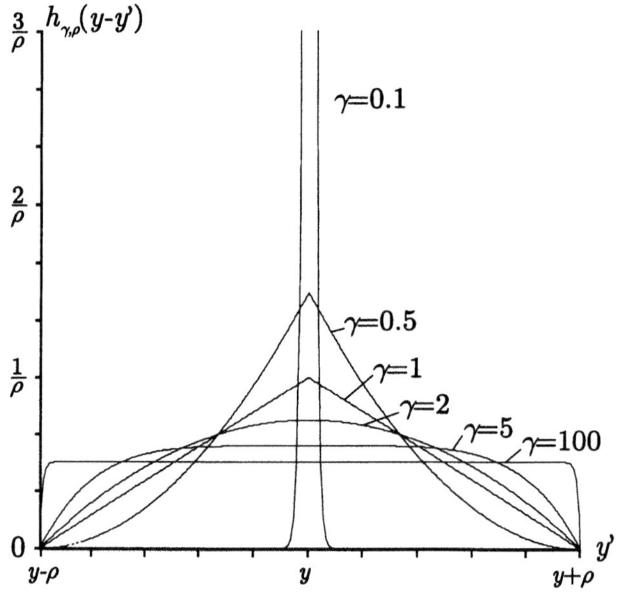

Fig. 2. Distance function $h_{\gamma,\rho}(y - y')$ for different values of γ.

The conditions

$$\int_{-\infty}^{\infty} h_{\gamma,\rho}(x)dx = 1 \quad \text{and} \quad 0 \leq \mu(y) \leq 1 \tag{6}$$

serve to ensure that the functional values of the filtered membership function $\hat{\mu}(y)$ do not exceed the value one.

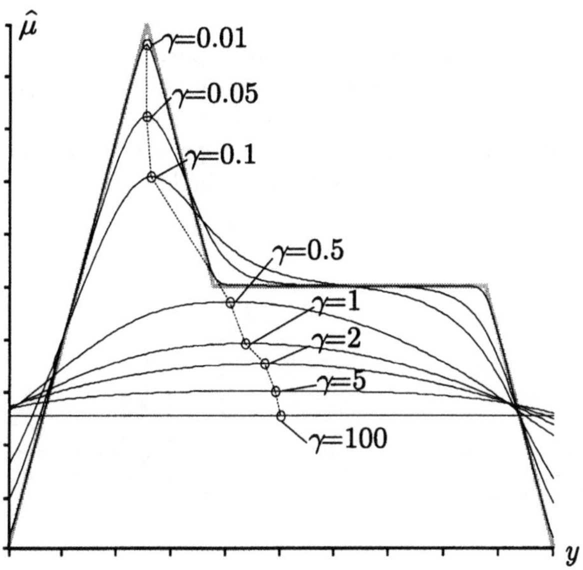

Fig. 3. Attractiveness function $\hat{\mu}$ resulting from an output membership function μ (grey line) and different distance functions $h_{\gamma,\rho}(y - y')$. The dots show the maxima of the attractiveness functions.

Fig. 3 shows the resulting attractiveness functions $\hat{\mu}(y)$ of a given output membership function $\mu(y)$ using different values for γ. The parameter ρ is set to $y_{max} - y_{min}$ ("full range").

In cases in which the linguistic values of the output variable of the fuzzy system are chosen in the form of singletons, the attractiveness function is given — instead of Eq. (3) — by the convolution sum

$$\mu(y) = \sum_i \mu(y_i)h(y - y_i) \quad . \tag{7}$$

It should be mentioned that the distance function (5) can be specialized without losing essential flexibility by selecting $\gamma \in \mathbb{N}$. If the output membership function $\mu(y)$ is either piecewise linear or consists of singletons, the filtered output membership function $\hat{\mu}(y)$ can be calculated analytically. In addition, if γ is small, the maximum of $\hat{\mu}(y)$ can be determined analytically. This is favorable for fast real-time applications [6].

3 The Advanced Inference Filter Concept

3.1 Motivation

The above convolution realization of the inference filter uses a constant distance effect, which can only be adjusted once before it is applied. This does not always result in the best-motivated output value. To obtain an output value that corresponds better to the decision of a human expert, a distance effect that could automatically be adapted to individual situations would be more appropriate. This paper presents a filtering approach that meets this idea.

To highlight the motivation for this new approach, we consider the following output membership functions in the form of singletons.

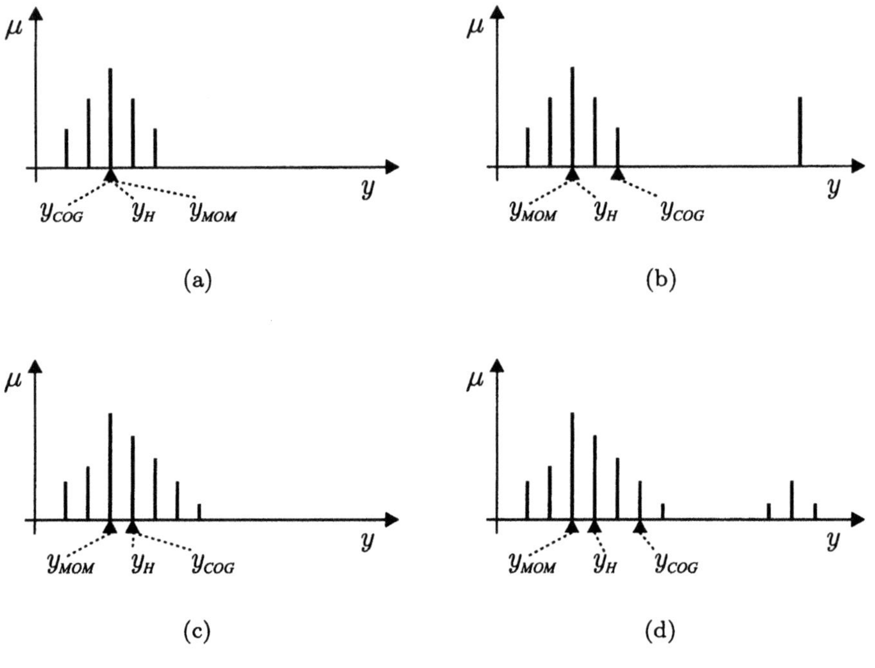

Fig. 4. Output membership functions and the resulting output values y_{COG}, y_{MOM} and y_H for COG defuzzification, MOM defuzzification and the value most likely to be chosen by a human expert, respectively.

Fig. 4(a) shows a clear situation where the center of gravity and mean of maximum defuzzification methods supply the same output value and are in agreement with the decision of a human expert. Fig. 4(b) shows an ambiguous situation: compared to the membership function of Fig. 4(a), an additional singleton is introduced that strongly contradicts the recommendations of the other singletons. As this additional singleton is activated to a smaller extent than the other singletons, the mean of maximum defuzzification method supplies the same output value as in Fig. 4(a), again in agreement with the decision of the human expert. In contrast, the center of gravity defuzzification method does not ignore the strongly contradictory recommendation. Consequently, the resulting compromise value differs from the decision of the human expert. Fig.

4(c) shows a situation that is not as clear as in Fig. 4(a), but much clearer than in Fig. 4(b). Here, the center of gravity defuzzification (but not the mean of maximum defuzzification) matches the human expert's decision. Fig. 4(d) shows a situation where all resulting values y_{COG}, y_{MOM} and y_H differ from each other. The human expert would choose the same output values as in Fig. 4(c); because the additional strongly contradictory recommendations are too weakly activated, they would be ignored. The center of gravity defuzzification again takes a compromise between all recommendations.

The above realization of the inference filter allows it to supply an output value y_H in all situations. However, we require different distance functions in different situations; for example, in Fig. 4(b) the δ–function, in Fig. 4(c) a quadratic function, and in Fig. 4(d)

$$h(y-y') = \begin{cases} \frac{3}{5}(1-|y-y'|^5) & |y-y'| \leq 1 \\ 0 & |y-y'| > 1 \end{cases} \quad . \tag{8}$$

3.2 The Advanced Inference Filter

In contrast, the realization of the inference filter proposed here supplies the desired output value y_H in all the above examples. The new approach is outlined here for membership functions in the form of singletons, but it can easily be extended to continuous membership functions.

We consider an output membership function of the form of n singletons Y_i, $i = 1, 2, \ldots, n$, where each singleton Y_i is described by a function

$$\mu_i(y) = \begin{cases} \mu_i & , \quad \text{if} \quad y = y_i \\ 0 & , \quad \text{else} \end{cases} \tag{9}$$

where y_i is the position of the singleton and μ_i is the corresponding activation. We assume that $y_1 < y_2 < \cdots < y_n$ holds. The output membership function can be written in the form

$$\mu(y) = \bigvee_{i=1}^{n} \mu_i(y) \quad . \tag{10}$$

For the filtering process, we consider all membership functions $\mu_{j,k}(y)$ that are induced by the output membership function $\mu(y)$ and defined by

$$\mu_{j,k}(y) = \bigvee_{i=j}^{k} \mu_i(y) \quad , \quad 1 \leq j \leq k \leq n \quad . \tag{11}$$

Each membership function $\mu_{j,k}(y)$ is defuzzified by the center of gravity method and a degree of attractiveness $\alpha_{j,k}$ is attached to the resulting value $\mathrm{COG}(\mu_{j,k}(y))$, imitating the way a human expert would defuzzify an output membership function. The expert considers a value to be more attractive if neighboring singletons have higher degrees of activation and the "mass" $M = \sum \mu_i$ of the activation is concentrated on neighboring singletons. Following this idea, we attach to each membership function $\mu_{j,k}(y)$ a singleton according to

$$\hat{\mu}_{j,k}(y) = \begin{cases} \alpha_{j,k} & , \quad \text{if} \quad y = \mathrm{COG}(\mu_{j,k}(y)) \\ 0 & , \quad \text{else} \end{cases} \tag{12}$$

with

$$\alpha_{j,k} = M_{j,k} \cdot e^{-q \cdot \sigma_{j,k}} \tag{13}$$

and where $M_{j,k}$ and $\sigma_{j,k}$ are given by

$$M_{j,k} = \sum_{i=j}^{k} \mu_i \quad \text{and} \quad \sigma_{j,k} = \sqrt{\frac{\sum_{i=j}^{k} (\mathrm{COG}(\mu_{j,k}(y)) - y_i)^2}{M_{j,k}}} \quad . \tag{14}$$

Finally, we define the resulting filtered function $\hat{\mu}(y)$ by

$$\hat{\mu}(y) = \bigvee_{\substack{j,k \\ 1 \leq j \leq k \leq n}} \mu_{j,k}(y) \tag{15}$$

where we realize the OR–operator by the maximum. The output value is given — as generally provided in the inference filter approach — by the value of y where $\hat{\mu}(y)$ takes its maximum.

The factor q introduced in Eq. (13) determines how strongly the neighboring singletons contribute to the attractiveness $\alpha_{j,k}$ of the family of singletons under consideration. If q is set to zero only the "mass" is crucial for the determination of the attractiveness. Thus, the family of singletons with the highest "mass" has the highest attractiveness, which is naturally $\mu_{1,n}(y)$. This means that this choice of q leads to the center of gravity defuzzification method. On the other hand, if q exceeds a certain limit the singleton with the highest activation becomes the most attractive membership function, which corresponds to the mean of maximum defuzzification method. Thus, by providing the range of $0 \leq q < \infty$ we obtain a model quality, which is at least as good as with the classical defuzzification methods. However, it has turned out in many examples that a compromise value of q improves the performance of the fuzzy system.

4 Examples

Before presenting some examples, we reconsider the example given in Fig. 4. As we stated there, it was not possible to find a way to imitate the defuzzification of the human expert. Now, with the advanced inference filter it is possible with a certain range of q to obtain the same output values as the human expert would suggest.

In the following section, we consider two examples, for which fuzzy models have been generated previously by a data–based fuzzy modeling method. The results for using conventional defuzzification methods and for applying the advanced inference filter defuzzification method are compared. The quality of a model is rated by the mean absolute error between the output of the model and the original values of the output of the validation data points.

4.1 The Model of a Heat Exchanger

In an industrial application, a model of a complex chemical process for the production of polyether is required. To achieve the global aim, the process is split into subprocesses. Then, for each subprocess a model is independently constructed. One such subprocess is the heat exchanger [11] (Fig. 5). The input variables for the model are the temperature ϑ_R of the reactor, the pressure p_R of the reactor and the manipulated variable u for the regulation of the heat exchanger. The output variable is the temperature ϑ_H of the cooling fluid in the heat exchanger. The fuzzy model of the heat exchanger was generated with the Fuzzy–ROSA–method [5, 8, 13].

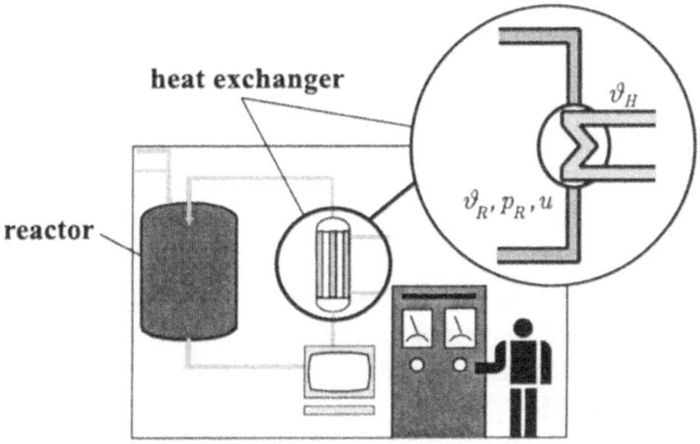

Fig. 5. Chemical process with a heat exchanger.

Fig. 6 shows the resulting error values of the model applying the advanced inference filter method for different values of q. For $q \to 0$ the same result as for using the center of gravity defuzzification method is obtained. Conversely, if q exceed a certain value the same result as for using the mean of maximum defuzzification method is obtained. However, by setting $q = 0.6$ the advanced inference filter method supplies a better model quality than either of the classical methods.

4.2 Experts' Prediction of Necessary Power Reserve

The operators of power generation and distribution networks must ensure that at any time a certain amount of power is held as a reserve. This reserve could be

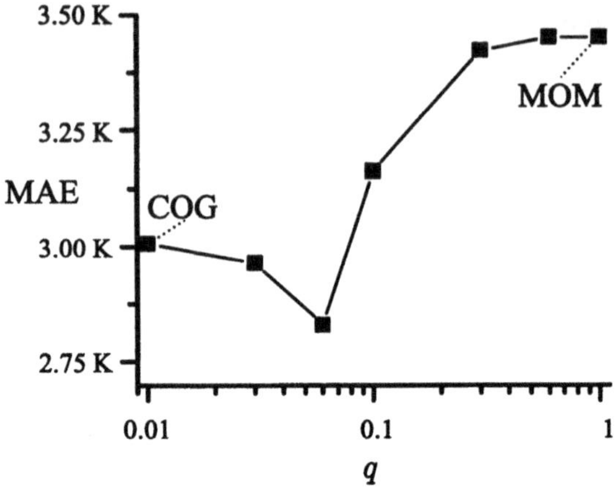

Fig. 6. Mean absolute temperature error in K of a fuzzy system using the advanced inference filter defuzzification method with different values of the factor q.

required if unexpected situations occur in the network. Therefore, 882 situations were presented to an expert who was required to decide what amount of power should be held as a reserve, depending on the situations. The situations were described by seven characteristic values. The task of modeling the decisions of the expert is complicated because experts do not make up their minds in identical situations in the same way. The modeling has been also performed with Fuzzy–ROSA–method.

Fig. 7 shows the error values of the model applying the advanced inference filter method for different values of q. The cases $q = 0$ and $q = 0.1$ correspond to the classical defuzzification methods. Compared to these, a better model quality is obtained with $q = 0.05$.

5 Conclusions

In this paper we recapitulate the inference filter concept and develop it further by introducing an advanced filtering approach instead of the filtering by convolution described previously.

Most parts of a fuzzy system — such as the inference mechanism — are well understood and founded on a theoretical basis. Yet defuzzification, which is the link between the fuzzy and real valued variables, is mainly based on heuristics. Classical defuzzification methods are not satisfactory if the output membership function shows strong ambiguities.

The first step to improve the defuzzification was the introduction of the inference filter concept and its realization by a convolution integral. This approach allows the reproduction of classical defuzzification methods and, moreover, exhibits additional degrees of freedom through the parameters of the distance

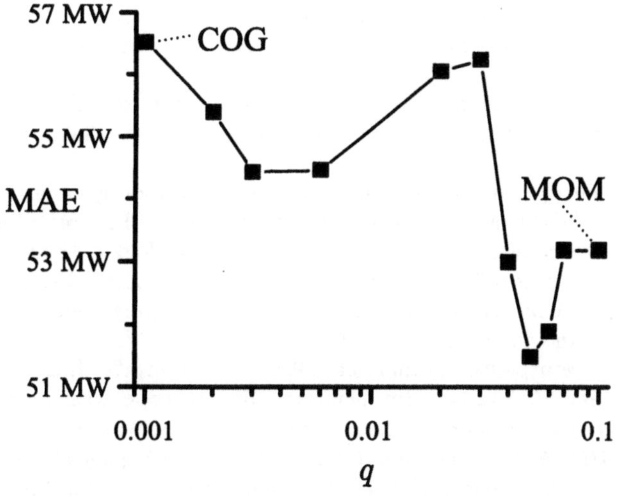

Fig. 7. Mean absolute error in MW of a fuzzy system using the advanced inference filter defuzzification method with different values of the factor q.

function of the convolution integral. By making use of these tuning parameters, in most cases a better performance can be achieved than with classical defuzzification methods. However, this approach is not free of drawbacks when there are explicit ambiguities in the output membership function.

Another step to improve the defuzzification is proposed here. To overcome the above drawbacks, we present the advanced inference filter. It maintains the properties of reproducing the classical defuzzification methods and gives additional degrees of freedom by adjusting the parameter q. The effectiveness of the advanced inference filter defuzzification method has been demonstrated with two examples.

In addition to its usefulness for conventional fuzzy systems, the advanced inference filter defuzzification is well suited to a new modeling method under research, implicit modeling [7]. This method aims especially at handling ambiguities in data and their resolution.

Acknowledgment

This research was sponsored by the Deutsche Forschungsgemeinschaft (DFG), as part of the Collaborative Research Center 'Computational Intelligence' (531) of the University of Dortmund

References

1. D. Driankov, H. Hellendoorn, and M. Reinfrank. *An Introduction to Fuzzy Control.* Springer, Berlin Heidelberg, 1993.

2. H. Kiendl. The inference filter. In *Second European Congress on Intelligent Techniques and Soft Computing (EUFIT '94)*, volume 1, pages 443–452, Aachen, 1994. Verlag Mainz.
3. H. Kiendl. *Verfahren zur Defuzzifizierung für signalverarbeitende Fuzzy–Baueinheiten und Filtereinrichtungen hierfür.* Kiendl, 1995. Patent DE 44 16 465.
4. H. Kiendl. *Fuzzy–Control methodenorientiert.* Oldenbourg Verlag, München, 1997.
5. H. Kiendl. Non-translation-invariant defuzzification. In *Proceedings of the Sixth IEEE International Conference on Fuzzy Systems (FUZZ–IEEE '97), Barcelona, Spanien, 1997*, pages 737–742, Piscataway, NJ, 1997. IEEE Press.
6. H. Kiendl, R. Knicker, and F. Niewels. Two way fuzzy controllers based on hyperinference and inference filter. In *Proceedings of World Automation Congress (WAC '96)*, Montpellier, Frankreich, 1996. TSI Enterprise Inc.
7. H. Kiendl and P. Krause. Implicit modeling to cope with ambigous data. In *Proceedings of the IASTED International Conference Modeling, Identification and Control (MIC 2001), Innsbruck, Österreich*, volume 1, pages 291–296, 2001.
8. A. Krone. *Datenbasierte Generierung von relevanten Fuzzy–Regeln zur Modellierung von Prozesszusammenhängen und Bedienstrategien.* Fortschritt–Berichte VDI, Reihe 10, Nr. 615. VDI Verlag, Düsseldorf, 1999.
9. M. Mackey and L. Glass. Oscillation and Chaos in Physiological Control Systems. *Science*, 197:287–289, 1977.
10. F. Niewels. Analytische formeln für das inferenzfilter zur berechnung gefilterter zugehörigkeitsfunktionen. pages 1–14. Forschungsbericht der Fakultät für Elektrotechnik, Nr. 0896, Universität Dortmund, 1996.
11. J. Praczyk. *Entwurf von Fuzzy–Gütemaßen zur Prozeßbewertung.* Fortschritt–Berichte VDI, Reihe 8, Nr. 796. VDI Verlag, Düsseldorf, 1999.
12. G. Reil and H. Jessen. Fuzzy contour modelling of roll bent components using inference filter. In *Proceedings of the Third European Congress on Intelligent Techniques and Soft Computing (EUFIT '95)*, pages 771–774, Aachen, 1995. Verlag Mainz.
13. T. Slawinski, A. Krone, P. Krause, and H. Kiendl. The fuzzy–rosa method: A statistically motivated fuzzy approach for data–based generation of small interpretable rule bases in high–dimensional search spaces. In M. Last, A. Kandel, and H. Bunke, editors, *Data Mining and Computational Intelligence*, pages 141–166. Physica-Verlag Heidelberg, 2001.

On Cardinality and Singular Fuzzy Sets

Krzysztof Dyczkowski, Maciej Wygralak

Adam Mickiewicz University, Faculty of Mathematics and Computer Science
Matejki 48/49, 60-769 Poznań, Poland
{chris, wygralak}@math.amu.edu.pl

Abstract. Fuzzy sets with triangular norms and their cardinalities understood as convex fuzzy sets of usual cardinal numbers are the subject of this paper. It appears that if nonstrict Archimedean triangular norms are involved, some fuzzy sets become totally dissimilar to any set of any cardinality. In other words, they become singular with respect to cardinality. We present a study of those singular fuzzy sets. In particular, we give their characterization and formulate their main properties.

1 Preliminaries

Cardinality belongs to the most important characteristics of a fuzzy set in exactly the same way as it belongs to the most fundamental characteristics of a set. However, the graduation of membership makes counting and cardinal calculus under fuzziness a task which is generally more complicated, advanced and enigmatic than in the case of ordinary sets. Various examples of applications of that calculus can be found in mathematics (e.g. in the theory of fuzzy topological spaces) as well as in computing with words, communication with data bases and information/intelligent systems, modeling the meaning of imprecise quantifiers in natural language statements, decision-making in a fuzzy environment, probabilities of fuzzy events, etc. (see [1, 2, 11-13]).

There are two main groups of constructive approaches to the question of the cardinality of a fuzzy set. In the so-called scalar approaches, that cardinality is a single usual cardinal number or a nonnegative real number (see [8, 11, 12]). In this paper, however, the other group will be discussed. The cardinality of a fuzzy set is then understood as a convex fuzzy set of usual cardinal numbers, i.e. as a weighted family of usual cardinals, which is called a *generalized cardinal number*. This "fuzzy" perception of cardinality is more advanced and complicated from both the conceptual and operational viewpoints, but at that price it gives us in comparison with any scalar approach a considerably more complete and adequate cardinal information about the fuzzy set. Generalized cardinals will be denoted by lowercase letters from the beginning of the Greek alphabet. If a generalized cardinal α does express the cardinality of a fuzzy set $A\colon \boldsymbol{M} \to [0,1]$, we write $|A| = \alpha$ and we say that the cardinality of A is equal to α. Reviews of generalized cardinals of fuzzy sets with the standard operations $\wedge := \min$ and $\vee := \max$ as well as further references can be found in [2, 6, 7]; as usual, the symbol $:=$ stands for "equals by definition". In [9, 10], generalized cardinals of fuzzy sets with triangular norm-based operations are constructed and investigated.

B. Reusch (Ed.): Fuzzy Days 2001, LNCS 2206, pp. 261–268, 2001.

Our further discussion will be restricted to finite fuzzy sets with triangular norm-based operations. They play the central role in applications. From now on, if not emphasized otherwise, the phrase "fuzzy set" will mean "finite fuzzy set". The universe $\boldsymbol{M}$, however, is assumed to be of a quite arbitrary power. The family of all (finite) fuzzy sets in $\boldsymbol{M}$ will be denoted by FFS. Their generalized cardinals are just some convex fuzzy sets in the set $\mathbb{N}$ of all nonnegative integers. A convenient vector notation will be used for those cardinals, namely

$$\alpha = (a_0, a_1, \dots, a_k, (a))$$

with $k \in \mathbb{N}$ does mean that $\alpha(i) = a_i$ for $i \le k$ and $\alpha(i) = a$ for $i > k$. Moreover,

$$(a_0, a_1, \dots, a_k) := (a_0, a_1, \dots, a_k, (0)) .$$

For instance, $\alpha = (0.1, 0.7, (1))$ is the notation for α with $\alpha(0) = 0.1$, $\alpha(1) = 0.7$ and $\alpha(i) = 1$ for $i > 1$. $\alpha = (0.2, 0.7, 0.3)$ means that $\alpha(0) = 0.2$, $\alpha(1) = 0.7$, $\alpha(2) = 0.3$ and $\alpha(i) = 0$ for $i > 2$. $\alpha = (0.9)$ denotes α with $\alpha(0) = 0.9$ and $\alpha(i) = 0$ for $i > 0$.

Let $A \in \text{FFS}$ and $[A]_i := \bigvee\{t\colon |A_t| \ge i\}$ with $i \in \mathbb{N}$, where A_t denotes the t-cut set of A, i.e. $A_t := \{x \in \boldsymbol{M}\colon A(x) \ge t\}$. The set $\{x \in \boldsymbol{M}\colon A(x) > t\}$, the sharp t-cut set of A, will be denoted by A^t. Throughout, we put $n := |\text{supp}(A)|$ and $m := |\text{core}(A)|$. Each $[A]_i$ with $0 < i \le n$ is thus the ith element in the nonincreasingly ordered sequence of all positive membership values $A(x)$, including their possible repetitions. Clearly, $[A]_0 = 1$ and $[A]_i = 0$ for $i > n$.

For completeness, let us recall a few definitions and facts from the theory of triangular norms (see e.g. [3, 5]). A binary operation $\boldsymbol{t}$: $[0,1] \times [0,1] \to [0,1]$ is a *triangular norm* (*t-norm*, in short) if $\boldsymbol{t}$ is commutative, associative, nondecreasing in the first and, hence, in each argument, and has 1 as unit. A continuous t-norm $\boldsymbol{t}$ is called *Archimedean* if $a\,\boldsymbol{t}\,a < a$ for each $a \in (0,1)$. If an Archimedean t-norm is strictly increasing over $(0,1) \times (0,1)$, it is called *strict*. A typical example of a strict t-norm is the *algebraic t-norm* $\boldsymbol{t}_a$ with $a\,\boldsymbol{t}_a\,b := ab$, whereas the *Łukasiewicz t-norm* $\boldsymbol{t}_Ł$ with $a\,\boldsymbol{t}_Ł\,b := 0 \vee (a+b-1)$ is a typical example of a nonstrict Archimedean t-norm. The class of Archimedean t-norms seems to be especially important and interesting because, for each continuous $\boldsymbol{t}$, either $\boldsymbol{t} = \wedge$ or $\boldsymbol{t}$ is Archimedean or $\boldsymbol{t}$ is the ordinal sum of a family of Archimedean t-norms; see e.g. [3] for the definition of ordinal sums. Also, the following nice representation theorem by Ling holds true for Archimedean t-norms (see [4, 5]).

Theorem 1.1. *A t-norm* $\boldsymbol{t}$ *is Archimedean iff there exists a strictly decreasing and continuous generator* g: $[0,1] \to [0,\infty]$ *such that* $g(1) = 0$ *and*

$$a\,\boldsymbol{t}\,b = g^{-1}(g(0) \wedge (g(a) + g(b)))$$

for each $a, b \in [0,1]$. *Moreover,* $\boldsymbol{t}$ *is strict iff* $g(0) = \infty$.

If $g(0)=1$, g is called *normed* generator. By mathematical induction, we get

$$a_1\, t\, a_2\, t \dots t\, a_k = g^{-1}(g(0) \wedge \Sigma_{i=1}^{k} g(a_i)) \tag{1}$$

for each Archimedean t with generator g and for each system $a_1, a_2, \dots a_k \in [0,1]$ with $k \in \mathbb{N}$; so, $a_1\, t\, a_2\, t \dots t\, a_k = 1$ for $k=0$. Consequently,

$$a_1\, t\, a_2\, t \dots t\, a_k > 0 \;\text{ iff }\; \Sigma_{i=1}^{k} g(a_i) < g(0)\,. \tag{2}$$

Thus, $a_1\, t\, a_2\, t \dots t\, a_k > 0 \Leftrightarrow a_1, a_2, \dots a_k > 0$ when t is strict, which cannot be extended to nonstrict Archimedean t-norms.

Each nonincreasing function $\nu\colon [0,1] \to [0,1]$ with $\nu(0)=1$ and $\nu(1)=0$ will be called a *negation*. So, ν_* such that $\nu_*(a) := 0$ for each $a>0$ is the smallest possible negation, whereas ν^* with $\nu^*(a) := 1$ for each $a<1$ is the largest one. Strictly decreasing and continuous negations are called *strict negations*. Their class will be denoted by *Sng*. Involutive strict negations ($\nu(\nu(a)) = a$ for each a) are said to be *strong negations*. An important example of a strong negation is the *Łukasiewicz negation* $\nu_Ł$ with $\nu_Ł(a) := 1-a$. Each t-norm t does generate a negation defined by

$$\nu_t(a) := \bigvee\{c \in [0,1]\colon\ a\, t\, c = 0\}$$

for each $a \in [0,1]$. Such a negation is strong if t is Archimedean and nonstrict. We then have

$$\nu_t(a) = g^{-1}(g(0) - g(a)) \tag{3}$$

whenever g does generate t. If t is strict, then $\nu_t = \nu_*$.

2 Generalized Cardinals of Fuzzy Sets with Triangular Norms

The following three types of generalized cardinals can be constructed for fuzzy sets with a t-norm t and a negation ν (see [9, 10]). Let $A \in \mathrm{FFS}$ and $\alpha = |A|$.

(a) $$\alpha(k) := [A]_1\, t\, [A]_2\, t \dots t\, [A]_k \;\text{ for each }\; k \in \mathbb{N}\,.$$

Applying the vector notation, one can write

$$\alpha = (1, [A]_1, [A]_1\, t\, [A]_2, \dots, [A]_1\, t\, [A]_2\, t \dots t\, [A]_n)\,.$$

This generalized cardinal does form a lower cardinal evaluation of A because $\alpha(k)$ is a degree to which A contains (at least) i elements for each $i \le k$; the quantification "for each $i \le k$" is here realized by means of t (see [3]). Putting $t=\wedge$, α collapses to the well-known FGCount of A (see [11]).

(b)
$$\alpha(k) := \nu([A]_{k+1})\, t\, \nu([A]_{k+2})\, t \dots = \nu([A]_{k+1})\, t\, \nu([A]_{k+2})\, t \dots t\, \nu([A]_n)\,,$$

i.e.
$$\alpha = (\nu([A]_1)\, t\, \nu([A]_2)\, t \dots t\, \nu([A]_n),\ \nu([A]_2)\, t\, \nu([A]_3)\, t \dots t\, \nu([A]_n),\ \vdots\ \nu([A]_{n-1})\, t\, \nu([A]_n),\ \nu([A]_n), (1))\,.$$

This α is an upper cardinal evaluation of A: $\alpha(k)$ is now a degree to which A contains at most i elements for each $i \geq k$. If $t = \wedge$ and $\nu = \nu_Ł$, α becomes the FLCount of A (see [11]).

(c)
$$\alpha(k) := [A]_1\, t\, [A]_2\, t \dots t\, [A]_k\, t\, \nu([A]_{k+1})\, t \dots t\, \nu([A]_n)\,,$$

i.e.
$$\alpha = (\nu([A]_1)\, t\, \nu([A]_2)\, t \dots t\, \nu([A]_n),\ [A]_1\, t\, \nu([A]_2)\, t \dots t\, \nu([A]_n),\ [A]_1\, t\, [A]_2\, t\, \nu([A]_3)\, t \dots t\, \nu([A]_n),\ \vdots\ [A]_1\, t\, [A]_2\, t \dots t\, [A]_{n-1}\, t\, \nu([A]_n),\ [A]_1\, t\, [A]_2\, t \dots t\, [A]_n)\,.$$

This intersection of the lower and upper cardinal evaluations of A is always convex and can be viewed as a "proper" generalized cardinal number of A. In the language of many-valued logic with t-norms, $\alpha(k)$ is a degree to which A has at least i elements for each $i \leq k$ and, simultaneously, A has at most j elements for each $j \geq k$. Worth mentioning are the following particular cases:

($\nu = \nu_*$) $\quad \alpha = (0, 0, \dots, 0, [A]_1\, t\, [A]_2\, t \dots t\, [A]_n)$ with n zeros.

($t = \wedge$, $\nu = \nu^*$) $\quad \alpha = (0, 0, \dots, 0, 1, [A]_{m+1}, [A]_{m+2}, \dots, [A]_n)$ with m zeros,

i.e. α becomes the generalized cardinal of A due to Dubois ([2]).

($t = \wedge$, $\nu = \nu_Ł$) $\quad \alpha(k) = [A]_k \wedge (1 - [A]_{k+1})\,,$

i.e. α collapses to the so-called FECount of A. $\alpha(k)$ is a degree to which A has exactly k elements (see [7, 11]).

In this paper, we restrict our discusison to the generalized cardinals of the type (c). They are called *generalized FECounts* (*induced by a t-norm* t *and a negation* ν). Their family will be denoted by $\mathrm{GFE}_{t,\nu}$. Special attention will be paid to the generalized FECounts induced by a t-norm from the class *NArch* of nonstrict Archimedean t-norms and a strict negation. They seem to be exceptional in a way.

3 Nonstrict Archimedean t-Norms and Singular Fuzzy Sets

Nonstrict Archimedean t-norms do have zero divisors, i.e. $a\,t\,b = 0$ is possible for $a, b > 0$ (recall that $a\,t\,0 = 0$). One can say that to have zero divisors does mean to have some inertia in attaining positive values by t: each positive argument of t is treated as zero whenever the other one is not sufficiently "large", e.g. $0.2\,t_Ł\,0.7 = 0$, but $0.2\,t_Ł\,0.9 > 0$. Clearly, this feature can be useful and desired in many applications.

Let $A \in \mathrm{FFS}$ and $|A| = \alpha \in \mathrm{GFE}_{t,\nu}$ with $t \in NArch$ and $\nu \in Sng$. Since t has zero divisors, it is possible that α is the zero function, i.e. $\alpha = (0)$. All the weights assigned to nonnegative integers are then equal to zero, which means that A is totally dissimilar to any set of any cardinality. Let us stress in this place that $\alpha = 1_{\{k\}}$ whenever A is a k-element set. The generalized cardinal (0) will be called the *flat generalized cardinal*. If $\alpha = (0)$, we will say that A is a *singular fuzzy set*. An instance of such a fuzzy set is $A = 1/x_1 + 0.8/x_2 + 0.5/x_3 + 0.5/x_4$ with $t = t_Ł$ and $\nu = \nu_Ł$. This simple example suggests that quite "usual" fuzzy sets may be singular with respect to cardinality (see further discussion).

It is clear that $\nu \in Sng$ has a unique fixed point $a \in (0, 1)$. For instance, $a = 0.5$ for $\nu = \nu_Ł$. Let $s := \max\{i \in \mathbb{N}\colon [A]_i > a\} = |A^a|$. So, $s = s(A, \nu)$ and $m \le s \le n$.

Theorem 3.1. (a) $\alpha(k) \le \alpha(s)$ *for each* $k \in \mathbb{N}$.
(b) $\alpha = (0)$ *iff* $\alpha(s) = 0$.

Proof. (a) The thesis is satisfied for $n \le 1$. Indeed, if $n = 0$, then $\alpha = (1)$ and $s = 0$. If $n = 1$, we have $\alpha = (\nu([A]_1), [A]_1)$ with $[A]_1 \in (0, 1]$. So, $s = 0$ gives $\alpha(0) \ge \alpha(1)$ because $[A]_1 \le a$, whereas $s = 1$ leads to $\alpha(0) < \alpha(1)$. Assume $n \ge 2$. The following chains of implications hold true:

$$k \le s \Rightarrow [A]_k > a \Rightarrow \nu([A]_k) < [A]_k$$

and

$$k > s \Rightarrow [A]_k \le a \Rightarrow \nu([A]_k) \ge [A]_k\,.$$

Hence

$$\begin{aligned}\alpha(k) &= [A]_1\,t \ldots t\,[A]_k\,t\,\nu([A]_{k+1})\,t \ldots t\,\nu([A]_n)\\ &\le [A]_1\,t \ldots t\,[A]_s\,t\,\nu([A]_{s+1})\,t \ldots t\,\nu([A]_n)\\ &= \alpha(s)\end{aligned}$$

for $k \ne s$. This completes the proof of (a). (b) is its immediate consequence. □

Generalized FECounts have a rather complex general form. By Theorem 3.1, the procedure of checking whether $\alpha(k) = 0$ for each k can be reduced to verifying the single condition $\alpha(s) = 0$.

Theorem 3.2. *For $t \in NArch$ with normed generator g, we have*

$$\alpha(s) = 0 \ \ \textit{iff} \ \ \Sigma_{i=1}^{s} g([A]_i) + \Sigma_{i=s+1}^{n} g(\nu([A]_i)) \geq 1 .$$

Proof. The thesis is a consequence of (2). □

So, A is nonsingular ($\alpha \neq (0)$) iff the $[A]_i$'s with $i \leq s$ are sufficiently near to 1 and the $[A]_i$'s with $s+1 \leq i \leq n$ are sufficiently smaller than the fixed point of ν. $\alpha \neq (0)$ is thus possible only if A is similar to an s-element set (see Example 3.4).

Corollary 3.3. If $\nu = \nu_t$, *then* $\alpha(s) = 0$ *iff* $\Sigma_{i=1}^{s} g([A]_i) - \Sigma_{i=s+1}^{n} g([A]_i) \geq s - n + 1$.

Proof. Indeed, by (3), $g(\nu_t([A]_i)) = 1 - g([A]_i)$. □

For instance, $g(x) = 1 - x$ and $\nu_t = \nu_Ł$ if $t = t_Ł$. So, the condition $\alpha(s) = 0$ holds iff $\Sigma_{i=1}^{s} [A]_i - \Sigma_{i=s+1}^{n} [A]_i \leq s - 1$. Consequently, the inequality $[A]_1 \leq (s-1)/s$, $s \geq 3$, implies that $\alpha = |A|$ is flat. Each fuzzy set $A \in$ FFS with $t = t_Ł$, $\nu = \nu_Ł$ and, say, $s = 10$ is thus singular whenever $[A]_1 \leq 0.9$. Take a more general example in which t is a Schweizer t-norm, i.e. $t = t_{S,p}$ with $p > 0$ and $a\, t_{S,p}\, b := [0 \vee (a^p + b^p - 1)]^{1/p}$; so, $t_Ł = t_{S,1}$. Its generator is of the form $g(x) = 1 - x^p$ and $\nu_t(x) = (1 - x^p)^{1/p}$. By Corollary 3.3, we obtain $\alpha(s) = 0$ iff $\Sigma_{i=1}^{s} ([A]_i)^p - \Sigma_{i=s+1}^{n} ([A]_i)^p \leq s - 1$.

Example 3.4. (i) Let $t = t_Ł$ and $\nu = \nu_Ł$, i.e. $a = 0.5$. If

$$A = 0.8/x_1 + 0.2/x_2 + 0.1/x_3 + 0.1/x_4 ,$$

then $s = 1$ and $\alpha(s) = [A]_1\, t\, \nu([A]_2)\, t\, \nu([A]_3)\, t\, \nu([A]_4) = 0.8\, t\, 0.8\, t\, 0.9\, t\, 0.9 > 0$, i.e. A is nonsingular. This is because A is similar to a 1-element set.
(ii) Using the t and ν from (i), let

$$A = 0.95/x_1 + 0.95/x_2 + 0.95/x_3 + 0.5/x_4 + 0.4/x_5 .$$

We now have $s = 3$ and $\alpha(s) = 0.95\, t\, 0.95\, t\, 0.95\, t\, 0.5\, t\, 0.6 = 0$. A is singular because its similarity to a 3-element set is low as the membership grades of x_4 and x_5 are near to a. A fuzzy subset of A, namely

$$B = 0.9/x_1 + 0.9/x_2 + 0.9/x_3 + 0.2/x_4 + 0.1/x_5$$

for which $s = 3$, too, is nonsingular. It is more similar to a 3-element set than A.
(iii) Let $t = t_{S,4}$ and $\nu = \nu_t$, i.e. $\nu(x) = (1 - x^4)^{1/4}$ and $a = 0.5^{1/4} \approx 0.84$. For the fuzzy set A from (ii), we again have $s = 3$, but this time $\alpha(s) > 0$ by Corollary 3.3. So, A is nonsingular. The reason is that a plays the role of a threshold point: the membership values $> a$ are considered to be more or less similar to the maximal membership value 1, whereas those $\leq a$ are considered to be more or less similar to the minimal membership value 0. In this example, we have $a \approx 0.84$. The membership values 0.5 and 0.4 are now considerably smaller than the threshold value and, consequently, A becomes much more similar to a 3-element set than in (ii).

Summarizing the above remarks, it seems that fuzzy sets which are nonsingular are fuzzy sets having a small value of the entropy measure of fuzziness. That measure, however, must be constructed in a special way: a maximal fuzziness should be assigned to the fuzzy set identically equal to $\boldsymbol{a}$. In the classical approach to fuzziness measures, which seems to be suitable only for fuzzy sets with the standard operations, that maximal fuzziness is assigned to the fuzzy set identically equal to 0.5. Let us notice that, for each $t \in NArch$ and $\nu \in Sng$, if A and B are such that $\boldsymbol{a} < B(x) \leq A(x)$ or $A(x) \leq B(x) \leq \boldsymbol{a}$ for each $x \in \boldsymbol{M}$, then the singularity of A implies the singularity of B, i.e. the nonsingularity of B implies the nonsingularity of A. If $\nu = \nu_t$ and A is such that $A(x) = \boldsymbol{a}$ for each $x \in \text{supp}(A)$, $n \geq 2$, then A is singular.

If $t \in NArch$, $\nu \in Sng$ and $A = a/x$ with $a \in (0, 1]$, then $|A| = (\nu(a), a)$. Thus, a singleton cannot be singular. This is no longer valid for fuzzy sets with larger supports. By the way, it is easy to find a singular fuzzy set A with $n \geq 3$. For $n = 2$, that task is more difficult. If, say, $t = t_Ł$, $\nu = \nu_Ł$ and $A = a/x + b/y$ with $a, b > 0$, then A is singular iff $a = b = 0.5$.

It is also easy to notice that each singular nonempty fuzzy set has a proper fuzzy subset which is nonsingular, at least the empty fuzzy set $1_\emptyset$. The reason is that a proper fuzzy subset of A may be more distinct with respect to cardinality, more similar to an s-element set than A itself (see e.g. Example 3.4(ii)). On the other hand, the following useful *absorption property* holds true for singular fuzzy sets. As usual, $\cap$ and $\cup$ denote the standard intersection and sum operations defined via $\wedge$ and $\vee$.

Theorem 3.5. *For each nonstrict Archimedean t-norm* $\boldsymbol{t}$ *and strict negation* ν, *if* A *is singular and* B *is such that* $A \cap B = 1_\emptyset$, *then* $A \cup B$ *is singular, too.*

Proof. Suppose that the assumptions of the theorem are satisfied. Put $c_i := [A \cup B]_i$ for each $i \geq 1$, $r := \max\{i\colon c_i > \boldsymbol{a}\}$, $n^* := |\text{supp}(B)|$ and $l := |\text{supp}(A \cup B)|$; as previously, $s := \max\{i\colon [A]_i > \boldsymbol{a}\}$, where $\boldsymbol{a}$ is the fixed point of ν. The sequence $c_1, c_2, \dots, c_l$ is a result of joining the sequences $[A]_1, [A]_2, \dots, [A]_n$ and $[B]_1, [B]_2, \dots, [B]_{n^*}$ in the nonincreasing way. Hence $r \geq s$ and, by Theorem 3.2,

$$\Sigma_{i=1}^{r} g(c_i) + \Sigma_{i=r+1}^{l} g(\nu(c_i)) \geq \Sigma_{i=1}^{s} g([A]_i) + \Sigma_{i=s+1}^{n} g(\nu([A]_i)) \geq 1 .$$

So, $A \cup B$ is singular. □

The absorption property is reflected in the results generated by the triangular norm-based extension principle saying that $(\alpha + \beta)(k) := \bigvee\{\alpha(i)\, \boldsymbol{t}\, \beta(j)\colon i + j = k\}$. Namely, we get $(0) + \beta = (0)$ for each $\beta \in \text{GFE}_{t,\nu}$.

The following simple criterion can be derived from Theorem 3.5: if a fuzzy set $a_{i1}/x_{i1} + a_{i2}/x_{i2} + \dots + a_{ik}/x_{ik}$ is singular for some indices $1 \leq i1 < i2 < \dots < ik \leq n$ with $2 \leq k < n$, then $A = a_1/x_1 + a_2/x_2 + \dots + a_n/x_n$ is singular, too. This sufficient condition for A to be singular is convenient in practice because it is easy to check if a fuzzy set with a small support having, say, 2 or 3 elements is singular. However, that A is singular does not generally imply the existence of a singular fuzzy subset $a_{i1}/x_{i1}+a_{i2}/x_{i2}+\ldots+a_{ik}/x_{ik}$ of A. For instance, $A=0.6/x_1+0.6/x_2+0.6/x_3$ with $t=t_Ł$ and $\nu=\nu_Ł$ is singular, whereas $|0.6/x_i+0.6/x_j|=(0,0,0.2)\neq(0)$.

Finally, the singularity of a fuzzy set A generally refers to and depends on a concrete t-norm and negation. Worth noticing in this context is the following property. Let $t,u\in NArch$, $\nu,\sigma\in Sng$, $|A|=\alpha\in \mathrm{GFE}_{t,\nu}$ and $|A|=\beta\in \mathrm{GFE}_{u,\sigma}$. If $\alpha=(0)$, $u\leq t$ and $\sigma\leq\nu$, then $\beta=(0)$. In Example 3.4(iii), we have $u<t_Ł$, but $\nu_Ł<\nu_u$ for $u=t_{S,4}$.

References

1. Delgado, M., Sánchez, D., Vila, M.A.: Fuzzy Cardinality Based Evaluation of Quantified Sentences. Int. Jour. Approx. Reasoning **23** (2000) 23-66
2. Dubois, D., Prade, H.: Fuzzy Cardinality and the Modeling of Imprecise Quantification. Fuzzy Sets and Systems **16** (1985) 199-230
3. Gottwald, S.: Many-Valued Logic and Fuzzy Set Theory. In: Höhle, U., Rodabaugh, S.E. (eds.): Mathematics of Fuzzy Sets. Logic, Topology, and Measure Theory. Kluwer Acad. Publ., Boston Dordrecht London (1999) 5-89
4. Ling, C.H.: Representation of Associative Functions. Publ. Math. Debrecen **12** (1965) 189-212
5. Weber, S.: A General Concept of Fuzzy Connectives, Negations and Implications Based on t-Norms and t-Conorms. Fuzzy Sets and Systems **11** (1983) 115-134
6. Wygralak, M.: Vaguely Defined Objects. Representations, Fuzzy Sets and Nonclassical Cardinality Theory. Kluwer Acad. Publ., Dordrecht Boston London (1996)
7. Wygralak, M.: Questions of Cardinality of Finite Fuzzy Sets. Fuzzy Sets and Systems **102** (1999) 185-210
8. Wygralak, M.: Triangular Operations, Negations, and Scalar Cardinality of a Fuzzy Set. In: Zadeh, L.A., Kacprzyk, J. (eds.): Computing with Words in Information/Intelligent Systems, Vol. 1 - Foundations. Physica-Verlag, Heidelberg New York (1999) 326-341
9. Wygralak, M., Pilarski, D.: FGCounts of Fuzzy Sets with Triangular Norms. In: Hampel, R., Wagenknecht, M., Chaker, N. (eds.): Fuzzy Control. Theory and Practice. Physica-Verlag, Heidelberg New York (2000) 121-131
10. Wygralak, M.: Fuzzy Sets with Triangular Norms and Their Cardinality Theory. Fuzzy Sets and Systems, to appear
11. Zadeh, L.A.: A Computational Approach to Fuzzy Quantifiers in Natural Languages. Comput. and Math. with Appl. **9** (1983) 149-184
12. Zadeh, L.A.: From Computing with Numbers to Computing with Words - From Manipulation of Measurements to Manipulation of Perceptions. IEEE Trans. on Circuits and Systems - I: Fundamental Theory and Appl. **45** (1999) 105-119
13. Zadeh, L.A.: Fuzzy logic = Computing with Words. In: Zadeh, L.A., Kacprzyk, J. (eds.): Computing with Words in Information/Intelligent Systems, Vol. 1 - Foundations. Physica-Verlag, Heidelberg New York (1999) 3-23

Fast Iris Detection for Personal Verification Using Modular Neural Nets

Hazem El-Bakry

Faculty of Computer Science &Information Systems
Mansoura University - Egypt
hazemelbakry@hotmail.com

Abstract. In this paper, a combination of fast and cooperative modular neural nets to enhance the performance of the detection process is introduced. I have applied such approach successfully to detect human faces in cluttered scenes [10]. Here, this technique is used to identify human irises automatically in a given image. In the detection phase, neural nets are used to test whether a window of 20x20 pixels contains an iris or not. The major difficulty in the learning process comes from the large database required for iris / non-iris images. A simple design for cooperative modular neural nets is presented to solve this problem by dividing these data into three groups. Such division results in reduction of computational complexity and thus decreasing the time and memory needed during the test of an image. Simulation results for the proposed algorithm show a good performance. Furthermore, faster iris detection is obtained through image decomposition into many sub-images and applying cross correlation in frequency domain between each sub-image and the weights of the hidden layer.

1 Introduction

Biometrics offers new solutions for personal identification are user friendly and yet secure access. They work by inspecting biological features of ourselves that distinguish us from others. Examples include fingerprints, hand geometry, and voice recognition. It would be preferable to base recognition decisions upon features which had very little genetic variations, and stable over lifetime of the individuals. It is suggested in recent years that human irises might be as distinct as fingerprints for different individuals, lending to the idea that iris patterns may contain unique identification features. The motivation for this endeavor stems from the observation that the human iris provides a particularly interesting structure as well as noninvasive biometric assessment. The prospect of using the human iris for identification was firstly suggested by ophthalmologists, who noted that the pattern of the iris remained unchanged in clinical photographs taken over several decades. Further, since the iris is an overt body, its appearance is amenable to remote examination with the aid of a machine vision system.

To have a more practical biometric system, here we concentrate on personal identification through human iris recognition, a relatively new biometric technology that has some key advantages in terms of speed, simplicity, accuracy, and applicability [9]. Of course, face recognition does not need any voluntary action and can be used for covert surveillance. But, the problem of classification of twins makes

B. Reusch (Ed.): Fuzzy Days 2001, LNCS 2206, pp. 269–283, 2001.

the identification of persons based on face recognition unreliable. Moreover, for a given person, the patterns of right and left irises are different from each other, although they are genetically identical as the four irises of two identical twins. Furthermore, the spatial patterns which apparent in the human iris are highly distinctive to an individual as shown in Fig. 1 [1, 9]. Also, the size of iris is stable throughout life. We may have noticed how large the eyes of a baby seem to be when their heads are so small. This is because even from birth, the human iris is about 10 millimeters in diameter. More importantly, the iris pattern itself is also stable throughout life, except that the color may change in the first months. As the pupil size is constantly changed, neural networks are used to compensate for this distortion in the iris patterns.

The goal of this paper is to solve the problem of requiring large database to build an automatic system in order to detect the location of irises in scenes. This paper explores the use of modular neural network (MNN) classifiers. Non-modular classifiers tend to introduce high internal interference because of the strong coupling among their hidden layer weights [2]. As a result of this, slow learning or over fitting can occur during the learning process. Sometimes, the network could not be learned for complex tasks. Such tasks tend to introduce a wide range of overlap which, in turn, causes a wide range of deviations from efficient learning in the different regions of input space [3]. High coupling among hidden nodes will then, result in over and under learning at different regions [8]. Enlarging the network, increasing the number and quality of training samples, and techniques for avoiding local minina, will not stretch the learning capabilities of the NN classifier beyond a certain limit as long as hidden nodes are tightly coupled, and hence cross talking during learning [7]. A MNN classifier attempts to reduce the effect of these problems via a divide and conquer approach. It generally, decomposes the large size / high complexity task into several sub-tasks, each one is handled by a simple, fast, and efficient module. Then, sub-solutions are integrated via a multi-module decision-making strategy. Hence, MNN classifiers, generally, proved to be more efficient than non-modular alternatives [5,6]. In section 2, we present a method for detection of human irises in photo images. Also, an algorithm during the searching procedure is described. A fast searching algorithm for iris detection which reduces the computational complexity of neural nets is presented in section 3. A faster searching algorithm for iris detection which reduces the number of computation steps through image decomposition is presented in section 4.

2 Human Iris Detection Based on Neural Networks

Image acquisition of the iris can not be expected to yield an image containing only the iris. Rather, image acquisition will capture the iris as part of a larger image that also contains data derived from the immediately surrounding eye region. Therefore, prior to iris pattern matching, it is important to localize precisely that portion of the acquired image corresponding to an iris. In particular, it is necessary to localize that portion of the image derived from inside the limbus (the border between the sclera and the iris) and outside the pupil. Further if the eyelids are occluding part of the iris, then only that portion of the image below the upper eyelid and above the lower eyelid should be included. Moreover, the detection algorithm must be invariant to changes in pupillary constriction and overall iris image size, and hence also invariant to camera zoom factor and distance to the eye. Previous work in iris detection is accomplished by using 2D Gabor wavletes. Then a doubly dimensionaless coordinate system is used to map the tissue [1]. But, these methods are complex and require large number of computations and consume more time. Here neural networks are used to detect the iris with all of these variations and compensate automatically for the stretching of the iris tissue as the pupil dilates.

The human iris is a complex pattern. Finding human irises automatically in a scene is a difficult yet significant problem. It is the first step in fully automatic human iris recognition system. Iris detection is the fundamental step before the iris recognition or identification procedure. Its reliability and time response have a major influence on the performance and usability of the whole iris recognition system. Previous authors [1,9] assume that only the human eye is presented in front of camera. But if any other non-iris (for example objects, texture, natural scenes, parts of the eye that have not any information about iris or any part of the face that may be appeared in front of camera) is presented what will be the decision?. Training a neural network for the iris detection task is challenging because of the difficulty in characterizing prototypical "non-iris" images. Unlike iris recognition, in which the classes to be discriminated are different irises, the two classes to be discriminated in the iris detection are "image containing irises" and "image not containing irises". It is easy to get a representative sample of images that contain irises, but much harder to get a representative sample of those which do not. Feature information needs to be stored in the database for the purpose of retrieval. Information retrieval can be done by using a neural network approach which has the potential to embody both numerical and structural iris data. However, neural network approaches have been demonstrated only on limited database. The use of huge samples of iris/non-iris images makes the learning process very difficult for the neural network.

2.1 A Proposed Algorithm For Iris Detection Using MNNs

First, in an attempt to equalize the intensity values of the iris image, the image histogram is equalized. This not only reduce the variability of pixels generated by illumination conditions, and enhance the image contrast but also increases the number of correct pixels that can be actually encountered [10]. The next component of our system is a classifier that receives an input of 20x20 pixel region of gray scale image and generates an output region ranging from 1 to -1, signifying the presence or absence of a iris, respectively. This classification must have some invariance to position, rotation, and scale. To detect irises anywhere in the input, the classifier is applied at every location in the image. To detect irises larger than the window size, the input image is repeatedly reduced in size. The classifier is applied at every pixel position in the image and scales the image down by a factor of 1.2 for each step as shown in Fig. 2. So, the classifier is invariant to translation and scaling. To have rotation invariant, the neural network is trained for images rotated from $0°$ to $355°$ by a step of $5°$. In order to train neural networks used in this stage, a large number of iris and non-iris images are needed. A sample of non-iris images, which are collected from the world wide web, is shown in Fig. 3. So, conventional neural nets are not capable of realizing such a searching problem. As a result of this, MNNs are used for detecting the presence or absence of human irises for a given image. Images (iris and non-iris) in the database are divided into three groups which result in three neural networks. More divisions can occur without any restrictions in case of adding more samples to the database. Each group consists of 400 patterns (200 for iris and 200 for non-iris). Each group is used to train one neural network. Each network consists of hidden layer containing 13 neurons, and an output layer which contains only one neuron. Here, two models of MNNs are used. The first is the ensemble majority voting which gives a result of 80% detection rate. The other is the average voting which gives a better result of 84% detection rate.

2.2 Enhancement of Detection Performance

To enhance the detection decision, the detection results of neighboring windows to confirm the decision at a given location can be used. This will reduce false detection as neighboring windows may reveal the non-iris characteristics of the data. For each location, the number of detections within a specified neighborhood of that location can be counted. If the number is above a threshold, then that location is classified as an iris. Among a number of windows, the location with the higher number of detections in range of one pixel is preserved, and locations with fewer detections are eliminated. In our case, a threshold of 4 is chosen. Such strategy increases the detection rate to 96% (average voting), as a result of reducing the false detections. It is clear that, the use of MNNs with this enhancement improves the results over those obtained in the previous section. As shown in Fig. 4, when the light beam is focused more to inside the pupil, the pupil itself is compressed. The iris contained in the eye of Fig. 4(a) is fed to the neural net during the learning process while the other (which found in the eye of Fig.4 (b)) could be detected correctly during the test phase.

3 Fast Neural Nets for Iris Detection

In subsection 2.1, modular neural network for object detection using a sliding window is presented to parse a given input image. In this section, a fast algorithm for object detection (used with each of the neural nets presented in section 2.1) based on two dimensional cross correlations that take place between the tested image and the sliding window. Such window is represented by the neural net weights situated between the input unit and the hidden layer. The convolution theorem in mathematical analysis says that a convolution of f with h is identical to the result of the following steps: let **F** and **H** be the results of the Fourier transformation of f and h in the frequency domain. Multiply **F** and **H** in the frequency domain point by point and then transform this product into spatial domain via the inverse Fourier transform [4]. As a result of this, these cross correlations can be represented by a product in frequency domain. By using the Fast Fourier Transform speed up in an order of magnitude can be gained.

During the detection step, a sub image **I** of size mxn (sliding window) is extracted from the test image of size **PxT** and fed to the neural network. Let w_i be the vector of weights needed to compute the activity of hidden neuron. This vector of size mn can be represented as mxn matrix $\mathbf{X}_i$. The output of hidden neurons **h(i)** can be given by:

$$h_i = g\left(\sum_{j=1}^{m}\sum_{k=1}^{n} X_i(j,k)I(j,k) + b_i\right) \tag{1}$$

where g is the activation function and **b(i)** is the bias of each hidden neuron **(i)**. The expression is obtained for a particular sub-image **I**. Equ. (1) can be extended to the global image **Z** as follows:

$$h_i(u,v) = g\left(\sum_{j=-m/2}^{m/2}\sum_{k=-n/2}^{n/2} X_i(j,k)\, Z(u+j, v+k) + b_i\right) \tag{2}$$

This operation denotes a cross correlation operation. For any two functions **f** and **d**, their cross correlation is given by:

$$f(x,y)\otimes d(x,y) = \left(\sum_{m=-\infty}^{\infty}\sum_{n=-\infty}^{\infty} f(m,n)d(x+m, y+n)\right) \tag{3}$$

We may reformulate equ. (2) as:

$$h_i = g\left(X_i \otimes Z + b_i\right) \tag{4}$$

where h_i is the activity of the hidden unit **(i)** and $\mathbf{h_i(u,v)}$ is the activity of the hidden unit **(i)** when the observation window (sliding window) is located at position **(u,v)** and $\mathbf{(u,v) \in [P-m+1, T-n+1]}$.

Now, the above given cross correlation can be given in terms of Fourier Transform:

$$Z \otimes X_i = F^{-1}\left(F(Z) \bullet F^*\left(X_i\right)\right) \tag{5}$$

Evaluating this cross correlation using FFT, an important speed up can be gained in comparison to a classic neural network based on sliding windows. Similarly the final output of the neural network can be expressed by linear combination of the hidden unit activity:

$$O(u,v) = g\left(\sum_{i=1}^{q} w_o(i)\, h_i(u,v) + b_o\right) \tag{6}$$

O(u,v) is the output of the observed window located at the position **(u,v)** in the input image **Z**.

For a tested image of NxN pixels, the 2D FFT requires $O(N^2\log_2 N^2)$ computation steps. For the weight matrix $\mathbf{X_i}$, the 2D FFT can be computed off line since these are constant parameters of the network independent of the tested image. A 2D FFT of the tested image has to be computed, therefore the total number of FFT to compute is q+1 (one forward and q backward transforms). In addition, we have to multiply the transforms of the weights and the input image in the frequency domain adding a further (qN^2) computation steps. This yields a total of $O((q+1)N^2\log_2 N^2 + qN^2)$ computation steps for the fast neural networks.

For a classic neural network, $O((N\text{-}n+1)^2n^2q)$ computation steps are required, when one considers the activity of q neurons in the hidden layer and a square tested image of size NxN and a sliding window of size nxn. Therefore, the theoretical speed up factor K is:

$$K = \frac{q(N - n + 1)^2 n^2}{(q + 1)N^2 \log_2 N^2 + qN^2} \tag{7}$$

For example, if N=200, n=20, q=13, we may achieve a speed up factor of 18.7595. The speed up factor introduced in [11] for object detection which is given by:

$$K = \frac{qn^2}{(q + 1)\log^2 N} \tag{8}$$

is not correct since the number of computation steps required for the 2D FFT is $O(N^2\log_2 N^2)$ and not $O(N^2\log^2 N)$. Also, this is not a typing error as the curve in Fig. 2 in [11] realizes equ.8. Moreover, the speed up ratio presented in [11] not only contains an error but is also not precise. This is because for fast neural nets, the term (qN^2) must be added. Such term has a great effect on the speed up ratio. Furthermore, for conventional neural nets, the number of operations is $(q(N\text{-}n+1)^2n^2)$ and not (qN^2n^2).

3.1 Sub-Image Normalization in Frequency Domain

In [12], the authors stated that image normalization to avoid week or strong illumination can not be done in frequency space. This is because the image normalization is local and not easily computed in the Fourier space of the whole image. Here, a simple method for normalization of sub-images in frequency domain is presented. Local image normalization can be obtained by normalizing the weights as follows:

Let $\overline{X}_{rc}$ be the zero-mean centered sub-image located at (r,c) in the input image ψ:

$$\overline{X}_{rc} = X_{rc} - \bar{x}_{rc} \tag{9}$$

where $\bar{x}_{rc}$ is the mean value of the sub image located at (r,c). We are interested in computing the cross correlation between the sub-image $\overline{X}_{rc}$ and the weights W_i that is:

$$\overline{X}_{rc} \otimes W_i = X_{rc} \otimes W_i - \bar{x}_{rc} \otimes W_i \tag{10}$$

where:

$$\bar{x}_{rc} = \frac{X_{rc}}{n^2} \tag{11}$$

Combining 10 and 11, we get the following expression:

$$\overline{X}_{rc} \otimes W_i = X_{rc} \otimes W_i - \frac{X_{rc}}{n^2} \otimes W_i \tag{12}$$

which is the same as:

$$\overline{X}_{rc} \otimes W_i = X_{rc} \otimes W_i - X_{rc} \otimes \frac{W_i}{n^2} \tag{13}$$

The centered zero mean weights are given by:

$$\overline{W}_i = W_i - \frac{W_i}{n^2} \tag{14}$$

Also, equ. 13 can be written as:

$$\overline{X}_{rc} \otimes W_i = X_{rc} \otimes \left(W_i - \frac{W_i}{n^2} \right) \tag{15}$$

So, we may conclude that:

$$\overline{X}_{rc} \otimes W_i = X_{rc} \otimes \overline{W}_i \tag{16}$$

which means that cross-correlating a normalized image with the weight matrix is equal to the cross-correlation of the n
weight matrix.

Normalization of sub-images in the spatial domain (in case of using traditional neural nets) requires $(N\text{-}n\text{+}1)^2n^2$ computation steps. On the other hand, normalization of sub-images in frequency domain through normalizing the weights of the neural nets requires $2n^2$ operations. So, this proves also that, local image normalization in frequency domain is faster than that in the spatial one.

3. 2 Image Testing at Multi Resolutions

Images are tested for the presence of a iris (object) at different scales by building a pyramid of the input image which generates a set of images at different resolutions. The iris detector is then applied at each resolution and this process takes much more time as the number of processing steps will be increased. In [7], the author stated that the Fourier transforms of the new scales do not need to be computed. This is due to a property of the Fourier transform. If f(x,y) is the original and g(x,y) is the sub-sampled by a factor of 2 in each direction image then:

$$g(x,y) = FT(2x,2y) \tag{17}$$

$$F(u,v) = FT(f(x,y)) \tag{18}$$

$$FT(g(x,y) = G(u,v) = \frac{1}{4}F(\frac{u}{2},\frac{v}{2}) \tag{19}$$

This implies that we do not need to recompute the Fourier transform of images, as it can be directly obtained from the original Fourier transform. Then, the author claimed that the processing needs $O((q+2)N^2\log^2N)$ additional number of computation steps. Thus the speed up ratio will be:

$$\eta = \frac{qn^2}{(q+2)\log^2 N} \tag{20}$$

Of course this is not correct, because the inverse of the Fourier transform is required to be computed at each neuron in the hidden layer (for the result of dot product between the Fourier transform of input image and the weights of the neuron, the inverse of the Fourier transform must be computed). So, the total number of computation steps needed for fast neural networks is $O((2q+1)N^2\log N^2+2qN^2)$ and not $O((q+2)N^2\log^2N)$ as stated in [7]. Thus, the speed up ratio becomes:

$$\eta = \frac{q(N-n+1)^2n^2}{(q+1)N^2\log_2 N^2 + qN^2} \tag{21}$$

4 A New Faster Algorithm for Human Iris Detection Based on Image Decomposition

In this section, a new faster algorithm for iris detection is presented. The number of computation steps required for fast neural nets with different image sizes is listed in table 1. From this table, we may notice that as the image size is increased, the number of computation steps required by fast neural networks is much increased. For example, the number of computation steps required for an image of size (50x50 pixels) is much less than that needed for an image of size (100x100 pixels). Also, the number of computation steps required for an image of size (500x500 pixels) is much less than that needed for an image of size (1000x1000 pixels). As a result of this, for example, If an image of size (100x100 pixels) is decomposed into 4 sub-images of size (50x50 pixels) and each sub-image is tested separately, then a speed up factor for iris detection can be achieved. The number of computation steps required by fast neural networks to test an image after decomposition may be calculated as follows:

1- Assume that the size of the image under test is (NxN pixels).

2- Such image is decomposed into α (LxL pixels) sub-images. So, α can be computed as:

$$\alpha=(N/L)^2 \quad (22)$$

3- Assume that, number of computation steps required for testing one (LxL pixels) sub-image is β. So, the total number of computation steps (T) required for testing these sub-images resulted after the decomposition process is:

$$T=\alpha\,\beta \quad (23)$$

To detect a iris of size 20x20 pixels in an image of any size by using fast neural networks after image decomposition into sub-images, the optimal size of these sub-images must be computed. From table 1, we may conclude that, the most suitable size for the sub-image which requires the smallest number of computation steps is 25x25 pixels. A comparison between the speed up ratio for fast neural networks and fast neural networks after image decomposition with different sizes of the tested images is listed in table 2 (q=30, n=20). A comparison between the number of computation steps required by conventional, fast neural networks and fast neural networks after image decomposition with different sizes of the tested images is listed in table 3 (q=30, n=20).

Table 1. The number of computation steps required for by fast neural networks (FNN) for images of sizes (25x25 - 1050x1050 pixels).

Image size	No. of computation steps in case of using FNN	Image size	No. of computation steps in case of using FNN
25x25	198699	550x550	179807163
50x50	949798	600x600	216787233
100x100	4419191	650x650	257448831
150x150	10759202	700x700	301827794
200x200	20156763	750x750	349957294
250x250	32742414	800x800	401868214
300x300	48616808	850x850	457589436
350x350	67861948	900x900	517148091
400x400	90547053	950x950	580569757
450x450	116732022	1000x1000	647878626
500x500	146469656	1050x1050	689330147

Table 2. The speed up ratio in case of using FNN and FNN after image decomposition into sub-images (25x25 pixels) for images of different sizes.

Image size	Speed up ratio in case of using (FNN)	Speed up ratio in case of using FNN after image decomposition
50x50	12.1415	14.5094
100x100	17.8159	24.7640
150x150	19.1400	28.7888
200x200	19.5037	30.9145
250x250	19.5566	32.2261
300x300	19.4898	33.1158
350x350	19.3736	33.7586
400x400	19.2379	34.2448
450x450	19.0961	34.6253
500x500	18.9550	34.9313
550x550	18.8176	35.1826
600x600	18.6853	35.3928
650x650	18.5588	35.5711
700x700	18.4381	35.7242
750x750	18.3232	35.8572
800x800	18.2138	35.9738
850x850	18.1095	36.0769
900x900	18.0102	36.1686
950x950	17.9154	36.2507
1000x1000	17.8248	36.3248

Table 3. The number of computation steps required for conventional, FNN and FNN after image decomposition into sub-images (25x25 pixels) for images of different sizes.

Image size	No. of computation steps in case of using conventional neural nets	No. of computation steps in case of using FNN	No. of computation steps in case of using FNN after image decomposition
50x50	11532000	949798	794798
100x100	78732000	4419191	3179191
150x150	205932000	10759202	7153179
200x200	393132000	20156763	12716763
250x250	640332000	32742414	19869942
300x300	947532000	48616808	28612717
350x350	1314732000	67861948	38945087
400x400	1741932000	90547053	50867053
450x450	2229132000	116732022	64378614
500x500	2776332000	146469656	79479770
550x550	3383532000	179807163	96170522
600x600	4050732000	216787233	114450000
650x650	4777932000	257448831	134320000
700x700	5565132000	301827794	155780000
750x750	6412332000	349957294	178829000
800x800	7319532000	401868214	203468000
850x850	8286732000	457589436	229696000
900x900	9313932000	517148091	257514000
950x950	10401132000	580569757	286921000
1000x1000	11548332000	647878626	317919000

4.1 Simulation of the Fast Iris Detection Process (after Image Decomposition) using Parallel Processing Techniques

In the previous section, a new algorithm for iris detection based on decomposing the image under test to many sub-images has been presented. Then, for each sub-image, a fast neural network has been used to detect the presence/absence of human irises.

Here, to further reduce the running time as well as increase the speed up ratio of the detection process, a parallel processing technique is used. Each sub-image is tested using a fast neural network simulated on a single processor or a separated node in a clustered system.

The number of operations (ω) performed by each processor / node (sub-images tested by one processor/node) =

$$\omega = \frac{\text{The total number of sub} - \text{images}}{\text{Number of Processors / nodes}}$$

$$\omega = \frac{\alpha}{\text{Pr}} \tag{19}$$

where Pr is the Number of Processors or nodes.

The total number of computation steps (γ) required to test an image by using this approach can be calculated as:

$$\gamma=\omega\beta \tag{20}$$

As shown in table 4, using a symmetric multiprocessing system with 16 parallel processors or 16 nodes in either a massively parallel processing system or a clustered system, the number of computation steps is reduced as well as the speed up ratio (with respect to conventional neural networks) for human iris detection is increased.

5 Conclusion

A fast modular neural network approach has been introduced to identify human irises. The technical problem associated with large database (iris/non-iris) required for training neural networks has been solved using MNNs. A simple algorithm for fast iris detection based on cross correlation in frequency domain has been presented in order to speed up the execution time. Moreover, by decomposing the input image into many sub-images and test each one by using the fast neural network, the speed up ratio has been increased. Simulation results have shown that the new algorithm is an efficient method for finding locations of irises when the size of the iris is unknown as well as rotated, occluded, noised, and mirrored irises are detected correctly at different illumination levels. The proposed approach may be applied to detect the presence / absence of any other object in an image.

Table 4 The number of computation steps and the speed up ratio in case of using FNN after image decomposition into sub-images (25x25 pixels) for images of different sizes using 16 parallel processors or 16 nodes.

Image size	No. of computation steps	Speed up ratio
50x50	198699.42	58.0374
100x100	198699.42	396.2240
150x150	596098.26	345.4656
200x200	794797.68	494.6312
250x250	1390895.94	358.0678
300x300	1788294.78	529.8520
350x350	2583092.46	508.9758
400x400	3179190.72	547.9168
450x450	4172687.82	534.2197
500x500	9934971.00	558.9008
550x550	6159682.02	549.3030
600x600	7153179.12	566.2841
650x650	8544075.06	559.2099
700x700	9736271.58	571.5875
750x750	11325866.94	566.1669
800x800	12716762.88	575.5814
850x850	14505057.66	571.2995
900x900	16094653.02	578.6973
950x950	18081647.22	575.2297
1000x1000	19869942.00	581.1960

References

1. R. P. Wildes, " Iris Recognition: An Emerging Biometric Technology; " Proceedings of IEEE, September, Vol. 85, No. .9, (1997) 1347-1363.
2. R. Jacobs, M. Jordan, and A. Barto, " Task Decomposition Through Competition in a Modular Connectionist Architecture: The what and where vision tasks, " Neural Computation 3, (1991) 79-87.
3. G. Auda, M. Kamel, H. Raafat, "Voting Schemes for cooperative neural network classifiers, " IEEE International Conference on Neural Networks ICNN95, Perth, Australia, November, Vol. 3, (1995) 1240-1243.
4. R. Klette, and Zamperon, *Handbook of image processing operators*, John Wiley & Sonsltd, 1996.
5. G. Auda, and M. Kamel "CMNN: Cooperative Modular Neural Networks for Pattern Recognition, " Pattern Recognition Letters, Vol. 18, (1997) 1391-1398.
6. E. Alpaydin, " Multiple Networks for Function Learning, " Int. Conf. on Neural Networks, Vol.1 CA, USA, (1993) 9-14.
7. A. Waibel, "Modular Construction of Time Delay Neural Networks for Speech Recognition, " Neural Computing 1, (1989) 39-46.

8. K. Joe, Y. Mori, and S. Miyake, "Construction of a large scale neural network: Simulation of handwritten Japanese Character Recognition, " on NCUBE Concurrency 2 (2), (1990) 79-107.
9. A. Jain, R. Bolle and S. Pankanti, *BIOMETRICS: Personal Identification in Networked Society*, Kluwer Academic Publishers, Chap5, (1998) 103-122.
10. H. M. El-Bakry, M. A. Abo-Elsoud, M. S. Kamel, "Fast Modular Neural Networks for Human Face Detection, " Proc. of IJCNN International Joint Conference on Neural Networks 2000, Como, Italy, 24-27 July, Vol. III, (2000) 320-324.
11. S. Ben-Yacoub, "Fast Object Detection using MLP and FFT, " IDIAP-RR 11, IDIAP, (1997).
12. R. Feraud, O. Bernier, J. E. Viallet, and M. Collobert, " A Fast and Accurate Face Detector for Indexation of Face Images," Fourth IEEE International Conference on Automatic Face and Gesture Recognition, Grenoble, France, 28-30 March, (2000).

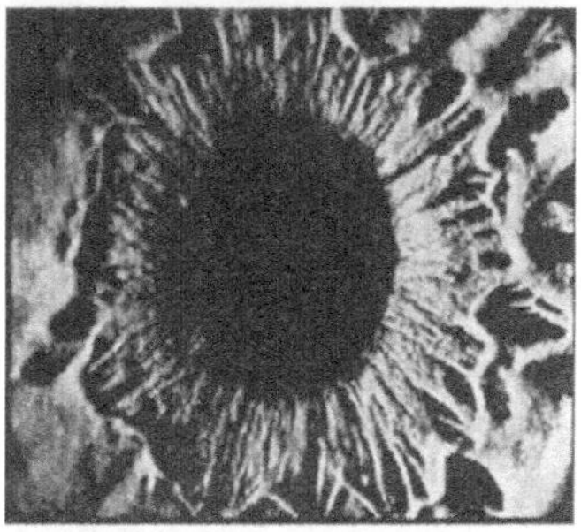

Fig. 1. The distinctiveness of the human iris.

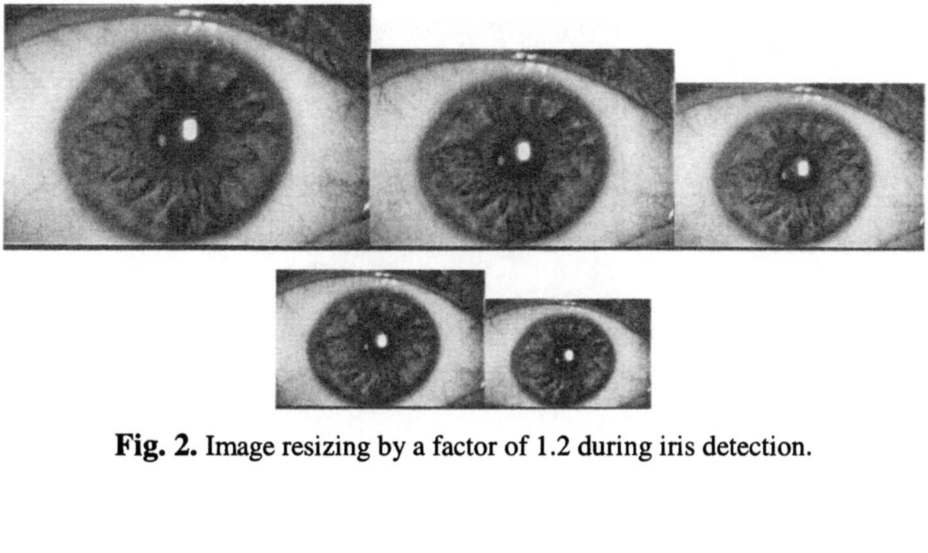

Fig. 2. Image resizing by a factor of 1.2 during iris detection.

Fig. 3. Examples of non-iris images.

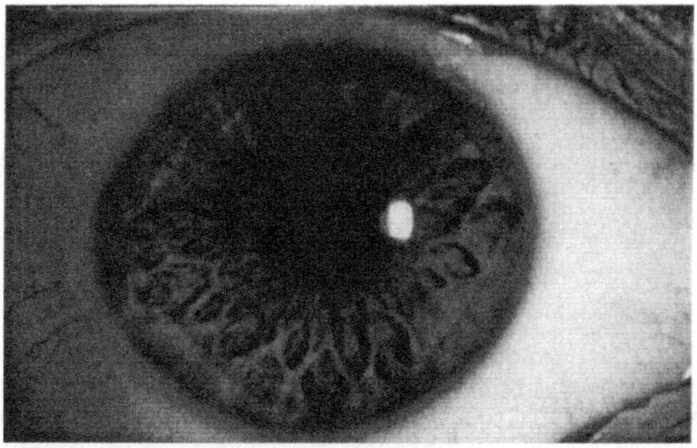

(a)

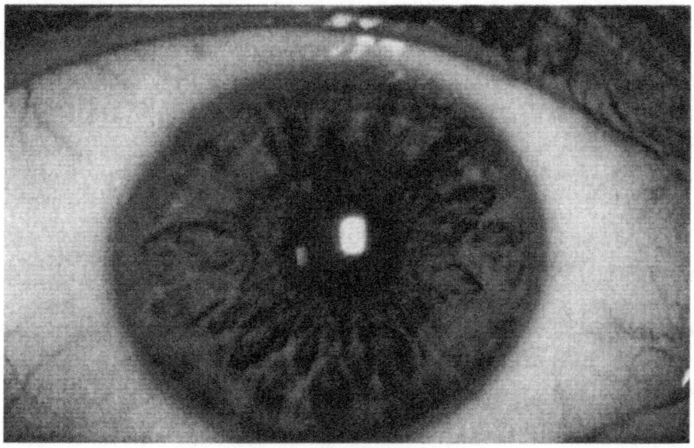

(b)

Fig. 4. Variation in pupil size when the light beam is more focused into the eye.

A Rotation Invariant Algorithm for Recognition

Hazem El-Bakry

Faculty of Computer Science &Information System
Mansoura University - Egypt
hazemelbakry@hotmail.com

Abstract. In this paper, a simple technique for human face classification using two transforms and neural nets is introduced. A new concept for rotation invariant based on Fourier descriptors and neural networks is presented. Although, the magnitude of the Fourier descriptors is translation invariant, there is no need for scaling or translation invariance. This is because the face sub-image here (20x20 pixels) is segmented from the whole image during the detection process as in my previous work [12]. The feature extraction algorithm based on Fourier descriptors is modified to reduce the number of neurons in the hidden layer. The second stage extracts wavelet coefficients of the resulted Fourier descriptors before application to neural network. Wavelet transforms have been shown to provide advantages in terms of better representation for a given data to be compressed. The final vector is fed to a neural net for face classification. Simulation results for the proposed algorithm show a good performance compared with previous results.

1. Introduction

Face recognition refers to the automatic identification of an individual, represented in a database, based on the information contained in a digital gray scale image. Automatic identification of the human faces has been an area of active research for the past 20 years. In addition to the importance of the face recognition task from a research point of view, it has a number of commercial and law enforcement applications. Personal identification from faces is performed by means of non contact measurement, which has advantages over contact techniques such as fingerprint, or close contact such as iris texture, or retina recognition.

Face recognition applications that are being implemented in practice include static matching of controlled photographs such as credit cards, photo Identification cards, driver licenses, and access control. A sharp increase in the number of services available over the internet has raised the problem of secure money transactions. At present, insurance for transactions costs 1.5 cents (1 pence) per transaction. In order to perform card holder identification in shops, a valid credit card number must be accompanied by the signature of the holder. Identification of the correct personality over the Internet cannot be done by conventional means, and visual face recognition using a camera provides, perhaps at the best, identification cue.

Traditional fixed template pattern matching techniques and geometrical model-based object recognition approaches tend to perform inadequately for recognizing faces [2,8]. Other approaches, such as eigenfaces do not provide invariance over changes in

B. Reusch (Ed.): Fuzzy Days 2001, LNCS 2206, pp. 284–290, 2001.

scale, head orientation, and lighting conditions. In another study, Blackwell et. al. [11] claimed that whole image preprocessing, such as PCA, can not solve the problems associated with learning large complicated data sets. These limitations arise due to processing of whole faces rather than their constituent parts [10].

My approach to the face recognition problem is to combine both Fourier and wavelet transforms with neural nets in order to enhance the recognition performance and reduce the response time during the test phase. In section 2, invariant Fourier descriptors are discussed and a modified algorithm is presented. Such modification allows to reduce the total number of neurons needed to classify the feature vectors and the time required during the learning / testing phase. In order to enhance the recognition performance as well as reduce the computation time, a combination between the FFT and DWT features is introduced in section 3.

2. Face Recognition Using Invariant Features

common and possibly unique to all face patterns, even under different imaging conditions [1]. Fourier descriptors are well suited for recognizing objects using the characteristics of the global shape. Although, the magnitude of the Fourier descriptors is translation invariant, there is no need for scaling or translation invariance. This is because, first the face image is located in the input image, and then segmented. During the detection phase, the image is resized by a step of 1.2, so the detection process is scaling invariant. Also, the detection is applied at every pixel position, and as a result the detection is translation invariant. The detection process is done by using fast modular neural networks [12]. Here, a new algorithm for rotation invariance based on Fourier descriptors and neural networks is presented. By computing the Fast Fourier Transform of the data, x[m], $1 \leq m \leq L$, we obtain the Fourier coefficients for $0 \leq m \leq L\text{-}1$ as follows:

$$a[k] = \sum_{m=1}^{L} x[m] e^{-jk(2\Pi/L)\,m} \tag{1}$$

For face recognition, we need a certain value, namely, a feature vector which represents the face. How to make a feature vector from a given graphic image data is an important issue in pattern recognition. As a feature, the Fourier descriptor are utilized because we can compress the data size significantly by eliminating one half of the Fourier components. The Fourier descriptors of the input face (20x20 pixels) are used as follows:

1-Discard one half of the data and the dc component a[0] since it depends only on the position of the image center.
2-Obtain the absolute value of the complex number a[k] which will make it invariant under rotation or translation as follows:

$$r[k] = \sqrt{Re(a[k])^2 + Im(a[k])^2} \tag{2}$$

3-Obtain the maximum component of the absolute values.
4-Dividing r[k] by the maximum absolute value of the other components.

Here, a backpropagation neural network is chosen for classification of feature vectors since it is known to have inherent robustness in classification and flexibility in its application. We construct a neural net for solving this classification problem. A network of three layers is trained to classify 120 patterns corresponding to ten human faces. A total number of 12 patterns are taken for each face through rotations from 0° to 165° by a step of 15°. This procedure not only achieves rotation invariance but also reduces the effect of background (for the same face) which may be changed from an image to another. The feature vector consists of 200 elements. Hence, it is common to select the input layer to have 200 nodes. The input nodes do not have any processing elements (neurons). The second layer containing 12 neurons while the output layer has 10 neurons to distinguish between 10 human faces.

In order to reduce the number of neurons in the hidden layer, we make a modification during the training process. Instead of training the neural net from 0° to 165°, we train on images rotated from 0° to 75°. This is done as follows:

1. First, the input (face) image is resized to 20x20.
2. Second, rotate the input (face) image by an angle of 90°.
3. Obtain FFT for the input (face) image and its rotated version.
4. Calculate the difference between the two vectors computed in the previous step.
5. Repeat the previous steps for the input image but with angles up to 75° by a step of 15°.
6. Do the above steps for all the faces to be stored in the database.
7. Train the neural networks with all of these patterns.

Using this procedure the neural network could be learned using only 8 neurons in the hidden layer instead of 12. For some examples of mirrored, noised, and occluded faces, experimental results have shown that the modified algorithm could recognize them correctly. Also, for the same face, this approach reduces the effect of background variation as the neural network is trained for rotated samples of each face.

3. Combining Fourier and Wavelet Transforms with Neural Nets for Face Recognition

Recently, wavelet transforms have been shown to provide certain advantages in terms of better data compaction for a given signal. But they are not invariant to rotation, translation, or scaling [7]. Using wavelet transform, we can analyze our signal in time for its frequency content. Unlike Fourier analysis, in which we analyze signals using sins and cosines, now we use wavelet function. Dilations and translation of the "mother function" or analyzing wavelet $\phi(x)$, defines an orthogonal basis, our wavelet basis:

$$\varphi(x) = 2^{\frac{s}{2}} \varphi(2^{-s} x - k) \tag{3}$$

The variables s and k are integers that scale and dilate the mother function ϕ to generate wavelets, such as a Daubechies wavelet family. The scale index s indicates the width of the wavelet, and the location k gives its position. The mother functions are rescaled, or dilated by powers of two, and translated by integers. What makes wavelet bases especially interesting is the self-similarity caused by the scales and dilations. Once we know about the mother functions, we know everything about the basis.

Compared with Fourier descriptor that uses global sinusoids as the basis functions, the wavelet descriptors is more efficient in representing and detecting local features due to the spatial and frequency localization property of wavelet bases [9]. The output vector in last section is then compressed using daubechies (s=3) wavelet. As a result of this, the new vector has a length of only 100 points. Thus, the number of neurons in the hidden layer is reduced to 10 neurons. So, over a sample of 30 images, the response time is reduced compared to using only Fourier coefficients as shown in Fig. 1. The complete learning algorithm is shown in Fig.2.

However, the input pattern is compressed from 200 to 100 elements when using (FFT+DWT), the recognition rate is increased as we feedback the negative examples (examples which could not be recognized correctly) to the trainer again but only in the case of using (FFT+DWT). When we add the negative examples to the trainer in case of using only FFT, the number of neurons in the hidden layer is increased above 8 neurons, so we fixed the training examples when using FFT and add the negative examples when using (FFT+DWT).

At the final test, as shown in table (1), the number of neurons in the hidden layer =5 for (FFT+DWT) and 8 neurons for FFT. So the recognition rate in case of (FFT+DWT) is better than using only FFT as the number of training examples is larger. But, of course the number of iterations is increased. However, during the test phase, it is cleared that using FFT+DWT is faster than using only FFT.

Table 1. A comparison between FFT only and FFT+DWT for Recognition for a sample of 100 faces for 10 Individuals.

Method	No. of neurons in hidden layer	No. of iterations	Recognition Rate
FFT	8	565	88%
FFT+DWT	5	2104	97%

As shown in Fig. 3, for some examples of Mirrored, noised, occluded, and horizontally/vertically skewed faces, the proposed algorithm has recognized them correctly with high recognition rate. Also, for a threshold value of 0.65 this combination is efficient in rejecting unknown faces.

4. Conclusion

A new rotation invariant approach has been introduced to identify frontal views of still human face images. Such approach has used a modified algorithm for obtaining Fourier descriptors and neural networks. In order to reduce the elapsed time, wavelet coefficients have been used to compress the Fourier feature vector. Simulation results have shown that the proposed algorithm is an efficient method for recognizing human faces even with occlusion, noise, or deformation in shape as well as rejecting unknown faces.

References

1. R. Chellappa, C. L. Wilson, and S. Sirohey, " Human and Machine Recognition of Faces: A Survey, " IEEE Proceedings., Vol. 83, No. 5, May, (1995) 707-740.
2. R. Brunelli, and T. Poggio, " Face Recognition: Features versus Templates, " IEEE Trans. on Pattern Analysis and Machine Intelligence, Vol. 15, No. 10, October, (1993) 1042- 1052.
3. H. Kim, and K. Nam, " Object recognition of one - DOF tools by a Back-Propagation Neural Net, " IEEE Trans. on Neural Networks, Vol. 6, No.2, March, (1995) 484-487.
4. M. Shridhar, and A. Badreldin, " High accuracy character recognition algorithm using Fourier and topological descriptors," Pattern Recognition, Vol. 17, No. 5, (1984) 515-524.
5. E. Persoon and K. S. Fu, " Shape discrimination using Fourier descriptors," IEEE Trans. Sys. Man, Cybern., Voil. SMC-7, No. 3, March (1977).
6. C. T. Zahn and R. Z. Roskies, " Fourier descriptors for plane closed curves, " IEEE Trans. Comput., Vol. C-21, No. 3, March (1972).
7. M. Y. Jaisimha, " Wavelet Features for similarity Based Retrieval of Logo Images, " SPIE Vol. 2660, (1989) 89-100.
8. N. Duta, and A. Jain, " Learning the human face concept from black and white pictures, " Proc. 14th Int'l. Conf. Pattern Recognition, Brisbane, August, (1998). 1365 1367.
9. G. C. H. Chuang and C.-C. Jay Kuo, "Wavelet Descriptor of Planar Curves: Theory and Applications," IEEE Trans. on Image Processing , Vol. 5, No.1, Jan., (1996) 56-70.
10. A. J. O Toole, H.ABDI, K. A. Deffenbacher, and D. Valentin, " A Perceptual Learning Theory of the Information in Faces", Cognitive and Computational Aspects of Face Recognition. Routledge: London, (1995) 159-182.
11. K. T. Blackwell, T. P. Vogl, S. D. Hyman, G. S. Barbour, and G. S. Alkon, "A New Approach to Hand Written Character Recognition, " IEEE Trans on Pattern Recognition Vol.25, 1992) 655-666.
12. H. M. El-Bakry, M. A. Abo-Elsoud, M. S. Kamel , "Fast Modular Neural Networks for Human Face Detection, " Proc. of IJCNN International Joint Conference on Neural Networks 2000, Como, Italy, Vol. III, 24-27 July, (2000) 320-323.

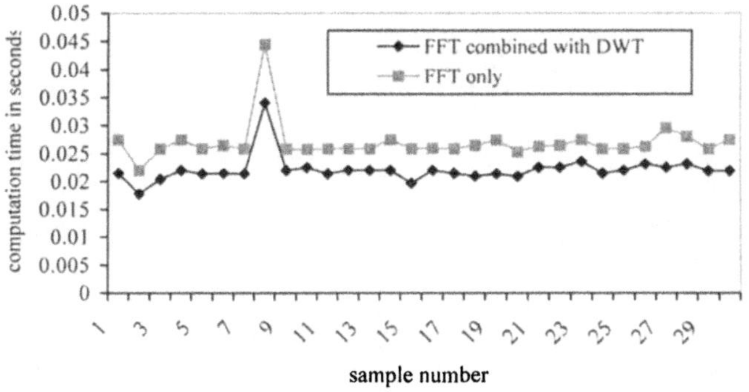

Fig. 1. A comparison between the elapsed time taken by FFT only and FFT combined with DWT.

11

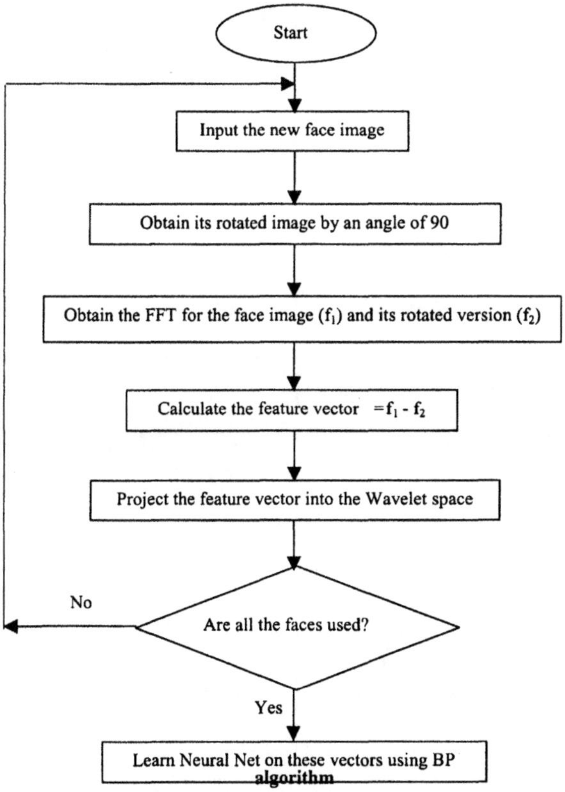

Fig.2. The learning algorithm for rotation invariant faces.

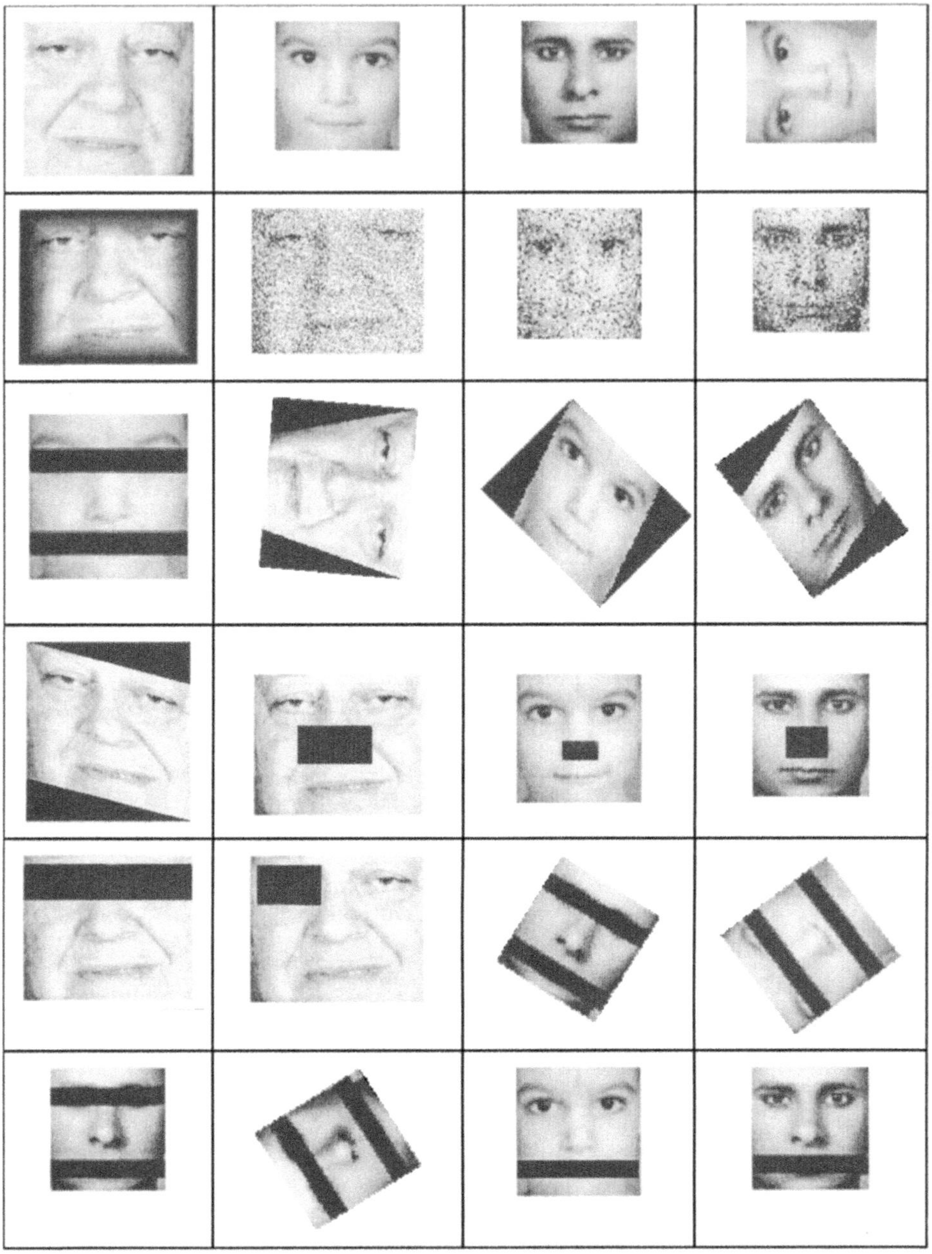

Fig.3. Some examples of mirrored, rotated, noised, deformed, occluded faces that could be recognized correctly using a combination of the modified FFT and DWT.

Application of the Neural Networks Based on Multi-valued Neurons to Classification of the Images of Gene Expression Patterns

Igor Aizenberg[1], Ekaterina Myasnikova[2], Maria Samsonova[2], and John Reinitz[3]

[1] Neural Networks Technologies (NNT) Ltd., 155 Bialik st., Ramat-Gan, 52523 Israel, igora@netvision.net.il;
[2] Institute of High Performance Computing and Data Bases ,118 Fontanka emb., St.Petersburg, 198005 Russia, {myasnikova,samson}@fn.csa.ru (MS),
[3] The Department of Applied Mathematics and Statistics, The University at Stony Brook, Stony Brook, NY 11794-3600, USA, reinitz@ams.sunysb.edu

Abstract. Multi-valued neurons (MVN) are the neural processing elements with complex-valued weights and high functionality. It is possible to implement an arbitrary mapping described by partial-defined multiple-valued function on the single MVN. The MVN-based neural networks are applied to temporal classification of images of gene expression patterns, obtained by confocal scanning microscopy. The classification results confirmed the efficiency of this method for image recognition. It was shown that frequency domain of the representation of images is highly effective for their description.

1 Introduction

The ability of neural networks to accumulate knowledge about objects and processes using learning algorithms makes their application in pattern recognition very promising and attractive [1]. In particular, different kinds of neural networks are successfully used for solving the image recognition problem [2].

Neural networks based on multi-valued neurons have been introduced in [3] and further developed in [4-7]. A comprehensive observation of multi-valued neurons theory, their learning and applications are given in [8]. Multi-valued neural element (MVN) is based on the ideas of multiple-valued threshold logic [8]. Its main properties are ability to implement arbitrary mapping between inputs and output described by partially defined multiple-valued function, quickly converging learning algorithms based on simple linear learning rules and complex-valued internal arithmetic. Several kinds of MVN-based neural networks have been proposed for solving the image recognition problems. Different models of associative memory have been considered in [4, 5, 8, 9, 10]. An approach to image recognition, which will be used here, has been introduced in [6] and further developed in [7, 8, 11, 12]. This approach is based on the following consideration. Since it is always difficult to formalize the image description so that it can be adapted for learning, a good solution to this challenge is an objectification of image presentation. Transformation of the

B. Reusch (Ed.): Fuzzy Days 2001, LNCS 2206, pp. 291–304, 2001.

image presentation from spatial to the frequency domain is an appropriate way to realize this objectification. Of course, in itself this approach is not new. Frequency domain representation of data for pattern recognition was proposed and successfully used in 70^{ths}, e.g., in [13]. The nature of this data presentation is clear: since in frequency domain the signal energy is concentrated in a small number of the low frequency part spectral coefficients, it is possible to use exactly these coefficients as an objective description of the signal.

In our study MVN-based neural network is applied to classification of images of gene expression patterns, obtained by confocal scanning microscopy [14]. This is a new promising approach for acquisition of quantitative data on gene expression at the resolution of a single cell. Gene expression data are of crucial importance for elucidation of mechanisms of cell functioning, as well as for the early diagnosis of many diseases.

The data are described in detail in Section 2. In our preliminary study we applied the discriminant function analysis for classification of one dimensional data with a fixed set of features. The results are given in Section 3. MVN, their learning and MVN-based neural network for image recognition are considered in Section 4. The results of classification obtained using MVN-based neural network are given in Section 5.

2 Description of the Data

We perform temporal classification of images of expression patterns of genes controlling segmentation in the fruit fly *Drosophila*, which is a model organism for molecular biology studies. Like all other insects, the body of the *Drosophila* is made up of repeated units called segments. During the process of segment determination a fly embryo consists of a roughly prolate spheroid of about 5000 nuclei. Genes that act to determine segments are expressed in patterns that become more spatially refined over time. One can view each gene's expression pattern as a collection of "domains" (stripes), each of which is a region of expression containing one maximum (Fig 1, 2). In our experiments gene expression was recorded by confocal scanning of embryos stained with fluorescence tagged antibodies. The obtained images were subjected to image segmentation procedure to obtain the data in terms of nuclear location [15] and then rescaled to remove a nonspecific background signal. In the processed image the nuclei are presented by single pixels with a fluorescence intensity proportional to the average value of gene expression in the respective nucleus.

Human observers classify the developmental stage of an embryo by careful study of its pattern, since each stripe possesses its own features at any stage of an embryo development. In such a way 809 embryos were subdivided into 8 temporal classes [16] (their representatives are shown in Fig.1). Each embryo was allocated to one of the temporal classes on the basis of thorough and extensive visual inspection of the expression pattern of the *eve* gene, which is highly dynamic. We selected embryos for scanning without regard for age, so we expect our dataset to be uniformly distributed in time. The 8 classes were approximately equally populated.

The evolution of the *eve* expression patterns during the segment determination shows the following tendency (see Fig. 1). Time classes 1, 2, and 3 do not have seven well-defined stripes and the number and location of stripes changes rapidly. The remaining groups (classes 4 to 8) do have seven well-defined stripes. After all the stripes are clearly visible their intensities increase in the posterior portion of the embryo. By the end of the period, all stripes have reached maximum and equal intensity and maximum sharpness. Time classes 1 to 3 could be grouped according to the number of individual stripes, which have formed. The remaining groups (classes 4, 5, 6, 7 and 8) all have 7 well defined stripes and were classified by features of the overall pattern.

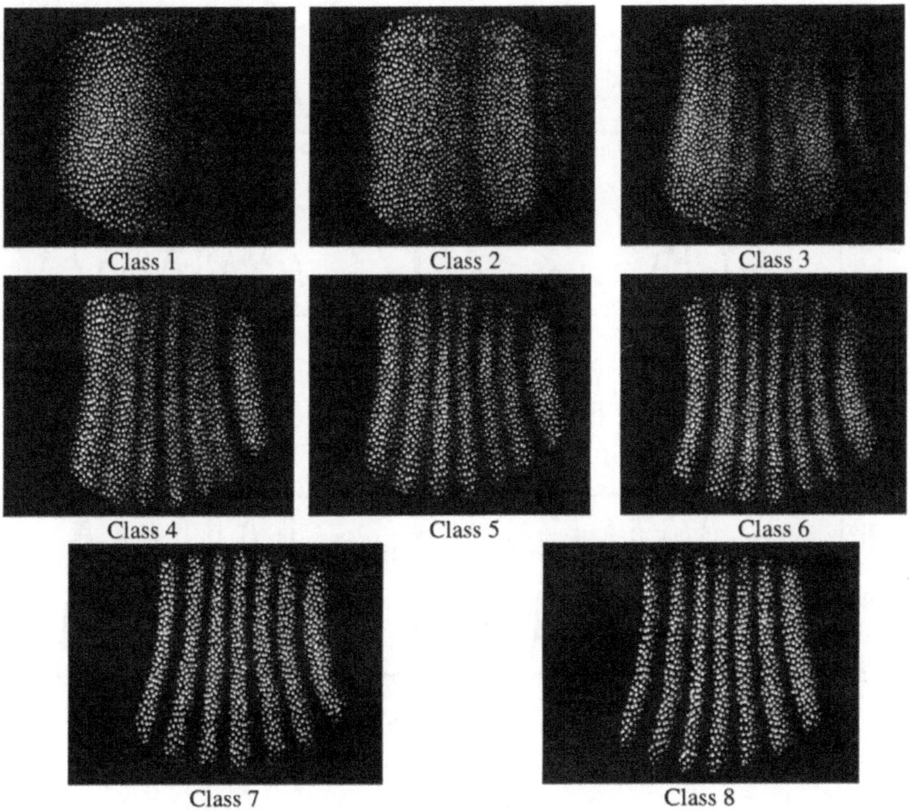

Fig. 1. Representative images of segmented and rescaled expression patterns belonging to the 8 temporal classes.

3 Preliminary Results: One Dimensional Classification by the Discriminant Function Analysis

The essential step for automated pattern classification in time is to characterize the patterns in terms of some observable parameters, and then use these parameters to read off the time. Expression of segmentation genes is largely a function of position along the anterior-posterior axis of the embryo body, and so can be well represented in one dimension. The time classification of expression patterns is performed for the data extracted from the central 10% of y values on the midline of an embryo in the x direction. The y values of these data are then ignored, and the patterns, demonstrating the variation of gene expression along the x axis, are presented as graphs. The stripes and interstripes then have the form of peaks and valleys respectively (see Fig.2) .

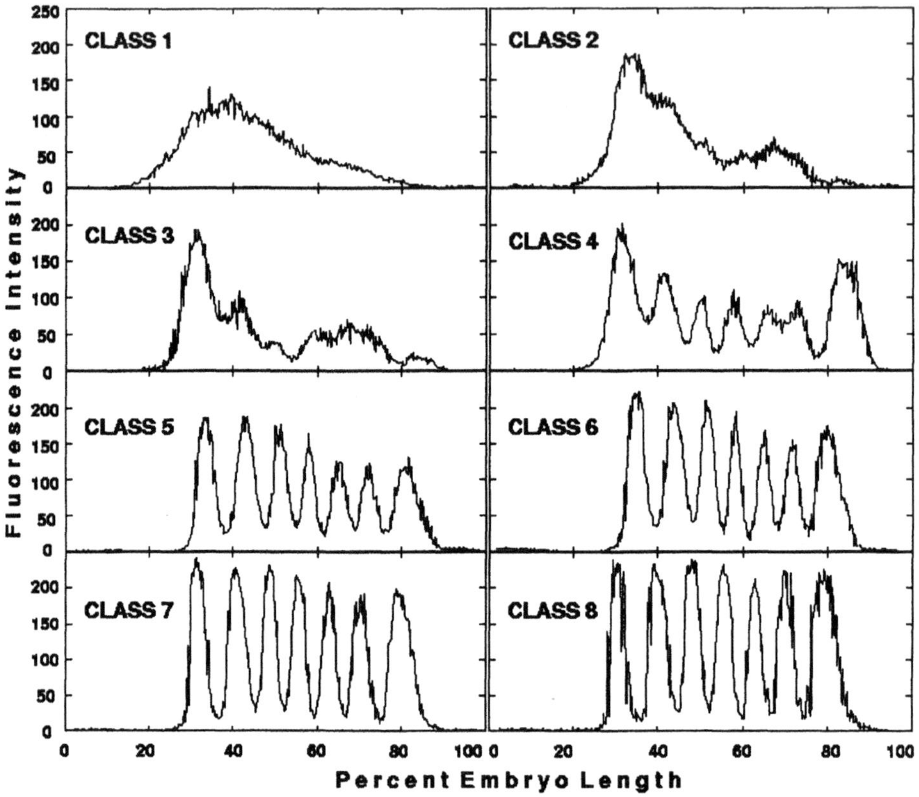

Fig. 2. Representative 1D expression patterns belonging to each of 8 temporal classes

As it is clear from the figure, it is natural to use extrema of expression domains as a set of variables for classification, but the noise on the peaks makes it rather difficult to find the extrema location directly from the raw data. There exist many methods based on different principles for extracting the features from an 1D signal of this type. We have employed quadratic spline approximation as well as the wavelet transform (with continuous basis) to extract extremal features of the pattern [16].

The classification of 1D images was performed by means of discriminant function analysis [17]. Discriminant analysis is mainly used to determine which variables discriminate between two or more naturally occurring groups. Another major purpose to which discriminant analysis is applied is the issue of predictive classification of objects. For each group in our sample, we can determine the location of the point that represents the means for all variables in the multivariate space defined by the variables in the model. These points are called group centroids. For each object we can then compute the Mahalanobis distances (of the respective object) from each of the group centroids. Again, we would classify the object as belonging to the group to which it is closest, that is, where the Mahalanobis distance is smallest.

Discriminant analysis will automatically compute the classification functions, which can be used to determine to which group each object most likely belongs. There are as many classification functions as there are groups. Each function allows us to compute classification scores for each object for each group, by applying the formula:

$$S_i = c_i + w_{i1}y_1 + w_{i2}y_2 + \cdots + w_{im}y_m.$$

In this formula, the subscript i denotes the respective group; the subscripts $1, 2, \cdots, m$ denote the m variables; c_i is a constant for the i th group, w_{ij} is the weight for the j th variable in the computation of the classification score for the i th group; y_j is the observed value of j th variable for the respective expression pattern. S_i is the resultant classification score. Now we classify the pattern as belonging to the group for which it has the highest classification score.

We estimate the discriminant functions that best discriminate between groups, based on some training data set, and then use them when predicting the other cases that were not used for the estimation of the discriminant function. As it has already been mentioned, the dataset is uniformly distributed in time and that means that assigning the borders between the classes is somewhat arbitrary. Therefore an embryo of the age intermediate between two adjacent classes may look quite atypical for the class to which it was assigned. To characterize a time class more definitely we use the set of most representative patterns, belonging to this class, as a training set for classification.

This method is appropriate for expression patterns at intermediate and late development stages when a full set of domains has formed. The recognition results are shown in Table 1. The class 4 images are best recognized (the attribution is 83.7%). Other embryo images are allocated to their classes with the accuracy of 49.4-67.5%. Thus the classification method is not very reliable (maximal attribution error is 50.6%). The reason for the high accuracy of recognition of embryos belonging to temporal class 4 may be the strong difference of the *eve* expression pattern at this age in comparison with other temporal classes.

To improve the quality of classification, we dropped the restriction to one dimensional data and used more powerful classifiers. Though expression of segmentation genes is largely an one dimensional function of position on the x axis, features of two dimensional images will supply more information to the classifier. Moreover, our choice of the characteristic features was based on a fixed set of features. We need to apply a more advanced classification method which makes an automated choice of the features responsible for the distribution of objects over the classes. In the next section we show that MVN-based neural networks achieve better temporal classification when applied to our whole dataset, including the very early embryos which were not considered in our previous classification.

Table 1. Recognition results for 1D data obtained by the discriminant analysis.

Class #	4	5	6	7	8
Number of Images (Training set)	33	36	30	30	36
Number of Images (Testing set)	43	77	86	62	40
Recognized	36 83.7%	38 49.4%	51 59.3%	38 61.3%	27 67.5%
Misclassified	7	39	35	24	13

4 Multi-valued Neuron and MVN-Based Neural Network for Image Recognition

4.1 Multi-valued Neuron

The concept of the multi-valued neuron (MVN) was introduced in [3]. The learning algorithms for MVN have been presented elsewhere [3, 5, 7]. The most comprehensive observation of MVN, its theoretical aspects, learning and properties are presented in [8]. Here we briefly review key mathematical properties of MVN and their learning algorithms.

An MVN [3, 8] performs a mapping between n inputs and single output. The mapping is described by multiple-valued (k-valued) function of n variables $f(x_1,...,x_n)$ characterized by $n+1$ complex-valued weights $w_0, w_1, ..., w_n$, so that

$$f(x_1,...,x_n) = P(w_0+w_1x_1 + ... + w_nx_n), \tag{1}$$

where $x_1, ..., x_n$ are neuron's inputs or in other words variables, on which the performed function depends. Values of the function and variables are also coded by complex numbers which are kth roots of unity: $\varepsilon^j = \exp(i2\pi\, j/k)$, $j \in \{0, k\text{-}1\}$, i is an imaginary unity. In other words, values of the k-valued logic are represented as kth roots of unity: $j \to \varepsilon^j$. P is the activation function of the neuron:

$$P(z)=\exp(i2\pi j/k), \quad if\ 2\pi j/k \le \arg(z) < 2\pi\,(j+1)/k, \tag{2}$$

where j=0, 1, ..., k-1 are the values of the k-valued logic, $z = w_0 + w_1x_1 + ... + w_nx_n$ is the weighted sum , arg(z) is the argument of the complex number z. Thus the complex plane is divided into k sectors by (2), and if z belongs to the jth sector, neuron's output is equal to ε^j (Fig. 3).

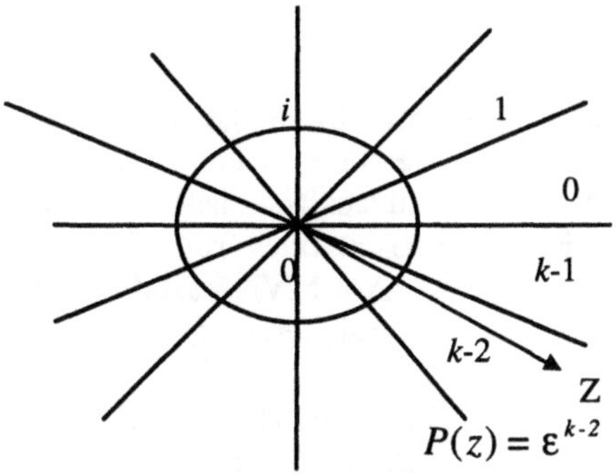

Fig. 3. Definition of the MVN activation function

MVN has some wonderful properties, which make them much more powerful than traditional artificial neurons. The representation (1) – (2) makes possible an implementation of the input/output mappings described by arbitrary partially defined multiple-valued functions. This makes it possible to develop more simple net-works adapted for solving complicated applied problems. Another important property of the MVN is the simplicity of its learning algorithms. Theoretical aspects of the learning, which are based on the motion within the unit circle, have been considered in [8]. Considering MVN learning as a generalization of perceptron learning, we obtain the following. If a perceptron output is incorrect for some element of the learning set (1 instead of -1, or -1 instead of 1) then the weights have to be corrected according to some rule to ensure an inversion of the weighted sum sign. Therefore, it is necessary to move the weighted sum to the opposite subdomain (respectively, from "positive" to "negative", or from "negative" to "positive"). For the MVN, which performs a mapping described by a *k*-valued function we have exactly *k* domains. Geometrically they are sectors on the complex plane (see Fig. 3). If the desired output of the MVN on some element from the learning set is equal to ε^q then the weighted sum has to be placed into the *q*th sector. But if the actual output is equal to ε^s then the weighted sum belongs to the sector number *s* (see Fig.4). The learning rule must correct the weights to move the weighted sum from sector *s* to sector *q*.

We use here a previously proposed correction rule for MVN learning [8]:

$$W_{m+1} = W_m + \frac{C_m}{(n+1)}(\varepsilon^q - \varepsilon^s)\,\overline{X}\,, \tag{3}$$

where W_m and W_{m+1} are the current and next weighting vectors, $\overline{X}$ is the complex conjugate of the vector of the neuron's input signals, ε is a primitive *k*th root of unity (*k* is chosen from (2)), C_m is a scale coefficient, *q* is a number of the desired sector on the complex plane, *s* is the sector into which the actual value of the weighted sum has fallen, *n* is the number of inputs to the neuron.

The learning rule (3) is a generalization of the perceptron error-correction rule, which has been considered, e.g, in [1]: Since the learning rule (3) is linear, its computing implementation is very simple. The learning algorithm based on the rule (3) is very quickly converging. It is always possible to find such a value of *k* in the representation (2) that (1) will hold for a given function *f* describing the mapping between neuron's inputs and output [8].

4.2 MVN-Based Neural Network for Pattern Recognition

Let us consider N classes of objects, which are presented by images of $n \times m$ pixels. The problem is formulated in the following way: we have to create a recognition system based on a neural network, which affords successful classification of the objects by fast learning on the minimal number of representatives from all the classes.

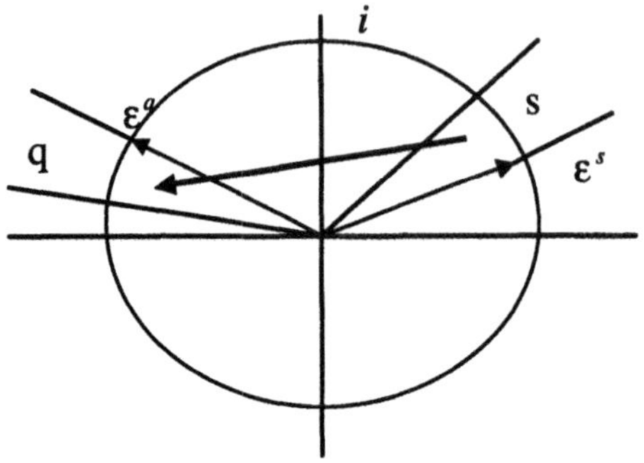

Fig. 4. Problem of the MVN learning

An MVN-based single-layer neural network, which contains the same number of neurons as the number of classes to be classified (Fig. 5), has been proposed in a form of such a system in [6].

This model has been further developed in [7, 8, 11, 12]. All 2D images are replaced by their Fourier spectra. Moreover, taking into account that a spectral phase contains more information about an object presented by a signal than the amplitude [10, 14], only the spectral phase is used for an object representation. Each neuron has to recognize pattern belonging only to its class and to reject any pattern from any other class. For different classes of discrete signals the sets of quarter to half low frequency spectral coefficients are very close to for the signals belonging to the same class from the standpoint of learning and analysis on the neural network [13]. To classify an object in terms of neural networks we have to train a neural network on a learning set containing the spectral phase of representatives of the classes. Then the weights obtained in the learning process will be used for the classification of unknown objects.

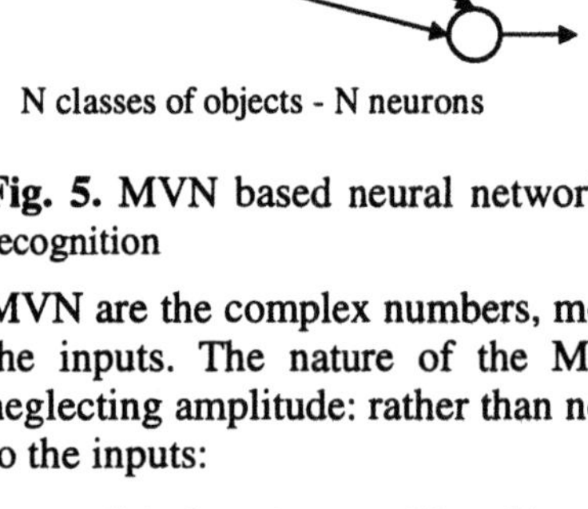

Fig. 5. MVN based neural network for image recognition

Representation of the recognized objects by phases of Fourier spectral coefficients is appropriate for an MVN-based neural network. Since the inputs and outputs of an MVN are the complex numbers, moreover roots of unity, it is natural to use phases as the inputs. The nature of the MVN suggests how to make use of phase while neglecting amplitude: rather than normalize, we apply the transformation based on (2) to the inputs:

$$P(phase) = \exp(i2\pi j/k), \quad if\ 2\pi j/k \le phaze < 2\pi(j+1)/k\,. \tag{4}$$

The recognition system has been successfully tested on an example of face recognition [11, 12]. The best results have been obtained experimentally, when for classification of the pattern as belonging to the given class, the first $l=k/2$ sectors on the complex plane (see (2)) have been reserved. The other $k/2$ sectors correspond to the rejected patterns (Fig. 6).

Extraction of the phases from Fourier spectral coefficients is organized according to frequency ordering. We start from the lowest ones and then we proceed according to the so-called "zigzag" rule (Fig. 7).

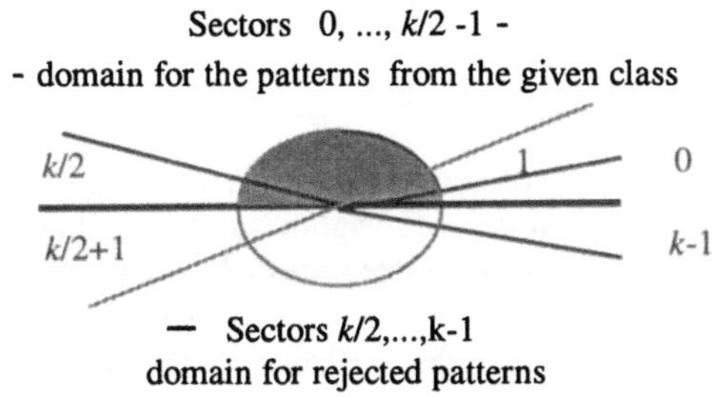

Fig. 6. Reservation of the domains for the recognition

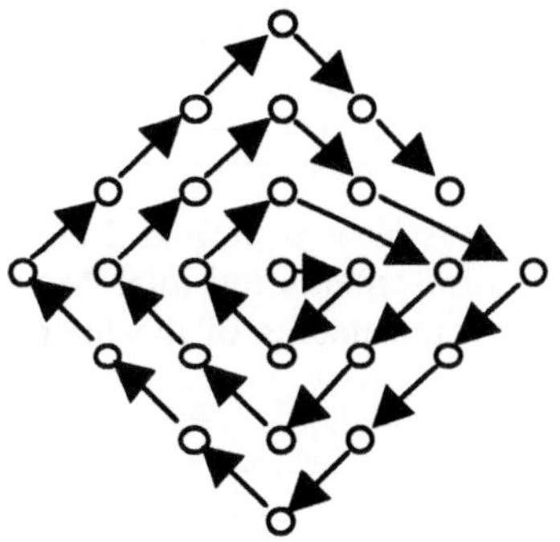

Fig. 7. Extraction of the phases

5 Recognition and Classification. Simulation Results.

To solve the classification problem we used the MVN-based neural network described above. Since the images from neighboring classes are often similar to each other, the architecture of the network has been modified in the following dichotomous way. A three-layered network was used. Two neurons of the first layer classify patterns into those belonging to classes 1-4 and 5-8, respectively. Four neurons of the second layer classify patterns into one of the classes 1-2, 3-4, 5-6 and 7-8, respectively. Finally, 8 neurons of the third layer classify a pattern into a particular class (Fig. 8), and the single neuron of output layer analyses the results coming from the neurons of the third layer.

To create representative learning sets for all the neurons, we used images that have been *a priori* correctly classified as representative for each particular class. This classification is based on the objective biological view described in section 2. Since it is natural to teach the neural network using maximally distinguished data, the definitely representative images were included in the learning set (from 28 to 35 images per class) and the other images were used for testing. All the neurons were taught using a learning algorithm based on the rule (3). We present here the final testing classification results obtained for the 3^d layer of the network. These results are given in Table 2. The results were obtained for k=2048 in (2) (2048-valued logic) and inputs taken from the Fourier phase coefficients corresponding to frequencies from 1 to 6 (84 neuron inputs). These parameters are chosen experimentally. A value k=2048 ensures a good precision for the representation of phases and provides the better results than those obtained with smaller values of k. A chosen number of frequencies is sufficient to achieve good results, which could not be obtained in case of a smaller number of frequencies. On the other hand the further increasing of this number does not lead to the recognition improvement.

Table 2. Classification results without preprocessing.

Class #	1	2	3	4	5	6	7	8
Number of Images	75	43	68	55	78	89	67	53
Recognized	61 81.3%	30 69.7%	49 72.0%	45 81.8%	59 75.6%	61 68.5%	46 68.7%	38 71.7 %
Unrecognized	3	1	1	1	1	3	1	1
Misclassified	11	12	18	9	18	25	20	14

It is clear from Fig. 2 that the images are of low contrast and many of important details remain invisible. Moreover, the small level of brightness jumps between important details of the images negatively affects the informational capacity of the Fourier representation of the image. Phases of the image spectrum corresponding to neighboring classes are sometimes very close to each other. As a result, the number of misclassified images still remains relatively high, especially for the neurons of the third layer.

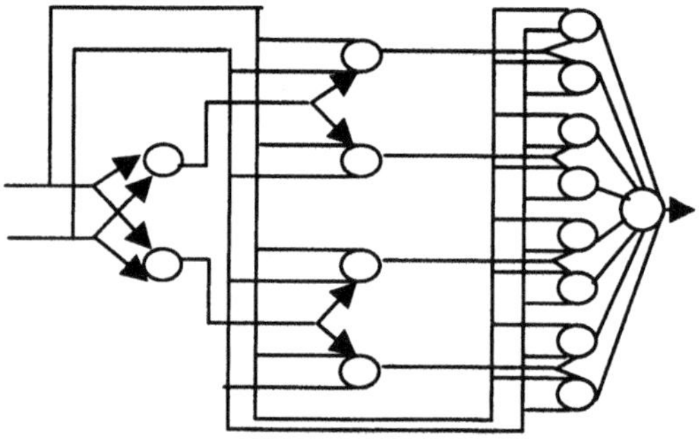

Fig. 8. The dichotomous neural network architecture used for classification.

To fix this problem it is natural to use a filter, which can amplify high spatial frequencies while preserving at the same time low and medium spatial frequencies. High frequency amplification makes it possible to enhance a contrast of the smallest objects against a residual image background. The application of spatial domain filter with the mentioned features significantly changes the spectrum modulo in the high frequency domain. At the same time such a filter changes the spectral phase in whole domain. The last property could be applied for the classification with the same parameters as those used before the proposed preprocessing procedure. We apply the following frequency correction spatial domain filter for the preprocessing:

$$\hat{B}_{ij}=(G_1+G_2)B_{ij}-G_2B_m+c\,. \tag{5}$$

where B_m is the local mean value in a 3×3 window surrounding the pixel B_{ij}; B_{ij}, and $\widehat{B}_{ij}$ are the signal values in ij^{th} pixel before and after processing; respectively, G_1 defines the correction of the low spatial frequencies, G_2 defines the correction of the high spatial frequencies, c is the mean value of the background after processing. The filter (5) has been considered, for example, in [19]. Note that if $G_1 = 1$, the filter (5) coincides with the "unsharp masking" filter (for example, see [20]). The significant improvement of image quality after applying the filter (5) is shown in Fig.9, which shows the same images, which are presented in Fig.1 processed with the filter (5).

The filter (5) was applied to all the images from all the classes. Then the recognition technique presented above was tested again using the images obtained after filtering. The learning and testing sets were composed of the same images as before. The learning process required approximately the same time, while the classification results were improved. Table 3 presents the classification results for the layers 1-3 of the neural network. Comparing the results for the *3^d* layer to the corresponding results presented in Table 2, one can clearly see that the proposed preprocessing technique is very important for the classification improvement.

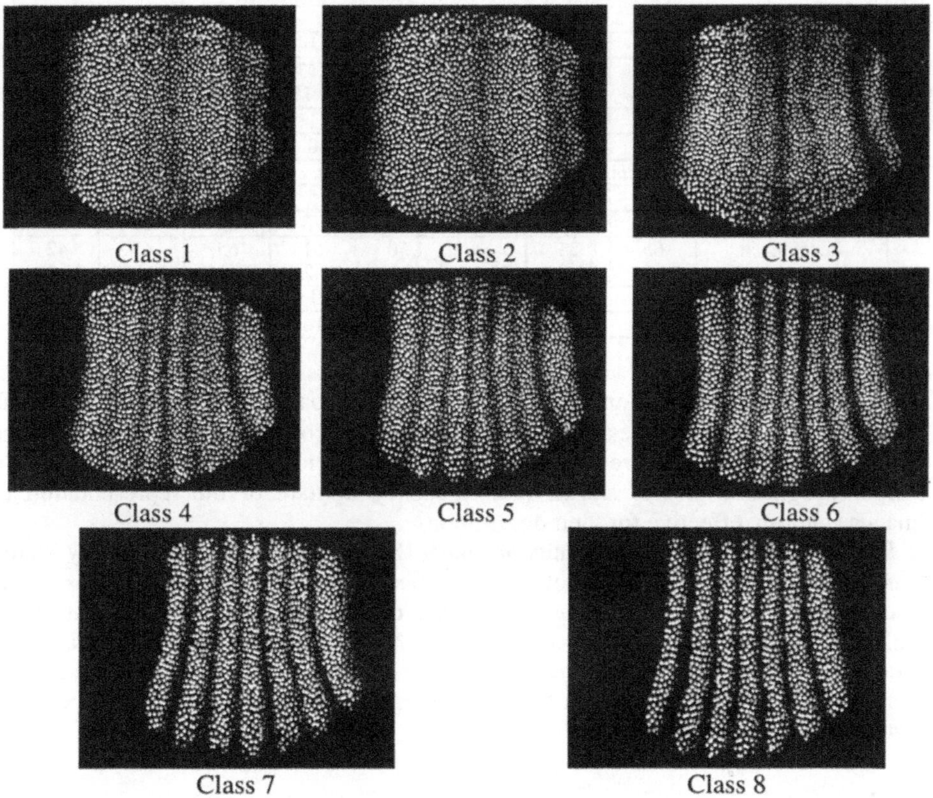

Fig. 9. Images presented in Fig.1 subjected to preprocessing with filter (5).

6 Conclusions

We have performed a temporal classification of *Drosophila* embryos on the basis of the knowledge about their gene expression patterns. In our preliminary study all embryos were subdivided into eight temporal classes by visual inspection. The 1D images were then automatically classified by means of the discriminant function analysis using a fixed set of features for classification. This method turned to be not very reliable and therefore the results are regarded just as preliminary.

Table 3. Classification results after preprocessing.

Class #	1	2	3	4	5	6	7	8
The results of testing for the 1st layer of neurons								
Classes	Classes 1-4				Classes 5-8			
Number of Images	75	72	99	53	83	123	99	85
Recognized	75 100%	68 94.4%	96 96.7%	41 77.4%	66 79.5%	120 97.5%	97 97.9%	85 100%
Unrecognized	0	1	3	0	4	3	2	0
Misclassified	0	3	0	12	13	0	0	0
The results of testing for the 2d layer of neurons								
Classes	1-2		3-4		5-6		7-8	
Number of Images	75	43	68	84	114	89	67	85
Recognized	71 94.6%	35 81.4%	54 79.4%	79 94.0%	113 99.1%	69 77.5%	61 91.0%	83 97.6%
Unrecognized	1	1	1	2	1	1	1	2
Misclassified	3	7	13	3	0	19	5	0
The results of testing for the 3d layer of neurons (final classification results)								
Classes	1	2	3	4	5	6	7	8
Number of Images	75	43	68	55	78	89	67	53
Recognized	65 86.6%	31 72.1%	51 75.0%	50 90.9%	61 78.2%	65 75.0%	47 70.2%	42 79.3%
Unrecognized	3	1	1	0	1	0	1	1
Misclassified	7	11	16	5	16	24	19	10

To improve classification we applied the MVN-based neural network to the recognition of the 2D images. We also used frequency correction filter to improve the images quality. The good results obtained by this method confirm its efficiency for image recognition. It was shown that frequency domain of the representation of images is highly effective for their description.

But while the recognition results are much better, than in the preliminary study, there still remain a number of misclassified and unrecognized images. This discrepancy may reflect inaccuracies not only of the recognition method but of the preliminary hand classification as well. Another reason may lie in the fact that subdivision of the dataset into a certain number of classes is somewhat arbitrary and therefore assigning the borders of these classes is also quite arbitrary. This can cause a certain number of erroneous attributions.

Acknowledgements

This work is supported by NIH grants 1 RO3 TW01147-01 and RO1 RR07801-08, NATO collaborate Linkage Grant LST 975850, GAP awards RBO-685 and RBO-895, the Ministry of Science and Technologies of the Russian Federation and the company Neural Networks Technologies (NNT-Israel) Ltd.

References

1. Haykin S. *Neural Networks. A Comprehensive Foundation. Second Edition.* Macmillan College Publishing Company, New York (1994).
2. Leondes C.T. (Ed.) *Image Processing and Pattern Recognition.* Academic Press, San Diego/London/Boston/New York/Sydney/Tokyo/Toronto (1997).
3. Aizenberg N.N. and Aizenberg I.N "CNN based on multi-valued neuron as a model of associative memory for gray-scale images", *Proc. of the 2-d IEEE International Workshop on Cellular Neural Networks and their Applications, Munich, October 12-14* (1992), 36-41.
4. Aizenberg N.N., Aizenberg I.N. and Krivosheev G.A. "Multi-Valued Neurons: Learning, Networks, Application to Image Recognition and Extrapolation of Temporal Series", *Lecture Notes in Computer Science*, **930**, (J.Mira, F.Sandoval - Eds.), Springer-Verlag, (1995), 389-395.
5. Aizenberg N.N., Aizenberg I.N. and Krivosheev G.A. "Multi-Valued Neurons: Mathematical model, Networks, Application to Pattern Recognition", *Proc. of the 13 Int.Conf. on Pattern Recognition, Vienna, August 25-30, 1996, Track D,* IEEE Computer Soc. Press, (1996), 185-189.
6. Aizenberg I.N. and Aizenberg N.N. "Application of the Neural Networks Based on Multi-Valued Neurons in Image Processing and Recognition", *SPIE Proceedings*, **3307** (1998) 88-97.
7. Aizenberg I.N. "Neural networks based on multi-valued and universal binary neurons: theory, application to image processing and recognition", *Lecture Notes in Computer Sciense*, **1625** (B. Reusch - Ed.), Springer-Verlag (1999), 306-316.
8. Aizenberg I.N., Aizenberg N.N. and Vandewalle J. *"Multi-valued and universal binary neurons: theory, learning, applications"*, Kluwer Academic Publishers, Boston/Dordrecht/London (2000).
9. Jankowski S., Lozowski A. and Zurada M. "Complex-Valued Multistate Neural Associative Memory", *IEEE Trans. on Neural Networks*, **7**, (1996) 1491-1496.
10. Aoki H. and Kosugi Y. "An Image Storage System Using Complex-Valued Associative Memory", *Proceedings of the 15th International Conference on Pattern Recognition, Barcelona, Spain September 3-8, 2000*, IEEE Computer Society Press, **2** (2000) 626-629, 2000.
11. Aizenberg I.N. and Aizenberg N.N. "Pattern recognition using neural network based on multi-valued neurons", *Lecture Notes in Computer Sciense*, **1607-II** (J.Mira, J.V.Sanches-Andres - Eds.), Springer-Verlag (1999) 383-392.
12. Aizenberg I., Aizenberg N., Butakoff C. and Farberov E. "Image Recognition on the Neural Network Based on Multi-Valued Neurons", *Proceedings of the 15th International Conference on Pattern Recognition, Barcelona, Spain September 3-8, 2000*, IEEE Computer Society Press, **2** (2000) 993-996.
13. Ahmed N. and Rao K.R. *"Orthogonal Transforms for Digital Signal Processing"*, Springer-Verlag (1975).
14. Kosman D., Reinitz J. and H.Sharp D. "Automated Assay of Gene Expression at Cellular Resolution", *Proceedings of the 1998 Pacific Symposium on Biocomputing,* World Scientific Press, Singapore, (1997) 6-17.

15. Kosman D., Small S. and Reinitz J."Rapid Preparation of a Panel of Polyclonal Antibodies to *Drosophila* Segmentation Proteins", *Development, Genes, and Evolution,* **208**, (1998) 290-294.
16. Myasnikova E., Samsonova A., Kozlov K., Samsonova M. and Reinitz J. "Registration of the Expression Patterns of Drosophila Segmentation Genes by Two Independent Methods", *Bioinformatics,* **17,** (2001), 3-12.
17. Press S.J. *"Applied Multivariate Analysis",* Holt, Rinehart & Winston (1972).
18. Oppenheim A.V. and. Lim S.J "The importance of phase in signals", *Proc. IEEE,* **69**, (1981) 529-541.
19. Aizenberg I.N."Processing of noisy and small-detailed gray-scale images using cellular neural networks", *Journal of Electronic Imaging*, **6**, No 3, (1997) 272-285.
20. Pratt W.K. *"Digital Image Processing"*, Second Edition, Wiley, New York (1992).

A Multi-purpose Visual Classification System

Gunther Heidemann

Universität Bielefeld, AG Neuroinformatik, Germany
gheidema@techfak.uni-bielefeld.de

Abstract. A computer vision system which can be trained to classification tasks from sample views is presented. It consists of several artificial neural networks which realize local PCA with subsequent expert nets as classifiers. The major benefit of the approach is that entirely different tasks can be solved with one and the same system without modifications or extensive parameter tuning. Therefore, the architecture is an example for the potential which lies in view based recognition: Making complicated tasks solvable with less and less expert knowledge.

1 Introduction

Artificial Neural Networks (ANN) have the advantage that they can acquire knowledge from examples. Therefore they are an ideal tool in image processing dealing with data that cannot easily be captured by explicit geometrical modeling. However, ANN cannot be applied directly to the raw pixel data because the "curse of dimensionality" would require nets of enormous size, and, even worse, huge sets of training images. Hence, a feature extraction stage that captures the interesting part of the image variance but removes redundancy has to preceed classification by ANN. The problem is, that once features have to be "designed" by a human expert, the advantages of ANN evaporate. Consequently, an ANN-based visual classification system has to provide both an adaptive feature extraction and a trainable feature classifier.

A well known solution to reduce dimensionality and exploit redundancy to stabilize the detected features is principal component analysis (PCA), as outlined e.g. in [18]. In most cases the image data are projected to the N_P principal components (PCs) with the largest eigenvalues, where N_P is much smaller than the original dimensionality (i.e. the number of pixels). Moreover, interpolation between different object views becomes much simpler in the projection space than in the original pixel space [16]. Being a linear method, the limitation of PCA is that it can capture only the "global" variance of the data, which is not necessarily the most relevant data variation with respect to a given computer vision task. A suitable non-linear extension of PCA is local PCA[22], which partitions the data into subsets for which PCA performs well. Because of the advantages of local over normal PCA, it is a consequent next step to look for a classification system which can exploit these advantages.

In this contribution, a three-stage system incorporating several ANN of different types is presented [6]. The first two stages realize a feature extraction

B. Reusch (Ed.): Fuzzy Days 2001, LNCS 2206, pp. 305–312, 2001.

by local principal component analysis (PCA), the last classifies the feature vectors using expert ANNs. This architecture involves a much greater variety of highly specific filters than normal PCA. This approach can deal with very high dimensional input (e.g. a whole image with three colour channels) as well as small image patches which are derived from prior processing stages. Hence, the system can be applied to a variety of visual classification tasks ranging from object recognition and pose estimation to region segmentation. Since the system is trained from sample views – instead of explicitly geometry modeling – it can be counted among the "view based" approaches. The system has been used stand alone in robotic tasks [9] and as the component of a larger hybrid system where it provides classification hypotheses to a semantic network [2].

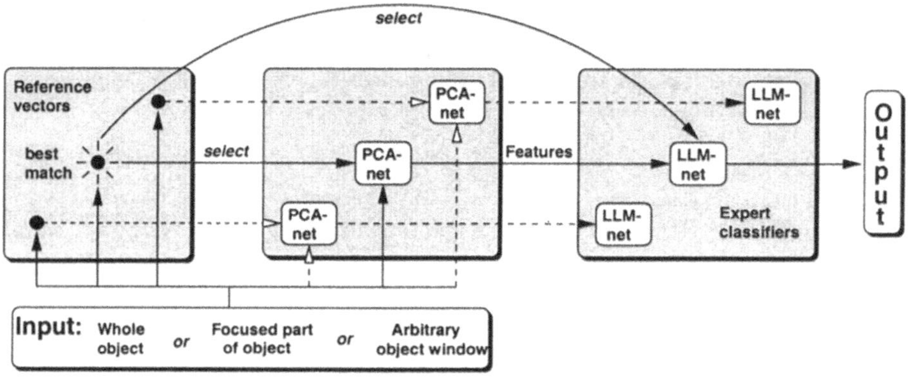

Fig. 1. The neural VPL-classification system maps an image patch or a complete image $\boldsymbol{x}$ to a vector valued output $\boldsymbol{y}$ in three steps: Left, the best match reference vector is determined. Middle: $\boldsymbol{x}$ is projected to the local PCs which were calculated by the attached PCA-net. Right: The projection is classified by the expert LLM-net.

2 The neural classification system

The trainable classifier performs a mapping $\boldsymbol{x} \to \boldsymbol{y}, \boldsymbol{x} \in \mathbb{R}^M, \boldsymbol{y} \in \mathbb{R}^N$ [6]. In computer vision, $\boldsymbol{x}$ consists of the pixels of an appropriately sized window of an image (grey level or colour). The window vector $\boldsymbol{x}$ can be mapped to a continuous valued output vector, e.g. for the estimation of pose parameters (angles). For classification tasks, the output is a (discrete valued) class k. In this case, one separate output channel for each of the N classes is used. Training is performed with labeled sample windows $\boldsymbol{x}_i^{Tr}$ and binary output vectors $(\boldsymbol{y}_i^{Tr})_j = \delta_{ij}, i, j = 1 \dots N$ coding class i. Classification of unknown windows $\boldsymbol{x}$ is carried out taking the class k of the output component with maximal value:
$k = \arg\max_j((\boldsymbol{y}(\boldsymbol{x}))_j)$.

The classifier combines visual feature extraction and classification in a three-stage architecture called VPL, which stands **V**ector quantization, **P**CA and

LLM-network. The vector quantization is carried out on the raw image windows to provide a first data partitioning using N_{VQ} reference vectors $\boldsymbol{r}_i \in \mathbb{R}^M, i = 1 \ldots N_{VQ}$. A suitable algorithm proved to be activity equalization vector quantization (AEV) [10] since one of the major difficulties in high dimensional spaces is codeword under-utilization [5]. Generation of a homogeneously used codebook using incremental algorithms like competitive learning (e.g. "winner takes all") is difficult because the huge volume of a high dimensional space usually cannot be "filled" by the provided data. In this case, only few reference vectors are attracted to the data whereas other reference vectors remain outliers.

Many solutions of this problem have been proposed like the neural gas [14] or the Kohonen self-organizing map [11, 13]. However, since these methods introduce interactions between the reference vectors, parameters have to be chosen appropriately which turns out to be difficult in cases where the designer has no intuition. Algorithms which are designed specially to provide a homogeneous codebook utilization are e.g. AEV [10] or frequency sensitive competitive learning [1, 3, 4] since they take into account the codeword access frequencies.

In the VPL-architecture, to each reference vector $\boldsymbol{r}_i$ of the primary vector quantization a single layer feed forward network is attached for the successive calculation of the principal components (PCs). The PCA-nets are trained as proposed by Sanger [20]. Currently, the number of PCs N_{PCA} is fixed for all reference vectors, in future extensions these numbers will be determined from the local eigenvalue spectra. The input $\boldsymbol{x}$ is projected to the N_{PCA} PCs with the largest eigenvalues: $\boldsymbol{x} \rightarrow \boldsymbol{p}_l(\boldsymbol{x}) \in \mathbb{R}^{N_{PCA}}, l = 1 \ldots N_{VQ}$. $\boldsymbol{p}_l(\boldsymbol{x})$ can be regarded as feature vector of input $\boldsymbol{x}$.

To each of the N_{VQ} PCA-nets one expert neural classifier of the Local Linear Map – type (LLM-net) is attached [19]. The LLM-nets perform the final mapping $\boldsymbol{p}_l(\boldsymbol{x}) \rightarrow \boldsymbol{y} \in \mathbb{R}^N$. The LLM-network is related to the self-organizing map [12] and the GRBF-approach [15]. The LLM-net performs a vector quantization using N_{LLM} nodes in the input space and can be trained to approximate a nonlinear function by a set of locally valid linear mappings which are attached to the reference vectors. For details see e.g. [19].

The three processing stages are trained successively, first vector quantization and PCA-nets (unsupervised), finally the LLM-nets (supervised). Classification of input $\boldsymbol{x}$ is carried out by finding the best match reference vector $\boldsymbol{r}_{n(\boldsymbol{x})}$, then mapping $\boldsymbol{x}$ to $\boldsymbol{p}_{n(\boldsymbol{x})}(\boldsymbol{x})$ by the attached PCA-net and finally mapping $\boldsymbol{p}_{n(\boldsymbol{x})}(\boldsymbol{x}) \rightarrow \boldsymbol{y}$. The major advantage of the VPL-classifier is its ability to form many highly specific feature detectors (the $N_{VQ} \cdot N_{PCA}$ local PCs), but needing to apply only $N_{VQ} + N_{PCA} - 1$ filter operations per classification. So a large set of filters storing implicit object knowledge has not to be paid by a high computational effort in application.

3 Application to computer vision tasks

A system that is to be used on a variety of tasks is not allowed to have a large parameter space that has to be sampled before successful application. For the

VPL-classifier, the crucial parameters are those of the local PCA (N_{VQ}, N_{PCA}) and the number of nodes in the expert LLM-nets N_{LLM}. Search in this parameter space is easy because the system performance is well behaved in all three parameters, i.e. the classification rate (rate of correct classifications) rises continuously when the parameter values are increased until saturation is reached. This could be shown in [8] for different sample tasks.

The bandwidth of the VPL-classifier will be demonstrated on tasks of different character: (1) classification of complete objects, (2) segmentation of partially occluded objects by classification of small image windows, (3) classification of hand gestures to show the capability to deal with nonrigid objects.

3.1 Object recognition

Fig. 2. Examples from the Columbia Object Image Library.

To evaluate the performance of the VPL-system for object classification the Columbia Object Image Library (COIL)[1] was used. A description of the database is given in [17]. It consists of single objects located on a turntable in a normalized position. Each object is rotated on a turntable at pose intervals of 5 degrees, so there are 72 images of each object (RGB). Fig. 2 shows some examples.

For the test, 60 objects were used. The resolution was subsampled by a factor 2 to 64×64. Since the objects are centered, the "where problem" had not to be solved here and the images could be directly used as input vectors $x \in \mathbb{R}^{3 \cdot 64^2}$. 18 images of each object at pose intervals of 20 degrees were used for training, the remaining 54 images for testing. The rate of correct classifications was about 93% for a VPL with $N_{VQ} = 9$, $N_{PCA} = 10$ and $N_{LLM} = 30$.

3.2 Object segmentation

Partial occlusion is a difficult problem in object recognition. A key ability in the construction of algorithms that can deal with occlusion is the segmentation

---
[1] Available at `http://www.cs.columbia.edu/CAVE`

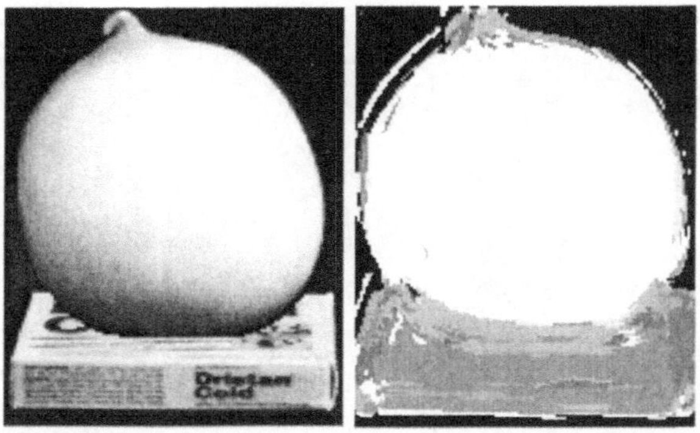

Fig. 3. Segmenting objects to detect partial occlusion using small windows as input (including colour) for the VPL-classifier. Left: Original, right: segmentation result using false colours. From [21].

of the different object regions as it allows to detect the object borders. This is a prerequisite for subsequent classification algorithms that can mask occluded object regions.

The VPL-classifier can be used to scan an entire image using an input window that is small compared to the objects but large enough to evaluate colour texture features [21, 7]. As in the previous section, the output is the object class. For the test, 20 objects of COIL were used (RGB). The system was trained with 18 images of each of the un-occluded objects, from which training windows of size 17×17 were sampled from the object area (image resolution 128×128). The dimensionality of the input vector $\boldsymbol{x}$ is $3 \cdot 17^2$.

The trained system was tested on images which were artificially generated from the remaining 54 images of each object. Each test image shows two different objects that are partly overlapping, see Fig. 3. The output classes were supplemented by an additional class for the background (in total 21). A segmentation example is shown in Fig. 3 (right), indicating the resulting class of each pixel using false colours. Of all pixels, 82% could be classified correctly. This rate is naturally below the results of the previous section as the input information is much smaller.

3.3 Gesture recognition

Gesture recognition is usually treated as a domain apart from object classification. However, since the major benefit of ANN-based approaches is the possibility to train a system just by providing labeled views, different hand poses can be handled in the same way as different objects. Fig. 4 shows six hand gestures of a human-machine interaction scenario which were evaluated for a test. The hand is over a mouse pad and indicates six pointing directions: Left, right, forwards, backwards, up and down. For the latter two "symbolic" hand postures were defined. The camera is located above the hand. The use of the mouse pad is to

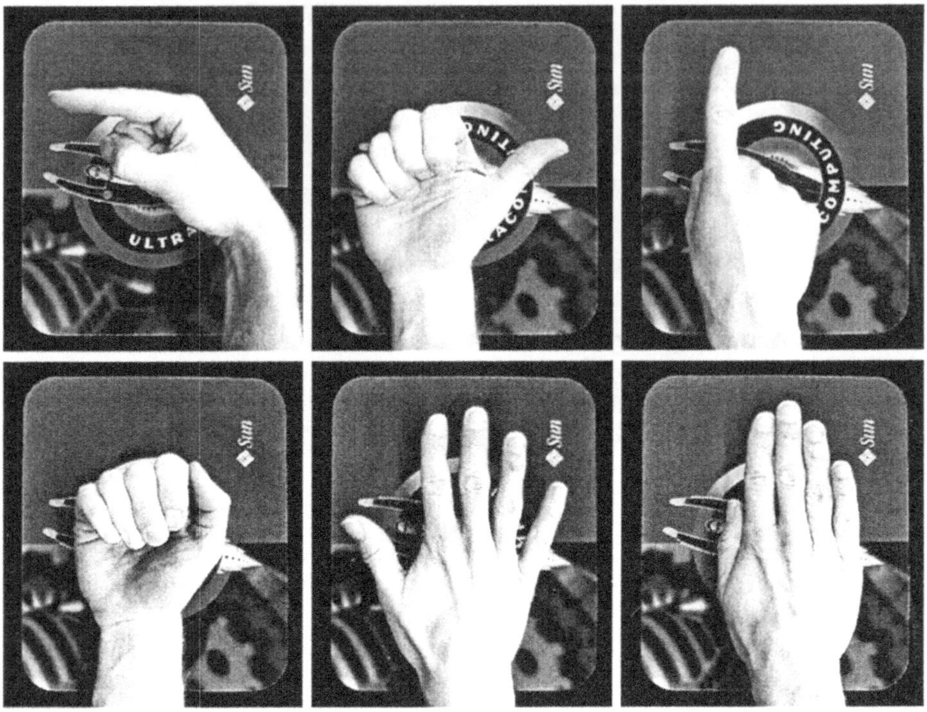

Fig. 4. Gesture classes (from upper left to lower right): "Left, right, away, towards instructor, up, down".

indicate to the user the area where the hand must be. Though the mouse pad is cluttered background, it does not disturb recognition as the system is adapted to its structure during the training phase.

The system was tested on a series of 48 images, 8 of each gesture. Using the "leaving one out" - method for evaluation, 89% correct classifications could be reached by a VPL with $N_{VQ} = 4$ reference vectors, $N_{PCA} = 6$ local PCs and $N_{LLM} = 20$ LLM-nodes.

4 Conclusion and Acknowledgement

Using three different scenarios, it could be shown that the VPL-classifier is well suited for view based solutions of computer vision tasks. The system is easy to adjust and has the potential for further development. The possible future development is worth some discussion.

It should not be overlooked that the view based approach still has its limitations. For example, the gesture recognition outlined in section 3.3 requires a fixed illumination. If the spectrum or direction of the illumination changes, the recognition results are much worse. Of course, different a priori known illuminations could be trained using sample images, but the system still has no inherent invariance properties of this kind. The same argument is true for other desirable, but not yet achievable robustness or invariance features such as independence of the camera position, scale or different skin colour.

There are promising approaches to overcome these limitations, e.g. the integration of view based recognition into hybrid systems involving high level knowledge. But even with the current restrictions, view based systems can do a lot: the major problem is to *find appropriate tasks*. Still, most of the work in the field of visual classification is aimed to discriminate classes which correspond to isolated objects. However, there are many other visual categories which can be classified without a decomposition into single objects. An example is given in [9], where the VPL-classifier is used to judge the stability of grasping positions of a robot hand on objects. In this scenario, a decomposition would be far too difficult, thus the problem is intractable by "classical" computer vision. Nevertheless, the view based system can be trained to judging the situations in the same way as classifying isolated objects. The conclusion can be drawn that the future potential of view based, trainable recognition systems lies not only in better methods, but also in the choice of the tasks.

I would like to thank Helge Ritter for initiating the research project and raising the money as well as for many helpful discussions. This work was supported by the Deutsche Forschungsgemeinschaft (DFG) within research project SFB 360 "Situated Artificial Communicators".

References

1. S.C. Ahalt, A.K. Krisnamurthy, P. Chen, and D.E. Melton. Competitive learning algorithms for vector quantization. *Neural Networks*, 3:277–290, 1990.
2. E. Braun, G. Heidemann, H. Ritter, and G. Sagerer. A multi-directional multiple path recognition scheme for complex objects. *Pattern Recognition Letters, special issue on Pattern Recognition in Practice VI, Vlieland, June 99*, 1999.
3. Aristides S. Galanopoulos and Stanley C. Ahalt. Codeword distribution for frequency sensitive competitive learning with one-dimensional input data. *IEEE Trans. On Neural Networks*, 7(3):752–756, 1996.
4. Aristides S. Galanopoulos, Randolph L. Moses, and Stanley C. Ahalt. Diffusion approximation of frequency sensitive competitive learning. *IEEE Trans. On Neural Networks*, 8(5):1026–1030, 1997.
5. S. Grossberg. Competitive learning: From interactive activation to adaptive resonance. *Cognitive Sci.*, 11:23–63, 1987.
6. G. Heidemann. *Ein flexibel einsetzbares Objekterkennungssystem auf der Basis neuronaler Netze.* PhD thesis, Univ. Bielefeld, Techn. Fak., 1998. Infix, DISKI 190.
7. G. Heidemann, D. Lücke, and H. Ritter. A system for various visual classification tasks based on neural networks. In A. Sanfeliu et al., editor, *Proc. 15th Int. Conf. on Pattern Recognition ICPR 2000, Barcelona*, volume I, pages 9–12, 2000.
8. G. Heidemann and H. Ritter. Combining multiple neural nets for visual feature selection and classification. In *ICANN 99, Ninth International Conference on Artificial Neural Networks*, pages 365–370, 1999.
9. G. Heidemann and H. Ritter. Visual checking of grasping positions of a three-fingered robot hand. In *ICANN 2001*, 2001. Accepted for publication.

10. G. Heidemann and H.J. Ritter. Efficient Vector Quantization using the WTA-rule with Activity Equalization. *Neural Processing Letters*, 13(1):17–30, 2001.
11. T. Kohonen. Self-organized formation of topologically correct feature maps. *Biol. Cybernetics*, 43:59–69, 1982.
12. T. Kohonen. Self-organization and associative memory. In *Springer Series in Information Sciences 8*. Springer-Verlag Heidelberg, 1984.
13. T. Kohonen. *Self-Organizing Maps*. Springer Verlag, 1995.
14. T. Martinetz, S.G. Berkovich, and K. Schulten. "Neural-gas" network for vector quantization and its application to time-series prediction. *IEEE Transactions on Neural Networks*, 4(4):558–569, 1993.
15. J. Moody and C. Darken. Learning with localized receptive fields. In *Proc. of the 1988 Connectionist Models Summer School*, pages 133–143. Morgan Kaufman Publishers, San Mateo, CA, 1988.
16. H. Murase and S.K. Nayar. Visual learning and recognition of 3-d objects from appearance. *Int. Journal of Computer Vision*, 14:5–24, 1995.
17. S.A. Nene, S.K. Nayar, and H. Murase. Columbia object image library: Coil-100. Technical Report CUCS-006-96, Department of Computer Science, Columbia University, 1996.
18. T. Poggio and S. Edelman. A network that learns to recognize three dimensional objects. *Nature*, pages 263–266, 1990.
19. H.J. Ritter, T.M. Martinetz, and K.J. Schulten. *Neuronale Netze*. Addison-Wesley, München, 1992.
20. T.D. Sanger. Optimal unsupervised learning in a single-layer linear feedforward neural network. *Neural Networks*, 2:459–473, 1989.
21. Dirk Strotmann. Lokale Klassifikation als Basis für die Erkennung teilverdeckter Objekte. Master's thesis, Universität Bielefeld, Technische Fakultät, 1999.
22. Michael E. Tipping and Christopher M. Bishop. Mixtures of probabilistic principal component analyzers. *Neural Computation*, 11(2):443–482, February 15 1999.

Problem Difficulty in Real-Valued Dynamic Problems

Karsten Weicker

University of Stuttgart, Institute of Computer Science,
Breitwiesenstr. 20–22, 70565 Stuttgart, Germany,
Karsten.Weicker@informatik.uni-stuttgart.de

Abstract. The article at hand identifies two kinds of problem difficulty in dynamic environments: discontinuities in the fitness landscape caused by moving optima and the discrepancy between tracking and optimization. These problem difficulties are supported by theoretical considerations and experimental findings. Various methods of resolution are discussed and a new adaptation method, namely the correction of strategy variables, is proposed.

1 Introduction

Recently there is an increasing interest in the application of evolutionary algorithms on dynamic, non-stationary problems. Despite of a growing number of published papers presenting new techniques for tackling those problems there is hardly any work available concerning the special requirements posed by dynamic behavior and the arising problem difficulty for certain evolutionary algorithms. Concerning the standard evolutionary algorithms, Bäck (1998) argues that self-adaptive evolution strategies (Rechenberg, 1994; Schwefel, 1995) are a particular good means to cope with dynamic problems. This work discusses the circumstances under which evolution strategies cannot cope with the dynamics and examines those circumstances empirically.

The various evolutionary standard algorithms have been extended by a wide range of techniques to tackle non-stationary problems. Examples are the addition of an external memory to remember past solution candidates (Branke, 1999b), a hypermutation operator which increases the diversity in the population (Cobb, 1990), also diploid concepts are used as an internal model to memorize partial previous solution candidates (Goldberg & Smith, 1987), ideas of thermodynamics to control selection (Mori, Kita, & Nishikawa, 1996), special local search operators to keep track of moving local optima in genetic algorithms Vavak, Fogarty, and Jukes (1996), and cultural algorithms to control the operators (Saleem & Reynolds, 2000). A more extensive overview on the whole field may be drawn from the survey by Branke (1999a). However, those techniques are out of the scope of this work which examines standard evolution strategies instead.

Preliminary experimental investigations on problem difficulties for evolution strategies in dynamic environments are contained in Weicker and Weicker (1999, 2000). There, specific problems are found to be rather difficult for evolution

B. Reusch (Ed.): Fuzzy Days 2001, LNCS 2206, pp. 313–325, 2001.

strategies. Those observations led to the more rigorous investigations in this paper.

2 Problem Difficulty in Dynamic Environments

As Bäck (1998) has shown empirically, evolution strategies (ES) are useful for tackling dynamic problems. Especially the self-adaptation enables the algorithm to reproduce slight dynamic modifications of the problem or even, in the case of bigger changes, to re-optimize the problem. However, the present article examines which characteristics of dynamic problems might hinder the optimization with evolution strategies and, thus, identifies a certain kind of ES-hardness. In particular, this paper is concerned with drifting landscapes where a static landscape moves slightly over time. As an exemplary application De Jong (2000) discusses the control of chemical production processes where a gradual movement of the optimal control point is caused by aging equipment or varying quality of raw material. Within such a problem framework, the following two subsections introduce two hypotheses which reflect properties of dynamic ES-hard problems.

2.1 Discontinuity Caused by Dynamics

In any evolutionary algorithm the operators have a certain underlying conception of how the next individuals should be created using previous individuals. These conceptions can be made explicit by operator-defined fitness landscapes (Jones, 1995) in the case of discrete search spaces and by probability density functions (pdf) in the case of continuous search spaces, which is the case with evolution strategies. In order to get good results there should be a high correlation between the characteristics of the problem and this conception of the operator. One example for a good operator-problem interaction are the Gaussian pdf of the evolution strategy mutation applied to real-valued, smooth, stationary problems. A smooth, partially monotonous search space (e.g. sphere model) guarantees that even a small step in the right direction is a good step. This fits perfectly to the internal model of the zero-mean Gaussian pdf. The additional self-adaptation mechanisms of the evolution strategies enable a quick adaptation to quite different problem spaces.

However, by introducing dynamic, time-varying aspects into a problem this simple correlation between operator and problem is disturbed. This may be explained for the evolution strategy mutation using Figure 1. The upper row shows three different one-dimensional fitness landscapes where the circled cross marks the current position of a candidate solution. The middle row shows how the fitness landscape is shifted from one generation to the next. And the lower row shows the arising discontinuity from this shift as the difference between the fitness values we could expect if the problem was stationary and the fitness values we encounter in the next generation. The smoothness we could actually expect gets disarranged from generation t to generation $t+1$. In the landscape in the middle row, a step to the right is necessary to reach the optimum, the mutation cannot do wrong at time t if a small step is chosen instead of a bigger step. But this is not true at time $t+1$ anymore since the small step reaches a worse fitness than at time t. Here a bigger step is desirable. Thus dynamics introduce a new difficulty in the optimization which is probably not met by a mutation preferring small steps as is summarized in the following hypothesis.

Hypothesis 1 *The Gaussian pdf does not match the character and the demands of tracking problems.*

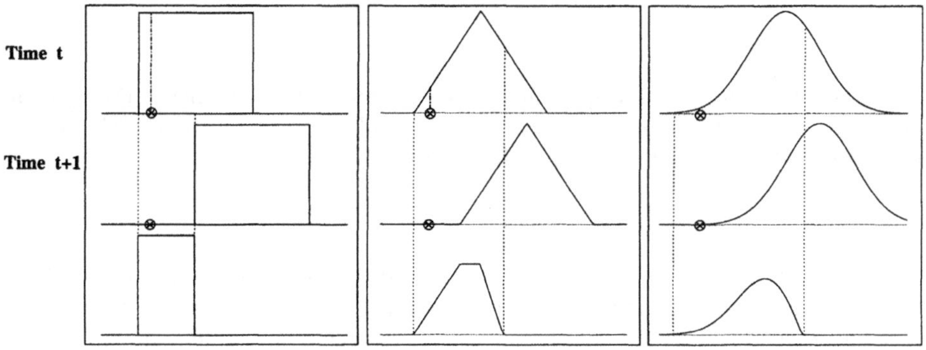

Fig. 1. This figure shows three different landscapes at time t (upper row) and $t+1$ (middle row). By the change of the landscape there arises a region of discontinuity shown in the lower row where the fitness should be equal or better to a point situated to the left but where the fitness decreases caused by the dynamics.

2.2 Combination of Tracking and Optimization

Another prerequisite in most research articles on dynamic optimization is the fact that usually only the tracking of an optimum is considered, i.e. the underlying optimization problem is assumed to be easy. However, if this is not the case, the question arises whether both the optimization task and the tracking task fit to the operator's pdf to generate new individuals. In particular, any self-adaptation mechanism has to support both tasks. The tracking of the current region of interest is a necessary requirement for the optimization – however, as it is explained in detail in Section 5, successful steps in the optimization task may disturb the settings of the self-adaptive strategy parameters in such a way that no tracking is possible anymore. This leads to the second hypothesis.

Hypothesis 2 *The ES self-adaptation mechanisms might fail in problem settings which require optimization within a tracking task.*

Such a combination of tracking and optimization may occur in a constrained optimization problem where the constraints are changing over time. In Section 4 a simple problem model for such a combination is given. A more realistic example can be found in (Weicker & Weicker, 1999).

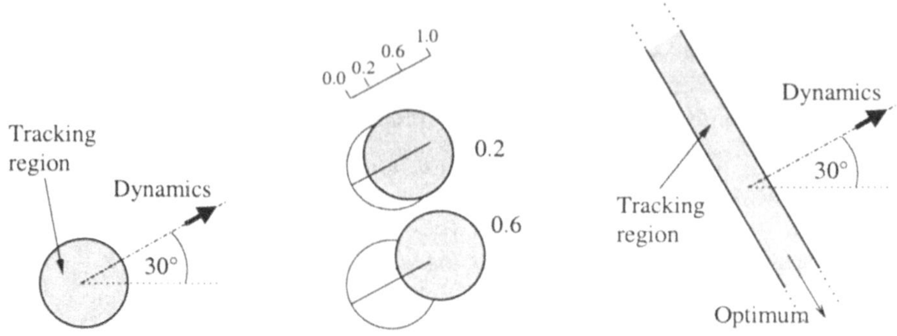

Fig. 2. On the left, the moving circle problem is shown, the meaning of the dynamics parameter is shown exemplary for the moving circle problem in the middle, and the moving corridor problem is shown on the right.

After a short introduction to the considered test problems in the next section, the hypotheses are supported by experiments in Sections 4 and 5. Also in both sections methods for the resolution of the problem are discussed and tested partially.

3 Problem Framework

In order to examine the hypotheses two very simple problems are defined resembling the underlying necessary characteristics of the hypotheses. For the sake of simplicity, a two-dimensional search space $\Sigma = \mathbb{R} \times \mathbb{R}$ is considered in which a tracking region with "good" fitness is defined. All other points are assigned a "bad" fitness. In more realistic optimization tasks the tracking region may be defined by points with near-optimal fitness or by dynamic constraints. This underlines the relevance of those simplified problems for dynamic optimization in general.

To study the first hypothesis, the *moving circle problem* as a pure tracking task is introduced. Here the tracking region consists of a circle which moves within the search space (cf. left part of Figure 2). The fitness within the tracking region is determined by the distance to the center of the circle where the optimum is located. All other points are assigned a constant bad fitness value. This problem is very similar to the moving hill or the moving sphere function in various previous examinations (e.g. Cobb, 1990; Collard, Escazut, & Gaspar, 1996; Branke, 1999b; Liles & De Jong, 1999; Morrison & De Jong, 1999). In order to avoid pathological behavior, the direction of the dynamics was chosen to be in a 30° angle from one axis. This prevents dependence on one axis only as well as the other symmetric extreme, an angle of 45°. The strength of the dynamics may vary in the range [0.0, 1.0] which corresponds to the fraction of new, non-overlapping parts of the tracking region (cf. middle part of Figure 2). After each generation the tracking region is moved according to the strength.

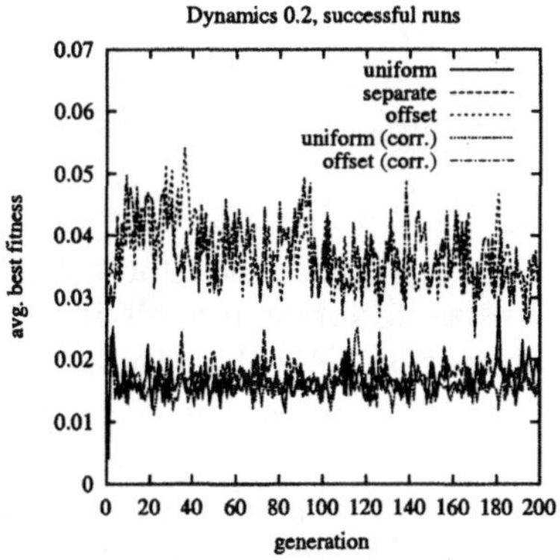

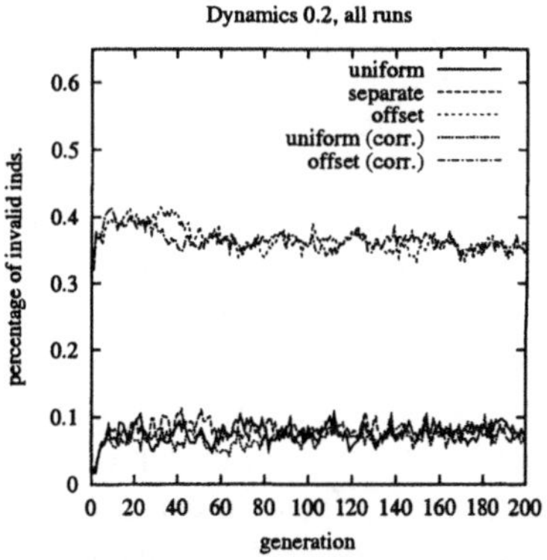

Fig. 3. Results for the moving circle problem with static strength of dynamics 0.2. The average best fitness of all successful experiments (left) shows better performance of all variants of standard step-size adaptation compared to both offset techniques. This behavior is also reflected in the percentage of invalid created individuals (right).

Strength 0.0 means that the problem is stationary and strength 1.0 imply that the circles do not overlap from one generation to the next but are still adjacent to each other.

In the examination of the second hypothesis, the focus is on a combination of tracking and optimization which might be unfavorable for the evolutionary algorithm. Thus a scenario was chosen where the direction of the tracking tasks is arranged orthogonal to the direction of the optimization task, the *moving corridor problem* (cf. the right part of Figure 2). The fitness value within the corridor is determined by how far we have moved towards the optimum. The strength of the dynamics is determined similarly to the moving circle model. Also, the corridor is arranged in a 30° angle from one axis. Such a model seems to be pathological in a two-dimensional problem setting, however, those orthogonal effects may occur easily in multi-dimensional search spaces with moving constraints.

In the following two sections various evolution strategies are applied to the two problems. As standard step-size adaptations, uniform self-adaptation with one strategy variable as standard deviation for all search space dimensions and separate self-adaptation with a different strategy variable for each search space dimension are compared. Also, two extensions are introduced in the next two sections. For each algorithm, problem, and strength of dynamics, 100 independent experiments are executed using different random seeds. All investigated algorithms use a $(1, 40)$-strategy without recombination.

4 Experiments: Moving Circle

The moving circle problem is used to examine Hypothesis 1 which states that the Gaussian pdf does not match the character and demands of tracking problems. Figure 1 sketches the underlying thoughts that a small step means worsening if the tracking region moves. This is empirically examined using the average best fitness over all experiments, the percentage of invalid individuals created, and the percentage of experiments which got completely lost, i.e. they could not track the tracking region.

The experiments are shown in Figures 3, 4, and 5. In uniform step-size adaptation the individual is extended by one strategy parameter for the standard deviation of the mutation, in separate step-size adaptation a distinct strategy parameter is used for each search space dimension. For the sake of completeness, the figures contain the performance of a technique, the offset mutation, introduced in the remainder of this section and a correction strategy which is combined with offset and uniform mutation and introduced in the next section.

From the experiments we can conclude that self-adaptive evolution strategies with either uniform step-size for all dimensions or separate step-sizes exhibit a substantially increasing number of invalid individuals with increasing dynamics. As a consequence the tracking accuracy measured by the average best fitness decreases with increasing dynamics. Also at the end of 200 generations, we see in Table 1 that for both uniform and separate step-size adaptation the fraction of complete lost runs increases considerable with increasing dynamics.

Experiments with the 1/5–success rule have been executed but are not included in the the results for reasons of very inferior performance. Covariance matrix adaptation was not used in the context of this work since preliminary studies (Weicker & Weicker, 1999) have shown that those adaptation mechanisms are often too inertial.

One approach to generate more effective new individuals in the face of dynamics is to shift the mean of the generating distribution away from zero, resulting in a directed mutation. One approach to reach such a behavior is the skewing of the underlying pdf as it was done by Hildebrand, Reusch, and Fathi (1999). Another approach by Ghozeil and Fogel (1996) uses an offset direction vector which determines the new position where a non-zero Gaussian mutation takes place. The second method was modified slightly for this work. In general the strategy parameters are an offset vector represented as an n-dimensional unit vector in Cartesian coordinates $\boldsymbol{v} \in \mathbb{R}^n$ and the standard deviation as $\sigma \in \mathbb{R}_+$. In this examination $n = 2$. The mutation on an individual $\boldsymbol{x}$ works as follows.

$$\begin{aligned} \sigma &= \sigma \exp\left(\frac{1.0}{\sqrt{1.0n}}\mathcal{N}(0,1)\right) \\ v_i &= v_i + \mathcal{N}(0,1) \\ x_i &= x_i + v_i + \mathcal{N}(0,\sigma) \end{aligned}$$

Concerning the adaptation mechanism for the offset vector this is an ad hoc approach. No fine tuning of the mechanism was done so far.

The experiments in Figures 3, 4, and 5 and Table 1 show that this approach is almost completely unaffected by the strength of the dynamics. The tracking accuracy is constant with respect to the dynamics. However, the standard adaptation mechanisms yield a higher tracking quality with slow dynamics.

Additional experiments with increasing (resp. decreasing) dynamics between 0.2 and 1.0 reflect exactly the behavior with static strength of dynamics and are

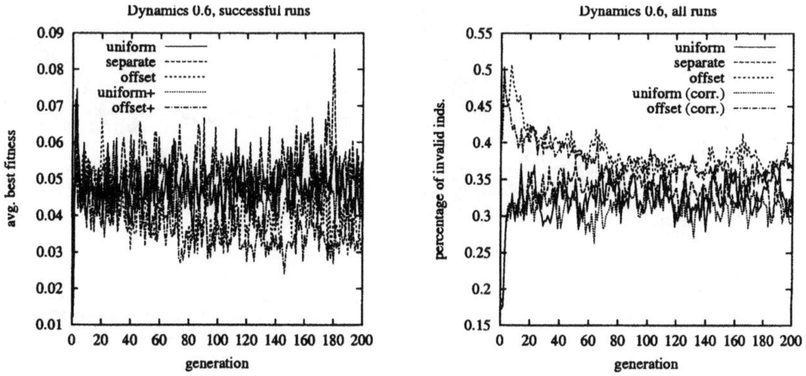

Fig. 4. Results for the moving circle problem with static strength of dynamics 0.6. Compared to the behavior with strength 0.2 in Figure 3 the performance of all techniques adjusts to each other. In case of the average best fitness there are even slight insignificant advantages for the offset techniques.

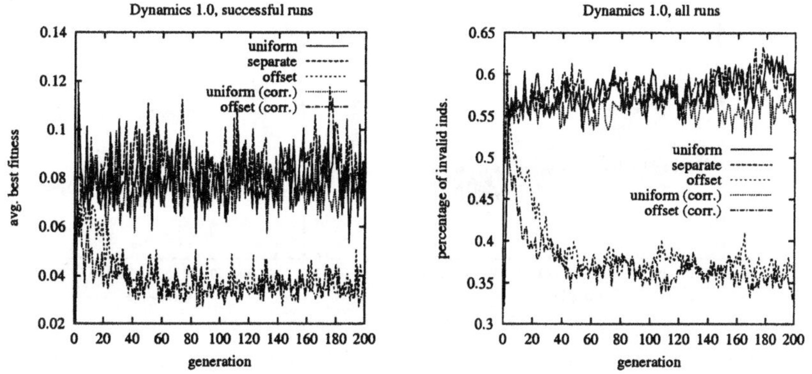

Fig. 5. Results for the moving circle problem with static strength of dynamics 1.0. The trend of Figures 3 and 4 continues. The offset technique preserves the same level of average fitness and percentage of invalid individuals where the standard step-size adaptation performs significantly worse.

Table 1. Fraction of completely lost experiments at the end of 200 generations for the moving circle problem.

	strength of dynamics		
	0.2	0.6	1.0
uniform	0.00	0.02	0.10
separate	0.00	0.02	0.09
offset	0.01	0.03	0.02

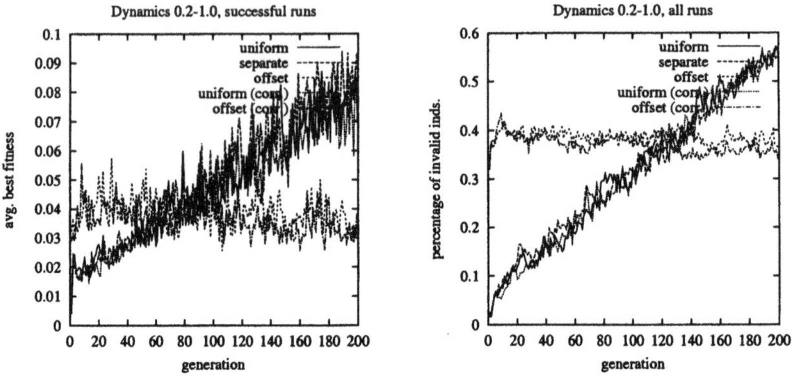

Fig. 6. The average best fitness (left) and the percentage of invalid created individuals (right) in experiments with the tracking circle model and strength of dynamics increasing linearly from 0.2 to 1.0 show an almost invariant behavior of the offset mutation and increasing problems of all other adaptation techniques.

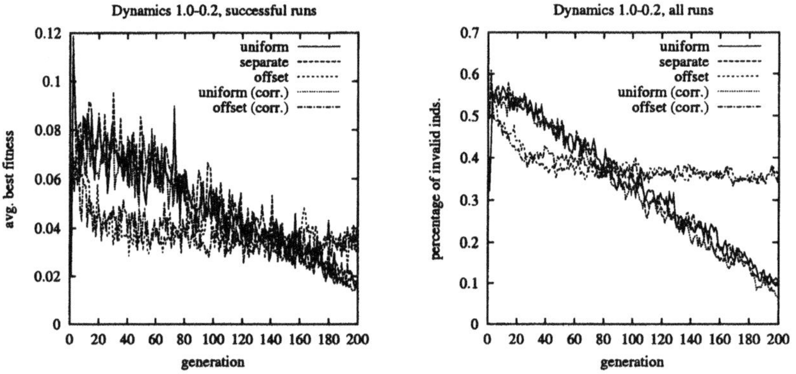

Fig. 7. Experiments with the tracking circle model and strength of dynamics decreasing linearly from 1.0 to 0.2 exhibit the same behavior as in the experiments in Figure 6.

shown in Figures 6 and 7. This underlines the adaptability of the introduced directed mutation.

Thus the first hypothesis is supported by the fact that there is a high dependency of the quality of standard adaptation mechanisms on the strength of dynamics and the existence of an adaptation mechanism which overcomes those limitations by breaking the principle of a zero-mean mutation favoring small steps.

5 Experiments: Moving Corridor

This section focuses on the second hypothesis which states that there are circumstances where optimization may be difficult in the presence of dynamics, i.e. if

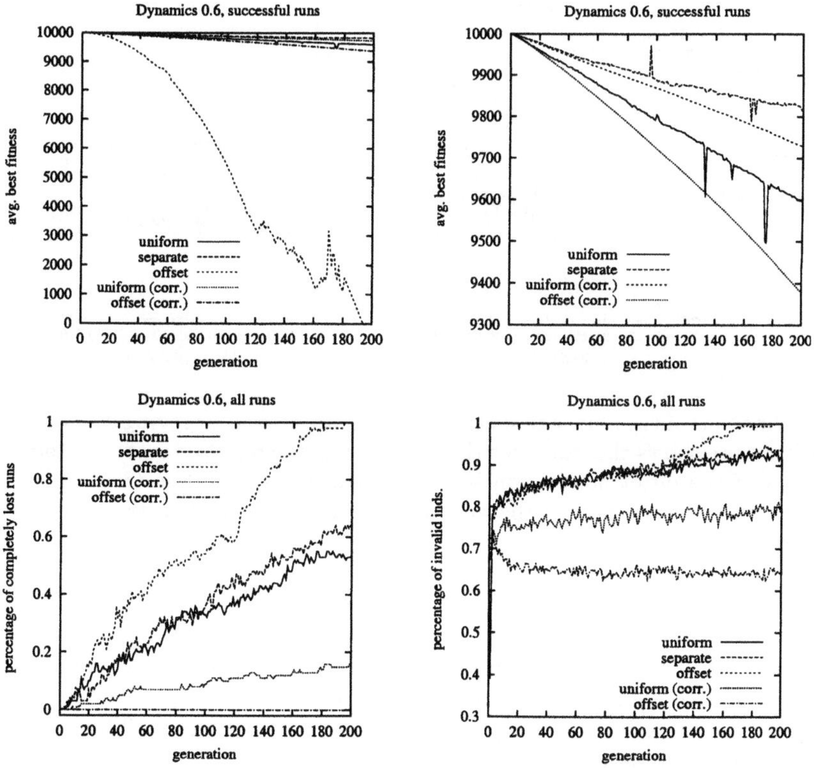

Fig. 8. Results for the moving corridor problem with strength of dynamics 0.6: the upper row shows the best fitness values of all successful experiments on average (with two differently scaled ordinates), the lower row shows the fraction of completely lost experiments (left) and the fraction of invalid generated individuals (right). Superior best fitness can be observed for offset mutation which comes along with high percentages of lost runs and invalid individuals. Only the offset mutation with strategy variable correction shows a stable performance concerning the invalid individuals and the lost runs.

successful tracking is a necessary prerequisite for optimization. This hypothesis is examined using the introduced moving corridor problem.

The results for the standard adaptation techniques as well as the directed offset mutation are shown in Figure 8 exemplary for dynamics 0.6. Other strength of dynamics reveal a similar behavior. The offset mutation shows in its successful runs a superior behavior (as would separate adaptation do if optimization was aligned with a coordinate axis.) However this is accompanied by a huge increase in completely lost runs and invalid individuals. At the end of 200 generations almost each created individual is invalid. Uniform and separate step-size adaptation show the same tendency. The reason of this effect can be clarified by looking at the evolution of the strategy parameters shown in Figure 9. Since the adaptation mechanism always accepts those individuals (and, thus, indirectly

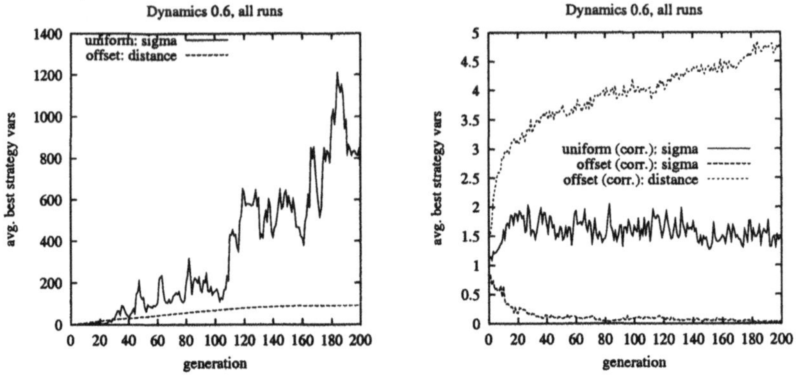

Fig. 9. Analysis of the adaptation of strategy variables for the moving corridor problem: the left picture shows the values for the uniform and the offset adaptation (the standard deviation of the offset adaptation is omitted since it stays close to 1) and the right picture shows the values for the adaptation techniques with correction of strategy variables.

the respective strategy values) that score the best fitness values, this results in accepting those settings which make huge jumps and hitting the tracking region by accident. However, there is only a very small probability that these strategy variables can reproduce the same effect in the next iterations. Thus the number of invalid individuals increases.

As a consequence, a technique is needed which takes care of the discrepancy between the greedy behavior concerning the optimization and the needs for successful tracking. We decided to correct the strategy variables of the accepted individual in such a way that tracking is still possible with a high success rate. The mutation of the strategy and object variables is not affected by this approach. But as soon as the surviving individual is selected (according to the best fitness) its strategy variables are modified in the following way. For all in the current generation created individuals with a higher fitness than the parent's fitness the distance $\sqrt{\sum_{i=1}^{n}(s_i^{child} - s_i^{parent})^2}$ of their strategy variables s^{child} and the parent's strategy variables s^{parent} are computed. The strategy variables of the individual with the minimal distance are used to replace the accepted individual's strategy variables. If there is none such individual no correction takes place.

The results for the moving corridor problem can be seen in Figures 8 and 9. The correction mechanism dampens the explosion of the strategy variables for both uniform and offset adaptation. As a consequence, the invalid created individuals can be kept on an almost constant level which leads to only few completely lost runs – in case of the offset adaptation even all runs are successful.

The results in Figures 10 and 11 with varying dynamics indicate that the correction strategy can handle non-stationary dynamics too. However diminishing dynamics are more problematic since there completely lost runs arise. In case of the uniform step-size adaptation with correction of strategy variables the

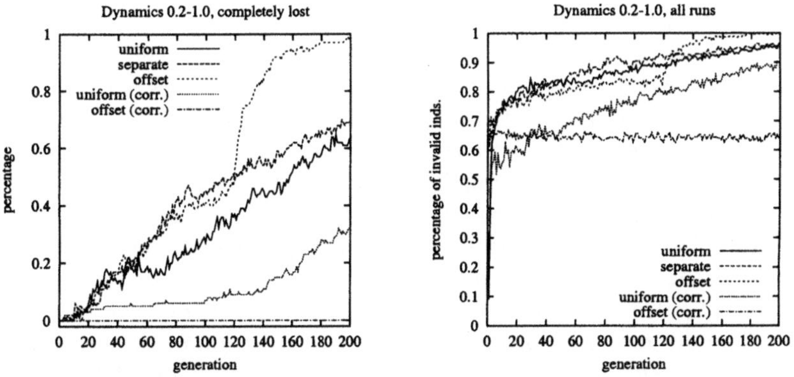

Fig. 10. Moving corridor problem with linearly increasing strength of dynamics from 0.2 to 1.0. The techniques without strategy variable correction show an almost chaotic behavior in generation 200. Uniform adaptation with correction performs better until generation 100, and, then, worsens considerably. Only offset mutation with correction shows a stable behavior and adapts without problems to the changing requirements.

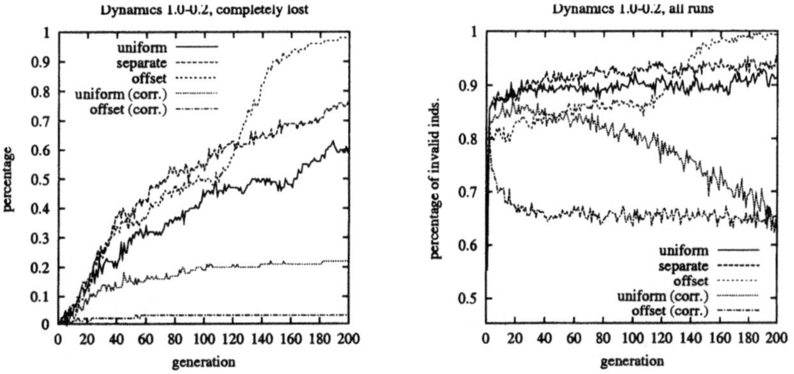

Fig. 11. Moving corridor problem with linearly decreasing strength of dynamics from 1.0 to 0.2. The results are similar to Figure 10. Only uniform adaptation with correction shows a better performance and offset mutation with correction exhibits a slightly higher number of completely lost runs.

increasing dynamics are more problematic since there seems to be a threshold where the percentage of invalid individuals does not suffice for the population size anymore.

The experiments with the moving circle problem in Figures 3, 4, and 5 show in addition that the mechanism does not hinder in tracking-only tasks.

All in all, the experiments show that there exist circumstances where standard adaptation techniques fail. The correction of strategy variables seems to be one possible remedy.

6 Conclusions

This work has shown empirically that two basic components of evolution strategies, the Gaussian pdf and the adaptation mechanisms, can have problems in dynamic environments that do not occur in stationary environments. By analyzing two simple dynamic problems indications to the reasons for those problems are found and guided the design of two new, proposed techniques. Directed mutations and the introduced correction of strategy variables seem to be at least feasible approaches to handle those problems. However, both extensions need further investigations on more general problems. Especially, the directed mutation must be evaluated on problems where the direction of the moving tracking region changes. Also it is clearly not a means to dynamic problems with a more chaotic behavior.

This clarifies that the goal of this work was not to replace the principles of standard ES mutation by new principles for dynamic problems. It rather illustrated a first attempt to tailor the mutation operator to a special class of dynamic problems. In the future, more theoretical considerations are necessary on the requirements posed by other classes of dynamic problems. This should lead in the long run to operators which are able to adapt best to different kinds of dynamics.

References

Bäck, T. (1998). On the behavior of evolutionary algorithms in dynamic environments. In *IEEE Int. Conf. on Evolutionary Computation* (pp. 446–451). IEEE Press.

Branke, J. (1999a). Evolutionary approaches to dynamic optimization problems: A survey. In J. Branke & T. Bäck (Eds.), *Evolutionary algorithms for dynamic optimization problems* (pp. 134–137). (part of GECCO Workshops, A. Wu (ed.))

Branke, J. (1999b). Memory enhanced evolutionary algorithms for changing optimization problems. In *1999 Congress on Evolutionary Computation* (pp. 1875–1882). Piscataway, NJ: IEEE Service Center.

Cobb, H. G. (1990). *An investigation into the use of hypermutation as an adaptive operator in genetic algorithms having continuous, time-dependent nonstationary environments* (Tech. Rep. No. 6760 (NLR Memorandum)). Washington, D.C.: Navy Center for Applied Research in Artificial Intelligence.

Collard, P., Escazut, C., & Gaspar, A. (1996). An evolutionary approach for time dependant optimization. In *Int. Conf. on Tools for Artificial Intelligence 96* (pp. 2–9). IEEE Computer Society Press.

De Jong, K. (2000). Evolving in a changing world. In Z. Ras & A. Skowron (Eds.), *Foundation of intelligent systems* (pp. 513–519). Berlin: Springer.

Ghozeil, A., & Fogel, D. B. (1996). A preliminary investigation into directed mutations in evolutionary algorithms. In H. Voigt, W. Ebeling, & I. Rechenberg (Eds.), *Parallel Problem Solving from Nature – PPSN IV (Berlin,*

1996) (Lecture Notes in Computer Science 1141) (pp. 329–335). Berlin: Springer.

Goldberg, D. E., & Smith, R. E. (1987). Nonstationary function optimization using genetic algorithms with dominance and diploidy. In J. J. Grefenstette (Ed.), *Proc. of the Second Int. Conf. on Genetic Algorithms* (pp. 59–68). Hillsdale, NJ: Lawrence Erlbaum Associates.

Hildebrand, L., Reusch, B., & Fathi, M. (1999). Directed mutation – A new self-adaptation for evolutionary strategies. In *1999 Congress on Evolutionary Computation* (pp. 1550–1557). Piscataway, NJ: IEEE Service Center.

Jones, T. (1995). *Evolutionary algorithms, fitness landscapes and search.* Unpublished doctoral dissertation, The University of New Mexico, Albuquerque, NM.

Liles, W., & De Jong, K. A. (1999). The usefulness of tag bits in changing environments. In *1999 Congress on Evolutionary Computation* (pp. 2054–2060). Piscataway, NJ: IEEE Service Center.

Mori, N., Kita, H., & Nishikawa, Y. (1996). Adaptation to a changing environment by means of the thermodynamical genetic algorithm. In H. Voigt, W. Ebeling, & I. Rechenberg (Eds.), *Parallel Problem Solving from Nature – PPSN IV (Berlin, 1996) (Lecture Notes in Computer Science 1141)* (pp. 513–522). Berlin: Springer.

Morrison, R. W., & De Jong, K. A. (1999). A test problem generator for non-stationary environments. In *1999 Congress on Evolutionary Computation* (pp. 2047–2053). Piscataway, NJ: IEEE Service Center.

Rechenberg, I. (1994). *Evolutionsstrategie '94.* Stuttgart: frommann-holzbog. (German)

Saleem, S., & Reynolds, R. (2000). Cultural algorithms in dynamic environments. In *Proc. of the 2000 Congress on Evolutionary Computation* (pp. 1513–1520). Piscataway, NJ: IEEE Service Center.

Schwefel, H.-P. (1995). *Evolution and optimum seeking.* New York: Wiley.

Vavak, F., Fogarty, T. C., & Jukes, K. (1996). A genetic algorithm with variable range of local search for tracking changing environments. In H.-M. Voigt, W. Ebeling, I. Rechenberg, & H.-P. Schwefel (Eds.), *Parallel problem solving from nature – PPSN IV* (pp. 376–385). Berlin: Springer.

Weicker, K., & Weicker, N. (1999). On evolution strategy optimization in dynamic environments. In *1999 Congress on Evolutionary Computation* (pp. 2039–2046). Piscataway, NJ: IEEE Service Center.

Weicker, K., & Weicker, N. (2000). Dynamic rotation and partial visibility. In *Proc. of the 2000 Congress on Evolutionary Computation* (pp. 1125–1131). Piscataway, NJ: IEEE Service Center.

GeLog - A System Combining Genetic Algorithm with Inductive Logic Programming

Gabriella Kókai[1]

Department of Computer Science, Programming Languages
Friedrich-Alexander University of Erlangen-Nürnberg
Martensstr. 3. D-91058 Erlangen, Germany
e-mail:kokai@informatik.uni-erlangen.de

Abstract. We have developed a genetic logic programming system (*GeLog*) which implements a combination of two different approaches for automatic programming: inductive logic programming and genetic algorithm. The paper presents the system and discusses its performance on a benchmark problem[1].

1 Introduction

This paper describes a genetic logic programming system. GeLog is an automatic programming system which combines the approaches of inductive logic programming (ILP, [9]) and genetic algorithm (EA, [4]). During conceptualization our major aims were focused on good and easy expansion in order to be able to deal with a wide range of problems, which often creates difficulties with already existing systems. The individuals in GeLog-System represent PROLOG programs, which can be regarded as a potential solution to a given problem. The first generation is generated by given basic knowledge. A favorable initial distribution can be reached through formulation of this basic knowledge and the algorithm for the generation of individuals. The course of evolution can therefore already be influenced at this point. For the evolutionary transformation of a population, various recombination and mutation operators are available, which can be adapted to different kinds of problems. The generated individuals are evaluated through positive and negative examples of the ILP problems. The best representatives of the newly generated individuals and in some cases also of the parental generation form the basis of the next step of evolution.

Both automatic programming methods (EA and ILP) have already proved their efficiency during work with different kinds of problems [14]. In order to combine the strengths of these two approaches, first Whigham and McKay ([16]) suggest the techniques of generalization and specialization deriving from inductive logic programming for development of recombination- and mutation operators in order to be able to realize a *directed search on the whole search space.* Several other system were developed [5], [2] and [17].

[1] This work is supported by the grants of the Bayerischer Staatsministerium für Wissenschaft, Forschung und Kunst, DAAD and Siemens

B. Reusch (Ed.): Fuzzy Days 2001, LNCS 2206, pp. 326–344, 2001.

In the remainder of this paper first a brief overview of the related works. Section 3 contains the description of GeLog, and discusses matters of its implementation. Some discussion about other existing systems can be found in section 4. In Section 5 two examples are given to demonstrate the learning ability. We evaluate the achieved results in Section 6. Finally in Section 7 a summary and outlook of future works will be made.

2 Theoretical Background

In this section the different learning methods used in this paper methods are introduced briefly.

2.1 Evolutionary Algorithms

Evolutionary algorithms are a model of natural evolution ([3]) that is often used to solve optimization problems ([4]). These algorithms are based on the collective adaptation and learning ability of *individuals*. Individuals form a *population* and each individual represents a possible solution of the problem. At the beginning, the individuals in the population are initialized randomly and they can be modified by the operators *selection*, *mutation* and *recombination*. These modifications lead to the evolution of the individuals; during the process better and better individuals appear. Thus in later generations individuals representing better solutions appear more likely. The flow chart of EA can be seen in the Figure 1.

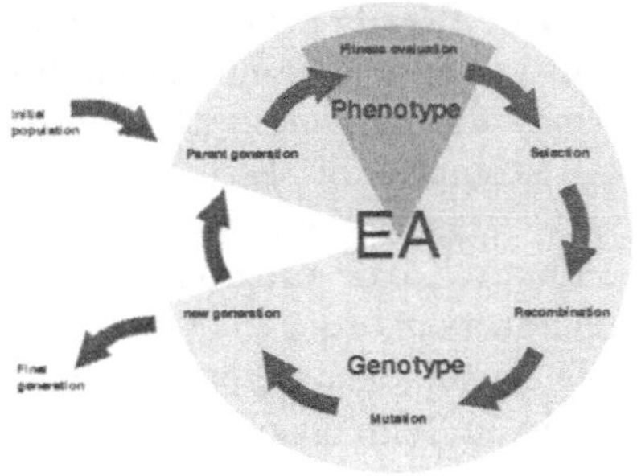

Fig. 1. The evolution process

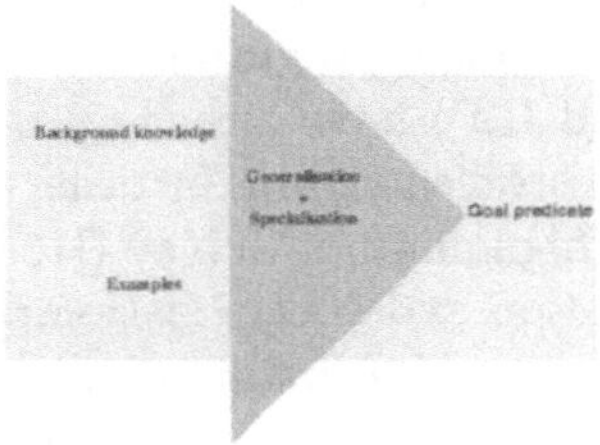

Fig. 2. The general construction of ILP

2.2 Inductive Logic Programming

Inductive logic programming (ILP, [9], [12]) is a research area lying at the intersection of inductive machine learning and logic programming. The general aim of ILP is to develop theories, techniques and applications of inductive learning from observations and background knowledge in a first order logical framework. ILP is perhaps the most general approach in the field of machine learning because of the complexity of the learned concepts. The use of programs from the logic of first order as underlying representation makes ILP systems more efficient and more useful than conventional empirical machine learning systems. ILP systems were used successfully for the solution of several problems of the real world ([10],[11]). Most of the representations in first order logic, which are used in ILP systems, are versions of the Horn clauses based on PROLOG.

The two most used operators (see Figure 2) with which *consistent* programs can be generated are *specialization* and *generalization*. Consistent is the fact that a program is *sufficient* and *necessary*. It means that the program executes a sufficient condition for positive training examples: all positive examples are derivable, and a necessary criterion for negative instances: none of those are generated by the program.

Specialization is a learning method where the required starting hypothesis stipulates that it must terminate for all positive and for some negative examples. Output is a modified starting hypothesis. The actual hypothesis is specialized as long as negative examples are no longer derivable. The *generalization* can easily regarded as the opposite direction of the specialization. There are positive examples, for which the program does not terminate. Frequently the starting hypothesis is the empty program or a program in which the errors are to be found.

3 Related Works

The following section gives a short overview of already existing work about genetic logic programming systems. Both automatic programming methods (EA and ILP) have already proved their efficiency during work with different kinds of problems [14]. In order to combine the strengths of these two approaches, Whigham and McKay ([16]) were the first to suggest the techniques of generalization and specialization deriving from inductive logic programming for development of recombination- and mutation operators in order to be able to realize a *directed search on the whole search space*. The available basic knowledge is represented in form of a context-free grammar. The individuals of the first generation are introduced from this grammar. The individuals are the derivation trees of words correspond to this grammar. The disadvantage of CFG-GP is the enormous effort for the user of the system. Beside the implementation of the fitness function it is necessary to put the crossover points manually into the attributed grammar.

In the work of Tveit ([15]) the representation form of the individuals is a vector. The length of these vectors depends directly on the number of variables used in the relation, which describes the background knowledge and the examples. The search space increases exponentially with the problem size. Another disadvantage, which results from the fact that the target clause is represented by a vector, is the missing possibility to replace the variables by terms. This system cannot be tested due to the missing implementation regarding the effectiveness.

The next system REGAL ([2]) is a distributed genetic algorithm-based system, designed for learning First Order Logic concept descriptions from examples. In REGAL the whole population is structured on three levels: population, species (subpopulations) and individuals, whereby a species concentrates on the solution of a sub-problem of the task. First the combination of selected representatives of several species of the population defines a solution for the problem.

Dogma was developed by Hekanaho ([6]). Dogma uses bit strings of fixed length as chromosomes of the genotype. An individual consists of the conjunction of predicates given in the background knowledge. The conjunction of all given predicates with all their result values represents the most general possible clause as the base of language (Language Template).

4 The GeLog System

GeLog is primarily a genetic logic programming system. The schematic structure of the GeLog system can be seen in Figure 3. In the following we give here a short introduction of the different part of the system.

4.1 Configureability of the system

As GeLog was not designed for certain types of problems we focussed on easy configureability. It is thus possible to adjust evolution without having to translate the system itself. Next to the parameters of the evolution strategy the importance of the evaluation criteria for the fitness function and the selection method can be determined in the configuration file (see Appendix 7). Newly developed operators can be integrated into the system by addition to the configuration file and unnecessary ones can be deleted. For initial creation of individuals the form of the phenotypes can be determined with the help of configuration parameters (maximal number of right hand sides per individual and maximal number of literals per right hand side of the individuals).

4.2 Phenotype and Genotype

During the course of evolution, the phenotype is the basis for the evaluation of an individual. The programming system must be capable to evaluate each individual of one generation in their "real environment" with regard to their suitability/ functionality. For this, the PROLOG programs are interpreted, the results are compared with the existing positive and negative examples. The genotype is the real representation of the individual and is stored as an object graph. The phenotype can be deduced from this and be evaluated on this basis. An individual consists of a predicate which can have various right hand sides containing logically connected basis predicates of background knowledge.

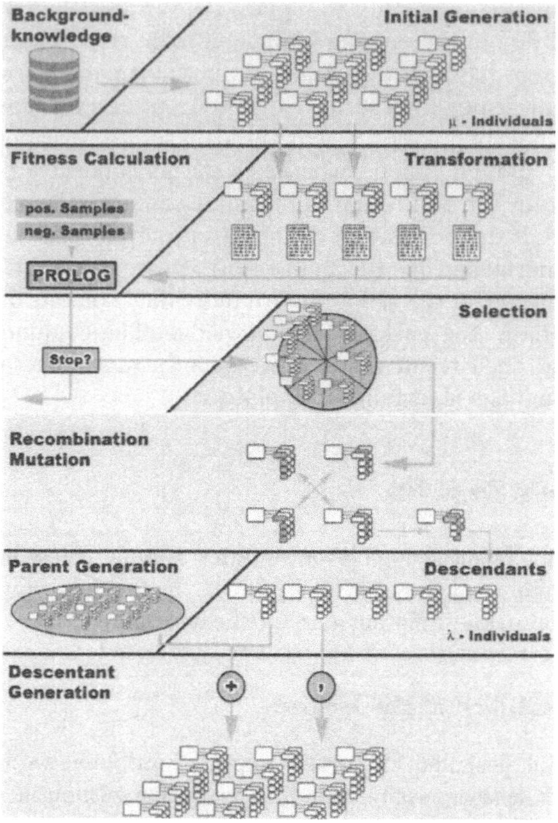

Fig. 3. Schematic structure of the GeLog system

GeLog processes three different groups of data: individuals, background knowledge and training instances. which build up the genotype and the phenotype. The following example taken from [9], can help to demonstrate the different constituents.

Example 1. Learning the family relationship
Background knowledge:

```
parent(ann,mary).        female(ann).
parent(ann,tom).         female(mary).
parent(tom,eve).         female(eve).
parent(tom,ian).
```

Positive examples:

```
daughter(mary,ann).
daughter(eve,tom).
```

Negative examples:

```
daughter(tom,ann).
daughter(eve,ann).
```

The following clause is a correct solution :

```
daughter(X,Y) :- female(X), parent(X,Y).
```

Depending on the background knowledge and the complexity of the target concept the target predicate definition may consist of a set of clauses such as:

```
daughter(X,Y) :- female(X), mother(X,Y).
daughter(X,Y) :- female(X), father(X,Y).
```

Let speaks first about the individuals. In connection with GeLog an individual represents a logic program, which is after the end of the evolution run in the sense of inductive logic programming complete and consistent. In order to generate such a program with the methods of the , it was clear that the code of the PROLOG program cannot be stored as an alphanumeric string since this gives no information about the structure of its contents. For the operators such a character string would only be controllable only with very high effort, because for each application a parsing of the string would be necessary to make meaningful and syntactically correct modifications.

By the analysis of the PROLOG clauses the following constituents which appear in similar form in the data structure were identified (see Figure 4):

- The *goal* represents the searched clause, which can be divided into a *head* and a *body*. The head has only got the goal literal. The body is formed by a disjunctive linked set of base literals.
- The base literal (BL) is defined by its arity and its name. The variables in the literals can be free or bound.
- Free variables are not bound to a value.
- A fixed value is assigned to bound variables.

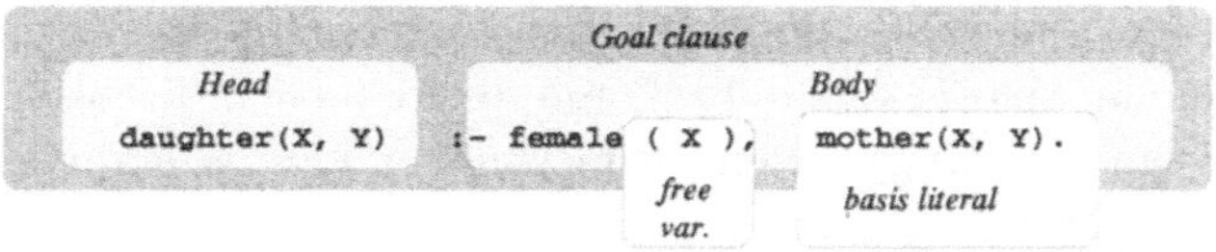

Fig. 4. Constituents of a PROLOG clause

On the assumption that the base literals which are created from the background knowledge are syntactically correct, any BL can be assigned to the body of the target. In the following conjunctions of base literals are called right side of the clause (RHS).

Concretely, it can be said that the genotype of a individual consists of so-called object graphs implemented in C++.

The partition of the constituents of a PROLOG clause was transferred directly to the class structure for the representation of the genotype. A target or an individual is an object, which differs from the other individuals within a population by a unique numeric identifier. All individuals together possess the same name and arity (name and arity of the target clause). Furthermore an individual contains one or more references on objects of the class RHS. A right hand side again contains references on any number of base-literals (see Figure 5).

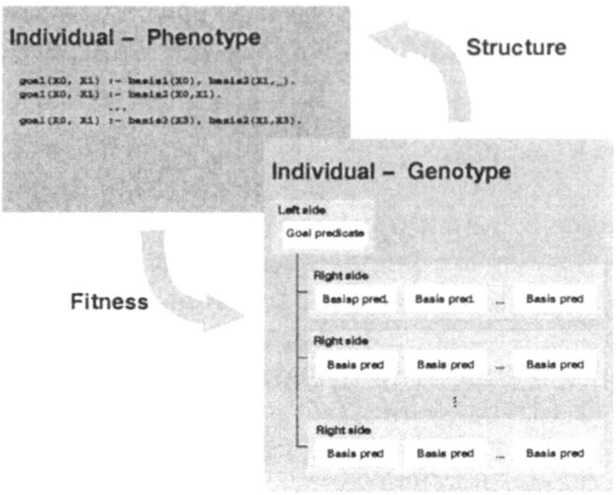

Fig. 5. The representation of the individual in Gelog

To formulate background knowledge all possibilities of PROLOG can be used. As we mentioned earlier it assumes that the clauses and facts are presented in syntactically correct form in the background knowledge, since they are used in the further process of the evolution also for the evaluation of the fitness.

As we mentioned GeLog uses a configuration file in order to define important parameters of the evolution cycle and the background knowledge. It can be the same for each population, but with different configuration files the evolution parameters and the set of the basic building elements for each population contained in it can be changed. Defined mechanism can be obtained by such a methodology as *seeding.*

If parts of the search area which promise a larger success of the search are already well-known before the evolution run, seeding mechanisms permit the production of individuals in these regions. By the use of more success promising base-literals and the specification of the build-in probability to each of the basic literals from the background knowledge the user can influence the seeding.

For the production of the configuration files the background knowledge must be converted into another form, so that it can be used during the generation of the initial population. For the recognition of a genotype, the constituents of clauses from the data structure used for the representation of an individual are necessary. to produce a genotype can be recognised. The name and the arity of the base literals flow into the configuration file from the background knowledge. It needs to be mentioned that no information about possible values that may be assigned to the parameters are stored here.

Clauses and facts contained in the background knowledge are generalized in this way since the bound variables are no more assigned to a value and in the representation if free variables appear. The allocation of values of variables takes place only in a later phase of the initialization process.

The training instances serve to test the quality of a produced program. It can happen that the phenotype of an individual is applied to an example and that the result is compared to the instance admitted one. GeLog differentiates only between two results. That also means it gives also only two classes of training instances, i.e. ***positive*** and ***negative***, which the program divides into ***true*** or *false* in each cases. In order to be able to identify a training instance, the positive and negative instances have a serial-number, in both cases beginning with zero. It can thus be said that an example in the training set consists of the class affiliation, the number and as many further values as the target descriptor has parameters. For the target predicate `daughter/2` two examples can be formulated as follows:

Example 2. Examples `daughter/2` in GeLog system

```
gelog_example(1, 0, 'marry', 'anne').
gelog_example(1, 1, 'eve', 'tom').
gelog_example(0, 0, 'tom', 'ann').
gelog_example(0, 1, 'eve', 'ann').
```

Because of the dualism in the representation of the individuals, it is necessary to transform the individuals as well as the background knowledge and the training examples. But it is never necessary to do it in both directions, since only the genotype of the individuals changes during the evolution run (see Table 1).

	Genotype → Phenotype	*Phenotype → Genotype*
Individuals	Necessary, occurs automatically during the evolution run	Never necessary because genotype is the primary representation
Background knowledge	Only necessary as part of an individual	Only necessary for basis predicates, occurs manually with the configuration
Training example	Never necessary, because the genotype is the primary representation	Never necessary, because it is not used in a genotype form

Table 1. Transformations between geno- und phenotype

The transformation of phenotype to genotype means that the PROLOG program code must be transformed into one `C++` object graph.

Table 1 makes clear that we only need it for the background knowledge. It must be available on the one hand to the fitness evaluation, on the other hand part of it serves as base predicate. These must be available as building block of the genotypes of an individual. There are several reasons against an automatic procedure, which led to the decision that the base predicate must be specified still once explicitly and central in the configuration file:

- Beside names and arities of the descriptor the a probability needed for the seeding must be indicated in each cases.

- In order to extract the base predicates automatically from the background knowledge available as PROLOG program codes, a PROLOG Parser would be necessary, if one does not want to introduce an unnecessary restriction for the formulation the background knowledge.
- Standard PROLOG predicates or such from program libraries can also serve as base descriptors. These are not however always present in the source text.
- The base predicate are usually a selected part of the whole set of predicates defined in the background knowledge. Therefore each modification in the set of the base predicates would make necessary a modification in the PROLOG program code.

The additional effort of the manual and double maintenance and error sources possibly resulting from it did not result problematic in practice.

The reverse direction of the transformation, of geno- to phenotype, means to transform a graph of C++ objects into a syntactically correct PROLOG program. In the Table 1 it can be seen that the only necessary transformation is the conversion of the individuals.

4.3 Generation and Population

Generally in evolutionary algorithms individuals are gathered to form a generation, the different generations are designated to a population. New individuals are created through evaluation of a generation, selection, recombination and mutation. These new individuals form the new generation within the population.

Concretely in GeLog the following evolutionary algorithm [8] was used: $\lambda_1(\mu_0/\rho_0 \overset{+}{,} \lambda_0||\Omega_0)^{\gamma_0}$ Each λ_1 population contains μ_0 individuals. For the reproduction cycle, ρ_0 multiplied with λ_0 parental individuals are selected and recombined with every new ρ_0. The descendants that are generated are modified by mutation operators and inserted into the next generation according to the plus or comma strategy chosen. Each population lives up to γ_0 generations. The parameter Ω_0 describes the way the operators work and can be modified during the evolution process.

4.4 Operators

Different recombination operators allow to interchange whole right hand sides as well as parts between different individuals (crossover). Mutation operators, on the contrary, only work on one individual. These operators can be divided into three categories, depending on if they add new knowledge (new program code), delete knowledge or modify existing knowledge. Concretely, basic predicates are added, deleted or their variables can be modified.

The definition of these operators bases on techniques developed in the field of ILP for generalization or specialization of the achieved knowledge. However, it is usually not possible to predict, which operator will be assigned to which one of both classes, as the renaming of a variable can be a generalization as well as a specialization.

The application probability and intensity of an operator as well as their modifications during evolution are determined at the beginning. The mutation's rate can therefore be regulated and adapted in accordance with the nature of the problem when the search area is scanned. For recombination two operators are available that operate on the levels of right sides and literals:

- *RecombineRHSOperator* executes a crossover from whole right sides. That means that in both lists of right sides of two individuals a crossover point is determined randomly in each case, and the two segments are exchanged according to the crossover points between the individuals. In the case of a universal crossover, several crossover points are determined and each second intermediate paragraph is exchanged (Figure 6).

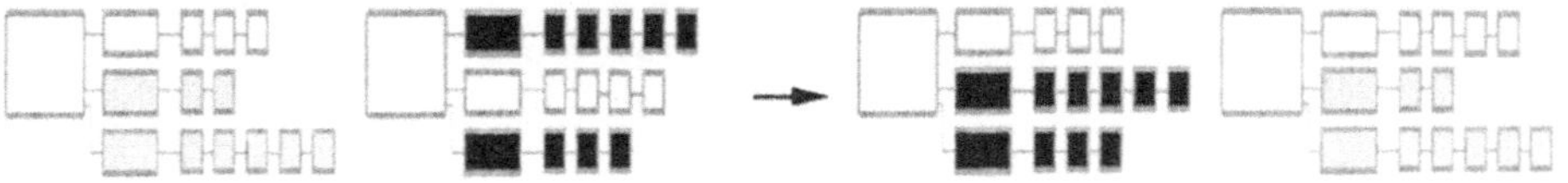

Fig. 6. Recombination of the right sides of two individuals

- *RecombineTwoRHSPredicateOperator* operates one level deeper and executes a crossover between literals of two right sides of different individuals. Universal crossover is also possible. This operator is applied randomly to several pairs of right sides. (Figure 7).

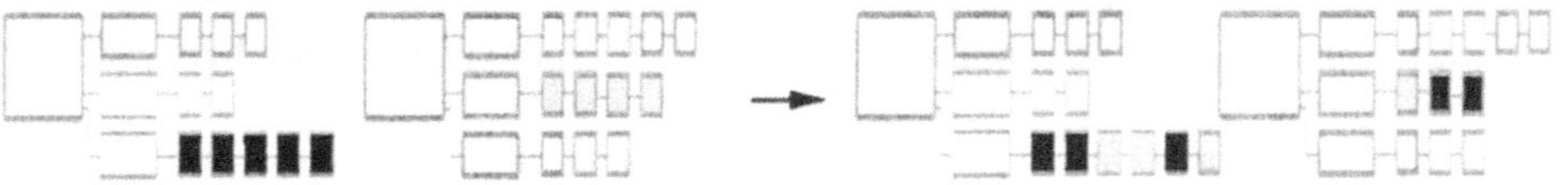

Fig. 7. Recombination of the right sides of two different individuals

For the mutation seven operators are available that operate in the levels of right sides, literals and variables. Here, the generalization and specialization methods from inductive logic programming appear. It cannot be forecasted for each operator whether its application on a concrete individual will lead to a specialization, generalization or to neither of them.

- *InsertNewRHSOperator* inserts new right side, which contains a random selected base predicate. It can work as generalizing by inserting a further disjunction.
- *DeleteRHSOperator* deletes a right side and can work as specializing.

- *InsertNewPredicateOperator* inserts a random selected base descriptor into a right side. It can cause a specialization by inserting a conjunction .
- *DeletePredicateOperator* deletes a base predicate from the right side. It can produce a generalization.
- *InsertNewLiteralOperator* inserts a new variable designator. It is assigned to a randomly selected parameter of a base descriptor. It can work as generalizing, if it replaces a once only used and thus free variable with a possibly bound one (Figure 8).

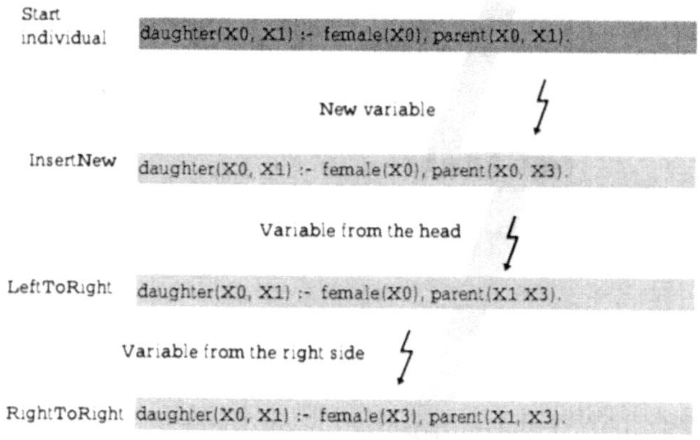

Fig. 8. Mutations on level of the variables

- *SubLeftToRightOperator* assigns a variable from the head of the clause to a random selected parameter of a base predicate. The effect always depends on the meaning of the replaced variable and can lead to specialization, or to the generalization of the clause.
- *SubRightToRightOperator* assigns any variable used in a base predicate to a random selected parameter of a base predicate. It can therefore behave as SubLeftToRightOperator by specializing or generalizing or it can also have no effect at all.

Example 3. Operators in GeLog
GeLog- Generalization

- Delete one right side:

```
daughter(X0, X1) :- female(X0), mother(X0,X1).
                 :- female(X1), father(X1, X0).
```

$\Downarrow$

```
daughter(X0, X1) :- female(X0), mother(X0,X1).
```

- Delete one literal:

```
daughter(X0, X1) :- female(X0), mother(X0,X1).
                 :- female(X1), father(X1, X0).
```

⇓

```
daughter(X0, X1) :- female(X0).
                 :- female(X1), father(X1, X0).
```

– Adding a new right side :

```
daughter(X0, X1) :- female(X0), mother(X0,X1).
```

⇓

```
daughter(X0, X1) :- female(X0), mother(X0,X1).
                 :- .
```

– Renaming a variable:

```
daughter(X0, X1) :- female(X1), father(X1, X0).
```

⇓

```
daughter(X0, X1) :- female(X1), father(X1, _).
```

GeLog-Specialization

– Adding a new literal from background knowledge:

```
daughter(X0, X1) :- female(X0).
                 :- female(X1), father(X1, X0).
```

⇓

```
daughter(X0, X1) :- female(X0), mother(X1, X0).
                 :- female(X1), father(X1, X0).
```

– Renaming a variable:

```
daughter(X0, X1) :- female(X0), mother(X0,_).
```

⇓

```
daughter(X0, X1) :- female(X0), mother(X0,X1).
```

4.5 Selection and fitness

For evaluation, the genotypes represented as object graphs are transformed into the corresponding phenotype and evaluated by the PROLOG interpreter. A criterion for the fitness of an individual is the number of positive solutions that where achieved. An individual is regarded as *ideal* for this criterion, if all positive and no negative example can be calculated through this criterion. Furthermore, the length of the program code as well as the time needed to calculate the solutions of the predicate are integrated into the fitness function.

In more detail we receive a linear fitness function through weighting of the values with a freely configurable factor and summing up:

$$\eta(s_i) = \sum_c a_c\, \eta_c(s_i) \qquad c \in \{pos, neg, rhs, lit, var\}$$

Where a_c is the weighting coefficient for the fitness function η_c. The index c means for one of the following environmental conditions:

- *pos* The completeness η_{pos} is determined as the proportion of the correctly classified positive training instances, related to their total number. Therefore value is situated in the interval of [0..1].
- *neg* The consistency η_{neg} is defined similarly to η_{pos}, the proportion of correct-proves negatively classified negative training instances.
- *rhs* The number of different right sides of the individual η_{rhs} represents the number of the outside disjunctions in the target predicate. It suggests how many comparisons a PROLOG interpreter at least needs in order to be able to declare evaluate that a descriptor must be regarded as false.
- *lit* On the other hand η_{lit} indicates the number of base predicates used in the entire individual. It represents an upper limit for the necessary comparisons during the analysis of the predicate.
- *var* The value of η_{var} indicates the number of all used different variables per right side. This should give, at least when homogeneous individuals are concerned, information on the necessary storage space during the analysis.

While the larger values for the completeness η_{pos} and the consistency η_{neg} are equivalent to higher fitness, η_{rhs}, η_{lit} and η_{var} behave differently. Thus the coefficients, a_{rhs} and a_{lit}, a_{var} are always negative so that the total η is reduced by higher complexity values. In the configuration file all coefficients are to be indicated as positive and are then accordingly negated. Now it is clear that even an ideal solution of the function will not achieve 100% of the theoretically possible fitness. This is justified by the fact that, dependent on the task, in the case of meaningful individuals values are to be expected to be at least greater than one in each case for the complexity mass and will therefore never have the optimal value zero.

A shift of the zero point for the complexity measure, so that an individual without right side is no more optimal, but one of expected size, would imply a more exact conception of the produced programs. In practice this quite simple pattern of a fitness function proved useful. Moreover, especially in case of adapted problems, the possibility of implementing fitness functions themselves also exists.

Gelog can distinguish between the selection of child individuals or the parental and child individuals respectively (comma, plus strategy). For selection of the individuals for the reproduction cycle, various selection mechanisms are at disposal: rank, proportional to fitness, elite, best. The first three methods are so-called probabilistic methods which are quite common for genetic algorithms. The selection of the best, however, is mainly used to select the surviving parental individuals when the plus strategy is applied.

5 Example

Below we want to comment the results GeLog delivered for a concrete example. The problem is to recognize certain kinds of mushrooms as poisonous due to certain characteristics. The example was borrowed from the Mushrooms Database of the UCI Machine Learning Repository [1]. We decided to choose this problem, because even after generalization there is still a relatively huge amount of background knowledge and a great number of positive and negative examples. The characteristics of a mushroom are described by a vector with 22 elements of nominal values. This example consists of 8124 different descriptions of mushrooms, 48, 2% of which are classified as being poisonous. There are various simple solutions to this problem which can recognize correctly nearly all examples ([13], [7]). As background knowledge, GeLog has 126 different basic predicates which test one specific characteristic of the mushroom. The goal predicate is `poisonous/1`. The parameter stands for a vector with the description of the mushroom. The evolutionary algorithm was applied with different selection mechanisms and comma and plus strategy. Table 2 shows the achieved fitness percentage. The results stem from five evolution runs per configuration. Beside strategy and selection all other parameters are kept constant.

configuration		best error rate		midle error rate	
strategy	selection	generation	individual	generation	individual
plus	prop	26.55%	3.10%	31.81%	3.72%
plus	rank	18.65%	1.03%	24.52%	3.04%
plus	elite	*11.94%*	*0.30%*	*17.55%*	*2.39%*
plus	best	29.53%	2.66%	33.36%	3.90%
comma	prop	37.28%	4.43%	43.39%	10.36%
comma	rank	28.32%	2.51%	33.91%	6.67%
comma	elite	13.68%	3.35%	18.47%	4.28%
comma	best	43.50%	3.05%	48.16%	13.22%
plus[1]	elite	25.14%	3.55%	27.98%	3.66%
plus[2]	elite	20.47%	1.77%	28.38%	2.90%

[1] μ_0=300 und λ_0=275, [2] extended background knowledge

Table 2. Result on the Mushroom Database

The diagrams 9 and 10 show in each case the averaged error rates for the best individual and the average value of the generation.

In none of the tests both a complete as well as a consistent solution can be found. However, with each configuration version at least one complete or consistent solution was determined. The differences between the configurations can therefore show, above all, how far the criterion (completeness or consistency) could still be achieved, if the other one has already been fulfilled.

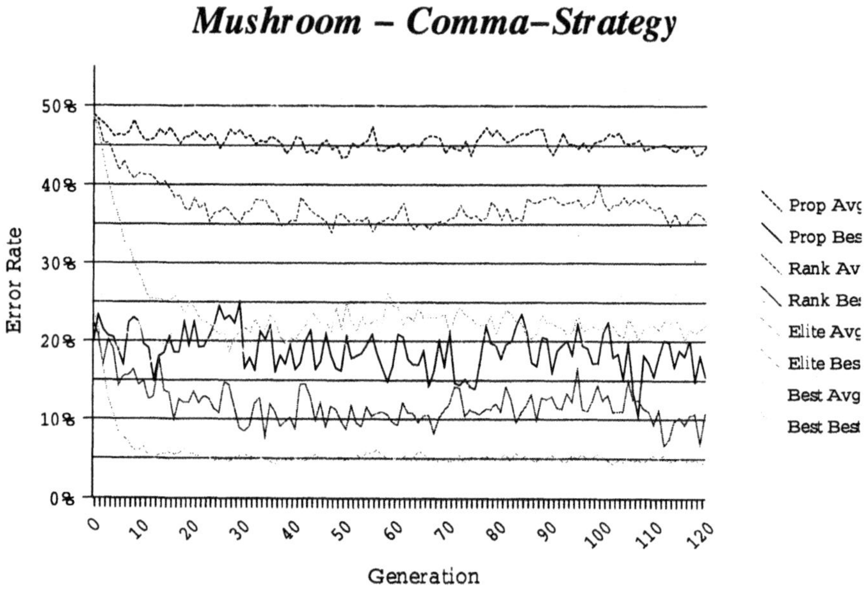

Fig. 9. The comma-strategy by Mushroom data set

The results achieved by the runs with comma-strategy (Figure 9) show as that expected result which reached a good solution can be lost again. The minimums of the error rate hold themselves only one generation and go lost then through recombination and mutation again. The individuals produced at these places fulfill usually one criterion to 100%. The other one lies in this case however only rarely over 97%. Obviously it is in this example an advantage if an individual which is either complete or consistent is as long as in the genepool available, until from that a better solution is developed. The optimization process runs sequential first for the one and then for the other criteria whereby it cannot say, which is first maximized. This can be recognized based on the result obtained with the plus-strategy (Figure 10). It is clearly to see that the error ratio of the best individuals is monotonously falling. One 100% reached criterion will retain and the value of the other criterion is subsequently improved.

It can be observed that the proportional and rank based selection lead to very similar results, whereby with progressive generation number the rank based one reaches somewhat littler error rate. The average error rate of the generation lies by rank based selection obviously lower, The best selection converges more slowly and obtain the worst result within the 120 generations. But the parameters are not advantageous selected for this selection procedure. The reason is that from 50 individuals always exactly the best 45 ones are selected two times for the recombination, good genes can only hard achieve and the algorithm works less precise. However the best-selection can compete with the other methods if the population or λ_0 are smaller. Both the fastest convergence as well as the best final results obtain the elite selection. Already at the beginning the optimization by the relatively high-value elite individuals can be very accomplished. On the one hand it provides the evolution more time for little improvement. On the other hand the best solutions do not dominate at the end as strongly as in proportional and rank based selection. Even in the comma strategy with the elite-selection acceptable solution are obtained. The curve for the best individual shows here a clearly smoother progress than by the other methods. Even the average of the generation with elite selection reaches error ratios in the area of the best individual of other selection procedures.

The incrementation of the size of the population μ_0 on 300 with a λ_0 of 275 has no advantages over the smaller population. Expanding the background knowledge with negated form of the previous basis predicates, enable to formulate negative solutions. Thus a certain value of a feature can be explicitly excluded. The test delivered similar results in this configuration as before, however no better individual with the help of the expanded background knowledge is located.

The duration of the evolution depends essentially on the amount of the generation cycles, the amount of individuals, and the amount of the example instances. The shown evolutions required respectively about 15 minutes computation time on an ATHLON™/600-System.

6 Conclusion and Future Work

This paper describes a genetic logic programming system. GeLog is an automatic programming system which combines the approaches of inductive logic programming and evolutionary algorithm. GeLog is a universal tool for work with problems in the ILP area. A independent representation of knowledge was therefore implemented, which can work with generally formulated genetic operators and selection functions. Experimental results in mushroom domain show the potential of this new approach.

Because of the various possibilities for parameterizing of GeLog a systematic logging and editing of the evolution flow would be useful in order to determine an optimal configuration. A statistical analysis and visualization of the evolution processes would be interesting. An automatic adaptation of the parameters would be desirable. This could take place similarly as with the evolution strate-

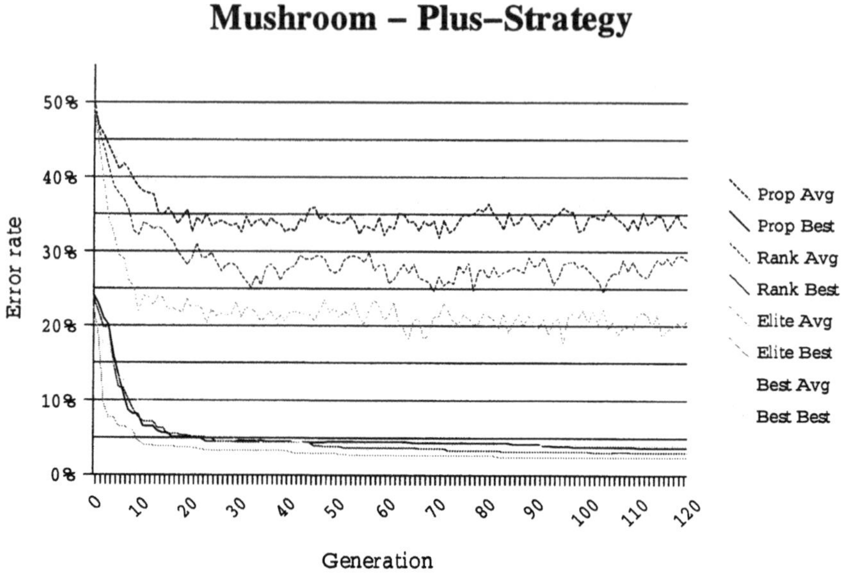

Fig. 10. Plus strategy by Mushroom data set

gies by means of the actual evolution run, as the parameter values themselves are parts of the genotypes. Alternatively, several evolution runs could be regarded again as individuals of a superordinate evolution. For the practice, some tools would be useful for the editing of the examples and the background knowledge, as well as for the systematic execution of evolution runs with different configurations. These could be implemented as components of GeLog or as independent tools. Furthermore, the possibility to include other phenotype areas such as LISP or JAVA instead of PROLOG and the resulting consequence for the efficiency of the system and the programs should be examined.

References

1. C. Blake and C. Merz. UCI Repository of machine learning Databases, University of California, Irvine, Dept. of Information and Computer Sciences, 1998. http://www.ics.uci.edu/~mlearn/MLRepository.html.
2. I. Bratko and M. Grobelnik. Inductive learning applied to program construction and verification. In *Proc of the 3rd International Workshop on Inductive Logic Programming*, pages 279– 292, 1993.
3. C. Darwin. *On the origin of species*. Murray, London, 1859.
4. S. Džeroski. Handling noise in inductive logic programming. Master's thesis, Faculty of Electrical Engineering and Computer Science, University of Ljubljana, Ljubljana, Slovenia, 1991.

5. J. Hekanaho. Dogma: A ga-based relational learner. In D. Page, editor, *Proceedings of the 8th International Conference on Inductive Logic Programming*, volume 1446 of *Lecture Notes in Artificial Intelligence*, pages 205–214. Springer-Verlag, 1998.
6. J. Hekanaho. *An Evolutionary Approach to Concept Learning*. Ph.d thesis, Department of Computer Science Abo Akademi University, 1998.
7. W. Iba, J. Wogulis, and P. Langley. 73-79. In *Proc. of the 5th Int. Conf. on Machine Learning*, page 1988, A. Arbor, Michigan, Trading off Simplicity and Coverage in Incremental Concept Learning. Morgan Kaufmann.
8. C. Jacob. *Principia Evolvica - Simulierte Evolution*. dpunkt Verlag, 1997.
9. N. Lavrac and S. Dzeroski. *Inductive Logic Programming Techniques and Application*. Ellis Horwood, 1995.
10. R. Mooney. Inductive logic programming for natural language processing. In S. Muggleton, editor, *Proceedings of the 6th International Workshop on Inductive Logic Programming*, volume 1314 of *Lecture Notes in Artificial Intelligence*, pages 3–24. Springer-Verlag, 1996.
11. S. Muggleton, A. Srinivasan, R. King, and M. Sternberg. Biochemical knowledge discovery using Inductive Logic Programming. In H. Motoda, editor, *Proc. of the first Conference on Discovery Science*, Berlin, 1998. Springer-Verlag.
12. S. H. Muggleton, editor. *Inductive Logic Programming*. Academic Press, New York, NY, 1992.
13. J. S. Schlimmer. Concept acquisition through representational adjustment. Technical Report 87-19, Department of Information and Computer Science, University of California, Irvine, CA, 1987.
14. L. R. Tang, M. E. Califf, and R. J. Mooney. An experimental comparison of genetic programming and inductive logic programming on learning recursive list functions. In *8th Int. Workshop on Inductive Logic Programming*, 1998.
15. A. Tveit. Genetic inductive logic programming. Diplomarbeit, Norwegian University of Science and Technology, NTNU, Dezember 1997.
16. P. A. Whigham and R. I. McKay. Genetic programming and inductive logic. Technical Report CS14/94, University College, University of New South Wales, Australia, 1994.
17. M. L. Wong and K. Leung. Genetic logic programming system: Inducing logic programs with genetic algorithms. *IEEE Expert/Intelligent Systems and Their Applications*, 10(5), 1995.

7 Appendix: The configuration file for the mushroom example

```
# Goal predicate:
search_for = poisonous;1

# Individuals of the initial generation 0 included max.:
max_pred_rhs = 3                    # Base predicate per right side
max_rhs = 5                         # right sides

# Strategy parameters:
pop_size = 100                      # μ0
strategy = plus                     # Strategy: plus / comma
descendants = 45                    # λ0
max_age = 120                       # γ0
```

```
# Weights of the fitness criterions:
fit_pos = 500                   # correct classified pos. example.
fit_neg = 500                   # wrongly classified neg. example.
fit_rhs = 1                     # number of the right sides
fit_var = 0                     # number of variables
fit_pred = 1                    # number of base predicate
# Selection function:
selection=rank                  # alt: prop, elite;n, best
# Mutation operators:
SubRightToRight
mutop = InsertNewPredicate;0.5;0.5
mutop = DeleteRHS;0.5;0.5
mutop = InsertNewRHS;0.5;0.5
mutop = DeletePredicate;0.5;0.5
# Recombination operators:
recop = RecombineTwoRHSPredicate;0.5;true
recop = RecombineRHS;0.5;true
# Training set - PROLOG Data:
examples = ../samples/Mushroom.Examples.pl
posExamplesCount = 3916         # Number of pos. examples.
negExamplesCount = 4208         # Number of neg. examples.
# Background knowledge - PROLOG Data :
background = ../samples/Mushroom.background.pl
# Base predicate:
sample = cap_shape_b;1;0.8
sample = cap_shape_c;1;0.8
[...]
sample = habitat_d;1;0.8
```

Decision Strategies in Evolutionary Optimization

Arita Takahashi[1] and Arkady Borisov[2]

[1] Institute of Information Technology, Technical University of Riga,
1, Kalkyu St., Riga LV-1658, Latvia
arita.takahashi@delfi.lv

[2] Institute of Information Technology, Technical University of Riga,
1, Kalkyu St., Riga LV-1658, Latvia
aborisov@egle.cs.rtu.lv

Abstract. This paper describes a sequel of the previous experiments on real number genetic algorithm behavior [2], [3]. A particular example of multi-criteria optimization is discussed. The behavior of the two previously explored genetic algorithms is compared with a simple evolutionary algorithm. The main idea of the experiments is to stimulate the algorithm to find the Pareto set without measuring dominance and non-dominance. The implication of maxi-min decision method is affecting optimization so that the final solutions lay closer to the Pareto set than those obtained without any decision method. This theoretical concept is tested and analyzed graphically by picturing populations after a certain number of generations. The differences in the algorithm behavior and causes of such differences are explained.

1 Introduction

As a rule, multi criteria optimization tasks are extremely complicated. The main reason for this is that usually no solution can be seen as supreme to others. But finding a region of good solutions raises another problem: what region exactly contains the best solutions? One may think the Pareto set is a very clear representation of good solutions for an algorithm to find. Although the indication of the Pareto set is very strict from the mathematical point of view, the physical borders of the Pareto region are gradual, not abrupt. This study of genetic algorithm behavior was started not to find reward mechanisms for non-dominated and punishments for dominated solutions. The research was made to clarify what part of the solution space the genetic algorithm will accept as preferable by itself.

1.1 Problem Formulation

Let us assume that two criteria have to be maximized simultaneously as shown in Fig.1, case a. An assumption is made in this study that both criteria have the same

B. Reusch (Ed.): Fuzzy Days 2001, LNCS 2206, pp. 345–356, 2001.

scale of measurement. They depend on two parameters, x_1 and x_2 (see the mathematical expressions (3), (4) and (5) at the end of this paper).

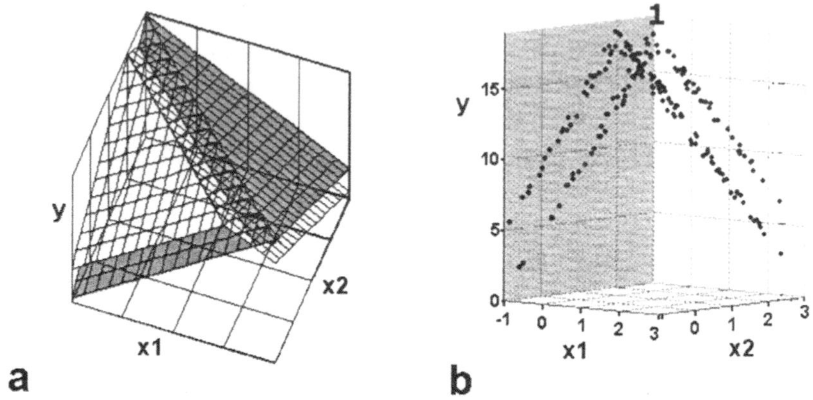

Fig. 1. The two criteria: (a) mathematical functions, (b) initial population cloud

Value Y is one of the criterion values. It will be identical to the first criterion's functional value $f_1(x_1,x_2)$ when a random number generated specifically for Y definition is $n_{rand}<0.5$, and it will be identical to the second criterion's functional value $f_2(x_1,x_2)$ when the random number is $n_{rand}\geq 0.5$. Fig. 1 (see case b) shows randomly generated initial population for maximization of the criteria displayed in Fig.1, case a.

Each individual of the population has only one fitness value Y. The problem is that the same individual will get a different fitness value each time when the individual's fitness switches from one criterion to another. Using of maxi-min strategy was the idea for the solution of the problem. The hypothesis should be approved or disapproved by experiments.

1.2 Plan of Experiments

To analyze the effect of the decision strategy more correctly, three different approaches were compared:

1. Maximization with no decision strategy used;
2. Maximization with maxi-min strategy;
3. Maximization with maxi-max strategy.

Three different algorithms were tested:

1. The classic genetic algorithm based on the crossover of real numbers, mutation, random selection and random coupling, described in [2];

2. The restricted genetic algorithm built on the crossover of real numbers, mutation, random selection and purposeful coupling, the specification of which is given in [3];

3. The evolutionary algorithm based on real number mutation operator and random selection operator. We will call this algorithm evolutionary because it has no crossover operator in it.

2 Decision Methods

There are three main decision strategies one may use in multi-attribute solution space [1]: dominance/non-dominance strategy, maxi-min strategy and maxi-max strategy. In the experiments described in this paper only maxi-min and maxi-max methods were used. The maxi-min method or the strategy of pessimist builds its outcome only on the worst attribute values, whereas the maxi-max method or the strategy of an optimist considers only the best attribute values.

3 Evolutionary Strategies

Wright [4] describes the nature of a crossover operator in the case when the phenotypes of binary strings are real numbers of parameter values. It can be seen that if two binary strings are crossed over in the place that separates the two representations of real values, the so-called real crossover occurs. It means the strings of real parameters may be used instead of those of binary symbols. For example, if the crossover had been performed between the first and the second parameter, then the two parents would have obtained the following children due to the crossover:

$$\begin{matrix} Parent\ 1 = (x_1, x_2, \ldots x_n) \\ Parent\ 2 = (x'_1, x'_2, \ldots x'_n) \end{matrix} \Rightarrow \begin{matrix} Child\ 1 = (x_1, x'_2, \ldots x'_n) \\ Child\ 2 = (x'_1, x_2, \ldots x_n) \end{matrix} \quad (1)$$

3.1 Strategies of Real Crossover

For better understanding, let us clarify the case of real crossover in two-parameter space. For example, if parameters of one parent are (2.5, 0.15) and parameters of the other parent are (3.7, 0.4) then these two children may be obtained:

$$\begin{matrix} Parent1 & (2.5, & \vert 0.15) \\ Parent2 & (3.7, & \vert 0.4) \end{matrix} \Rightarrow \begin{matrix} Child1 & (2.5, 0.4) \\ Child2 & (3.7, 0.15) \end{matrix} \quad (2)$$

In parameter space children are located on apices of parallelepiped. In the world of living beings, the offspring variety seems immeasurable. To make a model of the evolution of living beings the following definition of an offspring was introduced: For phenotypes $X=(x_1, x_2, x_3, \ldots x_n)$ and $X'=(x'_1, x'_2, x'_3, \ldots x'_n)$ a descendant is the phenotype whose parameters in the real numbers domain lay between the parameters of its parents. Therefore the first co-ordinate of the descendant is a random number lying between x_1 and x'_1, but the second co-ordinate is a random number between x_2 and x'_2 and so on. The potential identities of a child in a two-parameter case are uncountable as they occupy the whole space between the parameters' values of the parents (see Fig.2.). On the other hand both parents' coordinates determine limitations or boundaries for every parameter.

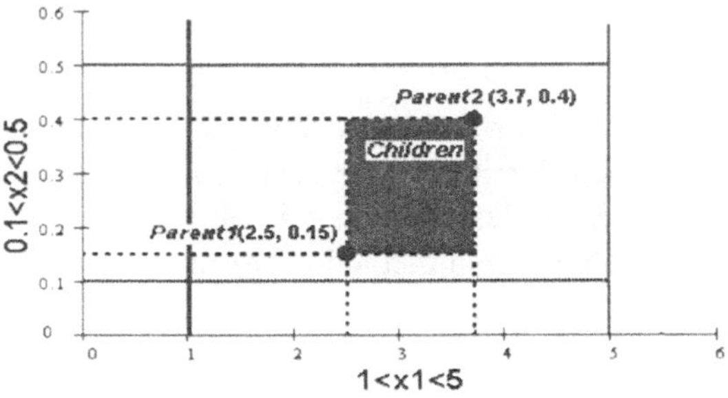

Fig. 2. The area of potential identities of a child

3.2 Mutation of Real Parameter

The mutation operator is executed by shifting the real number. The direction of the shift is random, there are no rules like "if parameter x_1<$value_1$ then its shift is positive, otherwise – negative". Mutation is uncontrollable, but random parameter values vary randomly. There are two aspects of mutation operator that have to be taken into account when creating a genetic algorithm based on genetic operators:

- By executing mutation operator a new step of search is done and a good solution may be obtained.
- Mutation operator may cause a loss of a very good solution.

In the case when the algorithm is based on the crossover operator depicted in Fig.2, there may be no need for mutation because the shift of parents' parameters is included in the definition of the crossover operator. On the other hand, there is a lower possibility of loosing a good solution without obtaining a better one or at least the one as valuable as the lost. The crossover operator is not an uncontrollable shift as it is limited by parents' parameter values. If all population individuals are crawling towards better solutions and if the space of better solutions occupy a settled place in the parameter space then the population shrink and the occupied area represent the surroundings of the best solution (see Fig.3).

Crawling in the right direction is achieved by using operators of selection and crossover. In the first approximation the surroundings of the best solution cover all the parameter space. In the next approximations it includes only the area occupied by the population. To represent this process more visually, the 1st, 8th, 16th and 50th generations of the population in two-parameter space are shown in Fig.3.

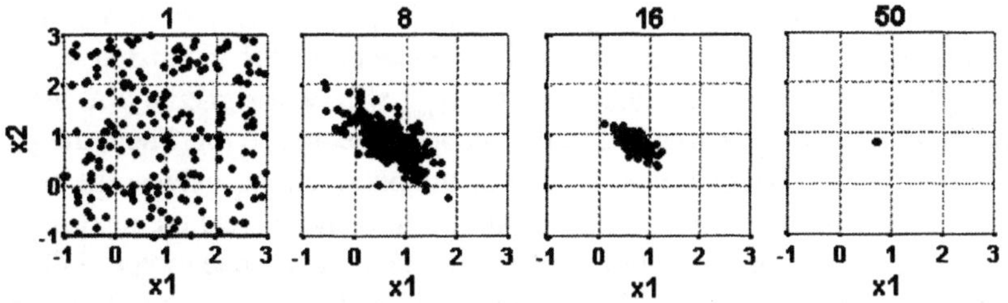

Fig. 3. Shrinking of the population towards the best solution (classic genetic algorithm, population size is 200, maxi-min strategy)

A special case may occur when both parents have identical parameters. This leads to an impasse because all the children produced by them are copies of the same solution. As a result of this, two negative aspects may appear:

- The stage of search for better solution is impossible.
- Assimilation process can take place.

Hence the following corrections can be recommended to improve the convergence of the algorithm:

1. Obtain a child by random initialization of a new individual independently from parents;
2. Shift by mutation one parameter of one parent.

Afterwards both parents and children participate in the selection of the next generation.

In an evolutionary algorithm, which does not use crossover, the only operator for producing a child is mutation. In two-parameter space one parameter is altered randomly while the other remains unchanged. Afterwards both parents and children have to compete for selection.

3.3 Purposeful Coupling

The analysis of marriage models in the human environment shows a variety of restrictions. For instance, in geographical terms, there will be a biological restriction imposed by the human race: there are individuals who are unable to get accustomed to very cold weather conditions because their body is adapted to life in a hot climate, and vice versa. In their turn, restrictions for mutual contacts and understanding may be imposed by the mother tongue or nationality. The principal idea here is that individuals may be crossed over only within the limits of one subpopulation, called a nation, whereas the selection is performed on the whole population.

The following model of this marriage in the restricted genetic algorithm is employed:

The example shown in Fig.1 relates to the two-parameter x_1, x_2 space (see Fig.4). Parameter x_1 varies from x_{1min} to x_{1max}, while parameter x_2 varies from x_{2min} to x_{2max}. First, phenotypes are ordered in the ascending sequence of the parameter x_1 values; the first n/5 phenotypes which are the nearest to x_{1min} will be assigned to the first race, and the next n/5 ones will be ascribed to the second race and so on. Then, the phenotypes are ordered in the ascending sequence of the parameter x_2 values; the first m/3 phenotypes of the first race will be assigned to the first nation, and the next m/3 ones will be ascribed to the second nation and so on. Thus, each phenotype will be assigned the "race" and "nation" ordinal numbers. Essentially, this algorithm may be repeated indefinitely.

The third restriction for marriage is represented by the so-called estate meaning that a person belonging to a specific estate marries a person of the same estate. It is in a sense a model of the Middle Ages relationships where upper class person has to marry equal position person by the conventional rules of the society. Although the poor have to marry the poor it seems that reproduction does not depend on an individual's welfare.

To model this process all phenotypes of a nation are arranged in the descending sequence of fitness values Y. After that individuals are coupled - the first best one with the second best one, the third with the fourth, and so on. The two parents have k (for example, 2) children. After that, the parents together with the children take part in the competitive activity for survival.

The classic genetic algorithm is a model of casual contacts between parents. One individual with high fitness can be a parent to all the children of the population if it is chosen for crossover as the first or the second parent. At the same time the other individual with low fitness is not participating in parenthood at all. Relationships "one individual with many others" may cause assimilation of population by one particular kind of the individuals (see [2]).

In contrast to the classic genetic algorithm, the restricted genetic algorithm is a model of devoted relationship between parents. The restrictions of coupling of parents lead to a more local search process (see [2] and [3]). This process can be seen in Fig.4.

While searching for good solutions inside each nation, the solutions that belong to the Pareto set are found. Comparing with 16th generation, depicted in Fig.3, the population of the restricted genetic algorithm concentrates around many different solutions, not around one single point.

3.4 Selection Operator

After the appearance of children, the size of the population is doubled. Because the crossover rate is 1.0, every couple after crossover in the genetic algorithms has two children. In the evolutionary algorithm each individual has a child obtained by mutation, therefore the population has double size as well. If the population size is 200 then the doubled population size is 400. By performing selection operator 200 times next generation is produced. In the experiments performed the roulette method of selection was used. To increase the frequency of choosing the best solutions, the relative fitness value $Y-Y_0$ was calculated and applied instead of the fitness value Y (Y_0 is the population's minimal value of Y, or "relative zero").

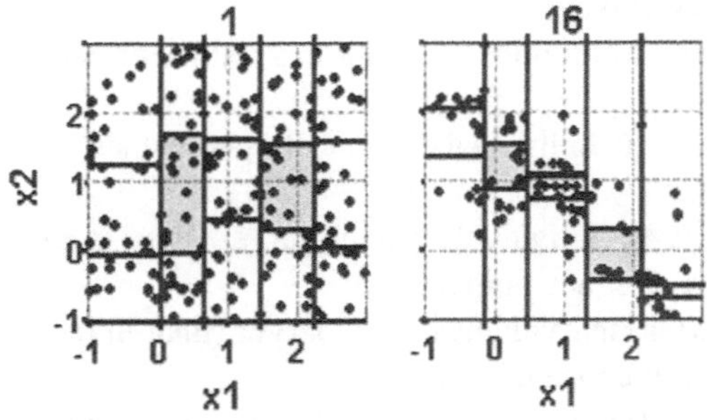

Fig. 4. Population of 15 nations (restricted genetic algorithm, population size is 200, maxi-min strategy): 1st generation and 16th generation

4 Description of Algorithms

There is a short description of all the algorithms.

As the experiments were not aimed to observe an influence of algorithms' parameter changes, the same parameters were used in all the experiments. Population size in all the experiments was 200. In every experiment 100 generations (iterations) were performed. The rate of crossover was 1,0 but the number of children obtained in one crossover operator was 2. In the evolutionary algorithm the rate of mutation was 1,0 (every individual was mutated) and the number of children obtained in one mutation operator was 1. In the restricted algorithm there were 5 races and 3 nations in every race (altogether 15 sub-populations).

4.1 The Classic Genetic Algorithm

The "classic genetic" algorithm is executed as follows:

1. Create the initial population by random generation of 200 individuals;
2. Calculate the fitness value Y and $Y-Y_0$ for each individual;
3. In loop 100 times: two times perform selection for choosing both parents, then cross over (operator, displayed in Fig.2) the parents and obtain 2 children (but if the parents coincide then initialize two new children and mutate the first parent by altering one parameter);
4. Calculate the fitness value Y and $Y-Y_0$ for each individual;
5. Collect the data for results;
6. Check the loop counter value: if it is 100 then finish, otherwise go back to step 3.

4.2 The Restricted Genetic Algorithm

The restricted genetic algorithm has the following steps:
1. Create an initial population by a random generation of 200 individuals;
2. Calculate the fitness value Y and $Y-Y_0$ for each individual;
3. Collect data for the results;
4. Assign the number of race:
   4.1. For the first 40 individuals having low values of parameter x_1 the race number is 1;
   4.2. For the next 40 individuals having larger values of parameter x_1 the race number is 2;
   4.3. For the next 40 individuals having larger (than in races 1 and 2) values of parameter x_1 the race number is 3;
   4.4. For the next 40 individuals having larger (than in races 1, 2 and 3) values of parameter x_1 the race number is 4;
   4.5. For the next 40 individuals having larger (than in races 1, 2, 3 and 4) values of parameter x_1 the race number is 5.
5. Divide each race into nations by assigning a nation number:
   5.1. For the first 13 individuals having low values of parameter x_2 the nation number is 1;
   5.2. For the next 13 individuals having larger values of parameter x_2 the nation number is 2;
   5.3. For the next 14 individuals having larger (than in nations 1 and 2) values of parameter x_2 the nation number is 3.
6. Arrange the individuals belonging to the same race and nation by Y value in the descending sequence. Assign couple numbers as follows: for the first two individuals number 1, for the second two number 2 and so on;
7. Cross over (operator, displayed in Fig.2) the parents (The individuals within the same race and the same nation having the same couple numbers) and obtain 2 children (but if the parents coincide then initialize two new children and mutate the first parent by altering one parameter);
8. Calculate the fitness value Y and $Y-Y_0$ for each individual;
9. Perform the selection operator 200 times;
10. Check the loop counter value: if it is 100 then finish, otherwise go back to step 3.

4.3 The Evolutionary Algorithm

This algorithm is very simple and includes the following steps:
1. Create the Initial population by random generation of 200 individuals;
2. Calculate the fitness value Y and $Y-Y_0$ for each individual;
3. Collect data for the results;
4. Mutate each individual by altering one of the parameter values;
5. Calculate the fitness value Y and $Y-Y_0$ for each individual;
6. Perform the selection operator 200 times;

7. Collect data for the results;
8. Check the loop counter value: if it is 100 then finish, otherwise go back to step 4.

5 Experiments

As the maximization criteria, two mathematical functions were used (see Fig.1, case a). They can be expressed as follows:

$$Y = 18.5 + \frac{1}{2} \cdot F(x_1, x_2) + \left[-1 + 4 \cdot \text{sgn}(1 - (1 + F(x_1, x_2))) \right] \cdot F(x_1, x_2) \tag{3}$$

where

$$F(x_1, x_2) = 1 - (x_1 + x_2) \quad or \quad F(x_1, x_2) = 1 - (x_1 + x_2 - 1) \tag{4}$$

$$and \quad \text{sgn}(x) = \begin{cases} +1 & \forall x > 0 \\ 0 & x = 0 \\ -1 & \forall x < 0 \end{cases} \tag{5}$$

The parameters intervals were the following:

$$-1 \le x_1 \le 3 \tag{6}$$

$$-1 \le x_2 \le 3 \tag{7}$$

As can be seen from the above expression the maximal value for both criteria is 18.5, but the fitness function's maximal value is 18.5 only under the maxi max strategy. However, as the difference Y-Y_0 is used for selection, the absolute function values do not play any role at all.

6 The Results

The most important traits of the experiments are shown in the appendix.

The results obtained were stored in the database and analyzed graphically by using SPSS graphical tools. The 1st, 8th, 16th, 24th, 50th and the 100th generations were saved. In patterns A, B, C one can see a visualization of the classic genetic algorithm, but D, E, F show the restricted algorithm behavior and G, H, L display the population of evolutionary algorithm. It can be seen that the root of the most marked analogy of behavior is the one built in the algorithm. Different ways of crossover operator, modifications in the coupling affect the solution space found by genetic or evolutionary algorithm. Therefore the classic genetic algorithm every time prefers only one point, usually in the middle of good solution area. The restricted genetic algorithm has the highest skill to cover all the area of good solutions although it has asymmetries in each subpopulation as well. The evolutionary algorithm has the largest variety of solutions although the strong migration caused by too high rate of mutation does not allow the algorithm to settle some part of population unchangeably.

The dark area is the target area. Although the size of this area is a disputed question (because the sensitivity of various algorithms is different and nobody can tell which precision is acceptable and which is not). The density of solutions inside the target areas is larger than the density outside of them. As the evolutionary algorithm is not sensitive enough it is difficult to determine if its concentration of solutions in target areas is larger.

The behavior of the classic genetic algorithm is illustrated in Fig. 5. There is depicted the 8th generation of the classic genetic algorithm, using maxi-min strategy. One can see that all individuals of the population are located inside the rectangle and the children individuals obtained from such population will also be located inside this rectangle. Therefore the algorithm is unable to create a child belonging to good solution area A or B.

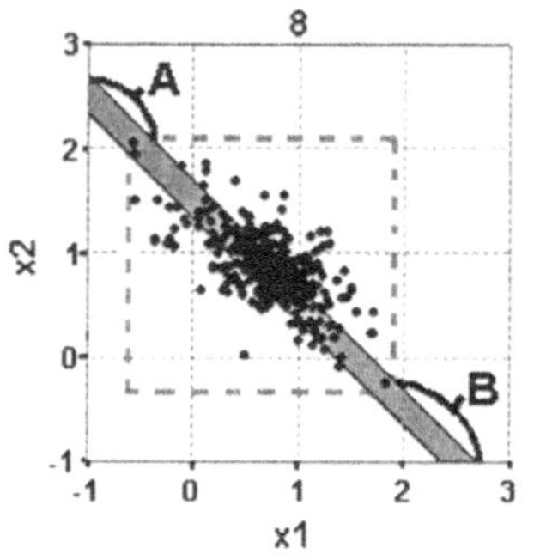

Fig. 5. The 8th generation of the classic genetic algorithm (maxi-min strategy)

As compared to the classic genetic algorithm, the restricted one has a different search tactic, because it arranges all individuals by their fitness values and because each individual can get only one partner for crossover and can not be selected for other individuals repeatedly.

7 Conclusions

The experiments showed the ability of classic and restricted genetic algorithm to response to different strategies by producing different final sets. The sensibility of the evolutionary algorithm was not strong enough. The use of maxi-min method increased the number of points belonging to the outcome of pessimist strategy. But as the number of points located outside the Pareto set depends to a high degree on the algorithm behavior, it is not possible to decrease it by trying the maxi-min decision strategy. Further research will be dealt with performing experiments aimed to measure the dominance and reward it by high fitness.

References

1. Hwang, C.L., Yoon, K.: Multiple attribute decision-making methods and application: A state of the art survey. Berlin, Springer Verl. (1981) 58–66

2. Takahashi, A.: Using Genetic Algorithms to Search for Maximum of the Multimodal Function. In: Proceedings of the 6th International Conference on Soft Computing MENDEL 2000, Brno University of Technology, Brno, Czech Republic (2000) 159–164
3. Takahashi A.: Genetic Algorithm Behaviour in the Multi-criteria Optimisation Task. In Proceedings of the Fourth International Conference on Knowledge-Based Intelligent Engineering Systems & Allied Technologies KES'2000, University of Brighton, Sussex, U.K. (2000) 514–520
4. Wright A. H.: Genetic Algorithms for Real Parameter Optimization. In: Foundations of GA. Gregory J.E. Rowlins, ed., Morgan Kaufmann Publishers (1991) 205–218

Appendix: Results - The 8th , 16th , 24th , and 50th generations

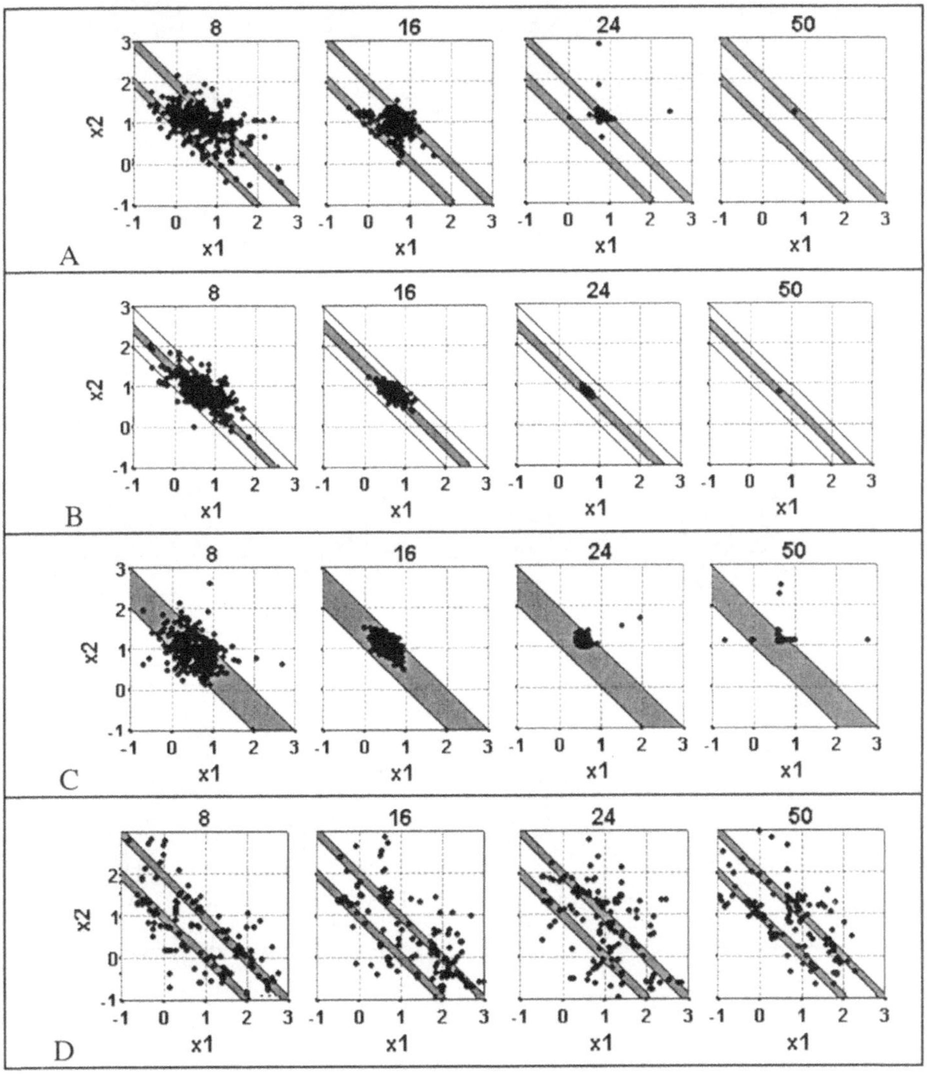

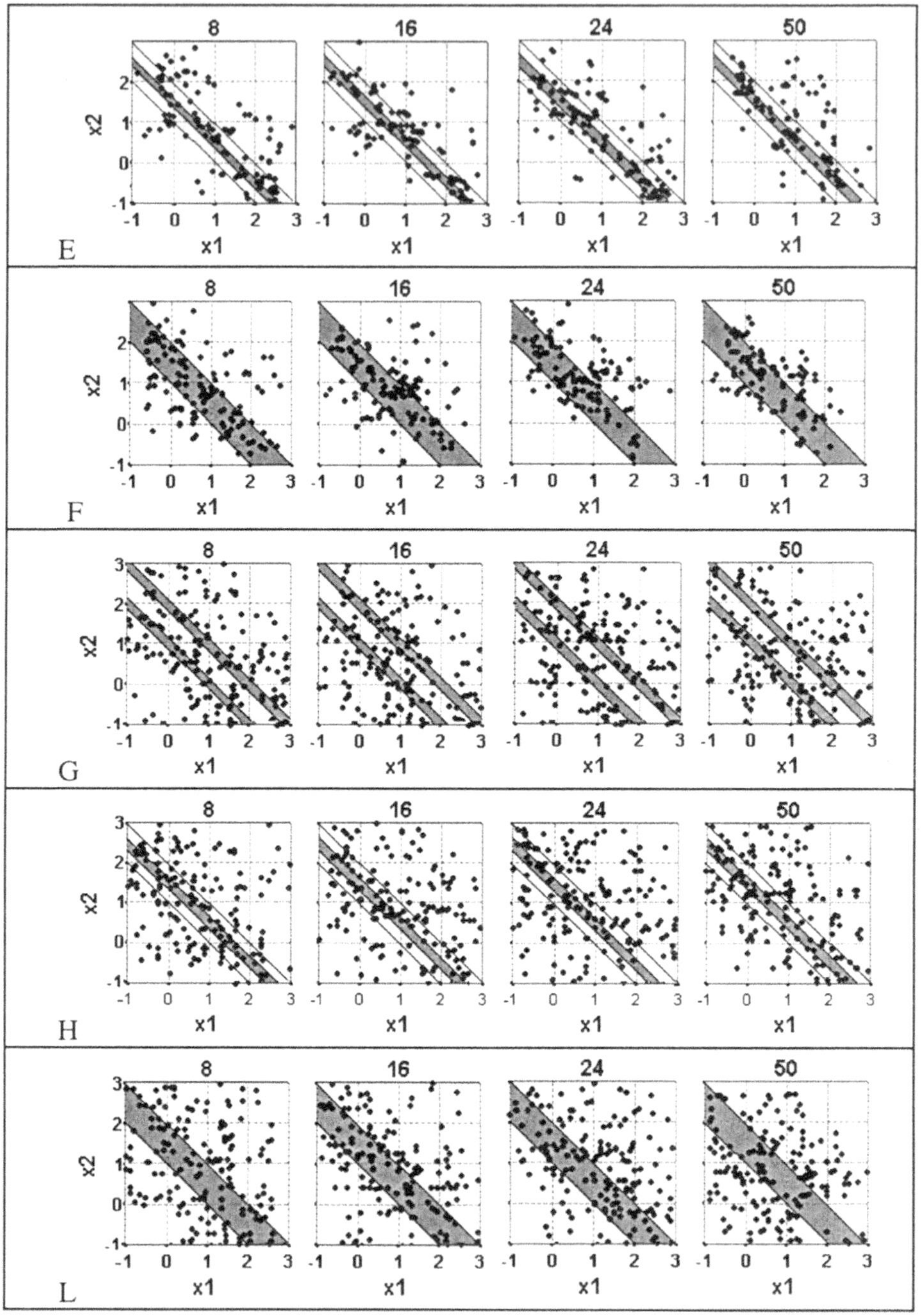
E
F
G
H
L
8
16
24
50
x1
x2

A Cluster-Based Evolutionary Algorithm for Multi-objective Optimization

István Borgulya

University of Pécs, Faculty of Business and Economics
H-7621 Pécs, Rákóczi út 80.
Borgulya@galileo.ktk.pte.hu

Abstract. In this paper a new evolutionary algorithm is described for multi-objective optimization. The new method handles non-linear objective functions and constraints and supports the decision-maker with an estimation of the Pareto set. This cluster-based method applies the Pareto-dominance principle. It approximates the Pareto set with the prototypes for each cluster and alternative prototypes as secondary population. The non-dominated set is continuously being up-dated: based on the Pareto ranking, the poorest clusters are regularly deleted, and the new ones are set.
The method solves the usual test problems with a satisfactory level of accuracy.

1. Introduction

Scientific and engineering problems often lead to multi-objective optimization problems (MOP). Several methods exist for solving these problems, and nowadays new, particularly evolutionary algorithm (EA)-based methods are developed (e.g. [8], [17]).

The solving of MOPs is a significantly more complex task than solving scalar optimization problems, because their solution is usually not a single value. The applicability of the best-known, classical methods for solving MOPs is mostly restricted to linear objective functions [15]. In order to solve the problem these methods reduce the objective functions to one function by means of weighing; for example, they use a lexicographic order or use linear programming procedures. In several cases the decision-maker's possibilities are widened by interactive facilities to allow the results to be monitored, to set parameters and to influence the directions of the search [15], [9].

In the case of non-linear objective functions the solving of the MOP is generally more difficult. We may choose from families of methods based on a fuzzy system (e.g. [12], [13]), based on methods applying parameter optimization (e.g. [7]), but primarily we choose methods incorporating EAs. Since 1985 several applications of the multi-objective EA (MOEA) technique were proposed and we can read the results in reviews (e.g. [5], [16]). Among others, van Veldhuizen and Lamont (2000) classify the MOEA variants in 3 major groups by the interaction of the decision-maker (DM) in the problem. Firstly, the problem may be reduced to a scalar optimization problem; secondly, the DM may take part – interactively - in the solving process; thirdly the result is given as a solution set (Pareto set) and the appropriate solution has to be

B. Reusch (Ed.): Fuzzy Days 2001, LNCS 2206, pp. 357–368, 2001.

chosen from this set. If we focus on the applied technique, another kind of classification is possible. For example, Fonseca and Fleming (1995) differentiate aggregation-based methods, population-based methods, Pareto-dominance principle based techniques and niche-method- based techniques. The majority of the methods use the Pareto-dominance principle.

Out of all the MOEA methods, several apply clustering. E.g. Rosenman and Gero (1985) proposed an algorithm to reduce the Pareto set based on clustering solutions, which are similar in all or some of the criteria. The algorithm by Cunha et al. (1997) is also based on the ideas of Rosenman and Gero; it reduces and ranks the points of the Pareto set. Zitzler and Thiele (2000) apply clustering for the purpose of reducing an externally stored non-dominated solution set.

The new MOEA method to be demonstrated - let us call it MOSCA (Multi-objective Optimum Search with Cluster-based Algorithm) - is a cluster-based EA also using the Pareto-dominance principle and which produces the approximation of the Pareto set with the DM. The MOSCA can handle non-linear objective functions and takes constraints into account. The new method has been developed from a scalar optimization method (the OSCA) which can solve constrained and also non-linear scalar optimization problems [1], [2]. The OSCA is a cluster-based EA, an evolutionary strategy (ES) whose structure and major steps can also be found in the new method.

MOSCA as a cluster based EA method uses some customary MOEA elements as well, such as: storing the non-dominated solutions found in an external set; using the concept of Pareto dominance to assign fitness values and to order them; and setting clusters to reduce the number of non-dominated solutions stored.

However the method of application, especially that of clustering, represents a new technique. That is:

- solutions are looked for as cluster prototypes, the method corresponds to a partially reduced storage method;
- non-prototypical solutions are sorted by clusters in an external, non-dominated set. The non-dominated set is continuously being up-dated: based on the Pareto ranking, the poorest clusters are regularly deleted, and the new ones are set.
- Finally, during clustering it is able to take the constraints into consideration.

By these means a new method is obtained, which e. g. compared to the NSGA method of Srinivas and Deb (1995): solves simpler tasks with less function evaluation; performs almost equally punctual solutions for a part of more difficult tasks; the method can be easily applied regardless of having a constrained or unconstrained problem, the handling of the objective functions and constraints is performed separately.

The remainder of the paper comprises of a description of the new method and a series of bi-objective test problems, which demonstrate the application possibilities.

2. Multi-objective Optimization Problem

We can examine a multi-objective optimization problem with the following form

Minimize $f(\mathbf{x}) = (f_1(\mathbf{x}), \ldots, f_k(\mathbf{x}))$

where $\mathbf{x}=(x_1, \ldots, x_m) \in F \subseteq S$

and where we have k objective functions $f_i\ \Re^m \to \Re$, $\mathbf{x}$ is the decision vector, S the decision space.

Let us suppose that the domain S of dimension m is

$$S= \{\mathbf{x} \in \Re^m \mid a_i \le x_i \le b_i,\ a_i,\ b_i,\ x_i \in \Re,\ i=1,2,\ldots,m\},$$

whereas the feasible region $F \subseteq S$ is defined by a set of r additional linear and/or nonlinear constraints ($r \ge 0$):

$$g_j(\mathbf{x}) \le 0,\ j= 0, 1, 2,\ldots, r.$$

Constraints in terms of equations e.g. $h(\mathbf{x})=0$, can be replaced by a pair of inequalities:

$$h(x)-\delta \le 0$$
$$- h(x)-\delta \le 0$$

where an additional parameter δ is used to define the tolerance of the algorithm.

Let us review some corresponding concepts and definitions ([16], [17]):

Definition 1.: (Pareto dominance) A solution $\mathbf{a} =(a_1, a_2, \ldots a_m) \in S$ is said to dominate a solution $\mathbf{b}=(b_1, b_2, \ldots, b_m) \in S$ (notation: $\mathbf{a} \cup \mathbf{b}$) if and only if

$$\forall i \in \{1,2,\ldots,m\}\ f(a_i) \le f(b_i) \wedge \exists j \in \{1,2,\ldots,m\}\ f(a_j) < f(b_j).$$

Definition 2.: (Pareto optimality) A solution $\mathbf{a} \in S$ is Pareto optimal if and only if

$$\neg\exists\ \mathbf{b} \in S: \mathbf{b} \prec \mathbf{a}.$$

Definition 3.: (Pareto optimal set) For a given MOP, the Pareto optimal set (PS) is defined as:

$$PS= \{\mathbf{a} \in S \mid \neg\exists\ \mathbf{b} \in S: \mathbf{b} \prec \mathbf{a}\ \}.$$

Definition 4.: (Pareto front) For a given MOP and Pareto optimal set PS, the Pareto front (PF) is defined as:

$$PF:= \{f(\mathbf{x})\ \Pi \mathbf{x} \in PS\}.$$

For the sake of simplicity the optimal attributes will not be used when mentioning the Pareto optimal set.

3. The New Algorithm

Let us examine first the MOSCA principle and then discuss the new model.

3.1 The MOSCA Principle

The fundamental principles of MOSCA are the following: let us cover S by t number of clusters, and let us approximate the points of the PS with the central points of clusters (with the prototypes).

As the PS is usually composed of an infinite number of points, the approximation might be refined by registering further 'alternative prototypes' for each cluster. These alternative prototypes are sets, which together with the prototype of the cluster, are non-dominated sets. This estimation allows that out of the elements of a cluster only the prototype and the set of alternative prototypes have to be stored.

The algorithm approximates to the PS with continuous modifications of the clusters. It continuously generates new points, which enables better prototype selection at a certain cluster. Concerning the process of generating new points, the operation of MOSCA can be divided into two stages. During the first stage it is divided into preliminary clusters by the S range: the algorithm forms clusters continuously from randomly generated points of S. During the second stage it already refines the convergence of the clusters by choosing the new points from a narrowing neighborhood of the prototype.

Adequate covering of the PS is secured by the occasional deletion of the least converging clusters. In the second stage at each kn^{th} iteration a given percentage of dominated prototypes (clusters) get deleted. For this, the prototypes are arranged, and the given percentage of the least dominated elements are deleted according to the Pareto ranking method by Goldberg (1989) and by the measure of the violation of the constraints [10]. The above percentage decreases parallel to the increasing number of iterations. The algorithm replaces the deleted clusters: the new prototypes are chosen from a wide neighborhood of an existing prototype at a given probability and this probability continuously decreases during the second stage.

The algorithm formed in this way is an EA. If we treat the characteristics of the clusters as an individual: I=(cluster$_1$, cluster$_2$,..., cluster$_t$). Then the generation and modification of clusters based on the randomly generated points can be defined as a mutation that modifies a single element, a single cluster of the individual. The estimation of the PS can be checked with a composite fitness function.

We describe the function of the MOSCA by means of two different (1+1)-ESs. These are as follows:

ES1: (The main algorithm) In the first stage it forms clusters using the ES2 procedure. In the second stage, continuing the formation of clusters, it improves the accuracy of the approximation of the PS with a local search (using the ES2 procedure), and it filters out the least converging clusters, and deletes them.

ES2: In the first stage it looks for non-dominated decision vectors at randomly selected points of S, and in the second stage it looks for non-dominated decision vectors in the vicinity of the prototypes with the normal distribution. The prototype (cluster) most similar to the result vector is modified, or a new cluster is generated.

Afterwards, the major steps of the algorithm are as follows:

ES1:

1. Setting the initial values
2. Mutation: modification of clusters (call ES2).
3. Ranking the prototype, a few dominated prototype deletions. (Stage 2)
4. Terminal condition (go to 2).

ES2:

Mutation: modification of clusters *kn*-time:

1. Decision vector generation.
2. The most similar prototype modification, or new cluster generation.

3.2 Introduced Notations

Let us introduce the following notations:

a) The Pareto dominance principle allows us to choose the non-dominated element as a prototype of the clusters. The choice of the prototype is controlled by a function, which assigns a rank number (oc_i) to each $\mathbf{p}_i$ prototype. Each cluster has its own rank number and so the rank numbers may have different values from cluster to cluster. Let us calculate the rank numbers for $\mathbf{x} \in S$ when $\mathbf{p}_i$ prototype and oc_i is known:

$$Rank(p_i, x) = \begin{cases} 0 & p_i \prec x \\ oc_i + 1 & p_i \succ x \\ oc_i & otherwise \end{cases}$$

b) We assign a set of alternative prototypes (Alt_i) to each cluster. This set contains the non-dominated decision vectors and the corresponding objective vectors having the same rank number as the prototype of the cluster (their quantity is maximized). The prototypes and the alternative prototypes of clusters form a non-dominated set (notation: NDSET, and NDSET=$\cup$ $\{\mathbf{p}_i \cup Alt_i\}$ (i=1,2,…,t)), which gives the improving approximation of the PS during the evolution.

c) The measure of similarity of two decision vectors $\mathbf{x}$ and $\mathbf{z}$ is denominated
   $H(\mathbf{x}, \mathbf{z}) = 1/(1 + d(\mathbf{x}, \mathbf{z}))$
   where $d(\mathbf{x}, \mathbf{z})$ is the Euclidean distance of the vectors.

d) Let $D(\mathbf{x})$ be the measure of violation of the constraints. (If $\mathbf{x} \in F$ or no constraints are specified $D(\mathbf{x})=0$):

$$D(x) = \left(\sum_{j=1}^{r} \max\{g_j(x), 0\}^2 \right)^{1/2}$$

   Let us utilize the value $D(\mathbf{x})$ in the optimum search for characterizing the decision vector $\mathbf{x}$ in the following way: vector $\mathbf{x}$ is better than vector $\mathbf{z}$ if $D(\mathbf{x})<D(\mathbf{z})$.
   In case $D(\mathbf{x}) = D(\mathbf{z})$ we call $\mathbf{x}$ better than $\mathbf{z}$ if $f(\mathbf{x})<f(\mathbf{z})$.

e) Let us designate the random number generator as rnd (uniform distribution on (0,1)).

f) Let us designate the normal random number generator as $N(0, \sigma)$ (where $N(0, \sigma)$ is a normal distribution with mean 0 and standard deviation σ).

g) The same population and individual is used with all two ESs. Let the labeling number, rank number, prototype, objective vector and the measure of violation of constraints of the prototypes corresponding to cluster number i be c_i, oc_i, $\mathbf{p}_i$, $f(\mathbf{p}_i)$ and $D(\mathbf{p}_i)$. If the ith cluster exists $c_i=1$ otherwise $c_i=0$. Let the population of the it^{th} generation be denoted by P(it), and the individual by
   $I=(c_1, oc_1, \mathbf{p}_1, f(\mathbf{p}_1), D(\mathbf{p}_1), \ldots, c_t, oc_t, \mathbf{p}_t, f(\mathbf{p}_t), D(\mathbf{p}_t))$.

As the (1+1)-ES uses only one parent and one descendent individual, the descendent individual is marked with stars:

$$I^*=(c_1^*,\ oc_1^*,\ \mathbf{p}_1^*,\ f(\mathbf{p}_1^*),\ D(\mathbf{p}_1^*),\ \ldots,\ c_t^*,\ oc_t^*,\ \mathbf{p}_t^*,\ f(\mathbf{p}_t^*),\ D(\mathbf{p}_t^*))$$

The fitness function can be composed of three different functions. These are the number of missing clusters (F1), the sum of the D distances of the prototypes (F2), and the reciprocals of the sum of the prototype rank numbers. (F3).

$$F1(I) = t - \sum_{j=1}^{t} c_j, \quad F2(I) = \sum_{j=1}^{t} D(p_j), \quad F3(I) = \frac{1}{\sum_{j=1}^{t} oc_i}$$

In the case of the MOSCA it is easier to use the three functions separately when evaluating or selecting them, than to formulate a common fitness function for them, and so the F1, F2 and F3 are applied separately.

3.3 The Steps of the MOSCA

9 parameters effect the run of the algorithm: *t, itt, itend, kn, ddp,* σ, *hmult, altt* and δ. Their roles one-by-one are:

- *t* - the number of the subsets of the Pareto optimal set to be found.
- *itt* - parameter of the local search. If the number of iterations (*it*) reaches *itt*, the local search begins in an ever-narrowing neighborhood of the prototypes.
- *itend* - parameter for the stop condition. The procedure is finished if the value of *it* is more than *itend*.
- *kn* - parameter which determines the timing of checks. Clusters are checked only in every *kn*th iteration.
- *ddp*- parameter of the local search. *ddp* percent of the dominated prototype of the individual will be deleted.
- σ - parameter of the local search. σ is the common standard deviation at the beginning of the local search.
- *hmult* - parameter of the local search. It is a multiplier affecting the extent of the standard deviation.
- *altt* – parameter of the non-dominated set. The maximal number of elements of each Alt_i set may be *altt*.
- δ - is used to define the tolerance of the algorithm by equation constraints.

The steps of MOSCA are as follows:
* ES1 *

1. Setting the initial values.
   Set the initial values of *t, itt, itend, kn,* σ, *hmult* , *ddp* , *altt* and δ. Let it=0, prot=*t*. Generate the I individual randomly: let c_i=1, $\mathbf{p}_i \in S$, oc_i=1, Alt_i=0 and num_i=0 (i=1,2,…,t). P(it) ← I.
2. The new population.
   2.1 Mutation (new clusters, or modification of clusters)
   if *it*<*itt* then call ES2 else $\sigma=\sigma*$ *hmult*, call ES2.
   2.2 Dominated prototype deletion. If it>*itt* then:
   Prototypes are ranked based on the criteria of Goldberg's Pareto ranking

method as well as on the measure of violation of the constraints, into an increasing sequence within the individual. Let Goldberg's method assign the rank number $grank_i$ to the $\mathbf{p}_i$ prototype.
Let j= Int(t * (0.9- ddp * (1-it / $itend$))).
For i=j, j+1, ...,t: If $\neg(grank_i=1 \wedge D(\mathbf{p}_i^*)=0)$ then
$c_i^*=0, oc_i^*=0$, prot=prot-1, $Alt_i=0$ and $num_i=0$.

2.3 P(it) ← I^*.

3. Terminal condition.
If (it> $itend$) is false, continue at step 2. Otherwise it's the end of the procedure.

*ES2 *

1. Evaluate F1(I), F2(I) and F3(I).
2. Mutations.
   2.1 Choosing a decision vector.
   it=it+1, $I^* \leftarrow I$,
   If it<itt then choose an **x** randomly from S.
   If it>itt then choose an **x** randomly from NDSET.
   $\forall i \in \{1,2, \ldots,n\}$:
   If (prot<t) ∧ (0.9-0.1*(1-it/$itend$)<rnd)
   then let $x_i=x_i$+ sign(N(0,1))*rnd else let $x_i=x_i + \sigma$*N(0,1)
   (and $\mathbf{x} \in S$).
   2.2 New cluster, or modification of a cluster.
   If $(\exists i)(c_i^*=0)$ then $\mathbf{p}_i^*=\mathbf{x}$, $c_i^*=1$, $oc_i^*=1$, $num_i=0$, prot=prot+1.
   else checking the analogies:
   let $H(\mathbf{x}, \mathbf{p}_i^*)=\max_q H(\mathbf{x}, \mathbf{p}_q^*)$; $q, i \in \{1,2,\ldots,t\}$.
   $\mathbf{p}_i^*=\mathbf{x}$, $num_i=0$, oc_i^*= Rank($\mathbf{p}_i$,$\mathbf{p}_i^*$).
3. Evaluate F1(I^*), F2(I^*) and F3(I^*).
4. Selection.
   If (F1(I) =F1(I^*)) ∧ (F2(I) =F2(I^*)) ∧ (F3(I) =F3(I^*)) ∧ (num_i<$altt$)
   then let $Alt_i=\mathbf{p}_i^* \cup Alt_i$, $num_i=num_i+1$.
   If (F1(I) >F1(I^*)) ∨ (F1(I) =F1(I^*)) ∧ (F2(I) >F2(I^*)) ∨
   (F1(I) =F1(I^*)) ∧ (F2(I) =F2(I^*)) ∧ (F3(I) >F3(I^*))
   then P(it) ← I^* else P(it) ← I.
5. Terminal condition.
   If mod(it, kn) =0 then exit, otherwise continue at step 1.

The results of the test examples show that the algorithm yields more accurate results if the elements are transformed to the [0, 1] interval (the $[0, 1]^m$ hypercube).

Determining Parameters

All parameters influence the speed of the convergence and the accuracy. Based on the test results we can assign values to some of the parameters, namely to *ddp*, δ , *hmult* and *kn*, where the algorithm provides a correct result (e.g. *ddp*=0.95, δ= 0.01, *hmult*=0.97, *kn*=500). With somewhat different values the accuracy of the result or the iteration number may change. However, the value of *itt* and σ are considerably task- dependent. Usually a local search can start after no more than 1500 iterations and σ=0.1.

The *t* and *altt* parameters affect the approximation of the PS fundamentally. They allow us to influence the partitioning of the PS. Usually the value of *t* is 30-40, and *altt*=2.

4. Test Results

The MOSCA is basically designed to handle non-linear, constrained MOP. The cluster- based approach enables the search for Pareto-fronts of various difficulty (e.g. convex, non-convex, or discontinuous fronts). To demonstrate the capabilities of MOSCA, we solved and discussed some test problems from other articles. Although all of these test problems are two-objective MOPs, the MOSCA is able to handle MOPs of more than two objectives.

The demonstrated test problems comprise of: the test problems of Laumanns et al. (1998) (g has locally Pareto-optimal solutions), the 3 test problems of Srinivas et al. (1995) (F1 is convex, F2 is discontinuous, F3 is a constrained MOP), 2 from Deb's (1999) test problems (test1 discontinuous, test2 with global and local Pareto fronts), and finally the first three test problems of Zitzler et al. (2000) (T1 convex, T2 non-convex and T3 discontinuous). The descriptions of the test problems can be found in the Appendix.

Problems	dim	*t*	*altt*	σ	*itend*
G	2	40	4	0.02	20000
F1	2	33	2	0.2	10000
F2	2	33	2	0.1	25000
F3	2	33	2	0.05	25000
test1	2	40	4	0.05	15000
test2	2	33	2	0.1	6000
T1	30	33	2	0.1	25000
T2	30	33	2	0.1	25000
T3	30	33	2	0.1	25000

Fig. 1. The parameter settings of the test problems (*kn*=500, *itt*=1400, *hmult*=0.97, *ddp*=0.95 and δ=0.01)

When setting the values of the parameters, our aim was to obtain results that were comparable to the results of other methods. Thus, we adopted the size of the population (under *t+t* altt*) together with the maximal number of function evolutions (*itend*) from other publications.

The parameter settings used for each test problem are summarized in Figure 1. The approximations of the PFs are shown in Figures 2 and 3. (Figure 3 additionally shows some NSGA results). As MOSCA is a stochastic method, each problem is run ten times with the same parameter values. Figures 2 and 3 display the best results of the runs. Deviation from these results was found only in the number of converging PS points; the algorithm usually generated 5-10% less PS points.

For the purpose of visual comparison we chose the NSGA method by Srinivas and Deb (1995). For the reason of size the results of MOSCA and NSGA (figure 3) are shown only at 3 tasks. Comparing the result (in case of populations of equal size) it was ascertained that:

- MOSCA solves simpler tasks with less function evaluation than NSGA. It solves e.g. F1 and F2 with as few as 10 thousand function evaluations instead of the 50 thousand, and solves test1 task with 15 thousand instead of 60 thousand. (See [14], [4]).
- The two systems solve a certain fraction of the more difficult tasks with approximately the same accuracy. E.g. in the cases of tasks T1 and T2, in addition to the 25 thousand allocated function evaluations, the two algorithms generated converging points in just about the same number and location. In task T3 the results of MOSCA are slightly poorer than those of the NSGA. On the other hand with task test2, MOSCA managed to find the global Pareto front in 100% of the runs, while NSGA performed correctly in 41%. (See [17], [4]).

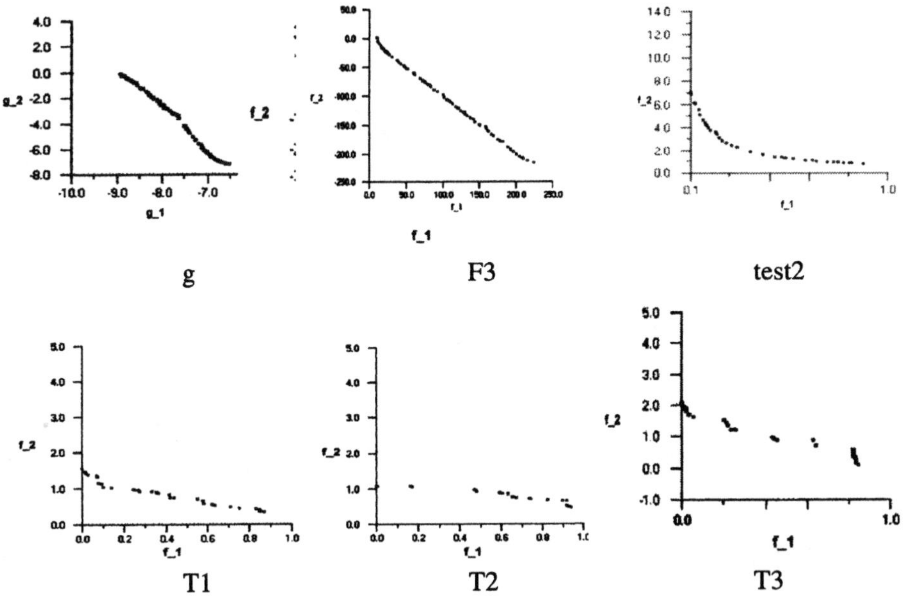

Fig. 2. The Pareto fronts of the g, F3, test2, T1, T2 and T3 problems

5 Conclusion

We can conclude that the MOSCA was successfully tested with different kinds of functions. The method solves the usual test problems at a satisfactory level of accuracy. Comparing the results with the solutions known from theory or from the respective publications it can be concluded that the solutions are correct.

The progression of the algorithm is influenced by 9 parameters, most of which can be set to standard values. Some parameters are problem-dependent: they influence the division of the Pareto set into subsets or the search for the Pareto front. In conclusion, the operation of MOSCA is simple; it is unnecessary to change any parameters other than the size of the population and standard deviation (σ).

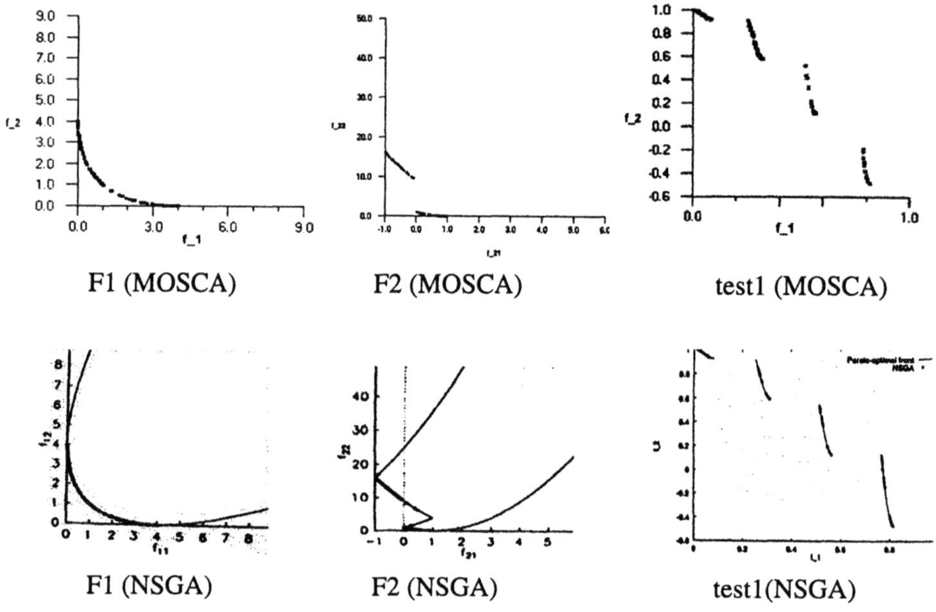

Fig. 3. The Pareto fronts of the F1, F2 and test1 problems

Although the method works properly, we can see two options for its further development. Firstly it is possible to refine PS convergence with the choice of an appropriate Pareto ranking method. Secondly the storage of the fixed number of alternative prototypes can also be excelled: by means of grouping and filtering the quality of the alternative prototypes can be improved.

Acknowledgments:

The study was supported by the Hungarian Research Program OTKA T 030861 and a DAAD grant at the University of Munich in 2000.

References

1. Borgulya I. (2000) Some polynomial programming problems solving with a clustering algorithm (in Hungarian) *Alkalmazott Matematikai Lapok* 20 19-34.
2. Borgulya I (2000): Constrained optimization using a clustering algorithm *Central European Journal of Operations Research.* 8 (1) 13-34.
3. Cunha A.G., Oliveria P., Covas J.A. (1997) Use of Genetic Algoritms in Multicriteria Optimization to Solve Industrial Problems. *Proc. Seventh Int. Conf. On Genetic Algorithms*, Morgan Kauffman, 682-688.
4. Deb K. (1999): Multi-objective Genetic Algorithms: Problem Difficulties and Construction of Test Problems. *Evolutionary Computing* Vol. 7. No. 3. 205-230.
5. Fonseca C. M, Fleming P. J. (1995): An Overview of Evolutionary Algorithms in Multiobjective Optimization. *Evolutionary Computing* Vol. 3. No. 1. 1-16.
6. Goldberg D.E. (1989). *Genetic Algorithm for Search, Optimization and Machine Learning*. Addison-Wesley, Reading, Massachusetts.
7. Gouljashki V. G, Kirilov L. M, Nerula S. C, Vassilev V. S. (1997): A reference Direction Interactive Algorithm of the Multiple Objective Nonlinear Integer Programming. In: Fandel G., Gal T. (ed.) *Multiple criteria Decision Making*, Springer Berlin. 308-317.
8. Laumann M , Rudolph G, Schwefel H-P. (1998): A Spatieel Predaton_Prey Approach to Multi-Objective Optimization: A preliminary Study. In Eihem A. E, Bäck T, Schoenauer M, Schwefel H-P (ed.) *Parallel Problem Solving from Nature –PPSN V.* Springer Berlin.
9. Miettiinen K, Mäkelä M. M. (2000): Interactive multiobjective optimization system WWW.NIMBUS on the Internet. *Computer & Operations Research* 27. 709-723.
10. Richardson J., Palmer M., Liepins G., Hilliard M. (1989): Some guidelines for genetic algorithms with penalty functions *Proc. of the 3th Int. Conf. on Genetic Algorithms,* Morgan Kaufmann, Los Altos, pp. 19-197.
11. Rosenman M.A., Gero J.S. (1985). Reducing the Pareto optimal set in multivriteria optimization. *Eng. Optim.* 8, 189-206.
12. Sakawa M. (1993) *Fuzzy sets and interactive multiobjective optimization.* Plenum Press, New York.
13. Sakawa M, Yauchi K. (1999): An interactive fuzzy satisficing method for multiobjective nonconvex programming problems through floating point genetic algorithm. *European Journal of Operation Researches* 117. 113-124.
14. Srinivas N, Deb K. (1995): Multiobjective Optimization Using Nondominated Sorting in Genetic Algorithm. *Evolutionary Computing* Vol. 2. No. 3. 221-248.
15. Steuer R. E (1989): *Multiple Criteria Optimization.* R. E. Klieger Pub. Com. Florida.
16. Van Veldhuizen D. A, Lamont G. B. (2000): Multiobjective Evolutionary Algorithms: Analyzing the State-of-the Art. *Evolutionary Computing* Vol. 8. No. 2. 125-147.
17. Zitzler E, Deb K, Thiele L. (2000) Comparison of Multiobjective Evolutionary Algorithms: Empirical Results. *Evolutionary Computing* Vol. 8. No. 2. 173-195.

Appendix. Test problems

g (Source [8].):

$$Min \quad g(x) = \begin{pmatrix} -10\exp(-0.2\sqrt{x_1^2 + x_2^2}) \\ |x_1|^{4/5} + |x_2|^{4/5} + 5(\sin^3 x_1 + \sin^3 x_2) \end{pmatrix}$$

where -3≤ x1, x2 ≤ 1.

F1 (Source [14]):

$$Min \quad F1(x) = \begin{pmatrix} x^2 \\ (x-2)^2 \end{pmatrix}$$

where -2≤ x ≤ 3.

F2 (Source [14]):

$$Min\ \ F2(x) = \left(\begin{cases} -x & if\ x \le 1 \\ -2+x & if\ 1 < x \le 3 \\ 4-x & if\ 3 < x \le 4 \\ -4+x & if\ x > 4 \end{cases} \atop (x-2)^2 \right)$$

where $-2 \le x \le 10$.

F3 (Source [14]):

$$Min\ \ F3(x) = \begin{pmatrix} (x_1-2)^2 + (x_2-1)^2 + 2 \\ 9x_1 - (x_2-1)^2 \end{pmatrix}$$

$$subject\ \ to\ \ x_1^2 + x_2^2 - 225 \le 0$$
$$x_1 - 3x_2 + 10 \le 0$$

where $-20 \le x1, x2 \le 20$.

Further on each of the test functions are defined in the same manner:

$$Min\ \ F(x) = \begin{pmatrix} f_1(x_1) \\ g(x_2,\ldots,x_m)\, h(f_1(x_1)\ , g(x_2,\ldots,x_m)) \end{pmatrix}$$

where $x = (x_1, x_2, \ldots, x_m)$

test1 (Source [3]):

$$f_1(x) = x_1$$
$$g(x_2) = 1 + 10x_2$$
$$h(f_1, g) = 1 - \left(\frac{f_1}{g}\right)^2 - \frac{f_1}{g}\sin(2\pi 4 f_1)$$

where $x_1, x_2 \in [0.0, 1.0]$.

test2 (Source [3]):

$$f_1(x) = x_1$$
$$g(x_2) = 2.0 - \exp\left(-\left(\frac{x_2 - 0.2}{0.04}\right)^2\right)$$
$$- 0.8\exp\left(-\left(\frac{x_2 - 0.6}{0.4}\right)^2\right)$$
$$h(f_1, g) = \frac{1}{f_1}$$

where $x_1, x_2 \in [0.1, 1.0]$.

T1 (Source [14]):

$$f_1(x) = x_1$$
$$g(x_2, x_3, \ldots, x_m) = 1 + 9\sum_{i=2}^{m} x_i / (m-1)$$
$$h(f_1, g) = 1 - \sqrt{\frac{f_1}{g}}$$

where m=30, and $x_i \in [0,1]$.

T2 (Source [14]):

$$f_1(x) = x_1$$
$$g(x_2, x_3, \ldots, x_m) = 1 + 9\sum_{i=2}^{m} x_i / (m-1)$$
$$h(f_1, g) = 1 - \left(\frac{f1}{g1}\right)^2$$

where m=30, and $x_i \in [0,1]$.

T3 (Source [14]):

$$f_1(x) = x_1$$
$$g(x_2, x_3, \ldots, x_m) = 1 + 9\sum_{i=2}^{m} x_i / (m-1)$$
$$h(f_1, g) = 1 - \sqrt{\frac{f_1}{g}} - \left(\frac{f_1}{g}\right)\sin(10\pi f_1)$$

where m=30, and $x_i \in [0,1]$.

Inherent Wave Estimation on Ultrasonic Non-destructive Testing Using Fuzzy Inference

Kouki Nagamune[1], Yutaka Hata[1], and Kazuhiko Taniguchi[2]

[1] Department of Computer Engineering, Himeji Institute of Technology, Japan
{nagamune, hata }@comp.eng.himeji-tech.ac.jp
http://wwwj3.comp.eng.himeji-tech.ac.jp
[2] Kinden Co., Ltd, Japan
k-tani@comp.eng.himeji-tech.ac.jp

Abstract. This paper proposes the new method for suppressing the inherent wave of the sensor system. The inherent wave often prevents from analyzing the signal of target. Previous method calculated the inherent wave to solve this problem using two assumptions. However, the calculated wave is differ from the original inherent wave in unsatisfying the assumptions. Therefore, the new method estimates the wave using fuzzy-logic without the assumptions. As a result, this method was able to detect with higher accuracy than previous method.

1 Introduction

The ultrasonic non-destructive testing (NDT) method is often used for disclosing the information in concrete members. The enormous progress to overcome the difficulties of testing concrete is mainly due to the development of new broadband transducers and high performance computers. The ultrasonic NDT of concrete members has been applied to evaluate crack depth [1] and internal defect [2]. However, the difficulty of understanding the reflected ultrasonic waves prevents us from solving problems for detecting and locating embedded objects [3].

One cause of those problems is that the received wave includes not only the echo from the inner concrete member but also the wave inherent in the measurement system such as the reflected wave in the probe, the surface echo and the back echo. These inherent waves reduce or cancel the echo from the embedded object. Therefore, it is necessary to eliminate the inherent wave. The previous method [4] estimates the inherent wave on two assumptions. First assumption was that the inherent wave is included in every measurement point. Second assumption was that the echo of the embedded object exists locally in comparison with the measurement range. However, the second assumption does not always hold. Then, since the estimated wave did not represent the real inherent wave, it is difficult to detect the embedded object.

This paper proposes a new method for estimating the inherent wave. The method solves the problem without considering the second assumption by using

B. Reusch (Ed.): Fuzzy Days 2001, LNCS 2206, pp. 369–379, 2001.

fuzzy inference [5]. Fuzzy inference can provide a powerful method for representing human knowledge. We use three types of knowledge for evaluating whether each wave is the inherent wave or not. We employ fuzzy if-then rules to express information about intensities of the back echo, and integration value of the absolute intensity between the time of front echo and the time of back echo. The degree of the inherent wave is calculated as an algebraic product of the degrees. The degree indicates the weight for each wave. After the inference result is calculated for each wave, the inherent wave is obtained with weighted average of all waves. Our method can estimate the inherent wave without considering the second assumption, because the higher weight is assigned to the wave whose property is more similar to the property of the inherent wave.

2 NDT System

We developed an NDT system using ultrasonic wave. The system consisted of an ultrasonic probe, an ultrasonic pulsar receiver, a biaxial scanner and a personal computer for data processing and driving the scanner. The system used a low frequency probe (Imaging Supersonic Laboratories Co., Ltd.) and a low frequency broadband pulsar receiver (Imaging Supersonic Laboratories Co., Ltd.). The reflected wave was transmitted and acquired via the pulsar receiver. The frequency range of the probe was 0.1 to 1.0 MHz. The sampling interval of the data was 0.2 μs. A coupling medium (sonicoto-E nichigo acetylene Co., Ltd.) was used for efficiently transferring the ultrasonic into the concrete between a concrete and the probe.

The overall of the developed system is shown in Fig. 1(a). A-scan waves are obtained by moving the probe on the surface of a concrete, and comprise a B-scan image. Because the probe operates in the frequency range of 0.1 to 1.0 MHz, we apply a band pass filter (BPF) to the wave. Figure 1 (c) shows a B-scan image generated A-scan waves shown in Fig. 1 (b). Intensity of the pixel in Fig. 1(c) represents amplitude of the A-scan wave. In our work, the image consists of 100 $\times$ 1000 pixels. The pixel dimension is 1.0mm $\times$ 0.4mm.

The NDT system we used here produced an A-scan wave for each measurement point on the concrete. Collecting A-scan waves forms a B-scan image for a line on the concrete. First, we subtract the inherent wave of the measurement NDT system from the A-scan wave, where the inherent wave is obtained using previous method and proposal method. The resultant waves are then collected and form a new B-scan image. In the obtained B-scan image, embedded objects will appear as parabola patterns. To emphasize the patterns, we apply the parabolic integration. The embedded objects can be detected by finding the local peak points of the intensity histogram on time-axis for the resultant B-scan image. We also present experimental results for a concrete member with one resin tube and one steel pipe. This method was able to detect with higher accuracy than previous method.

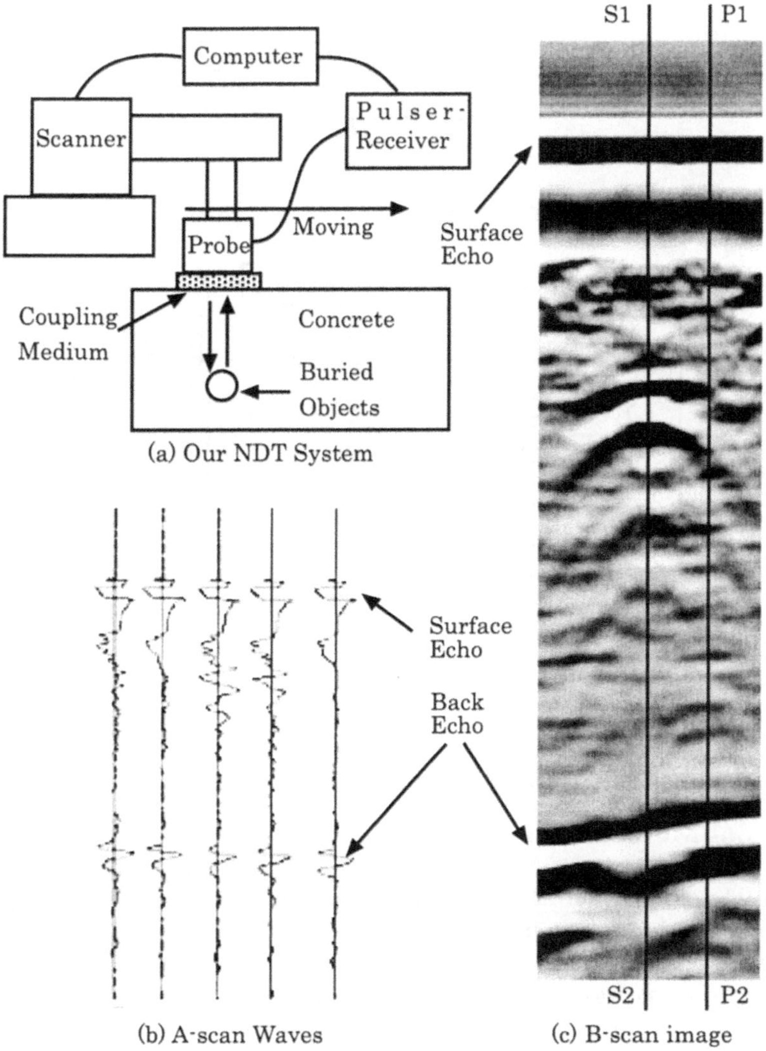

Fig. 1. An ultrasonic testing.

3 Method

3.1 Outline of the Procedure

This method consists of two stages as shown in Fig. 2. The first stage suppress the inherent wave from the received wave with estimating the inherent wave by using fuzzy-logic. Here, we explaine the new estimation method of inherent wave of the sensor system. The second stage detect automatically the location of embedded objects by analyzing the histogram of the intensity. The detail of each process is described in following subsection.

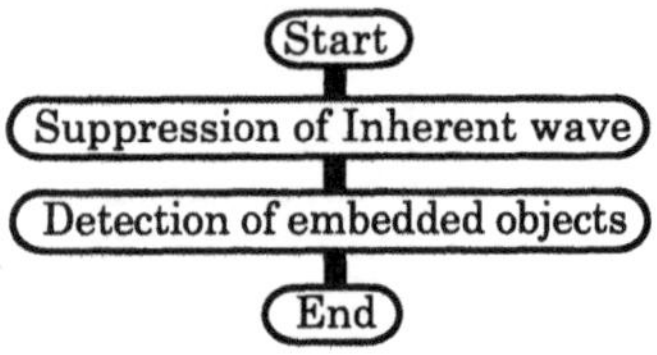

Fig. 2. The outline of procedure.

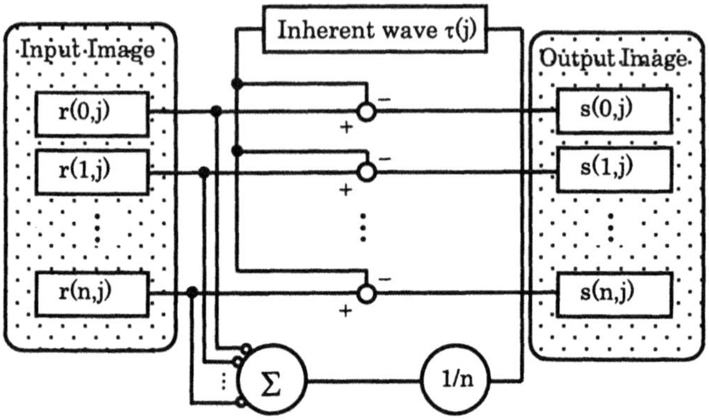

Fig. 3. The diagram of previous method.

3.2 Suppresstion of the Inherent Wave

Previous Method. A B-scan image includes the echo not only from embedded objects but also from the front surface of a concrete, the back surface of it, and so on. Here, we define the wave consists of the echo from every object except embedded objects. The inherent wave prevents us from finding the inner echo, because original B-scan image is the echo from embedded object superposed on the inherent wave. Therefore, a method for estimating the inherent wave is required to subtract the inherent wave from the original B-scan image. We already defined the inherent wave as a mean wave of all the reflected waves [4]. It is formulated as

$$\tau(j) = \frac{1}{N} \sum_{i=1}^{N} r(i, j) \tag{1}$$

where τ denotes the inherent wave, r denotes the reflected wave at the measurement point i, j denotes the time, N denotes the number of measurement points. We subtract the inherent wave from each reflected wave. Then only internal echo from embedded objects remains in the original image. The diagram of this procedure is shown in Fig. 3.

- assumption:1 the inherent wave is included in all measurement waves.
- assumption:2 the echo from the embedded object locally exists in comparison with measurement points.

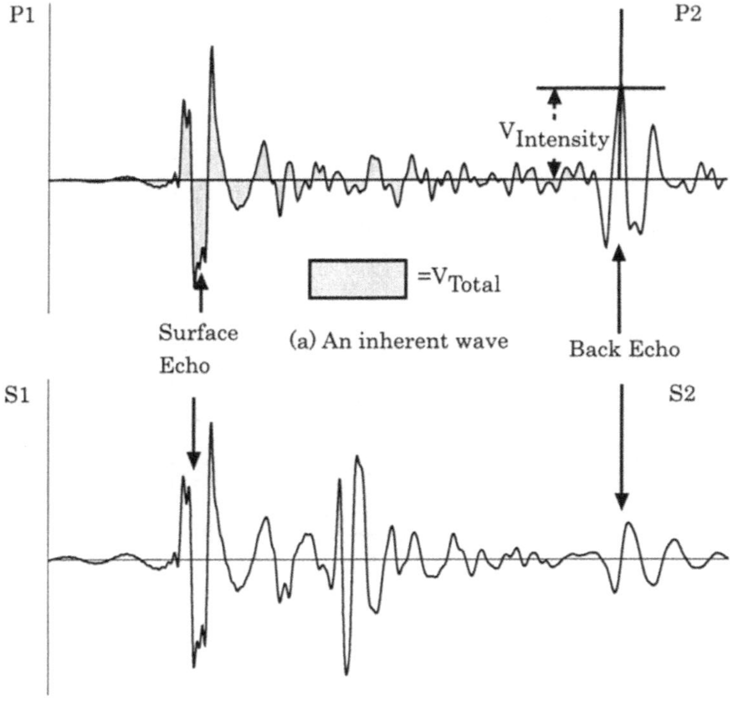

(b) An inherent wave added anothor echo

Fig. 4. Two types of received wave.

This method used two assumptions. The first assumption is that the inherent waves have similar pattern in all measurement waves. The second assumption is that the echo from the embedded object locally exists in comparison with measurement range. However, the second one does not always held. In this case, the suppression image provides some artifacts (Fig. 6(b)). This phenomenon prevents us from detecting the location of embedded objects. To solve this problem, we propose the new estimation method using fuzzy inference [5].

New method for inherent wave estimation. The reflected wave falls into two patterns: one is the pattern containing only inherent wave and another is the pattern containing echo from embedded object. Figure 4 shows the two types of receiving waves.

In this figure, $V_{Inherent}$, and V_{Total} denotes the intensity of the back echo, the integral intensity of the time series respectively. Estimating the inherent wave requires to distinguish the clean inherent wave from the additional one. Here, we use the fuzzy inference. It can provide a powerful method for representing human knowledge. Figure 5 shows the different two types of knowledge needed to make the inherent wave.

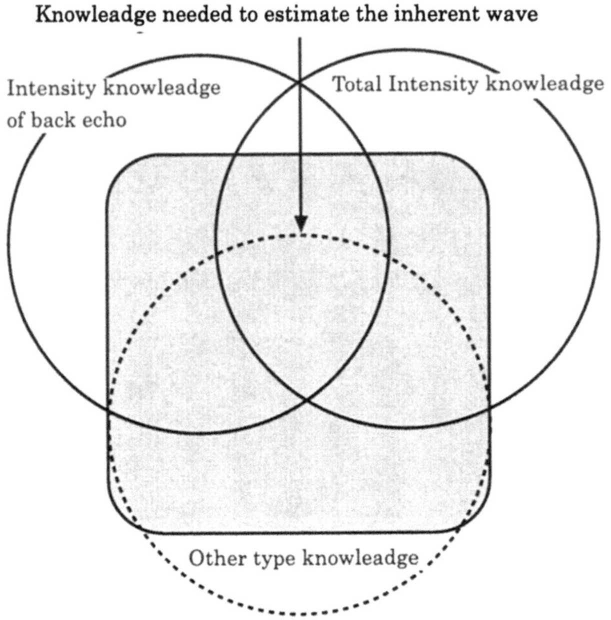

Fig. 5. Knowledge of inherent wave.

- Knowledge 1: the intensity of back echo is high.
- Knowledge 2: the integral value of the intensity between front echo and back echo is small.

The knowledge can be represented by membership functions $V_{Inherent}$, and V_{Total} as shown in Fig. 6.

The possibility of the inherent wave, $V_{Inherent}(i)$, is calculated as an algebraic product of the degrees.

$$\mu_{Inherent}(i) = \mu_{Intensity}(i) \cdot \mu_{Total}(i) \tag{2}$$

where i denotes a measurement point. By this degree we rewrite the previous Equation (1) as follows.

$$\tau(j) = \frac{\sum_{i=1}^{N} r(i,j) \cdot \mu_{Inherent}(i)}{\sum_{i=1}^{N} \mu_{Inherent}(i)}. \tag{3}$$

We can determine the inherent wave by using Equation (3), without considering the second assumption. The diagram of this procedure is described in Fig. 7. Moreover, artifacts do not appear in the resultant image (Fig. 8(c))

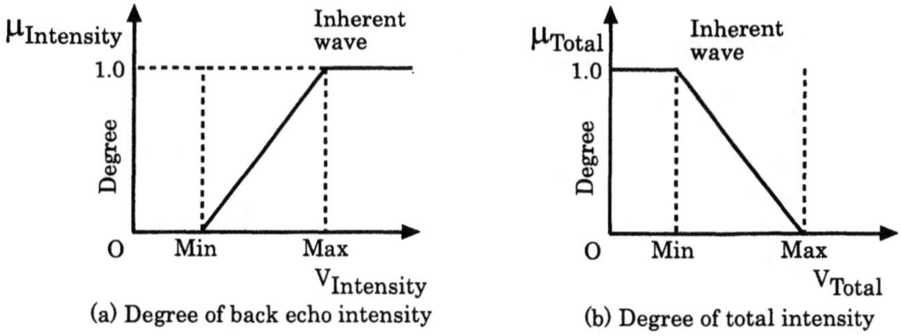

Fig. 6. Membership functions.

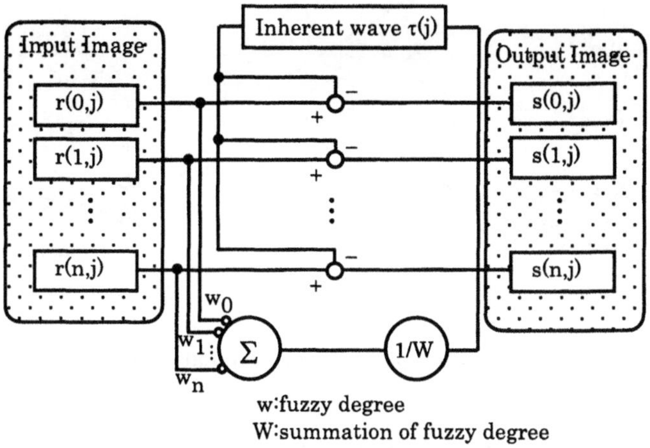

Fig. 7. The diagram of new method.

3.3 Detection of Embedded Object

The echo is generated at the place with differential acoustic impedance as the boundary between embedded objects and concrete. Analyzing this phenomenon, we attempt to detect the embedded objects with three steps.

First, we make a histogram of the summation of intensities between the front echo and the back echo for each measurement point in the suppression image, and then estimate the horizontal locations of candidate points by finding local maxima in the histogram.

Next, the incident wave of our system has two wavelengths. It causes the echo to four local extreme values. And, it has been proven that the third extrem value has max absolute intensity in the echo from the experience. Therefore, we detect the echo with the maximum absolute intensity as the echo of embedded objects in the wave of each estimated location. The depth of the embedded object is estimated at zero cross point before a wavelength from the maximum intensity point. This detection algorithm is summarized as follows.

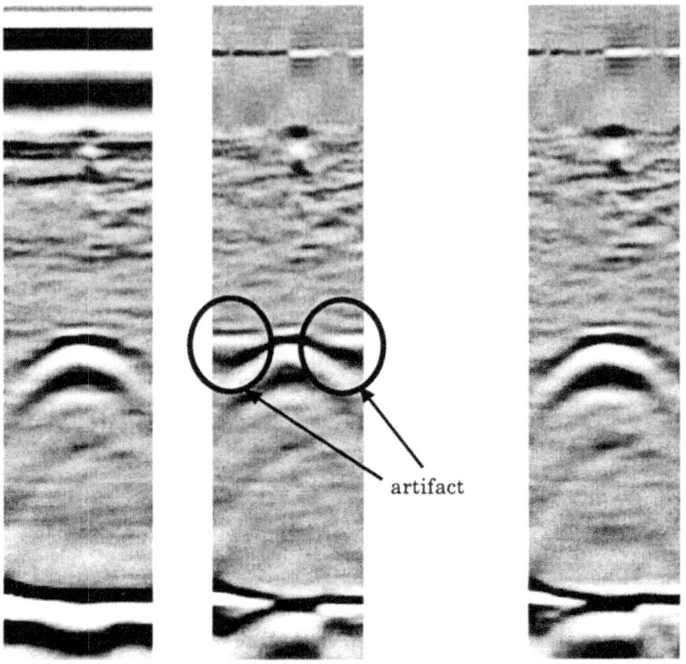

(a) Original (b) Previous method (c) Proposal method

Fig. 8. Suppressing the inherent wave.

- Step 1: Make a histogram of the summation of the absolute intensity between the time of front echo and the time of back echo for every measurement point (Fig. 9 step1).
- Step 2: Estimate the position of embedded objects by finding local maxima in the histogram of intensity (Fig. 9 step2).
- Step 3: For each wave that includes estimated point obtained in Step 2, we find a embedded point with maximum amplitude. The depth of the embedded objects are estimated at zero cross point before a wavelength from the maximum amplitude point (Fig. 9 step3).

4 Experimental Results and Conclusions

This method was applied to two cases. The details are tabulated in Table 1.

Figure 8 shows the histogram of the summation of absolute intensity between the time of the front echo and the time of the back echo for every measurement point. The histogram showed that the new method was able to assign higher intensity on the embedded object than the previous method. Table 2 tabulates the detection results for cases A and B. 10 trials were performed on each member. The results showed that the hit ratio of the embedded objects in the new method

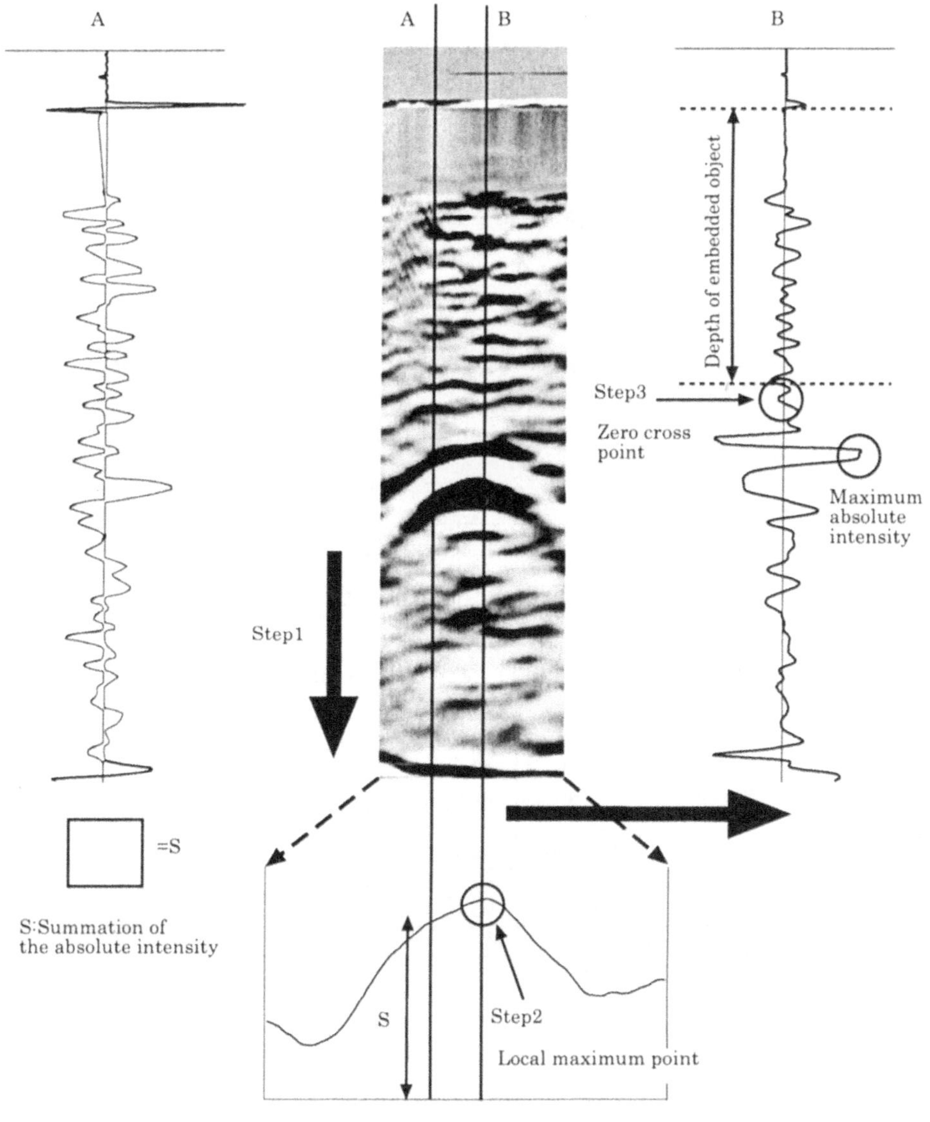

Fig. 9. Ditection process of embedded objects.

was higher than previous one. This also showed that our method was able to detect different types of embedded objects such as resin tube and steel pipe.

Consequently, this method would be of great interest in the field of ultrasonic NDT.

It reminds as a future work to develop a method of recognizing the material of embedded object.

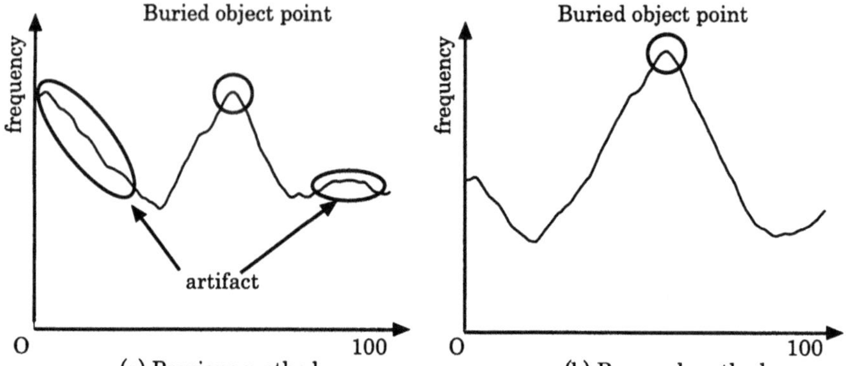

(a) Previous method (b) Proposal method

Fig. 10. Histgram of the intensity.

Table 1. The concrete members.

Case	Embedded object	Radius	X(mm)	Y(mm)
A	Steel pipe	20.0	50.0	80.0
B	Resin tube	33.0	50.0	50.0

Table 2. The detecting results.

Case		Object	X(mm)	Y(mm)	Number	Hit ratio (%)
A	Previous method	None	18.1	82.4	8	26.3
		None	20.8	80.9	4	
		None	32.5	76.8	2	
		Steel pipe	53.7	82.0	10	
		None	83.7	82.9	10	
		None	98.0	83.0	4	
	New method	Steel pipe	54.0	80.7	10	100.0
B	Previous method	None	15.3	51.0	10	40.0
		Resin tube	43.3	56.1	10	
		None	80.3	47.7	8	
	New method	Resin tube	43.5	56.0	10	52.6
		None	78.9	47.0	9	

References

1. J. Sanjie and M.A. Mohamed, "Ultrasonic Flaw Detection Based Mathematical Morphology," IEEE Trans. Ultrasonics Frequency Cont., vol. 41, pp. 150-160, Jan. 1994.
2. M. Karaoguz, N. Bilgutay, T. Akgul and S. Popovics, "Defect Detection in Concrete Using Split Spectrum Processing," IEEE Ultrasonics Sympo. Proc., pp. 843-846, 1998.
3. M. Krause, F. Mielents, B. Milman, H. Wiggenhauser, W. Muller and V. Schmitz, "Ultrasonic Imaging of Concrete Members using an Array System," Insight, vol. 42, pp. 447-450, Jun. 2000.

4. K. Nagamune, Y. Hata and K. Taniguchi, "Automated Extraction of Buried Pipes from 3D Ultrasonic Images by Fuzzy Inference," Int. Conf. Signal Processing Proc., pp. 877-881, 2000.
5. Y. Hata, S. Kobashi, S. Hirano, H. Kitagaki and E. Mori, "Automated Segmentation of Human Brain MR Images Aided by Fuzzy Information Granulation and Fuzzy Inference," IEEE Trans. Systems, Man, Cybernatics, vol. 30, pp. 381-395, 2000.

Fuzzy Association Rules: Semantic Issues and Quality Measures

Eyke Hüllermeier

Statistics and Decision Theory
University of Paderborn, Germany
eyke@upb.de

Abstract. Fuzzy association rules provide a data mining tool which is especially interesting from a knowledge-representational point of view since fuzzy attribute values allow for expressing rules in terms of natural language. So far, however, association rules of this type have not been investigated thoroughly from a semantical point of view. Particularly, the quality measures which have been proposed for assessing such rules are mostly "ad-hoc" generalizations of measures for classical rules. In this paper, we show that fuzzy association rules can be interpreted in different ways and that the interpretation has a strong influence on their assessment and, hence, on the process of rule mining. We motivate the use of multiple-valued implication operators in order to model fuzzy association rules and propose quality measures suitable for this type of rule.

1 Introduction

Data mining aims at discovering interesting patterns in (usually large) data sets. To this end, several techniques and representational frameworks have been developed, such as *association rules* which have recently gained considerable attraction. Association rules, syntactically written $A \Rightarrow B$, provide a means for representing dependencies between attributes. Typically, A and B denote sets of binary attributes, also called features or items. The intended meaning of a (binary) rule $A \Rightarrow B$ is that a transaction (a data record stored in the database) that contains the set of items A is likely to contain the items B as well. As a typical example consider the rule $\{eggs, butter\} \Rightarrow \{sugar\}$, suggesting that it is likely to find sugar in a transaction (which is here a purchase) which does already contain eggs and butter. Several efficient algorithms for mining association rules in large databases have been devised [1, 20, 23]. Typically, such algorithms perform by generating a set of candidate rules from selected itemsets which are then filtered according to several quality criteria.

Generally, a database does not only contain binary attributes but also attributes with values ranging on (completely) ordered scales, e.g. cardinal or ordinal attributes. This has motivated a corresponding generalization of (binary) association rules. Typically, a *quantitative association rule* specifies attribute

B. Reusch (Ed.): Fuzzy Days 2001, LNCS 2206, pp. 380–391, 2001.

values by means of intervals, as e.g. in the simple rule "Employees at the age of 30 to 40 have incomes between $50,000 and $70,000."

This paper investigates *fuzzy association rules*, which are basically obtained by replacing intervals in quantitative rules by fuzzy sets (intervals). The use of fuzzy sets in connection with association rules – as with data mining in general [21] – has recently been motivated by several authors (e.g. [2–7, 12–14, 16, 26, 28]). Among other aspects, fuzzy sets avoid an arbitrary determination of crisp boundaries for intervals. Furthermore, fuzzy association rules are very interesting from a knowledge representational point of view: The very idea of fuzzy sets is to act as an interface between a numeric scale and a symbolic scale which is usually composed of linguistic terms. Thus, the rules discovered in a database might be presented in a linguistic and hence comprehensible and user-friendly way. Example: "Middle-aged employees dispose of considerable incomes."

Even though fuzzy association rules have already been considered by some authors, the investigation of their semantics in the context of data mining has not received much attention as yet. This is somewhat surprising since a clear semantics is a necessary prerequisite, not only for the interpretation, but also for the rating and, hence, for the mining of fuzzy association rules.

The semantics of fuzzy association rules and their assessment by means of adequate quality measures constitute the main topics of the paper. In Section 2, we review the aforementioned types of association rules. Section 3 provides a thorough discussion of quality measures for fuzzy association rules. In this connection, it is shown that a generalization of (quantitative) association rules can proceed from different perspectives, which in turn suggest different types of measures. We especially motivate the use of multiple-valued implication operators in order to model fuzzy association rules. Besides, we present a semantic model which clarifies the meaning of fuzzy association rules and provides a sound theoretical basis. Finally, an alternative approach to the assessment of fuzzy association rules is discussed in Section 4. By making use of fuzzy quantifiers, this approach supports the evaluation of an association rule on a purely logical basis.

2 Association Rules

2.1 Binary Association Rules

Consider an association rule of the form $A \Rightarrow B$, where A and B denote subsets of an underlying set $\mathcal{A}$ of *items* (which can be considered as binary attributes). As already said above, the intended meaning of $A \Rightarrow B$ is that a *transaction* $T \subset \mathcal{A}$ which contains the items in A is likely to contain the items in B as well.

Of course, the development of algorithms for finding "interesting" association rules in a database D presupposes a formal definition of this qualification. A rule $A \Rightarrow B$ is generally rated according to several criteria, none of which should fall below a certain (user-defined) threshold. In common use are the following measures ($D_X \doteq \{T \in D \mid X \subset T\}$ denotes the transactions in the database D which contain the items $X \subset \mathcal{A}$, and $|D_X|$ is its cardinality):

- A measure of *support* defines the absolute number or the proportion of transactions in D containing $A \cup B$:

$$\mathsf{supp}(A \Rightarrow B) \doteq |D_{A \cup B}| \quad \text{or} \quad \mathsf{supp}(A \Rightarrow B) \doteq \frac{|D_{A \cup B}|}{|D|}. \tag{1}$$

- The *confidence* is the proportion of correct applications of the rule:

$$\mathsf{conf}(A \Rightarrow B) \doteq \frac{|D_{A \cup B}|}{|D_A|}. \tag{2}$$

- A rule $A \Rightarrow B$ should be interesting in the sense that it provides new information. That is, the occurrence of A should indeed have a positive influence on the occurrence of B. A common measure of the *interest* of a rule is

$$\mathsf{int}(A \Rightarrow B) \doteq \frac{|D_{A \cup B}|}{|D_A|} - \frac{|D_B|}{|D|}. \tag{3}$$

  This measure can be seen as an estimation of $\mathsf{Pr}(B \mid A) - \mathsf{Pr}(B)$, that is the increase in probability of B caused by the occurrence of A.

2.2 Quantitative Association Rules

In the above setting, a transaction T can be seen as a sequence $(x_1, \ldots, x_m)$ of values of binary variables $X_\imath$ with domain $\mathfrak{D}_{X_\imath} = \{0, 1\}$, where $x_\imath = T[X_\imath] = 1$ if the $\imath$th item $X_\imath$ is contained in T and $x_\imath = 0$ otherwise. Now, let X and Y be quantitative attributes (such as age or income) with completely ordered domains $\mathfrak{D}_X$ and $\mathfrak{D}_Y$, respectively. Without loss of generality we can assume that $\mathfrak{D}_X, \mathfrak{D}_Y \subset \mathfrak{R}$. A quantitative association rule involving the variables X and Y is then of the following form:

$$A \Rightarrow B: \quad \text{If } X \in A = [x_1, x_2] \text{ then } Y \in B = [y_1, y_2], \tag{4}$$

where $x_1, x_2 \in \mathfrak{D}_X$ and $y_1, y_2 \in \mathfrak{D}_Y$. This approach can simply be generalized to the case where X and Y are multi-dimensional vectors and, hence, A and B hyper-rectangles rather than intervals. Subsequently, we proceed from fixed variables X and Y, and consider the database D as a collection of data points $(x, y) = (T[X], T[Y])$, i.e. as a projection of the original database.

Note that the quality measures from Section 2.1 are applicable in the quantitative case as well:[1]

$$\mathsf{supp}(A \Rightarrow B) = |(\{(x, y) \in D \mid x \in A \wedge y \in B\}|, \tag{5}$$

$$\mathsf{conf}(A \Rightarrow B) = \frac{|\{(x, y) \in D \mid x \in A \wedge y \in B\}|}{|\{(x, y) \in D \mid x \in A\}|}. \tag{6}$$

[1] Subsequently we focus on *support* and *confidence* measures. The results can be transferred to other measures such as *interest* in a straightforward way.

In fact, each interval $[x_1, x_2]$ does again define a binary attribute $X_{x_1,x_2} = \mathbb{I}_{[x_1,x_2]}$. Thus, not only the rating but also the mining of quantitative rules can be reduced to the mining of binary association rules, by simply transforming the numerical data into binary data [19, 24]. Still, finding a useful transformation (binarization) of the data is a non-trivial problem by itself which affects both, the efficiency of subsequently applied mining algorithms and the potential quality of discovered rules. Apart from data transformation methods, clustering techniques can be applied which create intervals and rules at the same time [17, 27].

2.3 Fuzzy Association Rules

Replacing the sets (intervals) A and B in (4) by fuzzy sets (intervals) leads to fuzzy (quantitative) association rules. Thus, a fuzzy association rule is understood as a rule of the form $A \Rightarrow B$, where A and B are now fuzzy subsets rather than crisp subsets of the domains $\mathfrak{D}_X$ and $\mathfrak{D}_Y$ of variables X and Y, respectively. We shall use the same notation for ordinary sets and fuzzy sets. Moreover, we shall not distinguish between a fuzzy set and its membership function, that is, $A(x)$ denotes the degree of membership of the element x in the fuzzy set A. Note that an ordinary set A can be considered as a "degenerate" fuzzy set with membership degrees $A(x) = \mathbb{I}_A(x) \in \{0, 1\}$.

The standard approach to generalizing the quality measures for fuzzy association rules is to replace set-theoretic by fuzzy set-theoretic operations. The Cartesian product $A \times B$ of two fuzzy sets A and B is usually defined by the membership function $(x, y) \mapsto \min\{A(x), B(y)\}$. Moreover, the cardinality of a finite fuzzy set is simply the sum of its membership degrees [18]. Thus, (5) and (6) can be generalized as follows:

$$\mathsf{supp}(A \Rightarrow B) \doteq \sum_{(x,y)\in D} \min\{A(x), B(y)\}, \tag{7}$$

$$\mathsf{conf}(A \Rightarrow B) \doteq \frac{\sum_{(x,y)\in D} \min\{A(x), B(y)\}}{\sum_{(x,y)\in D} A(x)}. \tag{8}$$

Note that the support of $A \Rightarrow B$ corresponds to the sum of the *individual supports*, provided by tuples $(x, y) \in D$:

$$\mathsf{supp}_{[x,y]}(A \Rightarrow B) = \min\{A(x), B(y)\}. \tag{9}$$

According (9), (x, y) supports $A \Rightarrow B$ if both, $x \in A$ and $y \in B$.

3 Implication-Based Fuzzy Association Rules

3.1 Support

The fact that the antecedent A and the consequent B play symmetrical roles in (9) might appear strange. Indeed, a more logic-oriented approach to modeling a fuzzy rule "If X is A then Y is B" would use a generalized implication operator $\rightsquigarrow$, i.e. a mapping $[0,1] \times [0,1] \to [0,1]$ which generalizes the classical material implication (particularly, $\rightsquigarrow$ is non-increasing in the first and non-decreasing in the second argument). Thus, individual support can be defined as

$$\mathsf{supp}_{[x,y]}(A \Rightarrow B) \doteq A(x) \rightsquigarrow B(y) \tag{10}$$

and, hence, the overall support as

$$\mathsf{supp}(A \Rightarrow B) \doteq \sum_{(x,y)\in D} A(x) \rightsquigarrow B(y). \tag{11}$$

In order to realize the difference between (9) and (10), consider a simple rule "$A \Rightarrow B$: If X is approximately 10 then Y is almost 0" defined by two fuzzy subsets of the non-negative integers,

$$A : x \mapsto \begin{cases} 1 - \frac{|10-x|}{5} & \text{if } 6 \le x \le 14 \\ 0 & \text{otherwise} \end{cases}, \quad B : y \mapsto \begin{cases} 1 - \frac{x}{5} & \text{if } 0 \le x \le 4 \\ 0 & \text{otherwise} \end{cases}.$$

To which degree does the tuple $(x,y) = (8,3)$ support the above rule? According to (9), the support is $2/5$, namely the minimum of the membership of 8 in A and the membership of 3 in B. According to (10) with $\rightsquigarrow$ the Goguen implication

$$\alpha \rightsquigarrow \beta \doteq \begin{cases} 1 & \text{if } \alpha = 0 \\ \min\{1, \beta/\alpha\} & \text{if } \alpha > 0 \end{cases}$$

the individual support is larger, namely $2/3$. In fact, $(x,y) = (8,3)$ does hardly violate (and hence supports) the rule in the sense of (10): It is true that $y = 3$ does not fully satisfy the conclusion part of the rule; however, since $x = 8$ does not fully meet the condition part either, it is actually not expected to do so.

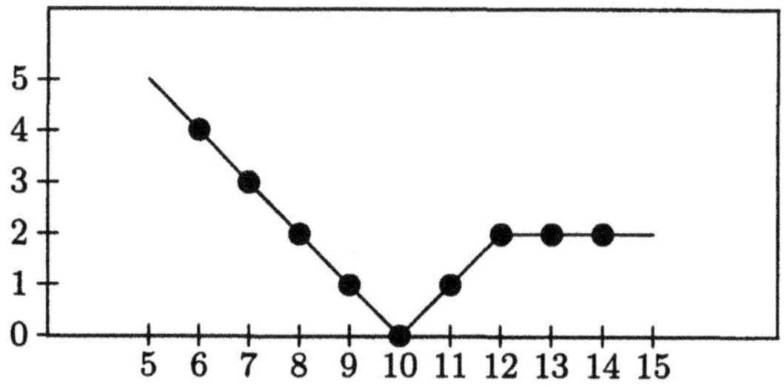

Fig. 1. A simple functional relation between two variables that can be described by means of a gradual fuzzy rule.

As can be seen, the definition of adequate quality measures for fuzzy association rules depends strongly on the interpretation of the rule. For further illustration, consider the nine observations shown in Figure 1.[2] In the sense of (10), each of these observations does fully support the rule "If X is approximately 10 then Y is almost 0." In fact, this rule is actually interpreted as "The closer X is to 10, the closer Y is to 0" or, more precisely, "The closer X is to 10, the more it is guaranteed that Y is close to 0." Therefore, (11) yields $\mathsf{supp}(A \Rightarrow B) = 9$. As opposed to this, the overall support is only 5 according to (7), since the individual support $\min\{A(x), B(y)\}$ which comes from a point (x, y) is bounded by $B(y)$, the closeness of y to 0. For instance, the support through $(6, 4)$ is only $1/5$ rather than 1. Note that, in the sense of (11), the support of the rule $A \Rightarrow B$ is larger than the support of $B \Rightarrow A$! This reflects the fact that the closeness of x to 10 is less guaranteed by the closeness of y to 0 than vice versa. For instance, $y = 2$ is "rather" close to 0, whereas $x = 14$ is only "more or less" close to 10, hence $\mathsf{supp}_{[2,14]}(B \Rightarrow A) = 1/3 < 1 = \mathsf{supp}_{[14,2]}(A \Rightarrow B)$.

The above example shows that the implication-based approach should be preferred whenever an association rule is thought of as expressing a gradual "the more ... the more ..."-relation between variables and all the more if this relation is not symmetric. For instance, the rule "Young people have low income" might actually be understood as "The younger a person, the lower the income," and this rule might well be distinguished from its inversion "The lower the income, the younger the person."

Concerning the adequacy of (11) as a support measure for association rules, two points deserve mentioning. Firstly, the concept of *support* in the context of association rules is actually intended as *non-trivial support*. Yet, the truth degree of a (generalized) implication $\alpha \rightsquigarrow \beta$ is 1 whenever $\alpha = 0$: From a logical point of view, a false premise entails any conclusion. That is, the rule $A \Rightarrow B$ would also be supported by those points (x, y) for which $x \notin A$. In order to avoid this effect, (10) can be modified as follows:

$$\mathsf{supp}_{[x,y]}(A \Rightarrow B) = \begin{cases} A(x) \rightsquigarrow B(y) & \text{if} \quad A(x) > 0 \\ 0 & \text{if} \quad A(x) = 0 \end{cases}. \tag{12}$$

According to (12), a point (x, y) supports a rule if both, it satisfies the rule (from a logical point of view) and it is non-trivial. Here, non-triviality is considered as a binary concept. However, it can also be quantified as a gradual property, namely as the degree to which x is in A. Combining satisfaction and non-triviality by means of a generalized logical conjunction $\top$, a so-called t-norm, then yields $\mathsf{supp}_{[x,y]}(A \Rightarrow B) = \top(A(x), A(x) \rightsquigarrow B(y))$. For example, by using the product operator as a special t-norm we obtain

$$\mathsf{supp}_{[x,y]}(A \Rightarrow B) = A(x) \cdot (A(x) \rightsquigarrow B(y)). \tag{13}$$

Note that (9), (12), and (13) are identical in the case where A and B are intervals, that is where $A(x), B(y) \in \{0, 1\}$.

The second point concerns the choice of the implication operator $\rightsquigarrow$. In fact, different types of implication operators exist which support different interpretations of a fuzzy rule [9]. The gradual "the more ... the more..."-interpretation

[2] Needless to say, this is a somewhat artificial example not at all typical of data mining.

discussed above is supported by so-called R(esiduated)-implications. An implication $\rightsquigarrow$ of this type can be derived from a t-norm $\top$ by residuation (hence the name):

$$\alpha \rightsquigarrow \beta \doteq \sup\{\gamma \in [0,1] \mid \top(\alpha,\gamma) \leq \beta\}.$$

A second important class is given by so-called S(trong)-implications, which are defined by $\alpha \rightsquigarrow \beta \doteq n(\alpha) \oplus \beta$, where $\oplus$ is a t-conorm (a generalized disjunction) and n a strong (hence the name) negation. For example, taking $n(\cdot) = 1-(\cdot)$ and $\oplus = \max$, one obtains the Kleene-Dienes implication $\alpha \rightsquigarrow \beta = \max\{1-\alpha, \beta\}$.

S-implications support a different type of fuzzy rule, often referred to as *certainty rules*. Basically, they attach a level of uncertainty to the conclusion part of the rule, in correspondence with the truth degree of the condition part. These rules, however, appear less reasonable in the context of association rules. This can be exemplified by the Kleene-Dienes implication. For this operator, the truth degree is *lower-bounded* by $1-A(x)$. Thus, (13) entails $\mathsf{supp}_{[x,y]}(A \Rightarrow B) \geq \min\{A(x), 1-A(x)\}$ regardless of the value $B(y)$. For example, if $A(x) = 1/2$, then $\mathsf{supp}_{[x,y]}(A \Rightarrow B) = 1/2$, no matter whether $B(y)$ is 1/2, 1/4, or even 0. This contrasts with R-implications, for which $\beta = 0$ generally implies $\alpha \rightsquigarrow \beta = 0$.

3.2 Confidence

A measure of confidence of a fuzzy association rule can be derived from a corresponding measure of support. Indeed, in the non-fuzzy case, the confidence of $A \Rightarrow B$ is nothing else than the support of $A \Rightarrow B$ over the support of A, that is, the support of $A \Rightarrow \mathfrak{D}_Y$. Interestingly enough, however, the minimal confidence condition $\mathsf{conf}(A \Rightarrow B) \geq \Delta$ can be interpreted in different ways, which in turn suggest different generalizations.

According to the aforementioned interpretation which relates the support of $A \Rightarrow B$ to the support of A, one obtains the generalized confidence measure

$$\mathsf{conf}(A \Rightarrow B) \doteq \frac{\sum_{(x,y)\in D} \mathsf{supp}_{[x,y]}(A \Rightarrow B)}{\sum_{(x,y)\in D} \mathsf{supp}_{[x,y]}(A \Rightarrow \mathfrak{D}_Y)}. \tag{14}$$

Note that $\alpha \rightsquigarrow 1 = 1$ for all α and, hence, the denominator in (14) simplifies to $\sum_{(x,y)\in D} \mathbb{I}_{(0,1]}(A(x))$ for (12) and $\sum_{(x,y)\in D} A(x)$ for (13).

A second possibility is to relate the support of $A \Rightarrow B$ to the support of $A \Rightarrow \neg B$. In this case, the minimal confidence condition means that the rule $A \Rightarrow B$ should be supported much better than $A \Rightarrow \neg B$:

$$\mathsf{conf}(A \Rightarrow B) = \frac{\sum_{(x,y)\in D} \mathsf{supp}_{[x,y]}(A \Rightarrow B)}{\sum_{(x,y)\in D} \mathsf{supp}_{[x,y]}(A \Rightarrow \neg B)}. \tag{15}$$

Interestingly enough, (14) and (15) are equivalent in the non-fuzzy case, since then $A(x) = \mathsf{supp}_{[x,y]}(A \Rightarrow \mathfrak{D}_Y) = \mathsf{supp}_{[x,y]}(A \Rightarrow B) + \mathsf{supp}_{[x,y]}(A \Rightarrow \neg B)$ for all (x,y) and, therefore,

$$\frac{\mathsf{supp}(A \Rightarrow B)}{\mathsf{supp}(A \Rightarrow \mathfrak{D}_Y)} \geq \Delta \Leftrightarrow \frac{\mathsf{supp}(A \Rightarrow B)}{\mathsf{supp}(A \Rightarrow \neg B)} \geq \frac{\Delta}{1-\Delta}.$$

3.3 Semantic Interpretation of Fuzzy Association Rules

A sound theoretical basis for the type of fuzzy association rule introduced in this section can be derived from the semantic model of implication-based fuzzy rules proposed in [8]. The idea is to represent an implication-based fuzzy rule $A \rightsquigarrow B$, i.e. the rule $A \Rightarrow B$ modeled by the implication operator $\rightsquigarrow$, as a convex combination

$$A \rightsquigarrow B \equiv \sum_{A \rightarrow_m B \in \mathcal{G}} p_m \cdot [A \rightarrow_m B], \tag{16}$$

where $\mathcal{G}$ is a class of (modified) pure gradual rules and the p_m are weights which might be interpreted as probability degrees. The rule $A \rightarrow_m B$ is the fuzzy rule $A \Rightarrow B$ modeled by the modified Rescher-Gaines implication $\rightarrow_m$ with $\alpha \rightarrow_m \beta = 1$ if $m(\alpha) \le \beta$ and 0 otherwise. As can be seen, a pure gradual rule induces a crisp relation of admissible tuples (x, y) and, hence, disposes of a very simple semantic interpretation. In fact, a tuple (x, y) satisfies the rule $A \rightarrow_m B$ iff it satisfies the constraints $x \in A_\lambda \Rightarrow y \in B_{m(\lambda)}$ for all level-cuts $\lambda \in \mathcal{L}$ ($\mathcal{L}$ is an underlying finite scale of membership degrees and A_λ is the λ-cut of A). Moreover, the degree to which (x, y) satisfies $A \rightsquigarrow B$ is given by the sum of the weights p_m of those rules $A \rightarrow_m B$ satisfied by (x, y). For details we refer to [8], where necessary and sufficient conditions for the existence and uniqueness of the representation (16) are given.

Above we have suggested to consider a fuzzy association rule $A \Rightarrow B$ as an implication-based (gradual) fuzzy rule. Proceeding from (16), a rule $A \Rightarrow B$ can hence be seen as a convex combination of simple or *pure gradual* association rules

$$A \rightarrow_m B. \tag{17}$$

Each of these gradual association rules in turn corresponds to a collection

$$A_\lambda \Rightarrow B_{m(\lambda)} \qquad (\lambda \in \mathcal{L}) \tag{18}$$

of ordinary association rules. In fact, if the level-cuts A_λ and $B_{m(\lambda)}$ of the fuzzy sets A and B are intervals, which holds true for commonly used membership functions, then (18) reduces to a class of interval-based association rules.

This interpretation assigns an association rule a concrete meaning and might hence be helpful in connection with the acquisition (mining) and interpretation of such rules. Apart from this, it provides a basis for justifying the quality measures for fuzzy association rules proposed in this section. In fact, proceeding from the convex combination of rules (17), it is obvious to define the support of $A \Rightarrow B$ as the convex combination of the supports of the rules $A \rightarrow_m B$:

$$\mathsf{supp}(A \Rightarrow B) = \sum_{A \rightarrow_m B \in \mathcal{G}} p_m \cdot \mathsf{supp}(A \rightarrow_m B). \tag{19}$$

Thus, it remains to define the (non-trivial) support of a pure gradual association rule $A \rightarrow_m B$, i.e. of a collection of ordinary association rules (18). To which degree does a point (x, y) support this class of constraints? One possibility is to say that (x, y) supports $A \rightarrow_m B$ if it satisfies *all* individual constraints and, furthermore, at least one of these constraints is non-trivial:

$$\mathsf{supp}_{[x,y]}(A \rightarrow_m B) = \begin{cases} 1 & \text{if } A(x) > 0 \wedge m(A(x)) \leq B(y) \\ 0 & \text{otherwise} \end{cases} . \tag{20}$$

A second possibility is to define $\mathsf{supp}_{[x,y]}(A \rightarrow_m B)$ as the sum of weights of those individual constraints which are indeed non-trivially satisfied:

$$\mathsf{supp}_{[x,y]}(A \rightarrow_m B) = \begin{cases} A(x) & \text{if } m(A(x)) \leq B(y) \\ 0 & \text{otherwise} \end{cases} . \tag{21}$$

It is readily verified that (20), in conjunction with (19), yields the support measure (12), and that (21) in place of (20) implies (13). This result provides a sound basis for these measures of (individual) support and, hence, for further quality measures derived from them.

4 Logic-oriented evaluation

The approach to evaluating fuzzy association rules pursued in Section 3 combines statistical and logical aspects: The individual support (13) of a rule $A \Rightarrow B$, provided by a tuple (x, y), takes the logical nature of the assumed implicative relation between A and B into account. Still, the overall support of $A \Rightarrow B$ is obtained by *summing* the individual supports, i.e. by means of a "counting procedure" which has a more statistical flavor.

Now, a fuzzy approach does also enable a purely logical evaluation of an association rule $A \Rightarrow B$. For example, a proposition such as "all points supporting A do also support B" can be translated directly into a fuzzy logical expression. The truth degree of this proposition, e.g.

$$\min_{(x,y)} A(x) \rightsquigarrow B(y), \tag{22}$$

can then be used as an evaluation measure of $A \Rightarrow B$. Of course, this type of evaluation is too strong and not in line with the purpose of data mining. To avoid this problem, the universal quantifier used in the above proposition, and modeled by the minimum operator in (22), can be replaced by a so-called *fuzzy quantifier*. This allows one to say, e.g., that "most of the points supporting A do also support B." As can be seen, a fuzzy quantifier can serve as an exception-tolerant relaxation of the universal quantifier. More generally, such quantifiers can be used for describing an approximate number ("about ten") or an approximate proportion ("around one-half").

Quantifiers such as "for most" or "for many" can be modeled by means of a fuzzy set [25, 10] with membership function $\mu : \{0, 1, \ldots, m\} \rightarrow [0, 1]$ such that $\mu(m) = 1$ and $\mu(\imath) \leq \mu(\jmath)$ for $0 \leq \imath < \jmath \leq m$.[3] Here, $\mu(\jmath)$ defines the truth

[3] Other possibilities of expressing a fuzzy quantifier exist as well, including the use of order-statistics [22] and an ordered weighted minimum or maximum [11].

degree of the proposition "most elements in the set X have property P" if P does actually apply to $\jmath$ among the m elements in X. We define an associated membership function $\overline{\mu}$ by $\overline{\mu}(0) = 0$ and $\overline{\mu}(\jmath) = 1 - \mu(\jmath - 1)$ for $1 \leq \jmath \leq m$. A membership degree $\overline{\mu}(\jmath)$ can then be interpreted as quantifying the importance that the property P is satisfied for exactly $\jmath$ (out of the m) elements.

Let $\mu, \overline{\mu} : \{0, 1, \ldots, N\} \rightarrow [0, 1]$ be membership functions modeling the "for many" concept, where N is a large enough constant. The truth degree of the proposition that "$A \Rightarrow B$ is supported by *many* points" –underlying the idea of the support of an association rule– can then be modeled as follows:

$$\mathsf{supp}(A \Rightarrow B) = \min_{0 \leq \jmath \leq |D|} \max\{1 - \overline{\mu}(\jmath), \gamma(\jmath)\}, \tag{23}$$

where

$$\gamma(\jmath) = \max_{D' \subset D : |D'| = \jmath} \min_{(x,y) \in D'} \mathsf{supp}_{[x,y]}(A \Rightarrow B)$$

defines the degree to which "$A \Rightarrow B$ is supported by $\jmath$ points." Moreover, $\mathsf{supp}_{[x,y]}(A \Rightarrow B)$ (as defined in (9) or (13)) denotes the individual support given by (x, y) to the rule $A \Rightarrow B$. The extent to which a (small) degree $\gamma(\jmath)$ decreases the overall support is upper-bounded by $1 - \overline{\mu}(\jmath)$, i.e. by the respective level of (un-)importance. Notice that we do not have to consider all subsets $D' \subset D$ of size $\jmath$ for deriving $\gamma(\jmath)$. If the $n = |D|$ values $v = \mathsf{supp}_{[x,y]}(A \Rightarrow B)$ are arranged in a non-increasing order $v_1 \geq v_2 \geq \ldots \geq v_n$, then (23) is equivalent to

$$\mathsf{supp}(A \Rightarrow B) = \min_{0 \leq \jmath \leq n} \max\{1 - \overline{\mu}(\jmath), v_\jmath\}, \tag{24}$$

where $v_0 = 1$ by definition. (24) can be interpreted by the fuzzy proposition that, for all $\jmath$, either it is not important that the rule $A \Rightarrow B$ is supported by $\jmath$ points or it is indeed supported by $\jmath$ points.

As a decisive difference between (24) and the measures in Sections 3 it should be observed that (24) does not "count" individual support values $\mathsf{supp}_{[x,y]}(A \Rightarrow B)$. For instance, let $\mu(k) = 1$ if $k \geq \Delta$ and $\mu(k) = 0$ otherwise, which corresponds to the condition that $A \Rightarrow B$ is supported by at least Δ points. We then have $\mathsf{supp}(A \Rightarrow B) = v_\Delta$ according to (24). That is, the overall support of $A \Rightarrow B$ is determined by one individual support alone, except the fact that the other individual support values have influence on the position of that value in the ordered sequence $v_1 \geq v_2 \geq \ldots \geq v_n$. Of course, in the crisp case where $v_\imath \in \{0, 1\}$ this is equivalent to the condition $|D_{A \cup B}| \geq \Delta$, that is, we obtain $\mathsf{supp}(A \Rightarrow B) = 1$ if $|D_{A \cup B}| \geq \Delta$ and $\mathsf{supp}(A \Rightarrow B) = 0$ otherwise.

According to (23), the support of a rule $A \Rightarrow B$ corresponds to the truth degree of the fuzzy proposition "$A \Rightarrow B$ is confirmed by many points." A similar evaluation can also be obtained for the confidence of a rule. To this end, let $\mu, \overline{\mu} : \{0, 1, \ldots, m\} \rightarrow [0, 1]$ be membership functions modeling the "for most" concept, where $m = |\{(x, y) \mid A(x) > 0\}|$. The truth degree of the statement that "most points supporting A do also support B" is then given by

$$\mathsf{conf}(A \Rightarrow B) = \min_{0 \leq \jmath \leq m} \max\{1 - \overline{\mu}(\jmath), v_\jmath\},$$

where $v_1 \geq v_2 \geq \ldots \geq v_m$ is the ordered sequence of values $\{\mathsf{supp}_{[x,y]}(A \Rightarrow B) \,|\, A(x) > 0\}$. In a similar way, it is possible to derive a fuzzy measure of the interest of a rule $A \Rightarrow B$. This can be done by using a fuzzy relation such as "considerably larger than" in order to translate a linguistically specified criterion such as "the support of $A \Rightarrow B$ is considerably larger than the support of the default rule $\top \Rightarrow B$."

5 Summary and Conclusions

The standard approach of replacing intervals by fuzzy sets and counting measures by fuzzy cardinality neglects important semantic issues of fuzzy association rules. In fact, different interpretations of the commonly used quality measures suggest different measures for fuzzy associations. This shows that semantic aspects become especially important in the fuzzy case, not only from a knowledge representational point of view.

We have proposed two types of evaluation for fuzzy association rules. The first one equals the standard approach in that the support of a rule is the sum of the individual supports through stored transactions. The individual support measure, however, employs a multiple-valued implication operator in order to reflect the implicative nature of associations. The second approach supports the evaluation of an association rule on a purely logical basis. In connection with the first approach, we have also outlined a semantic model which considers a fuzzy association rule as a convex combination of simple association rules. Thus, it clarifies the meaning and provides a sound basis of fuzzy associations.

The paper has mainly focused on theoretical foundations of fuzzy association rules. An important aspect of ongoing research is the practical realization of the results, that is the development of rule mining procedures. Our current implementation is presented in [15]. It extends the APRIORI algorithm [1] so as to cope with fuzzy attribute values and asymmetric support measures.

References

1. R. Agrawal and S. Srikant. Fast algorithms for mining association rules. In *Proceedings of the 20th Conference on* VLDB, Santiago, Chile, 1994.
2. Wai-Ho Au and K.C.C. Chan. An effective algorithm for discovering fuzzy rules in relational databases. In *Proceedings* IEEE *World Congress on Computational Intelligence*, pages 1314 –1319, 1998.
3. Wai-Ho Au and K.C.C. Chan. FARM: A data mining system for discovering fuzzy association rules. In *Proceedings* FUZZ–IEEE–99, pages 1217 –1222, 1999.
4. K.C.C. Chan and Wai-Ho Au. Mining fuzzy association rules. In *Proceedings* CIKM–97, *6th Int. Conf. on Inform. and Knowl. Management*, pages 10–14, 1997.
5. G. Chen, Q. Wei, and E.E. Kerre. Fuzzy data mining: Discovery of fuzzy generalized association rules. In G. Bordogna and G. Pasi, editors, *Recent Issues on Fuzzy Databases*. Springer-Verlag, 2000.
6. G. Chen, G. Wets, and K. Vanhoof. Representation and discovery of fuzzy association rules. Technical Report 00/01, ITEO Research Papers, 2000.

7. M. Delgado, D. Sanchez, and M.A. Vila. Acquisition of fuzzy association rules from medical data. In S. Barro and R. Marin, editors, *Fuzzy Logic in Medicine*. Physica Verlag, 2000.
8. D. Dubois, E. Hüllermeier, and H. Prade. Toward the representation of implicative fuzzy rules in terms of crisp rules. In *Proc.* IFSA/NAFIPS-2001, Vancouver, Canada, July 2001. To appear.
9. D. Dubois and H. Prade. What are fuzzy rules and how to use them. *Fuzzy Sets and Systems*, 84:169–185, 1996.
10. D. Dubois, H. Prade, and C. Testemale. Weighted fuzzy pattern matching. *Fuzzy Sets and Systems*, 28:313–331, 1988.
11. D. Dubois, H. Prade, and R.R. Yager. Merging fuzzy information. In J.C. Bezdek, D. Dubois, and H. Prade, editors, *Fuzzy Sets in Approximate Reasoning and Information Systems*, pages 335–401. Kluwer Academic Publishers, Boston, 1999.
12. A. Fu, M.H. Wong, S.C. Sze, W.C. Wong, W.L. Wong, and W.K Yu. Finding fuzzy sets for the mining of fuzzy association rules for numerical attributes. In IDEAL–98, *1st Int. Symp. on Intell. Data Engineering and Learning*, pages 263–268, 1998.
13. A. Gyenesei. A fuzzy approach for mining quantitative association rules. Technical Report 336, Turku Centre for Computer Science, 2000.
14. A. Gyenesei. Mining weighted association rules for fuzzy quantitative items. Technical Report 346, Turku Centre for Computer Science, 2000.
15. E. Hüllermeier. Mining fuzzy association rules in databases. Submitted.
16. C. Man Kuok, A. Fu, and M. Hon Wong. Mining fuzzy association rules in databases. *SIGMOD Record*, 27:41–46, 1998.
17. B. Lent, A. Swami, and J. Widom. Clustering association rules. In *Proceedings* ICDE–97, Birmingham, UK, 1997.
18. A. De Luca and S. Termini. Entropy of L-fuzzy sets. *Information and Control*, 24:55–73, 1974.
19. R.J. Miller and Y. Yang. Association rules over interval data. In *Proc.* ACM SIGMOD *Int. Conf. on Management of Data*, pages 452–461, 1997.
20. J.S. Park, M.S. Chen, and P.S. Yu. An efficient hash-based algorithm for mining association rules. In ACM SIGMOD *Int. Conf. on Management of Data*, 1995.
21. W. Pedrycz. Data mining and fuzzy modeling. In *Proc. of the Biennial Conference of the NAFIPS*, pages 263–267, Berkeley, CA, 1996.
22. H. Prade and R.R. Yager. Estimations of expectedness and potential surprize in possibility theory. *International Journal of Uncertainty, Fuzziness and Knowledge-Based Systems*, 2:417–428, 1994.
23. A. Savasere, E. Omiecinski, and S. Navathe. An efficient algorithm for mining association rules in large databases. In *Proceedings* VLDB–95, Zurich, 1995.
24. R. Skrikant and R. Agrawal. Mining quantitative association rules in large relational tables. In *Proceedings of the* ACM SIGMOD *International Conference on Management of Data*, pages 1–12, 1996.
25. R.R. Yager. Aggregating evidence using quantified statements. *Information Sciences*, 36:179–206, 1985.
26. S. Ben Yahia and A. Jaoua. Mining linguistic summaries of databases using Lukasiewicz implication fuzzy functional dependency. In *Proceedings* FUZZ–IEEE–99, pages 1246 –1250, 1999.
27. Y. Yang and M. Singhal. Fuzzy functional dependencies and fuzzy association rules. In *Data Warehousing and Knowledge Discovery, Proceedings* DAWAK–99, pages 229–240. Springer-Verlag, 1999.
28. W. Zhang. Mining fuzzy quantitative association rules. In *Proc. 11th* IEEE *Int. Conf. on Tools with Artificial Intelligence*, Chicago, Illinois, 1999.

Calculation vs. Subjective Assessment with Respect to Fuzzy Probability*

Chongfu Huang[1], Claudio Moraga[2], and Xingguang Yuan[2]

[1] Institute of Resources Science. Beijing Normal University. Beijing 100875. China
hchongfu@hotmail.com

[2] Department of Computer Science, Computer Engineering and Computing Education. University of Dortmund. 44221 Dortmund. Germany

Abstract. When a sample is drawn from a population with infinite elements, it is impossible to precisely get the probability distribution of the population from the sample. Particularly, when the size of the sample is small, the estimated values of the probabilities must be so imprecise that they would be represented by some fuzzy numbers. In that case, we can use the interior-outer-set model to calculate a fuzzy probability distribution, or invite some experts to review the sample and to subjectively assess. In this paper, with simulation experiments and inquiring experts, we prove that, the results from the calculation and the subjective assessment are very near in terms of the fuzzy expected value and the standard deviation. It implies that the interior-outer-set model can replace experts to give fuzzy probabilities.

1 Introduction

Knowledge acquisition is the so-called "bottle neck" problem of knowledge engineering. For example, whether an expert system is good or not is usually determined by whether or not the knowledge acquisition can be successfully carried. The fuzzy set methodology allows us to digitally code the knowledge elicited from experts, that is often incomplete, imprecise, fragmentary, not fully reliable, vague, contradictory, or deficient in some other way. However, in many cases, it is not easy to bring together domain experts to give reasonable knowledge. Machine learning may help to overcome the bottle-neck problem by learning from the materials from which experts get their knowledge.

When the materials are data associated with a particular type of frequency, experts would get the knowledge related to probabilities of occurrences. For example, studying the recent data from the New York Stock Exchange, an expert can get some knowledge for consulting services. This kind of knowledge often contains a probability component because a useful tool for options trading is to determine a variety of probabilities of where a stock price will be in the future. In general, from these data, it is impossible for an expert to precisely estimate

* Project supported by a Mercator Visiting Professorship of the German Research Society DFG, granted to Prof. Chongfu Huang at the University of Dortmund.

B. Reusch (Ed.): Fuzzy Days 2001, LNCS 2206, pp. 392–411, 2001.

the probabilities he needs. In his consciousness, the probabilities of the stock price changes are fuzzy concepts, such as "very small," " not large", and so on.

With current science and technology, it is practically impossible to replace all expert knowledge with machine learning. However, machine learning has the potential to learning from the data and convert them into some knowledge that people can use. So far, it has been used on fuzzy risk assessment [5] under small-sample condition. In 1998, Huang [7] studied the possibility to introduce the method of information distribution [6] [13] to calculate a fuzzy probability distribution without any expert knowledge. Huang and Shi improved the study in paper [12]. With the calculation case of the risk of crop flood, Huang and Bai [8] [10] proved that the new approach can give a better result to support risk management in crops avoiding flood than the traditional probability method. Huang and Moraga [11] mathematically formulated the approach as a general model, called interior-outer-set model, meanwhile, and gave a matrix algorithm to execute this model. In 2000, Huang [9] proposed two models based on Monte Carlo method to study the reliability of fuzzy risk calculated by information distribution. One is for the soft histogram estimate, another for the possibility-probability distribution.

In this paper, with simulation experiments and inquiring experts, we study the possibility of replacing the expert assessment with calculation.

The paper is organized as follows: Section 2 reviews some related basic definitions and terminology; Section 3 introduces the interior-outer-set model; Section 4 presents the experiments in calculation; Section 5 presents the experiments in subjective assessment; In section 6, we compare the results from the calculation and the subjective assessment in terms of the expected values and standard deviations. We conclude this paper with section 7.

2 Basic Definitions and Terminology

2.1 Fuzzy Probability

The earliest concept of fuzzy probability is the so-called probability of a fuzzy event, given by Zadeh in 1968 [16]. Before we calculate the probability of a fuzzy event, we have to know a basic probability distribution. In many cases, unfortunately, the focus point is just to find the basic distribution. We believe that, the most fuzzy probabilities connect to imprecise probabilities rather than to the probabilities of fuzzy events.

There is a whole set of other theories and models addressing imprecision, uncertainty and partial knowledge with respect to probability. They are referred to as imprecise probabilities. Imprecise probability is a generic term for many mathematical models which measure chance or uncertainty without sharp numerical probabilities. These models include belief functions, Choquet capacities, comparative probability orderings, convex sets of probability measure, fuzzy measures, interval-valued probabilities, possibility measures, plausibility measures and upper and lower coherent previsions and expectations. In statistical applications, imprecise probabilities usually come from subjectively assessed prior probabilities. The fuzzy set theory is applicable to the modeling of imprecise subjective probabilities, suggested by many researchers (for example, Freeling [4], Watson, Weiss and Donnell [15], and Dubois and Prade [3]). We consider only a particular case among imprecise probabilities, called possibility-probability distribution.

Definition 1. *Let $(\Omega, \mathcal{A}, P)$ be a probability space, and $x \in \Omega$. Let the possibility that the probability of x occurring is p be $\pi_x(p)$.*

$$\Pi_{\Omega,P} = \{\pi_x(p)|x \in \Omega, p \in P\} \tag{1}$$

is called a possibility-probability distribution (PPD).

For example [11], the Nasdaq composite index fell for a third on Thursday March 15, 2001. As all investors, Mr. Smith was nervous about his stock market investment risk. Although he analysed all collected information, it was impossible for him to estimate precisely the probability-risk for next week. He concluded the following fuzzy estimate:
"Probability of loss 2% investment is large;"
"Probability of loss 4% investment is small; "
"Probability of loss 6% investment is very small."
Suppose in his fuzzy terms

$$\text{Small probability} = 1/0 + 0.8/0.1 + 0.2/0.2,$$
$$\text{Large probability} = 0.2/0.8 + 0.8/0.9 + 1/1,$$

where, for a/b, a is membership, b is probability. Using the concentration operator $\mu_{con(A)}(p) = (\mu_A(p))^2$, we have

$$\text{Very small probability} = 1/0 + 0.64/0.1 + 0.04/0.2.$$

Then, the PPD given by Mr. Smith can be represented by Table 1.

Table 1. The fuzzy risk of Mr. Smith's stock market investment

	Probability Value p						
$\pi_l(p)$	0	0.1	0.2	⋯	0.8	0.9	1
l_1 – (loss 2%)	0	0	0	⋯	0.2	0.8	1
l_2 – (loss 4%)	1	0.8	0.2	⋯	0	0	0
l_3 – (loss 6%)	1	0.64	0.04	⋯	0	0	0

In fact, PPD is nothing else than a model of the second-order uncertainty [1] and together with the first-order uncertainty they form hierarchical models [2]. The reason why we redefine it is to avoid any confusion and to restrict ourselves in studying the imprecise probability that can be represented by a fuzzy relation.

2.2 Fuzzy Expected Value and Standard Deviation

Strictly speaking, the fuzzy expected value of a fuzzy probability distribution is a fuzzy set (or called fuzzy number). For comparing different fuzzy probability distributions, the simplest way is to defuzzify the fuzzy expected value into a crisp value. In this subsection, we give all relative formulae to obtain the crisp expected value from a fuzzy probability distribution, but we only give the final formula to calculate the standard deviation of a fuzzy probability distribution.

Unless stated otherwise, it is assumed that $\underset{\sim}{p}(x)$ denotes a general fuzzy probability distribution.

Definition 2. *Let Ω be the universe of discourse of random variable x, P be the universe of discourse of probability, and*

$$\underset{\sim}{p}(x) \triangleq \{\mu_x(p)|x \in \Omega, p \in P\} \tag{2}$$

be a fuzzy probability distribution. $\forall \alpha \in P$, *let*

$$\underline{p}_\alpha(x) = \min\{p|p \in P, \mu_x(p) \geq \alpha\}, \tag{3}$$
$$\overline{p}_\alpha(x) = \max\{p|p \in P, \mu_x(p) \geq \alpha\}. \tag{4}$$

$\underline{p}_\alpha(x)$ is called the minimum probability in α-cut with respect to x, and $\overline{p}_\alpha(x)$ is called the maximum probability in α-cut with respect to x.

The finite closed interval

$$p_\alpha(x) \triangleq [\underline{p}_\alpha(x), \overline{p}_\alpha(x)] \tag{5}$$

is called *α-cut* of $\underset{\sim}{p}(x)$ with respect to x.

Definition 3. *Let*

$$\underline{E}_\alpha(x) = \int_\Omega x\underline{p}_\alpha(x)dx, \quad \overline{E}_\alpha(x) = \int_\Omega x\overline{p}_\alpha(x)dx. \tag{6}$$

We call

$$E_\alpha(x) \triangleq [\underline{E}_\alpha(x), \overline{E}_\alpha(x)] \tag{7}$$

the expected interval of α-level of $\underset{\sim}{p}(x)$ with respect to x.

$$\underset{\sim}{E}(x) \triangleq \int_\Omega x\underset{\sim}{p}(x)dx \triangleq \bigcup_{\alpha\in]0,1]} \alpha E_\alpha(x) \tag{8}$$

is called the fuzzy expected value (FEV) of $\underset{\sim}{p}(x)$, also denoted as $FEV(\underset{\sim}{p}(x))$.

The simplest approach to defuzzify a FEV $\underset{\sim}{E}(x)$ into a crisp value is the so-called center-of-gravity method:

$$CG(\underset{\sim}{E}(x)) = \int_{-\infty}^{+\infty} x\mu_{E}(x)dx \Big/ \int_{-\infty}^{+\infty} \mu_{E}(x)dx. \tag{9}$$

Noticing the operator changing $\underset{\sim}{p}(x)$ into $\underset{\sim}{E}(x)$ is linear, we can avoid α-cuts computation to calculate the center of gravity.

$\forall x \in \Omega$, we let

$$c(x) = \int_P p\mu_x(p)dp \Big/ \int_P \mu_x(p)dp, \tag{10}$$

is called the average probability of that x occurring.

Then, the normalized average probability distribution of $\underset{\sim}{p}(x)$ is defined by

$$p(x) = c(x) \Big/ \int_\Omega c(x)dx, \quad x \in \Omega. \tag{11}$$

With it, without any α-cuts computation, we can obtain the center of gravity of FEV of $\underset{\sim}{p}(x)$

$$CG(FEV(\underset{\sim}{p}(x))) = \int_\Omega xp(x)dx. \tag{12}$$

Definition 4. *Let $\underset{\sim}{p}(x)$ be a fuzzy probability distribution and $p(x)$ be the normalized average probability distribution of $\underset{\sim}{p}(x)$.*

$$Var(\underset{\sim}{p}(x)) = \int_\Omega p(x)(CG(FEV(\underset{\sim}{p}(x))) - x)^2 dx \tag{13}$$

is called the variance (Var) of $\underset{\sim}{p}(x)$,

$$s = \sqrt{Var(\underset{\sim}{p}(x))} \tag{14}$$

is called the standard deviation of $\underset{\sim}{p}(x)$.

2.3 Method of Information Distribution

The base of the interior-outer-set model is the method of information distribution. To help the readers who never met earlier the concept, in this subsection, we outline this concept.

Roughly speaking, the method of information distribution assumes that an observation of a small sample is a piece of fuzzy information due to that a small sample cannot provide sufficient information for understanding its population. When we use the histogram method to estimate the probability distribution of a population, the membership degree of an observation belonging to an interval, called *histogram interval*, must be 1 or 0. In this conventional model, all observations falling into a same interval are considered to play a same role. It neglects the difference between them. In fact, the observations may occupy different positions in the interval. Neglecting the position difference implies that we throw away some information. When we use the method of information distribution to estimate the probability distribution of a population, the membership degree is determined by the distance between the observation and the center of an interval. If an observation is near the center of an interval, the membership degree is great (not larger than 1), otherwise it is small (not less than 0). That is, an observation can belong to more than one interval in nonzero membership degree. In other words, an observation is distributed to more than one interval. Because an observation is regarded as a piece of information, this kind of classification is called *information distribution*. By this way, we can pick up the information about the differences. The result from the method of information distribution is the same as that if we use fuzzy bounds instead of crisp bounds of the intervals in the histogram method.

The mathematical definition of the information distribution is a mapping from a Cartesian product to the unit interval [0,1].

Unless stated otherwise, it is always assumed $X = \{x_i | i = 1, 2, \cdots, n\}$ is a sample, $X \subset \mathbb{R}$ (set of real numbers), and $U = \{u_j | j = 1, 2, \cdots, m\} \subset \mathbb{R}$ is a discrete universe with step Δ, i.e., $\forall i,\ 1 < i < m,\ u_{i+1} - u_i = u_i - u_{i-1}$, $\Delta = u_2 - u_1$, where u_j, $j = 1, 2, \cdots, m$, are called *controlling points* and Δ called the *step* of controlling points.

Definition 5. *A mapping from $X \times U$ to [0,1]*

$$\begin{aligned} \mu : X \times U &\to [0,1] \\ (x, u) &\mapsto \mu(x, u) \end{aligned}$$

is called an information distribution of X on U, if $\mu(x, u)$ has the following properties :
(1) $\forall x \in X$, if $\exists u \in U$, such that $x = u$, then $\mu(x, u) = 1$, i.e., μ is reflexive;
(2) For $x \in X$, $\forall u', u'' \in U$, if $|u' - x| \le |u'' - x|$ then $\mu(x, u') \ge \mu(x, u'')$, i.e., μ is decreasing when $|x - u|$ is increasing;
(3) $\sum_{j=1}^{m} \mu(x_i, u_j) = 1$, $i = 1, 2, \cdots, n$, i.e., conservation of information.

This μ is called a *distribution function* of X on U. We say that observation x gives information, gain at $q = \mu(x, u)$, to the controlling point u. q is called *distributed information* on u from x. The simplest distribution function is the 1-dimension linear-information-distribution:

$$\mu(x, u) = \begin{cases} 1 - \mid x - u \mid / \Delta, & \text{if } |x - u| \le \Delta; \\ 0, & \text{otherwise.} \end{cases}$$

It is the same as the following formula

$$q_{ij} = \begin{cases} 1 - |x_i - u_j| / \Delta, & \text{if } |x_i - u_j| \le \Delta; \\ 0, & \text{if } |x_i - u_j| > \Delta. \end{cases} \tag{15}$$

3 Interior-Outer-Set Model

Using the information distribution formula (15), we can calculate a PPD from a given sample X and an universe U. The calculation is done on intervals

$$I_j = [u_j - \Delta/2, u_j + \Delta/2[, \quad u_j \in U.$$

and the universe of discourse of probability is

$$P = \{p_k | k = 0, 1, 2, \cdots, n\} = \{0, \frac{1}{n}, \frac{2}{n}, \cdots, 1\}$$

For interval I_j, we use $\pi_{I_j}(p_k)$, or $\pi_{u_j}(p_k)$, $k = 0, 1, 2, \cdots, n$, to represent its possibility-probabilities.

Definition 6. $X_{in-j} = X \cap I_j$ *is called the interior set* of interval I_j. The elements of X_{in-j} are called the *interior points* of I_j.

Obviously, we have

$$\forall i \neq j, \quad I_i \cap I_j = \emptyset,$$

and

$$X = X_{in-1} \cup X_{in-2} \cup \cdots \cup X_{in-m}.$$

Definition 7. *Let X_{in-j} be the interior set of interval I_j. $X_{out-j} = X \setminus X_{in-j}$ is called outer set* of interval I_j. The elements of X_{out-j} are called the *outer points* of I_j.

Let S_j be an index set such that $\forall s \in S_j$ then $x_s \in X_{in-j}$, and $\{x_s | s \in S_j\} = X_{in-j}$. Let T_j be one for X_{out-j}, that is, $\{x_t | t \in T_j\} = X_{out-j}$. S_j is called *interior index set*, and T_j *outer index set.*

$\forall x_i \in X$, if $x_i \in X_{in-j}$ we say that it loses information, by $1 - q_{ij}$, to other interval, we use $q_{ij}^- = 1 - q_{ij}$ to represent the loss; if $x_i \in X_{out-j}$ we say that it gives information, by q_{ij}, to I_j, we use q_{ij}^+ to represent the addition, i.e., when x_i is an outer point of I_j, $q_{ij}^+ = q_{ij}$. In short, q_{ij} means that x_i may leave I_j with possibility q_{ij}^- if $x_i \in X_{in-j}$, or x_i may join I_j with possibility q_{ij}^+ if $x_i \in X_{out-j}$.

q_{ij}^- is called *leaving possibility*, and q_{ij}^+ called *joining possibility*. The leaving possibility of an outer point is defined as 0 (it has gone). the joining possibility of an interior point is defined as 0 (it is in the interval).

When there are n_j observations $\{x_s | s \in S_j\}$ in interval I_j, we can obtain a formula to calculate a PPD as

$$\pi_{I_j}(p) = \begin{cases} \bigwedge_{s \in S_j} q_{sj}^-, & p = p_0; \\ \cdots & \cdots \\ \bigvee_{s_1, s_2, s_3 \in S_j, s_1 \neq s_2 \neq s_3} (q_{s_1 j}^- \wedge q_{s_2 j}^- \wedge q_{s_3 j}^-), & p = p_{n_j - 3}; \\ \bigvee_{s_1, s_2 \in S_j, s_1 \neq s_2} (q_{s_1 j}^- \wedge q_{s_2 j}^-), & p = p_{n_j - 2}; \\ \bigvee_{s \in S_j} q_{sj}^-, & p = p_{n_j - 1}; \\ 1, & p = p_{n_j}; \\ \bigvee_{t \in T_j} q_{tj}^+, & p = p_{n_j + 1}; \\ \bigvee_{t_1, t_2 \in T_j, t_1 \neq t_2} (q_{t_1 j}^+ \wedge q_{t_2 j}^+), & p = p_{n_j + 2}; \\ \bigvee_{t_1, t_2, t_3 \in T_j, t_1 \neq t_2 \neq t_3} (q_{t_1 j}^+ \wedge q_{t_2 j}^+ \wedge q_{t_3 j}^+), & p = p_{n_j + 3}; \\ \cdots & \cdots \\ \bigwedge_{t \in T_j} q_{tj}^+, & p = p_n. \end{cases} \tag{16}$$

where $j = 1, 2, \cdots, m$; $p_0 = 0, p_1 = 1/n, \cdots, p_{n_j} = n_j/n, \cdots, p_n = 1$. The formula (16) is called the *interior-outer-set model.*

4 Experiments in Calculation

The aim of experiments in calculation is to calculate PPDs from given samples and calculate their expected values and the standard deviations. In general, the samples can be drawn from any populations. However, in this paper, we focus our interest on the comparison of calculation and subjective assessment. Hence, the experiments involve only the two populations frequently used in a wide variety of applications, the normal distribution and exponential distribution. We employ the computer simulation model based on the Monte Carlo method to generate some samples.

When a given sample is $X = \{x_1, x_2, \cdots, x_n\}$, we let

$$b = \max_{1\le i\le n} \{x_i\} + 0.05, \quad a = \min_{1\le i\le n} \{x_i\} - 0.05. \tag{17}$$

Interval [a,b] must cover all observations and no observation is just at a boundary. With $\Delta = (b-a)/m$ being the width of an interval, we divide [a,b] into m histogram intervals, $I_1, I_2, \cdots, I_m$, whose center points are denoted by $u_1, u_2, \cdots, u_m$, respectively. For the sake of simplicity, we always use u_j to represent I_j.

For size n, the universe of discourse of probability is

$$P = \{p_0, p_1, p_2, \cdots, p_n\} = \{0, \frac{1}{n}, \frac{2}{n}, \cdots, \frac{n}{n}\}. \tag{18}$$

For this sample X, using formula (16), we can obtain a PPD, $\pi_{I_j}(p)$, $j = 1, 2, 3, 4$; $p \in P$, that is a fuzzy relation on the Cartesian product space $\{I_1, I_2, \cdots, I_m\} \times \{p_0, p_1, p_2, \cdots, p_n\}$, written as

$$R_X = (\pi_{I_j}(p_k))_{m\times(n+1)}$$
$$\begin{array}{c} \\ = \begin{array}{c} u_1 \\ u_2 \\ \cdots \\ u_m \end{array} \end{array} \begin{array}{c} \begin{array}{ccccc} p_0 & p_1 & p_3 & \cdots & p_n \end{array} \\ \begin{pmatrix} \pi_{I_1}(p_0) & \pi_{I_1}(p_1) & \pi_{I_1}(p_2) & \cdots & \pi_{I_1}(p_n) \\ \pi_{I_2}(p_0) & \pi_{I_2}(p_1) & \pi_{I_2}(p_2) & \cdots & \pi_{I_2}(p_n) \\ \cdots & \cdots & \cdots & \cdots & \cdots \\ \pi_{I_m}(p_0) & \pi_{I_m}(p_1) & \pi_{I_m}(p_2) & \cdots & \pi_{I_m}(p_n) \end{pmatrix} \end{array}. \tag{19}$$

4.1 Samples from Normal Distribution

A sample with size N from the normal distribution with expected value MU and standard deviation S can be generated by using Program 1 with a random seed number SEED.

Program 1. Generator of random numbers obeying the normal distribution

```
PROGRAM MAIN
INTEGER N,SEED
REAL X(100),MU,S
READ(*,*)N,MU,S,SEED
DO 10 I=1,N
U1=RANUN(SEED)
```

```
      U2=RANUN(SEED)
      Z=SQRT(-2.*ALOG(U2))*COS(6.283*U1)
      X(I)=MU+S*Z
10    CONTINUE
      WRITE(*,20)(X(I),I=1,N)
20    FORMAT(10(F5.2,','))
      STOP
      END

      FUNCTION RANUN(IX)
      K1=IX/60466
      IX=35515*(IX-K1*60466)-K1*33657
      IF(IX.LT.0)IX=IX+2147483647
      K1=IX/102657
      IX=20919*(IX-K1*102657)-K1*1864
      IF(IX.LT.0)IX=IX+2147483647
      RANUN=FLOAT(IX)/2.147483647e9
      RETURN
      END
```

Let N=20, MU=6.86, S=0.372. For SEED=6435435, 9643653, 54355633, 1934765, 27865365, respectively, we obtain the following 5 samples.

$$\begin{aligned}
X_1 &= \{6.58, 6.65, 7.10, 7.42, 7.46, 6.69, 6.58, 6.37, 6.85, 6.95,\\
&\qquad 6.87, 6.76, 6.72, 6.97, 7.33, 6.60, 7.07, 7.42, 6.55, 6.61\},\\
X_2 &= \{7.02, 7.34, 6.53, 7.03, 6.79, 7.04, 6.92, 7.02, 7.32, 7.72,\\
&\qquad 7.13, 6.98, 6.62, 6.44, 7.19, 7.37, 7.46, 7.21, 6.95, 6.62\},\\
X_3 &= \{6.86, 6.83, 6.75, 7.40, 6.94, 6.63, 7.40, 6.98, 6.82, 7.11,\\
&\qquad 7.06, 7.03, 6.50, 6.77, 7.25, 6.59, 6.78, 6.86, 6.83, 6.43\},\\
X_4 &= \{6.33, 6.35, 6.71, 6.89, 7.04, 6.61, 5.88, 6.93, 7.10, 6.53,\\
&\qquad 6.39, 6.99, 6.26, 7.16, 6.66, 6.65, 6.48, 7.89, 7.15, 6.64\},\\
X_5 &= \{6.87, 6.70, 7.41, 6.61, 7.38, 6.80, 6.35, 7.92, 7.20, 6.87,\\
&\qquad 6.69, 7.10, 6.69, 7.70, 7.29, 7.05, 7.55, 6.78, 7.10, 6.69\}.
\end{aligned} \tag{20}$$

In our case we want to emphasize that we only show two places following the decimal point for each elements. For example, in X_1, the first element x_1 is shown by 6.58, however, the real value of this element stored in the computer is $x_1 = 6.57538128$. In the following calculation, the numbers used are ones stored in computer.

Using formula (16) to, respectively, calculate these samples, we obtain the following fuzzy relations.

$$R_{X_1} = \begin{array}{c} \\ 6.47 \\ 6.77 \\ 7.06 \\ 7.36 \end{array}
\begin{array}{c}
\begin{array}{ccccccccccccccccccccc} .00 & .05 & .10 & .15 & .20 & .25 & .30 & .35 & .40 & .45 & .50 & .55 & .60 & .65 & .70 & .75 & .80 & .85 & .90 & .95 & 1 \end{array} \\
\begin{pmatrix}
.28 & .33 & .35 & .36 & .42 & .49 & 1 & .40 & .27 & .18 & .02 & .00 & .00 & .00 & .00 & .00 & .00 & .00 & .00 & .00 & .00 \\
.02 & .18 & .27 & .30 & .35 & .40 & 1 & .49 & .42 & .38 & .36 & .30 & .28 & .00 & .00 & .00 & .00 & .00 & .00 & .00 & .00 \\
.00 & .12 & .30 & .38 & 1 & .35 & .30 & .11 & .00 & .00 & .00 & .00 & .00 & .00 & .00 & .00 & .00 & .00 & .00 & .00 & .00 \\
.11 & .19 & .22 & .33 & 1 & .12 & .00 & .00 & .00 & .00 & .00 & .00 & .00 & .00 & .00 & .00 & .00 & .00 & .00 & .00 & .00
\end{pmatrix}
\end{array},$$

$$R_{X_2} = \begin{array}{c} \\ 6.56 \\ 6.91 \\ 7.25 \\ 7.60 \end{array} \begin{array}{c} \begin{array}{ccccccccccccccccccccc} .00 & .05 & .10 & .15 & .20 & .25 & .30 & .35 & .40 & .45 & .50 & .55 & .60 & .65 & .70 & .75 & .80 & .85 & .90 & .95 & 1 \end{array} \\ \begin{pmatrix} .09 & .16 & .17 & .36 & 1 & .33 & .00 & .00 & .00 & .00 & .00 & .00 & .00 & .00 & .00 & .00 & .00 & .00 & .00 & .00 & .00 \\ .03 & .14 & .21 & .33 & .33 & .33 & .36 & .40 & 1 & .35 & .18 & .17 & .16 & .12 & .00 & .00 & .00 & .00 & .00 & .00 & .00 \\ .12 & .18 & .19 & .26 & .35 & .35 & 1 & .40 & .39 & .36 & .33 & .33 & .14 & .03 & .00 & .03 & .00 & .00 & .00 & .00 & .00 \\ .36 & .39 & 1 & .35 & .26 & .19 & .00 & .00 & .00 & .26 & .26 & .26 & .26 & .00 & .00 & .00 & .00 & .00 & .00 & .00 & .00 \end{pmatrix} \end{array},$$

$$R_{X_3} = \begin{array}{c} \\ 6.52 \\ 6.79 \\ 7.05 \\ 7.32 \end{array} \begin{array}{c} \begin{array}{ccccccccccccccccccccc} .00 & .05 & .10 & .15 & .20 & .25 & .30 & .35 & .40 & .45 & .50 & .55 & .60 & .65 & .70 & .75 & .80 & .85 & .90 & .95 & 1 \end{array} \\ \begin{pmatrix} .05 & .25 & .31 & .42 & 1 & .13 & .04 & .02 & .00 & .00 & .00 & .00 & .00 & .00 & .00 & .00 & .00 & .00 & .00 & .00 & .00 \\ .02 & .04 & .13 & .13 & .16 & .16 & .28 & .29 & 1 & .42 & .42 & .28 & .25 & .08 & .00 & .00 & .00 & .00 & .00 & .00 & .00 \\ .03 & .08 & .20 & .28 & .42 & 1 & .29 & .28 & .27 & .16 & .00 & .00 & .00 & .00 & .00 & .00 & .00 & .00 & .00 & .00 & .00 \\ .27 & .31 & .31 & 1 & .20 & .03 & .00 & .00 & .00 & .00 & .00 & .00 & .00 & .00 & .00 & .00 & .00 & .00 & .00 & .00 & .00 \end{pmatrix} \end{array},$$

$$R_{X_4} = \begin{array}{c} \\ 6.10 \\ 6.62 \\ 7.15 \\ 7.68 \end{array} \begin{array}{c} \begin{array}{ccccccccccccccccccccc} .00 & .05 & .10 & .15 & .20 & .25 & .30 & .35 & .40 & .45 & .50 & .55 & .60 & .65 & .70 & .75 & .80 & .85 & .90 & .95 & 1 \end{array} \\ \begin{pmatrix} .32 & .41 & .44 & .47 & 1 & .44 & .27 & .17 & .02 & .00 & .00 & .00 & .00 & .00 & .00 & .00 & .00 & .00 & .00 & .00 & .00 \\ .02 & .02 & .05 & .07 & .16 & .17 & .27 & .44 & 1 & .49 & .47 & .44 & .42 & .32 & .31 & .22 & .10 & .00 & .00 & .00 & .00 \\ .00 & .01 & .10 & .22 & .31 & .42 & .49 & 1 & .16 & .07 & .05 & .02 & .00 & .00 & .00 & .00 & .00 & .00 & .00 & .00 & .00 \\ .41 & 1 & .01 & .00 & .00 & .00 & .00 & .00 & .00 & .00 & .00 & .00 & .00 & .00 & .00 & .00 & .00 & .00 & .00 & .00 & .00 \end{pmatrix} \end{array},$$

$$R_{X_5} = \begin{array}{c} \\ 6.51 \\ 6.92 \\ 7.34 \\ 7.76 \end{array} \begin{array}{c} \begin{array}{ccccccccccccccccccccc} .00 & .05 & .10 & .15 & .20 & .25 & .30 & .35 & .40 & .45 & .50 & .55 & .60 & .65 & .70 & .75 & .80 & .85 & .90 & .95 & 1 \end{array} \\ \begin{pmatrix} .24 & .38 & .43 & .44 & .45 & .47 & 1 & .35 & .31 & .14 & .13 & .00 & .00 & .00 & .00 & .00 & .00 & .00 & .00 & .00 & .00 \\ .13 & .14 & .31 & .31 & .35 & .42 & .42 & 1 & .47 & .45 & .44 & .43 & .34 & .13 & .00 & .00 & .00 & .00 & .00 & .00 & .00 \\ .10 & .13 & .16 & .34 & 1 & .50 & .42 & .42 & .31 & .14 & .00 & .00 & .00 & .00 & .00 & .00 & .00 & .00 & .00 & .00 & .00 \\ .14 & .38 & .50 & 1 & .16 & .10 & .00 & .00 & .00 & .00 & .00 & .00 & .00 & .00 & .00 & .00 & .00 & .00 & .00 & .00 & .00 \end{pmatrix} \end{array},$$

where we omit the digit before the decimal point when the digit is 0. And, we only show two places following the decimal point for each elements. For example, in R_{X_1}, the first element $\pi_{I_1}(p_0)$ is shown as .28, however, the real value of this element stored in the computer is $\pi_{I_1}(p_0) = 0.282436252$. When we calculate the expected values and the standard deviations, we use the values of R's stored in computer.

4.2 Samples from Exponential Distribution

A sample with size N from the exponential distribution with parameter LAMBDA can be generated by using Program 2 with a random seed number SEED. When LAMBDA=λ, the expected value of the exponential distribution is $1/\lambda$.

Program 2. Generator of random numbers obeying the exponential distribution

```
      PROGRAM MAIN
      INTEGER N,SEED
      REAL X(100),LAMBDA,MU
      READ(*,*)N,LAMBDA,SEED
      MU=1.0/LAMBDA
      DO 10 I=1,N
      Z=RANUN(SEED)
      X(I)=-MU*ALOG(Z)
10    CONTINUE
      WRITE(*,20)(X(I),I=1,N)
20    FORMAT(10(F5.2,','))
      STOP
      END
```

Let N=20, LAMBDA=3. For SEED=76543, 25235, 9463545, 1454235, 4654381, respectively, we obtain the following 5 samples.

$$
\begin{aligned}
X_6 &= \{0.15, 0.20, 0.02, 0.37, 0.35, 0.42, 0.26, 0.27, 0.83, 0.10,\\
&\qquad 0.15, 0.16, 0.04, 0.12, 0.46, 0.35, 0.24, 0.51, 0.27, 0.08\},\\
X_7 &= \{0.48, 0.01, 0.41, 0.17, 0.15, 0.57, 0.13, 0.11, 0.30, 0.61,\\
&\qquad 0.05, 0.04, 0.71, 0.14, 0.39, 1.10, 0.03, 0.14, 0.63, 0.24\},\\
X_8 &= \{0.40, 0.71, 0.02, 0.21, 0.39, 0.21, 0.33, 0.14, 0.19, 0.21,\\
&\qquad 0.34, 0.68, 0.30, 1.76, 0.02, 0.53, 0.26, 0.10, 1.53, 0.49\},\\
X_9 &= \{0.14, 0.67, 0.99, 0.82, 0.04, 0.03, 0.00, 0.57, 0.41, 0.11,\\
&\qquad 0.08, 0.07, 0.24, 0.06, 0.12, 0.04, 0.17, 0.57, 0.33, 0.03\},\\
X_{10} &= \{0.25, 0.23, 0.12, 0.04, 0.87, 0.36, 0.02, 0.28, 0.09, 0.37,\\
&\qquad 0.64, 0.51, 0.26, 0.46, 1.06, 0.17, 0.48, 0.13, 0.86, 0.46\}.
\end{aligned}
\tag{21}
$$

Here, we also only show two places following the decimal point for each elements. The real values of the elements of these samples stored in the computer are longer.

Using formula (16) to, respectively, calculate these samples, we obtain the following fuzzy relations.

$$
R_{X_6} = \begin{array}{c} \\ 0.13 \\ 0.35 \\ 0.56 \\ 0.77 \end{array}
\begin{array}{ccccccccccccccccccccc}
.00 & .05 & .10 & .15 & .20 & .25 & .30 & .35 & .40 & .45 & .50 & .55 & .60 & .65 & .70 & .75 & .80 & .85 & .90 & .95 & 1 \\
.06 & .06 & .09 & .15 & .15 & .22 & .31 & .40 & .50 & 1 & .50 & .42 & .37 & .36 & .00 & .00 & .00 & .00 & .00 & .00 & .00 \\
.00 & .02 & .10 & .34 & .36 & .37 & .42 & .50 & 1 & .46 & .31 & .22 & .15 & .09 & .06 & .00 & .00 & .00 & .00 & .00 & .00 \\
.22 & .46 & 1 & .34 & .10 & .02 & .00 & .00 & .00 & .00 & .00 & .00 & .00 & .00 & .00 & .00 & .00 & .00 & .00 & .00 & .00 \\
.27 & 1 & .00 & .00 & .00 & .00 & .00 & .00 & .00 & .00 & .00 & .00 & .00 & .00 & .00 & .00 & .00 & .00 & .00 & .00 & .00
\end{array},
$$

$$
R_{X_7} = \begin{array}{c} \\ 0.15 \\ 0.44 \\ 0.72 \\ 1.01 \end{array}
\begin{array}{ccccccccccccccccccccc}
.00 & .05 & .10 & .15 & .20 & .25 & .30 & .35 & .40 & .45 & .50 & .55 & .60 & .65 & .70 & .75 & .80 & .85 & .90 & .95 & 1 \\
.01 & .02 & .04 & .07 & .08 & .13 & .32 & .34 & .37 & .41 & .50 & 1 & .47 & .17 & .08 & .00 & .00 & .00 & .00 & .00 & .00 \\
.08 & .13 & .17 & .47 & .48 & 1 & .41 & .32 & .32 & .07 & .03 & .00 & .00 & .00 & .00 & .00 & .00 & .00 & .00 & .00 & .00 \\
.03 & .32 & .41 & 1 & .48 & .13 & .00 & .00 & .00 & .00 & .00 & .00 & .00 & .00 & .00 & .00 & .00 & .00 & .00 & .00 & .00 \\
.32 & 1 & .00 & .00 & .00 & .00 & .00 & .00 & .00 & .00 & .00 & .00 & .00 & .00 & .00 & .00 & .00 & .00 & .00 & .00 & .00
\end{array},
$$

$$
R_{X_8} = \begin{array}{c} \\ 0.25 \\ 0.69 \\ 1.14 \\ 1.59 \end{array}
\begin{array}{ccccccccccccccccccccc}
.00 & .05 & .10 & .15 & .20 & .25 & .30 & .35 & .40 & .45 & .50 & .55 & .60 & .65 & .70 & .75 & .80 & .85 & .90 & .95 & 1 \\
.03 & .07 & .08 & .08 & .11 & .13 & .18 & .20 & .24 & .32 & .32 & .35 & .49 & .50 & 1 & .46 & .37 & .02 & .00 & .00 & .00 \\
.02 & .03 & .37 & .46 & 1 & .35 & .32 & .20 & .18 & .11 & .03 & .00 & .00 & .00 & .00 & .00 & .00 & .00 & .00 & .00 & .00 \\
1 & .12 & .03 & .00 & .00 & .00 & .00 & .00 & .00 & .00 & .00 & .00 & .00 & .00 & .00 & .00 & .00 & .00 & .00 & .00 & .00 \\
.12 & .39 & 1 & .00 & .00 & .00 & .00 & .00 & .00 & .00 & .00 & .00 & .00 & .00 & .00 & .00 & .00 & .00 & .00 & .00 & .00
\end{array},
$$

$$
R_{X_9} = \begin{array}{c} \\ 0.14 \\ 0.39 \\ 0.65 \\ 0.91 \end{array}
\begin{array}{ccccccccccccccccccccc}
.00 & .05 & .10 & .15 & .20 & .25 & .30 & .35 & .40 & .45 & .50 & .55 & .60 & .65 & .70 & .75 & .80 & .85 & .90 & .95 & 1 \\
.03 & .05 & .08 & .14 & .22 & .22 & .26 & .34 & .36 & .39 & .39 & .40 & .50 & 1 & .24 & .00 & .00 & .00 & .00 & .00 & .00 \\
.06 & .24 & 1 & .40 & .33 & .32 & .14 & .03 & .00 & .00 & .00 & .00 & .00 & .00 & .00 & .00 & .00 & .00 & .00 & .00 & .00 \\
.06 & .32 & .33 & 1 & .34 & .06 & .00 & .00 & .00 & .00 & .00 & .00 & .00 & .00 & .00 & .00 & .00 & .00 & .00 & .00 & .00 \\
.31 & .34 & 1 & .06 & .00 & .00 & .00 & .00 & .00 & .00 & .00 & .00 & .00 & .00 & .00 & .00 & .00 & .00 & .00 & .00 & .00
\end{array},
$$

$$
R_{X_{10}} = \begin{array}{c} \\ 0.16 \\ 0.43 \\ 0.70 \\ 0.97 \end{array}
\begin{array}{ccccccccccccccccccccc}
.00 & .05 & .10 & .15 & .20 & .25 & .30 & .35 & .40 & .45 & .50 & .55 & .60 & .65 & .70 & .75 & .80 & .85 & .90 & .95 & 1 \\
.04 & .11 & .14 & .23 & .27 & .34 & .36 & .42 & .45 & .50 & 1 & .26 & .22 & .00 & .00 & .00 & .00 & .00 & .00 & .00 & .00 \\
.09 & .10 & .17 & .22 & .26 & .29 & 1 & .45 & .36 & .34 & .27 & .25 & .04 & .00 & .00 & .00 & .00 & .00 & .00 & .00 & .00 \\
.25 & 1 & .43 & .37 & .29 & .17 & .10 & .09 & .00 & .00 & .00 & .00 & .00 & .00 & .00 & .00 & .00 & .00 & .00 & .00 & .00 \\
.32 & .37 & .43 & 1 & .00 & .00 & .00 & .00 & .00 & .00 & .00 & .00 & .00 & .00 & .00 & .00 & .00 & .00 & .00 & .00 & .00
\end{array},
$$

where we also omit the digit before the decimal point when the digit is 0, and, we only show two places following the decimal point for each elements.

4.3 Calculation of Expected Values and Standard Deviations

According to formula (10), for a PPD $R = (\pi_{I_j}(p_k))_{m\times(n+1)}$, we have a formula as the following (22) to calculate the average probability of that event $x \in I_j$ occurring.

$$c(x \in I_j) = \sum_{k=0}^{n} p_k \pi_{I_j}(p_k) \Big/ \sum_{k=0}^{n} \pi_{I_j}(p_k), \quad j = 1, 2, \cdots, m. \tag{22}$$

The normalized average probability distribution of the R is

$$p(x \in I_j) = c(x \in I_j) \Big/ \sum_{j=1}^{m} c(x \in I_j), \quad j = 1, 2, \cdots, m. \tag{23}$$

Then, the center of gravity of FEV of the R is

$$CG(FEV(R)) = \sum_{j=1}^{m} u_j p(x \in I_j). \tag{24}$$

where u_j is the center point of the interval I_j.

And, the standard deviation of the R is

$$s(R) = \sqrt{Var(R)} = \sqrt{\sum_{j=1}^{m} p(x \in I_j)(CG(FEV(R)) - u_j)^2}. \tag{25}$$

Table 2 shows the expected values and standard deviations of $R_{X_1}, R_{X_2}, \cdots, R_{X_{10}}$.

Table 2. The expected values and standard deviations of PPDs

Sample	X_1	X_2	X_3	X_4	X_5	X_6	X_7	X_8	X_9	X_{10}
$CG(FEV(R_X))$	6.86	7.10	6.88	6.73	7.04	0.29	0.36	0.49	0.33	0.41
$s(R_X)$	0.30	0.34	0.24	0.41	0.40	0.17	0.26	0.40	0.26	0.27

5 Experiment in Subjective Assessment

5.1 Illustrating the Samples

For a give sample $X = \{x_1, x_2, \cdots, x_n\}$, we randomly sample n number y_1, $y_2, \cdots,$ y_n from the uniform distribution $U(0,1)$. Printing dots (x_1, y_1), $(x_2, y_2), \cdots, (x_n, y_n)$ as tiny circles on a piece of paper, we can illustrate the X by a visible image.

The area of these dots must be covered by a rectangle with width=$b - a$, height=1, where a, b are defined by formula (17). Diving the rectangle by the histogram intervals used for calculating a PPD, we can get some small rectangles.

Using this method, we make Fig. 1 to show $X_1, X_2, \cdots, X_{10}$ given in (20) and (21).

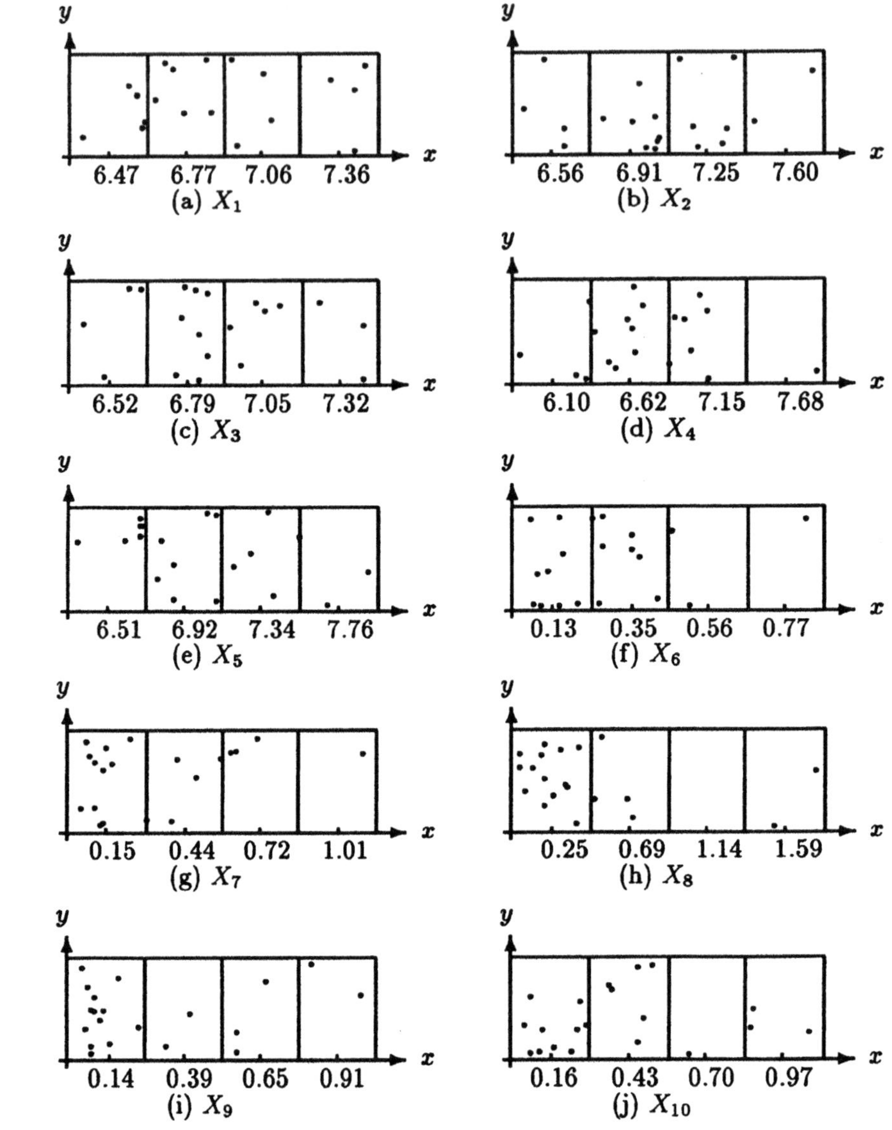

Fig. 1. The visible images of $X_1, X_2, \cdots, X_{10}$

5.2 Subjective Assessments

Because the dots of the 2-dimension observations can be illustrated by a visible image, it is more easy for an expert to review a sample consisting of 2-dimension observations and to subjectively assess.

A subjective assessment was acquired from an expert by the following inquiry:

Step 1. Telling the invited expert that the given sample was drawn form a population with infinite elements, and it is impossible to precisely estimate the probability distribution of the population from the sample;

Step 2. The expert reviews the dots shown by a figure with some small rectangles, and gives a subjective assessment for each small rectangle in an interval of probability that a random observation (x, y) would fall into the small rectangle.

An interval of probability given by the expert is called *assessment interval*, denoted by $I = [p_1, p_2]$. It is useful to notice that a histogram interval is defined on Ω, but, an assessment interval is defined on P.

In our case, the probability that a 2-dimension observation (x, y) falls into a small rectangle associated with histogram interval I_j is the same as the probability that a 1-dimension observation x falls into the corresponding interval I_j. Therefore, the assessments from the experts are just for the probabilities that x falls into the histogram intervals $I_1, I_2, \cdots, I_m$.

In this experiment we invited 5 experts to review the samples shown by Fig. 1. Their subjective assessments are shown in Table 3 to Table 12.

5.3 Calculation of Expected Values and Standard Deviations

For a given sample X, with respect to a histogram interval I_j, if there are t experts who give t assessment intervals as

$$I^{(1)} = [p_1^{(1)}, p_2^{(1)}],\ I^{(2)} = [p_1^{(2)}, p_2^{(2)}],\ \cdots,\ I^{(t)} = [p_1^{(t)}, p_2^{(t)}]. \tag{26}$$

These sets can be represented by their characteristic functions

$$\chi_k(p) = \begin{cases} 1, \text{ if } p \in I^{(k)}; \\ 0, \text{ if } p \notin I^{(k)}. \end{cases} \quad p \in P, \tag{27}$$

$k = 1, 2, \cdots, t;\ P = [0, 1]$.

Employing the falling-shadows formula [14], we can combine the assessment intervals into a membership function of the subjective fuzzy probability with respect to a histogram interval I_j, as

$$\mu_{u_j}(p) = \frac{1}{t}\sum_{k=1}^{t}\chi_k(p), \quad p \in P, \tag{28}$$

where u_j is the center point of the histogram interval I_j.

The final subjective assessment from X is a fuzzy set on the Cartesian product space $\{I_1, I_2, \cdots, I_m\} \times [0, 1]$, written as

$$S_X = \begin{pmatrix} \int_P \mu_{u_1}(p)/p \\ \int_P \mu_{u_2}(p)/p \\ \cdots \\ \int_P \mu_{u_m}(p)/p \end{pmatrix}. \tag{29}$$

Table 3. Assessment intervals for sample X_1

	u_1=6.47	u_2=6.77	u_3=7.06	u_4=7.36
Expert1	[0.10,0.20]	[0.20,0.25]	[0.20,0.25]	[0.15,0.20]
Expert2	[0.25,0.35]	[0.25,0.35]	[0.15,0.25]	[0.15,0.25]
Expert3	[0.10,0.25]	[0.20,0.30]	[0.10,0.22]	[0.15,0.22]
Expert4	[0.20,0.25]	[0.30,0.40]	[0.15,0.20]	[0.15,0.20]
Expert5	[0.25,0.30]	[0.25,0.30]	[0.15,0.20]	[0.15,0.18]

Table 4. Assessment intervals for sample X_2

	u_1=6.56	u_2=6.91	u_3=7.25	u_4=7.60
Expert1	[0.10,0.15]	[0.25,0.30]	[0.20,0.30]	[0.10,0.15]
Expert2	[0.15,0.20]	[0.35,0.45]	[0.15,0.25]	[0.05,0.10]
Expert3	[0.15,0.20]	[0.30,0.40]	[0.25,0.30]	[0.05,0.10]
Expert4	[0.15,0.20]	[0.40,0.50]	[0.40,0.45]	[0.05,0.10]
Expert5	[0.15,0.20]	[0.20,0.25]	[0.15,0.20]	[0.10,0.15]

Table 5. Assessment intervals for sample X_3

	u_1=6.52	u_2=6.79	u_3=7.05	u_4=7.32
Expert1	[0.05,0.25]	[0.25,0.50]	[0.25,0.45]	[0.10,0.15]
Expert2	[0.15,0.20]	[0.30,0.40]	[0.20,0.30]	[0.10,0.15]
Expert3	[0.10,0.25]	[0.25,0.35]	[0.20,0.30]	[0.05,0.10]
Expert4	[0.15,0.20]	[0.30,0.40]	[0.30,0.35]	[0.10,0.15]
Expert5	[0.15,0.20]	[0.25,0.30]	[0.20,0.25]	[0.10,0.15]

Table 6. Assessment intervals for sample X_4

	u_1=6.10	u_2=6.62	u_3=7.15	u_4=7.68
Expert1	[0.10,0.12]	[0.45,0.55]	[0.40,0.50]	[0.01,0.10]
Expert2	[0.10,0.20]	[0.50,0.60]	[0.30,0.40]	[0.00,0.05]
Expert3	[0.10,0.25]	[0.35,0.40]	[0.25,0.30]	[0.00,0.05]
Expert4	[0.05,0.10]	[0.40,0.50]	[0.30,0.40]	[0.01,0.05]
Expert5	[0.20,0.25]	[0.40,0.50]	[0.25,0.30]	[0.05,0.10]

Table 7. Assessment intervals for sample X_5

	u_1=6.51	u_2=6.92	u_3=7.34	u_4=7.76
Expert1	[0.25,0.35]	[0.25,0.35]	[0.20,0.30]	[0.10,0.20]
Expert2	[0.20,0.30]	[0.40,0.50]	[0.15,0.20]	[0.00,0.05]
Expert3	[0.20,0.30]	[0.25,0.35]	[0.15,0.30]	[0.00,0.05]
Expert4	[0.20,0.25]	[0.25,0.40]	[0.15,0.20]	[0.10,0.15]
Expert5	[0.20,0.25]	[0.30,0.35]	[0.20,0.25]	[0.05,0.10]

For a subjective assessment S, its average probability of that event $x \in I_j$ occurring can be directly calculated from the assessment intervals.

Let

$$c_{jk} = \frac{1}{2}\left(p_1^{(k)}(u_j) + p_2^{(k)}(u_j)\right), \quad j = 1, 2, \cdots, m; \ k = 1, 2, \cdots, t. \tag{30}$$

Table 8. Assessment intervals for sample X_6

	u_1=0.13	u_2=0.35	u_3=0.56	u_4=0.77
Expert1	[0.40,0.45]	[0.40,0.45]	[0.10,0.15]	[0.05,0.10]
Expert2	[0.45,0.60]	[0.40,0.50]	[0.05,0.10]	[0.00,0.05]
Expert3	[0.25,0.35]	[0.30,0.35]	[0.15,0.20]	[0.05,0.10]
Expert4	[0.45,0.50]	[0.30,0.40]	[0.05,0.10]	[0.03,0.05]
Expert5	[0.40,0.45]	[0.40,0.45]	[0.10,0.15]	[0.06,0.08]

Table 9. Assessment intervals for sample X_7

	u_1=0.15	u_2=0.44	u_3=0.72	u_4=1.01
Expert1	[0.55,0.65]	[0.20,0.45]	[0.25,0.30]	[0.10,0.15]
Expert2	[0.50,0.60]	[0.20,0.25]	[0.10,0.15]	[0.00,0.05]
Expert3	[0.35,0.45]	[0.10,0.20]	[0.15,0.20]	[0.05,0.15]
Expert4	[0.60,0.65]	[0.20,0.25]	[0.15,0.20]	[0.02,0.05]
Expert5	[0.40,0.45]	[0.25,0.30]	[0.20,0.25]	[0.05,0.10]

Table 10. Assessment intervals for sample X_8

	u_1=0.25	u_2=0.69	u_3=1.14	u_4=1.59
Expert1	[0.65,0.85]	[0.20,0.25]	[0.01,0.05]	[0.05,0.10]
Expert2	[0.60,0.70]	[0.20,0.30]	[0.00,0.05]	[0.05,0.10]
Expert3	[0.60,0.70]	[0.15,0.20]	[0.00,0.05]	[0.05,0.10]
Expert4	[0.70,0.90]	[0.10,0.15]	[0.00,0.05]	[0.05,0.10]
Expert5	[0.70,0.75]	[0.15,0.20]	[0.01,0.02]	[0.05,0.10]

Table 11. Assessment intervals for sample X_9

	u_1=0.14	u_2=0.39	u_3=0.65	u_4=1.91
Expert1	[0.65,0.85]	[0.20,0.30]	[0.10,0.20]	[0.10,0.15]
Expert2	[0.70,0.80]	[0.05,0.10]	[0.10,0.15]	[0.05,0.10]
Expert3	[0.50,0.60]	[0.05,0.10]	[0.15,0.25]	[0.02,0.05]
Expert4	[0.60,0.70]	[0.10,0.15]	[0.10,0.20]	[0.10,0.15]
Expert5	[0.45,0.50]	[0.15,0.20]	[0.15,0.20]	[0.10,0.15]

Table 12. Assessment intervals for sample X_{10}

	u_1=0.16	u_2=0.43	u_3=0.70	u_4=1.97
Expert1	[0.50,0.60]	[0.40,0.50]	[0.01,0.05]	[0.10,0.15]
Expert2	[0.40,0.50]	[0.30,0.40]	[0.00,0.05]	[0.10,0.15]
Expert3	[0.30,0.40]	[0.30,0.40]	[0.00,0.05]	[0.10,0.15]
Expert4	[0.40,0.55]	[0.40,0.45]	[0.10,0.15]	[0.20,0.30]
Expert5	[0.40,0.45]	[0.30,0.45]	[0.00,0.05]	[0.10,0.15]

c_{jk} is the center of $I^{(k)}(u_j) = [p_1^{(k)}(u_j), p_2^{(k)}(u_j)]$. Obviously, the average probability of $x \in I_j$ is

$$c(x \in I_j) = \frac{1}{t}\sum_{k=1}^{t} c_{jk}. \tag{31}$$

We normalize these probabilities to get $p(x \in I_j)$

$$p(x \in I_j) = c(x \in I_j) \Big/ \sum_{j=1}^{m} c(x \in I_j). \tag{32}$$

Then, the center of gravity of the fuzzy expected value is

$$CG(FEV(S)) = \sum_{j=1}^{m} u_j p(x \in I_j). \tag{33}$$

And, the standard deviation of the S is

$$s(S) = \sqrt{Var(S)} = \sqrt{\sum_{j=1}^{m} p(x \in I_j)(CG(FEV(S)) - u_j)^2}. \tag{34}$$

Table 13 shows the expected values and standard deviations of $S_{X_1}, S_{X_2}, \cdots, S_{X_{10}}$.

Table 13. The expected values and standard deviations from the subjective assessments

Sample	X_1	X_2	X_3	X_4	X_5	X_6	X_7	X_8	X_9	X_{10}
$CG(FEV(S_X))$	6.88	7.02	6.89	6.76	6.98	0.31	0.39	0.45	0.41	0.54
$s(S_X)$	0.32	0.31	0.25	0.40	0.39	0.19	0.28	0.37	0.51	0.59

6 Comparison of Calculation and Subjective Assessment

Comparing Table 2 and Table 13, we see that, from X_1 to X_8,

$$CG(FEV(R_X)) \approx CG(FEV(S_X)), \quad s(R_X) \approx s(S_X).$$

For X_9 and X_{10}, we see evident differences.

These differences mean that one of calculation and subjective assessment does not harmonize with the sample drawn from an exponential distribution population. Recalling that we used parameter LAMBDA=3 to generate X_9 and X_{10}. It means that the expected value of the population is 1/3=0.33. The standard deviation has the same value. Because

$$|CG(FEV(R_{X_9})) - 0.33| = 0, \quad |CG(FEV(S_{X_9})) - 0.33| = 0.08,$$

$$|CG(FEV(R_{X_{10}})) - 0.33| = 0.08, \quad |CG(FEV(S_{X_9})) - 0.33| = 0.21,$$

$$|s(R_{X_9}) - 0.33| = 0.07, \quad |s(S_{X_9}) - 0.33| = 0.18,$$

$$|s(R_{X_{10}}) - 0.33| = 0.06, \quad |s(S_{X_{10}}) - 0.33| = 0.26,$$

we conclude that the differences from the interior-outer-set model are smaller than from the expert assessment. It may imply that some inquired experts are not familiar with the exponential distribution.

It is interesting to notice that, theoretically the conventional method calculating expected value μ and standard deviation s is unbiased (the expected value of the error is zero). Using the conventional method to estimate the expected value and standard deviation of the population from which the X is drawn, we have

$$\hat{\mu} = \frac{1}{n}\sum_{i=1}^{n} x_i, \tag{35}$$

$$\hat{s} = \sqrt{\frac{1}{n}\sum_{i=1}^{n}(\hat{\mu} - x_i)^2}. \tag{36}$$

The unbiased property means

$$E(\mu - \hat{\mu}) = 0, \quad E(s - \hat{s}) = 0.$$

For $X_1, X_2, \cdots, X_{10}$, Table 14 shows the expected values and standard deviations calculated by using the conventional method.

Table 14. The expected values and standard deviations by the conventional method

Sample	X_1	X_2	X_3	X_4	X_5	X_6	X_7	X_8	X_9	X_{10}
$\hat{\mu}_X$	6.88	7.04	6.89	6.73	7.04	0.27	0.32	0.44	0.28	0.38
$\hat{s}_X$	0.32	0.32	0.26	0.42	0.40	0.19	0.28	0.44	0.29	0.28

Comparing Table 2 and Table 14, we notice that, for all Xs used in our experiments,

$$CG(FEV(R_X)) \approx \hat{\mu}_X, \quad s(R_X) \approx \hat{s}_X.$$

Then, a new question has arisen. It is unnecessary to calculate a PPD or inquire experts for subjective assessment because we can easily get unbiased estimators of the expected value and standard deviation by using formulae (35) and (36). Yes, it is right, if we can make decision only based on these statistical characteristics. However, when X is small, we not only want to know the statistical characteristics, but also want to know more information. The interior-outer-set model and subjective assessment can respectively give a fuzzy probability distribution whose performance is nearer the real situation: the estimated probabilities must be imprecise. Although the fuzzy representation of imprecise probability is only one of many alternatives, it is better than crisp representation of imprecise probability. A PPD and a subjective assessment are fuzzy representations of imprecise probability.

In the sense that the interior-outer-set model more precisely calculate the statistical characteristics than subjective assessment, the interior-outer-set model is better than subjective assessment. However, the membership function shapes of a PPD and a subjective assessment are different. A PPD R is a fuzzy relation matrix, but a subjective assessment S consists of j separating functions. Until now, we cannot judge which shape is much better. In spite of the outstanding question, it is logical to conclude that the interior-outer-set model can replace experts to give fuzzy probabilities because the calculation and the subjective assessment are very near in terms of the fuzzy expected value and the standard deviation.

7 Conclusion and Discussion

In many cases, it is impossible for an expert to precisely estimate the probabilities he needs. In his consciousness, the probabilities are fuzzy concepts, such as "very small," " not large", and so on. It is easy to acquire the fuzzy probabilities from experts. The practical troubles appear to bring together domain experts and guarantee that the inquired experts are familial with the population involved. If we are not able to overcome these difficulties, the knowledge acquisition is doomed to fail.

We can use the interior-outer-set model to calculate a fuzzy probability distribution, called possibility-probability distribution, that is very near a subjective assessment in terms of the fuzzy expected value and the standard deviation. On the basis of analysis for the method of information distribution, we conclude that, if the fuzziness of probability estimation comes from the fact that the size of the given sample is small, the interior-outer-set model can take the role of an expert to give fuzzy probabilities.

References

1. Cooman, G. de: Possibilistic second-order probability models. Advances in Cybernetic Modelling of Complex Systems (Part 5 of Proceedings of InterSymp'97, Baden-Baden, Germany, August 18-23, 1997), ed. G. E. Lasker, 6–10
2. Cooman, G. de: Lower desirability functions: a convenient imprecise hierarchical uncertainty model. Proc. 1st International Symposium on Imprecise Probabilities and Their Applications, eds. G. de Cooman, F. G. Cozman, S. Moral and P. Walley, Ghent, Belgium, 30 June - 2 July 1999, 111–120
3. Dubois, D., Prade, H.: Fuzzy sets, probability, and Measurement. European J. Operns. Res. **40** (1989) 135–154
4. Freeling, A. N. S.: Fuzzy sets and decision analysis. IEEE Trans. Sys. Man & Cyb. **SMC-10** (1980) 341–354
5. Huang, C. F.: Fuzzy risk assessment of urban natural hazards. Fuzzy Sets and Systems **83** (1996) 271–282
6. Huang, C. F.: Principle of information diffusion. Fuzzy Sets and Systems **91** (1997) 69–90

7. Huang, C. F.: Concepts and methods of fuzzy risk analysis. Risk Research and Management in Asian Perspective (Proceedings of the First China-Japan Conference on Risk Assessment and Management, November 23-26, 1998, Beijing, International Academic Publishers) 12–23
8. Huang, C. F.: An application of calculated fuzzy risk. Information Science (in press, accepted in 2001)
9. Huang, C. F.: Reliability of fuzzy risk assessment by information distribution. Proc. 19th International Conference of the North American Fuzzy Information Processing Society, Atlanta, on July 13-15, 2000, 278–282
10. Huang, C. F., Bai, H. L.: Calculation fuzzy risk with incomplete data. Proc. 4th International FLINS Conference on Intelligent Techniques and Soft Computing in Nuclear Science and Engineering, Bruges, Belgium, August 28-30, 2000, 180–187
11. Huang, C. F., Moraga, C.: Interior-outer-set model for fuzzy risk assessment and its matrix algorithm. International Journal of Uncertainty, Fuzziness and Knowledge-Based Systems (submitted in 2001)
12. Huang, C. F., Shi, P. J.: Fuzzy risk and calculation. Proc. 18th International Conference of the North American Fuzzy Information Processing Society, New York, June 10-12, 1999, 90–94
13. Liu, Z. R., Huang, C. F.: Information distribution method relevant in fuzzy information analysis. Fuzzy Sets and Systems **36** (1990) 67–76
14. Wang, P. Z.: Fuzzy Sets and Falling Shadows of Random Sets. Beijing Normal University Press, Beijing, 1985, (in Chinese)
15. Watson, S. R., Weiss, J. J. and Donnell, M. L.: Fuzzy decision analysis. IEEE Trans Sys. Man & Cyb. **SMC-9** (1979) 1–9
16. Zadeh, L.A.: Probability measures of fuzzy events. Journal of Mathematics Analysis and Applications **23** (1968) 421–427

Methodology for Optimizing Fuzzy Classifiers Based on Computational Intelligence

Ingo Renners, Adolf Grauel, and Ernesto Saavedra

Paderborn University, Soest Campus, Soest 59494, Germany
renners@ibm16.uni-paderborn.de

Abstract. In this paper a methodology using evolutionary algorithms is introduced for the optimization of fuzzy classifiers based on B-splines. The proposed algorithm maximizes the performance and minimizes the size of the classifier. On a well-known classification problem the algorithm performs an input selection over 9 observed characteristics yielding in a statement which attributes are important with respect to diagnose malignant or benign type of cancer.

1 Introduction

In the next section we shortly introduce the dataset which we used. Hence we implemented a mixed Evolutionary Algorithm (EA) consisting of Genetic Algorithms (GAs) and Evolutionary Strategies (ESs) a short introduction to evolutionary algorithms including GAs and ESs follows. The next section describes the fuzzy-classification algorithm and the encoding of the fuzzy-system into a genotype to perform structure optimization by the EA. By structure optimization we mean the selection of relevant input features out of a bulk of possible inputs, the covering of these relevant inputs with a certain number of linguistic terms, determine the shape of these linguistic terms and finding a rule-base for the fuzzy system consisting of a number of if-then-rules, where the conclusion is computed by a fuzzy-classification algorithm.

2 The Data

The Wisconsin breast cancer dataset was compiled by Dr. William H. Wolberg from the University of Wisconsin Hospitals, Madison [1, 2]. It contains $J = 699$ cases (458 benign, 241 malignant), where 16 cases have missing values, separated in $M = 2$ classes (sample diagnosis benign or malignant) and characterized by $N = 9$ ordinal discrete features (visually assessed characteristics on a scale from 1 to 10): Clump Thickness, Uniformity of Cell Size, Uniformity of Cell Shape, Marginal Adhesion, Single Epithelial Cell Size, Bare Nuclei, Bland Chromatin, Normal Nucleoli and Mitoses. 16 cases containing missing values were removed from the data set, leading to $J = 683$ cases used.

B. Reusch (Ed.): Fuzzy Days 2001, LNCS 2206, pp. 412–419, 2001.

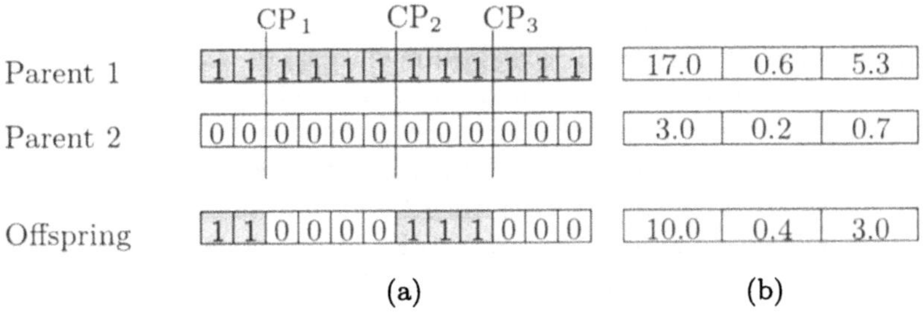

Fig. 1. Discrete 3-point crossover (a) and intermediate crossover (b).

3 Evolutionary Algorithms

Evolutionary algorithms are inspired by the principles of natural selection which adapt the genes of a species towards an optimal configuration concerning to their environment. Like the interaction of a biological individual with its environment, the interaction of a mathematical formalized chromosome with a fitness function provides a numerical valued measure of fitness. This fitness represents the quality of an individual and is used to bias the selection by preserving better individuals a higher probability to reproduce. Offspring share different chromosome parts of their parents through recombination and diversity is sustained by mutation. Thus all evolutionary algorithms share the same general proceeding of

- Step 0: initialize a start population of solutions
- Step 1: evaluate the solutions by a fitness function
- Step 2: create offsprings by using recombination
- Step 3: mutation
- Step 4: deterministic selection
- Step 5: goto step 1 until stop criteria

3.1 Genetic Algorithms

GAs were developed for optimizing binary (or at most integer) based solutions. The main operator in GAs is crossover which directs the search to new and better solutions areas. This simplest crossover schema determines by randomly chosen Crossover Points (CPs) which parts of the parent chromosomes are passed on the child (see Fig. 1a). By using only crossover a gene with identical parameter values in all individuals will never change. Therefore mutation is important to avoid stagnation of the search process in a subspace of the original search space. Mutation is implemented by inverting each gene with a priori given mutation probability.

3.2 Evolutionary Strategies

ESs were mainly developed to optimize real decision variables. Every ES-individual is equivalent to a vector of real numbers and contains values for all decision variables $x_j \in \mathbf{R}(j = 1, 2, \ldots, n)$ for the stated problem. Furthermore each individual contains $n_\sigma (1 \leq n_\sigma \leq n)$ standard deviations $\sigma_k \in \mathbf{R}^+(k = 1, 2, \ldots, n_\sigma)$ which are called (average) *rate of mutations*. These σ_k are strategy parameters which are adjusted self-adaptively during the optimization process. The decision variables of an offspring are inherited by one of the parents (same as in GAs), whereas the strategy parameters are inheritated by intermediate crossover (see Fig. 1b). Mutation is the main operator in ES and is done by changing the values of σ_j and x_j by two different methods. First the values σ_j are multiplied with a normal distributed random number. Then every decision variable (x_j) is changed by adding a normal distributed random number with expectation value 0 and standard deviation σ_j.

3.3 The Used Encoding Nomenclature

Assume the genotype which encodes the phenotype expressions by *decision variables* consists of i genes where each gene encodes an atomar value (real or integer). Starting with atomar (order zero) genes gene_i^0 $(0 \leq i \leq k)$, we can build gene compounds by introducing the nomenclature of higher order genes: $\text{gene}_i^{j-1} \subset \text{gene}_i^j$ $(j > 0, \text{gene}_i^{j-1} \neq \emptyset)$. Furthermore, let us introduce the concept of promoter genes which are used to activate sets of genes according to their value (e.g. 1 = active, 0 = inactive). Thereby the biological concept of introns is imitated.

4 Classification Based on Fuzzy Relations

By considering classification in a fuzzy environment [3] the ordinal discrete features are represented by fuzzy labels such as "if *clump thickness is big* and *uniformity of cell shape is low* then instance belongs to class C_i $(i = 0, 1)$". Therefore let $[a_n, b_n]^N$ be an N-dimensional continuous domain (in the form of a hyper-rectangle) of J real-valued vectors

$$\mathbf{x}_j = (x_{j1}, x_{j2}, \ldots, x_{jN}) \,\text{with}\, x_{jn} \in [a_n, b_n] \tag{1}$$

given as points for a pattern classification into one of M classes. Each fuzzy classification rule k has the following form:

IF x_{j1} is A_1^r **AND** ... **AND** x_{jN} is A_N^r **THEN** $\mathbf{x}_j$ belongs to class C^r,

where A_n^r are the linguistic terms in the input space with B-splines (section 4.2) as membership functions and $C^r \in \{1, 2, \ldots, M\}$ is the class assigned with a degree of certainty $CF^r \in [0.0, 1.0]$.

4.1 The Classification Algorithm

In previous papers [4] we have tested a fuzzy reasoning method based on the following described classification algorithm. To compute the activation $\alpha^k(\mathbf{x}_j)$ of rule r $(r \leq R)$ by point $\mathbf{x}_j$ use

$$\alpha^r(\mathbf{x}_j) = \mu_1^r(x_{j1}) + \mu_2^r(x_{j2}) + \cdots + \mu_N^r(x_{jN}) \tag{3}$$

with the degree of membership $\mu_n^r(x_{jn})$ of the vector component x_{jn} to the linguistic term A_n^r. The algebraic sum as a s-norm represents the AND-conjunction of the premises. Then let

$$\beta_m^r = \sum_{x_j \in C_m} \alpha^r(\mathbf{x}_j) \text{ with } m = 1, 2, \ldots, M. \tag{4}$$

Assign the class C^r such that

$$\beta_{C^r}^r = \max\{\beta_1^r, \beta_2^r, \ldots, \beta_M^r\}. \tag{5}$$

If there is no exclusive largest β_m^r then the consequent C^r of the fuzzy rule r cannot be determined uniquely and $C^r = \emptyset$ and $CF^r = 0$. Otherwise determine CF^r by

$$CF^r = \frac{\mid \beta_{C^r}^k - \beta \mid}{\sum_{m=1}^{M} \beta_m^r}, \tag{6}$$

where

$$\beta = \frac{\sum_{m \neq C^r} \beta_m^r}{M-1}. \tag{7}$$

To classify of a new point $\mathbf{x}_j$ the rule with the highest product of degree of membership and degree of certainty determines the class assigned C^r:

$$\alpha^r(\mathbf{x}_j) \cdot CF^r = \max_r\{\alpha^r(\mathbf{x}_j) \cdot CF^r\}. \tag{8}$$

The classification is invalid if several fuzzy rules r supply the same membership $\alpha^r(\mathbf{x}_j)$ but assign the point $\mathbf{x}_j$ to different classes C^r, or if no rule is activated by the input point $\mathbf{x}_j$, i.e. if $\alpha^r(\mathbf{x}_j) = 0 \ \forall r$.

4.2 The Linguistic Terms

Linguistic terms of fuzzy-systems can be formed by arbitrary basis functions defined over a knot-vector. Most common are triangular, trapezoidal or Gaussian functions. We used B-spline because of their inherent flexibility to form extremely different shapes simply by changing their order or their knot-positions. B-splines have been employed in surface-fitting algorithms for computer aided design tasks [5] and they are recursively defined by

$$\begin{aligned} B_j^{k+1}(x) &= \frac{x - \lambda_j}{\lambda_{j+k-1} - \lambda_j} B_j^k(x) + \frac{\lambda_{j+k} - x}{\lambda_{j+k} - \lambda_{j+1}} B_{j+1}^k(x) \\ B_j^1(x) &= 1, \text{if } x \in [\lambda_j, \lambda_{j+1}) \\ &= 0, \text{otherwise} \end{aligned} \tag{9}$$

with characteristics as positivity, local support and partition of unity.

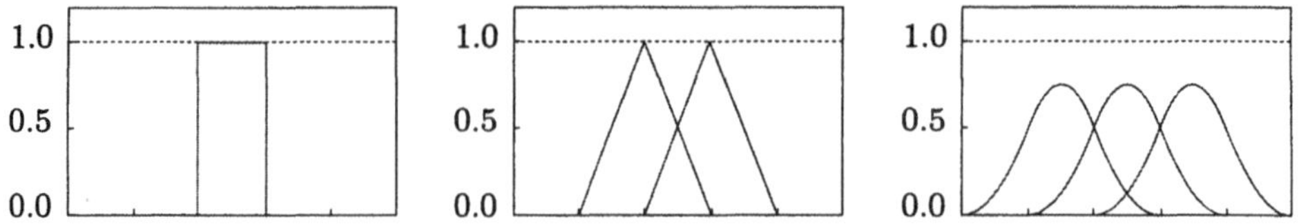

(a) Order 1 (b) Order 2 (c) Order 3

Fig. 2. Univariate B-splines of order 1, 2 and 3.

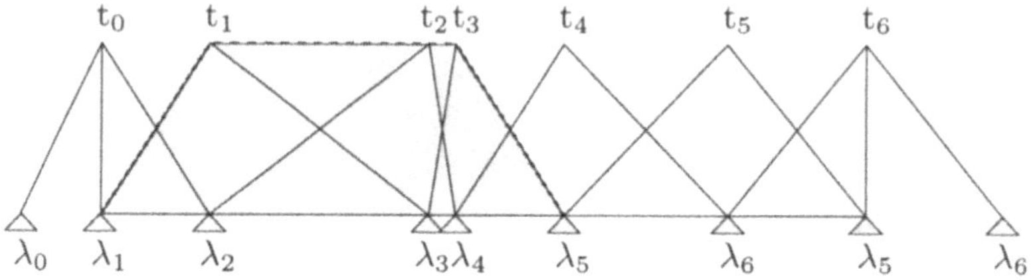

Fig. 3. Terms covering an input

4.3 Evolutionary Encoding of the Used Fuzzy-System

The important parameters for the determination of the basis functions covering an input are the order k, the number of knots (i.e. linguistic terms) and the knot-positions (i.e. appearance of the linguistic terms). This genotype assignment leads to a phenotype as exemplary depicted by the solid basis functions in Fig. 3 and is done as listed in Tab. 1. Furthermore a rule-base has to be encoded, which is done as listed in Tab. 2. Each encoded premise contain a promoter-gene which de/activates a premise and two genes which encode two position on the input interval. For both positions the nearest knot-index (i.e. λ_1 and λ_5) is evaluated yielding in a phenotype of a linguistic term as exemplary depicted by the grey-shaded basis function in Fig. 3.

4.4 The Fitness Function

The main concern is to build a system which classifies as much as possible patterns correctly. Furthermore, the system should use as few as possible rules and premises where the degree of certainty for each rule should be high. Because in this contribution we only want to demonstrate the feasibility of the approach we focus on the special case of using 2 rules (one for malignant and one for benign) to find a (sub)optimal joint face to separate the two classes. To handle (maybe) very small inclusions of pattern-clusters of one class in the hyperspace dominated by patterns of the other class some modifications of the algorithm has to be done as described in section 5. The fitness of each individual comprises of four specific fitness values (the smaller the better) which are listed according to their importance:

Table 1. Encoding of the first knot of basis functions covering an input.

gene	number type	task	phenotype
input 0			
$gene_0^0$	integer	promoter for input	—
$gene_1^0$	integer (GA)	decision variable	order
knot 0			
$gene_2^0$	integer	promoter for knot	—
$gene_3^0$	double (ES)	decision variable	position λ_0
$gene_4^0$	double (ES)	mutation rate $\sigma_{gene_3^0}$	—

Table 2. Encoding of the first premise of a rule.

gene	number type	task	phenotype
rule 0			
$gene_0^x$	integer	promoter for rule	—
premise 0			
$gene_1^x$	integer	promoter for premise	—
$gene_2^x$	integer (GA)	decision variable	knot number input 0
$gene_3^x$	double (ES)	mutation rate $\sigma_{gene_2^x}$	—
$gene_4^x$	integer (GA)	decision variable	knot number input 0
$gene_5^x$	double (ES)	mutation rate $\sigma_{gene_4^x}$	—

1. total number of Different Classes (DC) assigned: $\frac{1}{DC}$
2. Correct Classified Patterns (CCP): $\frac{1}{CCP+1}$
3. high confidence values and few Premises (P): $\sum_{r=1}^{R} \frac{CF_r}{P_r}$
4. total number of Different Input Features used (DIF): $\frac{1}{DIF}$

If two individuals are compared regarding to their fitness, the specific fitness values of each individual are compared until one one these values is smaller than the value of the other individual.

4.5 Evolutionary Parameters

The main parameters used by the evolutionary algorithm to achieve the results as shown in section 5 are: Population size = 200, tournament selection with 10 individuals (elite), stop criteria = 100th generation, used rules (fixed) = 2, encoded rule size = 10, number of premises = 1-9, number of terms = 2-30, encoded knot size = 35, possible orders = 1-5, different input features ≤ 9 and random but ≤ 10 crossover points. Mutation rates are set depending on the kind and number of encoded decision variables but all in $[0.1, 0.001]$.

5 Results and Discussion

We achieved a good classification rate of 0.987% (674 out of 683) by using only the following two rules (see also Fig. 4):

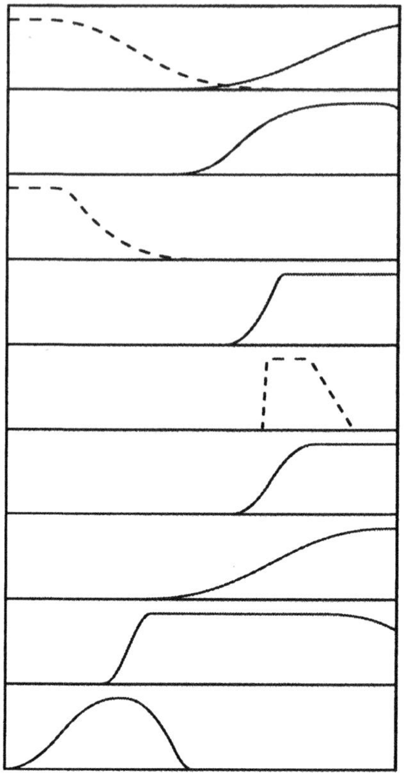

Fig. 4. Selected features and computed linguistic term distribution.

- **IF** *Clump Thickness is dashed term*
  AND *Uniformity of Cell Shape is dashed term*
  AND *Single Epithelial Cell Size is dashed term*
  THEN diagnosis is benign with degree of certainty 0.78.

- **IF** *Clump Thickness is solid term*
  AND *Uniformity of Cell Size is solid term*
  AND *Marginal Adhesion is solid term*
  AND *Bare Nuclei is solid term*
  AND *Bland Chromatin is solid term*
  AND *Normal Nucleoli is solid term*
  AND *Mitoses is solid term*
  THEN diagnosis is malignant with degree of certainty 0.92.

To classify the last 9 patterns correctly an extension to the here obtained algorithm must be performed. Since in the first generations the EA tries to find more general rules to classify the mass of the patterns correctly the maybe present rudiment potential to build rules for the very special cases (few pattern of one class nested in hyperspace domains of many patterns of the other class) is erased during evolving. An easy approach to classify such patterns correctly is to introduce specialized rules which are generated by covering each input feature of a misclassified pattern with very narrow linguistic terms. Unfortunately this approach gets more complicated if the general rule for the big cluster is still activated to the same degree as the specialized rule for the misclassified pattern and therefore in many cases the general rule has to be split up in several rules.

References

1. Mangasarin, O.L, Wolberg, W.H.: Cancer Diagnosis via linear programming. SIAM News 23 (5),(1990) 1–18.
2. Wolberg, W.H., Mangasarin, O.L.: Multisurface method of pattern separation for medical diagnosis applied to breast cytology. Proc. National Academy of Sciences U.S.A. 87 (1990) 9193–9196.
3. Grauel A.: Taschenbuch der Mathematik, Fuzzy Logic, Chapter 5.9 (Eds. Bronstein, Semendjajew, Musiol, Mühlig) Physica-Verlag, Ed. Springer-Verlag, Heidelberg-Berlin, Germany, (2002) 372–391.
4. Renners I., Saavedra E., Grauel A.: Data Mining and Classification Techniques based on Methods of Computational Intelligence. Springer-Verlag, Berlin-Heidelberg, Germany, (2001).
5. Cox, M.G.: Algorithms for Spline Curves and Surfaces. NPL Report DITC 166/90, (1990) 36 pages.

Analysing the Structure of Poly-crystalline Materials by 2-Dimensional DLS-Spectra and Neural Nets

Matthias Reuter

University of Dortmund, Chair of Computer Science I
Otto-Hahn Str. 16, D- 44221 Dortmund
Tel: 0231/755 6341, Fax: 0231/755 65555
reuter@ls1.cs.uni-dortmund.de

Abstract. A method to analyse the structure of poly-crystalline material by 2-dimensional DLS-Spectra and Backpropagation neural nets is introduced. It will be shown that DLS-Spectra "clean up" X-Ray-pictures such that the Radon-transformation of the Kikuchi-diagrams leads to better results. Further it will be shown, how these cleaned up Kikuchi-diagrams enable an automatic classification of poly-crystalline materials by neural hyper-classification systems based on the Backpropagation-algorithm.

Keywords: Kikuchi-diagrams, Radon-transformation, pattern recognition with neural nets.

1 Introduction

One of the essential features of poly-crystalline materials is that the distribution of the crystals grain-orientation (called texture of the crystal) can be detected by Kikuchi-diagrams in an automatic way. Rationale of Kikuchi-diagrams is the interaction of matter and electron beams which meet crystal probes. The interaction itself corresponds to inelastic scattering of the (wave-like) electrons by the grid system of the crystal probe, whereby the originated secondary electrons will be diffracted. This diffraction – described by Braggs law – creates a special pattern on a fluoroscopic, called Kikuchi-diagram. A Kikuchi-diagram shows straight bands – called Kikuchi-bands - and stars formed by the crosstalk of these bands. The structure of a Kikuchi-band is defined by its width which is reciprocal proportional to the shoal's distance of the of the diffracting net-planes belonging to the regarded band. The cutting angle of two bands corresponds to the angle of the according shoals of diffracting net-planes in the crystal. Thus it is depending on the orientation of the crystal which bands can be detected by Kikuchi-diagrams and how the poly-crystalline structures can be estimated. In Figure 1 a typical Kikuchi-diagram is shown; clearly the different bands and several crosstalks can be seen.

The analysis of the crystals structure is done in five steps: 1. Orientating the electron beam, 2. Recording and digitalizing the Kikuchi-diagrams, 3. Detecting the Kikuchi-bands, 4. Subscripting thc detected bands 5. Computing the texture of the crystal by using the orientation of the indexed bands.

B. Reusch (Ed.): Fuzzy Days 2001, LNCS 2206, pp. 420–429, 2001.

Fig. 1. Kikuchi-diagram

In this paper it will be shown, how an automatic detection of Kikuchi-bands (3.) can be done with the help of DLS-spectra and neural nets. Further it will be shown that the Radon-transformation can be observed as an adequate pre-processing step for such soft-computing systems, as Kikuchi-bands correspond with simple peaks in the Radon space.

2 Preprocessing Kikuchi-Diagrams

As the diffraction of the electrons results in a very poor yield of light, Kukichi-bands normally cannot be detected directly out of a fluoroscopic's activity pattern (or its record). Only a quasi dark, lightly structured background of the activity can be observed. This background results from electrons which have been scattered in an elastic way or from electron-waves which do not fulfil Braggs equation. To reduce or eliminate these disturbing noisy contributions of the picture we use 2-dimensional difference-autopower-spectra (2-dim DLS), as DLS-Spectra can be considered a band-pass-filter too [5]. The DLS-Operation itself consists of two calculation steps: 1. Calculating the average of a picture's background. 2. Subtracting the average from the original picture.

The first calculation step of the DLS-operation will be done by averaging a floating mask of the gray tones of the fluoroscopic's activity pattern. In detail the gray tone values of a $(m+1)$ x $(m+1)$ mask are added up whereby the point N denotes the epicentre of the mask. The calculated gray value is divided by the number of points added up and this value is assigned to the point N. Next the mask is displaced to the next point (pixel) of the picture and the calculation is replicated until the average has been calculated for all points of the picture. This operation can formally be regarded as convolution in the space of location for which

$$pict'_{y,x} = \sum_{i=-r}^{r} \sum_{i=-r}^{r} pict_{i,j} \cdot M_{y-i,x-j}$$

holds, whereby M denotes the mask of convolution. The entries of the mask are all "1" at the beginning. The parameter *m* denotes the width of the number of pixel points which will be added up by the DLS-operator. The parameters x, y denote coordinates of the epicentre of the mask M. The parameter r denotes the radius of the mask M, whereby for r

$$r = (m - 1)/2$$

holds.

After the 'average mask' of the background has been calculated, the dedicated DLS-values of all pixels are subtracted from the corresponding original gray values of the pixels. This "clean up" of a recorded x-ray picture done by a DLS- operation is shown in Figure 2.

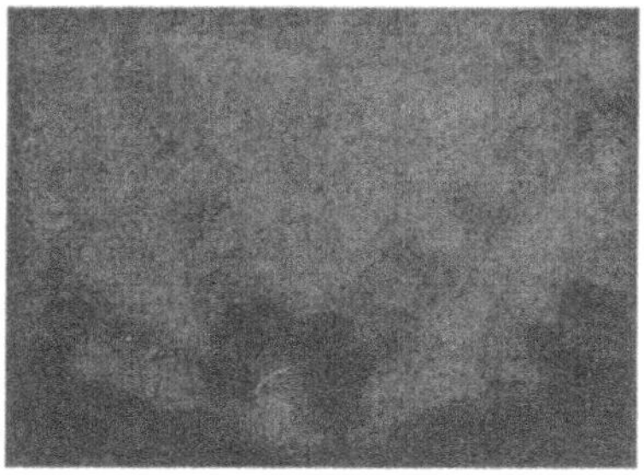

Original picture (left) and its average background (right)

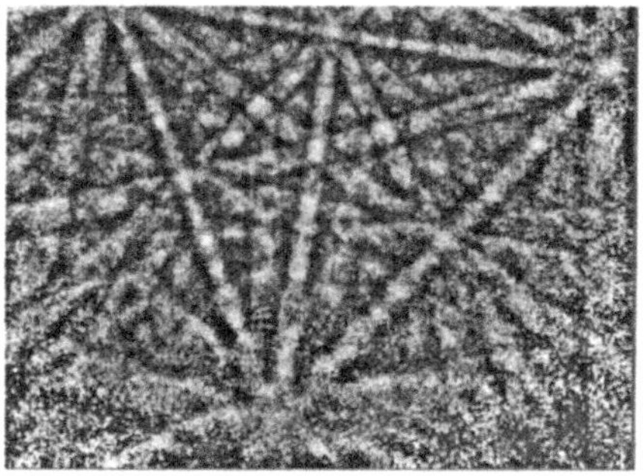

Fig. 2. Result of an amendment of an x-ray-picture using the DLS-Operation

3 The Radon-Transformation

In 1917 the Austrian mathematician Johann Radon published a paper about the integration of functions on manifolds[1] [6]. The Radon-transformation is defined by the formula

$$\Re(p_1,\ldots,p_m) = \int_{-\infty}^{+\infty} \int_{-\infty}^{+\infty} \ldots \int_{-\infty}^{+\infty} f(x_1,\ldots,x_n) \cdot \delta(h(p_1,\ldots,p_m,x_1,\ldots,x_n))\, dx_1 \ldots dx_n$$

[1] Manifolds are functions without any singularities like lines, circles, ellipses etc.

The given form of the Radon-transformation yields to a point in a so called 'Radon space'. This space has m+1 dimensions, like $\Re^{m+1}$. The expression $p_{m+1} = \Re(p_1, p_2, .., p_m)$ denotes a point in this space whose coordinates are given by the vector $(p_1, p_2, .., p_m)$. The function f: $\Re^n \rightarrow \Re$ specifies the intensities of the picture which have to be integrated along the manifold defined by the function h. It is important to mention that the function h: $\Re^{n+m} \rightarrow \Re$ has to be a canonical form of this manifold: This canonical form is set to zero if the parameter of the function fulfils the form and is set unequal zero otherwise.

Especially in image processing a scion of the Radon-transformation is used, which is called Hough-transformation [3]. The Hough-transforma-tion has the canonical form of the equation of a straight line v: $y = ax + b$, where a denotes the slope of the line and b denotes the actual value of the y–axis resp. y suffices the canonical Hessian form of a straight line, given by

$$\rho = x \cos(\phi) + y \sin(\phi)$$

Here ρ denotes the length of the perpendicula down to the straight line and ϕ denotes the angle between the x-axis and the perpendicular down to the straight line. Under these assumptions for the function h of the Radon-transformation

$$h = x\cos\phi + y\sin\phi - \rho$$

holds. In Figure 3 a straight line given by the polar-coordinates (ρ, ϕ) and the corresponding point in the Radon space (ρ, ϕ) is shown. Please note that the Radon-transformation has to be calculated for every point of the original picture to fill up the whole Radon space with values.

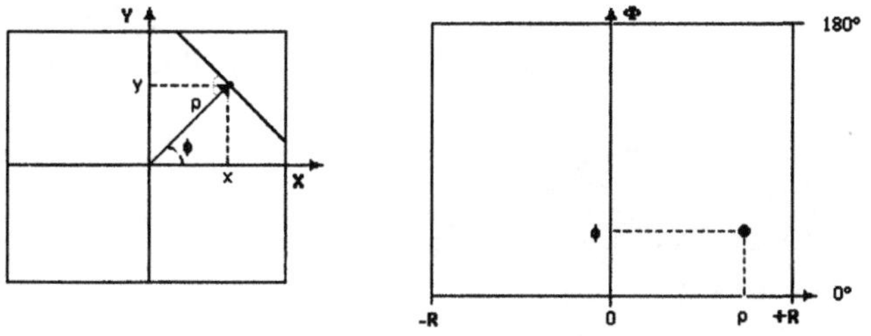

Fig. 3. Straight line (left) and the corresponding point in the Radon space (right)

The parameters p_1, p_2, x_1, x_2 are defined by: $p_1 = \phi$, $p_2 = \rho$, $x_1 = x$, $x_2 = y$. In this special case the Radon space has three dimensions as for m holds $m = 2$. If a pictures has 256 gray levels f is given by f: $|\Re^2 \rightarrow [0..255]$. This projection f combines the gray tones of the pixels with the coordinates (x, y). For the special case: 'manifold = straight line' and 'gray tone = intensity of the function' the Radon-transformation will be given by the form

$$R(\phi, \rho) = \int_{-\infty}^{+\infty} \int_{-\infty}^{+\infty} I(x, y) \cdot \delta(x \cdot \cos\phi + y \cdot \sin\phi - \rho) dxdy$$

Here $I(x,y)\ \delta(x\cos\phi + y\sin\phi - \rho)$ describes the intensities of the pixels placed on the straight line (ϕ, ρ). All other intensities are multiplied with zero. This way these intensities do not contribute to the integral given above. As we are dealing here with gray toned pixels, the discrete case of the theory has to be regarded and the integral will change to a sum. If a discrete Randon-transformation is used the resulting shape of the a Radon space will be similar to Figure 4. The white spikes define the positions of the Kikuchi-band in a coordinate system stretched by the parameters R and the angle ϕ.

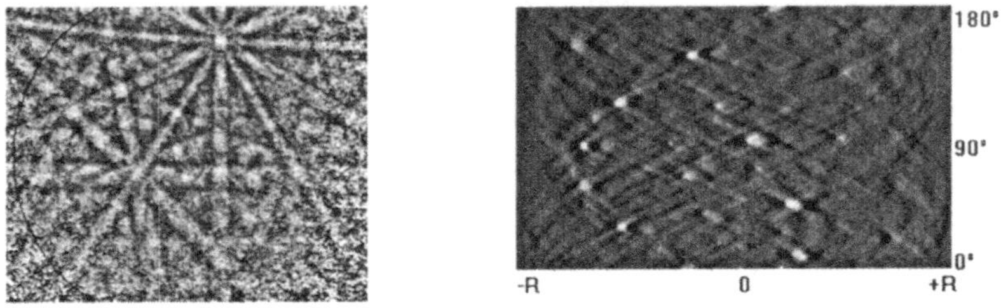

Fig. 4. Kikuchi-diagram (left) and its Radon space (right)

4 Peak-Detektion

Taking a closer look at the white spots in the Radon representation, it is striking that these white spots are hemmed by black spheres. The rectangle, which contains these two black spots and a white spot in the middle of both, should be regarded as a "peak". It is found that the white part of the peak marks the middle of the corresponding Kikuchi-band and the dark parts correspond to the edges of the band as shown in Figure 5. These features of the peaks can be used to detect Kikuchi-bands.

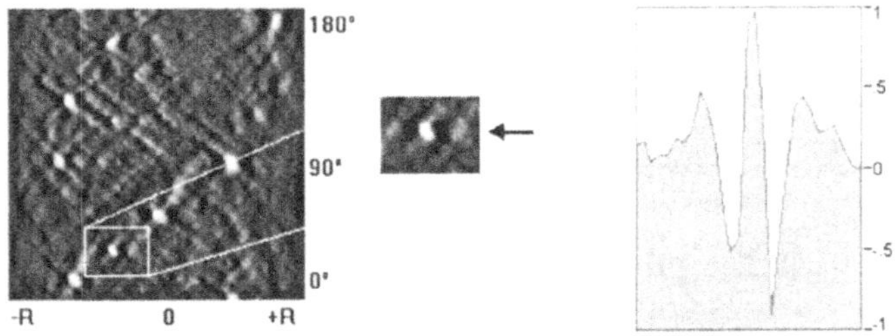

Fig. 5. Radon space with enlarged range of a peak (left) and corresponding distribution of the intensity of the Kikuchi-band (right)

On the other hand the shape of the peaks do vary in a very large range as shown in Figure 6.

Besides we can learn from Figure 6 that all peaks have the same features: a maximum value which is bounded by one minimum on the left hand side and one minimum on the right hand side. Furthermore the experiments show: if the Kikuchi-band is small the maximum values are positioned quite in the middle of the peak otherwise the maximum value differs from the middle. All these features can be used to detect Kikuchi-bands. It is important to mention that these features are only a necessary condition. With other words, even a maximum has a minimum on the left hand side and on the right hand side it is not necessarily sure that a band was detected that corresponds to a band in the Kikuchi-diagram. Therefore by using this method all right bands and some wrong ones can be detected by an algorithm.

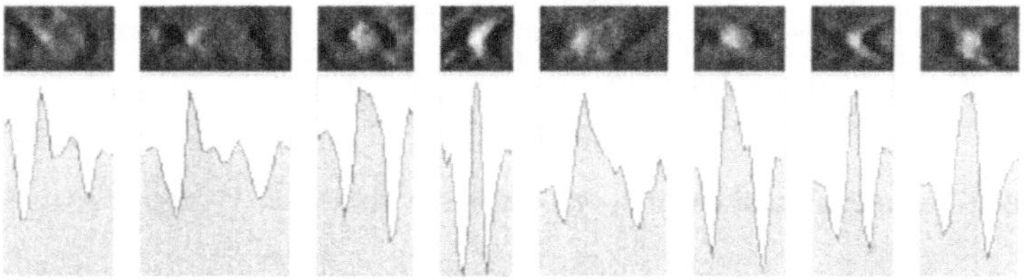

Fig. 6. Different peaks and the corresponding distribution of intensities

For detecting more 'real' Kikuchi-bands 3 steps have to be done: first the local minimums (the Radon points with an intensity value lower than the intensities of their vicinity) have to be detected. Now all local maximums (the Radon points with an intensity value larger than intensities of their vicinity) have to be detected. At least the quantum of the Radon peaks is identified as follows: for every Radon point *M* out of the quantum of all maximums two Radon points (L,R) out of the quantum of all minimums with the feature "being neighbors" are selected. These points are stored in a matrix. By this way the Radon space is divided in equal parts by a regular grid. Only these matrixes, called "cluster-array" are the objects which will be analysed subsequently. In Figure 7 such a cluster-array is shown.

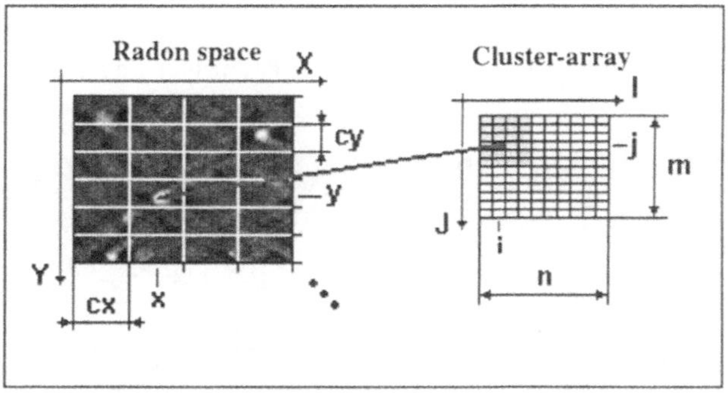

Fig. 7. Radon space (left) and corresponding cluster-array of a Radon spot (right)

Contingent on the subdivison of the Radon space it is possible that the limit of a cluster is positioned at another maximum or minimum. So a maximum or minimum can be assigned to two different Radon points of adjacent cluster-arrays (shown in Figure 8). This effect can be corrected by testing if detected Radon points are very close to each other. If this test grips, the better Radon point should be selected for further investigations.

If the detection of the peaks is done this way, normally (if the quality of the Kikuchi-diagram is good enough) between 7 and 10 Radon peaks per recorded picture can be detected.

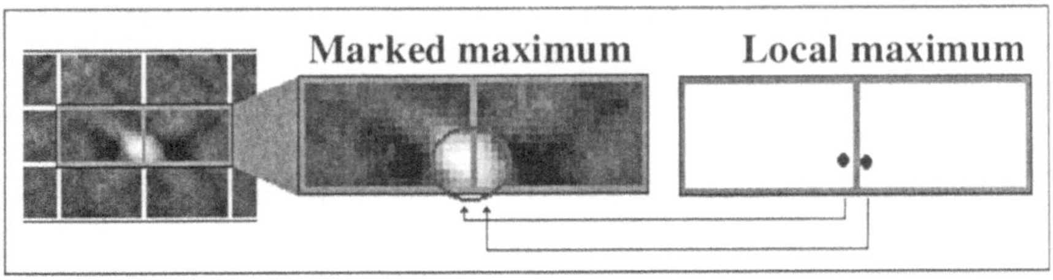

Fig. 8. Derangement at the limit of a cluster: a maximum is mapped to two clusters

Figure 9 shows an example of a DLS pre-processed x-ray picture and the corresponding result of the Radon-transformation. An example of the peak-detection is given in Figure 10, whereby three copies of the evaluated Radon space are shown: On the left hand side the detected minimums are shown. The second medially picture shows the detected maximums. The right hand picture shows the resulting peaks framed by quadrangles.

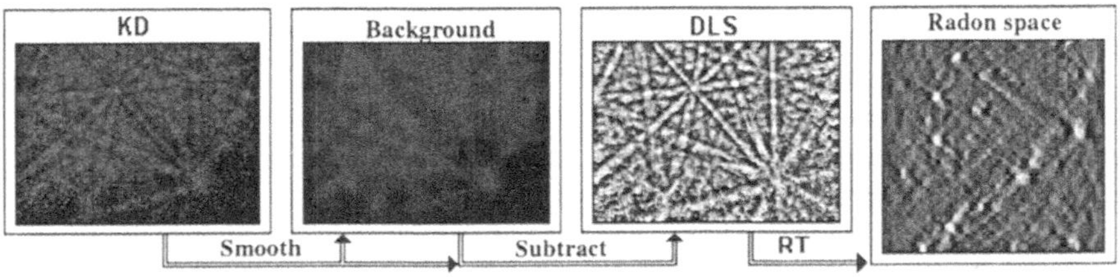

Fig. 9. Pre-processed Kikuchi-diagram and corresponding Radon space

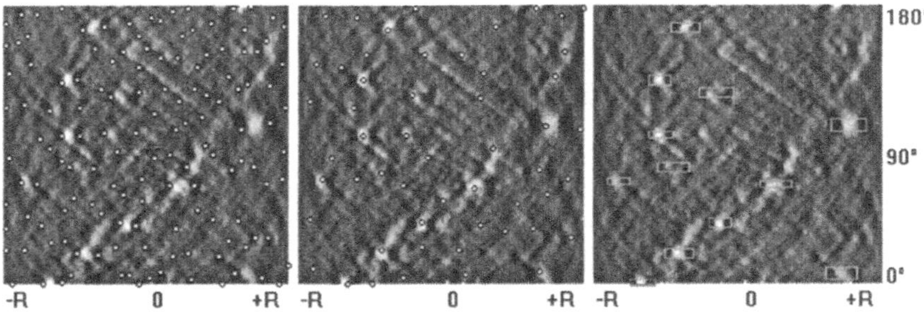

Fig. 10. Radon space with marked minimums (left), maximums (centre) and detected peaks (right)

5 Peak-Verification by Neuronal Nets

As the feature values of the given criteria can vary in a large range, common algorithms still detect a large number of peaks in the Radon space, which have no corresponding band in the Kikuchi-diagram. For that reason we used 3-layer Backpropagation-nets with the dimensions: 20 input neurons, 10 hidden layer neurons and 2 output neurons for verification. These nets have been trained to detect peaks as follows: if a corresponding Kikuchi-band for a peak in the Radon space exists, this peak is a real one and will be assigned to the neuron representing the class "correctly detected bands". Otherwise a peak is an unreal one and will be assigned to the neuron representing the class "incorrectly detected bands". The input-vectors m_p of the nets contain the intensities of the peaks and the profile (form) of the peaks, whereby the intensities have been normalised to a range [0..1]. The classifications "correctly detected bands" and "incorrectly detected bands" have been coded by the output-vectors (1,0) , (0,1).

It is important to mention that for getting better results the quantum of the data-set to be trained has to represent the two classification classes in a more or less homogeneous way. For the class "correct detected bands " this demand will be no problem, if a sufficient number of Kikuchi-diagrams is involved in the training set. But unfortunately the class "incorrect detected bands " has an uncountable number of representatives. So we decided to use the following predicate for this class: "a peak is incorrect, if it is not correct". The basic assumption of this definition is to calculate the incorrect peaks out of the characteristics of the correct peaks. So first the intensity of the correct peaks has been in verted and the inverted peaks have been flipped horizontal.

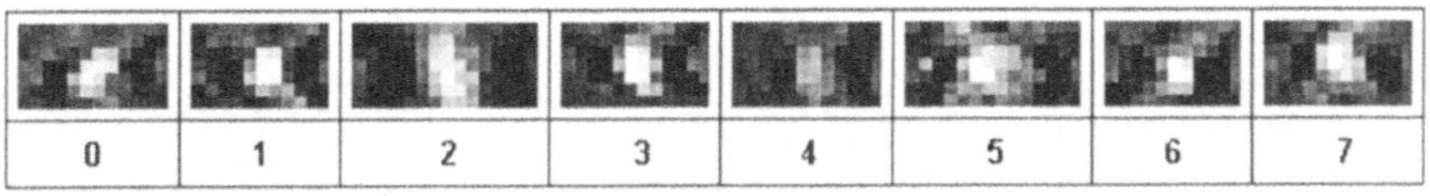

Fig. 11. Intensities of some cluster-array and their indices

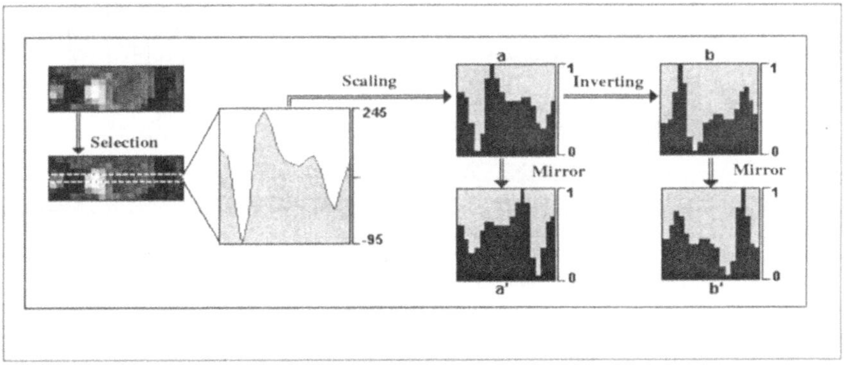

Fig. 12. Creating representatives for the classification classes "correct" and "incorrect"

The result of these operations was assigned to the class "incorrect detected bands ", as these peaks have a minimum of correlation with the correct peaks. Furthermore the correct peaks have been flipped horizontal at the x-axis and assigned to the class "correct detected bands ". Figure 11 shows a set of peaks which have been used to represent the class "correctly detected bands. In Figure 12 it is demonstrated how to create the cluster-arrays "correct peaks" and "incorrect peaks".

As common Backpropagation nets have a fixed number of neurons it is predestined to classify one cluster-array pattern only. As a matter of principle it may be possible to teach different patterns to one neural net if the dimension of the input-layer is chosen as follows: the maximum of a Radon peak (*maxB*) is coded by the number of input-neurons. On the other hand in this case all patterns of the training set which are smaller than *maxB* have to be coded by filling the "missing" width in the data string representing *maxB* with zeros. According to out tests this is not the way it works. Therefore we create a neural net for every possible width [*minB*, *maxB*] of the peaks. This way a hyper-classificator with a structure shown in the next Figure was created. On the left hand side the training patterns $TM_1,...,TM_n$ are sorted from the smallest to the largest width. With these sorted widths the Backpropagation nets $KNN_1,..., KNN_n$ are trained one by one. The terms of learning procedure are shown on the right hand side of the next Figure.

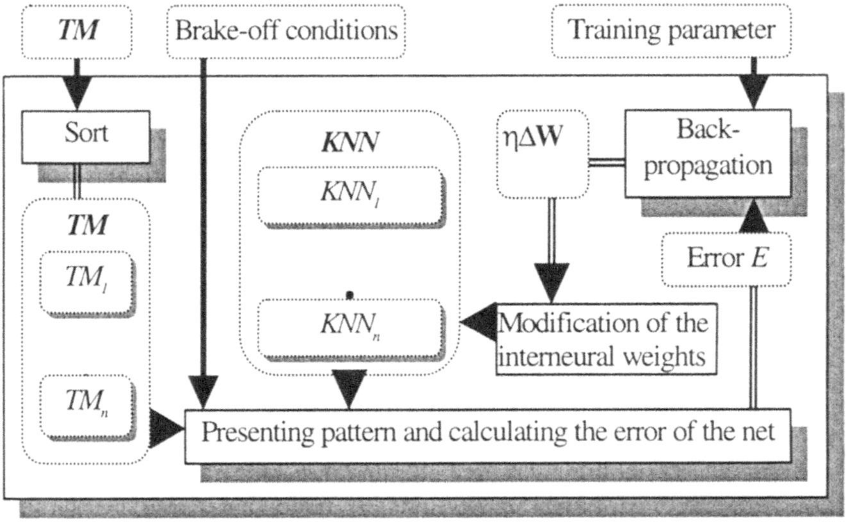

With this hyper-classificator 80 Kikuchi-diagrams have been analysed. All of them contain 8 to 10 Kikuchi-bands. As the results of Table 1 show the detection rate of the "correctly detected bands" can be improved by about 20 %, if neural nets are used instead of standard classification methods like fixed rule based system.

Bands sought in *one* picture	**all bands analysed**	**"real bands" detected**	**"imaginary bands" detected**	**rate of "correct" detection done by neural nets in %** (rounded)	**rate of "correct" detection done by standard methods in %** (rounded)
10	800	785	15	98.1	*71.3*
9	720	714	6	99.2	*76.4*
8	640	638	2	99.7	*77.1*

6 Summary

The combination of the Hough - transformation and neural Backpropagation-nets allows a quick and effective interpretation of diffraction patterns of poly-crystalline materials, if a DLS like pre-processing has been applied to the Kikuchi-diagrams.

References

1. Adams, B.L., Wright, S.I., Kunze, K.: Orientation imaging – The emergence of a new microscopy. In: Metallurgical Transaction A 24 (1993) 819-831
2. Bunge, H.J.: Texture Analysis in Materials Science. Butterworths Publ. London, 1982.
3. Hough, P.V.C.: A method and means for recognizing complex patterns. U.S. Patent 3.069.654, 1962.
4. Peterson, D.: Neural Nets. Prentice Hall, 1996, ISBN 3-8272-9531-9
5. Reuter, M.: Frequenzdifferenzspektren als Preprocessing Verfahren für neuronale Netze und klassische akustische Identifikatoren/Klassifikatoren. In: Fuzzy-Logik, Theorie und Praxis, Hrsg. B. Reusch, Informatik aktuell, pp. 111-118, Springer, Frankfurt, 1994
6. Radon, J.: Über die Bestimmung von Funktionen durch ihre Integralwerte längs gewisser Mannigfaltigkeiten. Berichte Sächsische Akademie der Wissenschaften, Seiten 262-276. Leipzig,1917.

A Neuron-MOS Threshold Element with Switching Capacitors

Victor Varshavsky[1] and Vyacheslav Marakhovsky[2]

[1] Neural Network Technologies Ltd., 3, Ha'Hashmonaim
Bnei-Brak, 51264, Israel
victor@nnt-group.com
[2] The University of Aizu, Hardware Department
Aizu-Wakamatsu City, 965-8580 Japan
marak@u-aizu.ac.jp

Abstract. In the paper a new approach to building a νMOS threshold element and an artificial neuron on its base is discussed. The implementability of the known νMOS threshold elements is restricted by the sum of input weights. It is shown that by switching the element capacitors in the pre-charge phase it is possible to reduce the implementability restriction to the maximum value of the threshold. It essentially expands the set of threshold functions realizable on one element.

1 Introduction

Among various ways of analog-digital implementation of threshold elements and artificial neurons, neuron-CMOS (νCMOS) threshold elements (containing a transistor pare with floating gates) are especially popular [1-5].

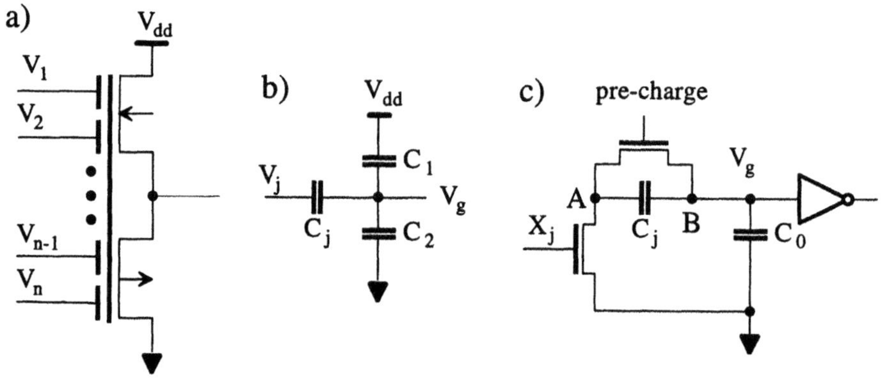

Fig. 1. νCMOS threshold element (a); Equivalent circuit for determining dependency $\Delta V_g(\Delta V_j)$; Modified νCMOS threshold element with one input (c).

The general idea of building νCMOS threshold elements (fig.1,a) is using a capacitor divider on the gates of a couple of CMOS transistors. The influence

B. Reusch (Ed.): Fuzzy Days 2001, LNCS 2206, pp. 430–435, 2001.

of ΔV_j (change of voltage V_j at the j-th input) on change of voltage V_g is determined by an equivalent circuit in fig.1,b where C_1 and C_2 are equivalent capacitors which depend on the state of the inputs ($C_1 + C_2 = \sum_{i \neq j} C_i + C_{gp} + C_{gn}$). Here is the formula of the dependency:

$$V_g = \frac{\Delta V_j C_j}{C_1 + C_2 + C_j} \tag{1}$$

If x_j is a Boolean variable such that $x_j \Delta V(j)$ is the voltage at the j-th input (when $x = 0$, the input voltage is 0), then

$$\Delta V_g = \frac{\sum_{j=0}^{n-1} x_j \Delta V_j C_j}{C_{gp} + C_{gn} + \sum_{j=0}^{n-1} C_j} \tag{2}$$

that makes possible to build the threshold element. The input weights can be determined by the values of the input capacitors C_j when all ΔV_j are equal (e.g. $\Delta V_j = V_{dd}$) or by the values of the input voltages ΔV_j when the input capacitor values are equal (e.g. in learnable artificial neurons).

One of the important problems in using *nu*CMOS threshold elements and artificial neurons on their base is their implementability, i.e. the problem of restrictions on the class of threshold functions that can be implemented by one element. These restrictions are caused mostly by the precision of fabricating capacitors C_j for the threshold elements, or by the precision of producing and holding the input voltages ΔV_j for learnable artificial neurons. As it is easy to see from (2), the precision of setting ΔV_g depends on the precision of the total capacitance $C = C_{gp} + C_{gn} + \sum_{j=0}^{n-1} C_j$, i.e. on the sum of the input weights.

On the other hand, when using changeable input voltages formed during the learning, the influence of the mistakes in capacitor fabrication is compensated during the learning and the precision of setting ΔV_g depends on the mistakes in setting only those ΔV_j for which $x_j = 1$. Since we are interested in the behavior of ΔV_g only in the area close to the threshold value, in this case the restrictions on the implementable threshold functions are determined not by the limit sum of the input weights but by the limit threshold value.

As we showed in [6], the maximum weights of the inputs, threshold and sum of the input weights grow with the increase of the number of variable not slower than the respective Fibonacci numbers. Namely, for threshold functions of n variables:

$$\begin{aligned} \max_{f,j} w_j &\geq \frac{1}{\sqrt{5}}[(\frac{1+\sqrt{5}}{2})^n - (\frac{1-\sqrt{5}}{2})^n] = \Phi_n; \\ \max_f T &\geq \frac{1}{\sqrt{5}}[(\frac{1+\sqrt{5}}{2})^{n+1} - (\frac{1-\sqrt{5}}{2})^{n+1}] = \Phi_{n+1}; \\ \max_f (T + \sum_{j=0}^{n-1} w_j) &\geq \frac{1}{\sqrt{5}}[(\frac{1+\sqrt{5}}{2})^{n+3} - (\frac{1-\sqrt{5}}{2})^{n+3}] - 1 = \Phi_{n+3} - 1; \end{aligned} \tag{3}$$

This means that passing from input weights sum restrictions to threshold restriction not only increases the number of variables for which all the threshold functions can be implemented on one element, but provides that the set of threshold functions implementable on one element is, generally speaking, open.

2 Threshold Element with Switching Capacitors

In the connection with the above, it looks very interesting to study the possibility of building a non-learnable νCMOS threshold element with a restriction on implementability associated not with the input weights sum but with the threshold. The solution can be based on the idea that comes from the above: only those capacitors C_j must take part in forming ΔV_g for which $x_j = 1$. Note that we used a similar approach when we built β-driven CMOS threshold elements [7-9].

The general idea of building such a threshold element is clear from fig.1,c. In the pre-charge phase, C_j is uncharged and cut from the ground ($x_j = 0$), voltage on the C_0 (in the point B and in the point A) is equal to a certain value $V_g = V_0$.[3] The charge on the capacitor C_0 is $Q_0 = V_0 C_0$ and on C_j the charge is $Q_j = 0$. In the working phase, the transistor shunting C_j is closed, and the voltage in the point A is equal to $\bar{x}_j V_0$. In the case of $x_j = 0$, the charge and voltage on C_0 are being kept. When $x_j = 1$, the general charge is kept but re-distributed between C_0 and C_j. Then

$$Q_0 = V_0 C_0 = V_g(C_0 + C_j), \; V_g = V_0 \frac{C_0}{C_0 + C_j}$$

and, if the circuit has n inputs, then

$$V_g = V_0 \frac{C_0}{C_0 + \sum_{j=0}^{n-1} x_j C_j}. \tag{4}$$

Now, let C_0 be proportional to the threshold and C_j proportional to the input weights, i.e. $C_0 = kT$; $C_j = kw_j$, and let the initial setting be $V_0 = 2V_{th}$ where V_{th} is the threshold of the output amplifier. Then it follows from (4) that

$$\begin{aligned} V_g \leq V_{th} \text{ if } \sum_{j=0}^{n-1} x_j C_j - C_0 = k(\sum_{j=0}^{n-1} w_j x_j - T) \geq 0, \; C_0 + \sum_{j=0}^{n-1} x_j C_j \geq 2C_0; \\ V_g > V_{th} \text{ if } \sum_{j=0}^{n-1} x_j C_j - C_0 = k(\sum_{j=0}^{n-1} w_j x_j - T) < 0, \; C_0 + \sum_{j=0}^{n-1} x_j C_j < 2C_0. \end{aligned} \tag{5}$$

It is easy to see that for V_g the condition of attaining the threshold value depends only on the threshold value and does not depend on the sum of the input weights,

[3] The procedure of setting V_0 will be discussed further.

because in the working point ($V_g = V_{th}$) the voltage V_g is formed only by those capacitors for which

$$\sum_{j=0}^{n-1} x_j C_j = C_0.$$

Now let us consider the pre-charge procedure that allows us to set $V_0 = 2V_{th}$. As a base for that, we will use the pre-charge procedure used in the νMOS implementation of threshold elements for threshold setting [5].

If in the circuit in fig.2,a the inverter's input and output are shorted (pre-

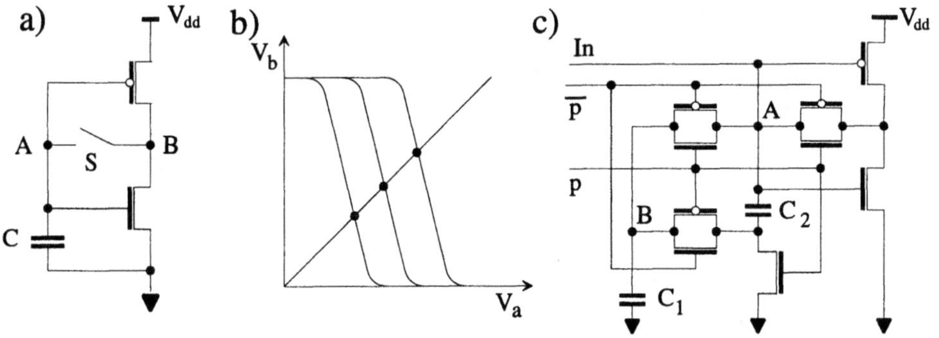

Fig. 2. Inverter with setting $V_0 = V_{th}$ in pre-charge phase (a); Input-output characteristics of dissymmetrical inverter (b); Inverter with possibility of setting $V_0 = 2V_{th}$ (c).

charge mode, the switch S is closed), the symmetrical inverter in points A and B has the threshold voltage of V_{th}. The inverter asymmetry shifts the voltage in respect to the threshold, leaving the pre-charge voltage on the line of the maximum steepness of inverter's input-output characteristic in a wide range of the inverter asymmetry variations (fig.2,b). In the working phase, the switch S is open and the threshold voltage V_{th} is stored on the capacitor C (fig.2,a).

To set in the working phase the initial voltage $V_0 = 2V_{th}$, let us use the standard scheme of doubling the voltage by commutation of the capacitors (fig.2,c). In the pre-charge mode ($p = 1$, $\bar{p} = 0$), C_1 and C_2 are put in parallel ($C_1 = C_2 = 2C_0$)[4] and for the symmetrical inverter the voltages in the points A and B are equal to the threshold voltage of the inverter ($V_A = V_B = V_{th}$). In the working mode ($p = 0$, $\bar{p} = 1$), capacitors C_1 and C_2 are put in sequence; their voltages are summed up so that the voltage on the gates of inverter's transistors is equal to the doubled threshold voltage of the inverter ($V_A = V_0 = 2V_{th}$) and the threshold capacitance is equal to C_0.

This exhausts the idea discussed in this paper.

[4] Taking into account the undergate capacitances of the inverter's transistors.

3 Conclusion

In conclusion, let us consider the results of SPICE-simulation of the νMOS threshold element circuit represented in fig.3.

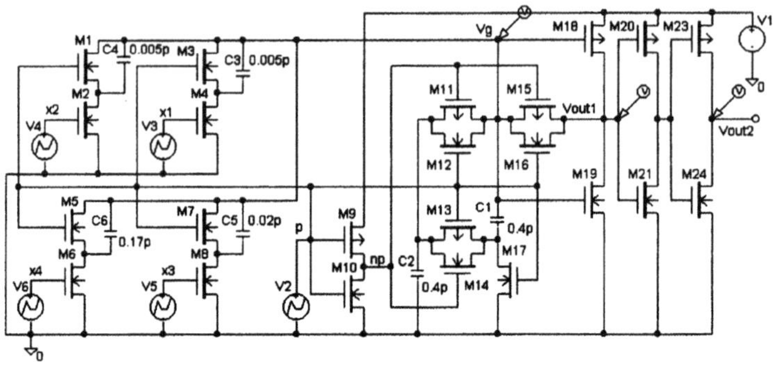

Fig. 3. Simulated circuit of the νMOS threshold element with switching capacitors.

This circuit has four logical inputs x_1–x_4 and an additional output amplifier that consists of two connected in series inverters (transistors M_{20}–M_{24}). Sources V_3–V_6 produce voltages corresponding to input variable values. The values of capacitors C_3–C_6 correspond to the input weights. The source of the voltage V_2 generates the precharge signal p. The inverter on transistors M_9, M_{10} produces the signal $\bar{p}$. Capacitors C_1 and C_2 are connected in parallel in pre-charge phase ($p = 1$) and in serial in evaluation phase ($p = 0$). Their values are chosen on the strength of the threshold ($C_1 = C_2 = 2C_0$, $T \approx C_0/min_j C_j$). The threshold value of the circuit is approximately equal to $0.2p/0.005p = 40$.

The behavior of the voltages $V_g(t)$, $V_{out1}(t)$ and $V_{out2}(t)$ getting by SPICE simulation is shown in fig.4. At the time instant $t = 0$s, $x_1 = x_2 = x_3 = x_4 = 0$, $p = 1$ and the circuit is in the precharge phase: $V_g = V_{out1} = V_{th}$, $V_{out1} = 5$V, V_{th} is the threshold voltage of the first inverter (M_{18}, M_{19}). When $t = 20$ns the control signal p becomes equal to zero starting the evaluation phase. In the begining of this phase all $x_j = 0$, the voltage $V_g = 2V_{th}$, and the output voltage $V_{out2} = 0$. After that 20ns later the input variables x_4, x_3, x_2, x_1 become equal to logical "1" by one in the corresponding order with the interval of 20ns.

Switching of each input variable leads to redistribution of capacitor charge storaged during pre-charge phase between the capacitors and we can see in fig.4 that the voltage V_g represents a ladder falling down to the threshold V_{th}. This ladder reaches to the threshold voltage when the last variable x_1 will be equal to logical "1" (the condition $\Sigma_{j=1}^{j=4} x_j C_j = C_0$ will become true). At this instant the output voltage V_{out2} of the threshold element goes to the state "1" (5V).

The first inverter (M_{18}, M_{19}) of the output amplifier amplifies the input ladder V_g with maximum gain factor on the last ladder step (V_{out1} in fig.4). The threshold of the next amplifier inverter (M_{20}, M_{21}) have to be fixed in the middle of the last ladder step of the voltage V_{out1} as it is shown in fig.4.

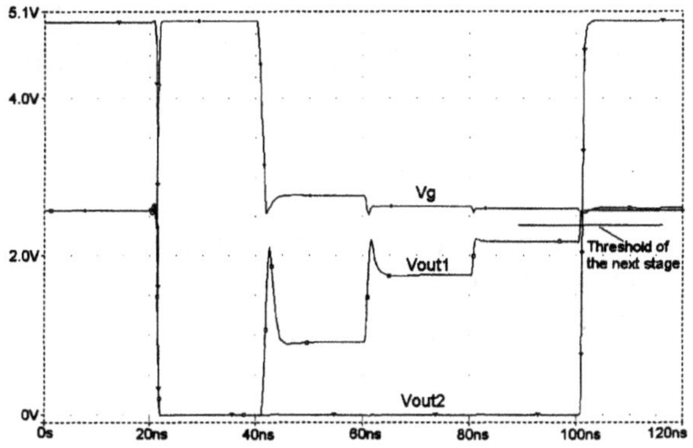

Fig. 4. SPICE simulation results for the circuit in fig.3.

Note, during transient processes of the charge redistribution the ladder V_g has glitches. These glitches can lead to wrong firing of the threshold element. To avoid this effect the output signal V_{out2} should be strobed at the time intervals when all transient processes are completed.

References

1. Shibata, T., Ohmi, T.: A functional MOS transistor featuring gate-level weighted sum and threshold operations. IEEE Transactions on Electron Devices, Vol.**39**, No.**6** (1992) 1444-1455.
2. Shibata, T., Ohmi, T.: Neuron MOS binary-logic integrated circuits: Part 1, Design fundamentals and soft-hardware logic circuit implementation. IEEE Trans. Electron Devices Vol.**40**, No.**5** (1993) 974–979.
3. Ohmi, T.: VLSI reliability through ultraclean processing. Proc. of the IEEE, Vol.**81**, No.**5** (1993) 716-729
4. Ohmi, T., Shibata, T., Kotani, K.: Four-Terminal Device Concept for Intelligence Soft Computing on Silicon Integrated Circuits. Proc. Of IIZUKA'96 (1996) 49–59.
5. Ozdemir, H., Kepkep, A., Pamir, B., Leblebici, Y., Cilingiroglu, U.: A Capacitive Threshold Logic Gate. IEEE Journal of Solid-State Circuits, Vol.**312**, No.**8** (1996) 1141-1150.
6. Varshavsky, V., Marakhovsky, V.: Learning Experiments with CMOS Artificial Neuron. Lecture Notes in Computer Science 1625, Computational Intelligence Theory and Applications, ed. by Bernard Reusch. Proceedings of the 6th Fuzzy Days International Conference on Computational Intelligence (Dortmund, Germany), Springer, 1999, pp. 706-707.
7. Varshavsky, V.: Beta-Driven Threshold Elements. Proceedings of the 8-th Great Lakes Symposium on VLSI, IEEE Computer Society, Feb. 19-21 (1998) 52–58
8. Varshavsky, V.: Simple CMOS Learnable Threshold Element. International ICSC/IFAC Symposium on Neural Computation, Vienna, Austria, (1998)
9. Varshavsky, V.: CMOS Artificial Neuron on the Base of Beta-Driven Threshold Elements. IEEE International Conference on Systems, Man and Cybernetics, San Diego, CA (1998) 1897–1861.

A Neural Approach to Compression of Hyperspectral Remote Sensing Imagery

Victor Neagoe

Polytechnic University of Bucharest,
Department of Applied Electronics and Information Engineering,
Splaiul Independentei 313, Bucharest 16, Romania 77206
vicneagoe@yahoo.com

Abstract. This paper presents an original research for hyperspectral satellite image compression using a fully neural system with the following processing stages: (1) a Hebbian network performing the principal component selection; (2) a system of "k" circular self-organizing maps for vector quantization of the previously extracted components. The software implementation of the above system has been trained and tested for a hyperspectral image segment of type AVIRIS with 16 bits/pixel/band (b/p/b). One obtains the peak-signal-to-quantization noise ratio of about 50 dB, for a bit rate of 0.07 b/p/b (a compression ratio of 228:1). We also extend the previous model for removal of the spectral redundancy (between the R, G, B channels) of color images as a particular case of multispectral image compression; we consider both the case of color still images and that of color image sequences.

1 Introduction

Over the next decade the volume of image data generated by airborne and spaceborne *remote sensing* missions will increase dramatically due to the commissioning and launching of sensors with high spatial and spectral resolution. The economics of transmission or storage of these *hyperspectral images* dictates that data *compression* is essential. A hyperspectral image comprises a number of bands, each of which represents the intensity of return from an image scene that is received by a sensor at a particular wavelength.

Hyperspectral imagery provides more information than multispectral imagery in the sense that the spectral resolution of the former is much better than that of the latter. While a multispectral image (for example, LANDSAT), generally requires only five to seven bands, a hyperspectral image of type *AVIRIS* (Airborne Visible/Infrared Imaging Spectrometer) [1], [17] simultaneously acquires *224 channels* (bands) of data in the range of 0.4 to 2.45 μm with an average spectral resolution of 10 nm. Channels of AVIRIS image are originally recorded with 12-bit resolution (compared with typically 8 bits for video) but, after radiometric correction, data is stored as 16-bit words. A property of fine-spectral-resolution imagery is interband correlation. The

B. Reusch (Ed.): Fuzzy Days 2001, LNCS 2206, pp. 436–449, 2001.

3-d correlation (two intraband (spatial) correlations as well as the third interband correlation) facilitates substantial reduction of the data required for storing and/or transmitting such imagery.

A well-known method for image compression is to extract the main directions of the input data set; this is equivalent to the computation of the Karhunen-Loeve Transform (KLT) [14]. The corresponding KLT matrix is obtained by computing the eigenvectors of the autocorrelation matrix of the input data. This problem is also called "Principal Component Analysis" PCA).

We have chosen *a neural solution of the PCA* by maximizing the information contained at the outputs of a special neural network called "Hebbian" [5], [14], [16]. If we use a specific training rule called Sanger rule [5], [14], [18], then we can prove that the weight vectors do not depend on the initial conditions and they will always converge to the eigenvectors of the autocorrelation matrix of the input data. Since then, *the Hebbian net may be considered as a neural equivalent of the KLT.*

Another common image compression method is *vector quantization*, which can achieve high compression ratios [9]. A vector quantizer makes use of the fact that a large number of possible blocks in an image look similar. These blocks are mapped to a single block (called *prototype* of the corresponding class), which is given a code that has fewer bits than the actual block representation. The image compression problem then becomes the task of finding the block in the codebook, which most closely represents an original block (namely, finding the *nearest prototype*).

Some advanced techniques for *vector quantization* belong to the field of *computational intelligence* using *neural* models. *Neural vector quantization* of images [11], [12], [13], [15] is based especially on the Kohonen Self-Organizing Map (SOM) [7]. Neighboring neurons in the above-unsupervised neural network develop adaptively into specific detectors of different vector patterns. The neurons become specifically tuned to various classes of patterns through a competitive, unsupervised or self-organizing learning. Only one cell (neuron) or group of cells at a time gives the active response to the current input. The spatial location of a cell in the network (given by its co-ordinates) corresponds to a particular input vector pattern.

First contribution of the present paper is the design, software implementation and evaluation of a fully neural model for compression of hyperspectral satellite imagery (instead of the conventional (non-neural) methods used in [1]). Our model consists of a Hebbian network (for principal component selection, that extracts the 3-d correlation of the hyperspectral image data) cascaded by a set of Kohonen network (for neural vector quantization). *The second contribution of the paper is to extend the present model based on interband correlation by considering a color image as a multispectral picture corresponding to the three R, G, B principal components.* For compression of color still images, the scheme remains the same as for hyperspectral satellite images, but the number of bands becomes three. For representation of color image sequences, the model includes a first processing stage consisting of a 4-dimensional orthogonal transform (instead of the 3-d transform used for hyperspectral imagery) for extraction of the principal component of the input color image sequence followed by a second processing stage of *neural* vector quantization. The experimental compression results are given both for the principal model (compression of hyperspectral satellite imagery) as well as for the special application (compression of color images).

2 A Fully Neural Model for Compression of Hyperspectral Imagery

2.1 Model Description

The proposed model (Fig. 1) contains the following processing cascade:

(a) The Hebbian network for *extraction of the principal components*;

(b) A set of self-organizing neural networks (Kohonen) for *vector quantization* of the principal components.

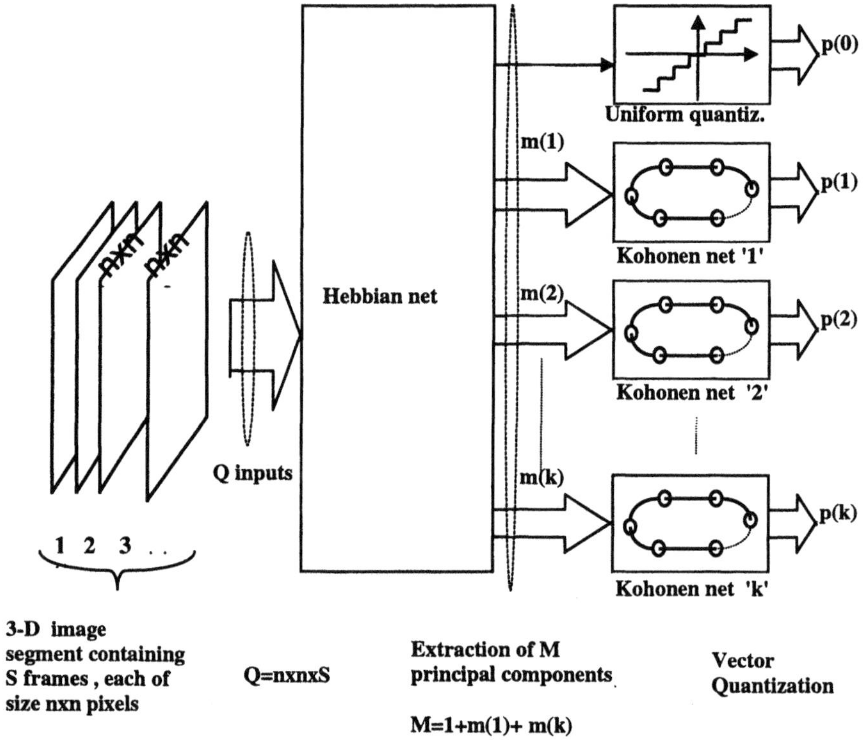

Fig. 1. Architecture of the neural system for compression of hyperspectral imagery.

(a) *The Hebbian network* processes the 3-d elementary blocks of "n x n x S" pixels of the input hyperspectral sequence (where "n x n" is the elementary analysis square in each "N x N" pixel band, and S is the number of bands. This network is a neural replica of the optimum 3-d Karhunen-Loeve transform.

To improve the convergence, we have chosen the initial weights of the network to be given by the elements of the matrix defining the 3-d Discrete Cosine Transform (3-d DCT). The network has Q=n x n x S inputs, corresponding to the above-mentioned 3-d multispectral segment and M outputs, corresponding to the principal components. The neural variant has the advantage (over the non-neural one=KLT) of deducing the optimum transformation by a simple and iterative technique instead of requiring a significant computational effort for evaluating the autocorrelation matrix, the eigenvalues and eigenvectors!

(b) *The system of "k" circular self-organizing maps* performs vector quantization of the M-1 (AC) principal components, given by the Hebbian network. These components are grouped into "k" subsets, so that m (1) + m (2)+...+m (k)= M-1, where m (h) is the number of inputs of the self-organizing map of index "h"; each network has 2exp[p (h)] neurons (outputs), where p (h) is the number of bits for encoding the segment "h" of the principal component set. First component is uniformly quantized with p (0) bits. Since then, the bit rate provided by the neural compression system is R = p (0)+p (1)+...+p (k)]/(n x n x S) bits/pixel/band (b/p/b). The radius of the neighborhood of each neuron decreases with the iteration. The *circular* architecture of the network implies a perfect symmetry.

2.2 Experimental Results for Compression of Hyperspectral Imagery

We have used hyperspectral images of the type AVIRIS (Airborne Visible/Infrared Imaging Spectrometer). The images are selected from a hypercube containing 128 spectral bands, each band with 128 x 128 pixels. The images are represented with a radiometric resolution of 16 bits/pixel/band and correspond to an urban area.

2.2.1 Training

a. Selection of the principal components (Hebbian Network)

- We have used S=8 spectral bands (avir_1.raw,....., avir_8.raw)
- The size of each band: N x N= 128 x 128 pixels
- Resolution: 16 bits/pixel
- The input multispectral image is segmented into 3-D blocks of 8 x 8 x 8 (namely, n=S=8)
- Number of inputs of the Hebbian network: n x n x S=512
- Number of outputs (selected components): M= 20

The training of the Hebbian network may be evaluated in Table 1.

b. Quantization of the Principal Components (Neural Self-Organizing System)

- The DC coefficient is scalar quantized with p (0)=9 bits.
- The set of M-1 = 19 AC coefficients are vectorially quantized by segmenting the set into k = 3 subsets of sizes: m (1)=7; m (2)=6; m (3)=6.
- Each of the three neural networks has a circular architecture with 512 neurons (it implies that a corresponding prototype is encoded with p (1)=p (2)=p (3)= 9 bits).
- The resulted bit rate is R= (9+27)/512= 0.07 bits/pixel/band (b/p/b), corresponding to the compression factor of F=16/0.07=228:1.

• The objective quality of the reconstructed bands of the hyperspectral training image after processing by the Hebbian network, *with* and *without* neural quantization may be evaluated from the Table 2.

Table 1. Peak signal-to-quantization noise ratios during the refinement of the Hebbian network for the hyperspectral image AVIRIS (8 bands: avir_1.raw, avir_2.raw,…,avir_8.raw); number of retained coefficients: M=20 (t= index of epoch)

t (epoch)	0	1	2	3	4	Frozen after t=4
(PSNR) dB (Global)	49.56	49.65	49.77	49.84	49.87	49.87
(PSNR) dB (Band1)	48.41	48.79	49.53	50.10	50.52	50.73
(PSNR) dB (Band 2)	51.37	51.35	51.31	51.21	51.07	51.01
(PSNR) dB (Band 3)	51.06	51.00	50.87	50.75	50.65	50.59
(PSNR) dB (Band 4)	48.72	48.80	48.94	49.05	49.16	49.20
(PSNR) dB (Band 5)	50.04	50.11	50.23	50.33	50.41	50.45
PSNR) dB (Band 6)	50.76	50.78	50.84	50.86	50.86	50.87
(PSNR) dB (Band 7)	51.05	51.21	51.43	51.53	51.55	51.54
(PSNR) dB (Band 8)	47.11	47.10	47.00	46.91	46.83	46.76

Table 2. Peak signal-to-quantization noise ratios of the hyperspectral *training* sequence AVIRIS (8 bands: avir_1.raw,…,avir_8.raw) processed firstly by the Hebbian network (after freezing the weights obtained during 4 epochs of training and retaining M=20 components) and then reconstructed *without* or *with* neural quantization

	Global	Band 1	Band 2	Band 3	Band 4	Band 5	Band 6	Band 7	Band 8
(PSNR) dB Reconstruction **without** quantization	49.87	50.73	51.01	50.59	49.20	50.45	50.87	51.54	46.76
(PSNR) dB Reconstruction **with.** quantization	49.69	50.53	50.79	50.39	49.05	50.25	50.65	51.28	46.64

•We can remark a high fidelity of the quantization (the global signal-to-quantization noise ratio decreases only from 49.87 dB to 49.69 dB as effect of quantization!).

• In Fig. 2 (a, b, c), we can subjectively evaluate the quality of the reconstructed image corresponding to Table 2 (band 3 of the considered training image). Visually, we cannot remark any difference between the input and the reconstructed image.

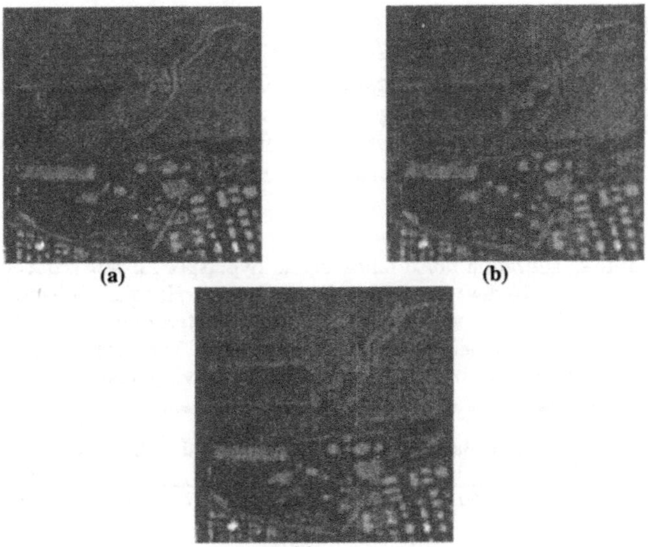

Fig. 2. (a) Original band 3 (avir_3.raw). (b) Band 3 reconstructed after Hebbian selection of the principal components without quantization (PSNR=50.59 dB). (c) Band 3 reconstructed after Hebbian selection of the principal components and vector quantization with a system of three self-organizing neural networks (PSNR=50.39 dB, R= 0.07 b/p/b; compression ratio=228:1).

2.2.2 Testing

• We have used the set of eight bands of the hyperspectral image (avir_9.raw, ...,avir_16.raw), different from those used for training, but corresponding to the same urban area.

• The parameters of the input sequence, those of the Hebbian network, as well as those of the Kohonen system are the same as for the training phase.

• In Table 3, we can evaluate the objective quality of the reconstructed picture (peak signal-to-quantization-noise ratio =PSNR) for the hyperspectral test image, while in Fig. 3 (a, b, c) we can visually evaluate the reconstructed test picture (band 10). *The signal-to-noise ratio (**about 50 dB!**) and the high compression ratio (of **228:1**) combine the high quality of reconstruction with an important coding efficiency.*

3 Removal of the Spectral Redundancy of Color Images as a Particular Case of Multispectral Image Compression

3.1 Compression of Still Color Images

We further extend the present model based on interband correlation by considering a color image as a multispectral picture corresponding to the three R, G, B component images! For compression of color still images, the scheme remains the same as for hyperspectral satellite images, but the number of bands becomes 3. Thus, *we build an original model for color image representation,* by considering in the same 3-d orthogonal transformation not only the 2-d spatial correlation ***but also the spectral correlation between the R, G, and B components***! One can approximate the Hebbian network by a suboptimum 3-d orthogonal transform like 3-d Discrete Cosine Transform (3-d DCT) with a small reduction of data compression performances but with a significant increasing of the computation speed.

Table 3. Peak signal-to-quantization noise ratios of the hyperspectral *test* sequence AVIRIS (8 bands: avir_9.raw,...,avir_16.raw) processed firstly by the Hebbian network, and then reconstructed *without* or *with* neural vector quantization (after freezing the weights obtained during 4 epochs of training and retaining M=20 components). The neural system has been trained on the multispectral sequence of 8 bands: avir_1.raw,...,avir_8.raw.

	Global	Band 9	Band 10	Band 11	Band 12	Band 13	Band 14	Band 15	Band 16
(PSNR) dB Reconstruction **without** quantization	**49.10**	46.16	50.46	50.15	48.62	49.57	50.73	50.33	48.83
(PSNR) dB Reconstruction **with** quantization	**47.95**	45.40	48.87	48.83	47.69	48.41	49.18	48.60	47.93

Experimental Results

We have trained and tested this special application for the color picture "Girl" of 512 x 512 pixels, represented in true-color bmp (24 bits/pixel).

General parameters

- The 3-d segment has the sizes: $n_1 = n_2 = 8$, $S = 3$, corresponding to the hyper-rectangle of 8 x 8 x 3.
- Number of retained coefficients (principal components) after 3-d DCT processing: M=25
- The first 3-d DCT coefficient (0,0,0) has been scalar quantized with p (0) = 8 bits.

Parameters of the neural system

- The set of M-1 = 24 AC coefficients are vectorially quantized by segmenting the set into k= 3 subsets of sizes: m (1)= m (2)= m (3)=8, each subset containing the inputs of a corresponding circular self-organizing map for vector quantization.
- Size of each ring network: 256 x 1 (p (1) = ... =p (3)= 8 bits)
- The resulted bit rate is R= (8+24)/(8 x 8 x 3) = 0.167 bits/pixel/channel (b/p/c), or, for other representation is R= (8+24)/(8 x 8)=0.5 bits/true-color pixel. It corresponds to the compression factor of F=24/0.5=48:1.

(a)

(b)

(c)

Fig. 3. (a) Band 10 original (avir_10.raw; 128 x 128 pixels). (b) Band 10 reconstructed after Hebbian selection of the principal components without quantization (PSNR=50.46 dB). (c) Band 10 reconstructed after Hebbian selection of the principal components and vector quantization with a system of 3 self-organizing neural networks (PSNR=48.87dB, R= 0.07 b/p/b; compression ratio=228:1). The neural system (Hebbian + Kohonen) has been trained on the multispectral sequence of 8 bands: avir_1.raw,…, avir_8.raw.

Table 4. Signal-to-quantization noise ratios for each color channel of the reconstructed color image "Girl" after 3-d DCT and neural vector quantization

(PSNR) red [dB]	(PSNR) green [dB]	(PSNR) blue [dB]
25.02	22.37	26.60

The *objective quality of the reconstructed "Girl" is given in Table 4 and the subjective* quality of the reconstructed color image after compression may be evaluated from Fig. 4.

3.2 Compression of Color Image Sequences

We extend for color image sequences the previous model of compression of hyperspectral images. Instead of separately processing the color image sequences (for each of the fundamental colors R, G, B), we have chosen a global processing for

(a)

(b)

Fig. 4. (a) Original "Girl". (b) Reconstruction of the "Girl" after 3-d DCT and neural quantization (compression ratio R=48:1).

redundancy removal taking into account in the same processing stage the 4-d correlation of the color image sequences: two dimensions of spatial correlation, one dimension for temporal correlation *and also one dimension for spectral correlation (corresponding to the R, G, B bands!). We choose a 4-dimensional orthogonal transform for color sequence representation,* instead of the well-known 3-d transform or hybrid coding (2-d transforms combined with prediction) for each color components. Thus, we consider in the same orthogonal transformation not only the spatial correlation and the temporal one, *but also the spectral correlation between the R, G, B channels. For example, we have chosen* a 4-dimensional Discrete Cosine Transform (4-d DCT), that is an approximation of the KLT (Hebbian net) that reduces the computational complexity. The 4-d DCT coefficients are given by the relation

$$C(k_1,k_2,k_3,k_4)=\frac{u(k_1)\cdot u(k_2)\cdot u(k_3)\cdot u(k_4)}{\sqrt{N_1N_2N_3N_4}}\cdot$$

$$\left(\sum_{n_1=0}^{N_1-1}\sum_{n_2=0}^{N_2-1}\sum_{n_3=0}^{N_3-1}\sum_{n_4=0}^{N_4-1} x(n_1,n_2,n_3,n_4)\cos\frac{\pi(2n_1+1)k_1}{2N_1}\cos\frac{\pi(2n_2+1)k_2}{2N_2}\cos\frac{\pi(2n_3+1)k_3}{2N_3}\cos\frac{\pi(2n_4+1)k_4}{2N_4}\right)$$

where $u(k_i)=\begin{cases}1, & k_i=0\\ \sqrt{2}, & k_i=1,\dots,N_i-1\end{cases}$

$i\in 1,2,3,4$, n_i and k_i belong respectively to the sets $\{0,\dots,N_1-1\}$, $\{0,\dots,N_2-1\}$, $\{0,\dots,N_3-1\}$ and $\{0,\dots,N_4-1\}$. Here, $N_1 = N_2 = n$ (elementary analysis square); N_3 is equal to S=frame number; N_4 =3 (number of channels). In the previous relation, we consider that a color image sequence segment expressed in the (R, G, B) format are represented by the corresponding 4-dimensional matrix x (i, j, k, h), for i = 0,..., N_1-1, j = 0,...,N_2-1, k = 0,...,N_3-1 and h = 0,...,N_4-1. The M retained 4-d DCT coefficients corresponding to the principal components of the input color image sequence segment are grouped in several sets (vectors) and each such a vector is applied to a corresponding *neural quantizer*.

Processing stages of the proposed model:

(a) a 4-d orthogonal transform of each input 4-d matrix of "N_1 x N_2 x N_3 x N_4" fundamental color pixels into a set of M selected components in the frequency domain (where N_1 x N_2 are the sizes of the square analysis segment of a certain frame, N_3 is the number of frames considered to be redundant, and a color pixel corresponds to N_4 =3 monochrome pixels, one for each of the fundamental colors R, G, B).

(b) a neural vector quantization system, consisting of "k" vector quantizers of the M-1 selected components (the AC ones) obtained as a result of the previous processing stage, where the selected components are grouped into "k" subsets

Remarks:

• All "k" neural vector quantizers are trained using one or several image sequences.

• **After training, we perform the processing of an input sequence according to the previous mentioned stages (a) and (b).**

Experimental Results

We have designed and implemented the software of the **neural** system shown in Fig. 1, where instead of the Hebbian network, we have used a 4-d DCT. For experimenting the corresponding system, we have used two image sequences: "Miss America" (misa) and "Salesman"; each experimental sequence had a length of maximum 64 frames of 256 x 256 pixels /frame with 24 bits/true-color pixel.

General parameters

• Number of retained coefficients (principal components): M=270

• The first 4-d DCT coefficient (0,0,0,0) has been scalar quantized with p (0) = 10 bits.

(a)

Fig. 5. (a) Original first frame of the color sequence "Miss America". (b) Reconstruction of the first frame of the color sequence "Miss America" (having 8 frames) using the proposed neural model (trained on the same sequence) (compression ratio R=150:1).

Parameters of the neural system

We have used a system of k=9 circular self-organizing maps (Kohonen networks). Each network has a number of 2exp [p (h)] neurons (outputs), where p (h) is the number of bits to encode the group ("h") of m (h) principal components. The neural system has the following parameters:

• *First six networks:*
  ♦ number of inputs: m (1) = ...=m (6)=36;
  ♦ size of the ring networks: 256 x 1 (p (1) = ... =p (6)= 8 bits)

• *Seventh and eight networks:*
  ♦ number of inputs: m (7) = m (8)=37
  ♦ size of the ring networks: 256 x 1; (p (7) = p (8)=8 bits)

• *Ninth network*
  ♦ number of inputs: m (9) = 39
  ♦ size of the ring network: 256 x 1 (p (9) = 8 bits).

• Resulted bit rate is BR=0.16 bits/true-color pixel (compression ratio R= 150:1)

• The peak signal-to-quantization noise ratios of the reconstructed first frame of the sequence "Miss America" (for the main bands R, G, B) are given in Table 5, while the *subjective* quality of the reconstructed color frame after compression may be evaluated in Fig. 5.

Table 5. Signal-to-quantization noise ratios for each color channel of the reconstructed first frame of the color sequence „Miss America".

(PSNR) red [dB]	(PSNR) green [dB]	(PSNR) blue [dB]
33.54	34.95	33.14

4 Concluding Remarks

1. This paper *presents a fully neural model for compression of hyperspectral satellite imagery* consisting of a Hebbian network (for principal component selection, that extracts the 3-D correlation of the hyperspectral image data) cascaded with a set of ring Self-Organizing Maps (for neural vector quantization of the previously extracted components).

2. By the proposed neural model, we point out the feasibility of applying an exciting technique of computational intelligence for compression of satellite imagery, instead of the conventional techniques.

3. If we compare the Hebbian network with the KLT, the neural variant has a significant advantage in reducing the computational effort, avoiding the necessity of deducing the autocorrelation matrix, its eigenvalues and eigenvectors and so on. The neural vector quantization proves also to be competitive with the classical (non-neural) vector quantization for the image compression task.

4. We give the experimental results of the software implementation of the previous model for compression of the hyperspectrral images AVIRIS. One obtains very good results: the peak-signal-to-quantization-noise-ratio of about 50 dB for each band, for a bit rate of 0.07 b/p/b (a compression ratio of 228:1). *This means a high quality of image reconstruction combined with a significant coding efficiency.*

5. As a special application, we extend the present model based on interband correlation by considering a color image as a multispectral picture corresponding to the three R, G, B channels. For compression of color still images, the scheme remains the same as for hyperspectral satellite images, but the number of bands becomes 3, corresponding to the R, G, B channels. Thus, *we obtained an original model for color image representation,* by considering in the same 3-d orthogonal transformation not only the 2-d spatial correlation *but also the spectral correlation between the R, G, B components*! To increase the computation speed we replace the Hebbian network by the 3-d DCT.

6. By extending the initial scheme to *the representation of color image sequences, we build a new model that includes a 4-dimensional orthogonal transform as a first processing stage* (instead of the 3-d transform for hyperspectral imagery) for extraction of the principal components. Thus, we consider in the same 4-d orthogonal transformation the redundancy removal corresponding to the following four correlation dimensions: the 2-d spatial correlation (the first two dimensions), the temporal one (the third dimension), and the spectral correlation between the R, G, B bands (the fourth dimension)! We have applied this 4-d orthogonal representation model for the particular case of the 4-d DCT, instead of the Hebbian net, to reduce the computational effort. The second processing stage (*neural* vector quantization) remains the same as for compression of hyperspectral images.

7. The very good experimental compression results are obtained both for color still images (compression ratio of 48:1) and also for color image sequences (compression ratio of 150:1).

References

1. Abousleman, G. P. , Marcellin, M. W., Hunt, B. R.: Compression of Hyperspectral Imagery Using the 3-D DCT and Hybrid DPCM/DCT. IEEE Trans. Geosci. Remote Sensing. 33 (1995) 26-34
2. Bishop, C.M.: Neural Networks for Pattern Recognition. Oxford University Press, New York (1995)
3. Chan, Y. -L., Siu, W. -C.: Variable Temporal - Length 3-D Discrete Cosine Transform Coding. IEEE Trans. Image Proc. 6 (1997) 758-763
4. Cramer, C., Gelenbe, E., Bakircioglu, H.: Low Bit-Rate Video Compression with Neural Networks and Temporal Subsampling. Proceedings IEEE. 84 (1996) 1529-1543
5. Hertz, J., Krogh, A., Palmer, R.: Introduction to the Theory of Neural Computation. Addison -Wesley Publishing Company, Redwood City California (1990)
6. Jain, A.K.: Fundamentals of Digital Image Processing, Prentice-Hall, Englewood Cliffs NJ (1989)
7. Kohonen, T.: The Self-Organizing Map. Proceedings IEEE. 78 1461-1480 (1990)
8. Li, H., Lundmark, A., Forchheimer, R.: Image Sequence Coding at Very Low Bit Rates: A Review. IEEE Trans. Image Proc. 3 (1994) 589-608
9. Nasrabadi, N.M., King, R.: Image Coding Using Vector Quantization: A Review. IEEE Trans. Commun. COM-36 (1998) 957-971
10. Neagoe, V. -E.: Predictive Ordering Technique and Feedback Transform Coding for Data Compression of Still Pictures. IEEE Trans Commun. COM-40 (1992) 386-396
11. Neagoe, V.-E.: A Circular Kohonen Network for Image Vector Quantization, In: D'Hollander, E.H., Joubert, G. R. Peters, F. J., Trystram, D. (eds.): Advances in Parallel Computing, Vol. 11. Elsevier, Amsterdam New York (1996) 677-680

12. Neagoe, V.-E., Szabo, F., Fratila, I.: A Fully Neural Approach to Color Image Compression. Proceedings of the International Symposium on Communications'96. Bucharest (1996) 476-481
13. Neagoe, V.–E., Georgescu, B.: A Neural Vector Quantization for Image Sequence Compression. In: Reusch, B., Dascalu, D. (eds.): Real World Applications of Intelligent Technologies. Part II. printed by National Institute for Research and Development in Microtechnologies, Bucharest (1998) 86-90
14. Neagoe, V.-E., Stanasila, O.: Recunoasterea formelor si retele neurale - algoritmi fundamentali (Pattern Recognition and Neural Networks-Fundamental Algorithms). Ed. Matrix Rom, Bucharest (1999)
15. Neagoe, V.-E.: A Neural Vector Quantization of 4-D Orthogonally Transformed Color Image Sequences. In: Borcoci, E., Dini, P., Vladeanu, C., Serbanescu, A. (eds.): Proceedings of the IEEE International Conference on Telecommunications, 4-7 June 2001, Bucharest, Romania, Vol. Special Sessions. Printed by Geoma, Bucharest (2001) 247-251
16. Oja, E.: A Simplified Neuron Model as a Principal Component Analyzer. Math. Biol. 15, 267-273 (1982) 267-273
17. Ryan, M. J. Arnold, J. F.: The Lossless Compression of AVIRIS Images by Vector Quantization. IEEE Trans. Geosci. Remote Sensing. 35 (1997) 546-550
18. Sanger, T. D.: Optimal Unsupervised Learning in a Single Layer Linear Feedforward Neural Network. Neural Networks. 2 (1989) 459-473

Multi-valued and Universal Binary Neurons: New Applications in Intelligent Image Processing

Igor Aizenberg

Neural Networks Technologies (NNT) Ltd., 155 Bialik st.,
Ramat-Gan, 52523 Israel
igora@netvision.net.il;

Abstract. Multi-valued neurons (MVN) and universal binary neurons (UBN) are neural elements with complex-valued weights and high functionality. It is possible to implement the arbitrary mapping described by partially defined multiple-valued function on the single MVN and the arbitrary mapping described by Boolean function (which may not be threshold) on the single UBN. In this paper we consider some applications carried out using these wonderful features of MVN and UBN. Conception of cellular neural networks based on MVN and UBN becomes a base for nonlinear cellular neural filtering. Application of the corresponding filters for edge detection and solving of the super-resolution problem are considered in the paper.

1 Introduction

The intensive development of the neural networks during last years makes their application to image processing, analysis and recognition very attractive. Since a processing within a local window around the pixel is not recursive, it may be organized simultaneously for all the pixels, independently of each other. So it is natural to organize this process using some appropriate kind of neural network. For example, many of the algorithms of linear and nonlinear filtering in spatial domain are reduced to the processing within a local window around each pixel of the image. We will concentrate here exactly on the nonlinear filtering algorithms that are reduced to the local processing within a window around each pixel and based on the nonlinear transformation of the result of linear convolution with the weighting kernel (template). The main operations in this case are noise reduction, frequency correction (extraction of image details) and edge detection.

The most appropriate neural network for solving of these problems is the Cellular Neural Network (CNN). CNN has been introduced in [1] as a special high-speed parallel neural structure for image processing and recognition. CNN conception has been intensively developed. Many results from simple filters for binary images [1-2] to algorithms for color image processing [3] were carried out. A main property of the CNN is their local connectivity. The neural networks considered before the paper [1] usually where the fully- connected networks (e.g., Hopfield network), or multi-layered neural networks with full connections between neurons of neighbor layers (e.g., multi-layered perceptron) [4]. A CNN conception supposes a cellular structure of the network: each neuron is connected only with the neurons from its nearest

B. Reusch (Ed.): Fuzzy Days 2001, LNCS 2206, pp. 450–463, 2001.

neighborhood (see Fig. 1). It means that the corresponding inputs of each neuron are connected with outputs of neurons from the nearest *rxr* neighborhood, and on the other hand, the output of each neuron is connected with the inputs of neurons from the same neighborhood. The neurons of such a network are also often called cells.

Depending on the type of neurons that are basic elements of the network, it is possible to distinguish continuous-time CNN (CCNN) [1], discrete-time CNN (DTCNN) [5] (oriented especially on binary image processing), CNN based on multi-valued neurons (CNN-MVN) [6, 7] and CNN based on universal binary neurons (CNN-UBN) [7-9]. CNN-MVN makes possible processing defined by multiple-valued threshold functions, and CNN-UBN makes possible processing defined not only by threshold, but also by arbitrary Boolean function. These properties of the CNN-MVN and CNN-UBN will be used here to develop a conception of the nonlinear cellular neural filtering (NCNF) [1, 10].

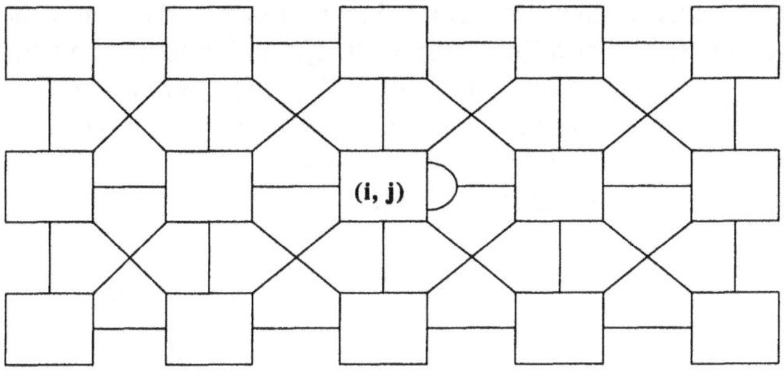

Fig. 1. CNN of a dimension 3x5 with the local connections in a 3x3 neighborhood: each neuron is connected with 8 neurons around it and with itself

We would like to consider below some effective image processing algorithms, which should be implemented using CNN-MVN and CNN-UBN. There are algorithms belonging to the family of nonlinear cellular neural filters. The first group of algorithms includes multi-valued filters (MVF) for noise reduction and frequency correction. The second one includes filters for the precise edge detection. We will also consider how MVF should be effectively used for solving the super resolution problem. All the problems to be considered here are very important for different tasks of applied image processing: medical imaging, satellite imaging, graphic art, etc.

2 A Problem of Nonlinear Filtering

Since a final purpose of this work is to present some new effective solutions in nonlinear image filtering we would like to make here a brief and concentrate observation of the considered problem. We will not present here a comprehensive observation of the nonlinear filtering. The reader, who is interested, should address to the latest monographs devoted to this problem [11-12]. The goal of the filtering operation is assumed to be an effective elimination or attenuation of the noise that is corrupting the signal. We will consider here 2D signals (images) and therefore 2D filters. Let us consider N x M image corrupted by noise. The simplest spatial domain filtering operation means filter defined as

$$\hat{B}_{ij} = w_0 + \sum_{\substack{i-n \le k \le i+n \\ j-m \le l \le j+m}} w_{kl} B_{kl}, \tag{1}$$

where B_{kl} are the values of the input signal, $\hat{B}_{ij}$ is the corrected value of the ij^{th} pixel, $w_0 \ge 0, w_{kl} > 0$. The range of indices k and l defines the filter window. It is evident that mean filter is a linear operation. But often it is impossible to obtain nice results with the filters like (1). They reduce noise, but not completely, and simultaneously they smooth the signal. A fundamental difference between the behavior of linear and nonlinear filters is the following. The impulse response (that is, the output signal when the input signal is equal to 1 at time instant 0 and zero otherwise) of a nontrivial time-invariant linear filter clearly cannot be a zero-valued sequence [11]. It means that a linear filter always will smooth the signal. A nonlinear filter preserves signal carefully, and usually is more effective for noise removal despite that it can be idle only for the partial signals. Different types of nonlinear filter were elaborated and investigated, e.g., *L*-filters [13], rank-order filters [13]. The reader who is interested may address also to weighted median filters [14], stack filters [11, 12], etc.

It is easy to see from Eq. (1) - (3) that a problem of the image filtering in the spatial domain is reduced to the replace of the current value of brightness B_{ij} in ij^{th} pixel by the value $\widehat{B}_{ij}$, which is better from the some point of view. Let us consider the linear filters (1) and (2). The arithmetic mean is a most popular, but not a single mean. Other widely used mean exist, e.g., geometric, harmonic, L_p. But obtaining of these means is connected with nonlinear operations. A general form of nonlinear mean filters has been considered recently [11]. It is (for 1D filter for simplicity):

$$B_i = g^{-1}\left(\frac{\sum_{k=-n}^{n} w_k g(B_k)}{\sum_{k=-n}^{n} w_k} \right),$$

where $g(x)$ is, for example, one of the following functions $g\colon R \to R$:

$$g(x) = \begin{cases} x & \textit{the arithmetic mean} \\ 1/x & \textit{the harmonic mean} \\ \ln x & \textit{the geometric mean} \\ x^p,\ p \in R \backslash \{-1,\ 0,\ 1\} & \textit{the } L_p \textit{ mean} \end{cases}.$$

Since function $g(x)$ is used for averaging of the signal, a following question is natural: is it possible to define other nonlinear means, and therefore, other types of averaging, which may be used for filtering? The main part of the paper gives a positive answer on this question. The issue of the approach, which is developed here, is application the non-linearities of the neurons activation functions. Introduction of the CNN based on MVN (CNN-MVN) [7, 15] and CNN based on UBN (CNN-UBN) [7, 8] opened a possibility to implement the original nonlinear spatial domain filters using this neural network. There are multi-valued filters (MVF) and cellular neural Boolean filters (CNBF). They should be considered as parts of the nonlinear cellular neural filters (NCNF) family. This family will be considered below.

3 Multi-valued and Universal Binary Neurons, CNN-UBN and CNN-MVN

Universal Binary Neuron (UBN) performs mappings described by arbitrary Boolean function of n variables. Multi-Valued Neuron (MVN) performs mappings described by full-defined threshold or partial-defined k-valued function (function of k-valued logic), where k is in general arbitrary integer>0.

Common basic mathematical background of the UBN and MVN is the following. An arbitrary Boolean function of n variables or k-valued function of n variables is represented by n+1 complex-valued weights $w_0, w_1, ..., w_n$ [7]:

$$f(x_1,...,x_n) = P(w_0 + w_1 x_1 + ... + w_n x_n), \tag{2}$$

where $x_1, ..., x_n$ are the variables, of which performed function depends (neuron inputs) and P is the activation function, which is defined in the following way.

1) For Multi-Valued Neuron:

$$P(z) = j, \text{if } 2\pi j/k \leq \arg(z) < 2\pi\,(j+1)/k, \tag{3}$$

where j=0, 1, ..., k-1 are values of the k-valued logic, i is imaginary unity, $z = w_0 + w_1 x_1 + w_n x_n$ is the weighted sum, arg(z) is the argument of the complex number z. (values of the function and of the variables are coded by the complex numbers, which are k-th roots of unity: $e^j = \exp(i2p\,j/k)$, $j \in \{0,1,...,k\text{-}1\}$, i is imaginary unity, in other words values of the k-valued logic are represented by k^{th} roots of unity: $j \rightarrow \varepsilon^j$);

2) For Universal Binary Neuron

$$P_B(z) = (-1)^j \text{ if } (2\pi j/m) \leq \arg(z) < (2\pi\,(j+1)/m) \tag{4}$$

where m is a positive integer, j is a non-negative integer $0 \leq j < m$, $z = w_0 + w_1 x_1 + w_n x_n$ is the weighted sum, arg(z) is the argument of the complex number z. Estimation for m is $m \geq 2n$ and it is considered in [7].

Let us consider a CNN with the obvious local connection structure (see Fig. 1) of a dimension N x M. Consider MVN or UBN as a basic neuron of such a network. Evidently, each cell of CNN-MVN (or CNN-UBN) will perform the following correspondence between neuron's inputs and output:

$$Y_{ij}(t+1) = F\left[w_0 + \sum_{k=-r}^{r}\sum_{l=-r}^{r} w_{kl}^{ij} x_{kl}^{ij}(t)\right] \tag{5}$$

where: ij are the two-dimensional coordinates (number) of a neuron; Y_{ij} is the neuron's output; w_{kl}^{ij} is the synaptic weight corresponding to the input of ij-th neuron, to which the signal from the output of kl^{th} neuron is transmitted; x_{kl}^{ij} is the input of ij-th neuron, to which the signal from the output of kl^{th} neuron is transmitted; F is the activation function of the neuron, which is defined by the equation (3) for MVN, and by the equation (4) for UBN; $\sum_{m=-r}^{r}$ means that each neuron is only connected with the neurons from its nearest r-neighborhood, as usual for CNN. Single weighting template of CNN-UBN and CNN-MVN may be written in the following form:

$$w_0;\ W = \begin{pmatrix} w_{i-r,j-r}^{ij} & \cdots & w_{i-r,j+r}^{ij} \\ \cdots & w_{ij}^{ij} & \cdots \\ w_{i+r,j-r}^{ij} & \cdots & w_{i+r,j+r}^{ij} \end{pmatrix},$$

where W is a rxr matrix.

4 Nonlinear Cellular Neural Filtering

Two-dimensional nonlinear cellular neural filter (NCNF) may be defined in the following way:

$$\widehat{B}_{ij} = F(w_0 + \sum_{\substack{i-n \le k \le i+n \\ j-m \le l \le j+m}} w_{kl} Y_{kl}) \tag{6}$$

Here m x n is a filter window, w_0, w_{kl} - filter weighting coefficients that are complex-valued, $\widehat{B}_{ij}$ - signal value in the ijth pixel after filtering, definition of Y_{kl} and F depends on a type of two NCNF subfamilies – multi-valued filters (MVF) and cellular neural Boolean filters (CNBF). Let us consider both subfamilies.

Multi-valued nonlinear filters (MVF) have been introduced recently [7, 8]. A specific complex non-linearity of the multi-valued neuron activation function [7] is a key for developing the new filters family. Let $0 \le B \le k-1$ is a dynamic range of 2D signal. Let consider a set of the kth roots of unity. We can put the root of unity

$$e^B = \exp(i2\pi B/k) = Y \tag{7}$$

to the correspondence with the real value B. Thus, we have the univalent mapping $B \leftrightarrow \varepsilon^B$, where ε is a primitive k^{th} root of unity. A two-dimensional *Multi-Valued Filter* (MVF) is a filter, which is defined by the following formula [9]:

$$\widehat{B}_{ij} = P(w_0 + \sum_{\substack{i-n \le k \le i+n \\ j-m \le l \le j+m}} w_{kl} Y_{kl}) \tag{8}$$

where P is the activation function (3) of the multi-valued neuron, Y_{kl} is obtained from B_{kl} by equation (7), i,j are the coordinates of the filtered pixel, $n \times m$ is a filter window, w_{kl} are the weighting coefficients (complex-valued in general). Evidently, the equality (8) defines a class of filters. Each of them is defined by the corresponding set of the weighting coefficients (weighting template).

One remark should be given before we will move to the applications of MVF. The new different families of nonlinear filters may be considered, if well-known nonlinear spatial domain filters will be connected with MVF. Let consider, e.g., a filter, also defined by equation (8), but simultaneously connected with the weighted rank-order filter. We will obtain a *rank-order multi-valued filter* (ROMVF). MVF may be also connected with other nonlinear filters in the same way.

Evidently, the most appropriate structure for the MVF implementation is CNN-MVN. The most simple, but very effective template for the reduction of Gaussian or uniform noise is the following:

$$W = \begin{pmatrix} 1 & 1 & 1 \\ 1 & G & 1 \\ 1 & 1 & 1 \end{pmatrix} \tag{9}$$

where G is a parameter. To remove a noise with the impulsive component we have to recommend $0 \le G \le 1$. To remove Gaussian or uniform noise, which is not mixed with the impulsive one our recommendation for G is the following: $1 \le G \le 16$.

(a) Input noisy image (b) MVF, 3x3, G=16, 2 iterations (c) rank-order, 3x3, rank=9, 2 iterations

Fig. 2. Multi-valued filter and its comparison to rank-order filter

Let us consider an example. It shows a reduction of zero-mean Gaussian noise with a dispersion equal to 0.3σ, and it is presented in Fig. 2. The results of rank-order filtering of the same images are given for comparison. Table 1 contains the objective statistical characteristics of the corrupted, and filtered images.

Table 1. Statistical characteristics for the example given in Fig. 3

Image	"Kremlin Noisy"	MVF, G=16	ROF
ND (σ)	445.7	157.4	420.0
PSNR	7.57	18.6	8.0

ND (σ) – estimation of white noise dispersion, PSNR – estimation of peak signal to noise ratio, SD – standard deviation from the original image.

Multi-valued filters are also a highly effective mean for the amplification of high and medium frequency. Solving this frequency correction problem leads to the image sharpening and extraction of image details against a preventing background. The extraction, localization, and detection of important features on the complex image background were proposed to consider [15] as a problem of localization of objects or details of a given type on the image. Such a type has been defined as sizes of objects. Then this idea has been developed [8], where the solution of the problem using traditional CNN has been proposed. The following linear filters have been proposed respectively, for the solution of global and high frequency correction problem

$$\widehat{B}_{ij}=G_1B_{ij}+G_2(B_{ij}-B_m)+G_3B_m+c \tag{10}$$

$$\hat{B}_{ij}=(G_1+G_2)B_{ij}-G_2B_m+c\,, \tag{11}$$

where B_m is a local mean value in a window around the pixel B_{ij}; B_{ij}, and $\widehat{B}_{ij}$ are the signal values in ij^{th} pixel, respectively, before and after processing; G_1 and G_3 define correction of the low spatial frequencies, G_2 defines correction of the high spatial frequencies, c is the mean value of the background after processing. By the way, if $G_1 = 1$ then filter (11) coincides with the filter called "unsharp masking" in different issues (see, e.g., [16]).

Despite the filters (10) and (11) give good results, they have some significant disadvantages. They are very sensitive to choice of the weighting coefficients. Even minimal change of any one may move a general dynamic range of the image to "white" or "black" side. As a result, the global image histogram may be dramatically changed. To break these disadvantages linear filters (10) and (11) should be transformed to the nonlinear ones. Since these filters operate with the arithmetic mean, they should be generalized using nonlinear means.

Taking into account (7) it is easy to obtain the following multi-valued generalizations, respectively for the filters (10) and (11):

$$\hat{B}_{ij} = P\left[c+(G_1+G_2)Y_{ij}+(G_3-G_2)\sum_{Y(k,l)\in R_{ij}}Y_{kl}\right], \tag{12}$$

$$\hat{B}_{ij}=P\left[c+(G_1+G_2)Y_{ij}-G_2\sum_{Y_{kl}\in R_{ij}}Y_{kl}\right], \tag{13}$$

where Y_{kl} is obtained from B_{kl} by equation (7), R_{ij} is a local window around pixel Y_{ij}; B_{ij} (Y_{ij}) and $\hat{B}_{ij}$ are signal values in ij^{th} pixel before and after processing, respectively, G_1, G_3 are the coefficients, which define correction of the low frequencies, G_2 defines correction of the high frequencies, c is the constant.

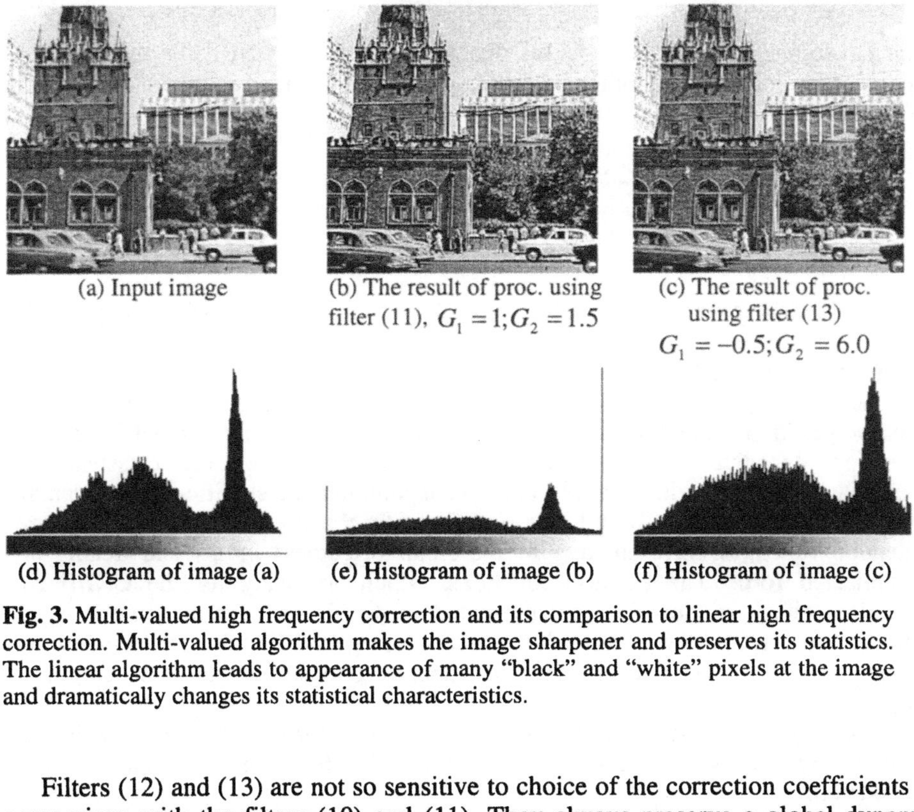

Fig. 3. Multi-valued high frequency correction and its comparison to linear high frequency correction. Multi-valued algorithm makes the image sharpener and preserves its statistics. The linear algorithm leads to appearance of many "black" and "white" pixels at the image and dramatically changes its statistical characteristics.

Filters (12) and (13) are not so sensitive to choice of the correction coefficients in comparison with the filters (10) and (11). They always preserve a global dynamic range of the image and its global histogram. Such a tolerance may be used for the processing of the images that contain many details of different sizes and different configuration. It is natural to implement both filters using CNN-MVN. The example of the image processing using the filter (13) and its comparison to the filter (11) is given in Fig. 3.

The next representative from the NCNF filter family is cellular neural Boolean filter (CNBF). The CNBF idea proposed in [8] is based on the separate processing of the image binary planes using a special spatial domain filter defined by some Boolean function and further integration of the resulting binary planes into the resulting image.

A key issue for the method was ability of CNN-UBN to implement a spatial domain filter described by arbitrary Boolean function [7, 8], not only by the threshold one.

Let B_{kl} be a signal value in the ij[th] pixel and $B \in \{0,1,...,2^k - 1\}$. Let us separate the image onto *k* binary planes directly, without any thresholding. Let B_{ij}^s be a binary value of the signal corresponding to s^{th} binary plane, Y_{ij}^s is the same value transformed to the Boolean alphabet {1, -1}. Let *F* in (6) be the UBN activation function (4). Taking into account (4), (6) is transformed to the following:

$$\hat{B}_{ij}^s = P_B(w_0 + \sum_{\substack{i-n \le k \le i+n \\ j-m \le l \le j+m}} w_{kl} Y_{kl}^s), \tag{14}$$

The equation (24) establishes CNBF for a binary plane. CNBF for the gray-scale image or for color channel of the color image should be written in the following form:

$$\hat{B}_{ij} = \bigoplus_{s=0}^{k-1} \left[P_B(w_0 + \sum_{\substack{i-n \le k \le i+n \\ j-m \le l \le j+m}} w_{kl} Y_{kl}^s) \right], \tag{15}$$

where $\bigoplus_{s=0}^{k-1}$ denotes integration of *k* binary planes from 0th until (*k*-1)[th] to the gray-scale image (color channel).

The most impressive application of CNBF is the precise edge detection, which has been proposed in [8] and then developed (see, e.g., [7, 9]). Many of the different edge detection algorithms are well known. Some of them are even classical (e.g., Laplacian, Sobel, Prewitt). All of them have a common and significant disadvantage: they are usually sensitive only to the significant brightness jumps and can't ensure a detection of edges corresponding to the small brightness jumps and details of a complicated form. The Boolean functions, which are using for the CNBF edge detection with a window 3x3, are presented in [8, 9]. At the same time it is very attractive to consider the precise edge detection using CNBF with a 5x5 window. It is important for the detection of edges corresponding to the details of a complicated form.

Evidently, it is possible to design many functions for edge detection with a 5x5 window by narrow direction. For example, the following functions may be considered:

Upward jumps, South-East

$$f\begin{bmatrix} x_1 & x_2 & x_3 & x_4 & x_5 \\ x_6 & x_7 & x_8 & x_9 & x_{10} \\ x_{11} & x_{12} & \boldsymbol{x_{13}} & x_{14} & x_{15} \\ x_{16} & x_{17} & x_{18} & \boldsymbol{x_{19}} & x_{20} \\ x_{21} & x_{22} & x_{23} & x_{24} & \boldsymbol{x_{25}} \end{bmatrix} = (x_{13} \,\&\, x_{19} \,\&\, x_{25}) \,\&\, (\overline{x_{14}} \vee \overline{x_{18}}) \,\&\, (\overline{x_{20}} \vee \overline{x_{24}})$$

Downward jumps, South-East

$$f\left[\begin{array}{c|ccc|c} x_1 & x_2 & x_3 & x_4 & x_5 \\ \hline x_6 & x_7 & x_8 & x_9 & x_{10} \\ x_{11} & x_{12} & X_{13} & x_{14} & x_{15} \\ x_{16} & x_{17} & x_{18} & X_{19} & x_{20} \\ \hline x_{21} & x_{22} & x_{23} & x_{24} & X_{25} \end{array}\right] =$$

$$(\overline{x_{13}} \,\&\, \overline{x_{19}} \,\&\, \overline{x_{25}}) \,\&\, (x_{14} \vee x_{18}) \,\&\, (x_{20} \vee x_{24})$$

Evidently, a function for the edge detection in all directions consists of the disjunction of the functions, which detect the edges in all the particular directions.

Fig. 4 presents the results of precise edge detection using CNBF with a 5x5 window. This example is a very nice illustration of how it is possible to detect the sea streams in the different directions using the presented technique.

5 Application of the Multi-valued Filters in Solving of the Super-resolution Problem

The approach to solving the super-resolution problem, which will be developed here, has been proposed in [17]. This approach is based on the iterative approximation of the Fourier, Cosine or Walsh image spectrum in the highest frequency domain.

Let $f(x, y)$ be a 2D signal or discrete image of an n x n sizes (without loss of generality), A - spatial domain, which signal is defined on, $\tilde{A} \subset A$ - its subdomain. The function $F(u,v) = \Phi[f(x, y)]; u,v \in \{0,1,...,n-1\}$, where Φ is Fourier, Cosine, or Walsh (ordered by Walsh) transform, is the spectrum of the signal $f(x, y)$. The super-resolution problem can be considered as two dual problems: extrapolation of spectrum to the domain $u,v \in \{0,1,...,n-1, n,...,2n-1\}$ and interpolation of signal on the whole domain A. The following iterative process has been proposed in [17].

To begin, we will need to extend our signal from domain $\tilde{A}$ to domain A. So let us build the first approximation to the resulting signal:

$$g(x,y) = \begin{cases} f(x, y), \text{if } (x,y) \in \tilde{A} \\ s(x, y), \text{if } (x,y) \in A \setminus \tilde{A} \end{cases} \tag{16}$$

where $s(x, y)$ is a uniform noise with a small dispersion.

It was suggested in [17] to consider $g(x, y)$ as $f(x, y)$, corrupted by additive uniform noise in the domain $A / \tilde{A}$. It was also suggested to take a uniform noise with a small dispersion and the same mean as $f(x, y)$ mean as $s(x, y)$. Then the problem has been transformed to the problem of noise reduction and correction of the highest frequencies. We would like to suggest here to use the result of some $f(x, y)$ interpolation as $s(x, y)$ instead of the noise (e.g., bicubic or spline interpolation should be used). Thus let us assume that $f_1(x, y) = g(x, y)$. The nth approximation $f_n(x, y)$ for $f(x, y)$ may be obtained in the following way:

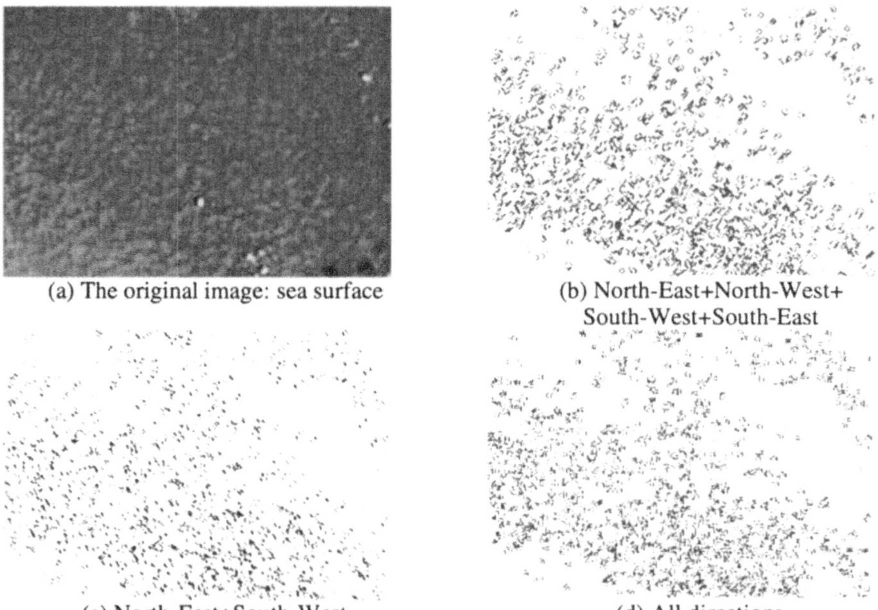

(a) The original image: sea surface

(b) North-East+North-West+South-West+South-East

(c) North-East+South-West

(d) All directions

Fig. 4. Precise edge detection using CNBF with a 5x5 window (upward brightness jumps)

$$F_{n-1}(u,v) = \begin{cases} F(u,v); \ u,v \in \{0,1,..., n-1\} \\ \Phi[f_{n-1}(x,y)]; \ u,v \in \{n,...,2n-1\} \end{cases} \tag{17}$$

$$g_{n-1}(x,y) = \Phi^{-1}[F_{n-1}(u,v)] \tag{18}$$

$$f_n(x,y) = \begin{cases} f(x,y), \text{if } (x,y) \in \tilde{A} \\ g_{n-1}(x,y), \text{if } (x,y) \in A \setminus \tilde{A} \end{cases} \tag{19}$$

The expressions (17) - (19) define the iterative process, which allows to obtain the best approximation for the signal $f(x, y)$ and its spectrum $F(u, v)$. This process can not be infinite, because the standard deviation is decreasing on each step. Experiments had shown, that the process defined by (17)-(19) is converged quickly. As a result we will obtain the following:

$$f_n(x,y) = \begin{cases} f(x,y), \text{if } (x,y) \in \tilde{A} \\ f(x,y) + \tilde{s}(x,y), \text{if } (x,y) \in A \setminus \tilde{A} \end{cases} \tag{20}$$

where $\tilde{s}(x, y)$ is an additive noise, and

$$f_n(x,y) - f_{n-1}(x,y) = \begin{cases} 0, \text{if } (x,y) \in \tilde{A} \\ \varepsilon, \text{if } (x,y) \in A \setminus \tilde{A} \end{cases} \tag{21}$$

where ε is close to zero. Experiments also show that less than 10 iterations are needed to obtain the best result.

According to the equation (20), $\tilde{f}(x,y) = f_n(x,y)$ contains an additive noise in $A \setminus \tilde{A}$ subdomain. It means, that to complete a process of spectrum and signal restoring, we have to remove the noise and correct the highest frequencies, which may be corrupted during the noise removal. There are many different filters, which may be used to solve this problem. The best filter to apply, both from the point of view of noise reduction quality and preservation of the image boundaries, is MVF's, defined by (7)-(9) and (7)-(11). So, the following final corrections should be performed:

$$S(x,y) = \begin{cases} f(x,y), \text{ if } (x,y) \in \tilde{A} \\ MVF\,[\tilde{f}(x,y)], \text{if } (x,y) \in A / \tilde{A} \end{cases} \tag{22}$$

$$\tilde{S}(x,y) = \begin{cases} f(x,y), \text{ if } (x,y) \in \tilde{A} \\ MVF_{fc}\,[S(x,y)], \text{if } (x,y) \in A / \tilde{A} \end{cases} \tag{23}$$

Quite a good result can be achieved using the described approach, but this process still can be developed to improve the quality of the resulting image.

It is possible to extrapolate the image spectrum without use the original one in (17). Let us take a closer look: The following correction has to be made in (17) - (19):

$$F_{n-1}(u,v) = \begin{cases} \Phi[f_{n-1}(x,y)];\, u,v \in \{0,1,\ldots,n-1\} \\ 0; \qquad u,v \in \{n,\ldots,2n-1\} \end{cases} \tag{24}$$

This technique requires a little bit more iterations for the convergence of the iterative process (17) – (19), but at the same time a quality of the resulting image often could be better. Another important improvement of the proposed algorithm is the following. It is well known that spatial limitation of the discrete images and therefore limitation of their discrete spectral representation leads to the appearance of so-called boundary effects at the results of frequency domain filtering. These boundary effects become apparent as series of stripes, which are often clearly visible at the regions of image boundaries. To eliminate this effect, it is possible to use a symmetrical extension of the original image in all directions (even extension of the image). Of course, this approach requires more resources and leads to increase of the processing time. At the same time a quality improvement of the resulting image should be gotten.

Let's now compare the new results to the old ones [17] for all transforms. The image shown in Fig.5a was taken for this purpose and resized to become 2 times smaller as shown in Fig.5b. The latter was then enlarged by previously described algorithm (Fig. 4c) and compared to the original image. Sinc 16 interpolation algorithm has been used to obtain a starting approximation of the resulting image.

(a) Original image (b) Downscaled image (c) Enlarged image (Cosine transform using (24) and corrections (22)-(23))

Fig. 5. Solution of the super-resolution problem

Table 2. Comparison of the improved super-resolution algorithm to the old one

Transform and algorithm	Standard deviation from the original	
	The old algorithm	The improved algorithm
Fourier with even extension	4.483	4.065
Cosine without even extension	4.531	4.065
Walsh without even extension	6.942	6.836

It is clear from the shown results that the best branch of the algorithm is one, which uses Cosine transform without even extension of the signal.

6 Conclusions

We have presented the cellular neural networks based on multi-valued and universal binary neurons (CNN-MVN and CNN-UBN). The results presented in this paper show high efficiency of CNN-MVN and CNN-UBN in image processing applications.

On the other hand, consideration of these neural networks involved development of a new nonlinear filters family - nonlinear cellular neural filters. As it was shown, these filters are very effective and useful for solving of such problems as noise reduction, frequency correction and edge detection. It is also shown that the described filters may be used for solving of the super-resolution problem, to overcome the results obtained by interpolation algorithms and the results obtained without use of multi-valued filters.

References

1. Chua L.O. and Yang, L. "Cellular neural networks: Theory", *IEEE Transactions on Circuits and Systems.* **35**, No 10, (1988) 1257-1290.
2. Roska T. and Vandewalle J. (ed.) *"Cellular Neural Networks"*, New York: John Wiley & Sons (1993).
3. Lee C.-C. and. Pineda de Gyvez J. "Color Image Processing in a Cellular Neural - Network Environment", IEEE *Trans. On Neural Networks,* **7**, No 5 (1996) 1086-1088.
4. Haykin S. *Neural Networks. A Comprehensive Foundation. Second Edition.* Macmillan College Publishing Company, New York (1999).
5. Harrer H. and Nossek J.A. "Discrete-time cellular neural networks", *International Journal of Circuit Theory and Applications,* **20** (1992) 453-467.
6. Aizenberg I. and Aizenberg N. "Application of the neural networks based on multi-valued neurons in image processing and recognition", *SPIE Proceedings,* **3307** (1998) 88-97.
7. Aizenberg I., Aizenberg N. and Vandewalle J. *"Multi-valued and universal binary neurons: theory, learning, applications"*. Kluwer Academic Publishers, Boston/Dordrecht/London: (2000).
8. Aizenberg I. "Processing of noisy and small-detailed gray-scale images using cellular neural networks" *Journal of Electronic Imaging,* **6**, No 3 (1997) 272-285.
9. Aizenberg I., Aizenberg N., Agaian S., Astola J. and Egiazarian K. "Nonlinear cellular neural filtering for noise reduction and extraction of image details", *SPIE Proceedings,* **3646** (1999), pp.100-111.
10. Rekeczky C., Roska T. and Ushida A. "CNN Based Self-Adjusting Nonlinear Filters" *Proc. of the Fourth IEEE International Workshop on Cellular Neural Networks and their Applications,* Seville, Spain, June 1996, pp. 309-314.
11. J.Astola, and P.Kuosmanen *Fundamentals of non-linear digital filtering.* CRC Press, Boca Raton, New York (1997).
12. Agaian S., Astola J. and Egiazarian K. *Binary polynomial transforms and nonlinear digital filters.* Marcel Dekker, Inc, New York/Boston/Hong-Kong, (1995).
13. Bovik A.C., Huang T.S., and Mudson D.C. A Generalization of Median Filtering Using Linear Combinations of Order Statistics', *IEEE Transactions on Acoustic, Speech, Signal Processing* **31** (1983) 1342-1349.
14. Justusson B.I. Median Filtering: Statistical Properties. *Topics in applied Physics - 43, Two-dimensional Digital Signal Processing II,* (Ed. T.S.Huang), Springer-Verlag, Berlin (1981) 161-196.
15. Belikova T.P. "Simulation of the linear filters in the problems of medical diagnostics", *Digital Image and Fields Processing in the Experimental Researches,* (I. Ovseevich - Ed.), Nauka Publisher House, Moscow (1990) 130-152.
16. W.KPratt "*Digital Image Processing. Second Edition*" John Wiley & Sons New York (1992).
17. Aizenberg I.N., Aizenberg N.N., Astola J., Bregin T., Butakov C. "New applications of the nonlinear cellular neural filters in image processing,", SPIE Proceedings, **3961** (2000), 45-56.

Hybrid Two-Population Genetic Algorithm

Julian Dorado, Antonino Santos, Juan R. Rabuñal,
Nieves Pedreira, and Alejandro Pazos

Univ. da Coruña, Facultad Informática, Campus Elviña, 15071 A Coruña, Spain
{julian,nino,juanra,nieves,ciapazos}@udc.es

Abstract. Genetic Algorithms are non-deterministic, stochastic-search adaptive methods which use the theories of natural evolution and selection in order to solve a problem within a complex range of possible solutions. The aim is to control the distribution of the search space by incorporating an exhaustive method in order to maintain a constant evolution of the population. The main goal is that of redesigning the algorithm in order to add to the classic genetic algorithm method those characteristics which favour exhaustive search methods. The method explained guarantees the achievement of reasonably satisfactory solutions in short time-spans and in a deterministic way, which entails that successive repetitions of the algorithm will achieve the same solutions in almost constant time-spans. We are, therefore, dealing with an evolutionary technique which makes the most of the characteristics of genetic algorithms and exhaustive methods.

1 Introduction

Genetic Algorithms (GA) [1] are systems which solve optimization problems [2][3] by using the mechanisms of natural evolution. GAs are part of evolutionary computation, together with other techniques such as Evolutionary Strategies and Genetic Programming [4]. They all share a common conceptual basis: they simulate the evolution of individual structures via selection, mutation and reproduction proceses.

Evolutionary mechanisms keep a population of structures which evolve according to the crossover, mutation and selection rules known as "genetic operators" [5]. In nature, survival is the main goal of every living being. In a GA, there has to be a fitness function [6] which establishes a classification among the different individuals in order to determine which ones shall survive and which ones shall perish. This is how the population evolves through the use of existing genetic combinations (crossover) and the exploration of new possibilities of genetic material (mutation). There is an intrinsic basis throughout this process, which is randomness.

2 State of the Art

Thus evolutionary computation is based on the randomness which lies in the genetic operators. This distributes the search space (in which there is, at least, one optimal solution) uniformly across the whole algorithm, while the solution is gradually improving towards the absolute minimum. Obviously, this method avoids the problem of the existence of local minima which make other search methods face a dead-end situation. The disadvantage would be that, being based on randomness, the search

B. Reusch (Ed.): Fuzzy Days 2001, LNCS 2206, pp. 464–470, 2001.

method is a non-deterministic one. The GA's behaviour has not been surveyed yet, because if we get rid of the random component, the method's advantages are eliminated. One of the problems caused by the random component is that no acceptable solution can be guaranteed in a given time. For that reason, a GA-based method is proposed which limits randomness and incorporates systematic procedures to genetic operations [7]. This gets acceptable solutions in reasonable time-spans. The algorithm explores the search space in a uniformly distributed manner throughout the simulation [8].

3 Description of the System

The proposed method divides the search space into subspaces, assigning one individual to each of them, and composing a population which shall be known as "genetic pool". Additionally, another population will be used (known as "ordinary population"), where individuals will evolve according to the classic GA, interacting with individuals from the genetic pool [8].

3.1 Genetic Pool

The individuals from the genetic pool will have as structure of the genetic material or chromosome a sequence of genes formed by three groups: the first one will be called G, and it will be made of the chain of genes which are needed in order to solve a problem. The second sequence, called B, will be made of Boolean values which indicate when a G-type gene is part of the local optimum individual. The third sequence, called I, will indicate the relative increment to the G gene to be applied to the mutation operation.

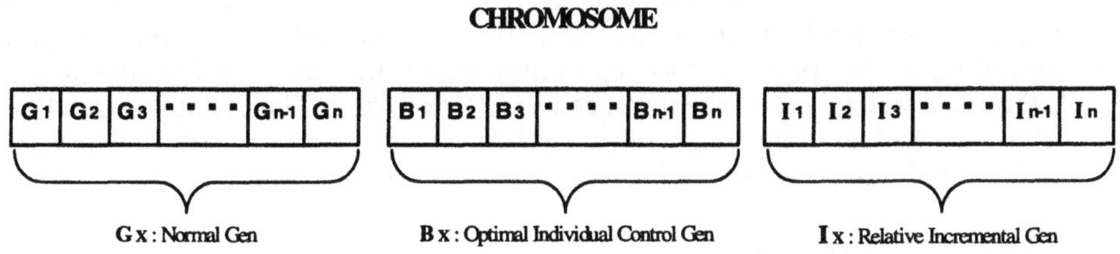

Fig. 1. Chromosome of individuals from genetic pool

These individual's genetic material is controlled by two basic rules:

- The range of values to be adopted by the genes is divided into as many parts as individuals there are in the genetic pool.
- One range will be assigned to each of them. This entails that the genes' values will be within a subrange.

The purpose of this procedure is the division of the search space into as many sections of identical size as individuals existing in the genetic pool. Thus the search space is uniformly distributed among all the individuals.

3.2 Ordinary Population

In this population, individuals evolve according to the rules of a classic GA, without the mutation operation and via the crossover operation with individuals from the ordinary population and the genetic pool.

3.3 Evolution of Both Populations

The idea of having an algorithm with two population in mutual iteration is that of keeping control on the individuals' evolution. This monitoring is done in various forms: 1) by controlling the genetic diversity existing among the population. A subsection of the search space is assigned to the individuals from the genetic pool, which cannot leave it. 2) by defining the mutation operator with increments which diminish through time.

The classic genetic algorithm may explore the same space areas several times, it may even explore several individuals simultaneously, causing an excessively homogeneous population. This approach achieves a more distributed and uniform performance in the search space, using as many individuals from the genetic pool as the portions or subsections into which the space may be divided (granularity). Thus, the mutation operation secures the systematic exploration of every area of the searching space, avoiding the homogeneity of the population.

4 Genetic Operations

An elitist strategy has been used both for the mutation and crossover genetic operations. There is, for this purpose, an external individual to both systems, known as local optimum superindividual, which has the best Fitness value in both populations. The genes of this superindividual are distributed among the individuals from the genetic pool, where every time the superindividual is improved, the corresponding genes are marked as supergenes inside the various individuals of the pool (the corresponding B gene is changed to TRUE value). The mutation operation cannot be applied to these genes, thus guaranteeing that those genes will remain within the population, constituting the local optimum superindividual.

4.1 Crossover Operation

The crossover operation used differentiates itself from the classic operation according to the following rules:

- A recombination of genetic material will take place between a randomly selected individual from the ordinary population and "n" genes randomly selected among the individuals from the genetic pool.
- The crossover type to be performed will be based on multiple crossover points, which secures a greater genetic recombination.

The individual selected from the ordinary population will be substituted by the best offspring individual, but only if it shows a better environmental adaptation capacity than the previous one from that population.

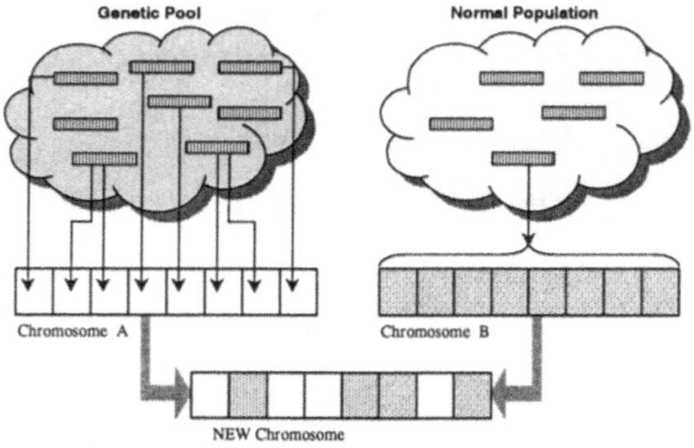

Fig. 2. Crossover operation

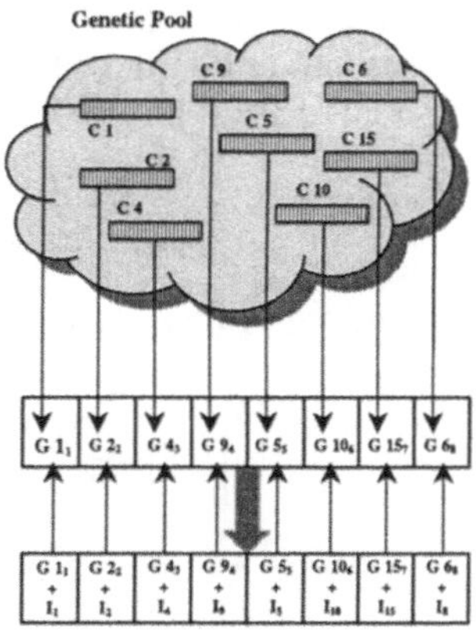

fig. 3. Mutation operation

4.2 Mutation Operation

The following rules are applied to the mutation genetic operation:

- Genetic operations will only be applied to individuals from the genetic pool.
- The genetic operation is based on the application of an established I_x increments to a randomly selected G_x gene.
- The increments (type I_x genes) are related to each of the genes. When they are applied, the increase is added to the gene's value and, if the end of the subrange is reached, the value is changed by the minimum of the subrange.
- If a gene belongs to the local optimum individual, the mutation operation cannot be applied to it (controlled by type B genes).

As a consequence of the application of crossover and mutation, the ordinary population may become homogeneous, which is a typical problem of the classic genetic algorithm. But this circumstance does not cause a problem in the proposed algorithm, given that the individuals from the genetic pool suffer a constant modification in their genes (increments). Therefore, when a better individual is found, after a fine adjustment of the genes' values in the mutation operation, it is included in the ordinary population, so that the genetic material of that population will be diversified in future generations. Consequently, it can be guaranteed that the system's global evolution will never stagnate.

5 Results

In order to compare the algorithm's correct functioning, an application has been developed using an implementation of the proposed GA and a classic GA. The tests carried out try to establish the level of convergence, the accountability, stability and level of adaptive and exhaustive search of the new algorithm compared to the classic GA. For that purpose, three minimum search problems are used, going from a less to a more complex solution. The three problems search for the minimum of the addition of 5 values of the same function *f(x)*. Individuals with 5-length chromosomes which codify values at floating point have been used. For chromosomes G_1, G_2, G_3, G_4, G_5 the calculation of the Fitness (*Fit*) in the 3 problems is the following:

$$Fit_1 = f_1(G_1) + f_1(G_2) + f_1(G_3) + f_1(G_4) + f_1(G_5)$$

$$Fit_2 = f_2(G_1) + f_2(G_2) + f_2(G_3) + f_2(G_4) + f_2(G_5)$$

where $f_1(x) = |\sin(x)|$ and $f_2(x) = \cos(x^2) / (1+x^2)$

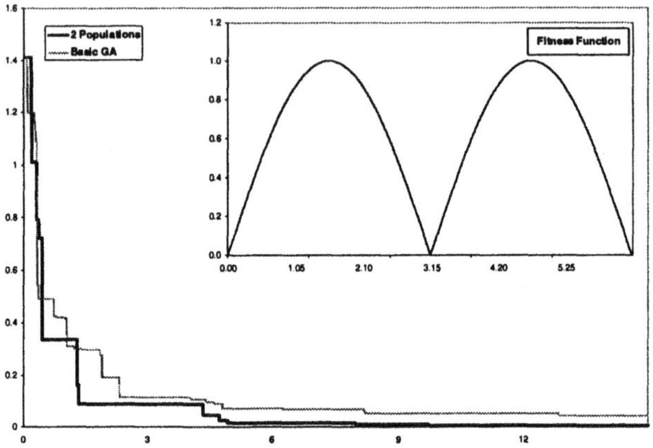

Fig. 4. Positive sin function

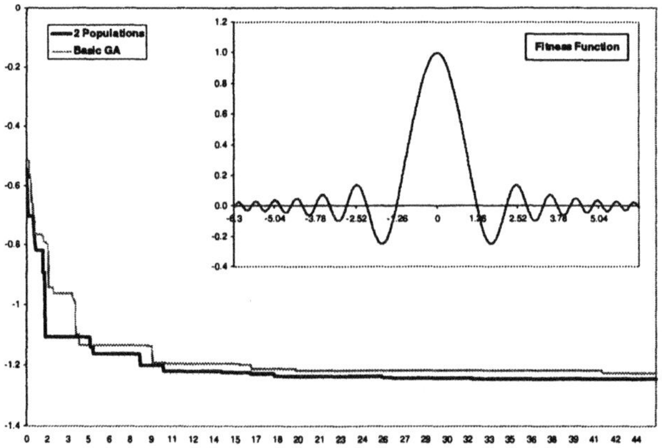

Fig. 5. Mexican hat function

The previous charts show the results obtained when solving the three problems and the evolution of the algorithms. The results shown in the charts have been achieved with populations made of 100 individuals, with 70% crossover and 30% mutation in both algorithms. As shown in the figures, in the three cases the algorithm has a greater convergence speed, particularly in the last case, where the absolute minimum is hard to reach.

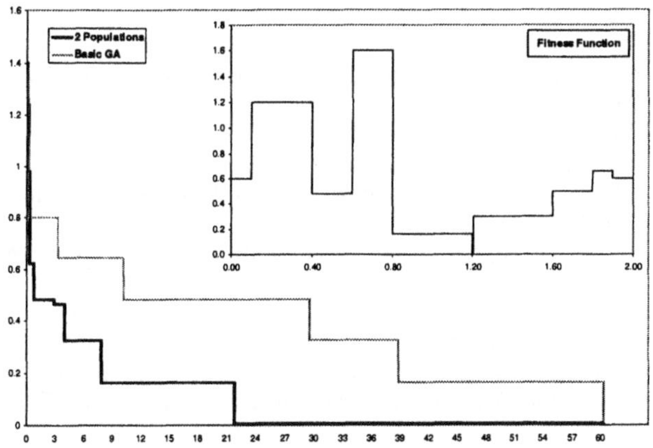

Fig. 6. Two wide local minima function

6 Conclusions

The system shows a considerable speed at finding solutions compared to the classic GA, it secures a more uniform exploration of the search space, and it guarantees the variety of the genetic material thanks to the genetic pool.

Moreover, the number of individuals in the populations is less influential than in the classic GA, achieving good results with few individuals. Tests have been carried out by reducing the number of individuals in the ordinary population to 10. The performance level of the simulation was average to the one reflected in the previous charts. This is due to the fact that the variety of the genetic material is constituted in the genetic pool, regardless of the number of individuals of the ordinary population, while in the classic GA, it is distributed among all the individuals in the population. This causes the usual problem of having to balance the crossover and mutation ratio if we want to avoid obtaining an excessively homogeneous population. The proposed algorithm does not entail such problem, since the crossover operator is carried out between both populations, and genetic variety is guaranteed in one of them (the pool). The fact that the ordinary population may become homogeneous is a lesser problem.

7 Future Works

The aim is to carry out the algorithm's parameter tests to deal with the number of subsections of the genetic pool, types of problems to be solved by the system, and adequate crossover and mutation ratios for each problem.

Another research line would be that of having several distributed systems interacting in a different way from those known at present. The development of the proposed algorithm allows the construction of new distributed genetic operators based on the interaction of solution-search areas within genetic pools. Thus, the interaction among populations does not have to be based exclusively on the exchange of individuals or operations, but a new possibility opens: the exchange of solution-search areas.

References

1. Holland J., "Adaptation in Natural and Artificial Systems", University of Michigan Press, Ann Arbor, 1975
2. Goldberg, D.E., "Genetic Algorithms in search, optimization and machine learning", Addison-Wesley, Reading, Massachusetts, 1989.
3. Morrow M., "Genetic Algorithms a new class of Searching Algorithms", Dr. Dobbb's Journal, April 1991.
4. Bäck, T., "Evolutionary Algorithms in Theory and Practice". Ed. Oxford University Press. 1996.
5. Bramlette M.F., "Initialization, Mutation and Selection Methods in Genetic Algorithms for Function Optimization". Proceedings of the Fourth International Conference on Genetic Algorithms. Ed. Morgan Kaufmann. Eds. R.K. Belew & L.B. Booker. Pp. 100-107. 1991.
6. Bethke, A.D., "Genetic Algorithms and Function Optimizers". Thesis. University of Michigan. 1981.
7. Kumar, A., Morris, A., "A hybrid evolutionary algorithm for global optimisation", GECCO-2000, Las Vegas, Nevada, July, 2000.
8. Liang K.-H., Yao X., Newton, C., "Evolutionary search of approximated N-dimensional landscapes" Int. Journal of Knowledge-Based Intelligent Engineering Systems, (3):172-183, July 2000.

Reducing Bloat in Genetic Programming

Patrick Monsieurs and Eddy Flerackers

Expertise Center for Digital Media, Wetenschapspark 2, 3590 Diepenbeek, Belgium
{patrick.monsieurs,eddy.flerackers}@luc.ac.be

Abstract. In this paper, several techniques will be presented to constrain the growth of solutions that are constructed by genetic programming. The most successful technique imposes a maximum size on the created individuals of the population that depends solely on the size of the best individual of the population. This method will be compared with other methods to reduce bloat, demonstrating that this method reduces bloat significantly better than the other methods.

1 Introduction

Genetic programming is an optimization technique inspired by the evolution theory of Darwin, and similar to genetic algorithms [1]. Typically, the solutions created by genetic programming have a variable length, and can be represented as a tree. When two parent individuals are selected to create two new children via crossover, the representations of both parents are split in two parts, and these parts are exchanged to form the new individuals. As a result, in most cases one of these children will be smaller then the parents, and the other will be larger. In most cases however, the larger individual will have a better performance than the smaller individual. This can cause the size of the individuals to grow very rapidly. Because of the large size of the individuals, the population will soon become unmanageable, because of memory problems and the increasing time to evaluate the solutions. Because of this, it is important to limit the growth of the individuals. How this can be accomplished will be described in the remainder of this paper.

In section 2, related work on the limitation of the size of individuals will be discussed. section 3 discusses the causes of bloat, and section 4 will present possible solutions. In section 5, these solutions will be tested and compared by running several experiments. Finally, some conclusions will be presented in section 6.

2 Related Work

Because the problem of bloat is known since the beginning of genetic programming, a lot of related work exists. Koza [1] uses the following techniques to fight bloat:

- The individuals that are created for the initial populations are restricted to depths between 2 and 6.
- Subtrees that are generated for subtree mutation are limited to a depth of 4, and non-terminal nodes are selected as mutation points with a probability of 90%.

B. Reusch (Ed.): Fuzzy Days 2001, LNCS 2206, pp. 471–478, 2001.

- The creation of new individuals is limited to a depth of 17.

Another often used technique to slow bloat is the use of parsimony pressure. In this case, the fitness function of an individual is a (linear) combination of its performance and its size, where a smaller size increases the fitness value. While this approach reduces the average size of the population, it can also reduce the speed at which better solutions are discovered. Furthermore, it is possible that the population will get trapped in a local minimum where the combined fitness function is dominated by the size metric and only individuals of a minimal size exist [2].

[3] describes a method where two populations of individuals are maintained. The first population contains individuals that are selected based only on their performance, while the second population contains individuals that have a small size. Crossover is then applied between individuals from both populations to create new individuals that are small and have a high performance.

Code editing is another way to reduce the size of individuals. In this approach, parts of individuals that have no effect on the result of the individual, called introns, are removed from the individual. However, introns are not the only way to introduce bloat, and by removing introns, other ways of bloating will begin to dominate the populations, as will be demonstrated in paragraph 5.

Finally, it is possible to introduce a form of hill climbing in the crossover process [4]. Using this approach, individuals created by the crossover operator are rejected if the result is not fitter or smaller than its parents. This approach restricts bloat considerably, but more time will be needed to find better solutions, as will be shown in paragraph 5.

3 Causes of Bloat

[5] describes four major theories of bloat:

- Hitchhiking: This theory, introduced by [6], says that introns are spread from individual to individual by crossover, coming along with the essential nodes of the individual.
- Defense against crossover: [7] argues that adding introns to partial solutions protects these partial solutions from being split up and destroyed because of crossover.
- Removal bias: [8] focuses on a special case of introns, namely unviable code, which is code which can never have any effect on the result of an individual, even when modified. This theory says that an individual with more unviable code has a better chance to survive crossover.
- Diffusion: [9] argue that because a solution can be represented in different ways, and there are more solutions with a greater size than simple solutions, it is only natural that more of the larger solutions are found.

However, introns are not the only way to increase the size of individuals. Another way to add unnecessary nodes to a solution is by adding some node to a solution, as well as another node that nullifies the first node. These nodes can even be located at great distances from each other in the representation, and are very difficult to detect. For example, in the expressions ($\underline{x+}$((x*x)$\underline{-x}$) and ($\underline{x+}$(A+(B$\underline{-x}$))), the underlined expressions cancel each other out.

A last cause of bloat is a form of parasite. In this case, a successful individual is appended with a subtree that only slightly reduces the fitness of the individual. In this case, the added subtree uses the high fitness of the better individual to ensure that the new individual also has a sufficiently high fitness to survive. For example, if the individual (x^5-2x^3) has a high fitness, the individual ((x^5-2x^3)+0.01) will also have a high fitness.

4 Possible Methods to Reduce Bloat

In this section, three methods to reduce bloat are presented: code editing, hill climbing and dynamic size limiting.

4.1 Code Editing

The experiments performed for this paper used symbolic regression as a test case. The primitives used to construct functions were the non-terminal nodes {+, -, *, /}, where "/" is safe division, meaning *a*/0 returns the value *a*. The set of terminal nodes consists of the variable *x*, but in some experiments also uses constant floating point values. The fitness of an expression is calculated by evaluating the expression over several values for *x*, and comparing each result with the desired function. The total error of an expression is the sum of the differences over all values of *x* that were evaluated.

Code editing for this domain can be performed at two levels. At a purely syntactical level, expressions can be reduced to more simple expressions. For example, the expression (1+(*x*-2)) can be reduced to (*x*-1), and constant expressions can be evaluated and replaced by a constant value.

At a higher level, it is also possible to remove nodes that have no influence on the result of the individual. For example, the expression (*x*/*x*)**x* can be reduced to *x*, and ((*x*-*x*)*(*x*+*x*))+*x* can be reduced to *x*.

It is not always obvious which expressions can be removed when looking only at the syntactical form of the expression. An easier method is to calculate the influence every child node has on the result of the parent node. When this influence is known, the child nodes that have very little or no influence on the result can be removed.

To determine the influence of the children of every node in the symbolic regression domain, the values of every child during every evaluation of the individual are examined. For example, the influence of children of the + and – operator can be determined as follows. If the operator *a*+*b* is evaluated, where *a* and *b* are subtrees that are evaluated to two constant values, the influence of subtree *a* is given by $|a|/(|a|+|b|)$. For example, in the expression 3+5, 3 is responsible for 3/8 of the result, while 5 is responsible for 5/8. By calculating the influences of the children over different values of x, and averaging the results, introns can be detected because they have an influence close to 0. Subtrees that have a low influence can now be removed. For * and /, the influence can be calculated in a similar way. In the implementation used for this paper, subtrees with an influence value less then 0.05 were removed.

4.2 Hill Climbing

This is essentially the approach used by [4]. The method was described in section 2.

4.3 Dynamic Size Limiting

In this approach, new individuals that are larger than a value MaxNewWeight are rejected. MaxNewWeight is set at the beginning of every generation, and is equal to C*CurrentBestWeight, where CurrentBestWeight is the weight of the current best individual of the population, and C is a constant. The best individual of a population is the individual with the highest fitness value. When several individuals have an identical fitness value, the smallest individual is the best one.

5 Experiments

To compare the different optimization techniques, the domain of symbolic regression was used. In all experiments, the objective was to find the function x^5-$2x^3$+x. The terminal set only contained the variable x, and the set of non-terminal nodes was {+, -, *, /}, where '/' is protected division. The population size in every experiment was 200, and a maximum of 100 generations were calculated. The three different optimization techniques (and combinations of those techniques) described in paragraph 4 were tested on this domain. For every experiment, the algorithm was run 50 times, and the averages of those runs are displayed. A run is ended when a correct solution is found, or after 100 generations. In the figures of the experiments, the average size of the individuals at every generation is shown, as well as the number of uncompleted runs at every generation.

5.1 No Optimizations

Figure 1 shows the result when no optimizations are used. As expected, the size of the individuals rises rapidly, up to 650 nodes at generation 100. In 74% of the runs, a solution is found within 100 generations.

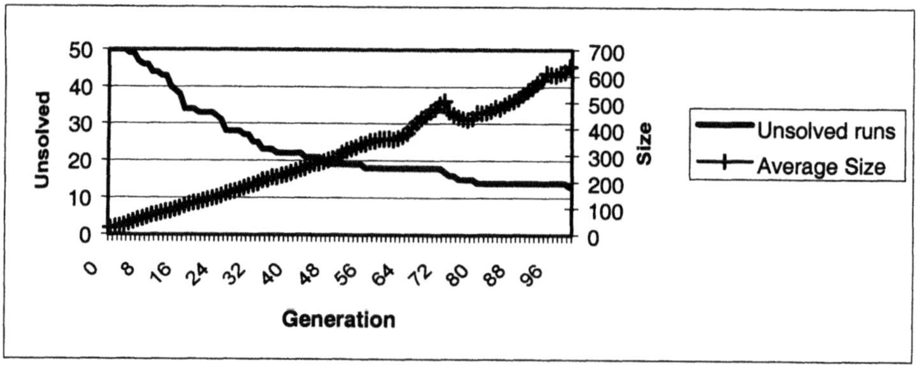

Fig. 1. Result without optimizations

5.2 Code Editing

In figure 2, the effect of removing introns is demonstrated. The results are only slightly better than the unoptimized case. Initially, the average size rises more slowly, but still rises to 625, and 74% of the runs find a solution. This demonstrates that introns are only a small part of the cause of bloating.

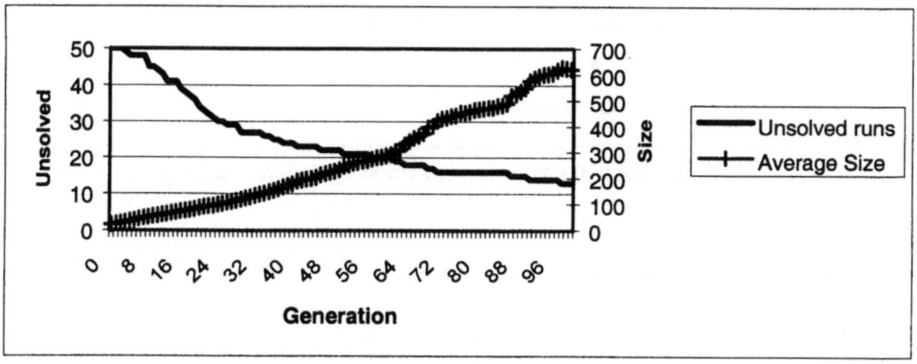

Fig. 2. Result of code editing

5.3 Hill Climbing

Figure 3 shows the result of using the hill climbing approach. The average size of the individuals is reduced significantly to 90 nodes after 100 generations, but in only 70% of the runs a solution is discovered.

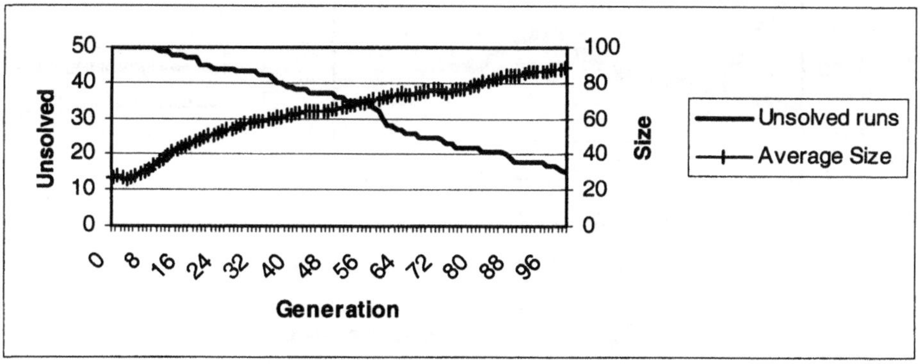

Fig. 3. Result of hill climbing optimization

5.4 Hill Climbing and Code Editing

When code editing is combined with hill climbing, the results shown in figure 4 are achieved. The average size is reduced significantly to 60, but the chance to find a solution decreases to 54%.

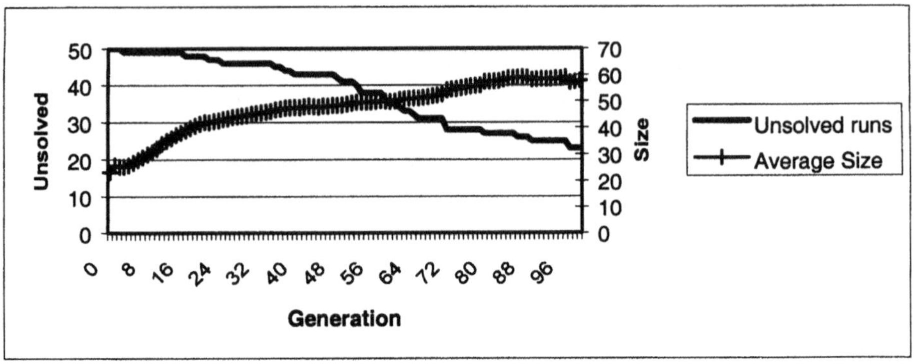

Fig. 4. Result of combining hill climbing and code editing

5.5 Dynamic Size Limiting

In this case, the size of new individuals is restricted to 1.333 times the size of the best individual of the population (or to a size of 20 for the initial population). For small sizes of the best individual, the new maximum weight is set to at least 2 more than the size of the best individual, to ensure that evolution is not stopped (because the minimum increase in size of a binary tree is 2). Figure 5 illustrates the result of this approach. The average size of the individuals is reduced even more compared to the hill climbing approach, to a maximum size of 32. Additionally, all of the 50 runs were completed successfully at generation 74.

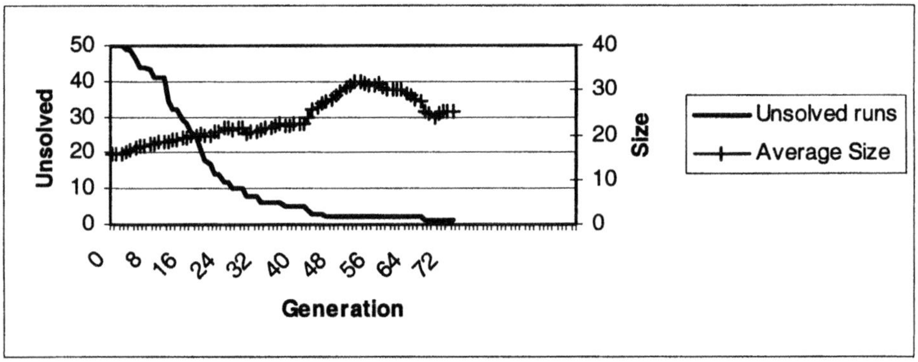

Fig. 5. Result of dynamic size limiting

5.6 Dynamic Size Limiting and Hill Climbing

By combining these two approaches, the average size of the individuals is reduced even more (to a size of 17), but the success rate decreases to 70%. These results seem to imply that hill climbing is a successful way to reduce the size, at the cost of finding solutions faster. These results are shown in figure 6.

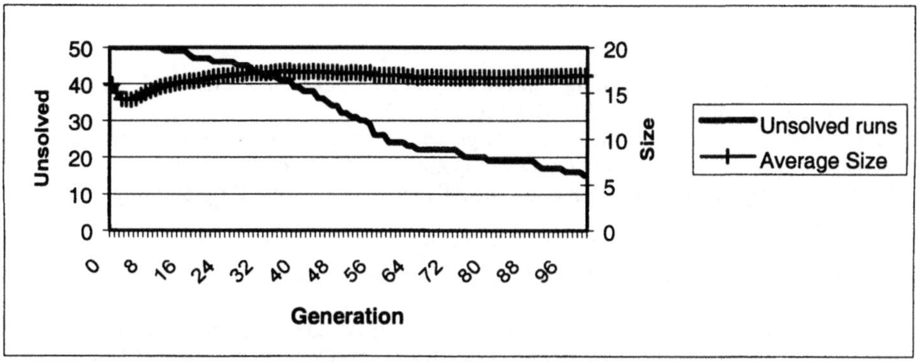

Fig. 6. Result of combining dynamic size limiting and hill climbing

5.7 Dynamic Size Limiting and Code Editing

Combining these two optimizations appears to be the best solution, as shown in figure 7: the average size is reduced to 28, and 98% of the runs are successful after 61 generations. Only a single run failed to find a solution.

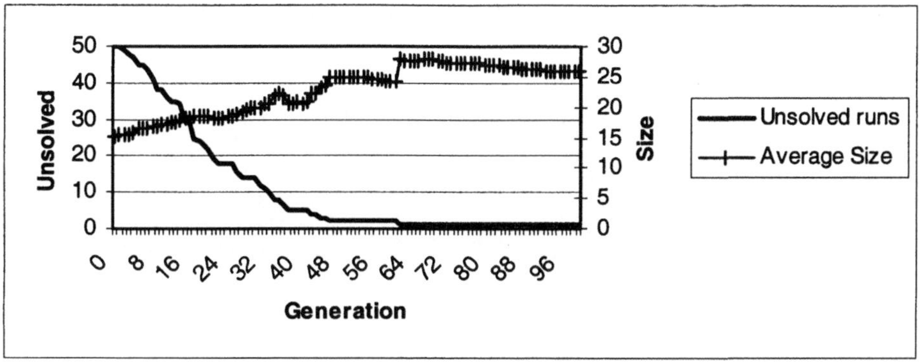

Fig. 7. Result of combining dynamic size limiting and code editing

6 Conclusions

In this paper, a new approach to reduce bloat in genetic programming was introduced. This approach restricts the size of new individuals to a constant C times the size of the current best solution. The approach is even more effective when combined with a code editing approach that removes introns from solutions. This approach has benefits over other approaches because it is very simple, improves the speed at which a solution is found.

7 References

1. Koza, J. R. Genetic Programming: On the Programming of Computers by Means of Natural Selection. Cambridge, MA, USA: MIT Press (1995)
2. T. Soule and J. A. Foster. Effects of code growth and parsimony pressure on populations in genetic programming. In Evolutionary Computation, 6(4) (1998) 293-309
3. Ryan, C. Pygmies and Civil Servants. In Advances in Genetic Programming. MIT Press (1994) 243-263
4. O'Reilly, U.-M. An Analysis of Genetic Programming. Ph.D. Dissertation, Carleton University, Ottawa-Carleton Institute for Computer Science, Ottawa, Ontario, Canada (1995)
5. Luke, Sean. Issues in Scaling Genetic Programming: Breeding Strategies, Tree Generation, and Code Bloat. Ph.D. Dissertation, Department of Computer Science, University of Maryland, College Park, Maryland (2000)
6. Tackett, W. A. Recombination, Selection, and the Genetic Construction of Computer Programs. Ph.D. Dissertation, University of Southern California, Department of Electrical Engineering Systems (1994)
7. Blickle, T. and Thiele, L. Genetic programming and redundancy. In Hopf, J., ed., Genetic Algorithms within the Framework of Evolutionary Computation (Workshop at KI-94, Saarbrücken). Saarbrücken,Germany: Max-Planck-Institut fur Informatik (1994) 33-38
8. Soule, T. and Foster, J. A. Removal bias: a new cause of code growth in tree based evolutionary programming. In 1998 IEEE International Conference on Evolutionary Computation. Anchorage, Alaska, USA: IEEE Press (1998) 781-186
9. Langdon, W. B. and Poli, R. Fitness causes bloat. In Chawdhry. Soft Computing in Engineering Design and Manufacturing. Springer-Verlag London (1997) 13-22

Too Much Knowledge Hurts: Acceleration of Genetic Programs for Learning Heuristics

Frank Schmiedle, Daniel Große, Rolf Drechsler, and Bernd Becker

Institute of Computer Science
Albert-Ludwigs-University
Georges-Köhler-Allee 51
79110 Freiburg, Germany
{schmiedl,grosse,drechsle,becker}@informatik.uni-freiburg.de

Abstract. Among many other applications, evolutionary methods have been used to develop heuristics for several optimization problems in VLSI CAD in recent years. Although learning is performed according to a set of training benchmarks, it is most important to generate heuristics that have a good generalization behaviour and hence are well suited to be applied to unknown examples. Besides large runtimes for learning, the major drawback of these approaches is that they are very sensitive to a variety of parameters for the learning process.

In this paper, we study the impact of different parameters, like e.g. stopping conditions, on the quality of the results for learning heuristics for BDD minimization. If learning takes too long, the developed heuristics become too specific for the set of training examples and in that case results of application to unknown problem instances deteriorate. It will be demonstrated here that runtime can be saved while even improving the generalization behaviour of the heuristics.

1 Introduction

Over the past 10 years, evolutionary approaches, e.g. *Genetic Algorithms (GAs)* and *Genetic Programming (GP)*, have become more and more popular [3,10, 11] and they have been successfully applied to many optimization problems, particularly in the area of VLSI CAD (e.g. [5]).

Instead of directly solving an optimization problem by GAs, in [6] a model has been proposed where GAs are used to learn good heuristics that can be applied to problem instances after the learning phase. Recently, this model has been generalized to the use of the more powerful and flexible GP instead of GAs in [4]. A set of *Basic Optimization Modules (BOMs)* is given and a heuristics includes a combination of some of the BOMs. In GAs, heuristics are always sequences of BOMs. In contrast to this, "conditional" BOM sequences can be represented by the heuristics as well if GP is used. The genetic program uses a set of training examples to learn good heuristics and during evolution, the *fitness*

B. Reusch (Ed.): Fuzzy Days 2001, LNCS 2206, pp. 479–491, 2001.

of the heuristics is determined by the results when being applied to the training set.

Since the fundamental component in *Heuristic Learning* is the incorporated GA or GP, respectively, the learning process is based on simulation. Therefore, often runtimes are very large. Moreover, the results are largely dependent on parameter settings of the evolutionary optimization method. Parameters like stopping conditions or the starting points for the evolution (=the initial population) can have strong impact on the runtimes and as well on the quality of the resulting heuristics.

Different from the fitness, the *quality* of a heuristics is determined by its generalization behaviour, i.e. the results obtained when being applied to problem instances that do not belong to the training set. A larger fitness value does not always imply a higher quality. If the GP is directed too much towards "exploitation" of the training set by parameter settings, in the final stage of evolution improving the fitness can result in a decrease of quality. The reason for this is that in such cases development of heuristics focuses only on the results obtained by application to training set examples, while generalization abilities cannot be taken into consideration.

In this paper, we analyze the impact of parameters like reproduction rate, multi-objective optimization type, start points and stopping conditions on the quality of the results for the problem of learning heuristics for BDD minimization by variable re-ordering. This problem is suitable for our study since it is of high practical importance. Moreover, it has been studied intensively and therefore in the meantime it is well understood.

The best results for BDD minimization so far have been obtained by using evolutionary approaches, but runtimes are very large. We demonstrate for heuristic learning based on GP that in many cases, slight learning improvements come with decreasing quality of the resulting heuristics at a late stage of evolution. Consequently, about 50% of the runtime can be saved by using better stopping criteria while at the same time, the quality of the heuristics can be increased.

2 Preliminaries

We assume that the reader is familiar with GA and GP concepts and refer to [3, 10] for details. A model for heuristic learning with GA has been proposed in [6]. It has been applied to various optimization problems in VLSI CAD. Recently, a generalization to the GP domain has been reported and and the GP-based heuristic learning has been successfully applied to BDD minimization for small benchmark circuits [4]. Since in this paper, various parameters and their impact on resulting heuristics are investigated in order to optimize the learning process, we give a brief review of heuristic learning and also describe how this approach is used for GP-based BDD minimization. By this, the paper becomes self-contained. Theoretic aspects are discussed for both GA and GP, but in all experiments, the GP-based method has been used.

2.1 Heuristic Learning

For learning heuristics in order to find good solutions to an optimization problem, it is necessary that several (non-optimal) strategies to solve the problem can be provided. Typically, for different classes of problem instances there are also different strategies that perform best. A strategy that behaves well on one problem class may return poor results when being applied to another problem instance that belongs to some different class. Thus it is promising to learn how to combine the strategies to heuristics that can be applied successfully to most or even all classes of problems.

The learning process is controlled either by GA or by GP. For a better understanding, some fundamental terms are introduced by the following

Definition 1. *Given an optimization problem P and a non-empty set of different non-optimal strategies $B = \{b_1, \ldots, b_{max}\}$ to solve the problem. Then the elements of B are called* BOMs (Basic Optimization Modules). *A* heuristics *for P is*

- *a non-empty sequence of BOMs in the case of learning by GA.*
- *an algorithm that can generate a sequence of BOMs if GP is used for heuristic learning.*

Furthermore, a flat sequence *of a heuristics is a concrete BOM sequence that has been calculated by the heuristics. If the heuristics has been generated by GA, its only flat sequence is the heuristics itself.*

Example 1. For an optimization problem P, a set of BOMs $B = \{A, B, C, D, E\}$ is given. A heuristics h which has been learned during a GP-driven evolutionary process is illustrated in Figure 1. Possible flat sequences of h are e.g. $ABDE$, $ACDE$ and $ABDDDDE$.

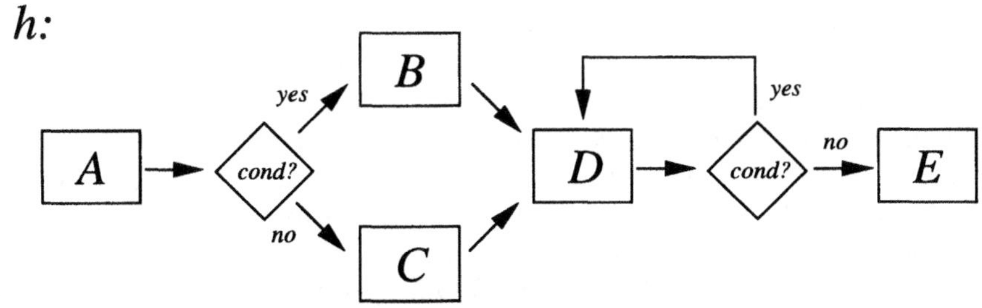

Fig. 1. Example for a heuristic generated during GP-based heuristic learning

Heuristics can be applied to p (an instance of P) by subsequently calling the BOMs included in the flat sequence to p. If GP is used, the flat sequence is generated on-the-fly during application of the heuristics (see Section 2.2).

During evolution, the strategies/BOMs are combined to generate heuristics that are the individuals in the population. (The fitness of) an individual can be evaluated by application of the heuristics to a training set of problem instances. In the learning phase, the target is that heuristics perform well according to some given optimization criteria when being applied to the training set. But for the success of the overall method, mainly a good generalization is important, i.e. the intention is that a heuristics that returns good results for the training set examples also performs well on unknown instances of P. Please note that for this, the handling of the different criteria, i.e. the special multi-objective optimization approach, plays a critical role for the success of the evolutionary process.

A scheme for the scenario of heuristic learning is outlined in Figure 2 where learning by GA and also learning by GP are included.

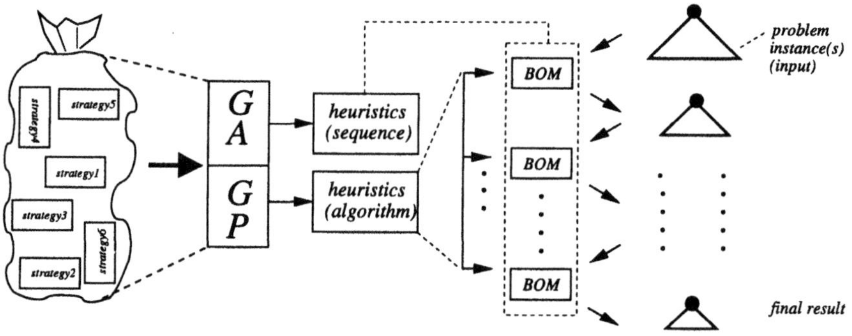

Fig. 2. Heuristic learning process

On the left, there is a set of problem solving strategies — this is the set of BOMs that are provided. Although GA *and* GP are included in the figure, only one of them is used for heuristic learning at the same time. In both cases, the result consists of one or several heuristics that can be applied to instances of P as illustrated at the right side of the figure. When a heuristics h is applied, the slots denoted by BOM are filled by elements of B. Problem instances are given as triangles[1] in the figure. Calling some BOM b_i returns an intermediate result that is passed to b_j which is the next one in the flat sequence. This is proceeded until the last BOM in the flat sequence of the heuristics terminates. As already mentioned, heuristics produced by GA are identical to their flat sequence. How flat sequences are generated for heuristics learned by GP is described in the following section.

[1] They can be considered e.g. as BDDs that have to be minimized.

2.2 BDD Minimization by Genetic Programming

Binary Decision Diagrams (BDDs) [2] are a state-of-the-art data structure often used in VLSI CAD for efficient representation and manipulation of Boolean functions. BDDs suffer from their size being strongly dependent on the variable ordering that is used. In fact, BDD sizes may vary from linear to exponential for different orderings. Optimization of variable orderings for BDDs is difficult [1], but nevertheless, strategies for BDD minimization that are based on dynamic variable ordering, i.e. *sifting* have been reported e.g. in [8, 13, 12].

For BDD minimization by heuristic learning, we use the GP-based approach that has been introduced in 2.1. The strategies sifting (SIFT), group sifting (GROUP), symmetric sifting (SYMM), window permutation of size 3 and 4, respectively, (WIN3, WIN4) serve as BOMs. For all these techniques there is an additional BOM that iterates the method until convergence is reached. These BOMs are denoted by the appendix CO, e.g. iterated sifting is denoted by SIFTCO. The "empty" operator called NOOP completes the set of BOMs B and hence $|B| = 11$.

The individuals in the GP consist of trees with leaf nodes labeled with BOMs and inner operator nodes that belong to the three different types CONCAT, IF and WHILE. During evaluation of the heuristics, the tree is traversed by a *depth-first-search*-based method in order to generate a flat sequence. The types of the inner nodes decide if

- both subtrees are evaluated subsequently (CONCAT),
- according to the truth value of a given condition either the left or the right son is considered (IF) or
- evaluation of the sons is iterated until a truncation condition is fulfilled (WHILE)

For recombination, two crossover operators are provided. While CAT concatenates two parents, the more sophisticated MERGE does the same for subtrees of the parents and by that, bloating can be prevented. In addition to this, there are four mutation operators that exchange BOMs (BMUT), node types (CIMUT, CWMUT) and modify conditions of IF-nodes (IFMUT), respectively. A probability table, i.e. Table 1, determines the frequencies for using the different recombination operators.

Table 1. Probabilities for recombination operators

Operator	CAT	MERGE	BMUT	CIMUT	CWMUT	IFMUT
Probability	5	45	20	10	10	5

An additional problem specific operator REDUCE is applied to each individual during evaluation. It eliminates redundant parts from the trees and hence speeds up the learning process. For evaluation of the heuristics, a training set of small

BDD benchmarks is used. Heuristics are subsequently applied to each of the benchmarks in that training set and the resulting sizes are the optimization criteria. To calculate the fitness of an individual, the multi-objective optimization techniques proposed in [14] are used.

3 Influence and Selection of Parameters

For the method introduced in Section 2, a standard algorithm for the evolutionary loop has been used [7, 9]. The algorithmic scheme of GA/GP leaves still some degrees of freedom to the user like e.g. generation of the initial population. In the following sections, such issues are addressed and the appropriate parameters and subroutines are discussed.

3.1 Initial Population

The initial population defines the point in the search space where evolutionary optimization is started. If by some problem specific method for generation of the initial population, a starting point can be found that is not so far from the optimum, it is likely that the optimum will be found early during the optimization process. Furthermore, the risk of getting stuck in a local optimum is also decreased in that case.

If on the other hand the starting point is close to some local optimum O_l, optimization will be directed towards O_l and with high probability it will not be possible to escape from O_l again. E.g. this happens if the individuals included in the initial population are too similar to each other.

As a consequence, for the problem of BDD minimization by GP-based heuristic learning, this means that at least the types IF and WHILE of inner nodes have to be part of some of the initial individuals. CONCAT nodes can be inserted, too, but it is not so essential that they occur in the beginning since such nodes are frequently inserted by the recombination operators during evolution. Besides the different node types, a large variety of BOMs should be included as well. By randomized insertion of BOMs into individuals, similarity can be avoided.

To determine the best suited structure of the trees that represent the initial individuals is a non-trivial problem and furthermore it is not obvious whether or not this structure should differ for different individuals of the initial population. To investigate this, we implemented the 3 approaches FIXED1, FIXED2 and VARIED that are explained in the following:

- FIXED1: The individuals in the initial population have a unique structure with only one level of inner nodes. All trees consist of 3 nodes: the root and its 2 sons that are leafs. The leaves are labeled with BOMs that are randomly selected out of B. The type of the root node is CONCAT in 50% of the cases, IF and WHILE are assigned to 25% of the root nodes each.
- FIXED2: All the individuals have 2 levels of inner nodes and 3 leaf nodes. Figure 3 illustrates 5 different classes of trees and each class is used for 20%

of the initial population. Analogously to the FIXED1 method, the BOMs are selected at random.

- VARIED: Here, the number of levels in the trees is arbitrary while only the *number of BOMs* in a heuristics is limited. A tree of infinite size with unlabeled nodes is traversed in a breadth first search manner, and each node is labeled with either IF (probability 20%), WHILE (15%) or BOM (65%). If a BOM is chosen, it is selected out of the set B at random and the sub-trees of the node are eliminated. When the BOM limit is reached, all following nodes are labeled with NOOP and their sub-trees are removed. The individual is complete when there are no unlabeled nodes any more.

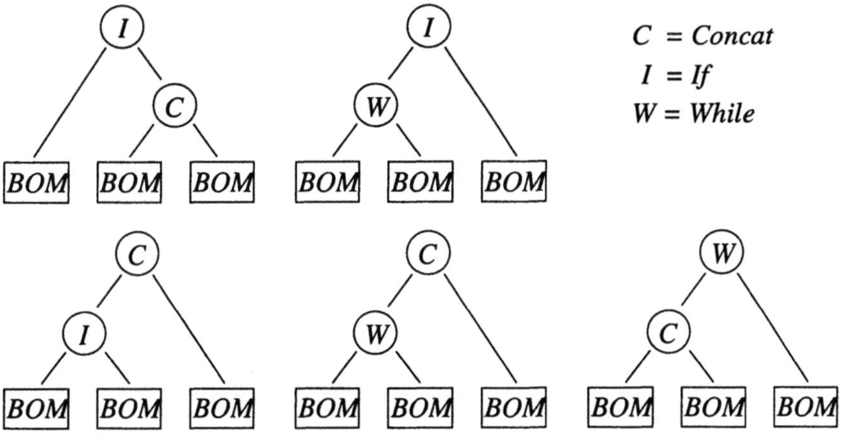

Fig. 3. Classes of trees for FIXED2

3.2 Training Set

The training set is needed during evolution for evaluation of the individuals' fitness. The heuristics represented by an individual is subsequently applied to each of the training set examples and based on the resulting BDD sizes, the fitness value of the heuristics is calculated. During the learning phase, search is directed towards individuals with good fitness values, i.e. optimization is done by trying to find heuristics that produce best results for the set of training examples.

Different from the fitness, the *quality* of a heuristics is determined by its generalization behaviour, i.e. the results obtained when being applied to benchmarks that the training set does not include. Since the overall task of heuristic learning is to find heuristics with good quality, it is essential that the training set is chosen in a way that keeps the average of the difference between fitness and quality as small as possible. As a consequence, a good training set has to be *representative* which means that the properties of each possible benchmark circuit have to be represented by one or several elements of the training set.

We use training sets that consist of 8 elements each. The training examples are benchmark circuits that are taken from LGSYNTH91 [15]. For all training example candidates, we measured the initial BDD size, i.e. the number of BDD nodes that was needed to construct the BDD without further minimization steps by one of the BOMs. Then we selected benchmarks with varying initial size and numbers of inputs/outputs in order to have more representative training sets. In total, 3 different training sets T_1, T_2 and T_3, respectively, have been used. Names, numbers of inputs/outputs and initial sizes of the benchmarks included in these sets are listed in Table 2.

Table 2. Training sets for heuristic learning

Training Set T_1				Training Set T_2				Training Set T_3			
Name	#in	#out	init size	Name	#in	#out	init size	Name	#in	#out	init size
bc0	26	11	589	*term1*	34	10	579	*alu4*	14	8	1181
ex7	16	5	172	*b9*	41	21	177	*s208.1*	18	9	1032
frg1	28	3	203	*count*	35	16	233	*s832*	23	24	2650
ibm	48	17	835	*x4*	94	71	890	*cm151a*	12	2	510
in2	19	10	2360	*s820*	23	24	2650	*i7*	199	67	504
s1196	32	31	2293	*i9*	88	63	2277	*apex6*	135	99	2759
ts10	22	16	4390	*vda*	17	39	4344	*example2*	85	66	468
x6dn	39	5	274	*s386*	13	13	280	*c432*	36	7	1732

3.3 Stopping Condition

During heuristic learning, a stopping condition is checked each time a new generation (of individuals) has been calculated. If the condition evaluates to *true*, then the evolutionary learning process is terminated.

Up to a certain number of generations g_{sat}, the computation of a larger number of generations leads to an increased fitness in most cases. But a saturation of fitness may be reached if g_{sat} is exceeded. Then the fitness of the best individual does not increase significantly in the following generations and therefore runtime is wasted if computation is continued. Even worse, a better fitness does not always imply a higher quality. Especially at a late stage of evolution, improvements of the fitness can result in a decrease of the quality which is the overall optimization task. The reason for this is that development of heuristics only focuses on the training set examples while the quality on other instances is not considered. Hence, overfitting on the training set can be observed in such cases.

In Figure 4, a progression of fitness and quality during evolution can be seen and in this case, overfitting can be observed. After g_{sat} is exceeded, only a slight increase of fitness can be obtained. Regardless of the fitness, evolution should be stopped no later than g_{opt}, since afterwards, the quality of the individuals in the population decreases.

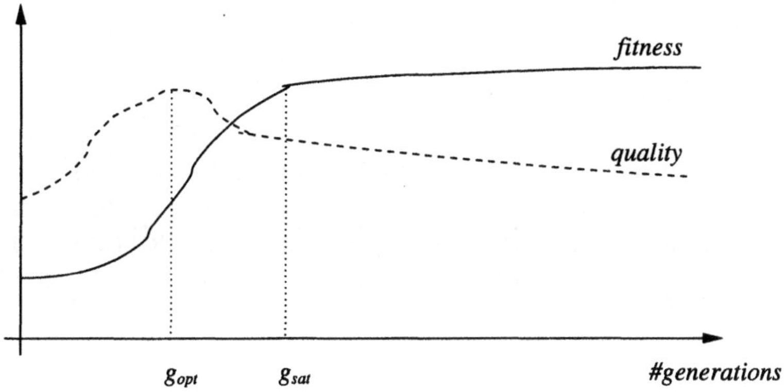

Fig. 4. Progression of quality and fitness during evolution

For BDD minimization by heuristic learning with GP, different absolute limits in numbers of generations have been used as stopping conditions for our method. By this, it is possible to investigate whether saturation and/or overfitting occur and if so, g_{sat} and g_{opt} can easily be determined as well.

4 Experimental Results

In this section, experimental results are presented. In the experiments that have been conducted, the population size has been set to 20 in each case and *tournament selection* has been used to select individuals for recombination. We have used reproduction rates of 50% (10 offsprings per generation) and 80% (16 offsprings per generation), respectively. The training sets T_1–T_3 have been utilized and for calculation of the fitness, we have applied the two different multi-objective approaches PREFERRED and PRIORITY. One of these techniques assumes that the importance of the benchmarks in the training set is unique, the other one deals with priorities (for details see [14]). Since the duration of computation for one generation is roughly constant during evolution, runtimes have been measured in number of generations.

The quality of heuristics learned by the GP is measured according to the results on a set C of the circuits *apex2, apex7, bcd, chkn, cps, in7, pdc, s1494, t1* and *vg2*. These benchmarks have not been considered during the evolutionary

Table 3. BDD sizes for heuristics learned by GP

Experiments with Training Set T_1											
set of bench.	rep. rate	MOO type	Sum of BDD sizes for #generations: 5	10	15	20	25	30	40	50	60
T_1	50%	Pref.	2307	**2093**	**2093**	**2093**	**2093**	**2093**	**2093**	**2093**	**2093**
C			**3923**	4145	4144	4144	4144	4144	4144	4144	4144
T_1	80%	Pref.	2097	2096	2260	2416	2084	2570	**2085**	2110	2377
C			3894	3849	3815	**3777**	3898	3881	3841	3938	3951
T_1	50%	Prio.	2245	**2095**	**2095**	**2095**	**2095**	**2095**	**2095**	**2095**	**2095**
C			3911	**3892**	**3892**	**3892**	**3892**	**3892**	**3892**	**3892**	**3892**
T_1	80%	Prio.	2095	2095	2095	**2093**	**2093**	**2093**	**2093**	**2093**	**2093**
C			3904	**3884**	**3884**	4146	4146	4146	4146	4146	4146

Experiments with Training Set T_2											
set of bench.	rep. rate	MOO type	Sum of BDD sizes for #generations: 5	10	15	20	25	30	40	50	60
T_2	50%	Pref.	2560	2578	2646	2205	2571	2698	2563	**2118**	2679
C			3999	4005	4055	**3848**	3946	4032	4022	3925	4048
T_2	80%	Pref.	2240	2510	2501	2511	**2106**	2458	2601	2599	2311
C			3879	4021	4008	4009	3949	**3861**	3974	4168	3884
T_2	50%	Prio.	2647	2419	**2331**	**2331**	**2331**	**2331**	**2331**	**2331**	**2331**
C			4122	3976	3975	3975	3975	3975	**3973**	**3973**	**3973**
T_2	80%	Prio.	2579	2537	**2101**	2610	2608	2144	2652	2490	2658
C			4156	4075	3796	4097	4029	**3786**	4277	4003	4175

Experiments with Training Set T_3											
set of bench.	rep. rate	MOO type	Sum of BDD sizes for #generations: 5	10	15	20	25	30	40	50	60
T_3	50%	Pref.	**2117**	2239	2239	2239	2239	2239	2239	2239	2239
C			3872	**3860**	**3860**	**3860**	**3860**	**3860**	**3860**	**3860**	**3860**
T_3	80%	Pref.	2243	**2123**	2124	2124	**2123**	**2123**	**2123**	**2123**	**2123**
C			3889	**3870**	3874	3874	**3870**	**3870**	**3870**	**3870**	**3870**
T_3	50%	Prio.	2450	2123	2093	**2113**	**2113**	**2113**	**2113**	**2113**	**2113**
C			3999	3870	4139	**3869**	**3869**	**3869**	**3869**	**3869**	**3869**
T_3	80%	Prio.	2410	**2243**	**2243**	**2243**	**2243**	**2243**	**2243**	**2243**	**2243**
C			3926	**3889**	**3889**	**3889**	**3889**	**3889**	**3889**	**3889**	3900

optimization process. The smaller the sum of BDD nodes of all elements in C after application of a heuristics, the higher is the quality of the heuristics.

In a first set of experiments, the initial population was generated by the VARIED approach and Table 3 illustrates the results for different numbers of generations computed during evolution. The table is divided into 3 parts whereas in the different parts, the use of training set T_1, T_2 and T_3, respectively, is considered. Each (double) row includes results on the training set and the set of unknown circuits C, respectively, for varying stopping conditions in numbers of generations but fixed reproduction rates and multi-objective optimization types. In the first column of each row, the appropriate set of benchmarks is given. The second and third column include the reproduction rate and the multi-objective optimization type, respectively. Sums of BDD sizes for all elements included in the benchmark set are given in columns 4–12 for a heuristics that has the best quality among the elements in the population after certain numbers of generations. The best quality and fitness values are illustrated in bold face.

It can be seen that the use of each of the training sets leads to the best quality result for at least one combination of reproduction rate and multi-objective optimization approach. Thus none of the training sets is superior to another one. The best quality values itself found during evolution only slightly differ if the set of training examples is exchanged. The largest deviation can be observed if the PRIORITY method and a reproduction rate of 50% are used: in that case, the best quality found by learning with T_3 is 2.6% better than if T_2 is used while optimization with training set T_1 returns a result in between. If PREFERRED is used instead of PRIORITY, the deviation even decreases to 1.9%. For the combination of the PRIORITY method and a reproduction rate of 50%, an early saturation can be observed regardless of the training set used. Nevertheless, in general the behaviour during evolutionary optimization is not unique for different training sets. This can be observed for the combination reproduction rate 50%/PREFERRED: if T_1 is used, then there is an early saturation while for T_2, no saturation occurs up to generation 60.

After a certain number of generations, slight improvements on the training set T often can only be achieved at the cost of a considerably decreased quality. E.g. this can be seen in the last rows of part 1 (training set T_1). However, the most important observation that can be made is that the heuristics with the best quality has been found within the first 30 generations in each of the 12 experiments unless one. Even by the exceptional setup (training set T_2/reproduction rate 50%/PRIORITY), the improvement of the quality after generation 30 is very small, i.e. $< 1\%$. As a consequence, by stopping the evolution after 30 generations, approximately 50% of the runtime can be saved while the quality of the resulting heuristics even increases on average.

The different methods for the construction of the initial population have been compared in further experiments. T_1 has been selected as training set and the reproduction rate was 50%. The fitness values have been computed by the PREFERRED approach for multi-objective optimization. Results are illustrated in Table 4, where analogously to Table 3, sums of BDD sizes for all benchmark examples are given.

Table 4. Comparison of different methods for generation of the initial population

generation of initial population	set of bench marks	Sum of BDD sizes for #generations: 5	10	15	20	25	30	40	50	60
VARIED	T_1	2307	**2093**	**2093**	**2093**	**2093**	**2093**	**2093**	**2093**	**2093**
	C	**3923**	4145	4144	4144	4144	4144	4144	4144	4144
FIXED1	T_1	2097	2096	2096	2096	2095	2093	**2091**	**2091**	**2091**
	C	4147	3892	**3891**	**3891**	4156	4144	4144	4144	4144
FIXED2	T_1	2096	2096	2094	2094	**2093**	**2093**	**2093**	**2093**	**2093**
	C	**3891**	**3891**	4144	4144	4144	4146	4146	4146	4146

By using FIXED1 and FIXED2, respectively, slight improvements in quality can be achieved. The saturation point g_{sat} can be "delayed", i.e. for FIXED1, and FIXED2, the best fitness is observed approximately 30 and 15 generations later than in the case of using VARIED. In each case, there is an overfitting behaviour during the learning process, but the point of time when the heuristics with the best quality occur is different. For the given setup, FIXED2 is the best method to generate the initial population since best qualities for heuristics that occur in any of the three experiments are obtained very early.

5 Conclusions

In this paper, GP-based heuristic learning for BDD minimization has been considered. Several parameters of the GP have been analyzed, i.e. generation of the initial population, selection of the training set examples and stopping conditions. Particularly, the generalization behaviour of the heuristics that are learned during evolution has been investigated. It can be seen that the quality of a heuristics does not always correspond directly to its fitness during the GP. Experimental results have been given that prove that by determination of suitable stopping conditions, approximately 50% of the runtime can be saved while the quality of heuristics learned is even improved.

References

1. B. Bollig and I. Wegener. Improving the variable ordering of OBDDs is NP-complete. *IEEE Trans. on Comp.*, 45(9):993–1002, 1996.
2. R.E. Bryant. Graph - based algorithms for Boolean function manipulation. *IEEE Trans. on Comp.*, 35(8):677–691, 1986.

3. L. Davis. *Handbook of Genetic Algorithms.* van Nostrand Reinhold, New York, 1991.
4. N. Drechsler, F. Schmiedle, D. Große, and R. Drechsler. Heuristic learning based on genetic programming. In *EuroGP*, volume 2038 of *LNCS*, pages 1–10. Springer Verlag, 2001.
5. R. Drechsler. *Evolutionary Algorithms for VLSI CAD.* Kluwer Academic Publisher, 1998.
6. R. Drechsler and B. Becker. Learning heuristics by genetic algorithms. In *ASP Design Automation Conf.*, pages 349–352, 1995.
7. R. Drechsler, B. Becker, and N. Göckel. Minimization of OKFDDs by genetic algorithms. In *Int'l Symposium on Soft Computing*, pages B:271–B:277, 1996.
8. M. Fujita, Y. Matsunaga, and T. Kakuda. On variable ordering of binary decision diagrams for the application of multi-level synthesis. In *European Conf. on Design Automation*, pages 50–54, 1991.
9. D.E. Goldberg. *Genetic Algorithms in Search, Optimization & Machine Learning.* Addision-Wesley Publisher Company, Inc., 1989.
10. J. Koza. *Genetic Programming - On the Programming of Computers by means of Natural Selection.* MIT Press, 1992.
11. Z. Michalewicz. *Genetic Algorithms + Data Structures = Evolution Programs.* Springer-Verlag, 1994.
12. S. Panda and F. Somenzi. Who are the variables in your neighborhood. In *Int'l Conf. on CAD*, pages 74–77, 1995.
13. R. Rudell. Dynamic variable ordering for ordered binary decision diagrams. In *Int'l Conf. on CAD*, pages 42–47, 1993.
14. F. Schmiedle, N. Drechsler, D. Große, and R. Drechsler. Priorities in multi-objective optimization for genetic programming. In *Genetic and Evolutionary Computation Conference*, 2001.
15. S. Yang. Logic synthesis and optimization benchmarks user guide. Technical Report 1/95, Microelectronic Center of North Carolina, 1991.

Gaussians-Based Hybrid System for Prediction and Classification

Ernesto Saavedra[1], Ingo Renners[1], Adolf Grauel[1], H.J. Convey[2], and A. Razak[2]

[1] University of Paderborn, Soest Campus, Soest 59494, Germany
grauel@ibm5.uni-paderborn.de
[2] Bolton Institute of Higher Education, Deane Campus, Bolton, BL3 5AB, England
hc1@bolton.ac

Abstract. We propose a hybrid model based on Genetic Algorithms (GA), Lattice Based Associative Memory Networks (LB-AMN) and Radial Basis Function Networks (RBFN) for the solution of prediction and classification problems. LB-AMN and RBFN have as basis in their structure a type of asymmetric radial basis function (RBF) which results from the combination of two Gaussian functions. In the first sections we describe the mathematical models used to build the hybrid system. Afterwards, we apply the model to the problem of breast cancer and toxicity prediction. In both cases, the obtained results were better than the ones obtained using other approaches. Finally, some conclusions are given.

1 Introduction

A main problem in science and industry is concerned with the establishing of functional relationships between sets of inputs and outputs which can be either linear or non-linear. Related to it, it is also the problem of finding the inputs on which the outputs most depend. Given a system from which we can take measures at its inputs and outputs, we want to find out which are the inputs that most affect the outputs and, to define a model that best approximates the systems behaviour. As an aim to solve this sort of problems, a model based on GAs and RBFs is proposed.

When RBFs are used, the criteria of distance used to evaluate them are either the Euclidean or Mahalanobis distance. If the Euclidean distance is applied, the input space is covered by hyper-spheres, whereas the Mahalanobis distance generates hyper-ellipses. The Euclidean-based approach usually result in a poor performance compared with models that use the Mahalanobis distance. However, a disadvantage in using Mahalanobis is that the numbers of parameters that define each basis function increases exponentially with the number of inputs. As aim to overcome this problem, we propose an asymmetric basis function resulting from the combination of two Gaussians.

The model is applied to the problem of breast cancer prediction, and to a problem of toxicity prediction through QSARs. The first one is a classification problem in which there are 9 possible inputs and one output with only two possible values that define whether it belongs to the malignant or benign group. The second problem is more difficult because the number of possible inputs is 150 approximately and the LB-AMN or RBFN must represent the dependency between the degree of toxicity and the found inputs.

B. Reusch (Ed.): Fuzzy Days 2001, LNCS 2206, pp. 492–499, 2001.

The models' performance is evaluated through the mean square error (MSE) and the correlation desired vs. computed output using the **Leave-one-out** technique. Using this method, the network topology is calculated with the N available samples. To compute the output for one sample, we leave it out and calculate the weights with the remaining N-1 samples, then we use such weights to compute the output using the left out sample as input. This method tests the topology robustness with all the available data and therefore the error obtained can be considered as prediction error.

2 Radial Basis Function (RBF)

Radial basis functions can be expressed as: $h(X) = f(\| X - c \|)$ where c is a n-dimensional vector denoting the *center* of the basis function, $\| X - c \|$ is the Euclidean norm, X denotes the n-dimensional vector and $f(\cdot)$ is an uni-variate from $\Re^+$. The output $f(\cdot)$ of the RBF depends only on the distance between the input X and the centres' vector c, and it is this proportion that gave them the name of RBF. They can have different shapes, e.g the thin plate spline function, the Gaussian function, etc. Gaussian functions can be written as a product of uni-variate functions represented by:

$$h(X) = e^{-\frac{1}{2\sigma^2}\|X-c\|} = e^{-\frac{1}{2}\frac{(X-c)^2}{\sigma^2}} = \prod_{i=1}^{n} e^{-\frac{1}{2}\frac{(x_i-c_i)^2}{\sigma^2}}, \tag{1}$$

where σ^2 is a constant of the function $h(X)$, i=1 ... n denotes the number of elements of vector X, c denotes the vector of centers, and c_i is the i-th. center of c that corresponds to x_i. If X is composed by only one element, c and σ^2 correspond to the mean and variance respectively. For the multi-dimensional case, the term $\frac{\|X-c\|}{\sigma^2}$ in Eq.1. can be replaced by: $\| M \cdot (X - c) \|$ known as the Mahalanobis distance which is invariant against linear transformations, i.e. scaling, rotation and translation [1]. The matrix **M** normalizes the input vector **X** in order to balance the different scaling influences of the origins and it is related to the variances matrix Ω by:$\Omega^{-1} = M^T \cdot M$. It can be calculated using:

$$\| M \cdot (X - c) \| = (X - c)^T \cdot M^T \cdot M \cdot (X - c). \tag{2}$$

Let us consider the cases when **M** is quadratic:

1. If **M** is the identity matrix multiplied by a certain constant, the receptive field generated by the basis function will be a n-dimensional sphere. For this case, the number of parameters, including the centers or elements of vector c, that define a n-dimensional Gaussian function will be:

$$Params(RBF)\,|_{Euclidean} = dim(X) + 1, \tag{3}$$

   where $dim(X)$ denotes the dimension of the independent variables **X**.
2. If **M** has different elements only in the main diagonal, the receptive field will be a n-dimensional ellipse with axes parallel to the co-ordinates and the number of parameters for each basis function will be defined by:

$$Params(RBF)\,|_{Mah.Simpler} = 2 \cdot dim(X). \tag{4}$$

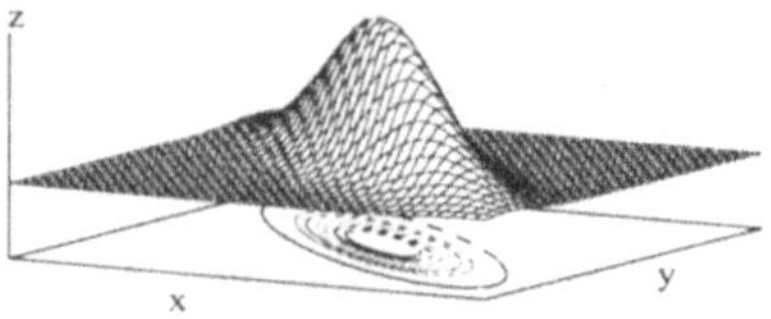

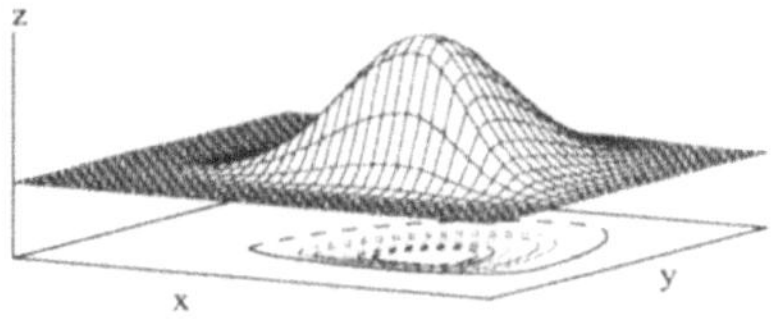

(a) Mahalanobis (b) Asymmetric

Fig. 1. Bi-dimensional Gaussian functions

3. For the other cases when the elements of **M** are different than zero, the Mahalanobis distance defines a n-dimensional ellipse with some rotational angle in relation to its co-ordinates (see Fig.1.a). The number of parameters that define such functions is:

$$Params(RBF)\mid_{Mahalanobis} = dim(X) \cdot (dim(X) + 1). \quad (5)$$

2.1 Asymmetric RBF

We propose another type of RBF based on Gaussian functions as well that, even though it can not generate arbitrary n-dimensional ellipses as receptive fields, it is more flexible than Mahalanobis for the case 2, and unlike case 3, its parameters increase linearly with the dimension of **X**. Let us define the following basis function for the uni-dimensional case:

$$h(x) = \begin{cases} e^{-\frac{1}{2}\frac{(x-c)^2}{\sigma_L^2}} & \text{if } x < c \\ e^{-\frac{1}{2}\frac{(x-c)^2}{\sigma_R^2}} & \text{if } x \geq c. \end{cases} \quad (6)$$

The function defined by Eq.6 decreases monotonically, it is a positive function for every x and it is not symmetric in relation to its center c. If we extend this concept to n-dimensional Gaussian functions evaluated by the Mahalanobis distance, the matrix **M** will take the form:

$$M = \begin{bmatrix} \sigma_{L_1}^2, \sigma_{R_1}^2 & 0 & \cdots & 0 \\ 0 & \sigma_{L_2}^2, \sigma_{R_2}^2 & 0 & 0 \\ \vdots & 0 & \ddots & \vdots \\ 0 & 0 & \cdots & \sigma_{L_n}^2, \sigma_{R_n}^2 \end{bmatrix}. \quad (7)$$

Using this type of function it is possible to generate receptive fields which are combinations of two elliptical functions with axes parallel to the co-ordinates (See Fig.1.b). The number of parameters that define a n-dimensional basis function of this type is:

$$Params(RBF)\mid_{Mah.Asymm} = 3 \cdot dim(X). \quad (8)$$

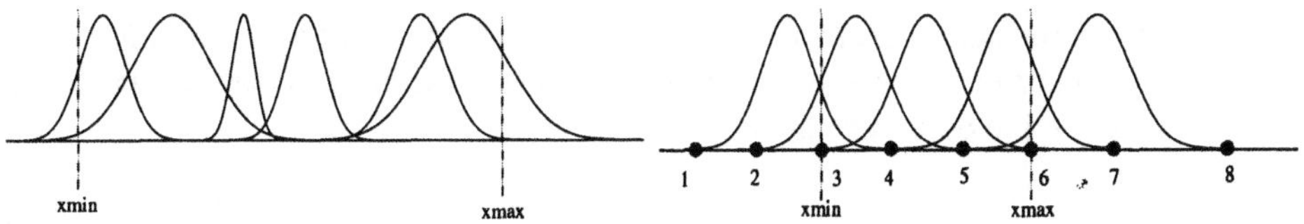

(a) RBFN (b) LB-AMN. knots=8, terms=5, order=3

Fig. 2. Functions distribution using RBFNs and LB-AMNs

2.2 Radial Basis Function Network (RBFN)

These type of Artificial Neural Network (ANN) is an instar-outstar model proposed initially to perform input-output mappings trained by examples [2]. The output $f(X)$ of the network is computed by:

$$f(X) = \sum_{j=1}^{m} w_j \cdot h_j(X), \tag{9}$$

where w_j is the function value, or weight of the *i-th* receptive field. Thus, each of the *n* components of the input vector **X** feeds forward to *m* radial basis functions or nodes whose outputs are linearly combined with weights $\{w_j\}_{j=1}^{m}$ into the network output $f(X)$. Unlike the LB-AMN model, the Gaussian functions used to cover the inputs might not be arranged in a regular fashion and each function covers the whole range of an input and not only part of it as in the case of LBAMNs. Figure 2.a shows an example of the Gaussians distribution over an input using a RBFN.

2.3 Lattice-Based Associative Memory Network (LB-AMN)

LB-AMNs are a class of ANN that normalizes the input space by a lattice formed by a certain number of knots on which the basis functions are defined. These networks have many desirable features: the address of the non-zero basis functions can be explicitly calculated, the transformed input vector is generally sparse which means that knowledge is stored and adapted locally [3]. They are also more transparent, which makes possible to interpret them as Tagaki-Sugeno fuzzy controllers. The output of the LB-AMN, as in the case of RBFNs, is formed from a linear combination of the outputs of the basis functions given by (9), where w_j, are adjustable weights, $h_j(X)$ are the activation values of the n-dimensional basis functions, and m is the number of basis functions in the lattice that the input vector **X** activates at time *t*. For LB-AMN, the basis functions are defined in such a way that they do not cover the whole range of an input but only sections of it, and they are created on a basis of a knots vector. The number of knots and basis functions are related by: $knots = terms + order$, where $terms$ is the number of basis functions used to cover an input, and $order$ is the number of functions that an input activates. Figure 2.b represents the Gaussians distribution over an input using LB-AMN models. In both cases, the determination of the weights vector w will lead to the solution of an over-determined system of linear equations written in a matrix form as $A(X) \cdot w = Y$, where $A(X)$ is the activation matrix and Y is the desired output of the RBFN or LB-AMN. Concerning the widhts and centers of the basis functions, Genetic Algorithms (GAs) can be used to determine them.

3 Hybrid Model for Prediction and Classification

The proposed model is based on genetic algorithms (GA) [4] and RBFs. The GA block will perform two tasks: **1.** Search the most relevant sub-set of inputs related to the output. **2.** Find the LB-AMNs or RBFNs topologies that best map the relationship between inputs and output. The model also includes a routine for calculating the ANNs' weights by solving an over-determined system of linear equations, and it evaluates de the ANNs' performance using the **Leave-one-out** technique.

3.1 GA Encoding

We define the following terminology to encode a possible solution that will imply inputs selection and topology optimization: Let us name $gene^0$ (read as gene of order 0) to each value or number which is encoded in the genome (the complete bit-string of one individual). Several genes of order zero can be merged to a compound defined as $gene^1$, and the members of this compound denoted as $gene_i^0$. This notation can be extended to an arbitrary order. We proceed as follows to encode the solutions:

$gene_0^0$ consisting of one bit. It defines whether the following genes are used or not. From the biological viewpoint, this gene is a promoter.

$gene_1^0$ encodes the column number of the input matrix. It performs the inputs selection.

$gene_2^0$ For the LB-AMN model, it encodes the *order* of the basis functions covering the selected input. In case of RBFN, it will encode the number of nodes.

$gene_3^0$ For the LB-AMN model, it encodes the number of knots or size of the knots vector which will permit to create a set of basis functions used to cover an input. It is related with the number of terms by: $knots = terms + order$. In case of RBFN, it encodes the size of a vector which will contain *m* centres plus $D \cdot m$ widths. $D = 1$ will define symmetric hyper-ellipses as receptive fields (Mahalanobis simpler in section 2), and $D = 2$ will define asymmetric receptive fields composed by two hyper ellipses (Mahalanobis asymmetric).

$gene_4^0$ to $gene_{p+4}^0$ encode the knots positions for the LB-AMN models where p = bits of $gene_3^0$. In case of RBFN, it encodes *m* centres positions and widths. We combine the parts consisting of $gene_0^0$ to $gene_{p+4}^0$ in a gene of order 1. A genome may have several genes of order 1, each one encoding a complete input.

4 Model Testing

The model is first applied to a problem of classification and afterwards to a more complicated of prediction. In both cases, we use the correlation coefficient between the measured and computed output from the models as parameter to define the model accuracy.

Table 1. Models' parameters and results for the breast cancer problem.

Optimisation parameters	wbc.txt: 683 samples			
Neural network	LB-AMN	LB-AMN	RBFN	RBFN
Topology dimension	3-3	3-3	5-5	5-5
Order size	2-4	2-4	—	—
Number of terms	2-6	1-6	—	—
Number of weights	3-50	3-50	—	—
Number of nodes	—	—	10-20	10-20
Number of generations	50	50	400	400
Genotype input size	20	20	30	30
N. of knots ($gene_3^0$)	100	40	60	60
Knots. Res. ($gene_4^0$)[bits]	15	15	14	14
Individuals	40	20	40	40
Combatants	10	5	10	10
Type of Gaussian	Symm.	Asymm.	—	—
Used distance	—	—	M. Simple	M. Asym
Results				
Found inputs	1,2,7	1,5,7	8,4,1,1,6	1,4,1,3,6
Generated functions	5,3,3	5,3,3	—	—
Found nodes	—	—	11	16
Patterns classified correctly	662	662	675	674
Percentage of success	96,92%	96.92%	98.82%	98.68%

4.1 Breast Cancer Classification

The goal in this problem is to find a model that permit us to define, based on certain descriptors, whether cells extracted from the body are benign or malignant. The data [7] consists of 683 samples with 9 cell descriptors (1. Clump thickness, 2. Uniformity of cell size, 3. Uniformity of cell shape, 4. Marginal adhesion, 5. Single epithelial cell size, 6. Bare nuclei, 7. Bland chromatin, 8. Normal nucleoli, 9. Mitosis). Table 1 shows the configuration parameters of the models and the obtained results.

4.2 Prediction of Toxicity in Chemical Compounds through QSAR

It is supported as a general assumption that macroscopic properties of chemicals like toxicity and ecotoxicity strongly depend on microscopic features and on the molecules structure. [5] and co-workers also assume that such structures can be characterized by molecular descriptors and molecular quantum similarity measures (MQSMs) which can be independently used as inputs to an ANN to model the QSARs. However, when QSAR methods are used for establishing the toxicity of chemicals, the relationship: molecular descriptors - degree of toxicity can be non-linear, and the number of descriptors and MQSMs can ascend to tens or even hundreds [5]. Therefore, in order to deal with this problem, it is necessary to use models which could represent non-linear relationships between inputs and output, and techniques for selecting sub-sets of most relevant descriptors to use them as inputs to the non-linear toxicity models. Following the approach proposed by [6], the toxicity prediction problem was split in 6 sub-problems each one corresponding to a chemical subgroup. We will obtain models for each chemical subgroup and, in order to evaluate the overall performance, we will reunify the predicted outputs and calculate an overall correlation between measured and computed toxicity. The data set was provided by [11] and it consists in sets of molecular descriptors of certain chemical compounds and their degree of toxicity measured in populations of

Table 2. Correlation measured vs. computed toxicity in Daphnias and Rats

Model	Anil.	Carba.	Hal.Arom.	Hetero.	Org.	Urea	Total Corr.
Daphnia (333 samples)							
LB-AMN Sym.	0.86	0.92	0.79	0.65	0.77	0.94	0.81
LB-AMN Asy.	0.81	1.00	0.68	0.71	0.77	1.00	0.82
RBFN Mah. Simpler	1.00	0.90	0.75	0.80	0.86	0.97	0.90
RBFN Mah. Asym.	0.89	0.98	0.81	0.84	0.93	0.99	0.92
Linear Reg.	0.86	0.87	0.70	0.76	0.80	0.93	0.86
LB B-splines	0.93	0.92	0.77	0.77	0.88	0.95	0.89
Rat (354 samples)							
LB-AMN Sym.	0.48	0.74	0.50	0.67	0.71	0.69	0.78
LB-AMN Asy.	0.63	0.81	0.54	0.62	0.81	0.79	0.80
RBFN Mah. Simpler	0.72	0.88	0.74	0.76	0.84	0.87	0.91
RBFN Mah. Asym.	0.78	0.95	0.71	0.74	0.76	0.90	0.89
Linear Reg.	0.70	0.90	0.58	0.68	0.64	0.78	0.83
LB B-splines	0.65	0.87	0.65	0.70	0.69	0.85	0.84

Daphnias, Trouts, Rats, Md, Bq and the parameter Acceptable Daily Intake (ADI). The molecular descriptors or inputs to the model are 150 approximately for each chemical subgroup.

Table 2 shows the results for each chemical sub-group as well as the total correlation for Daphnias, and Rats. We have also included the results obtained using B-splines LB-AMNs [8],[10] and Multilinear Regression models with 3 inputs found using search permutation techniques.

5 Conclusions

We have proposed a prediction-classification hybrid model using GAs, LB-AMNs and RBFNs with symmetric and asymmetric Gaussians as basis functions. In most of the cases, the obtained correlations have improved in comparison to other approaches based on B-Splines LB-AMNs and Multi-Linear Regression.

Using LB-AMN, learning is local (learning about one part of the input space does not affect knowledge stored in different area), and the knowledge is stored in a transparent fashion. A big disadvantage in using LB-AMNs is that the number of weights is exponentially dependent on the number of inputs, which is also the case with the number of parameters that define a n-dimensional function if the Mahalanobis distance in its original form is used. To overcome this problem, we used two other criteria of distance, named as Mahalanobis Simpler and Asymmetric.

On the other hand, we have avoided the usage of complementary to one functions for covering the input space. Such functions can only generate input-output mappings which are as complex as the functions [9]. Thus, linear functions will lead to linear input-output mappings, sinusoidal functions will lead to sinusoidal mappings and so on, whereas using non-complementary functions, as it is case with the models that we propose, it becomes possible to get more complex input-output mappings.

An important element in the models is the binary GA. We have used it for finding subsets of relevant inputs in high-dimensional problems instead of using permutation search techniques which require more computation. They have demonstrated to be much more faster and were able to find in most of the cases the same subsets of inputs regardless the cost function. GAs were also used to optimize the LB-AMN and RBFN topologies, avoiding the usage of clustering methods, non-linear gradient descent methods or orthogonal least squares algorithms. The usage of GAs as basic optimization tool permits to propose solutions based only on two sources: genetic algorithms and radial basis functions.

References

1. A. Grauel.Evolutionary Computation: Optimisation and Decision Making Proceeedings of the 5th International Conference on Control, Automation, Robotics and Vision. part1, pp. 111-115. Mandarin Singapore. 1998.
2. Chin-Teng Lin & C.S. George Lee.Neural Fuzzy System. pp. 328-331. Prentice Hall, 1995.
3. Martin Brown & Chris Harris. Neurofuzzy Adaptive Modelling Control. Prentice Hall, 1994.
4. Holland, John H.Adaptation in natural and artificial systems : an introductory analysis with applications to biology, control, and artificial intelligence. 5th. printing. Cambridge, Mass. [u.a.] : MIT Press, 1998.
5. E. Benfenati, S. Pelagatti. Instituto di Ricerche Farmacologiche Mario Negr, Milan, Italy. D. Robert, R. Carbó-Dorca, X. Gironés. University of Girona, Campus Montilivi, Institute of Computational Chemistry, Catalonia, Spain. A. Grauel, I. Renners, L.A. Ludwig. Paderborn University, Campus Soest, Department of Mathematics. Soest, Germany.
6. A. Grauel, I. Renners, L.A. Ludwig. "B-spline Neuro-Fuzzy Systems for Toxicity Prediction". EUFIt'99, proc. 7th. European Cong. Intelligent Techniques and Soft Computing, abstract pp. 206-207. Aachen, Germany. 1999
7. O. L. Mangasarian, R. Setiono, and W.H. Wolberg: "Pattern recognition via linear programming: Theory and application to medical diagnosis", in: "Large-scale numerical optimization", Thomas F. Coleman and Yuying Li, editors, SIAM Publications, Philadelphia 1990, pp 22-30.
8. A. Grauel, I. Renners, L.A. Ludwig, E. Benfenati, S. Pelegatti, R. Carbo-dorca, X. Gironés. "Modeling Toxicity with Molecular Descriptors and Similarity Measures via B-spline Networks". Proceedings IPMU'2000 8th. International Conference on Information Processing and Management of Uncertainty in Knowledge Based Systems. pp. 1021-1026. Madrid, Spain. 2000.
9. A. Grauel, H. Mackenberg.'Mathematical analysis of the Sugeno controller leading to general design rules' Fuzzy sets and systems, 85, pp.165-175. 1995.
10. I. Renners. Internal Reports on toxicity predicition using B-splines LB-AMNs. Department of Mathematics, Campus Soest. Paderborn University, Soest - Germany, 2000.
11. Benfenati E. Instituto di Richerche Farmacologiche Mario Negri, Milan Italy. 1999-2000.

A Color Problem for Fuzzy Graph

Leonid Bershtein[1] and Alexander Bozhenuk[2]

[1] Taganrog State University of Radioengineering,
Nekrasovsky, 44, 347900 Taganrog, Russia
lsb@pbox.ttn.ru
[2] Taganrog State University of Radioengineering,
Nekrasovsky, 44, 347900 Taganrog, Russia
Bozhenuk@aport.ru

Abstract. In this paper the questions of coloring of fuzzy graphs are observed. Definitions of separation degree and fuzzy chromatic set of fuzzy graphs are presented. Properties of fuzzy chromatic set are considered. The method for finding fuzzy chromatic set is suggested and substantiated. The example of founding fuzzy chromatic set is considered.

1 Introduction

The idea of fuzzy graphs has been introduced by Rosenfeld in a paper in [1], which has also been discussed in [2]. However, using of the fuzzy graphs as the models of various systems (social, economic systems, communication networks and others) leads to difficulties. It takes place because the majority of isomorphous transformations of the graph changes its external presentation and don't change its signature. Fuzzy chromatic set is invariant concerning the isomorphous transformations of the fuzzy graph and allows make its structural analysis.

2 A Coloring of Fuzzy Graph

Let a fuzzy graph $\tilde{G} = (X, \tilde{U})$ be given where $X = \{x_j\}, j \in J = \{1,2...,n\}$ is a set of vertices and $\tilde{U} = \{< \mu_G(x, y)/(x, y) >\}$ is a fuzzy set of edges with a membership function $\mu_G : X^2 \to [0,1]$ [2]. We color each vertex $x_j \in X$ of one of k colors $(1 \le k \le n)$ and consider a subgraph $\tilde{G}_i = (X_i, \tilde{U}_i)$. Here X_i is a subset of vertices, which have color i. Then the value $\alpha_i = 1 - \bigvee_{x,y \in X_i} \mu_G(x, y)$ defines the degree of internal stable of fussy subgraph $\tilde{G}_i = (X_i, \tilde{U}_i)$ [3].

B. Reusch (Ed.): Fuzzy Days 2001, LNCS 2206, pp. 500–505, 2001.

Definition 1. A value $L = \underset{i=1,k}{\&}\alpha_i = \underset{i=1,k}{\&}(1 - \underset{x,y\in X_i}{\vee}\mu_G(x,y))$ is called a separation degree of fuzzy graph $\tilde{G}$ with k colors.

We shall consider next tasks: 1. For a given fuzzy graph $\tilde{G}$ and number of colors k we define a maximal degree of separation L; 2. For a given fuzzy graph $\tilde{G}$ and degree of separation L we define a minimal number of colors k.

The fuzzy graph $\tilde{G}$ may be colored in a arbitrary number of colors. In this case the separation degree L depends of the number of colors. For the fuzzy graph $\tilde{G}$ we relate a family of fuzzy sets $\Re = \{\tilde{A}_G\}$, $\tilde{A}_G = \{< L_{\tilde{A}}(k)/k | k = \overline{1,n}\}$, where $L_{\tilde{A}}(k)$ defines a degree of separation of fuzzy graph $\tilde{G}$ with k colors.

Definition 2. A fuzzy set $\tilde{\gamma} = \{< L_{\tilde{\gamma}}(k)/k | k = \overline{1,n}\}$ is called a fuzzy chromatic set of graph $\tilde{G}$ if the condition $\tilde{A}_G \subseteq \tilde{\gamma}$ is performed for any set $\tilde{A}_G \in \Re$, or else: $(\forall \tilde{A}_G \in \Re)(\forall k = \overline{1,n})[L_A(k) \le L_{\tilde{\gamma}}(k)]$.

Otherwise, the fuzzy chromatic set defines a maximal separation degree of fuzzy graph $\tilde{G}$ with k= 1, 2,..., n colors.

There are some properties of the fuzzy chromatic set.

Property 1. $L_{\tilde{\gamma}}(n) = 1$. Otherwise, let the fuzzy graph $\tilde{G} = (X, \tilde{U}), |X| = n$ is colored in n colors. Then the degree of separation equals one.

Property 2. $L_{\tilde{\gamma}}(1) \ge 0$. Moreover, the following proposition is true:

$$(L_{\tilde{\gamma}}(1) = 0) \leftrightarrow (\exists x, y \in X | \mu_G(x,y) = 1).$$

Otherwise, let the fuzzy graph $\tilde{G}$ is colored in one color. In the case where the graph $\tilde{G}$ has a edge with the membership function equals one, then the degree of separation equals zero.

Property 3. The following proposition is true:

$$(\forall i, j = \overline{1,n})(i > j \rightarrow L_{\tilde{\gamma}}(i) \ge L_{\tilde{\gamma}}(j)).$$

In other words, the larger number of colors is used for fuzzy graph coloring $\tilde{G}$, the larger its optimal separation degree L.

Property 4. The following proposition is true:

$$(\forall i, j, k = \overline{1,n})[(i > k > j) \& (L_{\tilde{\gamma}}(i) = L_{\tilde{\gamma}}(j)) \rightarrow (L(k) = L(i) = L(j))].$$

In other words, if separation degrees of fuzzy graph $\tilde{G}$ coloring in i and j colors are equal, then separation degree is invariable for fuzzy graph $\tilde{G}$ coloring in different number of colors, laying between i and j numbers.

Property 5. The following proposition is true:

$$(L_{\tilde{\gamma}}(i) = 1 \& i \ne n) \rightarrow (\forall k \in \overline{i+1,n})(L_{\tilde{\gamma}}(k) = 1).$$

In other words, if for some numbers of colors i ($i \neq n$), the fuzzy graph $\tilde{G}$ coloring with separation degree equal to 1 exists, then for any other numbers of color, larger than i, the fuzzy graph $\tilde{G}$ coloring with separation degree equal to 1 exists too.
Property 6. The following proposition is true:

$$(L_{\tilde{\gamma}}(i) = 0 \,\&\, i \neq 1) \rightarrow (\forall k \in \overline{1, i-1})(L_{\tilde{\gamma}}(k) = 0).$$

In other words, if fuzzy graph $\tilde{G}$ coloring with separation degree different from 0 does not exist for some number of colors i ($i \neq 1$), then for other number of colors, the fuzzy graph $\tilde{G}$ coloring with separation degrees different from zero does not exists too.

If the fuzzy chromatic set $\tilde{\gamma}(\tilde{G})$ of graph $\tilde{G}$ is known, then the considered above tasks are solved automatically. That is why we consider a task of finding the fuzzy chromatic set.

Definition 3 [3]. We call the subset $\Psi' \subseteq X$ a graph $\tilde{G}$ maximal internal stable set with the degree $\alpha(\Psi')$, if the condition $\alpha(\Psi'') < \alpha(\Psi')$ is true for any subset $\Psi' \subset \Psi''$.
The following proposition is true.

Proposition. A fuzzy set $\tilde{\gamma} = \{< L_{\tilde{\gamma}}(k) / k | k = \overline{1,n}\}$ is a fuzzy chromatic set if and only if it is not more then k maximal internal stable sets Ψ_1, Ψ_2, ..., $\Psi_{k'}$ with the degrees of internal stability α_1, α_2, ..., $\alpha_{k'}$, $(k' \leq k)$ and moreover:

1) $\min\{\alpha_1, \alpha_2, ..., \alpha_{k'}\} = L_{\tilde{\gamma}}(k)$;
2) $\bigcup_{j=\overline{1,k'}} \Psi_j = X$;
3) do not exist another family $\{\Psi'_1, \Psi'_2, ..., \Psi'_{k''}\}$, $k'' \leq k$ for which $\min\{\alpha'_1, \alpha'_2, ..., \alpha'_{k''}\} > \min\{\alpha_1, \alpha_2, ..., \alpha_{k'}\}$ and the condition 2 is true.

The observed proposition proves the following algorithm of finding fuzzy chromatic set.

1^0. To determine the family of maximal internal stable sets T={Ψ_1, Ψ_2, ..., Ψ_t} with the degrees of internal stability α_1, α_2, ..., α_t respectively.

2^0. To determine the family $\{\Psi'_1, \Psi'_2, ..., \Psi'_k\} \subseteq T$ for which the condition 2) is true and $\min\{\alpha'_1, \alpha'_2, ..., \alpha'_k\}$ takes the biggest value. To determine $L_{\tilde{\gamma}}(k) = \min\{\alpha'_1, \alpha'_2, ..., \alpha'_k\}$.

3^0. Taking into account properties 1-6, the step 2^0 is repeated for all $k = \overline{1, n-1}$.

Example 1. For fuzzy graph $\tilde{G}$ in Fig.1, find fuzzy chromatic set.

We will consider a method for determination of all maximal internal stable sets with the highest degree of internal stable. This method is a generalisation of the Maghout's method for fuzzy graphs [3].

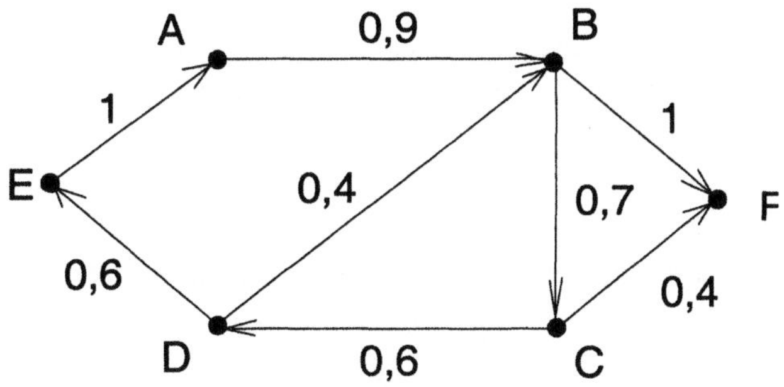

Fig. 1. Example of fuzzy graph

Let Ψ be a certain maximal internal stable set with the degree of internal stable $\alpha(\Psi)$. For arbitrary vertices $x_i, x_j \in X$, one of the following cases may be realised: a) $x_i \notin \Psi$; b) $x_j \notin \Psi$; c) $x_i \in \Psi$ and $x_j \in \Psi$. In the last case the degree $\alpha(\Psi) \leq 1-\mu_U(x_i, x_j)$.

In other words, the following expression is true:

$$(\forall x_i, x_j \in X)[x_i \notin \Psi \vee x_j \notin \Psi \vee (\alpha(\Psi) \leq 1-\mu_U(x_i, x_j))]. \tag{1}$$

We connect a Boolean variable p_i taking 1 when $x_i \in \Psi$ and 0 when $x_i \notin \Psi$, with each vertex $x_i \in X$. We associate the expression $\alpha(\Psi) \leq 1-\mu_U(x_i, x_j)$ with a fuzzy variable $\xi_{ij} = 1-\mu_U(x_i, x_j)$.

Considering the expression (1) for all possible values i and j we obtain the truth of the following expression:

$$\Phi_\Psi = \mathop{\&}_{i} \mathop{\&}_{j \neq i} (\overline{p}_i \vee \overline{p}_j \vee \xi_{ij}) = 1.$$

We open the parentheses and reduce the similar terms using the following rule:

$$\xi' \& a \vee \xi'' \& a \& b = \xi' \& a, \text{ for } \xi' \geq \xi''. \tag{2}$$

Here, $a, b \in \{0,1\}$ and $\xi', \xi'' \in [0,1]$.

Then for each disjunctive term, the totality of all vertices corresponding to the variables missing in the totality, gives a maximal internal stable set with the obtained degree of internal stable.

For the fuzzy graph $\tilde{G}$ presented in Fig.1, we obtain:

$$\Phi_\Psi = (\overline{p}_A \vee \overline{p}_B \vee 0{,}1) \& (\overline{p}_B \vee \overline{p}_C \vee 0{,}3) \& (\overline{p}_B \vee \overline{p}_F) \& (\overline{p}_C \vee \overline{p}_D \vee 0{,}4) \&$$
$$\& (\overline{p}_C \vee \overline{p}_F \vee 0{,}6) \& (\overline{p}_D \vee \overline{p}_B \vee 0{,}6) \& (\overline{p}_D \vee \overline{p}_E \vee 0{,}4) \& (\overline{p}_E \vee \overline{p}_A) = 1.$$

Completing the transformations of the fuzzy logical variables, we finally have:

$$\Phi_\Psi = \overline{p}_A\overline{p}_B\overline{p}_C\overline{p}_D \vee \overline{p}_B\overline{p}_C\overline{p}_E \vee \overline{p}_A\overline{p}_B\overline{p}_D\overline{p}_F \vee \overline{p}_B\overline{p}_D\overline{p}_E\overline{p}_F \vee \overline{p}_A\overline{p}_C\overline{p}_D\overline{p}_F \vee$$
$$\vee 0{,}6\overline{p}_A\overline{p}_B\overline{p}_D \vee 0{,}6\overline{p}_B\overline{p}_D\overline{p}_E \vee 0{,}6\overline{p}_A\overline{p}_C\overline{p}_F\overline{p}_E \vee 0{,}4\overline{p}_B\overline{p}_E \vee 0{,}4\overline{p}_A\overline{p}_B \vee$$
$$\vee 0{,}4\overline{p}_A\overline{p}_C\overline{p}_F \vee 0{,}3\overline{p}_A\overline{p}_F \vee \vee 0{,}1\overline{p}_E\overline{p}_F = 1.$$

It follows from the last expression that the considered fuzzy graph $\tilde{G}$ has 13 maximal internal stable sets: $\Psi_1 = \{E,F\}$, $\Psi_2 = \{A,D,F\}$, $\Psi_3 = \{C,E\}$, $\Psi_4 = \{A,C\}$, $\Psi_5 = \{B,E\}$ with the degrees of internal stability $\alpha_1 = \alpha_2 = \alpha_3 = \alpha_4 = \alpha_5 = 1$; $\Psi_6 = \{C,E,F\}$, $\Psi_7 = \{A,C,F\}$, $\Psi_8 = \{B,D\}$ with the degree of internal stability $\alpha_6 = \alpha_7 = \alpha_8 = 0,6$; $\Psi_9 = \{A,C,D,F\}$, $\Psi_{10} = \{C,D,E,F\}$, $\Psi_{11} = \{B,D,E\}$ with the degree of internal stability $\alpha_9 = \alpha_{10} = \alpha_{11} = 0,4$; $\Psi_{12} = \{B,C,D,E\}$ with the degree of internal stability $\alpha_{12} = 0,3$ and $\Psi_{13} = \{A,B,C,D\}$ with the degree of internal stability $\alpha_6 = 0,1$.

We construct the matrix $R = \|r_{ij}\|, i = \overline{1,n}, j = \overline{1,t}$ where the lines correspond to the vertices of graph $\tilde{G}$ and columns correspond to the maximal internal stable sets. If $x_i \in \Psi_j$, then the value r_{ij} has the value α_j, if $x_i \notin \Psi_j$, then the value r_{ij} has the value 0. For the fuzzy graph $\tilde{G}$ presented in Fig.1, we obtain:

	Ψ_1	Ψ_2	Ψ_3	Ψ_4	Ψ_5	Ψ_6	Ψ_7	Ψ_8	Ψ_9	Ψ_{10}	Ψ_{11}	Ψ_{12}	Ψ_{13}
A	0	1	0	1	0	0	0,6	0	0,4	0	0	0	0,1
B	0	0	0	0	1	0	0	0,6	0	0	0,4	0,3	0,1
C	0	0	1	1	0	0,6	0,6	0	0,4	0,4	0	0,3	0,1
D	0	1	0	0	0	0	0	0,6	0,4	0,4	0,4	0,3	0,1
E	1	0	1	0	1	0,6	0	0	0	0,4	0,4	0,3	0
F	1	1	0	0	0	0,6	0,6	0	0,4	0,4	0	0,3	0

The task of finding fuzzy chromatic set $\tilde{\gamma}$ is the task of finding the covering of all lines by k columns ($k = \overline{1, n-1}$) with the maximum of the volume $\min\{\alpha_{i_1}, \alpha_{i_2}, ..., \alpha_{i_k}\}$.

If k=2, then the covering defines the columns Ψ_2 and Ψ_{12} with the degree L(2)=1&0,3 = 0,3. If k=3, then the covering defines the columns Ψ_2, Ψ_3 and Ψ_5 with the degree L(1)=1&1&1 = 1.

So, for the fuzzy graph $\tilde{G}$ presented in Fig.1, the fuzzy chromatic set is $\tilde{\gamma}(\tilde{G}) = \{<0/1>, <0,3/2>, <1/3>, <1,4>, <1,5>, <1,6>\}$. Otherwise, the fuzzy graph $\tilde{G}$ may be colored by one color with the degree of separation 0; by 2 colors with the degree of separation 0,3 (vertices A, D, F - first color, vertices B, C, E - second color); by 3 colors with the degree of separation 1 (vertices A, D, F - first color, vertices C, E - second color, vertex B - third color).

3 Conclusion

The method of the finding fuzzy chromatic set of the fuzzy graph was considered. The definitions of separation degree and fuzzy chromatic set allow estimate the any fuzzy graph with the position of existence of fuzzy invariants. It is necessary to mark that the suggested method is the method of ordered full selection, because these tasks are reduced to the task of covering, i.e. these tasks are NP-complete tasks. However, this method is effective for the graphs, which have not homogeneous structure and not large dimensionality. Therefore this method can be used for the solving of practical problems, as the structurization of the fuzzy linguistic and social graphs.

References

1. Zadeh, L.: Fuzzy Sets and Their Application to Cognitive and Decision. Academic Press, New York (1975)
2. Kaufmann, A.: Introduction a la Theorie des Sous-Ensemles Flous. Masson, Paris (1977)
3. Bershtein, L., Bozhenuk, A.: Determination of Fuzzy Internal Stable, External Stable Sets, and Kernels of Fuzzy Oriented Graphs. Journal of Computer and Systems Sciences International. **38(1)** (1999) 153-157

On the Uniqueness of Interpretations for Fuzzy IF-THEN Rule Bases*

Helmut Thiele

University of Dortmund, Department of Computer Science I,
D-44221 Dortmund, Germany
Phone: +49 231 755 6152
Fax: +49 231 755 6555
E-Mail: thiele@ls1.cs.uni-dortmund.de
WWW: http://ls1-www.cs.uni-dortmund.de

Abstract. A fuzzy IF-THEN rule base is considered as a system of equations for a functional operator. We study several conditions so that such a system of functional equations has at most one solution. Several versions of compactness inspired by the compactness of consequence operators used in the theory of formal systems from mathematical logic play a fundamental role in formulating these conditions and in proving corresponding theorems.
Keywords: IF-THEN rule bases, uniqueness of interpretations, versions of compactness.

1 Fundamental Definitions

For definiteness we recall the following fundamental definitions.

Let U be a non-empty crisp set. By $\subseteq$, $\cup$, and $\cap$ we denote the subset relation, the union, and the intersection of crisp sets, respectively. The power set $\mathbb{P}U$ of U is defined by $\mathbb{P}U =_{def} \{X | X \subseteq U\}$. For $\mathcal{A} \subseteq \mathbb{P}U$, we define $\bigcup \mathcal{A} =_{def} \{a | \exists A(A \in \mathcal{A} \land a \in A)$. The empty set is denoted by $\emptyset$.

By $\mathbb{R}$ and $\langle 0, 1\rangle$ we denote the set of all real numbers and the set of all $r \in \mathbb{R}$ with $0 \leq r \leq 1$, respectively. Fuzzy sets on U are mappings $F : U \to \langle 0, 1\rangle$. The fuzzy power set $\mathbb{FP}U$ of U is defined by $\mathbb{FP}U =_{def} \{F | F : U \to \langle 0, 1\rangle\}$. For arbitrary fuzzy sets F, G on U and $\mathcal{F} \subseteq \mathbb{FP}U$ we define for $x \in U$

$$F \sqsubseteq G =_{def} \forall x(x \in U \to F(x) \leq G(x)),$$
$$(F \sqcup G)(x) =_{def} \max(F(x), G(x)),$$
$$(F \sqcap G)(x) =_{def} \min(F(x), G(x)),$$
$$\left(\bigsqcup \mathcal{F}\right)(x) =_{def} \mathrm{Sup}\{F(x) | F \in \mathcal{F}\}.$$

The support $\mathrm{SUPP}(F)$ of $F \in \mathbb{FP}U$ is defined as $\mathrm{SUPP}(F) =_{def} \{x | x \in U \land F(x) > 0\}$. F is said to be finite and a singleton if and only if $\mathrm{SUPP}(F)$ is finite and a singleton, respectively.

* This research was supported by the Deutsche Forschungsgemeinschaft as part of the Collaborative Research Center "Computational Intelligence" (SFB 531)

B. Reusch (Ed.): Fuzzy Days 2001, LNCS 2206, pp. 506–525, 2001.

2 Fuzzy IF-THEN Rule Bases and Their Interpretation using Concepts of Functional Analysis

Let n be an integer with $n \geq 1$. Furthermore let $F_1, \ldots, F_n, G_1, \ldots, G_n$ be fuzzy sets on U. A scheme of the form

$$\mathrm{RB} =_{def} \begin{array}{c} \text{IF } F_1 \text{ THEN } G_1 \\ \vdots \\ \text{IF } F_n \text{ THEN } G_n \end{array}$$

is said to be a fuzzy IF-THEN rule base in normal form.

In [26] we have developed an algorithm which transforms a fuzzy IF-THEN rule base used in practice into a normal form.

Let Ψ be a functional operator with $\Psi : \mathbb{FP}U \to \mathbb{FP}U$.

Definition 21 *Ψ is said to be an interpretation of the fuzzy IF-THEN rule base* $\mathrm{RB} =_{def}$

$$\begin{aligned} \Psi(F_1) &= G_1 \\ &\vdots \\ \Psi(F_n) &= G_n \end{aligned}$$

Now, for a given rule base RB we are faced with two problems:

- the **existence problem** of an interpretation for RB and
- the **uniqueness problem** of interpretations for RB.

At first glance we can state that the **existence problem** is trivially solvable if for RB for every $i, j \in \{1, \ldots, n\}$ the equation $F_i = F_j$ implies the equation $G_i = G_j$. In this case we define for $F \in \mathbb{FP}U$

$$\Psi(F) =_{def} \begin{cases} G_i & \text{if } F = F_i \\ G_0 & \text{if } F \notin \{F_1, \ldots, F_n\} \end{cases}$$

where G_0 is a fixed fuzzy set from $\mathbb{FP}U$.

But such interpretation is useless in practice, in general. For practical problems we need interpretations Ψ such that $\Psi(F)$ is "near" to G_i if F is "near" to F_i. Obviously, this condition is a intuitive formulation of the continuity of Ψ where a suitable topology is to be defined on $\mathbb{FP}U$.

Since publication of the seminal paper [31] by L.A.ZADEH, numerous methods for constructing interpretations have been developed and published.

As examples we only mention the algorithms FATI and FITA based on the COMPOSITIONAL RULE OF INFERENCE [2, 16, 19, 20, 24, 27, 30], the construction of interpretations by solving systems of relational equations [4–8], and several methods based on the so-called SIMILARITY BASED REASONING [17, 18, 23, 29, 32]. Further methods one can find in [9–15], for instance.

For definiteness here we are going to describe the algorithms FATI and FITA, but compared with [30] in a generalized form as we have defined then in [16, 18, 20, 24].

For making precise FATI we define

Definition 22 FAIB $=_{def}$ $[\pi_1, \dots \pi_n, \alpha, \kappa, Q]$ *is said to be a* FATI-*like interpreting base*

$=_{def}$ *1.* $\pi_1, \dots \pi_n : \langle 0,1\rangle \times \langle 0,1\rangle \to \langle 0,1\rangle$
2. $\alpha : \langle 0,1\rangle^n \to \langle 0,1\rangle$
3. $\kappa : \langle 0,1\rangle \times \langle 0,1\rangle \to \langle 0,1\rangle$
4. $Q : \mathbb{P}\langle 0,1\rangle \to \langle 0,1\rangle$

For a given fuzzy IF-THEN rule base RB (see above) by means of a given FATI-like interpreting base FAIB and the algorithm FATI, i. e. First Aggregation Then Inference, we construct an functional operator $\Psi^{RB}_{\text{FATI}} : \mathbb{FP}U \to \mathbb{FP}U$ by carring out the following three steps

Step 1 (FATI). Interpreting the rules of RB by constructing binary fuzzy relations $S_1, \dots, S_n$ by means of $\pi_1, \dots, \pi_n$, i. e. for $i \in \{1, \dots, n\}$ and $x, y \in U$ we define

$$S_i(x,y) =_{def} \pi_i(F_i(x), G_i(y))$$

Step 2 (FATI). Aggregation the relations defined by means of α, i. e. for $x, y \in U$ we define

$$S(x,y) =_{def} \alpha(S_1(x,y), \dots, S_n(x,y))$$

Step 3 (FATI). Inference of the functional operator Ψ^{RB}_{FAIB} from S by means of κ and Q based on the COMPOSITIONAL RULE OF INFERENCE, i. e. for $F \in \mathbb{FP}U$ we define

$$\Psi^{RB}_{\text{FAIB}}(F)(y) =_{def} Q\{\kappa(F(x), S(x,y)) | x \in U\}$$

For making precise FITA we define

Definition 23 FIIB $=_{def}$ $[\Psi_1, \dots, \Psi_n, \kappa_1, \dots, \kappa_n, Q_1, \dots, Q_n, \beta]$ *is said to be a* FITA-*like interpreting base*

$=_{def}$ *1.* $\pi_1, \dots \pi_n : \langle 0,1\rangle \times \langle 0,1\rangle \to \langle 0,1\rangle$
2. $\kappa_1, \dots \kappa_n : \langle 0,1\rangle \times \langle 0,1\rangle \to \langle 0,1\rangle$
3. $Q_1, \dots Q_n : \mathbb{P}\langle 0,1\rangle \to \langle 0,1\rangle$
4. $\beta : \langle 0,1\rangle^n \to \langle 0,1\rangle$

Again using the given fuzzy IF-THEN rule base RB by means of a given FITA-like interpreting base FIIB and the algorithm FITA, i. e. First Inference Then Aggregation, we construct a functional operator $\Psi^{RB}_{\text{FIIB}} : \mathbb{FP}U \to \mathbb{FP}U$ by carring out the following three steps

Step 1 (FITA). As step 1 from FATI, i. e. for $i \in \{1, \dots, n\}$ and $x, y \in U$

$$S_i(x,y) =_{def} \pi_i(F_i(x), G_i(y))$$

Step 2 (FITA). For each $i \in \{1, \dots, n\}$ we infer the functional operator Ψ_i from S_i by means of κ_i and Q_i based on the COMPOSITIONAL RULE OF INFERENCE, i. e. for $F \in \mathbb{FP}U$ and $y \in U$ we define

$$\Psi_i(F)(y) =_{def} Q_i\{\kappa_i(F(x), S_i(x,y)) | x \in U\}$$

Step 3 (FITA). Using the defined operators $\Psi_1, \dots, \Psi_n$ by means of the "aggregation" function β for $F \in \mathbb{FP}U$ and $y \in \mathbb{FP}U$ we define

$$\Psi^{RB}_{\text{FIIB}}(F)(y) =_{def} \beta(\Psi_1(F)(y), \dots, \Psi_n(F)(y))$$

Now, let Ψ be one of the operators Ψ^{RB}_{FAIB} and Ψ^{RB}_{FIIB}. The we are faced with the following problems.

1. Under which condition of RB and FAIB and FIIB, respectively, is Ψ an interpretation of RB, i. e. the equations

$$\Psi(F_1) = G_1$$
$$\vdots$$
$$\Psi(F_n) = G_n$$

hold? In [2, 16, 18, 20] one can find several theorems concerning these problems.
2. Analogously we ask whether and where Ψ is continous with respect to a given topology in $\mathbb{FP}U$.

With respect to the **uniqueness problem** one can hardly meet results in the literature.

Obviously, we can trivially restrict the variety of all interpretations for a given rule base RB by adding "new" rules.

Furthermore, assuming the continuity of Ψ with respect to a given topology on $\mathbb{FP}U$ implies a strong restriction of the set of all interpretations for RB.

"Axiomatic" approaches in literature, for instance in [1, 3], can be considered as conditions for restricting the set of all interpretations for RB.

However, we have to state that up to now there are no systematic investigations of the uniqueness problem.

The following concepts and investigations are inspired by Mathematical Logic, in particular, by the theory of consequence operators in formal systems.

3 Some Ideas from Mathematical Logic

For better understanding of the concepts and investigations presented in section 4 we shall recall some well-known notions and results from Mathematical Logic and shall develop some generalizations of them.

Let U be the set of formulas of a given formal logical system including a binary connector $\rightsquigarrow$ for generating formulas, called implication.

For a mapping $CONS : \mathbb{P}U \to \mathbb{P}U$, called consequence operator of the given formal system, we define

Definition 31

1. $CONS$ is said to be monotone on $\mathbb{P}U$
$=_{def} \forall X \forall Y (X \subseteq Y \subseteq U \to CONS(X) \subseteq CONS(Y))$.

2. *$CONS$ is said to be compact on $\mathbb{P}U$*
$=_{def} \forall y \forall X(X \subseteq U \wedge y \in CONS(X) \rightarrow \exists X_{fin}(X_{fin} \subseteq X \wedge X_{fin}$ *is finite* $\wedge y \in CONS(X_{fin})))$.
3. *$CONS$ fulfils the deduction property on $\mathbb{P}U$ with respect to* $\rightsquigarrow$
$=_{def} \forall x \forall y \forall X(x \in U \wedge X \subseteq U \wedge y \in CONS(X \cup \{x\}) \rightarrow x \rightsquigarrow y \in CONS(X))$.
4. *$CONS$ fulfils the inference property on $\mathbb{P}U$ with respect to* $\rightsquigarrow$
$=_{def} \forall x \forall y \forall X(x \in U \wedge X \subseteq U \wedge x \rightsquigarrow y \in CONS(X) \rightarrow y \in CONS(X \cup \{x\}))$.

Now, we recall the following well-known

Theorem 31 *If*

1. $CONS, CONS' : \mathbb{P}U \rightarrow \mathbb{P}U$
2. *$CONS$ and $CONS'$ are monotone on $\mathbb{P}U$*
3. *$CONS$ and $CONS'$ are compact on $\mathbb{P}U$*
4. *$CONS$ and $CONS'$ fulfil the deduction property on $\mathbb{P}U$ with respect to* $\rightsquigarrow$
5. *$CONS$ and $CONS'$ fulfil the inference property on $\mathbb{P}U$ with respect to* $\rightsquigarrow$
6. $CONS(\emptyset) = CONS'(\emptyset)$

then $\forall X(X \subseteq U \rightarrow CONS(X) = CONS'(X))$.

With respect to the investigations in section 4 we are going to generalize the theorem above to several versions.

First, we replace the empty set $\emptyset$ by a fixed system $\mathcal{A}$ of sets $A \subseteq U$ and replace the assumption

$$CONS(\emptyset) = CONS'(\emptyset)$$

by

$$\forall(A \in \mathcal{A} \rightarrow CONS(A) = CONS'(A)).$$

Then we can prove the following

Theorem 32 *If*

1. $CONS, CONS' : \mathbb{P}U \rightarrow \mathbb{P}U$
2. $\mathcal{A} \subseteq \mathbb{P}U$ *and* $X \subseteq U$
3. *$CONS$ and $CONS'$ are monotone on $\mathbb{P}U$*
4. *$CONS$ and $CONS'$ are compact on $\mathbb{P}U$*
5. *$CONS$ and $CONS'$ fulfil the deduction property on $\mathbb{P}U$ with respect to* $\rightsquigarrow$
6. *$CONS$ and $CONS'$ fulfil the inference property on $\mathbb{P}U$ with respect to* $\rightsquigarrow$
7. $\forall A(A \in \mathcal{A} \rightarrow CONS(A) = CONS'(A))$
8. $\bigcup \mathcal{A} \subseteq X$

then $CONS(X) = CONS'(X)$.

One could have the opinion that the assumption 8, i. e. $\bigcup \mathcal{A} \subseteq X$, is to strong. Hence, one should try to replace it by $\bigcup \mathcal{A} \subseteq CONS(X)$ and $\bigcup \mathcal{A} \subseteq CONS'(X)$. But in this case we have to replace the assumption 3, by the assumption that $CONS$ and $CONS'$ are closure operators defined as follows.

Definition 32

1. *$CONS$ is said to be embedding (or reflexive) on* $\mathbb{P}U$
   $=_{def} \forall X(X \subseteq U \rightarrow X \subseteq CONS(X))$.
2. *$CONS$ is said to be closed on* $\mathbb{P}U$
   $=_{def} \forall X(X \subseteq U \rightarrow CONS(CONS(X)) \subseteq CONS(X))$.
3. *$CONS$ is said to be a closure operator on* $\mathbb{P}U$
   $=_{def}$ *$CONS$ is monotone, embedding and closed on* $\mathbb{P}U$.

Theorem 33 *If*

1. $CONS, CONS' : \mathbb{P}U \rightarrow \mathbb{P}U$
2. $\mathcal{A} \subseteq \mathbb{P}U$ *and* $X \subseteq U$
3. $CONS$ *and* $CONS'$ *are closure operators on* $\mathbb{P}U$
4. $CONS$ *and* $CONS'$ *are compact on* $\mathbb{P}U$
5. $CONS$ *and* $CONS'$ *fulfil the deduction property on* $\mathbb{P}U$ *with respect to* $\rightsquigarrow$
6. $CONS$ *and* $CONS'$ *fulfil the inference property on* $\mathbb{P}U$ *with respect to* $\rightsquigarrow$
7. $\forall A(A \in \mathcal{A} \rightarrow CONS(A) = CONS'(A)$
8. $\forall A(A \in \mathcal{A} \rightarrow A \subseteq CONS(X) \wedge A \subseteq CONS'(X))$

then $CONS(X) = CONS'(X)$.

Now, we ask the question how can we formulate (modify) the theorems 31 – 33 if there does not exist a binary connector $\rightsquigarrow$ in the formal system given.

Therefore we formulate the following modification of the compactness where $\mathcal{A} \subseteq \mathbb{P}U$

Definition 33

1. *$CONS$ is said to be $\mathcal{A}$-compact on* $\mathbb{P}U$
   $=_{def} \forall y \forall X(X \subseteq U \wedge \bigcup \mathcal{A} \subseteq X \wedge y \in CONS(X) \rightarrow \exists A \exists Z(A \in \mathcal{A} \wedge Z \subseteq A \wedge y \in CONS(Z)))$.
2. *$CONS$ is said to be strongly $\mathcal{A}$-compact on* $\mathbb{P}U$
   $=_{def} \forall y \forall X(X \subseteq U \wedge \bigcup \mathcal{A} \subseteq CONS(X) \wedge y \in CONS(X) \rightarrow \exists A \exists Z(A \in \mathcal{A} \wedge Z \subseteq A \wedge y \in CONS(Z)))$.

Obviously, if $CONS$ is reflexive on $\mathbb{P}U$ then the $\mathcal{A}$-compactness and the strong $\mathcal{A}$-compactness are equivalent.

Theorem 34 *If*

1. $CONS, CONS' : \mathbb{P}U \rightarrow \mathbb{P}U$
2. $\mathcal{A} \subseteq \mathbb{P}U$ *and* $X \subseteq U$
3. $CONS$ *and* $CONS'$ *are monotone on* $\mathbb{P}U$
4. $CONS$ *and* $CONS'$ *are $\mathcal{A}$-compact on* $\mathbb{P}U$
5. $\forall A(A \in \mathcal{A} \rightarrow CONS(A) = CONS'(A))$
6. $\bigcup \mathcal{A} \subseteq X$

then $CONS(X) = CONS'(X)$.

Theorem 35 *If*

1. $CONS, CONS' : \mathbb{P}U \to \mathbb{P}U$
2. $\mathcal{A} \subseteq \mathbb{P}U$ *and* $X \subseteq U$
3. $CONS$ *and* $CONS'$ *are closure operators on* $\mathbb{P}U$
4. $CONS$ *and* $CONS'$ *are* $\mathcal{A}$*-compact on* $\mathbb{P}U$
5. $\forall A(A \in \mathcal{A} \to CONS(A) = CONS'(A))$
6. $\bigcup \mathcal{A} \subseteq CONS(X) \wedge \bigcup \mathcal{A} \subseteq CONS'(X)$

then $CONS(X) = CONS'(X)$.

The concepts of $\mathcal{A}$-compactness can be replaced by the following notion of inductiveness where $CONS, CONS' : \mathbb{P}U \to \mathbb{P}U$.

Definition 34 *The set* $\{CONS, CONS'\}$ *fulfils the induction property on* $\mathbb{P}U =_{def}$ $\forall X \forall Z(X, Z \subseteq U \wedge Z$ *is finite* $\wedge\, CONS(X) = CONS'(X) \to CONS(X \cup Z) = CONS'(X \cup Z))$.

Theorem 36 *If*

1. $CONS, CONS' : \mathbb{P}U \to \mathbb{P}U$
2. $\mathcal{A} \subseteq \mathbb{P}U$ *and* $X \subseteq U$
3. $CONS$ *and* $CONS'$ *are monotone on* $\mathbb{P}U$
4. $CONS$ *and* $CONS'$ *are* $\mathcal{A}$*-compact on* $\mathbb{P}U$
5. *The set* $\{CONS, CONS'$ *fulfils the induction property on* $\mathbb{P}U$.
6. $\forall A(A \in \mathcal{A} \to CONS(A) = CONS'(A))$
7. $\bigcup \mathcal{A} \subseteq X$

then $CONS(X) = CONS'(X)$.

Theorem 37 *If*

1. $CONS, CONS' : \mathbb{P}U \to \mathbb{P}U$
2. $\mathcal{A} \subseteq \mathbb{P}U$ *and* $X \subseteq U$
3. $CONS$ *and* $CONS'$ *are closure operators on* $\mathbb{P}U$
4. $CONS$ *and* $CONS'$ *are compact on* $\mathbb{P}U$
5. *The set* $\{CONS, CONS'\}$ *fulfils the induction property on* $\mathbb{P}U$
6. $\forall A(A \in \mathcal{A} \to CONS(A) = CONS'(A)$
7. $\bigcup \mathcal{A} \subseteq CONS(X) \wedge \bigcup \mathcal{A} \subseteq CONS'(X))$

then $CONS(X) = CONS'(X)$.

4 Fundamental Concepts and Uniqueness Results for Interpretations of Fuzzy IF-THEN Rule Bases

Assume $\Psi : \mathbb{FP}U \rightarrow \mathbb{FP}U$.

Definition 41

1. *Ψ is said to be monotone on $\mathbb{FP}U$*
   $=_{def} \forall F \forall G(F, G \in \mathbb{FP}U \wedge F \sqsubseteq G \rightarrow \Psi(F) \sqsubseteq \Psi(G))$
2. *Ψ is said to be reflexive (or embedding) on $\mathbb{FP}U$*
   $=_{def} \forall F(F \in \mathbb{FP}U \rightarrow F \sqsubseteq \Psi(F))$
3. *Ψ is said to be closed on $\mathbb{FP}U$*
   $=_{def} \forall F(F \in \mathbb{FP}U \rightarrow \Psi(\Psi(F)) \sqsubseteq \Psi(F))$
4. *Ψ is said to be closure operator on $\mathbb{FP}U$*
   $=_{def}$ *Ψ is monotone, reflexive, and closed on $\mathbb{FP}U$.*

In order to prove the following two theorems 41 and 42 we need the two versions of compactness formulated in definition 42. Assume $\mathcal{F} \subseteq \mathbb{FP}U$.

Definition 42

1. *Ψ is said to be $\mathcal{F}$-compact on $\mathbb{FP}U$*
   $=_{def} \forall x \forall G(x \in U \wedge G \in \mathbb{FP}U \wedge \bigsqcup \mathcal{F} \sqsubseteq G \rightarrow \exists F \exists H(F \in \mathcal{F} \wedge H \sqsubseteq F \wedge \Psi(G)(x) \leq \Psi(H)(x)))$
2. *Ψ is said to be modified $\mathcal{F}$-compact on $\mathbb{FP}U$*
   $=_{def} \forall x \forall G(x \in U \wedge G \in \mathbb{FP}U \wedge \bigsqcup \mathcal{F} \sqsubseteq \Psi(G) \rightarrow \exists F \exists H(F \in \mathcal{F} \wedge H \sqsubseteq F \wedge \Psi(G)(x) \leq \Psi(H)(x)))$

If Ψ is reflexive then the $\mathcal{F}$-compactness and the strong $\mathcal{F}$-compactness are equivalent.
Now we can prove the following two theorems

Theorem 41 *If*

1. *$\Psi, \Psi' : \mathbb{FP}U \rightarrow \mathbb{FP}U$*
2. *$\mathcal{F} \subseteq \mathbb{FP}U$ and $G \in \mathbb{FP}U$*
3. *Ψ and Ψ' are monotone on $\mathbb{FP}U$*
4. *Ψ and Ψ' are $\mathcal{F}$-compact on $\mathbb{FP}U$*
5. *$\forall F(F \in \mathcal{F} \rightarrow \Psi(F) = \Psi'(F))$*
6. *$\bigsqcup F \sqsubseteq G$*

then $\Psi(G) = \Psi'(G)$.

Theorem 42 *If*

1. *$\Psi, \Psi' : \mathbb{FP}U \rightarrow \mathbb{FP}U$*
2. *$\mathcal{F} \subseteq \mathbb{FP}U$ and $G \in \mathbb{FP}U$*
3. *Ψ and Ψ' are closure operators on $\mathbb{FP}U$*
4. *Ψ and Ψ' are $\mathcal{F}$-compact on $\mathbb{FP}U$*
5. *$\forall F(F \in \mathcal{F} \rightarrow \Psi(F) = \Psi'(F))$*
6. *$\bigsqcup F \sqsubseteq \Psi(G)$ and $\bigsqcup F \sqsubseteq \Psi'(G)$*

then $\Psi(G) = \Psi'(G)$.

If $\mathcal{F}$ is the set of all premisses of a given fuzzy IF-THEN rule base RB then condition 5 of the theorems above follows if Ψ and $\Psi^{'}$ are interpretations of RB. In order to prove the equation $\Psi(G) = \Psi^{'}(G)$ in theorem 41 we have the strong assumption $\bigsqcup F \sqsubseteq G$ which strongly restricts the range of sets $G \in \mathbb{FP}U$ fulfilling $\Psi(G) = \Psi^{'}(G)$. On the other hand we need only the monotonicity of Ψ and $\Psi^{'}$.

In theorem 42 the assumption $\bigsqcup F \sqsubseteq G$ is replaced by the weaker conditions (if Ψ and $\Psi^{'}$ are reflexive) $\bigsqcup F \sqsubseteq \Psi(G)$ and $\bigsqcup F \sqsubseteq \Psi^{'}(G)$, so that there are more $G \in \mathbb{FP}U$ fulfilling the equation $\Psi(G) = \Psi^{'}(G)$, in general. But for this advantage we have to pay by the stronger assumptions that Ψ and $\Psi^{'}$ are closure operators and strongly $\mathcal{F}$-compact.

Thus, in applications one has to check and to decide which version is more convenient.

Now we are going to develop concepts and theorems which are oriented towards the "classical" definition 31 and theorem 31.

Definition 43

1. *Ψ is said to be finitary $\bigsqcup \mathcal{F}$-compact on $\mathbb{FP}U$*
   $=_{def} \forall x \forall G(x \in U \wedge G \in \mathbb{FP}U \wedge \bigsqcup \mathcal{F} \sqsubseteq G \rightarrow \exists H(H \textit{ is finite } \wedge H \sqsubseteq G \wedge \Psi(G)(x) \leq \Psi(H)(x)))$
2. *Ψ is said to be strongly finitary $\bigsqcup \mathcal{F}$-compact on $\mathbb{FP}U$*
   $=_{def} \forall x \forall G(x \in U \wedge G \in \mathbb{FP}U \wedge \bigsqcup \mathcal{F} \sqsubseteq \Psi(G) \rightarrow \exists H(H \textit{ is finite } \wedge H \sqsubseteq G \wedge \Psi(G)(x) \leq \Psi(H)(x)))$

Obviously, if Ψ is reflexive on $\mathbb{FP}U$ then the strong finitary $\bigsqcup \mathcal{F}$-compactness implies the finitary $\bigsqcup \mathcal{F}$-compactness.

For formulating a suitable deduction and inference property we fix a mapping

$$\Rightarrow : \mathbb{FP}U \rightarrow \mathbb{FP}U$$

and put

Definition 44

1. *Ψ fulfils the $\Rightarrow$-deduction property on $\mathbb{FP}U$*
   $=_{def} \forall G \forall P \forall C(G, P, C \in \mathbb{FP}U \wedge C \sqsubseteq \Psi(G \sqcup P) \rightarrow P \Rightarrow C \sqsubseteq \Psi(G))$.
2. *Ψ fulfils the $\Rightarrow$-inference property on $\mathbb{FP}U$*
   $=_{def} \forall G \forall P \forall C(G, P, C \in \mathbb{FP}U \wedge P \Rightarrow C \sqsubseteq \Psi(G) \rightarrow C \sqsubseteq \Psi(G \sqcup P))$
3. *The properties defined above are called "weak" if the fuzzy sets $P, C \in \mathbb{FP}U$ considered are finite.*

Theorem 43 *If*

1. $\Psi, \Psi^{'} : \mathbb{FP}U \rightarrow \mathbb{FP}U$
2. *$\mathcal{F} \subseteq \mathbb{FP}U$ and $G \in \mathbb{FP}U$*
3. *Ψ and $\Psi^{'}$ are monotone on $\mathbb{FP}U$*
4. *Ψ and $\Psi^{'}$ are finitary $\bigsqcup \mathcal{F}$-compact on $\mathbb{FP}U$*
5. *Ψ and $\Psi^{'}$ fulfil the $\Rightarrow$-deduction property on $\mathbb{FP}U$*
6. *Ψ and $\Psi^{'}$ fulfil the $\Rightarrow$-inference property on $\mathbb{FP}U$*
7. $\forall F(F \in \mathcal{F} \rightarrow \Psi(F) = \Psi(F^{'}))$

8. $\bigsqcup\mathcal{F} \sqsubseteq G$

then $\Psi(G) = \Psi'(G)$.

Theorem 44 *If*

1. $\Psi, \Psi' : \mathbb{FP}U \to \mathbb{FP}U$
2. $\mathcal{F} \subseteq \mathbb{FP}U$ *and* $G \in \mathbb{FP}U$
3. Ψ *and* Ψ' *are closure operators on* $\mathbb{FP}U$
4. Ψ *and* Ψ' *are finitary* $\bigsqcup\mathcal{F}$*-compact on* $\mathbb{FP}U$
5. Ψ *and* Ψ' *fulfil the weak* $\Rightarrow$*-deduction property on* $\mathbb{FP}U$
6. Ψ *and* Ψ' *fulfil the weak* $\Rightarrow$*-inference property on* $\mathbb{FP}U$
7. $\forall F(F \in \mathcal{F} \to \Psi(F) = \Psi'(F))$
8. $\bigsqcup\mathcal{F} \sqsubseteq \Psi(G)$ *and* $\bigsqcup\mathcal{F} \sqsubseteq \Psi'(G)$

then $\Psi(G) = \Psi'(G)$.

Now, we are going to replace the deduction and inference property defined by definition 44 by the so-called induction property as follows.

Assume $\Psi, \Psi' : \mathbb{FP}U \to \mathbb{FP}U$

Definition 45 *The set* $\{\Psi, \Psi'\}$ *fulfils the* $\mathcal{F}$*-induction property on* $\mathbb{FP}U$
$=_{def} \forall F \forall H(F \in \mathcal{F} \wedge H \in \mathbb{FP}U \wedge H \text{ is finite } \wedge \Psi(F) = \Psi'(F)$
$\to \Psi(F \sqcup H) = \Psi'(F \sqcup H))$.

Theorem 45 *If*

1. $\Psi, \Psi' : \mathbb{FP}U \to \mathbb{FP}U$
2. $\mathcal{F} \subseteq \mathbb{FP}U$ *and* $G \in \mathbb{FP}U$
3. Ψ *and* Ψ' *are monotone on* $\mathbb{FP}U$
4. Ψ *and* Ψ' *are finitary* $\bigsqcup\mathcal{F}$*-compact on* $\mathbb{FP}U$
5. $\{\Psi, \Psi'\}$ *fulfil the* $\mathcal{F}$*-induction property on* $\mathbb{FP}U$
6. $\forall F(F \in \mathcal{F} \to \Psi(F) = \Psi'(F'))$
7. $\bigsqcup\mathcal{F} \sqsubseteq G$

then $\Psi(G) = \Psi'(G)$.

Theorem 46 *If*

1. $\Psi, \Psi' : \mathbb{FP}U \to \mathbb{FP}U$
2. $\mathcal{F} \subseteq \mathbb{FP}U$ *and* $G \in \mathbb{FP}U$
3. Ψ *and* Ψ' *are closure operators on* $\mathbb{FP}U$
4. Ψ *and* Ψ' *are finitary* $\bigsqcup\mathcal{F}$*-compact on* $\mathbb{FP}U$
5. *The set* $\{\Psi, \Psi'\}$ *fulfil the* $\mathcal{F}$*-induction property on* $\mathbb{FP}U$
6. $\forall F(F \in \mathcal{F} \to \Psi(F) = \Psi'(F))$
7. $\bigsqcup\mathcal{F} \sqsubseteq \Psi(G)$ *and* $\bigsqcup\mathcal{F} \sqsubseteq \Psi'(G)$

then $\Psi(G) = \Psi'(G)$.

If the set $\mathcal{F} \subseteq \mathbb{FP}U$ is given by the set of all premisses of a fuzzy IF-THEN rule base then for applications of such base very often it is assumed that for every $x \in U$ at least one rule is activated to certain extend, i. e. that there exists a rule with a premise F such that $F(x) \geq d$ where $d \in \langle 0, 1\rangle$ and $d > 0$. This assumption means, with other words, that $\mathcal{F}$ is a so-called d-covering of U (see [25]).

On the basis of this assumption we modify the compactness concept defined in definition 43 by replacing the condition "H is finite and $H \sqsubseteq G$" by "$\forall x(x \in U \rightarrow H(x) \leq d)$". Obviously, the last condition means that for every $x \in U$ the membership degree $H(x)$ of x with respect to H is not to high, i. e. $H(x) \leq d$, which can be interpreted as a compensation for the finiteness of H.

Furthermore, we add the assumption that $\mathcal{F}$ is finite and the assumption that Ψ and $\Psi^{'}$ are $\bigsqcup$-$\sqsubseteq$-topological (see [25]) as defined below.

We underline that the definitions and theorems below have no analogies in section 3.

Definition 46

1. *Ψ is said to be d-$\bigsqcup\mathcal{F}$-compact on $\mathbb{FP}U$*
   $=_{def} \forall x \forall G(x \in U \wedge G \in \mathbb{FP}U \wedge \bigsqcup\mathcal{F} \sqsubseteq G \rightarrow \exists H(H \in \mathbb{FP}U \wedge \forall y(y \in U \rightarrow H(y) \leq d) \wedge \Psi(G)(x) \leq \Psi(H)(x)))$.
2. *Ψ is said to be strongly d-$\bigsqcup\mathcal{F}$-compact on $\mathbb{FP}U$ if in the formulation above the condition $\bigsqcup\mathcal{F} \sqsubseteq G$ is replaced by $\bigsqcup\mathcal{F} \sqsubseteq \Psi(G)$.*

Definition 47 *Ψ is said to be d-$\bigsqcup\mathcal{F}$-topological on $\mathbb{FP}U$*
$=_{def} \forall F \forall G(F, G \in \mathbb{FP}U \rightarrow \Psi(F \sqcup G) \sqsubseteq \Psi(F) \sqcup \Psi(G))$.

Theorem 47 *If*

1. *$\Psi, \Psi^{'} : \mathbb{FP}U \rightarrow \mathbb{FP}U$*
2. *$\mathcal{F} \subseteq \mathbb{FP}U$, $G \in \mathbb{FP}U$, and $d \in \langle 0, 1\rangle$*
3. *Ψ and $\Psi^{'}$ are monotone and topological on $\mathbb{FP}U$*
4. *Ψ and $\Psi^{'}$ are d-$\bigsqcup\mathcal{F}$-compact on $\mathbb{FP}U$*
5. *$\forall F(F \in \mathcal{F} \rightarrow \Psi(F) = \Psi^{'}(F^{'}))$*
6. *$\bigsqcup\mathcal{F} \sqsubseteq G$*
7. *$\mathcal{F}$ is finite*

then $\Psi(G) = \Psi(G^{'})$.

Theorem 48 *If*

1. *$\Psi, \Psi^{'} : \mathbb{FP}U \rightarrow \mathbb{FP}U$*
2. *$\mathcal{F} \subseteq \mathbb{FP}U$, $G \in \mathbb{FP}U$, and $d \in \langle 0, 1\rangle$*
3. *Ψ and $\Psi^{'}$ are topological closure operators on $\mathbb{FP}U$*
4. *Ψ and $\Psi^{'}$ are strongly d-$\bigsqcup\mathcal{F}$-compact on $\mathbb{FP}U$*
5. *$\forall F(F \in \mathcal{F} \rightarrow \Psi(F) = \Psi^{'}(F))$*
6. *$\bigsqcup\mathcal{F} \sqsubseteq \Psi(G)$ and $\bigsqcup\mathcal{F} \sqsubseteq \Psi^{'}(G)$*
7. *$\mathcal{F}$ is finite*

then $\Psi(G) = \Psi^{'}(G)$.

5 Concluding Remarks

1. As we have remarked in section 2 there exists a lot of methods and algorithms in order to construct interpretations for a given fuzzy IF-THEN rule base. Because of lacking space in the paper presented we couldn't investigate the following important problem: Under which conditions do the interpreting operators which can be constructed by algorithms mentioned in section 2 fulfil the theorems formulated in in section 4? So in the positive case, we can state the uniqueness of the interpretation constructed. In a forthcoming paper we shall study this problem.

2. In a second forthcoming paper we shall investigate the problem whether uniqueness theorems from fuzzy analysis can give another approach to the uniqueness problem of fuzzy IF-THEN rule bases.

Acknowledgements

The author would like to thank STEPHAN LEHMKE for scientific discussions on the subject and VOLKHER KASCHLUN for his help in preparing the manuscript.

References

1. J.F. Baldwin and B.W. Pilsworth. *Axiomatic approach to implication for approximate reasoning with fuzzy logic.* Fuzzy Sets and Systems 3, 1980, pages 193–219.
2. M. Diehl. *Untersuchungen zum Kalkül der Fuzzy-IF-THEN-Regeln.* Diplomarbeit, Universität Dortmund, Fachbereich Informatik, Lehrstuhl I, Juli 1994.
3. S. Fukami, M. Mizumoto, and K. Tanaka. *Some considerations of fuzzy conditional inference.* Fuzzy Sets and Systems 4, 1980, pages 243–273.
4. S. Gottwald. *On the existence of solutions of systems of fuzzy equations.* Fuzzy Sets and Systems 12, 1984, pages 301–302.
5. S. Gottwald. *Characterizations of the solvability of fuzzy equations.* Elektron. Informationsverarb. Kybernet. 22, 1986, pages 67–91.
6. S. Gottwald. *On some theoretical results concerning the construction of fuzzy controllers.* In St.F. Bocklisch, et all (editors). *Fuzzy Sets Applications, Methodological Approaches, and Results.* Math. Research, volume 30, Akademie-Verlag, Berlin, 1986, pages 45–55.
7. S. Gottwald and W. Pedrycz. *Solvability of fuzzy relational equations and manipulation of fuzzy data.* Fuzzy Sets and Systems 18, 1986, pages 1–21.
8. S. Gottwald and W. Pedrycz. *On the methodology of solving fuzzy relational equations and its impact on fuzzy modelling.* In M.M. Gupta, and T. Yamakawa (editors). *Fuzzy Logic in Knowledge-Based Systems, Decision and Control.* North Holland Publ. Corp., Amsterdam, 1988, pages 197–210.
9. F. Klawonn and R. Kruse. *Equality relations as a basis for fuzzy control.* Fuzzy Sets and Systems 54, 1993, pages 147–156.
10. F. Klawonn and R. Kruse. *Fuzzy control as interpolation on the basis of equality relations.* Proceedings of the Second IEEE International Conference on Fuzzy Systems, San Francisco, California, March 28-April 1,1993.
11. L.T. Kóczy. *Approximate reasoning and control with sparse and/or inconsistent fuzzy rule bases.* In B. Reusch (editor). *Fuzzy Logic – Theorie und Praxis.* 2. Dortmunder Fuzzy-Tage, Dortmund, June 9-10, 1992, Springer-Verlag, 1993, pages 42–65.

12. L.T. Kóczy. *Compression of fuzzy rule bases by interpolation.* First Asian Fuzzy Systems Symposium, Singapore, November 23-26, 1993. Institute of Systems Science, National University of Singapore. Proceedings.
13. L.T. Kóczy. *On the calculus of fuzzy rules.* Fifth International Fuzzy Systems Association World Congress'93, Proceedings, volume 1, Seoul, Korea, July 4-9, 1993, pages 1–2.
14. L.T. Kóczy (editor). *Fuzzy Logic, Texts.* Volume 2, Fuzzy Systems II (Fuzzy Reasoning and Control), TEMPUS_JEP 07759-94-MODIFY, Budapest 1997.
15. R. Kruse, J. Gebhardt, and F. Klawonn. *Foundations of Fuzzy Systems.* Wiley, Chichester, 1994.
16. S. Lehmke, B. Reusch, K.-H. Temme, and H. Thiele. *On interpreting fuzzy IF-THEN rule bases by concepts of functional analysis.* Technical Report CI-19/98, University of Dortmund, Collaborative Research Center 531, February 1998.
17. H. Nakanishi, I.B. Türkşen, and M. Sugeno. *A review and comparison of six reasoning methods.* Fuzzy Sets and Systems 57, 1993, pages 257–294.
18. Bernd Reusch, and Helmut Thiele. *On semantic interpretations of fuzzy IF-THEN rule bases by concepts of Functional Analysis.* Paper. To appear in Fuzzy Sets and Systems.
19. K.-H. Temme, and H. Thiele. *On the correctness of the principles FATI and FITA and their equivalence.* IFSA'95 – Sixth International Fuzzy Systems Association World Congress, volume 2, São Paulo, Brazil, July 22-28, 1995, pages 475–478.
20. H. Thiele. *On the calculus of fuzzy IF-THEN rules.* In H.-J. Zimmermann, M.G. Negoiţă, and D. Dascălu (editors). *Real World Applications of Intelligent Technologies.* Mangalia, Romania, September 1995, Black Sea University, Editura Academiei Române, pages 91–111.
21. H. Thiele. *On similarity based fuzzy clusterings.* 17th Linz Seminar on Fuzzy Set Theory, Johannes Kepler Universität, Linz, Austria, February 20-24, 1996. Abstracts, pages 135–138.
22. H. Thiele. *A characterization of RUSPINI-partitions by similarity relations.* IPMU'96, International Conference on Information Processing and Management of Uncertainty in Knowledge-Based Systems, Universidad de Granada, Spain, July 1-5, 1996. Conference Proceedings, vol. I, pages 389–394.
23. H. Thiele. *Zur funktionalanalytischen Begründung des approximativen Schließens und von Fuzzy Control.* 16. Workshop "Interdiziplinäre Methoden in der Informatik", Meinerzhagen-Valbert, 16.-19. September, 1996. Universität Dortmund, Fachbereich Informatik, Lehrstuhl I, Forschungsbericht Nr. 639 des Fachbereichs Informatik, Seiten 1–20.
24. H. Thiele. *Investigating approximate reasoning and fuzzy control by concepts of functional analysis.* First International Conference on Conventional and Knowledge-Based Intelligent Electronic Systems, volume 2, Adelaide, Australia, May 21-23, 1997, pages 493–500.
25. H. Thiele. *On closure operators in fuzzy deductive systems and fuzzy algebras.* 28th International Symposium on Multiple-Valued Logic, Fukuoka, Japan, May 27-29, 1998. Proceedings, pages 304–309. Extended Version: *On Closure Operators in Fuzzy Algebras and Fuzzy Deductive Systems.* Technical Report CI-34/98, University of Dortmund, Collaborative Research Center 531 (Computational Intelligence), April 1998, 23 pages.
26. H. Thiele. *Investigating the logical structure of fuzzy IF-THEN rule bases using concepts of mathematical logic and of functional analysis.* In M. Jamshidi, et all (editors). *World Automation Congress 1998 – Advances in Robotics, Manufacturing, Automation, Control, Soft Computing, Multimedia, and Image Processing (WAC'98): Proceedings.* Anchorage, Alaska, USA, May 10-14, 1998, NASA, ISSCI, pages 132.1–132.6.
27. H. Thiele. *Funktionalanalytische Methoden zur Interpretation von Fuzzy Control Regelbasen (Methods of Functional Analysis for Interpreting Fuzzy Control Rule Bases.* 9. Workshop "Fuzzy Control" des GMD-Fachausschusses 5.22, Dortmund, Germany, 4.-5. November 1999, Forschungsbericht Nr. 0499 der Fakultät Elektrotechnik, Seiten 51–75.

28. H. Thiele. *On fuzzy clusterings with variable boundary conditions.* Fith International Conference on Fuzzy Sets Theory and its Applications (FSTA 2000), Liptovsky Mikulas, The Slovak Republic, January 30-February 4, 2000. To appear in "TATRA MOUNTAINS MATHEMATICAL PUBLICATIONS".
29. E.C.C. Tsang, J.W.T. Lee, and D.S. Yeung. *Similarity based fuzzy reasoning methods for fuzzy production rules.* IFSA'95 – Sixth International Fuzzy Systems Association World Congress, volume 1, São Paulo, Brazil, July 22-28, 1995, pages 157–160.
30. I.B. Türkşen, and Y. Tian. *Combination of rules or their consequences in fuzzy expert systems.* Fuzzy Sets and Systems 58, 1993, pages 3–40.
31. L.A. Zadeh. *Outline of a new approach to the analysis of complex systems and decision processes.* IEEE Trans. on Systems, Man and Cybernetics 3(1), 1973, pages 28–44.
32. Z. Zhong, and I.B. Türkşen. *An approximate reasoning schema based on similarity measures and interval-valued fuzzy sets.* Fuzzy Sets and Systems 34, 1990, pages 323–346.

Appendix

Proofs

Proof of Theorem 41. Assume

(1) $$\Psi, \Psi' : \mathbb{FP}U \to \mathbb{FP}U$$

and

(2) $$\mathcal{F} \subseteq \mathbb{FP}U \text{ and } G \in \mathbb{FP}U$$

We have to prove

(3) $$\Psi(G) = \Psi'(G).$$

To prove (3) it is sufficient to show

(4) $$\Psi(G) \sqsubseteq \Psi'(G)$$

and

(5) $$\Psi'(G) \sqsubseteq \Psi(G)$$

We show only (4) because (5) can be proved analogously.

To prove (4) it is sufficient to show

(6) $$\forall x(x \in U \to \Psi(G)(x) \leq \Psi'(G)(x))$$

Because Ψ is $\mathcal{F}$-compact on $\mathbb{FP}U$ and $\bigsqcup \mathcal{F} \sqsubseteq G$ (assumption 6) there exist $F, H \in \mathbb{FP}U$ such that

(7) $$F \in \mathcal{F}$$

and

(8) $$H \sqsubseteq F$$

and

(9) $$\Psi(G)(x) \leq \Psi(H)(x)$$

From (8) and the $\sqsubseteq$-monotonicity of Ψ on $\mathbb{FP}U$ we obtain

(10) $$\Psi(H) \sqsubseteq \Psi(F).$$

From (7) by assumption 5 we get

(11) $$\Psi(F) = \Psi^{'}(F).$$

Again, from (7) we obtain

(12) $$F \sqsubseteq \bigsqcup \mathcal{F},$$

hence by the $\sqsubseteq$-monotonicity of $\Psi^{'}$ on $\mathbb{FP}U$,

(13) $$\Psi^{'}(F) \sqsubseteq \Psi^{'}(\bigsqcup \mathcal{F}),$$

and from assumption 6 again by the $\sqsubseteq$-monotonicity of $\Psi^{'}$ on $\mathbb{FP}U$ we have

(14) $$\Psi^{'}(\bigsqcup \mathcal{F}) \sqsubseteq \Psi^{'}(G).$$

Thus, (9), (10), (11), (13), and (14) imply (6). ■

Proof of Theorem 42. By carring out the steps (1) - (13) of the proof given for theorem 41, for $x \in U$ we have

(15) $$\Psi(G)(x) \leq \Psi^{'}(\bigsqcup \mathcal{F})(x).$$

By $\sqsubseteq$-monoticity of $\Psi^{'}$ on $\mathbb{FP}U$ we get

(16) $$\Psi^{'}(\bigsqcup \mathcal{F}) \sqsubseteq \Psi^{'}(\bigsqcup \mathcal{F} \sqcup G).$$

Now, by assumption 6 of theorem 42 we have

(17) $$\bigsqcup \mathcal{F} \sqsubseteq \Psi^{'}(G),$$

furthermore by reflexivity of $\Psi^{'}$ we get

(18) $$G \sqsubseteq \Psi^{'}(G),$$

hence by (17) and (18),

(19) $$\bigsqcup \mathcal{F} \sqcup G \sqsubseteq \Psi^{'}(G),$$

hence by $\sqsubseteq$-monotonicity os $\Psi^{'}$ on $\mathbb{FP}U$

(20) $$\Psi^{'}(\bigsqcup F \sqcup G) \sqsubseteq \Psi^{'}(\Psi^{'}(G)).$$

By closedness of $\Psi^{'}$ on $\mathbb{FP}U$ it holds

(21) $$\Psi^{'}(\Psi^{'}(G)) \sqsubseteq \Psi^{'}(G).$$

Thus, (15), (16), (20), and (21) imply

(22) $$\Psi(G)(x) \leq \Psi^{'}(G)(x).$$

■

Proof of Theorem 43. By carring out the steps (1)–(6) of the proof given for the theorem 41 it is sufficient to prove

(6) $$\forall x(x \in U \to \Psi(G)(x) \leq \Psi^{'}(G)(x))$$

Because Ψ is finitary $\bigsqcup \mathcal{F}$-compact on $\mathbb{FP}U$ and $\bigsqcup \mathcal{F} \sqsubseteq G$ (assumption 8 of theorem 43) there exists an $H \in \mathbb{FP}U$ such that

(23) H is finite

and

(24) $$H \sqsubseteq G$$

and

(25) $$\Psi(G)(x) \leq \Psi(H)(x).$$

Assume $F \in \mathcal{F}$. Then by $\sqsubseteq$-monotonicity of Ψ on $\mathbb{FP}U$ we obtain

(26) $$\Psi(H) \sqsubseteq \Psi(F \sqcup H),$$

hence, because Ψ fulfils the $\Rightarrow$-deduction property on $\mathbb{FP}U$,

(27) $$H \Rightarrow \Psi(H) \sqsubseteq \Psi(F).$$

From $F \in \mathcal{F}$ we get

(28) $$\Psi(F) = \Psi^{'}(F),$$

hence by (27) and because $\Psi^{'}$ fulfils the $\Rightarrow$-inference property on $\mathbb{FP}U$,

(29) $$\Psi(H) \sqsubseteq \Psi^{'}(F \sqcup H).$$

By (24), $F \in \mathcal{F}$, and assumption 8 we have

(30) $$F \sqcup H \sqsubseteq \bigsqcup \mathcal{F} \sqcup G \sqsubseteq G,$$

hence by $\sqsubseteq$-monotonicity of Ψ' on $\mathbb{FP}U$,

(31) $$\Psi'(F \sqcup H) \sqsubseteq \Psi'(\bigsqcup \mathcal{F} \sqcup G) \sqsubseteq \Psi'(G).$$

Thus, (25), (29), and (31) imply (6). ■

Proof of Theorem 44. By carring out the steps (1)–(6) of the proof given for theorem 41 and the steps (23)–(29) of the proof given for theorem 43 we have

(29) $$\Psi(H) \sqsubseteq \Psi'(F \sqcup H)$$

where H is finite, $H \sqsubseteq G$, and $F \in \mathcal{F}$. Then we have

(32) $$F \sqcup H \sqsubseteq \bigsqcup \mathcal{F} \sqcup H \sqsubseteq \bigsqcup \mathcal{F} \sqcup G,$$

hence by the $\sqsubseteq$-monotonicity of Ψ' on $\mathbb{FP}U$,

(33) $$\Psi'(F \sqcup H) \sqsubseteq \Psi'(\bigsqcup \mathcal{F} \sqcup H) \sqsubseteq \Psi'(\bigsqcup \mathcal{F} \sqcup G).$$

Assumption 8 gives

(34) $$\bigsqcup \mathcal{F} \sqsubseteq \Psi(G).$$

Because Ψ' is reflexive on $\mathbb{FP}U$, we have

(35) $$G \sqsubseteq \Psi'(G),$$

hence (34) and (35) imply

(36) $$\bigsqcup \mathcal{F} \sqcup G \sqsubseteq \Psi'(G),$$

hence by $\sqsubseteq$-monotonicity and closedness of Ψ' on $\mathbb{FP}U$

(37) $$\Psi'(\bigsqcup \mathcal{F} \sqcup G) \sqsubseteq \Psi'(\Psi'(G)) \sqsubseteq \Psi'(G).$$

Thus, (25), (29), (33), and (37) implies (6). ■

Proof of Theorem 45. By carring out the steps (1)–(6) and the steps (23)–(26) of the proof given for the theorem 43 we have an $H \in \mathbb{FP}U$ such that

(23) $$H \text{ is finite}$$

and

(24) $$H \sqsubseteq G$$

and

(25) $$\Psi(G)(x) \leq \Psi(H)(x).$$

For $F \in \mathcal{F}$ by $\sqsubseteq$-monotonicity of $\Psi \mathbb{FP} U$ we get

(26) $$\Psi(H) \sqsubseteq \Psi(F \sqcup H),$$

Furthermore, assumption 6 implies

(28) $$\Psi(F) = \Psi'(F),$$

hence, because $\{\Psi, \Psi'\}$ fulfils the $\mathcal{F}$-induction property (see assumption 5) and by (23) H is finite, we obtain

(38) $$\Psi(F \sqcup H) = \Psi'(F \sqcup H).$$

Because of $F \sqsubseteq \bigsqcup \mathcal{F}, \bigsqcup \mathcal{F} \sqsubseteq G$ (see assumption 7),

(24), i. e. $H \sqsubseteq G$, and the $\sqsubseteq$-monotonicity of Ψ' on $\mathbb{FP}U$, we get

(39) $$\Psi'(F \sqcup H) \sqsubseteq \Psi'(G).$$

Thus, (25), (26), (38), and (39) imply (6), i. e. $\Psi(G)(x) \leq \Psi'(G)(x)$. ∎

Proof of Theorem 46. By carring out the steps (1)–(6) and the steps (23)–(26), and (28) of the proof given for the theorem 45 we can state

(25) $$\Psi(G)(x) \leq \Psi(H)(x).$$

where H is finite, $H \sqsubseteq G$, and

(40) $$\Psi(H) \sqsubseteq \Psi'(F \sqcup H)$$

hence by $F \sqsubseteq \bigsqcup \mathcal{F}$, $H \sqsubseteq G$, and the $\sqsubseteq$-monotonicity of Ψ' on $\mathbb{FP}U$,

(41) $$\Psi'(F \sqcup H) \sqsubseteq \Psi'(\bigsqcup \mathcal{F} \sqcup G).$$

Now, by assumption 7 we have

(42) $$\bigsqcup \mathcal{F} \sqsubseteq \Psi'(G),$$

furthermore by the reflexivity of Ψ' on $\mathbb{FP}U$,

(43) $$G \sqsubseteq \Psi'(G),$$

hence

(44) $$\bigsqcup \mathcal{F} \sqcup G \sqsubseteq \Psi'(G),$$

thus, by monotonicity and closedness of Ψ' on $\mathbb{FP}U$,

(45) $$\Psi'(\bigsqcup \mathcal{F} \sqcup G) \sqsubseteq \Psi'(\Psi'(G)) \sqsubseteq \Psi'(G).$$

Now, (25), (40), (41), and (45) implies (6). ■

Proof of Theorem 47. By carring out the steps (1)–(6) of the proof given for the theorem 41 we state for $x \in U$ it is sufficient to show

(46) $$\Psi(G)(x) \leq \Psi'(G)(x).$$

Because Ψ is d-$\bigsqcup\mathcal{F}$-compact on $\mathbb{FP}U$ and $\bigsqcup\mathcal{F} \sqsubseteq G$ there exists an $H \in \mathbb{FP}U$ such that

(47) $$H \text{ is finite}$$

and

(48) $$\forall y(y \in U \rightarrow H(y) \leq d)$$

and

(49) $$\Psi(G)(x) \leq \Psi(H)(x).$$

Because $\mathcal{F}$ is a d-covering of U, for $x \in U$ there is an $F \in \mathcal{F}$ such that $F(x) \geq d$, hence we get

(50) $$\forall y(y \in U \rightarrow (\bigsqcup \mathcal{F})(y) \geq d).$$

From (48) and (50) we obtain

(51) $$H \sqsubseteq \bigsqcup \mathcal{F},$$

hence by the $\sqsubseteq$-monotonicity of Ψ on $\mathbb{FP}U$,

(52) $$\Psi(H) \sqsubseteq \Psi(\bigsqcup \mathcal{F}).$$

Because $\mathcal{F}$ is finite and Ψ is $\bigsqcup$-$\sqsubseteq$ topological on $\mathbb{FP}U$, we obtain

(53) $$\Psi(\bigsqcup \mathcal{F}) \sqsubseteq \bigsqcup\{\Psi(F) | F \in \mathcal{F}\}.$$

By assumption 5 we have

(54) $$\forall F(F \in \mathcal{F} \rightarrow \Psi(F) = \Psi'(F)),$$

hence

(55) $$\bigsqcup\{\Psi(F) | F \in \mathcal{F}\} = \bigsqcup\{\Psi'(F) | F \in \mathcal{F}\}.$$

Furthermore, by $\sqsubseteq$-monotonicity of Ψ' on $\mathbb{FP}U$ we get

(56) $$\bigsqcup\{\Psi'(F) | F \in \mathcal{F}\} \sqsubseteq \Psi'(\bigsqcup \mathcal{F})$$

and again by $\sqsubseteq$-monotonicity of Ψ' on $\mathbb{FP}U$ from $\bigsqcup \mathcal{F} \sqsubseteq G$ we obtain

(57) $$\Psi'(\bigsqcup \mathcal{F}) \sqsubseteq \Psi'(G)$$

Thus, (49), (53), (55), (56), and (57) imply(46). ■

Proof of Theorem 48. By carring out the steps (1)–(6) and (47)–(56) of the proof given for the theorem 47 we can conclude for $x \in U$

(58) $$\Psi(G)(x) \leq \Psi'(\bigsqcup \mathcal{F})(x).$$

By assumption 6 we have

$$G \sqsubseteq \Psi'(G),$$

hence

$$\bigsqcup \mathcal{F} \sqcup G \sqsubseteq \Psi'(G),$$

thus, by $\sqsubseteq$-monotonicity and closedness of Ψ' on $\mathbb{FP}U$ we obtain

(59) $$\Psi'(\bigsqcup \mathcal{F} \sqcup G) \sqsubseteq \Psi'(\Psi'(G)) \sqsubseteq \Psi'(G).$$

Thus (58) and (59) imply (46). ■

Remark

The proof of theorem 31 is well-known from Mathematical Logic.

Proofs for the remaining theorems 32 – 37 can be obtained by adopting and modifying proofs for theorems of section 4, i. e. 32 by 43, 33 by 44, 34 by 41, 35 by 42, 36 by 45, and 37 by 46.

Quantum Computers as Fuzzy Computers

Jarosław Pykacz[1], Bart D'Hooghe[2], and Roman R. Zapatrin[3]

[1] Instytut Matematyki, Uniwersytet Gdański,
Wita Stwosza 57, 80-952, Gdańsk, Poland
pykacz@delta.math.univ.gda.pl
[2] Departement Wiskunde, Vrije Universiteit Brussel,
Pleinlaan 2, B-1050, Brussel, Belgium
bdhooghe@vub.ac.be
[3] Starlab nv/sa,
Engelandstraat 555, B-1180, Brussel, Belgium
romanycz@starlab.net
http://www.quantumtopology.com

Abstract. An implementation of digitised fuzzy numbers on quantum computers is suggested. It is shown that due to the famous quantum parallelism quantum computers can operate "globally" on whole membership functions of fuzzy numbers, not by calculating them "point by point" as classical computers do, which leads to the considerable decrease in the number of operations involved in storing and calculating fuzzy numbers. In particular, we show that the standard quantum adder is perfectly suited to realize Kaufmann-like addition of fuzzy numbers encoded in the form of suitably prepared superpositions of input qubits and that it does it in a single run. Although this computational gain is to some extent lost while reading the output, which has to be done statistically on the enough big sample of single runs of the adder, suitably chosen method of defuzzyfication allows to save a great deal of the original gain.

1 Introduction

The theory of quantum computations attracts nowadays a lot of attention in the physical and mathematical community. This interest is justified by well-founded hopes that quantum computers (when built) will outrange classical computers with respect to both speed and computational possibilities. For example, the famous Shor's factoring algorithm [6], [2], if implemented on a quantum computer, should allow to factorize numbers in a polynomial (with respect to the length of an input) time in contrary to the exponential time required by the best of the existing classical algorithms. As a consequence, it will make the safest of the existing encryption methods questionable.

It is an intrinsic feature of a quantum theory that it hardly ever allows to predict the outcomes of experiments/measurements *with certainty*. In all but exceptional cases quantum mechanics allows only to calculate *probabilities* of

B. Reusch (Ed.): Fuzzy Days 2001, LNCS 2206, pp. 526–535, 2001.

getting specific outcomes. This intrinsic "fuzziness" of a quantum theory suggested the idea that quantum computers might as well be "fuzzy" computers well suited to deal directly with vague or imprecise data [5].

The present paper is devoted to the development of the above-stated idea. We consider the standard quantum addition algorithm described in many papers on quantum computations (see, e.g., [7], [8]). The usual mode in which the standard quantum adder works consists of feeding in two numbers whose binary expansions are encoded in a form of suitable tensor products of basic vectors of a 2-dimensional complex Hilbert space. The obtained sum is also encoded in this way, therefore, it represents an ordinary ("crisp") number. The novelty of our approach consists in preparing the input to the adder in the form of specific *superpositions* of product states that encode membership functions of fuzzy numbers to be added. Then we show that by runing the adder *only once* we get the result also in the form of a superposition of product states that again represents a digitised fuzzy number – the result of addition. As we shall show, this yields a considerable gain in the number of operations required to store and add two digitised fuzzy numbers with respect to what can be achieved while performing such tasks on conventional digital computers.

The paper is organized as follows: Section 2 contains the most basic ideas of the standard fuzzy arithmetics while Section 3 describes how quantum computers work and explains their expected advantages over classical computers. In Section 4 we introduce quantum representation of digitised fuzzy numbers in a form of suitable superpositions of product states, and we show that already considered in the literature quantum adder is perfectly suited to perform Kaufman-like addition of digitised fuzzy numbers. Finally, in Section 5 we consider the problems encountered when one wants to read the output.

2 The Standard Fuzzy Arithmetics

2.1 Fuzzy Numbers

A *fuzzy number* is usually defined as a convex and normalized fuzzy subset A of the real line. We shall also assume that its membership function μ_A reaches its maximum in exactly one point – the crisp counterpart of a considered fuzzy number (otherwise one deals with fuzzy intervals rather than fuzzy numbers). In symbols, A is a fuzzy number if the following conditions hold:

convexity:

$$\forall x_1, x_2 \in \mathbb{R} \quad \forall \lambda \in [0,1] \qquad \mu_A(\lambda x_1 + (1-\lambda)x_2) \geq \min(\mu_A(x_1), \mu_A(x_2)) \quad (1)$$

normalization:

$$\exists! x_0 \in \mathbb{R} \quad \text{such that} \quad \mu_A(x_0) = 1. \quad (2)$$

In order to simplify the notation throughout the rest of the paper we adopt the popular convention of identifying fuzzy sets with their membership functions. Moreover, in the case of a fuzzy number we shall denote it using the unique value

x_0 in which it reaches its maximum overlined by a tilda. For example, $\tilde{2}$ denotes "fuzzy two" and $\tilde{2}(3)$ denotes the value of (the membership function of) "fuzzy two" in the point 3. Let us note that the normalization condition is not a very stringent one since we can always "normalize" arbitrary membership function that attains maximum dividing it by the value of this maximum.

The standard way of defining arithmetic operations on fuzzy numbers utilizes the sup-min extension principle. For example, addition of two fuzzy numbers is usually defined as follows:

$$(\tilde{x} + \tilde{y})(z) = \sup_{u+v=z} \min(\tilde{x}(u), \tilde{y}(v)). \tag{3}$$

The sup-min operations on fuzzy numbers, although the most popular, are not the only one considered in the literature (see, e.g., [1]). A very interesting (from the point of view of our purpose: implementation of addition of fuzzy numbers on a quantum computer) way of adding fuzzy numbers was considered by Kaufmann in his book [3] already in 1975. He used a probabilistic method of extending addition to fuzzy numbers by means of an ordinary convolution:

$$(\tilde{x} + \tilde{y})(z) = \int_0^z \tilde{x}(u)\tilde{y}(z-u)du \tag{4}$$

for some particular kinds of fuzzy numbers. We shall use this operations in a slightly modified way to perform addition of digitised fuzzy numbers on a quantum computer.

2.2 Addition of Fuzzy Numbers on a Classical Computer

The most general way in which operations on fuzzy sets can be emulated on a conventional digital computer consists in digitising all membership functions of an input and performing required operations pointwisely in order to get digitised membership function of an output. Since this is time and memory consuming, it is a common practice (see, e.g., numerous papers published in [4]) to restrict considerations to some types of membership functions (e.g., triangular or trapezoidal) only. Although this way of elaborating fuzzy data might be sufficient for many practical purposes, it is by no means a general one. It is also not "universal" in a sense that a procedure elaborated to treat triangular membership functions is usually not appropriate for membership functions of another shape, etc.

We shall show in the sequel that quantum computers are, by their very nature, much better suited to perform Kaufmann-like addition (4) of arbitrary fuzzy numbers than conventional digital computers are. Therefore, we shall estimate now the number of operations needed by a conventional digital computer to perform such addition.

Since we shall deal with digitised fuzzy numbers, we shall denote them using Zadeh [9] convention (which, by the way, occurs to be nearly identical to the most standard way of representing superpositions of quantum-mechanical vectors according to Dirac notation). Without big loss of generality we can also assume that a universe of discourse on which our digitised fuzzy numbers are defined consists of a finite number of integers: $\{0, 1, 2, ..., n\}$, so we deal with numbers of the form:

$$\tilde{a} = \tilde{a}(0)/0 + \tilde{a}(1)/1 + ... + \tilde{a}(n)/n = \sum_{i=0}^{n} \tilde{a}(i)/i, \tag{5}$$

where we admit that some of values $\tilde{a}(i)$ may equal zero. In this case Kaufmann addition (4) takes the discrete form:

$$(\tilde{a} + \tilde{b})(k) = \sum_{i+j=k} \tilde{a}(i)\tilde{b}(j). \tag{6}$$

Since both the number of values $(\tilde{a} + \tilde{b})(k)$ that a computer has to calculate and the number of products needed to calculate each of these values are of the order n, we see that the total number of arithmetic operations that a conventional digital computer has to perform in order to obtain a membership function of a sum of two fuzzy numbers (5) is of the order n^2. As we shall see in the sequel a quantum adder will get this membership function in the course of the only one run.

3 How Quantum Computer Works

We shall consider first a draft scheme of a classical computer from the physical point of view. It necessarily has a *register*, and a *processor*. Any bi-stable classical physical system, whose two possible states represent binary digits 0 and 1, can serve as a physical carrier of a classical bit (= binary unit) of information. The register consists of a number of such bi-stable physical systems. To store the input data means to prepare the register in a particular state. A classical l-bit register can be in 2^l states, denote them $|\, n\rangle$, where n is a number represented by a binary word of the length l. To carry out the calculation means that the register undergoes certain prescribed evolution performed by means of the processor.

Contrary to classical physical systems, quantum bi-stable systems (e.g., two-level atoms, spin-1/2 particles, etc.) can be in so called *superposition states* that, in a sense, encode simultaneously both 0 and 1. Mathematically, superposition states are expressed as linear combinations of states $|0\rangle$ and $|\,1\rangle$ that encode "classical" 0 and 1 with complex coefficients whose square moduli sum up to 1:

$$|\, s\rangle = c_0|\, 0\rangle + c_1|\, 1\rangle, \qquad c_0, c_1 \in \mathbb{C}, \qquad |c_0|^2 + |c_1|^2 = 1. \tag{7}$$

It should be stressed that although any measurement of $|\, s\rangle$ necessarily yields either 0 (with probability $|c_0|^2$) or 1 (with probability $|c_1|^2$), $|\, s\rangle$ cannot be interpreted as an unknown state that represents either 0 or 1 with respective probabilities $|c_0|^2$ and $|c_1|^2$. Because the coefficients c_0 and c_1 are complex, not real numbers, it does not represent a statistical mixture of $|0\rangle$ and $|\,1\rangle$. Neither it can be interpreted as representing some value "between" 0 and 1. It is an entirely new entity having no counterpart in classical physics and a unit of information it carries is customarily called *qubit* (= quantum bit). In the sequel, following the widespread custom, we shall not distinguish between qubits and their physical carriers.

The theory of quantum computations (the practice is still "in statu nascendi") tells how to encode an input to a quantum computer on a number of qubits that form a *quantum register*, and how to operate on them, with the aid of a *quantum processor* that works according to the laws of quantum mechanics, in order to get the desired output. Now we shall describe this process in a more detailed way.

If a classical system can be in a number of configurations (states) then the corresponding quantum system can be in any linear superposition (i.e., linear combination with complex coefficients whose square moduli sum up to 1) of such configurations. Thus, the possible quantum states form a subset of a complex linear space (a Hilbert space) spanned by classical configurations. A general state of a quantum register of l qubits is given by

$$\sum_{n=0}^{2^l-1} c_n \,|\, n\rangle, \qquad c_n \in \mathbb{C}, \tag{8}$$

where $|\, n\rangle$ denotes the state of the register that encodes the binary expansion of the number n, e.g., $|\, 9\rangle = |\, 1001\rangle$, and $\sum |c_n|^2 = 1$. These 2^l pure states of the form $|\, \sigma_1, \ldots, \sigma_l\rangle$ with $\sigma_i = 0, 1$ form a basis of the register's state space which is called *computational basis*.

The Hilbert space of a collection of quantum systems is the tensor product of the Hilbert spaces of these systems. Thus, the Hilbert space of an l-qubit quantum register can also be seen as a tensor product of l 2-dimensional spaces of the individual qubits. Therefore, if appropriate, we can write, e.g., $|1\rangle \otimes |0\rangle \otimes |0\rangle \otimes |1\rangle$ instead of $|1001\rangle$.

The time evolution of a quantum system is restricted to unitary transformations of the state vector. In the case of quantum computations usually one considers unitary transformations acting only on a few (say, 1, 2 or 3) quantum bits at a time, called *quantum gates*. It was proved (see, e.g., [7]) that this does not restrict the variety of arithmetic operations which can be performed.

One of the frequently used quantum gates is the *controlled-NOT* gate that operates on two qubits and changes the second bit iff the first bit is 1:

$$\begin{aligned} C_{not} : |00\rangle &\to |00\rangle \\ |01\rangle &\to |01\rangle \\ |10\rangle &\to |11\rangle \\ |11\rangle &\to |10\rangle \end{aligned} \tag{9}$$

The C_{not} gate is usually represented graphically by a following circuit:
where the circle represents the first (control) bit and the cross represents the conditional negation of the second bit.

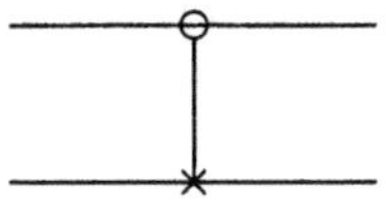

Fig. 1. Graphical representation of the controlled-NOT gate

The *controlled-controlled-NOT* gate, also called *Toffoli* gate, which operates on three qubits and negates the third bit iff the first two bits are 1 is represented by a circuit of the form:

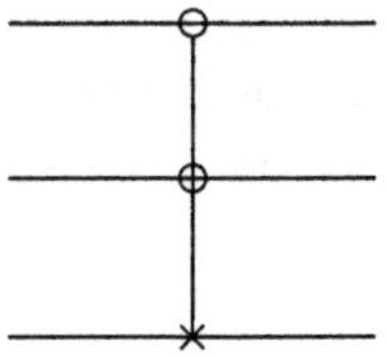

Fig. 2. Graphical representation of the Toffoli gate

These two quantum gates suffice to construct a 1-bit full quantum adder [7], [8]:

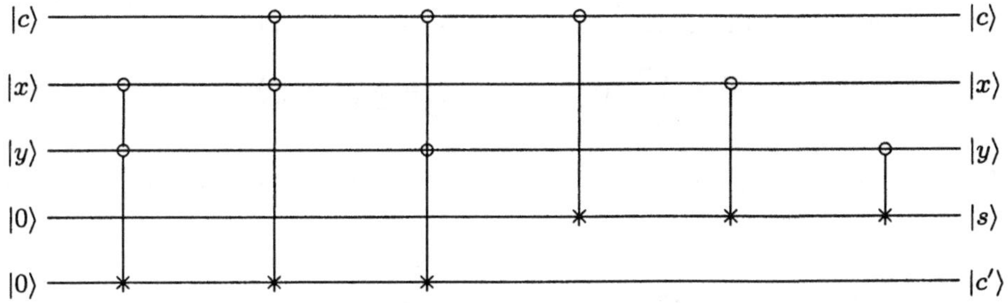

Fig. 3. The 1-bit full adder

It can be easily checked that it realizes addition modulo 2 of the incoming carry bit c and two data bits x, y yielding the sum bit s and the new carry bit c'. The input bits c, x, and y, are also rewritten to the output in order to preserve reversibility of the circuit which is a necessary requirement in quantum computations since without it the evolution could not be unitary. A number of such circuits can be combined to make a complete plain adder [7], [8] that realizes addition of two crisp numbers encoded in the binary way. We shall show in the next section that the same quantum adder is also perfectly suited to perform in the course of the only one run Kaufmann-like addition of suitably encoded fuzzy numbers.

4 Quantum Fuzzy Arithmetics

As it was anticipated in Section 2, we deal with discrete representation of fuzzy numbers, so membership functions of the fuzzy numbers are not continuous functions defined on $\mathbb{R}$ but they are defined on finite sets of integers. No "usual" computer actually deals with real numbers, they are always substituted by rationals of an appropriate length. In our paper we shall follow this model. Since we deal with fuzzy numbers, the results of the readout of the register's state associated to particular crisp numbers should be within $[0, 1]$.

Physically, we represent a fuzzy number by a state of the register (8) which is a superposition of tensor product states that form a computational basis. The computation is made by invoking the processor *only once* which, by the very nature of quantum interactions, simultaneously affects all pure states that enter the superposition (the famous *quantum parallelism*).

The basic scheme is the following. Given a fuzzy number with a membership function $\tilde{a}(i)$, we can represent it by a suitable superposition of states of the computational basis in the Hilbert space – the state space of the register. From the geometrical point of view, this amounts to perform a suitable rotation on the unit sphere in 2^l-dimensional Hilbert space which is a tensor product of l 2-dimensional state spaces of individual qubits, and can be always done by applying suitable unitary transformations [10]. Suppose now that we have to add two fuzzy numbers $\tilde{a} = \tilde{a}(i)$ and $\tilde{b} = \tilde{b}(j)$. They can be represented by the following superpositions of the states of the computational basis:

$$|a\rangle = \sum_i \varphi(i)\,|i\rangle\,, \qquad |b\rangle = \sum_j \psi(j)\,|j\rangle \tag{10}$$

where $|\varphi(i)|^2 = \tilde{a}(i)$ and $|\psi(j)|^2 = \tilde{b}(j)$. We feed the following input to the adder

$$|a\rangle \otimes |b\rangle \otimes |0\rangle = {}_{i,j}\sum \varphi(i)\,\psi(j)\,|i\rangle \otimes |j\rangle \otimes |0\rangle\,. \tag{11}$$

The adder acts on states of the computational basis as follows:

$$|i\rangle \otimes |j\rangle \otimes |0\rangle \xrightarrow{U} |i\rangle \otimes |j\rangle \otimes |i+j\rangle\,. \tag{12}$$

Substituting (12) in (11) and rearranging the sum with respect to the last factor of the tensor product we have

$$\begin{aligned} |i\rangle \otimes |j\rangle \otimes |0\rangle &\xrightarrow{U} {}_{i,j}\sum \varphi(i)\,\psi(j)\,|i\rangle \otimes |j\rangle \otimes |i+j\rangle \\ &= \sum_k \sum_{i+j=k} \varphi(i)\,\psi(j)\,|i\rangle \otimes |j\rangle \otimes |k\rangle\,. \end{aligned} \tag{13}$$

The probabilities to get the particular values of the result are obtained by representing the state (13) as a projector in the tensor product space $H \otimes H \otimes H$

and taking the partial trace over the two first spaces in the tensor product. Let us do it stepwise. In terms of projectors (13) becomes:

$$\sum_{k,k'} \sum_{i+j=k} \sum_{i'+j'=k'} \varphi(i)\overline{\varphi(i')}\psi(j)\overline{\psi(j')} \left|i\right\rangle \left\langle i'\right| \otimes \left|j\right\rangle \left\langle j'\right| \otimes \left|k\right\rangle \left\langle k'\right|. \tag{14}$$

Now taking the partial trace it over ii' and jj' one obtains:

$$\sum_{k,k'} \sum_{i+j=k} \sum_{i'+j'=k'} \delta_{ii'}\delta_{jj'}\varphi(i)\overline{\varphi(i)}\psi(j)\overline{\psi(j)} \left|i\right\rangle \left\langle i'\right| \otimes \left|j\right\rangle \left\langle j'\right| \otimes \left|k\right\rangle \left\langle k'\right|, \tag{15}$$

which yields the following vector:

$$\sum_{k} \sum_{i+j=k} \left|\varphi(i)\right|^2 \left|\psi(j)\right|^2 \left|k\right\rangle. \tag{16}$$

is associated with the density operator (15).

We can interpret the formula (16) directly in terms of membership functions, and we see that the addition of fuzzy numbers realized by the standard quantum adder is exactly the Kaufmann's discrete addition (6). So we have

Theorem 1. *If we have two digitised fuzzy numbers $\tilde{a} = \tilde{a}(i)$ and $\tilde{b} = \tilde{b}(j)$, and an input to the standard quantum adder is prepared in the form of superpositions $\left|a\right\rangle = \sum_i \varphi(i)\left|i\right\rangle$, $\left|b\right\rangle = \sum_j \psi(j)\left|j\right\rangle$, where $\left|\varphi(i)\right|^2 = \tilde{a}(i)$ and $\left|\psi(j)\right|^2 = \tilde{b}(j)$, then the quantum adder realizes Kaufmann's discrete addition:*

$$(\tilde{a}+\tilde{b})(k) = \sum_{i+j=k} \tilde{a}(i)\tilde{b}(j).$$

5 Getting the Output

As we saw in the previous Section when a quantum processor acts on an input data prepared in the form of suitable superpositions of states of the computational basis, it affects all components of these superpositions simultaneously, so it behaves as a family of classical processors that work in a parallel way. This famous "quantum parallelism" is actually a source of considerable computational speed-up which gave rise to hope that quantum computers can solve in reasonable time problems untractable by classical computers like factorization of very big numbers. Unfortunately, quantum computations have also a dark side: as an output we get a quantum state of a register that we can not access directly. Any measurement of the state of the register yields randomly only one particular value k (represented in the superposition by a state $\left|k\right\rangle$ of the computational basis) with probability $|c_k|^2$ which, in our case equals

$$|c_k|^2 = \sum_{i+j=k} \left|\varphi(i)\right|^2 \left|\psi(j)\right|^2 = \sum_{i+j=k} \tilde{a}(i)\cdot\tilde{b}(j). \tag{17}$$

Therefore, if our task is to reconstruct precisely the entire membership function of the result, one would have to repeat addition many times, loosing in this way a big part of the gain that follows from changing the classical adder to the quantum one. However, if the task is to determine some property of the output state of the register, the situation may change drastically. For example, the incredible efficiency of Shor's factoring algorithm [6] follows from the fact that it actually searches for a period of a function encoded in the output register, not for this function itself. Since most of computations performed with fuzzy numbers finishes with some defuzzyfication procedure, we may assume that our task is not to reconstruct the entire membership function of the result (cf. [11]), but, e.g., to find its barycenter. In our case, the x-coordinate of this barycenter is just the mean value of probability distribution of getting particular results of measurements performed on the register that achieved its final state after consecutive runs of the adder. Since the standard error of the mean value obtained in the course of k runs is proportional to $1/\sqrt{k}$, we see that if the desired accuracy of obtaining the "defuzzyfied result" is not too high, the number k may be much smaller than the number n^2 evaluated in Section 2 as the number of arithmetic operations that a classical computer has to perform in order to get the membership function of the output, which still would have to be augmented if the final task were to find the barycenter. Moreover, in the case of more complicated calculations involving many fuzzy numbers, the number of runs of a quantum machine required to obtain the defuzzyfied result would remain practically the same. On the contrary, the number of arithmetic operations of a classical computer increases drastically with every fuzzy number added to computations, so in this case the computational superiority of quantum machines over classical ones should be still more significant.

Acknowledgments

The main part of the paper was written when all the three authors met in the Center Leo Apostel (CLEA), Brussels in September 2000. The authors are deeply indebted to CLEA for providing them with the facilities that stimulated their work. BD'H and JP were supported by the grant under the Polish-Flemish Research Project 127/E-335/S/2000 and RRZ by the Starlab, Brussels, Q-ACTA Project and the research grant "Universities of Russia".

References

1. Dubois, D. Prade, H.: Fuzzy Sets and Systems. Theory and Applications, Academic Press, New York, (1980)
2. Ekert, A., Jozsa, R.: Quantum Computation and Shor's Factoring Algorithm, Rev. Mod. Phys. **68** (1996) 733-753
3. Kaufmann, A.: Introduction to the Theory of Fuzzy Subsets, Vol. I, Academic Press, New York, (1975)
4. Mareš, M., et al., (eds): Proc. 7th IFSA World Congress, Prague, July 1997, Academia, Prague, (1997)

5. Pykacz, J.: Quantum Logic as a Basis for Computations, International Journal of Theoretical Physics, **39** (2000)839-850
6. Shor, P. W.: Algorithms for Quantum Computation: Discrete Logarithms and Factoring, In: Goldwasser, S. (ed.): Proc. 35th Ann. Symp. Found. Comp. Sci. IEEE Computer Society Press, Los Alamitos, CA (1994) 124-134
7. Vedral, V., Barenco, A., Ekert, A.: Quantum Networks for Elementary Arithmetic Operations, Phys. Rev. A **54** (1996) 147-153
8. Vedral, V., Plenio, M. B.: Basics of Quantum Computation, Progress of Quantum Electronics **22** (1998) 1-40; see also e-print: quant-ph/9802065
9. Zadeh, L. A.: A Fuzzy Set Theoretic Interpretation of Linguistic Hedges, J. of Cybernetics **2** (1972) 4-34
10. Zalka, C.: Simulating Quantum Systems on a Quantum Computer, Proc. R. Soc. Lond. **A454** (1998) 313-32; see also e-print: quant-ph/9603026
11. Zapatrin, R.R., Logic Programming As Quantum Measurement, International Journal of Theoretical Physics, **34**, 1813–1821, (1995), see also e-print: quant-ph/9506041

Improvement of the Cluster Searching Algorithm in Sugeno and Yasukawa's Qualitative Modeling Approach

K.W. Wong[1], L.T. Kóczy[1,2], T.D. Gedeon[1], A. Chong[1], D. Tikk[2]

[1]School of Information Technology
Murdoch University
South St, Murdoch
Western Australia 6150
{ k.wong, t.gedeon, cchong } @murdoch.edu.au

[2]Department of Telecom. & Telematics
Budapest University of Technology and Economics
H-1117 Budapest, Pázmány sétány 1/d
Hungary
{ koczy, tikk } @ttt.bme.hu

Abstract. Fuzzy modeling has become very popular because of its main feature being the ability to assign meaningful linguistic labels to the fuzzy sets in the rule base. This paper examines Sugeno and Yasukawa's qualitative modeling approach, and addresses one of the remarks in the original paper. We propose a cluster search algorithm that can be used to provide a better projection of the output space to the input space. This algorithm can efficiently identify two or more fuzzy clusters in the input space that have the same output fuzzy cluster.

1 Introduction

Fuzzy modeling has become very popular because of the main feature of its ability to assign meaningful linguistic labels to the fuzzy sets [1] in the rule base [2,3]. Sugeno and Yasukawa's qualitative modeling (SY) method [4] has gained much attention in the fuzzy research field mainly due to its advantage of building fuzzy rule bases automatically from sample input-out data. The fuzzy rule bases extracted by the SY method are sparse fuzzy rule bases, i.e., where there are "gaps" among the rules, so that these regions can be interpolated from the remaining areas and rules [5,6,7]. In our approach, we intend to extend this method – with the necessary alterations and additional steps. The usual fuzzy controller identification methods generate dense fuzzy rule bases, so that the rule premises form a fuzzy partition of the input space. In a dense fuzzy rule base, the number of rules is very high, as it depends on the number of inputs k and the number of partitions per variable T in an exponential way. Assuming all the partitions are consistent in all premises and consequents, the total number of rules is $|R| = O(T^k)$. In order to avoid this exponential number of rules, the SY method puts emphasis on the rule consequents, i.e., the output space, and first finds a partition in Y. The determination of premises in the input space X is done by splitting appropriately the inverse images of the output clusters. Using this approach,

B. Reusch (Ed.): Fuzzy Days 2001, LNCS 2206, pp. 536–549, 2001.

the partitioning of the input space is done in a secondary manner, thus the number of fuzzy rules does not increase exponentially with the number of inputs.

One of the important remarks made in the paper by Sugeno and Yasukawa [4], is the condition under which more than one fuzzy cluster can be found in the input space, so that they correspond to the same output cluster. It is made clear that special care has to be taken to form two or more convex input clusters. However, the details of this particularly important step are not given in any detail in the paper. With regards to this remark, it has to be stated that two conditions have to be considered when proposing an algorithm that could efficiently handle the problem. First, the algorithm has to be able to identify the occurrence of several (at least two) rules in one fuzzy output cluster, i.e., the presence of more than one corresponding fuzzy cluster in the input space. Second, the algorithm should be able to decide whether these input fuzzy clusters identified can be merged or any of them can be discarded. This paper shows a detailed analysis of these two conditions and proposes a cluster search algorithm that can address this problem efficiently.

2 The Sugeno and Yasukawa's Qualitative Modeling Method

In a given data set, with k inputs, the given input-output data pairs for n patterns are:

$$(x_1^i, x_2^i, x_3^i, \ldots, x_k^i; y^i)$$
where $i = 1, 2, 3, \ldots n$

The SY method [4] is able to perform two main steps, *identification* and *qualitative modeling* in obtaining a fuzzy model as:

$$R^i : \text{if } x_1 \text{ is } A_1^i \text{ and } x_2 \text{ is } A_2^i \ldots \text{ and } x_n \text{ is } A_n^i \text{ then } y \text{ is } B^i \quad (1)$$

In the identification step, it can be subdivided into *structure identification I and II*, and *parameter identification*. In structure identification I, the main purpose is to find appropriate input candidates and input variables in building the model. Structure identification II is concerned with the input-output relations by concentrating on the number of rules and partitions of the input space. The parameter identification step is basically used to tune the parameters in the membership functions of the fuzzy sets. Finally, linguistic labels can be assigned to the finalized fuzzy sets in the rule base. In this paper, we are mainly concerned with the structure identification II stage.

In order to extract a fuzzy rule base, the SY method uses two distinctive characteristics. First, it partitions the consequents of the rule and then finds a relationship concerning the premises. It is important that it does not use an ordinary fuzzy partition of the input space as shown in Figure 1. In order to satisfy the first characteristic, it makes use of the fuzzy c-means method (FCM) [8] to search for the fuzzy clusters in the output space using all available data. When determining the best clusters in the output space of the following selecting criterion [4] is used:

$$S(c)=\sum_{k=1}^{n}\sum_{i=1}^{c}(\mu_{ik})^m(\|x_k-v_i\|^2-\|v_i-\bar{x}\|^2) \quad (2)$$

where

n:	number of data to be clustered;
c:	number of clusters, $c \geq 2$;
x_k:	kth data;
$\bar{x}$:	average of data;
v_i:	vector expressing the center of the ith cluster;
$\|\bullet\|$:	norm;
μ_{ik}:	grade of the kth data belonging to the ith cluster
m:	adjustable weight

The best cluster is determined when the selecting criterion *S(c)* reaches a minimum when the number of clusters, *c*, increases. The final results of the clustering have the relation between the grade of membership belonging to a fuzzy cluster *B* for every output *y*:

$$y^i \ in \ B^j (1 \leq j \leq c):$$

$$(x^i, y^i), B^1(y^i), B^2(y^i), \ldots, B^c(y^i) \quad (3)$$

From the output fuzzy cluster *B*, a fuzzy cluster *A* is induced in the input space as shown in Figure 2. If there are two input dimensions x_1 and x_2, the fuzzy cluster *A* can then be projected to the axes x_1 and x_2 as shown in Figure 3.

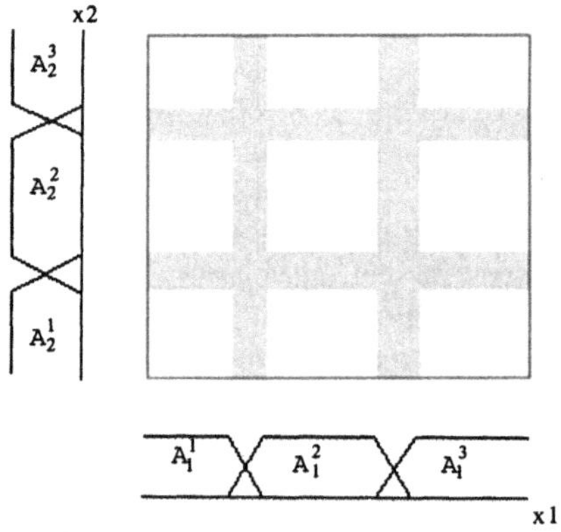

Fig. 1. Ordinary fuzzy partition of the input space

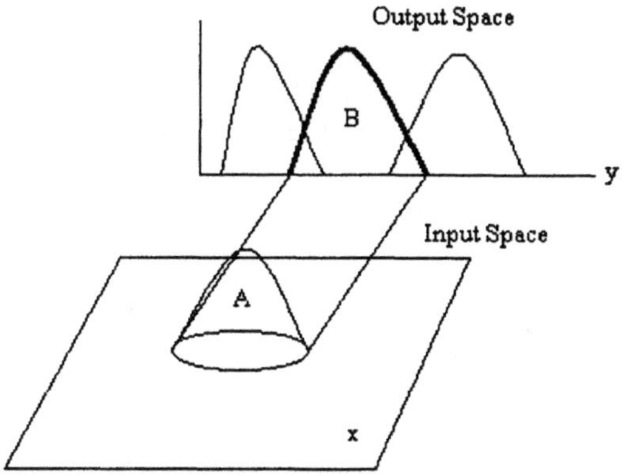

Fig. 2. Fuzzy cluster A from output cluster B

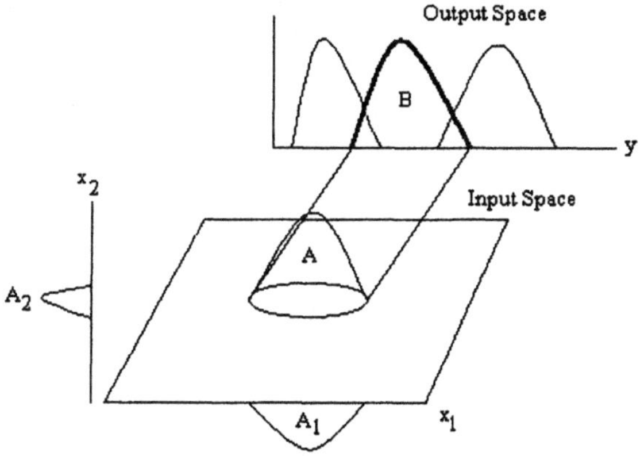

Fig. 3. Fuzzy cluster A for two input dimensions

From these cluster relations, this can be formulated as follows:

$$A_1(x_1^i) = A_2(x_2^i) = B(y^i) \tag{4}$$

From this output cluster, a fuzzy rule is generated as:

if x_1 is A_1 and x_2 is A_2, then y is B.

3 The Problem of Clustering Input Space

In the original approach, the output cluster is typically used to obtain the corresponding (single) input cluster as well, by projecting the output cluster onto the input space. However, in cases where there is more than one cluster in the input space corresponding to the same output cluster, e.g. A^l to A^j corresponding to the same output cluster *B*, a further step has to be done in order to separate them. The SY method suggests very generally that in the case of two (or more) input fuzzy clusters, two (or more) convex fuzzy clusters should be "formed carefully" in the input space as shown in Figure 4 and Figure 5. However, the details of the method are not shown in the paper as to how this can be done "carefully". The first problem that we address here is how to separate two or more input clusters and corresponding antecedents, especially in multi-dimensional cases.

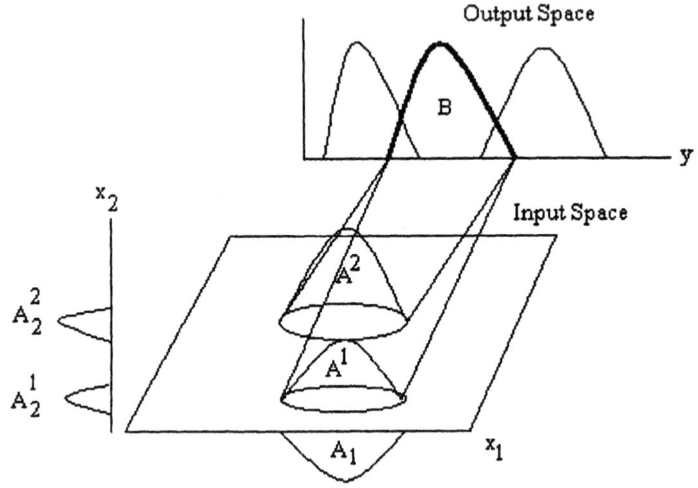

Fig. 4. Two fuzzy clusters for one output cluster.

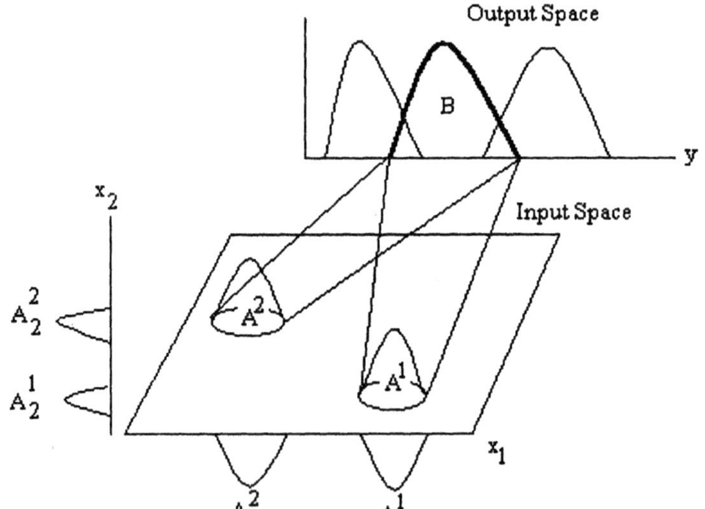

Fig. 5. Another example of two fuzzy clusters for one output cluster.

Besides identifying the fuzzy clusters that may correspond in the input space to a specific output fuzzy cluster, some other factors need to be considered. They can be classified in two main areas as follows:

(1) Merging of Input Clusters

After identifying the possible clusters in the input space, the next question is whether all the identified clusters are necessary to produce a reasonable model. Depending on the distribution of the sample data as well as the nature of the problem, the seemingly separate two or more input clusters may appear just because there is a slight decrease between two peaks, indicating a lower frequency of data there, due to the uneven distribution of the available data. In this case, separating the clusters may not really improve the model, but may increase the number of fuzzy rules constructed.

When sampling the data, noise may accidentally be included in the given input-output sample. Therefore it is not possible to define an exact model that describes the relationship between X and Y when noise exists in the input-output pair. However, a probabilistic relationship governed by a joint probability law $P(v)$ can be used to describe the relative frequency of occurrence of input-output pairs (x^i, y^i) for n training patterns. The joint probability law $P(v)$ can be further separated into an environmental probability law $P(\mu)$ and a conditional probability law $P(\gamma)$. For notation expression, the probability law is expressed as:

$$P(v) = P(\mu)P(\gamma) \tag{5}$$

The environmental probability law $P(\mu)$ describes the occurrence of the input X. The conditional probability law $P(\gamma)$ describes the occurrence of the Y based on the given input X. An input-output pair (X, Y) is considered to be noise if X does not follow the environmental probability law $P(\mu)$, or the output Y based on the given X does not follow the conditional probability law $P(\gamma)$. In this case, even when the input clusters have been identified, it may be a noise motivated clustering result. In this case, using all the clusters may not really improve the model, but may increase the number of fuzzy rules constructed and thus the computational complexity.

(2) Computational Efficiency

The main advantage of the SY method as compared to traditional methods is that it works on the consequents rather than the premises. This is very computationally efficient. Besides, after the fuzzy model; which is a sparse fuzzy rule base; has been built, the model does not require intensive computation to generate an answer, as the number of fuzzy rules is small for most cases. Care must be taken when constructing an algorithm to identify the input fuzzy clusters within the output cluster. The algorithm has to be computationally efficient and at the same time does not increase the number of fuzzy rules generated tremendously.

4 Identification of the Clustering in the Input Space

Cluster the output as per the SY method using fuzzy c-means. In all our analysis here, we are assuming that the distribution in each cluster can be approximated by a normal distribution.

For cluster C_i:

(1) Random selection

In each cluster, the population (U_i) of that cluster consists of data with membership not smaller than 0.2:

$$U_i = \{x_j \mid \mu_{ij} \geq 0.2, \ where \ j = 1,\ldots, N\} \tag{6}$$

where N is the number of data in cluster i.

Those data that have membership grades smaller than 0.2 are left out, and are considered to be insignificant in contributing to the features of the cluster. When dealing with very large number of data, which is common in most real world applications, if we are to project all the output points back to the input space, this may be very time consuming. The total projections, p required are:

$$p = \sum_{i=1}^{c}(N_i) \tag{7}$$

and $p > n$

where N_i is the number of data in cluster i

We propose here to project a subset of N by using sampling distribution theory, being s. Regardless the distribution of the data in the cluster, s is always smaller than N [9].

When performing random selection, it is always necessary to ensure that the sample contains the major information in the population, U_i. The standard of measure is by using sampling distribution theory [10,11]. The basic idea is to ensure that the selected data can approximate the population in the cluster (U_i), by investigating the mean ($\overline{u}_i$), variance ($\sigma_{\overline{u}}^2$) and standard deviation ($\sigma_{\overline{u}}$) of the population.

For random sampling without replacement in a finite population, the variance ($\sigma_{s_i}^2$) and standard deviation (σ_{s_i}) of the sample are as follows (with the assumption that $s > 0.05N$):

$$\sigma_{s_i}^2 = \frac{\sigma_{\overline{u}}^2}{s} \bullet \frac{N-s}{N-1} \tag{8}$$

$$\sigma_{s_i} = \frac{\sigma_{\bar{u}}}{\sqrt{s}} \bullet \sqrt{\frac{N-s}{N-1}} \tag{9}$$

where $\frac{N-s}{N-1}$ is the finite population correction factor.

However, according to the Central Limit Theorem, the finite population correction factor can be ignored if and only if $s \leq 0.05N$.

If the sample (S_i) we have selected is a good representation of the population (U_i) in the cluster, then the sampling error should be small. Sampling error (E) is the measure of the difference between the sample average and the population average:

$$E = \bar{u}_i - \bar{s}_i \tag{10}$$

In order to obtain a better random sample for use in the projection, three random data sets are obtained and their sampling errors are compared. The best sampling error of the three are then selected for use in the next step. However, the next issue we have to deal with is the size of s. In this paper, we make use of the mean as a guideline. The size of sample, s, required for accuracy of $\sigma_{\bar{u}}$, A, for a confidence level, l are:

$$s = \frac{\sigma_{\bar{u}}^2}{\frac{A^2}{Z^2} + \frac{\sigma_{\bar{u}}^2}{N}} \tag{11}$$

where Z is the z-value corresponding to the desired confidence level, l

As we are using 5% accuracy and a confidence level of 95%, s can be simplified to:

$$s = \frac{\sigma_{\bar{u}}^2 N}{0.00065N + \sigma_{\bar{u}}^2} \tag{12}$$

As the denominator is a factor of 0.00065, s will be very small compared to N if N is very large, thus the computational complexity of projecting p will be reduced.

(2) Partitioning of input

After the random sample s has been selected, partitioning of the input variables into intervals needs to be carried out in order to identify the clusters in the input space. In this paper, we make use of the *partial completeness measure* K to account for the information lost by partitioning [12]. The decisions on how to partition an input variable and the number of partitions required depends on how this partial completeness measure reflects the information lost. Information lost while partitioning mainly occurs in the next stage. The main reason for information loss is due to combining the adjacent intervals in searching for the normal distributed clusters. Besides, the weighting function in discarding any rules as noise is also implied by the partitions, this will in turn cause information to be lost. Let *Ru* be the set of rules obtained by considering all ranges over the variables, then *Ru'* be the set of rules obtained by considering all ranges from the partitioning of the variables. When *Ru* is transformed into *Ru'*, information is lost. The partial completeness measure is basically used to measure how far or close is *Ru'* is to *Ru*.

The partial completeness measure used to identify the number of intervals has direct effect from the human's decision on the confidence level and support level for each rule. It is known that in order to guarantee close rules will be generated, the minimum confidence must be set to $1/K$ times the desired level [12]. It is also shown that equi-depth partitioning minimizes the partial completeness level for any specified number of intervals.

After the analysis, the number of partitions I required can be calculated by the formula:

$$No.\ of\ \ Partitions\ \ (I) = \frac{2k}{\min \sup(K-1)} \tag{13}$$

where k is the number of input variables,
min sup is the minimum support (as a fraction) specified by the user,
and K is the partial completeness level.

It is important to have the partitions as small as possible in order to obtain a constructed rule base *Ru'* that is as close to *Ru* as possible, but this will effectively increase the number of partitions. Care needs to be taken while selecting the appropriate partial completeness level, which is normally chosen to be between 1.5 and 5 [12]. If the number of partitions for an input variable is large, the support for a partition may be low. If the support for a partition is too low and just happens to be away from the searched distribution as shown in Figure 6, it will be treated as an outlier and be discarded while forming the cluster. In order to solve this problem, we introduce an algorithm for combining adjacent partitions while searching for input clusters. The details of the combination algorithm will be discussed in the next section.

(3) Identifying input clusters

In this stage, the s projections are made and the counter $Count_{Bin_j}$ for each bin Bin_j are also set when it is hit by the projection. Bins refer to the intervals of the partitions. Figure 7 shows the bins and counters for a one-dimensional input space and the projections from an output cluster.

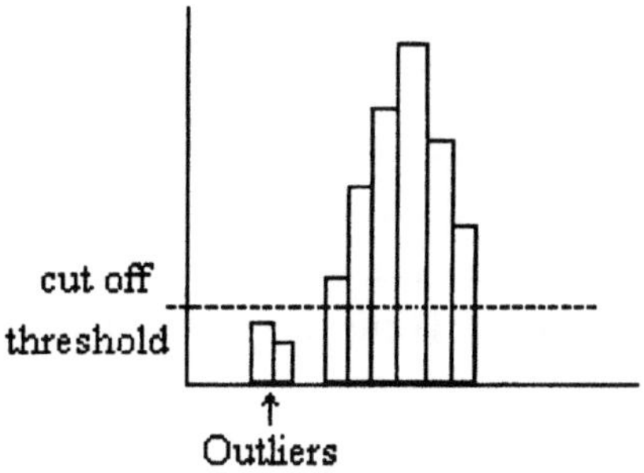

Fig. 6. Partitions with support less than cut off threshold are treated as outlier.

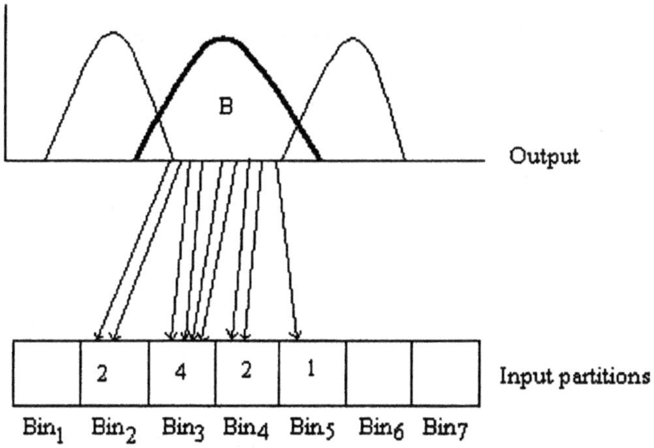

Fig. 7. Example of the projections and counters in the bins.

Beside the counters, a relation is formed by all the input variables and the output. Therefore, a total of s relations are formed. The relations are constructed as:

$$\{[x_1^i(Bin_{1j}), x_2^i(Bin_{2j}), x_3^i(Bin_{3j}), \ldots, x_k^i(Bin_{kj}); C_i]\} \qquad (14)$$

where $i = 1, 2, 3, \ldots s$
and $j = 1, 2, 3, \ldots I$

Based on the counter value in each bin, they are recombined in order to search for fuzzy sets in the input space. The combination algorithm is as follow:

A. Identifying the centre of distribution

By moving from left to right along the bins, the counter values are compared. If $max(Count_{Binj-1}, Count_{Binj+1}) < Count_{Bin}$, then Bin_j is the centre of the distribution. Figure 8 shows an example of the centres of the distribution with the circled bins with highest hits, i.e. centres of the distribution.

6	8	(13)	9	6	(8)	4	2	(7)	4

Fig. 8. Illustration of centre identification.

B. Distance measure.

Find the distance between the left and the centre of the distribution:

$$dist_L = Count_{Binj} - Count_{Binj-1} \qquad (15)$$

If $dist_L$ is small (below a threshold) the two bins are combined into one.

The distance between the right and the centre of the distribution:

$$dist_R = Count_{Binj+1} - Count_{Binj} \qquad (16)$$

Similar to the left hand side, if $dist_R$ is below the threshold, the two bins are combined. Figure 9 shows how they are combined. The above two steps are repeated until the two neighbouring distance measures are above the threshold.

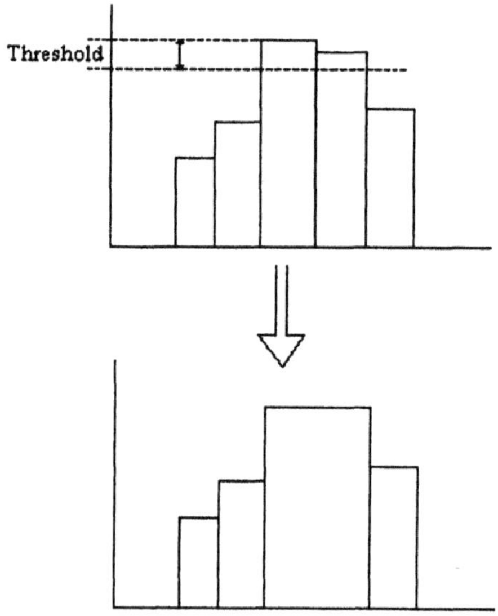

Fig. 9. Combination of bins.

With these normal distributions found in the input space, input clusters are constructed. Trapezoidal approximations based on Ruspini partitions [13] are then used to convert the input clusters to trapezoidal fuzzy membership functions. Basically, the condition below is met when performing the trapezoidal approximations:

$$\sum_{i=1}^{T} A_i(x) = 1 \qquad (17)$$

Trapezoidal approximations based on Ruspini partitions are illustrated in Figure 10.

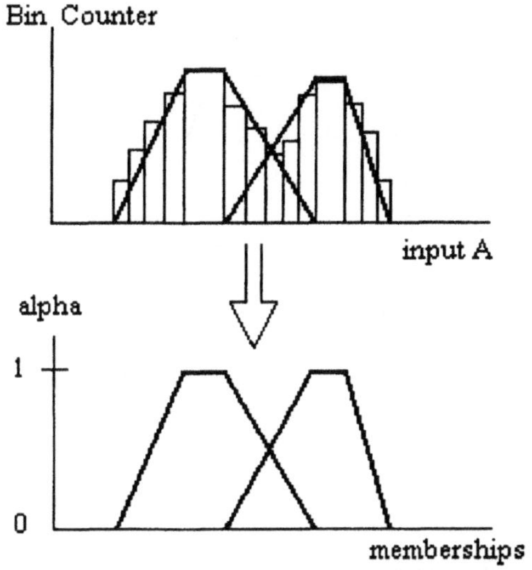

Fig. 10. Approximating trapezoidal fuzzy membership functions.

The bins information in the relation is then substituted with their respective fuzzy membership information. Any repetition relations are then discarded. Relations with small weighting functions or support level are treated as noise and are also discarded.

(4) Constructing fuzzy rules

For this output cluster with the relation information of the input space, fuzzy rules can be constructed. If any fuzzy memberships are found to be next to each other, they are merged into one fuzzy membership as shown in Figure 11. A prediction index (PI) [4] is then calculated based on the predicted output and observed output. The prediction index (PI) [4] is:

$$PI = \sum_{i=1}^{n} \left[\frac{(y^i - o^i)^2}{n} \right] \qquad (18)$$

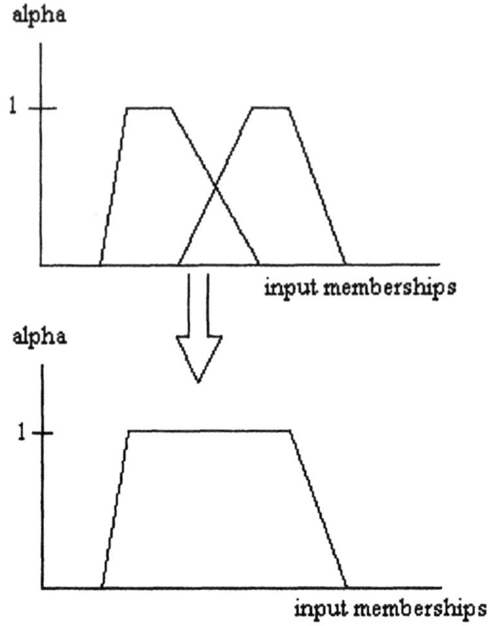

Fig. 11. Merging of fuzzy membership functions.

If it does not affect the PI, they will remain merged. This merging of fuzzy rules will help to reduce the number of fuzzy rules extracted.

5 Conclusion

This paper has examined the Sugeno and Yasukawa's qualitative modeling (SY) method [4]. The popularity of the SY method is mainly due to its rule extraction algorithm. SY method finds the partition in the output space and then projects back to the input space in searching for the input partitions. This results in a sparse fuzzy rule base. This method also avoids the possibility of exponential growth of the fuzzy rule base. One of the important remarks made in the paper [4] is the condition under which more than one fuzzy cluster can be found in the input space that corresponds to the same output cluster. This paper has examined this issue and formulated an algorithm to address it. There are two conditions that are taken into consideration while formulating the approach. First, the algorithm has to be able to identify the occurrence of several rules in one fuzzy output cluster, i.e., the presence of more than one corresponding fuzzy cluster in the input space. Second, the algorithm should be able to decide whether these input fuzzy clusters identified can be merged or any of them can be discarded. The main objective of our approach was to formulate an algorithm that preserves the advantages of the original SY method, which are computational efficiency and small number of fuzzy rules produced.

References

[1] Zadeh, L.A. (1968) "Fuzzy Algorithm", *Information and Control*, vol. 12, pp. 94-102.

[2] Sugeno, M., and Takagi, T. (1983) "A New Approach to Design of Fuzzy Controller," *Advances in Fuzzy Sets, Possibility Theory and Applications*, pp. 325-334.

[3] Nguyen, H.T., and Sugeno, M. (Eds) (1998) *Fuzzy Systems: Modeling and Control*, The Handbook of Fuzzy Sets Series, Kluwer Academic Publishers.

[4] Sugeno, M., and Yasukawa, T. (1993) "A Fuzzy Logic Based Approach to Qualitative Modeling," *IEEE Transactions on Fuzzy System*, vol. 1 no. 1, pp. 7-31.

[5] Kóczy, L.T. and Hirota, K., (1993) "Approximate reasoning by linear rule interpolation and general approximation," *Int. J. Approx. Reason*, Vol. 9, pp.197-225.

[6] Gedeon, T.D. and Kóczy, L.T., (1996) "Conservation of fuzziness in rule interpolation," *Intelligent Technologies, International Symposium on New Trends in Control of Large Scale Systems*, vol. 1, Herlany, pp. 13-19.

[7] Tikk, D., and Baranyi, P., (2000) "Comprehensive analysis of a new fuzzy rule interpolation method," *IEEE Trans. on Fuzzy Systems*, vol.8 no. 3, pp. 281-296.

[8] Bezdek, J.C. (1981) *Pattern Recognition with Fuzzy Objective Function Algorithm*, Plenum Press.

[9] Wadsworth, G.P., and Bryan, J.G., (1974) *Applications of Probability and Random Variables*, Second Edition, McGraw-Hill.

[10] Cochran, W.G. (1977) *Sampling Techniques*, Wiley.

[11] Anderson, T.W. (1996) *The new statistical analysis of data*, Springer

[12] Srikant, R., and Agrawal, R., (1996) "Mining Quantitative Association Rules in Large Relational Tables," *Proceedings of ACM SIGMOD Conference on Management of Data*, Montreal, Canada, pp. 1-12.

[13] Ruspini, E.H., (1969) "A new approach to clustering," *Information and Control*, vol. 15, pp. 22-32.

Can Perceptrons Identify Users of Body Lotion?

B.-J. Falkowski and J. Nietzschmann

University of Applied Sciences Stralsund
Fachbereich Wirtschaft
Zur Schwedenschanze 15
D-18435 Stralsund
Germany
{Bernd.Falkowski,Joerg.Nietzschmann}@fh-stralsund.de

Abstract. A real world problem (identifying users of body lotion from the answers given in a questionaire) is treated using a linear discriminant and a modified fault tolerant perceptron learning algorithm (pocket algorithm). As was to be expected, a fairly large error rate, which may nevertheless be acceptable for practical purposes, is obtained. A comparison with classical statistical methods (binary logistic regression) shows that the pocket algorithm does not perform as well as was previously predicted in general.

1 The Problem

Manufacturers of a large variety of products are very interested in identifying the users of their products using knowledge about certain individual characteristics. They usually attempt to extract the relevant information from data derived from face-to-face interviews conducted by market research institutes. In Germany such data sets may be obtained, for example, from the so-called "Verbraucher-Analyse", a questioning of consumers. In 1999 within the frame of this research 31337 consumers were interviewed. Since a manufacturer of body lotion was interested in identifying the users of his (and similar) products, the interviews included questions concerning age, marital status, profession, personal net income, education, available household net income, address, and skintype. Persons interviewed were also asked, whether they used body lotion or not. Out of these there were 16501 data sets (concerning female persons only; about half of which were body lotion users) available for analysis. In order to make the problem more amenable to mathematical analysis, the attribute values were grouped together. The new attributes (e.g. age between 20 and 30 etc.) were only allowed to take the values 1 respectively 0 for fulfilled respectively not fulfilled. Thus mathematically speaking the aim was to construct a function f: $B^n \rightarrow B$,where B:= {0, 1}, that mirrors the user /non-user property (0/1) of the data-sets provided as accurately as possible (where a potential user is characterized by an element from B^n containing his coded answers to the questions asked) .

Of course, there are various ways to tackle this problem. In particular one could attempt to apply classical statistics methods such as binary logistic regression. However, this seemed somewhat dubious since nothing was known about the underlying distributions and in particular about independence conditions usually

B. Reusch (Ed.): Fuzzy Days 2001, LNCS 2206, pp. 550–553, 2001.

required, cf. e.g. [6], p.109. Hence it seemed advisable to employ somewhat less conservative methods and use classical statistics to evaluate their comparative value.

2 Systems Design

Although it could not reasonably be hoped, that the function f mentioned above would be completely linear, the results described in [4] seemed to indicate that the choice of such an "Ansatz" might lead to an acceptable rate of errors for practical purposes. This seemed all the more attractive since the so-called pocket algorithm (a fault tolerant perceptron learning algorithm, cf. [4]) is easy to implement in Visual Basic for Applications. Seeing that the data sets were also easily transferred into Excel spreadsheets this meant that there were only minimal Software requirements. In view of the probabilistic nature of the algorithm there did, however, remain some doubts concerning the run times required. Moreover it was not clear to what extent optimal learning (guaranteed for infinitely long run times) could be achieved, since very little is known about the rate of convergence of the algorithm. Here optimal means, of course, that the coefficients for the linear discriminant are chosen in such a way that the number of errors on the learning data is minimal. Nevertheless, in view of the fact that in 1990 good results (when compared to standard statistical methods) were reported in [4] and that personal computers available today are very fast indeed when compared to those of 10 years ago, it appeared reasonable, to conduct some experiments.

Another reason for utilizing this "Ansatz" was that it can fairly easily be generalized. Indeed, by explicitly embedding the original data space in a higher dimensional space along the lines indicated in [3], one can increase the capacity of the simple perceptron, cf. e.g. [6], p. 119 significantly. Moreover it is even possible to employ "support vector machines" using kernel methods, cf. e.g. [2], p. 26. As a first step, however, this seemed difficult: For the former method more programming effort and, most probably, considerably more complex hardware would have been needed, whilst for the latter method apart from the increased programming effort the choice of suitable kernels represented a problem.

3 Implementation

The above considerations naturally lead to a system implementation using Excel and VBA. The grouping of attribute values lead to between 39 and 99 binary attributes.

Hence the learning algorithm had to compute weights w_1, w_2, …, w_n for the binary values of the characteristics c_1, c_2, …, c_n and a cut-off c (where $39 \leq n \leq 99$), such that

$w_1*c_1 + w_2*c_2 + \ldots + w_n*c_n > c \leftrightarrow$ the consumer uses body lotion.

Moreover the number of mistakes of this function on the 8250 learning data sets had to be minimized.

This gave rise to rather large run times. Hence a slight variation of the original pocket algorithm was implemented: Essentially it employed sampling without replacement as opposed to sampling with replacement. Due to the probabilistic nature of the algorithm it was of course not possible to evaluate the improvements very precisely. However, during some experimental runs this resulted in runtime improvements by up to a factor of 15.

The experiments were conducted on a Pentium III PC with 256 MB RAM and 500 MHz. With this kind of hardware, which is of course not unusual nowadays, between 5 and 25 million iterations of the main loop of the algorithm could be effected using up CPU times of about 24 hours. This showed quite a marked improvement over the 50 000 iterations mentioned in [4].

4 Experimental Results

As hinted at above, of the available 16 501 data sets only about half were used for learning whilst the other half served as test data set concerning the generalization ability of the system. Both data sets contained about 50% users of body lotion.

Using 5 million iterations of the main loop of the algorithm the overall error rate was 36,67%. The rate of errors concerning the users of body lotion was only 32,76%, with a correspondingly worse rate of errors for non-users of body lotion (40,93%). Both rates refer to the learning data set. The situation appeared rather stable since on the test data set the overall error rate was also about 36%, with about 30% of the errors occurring for users of body lotion. This was in fact the best test run, and only 39 (carefully selected) attributes were used. The test runs concerning the complete set of 99 attributes were slightly inferior and hence discarded (Note here also, that it is advantageous to use as few attributes as possible, since the VC-Bound, see below, is smaller).

The results did not appear altogether satisfactory and hence another test run using only 200 learning data and 99 attributes was performed in order to test the correctness of the algorithm. Here, as was to be expected, the error rate was very much smaller (overall 8,5%, 3,5% for users of body lotion). However, as was also to be expected, the relevant weights lead to excessively large error rates when applied to the whole data set (The VC-Bound is increased significantly, see below).

To get some idea of the quality of approximation, similar tests were conducted using binary logistic regression (BLR) as implemented in SPSS, version 9.0, cf. [1], p. 17. For the former test situation quite similar results were obtained as far as the overall error rate was concerned. The only difference being that the error rate for users of body lotion was slightly better (≈2%), whilst the rate for non-users was correspondinly worse. In the latter test situation, however, the overall error rate using BLR was significantly worse (≈13%)

5 Conclusions

Unfortunately the rather propitious results described in [4] do not seem to apply in the situation described above. In particular the overall error rate seems almost identical with the error rate achieved using BLR, if a large number of data sets is used. This seems to indicate that in this case the number of iterations in the main loop of the (modified) pocket algorithm needed to achieve the (theoretically guaranteed) minimal error rate is prohibitively large. In contrast to this the (modified) pocket algorithm performs rather well, when compared with BLR, if only a small number of learning data is used. Unfortunately though, the tested small set of learning data did not generate weights, that could be successfully used in the general situation. This could, of course have been predicted using some results concerning the Vapnik-Chervonenkis bound, cf. [5], p.99. Since it is generally thought that these results are overly pessimistic, cf. e.g. [2], p. 54, it still seems useful to have run the test. This is all the more true since the generalization of the results obtained using the 8250 data seems far better than predicted by using the (distribution-free) VC-bound.

It should be pointed out that we were puzzled at first by the rather good predictions concerning users of body lotion as opposed to non-users and the (small) differences to results obtained using BLR. We suspect, however, that this may be due to the fact that the available data contained an unrealistically high number of users, whose characteristics could consequently be learned more accurately. If this were true then we could also reasonably conjecture that the (modified) pocket algorithm performs better than BLR if fewer learning data are available: This would be certainly be supported by our experiments in the case of 200 learning data as well as in the case of non-users of body lotion.

Finally we would like to point out that it is not yet clear whether these results are immediately useful from a practical point of view since the problem obviously requires a higher order separating hypersurface as opposed to a hxperplane. At any rate it seems that they are better than pure guesswork. If more accurate predictions are required, then one has to think again, whether the expense required for more sophisticated methods, see section 2. above, can be justified.

References

1. Bühl, A.; Zöfel, P.: Einführung in die moderne Datenanalyse unter Windows. Addison-Wesley, (2000)
2. Christianini, N.; Shawe-Taylor, J.: An Introduction to Support Vector Machines and other Kernel-Based Learning Methods. Cambridge University Press, (2000)
3. Cover, T.M.: Geometrical and Statistical Properties of Systems of Linear Inequalities with Applications in Pattern Recognition. IEEE Transactions on Electronic Computers, 14, (1965), pp. 326-334
4. Gallant; S.I.: Perceptron-based Learning Algorithms. IEEE Transactions on Neural Networks, Vol. I, No. 2, (1990), pp. 179-191
5. Haykin, S.: Neural Networks,.2nd edition, Prentice Hall, (1999)
6. Ripley, B.D.: Pattern Recognition and Neural Networks. Cambridge University Press, (1997)

Reinforcement Learning for Control of Traffic and Access Points in Intelligent Wireless ATM Networks

Jerzy Martyna

Jagiellonian University, Dept. of Computer Science
ul. Nawojki 11, 30-072 Cracow, Poland

Abstract. In this paper, we introduce the scheme for bandwidth allocation and congestion avoidance in wireless intelligent ATM networks which is based on reinforcement learning (RL). Our solution guarantees a desired bandwidth to connections which require a fixed wide bandwidth according to QoS constraints. Any unused bandwidth is momentarily backed up (returned) to Virtual Circuits. Proposed RL method is also suitable for supporting the mapping between a single ATM switch port and a wireless access points. It can control the access points in wireless intelligent ATM networks.

1 Introduction

Reinforcement learning (RL) was introduced to the literature by A. G. Barto and R. S. Sutton in their works [1], [15]. The work of Sutton presents the temporal credit assignment problem in which the ultimate results of the system act is in relation of merit and faulty earlier steps. Presently, most attention was paid to the temporal differences [16] in which credit must propagate back through the temporal sequence of decisions that led to success. This is known as Q-learning algorithm.

Some papers were devoted to applications of reinforcement learning to packet routing in dynamically changing networks [6], channel allocation in cellular telephone system [14], call admission control in telecommunication networks [9], etc. Recent works on the use of reinforcement learning in ATM networks [10], [18] has shown that learning can be used to work with such methods in the real broadband networks. The iteration of learning with peception is a crucial factor of this approach.

Wireless ATM (WATM) are recognized as a general solution for next-generation communication networks. These networks will be required to support multimedia traffic [2], [11], [5], [20]. Intelligent WATM belong to an architectural concept that will provide new services [4], [7], such as creation of *ad hoc* local area networks.

Integration of WATM with the intelligent functions make possible the mobile networking. In these intelligent networks it will be possible to provide bandwidth-on-demand to portable devices. Hence, all advanced multimedia services in mo-

B. Reusch (Ed.): Fuzzy Days 2001, LNCS 2206, pp. 554–564, 2001.

bile environment are performed. Moreover, in comparison with current networks in WATM networks more efficient use of network resources can be achieved.

In this paper, we will use the RL method that does not require iterative behaviour in learning process of the designed perception system. We will use the so called Lion algorithm introduced to the literature in the paper of Whitehead and Ballard [21] and modified by Piggott and Sattar [12]. In this approach, a perception system's total state reduces the complexity of learning. It allows the time reduction of learning process and also the simplification of implementation of perception systems in wireless intelligent ATM networks.

The rest of this paper is organized as follows. In section 2, we describe the problem of building of wireless intelligent ATM networks. Section 3 gives the solution of the problem by the use of the reinforcement learning method. In section 4 we give some experimental results obtained through simulation of our model. Section 5 concludes the paper.

2 Problem Description

In this section we formulate the problem of bandwidth allocation in intelligent wireless ATM networks (WATM networks). The typical intelligent WATM network's architecture is given on Fig. 1 (see [4]).

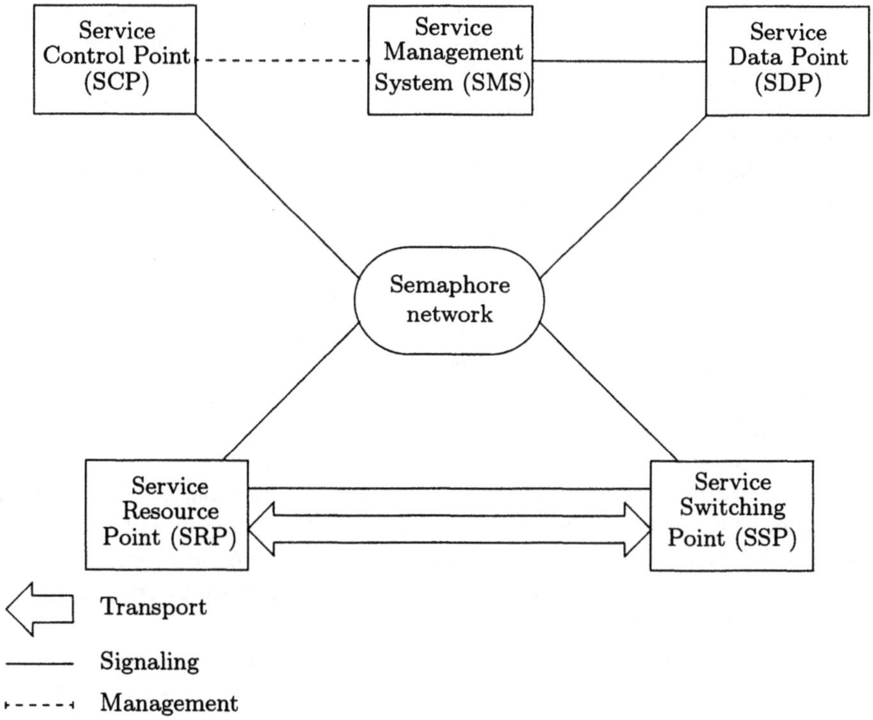

Fig. 1. An architecture of wireless intelligent ATM network

We assume that our intelligent Wireless ATM network performs as follows. When a subscriber asks for an intelligent service, the terminal starts its execution by sending a message to the Service Control Point (SCP). The SCP executes the intelligent service program based on information stored at the Service Data Point (SDP) and interactions over Signaling Network (SN) with the Service Switching Point (SSP). The SSP is able to command the connection processor, since it is placed at the switching node.

It is obvious that SSP anf Service Resource Point (SRP) form data streams from SRP or from users to other users according to the required service type. Therefore, one of the SRP and SSP must be adopted to all AAL types in ATM networks.

In our approach we use two categories of complementary services: switch changes and CAC services [4]. In the first service we admitted three possibilities:

a) dynamic busy server change,
b) dynamic change of mobile's server,
c) economic connection services.

Ad a) In *dynamic busy server change* services are activated either by the client or by the basic service supplier, if it is impossible to connect the user to a given destination server. In this service it is possible to change the requested connection rate and the servers' utilization.

Ad b) We assumed that in the *dynamic change of mobile's server* a mobile station is a base support station (BSS) and there is a connection to a special service. Hence, the typical mobile station is serviced by the nearest server providing the considered service. As the mobile station moves to another BSS, which supports the same application that needs fewer connections or has smaller cost of connection than the first one, the service will have to switch the connection to the second server. The dynamic change of mobile's server can decrease response time, reduce the effect of the jitter phenomena at the same quality of service (QoS) parameters of communication.

Ad c) In economic connection service the lowest connection by the same parameters of QoS is determined. It is possible, among others, thanks to the finding of the shortest way to the mobile's server.

In the CAC complementary service several options are possible. As first, we introduce *"additional connection"*. It is needed when a sender want to increase its output. A new additional connection will automatically transport the exceeding ATM cells according to the previously assumed QoS parameters. To another CAC complementary service belongs the *"call me back when possible"* option. This service is activated when the network rejects its connection request due to a lack of resources. In this complementary service we can automatically establish such criteria as priority, resource utilization and optimization.

In our approach we can establish networks the following intelligent services introduced above for wireless ATM networks. These services can manage the communication resources, placed at the SSP and SRP points in wireless ATM networks, etc. These intelligent services include:

1) *Dynamic change of unmobile, busy switch* - used in all situations in which the SCP point is idle or out of order state. Hence, the SSP point proposes the new connection to another switch. Any time there are signaling messages send to a client and to service point to compose the transmission connection. The SCP must know which SSP point to choose in order to satisfy a client request. It needs the whole intelligence and knowledge to make the best decision.

2) *Dynamic change of mobile's switch* - used in all situations in which the continuation of service's execution must be ensured. For the most part it is caused by the movement of the SSP which requires the new transmission connection with the SRP point. The client/server communication session is implemented as in the previous intelligent service.

3) *Intelligent traffic control service* handles all QoS parameters needed for the connection in traffic contract. In our approach, we include the peak cell rate, cell delay variance tolerance and the conformance definition. The conformance definition may be based on the generic cell rate algorithm (GCRA) to check the conformance to the traffic contract at a user-network interface.

The intelligence implemented in CAC algorithm determines whether the WATM can accept a request as the service with needed QoS parameters without degradations of existing connections.

3 Reinforcement Learning and Its Application to Control of Traffic and Access Points in Intelligent Wireless ATM Networks

In this section, we would refer to some results of reinforcement learning (RL). We recall that in the general formulation of the bandwidth allocation problem in intelligent WATM networks interaction between learning system and environment in time t is described as $< s_t, a_t, r_t >$, where s_t, a_t, r_t are descriptors in time t of state, action and reinforcement value, respectively. The fourth defined by the state, action, reinforcement is obtained after their performing. The next state is defined as the system experience. By the use of time steps t we can define our learning system as $< s_t, a_t, r_t, s_{t+1} >$.

We can describe the general scenario of reinforcement learning as the sequence of operations performed at each time step t (see Fig. 2). Step 5 is crucial for the

At each time step t:

1. access the current state s_t:
2. choose an action a_t to perform in the state s_t according to the actual decision strategy;
3. perform the action a_t;
4. obtain a reward r_{t+1} and access the new state s_{t+1};
5. analyse the decision strategy according to the system experience $< s_t, a_t, r_t, s_{t+1} >$

Fig. 2. General algorithm for reinforcement learning

At each time step t:
1. access the current state t;
2. choose an action a_t and perform it;
3. obtain a reward r_{t+1};
4. access the new state s_{t+1};
5. update $Q(s_t, a_t)$ using a learning factor α and a discount factor γ as follows:

$$Q_{t+1}(s_t, a_t) \leftarrow Q_t(s_t, a_t) + \alpha \Delta Q_t$$

where

$$\Delta Q_t \leftarrow (r_{t+1} + \gamma \max_a Q_t(s_{t+1}, a)) - Q_t(s_t, a_t)$$

Fig. 3. Q-learning algorithm estimates new state obtained by performing the chosen action at each time step

realization of reinforcement learning in the environment that can change its own states. In general, these changes are stochastic. The mathematical model for reinforcement learning for this environment is known as the Markov decision process. By using the stochastic dynamical programming [3] we can find the optimal decision strategy on the condition that all probabilities of state change and expected value of reinforcement for each pair state-action are known. Otherwise, we can use the method of reinforcement learning.

For the realization of reinforcement learning, the best means is the *temporal credit assignment* [15]. In this approach each action is assigned "merit" and "feel" for the whole action of the system given earlier which may allow to observe the results. The solution of this problem includes the *temporal differences* given by Sutton [16]; the best known algorithm takes the Q-algorithm.

In Q-algorithm we evaluate Q-function which estimates all pairs state-action in regard of expected rewards. Precisely, with each pair state-action the current *estimate* of the expected discounted rewards for the actions is associated, under the assumption that each forms part of an optimal policy. At each time step t, the value returned by Q for the action chosen is updated by an error factor ΔQ_t. This is the difference between the old value and the sum of the immediate reward and the future reward expected from continuing the policy, discounted by a constant factor γ. In each step Q-function (value) is modified according to the algorithm given in Fig. 3. It was showed by Watkins and Dayan [19] that Q-value (function) converges towards unique value, if the expected future rewards are represented discretely.

It is obvious that in Q-algorithm we have the exchange between exploration (executed actions that are not reckoned as the best goal to improve our knowledge) and exploit (use of actual accessed knowledge to obtain the reward).

The use of Q-algorithm is inaccurate in some situations. In the situations that are structurally similar, the same action will be taken in all of them. To the better classification of system to distinguish between some structural situations the Lion algorithm was introduced by Whitehead and Ballar [21]. The Lion algorithm (LA) identifies the so called *consistent* representations. It relies on the fact

5. Update $Q_{t+1}^{(i)}(v_t^{(i)}, a_t) \quad \leftarrow$

$$\begin{cases} 0 & \text{if } i \in \text{lions}_t \text{ and } \Delta Q_t < 0 \\ Q_t^{(i)}(v_t^{(i)}, a_t) + \alpha \Delta Q_t & \text{if } i \in \text{lions}_t \text{ and } \Delta Q_t \geq 0 \\ Q_t^{(i)}(v_t^{(i)}, a_t) + \alpha' \Delta Q_t & \text{otherwise} \end{cases}$$

where $\text{lions}_t \quad \leftarrow \quad \{i \mid Q_t^{(i)}(v_t^{(i)}, a_t) = \max_j Q_t^{(j)}(v_t^{(j)}, a_t)\}$

and

$$\Delta Q_t \leftarrow c_{t+1} - Q_t^{(i)}(v_t^{(i)}, a_t), \quad i \in \text{lions}_t$$

$$c_{t+1} \leftarrow \max[r_{t+1}, \gamma \max_{i,j} Q_t^{(i)}(v_t^{(i)}, a_j)]$$

Fig. 4. Q value update used by the Lion algorithm

that in a deterministic world only *inconsistent* states will ever cause the *negative* update of a Q value. An action decision that leads to an inconsistent state will require a negative update of a Q value. The decision system thus learns to use only consistent internal state.

In other words, this approach reduces the complexity of learning by removing the need to learn a separate rule for each successive state where the action required is the same. It requires the repeated execution of a loop. These iterations are significated limitation of the Lion algorithm. These limitations are omitted in the modified Lion algorithm proposed by Piggot and Sattar [12]. In this algorithm the fundamental unit called a *rule* is defined. A rule (v_i, a) pairs one view, or representation state of receptor i with one action, and has a Q value associated with it. At time t the rule $(v_t^{(i)}, a_t)$ is applicable when receptor i is viewed and has a v_i view. Among many of rules (if they are) that correspond to the action, a dominant active rule is a *lion*. The Fig. 4 shows the calculation of a new Q value in LA algorithm.

Figure 5 outlines the calculation of a new Q value in the Lion algorithm. It is expressed in terms of our problem solution. The Q value is updated in another

$$Q_{t+1}^{(i)} v_t^{(i)}, a_t) \leftarrow \begin{cases} 0 & \text{if } i \notin \text{significant} \\ Q_t^{(i)}(v_t^{(i)}, a_t) \ + \ \alpha \Delta Q_t & \text{if } i \in \text{lions}_t \\ Q_t^{(i)}(v_t^{(i)}, a_t) \ + \ \alpha' \Delta Q_t & \text{otherwise} \end{cases}$$

where

$\text{significant}_t^{(i)}$ if i $\in$ neighbouring cells

or $\text{significant}_t^{(i)} \ \leftarrow \{ \text{ i} \mid Q_t^{(i)}(v_t, a_t) \leq c_{t+1} \text{ or } v_t^{(i)} = v_{t+1}^{(i)}\}$

$\text{lions}_t \leftarrow \{i \mid Q_t^{(i)}(v_t^{(i)}, a_t) = \max_j (Q_t^{(j)}(v_t^{(j)}, a_t)\}$

and

$\Delta Q_t \leftarrow \max\{r_{t+1}, c_{t+1} - Q_t^{(i)}(v_t^{(i)}, a_t^{(i)}), \text{ where } i \in \text{lions}_t\}$

Fig. 5. Q value update used in our modification to the Lion algorithm

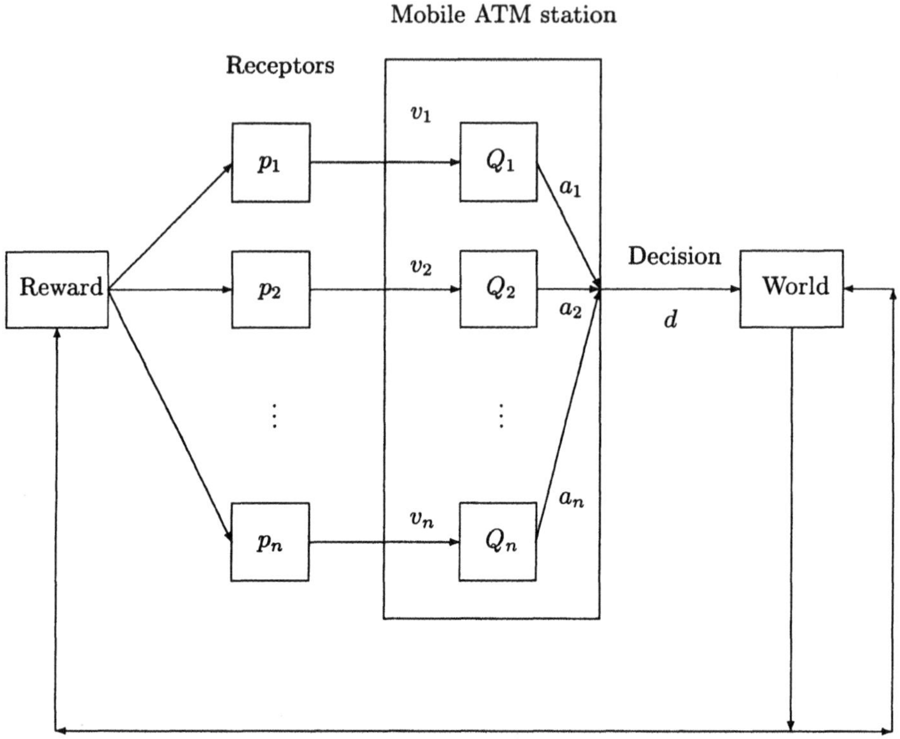

Fig. 6. A formal model of a mobile ATM station with n receptors

way. In our solution we addmitted new states known as "significant". To these states belong all possible situations described here by the receptors installed in the mobile station. In our approach, these situations are limited only to view in cell, in which our mobile station stays is viewed from the neighbouring cells. The credit value, c, is the same as in the previous given algorithm. We additionally obtain some informations by discrediting "significant" rules, not just belonging to the lion. As in the former presented algorithm, a lion is a significant active rule with the largest value of Q. Our ΔQ_t takes into consideration the current state given by the Q_t and the next state hidden in the reiforcement factor. It allows to select the value of the reward, r.

4 Computational Experiment

It was assumed that the mobile ATM staton is under an influence of the views $(v_i, i = 1, \ldots, n)$. They are taken away by the receptors $p_i, i = 1, \ldots, n)$. Our mobile ATM station undertakes the decision $d = v_1 \times \ldots \times v_n$, which concerns the control points and the addmissible traffic rate. The decision is not in association with the movement. A formal model of our mobile ATM station is given in Fig. 6.

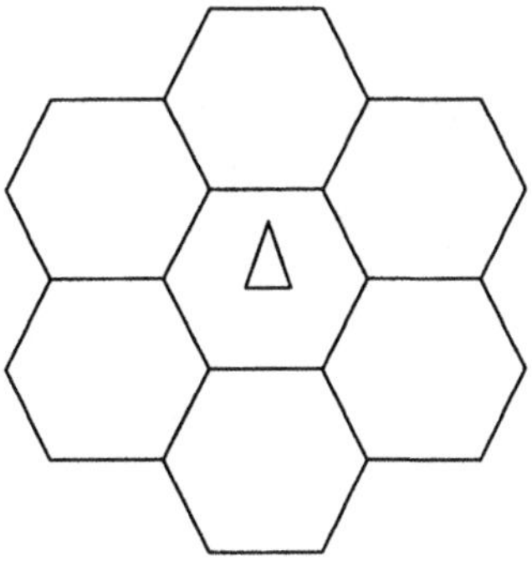

Fig. 7. Cell layout

The activity of our mobile ATM station is as follows. Each receptor p_i transforms the current world state to an internal representation view from the set v_i and has a corresponding Q function in the decision system. Our modification to the Lion algorithm discredits the Q values of the significant views. The decision system learns to choose the action that will maximize the return from the reward function r.

Our cellular concept of the intelligent wireless ATM network assumes the ability to use a radio channel from several directions in the geographic area simultaneously. This idea is illustrated in Fig. 7. The cell site at the center of each cells provides radio coverage for that cell. In our approach, each mobile ATM station contains seven receptors connected with antennas for the radio coverage. One of them is intended for the cell in which it stays. The other six receptors are intended for the six neighbouring cells. Crossing over the border between cells by the mobile ATM station is associated with switching to another receptor. An appreciation of possible traffic is here realized through two additional receptors used in current data transmission.

The first is devoted to measuring the peak cell rate (PCR) parameter and the second to the cell delay variation (CDV) parameter. Both parameters allow to determine the class of traffic rate, namely: CBR, rt-VBR, nrt-VBR or ABR. Changing of one of negotiated parameters brings about the switching to other cell, which is determined from among the neighbouring cells.

Three scenarios were tested in the computational experiment. For each scenario the intelligence of the mobile ATM station moved from point A to point B in the given topology of the wireless ATM network was studied. In the first experiment, (Fig. 8a) a dynamic change of an unmobile, base support station (BSS) was studied. In the second, (Fig. 8b) the dynamic change of the mobile's switch was tested. We assumed that the movement of the mobile's switch is in opposite direction to the same speed as the spped of mobile ATM station. The third experiment (Fig. 8c) tested for the change of PCR and CDV parameters.

The change was tested to show the possibilities of switching between two BSS to keep the PCR and CDV parameters for the assumed traffic rate.

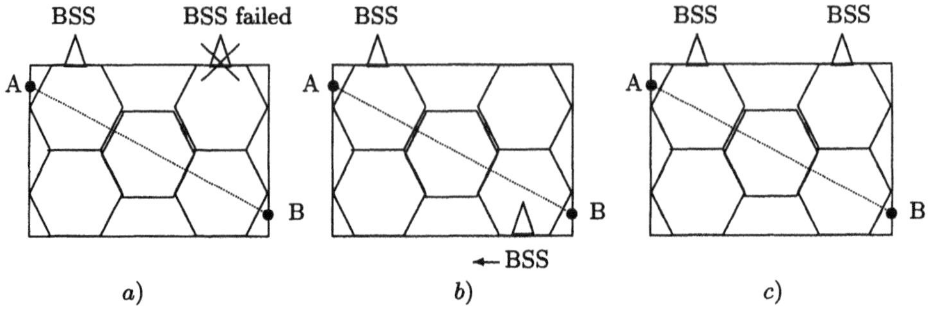

Fig. 8. Three scenarios used in the computational experiment. Mobile ATM station moves from point A to B

In our computational experimet we verified the effectiveness of the reinforcement learning method based on our modification of Lion algorithm. The assumed values of all the parameters are given in Table 1. In our implementation of the reinforcement learned algorithms, integers to represent the value of Q were used. It improved space and time performance. We observed that the reward is varied between 15 and 5693. The number of steps required for each change of access point or traffic rate control was averaged over 200 trials.

We found that our modified Lion algorithm had limited properties to learning in comparison with the typical reinforcement algorithm [16]. It was observed in the first switching. The average number of errors (see Table 2) was 10 percent greater than in the steady switching.

Table 1. Parameter settings using in simulation and learning

Parameter	Value
Reward	5693
α	0.6
α'	0.3
γ	0.5

5 Conclusions

The study has shown that the reinforcement learning method can be used to control of access ponts and traffic rate in intelligent wireless ATM networks. The model parameters evaluated included three situations: dynamic busy server

Table 2. Performance of computations

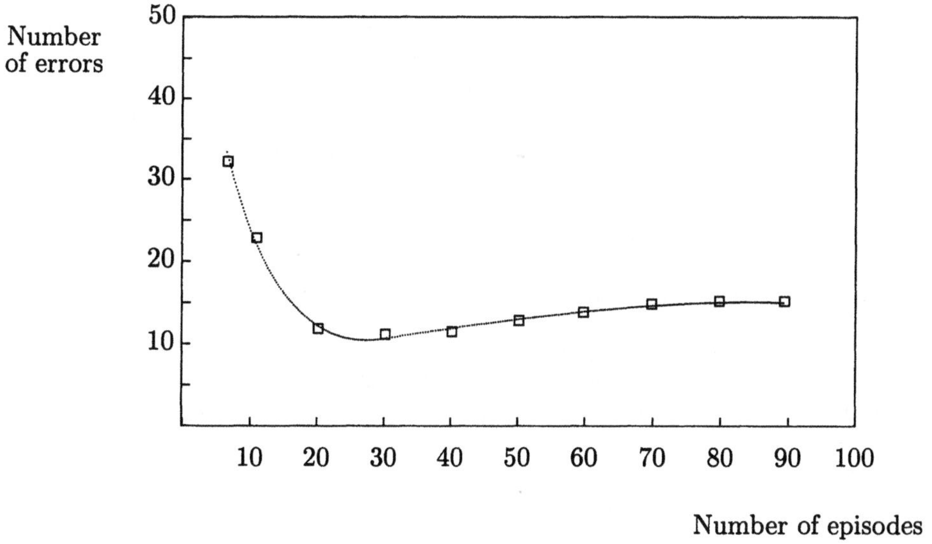

change, dynamic change of mobile's server, economic connection services. The study shown that the system is also flexible and easily to modified to include additional variables as well as increase its accuracy.

The reinforcement learning method approach is not based on mathematical model and requires no analytical or theoretical knowledge. This approach is feasible and lends itself to modification. While the traditional approach yields to fixed values for the change, the reinforcement learning method gives dynamic values depending on the various traffic rates and geometric conditions.

References

1. A.G. Barto, R.S. Sutton, C.W. Anderson, *Neuronlike Adaptive Elements That Can Solve Difficult Learning Control Problems*, IEEE Transactions on Systems, Man, and Cybernetics, Vol. 13, 1983, pp. 835 - 846.
2. R.R. Bhat, *Draft Baseline Text for Wireless ATM Capability Set 1 Specification*, ATM Forum BTD-WATM-01, 10 Dec. 1998.
3. R.E. Bellman, *Dynamic Programming*, Princeton University Press, Princeton, NJ, 1957.
4. C. Ben Ahmed, N. A. Boudriga, M. S. Obaidat, *A Performance Analysis Scheme for Intelligent Wireless ATM Networks*, Int. Journal of Communication Systems, Vo. 12, 1999, pp. 405 - 426.
5. S.K. Biswas, B. Sengupta, *Call Admissibility for Multirate Traffic in Wireless ATM Networks*, IEEE INFOCOM, 1997, pp. 649 - 657.
6. J.A. Boyan, M. Littmann, *Packet Routing in Dynamically Changing Networks: A Reinforcement Learning Approach*, in: J.D. Cowan et al. (Eds.), *Advances in Neural Information Processing Systems - 6*, Morgan Kauffman, San Francisco, 1994, pp. 671 - 678.

7. G. Christensen, R. Duncan, *Wireless Intelligent Networking*, Artech House, London, 2000.
8. S. Mahadevan, J. Connel, *Automatic Programming of Behaviour-Based Robots Using Reinforcement Learning*, Artificial Intelligence, Vol. 55, 1992, pp. 311 - 365.
9. P. Marbach, O. Mihatsch, M. Schulte, J.N. Tsitsiklis, *Reinforcement Learning for Call Admission Control and Routing in Integrated Service Networks*, M. Jordan, M. Kearns, S. Solla (Eds.), *Advances in NIPS - 10*, MIT Press 1998, Cambridge 1998.
10. P. Marbach, O. Mihatsch, J.N. Tsitsiklis, *Call Admission Control and Routing in Integrated Services Networks Using Neuro-Dynamic Programming*, IEEE Journal on Selected Areas in Comm., Vol. 18, No. 2, Feb. 2000, pp. 197 - 208.
11. M. Niemi, *Application Requirements for WATM*, ATM Forum, Document No. 96-1058, 1996.
12. P. Piggott, A. Sattar, *Reinforcement Learning of Iterative Behaviour with Multiple Sensors*, Journal of Applied Intelligence, 4, 1994, pp. 351 - 365.
13. S. P. Singh, *Reinforcement Learning Algorithms for Average-Payoff Markovian Decision Processes*, AAAI '94, pp. 700 - 705.
14. S.P. Singh, D. P. Bartsekas, *Reinforcements Learning for Dynamic Channel Allocation in Cellular Telephone Systems*, in: M. Mozer et al. (Eds.), *Advances in NIPS 9*, MIT Press, Cambridge, 1997, pp. 974 - 980.
15. R.S Sutton, *Temporal Credit Assignment in Reinforcement Learning*, Ph. D. Thesis, Dept. of Computer and Information Science, University of Massachusetts, 1984.
16. R.S. Sutton, *Learning to Predict by the Methods of Temporal Differences*, Machine Learning, Vol. 3, 1988, pp. 9 - 44.
17. R.S. Sutton, A.G. Barto, *Reinforcement Learning, An Introduction*, MIT Press, Cambridge, 1998.
18. H. Tong, T. X. Brown, *Adaptive Call Admission Control Under Quality of Service Constraints: A Reinforcement Learning Solution*, IEEE Journal on Selected Areas in Comm., Vol. 18, No. 2, Feb. 2000, pp. 209 - 221.
19. C.J.C.H. Watkins, P. Dayan, *Technical Note: Q-Learning*, Machine Learning, Vol. 8, 1992, pp. 279 - 292.
20. E. K. Wesel, *Wireless Multimedia Communication. Networking, Video, Voice, and Data*, Addison-Wesley, Reading, 1998.
21. S.D. Whitehead, D.H. Ballard, *Learning to Perceive and Act by Trial and Error*, Machine Learning, Vol. 7, No. 1, July 1991, pp. 45 - 83.

Categorization of Time Delay by Means of Neural Networks

Sabine Bostelmann

University of Dortmund
Chair of Computer Science I
Otto-Hahn-Str. 16, D-44221 Dortmund

Abstract. This paper presents a method to categorize and supervise time delays by means of Kohonen maps. In detail it will be shown that the evaluation of Weibull and Erlang distributions of the delay data of a company is an essential preprocessing step which enables a Kohonen map to re-evaluate how many classes of delay are coded in the data sets. Further on it can be shown that the knowledge of this topological mapping of time delay classes makes it possible to compute the severity of each delay class by using appropriate weighting factors, such that difficult problems can be remedied first and will not delay the course of the system any further.

1 Introduction

A regarded feature 'time delay' can be originated from any system, but in this paper the term delay will be applied to a company that transports goods between different places. Each good has an appointed time to be picked up and another one to be delivered, likewise each vehicle transporting these goods has a departure time and an arrival time. Every overstay of these times is regarded as a delay.

The input form of the categorizing and supervising program requires information on the underlying data and on the desired kind of output. The input data can contain pure delay times (in minutes, as plain text) or a class-related representation of the data set. The output of the result given by the analyzing program can be chosen to be made up of tabulars, graphics or in textual form.

1.1 The Evaluation of Time Delays

The categorization of time delay data consists of its interpretation generated by neural networks. In order to get good results, the delay data has to be preprocessed first.

The Preprocessing Strategy. To obtain convincing results by means of classifying data with neural networks, the delay data needs to be preprocessed. On the one hand the preprocessing has to guarantee an adequate problem oriented

B. Reusch (Ed.): Fuzzy Days 2001, LNCS 2206, pp. 565–575, 2001.

representation of the delay data, on the other hand it has to correct the data set by disregarding any disturbances. To achieve this goal the data is evaluated statistically first. Due to the mostly relative small number of delay data, class related statistics are applied, building the starting point of the whole preprocessing. Depending on the operator's selection the class related statistics are read directly from preprocessed data by the program or will be calculated in the first step of the data mining process.

By means of the disposition in delay classes the empirical distribution function of the input data is computed. Subsequent to this the theoretical distribution function is searched for which approaches the empirical distribution function best. At this, two functions can be selected. The WEIBULL distribution

$$F(t) = 1 - e^{-(\frac{t-t_0}{\alpha})^\beta}, \qquad t > t_0 \ ,$$

adapts very easily to different distribution functions due to its three parameters α, β and t_0, thus it is suitable to be approached to the empirical distribution function. For example: if $\beta = 1$ and $t_0 = 0$ the WEIBULL distribution equals the *exponential* distribution

$$F(t) = 1 - e^{-\frac{t}{\alpha}}$$

with the parameter $\frac{1}{\alpha}$; setting $\beta = 2$ and $t_0 = 0$ the WEIBULL distribution resembles the RAYLEIGH distribution

$$F(t) = 1 - e^{-(\frac{t}{\alpha})^2} \ .$$

Due to its distinct ability to adaption, positive as well as negative skewness can be shown.

The disadvantage of the WEIBULL distribution is that it is hard to interpret with a model because of its variability. For this reason the ERLANG-n distribution

$$F_n(t) = 1 - e^{-\lambda t} \sum_{i=0}^{n-1} \frac{(\lambda t)^i}{i!}$$

is used additionally. It is based on the Queuing Theory and is suitable to be applied to the actual problem [4]. Alike the WEIBULL distribution the ERLANG-n distribution with parameter $n = 1$ resembles the exponential distribution.

For further procedure the WEIBULL parameters α and β as well as the ERLANG parameter λ and n are calculated by a fitting algorithm such that the occurring squared deviation between theoretical and empirical distribution function is minimized. The parameter of position t_0 here is set to 0 as the minimum delay will always be greater than 0.

The resulting parameters of the WEIBULL and ERLANG distribution are coded in one vector per record, these vectors are categorized by the neural network afterwards.

2 Categorization of Statistical Data by Neural Networks

If a record – describing a so far unknown system state – should be categorized, the technique of vector quantization can be regarded as an appropriate method.

This technique goes back to the idea to represent a set of records that needs to be classified as an n-dimensional feature space. Each record corresponds to one feature vector in this space. During conditioning the net maps the training vectors – a subset of all records that is used to train the net – to vectors that represent the corresponding features. These vectors serve as reference vectors and correspond to the neurons of unsupervised conditioned nets. Thus a set M of feature vectors can be described by a finite set A of reference vectors $\boldsymbol{w}_j \in \mathbb{R}^n$, $j = 1, ..., |A|$, e.g. a subspace $M \subseteq \mathbb{R}^n$. After training the net it can classify any input signal due to its associativity by assigning the particular signal to the reference vector that is most similar to it. The neuron that is activated the most by a vector is called the 'winner neuron'. The quality of the classification depends on how good the input and the reference vector match. Assigning an input vector $v \in M$ to the reference vector that resembles it most, the feature space M is divided in different regions M_j which are called *Voronoi polygons* or *areas of reception.* Regarding any norm, the following is valid

$$M_j = \{\boldsymbol{v} \in M \mid \|\boldsymbol{v} - \boldsymbol{w}_j\| \leq \|\boldsymbol{v} - \boldsymbol{w}_i\| \, \forall i\} \ .$$

Figure 1 shows four neurons with their gray highlighted areas of reception tough only the interstice of the neurons is regarded.

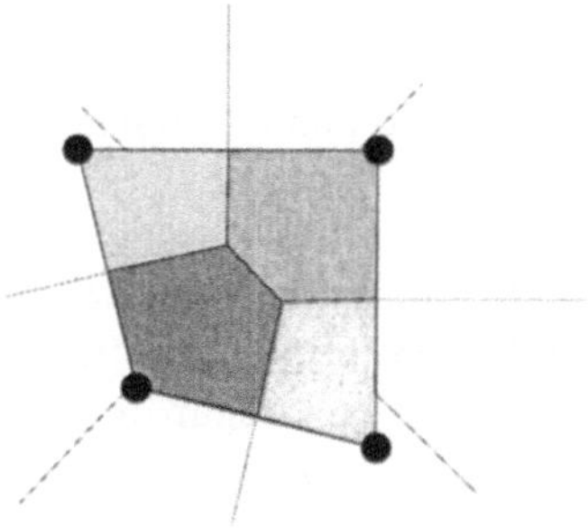

Fig. 1. Voronoi polygons [3]

A special vector quantization in which the input vectors are placed on a squared or hexagonal grid is done by Kohonen maps.

A common Kohonen map is a neural network consisting of two layers, the input layer and the so called Kohonen layer. As indicated in Fig. 2 all input neurons are connected to all output neurons. The weighting factors w_{ij} of the incoming connections build the weighting vector $\boldsymbol{w}_j$ of the neuron. Thus the dimension of the weighting vector is given by the number of input neurons.

The object of conditioning such a net is to distribute the neurons as even as possible in the input space. This is shown exemplary for a map of the dimension 2 in Fig. 3. This demand is put into practice by assigning a local neighborhood function to the neurons. In the first place this function allows the winner neuron

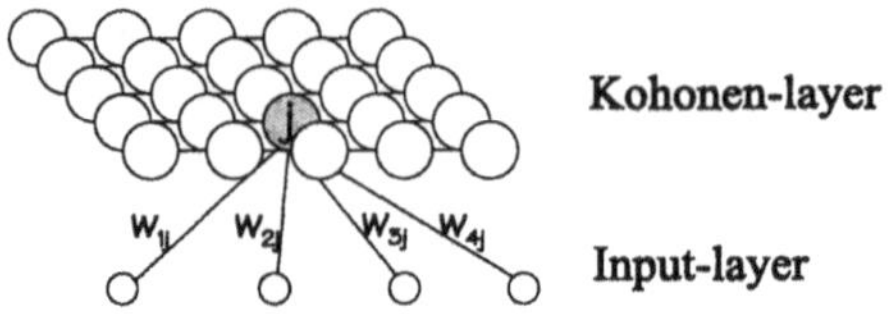

Fig. 2. Scheme of a Kohonen map [3]

to attract the whole activity of the map, in the second place it enables the winner to influence its neighbors. The interference always depends on the chosen neighborhood function. In general, the larger the distance from a neuron to the actual winner is the smaller the influence on the neuron will be.

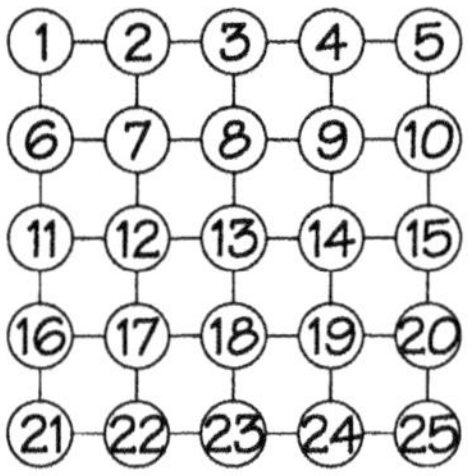

Fig. 3. Neighborhood on a 2-dimensional map [3]

During the iterative conditioning procedure the neighboring neurons of the map will become neighbors in the input space as the actual input vector $\boldsymbol{E}$ is compared with all weighting vectors $\boldsymbol{w}_j$ regarding any norm. To let the neuron j win that resembles the input vector the most, the euclidic distance is used as norm and is minimized

$$win_{\text{euclid}}(\boldsymbol{E}) = j \text{ with } \|\boldsymbol{E} - \boldsymbol{w}_j\| = \sum_{k=1}^{n} |e_k - w_{kj}| \stackrel{!}{=} \min$$

where at $win_{\text{euclid}}(\boldsymbol{E})$ represents the winner neuron.

To reward the neuron j that is energized the most by the actual input vector, the scalar product is applied as norm and is maximized

$$win_{\text{scalar}}(\boldsymbol{E}) = j \text{ with } <\boldsymbol{E}|\boldsymbol{w}_j> = \sum_{k=1}^{n} e_k \cdot w_{kj} = a_j \stackrel{!}{=} \max \ .$$

After the winner is appointed its weighting factors are changed such that it fulfills the principle of optimization even better, its neighbors' activities are influenced too. The changing of the weighting factors follows the *Hebb's learning rule.* This rule states that the connection weighting factors w_{ij} are always augmented when neuron j is highly activated by an input signal of the highly energized neuron i. The changing of the weighting factors Δw_{ij} follows the rule

$$\Delta w_{ij} = \nu \cdot A_i \cdot a_j \ .$$

Thus the change depends on the output signal of neuron i and the resulting activity of neuron j. This product is weighted with the constant learning rate ν.

Adapted to the Kohonen map the evaluation of the weighting factors in the $k+1$ step results in

$$\boldsymbol{w}_j(k+1) = \boldsymbol{w}_j(k) + \nu(k) \cdot h_{cj}(k) \cdot (\boldsymbol{E}(k) - \boldsymbol{w}_j(k)) \ .$$

Disaccording to the Hebb's learning rule, the learning rate $\nu(k)$ is not constant any more but mostly a monotonic decreasing function on the interval $[0, 1]$.

Thus a Kohonen map defines a topology-preserving mapping of the n-dimensional input vectors to the m-dimensional grid of the map. To obtain a clear man-machine-interface $m = 2$ is used most of the time.

The distance of the neurons c and j on the Kohonen map is coded in the neighborhood function $h_{cj}(k)$ which returns smaller values for neurons that are farther away from each other. The influenceable neighborhood decreases during the conditioning. In general it is useful to permit large values of the neighborhood function at the beginning of the conditioning in order to let the winner neuron influence even the most distant neurons. In the progress the values should decrease to make a stabilization of the evaluated topology possible.

The choice of the neighborhood function depends on the actual problem, but our experimental results show that for categorizing time delays the normalized Gaussian function is best. This neighborhood is defined by

$$h_{cj,\text{gauss}}(k) = \frac{1}{\sigma(k)\sqrt{2\pi}} \cdot e^{\frac{-\|\boldsymbol{r}_c - \boldsymbol{r}_j\|^2}{2\sigma^2(k)}} \ ,$$

where at the standard deviation $\sigma(k)$ defines the width of the kernel of neighborhood and $\|\boldsymbol{r}_c - \boldsymbol{r}_j\|$ describes the distance of two neurons c and j on the grid.

In addition to the neighborhood functions there exist other functions that describe direct net neighborhoods. They can be divided into neighborhoods with four, six or eight neighbors.

Applying these functions, only the winner neuron and its direct neighbors are influenced according to the formula

$$h_{c,\text{direkt}_d}(j) = \begin{cases} 1 & \text{if } c = j \ , \\ \frac{1}{10} & \text{if } j \in N_d(c) \ , \\ 0 & \text{else} \ , \end{cases}$$

where at c relates to the winner neuron, j to the neuron that is to change and $d \in \{4, 6, 8\}$ specifies the kind of the regarded neighborhood. The neighborhood with four neighbors is shown on the left hand-side of Fig. 4, the one with eight neighbors is shown on the right hand-side.

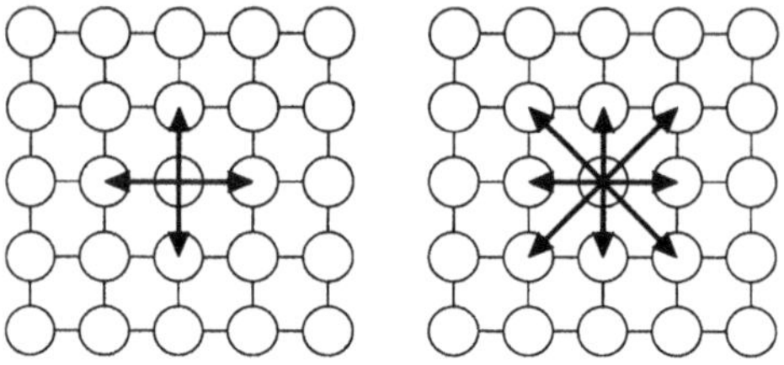

Fig. 4. Neighborhood with four and eight neighbors [3]

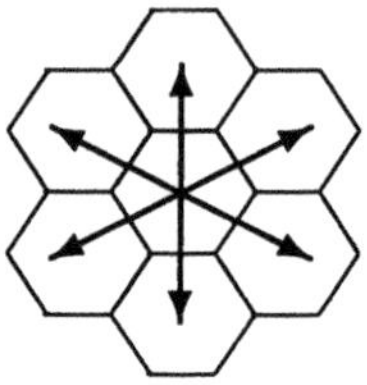

Fig. 5. Neighborhood on the hexagonal grid

Besides the squared placed neurons, the Kohonen map can be build hexagonal as exemplary shown in Fig. 5. The neurons on this map have six neighbors – as long as they are not located on the border of the map.

Vector Categorization of Time Delay Data. As experience shows delay vectors are categorized best by a Kohonen map on which the neurons are placed hexagonally. Furthermore, besides the representative real data, the training vector should contain the two extreme cases 'no delay occurred' and 'all "indefinitely" delayed'. This way the net was enabled to estimate the seasonal evolution of time delays.

To simplify working with the program, an additional interpretation of the resulting Kohonen map is given with the general output. This interpretation informs the operator which categories the actual record was assigned to. If more than a given number of categories is needed to represent the actual record, it is not possible to classify the vector without further information, because too many classification concepts are active and it is unknown which of these concepts will dominate in evaluations with succeeding data of delay. The operator gets a warning message that the actual vector could not clearly be classified.

For interpreting the classification results two strategies are used:

- On the one hand the neighborhood relations of the neurons on the Kohonen map can be regarded.
- On the other hand the error resulting in assigning records to categories can be minimized.

The choice of these possibilities is left to the operator because the kind of interpretation to be used depends on the actual problem.

Interpretation of the Classification Result Using Neighborhood Relations. Using neighborhood relations to interpret the classification behavior of the net, the assignment of an input vector to the given categories will be realized in two steps: First the number of activity centers of the resulting map has to be detected. The basic idea is that every neuron is either an activity center by itself – using the euclidic distance, this equals an activity minimum – or one of its neighbors' activity is smaller than its own. This is examined for every neuron and the overall number of centers is counted. The centers of the actual record are afterwards assigned to the activity centers of the categories made up by the training vectors.

The examination of the neighborhood activity of a neuron requires more energy than expected at first sight; due to the hexagonal placement the neurons are alternately displaced up and down as shown in Fig. 6. They are numbered from 1 on the upper left to *width* · *height* on the lower right of the map. To analyze the neighborhood of a neuron, the position of that neuron needs to be determined – based on its number – first, because neurons that touch the border of the map do not have six neighbors but only four, respectively two or three if they are situated in the corner of the map. Furthermore it is important to know if the neuron is displaced to the top or the bottom because two of its sideways neighbors are not determined by 'number of the actual neuron ±1'. This problem is illustrated in Fig. 6 where at M and N resemble the number of the actual neuron and b is set to the width of the map. The abbreviations $x = b - 1$ and $y = b + 1$ are used.

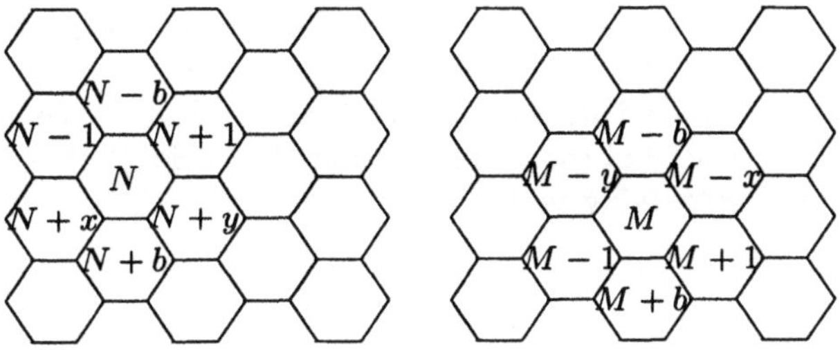

Fig. 6. Determination of the neighbors of a neuron

Even if it is possible to ascertain the displaced position (up or down) in the case of an even width of the map just by knowing the number of a neuron, this becomes more complicated if the width is uneven. It would be necessary to know if the neuron is part of an even or an uneven row to assign the correct neighbors to it, this is indicated in Fig. 7. To reduce the required energy, the position of a neuron within its row is determined – even positions are displaced to the top, uneven positions to the bottom.

After determining the activity centers of the evaluated map the second step can be done by assigning the activity centers to the given categories. Thereto each center is represented by its winner neuron with its activity. The activity

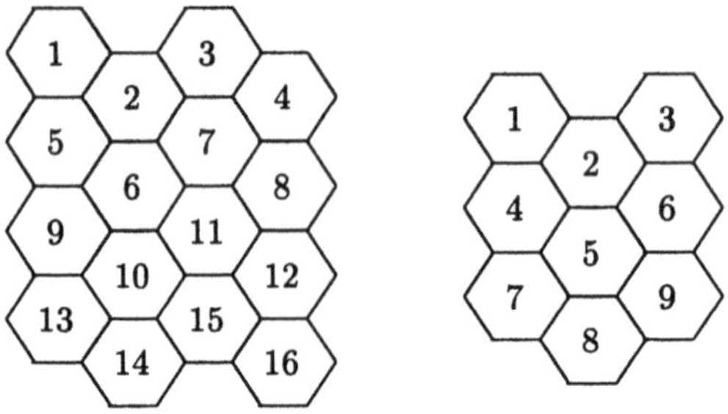

Fig. 7. Neighbor determination depending on the width of the map

patterns of the map resulting from the training vectors are compared in the place of the winner neuron. The regarded activity center is assigned to the map that has the minimum activity in the place of the winner neuron (using the euclidic distance, the minimum activity resembles the largest conformity). If the smallest activity in this place is larger than a given level, the record is not assigned to any category because the distance between the actual record and the training vector is too big. In this case a message appears that a new activity center was found. Thus the neural network has the possibility to point out that the actual data could not be classified properly.

In addition to the assignment, the activity center of the representing category is shown on the Kohonen map. This enables the operator of the program to estimate how well the calculated category of delay represents the actual input vector. Furthermore the value of the resulting error of the assignment is shown. In case of minimizing the distance between input and reference vector, the error is received by the activity of the actual map at the position of the winner neuron of the corresponding training vector.

Interpretation of the Classification Result Using Error Minimization. Using error minimization to interpret the classification of the Kohonen map, only the activities of the neurons are relevant – independent of the total number of existing activity centers. This allows a more intuitive assignment of the actual maps to the given categories but it does not contain any information on how many categories were found during the evaluation.

Working with error minimization the procedure is a lot simpler than the one described above. After the Kohonen map has classified a record, the activities of the neurons that are situated on the places of the winner neurons of the training vectors are examined. The actual record is assigned to a category if the activity is smaller than a given level. The selected categories are visualized adequately with their resulting errors. In case that more than a given number of categories is needed to classify the vector, the operator is warned again that it is momentarily not possible to calculate a generalized categorization because there are too many classification concepts active. Nevertheless in both cases a tabular containing the

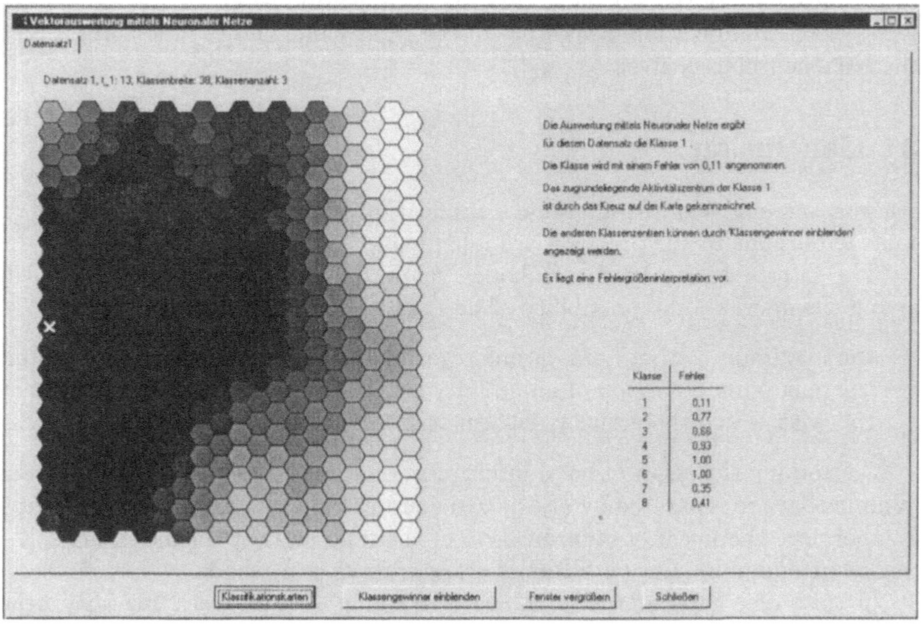

Fig. 8. The categorization of the Kohonen map

categories resembling the training vectors and the resulting matching-errors is displayed.

To give the operator the possibility for his own interpretation of the classification result, it is possible to project the maps evoked by the training vectors or their activity centers on the actual map.

2.1 The Output of the Categorization Result

The result of the data mining is presented in different output windows, whereby the desired kind of output can be chosen. Generally two groups can be selected: The first group contains very general information on the evaluation process – such as the parameters of the WEIBULL and ERLANG distribution, resulting of the statistical preprocessing and the errors, made by approaching the empirical function with the theoretical functions. The second group is made up of more detailed information, statistical means of the different distributions such as median, variance, skewness and excess are listed. Disaccording to the other kind of output, the presentation of the classification result of the Kohonen map can not be clearly assigned to one of these groups. On the one hand it contains the interpretation of the classification result – either using neighborhood relations or minimizing the error, on the other hand it provides the operator with the possibility to interpret the underlying Kohonen map by himself. Figure 8 shows a screenshot of the interface of the program being applied to delay data of a german customer. On the left hand side the classification result of a Kohonen map using error minimization is shown. On the right hand side the interpretation of the Kohonen map is given.

3 Conclusion

For guaranteeing a more individual classification of arbitrary data of time delays, some extensions to the program are imaginable.

The interpretation of the Kohonen map could be made more variable by giving the operator the possibility of adjusting the given levels such as:

- the maximum activity of neurons regarding the creating of activity centers,
- the maximum deviation of an activity center and the belonging winner neuron with which the activity pattern still belongs to the category.

Concerning the neighborhood interpretation it would be possible to join local minima that are separated by one or two neurons only to a single activity center. Furthermore the linear standardization of the map activity on the interval [0,1] can be extended by using additional standardization methods.

In spite of a representative choice of training vectors test runs with other records could not always be classified but sometimes built out new categories. These categories can be interpreted as 'question classes' that the neural network uses to tell the operator that the actual record could not be evaluated clearly.

Regarding the ideal vector which refers to 'no delay', it is to notice that this is the *mathematical* ideal case. This vector should be adapted to the objective target of the specific company because there is a big difference between the target 'as many deliveries as possible without or with very small delay' (this resembles target 1 in Fig. 9) and 'at the latest all deliveries have taken place after a given period of time' (target 2 in the figure).

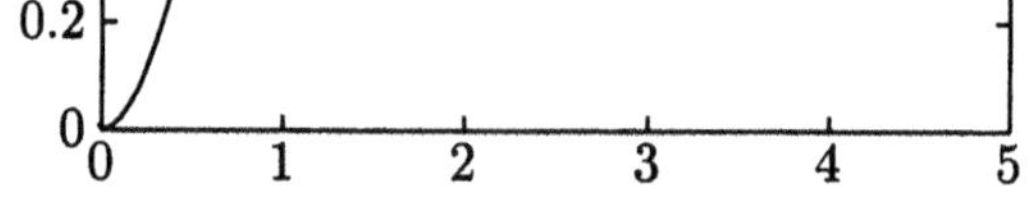

Fig. 9. Different meanings of punctuality

Thus concept 1 has a large number of (almost) undelayed deliveries but there can be deliveries with very long delay because the ideal function approaches 1 very slowly. On the other hand concept 2 has only a few undelayed transports but it is assured that all deliveries reach their destination within a relative small period of time.

References

1. Bostelmann, S.: Kategorisierung von Ablaufversp"atungen mittels Neuronaler Netze. Master's thesis, TU Clausthal (2000)
2. Kohonen, T.: Self-Organizing Maps. Springer (1995)
3. Reuter, M.: Die potentialorientierte Theorie der Neuronalen Netze. TU Clausthal (1998)
4. Schassberger, R.: Warteschlangen. Springer-Verlag (1973)
5. Stange, K.: Angewandte Statistik – Erster Teil: Eindimensionale Probleme. Springer (1970)
6. Zell, A.: Simulation Neuronaler Netze. Addison-Wesley (1990)

Neural Networks as a Tool for Gray Box Modelling in Reactive Distillation

Kai Dadhe[1], Volker Roßmann[2], Kazim Durmus[1], and Sebastian Engell[1]

[1] Process Control Laboratory, University of Dortmund, 44221 Dortmund, Germany
{dadhe, durmus, engell}@ast.chemietechnik.uni-dortmund.de
http://astwww.chemietechnik.uni-dortmund.de
[2] Bayer AG, PLT, 51368 Leverkusen, Germany
volker.rossman.vr@bayer-ag.de

Abstract. In this paper we discuss the use of neural networks as a tool for gray box modelling of the reactive distillation column. The basic idea is to replace certain correlations for the calculation of physical properties by neural networks. Different architectures as radial basis function networks and feedforward networks are compared and their approximation abilities are demonstrated.

1 Introduction

The rigorous dynamic modelling of chemical processes gives rise to large differential-algebraic equation systems (DAE), containing thousands of equations and variables. These models are very useful for simulation studies and offline optimization, but due to their size and complexity they cannot be used for online optimization and nonlinear model predictive control.

For online applications the computational expense has to be reduced. One option is to use better numerical algorithms and more powerful computers. But even if the online requirements are fulfilled, the effort may be prohibitive.

Another approach is the simplification of the model. As chemical processes and the physical properties therein often exhibit strongly nonlinear behaviour, the reduced order models must also be nonlinear if a large range of operation is to be covered. Neural networks with nonlinear activation functions are powerful nonlinear models and are hence studied in this paper.

The major bottleneck of model simplification is the reduced region of validity. As a trade-off, certain parts of the original model are kept whilst other or previously unknown parts are replaced by neural networks. This method is known as gray box or hybrid modelling.

The paper is structured as follows. First, the physical property correlations to be replaced by neural networks are described. A short discussion of the specific networks, their structure and mathematical formulation motivates the training procedure used here. Simulations prove the success of the approximations, and conclusions are given.

B. Reusch (Ed.): Fuzzy Days 2001, LNCS 2206, pp. 576–588, 2001.

2 Modelling Issues

The process under consideration is the heterogeneously catalyzed production of methyl acetate (MeAc) from acetic acid (HAc) and methanol (MeOH). A detailed process description can be found in [1], a general review on reactive distillation and its modelling aspects was recently published by [2].

The plant is depicted in Fig. 1. It consists of three parts, the reboiler, the condenser and reflux, and the column itself. Within the column there are three sections of structured packings, two catalytic ones at the bottom and a separating section at the top of the column.

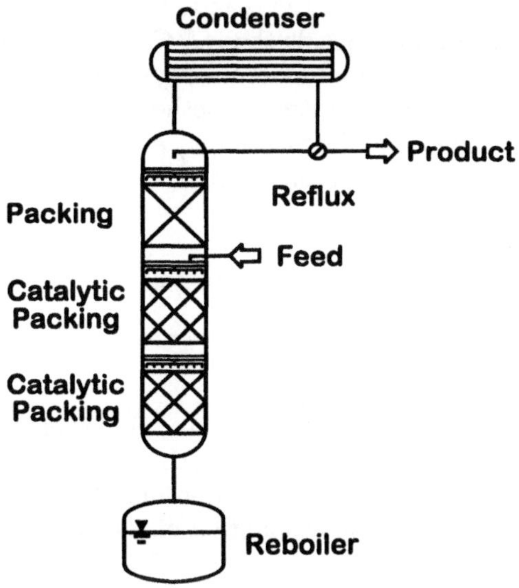

Fig. 1. Scheme of a reactive distillation column

Operating the plant in batch mode means that one has to supply all raw materials at startup in the reboiler. Since acetic acid is the highest boiling component, its concentration would already drop strongly in the lower part of the column causing vanishing reaction rates in the middle and upper part of the catalytic section. Therefore a semibatch mode is used in which the acetic acid is fed continuously above the catalytic packing.

2.1 Model of a theoretical stage

Although continuous in the spatial domain the packing sections are divided up into several theoretical stages, which are the core of the dynamic model.

The mass and energy balances for the ith catalytic stage are, according to Fig. 2:

$$\frac{d}{dt}\left(N_{tot}^i x_j^i\right) = L^{i+1}x_j^{i+1} - L^i x_j^i + \\ +V^i y_j^i - V^{i+1}y_j^{i+1} + dn_{j,reaction}^i, \; j = 1, \ldots, NC, \tag{1}$$

$$\frac{d}{dt}\left(N_{tot}^i h_L\right) = L^{i+1}h_L^{i+1} - L^i h_L^i + \\ +V^i h_V^i - V^{i+1}h_V^{i+1} + dH_{reaction}^i, \tag{2}$$

where L, V and x, y are the liquid and vapour flow rates and mole fractions, respectively, h_L, h_V, are the liquid and vapour enthalpies and N_{tot}^i the total molar holdup. For the separation stages, the reaction terms $dn_{reaction}^i$ and $dH_{reaction}^i$ in eq. (1) and (2) are omitted. Beside these $NC+1$ differential equations (NC: number of components) there are numerous algebraic equations for the calculation of hydrodynamics, reaction rates and other physical properties like liquid and vapour enthalpies and liquid densities. x_j^i and y_j^i are related by the vapour liquid equilibrium (VLE). Its computation is based on Wilson activity coefficients:

$$\varphi_j y_j p_{tot,j} = p_j^0 \gamma_j x_j, \tag{3}$$

$$\sum_j y_j = 1, \tag{4}$$

with the fugacity coefficients φ_j, the activity coefficients γ_j and the vapour pressures of the pure components p_j^0 of the jth component. This set of implicit equations gives rise to one third of all equations due to the large number of auxiliary variables needed for the calculation of the activity coefficients.

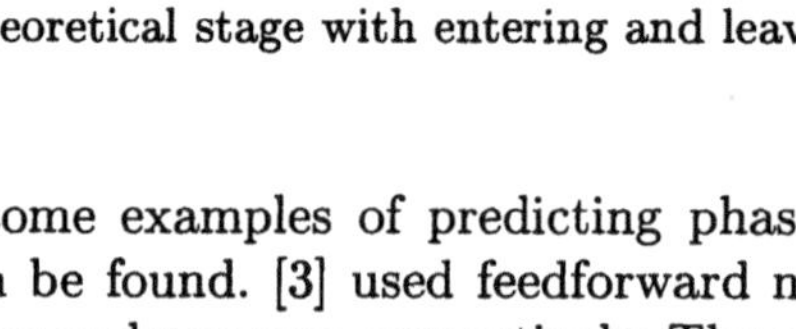

Fig. 2. A theoretical stage with entering and leaving streams

In the literature, some examples of predicting phase equilibria by means of neural networks can be found. [3] used feedforward networks with different numbers of hidden layers and neurons, respectively. They compared the network predictions with the UNIFAC method for three different binary mixtures. [4] used the equation-of-state (EOS) as basis of the comparison instead. [5] used a neural network as a decision-maker which VLE method to be chosen.

3 Neural Network Architectures

3.1 Feedforward Networks

It is well known that general feedforward networks with sigmoidal activation function can approximate any nonlinear function [6]. The number of hidden layers and the number of neurons therein as well as the activation function have been determined during design. The Matlab Neural Network Toolbox was used, information about the training algorithms can be found in [7]. The MISO[1] mapping with one hidden layer and linear output layer is

$$\hat{y}_k = v_0 + \sum_{j=1}^{m_k} w_{2,j} \tanh\left(w_{1,j}^T x + b_j\right) . \tag{5}$$

3.2 Radial Basis Function Networks

As a second architecture radial basis function (RBF) networks with a linear part that perform the following nonlinear mapping

$$\hat{y}_k = v_0 + v^T x + \sum_{j=1}^{m_k} w_j h_j(x), \tag{6}$$

and the basis functions h_j being of the Gaussian type

$$h_j(x) = e^{-\frac{||x - c_j||^2}{\sigma_j^2}} \tag{7}$$

were used. Almost the same approximation accuracy as feedforward networks faster training is possible as the weights to be determined can be found by linear optimization.

Again, the Matlab Toolbox[2] was used. As there are no possibilities to influence the positioning of the centres c_j of the radial basis function unless the training data set is changed, a toolbox developed at the Process Control Laboratory was applied. Fig. 3 shows the structure of the RBF networks.

The training procedure which was used in this work is divided into two steps. First the centres c_j and the widths σ_j of the basis functions are determined by unsupervised learning using only the input samples of the training data set. The centres are calculated by k-means-clustering, see [8] and [9]. This algorithm finds a set of cluster centres and a partition of the input samples into subsets. It uses m different randomly chosen training samples as cluster centres. Each sample is assigned to the centre next to it. Then the average position of the samples belonging to each cluster is found and the corresponding centre is moved to that point. Once all clusters have been updated, the procedure is repeated until it

[1] multiple input single output

[2] see [7] for details on design and training

converges. The radii σ_j are determined from the p-nearest neighbour heuristics [9]:

$$\sigma_j = q\sqrt{\frac{1}{p}\sum_{k=1}^{p}||c_j - c_k||^2}. \tag{8}$$

In (8) c_k are the p nearest neighbours of c_j and the factor q controls the degree of overlap of the basis functions. For given c_j and σ_j the output weights v_j and w_j can be calculated in a supervised manner by solving the quadratic program

$$\min_{v,w}\sum_{i=1}^{N}\left[y_i - \hat{y}_i(v,w)\right]^2, \tag{9}$$

where y_i are the target values of the model output from the training data set. To improve the extrapolation accuracy, the linear weights v_j are calculated first. The RBF weights w_j are chosen such that the residual error is minimized.

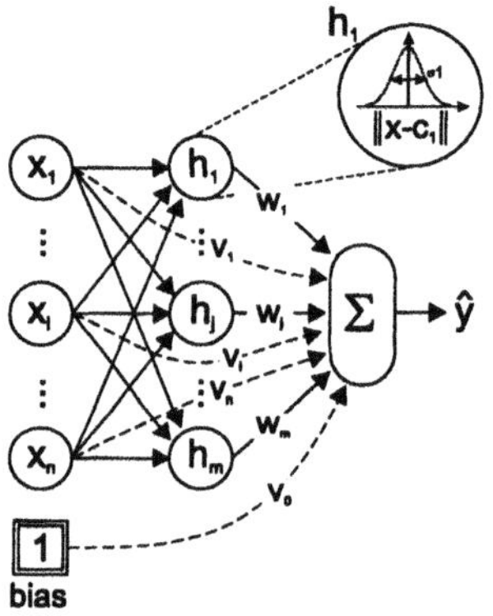

Fig. 3. Radial basis function network

4 Results

4.1 Training

The system of the vapour liquid equilibrium in the four component mixture has five inputs (four liquid mole fractions and the pressure) and five outputs (four vapour mole fractions and the temperature), therefore five MISO networks have been designed. The liquid and vapour mole fractions are between 0 and 1 by definition, therefore no scaling on these variables is necessary. As the column is operated at atmospheric pressure and $p_{tot,j}$ from Eq. (3) varies around 1.013 bar, it is scaled with

$$\bar{p}_{tot} = 100 \cdot (p_{tot} - 1.01) \tag{10}$$

and the temperature, which lies in the interval between 300K and 400K, is scaled with

$$\bar{T} = 0.01 \cdot (T - 300) \,. \tag{11}$$

Training as well as validation and the RMSE calculation therein are done with the scaled variables. The training data were generated by solving the set of nonlinear implicit functions from Eq. (3)-(4) for different liquid mole fractions x_i and pressures p_{tot}.

The measurement for training as well as for validation is the root mean squared error (RMSE) of the kth output:

$$RMSE^k = \sqrt{\frac{\sum_{i=1}^{N}(y_i^k - \hat{y}_i^k)^2}{N}} \tag{12}$$

The feedforward network was designed with two hidden layers consisting of ten neurons each, resulting in a summed RMSE of $\approx 2 \cdot 10^{-7}$. The training method applied here is a quasi Newton algorithm.

The first RBF networks were trained with the Matlab Toolbox, a desired RMSE for each output of 0.01 results in five nets with $m_k = \begin{bmatrix} 52\ 123\ 179\ 161\ 117 \end{bmatrix}$ neurons on the hidden layers. The radii are fixed to the value of $\sigma = (2\sqrt{\ln 2})^{-1}$ which turned out to give good overlapping of the radial basis functions as for a distance of $||x - c_j|| = 0.5$ the output of the neuron is 0.5.

Using the k-means-clustering and the p-nearest neighbour heuristic the parameters p (number of neighbours) and q (overlapping factor) from Eq. (8) were determined. Twenty neural networks for the first output ($k = 1$) with a fixed number of 30 neurons on the hidden layer were trained and the one with minimal RMSE was taken to evaluate the dependence on the above parameters. Varying p and q from 1 to 10 yielded $p = 1$ and $q = 7$ which was compared with the combination $p = 4$, $q = 3$, which was used in former applications, see [10].

To eliminate the influence of different random initialisations of the k-means-clustering, twenty neural networks were designed and the one with minimal RMSE was taken. If the actual RMSE was still higher than the desired 0.01, the number of neurons, starting from ten, was increased by five until the desired RMSE was met. The resulting numbers of neurons in the hidden layer were $m_k^{1,7} = \begin{bmatrix} 20\ 80\ 110\ 100\ 60 \end{bmatrix}$ for $p = 1$ and $q = 7$ and $m_k^{3,4} = \begin{bmatrix} 20\ 80\ 115\ 100\ 60 \end{bmatrix}$ for the $p = 3$, $q = 4$ combination, using the RBF networks with the first parameter setting in the following validation section. It is worth mentioning that the calculation of the linear weights in Eq. (6) and the subsequent training with the residuals $y_i - v^T x_i$ did not lead to a reduction of the networks' size.

4.2 Validation

The validation data set was generated by simulation of the rigorous dynamic model along an offline determined optimal trajectory. A control vector parametrization strategy was used and the optimization yielded piecewise constant manipulated variables and a maximal process productivity (see [11] for details of the optimal column operation). The data for the calculation of the VLE are sampled with a sampling interval of six minutes and are taken from all twelve theoretical stages as well as from the reboiler. All together the validation data set consisted of 2340 points. The root mean squared errors of the validation data set are given in table 1.

Table 1. Comparison of the different RMSE for the validation data set. The $\hat{y}_1$ to $\hat{y}_4$ are the vapour mole fractions, $\hat{T}$ is the scaled temperature.

Neural Network	$\hat{y}_1$	$\hat{y}_2$	$\hat{y}_3$	$\hat{y}_4$	$\hat{T}$
Feedforward networks within gPROMS	0.0070	0.0140	0.0181	0.0178	0.0132
Feedforward networks	0.0078	0.0133	0.0273	0.0076	0.0080
RBF networks with k-means clustering and p-nearest neighbour	0.0068	0.0166	0.0224	0.0160	0.0121
RBF with Matlab training	0.0029	0.0075	0.0125	0.0114	0.0092

Using the feedforward networks for the calculation of the VLE – just replacing the set of Eq. (3)-(4) with the neural networks – yields reasonable values for the RMSE. Implementing the feedforward networks into the simulation environment and simulating the optimal trajectory shifts the results to a better approximation of $\hat{y}_1$ and $\hat{y}_3$ and a worse approximation of the other variables.

The duration of the simulations with the rigorous model and with the neural networks integrated are almost the same. A redesign and optimization of the neural network code and a thorough optimization of the networks' size may lead to a further saving in computational expenses.

In order to assess the approximation abilities, the following figures (Fig. 4-9) show simulations of the vapour mole fractions within the column. The line styles are solid black for the simulation of the rigorous dynamic model, solid gray for the simulation of the dynamic model with the feedforward networks integrated in the simulation environment gPROMS, dash-dotted for the calculation of the VLE with the feedforward networks, dotted for the calculation with the RBF networks trained with the k-means clustering and p-nearest neighbour heuristic and dashed for the calculation with the RBF networks trained with the Matlab Toolbox. The stages are numbered from the top to the bottom, stages 1:4 are the separating unit, stages 5:8 and 9:12 are the upper and lower catalytic ones, respectively, and stage 13 represents the reboiler.

Fig. 4 shows the product concentration on the topmost stage. The same mole fractions can be found in the reflux of the column, which are measured and used for composition control of the column. Hence, it is important that these controlled variables are approximated properly as the dynamic model with the neural networks is potentially used within a nonlinear model predictive control

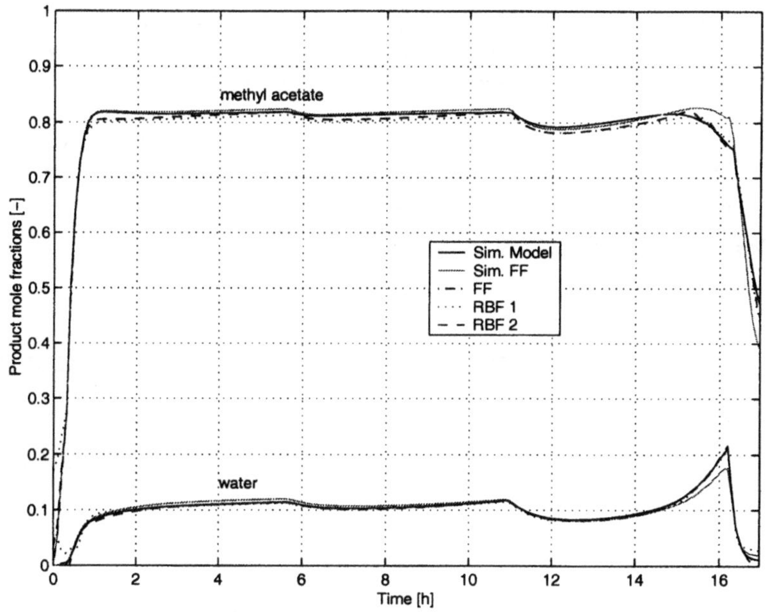

Fig. 4. Product mole fractions during optimal operation at the topmost stage

scheme. Beside some larger deviations at the end of the batch the approximation accuracy is good and satisfactory for control applications.

The vapour mole fraction profiles along the column are depicted in Fig. 5-8 and represent the process after 12 minutes, six, twelve and 16 hours of operation, respectively.

At the beginning of the operation the column is filled with boiling methanol and the acetic acid feed starts at $t = 0$. It takes approximately one hour to establish quasi stationary profiles within the column. The methanol and methyl acetate profiles change significantly during this period which is a typical startup scenario of the process.

The neural networks are able to reflect this behaviour and show good approximation abilities during the rest of the batch. Only at the end of the batch after 16 hours the profiles calculated by the neural networks differ from the simulation of the rigorous dynamic model which coincides with the results from Fig. 4. This can be tolerated as deviations in the process behaviour at the end of the batch do not harm the process objectives of the optimal operation. See table 2 for a comparison.

5 Conclusions

The results show the neural networks capabilities of predicting the vapour liquid equilibrium. Simulations with the neural networks integrated into the simulation

Table 2. Comparison of the process objectives that characterize the optimal operation between the rigorous dynamic model and the model with the feedforward neural networks integrated into the simulation environment gPROMS.

Process objectives	Rigorous dynamic model	Dynamic model with feedforward networks
Productivity [mmol/s]	26,94	27,04
Conversion acetic acid [%]	87,50	87,82
Conversion methanol [%]	83,79	84,10
Product purity [%]	79,68	79,86

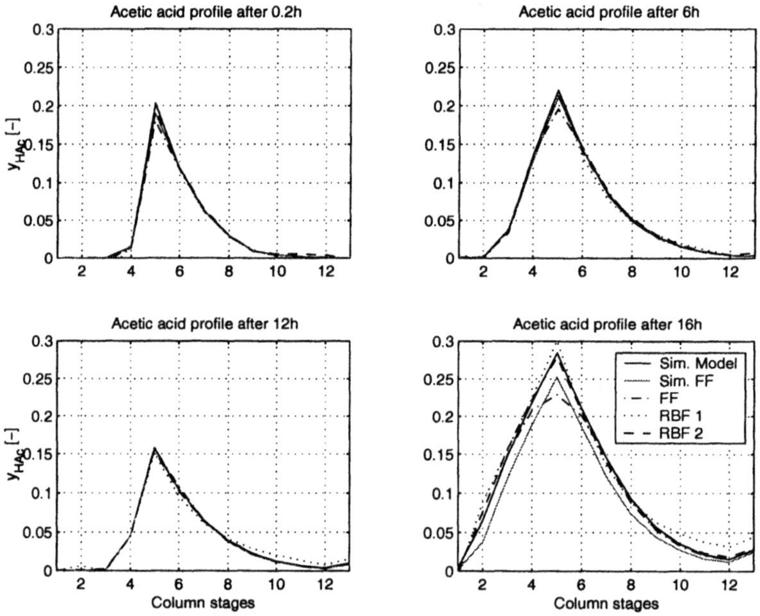

Fig. 5. Acetic acid profiles

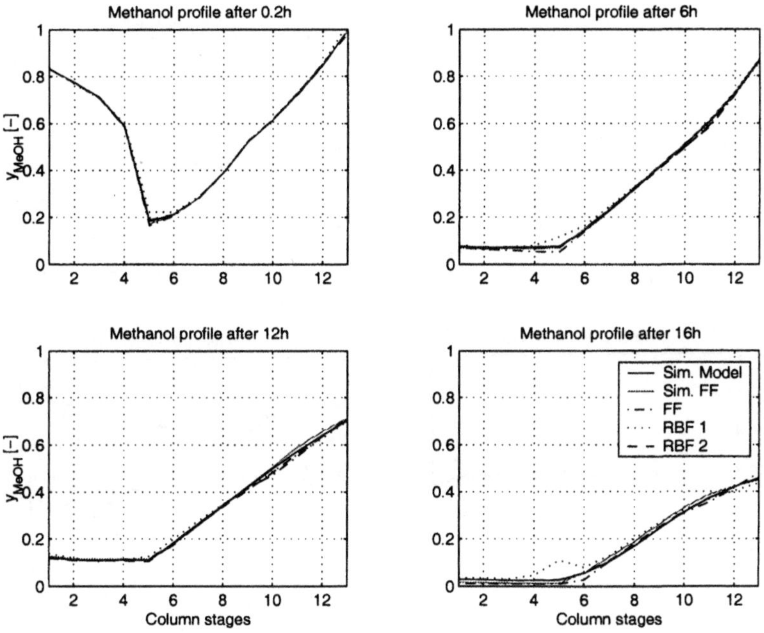

Fig. 6. Methanol profiles

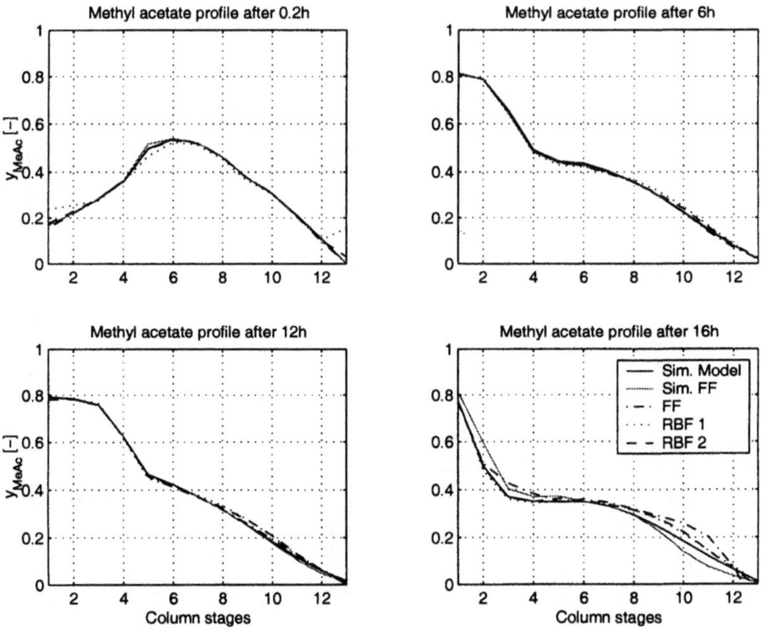

Fig. 7. Methyl acetate profiles

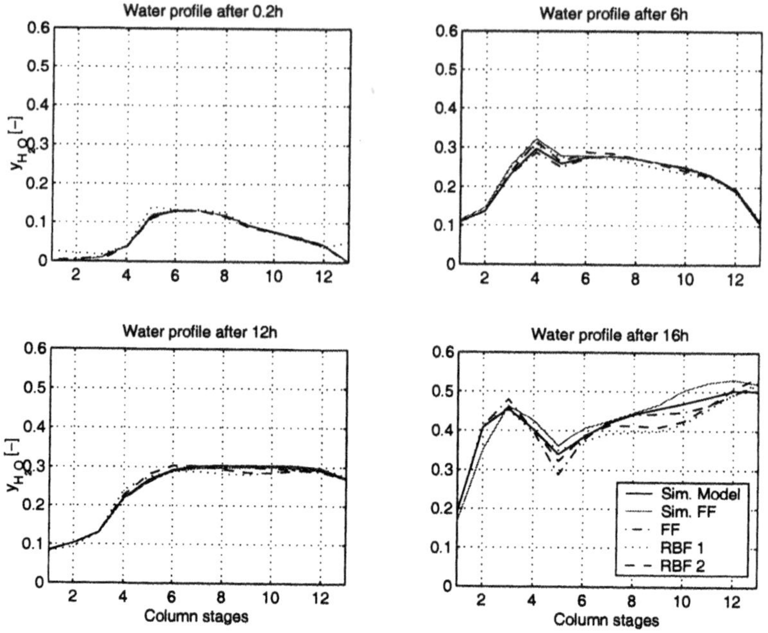

Fig. 8. Water profiles

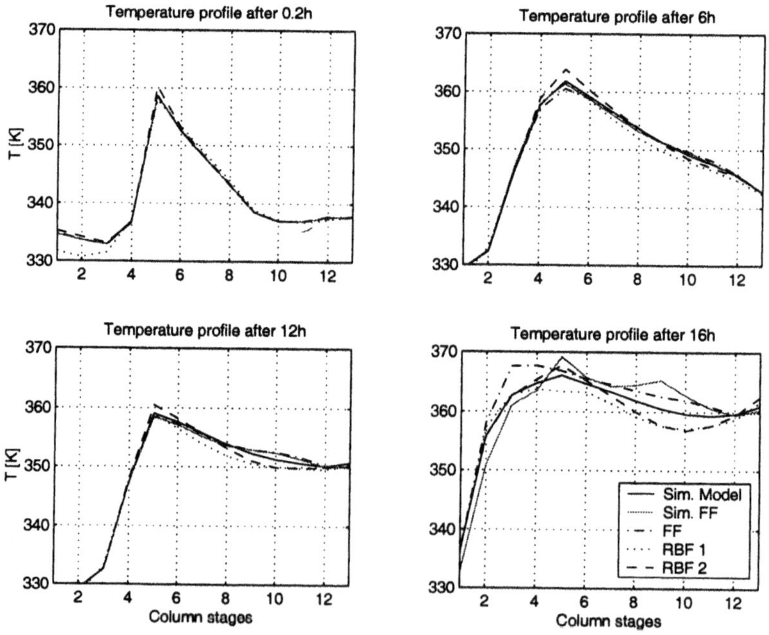

Fig. 9. Temperature profiles

environment gPROMS as well as VLE calculations with the neural networks reveal good approximation during the batch operation in which the variables pass through a wide operating range.

The training and validation data were generated by simulation of the rigorous dynamic model with existing formulas for the VLE. Using (online) measurement data instead the learning abilities of neural networks can be used for model adaption during operation. This is also possible for other physical properties.

It turned out that the neural networks are relatively complex with a large number of neurons in order to achieve a certain root mean squared error. Another possible approach could be the use of dynamic neural networks as approximation of the dynamic behaviour of a certain stage or of a complete packing. Further optimization of the neural networks size could lead to a reduced complexity and a significant benefit in computational expenses.

Acknowledgement

Part of this work was sponsored by the DFG (Deutsche Forschungsgemeinschaft). We want to thank our colleagues at the Thermal Separation Laboratory, University of Dortmund for the fruitful cooperation in the field of reactive distillation.

References

1. J. Krafczyk and J. Gmehling. Einsatz von Katalysatorpackungen für die Herstellung von Methylacetat durch reaktive Rektifikation (Use of catalytic packings for the production of methyl acetate by reactive distillation). *Chemie Ingenieur Technik*, 66(10):1372–1374 (in German), 1994.
2. R. Taylor and R. Krishna. Modelling reactive distillation. *Chem. Eng. Sci.*, 55:5183–5229, 2000.
3. R. Petersen, A. Fredenslund, and P. Rasmussen. Artificial neural networks as a predictive tool for vapor-liquid equilibrium. *Comp. & Chem. Engng.*, 18:S63–S67, 1994.
4. R. Sharma, D. Singhal, R. Ghosh, and A. Dwivedi. Potential application of artificial neural networks to thermodynamics: Vapor liquid equilibrium predictions. *Comp. & Chem. Engng.*, 23:385–390, 1999.
5. E. Alvarez, C. Riverol, J. M. Correa, and J. M. Navaza. Design of a combined mixing rule for the prediction of vapor-liquid equilibria using neural networks. *Ind. Eng. Chem. Res.*, 38:1706–1711, 1999.
6. K. Hornik. Approximation capabilities of multilayer feedforward neural networks. *Neural Networks*, 4:251–257, 1990.
7. H. Demuth and M. Beale. *Neural Network Toolbox*. The MathWorks Inc., 1998.
8. J. MacQueen. Some methods for classification and analysis of multivariate observations. In L. M. LeCam and J. Neyman, editors, *Proc. Fifth Berkeley Symp. Math. Stat. and Prop.*, pages 281–297. Berkeley Press, 1967.
9. J. A. Leonard and M. A. Kramer. Radial basis function networks for classifying process faults. *IEEE Control Systems*, 4, 1991.

10. G. Fernholz, V. Roßmann, S. Engell, and J.-P. Bredehöft. System identification using radial basis function nets for nonlinear model predictive control of a semi-batch reactive distillation column. In *Proc. IEE Conference Control 2000*, Cambridge, 2000.
11. G. Fernholz, S. Engell, L.-U. Kreul, and A. Górak. Optimal operation of a semibatch reactive distillation column. *Comp. & Chem. Engng.*, 24:1569 – 1575, 2000.

On Step Width Adaptation in Simulated Annealing for Continuous Parameter Optimisation

Lars Nolle[1], Alec Goodyear[1], Adrian A Hopgood[2], Phil D Picton[3], and Nicholas StJ Braithwaite[1]

[1]Oxford Research Unit, The Open University, Berkley Road, Oxford, OX1 5HR, UK
{L.Nolle, A.Goodyear, N.S.Braithwaite}@open.ac.uk
[2]Department of Computing, The Nottingham Trent University, Burton Street, Nottingham, NG1 4BU, UK
Adrian.Hopgood@ntu.ac.uk
[3]School of Technology and Design, University College Northampton, Northampton, NN2 6JD, UK
Phil.Picton@northampton.ac.uk

Abstract. Simulated annealing is a general optimisation algorithm, based on hill-climbing. As in hill-climbing, new candidate solutions are selected from the 'neighbourhood' of the current solution. For continuous parameter optimisation, it is practically impossible to choose direct neighbours, because of the vast number of points in the search space. In this case, it is necessary to choose new candidate solutions from a wider neighbourhood, i.e. from some distance of the current solution, for performance reasons. The right choice of this distance is often crucial for the success of the algorithm, especially in real-world application where the number of fitness evaluations is limited. This paper explains how in such a case the use of a variable radius of this neighbourhood, refereed to as maximum step width, can increase the over-all performance of simulated annealing. A real-world example demonstrates the increased performance of the modified algorithm.

1 Introduction

Simulated annealing (SA) is a robust general optimisation method that was first introduced by Kirkpatrick et. al. [1], based on the work of Metropolis et. al. [2]. It simulates the annealing of a metal, in which the metal is heated-up to a temperature near its melting point and then slowly cooled to allow the particles to move towards an optimum energy state. This results in a more uniform crystalline structure and so the process allows some control over the microstructure. SA has been demonstrated to be robust and capable of dealing with noisy and incomplete real-world data [3,4].

Simulated annealing is a variation of the hill-climbing algorithm. Both start off from a randomly selected point within the search space. Unlike in hill climbing, if the fitness of a new candidate solution is less than the fitness of the current solution, the new candidate solution is not automatically rejected. Instead it becomes the current solution with a certain transition probability $p(T)$. This transition probability depends on the difference in fitness ΔE and the temperature T. Here, 'temperature' is an abstract control parameter for the algorithm rather than a real physical measure.

B. Reusch (Ed.): Fuzzy Days 2001, LNCS 2206, pp. 589–598, 2001.

Equation 1 gives a common transition function p(*T*) for a given temperature and a given difference in fitness:

$$p(T) = \frac{1}{1 + e^{\frac{\Delta E}{T}}} \tag{1}$$

Where:

T	temperature – control parameter for cooling schedule
p(T)	probability of transition for temperature T
ΔE	difference between previous candidate solution and current one

The algorithm starts with a high temperature, which is subsequently reduced slowly, usually in steps. Equation 2 shows the standard cooling function introduced by Kirkpatrick [1]. Many others can be found in the literature [5,6].

$$T_{n+1} = \alpha T_n \tag{2}$$

Where:

T_n	temperature at step n
α	cooling coefficient ($\alpha < 1$)

On each step, the temperature must be held constant for an appropriate period of time (i.e. number of iterations) in order to allow the algorithm to settle in a 'thermal equilibrium', i.e. in a balanced state. If this time is too short, the algorithm is likely to converge to a local minimum. The combination of temperature steps and cooling times is known as the annealing schedule, which is usually selected empirically. Fig. 1 shows the flowchart of the basic simulated annealing algorithm.

Fig. 2 shows a typical run of a SA algorithm, optimising a real-world system [4]. The dotted line represents the best solution found so far, the thick line represents the fitness on each iteration. It can be observed that the algorithm initially explores the search space (large differences in fitness), while later on (after approximately 400 iterations in this case) it exploits the most promising region.

Hence, the behaviour of the algorithm at the beginning of the search is similar to a random walk, while towards the end it performs like ordinary hill climbing.

2 Need for step width variation

For continuous parameter optimisation, it is practically impossible to choose direct neighbours of the current solution as new candidate solutions, simply because of the vast number of points in the search space. In this case, it is necessary to choose new candidate solutions from some distance in a random direction of the current solution in order to travel in an acceptable time through the search space. This distance could either be a fixed step width s or it could have an upper limit s_{max}. In the first case, the neighbourhood would be defined as the surface of a hypersphere around the current solution, in the second case the neighbourhood would be defined as the volume of the hypersphere.

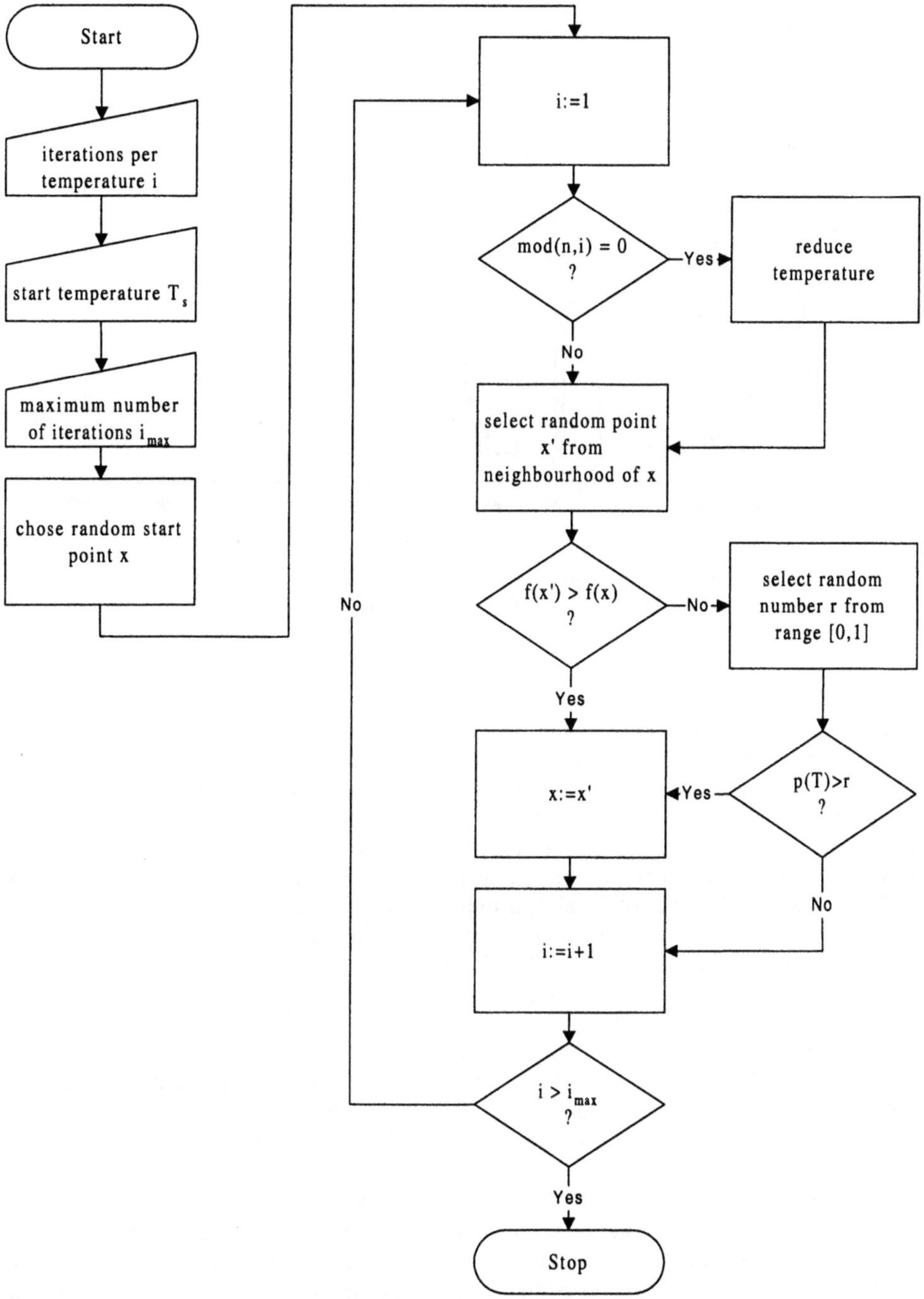

Fig. 1. Flowchart of the basic SA algorithm

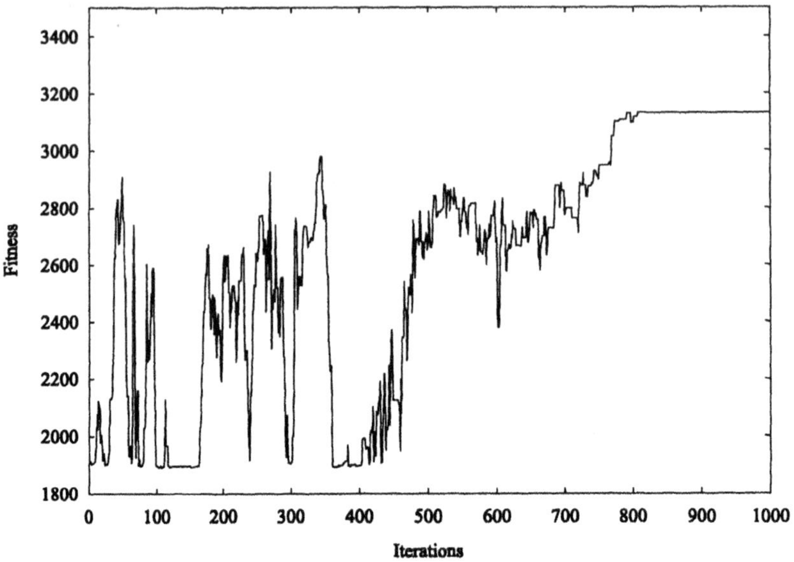

Fig. 2. Typical run of a SA algorithm; at the beginning of the search the algorithm explores the search space, while it is exploiting the most promising region of the search space after the algorithm has been 'frozen'

In the latter case, new candidate solutions might be generated by adding small, equally distributed random numbers from the interval [$-s_{max}$, s_{max}] to each component of the current solution vector. The maximum step width s_{max} is crucial to the success of SA. If s_{max} is chosen to be too small and the start point for a search run is too far away from the global optimum, the algorithm might not be able to get near that optimum before the algorithm 'freezes', i.e. the temperature becomes so small that p(T) is virtually zero and the algorithm starts to perform only hill climbing. In that case, it will get stuck in the nearest local optimum. If, on the other hand, the step width has been chosen to be too large, and the peak of the optimum is very narrow, the algorithm might well get near the global optimum before the algorithm 'freezes', but never reaches the top, because most of the steps are simply too large so that new candidate solutions 'fall off' the peak. Hence, there is always a trade-off between accuracy and robustness in selecting an appropriate maximum step width. If s_{max} is too small, SA has the potential to reach the peak of the 'frozen-in' optimum, but it cannot be guaranteed that this optimum is the global one. On the other hand, if s_{max} is too large, SA has the potential to get near the global optimum, but it might never reach the top of it.

A solution could be to use small steps and to adjust the cooling schedule to increase the length of the Markov chains. This is not always be possible in real-world optimisation problems with time constrains, i.e. a limited number of possible fitness evaluations. The solution proposed here is to adapt the maximum step width s_{max} to the search process itself. In general, there are two possible approaches. Firstly, s_{max} may be adapted to an SA parameter, like current iteration, temperature, etc. The search starts with a large value for s_{max} and is subsequently decreased towards the end. Secondly, s_{max} can be adapted to the fitness landscape, e.g. ΔE or p(T).

3 Real-world example

In this example, SA has been employed to deduce fourteen Fourier terms in a radio-frequency (RF) waveform for a Langmuir probe compensation system [4].

Langmuir probes [7] are important electrostatic diagnostics for RF-driven gas discharge plasmas [8]. These plasmas are inherently non-linear, and many harmonics of the fundamental are generated in the plasma.

RF components across the probe sheath distort the measurements made by the probes. To improve the accuracy of the measurements, these RF components must be removed. This has been achieved by active compensation, i.e. by applying an RF signal to the probe tip. Not only amplitude and phase of the applied signal have to match that of the exciting RF, also its waveform has to match that of the harmonics generated in the plasma. The active compensation system uses seven harmonics to generate the required waveform. Therefore, fourteen heavily interacting parameters (seven amplitudes and seven phases) need to be tuned before measurements can be taken. Fig. 3 shows the closed optimisation loop.

The goal is to maximise the parameter 'floating potential' of the Langmuir probe system, which acts as a fitness measure. Previous work has shown that this is a hard optimisation problem, because the inputs heavily influence each other. Also, the fitness at the global optimum is not known a priori.

Fig. 4 shows the set-up of the experiment and Fig. 5 shows the Langmuir probe inserted into an Argon plasma.

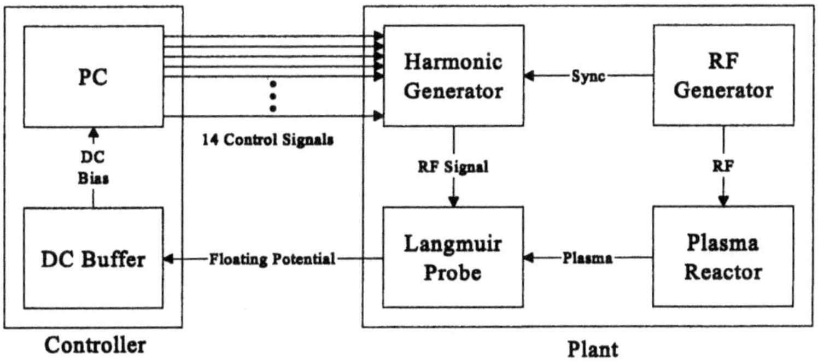

Fig. 3. Closed optimisation loop

Fig. 4. Experimental set-up

Fig. 5. Langmuir probe in Argon plasma

Each of the digital inputs and the digital output of the system to be optimised had a resolution of 12 bits, hence the search space consisted of approximately 3.7×10^{50} search points.

During the experiments, it became evident that the selection of s_{max} was crucial to the solutions found by SA. Fig. 6 shows a typical run with a small maximum step width (s_{max}=50). It can be observed that the algorithm has difficulties in getting out of the 'flat' region of the search space from where it started. Only after approx. 800 iterations did it reach the next optimum, where it performed hill climbing.

In Fig. 7, a typical run with a large step width can be seen (s_{max}=4000). While the algorithm seems to find the region of the global optimum easily, it failed to exploit this region.

Even with a carefully selected value for s_{max}, it was not possible to ensure stable and reproducible solutions. While most of the search runs resulted in fitness values of around 2800, in a few runs values of around 2900 have been found. In order to overcome this behaviour, a scaling function has been empirically designed to adapt the maximum step width to the current iteration (equation 3).

$$s_{\max}(n)=\frac{2s_0}{1+e^{\frac{\beta n}{n_{\max}}}} \quad (3)$$

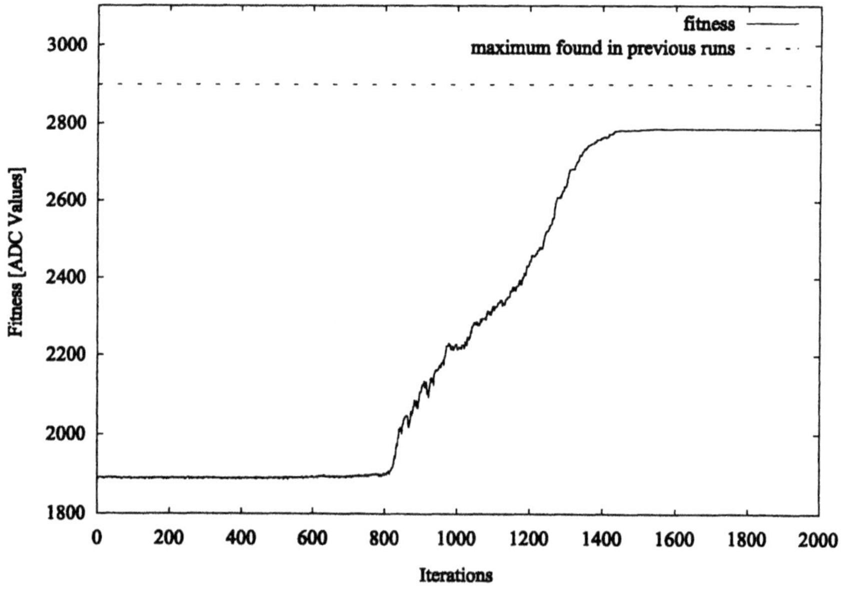

Fig. 6. The step width has been chosen too small; the algorithm failed to reach the region containing the maximum peak during the exploration phase of the search

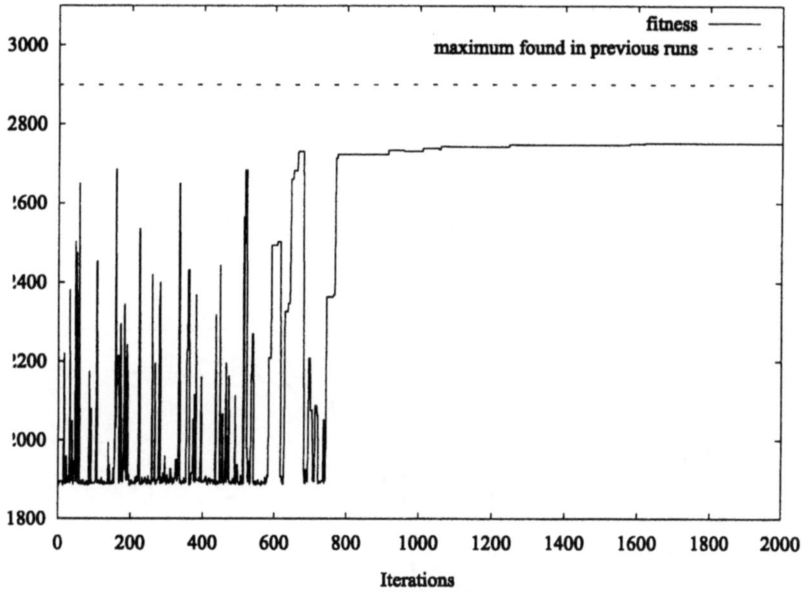

Fig. 7. The maximum step width has been chosen too large; the algorithm failed to find the maximum peak during the exploitation phase of the search run

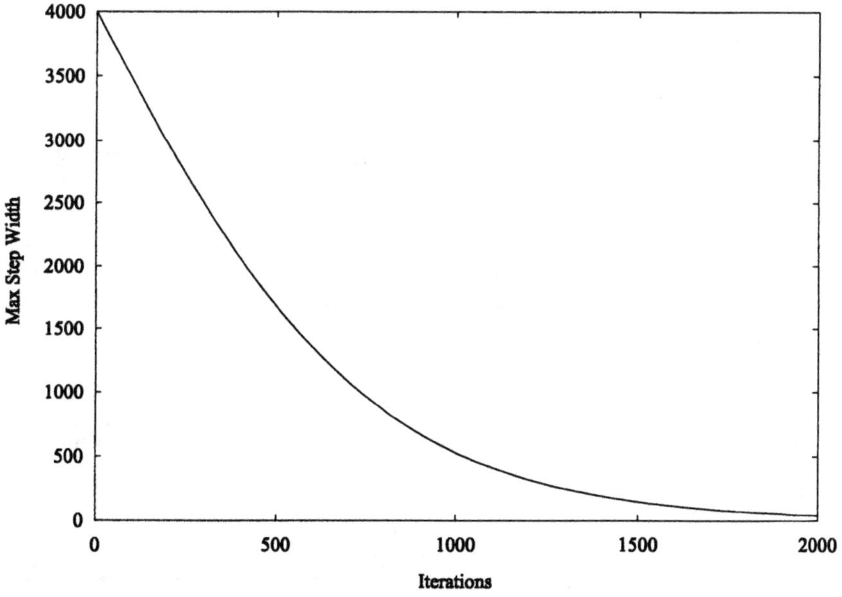

Fig. 8. Step width scaling function used in this research

Where:

$s_{max}(n)$	maximum step width at iteration n
s_0	initial maximum step width
n	current iteration
n_{max}	maximum number of iterations
β	adaptation constant; a value of 5.3 was chosen

Fig. 8 shows the graphical representation of the scaling function while Fig. 9 shows a typical run with the improved SA.

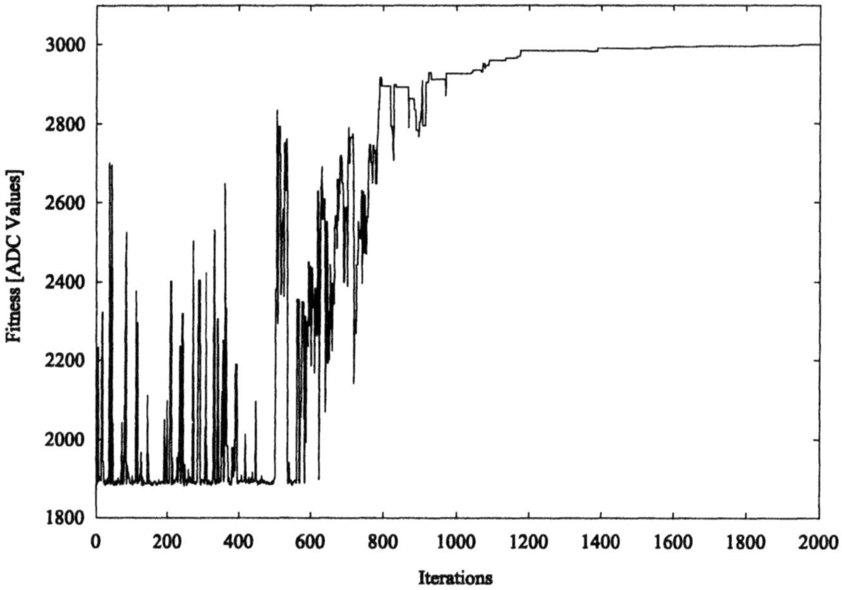

Fig. 9. Search run with step width adaptation; the algorithm explores the search space at the beginning of the search and transforms smoothly into hill-climbing

For evaluation, three different methods have been compared: method 1 used a static maximum step width of 50, method 2 used a static maximum step width of 4000 and finally, method 3 used the adaptive step width with an initial value of 4000. Table 1 shows the results of the experiments.

Table 1. Average Fitness and deviation of fitness for methods used during this research

Method No	Average Fitness [Volts]	Deviation of Fitness [Volts]
1	12.20	1.82
2	11.93	1.05
3	17.84	0.44

Fig. 10 gives a graphical representation, where the dots represent the average fitness gained and the error bars represent the standard deviation for the results.

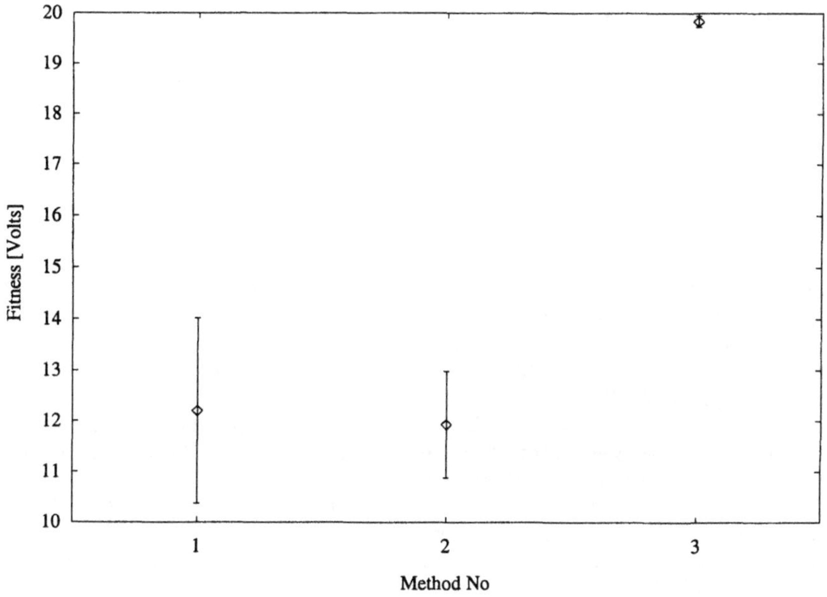

Fig. 10. Comparison of methods; method 1 is a standard SA algorithm with small maximum step width, method 2 is a standard SA algorithm with large maximum step width, and method 3 is a SA algorithm with step width adaptation

It can be observed that not only can the average fitness be improved significantly, but also the robustness, i.e. the reproducibility, has been improved dramatically (smaller error bars). That means the algorithm is more likely to reach the peak of the global optimum. This makes it suitable for industrial applications, which usually require a high degree of robustness and reproducibility.

4 Conclusions

For continuous parameter optimisation, selecting an appropriate maximum step width s_{max} for simulated annealing is always a compromise between accuracy and robustness. If it is too small, SA has good exploitation capabilities but reduced chances of reaching the global optimum to exploit it. If s_{max} is too large, the algorithm has good exploration capabilities, and hence is likely to find the region of the search space containing the global optimum, but it then performs badly in the exploitation of this region.

A solution to this problem is the on-line adaptation of the maximum step width to either a SA parameter, like temperature or iteration, or to the fitness landscape. In this research, the performance of SA in the optimisation of a Langmuir probe system has been dramatically improved by adapting the maximum step width to the current iteration. This has not only improved the average fitness for the system, it has also improved the reproducibility of the results, which is of great importance for industrial applications.

5 Acknowledgements

This work was funded by the Engineering and Physical Sciences Research Council (EPSRC) under grant reference GR/M71039/01.

6 References

1. Kirkpatrick, S., Gelatt Jr, C. D., Vecchi, M. P.: "Optimization by Simulated Annealing", Science, 13 May 1983, Vol. 220, No. 4598, pp 671-680
2. Metropolis, A., Rosenbluth, W., Rosenbluth, M. N., Teller, H., Teller, E.: "Equation of State Calculations by Fast Computing Machines", The Journal of Chemical Physics, Vol. 21, No. 6, June 1953, pp 1087-1092
3. Nolle, L., Walters, M., Armstrong, D.A., Ware, J.A.: "Optimum Finisher Mill Set-up in the Hot Rolling of Wide Steel Strip Using Simulated Annealing", Proceedings Of the 5th International Mendel Conference, Brno, 1999, pp 100-105
4. Nolle, L., Goodyear, A., Hopgood, A.A., Picton, P.D., Braithwaite N.StJ.: "Automated Control of an Actively Compensated Langmuir Probe System Using Simulated Annealing", submitted for publication at The 5th World Multi-Conference on Systemics, Cybernetics and Informatics, SCI 2001, Orlando, Florida USA, July 22-25, 2001
5. Geman, S., Geman, D.: "Stochastic Relaxation, Gibbs distribution and the Bayesian restoration of images", IEEE Trans. PAMI, 5 (1984), pp 721-741
6. Huang, M.D., Romeo, F., Sangiovanni-Vincentalli, A.: "An efficient general cooling schedule for simulated annealing", Proceedings of the IEEE International Conference on Computer Aided Design, Santa Clara, 1986, pp 381-384
7. Swift, J. D. S., Schwar, M. J. R.: "Electrical Probes for Plasma Diagnostics", Ilitte, London, 1970
8. Lieberman, M. A., Lichtenberg, A. J.: "Principles of Plasma Discharges for Materials Processing", Wiley Interscience, New York, 1994

GA-Based Search for Paths with Minimum Energy Consumption for Mobile Robot Systems

Marcus Gemeinder* and Michael Gerke

Fernuniversität Hagen, Faculty of Electrical and Computer Engineering,
58084 Hagen, Germany

Abstract. A GA based path planning software for mobile robot systems focusing on energy consumption is presented. One special feature of this software is the consideration of changing textures – and therefore changing energy requirements when moving – within given environments. Problem specific genetic operators are introduced, and especially the handling of exceptional situations is described in detail.

1 Introduction

A research area with special importance to mobile robot systems is devising suitable methods to plan optimum moving trajectories. Possible optimisation criteria could be path length, energy consumption, or time needed for movement.

An operational area with the crucial aspect being energy consumption are missions on other planets in our solar system. The little rover *Soujourner* has driven on the surface of Mars, and further similar missions will follow. To use the full potential of such vehicles, they should always move in an energy-saving manner. The typical task could be described as follows: for two arbitrary points in an area, find the drivable path between these points minimising the energy required.

There exist many approaches within the area of Computational Intelligence to solve the problem of finding collision-free trajectories in unknown environments (see [2], [3], [5], [6]). Most of them have in common that the aspect of energy consumption is considered only with respect to the length of the paths. Environments are modelled in a binary manner (obstacle/no obstacle), and further information about the texture of the ground is not taken into account.

Here a path planning software based on a Genetic Algorithm is presented, which considers changes in the texture of the ground. This is a crucial property allowing to find paths within an inhomogeneous area characterised by minimum energy consumption.

2 GA-Based Path Planning

In this application the genetic pool is a set of individuals, each representing a path from a given start point to a given goal. The paths are totally independent

* Corresponding author. E-mail: marcus.gemeinder@fernuni-hagen.de

B. Reusch (Ed.): Fuzzy Days 2001, LNCS 2206, pp. 599–607, 2001.

of the presence or distribution of obstacles. Information about the environment is considered later on when the fitness of the individuals is evaluated.

2.1 Genetic Operators

In the sequel, the genetic operators are described on an abstract level, where implementation details are not of interest, yet. Each path shall be seen as a continuous curve in a plane. On this level of abstraction, each operator is a tool for a geometrical manipulation of curves.

Mutation. First, two arbitrary points on a path are chosen. The sub-path between these points is replaced by two straight lines, which connect the start and the end point of the sub-path with a via point chosen at random (see Fig. 1(a)).

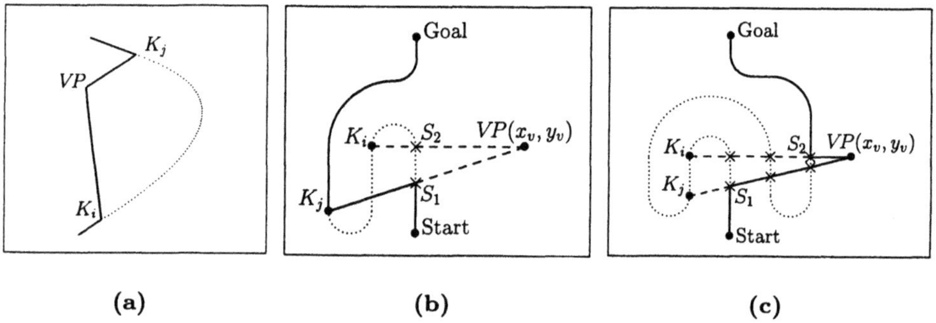

Fig. 1. Operator Mutation. (a) Mode of action. (b) Avoidance of closed loops. (c) Reversal of the new sub-path's direction.

The situation shown in Fig. 1(a) is typical, but cannot be taken for granted: there are no points of intersection between the new sub-path and the remaining sub-paths. Figure 1(b) shows another situation: as before, the dotted sub-path between the points K_i and K_j is to be replaced by the two straight lines connecting these points with the via point *VP*. In this special case one obtains intersections S_1 and S_2 between the straight lines and the old sub-path from *Start* to K_i, causing two closed loops which are, of course, undesired. An algorithm to handle such exceptional situations is given as follows:

1. Moving forward on the sub-path beginning in *Start*, the first intersection K_{first} between this sub-path and the straight lines is determined. This could be K_i.
2. Moving backward on the sub-path ending in *Goal*, the last intersection K_{last} between this sub-path and the straight lines is determined. This could be K_j.

3. The new path results as the connection of the path sections *Start* $\rightarrow K_{\text{first}}$ (start section), $K_{\text{first}} \rightarrow K_{\text{last}}$ (straight lines), and $K_{\text{last}} \rightarrow$ *Goal* (end section).

The operation of this algorithm is illustrated in Fig. 1(b): one obtains firstly S_1 as K_{first}, secondly K_j as K_{last} and, finally, the solid line connecting *Start* and *Goal.* Note that the via point is not part of the new path, anymore.

Another aspect has to be taken in account: in Fig. 1(c) a situation is shown, in which the natural direction of the inserted sub-path has to be reversed. Normally, the straight lines are traversed in the direction from K_i to K_j. In this special situation, the inserted sub-path has to be traversed in the direction from K_j to K_i (more precisely, from S_1 to S_2). From a merely geometrical point of view, there is no difference. On the level of implementation, where each path is a chain of adjacent entries in a matrix which models the environment (this aspect will be discussed later on), this differentiation is important.

Crossover. Having chosen two different individuals representing two paths, all intersection points between these paths are determined. After that, one intersection point is chosen at random, and the sub-paths following this point are exchanged (see Fig. 2(a)).

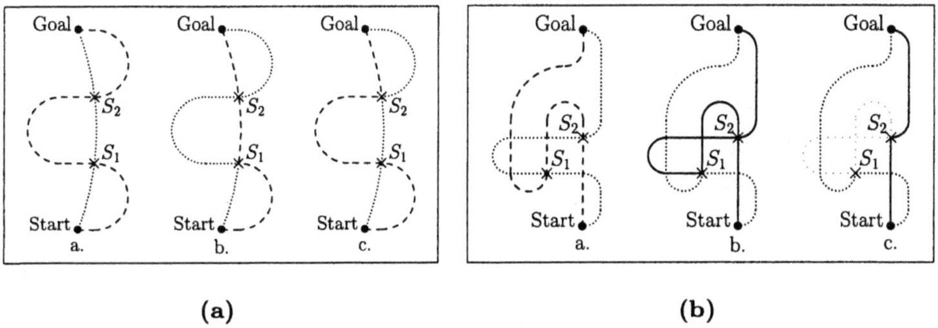

Fig. 2. Operator Crossover. (a) Mode of action (exchange after S_1 resp. S_2). (b) Avoidance of closed loops (K_i coincides with S_2).

As in the case of applying *mutation*, closed loops can be the result when the exchanged sections are connected. In Fig. 2(b) it can be seen, in which way these undesired loops are avoided; the algorithm achieving this is similar to the one described above:

1. Moving forward on the sub-path beginning in *Start*, the first intersection K_i between this sub-path and the other sub-path is determined.
2. The new path results as the connection of the path sections *Start* $\rightarrow K_i$ (first sub-path) and $K_i \rightarrow$ *Goal* (second sub-path).

Tighten. *Tighten* is a very specific operator for this application. It has two different effects: on the one hand it increases the variety within the genetic pool, and on the other hand it is suitable for optimising paths with respect to their length.

First, two arbitrary points K_i and K_j within a path are chosen. After that, every point K_l of the sub-path between these points is shifted towards the straight line g_h connecting K_i and K_j (see Fig. 3(a)). For the shifting vector $\overrightarrow{\Delta k_l}$ holds

$$\overrightarrow{\Delta k_l} = \begin{pmatrix} \Delta k_x \\ \Delta k_y \end{pmatrix} = r \cdot d(g_h, K_l) \begin{pmatrix} n_{0x} \\ n_{0y} \end{pmatrix} , \tag{1}$$

respectively. Thereby $d(g_h, K_l)$ is the distance between K_l and g_h; the values n_{0x} and n_{0y} are the components of the normal vector to g_h, which is directed from K_l towards g_h; and r is a real number chosen at random, fulfilling the inequation $0 \leq r \leq 1$ and determining the effect of *tighten.* If r is equal to 0, the path is not changed, and for $r = 1$ the resulting sub-path is identical to the straight line.

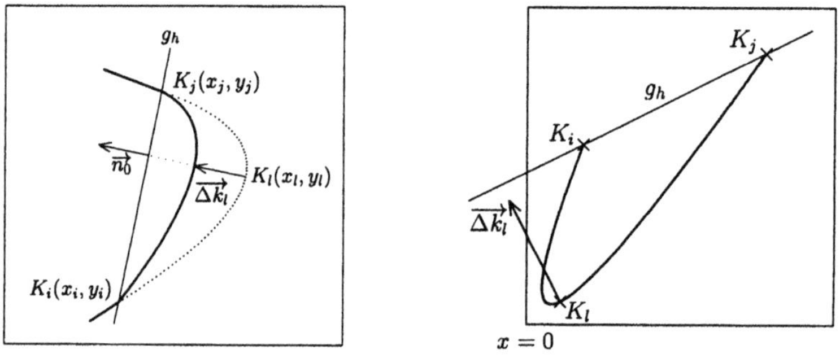

(a) (b)

Fig. 3. Operator Tighten. (a) Mode of action. (b) Exceeding boundaries.

Sometimes a point would be shifted beyond the boundaries of the given environment (see Fig. 3(b)). In this case, the affected points are only shifted until the boundary is reached. Closed loops are avoided in the same way as described above illustrating the operator *mutation*.

2.2 The Genetic Algorithm

The implemented GA is a very simple one. Each individual within the genetic pool represents a path from *Start* to *Goal.* The individuals are chosen from the pool in dependence on their fitness, and each time one of the genetic operators is applied. The operators are selected randomly, with probabilities P(Mutation),

P(Crossover) and P(Tighten) determining their frequencies of occurrence, respectively. The new individuals form a new pool, and the procedure is repeated until a formerly defined termination criterion is fulfilled.

Fitness evaluation and selection will be discussed within the following section.

3 Implementation Details

After the more abstract considerations above, in this section the modelling of the environment, the internal representation of paths, the process of fitness evaluation and selection, and the handling of paths will be described in detail.

3.1 Environments and Paths

Every path connects *Start* and *Goal* within an existing environment. This environment has to be made available in form of a matrix. The sensory data gathering and preparation will not be discussed in this paper. Robot dimensions will not be considered, either.

The regarded environment has to be subdivided into quadratic regions. Each region is well-defined by a pair of integer-numbered co-ordinates. Now, a path within this area is a sequence of steps from the centre of one region to the centre of an adjacent region, connecting the start and the end region. Thus, a path is internally represented as an array of pairs of integer co-ordinates.

Each region corresponds to an entry of the matrix mentioned above. In this matrix the environment information is stored as follows: each entry is a real value, indicating the texture dependent amount of energy that would be needed, if the robot moved through the corresponding region. This energy factor has to lie in the interval between 0 and 1. Weighting to extend this interval is possible. If the value is set equal to 1, the region cannot be used for the movement. This way obstacles are modelled.

3.2 Fitness Evaluation and Selection

In this application the fitness is evaluated according to energy consumption. This value is determined by adding up the energy costs for the single movements from one region's centre to the next region's centre. These costs result as product of the relative length of the movement and the average of the energy factors of both regions. Thereby the relative length is equal to 1 if the regions concerned differ in only one co-ordinate, and $\sqrt{2}$ otherwise. If one or more corresponding energy factors are equal to 1, the path is not drivable.

Selection is performed in a very easy manner: all paths are sorted, beginning with the drivable ones according to their energy consumption, followed by the non-drivable ones according to the number of obstacles touched. This number is not necessarily equal to the number of 1-regions belonging to a path. In fact, immediately adjacent 1-regions are considered as one obstacle contact, only. This modus operandi has proved to be much more efficient than punishing every single 1-region.

Having the complete ranking, each time a random number from the interval [0, 1] is squared and multiplied by the number of individuals, determining the rank of the individual to be selected. This rather inefficient selection method will be replaced by stochastic universal sampling (see [1]) soon. In addition, best-1-elitism was implemented to prevent losing the best path.

3.3 Handling of Paths

As mentioned above, closed loops within paths are undesired. This is a macroscopic requirement. On the level of path representation, other, microscopic requirements arise: Figure 4 shows two situations to be avoided. In the first one, the more direct diagonal movement is better than the two orthogonal steps (except for the rather unlikely setting, that the energy factor of the centre region is small enough to compensate for the distinctly longer way). In the second situation, the straight step is better in any case.

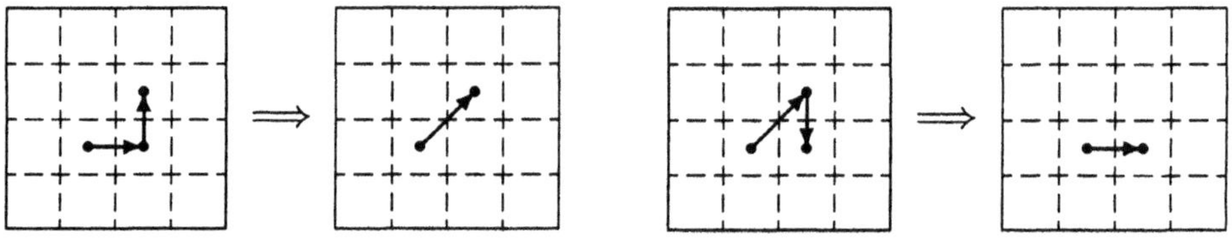

Fig. 4. Elimination of superfluous nodes in order to avoid undesired node sequences

The prevention of such undesired node sequences is ensured by a recursive function: each time a path section has to be extended by an additional node (a centre of a region), it is recursively tested whether preceding nodes have to be eliminated. Figure 5 illustrates this showing the addition of a further node to an existing path section.

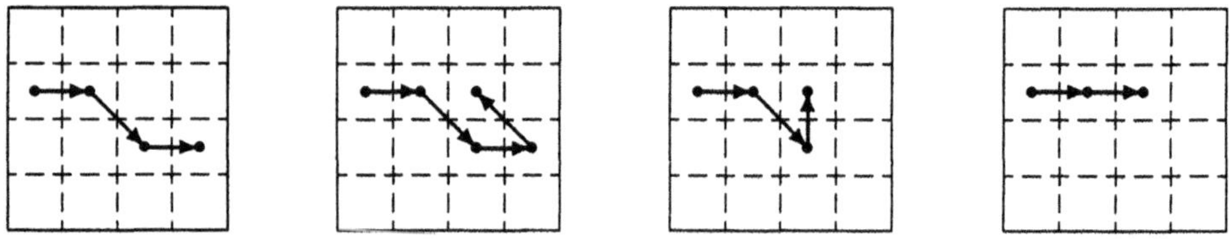

Fig. 5. Recursive addition of a fifth node to a path consisting of four nodes eliminating two superfluous nodes

If two sub-paths are to be connected, the situation must be considered in detail, too. Figure 6(a) shows two sub-paths – one labelled with circles, the other one with squares – which have one node in common (node 9, resp. 7). The shown concatenation is not optimum; actually the resulting path should be reduced to the one depicted in Fig. 6(b).

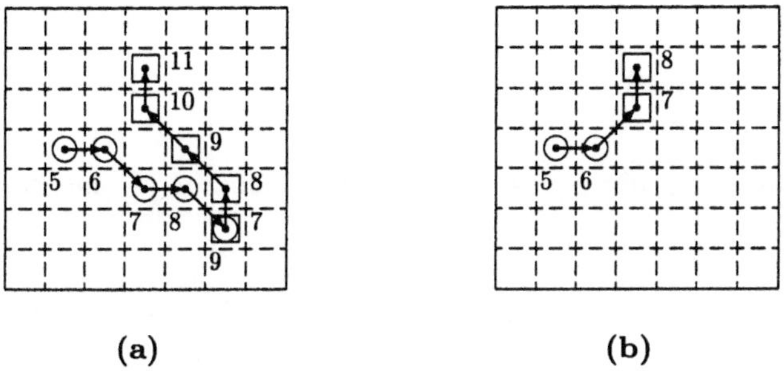

Fig. 6. Concatenation of two sub-paths with elimination of superfluous nodes

An algorithm achieving this reduction is a refined version of the algorithm described above when illustrating the operator *crossover*:

1. Moving forward on the first sub-path beginning in *Start*, the first node K_i is sought which has an immediate neighbour within the second sub-path.
2. In case of K_i having more than one immediate neighbour in the second sub-path, choose the one which lies nearest to *Goal*, calling it K_j.
3. The new path results as the connection of the path sections *Start* $\rightarrow K_i$ (first sub-path) and $K_j \rightarrow$ *Goal* (second sub-path).

A last process to be considered in detail is replacing a path section by a new one, as it is the case when applying the operators *mutation* and *tighten.* Figure 7(a) shows a path, labelled with circles, whose interior section between the nodes 2 and 11 is to be replaced by the sub-path labelled with squares.

The following algorithm – a refinement of the algorithm described above when illustrating the operator *mutation* – guarantees that the resulting path contains no undesired node sequences:

1. Moving forward on the sub-path beginning in *Start*, the first node K_{first} is sought which has an immediate neighbour within the new sub-path.
2. In case of K_{first} having more than one immediate neighbour in the new sub-path, choose the last one in the path's direction, calling it K_{insert1}.
3. Moving backward on the sub-path ending in *Goal*, the first node K_{last} is sought which has an immediate neighbour within the new sub-path.
4. In case of K_{last} having more than one immediate neighbour in the new sub-path, choose the first one in the path's direction, calling it K_{insert2}.

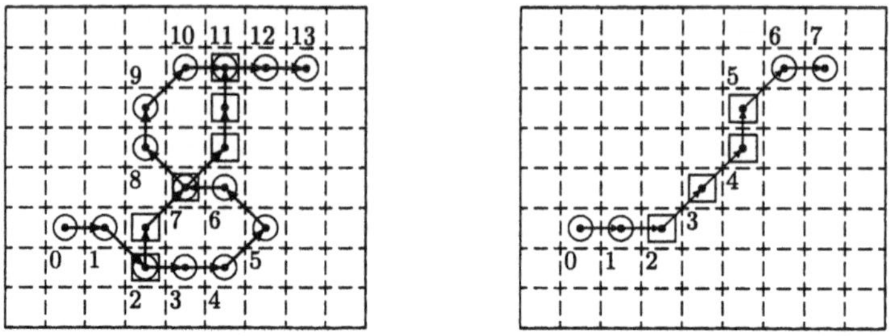

(a) (b)

Fig. 7. Insertion of a new sub-path with elimination of superfluous nodes

5. For insert1 $\leq$ insert2, the new path results as the connection of the path sections $Start \rightarrow K_{\mathbf{first}}$ (start section), $K_{\mathbf{insert1}} \rightarrow K_{\mathbf{insert2}}$ (new sub-path), and $K_{\mathbf{last}} \rightarrow Goal$ (end section). Otherwise, the new section has to be inserted in reverse direction.

The result of applying this algorithm can be seen in Fig. 7(b).

4 Operator Probabilities and Experimental Results

To find an optimum set of probabilities for the different genetic operators various sets were investigated. For each set of probabilities the GA was applied to three different environments (see Fig. 8), one hundred times each, employing 50 individuals. Two interesting results were obtained. Firstly, a probability greater than 0 for pure reproduction reduces the performance of the algorithm. Thus, the sum of the other probabilities should be set equal to 1. Secondly, the use of different sets for modifying drivable or non-drivable paths has been investigated. As the performance increased only slightly, this approach can be abandoned.

The best set of probabilities found was $P(\text{Mutation}) = 0.55$, $P(\text{Crossover}) = 0.35$, and $P(\text{Tighten}) = 0.10$. Using these probabilities, a first drivable path within the environments depicted in Fig. 8 was found after an average of about 182, 80 resp. 52 generations. Thereby the probability of mutation consists of two parts: 0.30 for normal mutation and 0.25 for a short mutation, which was restricted to modifications of path sections with a length smaller than a quarter of the overall path length. Furthermore, the vertical and horizontal distances between the start node of the replaced section and the via point had to be smaller than half of the environment's width or length, respectively. Having discovered a first drivable path, after some dozens of further generations an optimum path has been found in more than 90 percent of all cases.

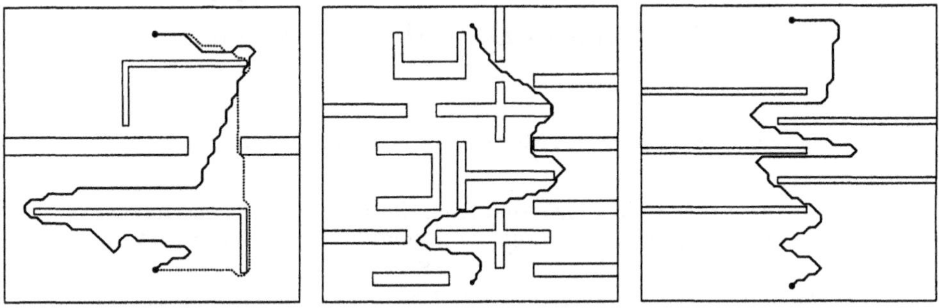

Fig. 8. Example environments, each with a not further optimised drivable path found by the GA. The first sub-figure shows in addition a dotted optimum path. The environments were subdivided into 70 × 70 squared regions.

5 Conclusions and Outlook

A new GA based path planning method was described, emphasising the handling of path modifications when applying the genetic operators. First applications of the algorithm showed very promising results, and particularly the operator *tighten* turned out to be very suitable for the given optimisation problem. Furthermore, this path planning method is successfully being utilised in a small size experimental environment employing Khepera robots (for more information about that robot see [4]). Nevertheless, there are some drawbacks: in certain environments the GA does not find drivable paths. This is the case, for instance, if a way out of a spiral with several windings is sought. Therefore, current considerations concentrate on developing an operator which searches for path sections circumventing obstacles.

References

[1] Baker, J.E.: Reducing Bias and Inefficiency in the Selection Algorithm. *Genetic Algorithms and their Applications: Proc. 2nd Intl. Conference on Genetic Algorithms.* Lawrence Erlbaum Associates, Hillsdale, NJ, 1987

[2] Gerke, M.: Genetische Bahnplanung für mobile Roboter. *8. Workshop Fuzzy-Control des GMA-FA 5.22.* Dortmund, 1998

[3] Lee, S., Kardaras G.: Collision-Free Path Planning with Neural Networks. *Proc. 1997 IEEE Intl. Conference on Robotics and Automation.* Albuquerque, NM, 1997

[4] Löffler, A., Mondada, F., Rieckert, U., editors: *Experiments with the Mini-Robot Khepera. Proc. 1st International Khepera Workshop.* Paderborn, 1999

[5] Xiao, J., Michalewicz, Z., Zhang, L., Trojanowski, K.: Adaptive Evolutionary Planner/Navigator for Mobile Robots. *IEEE Transactions on Evolutionary Computation,* 1(1), 1997

[6] Zhao, M., Ansari, N., Hou, E.S.H.: Mobile Manipulator Path Planning by a Genetic Algorithm. *Journal of Robotic Systems,* 11(3), 1994

A New Modified Genetic Algorithm for Multiuser Detection in DS/CDMA Systems

Mahrokh G. Shayesteh [1,3], Mohammad B. Menhaj [1,2], Babak G. Nobary [1]

[1] Dep. of Electrical Eng., Amir Kabir University of Tech., Tehran, Iran
g7523901@cic.aku.ac.ir, menhaj@cic.aku.ac.ir,
nobary@modares.ac.ir

[2] Dep. of Electrical & Computer Eng., Oklahoma State University, Oklahoma, USA
menhaj@ceat.okstate.edu

[3] Faculty of Engineering, Ourmieh University, Ourmieh, Iran
m.shayesteh@mail.urmia.ac.ir

Abstract. The conventional single user detector in DS/CDMA (Direct Sequence Code Division Multiple Access) systems involves multiple access interference and the near-far effect which cause the limitation of capacity. On the other hand, the computational complexity of the optimum multiuser detector grows exponentially with the number of users. There has been a lot of interest in suboptimal multiuser detectors with less complexity and reasonable performance. In this paper we apply the classic and a new modified genetic algorithm for multiuser detection of DS/CDMA signals. It is shown that the classic genetic algorithm (GA) reaches an error floor at high signal to noise ratios while the proposed method has higher performance than the classic one and comparable to the optimum single user detector with much less complexity than the optimum multiuser detector.

1 Introduction

DS/CDMA is considered as the third generation of cellular mobile, indoor wireless and personal communication systems. CDMA offers attractive features such as frequency reuse, soft handoff, increased capacity, multipath combating, and etc [1]. In a CDMA system, several users simultaneously transmit information over a common channel using pre-assigned signature codes.

The conventional single user detector in a DS/CDMA system consists of a bank of filters matched to the spreading codes and then deciding on the sign of the outputs. Ideally the spreading codes for DS/CDMA should be orthogonal. However, in practice, the codes are non-orthogonal, which creates interference in conventional matched filter demodulation. Multiple access interference (MAI) produced by the other co-channel users is a significant limitation to the capacity of this detector. Another limitation is the near-far problem which occurs when the relative received power of interfering signals becomes larger. These two problems severely limit the

B. Reusch (Ed.): Fuzzy Days 2001, LNCS 2206, pp. 608–618, 2001.

multi access capability of DS/CDMA. A potential solution is multiuser detection that exploits the information of the signals of interfering users [2].

The optimum multiuser detector [2] evaluates a log-likelihood function over the set of all possible information sequences. It consists of a bank of matched filters front end followed by Viterbi algorithm. The optimum receiver achieves low error probability at the expense of high computational complexity. The computational complexity of this detector increases exponentially with the number of users. So this method is extremely time consuming for a realistic number of users. Consequently, there has been considerable research into suboptimal detectors. These detectors achieve significant performance gains over the conventional detector without the exponential increase in complexity.

Genetic algorithms (GAs) have been employed for solving many optimization problems. They were first introduced by Holland [3] and have received considerable attention in DSP research [4]. GAs model natural selection and survival of the fittest [5]. While GAs are not perfect, i.e. they do not always find the optimal solution, they are very efficient in attaining a near-optimum solution in a much shorter time compared to the conventional point-by-point exhaustive search, especially in large solution spaces. GAs solve optimization problem iteratively. At each iteration (generation) they use a set (a population) of possible bit vectors (strings), from which a new set of bit vectors is generated based on the value (fitness) of the current bit vectors by applying genetic operators. This process continues until the terminating condition is satisfied.

GA-based multiuser detection has been proposed in [6-8]. In [6] a pure GA and a hybrid GA-multistage detector are used. In [7] a GA in conjunction with a modified Viterbi algorithm is used to detect the users bits sequentially in the asynchronous AWGN channel. However these papers does not examine the effect of high SNRs (Signal to Noise Ratio). In [8], good initial guesses concerning the possible user bit sequences obtained from the other detectors such as the matched filter and decorrelator [9] were required in order to attain an improved performance.

In this paper, we propose a new modified genetic algorithm. We apply the classic and modified GAs in DS/CDMA systems. The likelihood function is used as the objective function in GAs. It is shown that the bit error rate saturates at high signal to noise ratios using the classic GA. We show that the proposed method has much better performance than the classic one and is near to the optimum detector with much less complexity. The proposed algorithm finds the solution without using the other detectors and without reaching the error floor.

This paper is organized as follows. In section 2 we explain the classic genetic algorithm and then describe our proposed modified method. Section 3 describes the model of a typical DS/CDMA system in the AWGN. In section 4 simulations and results are discussed. Finally section 5 presents the conclusion.

2 Classic and New Modified GAs

A genetic algorithm (GA) is a guided random technique which mimics the process of natural evolution and is an efficient tool in searching for global optima. In this section, at first we describe the classic GA and then explain the modifications that leads to the proposed method.

2.1 The Classic GA

The classic GA encodes potential solutions to an optimization problem in the form of a generation containing *P* individuals called the population [3], [5]. By modifying the current generation, the next generation is obtained. After a certain number of iterations, the best individuals of the final population will be the solutions.

In the classic GA, the samples in the solution space are usually encoded as binary bit strings, and a finite number of them are selected randomly to form the initial population. The process of obtaining the next population from the current one involves applying two types of genetic operators to the current population. These operators are called crossover and mutation. A crossover operator requires two operands or so called parents and produces two outputs (offspring). Parent selection makes extensive use of the fitness value of the individuals. The fitness of each individual is its figure of merit and normally is the objective function. Selecting the parents for a crossover operator can be done using a so-called parent selection algorithm such as the Roulette wheel or Tournament selection methods. A crossover operator exchanges certain bits of the two parents according to a specific rule and produces two offspring. Some of the most common crossover operators are one-point, two-point and uniform. The mutation operator simply complements one or more bit of the individual it is applied to. In a given generation, the crossover and mutation operators are applied to the individuals with predefined probability p_c and p_m. Normally p_c is higher than p_m. The terminating condition for a genetic algorithm can be either the average or minimum fitness of the population reaching a certain value or the total number of generations exceeding a certain number (*G*).

According to the above, below a pseudo code form of the classic GA is given.

a) Set the required parameters:
- *P*: Population length
- p_c: Crossover probability
- p_m: Mutation probability
- *G*: The number of generations
- Select a parent selection method
- Select a crossover operator
- Select a mutation operator

b) Generate a random population. This will be the initial population and the starting point for the GA.

c) Apply the selected crossover operator to $P.\,p_c$ of the individuals and store the offspring.

d) Apply the selected mutation operator to $P.\,p_m$ of the individuals and store the resulting mutated individuals.

e) Replace the current population with the individuals resulting from crossover and mutation.
f) Repeat steps **c** to **e** until the termination criterion is satisfied.

2.1.1 Parent Selection Methods

Parent selection is analogous to the phenomenon of natural selection. Usually members are selected for mating with a selection probability proportional to their fitness values. Popular parent selection schemes include tournament selection and roulette-wheel selection. Here we describe the roulette-wheel method. In principle we construct a roulette wheel on which each member of the population is given a sector whose size is proportional to the individual's fitness. Then we spin the wheel and whichever individual comes up becomes a parent.

We can describe the implementation of the procedure follows:

a) Construct a list of fitness of all individuals in the population.
b) Generate a random number between zero and the total fitness in the population.
c) Return the first individual whose fitness, added to the fitness of all other elements before it, from the list in step **a**, is greater than or equal to the random number from step **b**.

2.1.2 Crossover Methods

To exploit the potential of the current population, we use crossover operators to generate new individuals that we hope will retain good features from the previous generation. Crossover is usually applied to the selected pairs of parents with a probability equal to a given crossover rate.

One-point crossover is the most basic crossover operator, where a crossover point on both parents is selected at random and two parents are interchanged at this point.

In two-point crossover, two crossover points are selected and the part of the individual between these two points is then swapped to generate two offspring. We can define n-point crossover similarly. In general, (n-1) point crossover is a special case of n-point crossover.

Another technique that is very common is uniform crossover. Uniform crossover is radically different from one-point and two-point crossovers. Copying the corresponding bit from one or the other parent, chosen according to a randomly generated crossover mask creates each bit in the offspring. Where there is a 1 in the mask, the bit is copied from the first parent, and where there is a 0 in the mask, the gene is copied from the second parent. The process is repeated with the parents exchanged to produce the second offspring. A new crossover mask is randomly generated for each pair of parents.

The crossover rate can be variable through time. As the population starts to converge the crossover rate can be increased so that the area in which the search is converging in is searched more thoroughly. The effect of crossover is similar to that of mating in the natural evolutionary process, in which parents pass segments of their own chromosomes on to their children. Therefore, some children are able to outperform their parents if they get good bits from both parents.

2.1.3 Mutation

Crossover exploits current bits potentials, but if the population does not contain all the encoded information needed to solve a particular problem, no amount of bit mixing can produce a satisfactory solution. For this reason, a mutation operator capable of spontaneously generating new individuals is included. The most common way of implementing mutation is to flip one bit of an individual with a probability equal to a very low given mutation rate. A mutation operator can prevent the population from converging at any local optima. The mutation rate is kept low so good individuals obtained from crossover are not lost. If the mutation rate is high (above 0.1), GA performance will approach that of a primitive random search.

2.2 Proposed Modified GA

The modified GA is obtained by applying the following modifications to the classic GA.

A) Parent selection is based on choosing individuals with better fitness. The mutation operation is performed on the individuals that have been involved in the crossover operation. The crossover and mutation rates are chosen in such a way that the total number of offspring and mutated individuals is less than the population length. The rest of the population is generated randomly. This added randomness prevents the algorithm from being confined to a specific area of the search space.

B) After applying the crossover and mutation operators to the current population, i.e. producing the offspring and the mutated individuals, the next generation is obtained by selecting the best P individuals from the current population and the generated individuals from A.

C) It is seen in [8] that good initial guesses (the first population) can greatly improve the performance of the GA. These initial guesses were obtained from the other detectors such as the decorrelator [9]. Here, we let the algorithm run for a certain number of generations and thus eliminate the randomness of the individuals. Next, with the exception of a number of fittest individuals, the rest of the population is randomly generated. With these fittest as initial guesses, the algorithm finds the solution in much fewer generations without making use of the other detectors and without reaching the error floor.

D) As we saw in the classic GA, the next generation is constructed by replacing the current generation with the individuals resulting from crossover and mutation. If the mutation rate increases, the number of high fitness individuals will decrease. On the other hand, decreasing the mutation rate too much will narrow the search preventing the search space to be searched effectively. So we have to make a trade-off between keeping the high fitness individuals (by decreasing the mutation rate and increasing the crossover rate) and expanding the search. Because of **B** the fittest individuals are never lost. As a result we need not worry about losing the fit individuals by increasing the mutation rate and can increase the mutation rate and even make it larger than the crossover rate.

3 System Model

The system model consists of K simultaneous users. The kth user's signature waveform of duration T and normalized power is composed of a spreading sequence of N chips:

$$s_k(t) = \sum_{n=0}^{N-1} a_n^k(t) p(t - nT_c) \tag{1}$$

where $a_n^k \in (-1,1)$ is the spreading sequence and $p(t)$ is the pulse waveform of duration T_c. For BPSK data modulation, the received baseband signal in an asynchronous AWGN channel is:

$$r(t) = \sum_{i=-P}^{P} \sum_{k=1}^{K} \sqrt{w_k} b_k(i) s_k(t - iT - \tau_k) + n(t) \tag{2}$$

where w_k is the received energy of the kth user, $b_k(i)$ is the data bit of the kth user during the ith interval with amplitude ± 1 and period $T = NT_c$, τ_k is the time delay of the kth user, $2P+1$ is number of bits in each transmission, and $n(t)$ is additive white Gaussian noise with two-sided spectral density equal to σ^2.

For the synchronous channel $\tau_k = 0 \ \forall k$, so we can drop the index i. In this case, the received signal can be written as:

$$r(t) = \sum_{k=1}^{K} \sqrt{w_k} b_k s_k(t) + n(t) = S(t,b) + n(t). \tag{3}$$

The conventional single user detector consists of a bank of filters matched to the spreading codes of users and a threshold device. The outputs of the filters can be written in matrix form as:

$$y = RWb + n \tag{4}$$

where for the synchronous case, we have

$$y = (y_1, \ldots, y_K)^T \tag{5}$$

$$b = (b_1, \ldots, b_K)^T \tag{6}$$

$$n = (n_1,\ldots,n_K)^T \tag{7}$$

$$W = diag(\sqrt{w_1},\ldots,\sqrt{w_K}) \tag{8}$$

$$R_{jk} = \int_0^T S_j(t)S_k(t)\,dt \tag{9}$$

In the above equations, y_k is the output of the matched filter for user *k*, b_k is the transmitted bit of user *k*, *R* is the correlation matrix with elements R_{ij} which shows the correlation between the spreading codes of users *i* and *j*, *W* is a diagonal matrix containing the users energy, and *n* is Gaussian noise vector with the covariance matrix $\sigma^2 R$.

Based on the sign of the outputs of filters, an estimate of the transmitted bits is obtained as

$$\hat{b}_k = sign(y_k) \qquad k = 1,\ldots,K \tag{10}$$

where *sign*(*x*) is signom function.

The conventional detector is optimum in the presence of only one single user and white Gaussian noise. However when there are other users, this detector suffers from two problems. First, as the number of users increases, it can become multiple access interference limit. Another disadvantage is the near-far problem, which refers to the phenomenon of high power interferes completely destroying communications from low power transmitters. One solution to these problems is multiuser detection where the receiver exploits the knowledge of the signals of interfering users [2]. The optimum multiuser detector maximizes the joint probability density function (maximum likelihood). It selects the data sequence *b* such that minimizes

$$\Lambda(b) = \int [r(t) - S(t,b)]^2\,dt \tag{11}$$

The minimization of the above equation results in the following log-likelihood detection

$$\hat{b}_{OMD} = \arg\left\{\max_{b\in\{-1,1\}} \left(2y^T b - b^T WRWb\right)\right\} \tag{12}$$

The optimum detector must evaluate the above metrics over all 2^K (synchronous) or $2^{K(2P+1)}$ (asynchronous) possible vectors in order to find the optimum *b*. Its computational complexity grows exponentially with the number of users and makes it impractical for a realistic environment. As a result, there has been a lot of interest into the suboptimum multiuser detectors with less complexity and near optimum performance.

Two types of receiver whose complexity is proportional to the number of users are decorrelating detector [9] and multistage detector [10]. The decorrelating detector applies the inverse of the correlation matrix to the matched filter outputs in order to decouple the received signal. This detector is linear and near-far resistance. It provides significant improvement over the conventional detector at conditions where MAI is dominant, without requiring knowledge of the user power. However the disadvantages of this detector is noise enhancement which causes poor performance at low signal to noise ratios. The multistage detector is nonlinear. It utilizes previously made decisions of the other users to cancel interference present in the signal of the desired user. This process is repeated for several stages. Multistage detector requires exact knowledge of the users power. Its performance depends heavily on the initial estimation.

In this experiment, we apply genetic algorithm for multiuser detection. We use the likelihood function as the fitness function which is always positive:

$$C(b) = \exp\left(2y^T b - b^T WRWb\right). \tag{13}$$

4 Simulations and Results

The spreading sequences for all users were chosen from a set of Gold codes of length N=31. The rectangular chip waveform was used for the spreading codes. The number of users is K=20. Monte Carlo simulation is used for computing the error rate. The effect of MAI is examined by keeping constant the relative powers of users and varying noise variance. The near-far effect is simulated by changing the relative power of users with constant noise variance. A lower bound for the BER (Bit Error Rate) of the optimum multiuser detector can be found as the BER of a single user receiver in the absence of other users.

Uniform crossover is used both for the classic and proposed GAs. In the classic GA, individuals obtained from crossover and mutation operators form the next generation. For the modified GA, in each generation, a number of random individuals is produced according to the modified algorithm. Also, after a certain number of generations, we retain some individuals and generate the rest randomly.

Fig. 1 shows the performance of the mentioned GAs for a 20 user system in the synchronous AWGN channel. The BER of the first user versus the signal to noise ratio (SNR) is obtained with $w_i / w_1 = 6\,dB$ for different values of population size (P) and generation number (G). It can be seen that in the classic GA, the BER saturates at high SNRs, while the proposed modified GA has a performance comparable to the single user lower band.

Fig. 2 shows the BER of user 1 versus the relative energy of the interfering users for $SNR1 = w_1 / \sigma^2 = 10\ dB$. It is seen that the classic GA is not robust to the near-far effect, while the modified one is resistant to the variation of relative energies.

Figs. 3 shows the effect of population size (P) and the number of generations (G) on the BER for $SNR1 = 10\ dB$ and $w_i / w_1 = 6\ dB$. It is observed that for a high population size P=40, the modified GA is capable of reaching the optimal value within about 35 generations. On the contrary, a population size of P=20 requires more generations before the residual error floor is reached; the amount of the error for this case is higher than that of P=40.

The choice of the crossover and mutation probabilities also affects the convergence rate and the residual BER. This is illustrated in Fig. 4 for different values of p_c and p_m. It is observed that the higher values of mutation rate and lower values of crossover probability improve the performance of the modified GA.

GAs are more efficient than the optimum detector in terms of the number of fitness function evaluations required. In GA, the objective function is evaluated $P \times G$ times which is much less than the 2^K required for optimum detection.

The above simulations can also be applied to the asynchronous AWGN case and multipath fading channel with L paths. Thus, the length of individuals are $(2P+1)K$ and $(2P+1)KL$, respectively. Of course, the computational complexity grows in comparison with the synchronous channel, but GAs still are more efficient than the optimum detectors which need $2^{K(2P+1)}$ evaluations.

5 Conclusions

In this paper, we examined the classic and a new modified genetic algorithms as multiuser detectors in DS/CDMA systems. The likelihood function was used as the fitness function. It was observed that for the classic GA, the bit error rate saturates at high signal to noise ratios. We proposed a modified genetic algorithm to solve this problem. Its performance is much better than the classic GA and is close to the single user lower band. The modified GA is also resistant to the near-far effect while the classic one does not have his property. The complexity of GAs is $P \times G$ objective function evaluations which is much less than 2^K required for optimum multiuser detection. The same results can be obtained for asynchronous AWGN and multipath fading channels.

References

1. Lee, W.C.Y.: Overview of Cellular CDMA. IEEE Trans. Vehicular Techn., Vt-40, No. 2 (1991) 291-302
2. Verdu, S.: Minimum Probability of Error for Asynchronous Gaussian Multiple-Access Channels. IEEE Trans. Information Theory, Vol. IT-32, No. 1, (1986) 85-96
3. Holland, J.H.: Adaptation in Natural and Artificial Systems. University of Michigan Press (1975)
4. Tang, K.S.K., Man, F., Kwong, He, S.Q.: Genetic Algorithms and Their Applications. IEEE Signal Processing Mag., Vol. 34, (1996) 22-37
5. Goldberg, D.E.: Genetic Algorithms in Search, Optimization, and Machine Learning. Addison-Wisely (1989)
6. Ergun, C., Hacioglu, K.: Multiuser Detection Using a Genetic Algorithm in CDMA Communications Systems. IEEE Trans. Commu., Vol. 48, No. 8, (2000) 1374-1383
7. Wang, X.F., Lu, W.S., Antonio, A.: A Genetic Algorithm-Based Multiuser Detector for Multiple Access Communications. ISCAS'98, Vol. 4, (1998) 534-537
8. Juntti, M.J., Schlosser, T., Lilleberg, J.O.: Genetic Algorithms for Multiuser Detection in Synchronous CDMA. IEEE International Symposium on Inform. Theory, (1997) 492

9. Lupas, R., Verdu, S.: Linear Multi-User Detectors for Synchronous Code-Division Multiple Access Channels. IEEE Trans. Information Theory, Vol. 38, No. 4, (1989) 496-508
10. Varansi, M.K., Aazhang, B.: Near Optimum Detection in Synchronous Code Division Multiple Access Systems. IEEE Trans. Commu., Vol. 39, (1991) 725-735

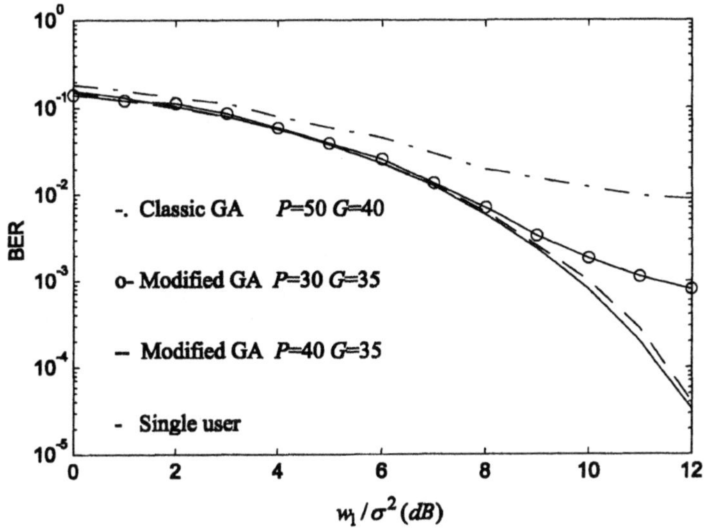

Fig. 1. The BER of user 1 versus *SNR*1 with $w_i/w_1 = 6dB$, *K*=20, *N*=31, p_c=0.3, p_m=0.5. For the classic GA: p_c=0.9, p_m=0.1 (the best case)

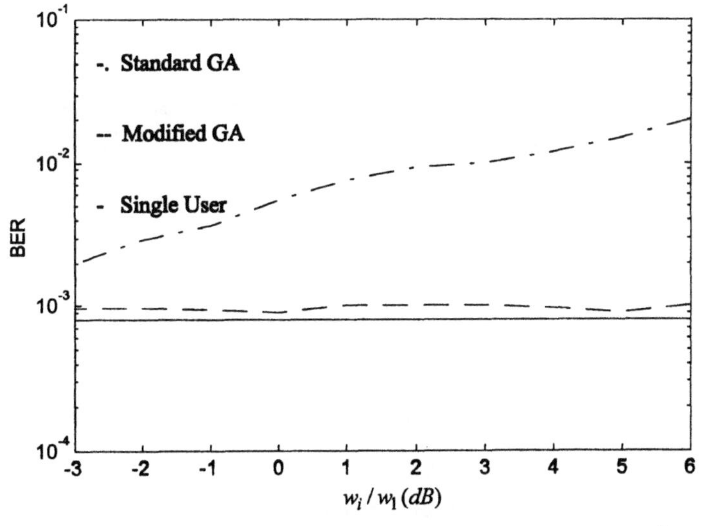

Fig. 2. The BER as a function of w_i/w_1 (near-far effect) with w_1/σ^2 =10 *dB*, *K*=20, *N*=31, p_c=0.3, p_m=0.5, *G*=35, and *P*=40. For the classic GA: p_c=0.9, p_m=0.1

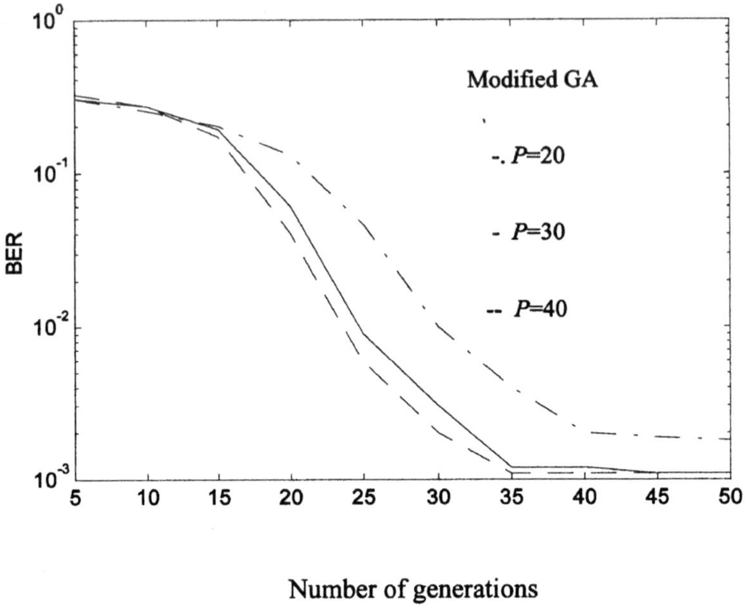

Fig. 3. The BER of user 1 as a function of the number of generations with $w_i / w_1 = 6\,dB$, $w_1/\sigma^2 = 10\ dB$, $p_c = 0.3$, $p_m = 0.5$, $K = 20$, $N = 31$

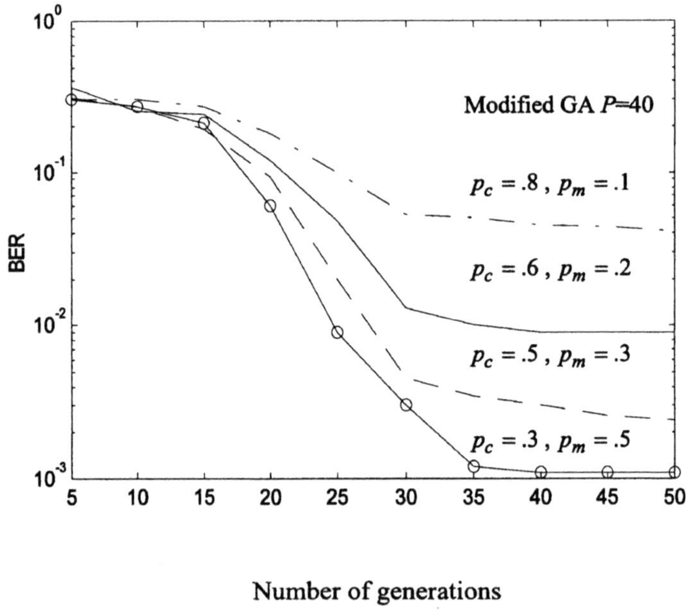

Fig. 4. The BER of user 1 as a function of the number of generations for various probabilities of crossover and mutation, $w_i / w_1 = 6\,dB$, $w_1/\sigma^2 = 10\ dB$, $K = 20$, $N = 31$

Clustering of Categoric Data in Medicine – Application of Evolutionary Algorithms

Thomas Villmann and Conny Albani

Universität Leipzip, Germany
Klinik für Psychotherapie, 04107 Leipzig, K.-Tauchnitz-Str. 25
villmann@informatik.uni-leipzig.de

Abstract Clustering of non-metric data sets often occurs in investigations in medicine and social science. The problem is to find suitable measures which describe similarities and, hence, are applicable to the clustering algorithm. In the present contribution we use evolutionary algorithms EA for clustering. Thereby, the similarity measures determine the respective fitness function for the EA. We consider several fitness functions and derive a new one which allows, additionally, the determination of a useful cluster number. — For the EA we use a new selection strategy combining the advantages of both the (μ, λ)- and $(\mu + \lambda)$-strategy and a multiple subpopulation approach with a migration scheme following the collective learning dynamic in self-organizing maps.

keywords: *evolutionary algorithms, clustering, medicine, self-organizing maps*

1 Introduction

Modern medicine research uses all the standard methods of mathematical statistics to extract relevant features, structural information and other data obtained medical investigations. However, a great number of informations are not available as metric data. Examples are verbal diagnostics (some times respective codes), categorical data as well as data received from questionnaires and other. The information extraction uses such databases usually is very difficult. In case of ordinal scales some times the ordinate is used as a metric scale. However, this approach is dangerous because of the implicit infiltration of (unknown) assumptions. In this contribution we want concentrate on the problem of clustering categoric data.

In the last years, several approaches were developed to overcome the problems for clustering of ordinal-scaled data [1], [2], [3], [4], [5], [6]. Thereby, new ideas have significantly improved the clustering abilities. In particular the approaches by Hofmann & Buhmann [4] and Graepel & Obermayer [5] which are dealing with the application of entropy maximization as cluster optimization criterion, provide powerful new methods. On the other hand, Hasenjäger

B. Reusch (Ed.): Fuzzy Days 2001, LNCS 2206, pp. 619–627, 2001.

ET AL. introduced a approach for clustering in case of missing data [3]. However, the used algorithms are advanced schemes such as neural network structures, simulated annealing and Monte-Carlo-Simulations which require a careful treatment and expert knowledge. We suggest in application of *evolutionary algorithms* (EAs) for solving the clustering tasks. EAs are robust algorithms and easily to implement in comparison to the above algorithms [7]. Moreover, by the strong decision between manipulating the objects and its assessment by the fitness function a wide class of optimization functions can serve as optimization goal for partitioning [8]. In the following we demonstrate the application of EAs for clustering of categoric data in psychotherapy research. At the same time we suggest some extensions of the basic EA schemes to improve the performance which are an *advanced selection strategy* and a *multiple population approach with migration* inspired by the learning dynamic in neural maps.

2 Clustering of Categoric Data in Psychotherapy - Fitness Function for Dissimilarity Data

One of the mostly used method for acquisition of structures of interpersonal relationships in the area of psycho–dynamic psychotherapy research is the method of the '*Core Conflictual Relationship Theme*' (CCRT) developed by LUBORSKY, [9]. The method investigates short stories about relationships, so–called *relationship–episodes*, which are often reported by the patients in their therapeutical sessions. In each of these episodes the components *wish of the subject*, *response of the object* and *response of the subject* were encoded which are used to perform the CCRT using a system $\mathcal{S}^w$ of $s^w_{\max} = 34$ so-called *standard categories* S^w_j to classify the wishes [10]. Each standard category describes an aspect in verbal manner. However, the categories are often correlated in meaning because of they are describing not only contrary but also the same or similar psychological topics. Therefore, they are collected in a set $\mathcal{C}^w$ of $c^w_{\max} = 8$ *clusters* C^w_k [11] which are then used in further considerations instead of the standard categories. Analogously, we have $s^{ro}_{\max} = 30$ categories $S^{ro}_j \in \mathcal{S}^{ro}$ for encoding the response of both the object and subject [10] which are collected in $c^{ro}_{\max} = 8$ clusters $C^{ro}_k \in \mathcal{C}^{ro}$ and $c^{ro}_{\max} = 8$ clusters $C^{rs}_k \in \mathcal{C}^{rs}$, respectively. However, the clusters also are still correlated again. Furthermore, several considerations have shown that *the used scheme of assignment still leads to unsatisfactory reliability rates* and interpretation problems in investigations based on this scheme. For further and more detailed critical remarks we refer to [12].

For solving the re-clustering problem of standard categories at first we determined the similarities between the standard categories S^w_i, S^w_j on the basis of their (symmetric) conditional probabilities $p^w(i \mid j)$ (in analogy $p^{ro}(i \mid j)$, $p^{rs}(i \mid j)$ for the responses of object and subject). Now, to perform the clustering, it has to be done as a clustering of proximity data. This problem we can apply the approach of vector quantization of proximity data introduced by HOFMANN & BUHMANN [4] and GREAPEL & OBERMAYER [5]:

$$\mathcal{H} = \frac{1}{2} \sum_{i=1}^{M} \sum_{k=1}^{M} \frac{D_{i,k}}{M} \left(\sum_{v=1}^{c_{\max}} \frac{u_{i,v} u_{k,v}}{p_v} - 1 \right) \tag{2.1}$$

with $p_v = \sum_{i=1}^{M} u_{i,v}/M$ the normalized percentage of the data in that cluster and M the number of data. The values $D_{i,k}$ are the *dissimilarities* between the data points which we can define as difference of the conditional probabilities from the unit. $\mathcal{H}$ serves as a cost function and has to be minimized. As pointed out in [4] $\mathcal{H}$ is independent on non-symmetric dissimilarities and permutations of the cluster indices. Moreover, the cluster assignments $u_{i,v}$ are distributed according to the Gibbs distribution

$$\mathbf{P}(\mathcal{H}) = \exp\left(-\frac{\mathcal{H} - \mathbf{F}(\mathcal{H})}{T} \right) \tag{2.2}$$

whereby T plays the role of a temperature, and $\mathbf{F}(\mathcal{H})$ is the free energy. Yet, because of the statistical dependence of the assignments the Gibbs distribution $\mathbf{P}(\mathcal{H})$ can not be exactly rewritten in a factorized form, however a mean field approximation and corrections to the assignment correlations has been derived [4]. Yet, $\mathcal{H}$ requires a *priori known value* $c_{\max}$, which is, in fact, a hard restriction and difficult to overcome [1]. For this purpose, the so-called *partition entropy* was introduced which *explicitly takes the number of clusters into account* [13]:

$$\mathcal{E}_{\mathcal{P}}(c_{\max}) = -\frac{\frac{1}{M \log(c_{\max})} \sum_{k=1}^{M} \sum_{i=1}^{c_{\max}} u_{k,i} \log(u_{k,i})}{1 - \frac{c_{\max}}{M}} \, . \tag{2.3}$$

$\mathcal{E}_{\mathcal{P}}(c_{\max})$ is often applied in medical applications [2]. Yet, these approaches require metric spaces as data in contrast to the topographic vector quantizer. Therefore, if we combining the partition entropy $\mathcal{E}_{\mathcal{P}}(c_{\max})$ (2.3) and $\mathcal{H}$ (2.1) we obtain as new fitness function:

$$\mathcal{F} = \frac{\frac{1}{M \ln(c_{\max})} \mathcal{H}}{1 - \frac{c_{\max}}{M}} \tag{2.4}$$

which explicitly takes the number of clusters into account and is also applicable to non-metric data. Hence, $\mathcal{F}$ incorporates both demanded features. As it shown in [5] it is possible to minimize $\mathcal{H}$ by simulated annealing, and, hence this approach could be applied to optimize $\mathcal{F}$.

Yet, it is difficult to generate a careful annealing scheme in the above approach, because the of the large number of local minima ant the occurring phase transitions in the dynamic of the annealing process [5]. A possibility for convergence improvement is the application of evolutionary algorithms to optimize $\mathcal{H}$ or $\mathcal{F}$.

For the application of EAs for clustering as presented below it is useful to interpret clustering as a *partitioning problem* under constraints. Thereby, a *partitioning* of a nonempty set $\mathcal{S}$ related to a nonempty set $\mathcal{C}$ is an unique and surjective mapping $\Phi : \mathcal{S} \to \mathcal{C}$. Then a *partition* Ψ_Φ of $\mathcal{S}$ with respect to the partitioning Φ is given by $\Psi_\Phi = \left\{ \Phi^{-1}(C) \mid C \in \operatorname{cod}(\Phi) \right\}$ whereby $\operatorname{cod}(\Phi)$ is the range of Φ. These constraints are collected in a fitness function which may be identified as the above $\mathcal{F}$.

For solving the partitioning task each individual in a generation of an EA describes a certain partition. In the present paper we assume S to be discrete containing $s_{\max}$ elements. Then we can take each individual as a string of length $s_{\max}$ the components of which contain the cluster index onto which the respective component has to be mapped. Mutation of an individual is defined as a random change of the mapping for a randomly selected individual component, and the crossover is the usual two-point crossing.

3 Advanced EA schemes

3.1 Multiple Population EAs with Toplogical Order and Neighborhood Related Migration Strategies

Several approaches were developed to improve the above basic EA-scheme. Especially, multiple subpopulation approaches are widely considered [14, 15]. Thereby the basic population $\mathbf{\Pi}$ is divided into subpopulations Π_i which have more or less communication during the evolution which may be realized as migration. To improve the performance of the basic subpopulation approach we arranged the subpopulations Π_i in a topological order usually chosen to be a regular lattice, for instance a ring or a quadratic lattice [1]. Between these subpopulations a *migration scheme* was introduced which is adapted from the SOM-learning: a visit (migration) from individuals between neighboring subpopulations is allowed for a short time *regarding to the topological order* Ω in $\mathbf{\Pi}$ and with respect to a time-dependent neighborhood function

$$h_{i^*}(t,k) = (1.0 - \epsilon_h) \cdot \exp\left(-\frac{r_{i^*,k}}{2\left(\sigma_h(t)\right)^2}\right) + \epsilon_h \tag{3.1}$$

with a small positive number ϵ_h [16]. During the evaluation of a certain subpopulation Π_{i^*} the neighborhood function h is applied to determine the number of visiting individuals from each other subpopulation. Because of $h_{i^*}(t,k) \in (0,1]$ we can interpret $h_{i^*}(t,k)$ as a *probability for migration* of an individual of the subpopulation Π_k into the actual evaluated Π_{i^*}. The value $r_{i^*,k}$ is defined as the rank of neighborhood $r_{i^*,k} = \operatorname{rank}(\Pi_{i^*}, \Pi_k; \Omega)$ with respect to the topological order Ω between the actually evaluated subpopulation Π_{i^*} and another Π_k. The defined migration scheme follows the idea of neighborhood cooperation in learning in neural maps. In fact, neural maps learning is indicated by a large cooperativity between topologically ordered neurons determined by a neighborhood function at the beginning of the adaptation process to implement a rough but fast learning. During the progress the cooperativity is shrinked to nearest neighbor learning which corresponds to a fine tuning of the neurons. As shown in [16, 17] this approach also in EA environments drastically improves the adaptation rate.

[1] In general, other arrangements are also admissible. Then the lattice can be defined by a connection matrix which describes the neighborhood relations.

Table 1: Values for the weighted concordance coefficient $\tilde{\kappa}$ and the fitness function $\mathcal{F}$ according to the original clusters and the EA generated solutions.

database	$\tilde{\kappa}$ original clusters	$\mathcal{F}$ original clusters	$\tilde{\kappa}$ EA clusters	$\mathcal{F}$ EA clusters	c_{max} EA clusters
P^w	0.334	6.88	0.435	6.24	10
P^{ro}	0.323	6.28	0.421	6.16	9
P^{rs}	0.479	6.16	0.504	6.09	9

Up to now we only implemented the neighborhood oriented cooperation scheme. However, additionally one can take the neighborhood knowledge into account to improve the genetic operators too. This work should be done in the next step and show some parallelism to the work of J. HUHSE [18, 19].

3.2 The $(\mu * \lambda)$-selection strategy

For selection of the offspring generation we have used a mixture of the well-known (μ, λ)– and the $(\mu + \lambda)$–strategy (in the notation of SCHWEFEL, [20]). While in the (μ, λ)–strategy where μ individuals produce λ children with $\mu < \lambda$ only the μ best of the λ children form the new population, in the $(\mu + \lambda)$–strategy all $\mu + \lambda$ individuals are allowed for the selection process. Whereas in the second strategy the best solution is preserved but the evolution tends to stagnate into a local minimum, in the first one the convergence is decelerated to allow reaching deeper minima (near the global minimum) but good solutions may be lost during the evolution. Balancing the advantages of both strategies [7] in the $(\mu * \lambda)$–approach again μ individuals produce the λ preliminary offsprings [21]. However, in the selection step the μ_t best individuals of the old generation and the λ new ones are allowed for comparison with respect to their fitness to generate the final offspring generation of μ individuals. Thereby μ_t depends on time t of evolution appearing as the number of generations performed:

$$\mu_t = \text{int}\left[(\mu - \mu_\tau) \cdot \gamma(t)\right] + \mu_\tau \quad , \tag{3.2}$$

with int $[x]$ stands for the integer value of x. The function $\gamma(t)$ is of decreasing sigmoid type with $0 \leq \gamma(t) \leq 1$ here chosen as the Fermi function $\gamma(t) = \frac{1}{1+\exp\left(\frac{(t-t_a)}{t_b}\right)}$ to switch near the t_a-*th* generation from the $(\mu + \lambda)$–strategy to the (μ, λ)–strategy in a definite range of generation steps ($\approx 4t_b$). We have $\mu_0 = \mu$ for the initial value and $\lim_{t \to \infty} \mu_t = \mu_\tau$ ($\mu_\tau \ll \mu$) coding a minimal survival probability for the parent individuals. In this way we get a smoothed switch from the $(\mu + \lambda)$– to the (μ, λ)–strategy, what we call $(\mu * \lambda)$–strategy, combining the advantages of both strategies and, additionally, always preserving the best μ_τ individuals (slightly different from the original (μ, λ)–strategy).

Table 2: Different values for the weighted concordance coefficient $\tilde{\kappa}$ and the respective meaning for intra cluster agreements of the considered observables

$\tilde{\kappa}$–coefficient	meaning
$\tilde{\kappa} < 0.1$	no agreement
$0.1 \leq \tilde{\kappa} < 0.4$	weak agreement
$0.4 \leq \tilde{\kappa} < 0.6$	clear agreement
$0.6 \leq \tilde{\kappa} < 0.8$	strong agreement
$0.8 \leq \tilde{\kappa}$	nearly complete agreement.

4 Results

At first we computed $\mathcal{F}$ according to (2.4) for the original clusters $\mathcal{C}^{w}$, $\mathcal{C}^{ro}$ and $\mathcal{C}^{rs}$ of the respective standard categories as defined in [11]. Additionally, we computed the weighted concordance coefficient $\tilde{\kappa}$ [12] for comparison with other investigations in the area of psychotherapy research (see Tab. 1). The $\tilde{\kappa}$–coefficients may be interpreted according to Tab. 2 [22]. Applying $\mathcal{F}$ from (2.4) as fitness measure in the EA we result cluster solutions with higher number of clusters as depicted in Tab. (1): However, the increased $\tilde{\kappa}$-values as well as the better (decreased) $\mathcal{F}$-values refer to a better intra-cluster agreement which correspond to psychological considerations: From a psychological point of view the new clusters show a better coherence which allows a more clear interpretation. Moreover, the increased cluster number allows a more detailed description [12, 23].

From technical point of view we used in our computations $\mu_{\text{all}} = 400$ individuals which were evenly distributed onto 10 subpopulations. The subpopulations were arranged on a ring as topological structure, whereby $\mu_{\mathcal{T}}$ in (3.2) was given as $\mu_{\mathcal{T}} = 1$. We trained the ensemble during $t_{\text{max}} = 5000$ time steps. The characteristic time scale for decreasing the neighborhood between the subpopulation was defined as linear shrinking of $\sigma_h(t)$ in (3.1) with $\sigma_h(0) = 3$ and $\sigma_h(t_{\text{max}}) = 0.2$.

The choice of the new fitness function $\mathcal{F}$ gives the ability to compare cluster solutions with different cluster number as well as optimization of the cluster number during the evolutionary process. Moreover, the SOM-like migration scheme generated in neighbored subpopulations cluster solutions which are judged by psychotherapist as similar. However, a mathematical proof of this fact is difficult because of the complicate structure of the fitness function but should subject of further considerations as well as an improvement of the fitness function. Moreover, Fuzzy-clustering should be investigated in the future to understand the complicate data structure in a better way.

5 Conclusion

In the contribution we suggest the application of Evolutionary algorithms for solving clustering tasks for categoric data as often occur in medical research. EAs as easy treatable and robust tools can incorporate several special optimization goals (fitness functions) for clustering, because the manipulation of the objects to be optimized is not restricted by the choice of the assessment. Hence, EAs provide a powerful method for realization which, in addition, allow the integration of expert knowledge. Moreover, the presented approach can be extended to *fuzzy clustering* straight forward.

References

[1] J. Buhmann, H. Kühnel, Vector quantization with complexity costs, IEEE Transactions on Information Theory 39 (1993) 1133–1145.

[2] F. Ermini, C. Marchesi, Intelligent data retrieval from multidimensional clinical archives, in: E. Ifeachor, A. Sperduti, A. Starita (Eds.), Neural Networks and Expert Systems in Medicine and Healthcare, World Scientific, Singapore, New Jersey, London, Hong Kong, 1998, pp. 295–303.

[3] M. Hasenjäger, H. Ritter, Active learning with local models, Neural Processing Letters 7 (1998) 107–117.

[4] T. Hofmann, J. Buhmann, Pairwise data clustering by deterministic annealing, IEEE Transactions on Pattern Analysis and Machine Intelligence 19 (1) (1997) 1–14.

[5] T. Greapel, K. Obermayer, A stochastic self-organizing map for proximity data, Neural Computation 11 (1) (1999) 139–155.

[6] T. Villmann, Application of evolutionary algorithms for clustering of nonmetric data in medicine, in: G. Brewka, R. Der, S. Gottwald, A. Schierwagen (Eds.), Fuzzy-Neuro-Systems'99 (FNS'99), Proc. Of the FNS-Workshop, Leipziger Universitätsverlag, Leipzig, 1999, pp. 155–165.

[7] Z. Michalewicz, Genetic Algorithms + Data Structures = Evolution Programs, third, revised and extended Edition, Springer–Verlag Berlin Heidelberg New York, 1996.

[8] T. Villmann, Neural networks approaches in medicine – a review of actual developments, in: Proc. Of European Symposium on Artificial Neural Networks (ESANN'2000), D facto publications, Brussels, Belgium, 2000, pp. 165–176.

[9] L. Luborsky, The core conflictual relationship scheme, in: N. Freedman, S. Grand (Eds.), Communicative Structure and Psychic Structures, Plenum Press New York, 1977.

[10] P. Crits-Christoph, A. Demorest, List of standard categories (edition 2), in: L. Luborsky, H. Kächele (Eds.), Der Zentrale Beziehungskonflikt - ein Arbeitsbuch, PSZ Verlag, Ulm, Germany, 1988.

[11] J. Barber, P. Crits-Christoph, L. Luborsky, A guide to the CCRT Standard Categories and their classification, in: L. Luborsky, P. Crits-Chrostoph (Eds.), Understanding Transference, Basic Books New York, 1990, pp. 37–50.

[12] C. Albani, T. Villmann, B. Villmann, A. Körner, M. Geyer, D. Pokorny, G. Blaser, H. Kächele, Kritik und erste Reformulierung der kategorialen Strukturen der Methode des Zentralen Beziehungs–Konflikt–Themas (ZBKT), Psychotherapie, Psychosomatik und Medizinische Psychologie 49 (11) (1999) 408–421.

[13] L. Kaufmann, P. Rousseuw, Finding Groups in Data - A Introduction to Cluster Analysis, John Wiley , Sons, 1990.

[14] D. B. Fogel, Evolutionary Computation: Towards a New Philosophy of Machine Intelligence, IEEE Press, Piscataway, NJ, 1995.

[15] H. Mühlenbein, M. Gorges-Schleuter, O. Krämer, Evolution Algorithm in Combinatorial Optimization, Parallel Computing (7) (1988) 65–88.

[16] T. Villmann, R. Haupt, K. Hering, H. Schulze, Parallel evolutionary algorithms with som-like migration, in: A. Dobnikar, N. Steele, D. W. Pearson, R. Albrecht (Eds.), Artificial Neural Networks and Genetic Algorithms (Proc. Of ICANNGA'99), Springer-Verlag, Wien - New York, 1999, pp. 274–279.

[17] T. Villmann, Evolutionary algorithms with subpopulations using a neural network like migration scheme and its application to real world problems, Integrated Computer- Aided Engineering (2001) to appear.

[18] J. Huhse, A. Zell, Evolutionary strategy with neighborhood attraction, in: H. Bothe, R.Rojas (Eds.), Neural Computation 2000, ICSC Academic Press, Zürich, 2000, pp. 363–369.

[19] J. Huhse, A. Zell, Investigating the influence of the neighborhood attraction factor to the evolution strategies with neighborhood attraction, in: M. Verleysen (Ed.), Proc. Of European Symposium on Artificial Neural Networks (ESANN'2001), D facto publications, Brussels, Belgium, 2001, pp. 179–184.

[20] H.-P. Schwefel, Numerical Optimization of Computer Models, Wiley and Sons, 1981.

[21] K. Hering, R. Haupt, T. Villmann, Hierarchical Strategy of Model Partitioning for VLSI–Design Using an Improved Mixture of Experts Approach, in: Proc. Of the Conference on Parallel and Distributed Simulation (PADS'96), IEEE Computer Society Press, Los Alamitos, 1996, pp. 106–113.

[22] L. Sachs, Angewandte Statistik, 7th Edition, Springer Verlag, 1992.

[23] A. Körner, Kategorisierung von Beziehungsschemata mit der Methode "Das Zentrale Beziehungskonfliktthema", Ph.D. thesis, University Leipzig, Germany (2000).

A New Direction in AI – Toward a Computational Theory of Perceptions

Professor Lotfi A. Zadeh*

Humans have a remarkable capability to perform a wide variety of physical and mental tasks without any measurements and any computations. Familar examples are: parking a car, driving in city traffic, playing golf, cooking a meal, and summarizing a story. In performing such tasks, humans employ perceptions of time, direction, speed, shape, possibility, likelihood, truth, and other attributes of physical and mental objects. Reflecting the bounded ability of the human brain to resolve detail, perceptions are intrinsically imprecise. In more concrete terms, perceptions are f-granular, meaning that (a) the boundaries of perceived classes are unsharp; and (b) the values of attributes are granulated, with a granule being a clump of values (points, objects) drawn together by indistinguishability, similarity, proximity, and functionality. For example, the granules of age might be labeled very young, young, middle-age, old, very old, etc.

F-granularity of perceptions puts them well beyond the reach of traditional methods of analysis based on predicate logic and/or probability theory. The computational theory of perceptions (CTP) that is outlined here adds to the armamentarium of AI a capability to compute and reason with perception-based information. The point of departure in CTP is the assumption that perceptions are described by propositions drawn from a natural language, e.g., it is very unlikely that there will be a significant increase in the price of oil in the near future.

In CTP, a proposition, p, is viewed as an answer to a question and the meaning of p is represented as a generalized constraint. To compute with perceptions, their descriptors are translated into what is called the Generalized Constraint Language (GCL). Then, a goal-directed constraint propagation is employed to answer a give query. A concept that plays a key role in CTP is that of precisiated natural language (PNL).

The computational theory of perceptions suggests a new direction in AI – a direction that may enhance the ability of AI to deal with real-world problems in which decision-relevant information is a mixture of measurements and perceptions. What is not widely recognized is that many important problems in AI fall into this category.

---

* Research supported in part by ONR Contract N00014-99-C-0298, NASAContract NCC2-1006, NASA Grant NAC-117, ONR Grant N00014-96-1-0556, ONR Grant FDN0014991035, ARO Grant DAAH 04-961-0341 and the BISC Programm of UC Berkeley. Lotfi A. Zadeh is Professor in the Graduate School and director, Berkeley initiative in Soft Computing (BISC), Computer Science Division and the Electronics Research Laboratory, Department of EECs, University of California, Berkeley, CA 94720-1776; Telephone: 510-642-4959; Fax: 510-642-1712; email: zadeh@cs.berkeley.edu.

B. Reusch (Ed.): Fuzzy Days 2001, LNCS 2206, p. 628, 2001.

Personalization of Newsletters Using Multistage Fuzzy Inference

Dipl.-Inform. Gero Presser

University of Dortmund, Department of Computer Science I,
D-44221 Dortmund, Germany.
Phone: +49 231 755 6309, Fax: +49 231 755 6555
presser@ls1.cs.uni-dortmund.de

Abstract. Personalization is becoming more and more important in any e-commerce application. However, nowadays it is primarily used to personalize web-pages whereas newsletters are in most cases static, i.e., independent of the customer's profile.
In this paper we introduce a method for personalizing newsletters based on fuzzy IF-THEN rules. The key intention for using *fuzzy* rules (instead of crisp rules) is to provide a framework that can easily be used to express *vague* concepts as they are frequently used in marketing issues. In particular, fuzzy rules are easy to understand (and define) for human operators like marketing experts.
Our approach makes use of multistage inference. This guarantees that the definition of the rulebase can be done in a modular fashion and hence the reuse of rules is directly supported.

1 Motivation

Without any doubt, e-commerce is becoming an important fundament of modern economies. Forrester Research (see `http://www.forrester.com`), for example, predicts that in 2003 the sales in Western Europe will exceed 700 Billions Euro in the segment business-to-business (B2B) and 125 Billions Euro in business-to-customer (B2C).

Most of the B2C-sales are generated via shops in the Internet (e-shops) like amazon.com. According to the most recent study of the Internet Software Consortium (isc, see `http://www.isc.org`), approximately 110,000,000 Hosts have been connected to the Internet in January 2001. Since for almost any need there is a plenty of web-sites fulfilling this specific need, is has become a serious problem to attract customers and to keep them coming back.

In contrast to the beginning of the "e-age", nowadays most commercial web-sites are no longer static but dynamically change their appearance (e.g., the products presented on the starting page of an e-shop) based on the profile of the respective customer. Personalization has become one of the most outstanding methods for increasing customer loyalty and its significance is still growing.

Another trend, that can clearly be observed, is the changing of the Internet from a pure pull-medium (where content is "pulled" by the customer) to a mixture of a push- and a pull-medium (cf. [2]). Almost any commercial site makes use

B. Reusch (Ed.): Fuzzy Days 2001, LNCS 2206, pp. 629–636, 2001.

of e-mail based services like newsletters or order-tracking messages to increase customer loyalty. Newsletter are very frequently used to inform customers of new products and related information.

However, most of these newsletters are *not* personalized. That is, any customer receives an identical copy of the newsletter independent of her specific profile. This is quite astonishing since even sites that massively make use of personalization techniques for tailoring the web-content still rely on static newsletters.

In this paper we introduce a framework that allows the personalization of newsletters in a convenient manner. This framework is based on multistage fuzzy inference where the "personalization logic" is encoded by means of fuzzy IF-THEN rules (called *business rules* in this context). The personalization of newsletters seems to be a fruitful application of fuzzy logic since there are many vague aspects involved. Because the concepts used from fuzzy logic are quite standard, the main focus in this paper lies on the explanation of the application as well as the model – we do *not* investigate aspects of fuzzy logic in more detail and even omit a discussion which fuzzy operators should be used.

The paper is organized as follows. In Section 2 we briefly discuss the term personalization as well as the related concept of user modelling. We continue by introducing the personalization of newsletters (Section 3) and the method that stands behind our framework (Section 4). Next, we describe the key element of this method in more detail (Section 5). The paper closes with a short outlook (Section 6) in which we address topics that are still under research.

2 Personalization and User Modelling

The term *personalization* is frequently used in a quite broad sense which is sometimes misleading: Its meaning ranges from simple personal greetings (as can be found on almost any web-site) and manually adjustable content (e.g., see MyYahoo!, `http://www.yahoo.com`[1]) to automatically tailored content (e.g., special offerings selected using some recommender engine based on user data as can be found on the web-site of amazon.com, cf. [11]). A mixture of different concepts is implemented on the Ziff Davis web-site `http://www.myzdnet.com`. Gartner Group predicts, that in 2003 more than 80% of all major web-sites will make use of personalization techniques to address users in a more efficient way (cf. [5]).

Numerous techniques are being used such as demographic profiling, collaborative filtering, and click-stream analysis (see [12] for example). One of the most important techniques is the rule-based approach where the "personalization

[1] Though this is quite frequently called personalization, it is actually *customization*. Personalization means that user data (like attributes, habits, ...) is used to retrieve content that is relevant to the respective user; this selection is done by the *server*. Customization, on the other hand, means that the *user* directly modifies the look and feel of content offered on a site. In practical applications the difference between the two concepts definitely matters (see [9]), hence the discrimination in not academical.

logic" is encoded by means of business IF-THEN rules (see [1]).[2] Our framework is an extension of this "classical" rule-based approach.

Basically, personalization means that content – which is delivered to some user – is tailored based on the user's profile. Hence, user modelling is a necessary prerequisite for any personalization method.

The purpose of a *user profiling* component is to acquire and manage a model of relevant user characteristics like interests, preferences and knowledge. In this paper we simply assume that there is some suitable user profiling component which can be employed for queries concerning user data. For details with regard to user profiling as well as best-practice examples see [3].

3 Personalization of Newsletters

Newsletters are among the most efficient marketing instruments reaching response rates up to 30% (see [13]). Nevertheless, there is one "dangerous" aspect about this marketing instrument which has strong implications: The user always has the possibility to ultimately unsubscribe, usually simply by clicking on a hyperlink at the end of the newsletter (or by replying in some special way). Hence the selection of the newsletter's content is of particular importance in order to keep the "opt-out" (i.e., the number of users unsubscribing) as small as possible.

Experience tells that the content should be desired, personal, and relevant (cf. [4]). Consequently, personalization is an outstanding method for retrieving the "right" content for a given customer since it exactly fits these needs.[3]

Most newsletters are merely personalized in a trivial sense (if ever): The customer is directly addressed by her or his name (e.g., "Hello Peter Stone"). In some cases, the user can choose in which categories she is interested in (e.g., see internet.com, `http://www.internet.com`). This information is used to binary decide which news are send to the customer (and which are *not* send). This method is quite useful if the content can clearly be divided into distinct categories and if there is a significant amount of news items. Therefore this kind of tailoring the content of newsletters is applied quite often by information sites.[4]

In our approach we go one step further and *dynamically* assemble the content for the newsletter based on the user's profile: A newsletter is split into a collection of *text modules* (e.g., any recommendation or news topic is a text module of its own). These text modules can be bundled in different ways depending on the profile of the respective recipient, thus leading to massively personalized newsletters (cf. Figure 1 for a schematic overview). Compare this to the classical

[2] Rule-based personalization is implemented, for example, in the Dynamo Web Server (art technology group, `http://www.atg.com`) or the One-to-One product suite (BroadVision, `http://www.broadvision.com`). Business rules are frequently used for up-selling, e.g. "IF Customer bought a Siemens Phone last week THEN show all Siemens accessories".

[3] Personalization increases the chance of retrieving both personal and relevant content. "Desired" means (briefly), that the user has voluntarily subscribed to the newsletter.

[4] However, it is rather *customization* of newsletters than personalization.

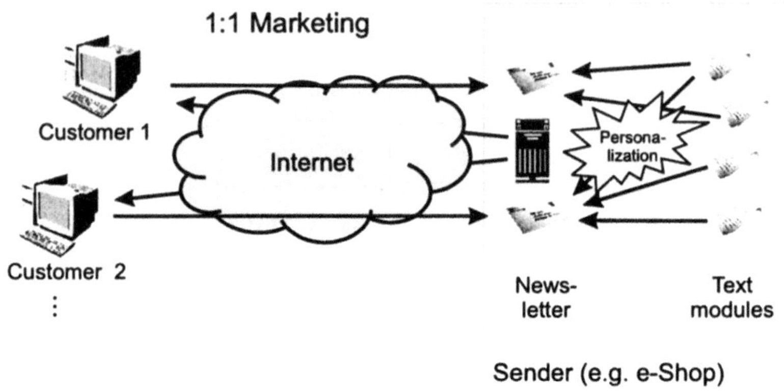

Fig. 1. Scheme of the newsletter personalization

method where the static bundle of *all* modules is sent to *any* recipient! Personalization massively helps in making the content of the newsletter relevant and hence in reducing the opt-out.

In addition, the *order* of the text modules that are sent to a given recipient can (and even should) depend on her profile. We suggest that the most important modules (for the given recipient) shall be put at the beginning of the newsletter.[5]

Thus, a method for estimating the degree of importance of a given text module for a given recipient (customer) is needed. In this context we suggest an approach that is based on a rulebase consisting of fuzzy IF-THEN rules. The key advantage of using *fuzzy* rules is that they seem to be very natural to human operators (e.g., marketing experts) and therefore are quite easy to understand as well as to define. In particular, fuzzy rules are capable of expressing vague concepts as they are frequently used in marketing issues.

There is one step that we have not mentioned yet but that turns out to be extremely important. We try to illustrate this point by means of an example. Assume that a newsletter has to be implemented for an existing online sports outfitter (e-shop). The problem with personalization in this example is that (in general) the operator of the e-shop does *not* really know his customers, making it almost impossible for him to say anything about the relevance of a given text module for a given customer.[6] But, on the other hand, the e-shop operator *can* easily define different target groups. In the sketched example, there might be the (target) group of football players and the group of skier.

Now, based on her activities in the online shop, for a given customer her membership to different target groups can be estimated. On the other hand, for

[5] This is known as the *"Reverse Pyramid"* concept, the most important topics are presented first. It is of particular importance since many users only read the first part of a newsletter.

[6] For a given customer all the operator knows might be that she bought a tennis court seven weeks ago. Does this customer like skiing?

a given target group and a given text module it turns out to be very easy (for the domain expert) to define the corresponding relevance. Putting both components together, a personalized newsletter can be assembled. In the following sections we formalize this approach.

4 Approach

Let $\mathcal{C}$ be the set of customers, $\mathcal{T}$ the set of text modules and $\mathcal{G}$ the set of target groups. We "need" a mapping

$$\mathit{group} : \mathcal{C} \times \mathcal{G} \rightarrow [0,1]$$

such that, for any customer $C \in \mathcal{C}$ and any target group $G \in \mathcal{G}$, $\mathit{group}(C, G)$ estimates the degree of membership of customer C to target group G.[7] Besides, a function

$$\mathit{rel} : \mathcal{G} \times \mathcal{T} \rightarrow [0,1]$$

has to be defined, where $\mathit{rel}(G, T)$ (for $G \in \mathcal{G}$ and $T \in \mathcal{T}$) is the degree of relevance of text module T for (a prototypical member of) target group G.

Given these two functions, it is possible to assemble a personalized newsletter (based on the set of text modules $\mathcal{T}$) for a given customer $C \in \mathcal{C}$.[8]

For defining function *group* we suggest an approach based on fuzzy IF-THEN rules. The key idea is that for a given customer $C \in \mathcal{C}$ there are some attributes (delivered by the user profiling component) which can be used for estimating $\mathit{group}(C, G)$ (for any $G \in \mathcal{G}$). Examples of suitable attributes are the amount of money spent for products of a certain category, whether the customer owns a Siemens phone, the customer's age, the overall amount of money spent or the web-pages (of the e-shop) visited. Though in most cases it is almost impossible to precisely define the function *group* based on these variables, it is quite easy to define rules like

> IF money-spent(C) is high AND money-spent-on(C,football) is low AND sites-visited(C,football) is low THEN interest(C,football) is low.[9]

[7] A possible extension of our model is to use fuzzy sets instead of crisp values here. Following this idea, $\mathit{group}(C, G)$ would be a fuzzy set on the unit interval $[0,1]$.

[8] However, this construction is by far not trivial since the customer belongs to more than one target group (in general). Our solution to this problem is straight-forward and the practical experience is promising: We interpret $\mathit{rel}(G, T)$ as kind of utility and estimate the relevance of text module T for a given customer C by

$$\sum_{G \in \mathcal{G}} \mathit{group}(C, G)\, \mathit{rel}(G, T).$$

Alternatively, $\mathit{rel}(G^*, T)$ where $\mathit{group}(C, G^*) = \max\{\mathit{group}(C, G) \mid G \in \mathcal{G}\}$ could be used. Though there are crucial differences between these two approaches, we do not want to go into details here. Anyway, both approaches can be justified and both approaches work well in practice.

[9] We omit a discussion of the rule's components since the names used should be self-explaining and we do not need more details here.

or

IF owns(C,Siemens-phone) is yes AND age(C) is young
THEN relevance(C,Siemens-sounds) is high.

In particular, rules of this type can be used to compute intermediate attributes which simplify the definition of the premises of other rules and which can be reused without any modification. E.g., in the above example interest(C,football) could be used as an input in another rule's premise.

Hence, a multistage inference has to be used. We investigate this topic (i.e., the representation of *group*) in more detail in the following section.

5 Function *group*

As pointed out before, we use a multistage fuzzy inference mechanism to define the function *group* which represents the "personalization logic". To keep things simple we use a hierarchical approach (cf. Figure 2).

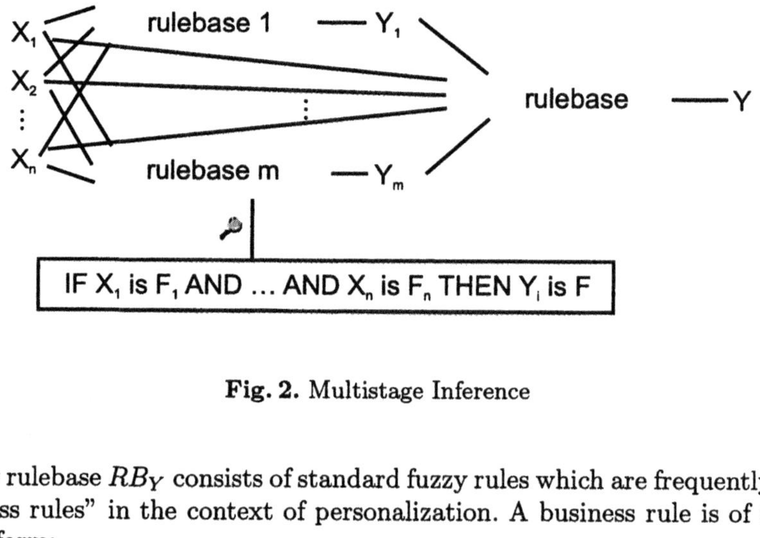

Fig. 2. Multistage Inference

Any rulebase RB_Y consists of standard fuzzy rules which are frequently called "business rules" in the context of personalization. A business rule is of the following form:

IF X_1 is L_1 AND ... AND X_n is L_n THEN Y is L,

where $X_1, \ldots, X_n, Y$ are fuzzy variables and $L_1, \ldots, L_n, L$ are fuzzy set membership functions (representing linguistic states of the corresponding variables). If necessary, the used variables have to be fuzzified first (e.g., some or even all variables used in the user profiling interface might be crisp variables). All rules in rulebase RB_Y have the output variable Y which can be used as an input in another rulebase.

As stated before, the rulebases are organized hierarchically. Hence, the inference can be done step-by-step. Given the values of the input variables of rulebase RB_Y and using some (fixed) inference mechanism, the result for the output variable Y can be computed.

Eventually, after defuzzification, for some given customer $C \in \mathcal{C}$ and any (target) group $G \in \mathcal{G}$, the value *group*(C, G) has been computed. Hence, the hierarchical fuzzy system realizes function *group* which allows – together with function *rel* and an ordering-scheme – to assemble a personalized newsletter.

6 Outlook

In this paper we have described the relevance of personalization of newsletters and introduced an approach based on fuzzy IF-THEN rules.

Though the approach was only sketched in this paper, there is nothing mystic about it and it can be implemented almost straight-forward. In fact, the author has developed a Java-based prototype which will continuously be improved. Nevertheless there are some points that need to be investigated in more detail, e.g. the ranking of the text modules, the fuzzy operators which are applied, and the inference mechanism used.

For real-world applications it is essential to have a graphical tool for editing the newsletter's content, i.e., the text modules and – in particular – the rulebase. This allows domain experts to easily give in the relevant information and hence produce personalized newsletters.

The rule-based approach to personalization originally stems from web-content personalization. It seems promising to employ *fuzzy* IF-THEN rules in this application as well, since the general question is identical. Again, fuzzy logic would have the advantage that the personalization logic – which is represented by business rules – becomes far easier to formalize and to maintain since vague concepts can be used.

References

1. Cliff Allen. Personalization – yesterday, today, and tomorrow, 2000. personalization.com, Soapbox.
`http://www.personalization.com/sopabox/columns/allen-column-1.asp`.
2. Daniel Amor. *The E-business (R)evolution.* Prentice Hall, Upper Saddle River, NJ, 1999.
3. Josef Fink and Alfred Kobsa. A review and analysis of commercial user modeling servers for personalization on the world wide web. *User Modeling and User-Adapted Interaction*, 10:209–249, 2000.
4. Seth Godin. *Permission Marketing.* Simon & Schuster, New York, 1999.
5. Gartner Group. Do you know what 'personalization' means?, 2000. Research Note.
6. B. Joseph Pine II. *Mass Customization.* Havard Business School, Boston, 1993.
7. Abraham Kandel, editor. *Fuzzy Expert System.* CRC Press, Boca Raton, FL, 1991.
8. George J. Klir and Bo Yuan. *Fuzzy Sets and Fuzzy Logic.* Prentice Hall, Upper Saddle River, NJ, 1995.
9. Paul F. Nunes and Ajit Kambil. Personalization? No Thanks. *Havard Business Review,* (4):32–34, 2001.

10. David M. Pennock and Eric Horvitz. Collaborative filtering by personality diagnosis: A hybrid memory- and model-based approach. In *Proceedings of the IJCAI Workshop on Machine Learning for Information Filtering*, 1999.
11. Badrul Sarwar, George Karypis, Joseph Konstan, and John Riedl. Analysis of recommendation algorithms for e-commerce. In *ACM Conference on Electronic Commerce*, pages 158–167, 2000.
12. J. Schafer, J. Konstan, and J. Riedl. Recommender systems in electronic commerce. In *Processings of the ACM Conference on Electronic Commerce (EC-99)*, pages 158–166, 1999.
13. Torsten Schwarz. *Permission Marketing – macht Kunden süchtig.* Max Schimmel, Würzburg, 2000.

Fuzzy Techniques for XML Data Smushing

Ernesto Damiani[1], Barbara Oliboni[2], and Letizia Tanca[2]

[1] Università di Milano, Dipartimento di Tecnologie dell'Informazione
Via Bramante 65, 26013 Crema, Italy
edamiani@crema.unimi.it

[2] Politecnico di Milano, Dipartimento di Elettronica e Informazione
Via Ponzio 1, 20100 Milano, Italy
oliboni,tanca@elet.polimi.it

Abstract. The recently proposed notion of a *Semantic Web* requires XML/RDF processing techniques able to locate, extract and organise heterogeneous information contained in XML documents coming from different sites, dealing flexibly with differences in structure and tag vocabulary. Such techniques should operate even when tagging is done in accordance with non-informative schemata, and even when no schema is available at all. In this paper, we review the main problems related to the processing and restructuring of large amounts of XML-based data, and propose some solutions in the framework of a flexible query and processing model for *well-formed* XML documents.

1 Introduction

The eXtensible Mark-up Language (XML) was initially proposed as a standard way to design mark-up languages for representing, exchanging and publishing information on the Web. Its usage has recently spread to many other application fields: to name but a few, XML markup is currently used for publishing legacy data, for enriching network resources with general-purpose metadata according to the *Resource Description Format* (RDF) standard, for storing data that cannot be represented with traditional data models and for ensuring interoperability among software systems. For the sake of brevity, we shall not review XML and RDF in this paper; the interested reader is referred to [W3C98b], [W3C99a]. XML mark-up and RDF metadata are used together to make documents self-describing, enabling applications to deal intelligently with document content. *Valid* XML documents are associated with a *Document Type Definition* (DTD) or a Schema, which defines their common syntactic structure. DTDs and Schemata, when available, further widen the range of applicable processing techniques. However, XML schemata and DTDs are often non-informative (for instance, the standard DTD for RDF metadata does not carry any information about the metadata content) or simply not available.

In this paper, we focus on techniques aimed at extracting and organizing information from heterogeneous *well-formed* (i.e., not valid) XML documents, dealing flexibly with differences in structure and tag vocabulary. The well-known

B. Reusch (Ed.): Fuzzy Days 2001, LNCS 2206, pp. 637–652, 2001.

XSLT language [W3C99b] provides powerful transformation and restructuring operators acting on XML data. A transformation expressed in XSLT allows to use declarative rules for transforming a source document into a result document. However, XSLT's `if match then action` paradigm makes its usage awkward when the structure of the target XML documents is heterogeneous and not known in advance. Moreover, XML processing techniques should be provided via languages easy to write and understand, as also non-technical persons are expected to access the large XML information bases supporting their businesses. In this paper we propose a fuzzy approach to the smushing problem that can be used in the framework of the fuzzy querying technique for XML documents, presented in [Dam00a]. Finally, we mention some possible commercial applications of our approach in the framework of e-business and knowledge-management systems.

Throughout the paper, we shall model XML documents as *multi-sorted trees* containing *typed nodes*. We loosely base our work on the XML Infoset data model which defines seven types of nodes, namely root, element, text, attribute, namespace, processing instruction and comment nodes. For the sake of conciseness, however, here we consider only three sorts: *element (E)*, *text (T)* and *attribute (A)* nodes. Other than that, the main difference between our data model and the XML Infoset is that in our approach a text node always contains at least one character of data. In other words, if the text node contains only whitespace, then the space is considered insignificant and stripped. This means that in our representation we have only nodes which have a content (*stripped trees*). For the sake of simplicity, we shall not elaborate further on the formal graph-theoretical definition of *well-formed* XML trees. Rather, we shall use a graphical representation of XML trees, outlining the properties they satisfy if and when needed. We represent element nodes as no-filled circles, attribute nodes as filled circles and a text nodes as rectangles[1].

The paper is organised as follows: after stating the problem in Section 2, we outline the proposed approximate matching technique for XML subtrees in Section 3, which also shows a worked out example. Section 4 sketches our prototype architecture while Section 5 draws the conclusions.

2 The Problem

Our research is aimed at a flexible search and processing technique, capable to extract relevant information from a (possibly huge) set of XML documents. Such XML documents may come from a number of different sources, each employing its own mark-up; this corresponds to a high degree of variability about the documents' structure and tag vocabulary. The relevant information is selected as follows [Dam00a]: first, a *XML pattern*, i.e. a partially specified subtree, is provided by the user. Then, the XML documents of the target dataset are scanned;

[1] Again for simplicity, in this work we assume that the elements of the document share the same namespace.

XML fragments are located and sorted according to their similarity to the pattern. This pattern-based approach is a step forward with respect to keyword-based techniques currently used by WWW search engines, as the latter cannot exploit XML mark-up, and therefore ignore potentially useful information about document structure.

In [Dam00a] we described a fuzzy technique for extending the structure of target document trees and weighing their arcs, in order to increase pattern recall. The problem of matching the pattern to the extended XML tree structure was described as a *fuzzy sub-graph matching* problem. Besides dealing with structure, however, any approximate matching technique needs to assess the degree of equivalence of the information carried by two nodes or subtrees. Such hidden equivalence should be reconciled or tolerated by means of flexible and approximate *smushing* techniques [2].

Suppose that two nodes having different tags contain "more or less" the same content (Fig. 1). For example, let A and B be two RDF nodes in the Semantic Web that do not have URIs, as follows:

```
node A:
  --rdf:type --> person
  --email --> "edamiani@crema.unimi.it"
  --firstName --> "Ernesto"
  --lastName --> "Damiani"
  --authorOf --> http://www.dreamwater.net/damiani


node B:
  --rdf:type --> person
  --email --> "edamiani@crema.unimi.it"
  --name --> "Ernesto Damiani"
  --homePage --> http://www.dreamwater.net/damiani
```

Intuitively, these two nodes should *match* to a degree, regardless of their tag name. Obviously, such node pairs are not crisply equivalent, but still they should be treated as equivalent to a degree, i.e. fuzzily. Mathematically the problem can be stated classically as finding a *distance* between points (the nodes) in a multidimensional space.

This approach has been proposed many times by information retrieval research, and the properties of the chosen distance metric have been found to be a prime determinant of how well an approach works. However, our scenario is quite different, as XML markup provides meta-information about field types and their value domains that can be used to perform some sort of smart reasoning about identity and similarity between nodes.

[2] For the original definition of the problem by Dan Brinkley, see `http://lists.w3.org/Archives/Public/www-rdf-interest/2000Dec/0191.html`

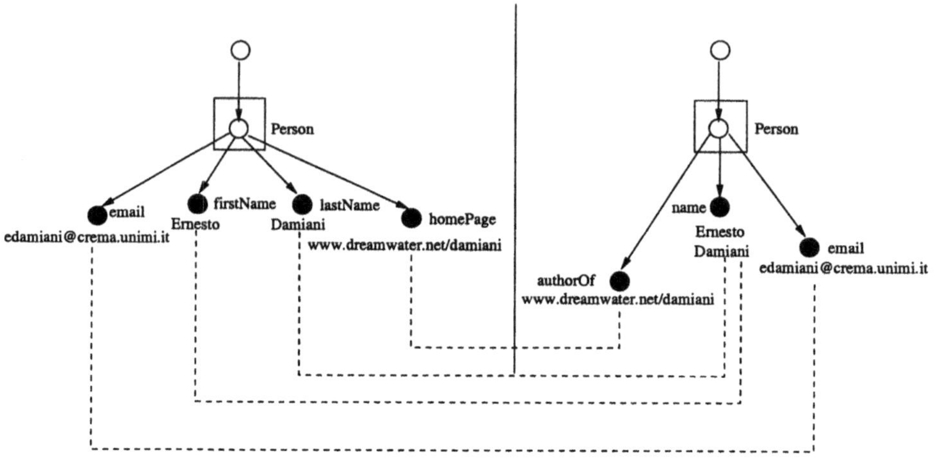

Fig. 1. Content similarity as a cause of equivalence between XML nodes

In the sequel, we rely on the notion of a *weighted XML tree*. Weighting techniques and fuzzy sub-graph matching of XML data were extensively discussed in [Dam00a] and [Dam00b].

In Fig.2 we show a XML graph with fuzzy weights. Note that the weight on each arc is referred to the correspondent attribute node or text node.

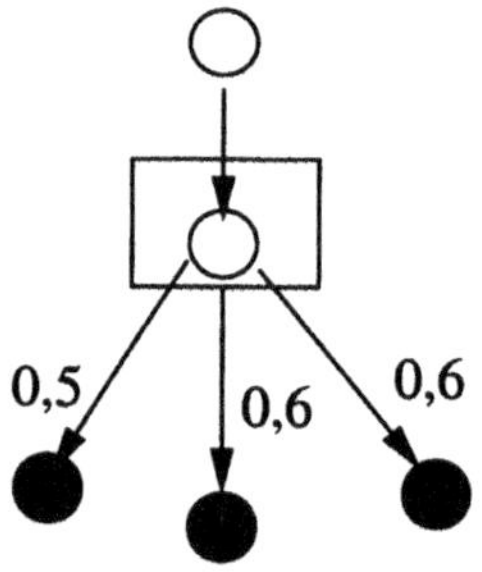

Fig. 2. A weighted XML tree

We are now ready to deal with the problem of computing a "degree of similarity" between two XML nodes and to decide whether it is sufficient to justify a match between them. In general, such problem can be stated as one of approximate subtree matching.

3 Approximate matching of XML subtrees

In order to compute the "degree of similarity" between a pattern and a document, we use a subgraph matching procedure to evaluate their structure similarity and a smushing technique to compute the degree of equivalence of the information carried by the leaf nodes, i.e. the nodes whose content includes attributes and data, but no sub-elements. In terms of the XML Infoset, those nodes are defined as having a *data-only content model.* We shall use the smushing technique also for the element nodes (non-leaf) which have a *mixed content model*, i.e. whose content includes sub-elements as well as attributes.

3.1 XML Thesauri

Depending on the specific application, XML tag names may carry a considerable amount of semantics that should be taken into account when computing similarity between subtrees.

Generally speaking, information about synonymy between XML tag names might be made available by providing suitable metadata. For instance, in a domain where multiple schemata are in use, RDF statements may be used to tie element definitions of different schemata together to obtain a formal standardized Thesaurus on all namespaces. Such an XML Thesaurus could then be used to assess synonymy between tag names even when well-formed (i.e., unvalidated) documents are involved.

The advent of the XML *namespace* technology [W3C99c] means that element types (and therefore, their names) can be unambiguously shared across documents as desired, increasing the potential usefulness of a Thesaurus.

Our approach does not rely on the availability of an XML Thesaurus; rather, we exploit it if available. We make a minimal set of assumptions about such Thesaurus; namely, we assume it to consist of a set of *pairs* of synonyms $\langle$ *TAGname(i), TAGname(j)* $\rangle$. Here, we do not deal with any synonymy metrics ([Dam99]): the presence of a pair of tag names in the set means that, according to the Thesaurus, these names are full synonyms.

Note that, in principle, each element of a pair is a sequence of a namespace name followed by a tag name; however, for the sake of conciseness, in the remainder of the paper we shall omit namespace prefixes before tag names.

3.2 Elementary type matching

To compute the "degree of similarity" between a pattern p_i and a document d_j we use a similarity $MATCH(p_i, d_j)$ considering the tag names of the nodes to evaluate, their types and contents.

$$MATCH(p_i, d_j) = \bigwedge_i \theta_{ij} \wedge (\bigvee_j (f_T(n_{p_i}, n_{d_j}))) \tag{1}$$

where $\bigwedge_i$ is one of the possible t-norms represented in Tab. 1 and $\bigvee_j$ is the corresponding t-conorm.

	t-norms
1	$min(x,y)$
2	xy
3	$max(x+y-1,0)$
4	$\frac{xy}{x+y-xy}$
5	$x(y=1), y(x=1)$, 0 o/w

Table 1. ***t*-norms**

Moreover, f_T is one of the possible functions $P \times D \rightarrow [0,1]$ to compute the type similarity between one node of the pattern (n_{p_i}) and one of the document (n_{d_j}) taking into account their estimated type T (Tab. 2).

As far as coefficient θ_{ij} is concerned, its role is to exploit the XML Thesaurus content if available[3]:

- If both tag names are contained in the Thesaurus, and belong to the same pair (i.e., according to the Thesaurus they are synonyms), $\theta_{ij} = 1$.
- When the Thesaurus contains only one of the considered tags, we get $\theta_{ij} = \frac{1}{4}$;
- if both the tags are not in the Thesaurus, $\theta_{ij} = \frac{1}{2}$.
- finally, if the tag names associated to nodes n_{p_i}, n_{d_j} appear in the Thesaurus, but do not belong to the same pair (i.e., according to the Thesaurus they are *not* synonyms), $\theta_{ij} = 0$.

The rationale for these choices is intuitive, as it relies on a conservative estimate about the Thesaurus completeness.

For an even more complete consideration of possible synonymies, we can apply the same conservative approach to the case of *transitive* synonymy, for instance *((character, letter), (letter, epistle))* or *((publication, work), (work, book))*. Thus, when two tags are related by a synonymy chain of length k, we can impose:

$$\theta_{ij} = \frac{1}{2} + \frac{1}{k+1} \tag{2}$$

Table 2 reports some possible functions f_T that compute the type similarity between one node of the pattern (n_{p_i}) and one of the document (n_{d_j}) taking into account their estimated type T.

Suppose that n_{p_i} and n_{d_j} are "URI" nodes, the content of n_{p_i} is `www.elet.polimi.it` and the content of n_{d_j} is `www.polimi.it`: in this case $l_{max} = 4$, $l_{min} = 3$. Note that k indicates the position of the component fields of the URIs from right to left and C will be 0 when the two parts are different, 1 otherwise.

[3] When no Thesaurus is available, Eq.1 can still be used, provided that we set $\bigwedge_i = min_i$ and $\theta_{ij} = 1 \; \forall i, j$

Type	Function f_T
URI	$\frac{1}{l_{max}} \sum_{k=1}^{l_{min}} \frac{1}{k} * C$ l_{max} = length of the longest URI l_{min} = length of the shortest URI $C(a,b) = \begin{cases} 0 & \text{if } a \neq b \\ 1 & \text{if } a = b \end{cases}$
DATE	$x = \frac{\alpha}{(\alpha+k)}$ k = distance between the dates α = parameter to customize function slope
STRING	*edit or cut-and-paste string matching metrics*

Table 2. Type Similarity

Thus with:

```
www.elet.polimi.it
     www.polimi.it
```

we have the following computation:

$$f_T(n_{p_i}, n_{d_j}) = \frac{1}{4}\sum_{k=1}^{3}\frac{1}{k} \; * \; C$$
$$= \frac{1}{4} * [(1*1) + (\frac{1}{2}*1) + (\frac{1}{3}*0) + (\frac{1}{4}*0)] = 0.375$$

In fact, the URIs have the same part in position $k = 1$ and $k = 2$.

With:

```
www.google.com
 www.polimi.it
```

we have $l_{max} = 3$, $l_{min} = 3$ and the computation is:

$$f_T(n_{p_i}, n_{d_j}) = \frac{1}{3}\sum_{k=1}^{3}\frac{1}{k} \; * \; C$$
$$= \frac{1}{4} * [(1*0) + (\frac{1}{2}*0) + (\frac{1}{3}*1)] = 0.1$$

With:

```
citeseer.nj.nec.com
 www.elet.polimi.it
```

we have $l_{max} = 4$, $l_{min} = 4$ and the computation is:

$$f_T(n_{p_i}, n_{d_j}) = \frac{1}{4}\sum_{k=1}^{4}\frac{1}{k} \ * \ C$$
$$= \frac{1}{4} * [(1 * 0) + (\frac{1}{2} * 0) + (\frac{1}{3} * 0) + (\frac{1}{4} * 0)] = 0$$

In fact `www.elet.polimi.it` and `citeseer.nj.nec.com` are different in all parts[4].

Moreover, suppose that n_{p_i} and n_{d_j} are "DATE" nodes: advantages in using a `Date` type include the possibility of quantifying the match in a subtler way than when considering simple strings.

Our method for matching two dates considers the match value as inversely proportional to the distance between the dates, measured in days. This difference can be automatically calculated by simple library functions, included in most high-level programming languages, and so we are not dealing with it.

The technique for match value computation exploits a hyperbolic, decreasing function of the distance between the dates k, calculating the value as

$$x = \frac{\alpha}{(\alpha + k)}$$

. The parameter α can be tuned for a customized function slope, in order to obtain a faster or slower convergence to zero w.r.t. the distance. Note that all values of α clearly return 1 for two identical dates, and make the function value decrease as the distance grows. For instance, a $\alpha = 1/2$ returns $x = 1/3$ for a 1-day distance, 1/15 for a 1-week distance, and 1/61 for a 1-month distance.

This result value could be corrected in particular cases: suppose for instance that we find exactly a 1-year difference between tow dates. This value could be considered as more significant than others, for it could represent a date of a periodic event.

With reference to the last row of Tab. 2, we observe that a wealth of metrics are available for approximate string matching, some of which are particularly suited for XML content obtained via specific authoring processes. For instance, cut-and-paste metrics [Bur00] has been proposed for document content known to have been produced manually or with the help of editor programs.

---

[4] Note that in this example we are also taking into account the "www" field, which could be otherwise considered non significant and thus excluded from the similarity computation.

3.3 A Worked-out Example

In [Dam00a] we described a fuzzy technique for extending the structure of target document trees and weighing their arcs, in order to increase pattern recall. In Fig. 3 we show a graph-based representation of an XML document containing information about students and their Master's thesis. The students are grouped by considering the topic of their thesis (Databases or Computer Networks). For each student the document contains information about the degree and other personal data. The document in Fig. 3 is a weighted XML document; in fact, we associate a weight to each arc by means of a *structure-related* weighing technique. This kind of weighing technique weighs the arcs of an XML graph by using topological parameters related to the position of elements.

In this example, to associate a fuzzy value to each arc (n_i, n_j), we use the *normalized distance* from root, defined as:

$$w_{arc}(n_i, n_j) = \frac{d_{max} - l}{d_{max}}$$

where d_{max} is the length of the longest path starting from root node and l is the distance from n_j to the root. Note that the arcs from an element to its attributes have the same weights as the ingoing arc of that element; indeed, to compute the longest path we considered only element nodes. The rationale for this choice is that in a XML subtree external elements are normally used to specify general information and therefore carry much of the subtree semantics.

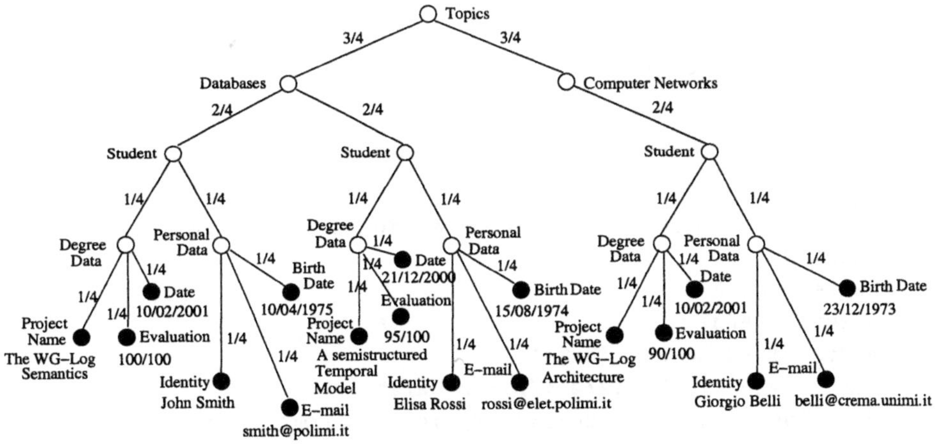

Fig. 3. A weighted XML document

In Fig. 4 we show the XML document of Fig. 3 after the computation of the *fuzzy transitive closure* which adds a new arc between two nodes if they are connected via a path of any length in the original graph.

The weights of the closure arcs can be computed by aggregating the weights of the original arcs by means of one of the $t-norms$ of Tab. 1. For example to compute the closure on the document of Fig. 3 we use the *min* function. For the sake of clarity, the closure arcs are depicted with dashed lines (Fig. 4).

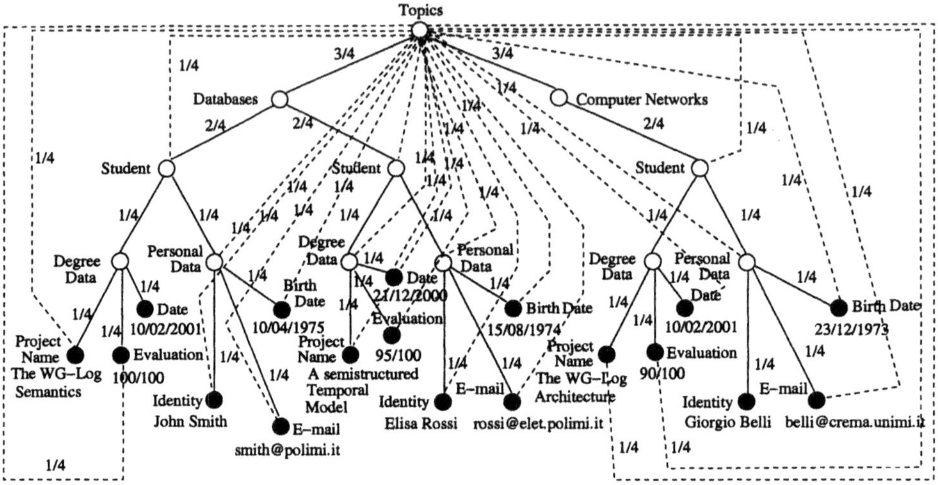

Fig. 4. Closure of the XML document of Fig. 3

The following fragment:

```
<Pattern>
 <Student>
  <Degree Data>
   <Thesis> A Semistructured Temporal Model </Thesis>
   <Mark> 95/100 </Mark>
  </Degree Data>
  <Personal Data>
   <Name> Elisa Rossi </Name>
   <Birth Date> 15/08/1974 </Birth Date>
   <Home Page> www.elet.polimi.it/~rossi </Home Page>
  </Personal Data>
 </Student>
</Pattern>
```

corresponds to the DOM tree of Figure 5 and requires information about student Elisa Rossi, whose birthdate is 15/08/1974; the title of this student's thesis (graduated with 95/100) could be something like "A Semistructured Temporal Model" and her home page "www.elet.polimi.it/∼rossi".

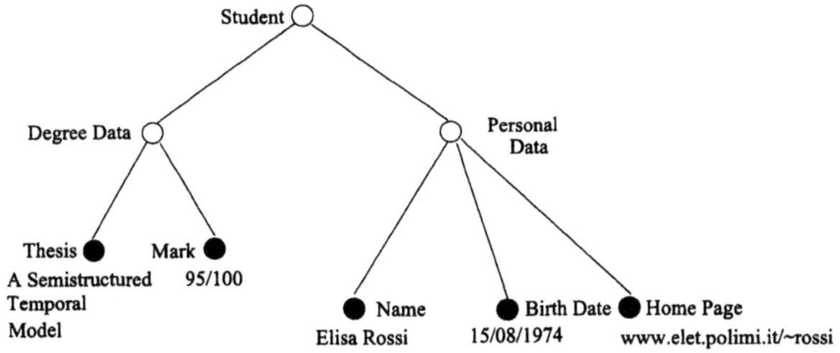

Fig. 5. A pattern

Our matching procedure searches a subtree having a structure similar to the DOM representation of the pattern (Fig. 5) inside an extended DOM representation of the document (Fig. 3). In order to obtain a more flexible pattern representation, *datatypes* can be taken into account by means of *pattern schemata* as follows:

```
<pattern xmlns:ipo="http://oursystem/IPO"
xmlns:xsi="http://www.w3.org/2000/10/XMLSchema-instance"
xsi:schemaLocation="http://oursystem/patternschema/thispattern.xsd">
<ipo:Student>
<ipo:Degree Data>
<ipo:Thesis> A Semistructured Temporal Model</ipo:Thesis>
<ipo:Mark ipo:XML-type="number"> 95/100 </ipo:Mark>
</ipo:Degree Data>
<ipo:Personal Data>
<ipo:Name> Elisa Rossi </ipo:Name>
<ipo:Birth Date ipo:XML-type="date"> 15/08/1974 </ipo:Birth Date>
<ipo:Home Page ipo:XML-type="uri"> www.elet.polimi.it/~rossi </ipo:Home Page>
</ipo:Personal Data>
</ipo:Student>
</pattern>
```

This technique allows us to *label* the pattern tree's nodes with type information. Information about types provided by the user is exploited by the `Smusher` module when estimating similarity between pattern and document nodes. Note that types' definitions are stored in a *pattern schema* (called `thispattern.xsd` in the previous example). Besides standard definitions of elementary types, this schema may contain domain-specific complex type definition.

The similarity degree between the pattern in Fig. 5 and the document of Fig. 3 is computed via function $MATCH(p_i, d_j)$ as described above. The similarity degree between the pattern node "Home Page" and the document node "E-mail" is computed in the following way:

```
www.elet.polimi.it
    elet.polimi.it
```

we have the following computation:

$$f_T(n_{p_i}, n_{d_j}) = \frac{1}{4}\sum_{k=1}^{3}\frac{1}{k} * C$$
$$= \frac{1}{4} * [(1 * 1) + (\frac{1}{2} * 1) + (\frac{1}{3} * 1) + (\frac{1}{4} * 0)] = 0.61$$

This value is also the result of $MATCH(p_i, d_j)$ because the other pairs of nodes have as similarity degree $f_T = 1$.

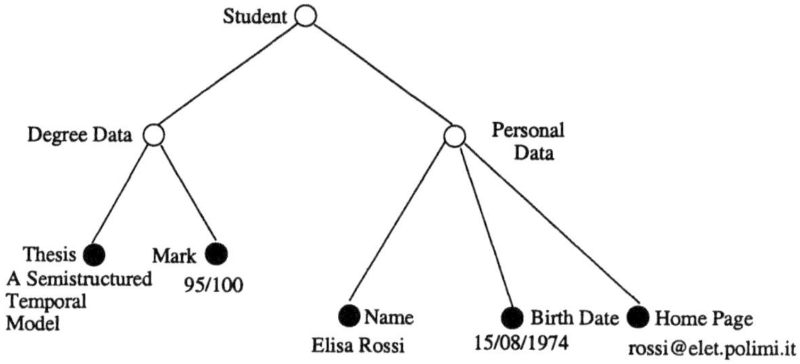

Fig. 6. Result

In Fig. 6 we show the result obtained by applying the pattern of Fig. 5 to the document of Fig. 3. The result is composed by *smushed* nodes, in fact the obtained nodes are constructed by considering both the document and the pattern. A smushed node has the tag name of the document node, but the attribute name indicated in the pattern. Thus, the information represented in the result is the information contained in the document and with the highest similarity degree. Note that smushing can be made reversible by allowing a "tentative" merge, where the nodes are matched but the matching can later be undone if further evidence or user input justifies it. The result visualization can be *restricted* or *total*; in our example we show only a *restricted visualization* of the result, in fact we report only information carried by the nodes indicated in the pattern, i.e. we show only information explicitly required. So, we do not report the Degree Date. The *total visualization* reports also information carried by all the nodes of the document related to the required information, highlighting the information contained in the nodes indicated in the pattern. In this way the user can obtain data not explicitly required, thus this kind of visualization can be useful when the request is general or the user does not know in advanced precisely what she is looking for.

3.4 Complex type matching

We are now ready to outline how our approach could be extended to internal nodes whose content model is *mixed*. In order to deal with this problem, we rely on the notion of *fuzzy subsethood* between the *states* of such nodes. In O-O data models, objects are usually characterized by *identity* and *state*. Identity distinguishes one object from another, while the state of an object consists of the values of its attributes. In fuzzy extensions of the O-O data model [Bor94], "similar" objects are grouped together to form a *fuzzy class*, according to some notion of *conceptual distance*. In the XML context, we associate to each node an identity by considering the unique path from the root to that node. Then, we represent the state of a node using the XML subtree rooted in it. For a leaf element node e_j, the state is readily defined as a pair of fuzzy sets:

$$S_{e_j} = (S_1, S_2) = (\{t_1/w_1, t_2/w_2, ..., t_k/w_k\}, \{a_1/w_1, a_2/w_2, ..., a_h/w_h\}) \quad (3)$$

These sets are defined respectively by the text and attribute nodes connected to them and their importance weights ([Dam00a]). Note that the elements of the state are values of text nodes and either value or name-value pairs for attribute nodes, the latter choice leading to a more conservative smushing. For non-leaf element nodes, the state can be obtained by recursively applying the fuzzy union operator to the states of its descendants. In other words, for a non-leaf element node e_i we have

$$S_{e_j} = \cup S_{e_t} \quad (4)$$

where e_t are the descendants of e_i in the tree and the fuzzy union is applied separately to the two fuzzy sets composing each state. When evaluating the equivalence between a node of the user-provided pattern and one of a candidate XML document, we compute the fuzzy subsethood between their states. If the overall degree of subsethood is "sufficient" with respect to a dataset-dependent threshold α [5], then we merge the two nodes, generating a *smushed* node, i.e. a merged information granule that is copied to the output. In other words, we compute:

$$\phi(S_i, S_j) = T(Incl(S_{i,1}, S_{j,1}), Incl(S_{i,2}, S_{j,2})) \quad (5)$$

Where *Incl* is classical cardinality-based fuzzy subsethood and T is, as usual, an aggregation operator. The choice of T is application and even dataset dependent, and therefore should be done semi-automatically based on user feedback ([Dam00b]).

4 Smushing in a framework of an XML search engine

Our system receives as input an XML information base whose mark-up structure is variable, while hopefully retaining some commonalities in tag repertoire and vocabulary. Another source of input is provided by the users, who express

[5] and, of course, provided the overall procedure of subgraph matching is successful

their interests to the system by writing a *search pattern*. Patterns comply to a graph-based syntax (straightforwardly expressed in XML) which is both easy to understand and suitable for visual representation. The system output is a (set of) well-formed XML fragments, whose content is taken from the information base and is as close as possible to the user-specified pattern. The system's software architecture is composed of two main modules, the `Pattern Locator` and the `Smusher`, corresponding to operations at two different levels of granularity. The `Pattern Locator` module is the core of our design. First, it parses and pre-processes the target document tree. Then, it uses a `Match` function to look for fragments of the target document having a *topological similarity* with the user pattern. The `Smusher` is a service module, which is called by the `Match` function of the Locator to perform XML *node smushing*, i.e. to evaluate similarity between elementary granules of information (such as XML nodes with their content and attributes) and create result nodes more suitable for user output. The final result of a `Pattern Locator` execution is a list of smushed XML fragments, ordered according to their similarity with the pattern; this list is sent to a final `Post-Processor` module that organizes it in a catalog suitable for user consultation or further processing. Figure 7 depicts our architectural design.

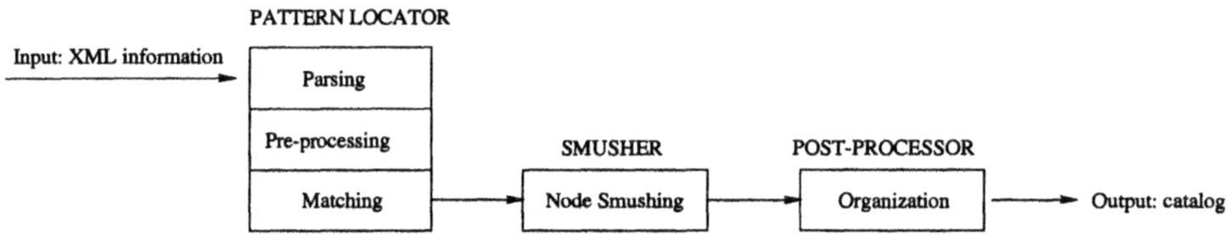

Fig. 7. The Architectural Design

5 Applications and Future Developments

Flexible querying and restructuring of XML data, including intelligent smushing, is needed for a number of new WWW-based services, such as *Web-mining systems* or *Corporate Vertical Portals*. The term "Web mining system" indicates a wide class of applications collecting domain-specific information from the Web. Such information is organized and presented to users in order to allow them to keep under control the information flow about their field of interest, possibly applying data-mining techniques. Current systems use Web spiders for data-collection and keyword-based clustering to organize and query HTML data. The availability of such an information flow in XML format is an occasion to provide high quality Web mining services via advanced reorganization tools specific for XML. Intelligent smushing is a basic requirement of these systems. Finally, we remark that from the user's point of view, smushing is *conservative* or *liberal* according to how likely it is to merge two nodes incorrectly or conversely fail to merge two nodes that represent the same thing. While one cannot expect to fully avoid both types of errors unless XML data are extremely well behaved, we are currently experimenting *precision* and *recall* rather precisely for given datasets, by means of sample queries.

References

[Bor94] G. Bordogna, D. Lucarella, G. Pasi, "A Fuzzy Object Oriented Data Model", IEEE International Conference on Fuzzy Systems, Vol.1, 1994, pp.313-317.

[Bou96] B. Bouchon-Meunier, M. Rifqi, S. Bothorel, "Towards General Measures of Comparison of Objects", *Fuzzy Sets and Systems, vol.84*, 1996.

[Bur00] P. Burchard, "Cut-Paste Metrics: A More Natural Approach to Approximate String Matching", `http://www.pobox.com/ burchard/bio/fuzzymatch/`, 2000

[Cer99b] S. Ceri, A. Bonifati. "Comparison of XML Query Languages". SIGMOD Record 29(1) (2000).

[Coh93] R. Cohen, G. Di Battista, A. Kanevsky, R. Tamassia. "Reinventing the Wheel: An Optimal Data Structure for Connectivity Queries". *Proc. of ACM-TOC Symp. on the Theory of Computing*, S.Diego, CA, (US) (1993).

[Com98] S. Comai, E. Damiani, R. Posenato, L. Tanca. "A Schema-Based Approach to Modeling and Querying WWW Data". In H. Cristiansen, ed., *Proceedings of Flexible Query Answering Systems* (FQAS '98), Roskilde (Denmark), *Lecture Notes in Artificial Intelligence* 1495, Springer (1998).

[Com99] S. Ceri, S. Comai, E. Damiani, P. Fraternali, S. Paraboschi, L. Tanca. "XML-GL: a Graphical Language for Querying and Restructuring XML Documents", *Computer Networks, Vol. 31 (1999), pp. 1171-1187*

[Coh00] S. Cohen, Y. Kogan, W. Nutt, Y. Sagiv, A. Serebrenik. "EquiX: Easy Querying in XML Databases" *WebDB (Informal Proceedings), 2000.*

[Dam99] E. Damiani, M.G. Fugini, C. Bellettini "A Hierarchy Aware Approach to Faceted Classification of Object-Oriented Components". ACM Transactions on Software Engineering Methodologies, vol.8 n.3 pp.215-262, 1999.

[Dam00a] E. Damiani, L. Tanca. "Blind Queries to XML Data". Proceedings of DEXA 2000, London, UK, September 4-8, 2000. *Lecture Notes in Computer Science, Vol. 1873, Springer, 2000, Pages: 345-356.*

[Dam00b] E. Damiani, L. Tanca, F. Arcelli Fontana. "Fuzzy XML Queries via Context-based Choice of Aggregations", Kybernetika n.16 vol.4, 2000.

[Dub98] D. Dubois, F. Esteva, P. Garcia, L. Godo, R. Lopez de Mantaras, H. Prade. "Fuzzy Set Modelling in Case-Based Reasoning". *Int. Jour. of Intelligent Systems* 13(1) (1998).

[Dub99] D. Dubois, H. Prade, F. Sedes. "Fuzzy Logic Techniques in Multimedia Database Querying: A Preliminary Investigations of the Potentials". In R. Meersman, Z. Tari and S. Stevens, eds., *Database Semantics: Semantic Issues in Multimedia Systems*, Kluwer Academic Publisher (1999).

[W3C98a] W3C. Extensible Stylesheet Language (XSL) Version 1.0. October 2000. `http://www.w3C.org/TR/xsl/`

[W3C98b] W3C. Extensible Markup Language (XML) 1.0. Feb. 1998. `http://www.w3C.org/TR/REC-xml/`

[W3C99a] W3C. Resource Description Framework (RDF) Model and Syntax Specification. W3C Recommendation 22 February 1999. http://www.w3.org/TR/REC-rdf-syntax/

[W3C99b] W3C. XSL Transformations (XSLT) Version 1.0 W3C Recommendation 16 November 1999. http://www.w3.org/TR/xslt

[W3C99c] W3C. Namespaces in XML W3C Recommendation 14 January 1999. http://www.w3.org/TR/1999/REC-xml-names-19990114/

A New Fuzzy-Based Multi-objective Optimization Concept for Process Control Systems

Thomas Bernard, Markoto Sajidman, and Helge-Björn Kuntze

Fraunhofer-Institut für Informations-
und Datenverarbeitung IITB
Fraunhoferstr. 1, 76131 Karlsruhe
Fax +49 (0)721/6091-413
bnd@iitb.fhg.de

Abstract. Control system design problems are generally multi-objective since they require several, generally conflicting requirements to be simultaneously met. In order to consider all individual objectives different vectorized solution concepts have been proposed in the past in which the problem is expressed as a set of algebraic inequalities. In this paper a new concept based on the theory of Fuzzy Decision Making will be introduced. According to the natural blurred character of the objectives and the human way of evaluating control performance the individual target functions will be expressed by Fuzzy membership functions which are aggregated to a common performance index by logical AND connection. The proposed Fuzzy-concept has been successfully applied to the control design of a complex industrial batch process which is characterized by heavy stochastic disturbances and large sensor time delay. By the employment of the developed optimization concept the implementing time of new controllers has been reduced and the product quality has been considerably improved.

1 Motivation

Most control design problems are multi-objective, in that there are several conflicting design aims which need to be simultaneously achieved. Typical objectives of industrial plants are e.g. (cf. Fig.1.):

- a command response providing stationary accuracy and a tight dynamics without overshoot,
- an almost complete rejection of stochastic and deterministic disturbances
- complete stability in all optional set points
- low actuation effort as regards required power and bandwidth and
- robustness with respect to altering plant parameters

The designer formulates his design goals as a set of positive-defined computational criteria, which can be a function of stability characteristics (e.g. eigenvalues), and time- and frequency response characteristics (e.g.step-response, overshoot and settling time, control rates, bandwidth, stability margins etc). In most cases, the objective functions are in conflict, so the reduction of one objective function leads to the increase in another. Subsequently, the result of the multi-

B. Reusch (Ed.): Fuzzy Days 2001, LNCS 2206, pp. 653–670, 2001.

objective optimization is known as a Pareto-optimal solution which has the property that it is not possible to reduce any of the objective functions without increasing at least one of the other objective functions. The problem with multi-objective optimization consists in representing the suitable set of Pareto-optimal solutions and in choosing the solution which provides the best design.

The most simple solution concept is to replace the difficult multi-objective design problem by a more easy mono-objective problem by summarizing all partial objectives in a substitute performance specification [1, 2]. Obviously, the shortcoming of this approach is that conflicting objectives can compensate each other. Thus, in order consider all individual objectives different vectorized solution concepts have been proposed in the past in which the problem is expressed as a set of algebraic inequalities [2, 3, 4, 5, 6, 7]. The aim of the design is to find control parameters that simultaneously satisfies the set of inequalities and provide the largest tolerable values of the objective functions.

As regards the actual numerical solution various algorithms have been proposed. Zakian and Al-Naib [7] developed a numerical search algorithm known as the moving boundaries process (MBP) which is based on Rosenbrock search method. Whidborne, Postlethwaite and Gu [6] prefered a mixed optimization approach. Ng [5] has proposed another algorithm based on hill-climbing. The multiobjective genetic algorithms (MOGA), developed by Fonseca and Fleming [4], provide several advantages especially global convergency.

In this paper a new concept based on the theory of Fuzzy Decision Making [8, 9] will be introduced. According to the natural blurred character of the objectives and the human way of evaluating control performance the individual target functions will be expressed by Fuzzy membership functions. The common performance index will be obtained by the logical AND connection of these individual membership functions. The proposed Fuzzy-based multi-objective control design concept has been successfully applied to a complex industrial batch process which is characterized by heavy stochastic disturbances and large sensor time delay [10].

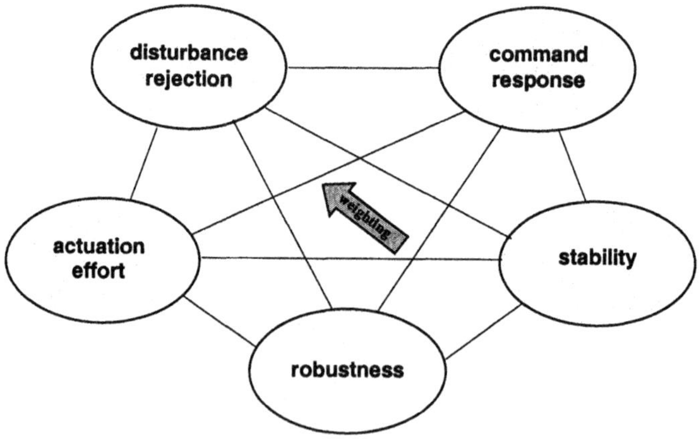

Fig. 1. Typical partly conflicting objectives of industrial processes

2 Multi-objective Fuzzy Control Design Concept

2.1 Basic Idea

The approach presented in this paper uses the transparent aggregation and weighting mechanisms of the criteria provided by the theory of Fuzzy Decision Making [8, 9]. A generic advantage of this is the transparent and smooth Fuzzy-AND aggregation of the performance criteria which is similar to human manner of evaluation. The algorithm of Fuzzy Decision Making guarantees that the obtained solution satisfies all criteria with nearly the same weighting degree. Furthermore, it can be shown that the solution is pareto--optimal. In the algorithm weighting parameters are introduced which allow a transparent weighting of individual criteria according to their specific importance. Similar approaches of the theory of Fuzzy Decision Making have been successfully applied at the IITB in the field of climate control with respect to comfort and economy criteria [11, 12].

The proposed concept can be subdivided into following three steps:

1. Definition of performance criteria
   Several performance criteria (e.g. an integral or a function) are defined which correspond to the requirements of the optimization problem. In the considered problem the criteria are ***disturbance rejection***, ***command response***, ***stability*** and *actuation effort* represented by functions or integrals J_{dist}, J_{com}, J_{stab}, J_{act}. The criteria are functions of the parameters $\mathbf{p}$ which have to be optimized.
2. Transformation in Fuzzy membership functions
   The performance criteria $J_{dist}(\mathbf{p})$, $J_{com}(\mathbf{p})$, $J_{stab}(\mathbf{p})$, $J_{act}(\mathbf{p})$ are transferred into Fuzzy membership functions $\mu_{dist}(\mathbf{p})$, $\mu_{com}(\mathbf{p})$, $\mu_{stab}(\mathbf{p})$, $\mu_{act}(\mathbf{p})$. The membership degree equal one corresponds with the optimal fulfillment of the objective while membership degree equal zero corresponds with an unacceptable performance.
3. Fuzzy Decision Making by weighting of the performance criteria
   The optimal parameters $\mathbf{p}^*$ are determined by an algorithm of Fuzzy Decision Making. A weighting of individual criteria is possible according their specific importance. An important property of the algorithm of Fuzzy Decision Making is the exclusion of all values $\mathbf{p}$ with membership degree zero. Hence it is guaranteed that the resulting solution is acceptable with respect to all criteria.

Robustness is considered implicitly in the algorithm a well. In the computation of the performance criteria (step 1) in general some model equations are involved. For the considered problem there are transfer functions with a proportional gain K_P and a time constant T_P of the plant. These parameters of the plant vary in a range $[K_0-\Delta K_1, K_0+\Delta K_2]$ and $[T_0-\Delta T_1, T_0+\Delta T_2]$. In the computation of Fuzzy membership functions (step 2) some typical values of K and T are used (e. g. $K_P \in \{K_0-\Delta K_1, K_0, K_0+\Delta K_2\}$ and $T_P \in \{T_0-\Delta T_1, T_0, T_0+\Delta T_2\}$). The obtained membership functions are aggregated by a Fuzzy-AND-operator. By this measure the membership degree of μ_{dist}, μ_{com}, μ_{stab}, μ_{act} corresponds to the worst case which was caused by the variation of the plant parameters.

2.2 Optimization by Fuzzy Decision Making

In the concept of Fuzzy Decision Making performance criteria and constraints are described as Fuzzy membership functions [8, 9]. The Fuzzy performance criteria are described *Fuzzy goals* $\mu_{G1}, \ldots, \mu_{GN}$ and by the *Fuzzy constraints* are denoted by $\mu_{C1}, \ldots, \mu_{CM}$. Fuzzy goals and Fuzzy constraints are desired to have maximum degree of membership. Consequently the Fuzzy goals and Fuzzy constraints are connected by Fuzzy-AND-operators (e.g. min-operator) to get the *Fuzzy decision* μ_D. In the simplest case with only one Fuzzy goal μ_G and one Fuzzy constraint μ_C and with the min-operator as Fuzzy-AND the *Fuzzy decision* μ_D is obtained by eq. (1)

$$\mu_D(x) = \mu_G(x) \wedge \mu_C(x), \qquad \wedge = \min \tag{1}$$

Eq. (1) can be easily extended for the case of N Fuzzy goals $\mu_{G1}, \ldots, \mu_{GN}$ and M Fuzzy constraints $\mu_{C1}, \ldots, \mu_{CM}$ (eq. (2)).

$$\mu_D(x) = \mu_{G1}(x) \wedge \ldots \wedge \mu_{GN}(x) \wedge \mu_{C1}(x) \wedge \ldots \wedge \mu_{CM}(x) \tag{2}$$

The *optimal decision* x^* is calculated by maximizing $\mu_D(x)$, where x is an element in the set of possible solutions X (eq. (3)). The algorithm is illustrated assuming for simplicity one Fuzzy goal and one Fuzzy constraint in Fig. 2..

$$\mu_D(x^*) = \max_{x \in X} \mu_D(x) \tag{3}$$

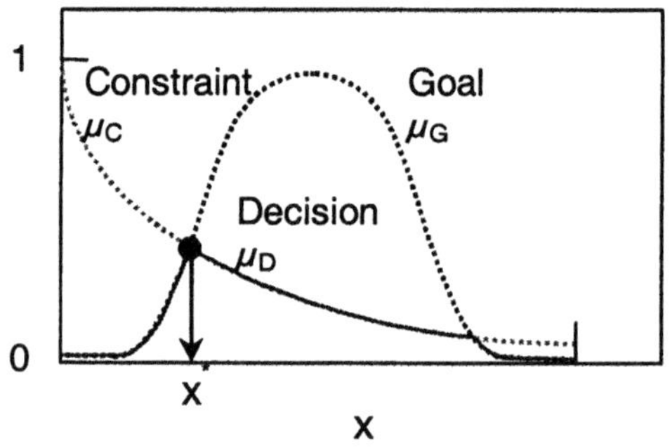

Fig. 2. Illustration of the principle of Fuzzy Decision Making with min-operator as Fuzzy-AND-operator

The method of Fuzzy Decision Making is characterized by four properties which are of importance for a multi-objective control design:

1. Constraints and goals are treated in an absolutely symmetrical way.
2. As shown by equation (2) the algorithm is well suited for multi-criteria optimization problems. Further performance criteria can be added easily.
3. It can easily be shown that values x which yield zero in *any* performance criteria μ_G or μ_C can not be the optimal decision x^*. This property can be used to define "forbidden ranges" of x which corresponds to hard restrictions in non-Fuzzy optimization problems.
4. The result will be pareto optimal.

2.3 Weighting of Different Performance Criteria

In many cases the performance criteria are not of equal importance. Thus, a transparent weighting of the individual performance criteria seems to be useful. For the Fuzzy Decision Making algorithm (eq. (2), (3)) there are many ways to do this [8, 9, 11]. It turned out that the weighting with multipliers λ_i, $0 < \lambda_i \leq 1$ in connection with the min-operator is an especially transparent way to implement the weighted AND-connection [11]. In eq. (7) we do not distinguish between Fuzzy goals and Fuzzy restrictions, hence the performance criteria are denoted by $\mu_i(x)$. The smaller λ_i is chosen, the *more* the i-th performance criteria $\mu_i(x)$ will be emphasized. The relative importance of two performance criteria μ_i, μ_k is represented by the ratio λ_i/λ_k. For $\lambda_i/\lambda_k \to 0$ the criterion μ_i is much more emphasized than μ_k.

$$\mu_D(x) = \lambda_1\mu_1(x) \wedge \ldots \wedge \lambda_N\mu_N(x), \; 0 < \lambda_i \leq 1, \wedge = \min \tag{4}$$

With eq. (4) the optimal decision x^* (cf. eq. (3)) could be ambiguous if there are ranges with constant degree of membership in one or several performance criteria μ_i. A non-ambiguous optimal decision is guaranteed if a correction term is added in eq. (4) which yields eq. (5) [11].

$$\mu_D(x) = \lambda_1\mu_1(x) \wedge \ldots \wedge \lambda_N\mu_N(x) + \underbrace{\varepsilon\mu_1(x)\cdot\ldots\cdot\mu_N(x)}_{\text{correctiorterm}}, \; 0 < \lambda_i \leq 1, \wedge = \min, 0 < \varepsilon << 1 \tag{5}$$

3 Application to an Industrial Batch Process

3.1 Plant Behaviour

The proposed multi-objective control design concept has been implemented and applied to an industrial automatically controlled glass tube manufacturing plant [10] shown in Fig. 3. (left). A glass cylinder is fed with constant speed into a furnace. It will be heated up to its softening temperature ϑ, and is pulled out with a controllable speed v. Simultaneously an inert gas is blown into the cylinder in order to create a difference pressure p. The controlled geometrical variables are the tube diameter D and the wall thickness W. A typical time plot of a production is shown in Fig. 3. (right).

The process is characterized by the following difficulties:

- high complexity of system dynamics due to a highly nonlinear rheology,
- nonlinear coupled multi-variable system behavior with time varying, distributed parameters,
- large time varying dead-times caused by large sensor distances and changing pulling speed,
- stochastic disturbances caused by inhomogeneous material (cf. Fig. 3. , right),
- large drift disturbances caused by thermic effects during start, transient and end phases of the process (cf. Fig. 3. , right).

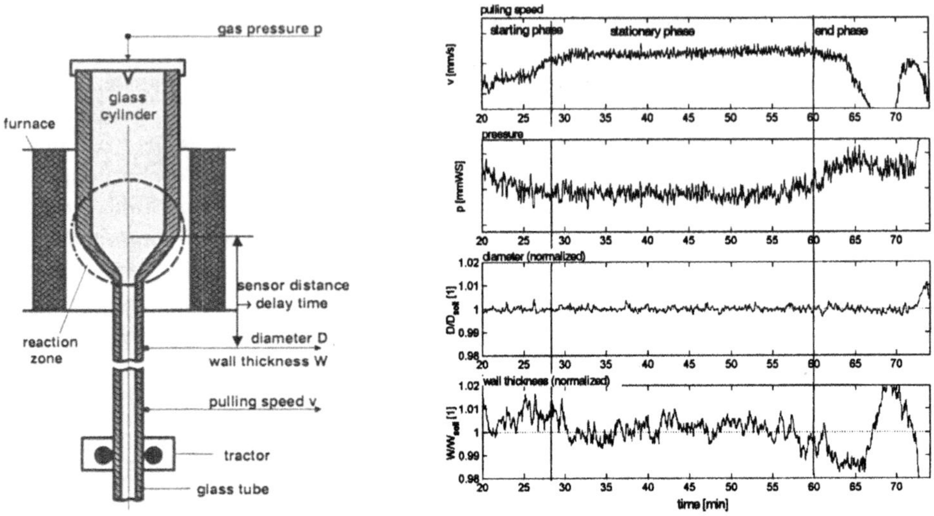

Fig. 3. Scheme of a glass manufacturing plant (left) and time plot of a production (right)

3.2 Process Model

The process can be approximately descrbed by a hybrid model structure according to Wiener [1], which is characterized by a nonlinear static transfer function in sequence with a linear dynamic transfer function. In the following a linearized partial plant model is assumed which represents in the considered operating point a very realistic process response. Deviations of the transfer gain k_M and of the time constant T_M with respect to the real plant (k_p, T_p) are considered by error parameters Δk_M and ΔT_M which denote the relative deviation (cf. Table 1.). For $\Delta k_M = \Delta T_M = 0$ model and plant response are identical. Since transfer gain k_p and time constant T_P can vary considerably the value sets $\Delta k_M = \Delta T_M = \{-0.3, 0, 0.3\}$ are used in the investigation. Thus, a model uncertainty of $\pm$ 30 % is considered as regards the gain and time constant respectively. The error parameters Δk_M, ΔT_M will be used to calculate robust control parameters (cf. section 3.5).

3.3 Control Concept

The control concept of the considered partial plant is based on a modified Smith predictor controller [13]. The control parameter optimization refers to the different control structures:

1 Conventional Smith predictor for linear plant model and PI controller
2 Control system with selective P loop and I loop. Thus, the low pass response of the I part is used for filtering stochastic disturbances while the P part fed back from the model output has mainly to guarantee stability. This structure is shown in, Fig. 4 the transfer functions are listed in Table 2..

3 Smith predictor with a splitted model. In this case the static model part is used for feedback stabilization while the dynamic model part has mainly to minimize the control error. In this approach only a I controller is used.

In chapter 4 only results of the optimization of structure 2 are presented.

Table 1. Transfer functions and parameters of the plants and the models

	transfer function	Parameters
VZ1-model	$G_M(s) = \frac{k_M}{1+T_M s}$	k_M=1, T_M=28 sec
VZ1-plant	$G_P(s) = \frac{k_P}{1+T_P s}$	$T_P = (1+\Delta T_M)T_M$ $k_P = (1+\Delta k_M)k_M$ $\Delta k_M \in \{-0.3, 1,1.3\}$, $\Delta T_M \in \{-0.5, 1,1.5\}$
VZ2-model	$G_M(s) = \frac{k_M}{(1+T_M s)^2}$	k_M=1, T_M=32 sec
VZ2-plant	$G_P(s) = \frac{k_P}{(1+T_P s)^2}$	$T_P = (1+\Delta T_M)T_M$ $k_P = (1+\Delta k_M)k_M$ $\Delta k_M \in \{-0.3, 1,1.3\}$, $\Delta T_M \in \{-0.5, 1,1.5\}$
time delay	$G_D(s) = \exp(-T_D s)$	$T_D=L_D/v$

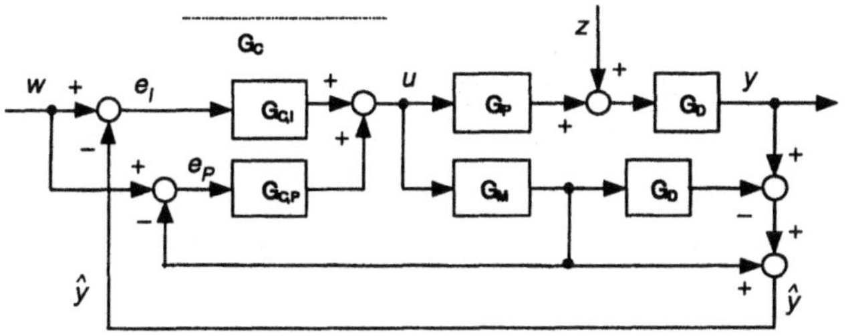

Fig. 4. Modified Smith predictor with selective P loop and I loop (control structure 2)

3.4 Fuzzy Performance Criteria

According to step 1 and 2 of the basic concept (cf. chapter 2.1) in the sequel the performance criteria are defined which correspond to the requirements of the optimization problem. In the treated problem these are the criteria *disturbance rejection*, *command response*, *stability* and *actuation effort* which are represented by functions or integrals J_{dist}, J_{com}, J_{stab}, J_{act}. These functions are then transformed into Fuzzy goals μ_{dist}, μ_{com}, μ_{stab}, μ_{act} which will be described in detail in the following sections.

Table 2. Transfer functions (TF) of the modified Smith predictor with selective P loop and I loop (control structure 2)

Basic Controllers	$G_{C,P}(s) = k_P,\ \ G_{C,I}(s) = \frac{k_I}{s},\ \ G_C(s) = G_{C,P}(s) + G_{C,I}(s)$
Command TF	$G_w = \frac{y}{w} = \frac{G_C G_P G_D}{1 + G_C G_M + G_{C,I}(G_P - G_M)G_D}$
Disturbance TF	$G_z = \frac{y}{z} = \frac{1 + G_C G_M - G_{C,I} G_M G_D}{1 + G_C G_M + G_{C,I}(G_P - G_M)G_D} G_D$

3.4.1 Disturbance Rejection

In order to obtain a good disturbance rejection of the control system the disturbance transfer function $G_z(\omega)$ must have a good damping in the required frequency intervall $[\omega_1, \omega_2]$. In Fig. 5. a typical disturbance transfer function of the smith predictor is shown. It is characterized by well-damped band-ranges for $|G_z(\omega)| < 1$ and disturbance amplifying band-ranges for $|G_z(\omega)| > 1$.

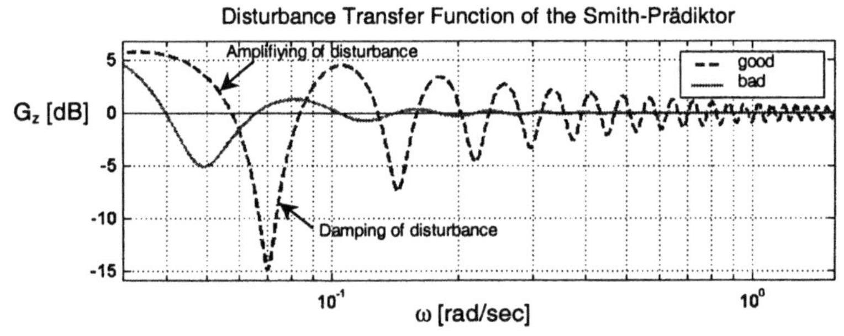

Fig. 5. Typical disturbance transfer function of the smith predictor

Since it is desired to shape $|G_z(\omega)|$ as negative and smooth as possible the performance index

$$J_{dist}(k_p,k_i) = \int_{\omega_1}^{\omega_g} \left(\log(|G_z(\omega,k_p,k_i)|)_{|G_z|<1} + \underbrace{\log(|G_z(\omega,k_p,k_i)|)^2_{|G_z|>1}}_{\text{penalty term for } |Gz|>1} + \alpha \cdot \underbrace{\left| \frac{\partial^2 |G_z(\omega,k_p,k_i)|}{\partial \omega^2} \right|}_{\text{penalty term for curvature of } Gz} \right) d\omega \quad (6)$$

is introduced. The 1st term of the integral function (6) describes the amplitude for $|G_z| < 1$. Positive amplitudes $|G_z| > 1$ are punished by the 2nd quadratic term. The 3rd penalty term refers to the smoothness of $|G_z(\omega)|$. The integration limits are set to $\omega_1 = 2\pi/T_1$ and $\omega_g = 2\pi/T_g$ with $T_1 = 100 \cdot T_s$ and $T_g = 2 \cdot T_s$ where a sampling time

$T_s = 2$ sec is chosen (Shannon's limit frequency is denoted by ω_g). This limits correspond to the relevant frequency regime. The integration ist carried out numerically.

J_{dist} can be interpreted as a measure for the standard deviation σ_y of the process output signal y which will complys with the well known relation [14]:

$$\sigma_y^2 = \Theta_{yy}(0) - \bar{y}^2, \tag{7}$$

where $\Theta_{yy}(\tau)$ denotes the autocorrelation function und $\bar{y}$ the mean value of y. $\Theta_{yy}(0)$ results from the transformation

$$\Theta_{yy}(0) = \frac{1}{\pi}\int_0^{\infty} \left|G_z(\omega)\right|^2 S_{zz}(\omega)d\omega, \tag{8}$$

where $S_{zz}(\omega)$ denotes the power density spectrum of the disturbance signal [14]. For white noise, this means $S_{zz}(\omega)$=const., eq. (8) has a similar structure as the 2nd term of eq. (6).

As for big negative values of J_{dist} the disturbance rejection can be regarded as optimal and for for big positive values of J_{dist} as not accaptable, the Fuzzy criterion μ_{dist} is defined by eq. (9) (cf. also Fig. 6.):

$$\mu_{dist}(k_p, k_i) = \min[1, \exp(-\frac{J_{dist}(k_p, k_i) - J_1}{J_2})] \tag{9}$$

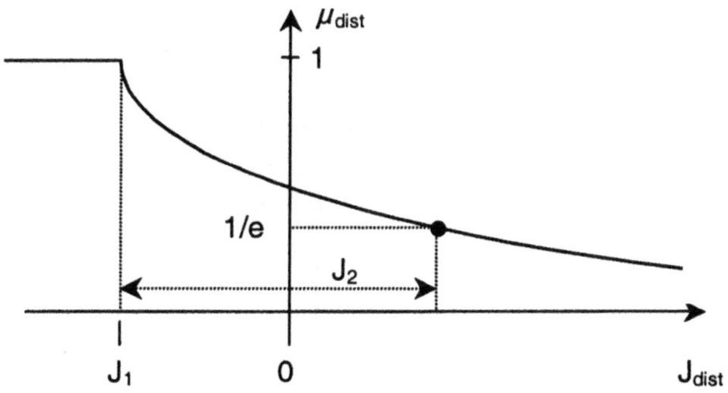

Fig. 6. Fuzzy criterion for good disturbance rejection μ_{dist}

The parameters J_1, J_2 have been choosen to $J_1 = -3$, $J_2 = -6$. It is easy to interpret the parameter values as J_{dist} corresponds with the mean value of $|G_z|$ in [dB] (cf. eq. (6)).

3.4.2 Command Response

In order to obtain good command response the amplitude of the command transfer function $G_w(\omega)$ for small frequencies has be approximately one, i.e. equal zero in logarithmic scale. Values $\log(|G_w(\omega)|) > 0$ correspond with overshooting and are undesired as well as $\log(|G_w(\omega)|) < 0$ which corresponds with a slowly reaching of the setpoint. Hence the performance criterion J_{com} is defined as the mean value of $|\log(|G_w(\omega)|)|$ for small frequencies and has to be minimized:

$$J_{com}(k_p, k_i) = \frac{1}{\omega_2 - \omega_1} \int_{\omega_1}^{\omega_2} |\log(|G_w(\omega, k_p, k_i)|)| d\omega \rightarrow \min, \quad \omega_2 \approx \omega_1 \tag{10}$$

The integration limits $\omega_1 = 2\pi/T_1$ and $\omega_2 = 2\pi/T_2$ correspond with $T_1 = 100 \cdot T_s = 200$ sec und $T_2 = 97.5 \cdot T_s = 195$ sec.

The command response can be regarded as optimal if J_{com} (in dB-scale) is equal zero and it is not acceptable if J_{com} is large. Hence the the Fuzzy criterion μ_{com} is defined by eq. (11) (cf. also Fig. 7.). The parameter J_1 has been chosen to $J_1 = 5$ dB.

$$\mu_{com}(k_p, k_i) = \exp\left(-\frac{J_{com}(k_p, k_i)}{J_1}\right), \tag{11}$$

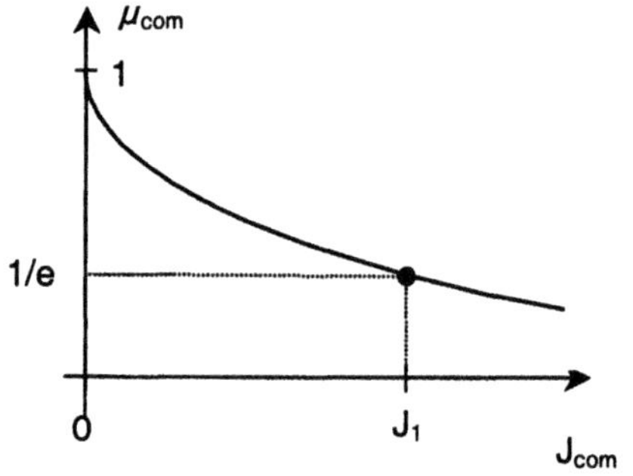

Fig. 7. Fuzzy criterion for good command response μ_{com}

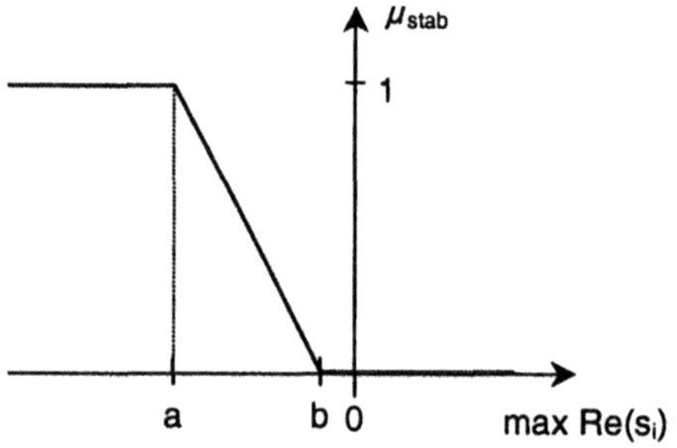

Fig. 8. Fuzzy criterion with respect to stability μ_{stab}

3.4.3 Stability

For evaluating the stability the poles s_i $(i = 1 \ldots N)$ of the closed control loop are of dominant importance. If all poles have a negative real part ($\mathrm{Re}(s_i) < 0$, $i = 1 \ldots N$), the control loop is stable. If one pole s_k exists with $Re(s_k) = 0$ and all other poles have a negative real part the control loop is at the stability border. If there exists one pole s_k with $Re(s_k) > 0$ the control loop is instable [14].

Hence in the definition of a Fuzzy criterion with respect to stability μ_{stab} the maximal real part of the poles (max $Re(s_i)$) is considered. An optimal value $a<0$ is defined with $\mu_{stab} = 1$ if max $Re(s_i) < a$. Also a minmal value b is defined with $\mu_{stab} = 0$ if max $Re(s_i) > b$. In between a und b the membership funktion is linear interpolated. Thus μ_{stab} is definded by eq. (12) (cf. also Fig. 8.).

$$\mu_{stab}(k_P, k_P) = \begin{cases} 1 & \text{if } \max Re(s(k_P, k_P)) < a < 0 \\ \dfrac{\max Re(s(k_P, k_P)) - b}{a - b} & \text{if } a < \max Re(s(k_P, k_P)) < b \\ 0 & \text{if } b < \max Re(s(k_P, k_P)) \end{cases} \tag{12}$$

The parameters a and have been chosen according the reachable stability as $a = -0.06/\text{sec}$ and $b = -0.02/\text{sec}$.

In the design of μ_{stab} it is of spezial importance that $\mu_{stab} = 0$ if max $Re(s_i) > b$. As it was discussed above the algorithm of Fuzzy Decision Making guarantees the solutions k_p, k_i which yield max $Re(s_i) > b$ and hence $\mu_{stab} = 0$ are excluded. Hence a stable solution is provided. This transparent mechanism is an important advantage of the proposed aproach.

3.4.4 Actuation Effort

Due to reduce energy consumption and wear of the actuators a further aim is to achieve low effort of the actuators. Since in the given plant in the stationary phase the error signal is mainly of stochastic character the standard deviation of the actuator signal u has to be minimized. Corresponding to the discussion in section 3.4.1 this is achieved by the minimization of the integral of the transfer function of the substitue controller G_C^* (eq. (13)).

$$J_{act}(k_p, k_i) = \int_{\omega_1}^{\omega_2} \log(|G_C^*(\omega, k_p, k_i)|) d\omega \rightarrow \min . \tag{13}$$

G_C^* is defined by

$$G_C^*(s) = \frac{u(s)}{\hat{e}(s)} \tag{14}$$

(cf. Table 2.). The minimization of J_{act} leads to small mean values of $\log(|G_C^*(\omega)|)$ which corresponds with low effort of the actuators.

The Fuzzy criterion with respect to low effort of the actuators μ_{act} is defined in an analogous way as μ_{dist} (cf. section 3.4.1). If $J_{act} << 0$ the effort of the actuators can be regarded as optimal, this corresponds with $\mu_{act} \rightarrow 1$. If $J_{act} >> 0$ the effort of the actuators is high which coresponds with $\mu_{act} \approx 0$. Hence the definition of μ_{act} which is provided by (15) resp. Fig. 9. is plausible.

$$\mu_{act}(k_p, k_i) = \min[1, \exp(-\frac{J_{act}(k_p, k_i) - J_1}{J_2})] \tag{15}$$

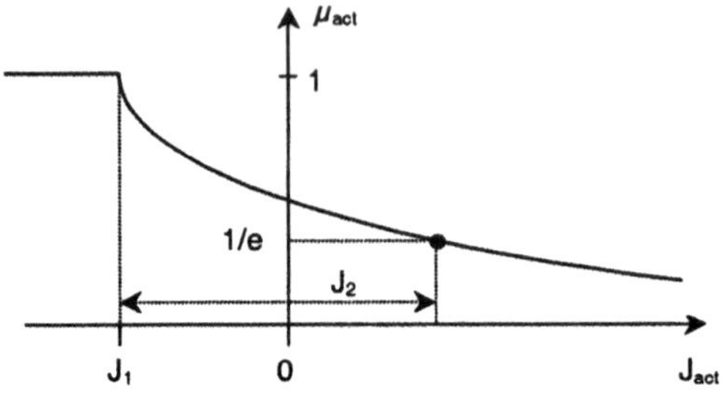

Fig. 9. Fuzzy criterion with respect to low effort of the actuators μ_{act}

The parameters J_1, J_2 have been chosen as $J_1 = -5$, $J_2 = 5$. The parameter values are not easy to interpret for this criterion. Hence in order to determine J_1, J_2 first the reachable range of J_{act} (i.e. inf J_{act}, sup J_{act}) has to be found out.

3.5 Robustness

Model based control structures naturally have the best performance if the model G_M exactly represents the plant G_P, i.e. $G_M = G_P$. E.g. it holds for the command transfer function of the modified Smith predictor with selective P loop and I loop (c.f. Table 2.):

$$G_w = \frac{y}{w} = \frac{G_C G_P G_D}{1 + G_C G_M + G_{C,I}(G_P - G_M)G_D} \tag{16}$$

The deviation of model and plant is described by the term $(G_P - G_M)$. Only in the ideal case $G_P = G_M$ the time delay is compensated.

In the proposed optimization scheme the deviation of model and plant are modeled by relative error parameters Δk_M, ΔT_M which describe the maximum and unknown deviation of gain and time constant of the plant (cf. Table 1.).

$$\begin{aligned} k_P &= (1 + \Delta k_M) \cdot k_M \\ T_P &= (1 + \Delta T_M) \cdot T_M \end{aligned} \tag{17}$$

E.g. for a system of first order the plant with its uncertainties with respect to k_P and T_P is modeled by by eq. (18) (c.f. Table 1.).

$$G_P(s) = \frac{k_P}{1 + T_P s} = \frac{(1 + \Delta k_M) k_M}{1 + (1 + \Delta T_M) T_M s} \equiv G_P(\Delta k_M, \Delta T_M, s)\,. \tag{18}$$

It would also be possible to include unknown model deviations of higher order, but for the given plant the order of the partial models of first and second order are good approximations.

With G_P also the transfer functions G_z, G_w, G_u depend on Δk_M and ΔT_M (cf. Table 2.) and hence also the performance criteria μ_{dist}, μ_{com},. μ_{stab}, μ_{act}. Thus μ_{dist}, μ_{com},. μ_{stab}, μ_{act} are functions of the controller parameters k_P, k_I and the unknown model deviations Δk_M, ΔT_M (eq. (19)).

$$\begin{aligned} \mu_{dist} &= \alpha(\Delta k_M, \Delta T_M, k_P, k_I) \\ \mu_{track} &= \beta(\Delta k_M, \Delta T_M, k_P, k_I) \\ \mu_{stab} &= \gamma(\Delta k_M, \Delta T_M, k_P, k_I) \\ \mu_{act} &= \delta(\Delta k_M, \Delta T_M, k_P, k_I) \end{aligned} \tag{19}$$

The unknown model deviations Δk_M, ΔT_M are considered by a Fuzzy-AND combination with representative values $\Delta k_M \in \{\Delta k_{M,i}\}$, $i=1,\ldots N$, $\Delta T_M \in \{\Delta T_{M,j}\}$, $j=1,\ldots,M$ (eq. (20)). For the considered application $\{\Delta k_{M,i}\}$, $\{\Delta T_{M,j}\}$ have been chosen as $\{\Delta k_{M,i}\} = \{\Delta T_{M,j}\} = \{-0.3, 0, 0.3\}$.

$$\begin{aligned} \tilde{\mu}_{dist}(k_P, k_I) &= \bigwedge_{\Delta k_{M,i}} \bigwedge_{\Delta T_{M,j}} \mu_{dist}(\Delta k_M, \Delta T_M, k_P, k_I), \quad \wedge = \min \\ \tilde{\mu}_{track}(k_P, k_I) &= \bigwedge_{\Delta k_{M,i}} \bigwedge_{\Delta T_{M,j}} \mu_{track}(\Delta k_M, \Delta T_M, k_P, k_I) \\ \tilde{\mu}_{stab}(k_P, k_I) &= \bigwedge_{\Delta k_{M,i}} \bigwedge_{\Delta T_{M,j}} \mu_{stab}(\Delta k_M, \Delta T_M, k_P, k_I) \\ \tilde{\mu}_{act}(k_P, k_I) &= \bigwedge_{\Delta k_{M,i}} \bigwedge_{\Delta T_{M,j}} \mu_{act}(\Delta k_M, \Delta T_M, k_P, k_I) \end{aligned} \tag{20}$$

By the Fuzzy AND combination of the performance criteria it is achieved that $\tilde{\mu}_{dist}$, $\tilde{\mu}_{track}$, $\tilde{\mu}_{stab}$, $\tilde{\mu}_{act}$ are dominated by the worst effect of a model deviation, i.e. the smallest value of μ_{dist}, μ_{com}, μ_{stab}, μ_{act}. With this "robust" performance criteria the Fuzzy decision μ_D (cf. eq. (4)) is calculated by eq. (21).

$$\begin{aligned} &\mu_D(k_P, k_I) = \\ &\lambda_{dist}\mu_{dist}(k_P, k_I) \wedge \lambda_{track}\mu_{track}(k_P, k_I) \wedge \lambda_{stab}\mu_{stab}(k_P, k_I) \wedge \lambda_{act}\mu_{act}(k_P, k_I), \\ &0 < \lambda_i \le 1 \end{aligned} \tag{21}$$

The smaller the weighting parameters λ_{dist}, λ_{com}, λ_{stab}, λ_{act} are chosen the more is the corresponding performance criterion emphasized. For equal importance of the criteria, this means $\lambda_{dist} = \lambda_{com} = \lambda_{stab} = \lambda_{act} = 1$, eq. (21) reduces to eq. (22).

$$\mu_D(k_P, k_I) = \mu_{dist}(k_P, k_I) \wedge \mu_{track}(k_P, k_I) \wedge \mu_{stab}(k_P, k_I) \wedge \mu_{act}(k_P, k_I) \tag{22}$$

The optimal parameters k_P^*, k_I^* are computed corresponding eq. (3) by eq. (23)

$$\mu_D(k_P^*, k_I^*) = \max_{k_P, k_I} \mu_D(k_P, k_I) \tag{23}$$

4 Results

In this chapter different results are discussed. In section 4.1 the influence of different weighting of the performance criteria is investigated. Section 4.2 discusses results with different requirements to robustness. In Section 4.3 the performance of the concept is shown by time plots of simulations with optimized parameters and heuristical chosen parameters.

4.1 Weighting of the Performance Criteria

One result of the optimization is shown in Fig. 10.. In the figures of the first row the Fuzzy performance criteria μ_{dist}, μ_{com}, μ_{stab}, μ_{act} and the resulting Fuzzy decision μ_D are shown in dependence of the parameters k_P, k_I. In the second row the same plots are shown as a two-dimensional projection. In these figures the optimal parameters are marked with circles. In order to demonstrate the influence of the different emphsize of one performance criterion the weighting parameter regarding the criterion *command response* μ_{com} has been varied in the range $\lambda_{com} = 1 \ldots 0.1$. The smaller λ_{com}, the stronger is the emphasis on a good command response of the control loop. The result can be clearly seen as K_P and K_I increase with smaller values of λ_{com}.

From Fig. 10. it can also be noticed that the performance criteria are partly contradictory. Given a stronger weighting on a good command response (i.e. increasing of μ_{com}) a degradation of disturbance rejection and effort of the actuators (i.e. decreasing of μ_{dist} and μ_{act}) is the result. In this example the degree of stability is not contradictory to a good command response, so both membership functions μ_{stab} and μ_{com} are increasing.

4.2 Different Robustness Requirements

The parameters k_p and k_I of the of the modified Smith predictor with selective P loop and I loop (c.f. Table 2.) have been optimized for a large spectrum of time delay T_D, which varies from $T_D = 0 \ldots 14 \cdot T_P$, where T_P is the time constant of the plant. The results for a VZ1 model with different unknown model errors $\Delta k_M = \Delta T_M \equiv \Delta$ are shown in Fig. 11.. For ideal correspondence of model and plant, i.e. $\Delta = 0$, K_P and K_I do not depend on T_D because of the ideal compensation of the time delay (cf. transfer functions in Table 2.). With increasing demand to robustness, i.e. increasing unknown model error Δ both k_P and k_I are increasing and the Fuzzy performance index μ_D is decreasing. This result is caused by the fact that with small model error the model based controller has a better performance and hence needs lower gain.

4.3 Process Performance

Since in most industrial applications for safety and cost reasons it is almost impossible to compare the performance of different controller with exactly the same parameters and disturbances of the plant, some representative simulations based on a very realistic plant model are presented. In Fig. 12. results are shown with heuristic parameters (left) and optimal parameters (right). An VZ1 model was assumed with time constant $T_P = 26$ sec and time delay $T_D = 130$ sec. The relative model error due to the internal model of the controller have been set to $\Delta k_M = \Delta T_M = 30\%$ which corrresponds to a big model deviation. An realistic noise disturbance has been added to the system input. The heuristic controller parameters are $k_P = 2.59$, $k_I = 0.1$, the optimal parameters $k_P = 0.36$, $k_I = 0.014$.

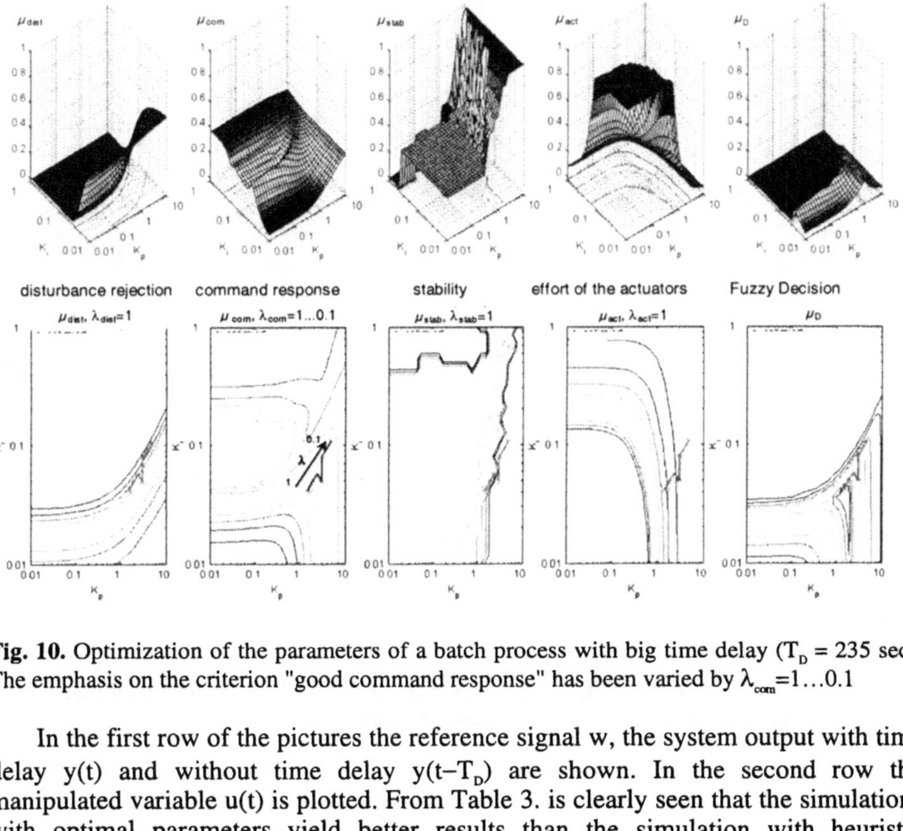

Fig. 10. Optimization of the parameters of a batch process with big time delay ($T_D = 235$ sec). The emphasis on the criterion "good command response" has been varied by λ_{com}=1...0.1

In the first row of the pictures the reference signal w, the system output with time delay y(t) and without time delay $y(t-T_D)$ are shown. In the second row the manipulated variable u(t) is plotted. From Table 3. is clearly seen that the simulations with optimal parameters yield better results than the simulation with heuristic parameters as the standard deviations of the error signal e (σ_e) and the manipulated variable u (σ_u) are much smaller.

Table 3. Standard deviations of the error signal e and the manipulated variable u of the simulations of Fig. 12. (calculated from t = 500 ... 1500 sec)

	heuristic	optimal
σ_e	0.18	0.15
σ_u	0.110	0.068

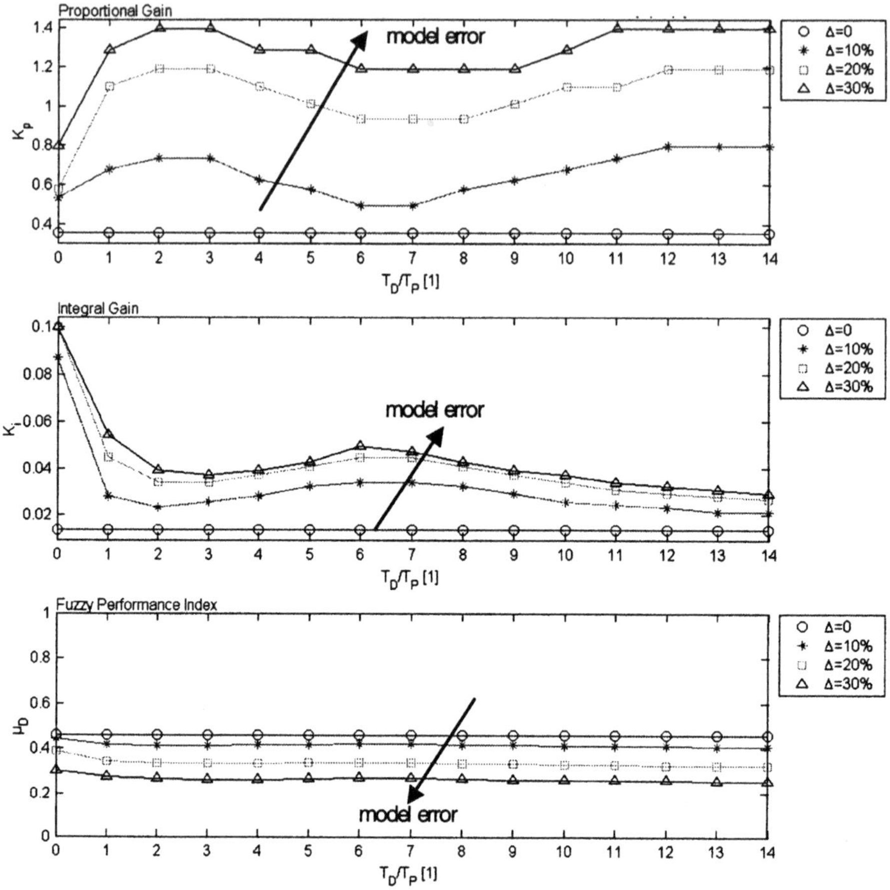

Fig. 11. Optimal Parameters k_P, k_I in dependence of T_D/T_P with different unknown model errors $\Delta k_M = \Delta T_M \equiv \Delta$ (T_D: time delay, T_P: time constantof the plant)

5 Conclusions

In the industrial process control system several, generally conflicting, requirements have simultaneously to be considered. In this paper a new concept based on the theory of Fuzzy Decision Making will be introduced in which the individual target functions will be expressed by Fuzzy membership functions which will be logically connected by an Fuzzy AND operator. The proposed Fuzzy-concept has been successfully applied to the optimal control design of a complex industrial glass manufacturing process. It has been shown that the implementing time of new controllers has been reduced and the product quality has been considerably improved.

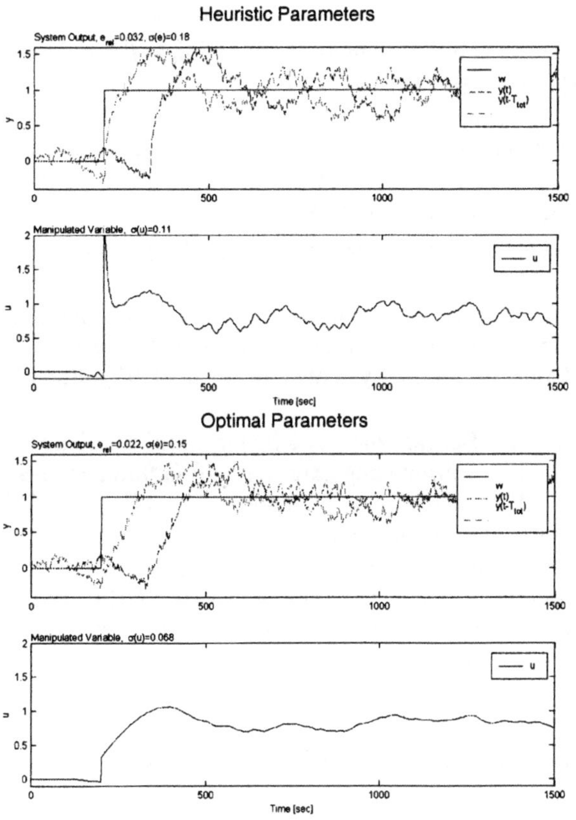

Fig. 12. Simulation results for a VZ1 model with heuristic parameters (left) and optimal parameters (right); $T_P = 26$ sec, $T_D = 130$ sec

References

[1] Ackermann, J.: Robuste Regelung. Springer, Heidelberg 1993

[2] Kahlert, J.: Vektorielle Optimierung mit Evolutionsstrategien und Anwendungen in der Regelungstechnik. VDI-Verlag, Düsseldorf, 1991

[3] Kreisselmeier, G; Steinhauser, R.: Systematische Auslegung von Reglern durch Optimierung eines vektoriellen Gütekriteriums, Regelungstechnik 27, pp. 76-79, 1979

[4] Fonseca, C.M.; Fleming, P.J.: Multiobjective optimization and multiple constraint handling with evolutionary algorithms - part I: a unified formulation. IEEE Trans. Syst. Man & Cybernetics A, 28 (1), pp. 26-37, 1998

[5] Ng, W.Y: Interactive Multi-Objective Programming as a Framework for Computer-Aided Control System Design, volume 132 of Lect. Notes Control & Inf. Sci. Springer-Verlag, Berlin, 1989

[6] Whidborne, J.F.; Postlethwaite, I.; Gu, D.-W.: Robust controller design using H_∞ loop-shaping and the method of inequalities. IEEE Trans. on Contr. Syst. Technology, 2(4), pp. 455-461, 1994

[7] Zakian, V.; Al-Naib, U.: Design of dynamical and control systems by the method of inequalities. Proc. IEE, 120(11), pp. 1421-1427, 1973

[8] Bellman, R.E.; Zadeh, L.A.: Decision Making In A Fuzzy Environment, Management Science, 17 (1970), S. 141-163

[9] Rommelfanger, H.: Fuzzy Decision Support Systeme, Springer, Heidelberg 1994

[10] Sajidman, M.; Kuntze, H.-B.: Integration of Fuzzy Control and Model Based Concepts for Disturbed Industrial Plants with Large Dead-Times. Proc. 6th IEEE Int. Conf. on Fuzzy Systems (FUZZ IEEE'97), Barcelona (Spain), July 1-5, 1997

[11] Bernard, T.: Ein Beitrag zur gewichteten multikriteriellen Optimierung von Heizungs- und Lüftungsregelkreisen auf Grundlage des Fuzzy Decision Making, Universität Karlsruhe (TH), 2000, [online: http://thbernard.leute.server.de/diss]

[12] Bernard, T; Kuntze, H.-B.: Multi-objective Optimization of Building Climate Control Systems Using Fuzzy-Logic, VDI/VDE-Gesellschaft Meß- und Automatisierungstechnik (GMA): Proc. European Control Conference ECC'99, Sept. 1999, Karlsruhe/Germany

[13] Smith, O. J.: A Controller to Overcome Deadtime, ISA Journal, 6 (1959) 2, pp. 28-33

[14] Isermann, R.: Digitale Regelsysteme, Springer, Berlin 1977

An Improved Shape-Preserving Affine Takagi-Sugeno Controller Using Splines

Felipe Fernández and Julio Gutiérrez

Dep. Tecnología Fotónica, Facultad de Informática U.P.M.
Campus de Montegancedo
28660 Madrid, Spain
felipe.fernandez@es.bosch.com, jgr@dtf.fi.upm.es

Abstract. First-order Takagi-Sugeno (TS) fuzzy systems are characterized by a multivariate fuzzy partition in the operating domain and a local affine function within each fuzzy region. This fuzzy inference method presents some undesirable properties in the regions where the membership functions of the antecedent fuzzy partitions overlap. The output function of a SISO affine TS system is in general a non-convex combination of the corresponding piecewise affine functions and the local derivative of the output is not bounded by the derivatives of the corresponding consequent affine functions. These characteristics can influence negatively the stability and robustness of the corresponding fuzzy system. In this paper, to overcome these drawbacks, SISO affine TS models defined on trapezoidal fuzzy partitions are transformed into equivalent zero-order TS models that are defined on triangular fuzzy partitions. Moreover, to improve the smoothness and continuity order of the output function, a B-spline convolution filter is applied to transform the corresponding C^0 triangular fuzzy partition into a C^m spline fuzzy partition (φ-spline fuzzy partition). The obtained spline fuzzy partition preserves the output function in the cores of the initial fuzzy intervals. This technique is directly generalized to a MISO system by the corresponding tensor product. Some examples are given to practically demonstrate the capability of the approach.

1 Introduction

Affine and linear functions are fundamental ingredients in the endeavours of scientists and engineers. They constitute a powerful tool for the simple analysis and design of dynamical systems. Unfortunately, most of the real systems are nonlinear and cannot be modelled using a single affine model for all operating conditions. The development of new general systematic procedures for modelling non-linear systems will probably be a key factor in the coming future of science and engineering.

A remarkable approach to identify non-linear systems is via composite local modelling. A non-linear system can be represented by means of piecewise-affine models. In order to ensure smooth transitions between the different regions, a set of suitable fuzzy blending functions can be used.

The reference basic model selected in this paper is a MISO product-sum affine Takagi-Sugeno model defined on trapezoidal fuzzy partitions (Ruspini partitions of unity). This approach is very simple, intuitively appealing and optimal in relation to weighted mean-square error-like integral [7]. Although, some undesirable properties

B. Reusch (Ed.): Fuzzy Days 2001, LNCS 2206, pp. 671–687, 2001.

(shape-failures) of the global output function are obtained when the standard computation method is applied.

Shape-preserving approximation schemes [4][9] usually means that where the initial data increases or decreases monotonically, the approximation function f has the same behaviour, f is convex (concave) over intervals where the data set is convex (concave), and f has no more intersection points with an arbitrary straight line more often than the corresponding control polygon. Last variation diminishing property also implies that a convex polygon generates a convex curve.

Most of approximation methods try to preserve shape features inherent in the given data such that convexity, monotony or variation diminishing properties. However, standard affine TS system has no convex hull property and no variation diminishing property. Since linear interpolation between functions may not be as intuitive as might be expected [4], and in general, the average of two convex functions is not necessarily convex and linear interpolation between functions is thus not shape preserving.

The quadratic output function of a univariate standard affine TS model, defined on trapezoidal fuzzy partition, has no C^1 continuity and it is a non-convex combination of the corresponding piecewise affine functions [1].

R. Babuska et al. [1][2] have introduced an alternative interpolation scheme with convexity and smoothness properties. The output function generated is smooth with continuous derivatives, and the output function derivative is bounded by the slopes of the corresponding affine functions. However, this approach requires the computation of a piecewise rational smooth approximation function and the method can only be applied to C^0 convex or concave piecewise affine functions. Moreover, the multivariate extension is not straightforward.

This paper presents a more general and easier approach based on an initial transformation of a first-order TS system, defined on a tensor product of trapezoidal fuzzy partitions, into a zero-order TS system, defined on an equivalent tensor product of triangular fuzzy partitions. The obtained output is a C^0 piecewise affine function.

To improve the smoothness and continuity order of the output function, an additional even spline convolution filter is applied onto the corresponding triangular partition of each input variable. The resultant fuzzy partition is piecewise polynomial spline, called in this paper φ-spline.

An additional enlargement of the cores of the initial trapezoidal membership function is also carried out to preserve the multilinear output function in the initial cores of the original multidimensional trapezoidal fuzzy partition (obtained by the tensor product of the corresponding trapezoidal membership functions). This way the ulterior spline convolution filter applied (which reduces the antecedent membership cores) restores the original cores of the initial multidimensional fuzzy partition.

These two transformations allow overcoming the mentioned drawbacks of the conventional affine TS model.

The rest of the final paper is organized as follows: Section 2 briefly reviews some general definitions and terminology on fuzzy numbers and fuzzy partitions; Section 3 presents a trapezoidal/triangular transformation of multivariate affine T-S fuzzy models; Section 4 reviews the fuzzification convolution; Section 5 describes the main properties of φ-spline partitions; Section 6 describes an smooth shape-preserving affine T-S model with continuous derivatives; Section 7 gives two application examples and Section 8 adds some concluding remarks.

2 Fuzzy Numbers and Fuzzy Partitions

To simplify the specification of a MISO fuzzy system it is advantageous to consider multidimensional fuzzy sets defined by the corresponding multidimensional membership functions $R^n \rightarrow [0,1]$. For convenience of ulterior development, we also introduce the non-standard concept of multidimensional S-fuzzy set (signal fuzzy set) defined as function: $R^n \rightarrow [0,\infty)$.

One common way to extend one-dimensional membership functions to multidimensional ones is via the corresponding tensor product:

$$A(u_1,u_2,...,u_n) = A(u_1) \times A(u_2) \times ... \times A(u_n)$$

A multidimensional fuzzy set $\boldsymbol{A}$ is normal when Height($\boldsymbol{A}$)=1. A multidimensional normal convex fuzzy set is usually called in the literature multidimensional fuzzy number or multidimensional fuzzy interval [8].

In this paper, for a more general consideration of multidimensional fuzzy numbers we relax the normality condition and only maintain the convexity constraint.

To accomplish general functional transformations, we introduce the nonnormal class (into the multidimensional S-fuzzy sets and S-fuzzy numbers) to designate those ones with Height($\boldsymbol{A}$)≠1 (subnormal when 0≤Height($\boldsymbol{A}$)<1 and supernormal when Height($\boldsymbol{A}$)>1).

For interest of ulterior development, we recall the following standard concepts [8]:

- Cardinality of fuzzy set $\boldsymbol{A}$, on a multidimensional discrete domain $\boldsymbol{U}$, is defined by

$$|A| = \sum_{u \in U} A(u)$$

- Volume of fuzzy set $\boldsymbol{A}$, on a continuous multidimensional domain U, is defined by

$$|A| = \int_U A(u)\, du$$

In this paper we name a S-fuzzy set C-normal when its volume or cardinality is $|\boldsymbol{A}|=1$.

Generally a set of linguistic terms is used in the antecedent description of a fuzzy controller. These linguistic terms can be seen as qualitative values used to describe fuzzy rules.

A set of M linguistic terms $\{A_1, A_2, ... A_M\}$ is defined in each one-dimensional domain U of a given scalar variable u. It is generally required that the linguistic terms satisfy the property of coverage, i.e. for each domain element is assigned at least one fuzzy set A_j with nonzero membership degree.

For practical reasons, the linguistic terms are usually constrained to be normal or subnormal fuzzy numbers. Also the model here considered, and presently in many other fuzzy control models [2][5][10][11], imposes stronger conditions on the linguistic terms used.

Ruspini-partition of unity $\{Aj\}$ of a domain U is defined by the following constraints

$$0 \le Aj(u) \le 1 \qquad \forall u \in U \quad 1 \le j \le M$$

$$\sum_{j=1}^{M} A_j(u) = 1 \qquad \forall u \in U$$

$$\sum_{u \in U} A_j(u) > 0 \quad 1 \leq j \leq M$$

First equation is the standard membership constraint. Second equation means that for each u, the sum of all membership degrees $A_j(u)$ equals one. Third equation implies that none of the fuzzy sets $A_j(u)$ is empty. This fuzzy segmentation of domain U is also called *fuzzy partition* [1][8].

Basic univariate fuzzy partitions on a domain U are usually referred to the corresponding sequence of *knot points*

$$a=t_0< t_1< \ldots < t_M=b \text{ with } [a,b]\in \mathrm{U}$$

- This sequence $T=(t_0,t_1,\ldots,t_M)$ can specify a basic crisp partition, determined by a set of M crisp intervals $\Pi=\{C_0,C_1,\ldots, C_{M-1}\} =\{[t_0,t_1),[t_1,t_2),\ldots,[t_{M-1},t_M),\}$ (Figure 1.a).
- An analogous knots sequence T can be used to specify a *trapezoidal fuzzy partition* $\Pi=\{C_0,L_1,C_2,L_3,\ldots,L_{M-2},C_{M-1}\}$ (Figure 1.b) composed by two interleaved sets of membership functions:
  1. The set of piecewise constant functions $\{C_j(u)=1\}$ defined in the corresponding even intervals (*interpolation intervals*) $C=\{[t_0,t_1), [t_2,t_3), [t_4,t_5), \ldots ,[t_{M-1},t_M)\}$, and
  2. The set of couples of piecewise linear affine functions $\{LU_j(t)=t,\ LD_j(t)=1-t \mid t=(t_{j+1}-u)/(t_{j+1}-t_j);\ (0\leq t\leq 1)\}$ defined in the corresponding odd intervals (*approximation intervals*) $L=\{[t_1,t_2), [t_3,t_4), [t_5,t_6),\ldots ,[t_{M-2},t_{M-1}),\}$
- The same previous sequence T can be used to define a *triangular fuzzy partition* $\Pi=\{L_0,L_1,L_2,L_3,\ldots,L_{M-2},L_{M-1}\}$ composed by a set of couples of piecewise linear affine functions $\{LU_j(t)=t,\ LD_j(t)=1-t \mid t=(t_{j+1}-u)/(t_{j+1}-t_j);\ (0\leq t\leq 1)\}$ defined in the corresponding intervals $\{[t_0,t_1),[t_1,t_2),[t_3,t_4),\ldots,[t_{M-1},t_M)\}$.

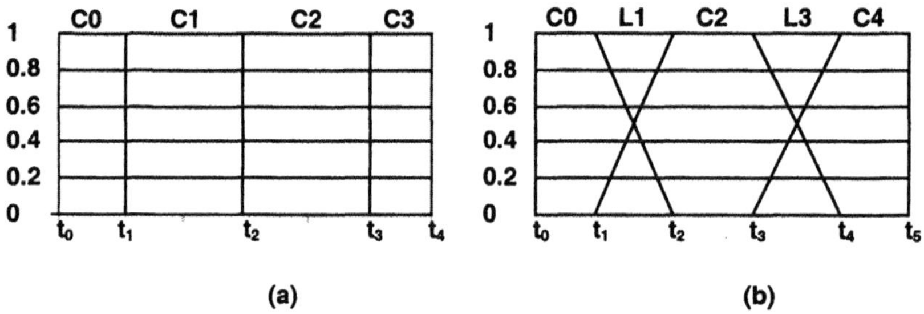

Fig. 1. Fuzzy partitions: (a) Crisp partition. (b) Trapezoidal fuzzy partition.

An important parameter of a fuzzy partition is the overlapping factor *ov*, which is the maximum number of membership functions with non-null membership degrees among all the elements of domain U.

$$\mathrm{ov}(\{A_j\}) = \max_{u \in U} \Big(\sum_{1\leq j \leq M} \lceil A_j(u) \rceil \Big)$$

where $\lceil \alpha \rceil$ denotes the ceiling function (smallest integer $\geq \alpha$). Usually the fuzzy partitions considered must have a low overlapping factor, in order to limit the computational complexity of the corresponding fuzzy system

Finally in this section we review one type of uncertainty measurement of special interest for fuzzy partition characterization: degree of fuzziness.

Any set that is not crisp has some degree of fuzziness that results from the imprecision of its boundaries [8][12]. For measuring fuzziness of a fuzzy set, we can use the function f to express the lack of distinction between the membership $A(u)$ and its complement 1-$A(u)$:

$$f(A) = \int_{Supp(A)} (1 - 2\,|\,A(u) - 0.5\,|)\,du$$

Analogously, it is possible to define the right and left fuzziness of a fuzzy number, respectively delimited on the left and right bounds. For example, the corresponding fuzziness of a normal trapezoidal fuzzy number B are [12]:

$$f(B)=(\alpha+\beta)/2; \quad f_l(B)=\alpha/2; \quad f_r(B)=\beta/2$$

where α and β are the bases length of the corresponding left and right rectangular triangles.

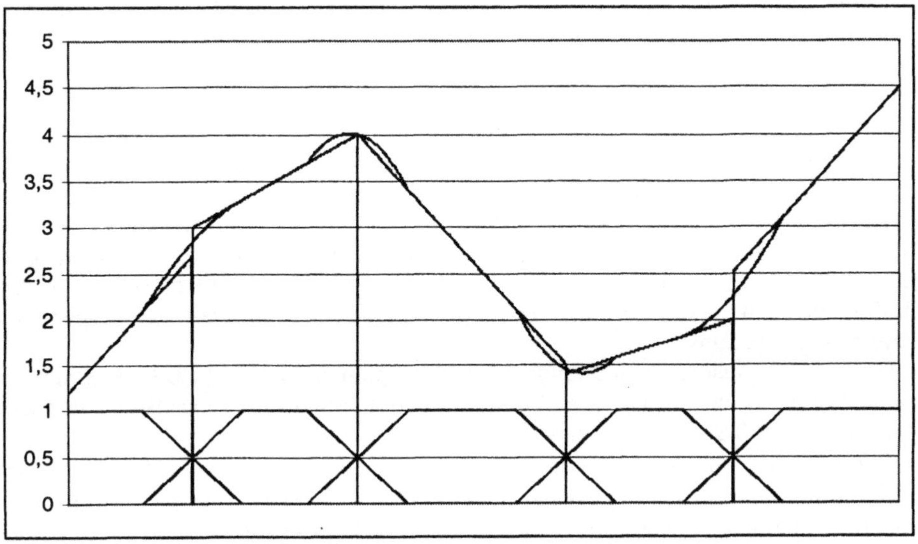

Fig. 2. Univariate output function of a standard affine TS controller

Therefore the global fuzziness of f_T of a trapezoidal fuzzy partition $\Pi=\{C_0,L_1,C_2,L_3,\ldots,L_{M-2},C_{M-1}\}$ can be easily computed by the addition of lengths of the corresponding approximation intervals L_j:

$$f_T(\{C_0,L_1,C_2,L_3,\ldots,L_{M-2},C_{M-1}\})=(|L_1|+|L_3|+\ldots+|L_{M-2}|)/2$$

In Section 4 a general convolution operator on a fuzzy partition that increases its global fuzziness is introduced. This operator allows to easily specify and control the related smoothness of the obtained fuzzy control surface.

3 Transformation of Affine T-S Fuzzy Models

A multivariate affine Takagi-Sugeno system rule, defined on a trapezoidal tensor product fuzzy partition, can be specified (using homogenous coordinates) by a set of r_f first-order canonical rules $\{R_i\}$:

$$\{R_i\text{: If } \mathbf{x} \text{ is } \mathbf{A}_i \text{ then } z = [\mathbf{a}_i\ b_i]\,[\mathbf{x}\ 1]^T\}$$

where $\mathbf{x}=(x_1,\ldots,x_n)$ is the multivariate input variable $\mathbf{A}_i=A_{i\,1}(x_1)\times \ldots \times A_{i\,n}(x_n)$ is the multivariate trapezoidal fuzzy number of the i-th rule defined by means of the corresponding tensor product of trapezoidal membership functions; z is the output variable and $[\mathbf{a}_i\ b_i]$ are the affine coefficients n+1-vector of the corresponding local hyperplane ($z = a_{i1}\,x_1 + a_{i2}\,x_2 + \ldots + a_{in}\,x_n + b_i$).

Using product-sum inference, this standard aggregated MISO fuzzy model with antecedent fuzzy partitions can be computed as

$$z(\mathbf{x}) = \sum_{i\in R_{active}} \alpha_i(\mathbf{x})\,[\mathbf{a}_i \quad b_i][\mathbf{x} \quad 1]^T$$

where $\alpha_i(\mathbf{x}) = A_{i\,1}(x_1)\times A_{i\,2}(x_2)\times\ldots\times A_{i\,n}(x_n)$ is the firing degree of the corresponding tensor product of trapezoidal membership functions of i-th rule, and R_{active} is the set of active rules. This model covers the domain of a non-linear plant with a set of local affine models, and the whole description of plant is obtained as a weighed average of the corresponding local models. The local affine functions are preserved in the multidimensional cores of the corresponding tensor product of trapezoidal membership functions.

This model presents shape-failures, as it is pointed out in [1][2]. A graphical example of a piecewise quadratic C^0 output function, obtained by means of a standard univariate affine TS system, is shown in Figure 2.

In this section an equivalent triangular zero-order TS model is considered, in order to obtain a shape preserving approximation of the corresponding piecewise affine functions. The new multivariate system is specified by a set of $r_z = r_f \times 2^n$ zero-order canonical rules $\{R_j\}$ corresponding to 2^n corners of each core of the original multivariate trapezoidal fuzzy numbers $\mathbf{A}_i$:

$$\{R_j\text{: If } \mathbf{x} \text{ is } \mathbf{B}_j \text{ then } z = c_j\}$$

where $\mathbf{x}=(x_1,\ldots,x_n)$ is the original multivariate input variable, $\mathbf{B}_j=B_{j\,1}(x_1)\times \ldots \times B_{j\,n}(x_n)$ is a new multivariate fuzzy number of the j-th rule defined by means of the corresponding tensor product of new triangular membership functions; z is the original output variable and c_j is a constant equal to the value of the original affine function $z = a_{i1}\,x_1 + a_{i2}\,x_2 + \ldots + a_{in}\,x_n + b_i$ in the core of the corresponding triangular tensor product.

To obtain the new univariate equivalent triangular fuzzy partition for each input variable, two triangles are derived for each initial trapezium. The cores of these triangles are placed in the corners of the corresponding trapezium cores and the support corners of these triangles are placed on the adjacent knots of the original fuzzy partition (Figure 3).

The output function c_j in the core of each tensor product of triangular membership functions preserves the value of the original affine function. This scheme also provides an additional multilinear interpolation function in the region where the original trapeziums overlap. The output function of this equivalent C^0 zero-order TS model is given by

$$z(\mathbf{x}) = \sum_{j \in R_{active}} \beta_j(\mathbf{x})\, c_j$$

where $\beta_j(\mathbf{x}) = B_{j\,1}(x_1) \times B_{j\,2}(x_2) \times \ldots \times B_{j\,n}(x_n)$ is the firing degree of the corresponding tensor product of triangular membership functions of j-th rule, and R_{active} is the set of active rules.

The C^0 multivariate output function computed by this scheme is a interleaved piecewise function composed by two kind of functions (Figure 4):

- Affine functions in the cores of the initial multivariate trapezoidal fuzzy partition (interpolation regions), and
- Multilinear interpolation functions in the rest of the domain regions (approximation regions).

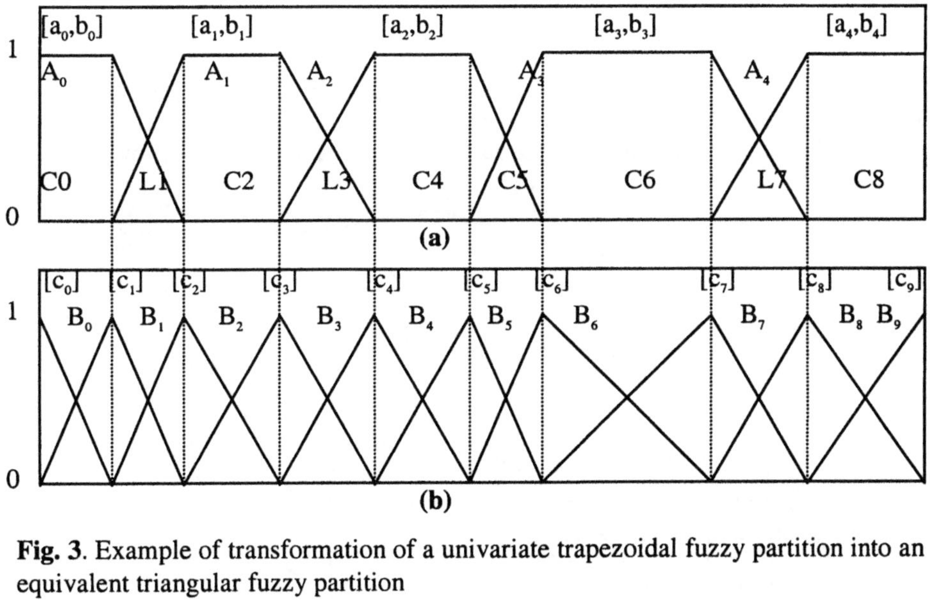

Fig. 3. Example of transformation of a univariate trapezoidal fuzzy partition into an equivalent triangular fuzzy partition

In this paper, in order to increase the continuity order and smoothness of this piecewise output function, a suitable B-spline convolution filter is considered. Next section reviews the general properties and interpretation of a general fuzzification convolution.

4 Fuzzification Convolution

This section introduces the concept of multivariate fuzzification convolution, which is a generalization of the concept of fuzzification function [5][6][8]. This transformation allows a global consideration of the fuzziness associated to the fuzzy input variables and to the fuzzy antecedent terms of a TS system, and can also model an additional smoothing output filter of a fuzzy controller.

The multivariate uniform fuzzification transform $\boldsymbol{F}$ is a spatially or shift invariant transformation that applies a multivariate even fuzzification function Φ (even C-normal S-fuzzy number) onto a multivariate fuzzy number f (normal or subnormal), defined on a multidimensional continuous domain $\boldsymbol{U}$, and gives a new multivariate fuzzy number f^{Φ} defined by means of the following cross-correlation/convolution operation

$$\boldsymbol{F}_{\Phi}(\boldsymbol{f})\ (\boldsymbol{v}) = \Phi * \boldsymbol{f}\ (\boldsymbol{v}) = \int_U \boldsymbol{f}(\boldsymbol{u})\cdot\Phi(\boldsymbol{u}-\boldsymbol{v})\ \mathrm{d}\boldsymbol{u} = \int_U \boldsymbol{f}(\boldsymbol{u})\cdot\Phi(\boldsymbol{v}-\boldsymbol{u})\ \mathrm{d}\boldsymbol{u} = \boldsymbol{f}^{\Phi}(\boldsymbol{v})$$

A Particular interesting case of fuzzification convolution is the multidimensional Dirac impulse fuzzification transform $\boldsymbol{F}_{\delta}$, normally called in the literature [5] singleton fuzzification transform:

$$\boldsymbol{F}_{\delta}\ (\mathbf{f}) = \delta * \mathbf{f} = \mathbf{f}$$

It is convenient to remark that, in general, the multivariate fuzzification transform modifies, in general, the support and core of the corresponding fuzzy number, but it preserves its volume and maintains the fuzzy partition as it is shown bellow.

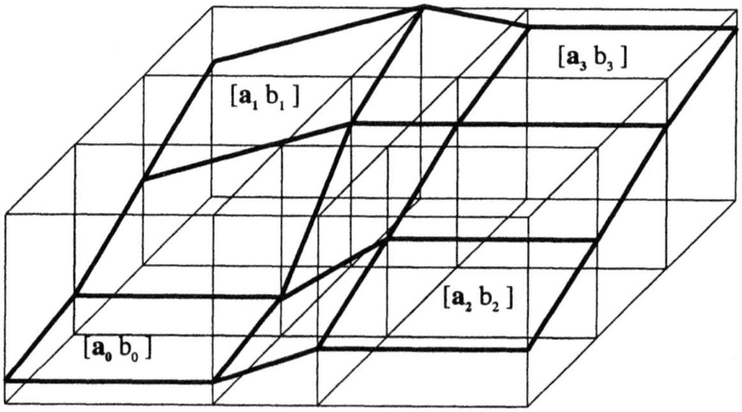

Fig- 4. A transformed affine TS fuzzy system computed by a bivariate C^0 piecewise output function composed by four affine cells and five bilinear cells.

On the contrary, if a max-min operation [5] [8][12] is used to define an analogous fuzzification transform (instead of sum-product), the corresponding operation modifies the support but does not change the core of the corresponding fuzzy number, and therefore it does not preserve the corresponding fuzzy partition.

Some important properties of the multidimensional fuzzification transform are:

- **Partition of unity conservation.** The fuzzification transform of a multidimensional fuzzy partition ($\{Aj\}$ $j=1..M$) is another fuzzy partition, i. e. partitions of unity are closed under fuzzification transform:

  $$\mathcal{F}_\Phi(\{A_j\}) = \Phi * \{A_j\} = \{\Phi * A_j\} = \{A^\Phi{}_j\}$$

  Proof. It is a consequence of the distributive property of convolution operation

  $$\sum_{j=1}^{M} \Phi * \{A_j\} = \Phi * 1 = 1 \qquad \forall u \in U$$

  •

- **Area preservation.** The fuzzification transform preserves the volume of a multidimensional fuzzy interval A

  $$|A * \Phi| = |A|$$

  Proof. It is a consequence of the commutative and distributive property of integral operator

  $$\int_U \mathcal{F}_\Phi(A)dv = \int_U \int_U \Phi(u-v)A(\mathbf{u})\, du\, dv = \int_U \int_U \Phi(u-v)\, dv\, A(\mathbf{u})\, du = 1 \cdot |\mathbf{A}| \qquad \forall u \in U$$

  •

- **S-Fuzzy numbers are closed with respect to uniform fuzzification transform.** The set of multidimensional S-fuzzy intervals is closed under the uniform fuzzification convolution.
  Proof. Since spatially invariant fuzzification filtering preserves the convexity property of a fuzzy interval.

  •

- **Smoothness of a uniform fuzzification transform**. Uniform fuzzification convolution is a smoothing transform: If Φ is spatially invariant and f and Φ have smoothness of order m and n respectively ($f \in C^{m-2}$ and $\Phi \in C^{n-2}$), then $f*\Phi$ has smoothness of order m+n ($f*\Phi \in C^{m+n-2}$). Smoothness of order k means [2] that the derivative of order k of the function becomes impulsive, and the differentiability class C^k is the space of functions that are k times continuously differentiable.
  Proof. Since the order of Fourier transforms F of f, Φ and $f * \Phi$ are respectively

  $$F(f) = o(|s|^{-m}); \quad F(\Phi) = o(|s|^{-n}); \quad F(f * \Phi) = o(|s|^{-(m+n)})$$

  hence the smoothness order of $f * \Phi$ is $m+n$ (its $(m+n)$-th derivative is impulsive) and consequently $f * \Phi$ belongs to the class C^{m+n-2}.

  •

- **Affine invariance.** Affine functions are fixed points of fuzzification transform:

  $$\mathcal{F}_\Phi(\mathbf{a} \cdot \mathbf{u} + \mathbf{b}) = \Phi(\mathbf{u}) * (\mathbf{a} \cdot \mathbf{u} + \mathbf{b}) = \mathbf{a} \cdot \mathbf{u} + \mathbf{b}$$

Proof. It is a consequence of the even and C-normal property of the fuzzification function $\Phi(\mathbf{u})$ used. In each position of the corresponding filter window, the corresponding average value, for two symmetrical points of this window, is equal to the corresponding central value. Note that this property is also maintained for a general nonuniform or spatially variant fuzzification transform.

•

In this paper, to solve the bound problem associated to every convolution operation, a progressive linear reduction of the kernel support of the fuzzification function is applied in the proximity of the limits of the domain.

A null window width on the limits of the domain interval eliminates the boundary problems of the average window used. This is a natural specification in a fuzzy context, because usually in practical applications, the extreme values of a variable can be fixed without imprecision, and generally we have no continuity constraints in the corresponding points of output control surface.

To conclude this section, we give three possible interpretations of multidimensional fuzzification convolution:

1. As a cross-correlation between a multidimensional linguistic term $A(.)$ and a multidimensional fuzzy input variable $X(.)$. This provides a measure of matching (X is A) between the fuzzy term A and the fuzzy variable X.

$$X * A \;(v) = \langle X(v), A(v) \rangle$$

The computation of a multidimensional fuzzy proposition (X is A) is greatly simplified during the execution time of the fuzzy algorithm when the fuzzification transform is applied to the corresponding linguistic label during the compilation time of the related fuzzy algorithm:

$$X * A \;(v) = (\,\delta_x * \Phi\mathbf{X}\,) * A \;(v) = \delta_x * (\Phi\mathbf{X} * A\,)\,(v) = \delta_{\mathbf{x}} * A^{\Phi\mathbf{x}} \;(v)$$

where $\delta_x\,(v) = \delta(v-x)$ and x is the centre of the fuzzy input $X = \Phi X = \delta_x * \Phi$.

2. As a multidimensional fuzzification operator with respect to a fuzzification function $\Phi\mathbf{A}(.)$ on a fuzzy number $A(.)$.

$$\Phi\mathbf{A} * A \;(v) = \langle \Phi\mathbf{A}\;(u-v), A\;(u) \rangle$$

Thus, it is possible to control the fuzziness degree of a fuzzy partition by means of the corresponding fuzzification operator.

3. As a spatial convolution filtering (multivariate low-pass filter) of a multidimensional signal z with the multidimensional smoothing function $\Phi\mathbf{Z}$.

$$\Phi\mathbf{Z} * z\;(v) = \langle \Phi\mathbf{Z}(v-u), z(u) \rangle$$

Thus, the low-pass multidimensional window function Φ can be used for filtering the output z of fuzzy controller. This local averaging has the effect of suppressing the high-frequency variations while preserving the basic shape of z.

All previous fuzzification functions (ΦX,ΦA,ΦZ) can be combined in a global fuzzification function Φ

$$\Phi = \Phi\mathbf{X} * \Phi\mathbf{A} * \Phi\mathbf{Z}$$

taking into account the additive nature of TS controllers and fuzzification transform considered.

5 Univariate φ-Spline Partitions

Nonuniform univariate normal triangular fuzzy partitions are determined by a set of knots, which defines the cores of the corresponding triangular membership functions. These control points are distributed throughout each univariate input domain $[a,b] \in U$, and define the ordered set: $a=t0= t_1 < t_2 < \ldots < t_M = t_{M+1} = b$. A normal triangular fuzzy number B_j^2 relative to the knot t_j ($1 \leq j \leq M$) is defined by:

$$B_j^2(u) = \{ \max(\min((u-t_{j-1})/(t_j-t_{j-1});(t_{j+1}-u)/(t_{j+1}-t_j)) ; 0) \}$$

This triangular fuzzy partition is also named nonuniform B-spline partition of order 2 (or degree 1) [3][4]. To add a uniform filter to the corresponding output function, a uniform spline convolution filter can be applied on this triangular partition of unity.

In general, if an even B-spline filter $N^m[\sigma]$ ($m \geq 1$) of order m and support length $\rho = m \cdot \sigma$ is convolved with a nonuniform normal triangular fuzzy partition $\{B_j^2\}$:

$$N^m[\sigma] * \{B_j^2\} = \{N^m[\sigma] * B_j^2\} = \{B_j^{m+2}[\sigma]\}$$

it gives a new fuzzy partition $\{B_j^{m+2}[\sigma]\}$ that is called in this paper, *φ-spline partition* of order m+2 (Figure 5).

To solve the mentioned domain bound problem associated to every convolution operation, a varying (trapezoidal) kernel support length ρ_u can be used:

$$\rho_u/2 = \min((u-a), -(u-b), \rho/2)$$

This function of convolution kernel support length is zero in the corners of the corresponding domain interval.

The B-spline convolution filter used $N^m[\sigma]$ determines the smoothness of the related output function. This way the designer can initially obtain a normal triangular fuzzy partition, which gives a non-smooth piecewise multilinear output function. Furthermore, it is possible to add an even B-spline filter to provide a suitable smoothness for the corresponding output function.

Properties . For each $m \geq 0$, regular φ-splines $B_j^{m+2}[\sigma]$ of degree (m+1) and (m+2)-th order have the following properties:

1. Support width: $\|\text{supp}(B_j^{m+2}[\sigma])\| = \|\text{supp}(B_j^2)\| + m*\sigma$
2. Core width: $\|\text{core}(B_j^{m+2}[\sigma])\| = 0$;
3. Continuity class: $\{B_j^{m+2}[\sigma]\} \in C^m$
4. Polynomial form: $B_j^{m+2}[\sigma]$ are formed by piecewise polynomial functions of degree m+1
5. Linearity preservation: If $B_j^2(u)$ is defined by the knots (t_{j-1}, t_j, t_{j+1}), then $B_j^{m+2}[\sigma]\,(u)$ is an affine function in the intervals: $(t_{j-1}+m\sigma/2\,,\, t_j-m\sigma/2)$ and $(t_j+m\sigma/2\,,\, t_{j+1}-m\sigma/2)$

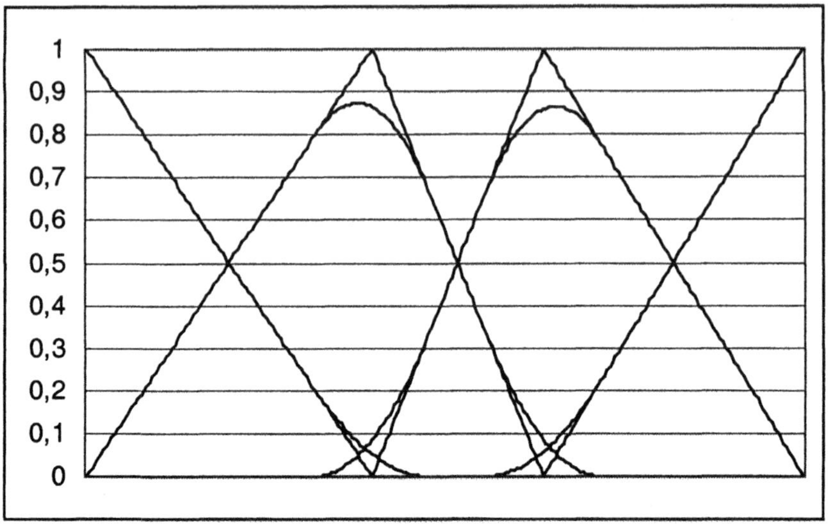

Fig. 5. A nonuniform triangular φ-spline fuzzy partition $\{B_j^2\}$ and a derived nonuniform quadratic φ-spline fuzzy partition $\{B_j^3\}$

Sketch of proof. These properties are a direct consequence of general convolution properties applied to nonuniform triangular fuzzy partitions.

6 Shape-Preserving Affine Takagi-Sugeno Model

If a smoother output function C^m is needed, an even B-spline fuzzification filter of order *m* with support $\rho = m \cdot \sigma$ can be applied to each univariate triangular partition (Figure 6). In this case a previous optional enlargement of each trapezium core can be advisable in order to preserve the original affine functions in the corresponding initial cores of the multivariate trapezoidal fuzzy numbers.

The improved shape-preserving affine Takagi-Sugeno system rule here considered, uses the same original multivariate affine T-S rules

$$\{R_i\text{: If } \mathbf{x} \text{ is } \mathbf{A}_i \text{ then } z = [\mathbf{a}_i\ b_i]\ [\mathbf{x}\ 1]^T\ \}$$

and transforms each univariate approximation interval $L_{k\,j} = [t_{k\,j}, t_{k\,j+1})$ of each univariate trapezoidal partition $\Pi_k = \{C_{k\,0}, L_{k\,1}, C_{k\,2}, L_{k\,3}, \ldots, L_{k\,M-2}, C_{k\,M-1}\}$ (defined in Section 2) into a narrower one $L'_{k\,j} = [t_{k\,j} + \rho_k/2\ , t_{k\,j+1} - \rho_k/2)$, where ρ_k is the kernel support length of the posterior B-spline filter applied on each univariate domain U_k (Figure 7). Consequently, the cores of each univariate trapezoidal membership function are originally enlarged.

The following hypothesis is assumed: $(t_{k,\,j} + \rho_k/2 \leq t_{k,\,j+1} - \rho_k/2)$ for each original univariate approximation interval L_{kj}. Under this assumption the global transformation (core enlargement and fuzzification convolution) preserves the affine functions in the cores of the initial trapezoidal tensor products.

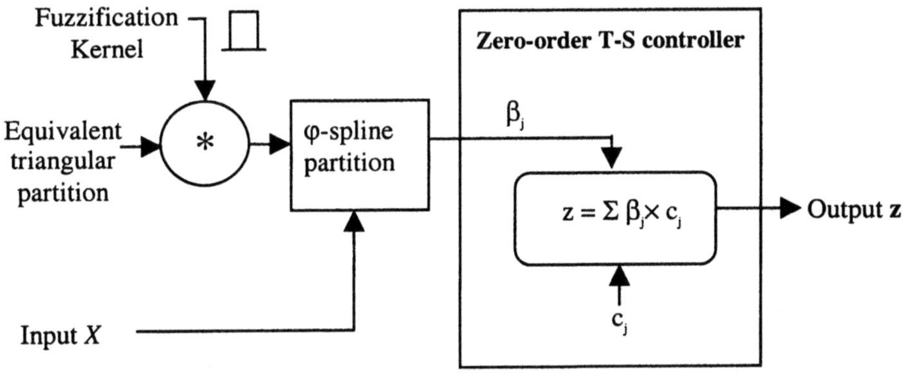

Fig. 6. Equivalent fuzzified triangular zero-order Takagi-Sugeno model.

The new multivariate fuzzified triangular TS system derived from the previous modified trapezoidal partition is specified by the corresponding zero-order canonical rules $\{R_j\}$:

$$\{R_j\text{: If } \mathbf{x} \text{ is } \mathbf{B}'_j \text{ then } z = c_j\}$$

where $\mathbf{x}=(x_1,\ldots,x_n)$ is the original multivariate input variable, $\mathbf{B}'_j=B'_{j1}(x_1)\times \ldots \times B'_{jn}(x_n)$ is the new multivariate φ-spline fuzzy number of the j-th rule defined by means of the corresponding tensor product of fuzzified triangular membership functions

$$\mathbf{B}'_j= \mathbf{B}_j * \Phi = B'_{j1}(x_1) * \Phi(x_1) \times \ldots \times B'_{jn}(x_n) * \Phi(x_1)$$

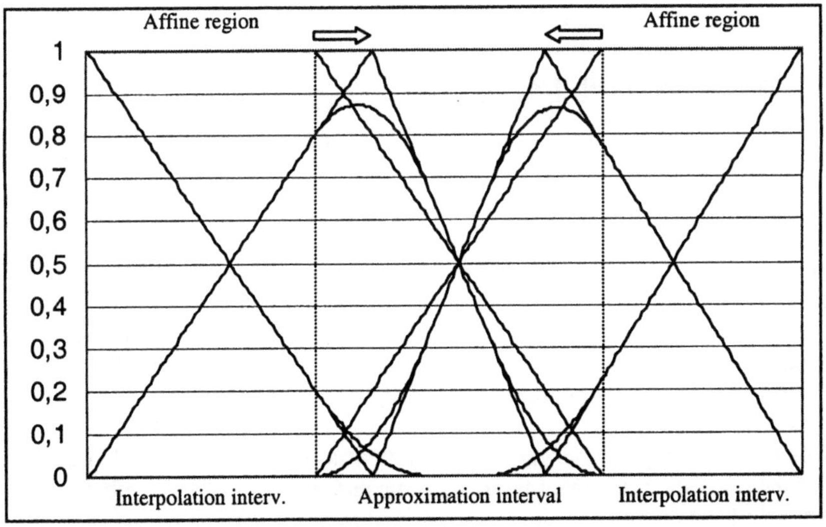

Fig. 7. Construction of a univariate φ-spline fuzzy partition from the original univariate trapezoidal fuzzy partition by means of core enlargement, trapezoidal-triangular transformation and additional B-spline fuzzification

The output z is the same original output variable and c_j is also a constant equal to the value of the original affine function $z = a_{i1} x_1 + a_{i2} x_2 + \ldots + a_{in} x_n + b_i$ in the core of the corresponding triangular tensor product.

7 Examples

This section describes two particular TS systems, to show the characteristics of the model considered. Both systems are specified by means of a set of affine functions that are defined on the corresponding trapezoidal partition. The fuzzification function applied is an even B-spline of order 1. Therefore the output functions obtained are piecewise affine-quadratic function, which have a relative low computational complexity.

Example of a SISO System

The first univariate affine T-S system considered is specified on a trapezoidal fuzzy partition $\{A_0, A1, \ldots, A_5\}$ defined by the knot sequence T

$$T=(t_0, t_1, \ldots, t_9) = (0, 9, 21, 29, 41, 54, 66, 74, 86, 100)$$

The corresponding affine fuzzy rules are

R0: **IF** x IS A0 **then** z = 0.10 (x-0) + 1.2

R1: **IF** x IS A1 **then** z = 0.05 (x-15) + 3

R2: **IF** x IS A2 **then** z = -0.10 (x-35) + 4

R3: **IF** x IS A2 **then** z = 0.03 (x-60) + 1.4

R4: **IF** x IS A2 **then** z = 0.10 (x-80) + 2.5

The fuzzification function applied is a univariate B-spline $B^1[9]$ of order 1 and support length 9.

Figure 8 depicts the original trapezoidal fuzzy partition, the crisp piecewise affine function, the output function C^0 obtained using the standard affine TS procedure and the output function C^1 computed using the shape-preserving scheme previously described.

Note that the shape-preserving output function has a local affine behaviour in the core of the corresponding original trapeziums and gives a good smooth approximation tin the rest of the domain.

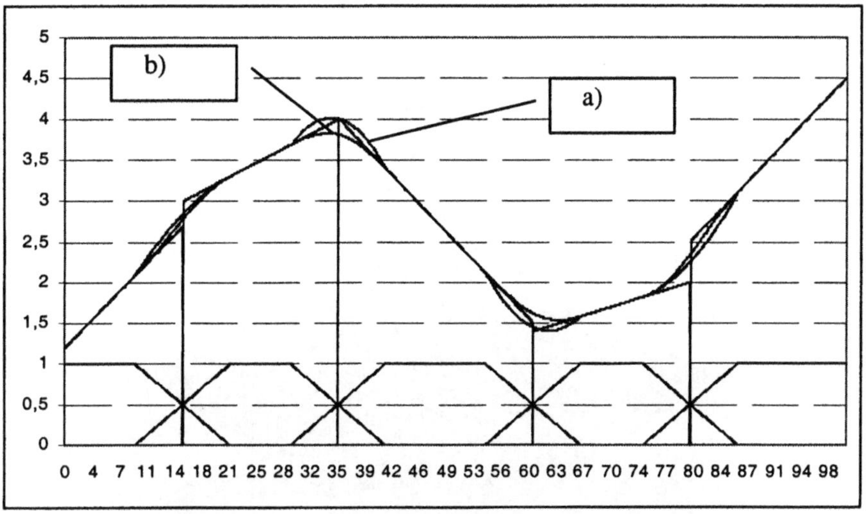

Fig. 8. Comparison between the output functions of a SISO affine TS controller: a) Standard method b) Shape-preserving method

Example of a MISO System

The second system analyzed is a bivariate T-S system specified on a tensor product of trapezoidal fuzzy partitions: $\{A_{10},A_{11}\}\times\{A_{20},A_{21}\}$ that are specified by the following knots sequences:

$$T_1 = (t_{10}, t_{11}, t_{12}, t_{13}) = (0, 80, 180, 250)$$

$$T_2 = (t_{20}, t_{21}, t_{22}, t_{23}) = (0, 80, 180, 250)$$

The corresponding affine fuzzy rules are:

R0: **IF** x_1 IS A_{10} **and** x_2 IS A_{20} **then** $z = 2.5$

R1: **IF** x_1 IS A_{10} **and** x_2 IS A_{21} **then** $z = 2.5$

R2: **IF** x_1 IS A_{11} **and** x_2 IS A_{20} **then** $z = 2.5$

R3: **IF** x_1 IS A_{11} **and** x_2 IS A_{21} **then** $z = 2.5 + 0.02(x_1\text{-}150)+0.02(x_2\text{-}150)$

The fuzzification function applied is a bivariate B-spline tensor-product $B^{1,1}[40,40]$ of order (1, 1) and support length (40,40).

A critical zone of the corresponding control surface is depicted in Figure 9.a, where it is possible to appreciate some shape failures when the output function is computed using the standard affine TS method. Figure 9.b shows the corresponding output surface zone using the shape-preserving method proposed.

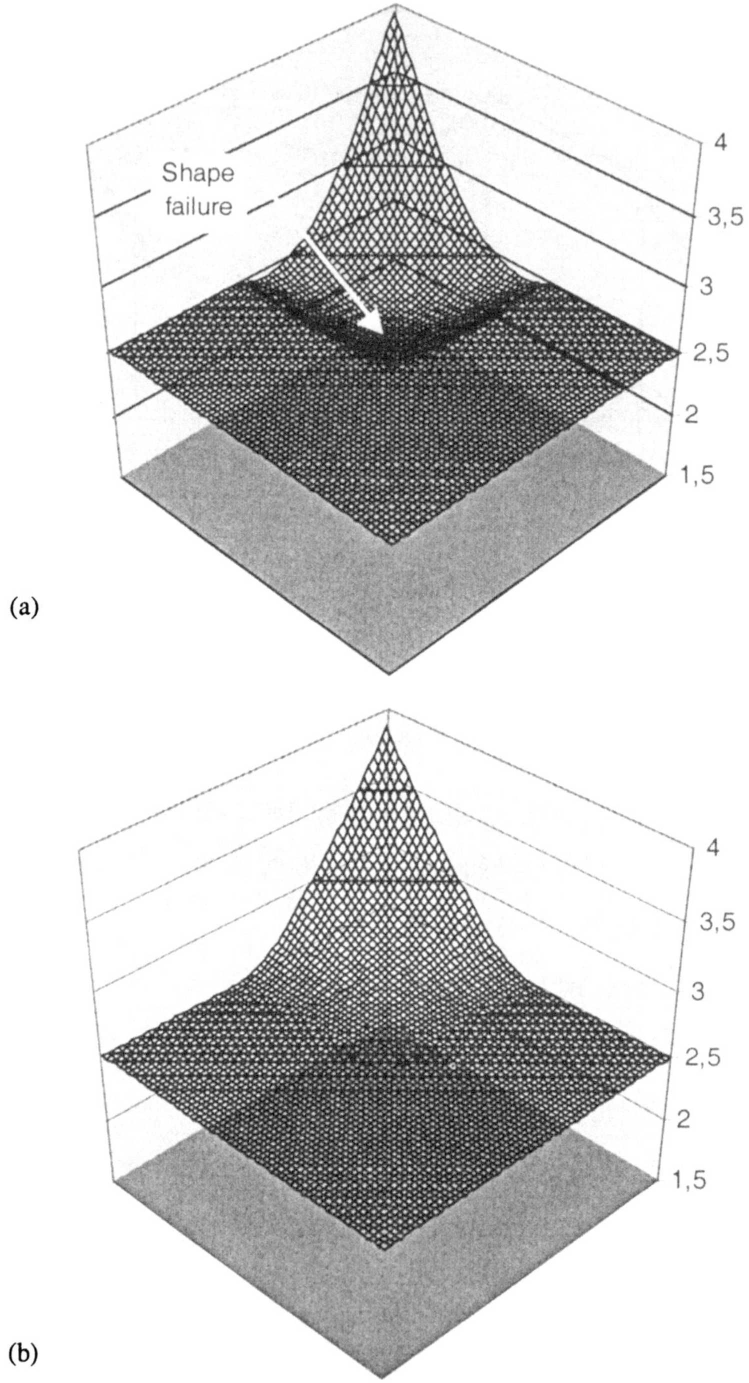

Fig. 9. Comparison between the output functions of a bivariate affine TS controller: a) Standard method b) Shape-preserving method

8 Conclusions

This paper has introduced a shape-preserving affine Takagi-Sugeno controller that facilitates the analysis and implementation of fuzzy systems. The described model also allows to specify different fuzziness of a fuzzy system by means of the B-spline fuzzification convolution considered.

Consequently a multivariate shape-preserving piecewise polynomial fuzzy modelling can be built up on the basis of multidimensional trapezoidal partition and B-spline fuzzification tools.

References

1. R. Babuska, C. Fantuzzi, U. Kaymak, and H. B. Verbrunggen, Improved inference for Takagi-Sugeno models, In Proc. Fifth IEEE International Conference on Fuzzy Systems, New Orleans, USA, pp. 701-706, 1996.
2. R. Babuska, *Fuzzy Modeling for Control*, Kluwer Academic Publisher, 1998.
3. C. K. Chui, Wavelets: A mathematical Tool for Signal Analysis, SIAM, 1997.
4. G. Farin, Curves and Surfaces for Computer-Aided Geometric design, Fourth edition, Academic Press, 1998.
5. D. Driankov, H. Hellendoorn, M. Reinfrank, An Introduction to Fuzzy Control, Springer-Verlag, 1993.
6. H. Geering, Introduction to Fuzzy Control, Institut fuer Mess- und Regeltechnik. IMRT-Bericht Nr. 24, October 1993.
7. T. A. Johansen, On the Optimality of Takagi-Sugeno-Kang Fuzzy Inference Mechanism, Tech. Report of Norwegian Institute of Technology, University of Trondheim, Norway, November 1994.
8. G. Klir, B. Yuan, Fuzzy Sets and Fuzzy Logic, Prentice Hall, 1995.
9. B. I. Kvasov, Methods of Shape-Preserving spline approximation, Word Scientific, 2000.
10. J. Zhang and A. Knoll, Constructing Fuzzy Controllers with B-spline Models, IEEE International Conference on Fuzzy Systems, 1996.
11. J. Zhang and A. Knoll, Unsupervised Learning of Control Surfaces Based on B-spline Models, IEEE International Conference on Fuzzy Systems, 1997, pp. 1725-1730.
12. F. Fernández and J. Gutiérrez, "A Transformational Approach to Fuzzy Matching", Proc. of 7th International Conference IPMU, 1998, Vol. II, pp. 1317-1324.

METALA: A Meta-learning Architecture*

Juan A. Botía, Antonio F. Gómez-Skarmeta,
Mercedes Valdés, and Antonio Padilla

Departamento de Ingeniería de la Información y las Comunicaciones,
Universidad de Murcia. Campus de Espinardo. E-30071. Murcia, Spain
{juanbot,skarmeta,mvaldes,apadilla}@um.es

Abstract. Meta-learning has been accepted, in the last five years, as a proper machine learning research field. In this concrete area of interest, the way in which different theories, each one produced either with the same algorithm or with many of them, are merged to produce a more accurate model has been the main topic. Now, new emerging techniques got more to do with inductive meta-learning. It is the process of learning from others learning experiences. This kind of learning imposes severe requisites, from the point of view of the software system that would support it. The purpose of this work is to show a software architecture for this type of learning. The architecture will give recommendations for building a system of this kind, that has to tackle with very precise but difficult problems at a time.

1 Introduction

This section is devoted to offer, very briefly, an overview of meta-learning, i.e. all the research work that has been or is being devoted to use the process of automatic learning to learn from it. Surfing through the literature, we have envisaged six different types of meta-learning in the whole machine learning research community. **Hybridization** consist in that different learning techniques are combined in a way in which each technique would attack the part of the problem that best suits to its own features. A very good introduction to this technique can be read in [7]. In **Composite meta-learning** local data theories are produced but a global theory of data is needed. This is done by a few kinds of models merging. The paradigmical example of this approach is the JAM system [10]. Then comes **Constructive meta-learning**. It is called in this way in a sense that it uses chuncks of all learning paradigms to design itself on-line, depending on the learning task and the learning data set. This is precissely the approach used in CAMLET [11]. **Boosting and Bagging** differ from composite meta-learning in that in these techniques, the same classifier is applied repeatly over the same data set. Boosting [5] works with a single type of classification algorithm. Basically, it is an iterative process in which in each iteration a new classifier is produced. In baggin [3] new classifiers are compound in parallel.

* Work partially supported by the European Comission through the project FEDER 1FD97-0255-C03-01

B. Reusch (Ed.): Fuzzy Days 2001, LNCS 2206, pp. 688–698, 2001.

Each of them is built from a different bootstrap training data constructed from the original data set. Then, as it is done in boosting, the final classifier is an aggregation of all formerly constructed base classifiers. In **Ranking methods** an ordered list of classifiers is created. This ranking will be used then to decide which algorithm should be used to produce a hopefully reasonably good model. In the work of [2], we can find a comparision on different ranking methods. **Inductive meta-learning** uses as the main criteria for selecting a model the predicted accuracy of the candidate model. There are many applications that have proben to be successful in this new area. A good example is the work of [1]. All these techniques require very special requisites, in order to develop a software system that holds this kind of learning. In the next section, the software architecture propossed to implement such kind of system will be introduced.

2 The Architecture

METALA (META-Learning Architecture) is a set of recommendations to build a software system which affords massive production of learning experiments, and the interaction of different learning paradigms, at the functional level. Its design is based on four different layers [6], as is depicted in figure 1. Its main features are the following:

- It is distributed: learning algorithms, as well as data sources, are transparently located in any machine through the Internet. There is a middleware that offers distribution services and holds communication between all software entities. This will depend upon the software technology used to implement these architecture. However, three prototypes are being developed. The first two are finished now, and are built using CORBA (Common Object Request Broker Architecture) and RMI (Remote Method Invocation) respectively. The third one is based on the MAS (Multi-Agent System) paradigm and is currently being developed, in the context of this work.
- It is totally configurable: i.e. it is capable of designing the learning task, in terms of the learning agents that will take part on it, and the learning experiments each one will perform.
- It is extensible: there are software templates, that can be used to integrate a new learning algorithm in the system, with a minimun programming effort.
- It is modular: in a sense that previously existing algorithms in the platform can be used (even if they are in another remote agent) to build a new (meta)learning algorithm, for example for boosting (see the section above).

In the following sections, we will describe each one of the layers appearing in figure 1.

2.1 The O.O. Layer

The design of the O.O. (Object Oriented) layer is based on the UML (Unified Modelling Language) notation. The use of METALA for implementing a software

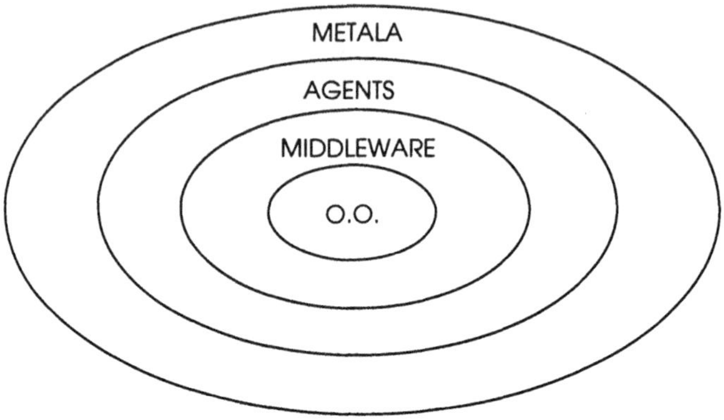

Fig. 1. The four levels of the METALA architecture.

system should consist only in the understanding and implementation of interfaces and interaction protocols included. The O.O. level covers the following design topics:

- Data sources: in METALA, we have defined all that is related with data. We give interfaces for generic data sources, different type of accesses to data (secuential, random, n-fold), etc. We also give a complete design of what a repository of sources should be.
- Learning algorithms: algorithms are seen in METALA as servicies. All inductive paradigms that have been taken into account in the architecture (i.e. neural network learning, decission tree and rules learning, inductive logic programming, and genetic algorithms based learning) are organized in a hierarchical fashion, and all of them must implement a concrete interface: the so called `LearningService` interface. Once all different algorithms can be managed in the same way, by means of that interface, they can be put together into a general repository. Designs for building such a repository are also given.
- Theories: The idea of contemplating learning algorithms and theories induced from data separately, is a central point of this architecture. That is why, at the same time it has a hierarchy for algorithms, it also has another hierarchy for models. Models are seen generically, in the context of this work, as pieces of knowledge that are obtained from the combination (*algorithm + parameters + data*) This pieces of knowledge can be shown to the user, in a friendly way, or they can be made permanently saved, as results of learning experiments. Of course, inferences can be made with the obtained models.
- Learning evolution: depending on the amount of data, and on the algorithm, an induction process can take several hours, or even days to finish. It is important, from both the perspective of the user and coopeating learning objects, to have the capability of following the evolution of a learning experiments. Different types of learning evolution have been defined and designed in METALA for the four learning paradigms afore mentioned.

- Learning experiments stuff: a learning experiment is a description of a learning session. It includes detailed description of data used, and how it has to be accessed, the learning algorithm to use for that data, the values for the corresponding configuration parameters of the algorithm, and so. All of this is specified in the architecture.

The O.O. layer is the base for the building of this kind of system. This layer is used then to build the following one, the Middleware layer.

2.2 The Middleware Layer

The architecture also pretends to follow a distributed computation mechanism (i.e. a middleware) for all techniques mentioned in the former section.

In this layer, all topics covered are related with distributed computation. And lessons have been learned from the use of CORBA and RMI for the implementation of two different prototypes. The two different middleware technologies show a very important difference, from the perspective of METALA, and that is that CORBA does not allow neither to pass objects by value as arguments in function calls nor to return objects by value as results from function calls. This is obviously caused by the mannner in which CORBA works. This technology integrates code from different high level programming languages. However, RMI does not (it only works with Java code) This issue is important in METALA as it influence the following issues:

- Trainning data access: in one hand, if we use technology like CORBA for developing a METALA based system, trainning data must be allways accessed remotely. This is good from a memory consumption perspective, in the side of the client as it does not have to have all trainning data in the machine it is located. However, if accesses to data are frecuent, or the chuck of data to induce is very big, thit will lead to a poor performance. By using RMI, this problem can be adressed more efficiently depending on the situation.
- Mobile code: it should possible in METALA to include learning algorithms that can migrate from one repository to another. This feature is very important in situations where we are talking about many megabytes of information to analyse. Or even when the data transport network we have is not trusted by us. In this situation, it is better for the algorithm to move to where the data is in. This can only be adressed with RMI. Nowadays, this feature is not included in METALA.
- Reuse of inductive algorithms implementation: METALA is an open architecture, in the sense that is should allow the integration of any algorithm pertainning to one of the considered learning paradigms. If we use RMI for that, only Java coded algorithm implementations can be integrated. Using CORBA, it is possible to integrate implementations done in virtually all most popular programming languages.

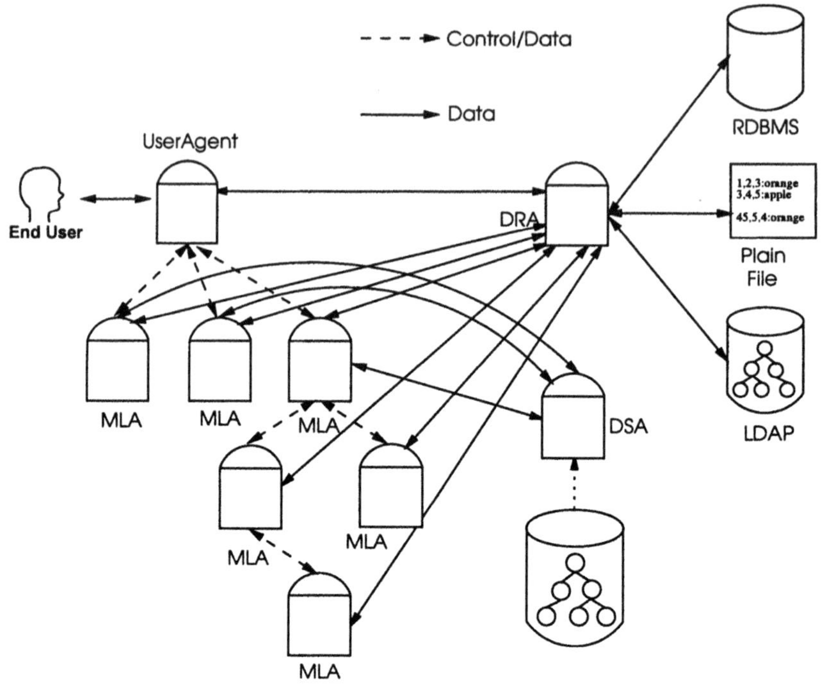

Fig. 2. Agents architecture for METALA.

2.3 The Agents Layer

The agents level architecture is dedicated to the study of high level problems related to (1) the management of tens of lerning servers and repositories, distributed across tens of computing machines, (2) the distribution of load in learning servers, to get the best possible response time for a learning request and (3) management of multiple users, working at the same time in METALA. Agents architecture can be seen in figure 2. In this figure, the following agents appear:

- DSA (*Directory Service Agent*): this agent is in charge of offering directory services to the system. Through this service, an agent can subscribe to the system. In this way, it is known for the rest of living agents. Once its execution is finished, it can unsubscribe itself from the directory. As well as directory related services, this agent acts as results repository (i.e. theories plus theories related values like generalization errors and such). This agent is the interface for the blackboard mechanism that support organization and coordination for METALA agents. It should only exists one DSA for the whole METALA system.
- DRA (*Data Repository Agent*): this agent manages access to learning data. It offers services defined by means of a concrete UML interface -`DataRepository`-. A DRA should exists in each machine in which interesting data can be found. This agent must be told about which sources it

has to manage, at launching time. Once all sources are validated by it, it can subscribe to the directory, along with details about all sources that can be accessed from it.
- MLA (*Machine Learning Agent*): offers learning services, and is located at a single machine (this detail is important when load balancing problem is addressed). This agent is the one that shows the highest level of autonomy and reflection. The MLA is the agent that demands more CPU power and primary memory availability. That is the main reason for incorporating a mechanism by means of which evaluate its own capacity at a given moment. Capacity must be understood in terms of new experiments that it is willing to afford. Once a MLA is launched, it subscribes to the system, through the DSA, along with its different learning services and features of that services. The DSA maintains a load estimation that reflects its load, and it can be used for load balancing. This is stored by the DSA, but updated by the corresponding MLA.
- UA (*User Agent*): this is, basically, a collection of graphical interfaces and functionalities. It interacts directly with the final user and acts in his behalf at the system. An UA will exists for each connected user. From this UA, all DRAs and MLAs that exist can be visualized and their features can be seen. This agent offers the user the possibility to define new experiments, edit already defined experiments, save them and executing then meanwhile it offers the evolution of all experiments launched, and still running, by the user. Once there are results available, the UA is capable of searching for the best model found (if a single learning technique was used) or for the best technique (if more than one technique was used in the same conditions).

The underlying organization and coordination mechanism is based on the blackboard paradigm [4]. A LDAP based directory [8] has been used to implement the directory infraestructure that the DSA needs, to serve all its services.

Design of a MLA

In the architecture, the most important entity is the MLA, briefly described in the section above. An

MLA can carry many learning algorithms with it (even only one). In this manner the architecture can observate hierarchies of specialized agents depending on the inductive paradigms they incorporate for learning. There are two differentiated software parts in a MLA. The first one is related with the middleware, and the second one is related with how the different algorithms are integrated and managed, in the same MLA. This division appears clearly differentiated in figure 3.

The following paragraph explains the main execution of a MLA. Once an experiment is received, it is put into the `Experiments Queue`. In this structure, experiments wait to be launched by the `Dispatcher`. When this module gets the next experiment in the queue, it (1) obtains the access to the data, (2) creates the learning object corresponding to the dessired learning algorithm, (3)

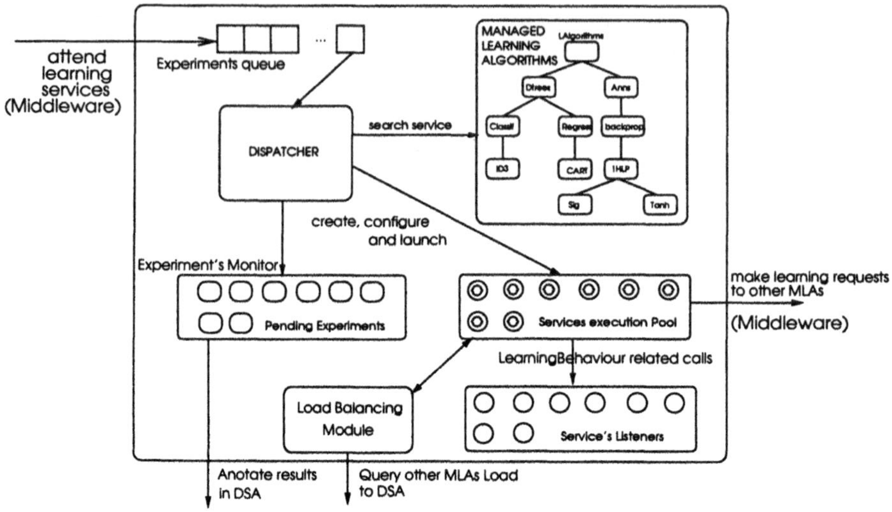

Fig. 3. Architecture for a MLA.

passes the access to the algorithm, (4) passes the corresponding listeners to it and creates a internal monitorization object for the experiment that is put into the `Experiments Monitor`, (5) sets the algorithm parameters, and finally (6) launches it into the `Services Execution Pool` in which the experiment starts learning concurrently.

In section 2.1 experiments evolution was firstly mentioned. When a learning request is done to a MLA, this request can not be blocking (i.e. it should finish inmediatly). This must be done asynchronously. That is why the MLA uses listeners mentioned in step (4) of the MLA's loop. Listeners are located in the client, and the object doing learning is in charge of making calls to the listener object giving information about the evolution of the experiment. In this manner, this information is received by the client that can follow all dessired experiments.

For some kind of experiments, it is possible that the MLA would need the help of other MLAs. Think, for example, on an hybridization learning problem (see section 1) in which different learning algorithms interact to arrange a theory from data. This procedural knowledge must be coded inside the corresponding meta-learning algorithm. It knows what learning services needs, but does not know who to ask for them. Then it is when the `Load Balancing Module` comes into play. The hybrid algorithm passes its request to that module, and it returns the most suitable agents for that (in terms of best load balance). After that, the algorithm can make the corresponding requests.

3 A Service for Massive Storage

Typical inductive meta-learning experiments do need to manage lots of data, coming from realized experiments. Besides, there are typical services, related with distributed computing, as a white and yellow pages service that also need mechanism to manage information about agents and services available in the system, where are and how to access them. The architecture uses a hierchical information model to represent all this data, from the conceptual point of view. Talking about engineering details, an LDAP (Lightweight Directory Access Protocol) type of protocol has been used in the development of the prototypes.

A directory based data repository is more suitable for a system as METALA, than a simple relational database management system. In a directory, entities are organized in a hierarchical way. A hierarchical organization is the natural manner in which entities are arranged in METALA.

4 On Performance in the Architecture

In the `Load balancing` module of the MLA (see figure 3) there has to be a concrete policy that rules the criteria by means of which an algorithm decides the next agent that will execute an experiment. When a set of experiments are ready to execute, the objetive is to finish them as soon as possible. To achive that requeriment, we have all MLAs at our disposition. However, each MLA will be situated on a different machine, and all these machines will show different computational power and working load. We have here a load balancing problem [9].

We have studied, and included, three different load balancing strategies in the architecture. Those are **a round-robin criteria, a statically weighted workload based strategy** and a **dinamically weighted workload based strategy**.

The round-robin based criteria is based on the assumption that all machines are of the same type and have the same working load. It works by obtainning, from the DSA, the list of available MLAs, and the next experiment is assigned to the next MLA in the list.

The second criteria assumes that hosts executing MLAs show different computational power, but fixed working load. It works assigning a weight to that server, that reflects, in some way, the computational power. This guides the experiments assignment. Let a MLA be labeled as h. Then its weight shall be w_h and the workload of h, W would be obtained with

$$W_h = w_h \times e_h,$$

being e_h the number of experiments currently performed by h. Clearly, the more computational power a host shows, the more smaller w_h it should have. A possible set of w_h values for the hosts we have used for experimentation appears in table 1, in the rightmost column.

This computational power model can be visualiced with the figure 4. It can be easily interpreted. For example, suposse that there are only two MLAs alive in the system, in a concrete moment. Let these MLAs be the ones with $w_6 = 0.7$ and $w_3 = 0.1$. The two of them are idle. Suposse that a new experiment comes. Notice that, in the x axis, at the point with $x = 1$, the workload for the MLA

Table 1. Hosts running the different MLAs for time experiments.

Order	Architecture	O.S.	RAM	w_h
1	SPARC II (Ultra 5)	Solaris 7	128MB	0.4
2	Intel Pentium III	Linux Red Hat 6.2	128MB	0.9
3	SPARC II (2Processors)	Solaris 7	256MB	0.1
4	SPARC II (Ultra 10)	Solaris 7	256MB	0.2
5	SPARC II (2Processors)	Solaris 7	256MB	0.1
6	SPARC I (Ultral)	Solaris 7	64MB	0.7
7	Intel Pentium III	Linux Mandrake 7.2	128MB	0.8

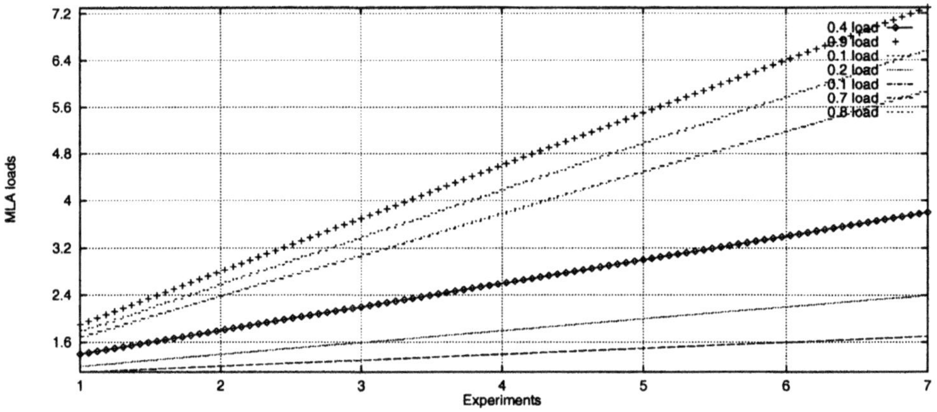

Fig. 4. Evolution of all loads, depending on the experiments each MLA is working on for the statically weighted strategy for load balancing.

labeled as 3 is smaller. Hence, the experiment would be assigned to it. Now, the workload for the agent labeled as 3 is the one meassured at the point where $x =$ 2 (as it is executing a new experiment). However, the following six experiments would be assigned to it, because in the whole x range [1; 7], the workload of the MLA labeled as 6 is never higher than the one of the MLA labeled as 3.

In the third criteria, all learning agents will launch in the system with the same initial w_h . Once a request of experiments is sent to the system, those are distributed, in a round-robin fashion (i.e. all w_h are the same). When all experiments are done, times for each one of the experiments are requested from the DSA and the w_h are adjusted accordingly to the differences between them. Formally, let S be defined as the set $S = \{e_1, e_2, \ldots, e_n\}$. S represents a complete learning session. The labels e_i are the experiments that must be executed in that session. Once S is defined by the user and launched into the user agent, and after all experiments are done, $t_i, 1 \leq i \leq n$ are obtained, where t_i is the elapsed time since the experiment i has been sended to the pertinent MLA to the time in which the MLA has finished with it.

Then, each weight w_h would be adjusted, for each experiment e_j by means of the expression

$$w'_h = w_h + w_h * \left(-\alpha + \frac{2\alpha(t_j - t_{min})}{t_{max} - t_{min}}\right)$$

where e_j is an experiment executed by the server h, t_{min} and t_{max} are the minimum and maximum elapsed time for all experiments in S, and α is a parameter that conditions the convergency speed for the adjustment.

In figure 5, concretly in plot (a) different elapsed times can be seen for different loads of experiments and load balancing strategies (both round robbin -RR- and statically weighted one -SW-). It can be noticed that the SW technique takes more profit from new added servers than RR.

In plot (b) of the same figure, the evolution of elapsed time is depicted as different learning sessions (of the same type) are executed, for the dinamically weighted (DW) workload based strategy. After about 12 sessions are executed, the system becomes almost stable at the minimun possible elapsed time. Parameter ff refers to the sensibility of maladjusted server time. Many experiments have been produced, that show the convenience of applying each one of the techniques depending on the system conditions. This has made possible a complete study of behaviour, in respect to resources consumption, of cooperating MLAs.

5 Conclusions and Future Works

METALA is a software architecture that offers basic recommendations to develop a software system, based on the MAS paradigm, to do meta-learning experiments. It is extensible, as it allows an easy integration of new learning algorithms, or new implementations of already integrated ones, into the system. METALA specifications include machine learning experiments monitorization and analysis of results, and it is defined as a multi-user environment. METALA provides with different load balancing strategies that allows an optimal distribution of experiments, to exploit all servers available.

However, the load balancing issue is not enough studied yet. Experiments must be done, simulating controlled workloads for each type of server (normal, high end and desktop servers). Cost should be taken into account as well, in situations in which CPU usage of concrete servers is more expensive than in the rest.

Scalability issues, related with managing hundreds or thousands of agents have not yet addressed. This problem should be addressed taking benefit of the distributed nature of a directory, definning federations of agents and communities of users working in these federations.

METALA agents are not so much *intelligent* yet. Currently, we are now developing a new revision of the architecture, by means of a prototype that will be developed on the basis of a MAS developing platform.

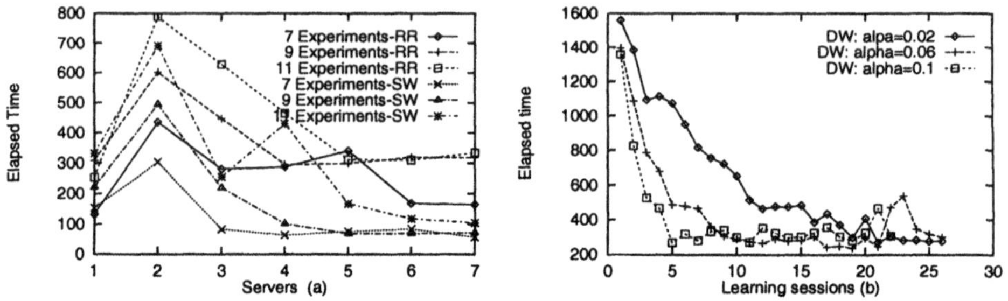

Fig. 5. Evolution of elapsed time for the execution of the whole request of experiments, depending on servers available (left, corresponding to RR and SW strategies) and learning sessions performed (right, corresponding to DW strategy).

References

1. Hilan Bensusan. Odd bites into bananas don't make you blind - learning about simplicity and attribute addition. In *Upgrading Lerning to the Meta-level: Model Selection and Data Transformation. ECML-98 Workshop Summary.*, pages 30–42.
2. Pavel B. Brazdil and C. Soares. A comparison of ranking methods for classification algorithm selection. In *ECML-99.*
3. Leo Breiman. Bagging predictors. Technical report, Department of Statistics. University of California., Berkeley, California 94720., September 1994.
4. Norman Carver and Victor Lesser. The evolution of blackboard control architectures. Technical Report 92-71, CMPSCI, October 1992.
5. Yoav Freund and Robert E. Schapire. A short introduction to boosting. *Journal of Japanese Society for Artificial Intelligence*, 14(5):771–780, September 1999.
6. David Garlan and Mary Shaw. An introduction to software architecture. In V. Ambriola and G. Tortora, editors, *Advances in Software Engineering and Knowledge Engineering*, pages 1–39, Singapore, 1993. World Scientific Publishing Company.
7. M. Hilario. An overview of strategies for neurosymbolic integration, 1995.
8. W. YeongT. Howes S. Kille. Lightweight directory access protocol. request for comments: 1777. Technical report, Performance Systems International, University of Michigan and ISODE Consortium, March 1995.
9. Luis Paulo Peixoto. Load distribution: a survey. Technical Report UM/DI/TR/96/03, Departamento de Inform'atica. Escola de Engenharia. Universidade do Minho, 1996.
10. Salvatore Stolfo, Andreas L. Prodromidis, Shelley Tselepis, Wenke Lee, and Dave W. Fan. JAM:Java Agents for Meta-Learning over Distributed Databases. In David Heckerman, Heikki Mannila, Daryl Pregibon, and Ramasamy Uthurusamy, editors, *The Third International Conference on Knowledge Discovery & Data Mining.* AAAI Press, August 1997.
11. A. Suyama, N. Negishi, and T. Yamaguchi. Camlet: A platform for automatic composition of inductive applications using ontologies. In *Recent Advances in Metalearning and Future Work. ICML-99 Workshop Summary.*

On-Line Signature Verification Using a Computational Intelligence Approach

W. Sardha Wijesoma, Mingming Ma, and K.W. Yue

School of Electrical and Electronics Engineering,
Nanyang Technological University,
Singapore 639798
eswwijesoma@ntu.edu.sg

Abstract. For signature verification, there can be a large number of features available in a signature and not all these features are of use as some even could be unfavorable for verification of particular signatures. Also, the large number of features imposes difficulty in comparing instances. This may be worsened if observed peculiarities between features could not be drawn to benefit the comparison process. Hence, to determine a personalized subset of features and allowing for the inclusion of heuristics in verification are prime factors to consider in the design of the verification model. In this novel approach, genetic algorithm determines the optimal personalized features for each subject and fuzzy logic transforms the heuristics conceived by a human verifier to rules that govern the behavior of the automated verification system. Experimental results are presented to demonstrate the effectiveness of this scheme.

1 Introduction

Need for automated signature verification is fast becoming an important element in the fields of finance, commerce and security [1]. The proliferation of sophisticated handheld communication devices like personal digital assistance (PDAs) and digital cellular phones has supported the development of these modern applications and thus brought forth the possibility of having the task of signature verification automated.

On-line signature verification is defined as the process that captures the actual production of the signature, its pen trajectory or dynamics, and then verifies its authenticity automatically. Traditional approaches to signature verification can be broadly classified into either a functional approach or a parametric approach. In the functional or time signal comparison approaches, signals captured from an instrumented pen or other devices are considered as mathematical functions whose values directly or indirectly constitute the feature set. In the parametric approach, specific parameters are computed from these time signals or (x,y) coordinates sequences to extract features and these features are used to construct reference class models of the signatures [2]. In our approach to signature verification, the parametric approach was adopted because of its algorithmic simplicity and less memory storage requirements.

There are two main aspects that must be tackled in the parametric approach to signature verification. One, to select the relevant features to constitute the final feature set for discrimination and two, to determine an effective way to compare a signature with its class model.

B. Reusch (Ed.): Fuzzy Days 2001, LNCS 2206, pp. 699–711, 2001.

The selection of features is simply done by measuring statistical quantities from the alleged class model. Features with small deviations from its average value can be reckoned as prime features for verification [3]. However, this is not reliable as they may not necessarily be good candidate features. On the contrary it can be even detrimental to verification as it may be a feature that can be imitated with ease. The existence of intra-personal variability complicates the matter even further. Thus, a system is considered to be feasible only if the system is insensitive to intra-personal variability, but sensitive to inter-personal variability [4], [5]. Earlier work has proposed a personalized feature selection algorithm based on a kind of discrimination power (DP) function of a feature [6], [5]. The methodology in [6] considered not only the consistency of a feature but also the discrimination power of inter-personal variations of the feature. To formulate this DP function, the mean and the deviation of both a particular signer's signatures and the general persons' handwritings are used. Experimental results yielded an average error rate of 4.82% for our own signature database comprising of 1230 genuine signatures of 41 persons and 410 skilled forgeries from 6 forgers. Major shortcomings of such methodologies are that the personalized feature vector's dimension is predetermined and fixed, and further a feature is included based on a figure of merit (discriminating power) calculated in isolation.

Conventionally, a class of signature instances that is characterized by a fixed number of features may be represented in a multidimensional representation space. Statistical analysis generates the probability distribution of the instance vectors around its class centroid. A distance threshold is used to discriminate a genuine instance from a forgery. An instance is genuine if its distance measure falls within the threshold. This distance may be Euclidean or other distance measures [7], [8].

In the proposed approach to signature verification, the problem of feature selection is formulated as an optimization problem. The object of optimization is to determine the type and the number of features constituting the feature subset for the signer that will maximize a suitably chosen discriminating function. Genetic algorithms (GA) can provide an effective solution for this kind of problem [9]. For comparison, fuzzy logic creates the possibility of including heuristic knowledge from the human expert in the verification of a signature. For example, a possible heuristic is to seek for observed consistencies in both the shape and the pen-dynamics when analyzing a signature instance. The reason being that it must be more difficult for forgers to imitate both shape and dynamics of the original signature simultaneously as compared to imitating either of the two aspects in isolation. Although similar in the way the logic 'AND' functions, fuzzy logic does this in a non-linear way. Taking a statistical view, this can be seen as a non-linear alteration in the distribution of the instances in the multidimensional space; and this sophistication is not possible for the straightforward distance-measure approaches mentioned. As a whole, the synergies between genetic algorithm and fuzzy logic make up the computational intelligence in verification system.

2 Data Acquisition and Preprocessing

On-line signatures may be captured using a variety of input devices. When one signs on a tablet, three types of information are captured: (i) the absolute coordinates $x(n)$,

$y(n)$, $n=1,\ldots,n_k$, (ii) the end point coordinate corresponding to each stroke, (iii) the timing information tagged to each pair of (x, y) coordinates.

The raw signature data hence captured is initially preprocessed so as to compensate for rotation and translation variances between different signature instances of a signer. The translation simply shifts the entire signature curve so that the centroid of the signature curve coincides with some reference point such as the origin of the coordinate frame [10]. The rotation is based on the idea of orienting the signature curve so that some reference line of the signature is parallel with the x-axis of the coordinate frame. The reference line chosen is the line connecting the starting point of the signature curve with the centroid of the signature.

3 Feature Extraction

The repertoire of features constitutes a total of 10 shape-related features and 14 dynamics-related features as shown in Table I in the appendix. The No.1 to No.10 are shape-related features, and the No.11 to No.24 are dynamics-related features. They are selected from a larger set of candidate features based on their generally good potential for signature verification.

4 Analysis of a Test Signature for Authenticity

Figure 1 shows the overall verification process of the proposed system. The process is divided into three stages.

Stage 1: The two sets of personalized features of a test signature are calculated and each feature is compared with its reference by a single-input, single-output and single rule system. The input is the normalized difference in feature value between a test signature and its reference and is given by,

$$Dist(i) = \frac{f_i - m_i'}{\sigma_i'}. \tag{1}$$

f_i is the value of the ith feature, m_i' and σ_i' are the mean and standard deviation of this feature across the genuine training samples. The output of the system is a measure of the degree of similarity of the test signature's specific feature value against the reference expressed as a percentage. This output is defined as the degree of match (DOM) of the feature.

Stage 2: Since, not all the personalized features contribute equally in discriminating a genuine signature from a forgery; every personalized dynamic and shape feature is weighted accordingly to reflect its relative importance. In stage 2 the DOMs of the individual dynamic and shape features are appropriately weighted and combined to separately yield two quantities. Each of these two outputs, defined as the total degree of match (TDOM), separately quantifies the sum total weighted feature vector match of a test signature to the reference shape or dynamics personalized feature vectors.

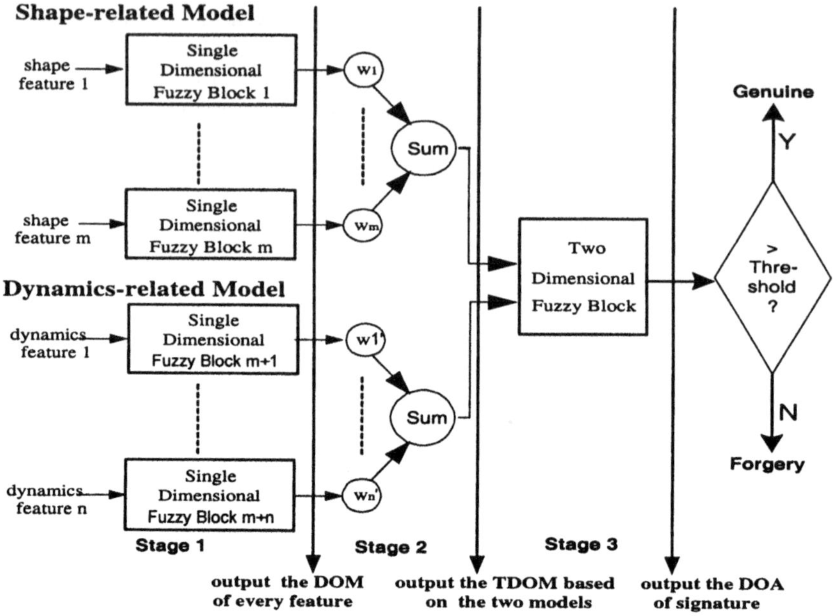

Fig. 1. Block diagram of signature verification process

In determining the weighting factors to reflect the relative importance of features, two aspects of a feature are considered. These two aspects are the ease with which the feature can be emulated and the other is the consistency or stability of the feature. As for ease of reproduction consider the velocity feature. In attempting to forge the signature of another we may feel that to forge a signature written faster is harder than a signature written slowly. By analyzing signatures written by a large number of people we may obtain an estimate of the general distribution of velocity for general handwriting for the group. If the location of the velocity feature for a specific signer obtained by sampling a number of signature instances is relatively higher than the velocity distribution obtained for the population of people (general persons' velocity distribution) we postulate that it is relatively difficult to emulate this velocity feature. A feature such as velocity is known as a type II feature, as the degree of difficulty emulating this feature is dependant on how far the individual's feature distribution lies to the right of the "general person's" feature distribution. For type I features what matter is the relative distance between the two feature distributions. The 10 shape features we extract are all of type I, and out of the 14 dynamics features, numbers 1 to 3, and 5 to 10 are type I features, 4, and 11 to 14 are type II.

The other aspect of a feature is its stability, which is a measure of the consistency of the feature values obtained over a sample of genuine signature instances of a signer. In particular, the smaller the standard deviation of the specific feature the better the feature is for verification. Now a type II feature's relative importance (RIM) is determined to reflect the above two aspects by:

$$RIM_i = \frac{(m_i' - m_i)}{\sqrt{\sigma_i'^2 + \sigma_i^2}} \; . \qquad (2)$$

Subscript i refers to the feature number and

σ_i: *feature standard deviation of general handwritings,*
σ_i': *feature standard deviation of a signer's signatures,*
m_i: *feature mean of general handwritings, and*
m_i': *feature mean of a signer's signatures.*

For a type I feature RIM is given by the absolute value of equation (2). The variance between RIMs of different features is very large and not suitable for direct inclusion as a weighting factor. A sigmoid function is used to map and hence normalize the RIM to a value in the range from 0 to 1, to obtain a measure of a feature's normalized relative importance (NRIM) as:

$$NRIM_i = 1/1 + \exp(-RIM_i) \; . \qquad (3)$$

Finally, the weight of the ith feature is obtained by:

$$W(i) = NRIM_i / \sum_{i=1}^{k} NRIM_i \; . \qquad (4)$$

k is the dimension of the optimal personalized shape or dynamic feature vector as the case may be.

Stage 3: TDOM of shape aspect and the TDOM of dynamics aspect of the test signature are analyzed by a two-input, single-output and multiple rules standard fuzzy system to give an overall measure of authenticity for the test signature. This measure is defined as the degree of authenticity (DOA) of the test signature and is output as a percentage.

5 The Computational Intelligence in the Verification System

Given an entire set of features, a verification system must determine how the authenticity of a signature may be deduced. The important considerations are, how to select the features that enhances the discriminative power of a signer; and with the reduced set of features, how can a legitimate analysis be made. In our parametric approach to signature verification, the 'intelligence' in the verification model is made up of fuzzy logic and genetic algorithm.

Fuzzy logic is a powerful tool for solving complex problems because of its ability to determine outputs for a given set of inputs without using conventional mathematical models. As Jain puts it [11], the basic motivation behind fuzzy set theory is the fact that the conventional methods had become so complex that researchers trying to apply them had to make a choice between a complex system and a complex tool. In our application, fuzzy logic gives the flexibility to include heuristics concocted by the human expert to affect the outcome from varying combination of input conditions. The observed relationships that exist between features are transformed into fuzzy rules. These relationships may bring about a greater discriminative power as compared to their isolated evaluation. In selecting discriminative features, genetic algorithm can determine the type and the number of features to constitute the feature set for the signer that will maximize a suitably chosen discriminating function. In the next two subsections, the process and the rationalizations in the use of each of these computational methods is exemplified.

5.1 Fuzzy Inference Systems

In fig 1., the fuzzy systems have been implemented primarily in stages 1 and 3. In stage 1, the fuzzy system is a single input, single output and single rule Takagi-Sugeno model shown in fig. 2. It effectively transforms a distance measure into an isolated percentage degree of match for different features.

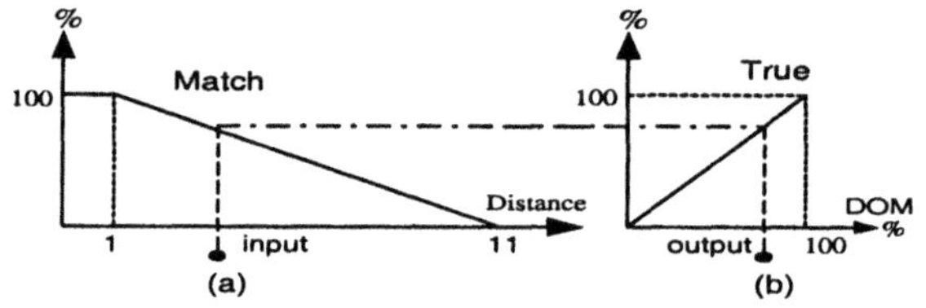

Fig. 2. Single input, single output, and single rule fuzzy system.

In stage 3, the two dimensional (2D) fuzzy system takes in the 'TDOM of shape' and 'TDOM of dynamics' as its inputs. The output linguistic variable of the fuzzy block is the 'DOA of signature'. The shape TDOM and dynamics TDOM of the previous stage (stage 2) are the inputs to this system.

The number of linguistic terms partitioning the input and output spaces is limited to the range three to nine. Too small a number would be insufficient to formulate sufficient rules, and a larger number computationally cumbersome [12]. The input linguistic variables 'TDOM of shape', and 'TDOM of dynamics' and the output linguistic variable 'DOA of signature' comprises of 5, 6 and 7 linguistic values respectively and their associated membership functions are as shown in Figure 3.

It is reasonable to assume that a signature's shape or image can be forged more easily than the pen dynamics of it. This is because of the ready accessibility of a person's signature image. To emulate or forge pen dynamics the forger may have to witness the act of signing and to some extent be aware of the dynamic features extracted of the signature for verification. This domain knowledge is embedded in the choice of the scales chosen for the respective linguistic variables. In particular, it may be noted that the 'False' linguistic term maps a TDOM of 70% or lower for the shape TDOM input and 60% or lower for the dynamics TDOM input. In other words the shape aspect of a signature is dealt with more strictly than the dynamics.

The fuzzy rule base used for inferring the overall DOA of a test signature based on the shape and dynamic features match is given in Figure 4. The assignment of the fuzzy rules embodies the heuristic that the lesser the degree of match of shape and/or dynamic aspect of a test signature the lesser the authenticity of the signature. Depending on the rules assigned, the final result could be biased or weighted more towards either the shape or the dynamics aspect. But it is suggested that it is not prudent to accentuate either one over the other too much, as it is more difficult for forgers to imitate both the shape and dynamics of the genuine signature simultaneously than just imitate one aspect. If one aspect is accentuated too much, forgers could deliberately imitate this aspect making it easier for them. Hence the rules are crafted not to accentuate either of the two aspects over the other by a large margin. Effective non-linear weighting of the two feature aspects can be obtained through careful calibration of the fuzzy rules.

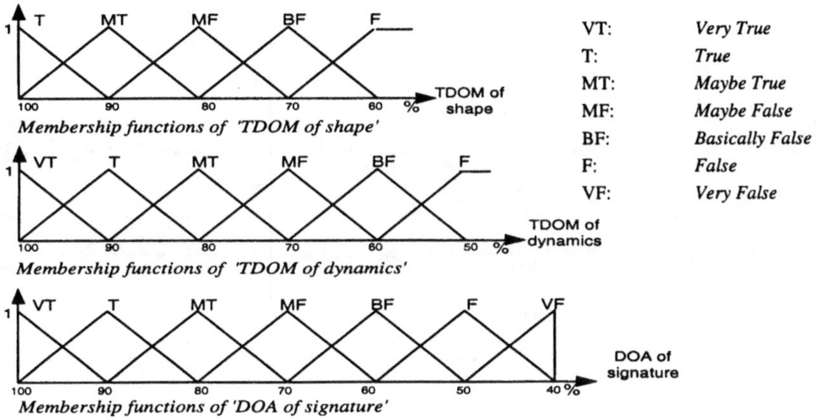

Fig. 3. Input/output membership functions of fuzzy 2D fuzzy system

Fuzzy inference is based on Larson's method, as illustrated in Figure 5. Singleton fuzzification and mean of maximum (MOM) defuzzification is used. For the scenario illustrated in Figure 5 the defuzzified percentage DOA of test signature is:

$$DOA\ \ of\ \ signature\,(\%) = \frac{0.2\times 90 + 0.2\times 80 + 0.7\times 80 + 0.3\times 70}{0.2 + 0.2 + 0.3 + 0.7}\times 100\% \ . \qquad \textbf{(5)}$$

The percentage DOA of the test signature hence obtained is classified as true or false by a simple threshold operation.

TDOM of dynamics		TDOM of shape				
		T	***MT***	***MF***	***BF***	***F***
	VT	VT	T	MT	BF	F
	T	T	MT	MF	BF	F
	MT	MT	MF	BF	F	VF
	MF	MF	BF	F	VF	VF
	BF	BF	F	VF	VF	VF
	F	F	VF	VF	VF	VF

Fig. 4. Fuzzy rule base for stage 3

With the fuzzy architecture in Figure 1 arranged in a hierarchical manner, the need for establishing a large rule-base is diminished. This in turn minimizes the risk of having unpredictable fuzzy outcome. The architecture also improves the robustness of the system as a whole. The reason being that if a different set of rules are to be established for different feature sets, the number of rule sets increases sharply with the number of features; and the requirements to store them can become impractical. In essence, the way in which the fuzzy architecture is constructed gives a valid rule set that generalizes well for different feature subsets.

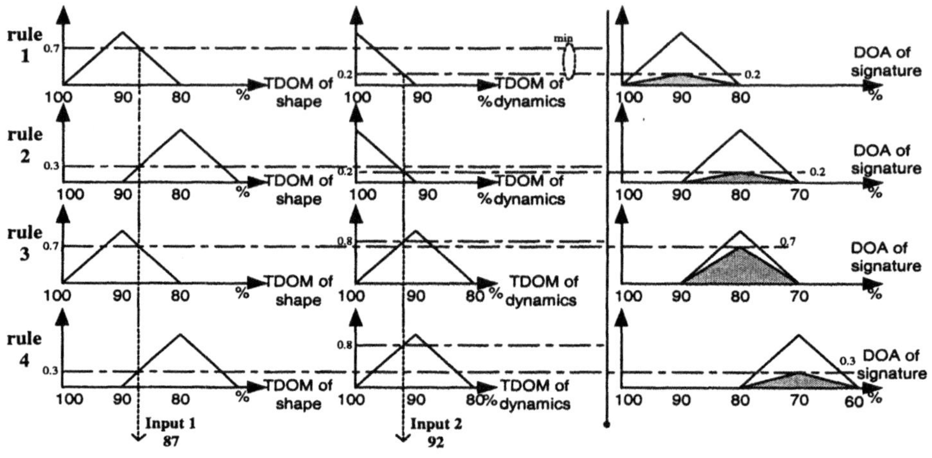

Fig. 5. An example of fuzzy inference process for determining the degree of authenticity of a test signature

5.2 Genetic Algorithms for Selection of Optimal Personalized Features

A genetic algorithmic process determines the optimal personalized shape and dynamic feature vectors of a signer. The personalized shape and dynamic reference parameter models are formulated by calculating the average of the feature values of the personalized feature types of both shape and dynamics as determined by the genetic algorithm for genuine instances of a signature.

1	2	3	4	5	6	7	8	9	10
1	1	1	0	0	1	1	0	0	0

Fig. 6. bit coding of shape features onto a chromosome

5.2.1 Genetic Coding of Feature Vectors

GA is used to select both the optimal personalized shape and dynamic feature subsets. In each case the feature vectors are binary encoded. Figure 6 shows the structure of the chromosome representing the feature vector catering to a maximum of 10 shape features. A binary '1' or a '0' in a cell indicates selection or de-selection of the feature corresponding to the cell position. For example, the features defined by the shape chromosome in Figure 6 are the shape feature numbers 1, 2, 3, 6, and 7. A similar structure and encoding is used for the dynamics related feature chromosome, however with a size of 14. This provides for a maximum of 14 dynamic features.

1	2	3	4	5	6	7	8	9	10
1001	0010	0000	0101	0100	1010	1011	0001	0000	0001

Fig. 7. 4-bit coding of shape features onto a chromosome

In the 1-bit coding of features, GA determines the optimal number of features for final discrimination. However we hypothesize that the exclusion of a feature entirely might cause the system to be vulnerable to illicit penetration by expert forgers due to the possibility of insufficient features. Thus, as an alternative, the 1-bit coding scheme may be replaced by a 4-bits coding scheme for shape features shown in Figure 7. These 4-bits numbers then become the weights that measure the relative importance of each feature. A weight of '1111' would correspond to the maximum weight attainable for any feature while '0000' corresponds to the minimum weight. Again, the dynamic features will have the same chromosome with the size of 14.

5.2.2 Genetic Algorithm

After the initial chromosome population (shape or dynamics) is randomly generated, genetic operations are applied to the current generation to evolve the next generation of chromosome as follows:

Selection: Upon evaluation of fitness of each chromosome tournament selection is performed. A number (usually called the tournament size, Tour) of individuals are chosen randomly from the population and the individual with highest fitness from this group is selected and reproduced as parent. This selection process is repeated N (population size) times. Some chromosomes with lower fitness values would be deleted by this operation.

Crossover: Single-point crossover is performed with a given crossover probability on the parent population hence selected to produce offspring. A crossover point is selected randomly and the genes are exchanged between the parents about this point.

Mutation: In order to introduce variation to the population mutation is introduced as the final step of reproduction. This involves the flipping of one or more binary bits (gene states) with a low probability.

Reinsertion: The chromosome or individual with the best fitness value is reinserted in the next generation. This process known as elitist selection guarantees that the best feature vector survives, ensuring that the population never loose ground on fitness.

These genetic operations are repeated until a stop condition is met. Then the features defined by the best chromosome of the last generation are the ones belonging to the optimal personalized feature vector, dynamic or shape as the case may be. It is to be noted that the dynamic and shape feature vectors are determined separately.

5.2.3 Fitness Function and Evaluation

To evaluate the goodness of a chromosome, verification is carried out on n_1 samples of genuine and n_2 samples of forged signatures of a person using the set of features as defined by the chromosome. Now a chromosome's fitness is calculated using the following fitness function:

$$F = (1 - FR) \times \frac{1}{n_1} \sum_{i=1}^{n_1} D1_i + (1 - FA) \times \frac{1}{n_2} \sum_{j=1}^{n_2} (100 - D2_j) \ . \qquad (6)$$

$D1_i$, is the TDOM for the ith genuine sample, and $D2_j$ the TDOM for the jth forgery sample. *FR* is the false rejection rate and *FA* the false acceptance rate obtained for the total signature samples (n_1+n_2) based on the shape and dynamic feature vectors alone.

It may be noted that the fitness value of a chromosome is based on how well it contributes to the successful verification on two fronts. One is magnitude of the TDOMs the chromosome gives rise to for both genuine and forgery samples. It is conjectured that larger the TDOM that results for genuine samples and smaller the TDOM for forgery samples, better the chromosome is. The other is that, obviously, smaller the false rejection and acceptance rates (type I and II errors) that result, better the chromosome is. Hence the fitness evaluation function is defined by adding the average of the TDOM for genuine samples to the average of (100-TDOM) for forgery samples, each weighted by acceptance rate and rejection rate respectively.

In the actual implementation a total of 10 genuine samples of each signature are used for fitness evaluation. In view of the fact that skillful-forged signature samples are not easily obtainable as compared to genuine ones in practice it is suggested that skillful-forged signatures should not be used for fitness evaluation. In the implementation random forgeries or other signers' signature samples (80 in this case) are used as forgeries during fitness evaluation of a chromosome.

6 Signature Database and Results

6.1 Database

The database consists of 1230 genuine signatures from 41 persons and 410 forgeries from six forgers. For each signer, there are 30 genuine samples and 10 forgery samples generated by two different forgers. In collecting genuine signature samples the signer was allowed to get accustomed to signing on the tablet using the stylus before providing 30 genuine samples over three separate sessions. The sessions were spaced a few days apart and each time, 10 samples were collected. This approach of collection prevents the signer from being exhausted from continual signature production and at the same time takes into consideration the variance in signature production over time. At the end of each signature sample production, each signer was allowed to delete and redo the sample, if in the opinion of the signer the signature was not "OK".

As for forgeries only skilled forgeries were taken into account. Each forger was shown all copies of the genuine signature samples of the 41 signers and allowed to forge 10 to 20 of these persons' signatures they thought they could forge. After permitting the forgers to practice as long as necessary, forgery samples were collected for each signature. Cash rewards were used to motivate the forgers. In the resulting database of signatures, the first 10 genuine samples of each signature were used to determine the optimal personalized shape and dynamic feature sets and the other remaining 20 genuine samples plus the 10 forgeries were used for testing. At any stage there was no deletion of "problem signatures".

6.2 Experimental Results

Due to the unavailability of a common published signature database a direct comparison of various signature verification algorithms reported in the literature is of questionable value [2]. Further results reported are produced under disparate conditions [4]. Here we compare the proposed optimal personalized features selection algorithm using GA to a personalized features selection algorithm based on a kind of DP function proposed by the authors [6]. The same database and the same modeling and testing parts of it are used for the comparison of the two approaches.

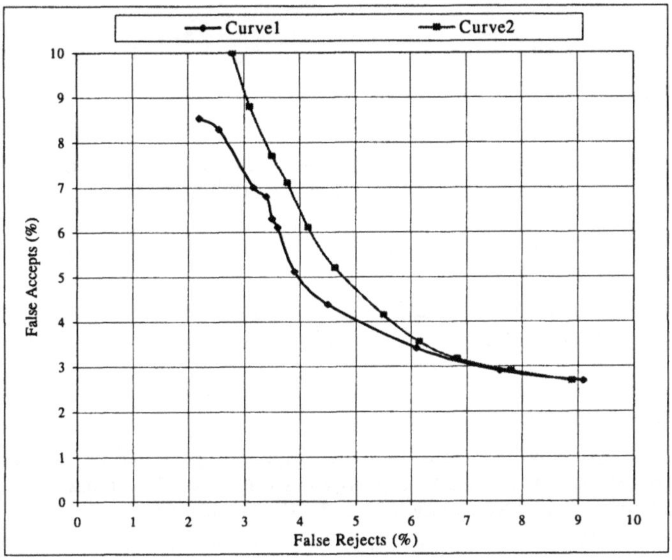

Fig. 8. Error trade-off curves

The parameters chosen for the GA algorithm are, tournament size of 3, crossover probability of 0.95, mutation probability of 0.09, and a stop condition of 50 generations or a maximum fitness of 195. The experimental results on these two systems are shown in Figure 8. Curve 1 in Figure 8 is the best error trade-off curve between FR (False Rejects) rate and FA (False Accepts) rate for different thresholds for the proposed verification system. The lowest average error-rate is 4.45%. Curve 2 in Figure 8 is the best error trade-off curve of the system based on the personalized features selection algorithm using a kind of DP function [6]. The lowest average error-rate of this system is 4.82%. It can also be seen that for any given threshold the proposed system outperforms the other.

7 Conclusion

The hierarchical architecture that encompasses the computational logic of genetic algorithms and fuzzy logic is an effective parametric approach to signature verification. Genetic algorithms provide for the selection of a personalized feature vector based on the discriminating ability of the feature vector as a whole rather than the individual features. Use of fuzzy logic for determination of the authenticity of a signature as opposed to distance measurements enables the inclusion of heuristics in signature verification. Together, they personify the parameters of the verification model to yield good results with moderate complexity.

In the future, it is envisaged to further improve on the performance by segmenting the signature into optimal segments and then determining the optimal feature subsets of these curve segments.

References

1. Plamondon, R., and Lorette, G.: Automatic signature verification and writer identification - The state of the art, *Pattern Recognition*, Vol. 22, (1989) 107-131
2. Plamondon, R.: The design of an on-line signature verification system: from theory to practice, *Progress in Automatic Signature Verification,* World Scientific Publishing Co. Pte. Ltd. (1994) 795-811
3. Nelson, W., Turin, W., and Hastie, T.: Statistical methods for on-line signature verification, Progress in Automatic Signature Verification, World Scientific Publishing Co. Pte Ltd., (1994) 749-769.
4. Nalwa, V. S.: Automatic on-line signature verification, Proceedings of the IEEE, Vol. 85, (February 1997) 213-239
5. Seong, H. K., Myoung, S. P., and Jaihie, K.: Applying personalized weights to a feature set for on-line signature verification, Proceedings of the 3rd International Conference on Document Analysis and Recognition, Vol. 2, (1995) 882-885.
6. Ma Mingming, Wijesoma, W.S., and Eric Sung,: An automatic on-line signature verification system based on three models, Proceedings of the 2000 Canadian Conference on Electrical and Computer Engineering, (2000) 890-894
7. Kashi, R. S., Turin, W., and Nelson, W. L.: On-line handwritten signature verification using stroke direction coding, Optical Eng. Vol. 35, No. 9, (1996) 2,526-2,533
8. Matsuura, T., and Yu, T.S.: On-line signature verification by IIR system, Proc. fifth Int'l Workshop Frontiers in Handwritting Recognition, Colchester, England, (Sept 1996) 413-416
9. Goldberg, D.: Genetic Algorithms in Search, Optimization, and Machine Learning. Addison-Wesley, Reading, MA (1989)
10. Wilfong, G., Sinden, F., and Ruedisueli, L.: On-line recognition of handwritten symbols, IEEE Trans. Pattern Analysis and Machine Intelligence, Vol.18, (1996) 935-940
11. Jain R.: Fuzzyism and real world problems, in Fuzzy Sets: Theory and Applications to Policy Analysis and Information Systems, Wang, P. P, and Chang, S., K., Eds. New York: Plenum (1980)
12. Dou Zhenzhong: Fuzzy Logic Control Technique and Application., BUAA Publishing House, (1997)

Appendix

Table 1. Shape and dynamics related feature set

No.	Feature Name	Method of Estimation
1	Height-to-width ratio	$Hw = Height / Width$
2	Length-to-width ratio	$Lw = L / Width$
3	Horizontal mean-min-difference	$Xcn = Xm - min(x(k))$
4	Vertical mean-min-difference	$Ycn = Ym - min(y(k))$
5	Left-to-right side ratio	$LR = (Xa-min(x(k)))/(max(x(k))-Xa)$
6	Upper-to-lower side ratio	$UL = (max(y(k))-Ya)/(Ya-min(y(k)))$
7 ~10	Direction histogram	$S_l = card\{\theta_k:(l-1)\pi/2<\theta_k\leq l\pi/2\}/(K-1)$, $k=2,\ldots,K$, $l=1,\ldots,4$
11	Total signature time	$T = t_k - t_1$
12	Pen-down time	$Td = \sum_{i=1}^{nums}(t_{end_i} - t_{start_i})$
13	Pen-down time ratio	$Tdr = Td / T$
14	RMS speed	$V = \left(\frac{1}{K-nums}\sum_{k=1}^{K-nums} v(k)^2\right)^{1/2}$
15	RMS-min-difference of speed	$V_{m\text{-}min} = V - min(v(k))$
16	Max-RMS-difference of speed	$V_{max\text{-}m} = max(v(k))-V$
17	Numr of positive x velocity	$NV_{xp} = card\{v_x(k) : v_x(k) > 0\}$
18	Num of negative x velocity	$NV_{xn} = card\{v_x(k) : v_x(k) < 0\}$
19	Num of positive y velocity	$NV_{yp} = card\{v_y(k) : v_y(k) > 0\}$
20	Num of negative y velocity	$NV_{yn} = card\{v_y(k) : v_y(k) < 0\}$
21	RMS centripetal acceleration	$A_c = \left(\frac{1}{K-2\times nums}\sum_{k=1}^{K-2\times nums} a_c(k)^2\right)^{1/2}$
22	RMS tangential acceleration	$A_t = \left(\frac{1}{K-2\times nums}\sum_{k=1}^{K-2\times nums} a_t(k)^2\right)^{1/2}$
23	RMS acceleration	$A = \left(\frac{1}{K-2\times nums}\sum_{k=1}^{K-2\times nums} a(k)^2\right)^{1/2}$
24	Maximum acceleration	$Max_a = max(A(k))$

K is the total number of sample points of a signature instance; nums is the number of strokes; Xm, and Ym are the mean values of the x, y data respectively; (Xa,Ya) is the centroid of the signature curve; L is the curve-length of signature; card{A} denotes the cardinality (i.e. the number of elements) of the set A.

Autonomous Driving through Intelligent Image Processing and Machine Learning

Michael Krödel and Klaus-Dieter Kuhnert

Universität – GH Siegen, Institute for Real-Time-Systems,
Hölderlinstrasse 3, D-57068 Siegen/Germany

Abstract. This abstract describes the current research in the area of autonomously driving of a vehicle along different road courses [1]. The focus of this paper are two main aspects: firstly, parameters of the environment are being extracted from a video image coming from one single camera which is installed in or in front of the vehicle which is to drive along the road course; secondly, the incoming images from the camera need to be processed by a computer system that way, that not only Steering Commands for the vehicle are being generated (for accelerator / brake as well as the steering wheel) but the appropriateness of those Steering Commands is being constantly weighed and continuously improved over time. Consequently, the current work focuses on a system which is able to learn and to develop completely on its own the ability to steer different vehicles in different environments and combines research in the areas of Intelligent Image Processing, Machine Learning and Robotics.

1 Introduction

The current research as described above is split into two subsystems: Intelligent Image Processing and Machine Learning. Both subsystems are being described in detail further below, followed by the presentation of the experimental results so far.

2 Intelligent Image Processing

The first main aspect, as mentioned above, is the extraction of parameters describing the environment which are being derived purely out of each Single Image of a single installed camera. Consequently, no ultrasonic sensors, no information from a speedometer or a compass system, no stereo cameras or no information from a GPS system are being required as utilised in projects like [2].

The relevant information to be extracted out of any Single Image are the Road Mark Edges of the currently visible part of the road course based on the point of view from the vehicle. Several approaches have been proposed for finding the roadway in video-images. We have chosen an edge based approach because the task-relevant information is very concentrated in this image region; i.e. it can be extracted very rapidly. Using purely standard techniques like horizontal difference operators with

B. Reusch (Ed.): Fuzzy Days 2001, LNCS 2206, pp. 712–718, 2001.

threshold would result in a Binary Image representing the shape of any object in the image; not only road marks but also objects next to the road like trees, bushes, etc. Additionally, any interconnection of the resulting pixels in a Binary Image to meaningful (i.e. task-relevant) objects leads to problems and the description of the current scene becomes typically very distorted. Therefore, we present a newly developed technique which takes into account the statistical properties of the object which has to be detected.

In our case the method is tuned to describe the properties of road borders. The current system determines, whether any contrast differences in the incoming image may be representing Road Mark Edges and builds up information on where, within the Single Image, Road Mark Edges may be found and in which direction, in dependency of the position within the Single Image, the Road Mark Edge may point. The knowledge is supervisedly learned from a training set of images. After the learning phase this information is utilised to connect points which could belong to a Road Mark into extended chains in a speed optimal way by filtering out disturbances at the same time.

To implement this intelligence, different Single Images are being analysed and Chaining Vectors, (vectors indicating in which direction Road Mark Edges are most probably oriented at that point) created and stored, resulting in a Vector Database. Since the Chaining Vectors vary from position to position within the Single Image, the Single Image is being divided in sub-areas called Tiles and each Tile is being processed individually. This allows to use different primary interconnection directions based on the actual location within the Single Image.

For each Tile, a set of Chaining Vectors is being used which do represent several expected orientations of Road Mark Edges. Such Chaining Vectors are being displayed in Figure 1 and created prior to the runtime as follows:

- Road Marks are marked manually in the Single Images of the training set.
- Each image is split into Tiles.
- The Tile of a Single Image is being split into its individual pixels with the dimension $n \times n$ with n being an odd number.
- An algorithm now moves from pixel to pixel within the Tile and determines the local direction and probability for an Road Mark. After generating this data the chaining directions of each Tile are sorted by their probability. Thus, when using this data structure for searching the next pixel within a Road Mark in the action phase the most probable connections will be tested and found first. These potential interconnections between pixels of a Road Mark are called Chaining Vectors.

In order to apply the Chaining Vectors, the Single Image is being scanned horizontally for differences in contrast. In order to filter small distortions and based on the fact that Road Mark Edges are being expected in a more or less vertical orientation, an average is computed over several horizontal lines of the Single Image. This leads to a Binary image on which the Chaining Vectors are applied to connect edge pixels to Road Mark contours. This way Road Marks which obtain the properties of the training set are most rapidly found. Those interconnections are being stored as being part of a Road Mark Edge and the result of such a processed Single Image is being referred to as a Situation Description.

Other aspects being handled within the area of Intelligent Image Processing are the compensation of the perspective distortion which comes from the reduction from a three dimensional scene to a two dimensional Single Image as well as the reduction of redundancy when one Road Mark Edge is being detected several times by the image processing algorithm.

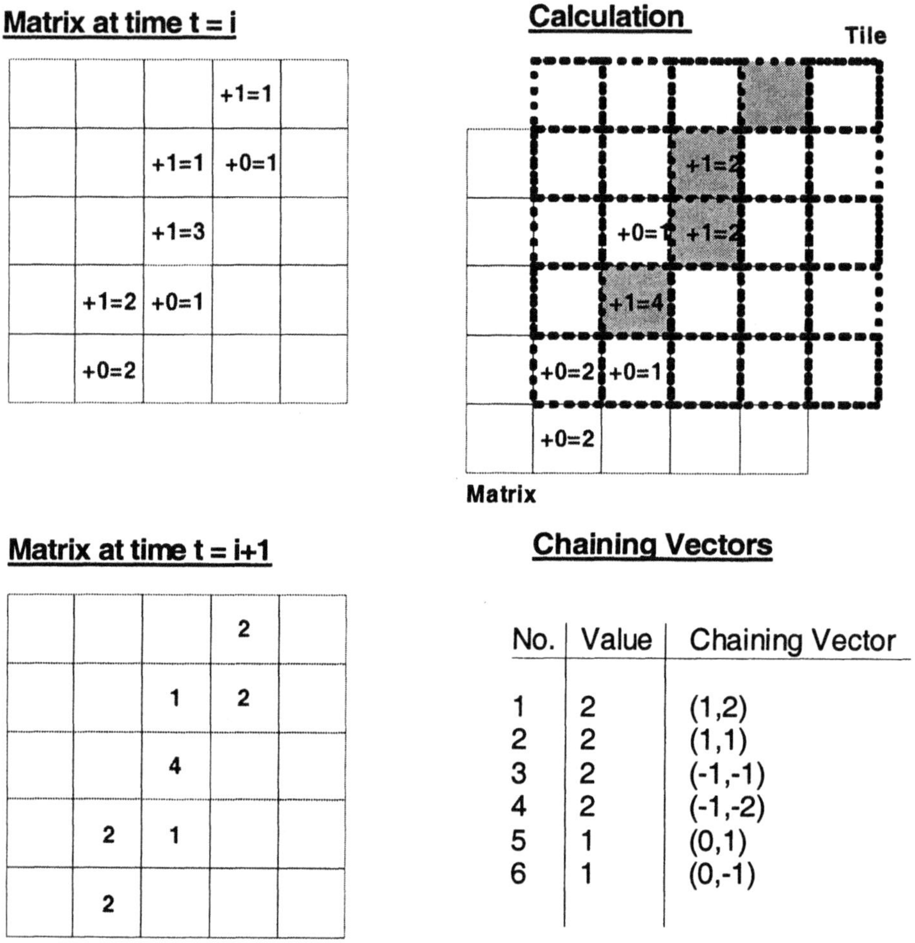

Fig. 1. Matrix elements of one Tile and Chaining Vectors

3 Machine Learning

The second main aspect, as mentioned above, is the capability of our system to improve quality of the issued Steering Commands over time. Several approaches have been undertaken to build a car which can drive autonomously. In order to cope with such requirement, other research projects use a direct (short cut) connection between image processing and a soft computing learning method e.g. a neural network. Such approaches provide good results but only for as long as input pictures of the scene are similar to the training pattern. In one project, this approach got enhanced by a multiple neural network. But the dependency problem of the taught training patterns is inherent to the neural network approach. Further developments like [3] and [4] are being done with different variations but still suffer from those aspects of neural networks.

A completely different approach is being followed in research projects which use explicit modelling; but those systems are always in need for loading them with many parameters for the modelling process [5].

The paradigm of this research is to avoid implicit learning (e.g. by a neuronal network) as well as explicit modelling. Instead, the system of this research learns how to act in certain situations by trial and error and building itself a set of Behavioural Patterns over time. A Behavioural Pattern is constituted by the description of the recorded situation, the generated Steering Commands from that time and a factor indicating the success of failure of this combination. Thus, we try to learn explicitly the relation between certain measurements and the reaction appropriate for the task. The system works in real time with continuous feedback from the environment and any such Behavioural Pattern is being continuously optimised. A complete protocol of all the Behavioural Pattern generated during a road course delivers the knowledge base for further actions.

The difficulty of optimising such Behavioural Pattern stems from the fact, that the success or failure of any generated Steering Command may not be able to be calculated immediately. Only few situations in the environment deliver directly a response about the value of the actual behaviour. Our optimisation criterion is the time needed to drive a specific route i.e. we try to learn to drive as fast as possible. With this criterion in mind only after a completed round or time stop an evaluation of the Behavioural Pattern used in between can be performed. Such characteristic of delayed rewarding or punishing makes learning much more difficult. Speaking in general terms, the learning methodology being used for this research can be referred to as modified Reinforcement Learning [6].

One crucial factor for any machine learning process is that input data are being prepared that way that it can be further processed. Consequently, the results from the Intelligent Image Processing which are so far the contours of the Road Mark Edges are being intersected with a grid of a certain size, resulting in position and angle of intersections between the contours and gridlines and further referred to as Abstract Complete Situation Description (ACSD).

Based on the those ACSD's a modified Nearest Neighbour (NN) Algorithm tries to find previously recorded ACSD's which correspond to the ACSD of the actual situation. This NN-algorithm has a O(n log n) characteristic which allows to search within approximate 10000 Patterns in several milliseconds. Without such a rapid similarity recognition method a real-time realisation of our approach would not be possible.

The result of the NN-algorithm is being used by a Position Estimation algorithm to estimate the current position within a road course. The requirement for position estimation is defined as the need for information on how the ACSD's of the further road might look like in order to gather as much information as possible for the machine learning algorithm and also as the need for the possibility to measure times for road sections as a basis for any performance determination.

The elements described so far result in a set of key information for a machine learning algorithm: a parametric description of the actual situation, an estimated position within the road course, and the records of previously driven road courses similar to the actual road course. Additionally, the Steering Commands as well as the time stamps available for each recorded ACSD implicitly contain the information on the quality of any set of Steering Commands. The difficulty at this point is that there is no direct link between the quality of a sequence of Steering Commands and any single Steering Command; also the ACSD's based on the actual Single Image and the recorded ACSD's and will usually not match identically but only match to some extend. Those information on one hand but the leeway for interpretation on the other hand spans the area for machine learning. As mentioned above, the current research is oriented towards the area of Reinforcement Learning.

Due to the volume of information available for the machine learning algorithm, a further clustering of those data is being done. At this point, the basic concept is that the Steering Commands are being generated depending on the situation the vehicle is in. A true situation, however, is not only captured by the information provided by the ACSD's but also by the Steering Commands active at each ACSD as well as a velocity factor. Our definition of similarity is behaviour oriented i.e. similarity among situations is defined that way in this project that the Steering Commands need to be similar. This allows the clustering into a two-dimensional space based on the values of direction (steering left, steering right) and acceleration (speed acceleration, speed reduction).

The integration of the machine learning aspects into the subsystem of Learning System is planned that way that the clustering of situations is being done for the previously recorded ACSD's at and around the identified location within the road course. Each situation within such clustering is then enriched with the information of the performance rating which has been done for the road section the situation was part of. Equally, the velocity factor valid at the time of recording is stored along with it's corresponding ACSD. Such combination is then being referred to as Behavioural Patterns (i.e. policy). The optimisation of such Behavioural Patterns, therefore to calculate new Steering Commands for the current ACSD and velocity factor and also, even if only delayed, enriching it with a performance value, is being expected to be done by means of Reinforcement Learning. This part of the concept, however, is currently under investigation and also still to be implemented

4 Experimental Results

The system is implemented in a simulation environment which allows very realistic generation of the images along several routes. A second computer serves as an autonomous driver. On this machine the described software system is being implemented.

Up to this point, the image processing section including the determination of Situation Descriptions for each Single Image is completed. Figure 2 shows the calculation of two different Situation Descriptions based on the same Single Image – one time with and one time without the usage of the described Chaining Vectors. It can be clearly seen that the usage of Chaining Vectors allows a less ragged approximation of the Road Mark Edges.

Furthermore, the conversion into ACSD's is also completed and several experimental runs with the driving learning system have been performed, also experimenting with different grid sizes (Figure 3). First experiments for estimating the position within a route show promising results (Figure 4).

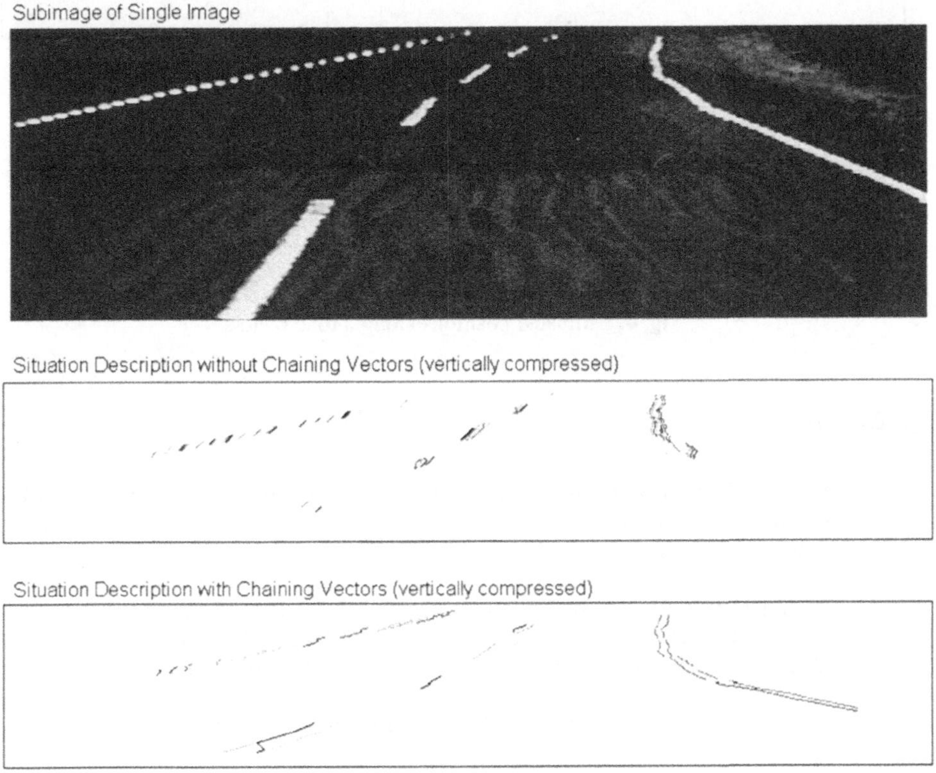

Fig. 2. Situation Descriptions calculated with and without Chaining Vectors

The interpolation of the Behavioural Patterns to cope with unknown routes and the calculation of any success of issued Steering Commands of a certain Single Image are currently still under study and subject to further research.

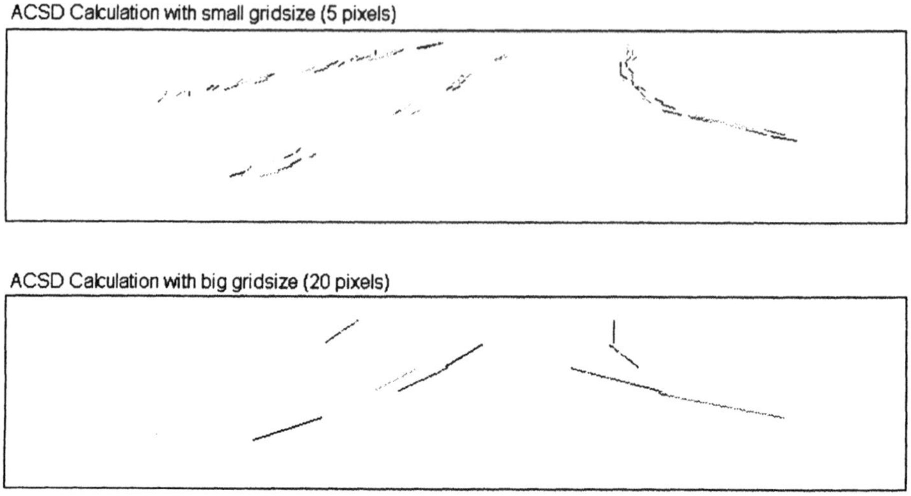

Fig. 3. ACSD's of one Single Image based on different grid sizes

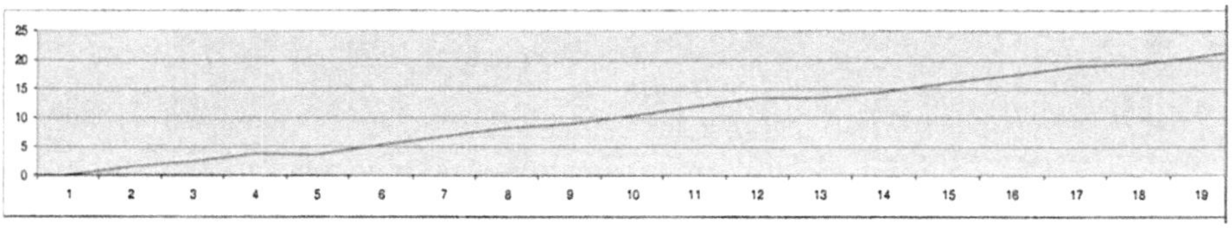

Fig. 4. Estimated positions along a road course

References

[1] M. Krödel, K.-D. Kuhnert, Towards a Learning Autonomous Driver System, IEEE International Conference on Industrial Electronics, Control and Instrumentation, October 22-28, 2000, Nagoya, Japan

[2] T.M.Jochem, D.A.Pomerleau, C.E.Thorpe, Vision Guided Lane Transition, Intelligent Vehicles '95 Symposium, September 25-26, 1995, Detroit/MI, USA

[3] T.M. Jochem, D.A. Pomerleau, C.E. Thorpe. MANIAC: A Next Generation Neurally Based Autonomous Road Follower, IAS-3, Int. Conference on Intelligent autonomous Systems, Feb. 15-18, 1993, Pittsburgh/PA, USA, F.C.A. Groen, S.Hirose, C.E.Thorpe (eds), IOS Press, Washington, Oxford, Amsterdam, Tokyo, 1993

[4] Expectation-based selective attention for visual monitoring and control of a robot vehicle, S.Baluja, D.A.Pomerleau, Robotics and Autonomous System, Vol.22, No.3-4, December 1997

[5] E.D.Dickmanns, A.Zapp, Autonomous High Speed Road Vehicle Guidance by Computer Vision, Preprints of the 10th World Congress on Automatic Control, Vol.4, International Federation of Automatic Control, Munich, Germany, July 27-31, 1987

[6] L. Baird III, Reinforcement Learning through Gradient Descent, Dissertation, Carnegie Mellon University, 1999 Pittsburgh, USA

SVD Reduction in Continuos Environment Reinforcement Learning

Szilveszter Kovács[1]

Research Institute of Manufacturing Information Technology Gifu Prefecture
4-179-19 Sue Kakamigahara Gifu 509-0108 Japan
On research leave from: Department of Information Technology,
University of Miskolc, Miskolc-Egyetemváros, Miskolc, H-3515, Hungary
szkszilv@gold.uni-miskolc.hu,
http://www.iit.uni-miskolc.hu/~szkovacs

Abstract. Reinforcement learning methods, surviving the control difficulties of the unknown environment, are gaining more and more popularity recently in the autonomous robotics community. One of the possible difficulties of the reinforcement learning applications in complex situations is the huge size of the state-value- or action-value-function representation [2]. The case of continuous environment (continuous valued) reinforcement learning could be even complicated, as the state-value- or action-value-functions are turning into continuous functions. In this paper we suggest a way for tackling these difficulties by the application of SVD (Singular Value Decomposition) methods [3], [4], [15], [26].

1 Introduction

Reinforcement learning methods are trial-and-error stile learning methods adapting dynamic environment through incremental iteration. The principal ideas of reinforcement learning methods, the dynamical system state and the idea of „optimal return“ or „value“ function are inherited from optimal control and dynamic programming [1]. One common goal of the reinforcement learning strategies is to find an optimal policy by building the *state-value-* or *action-value-function* [2]. The *state-value-function* $V^{\pi}(s)$, is a function of the expected return (a function of the cumulative reinforcements), related to a given state $s \in S$ as a starting point, following a given policy π. Where the states of the learning agent are observable and the reinforcements (or rewards) are given by the environment. These rewards are the expression of the goal of the learning agent as a kind of evaluation follows the recent action (in spite of the instructive manner of error feedback based approximation techniques, like the gradient descent training). The policy is the description of the agent behavior, in the form of mapping between the agent states and the corresponding suitable actions. The *action-value function* $Q^{\pi}(s,a)$, is a function of

[1] This research was partly supported by the Hungarian National Scientific Research Fund grant no: F 029904.

B. Reusch (Ed.): Fuzzy Days 2001, LNCS 2206, pp. 719–738, 2001.

the expected return, in case of taking action $a \in A_s$ in a given state s, and then following a given policy π. Having the action-value-function, the optimal (greedy) policy, which always take the optimal (the greatest estimated value) action in every states, can be constructed as [2]:

$$\pi(s) = \arg\max_{a \in A_s} Q^{\pi}(s,a). \qquad (1)$$

Namely for estimating the optimal policy, the action-value function $Q^{\pi}(s,a)$ is needed to be approximated. In discrete environment (discrete states and discrete actions) it means, that at least $\sum_{s \in S} \|A_s\|$ element must be handled. (Where $\|A_s\|$ is the cardinality of the set of possible actions in state s.) Having a complex task to adapt, both the number of possible states and the number of the possible actions could be an extremely high value.

1.1 Reinforcement Learning in Continuous Environment

To implement reinforcement learning in continuous environment (continuous valued states and actions), function approximation methods are widely used. Many of these methods are applying tailing or partitioning strategies to handle the continuous state and action spaces in the similar manner as it was done in the discrete case [2]. One of the difficulties of building an appropriate partition structure is the anonymity of the action-value-function structure. Applying fine resolution in the partition leads to high number of states, while coarse partitions could yield imprecise or unadaptable system. Handling high number of states also leads to high computational costs, which could be also unacceptable in many real time applications.

1.2 Fuzzy Techniques in Continuos Environment Reinforcement Learning

There are many methods in the literature for applying fuzzy techniques in reinforcement learning (e.g. for „Fuzzy Q-Learning“ [9], [10], [11], [12], [6]). One of the main reason of their application beyond the simplicity of expressing priory knowledge in the form of fuzzy rules, is the universal approximator property [7], [8] of the fuzzy inference. It means that any kind of function can be approximated in an acceptable level, even if the analytic structure of the function is unknown. Despite of this useful property, the use of fuzzy inference could be strictly limited in time-consuming reinforcement learning by its complexity problems [16], because of the exponential complexity problem of fuzzy rule bases [13], [3], [4]. Fuzzy logic inference systems are suffering from exponentially growing computational complexity in respect to their approximation property. This difficulty comes from two inevitable facts. The first is that the most adopted fuzzy inference techniques do not hold the universal approximation property, if the numbers of antecedent sets are limited, as stated by Tikk in [17]. Furthermore, their explicit functions are sparse in the approximation function space. This fact inspires to increase the density, the number of antecedents in pursuit of gaining a good approximation, which, however, may soon lead to a conflict with the computational capacity available for the implementation, since the increasing number of antecedents explodes the computational requirement. The latter is the second fact and stated by Kóczy et al. in [16]. The effect of this contradiction is gained by the lack of a mathematical framework capable of estimating the necessary minimal number of antecedent sets. Therefore a heuristic setting of the number of antecedent sets is applied, which usually overestimates, in order to be on the safe side, the necessary number of antecedents resulting in an unnecessarily high computational cost. E.g. the structurally different Fuzzy Q-Learning method implementations introduced [9], [10], [11] and [12] are sharing the same concept of fixed, predefined fuzzy antecedent partitions, for state representation. One possible solution for this problem is suggested in [6]. By introducing „Adaptive State Partitions", an incremental fuzzy clustering of the observed state transitions. This method can lead to a better partition than the simple heuristic, by finding the best fitting one in respect to the minimal squared error, but still has the problem of limited approximation property inherited from the limited number of antecedent fuzzy sets.

Another promising solution, as a new topic in fuzzy theory, is the application of fuzzy rule base complexity reduction techniques.

1.3 Fuzzy Rule Base Complexity Reduction

The main idea of application fuzzy rule base complexity reduction techniques for reinforcement learning is enhancing the universal approximator property of the fuzzy inference by extending the number of antecedent sets while the computational complexity is kept relatively low.

Some reduction techniques are classified regarding their concept in [14] and [4]. A fuzzy rule importance based technique is proposed by Song et al. in [20]. Another recent method proposed by Sudkamp et al. [22] combines rule learning with a region merging strategy.

Recently, several publications have applied orthogonal transformation methods for selecting important rules from a given rule base, for instance, in 1999 Yen and Wang investigated various techniques in [14] for possible fuzzy rule base simplification techniques such as orthogonal least-squares, eigenvalue decomposition, SVD-QR with column pivoting method, total least square method and direct SVD method. [21] also proposes an SVD based technique with examples.

SVD based fuzzy approximation technique was initialized in 1997 by Yam [15], which directly finds a minimal rule-base from sampled values. Shortly after, this concept was introduced as SVD fuzzy rule base reduction and structure decomposition in [3], [24], [25]. Its key idea is conducting SVD of the consequents and generating proper linear combinations of the original membership functions to form new ones for the reduced set. [3], [15] characterizes fuzzy functions by the conditions of sum-normalization (SN), nonnegativenes (NN) and normality (NO), and extends SVD reduction with further tools to preserve SN and NN conditions of the new membership functions. It may have significant role if the purpose is not only saving computational cost, but maintaining the fuzzy concept and having a theoretical study of the reduced rule's features.

An extension of [14] to multi-dimensional cases may also be conducted in a similar fashion as the higher order SVD reduction technique proposed in [3], [13], [15]. Further developments of SVD based fuzzy reduction [3] [15] are proposed in [13], [18], [19], [23].

The key idea of using SVD in complexity reduction is that the singular values can be applied to decompose a given system and indicate the degree of significance of the decomposed parts. Reduction is conceptually obtained by the truncation of those parts, which have weak or no contribution at all to the output, according to the assigned singular values. This advantageous feature of SVD is used in this paper for enhancing the universal approximator property of the fuzzy inference by extending the number of antecedent sets while the computational complexity is kept relatively low. The complexity and its reduction is discussed in regard of the number of rules, which result simplicity in operating with the rules, in reinforcement learning methods.

On the other hand, as one of the natural problems of any complexity reduction technique, the adaptivity property of the reduced approximation algorithm becomes highly restricted. Since the crucial concept of the Fuzzy Q-learning is based on the adaptivity of the action-value function this paper is aimed propose to adopt an algorithm [26] capable of embedding new approximation points into the reduced approximation while the calculation cost is kept.

This paper is organized as follows. Section 2 briefly summarizes the concept of Fuzzy Q-learning. Section 3 introduces the proposed Fuzzy Q-learning and dynamic partition allocation method. Section 4 examines various concepts of adaptation SVD based techniques, e.g. complexity reduction and approximation adaptation [26], for reinforcement learning. Section 5 gives two simple examples for the practical use of the proposed method.

2 Reinforcement Learning

For introducing a possible way of application of SVD complexity reduction techniques in Fuzzy Reinforcement Learning, a simple direct (model free) reinforcement learning method, the Q-Learning [5], was chosen.

The goal of the Q-learning is to find the fixed-point solution Q of the Bellman Equation [1] through iteration. In discrete environment *Q-Learning* [5], the action-value-function is approximated by the following iteration:

$$
\begin{aligned}
&Q_{i,u} \approx \tilde{Q}_{i,u}^{k+1} = \tilde{Q}_{i,u}^{k} + \Delta\tilde{Q}_{i,u}^{k+1} = \\
&\tilde{Q}_{i,u}^{k+1} = \tilde{Q}_{i,u}^{k} + \alpha_{i,u}^{k} \cdot \left(g_{i,u,j} + \gamma \cdot \max_{v \in U} \tilde{Q}_{j,v}^{k+1} - \tilde{Q}_{i,u}^{k} \right) \qquad \forall i \in I, \forall u \in U
\end{aligned}
\tag{2}
$$

where $\tilde{Q}_{i,u}^{k+1}$ is the $k+1$ iteration of the action-value taking action A_u in state S_i, S_j is the new observed state, $g_{i,u,j}$ is the observed reward completing the $S_i \rightarrow S_j$ state-transition, γ is the discount factor and $\alpha_{i,u}^{k} \in [0,1]$ is the step size parameter (which can change during the iteration steps).

For applying this iteration to continuous environment by adopting fuzzy inference (Fuzzy Q-Learning), there are many solutions exist in the literature [6], [9], [10], [11], [12].

Having only demonstrational purposes, in this paper one of the simplest one, the order-0 Takagi-Sugeno Fuzzy Inference based Fuzzy Q-Learning is studied (a slightly modified, simplified version of the Fuzzy Q-Learning introduced in [9] and [6]). This case, for characterising the value function $Q(s,a)$ in continuous state-action space, the order-0 Takagi-Sugeno Fuzzy Inference System approximation $\tilde{Q}(s,a)$ is adapted in the following manner:

$$\textbf{If } s \textbf{ is } S_i \textbf{ And } a \textbf{ is } A_u \textbf{ Then } \tilde{Q}(s,a)=Q_{i,u},\ i\in I, u\in U, \tag{3}$$

where S_i is the label of the ith membership function of the *n* dimensional state space, A_u is the label of the uth membership function of the one dimensional action space, $Q_{i,u}$ is the singleton conclusion and $\tilde{Q}(s,a)$ is the approximated continuous state-action-value function. Having the approximated state-action-value function $\tilde{Q}(s,a)$, the optimal policy can be constructed by function (1):

$$\tilde{\pi}(i)=\arg\max_{u\in U} Q(i,u), \tag{4}$$

Setting up the antecedent fuzzy partitions to be Ruspini partitions, the order-0 Takagi-Sugeno Fuzzy Inference forms the following approximation function:

$$f(x_1,x_2,\ldots,x_N)=\sum_{j_1,j_2,\cdots,j_N}^{J_1,J_2,\ldots,J_N}\prod_{n=1}^{N}\mu_{j_n,n}(x_n)b_{j_1j_2\ldots j_N}\,. \tag{5}$$

where $\mu_{j_n,n}(x_n)$ is the membership value of the j_n [th] antecedent fuzzy set at the n [th] dimension of the N dimensional antecedent universe X_n at the state-action observation x_n and $b_{j_1j_2\ldots j_N}$ is the value of the singleton conclusion of the $j_1j_2\ldots j_N$ [th] fuzzy rule. In this notation all combination of the antecedents corresponds to one consequent fuzzy set defined these relations are expressed by rules as: IF $\mu_{j_1,1}(x_1)$ and $\mu_{j_2,n}(x_2)$ and ... and $\mu_{j_N,N}(x_N)$ THEN $\beta_{j_1j_2\ldots j_N}$. Singleton consequent fuzzy sets $\beta_{j_1j_2\ldots j_N}$ are defined by their location $b_{j_1j_2\ldots j_N}$. on output universe Y.

Applying the notation introduced in (3), equation (5) turns to the following:

$$\tilde{Q}(s,a)=\sum_{i_1,i_2,\cdots,i_N,u}^{I_1,I_2,\ldots,I_N,U}\prod_{n=1}^{N}\mu_{i_n,n}(s_n)\cdot\mu_u(a)\cdot q_{i_1i_2\ldots i_N u} \tag{6}$$

where $\tilde{Q}(s,a)$ is the approximated state-action-value function $\mu_{i_n,n}(s_n)$ is the membership value of the i_n th state antecedent fuzzy set at the n th dimension of the N dimensional state antecedent universe at the state observation s_n, $\mu_u(a)$ is the membership value of the u th action antecedent fuzzy set of the one dimensional action antecedent universe at the action selection a and $q_{i_1 i_2 \dots i_N u}$ is the value of the singleton conclusion of the $i_1 i_2 \dots i_N u$ th fuzzy rule.

Applying the approximation formula of the Q-learning (2) for adjusting the singleton conclusions in (5), leads to the following function:

$$q^{k+1}_{i_1 i_2 \dots i_N u} = q^{k}_{i_1 i_2 \dots i_N u} + \prod_{n=1}^{N} \mu_{i_n,n}(s_n) \cdot \mu_u(a) \cdot \Delta\tilde{Q}^{k+1}_{i,u} \tag{7}$$

$$q^{k+1}_{i_1 i_2 \dots i_N u} = q^{k}_{i_1 i_2 \dots i_N u} + \prod_{n=1}^{N} \mu_{i_n,n}(s_n) \cdot \mu_u(a) \cdot \alpha^{k}_{i,u} \cdot \left(g_{i,u,j} + \gamma \cdot \max_{v \in U} \tilde{Q}^{k+1}_{j,v} - \tilde{Q}^{k}_{i,u} \right)$$

where $q^{k+1}_{i_1 i_2 \dots i_N u}$ is the $k+1$ iteration of the singleton conclusion of the $i_1 i_2 \dots i_N u$ th fuzzy rule taking action A_u in state S_i, S_j is the new observed state, $g_{i,u,j}$ is the observed reward completing the $S_i \to S_j$ state-transition, γ is the discount factor and $\alpha^{k}_{i,u} \in [0,1]$ is the step size parameter. The $\max_{v \in U} \tilde{Q}^{k+1}_{j,v}$ and $\tilde{Q}^{k}_{i,u}$ action-values can be approximated by equation (6).

The next problematic question of the Fuzzy Reinforcement Learning, as it was introduced in section 1.2, is the proper way of building the fuzzy partitions. The methods sharing the concept of fixed, predefined fuzzy partitions, like [9], [10], [11] and [12] are facing the following question: More detailed partitions are yielding exponentially growing state spaces (rule base sizes), elongating the adaptation time, and dramatically increasing the computational resource demand, while sparse partitions could cause high approximation error, or unadaptable situation. One possible solution for this problem is suggested in [6]. By introducing „Adaptive State Partitions", an incremental fuzzy clustering of the observed state transitions. This method can lead to a better partition than the simple heuristic, by finding the best fitting one in respect to the minimal squared error, but still has the problem of limited approximation property inherited from the limited number of antecedent fuzzy sets.

In this paper another dynamic partition allocation method is suggested, which is instead of adjusting the sets of the fuzzy partition, simply increase the number of the fuzzy sets by inserting new sets in the required positions.

3 The Proposed Method

The reinforcement learning method proposed in this paper can be divided to two main parts. The first is the reinforcement method itself. It is the direct (model free) Fuzzy

Q-Learning method as it was introduced in section 2 (order-0 Takagi-Sugeno Fuzzy Inference based modification of the Fuzzy Q-Learning introduced in [9] and [6]).

The second is the dynamic partition allocation method proposed in this paper. The main idea is very simple (see Fig.1. for an example). Initially a minimal sized (e.g. 2-3 sets only) Ruspini partition built up triangular shaped fuzzy sets on all the antecedent universes (see Fig.1.a.). In case if the action-value function update is high ($\Delta\tilde{Q} > \varepsilon_Q$), and the partition is not too dense already ($d_s > \varepsilon_s$), and the actual state-action point is far from the existing partition members (see Fig.1.b.), than a new fuzzy state is inserted to increase the resolution (see Fig.1.d.). If the update value is relatively low (see e.g. Fig.2.), or the actual state-action point is close to the existing partition members (Fig.3.), than the partition is staying unchanged. The state insertion is done in every state dimensions separately (in multidimensional case it means an insertion of a hyperplane), by interpolating the inserted values from the neighbouring ones (see Fig.1.e. and Fig.4. as a two dimensional example). Having the new state plane inserted in every required dimension, the value update is done regarding to the Fuzzy Q-Learning method as it was introduced in section 2, by the equation (7). (See e.g. on Fig.1.c,d, or Fig.4.d.)

The proposed dynamic partition allocation method has the property of local step-by-step refinement in a manner very similar to the binary search. It can locate the radical positions of the value action function with the precision of $d_s^{i+k} = \frac{d_s^i}{2^k}$ in k steps (where d_s^i is the starting precision).

The main problem of the proposed simple dynamic partition allocation method is the nondecreasing adaptation manner of the antecedent fuzzy partitions. In some situation, it could mean rapidly increasing partition sizes. Moreover, these cases also lead rapidly growing, or at least nondecreasing computational resource demand.

For retaining the benefits of the dynamic partition allocation and maintaining the overall computational resource demand low, in this paper, the adoption of Higher Order SVD [13] based fuzzy rule base complexity reduction techniques and its fast adaptation method (see in section 4.2 and 4.3) is suggested. The application of the fast adaptation method [26] gives a simple way for increasing the rule density of a rule base stored in a compressed form directly (see section 4.2 and 4.3, equation (9) and [26]). Providing an economic sized structure for handling continuously increasing and varying rule bases, which is so typical in reinforcement learning.

4 SVD Based Complexity Reduction

The essential idea of using SVD in complexity reduction is that the singular values can be applied to decompose a given system and indicate the degree of significance of the decomposed parts. Reduction is conceptually obtained by the truncation of those parts, which have weak or no contribution at all to the output, according to the assigned singular values. This advantageous feature of SVD is used in this paper for enhancing the universal approximator property of the fuzzy inference by extending the number of antecedent sets while the computational complexity is kept relatively low. The complexity and its reduction is discussed in regard of the number of rules, which result simplicity in operating with the rules, in reinforcement learning methods.

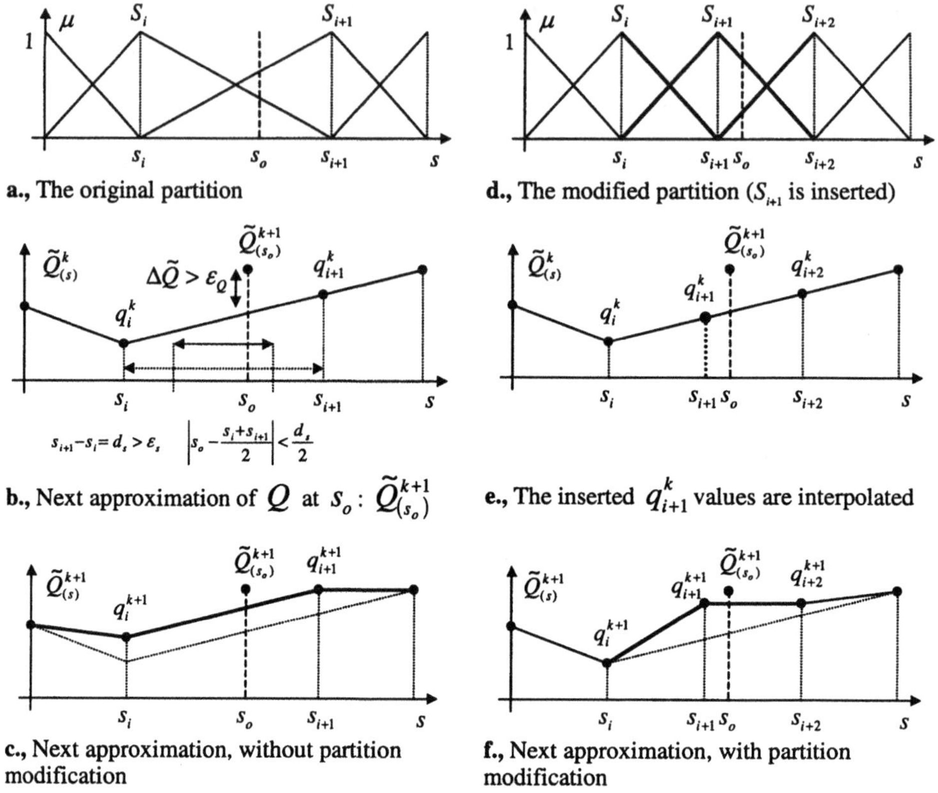

Fig. 1. The proposed dynamic partition allocation.

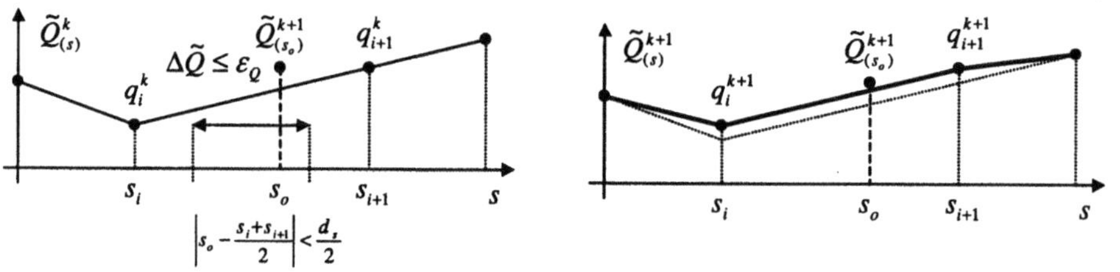

Fig. 2. The action-value function update is relatively low.

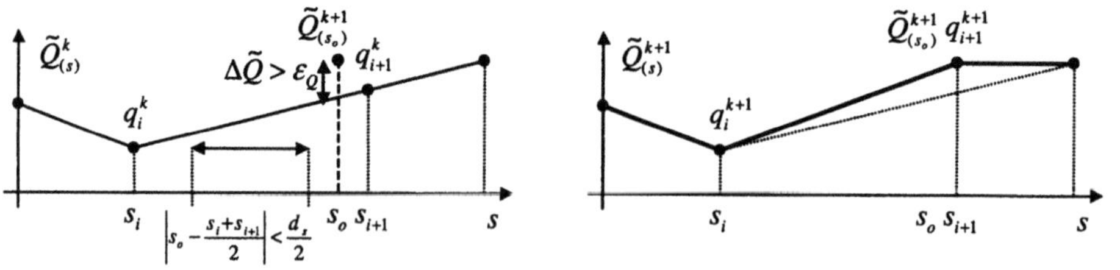

Fig. 3. The actual state-action point is close to the existing partition members.

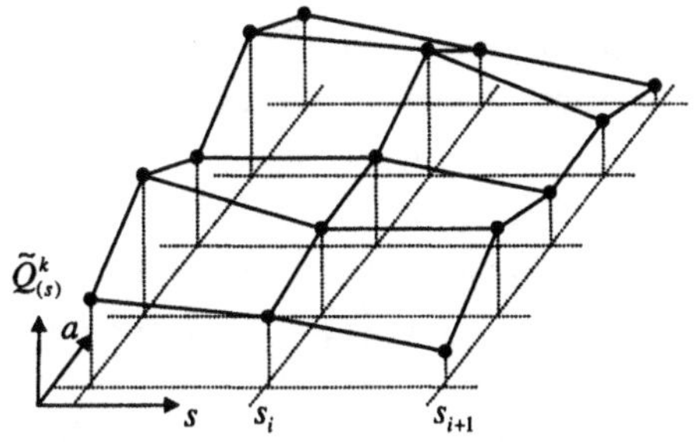

a., The original approximation of $\tilde{Q}^k_{(s)}$

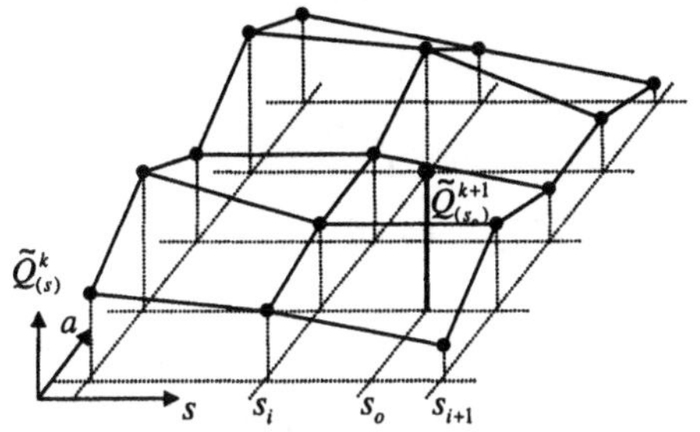

b., Next approximation of Q at s_o : $\tilde{Q}^{k+1}_{(s_o)}$

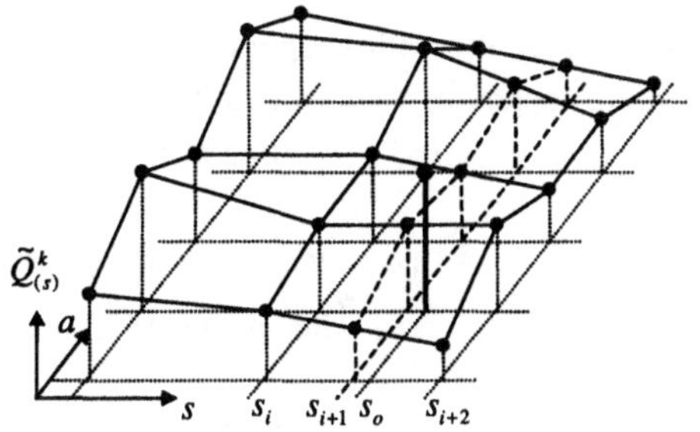

c., State insertion s_{i+1}, by interpolating the inserted action values $q^k_{i+1,j}$, $\forall j \in [1, J]$, from the neighbouring $q^k_{i,j}$ and $q^k_{i+2,j}$ ones.

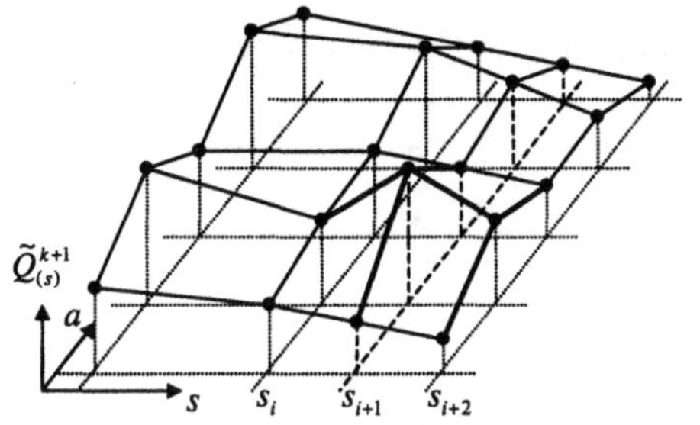

d., Next approximation, with state insertion s_{i+1}, and value update regarding to (7)

Fig. 4. The proposed dynamic partition allocation in two-dimensional (single state and action) antecedent case.

4.1 Definitions

In this section some elementary definitions and concepts utilized in the further sections will be introduced. With respect to the notation, to facilitate the distinction between the types of given quantities, the notation will be reflected by their representation: scalar values are denoted by lower-case letters $\{a, b, ...; \alpha, \beta, ...\}$; column vectors and matrices are given by bold-face letters as $\{\mathbf{a}, \mathbf{b}, ...\}$ and $\{\mathbf{A}, \mathbf{B}, ...\}$ respectively, matrix $\mathbf{0}$ contains zero values only; tensors correspond to capital letters as $\{A, B, ...\}$. The transpose of matrix $\mathbf{A}$ is denoted as $\mathbf{A}^{\mathrm{T}}$. Subscript is consistently used for a lower order of a given structure. E.g. and element of matrix $\mathbf{A}$ is defined by row-column number i, j symbolized as $(\mathbf{A})_{i,j} = a_{i,j}$. Systematically, the i-th column vector of $\mathbf{A}$ is denoted as $\mathbf{a}_i$, i.e. $\mathbf{A} = [\mathbf{a}_1 \quad \mathbf{a}_2 \quad \cdots]$. To enhance the overall readability characters $i, j, ...$ in the meaning of indices (counters), $I, J, ...$ are reserved to denote the index upper bounds, unless stated otherwise. $\Re^{I_1 \times I_2 \times \ldots \times I_N}$ is the vector space of real valued $(I_1 \times I_2 \times \ldots \times I_N)$-tensors. Letter N serves to denote the number of variables. Letter k has special role and it is: $k = 1 \ldots N$, $k \neq n$.

Definition 1. (n-mode matrix of tensor A) Assume an N-th order tensor $A \in \Re^{I_1 \times I_2 \times \ldots \times I_N}$. The n-mode matrix $A_{(n)} \in \Re^{I_n \times J}$, $J = \prod_k I_l$ contains all the vectors in the n-th dimension of tensor A. The ordering of the vectors is arbitrary, this ordering shall, however, be consistently used later on. $(\mathbf{A}_{(n)})_j$ is called an j-th n-mode vector.

Note that any matrix of which the columns are given by n-mode vectors $(\mathbf{A}_{(n)})_j$ can evidently be restored to be tensor A.

Definition 2. (tensor interval) Assume N-th order tensors $A, B, C \in \Re^{I_1 \times I_2 \times \ldots \times I_N}$. $C \in_t [A, B] \Leftrightarrow A \leq_t C \leq_t B$, where $A \leq_t B \Leftrightarrow \forall i_1 i_2 \ldots i_N : a_{i_1 i_2 \ldots i_N} \leq b_{i_1 i_2 \ldots i_N}$

Definition 3. (n-mode sub-tensor of tensor A) Assume an N-th order tensor $A \in \Re^{I_1 \times I_2 \times \ldots \times I_N}$. The n-mode sub-tensor $A_{i_n = \alpha}$ contains elements $a_{i_1, i_2, \ldots, i_{n-1}, \alpha, i_{n+1}, \ldots, i_N}$.

Definition 4. (n-mode matrix-tensor product) The n-mode product of tensor $A \in \Re^{I_1 \times I_2 \times \ldots \times I_N}$ by a matrix $\mathbf{U} \in \Re^{J \times I_n}$, denoted by $A \times_n \mathbf{U}$ is an $(I_1 \times I_2 \times \ldots \times I_{n-1} \times J \times I_{n+1} \times \ldots \times I_N)$-tensor of which the entries are given by $A \times_n \mathbf{U} = B$, where $B_{(n)} = \mathbf{U} \cdot A_{(n)}$. Let $A \times_1 \mathbf{U}_1 \times_2 \mathbf{U}_2 \ldots \times_N \mathbf{U}_N$ be noted for brevity as $A \bigotimes_{n=1}^{N} \mathbf{U}_n$.

There are major differences between matrices and higher-order tensors when rank properties are concerned. These differences directly affect the way an SVD generalization could look like. As a matter of fact, there is no unique way to generalize the rank concept. In this paper the description is restricted to n mode rank only.

Definition 5. (n mode rank of tensor) The n mode rank of A, denoted by $R_n = rank_n(A)$, is the dimension of the vector space spanned by the n mode vectors as $rank_n(A) = rank(A_{(n)})$.

Theorem (N-th Order SVD or HOSVD) Every tensor $A \in \Re^{I_1 \times I_2 \times .. \times I_N}$ can be written as the product $A = S \overset{N}{\underset{n=1}{\otimes}} \mathbf{U}_n$, in which $\mathbf{U}_n = [\mathbf{u}_{1,n} \quad \mathbf{u}_{2,n} \quad \dots \quad \mathbf{u}_{I_N,n}]$ is a unitary $(I_N \times I_N)$-matrix called n-mode singular matrix. Tensor $S \in \Re^{I_1 \times I_2 \times .. \times I_N}$ of which the subtensors $S_{i_n=\alpha}$ have the properties of all-orthogonality (two subtensors $S_{i_n=\alpha}$ and $S_{i_n=\beta}$ are orthogonal for all possible values of n, α and β, when $\alpha \neq \beta$) and ordering: $\left\|S_{i_n=1}\right\| \geq \left\|S_{i_n=2}\right\| \geq \dots \geq \left\|S_{i_n=I_n}\right\| \geq 0$ for all possible values of n.

See detailed discussion and notation of matrix SVD and Higher Order SVD (HOSVD) in [13].

4.2 SVD Based Fuzzy Rule Base Complexity Reduction

Since the state action value function is approximated by an order-0 Takagi-Sugeno Fuzzy Inference method this section is intended to provide a brief survey of the fundamentals in SVD based fuzzy rule base reduction techniques, which are proposed in [3], [4], [13], [15].

The calculation complexity of (5) explodes with values $J_1, J_2, \dots J_N$, in this regards, for comprehensive analysis and exact theorems, see [16]. Decreasing the upper bound of the indices in the sum operator of (5), namely the number of antecedent sets, leads to the initial idea of calculation reduction. Formula (5) can be equivalently written in tensor product form as: $f(x_1, x_2, \dots, x_N) = B \overset{N}{\underset{n=1}{\otimes}} \mathbf{m}_n$, where tensor $B \in \Re^{J_1 \times J_2 \dots \times J_N}$ and vector $\mathbf{m}_n$ respectively contain elements $b_{j_1 j_2 \dots j_N}$ and $\mu_{j_n,n}(x_n)$. This reduction can be conceptually obtained by reducing the size of tensor B via Higher Order SVD (HOSVD). According to the special terms in this topic, the following notation has emerged [3], [4], [13]:

Definition 6. (Exact / non-exact reduction) Assume an N-th order tensor $A \in \Re^{I_1 \times I_2 \times .. \times I_N}$. **Exact** reduced form $A = A^r \overset{N}{\underset{n=1}{\otimes}} \mathbf{U}_n$, where „r" denotes „reduced", is defined by tensor $A^r \in \Re^{I_1^r \times I_2^r \times .. \times I_N^r}$ and basis matrices $\mathbf{U}_n \in \Re^{I_n \times I_n^r}$, $\forall n : I_n^r \leq I_n$ which are the result of HOSVD, where only zero singular values and the corresponding singular vectors are discarded. **Non-exact** reduced form $\hat{A} = A^r \overset{N}{\underset{n=1}{\otimes}} \mathbf{U}_n$, is obtained if not only zero singular values and the corresponding singular vectors are discarded.

The above properties directly lead to the following fundamental concept of.

Method 1. (exact SVD based fuzzy rule base reduction) The SVD based fuzzy rule base reduction transforms equation (5) to the form of:

$$f(x_1,x_2,\ldots,x_N)=\sum_{j_1,j_2,\ldots,j_N}^{J_1^r,J_2^r,\ldots,J_3^r}\prod_{n=1}^{N}\mu_{j_n,n}^r(x_n)b_{j_1j_2\ldots j_N}^r, \tag{8}$$

where $\forall n: J_n^r \le J_n$ is obtained as the main essence of the reduction.

The reduced form is obtained via HOSVD capable of decomposing B into $B = B^r \bigotimes_{n=1}^{N} \mathbf{U}_n$. Having $B^r \in \Re^{J_1^r \times J_2^r \ldots \times J_N^r}$ and its singular vectors the reduced form is determined as: $f(x_1,x_2,\ldots,x_N) = B^r \bigotimes_{n=1}^{N} \mathbf{m}_n^r$, where $\mathbf{m}_n^r = \mathbf{m}_n \mathbf{U}_n$. Equation (5) is an equivalent of (8) that is the starting point for theoretical developments of this topic.

Remark 1. Note that, the obtained functions may not be interpretable as antecedent fuzzy sets. In order to obtain functions which can be antecedent fuzzy stets, further to have *Ruspini* partition, sum-normalization (SN), nonnegativenes (NN) and normality (NO) transformation techniques are developed to HOSVD algorithm in [3], [4], [13].

Remark 2. The error controllable advantage of the reduction technique is conceptually obtained by the truncation of non-zero singular values. The error bound of $\hat{f}(x_1,x_2,\ldots,x_N)$ can be estimated during the execution of the SVD reduction algorithm. Note that, the final error of $\hat{f}(x_1,x_2,\ldots,x_N)$ depends on the type of the antecedent functions applied. Typical practical cases are analysed in [4].

4.3 Adaptation of SVD Based Approximation

One of the natural problems of any complexity reduction technique is that the adaptivity property of the reduced approximation algorithm becomes highly restricted. Since the crucial concept of the reinforcement learning is based on the adaptivity of the action-value function, in this paper the „fast adaptation of SVD based approximation" (introduced in [26]) is suggested to adopt for reinforcement learning. This fast adaptation method, directly adapts the reduced approximation by replacing, or embedding new approximation points. The ability of embedding new approximation points provides the practical applicability of the proposed dynamic partition allocation method.

Therefore, the application of the fast adaptation method in the proposed reinforcement learning structure is twofold. On one side, it helps the dynamic partition allocation by increasing the rule density. On the other side, by the replacement of the previously fetched and modified values serves the adaptation of the approximated action-value function.

Consequently the key idea of the proposed reinforcement algorithm is to insert a set of new rules, for instance A, into the existing rule base B. As it is already discussed, in order to avoid complexity problems, the reduced form of B (namely B^r) is utilised. This, hence, means that the embedding of the new rules contained in A should directly performed on B^r. One more important constraint should be emphasised here. In order to fix the complexity of the rule base only those in formation of the new rules should be inserted into B^r which do not increase the size of B^r. Actually this is equivalent to the key idea of the fast adaptation introduced in [26]. More precisely in regard of the following algorithm, only those sub-tensors of A are embedded into B^r, which are linearly dependent from B^r [26]. Since the elements in B^r are fixed, no SVD is needed during embedding, which offers a chance to develop a fast algorithm to adapt HOSVD. Here an elementary step of the idea is discussed when the rule base is being increased in an arbitrary dimension n.

Method 2. (n mode fast adaptation [26]) Assume a reduced rule base defined by tensor $B^r \in \Re^{J_1^r \times J_2^r \times \ldots \times J_N^r}$ and its corresponding matrices $\mathbf{Z}_n \in \Re^{J_n \times J_n^r}$ resulted from B by HOSVD. Furthermore, let $A \in \Re^{J_1 \times J_2 \times \ldots J_{n-1} \times I \times J_{n+1} \times \ldots \times J_N}$ be given, that has the same size as B except in the n-th dimension where I may differ from J_n. The localized error threshold of the adaptation is defined by ∇.

The goal is to determine the reduced form E^r of extended rule base E defined by tensor $E' = \begin{bmatrix} B & A' \end{bmatrix}_n$, where E' contains the selected n mode sub-tensors of E according to the given error threshold ∇ as

$$\hat{E}' = \left(B^r \bigotimes_{k=1}^{N} \mathbf{Z}_k \right) \times_n \mathbf{U}, \tag{9}$$

and $A' \in \Re^{J_1 \times J_2 \times . J_{n-1} \times I' \times J_{n+1} \times . \times J_N}$ contains the selected n mode sub-tensors of A and lets the corresponding sub-tensors $T'_{\text{min/max}}$ selected from the corresponding $T_{\text{min/max}}$. For brevity let $\nabla' = [T'_{\text{min}} \quad T'_{\text{max}}]$. $\mathbf{U} = \begin{bmatrix} \mathbf{Z}_n & \mathbf{V} \end{bmatrix} \in \Re^{(J_n + I') \times J_n^r}$, $I' \le I$, where $\mathbf{V}$ is determined to fulfil (9) subject to $\hat{E}' - E' \in_t \nabla'$.

Method 2 has built-up from the following two algorithms:

Algorithm: (n-mode high order d way defective projection to a given basis [26]) This algorithm defectively projects, according to threshold ∇, a given tensor $A \in \Re^{J_1 \times J_2 \times \ldots J_{n-1} \times I \times J_{n+1} \times \ldots \times J_N}$ in dimension n to basis $\mathbf{Z}_n \in \Re^{I \times J_n^r}$. The result is $\hat{A}' = A^p \times_n \mathbf{Z}_n$, where $A' \in \Re^{J_1 \times J_2 \times \ldots J_{d-1} \times J'_d \times J_{d+1} \times \ldots \times J_N}$, which may be defective in the specified dimension d, $J'_d \le J_d$, and consists of selected d-mode sub-tensors of A. The projection is done by the above defined defective matrix projection to yield: $\mathbf{Z}_n A^p_{(n)} - A'_{(n)} \in_t \nabla'$ under the condition of projection as $\mathbf{P}_n A'_{(n)} - A'_{(n)} \in_t \nabla'_{(n)}$, where $\mathbf{P}_n = \mathbf{Z}_n \cdot \mathbf{Z}_n^+$. From the point of calculation, the condition is actually checked vector by vector as: $\mathbf{P}_n (A'_{(n)})_{i'} - (A'_{(n)})_{i'} \in_t (\nabla'_{(n)})_{i'}$. Important step is here, in high order case, that not only those vectors are ignored which do not satisfy the above condition of projection, but all vectors contained in a d-mode sub-tensor of A, where at least one vector exists not holding the condition of projection. The resulted $A^p_{(n)}$ projected from the remaining vectors can be restored to be tensor A^p. The size of A' may differ from A only in dimension d. The size of A' can be defined by removing the cancelled sub-tensors from A. $\mathbf{P}_n (A'_{(n)})_{i'} - (A'_{(n)})_{i'} = (H'_{(n)})_{i'}$ can be restored to be tensor H', like in the case of A'. Let tensor H_n with the size of A be generated via extending H' with zeros and be called as n mode projection error.

Having the above Algorithm, the first step of Method 2 is the following: Repeating the k-mode high order n-way defective projection to basis $\mathbf{Z}_k \in \Re^{J_k \times J_k^r}$ which results in $\hat{A}' = A^p \overset{N}{\underset{k=1}{\otimes}} \mathbf{Z}_k$, where $A' \in \Re^{J_1 \times J_2 \times \ldots J_{n-1} \times I' \times J_{n+1} \times \ldots \times J_N}$ and n-mode sub-tensors of A' are selected from among the n-mode sub-tensors of A. The cumulated error obtained by the defective matrix projection in each k-th step is $S'_k = \sum_k \left(H'_k \overset{k-1}{\underset{h=1,h\neq n}{\otimes}} \mathbf{Z}_k \right)$. As a matter of fact, the tolerance criteria of the condition of projection should be corrected in each step by S_k. The last step of Method 2 is the inverse of the previous step:

Algorithm: (*n-mode high order defective basis [26]*) Assume given tensor B^r with $\mathbf{Z}_k \in \Re^{J_k \times J_k^r}$ as the reduced form of B by HOSVD; corrected error threshold ∇ and $A^p \in \Re^{J_1^r \times J_2^r \times \ldots J_{n-1}^r \times I' \times J_{n+1}^r \times \ldots \times J_N^r}$ as a result of the first step of Method 2. This algorithm determines defective basis $\mathbf{U} \in \Re^{(J_n + I'') \times J_n^r}$ in such a way that $\left[\left(B^r \overset{N}{\underset{k=1}{\otimes}} \mathbf{Z}_k \right) \times_n \mathbf{U} - E' \right] \in_t \nabla''$, where $E'' = [B \quad A'']_n$. In other words $\mathbf{V} \in \Re^{I' \times J_n^r}$ is determined here which fulfills: $\left(\left(B^r \overset{N}{\underset{k=1}{\otimes}} \mathbf{Z}_k \right) \times_n \mathbf{V} - A'' \right) \in_t \nabla''$, where A'' is created from A' according to the condition of projection like A' is created from A. Applying n-order defective projection of $(A^r_{(n)})^{\mathrm{T}}$ to basis $(B^r_{(n)})^{\mathrm{T}}$ accordingly to error of $T'-S'_n$ to find $(\hat{A}^r_{(n)})^{\mathrm{T}} = (B^r_{(n)})^{\mathrm{T}} \cdot \mathbf{V}^{\mathrm{T}}$ leads to the solution where the rows of $A^r_{(n)}$ are selected from among the rows of $A^P_{(n)}$. Consequently, $\mathbf{V}$ is obtained that shows the linear dependence between the rest of A^P and B^r. The final step is, hence, to fit the new basis as: $\mathbf{U} = \begin{bmatrix} \mathbf{Z}_n \\ \mathbf{V} \end{bmatrix}$. The required antecedent sets are defined as: $\mathbf{s}^r_k = \mathbf{s}_k \mathbf{Z}_k$ and the new basis is $\mathbf{s}^r_n = \mathbf{s}_n \mathbf{U}$.

The more detailed description of the fast adaptation algorithm is given in [26]

5 Practical Use of the Proposed Reinforcement Technique

For introducing the proposed application way of SVD based fuzzy rule based approximation techniques in reinforcement learning, two simple application examples were chosen.

The first is a simple, for the sake of visualisation of the action value (Q) function, a one dimensional state-space system characterised by the following state-transition function (10):

$$s^{k+1} = 2 \cdot \left(s^k + a^k\right), \tag{10}$$

where $s \in S = [-1,1]$ is the one dimensional state and $a \in A = [-0.2, 0.2]$ is the action.

The reward is calculated in the following manner:

$$r = 1 \;\; iff \;\; s \in [-0.1, 0.1] \;\; else \;\; r = 0 \tag{11}$$

The second example is the well known cart-pole balancing application characterised by the following state-transition functions (12,13):

$$\ddot{x} = \frac{f - b_C \cdot \dot{x} + m \cdot l \cdot \sin(\varphi) \cdot \dot{\varphi}^2 - m \cdot g \cdot \cos(\varphi) \cdot \sin(\varphi) + b_P \cdot \dot{\varphi} \cdot \cos(\varphi)/l}{M + m - m \cdot \cos(\varphi)^2} \tag{12}$$

$$\ddot{\varphi} = \frac{(f - b_C \cdot \dot{x}) \cdot \cos(\varphi) - (M + m) \cdot (g \cdot \sin(\varphi) - b_P \cdot \dot{\varphi} / (l \cdot m)) + m \cdot l \cdot \cos(\varphi) \cdot \sin(\varphi) \cdot \dot{\varphi}^2}{m \cdot l \cdot \cos(\varphi)^2 - (M + m) \cdot l} \tag{13}$$

where x is the position of the cart, φ is the angle of the pole, f is the actuating force, m is the mass of the pole, l is the length of the pole, M is the mass of the cart, g is the gravity acceleration, b_C is the friction coefficient of the cart, b_P is the friction coefficient of the pole.

The reward in the cart-pole balancing example is calculated in the following manner:

$$r = 1 \ \textit{iff} \ \varphi \in [-0.0025 \cdot \pi, 0.0025 \cdot \pi] \ \textit{else} \ r = 0 \tag{14}$$

The first experiment is related to the efficiency of the proposed dynamic partition allocation method, and based on the first application example (see results on fig.5). Fig.5.b is introducing the two basic problems of fixed partition. The lack of universal approximator property in case of rough partition and the difficulties of adaptivity.

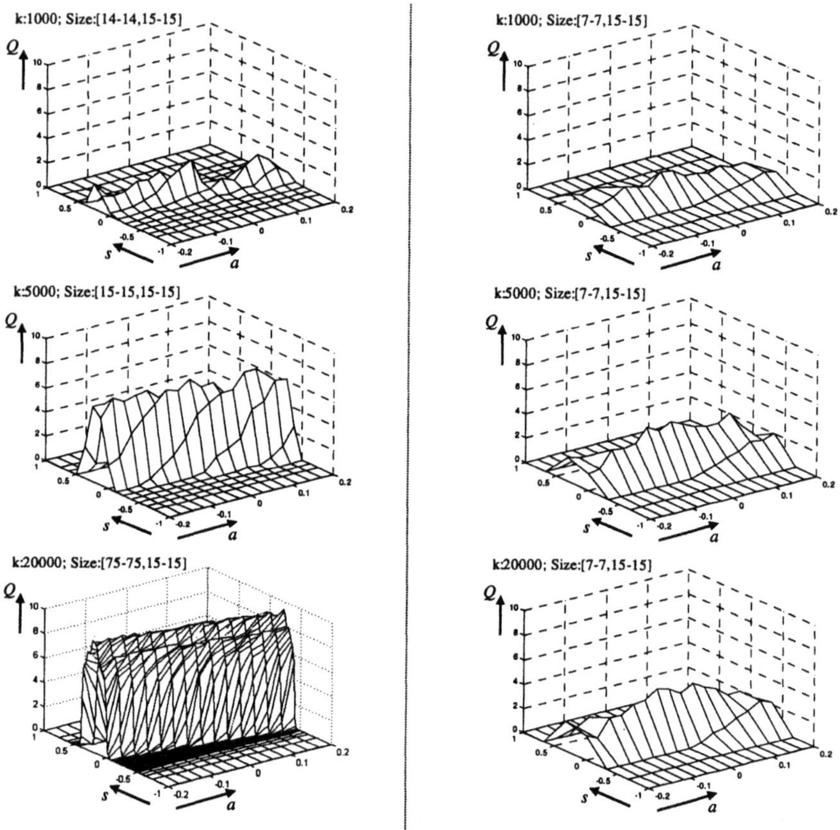

a., Dynamic partition allocation

b., Fixed, 7 equidistant set partition

Fig. 5. First application; dynamic and fixed partition allocation (k is the iteration number, Size: [S,A] set numbers)

The second experiment is related to the efficiency of the proposed SVD based complexity reduction and approximation adaptation (fast adaptation method). This experiment is based on the first and second application example (see results on fig.6. and fig.7.). In case of the first example application fig.5.a. introduces three stages of a 20000 step iteration. On fig.6.a. the same iteration process turns the action-value rule base to reduced form at the iteration step 5000, by applying the *Method 1* of section 4.2. From this step the iteration is continuing up to 20000 iterations using the fast adaptation method (*Method 2* of section 4.3.). Fig.6.b. is the same experiment as fig.6.a., except the turning the reduction is done earlier at the step 1000.

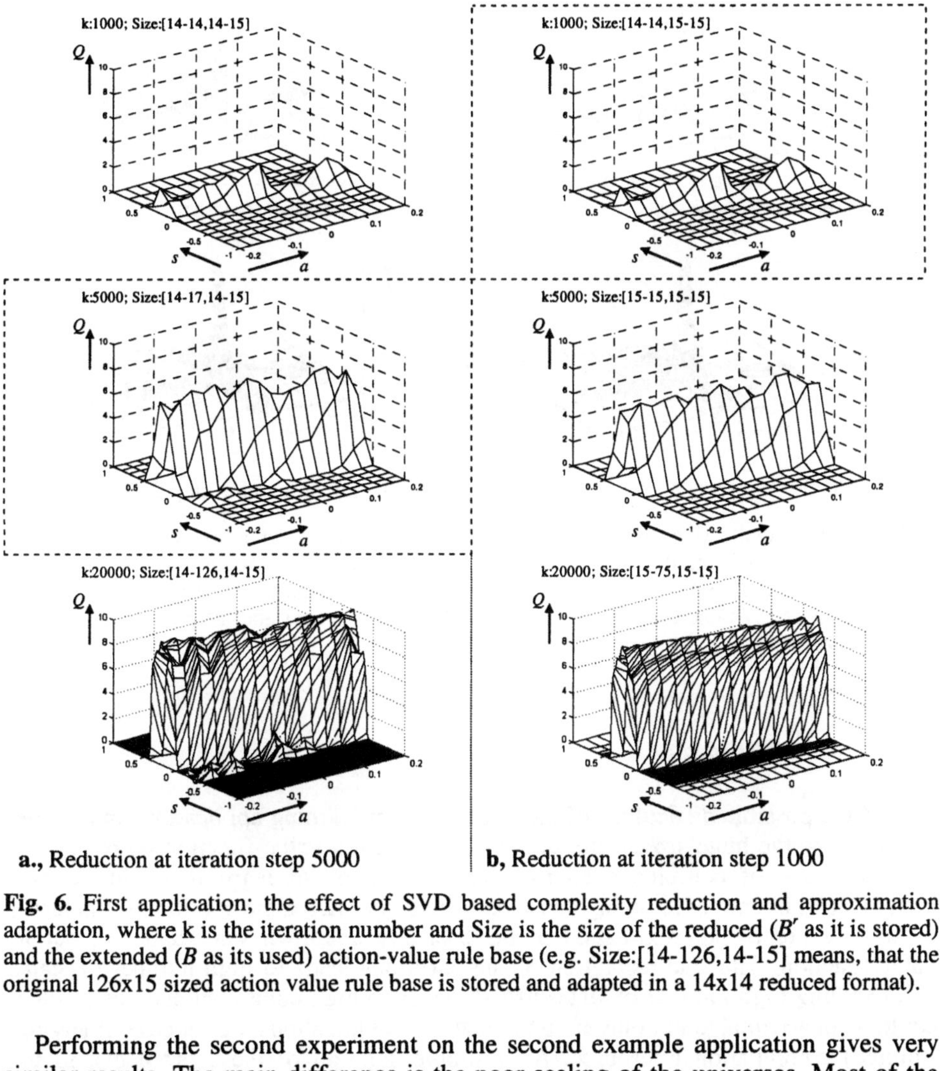

a., Reduction at iteration step 5000 **b,** Reduction at iteration step 1000

Fig. 6. First application; the effect of SVD based complexity reduction and approximation adaptation, where k is the iteration number and Size is the size of the reduced (*B′* as it is stored) and the extended (*B* as its used) action-value rule base (e.g. Size:[14-126,14-15] means, that the original 126x15 sized action value rule base is stored and adapted in a 14x14 reduced format).

Performing the second experiment on the second example application gives very similar results. The main difference is the poor scaling of the universes. Most of the reinforcements are gained in a very small area of the state space. Fixed partition state descriptions could have difficulties in case of inappropriate universe scaling, while the proposed dynamic partition allocation method simply overcome the situation.

As the main conclusion of the second experiment, it seems that in many cases the action-value function is considerably reducible. Moreover due to the fast adaptation method this reduction can be performed in an early stage of the adaptation and the iteration steps can be continued on an economic sized structure.

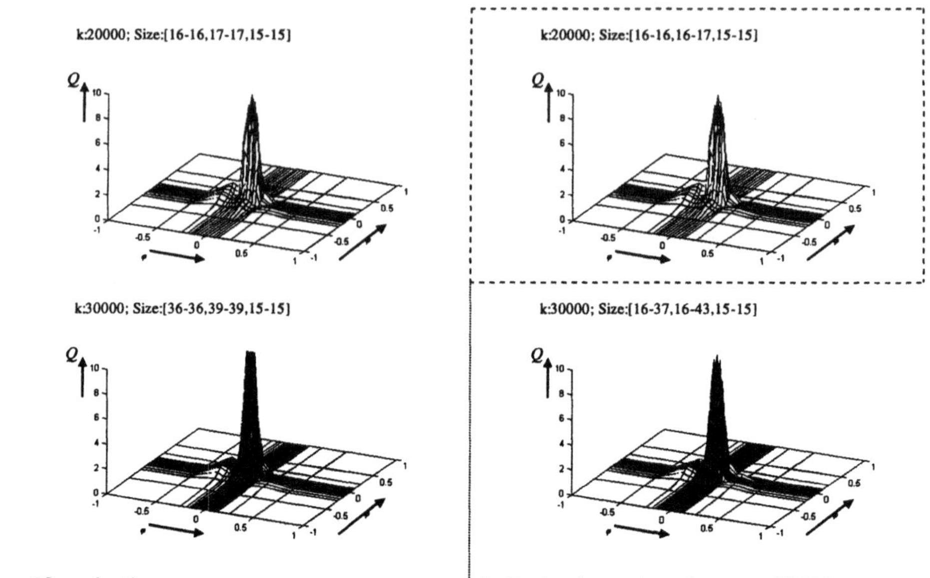

a., No reduction

b, Reduction at iteration step 20000

Fig. 7. Second application; the effect of SVD based complexity reduction and approximation adaptation, where k is the iteration number and Size is the size of the reduced (B' as it is stored) and the extended (B as its used) action-value rule base (e.g. Size:[16-37,16-43,15-15] means, that the original 37x43x15 sized action value rule base is stored and adapted in a 16x16x15 reduced format).

6 Conclusions

One of the possible difficulties of the reinforcement learning applications in complex situations, is the huge size of the state-value- or action-value-function representation [2]. The case of continuous environment reinforcement learning could be even complicated, in case of applying dense partitions to describe the continuous universes, to achieve precise approximation of the basically unknown state-value- or action-value-function. The fine resolution of the partitions leads to high number of states, and handling high number of states usually leads to high computational costs, which could be unacceptable not only in many real time applications, but in case of any real (limited) computational resource. As a simple solution of these problems, in this paper the adoption of Higher Order SVD [13] based fuzzy rule base complexity reduction techniques and its fast adaptation method [26] is suggested. The application of the fast adaptation method [26] gives a simple way for increasing the rule density of a rule base stored in a compressed form directly. To fully exploit this feature, a dynamic partition allocation method is also suggested.

Based on the application examples, the main conclusion of this paper is the reducibility of action-value function. It seems that in many cases the representation of the action-value function is considerably reducible. Moreover due to the fast adaptation method this reduction can be performed in an early stage of the adaptation and the iteration steps can be continued on an economic sized action-value function representation.

References

1. Bellman, R. E.: Dynamic Programming. Princeton University Press, Princeton, NJ (1957)
2. Sutton, R. S., A. G. Barto: Reinforcement Learning: An Introduction, MIT Press, Cambridge (1998)
3. Yam, Y., Baranyi, P., Yang, C. T.: Reduction of Fuzzy Rule Base Via Singular Value Decomposition. IEEE Transaction on Fuzzy Systems. Vol.: 7, No. 2 (1999) 120-131.
4. Baranyi, P., Yam, Y.: Fuzzy rule base reduction. Chapter 7 of Fuzzy IF-THEN Rules in Computational Intelligence: Theory and Applications Eds., D. Ruan and E.E. Kerre, Kluwer (2000) 135-160
5. Watkins, C. J. C. H.: Learning from Delayed Rewards. Ph.D. thesis, Cambridge University, Cambridge, England (1989)
6. Appl, M.: Model-based Reinforcement Learning in Continuous Environments. Ph.D. thesis, Technical University of München, München, Germany, dissertation.de, Verlag im Internet (2000)
7. Wang, L.X.: Fuzzy Systems are Universal Approximators. Proceedings of the First IEEE Conference on Fuzzy Systems, San Diego (1992) 1163-1169
8. Castro, J.L.: Fuzzy Logic Controllers are Universal Approximators. IEEE Transaction on SMC, Vol.25, 4 (1995)
9. Horiuchi, T., Fujino, A., Katai, O., Sawaragi, T.: Fuzzy Interpolation-Based Q-learning with Continuous States and Actions. Proc. of the 5th IEEE International Conference on Fuzzy Systems, Vol.1 (1996) 594-600
10. Glorennec, P.Y., Jouffe, L.: Fuzzy Q-Learning. Proc. of the 6th IEEE International Conference on Fuzzy Systems (1997) 659-662
11. Berenji, H.R.: Fuzzy Q-Learning for Generalization of Reinforcement Learning. Proc. of the 5th IEEE International Conference on Fuzzy Systems (1996) 2208-2214
12. Bonarini, A.: Delayed Reinforcement, Fuzzy Q-Learning and Fuzzy Logic Controllers. In Herrera, F., Verdegay, J. L. (Eds.) Genetic Algorithms and Soft Computing, (Studies in Fuzziness, 8), Physica-Verlag, Berlin, D, (1996) 447-466
13. Baranyi, P., Várkonyi-Kóczy, A., Yam, Y., Patton, R.J., Michelberger, P., Sugiyama M.:SVD Based Reduction of TS Models. IEEE Trans. Industrial Electronics (accepted with minor revision)
14. Yen, J., Wang, L.: Simplifying Fuzzy Rule-based Models Using Orthogonal Transformation Methods. IEEE Trans. SMC, Vol 29: Part B, No. 1 (1999) 13-24
15. Yam, Y.: Fuzzy approximation via grid point sampling and singular value decomposition. IEEE Trans. SMC, Vol. 27 (1997) 933-951
16. Kóczy, L.T., Hirota, K.: Size Reduction by Interpolation in Fuzzy Rule Bases. IEEE Trans. SMC, vol. 27 (1997) 14-25
17. Tikk, D.: On nowhere denseness of certain fuzzy controllers containing prerestricted number of rules. Tatra Mountains Mathematical Publications vol. 16. (1999) 369-377
18. Lei, K., Baranyi, P., Yam, Y.: Complexity Minimalisation of Non-singleton Based Fuzzy-Neural Network. International Journal of Advanced Computational Intelligence, vol.4, no.4 (2000) 1-8
19. Baranyi, P., Yam, Y., Várlaki, P., Michelberger, P.: Singular Value Decomposition of Linguistically Defined Relations. Int. Jour. Fuzzy Systems, Vol. 2, No. 2, June (2000) 108-116

20. Song, F., Smith, S.M.: A Simple Based Fuzzy Logic Controller Rule Base Reduction Method. IEEE Int. Conf. System Man and Cybernetics (IEEE SMC' 2000), Nashville, Tennessee, USA (2000) 3794-3798
21. Setnes, M., Hellendoorn, H.: Orthogonal Transforms for Ordering and Reduction of Fuzzy Rules. 9th IEEE Int. Conf. on Fuzzy Systems (FUZZ-IEEE 2000), San Antonio, Texas (2000) 700-705
22. Sudkamp, T., Knapp, A., Knapp J.: A Greedy Approach to Rule Reduction in Fuzzy Models. IEEE Int. Conf. System Man and Cybernetics (IEEE SMC'2000), Nashville, Tennessee, USA (2000) 3716-3721
23. Baranyi, P., Yam, Y., Yang, C.T., Várlaki, P., Michelberger, P.: Generalised SVD Fuzzy Rule Base Complexity Reduction. International Journal of Advanced Computational Intelligence (accepted, to be printed in 2001)
24. Baranyi, P., Yam, Y.: Singular Value-Based Approximation with Non-Singleton Fuzzy Rule Base. 7th Int. Fuzzy Systems Association World Congress (IFSA'97) Prague (1997) 127-132
25. Baranyi, P., Yam, Y.: Singular Value-Based Approximation with Takagi-Sugeno Type Fuzzy Rule Base. 6th IEEE Int. Conf. on Fuzzy Systems (FUZZ-IEEE'97) Barcelona, Spain (1997) 265-270
26. Baranyi, P., Várkonyi-Kóczy, A.R., Yam, Y., Várlaki, P., Michelberger, P.: An Adaption Technique to SVD Reduced Rule Bases. IFSA 2001, Vancouver (accepted for presentation)

Analyzing Directed Acyclic Graph Recombination

Carlos Cotta and José M. Troya

Dept. Lenguajes y Ciencias de la Computación,
ETSI Informática (Office 3.2.49), University of Málaga,
Campus de Teatinos, 29071 - Málaga - SPAIN
ccottap@lcc.uma.es

Abstract. This work studies the edge-based representation of directed acyclic graphs, as well as the properties of recombination operators working on it. It is shown that this representation is not separable, and the structure of the basic information units that must be processed in order to maintain feasibility of the solutions is described. As an experimental analysis indicates, a recombination operator using these units has subquadratic complexity in the graph size. It is also shown that a standard gene-transmission recombination operator is biased to produce solutions of lower edge-density than the parents' average. An unbiased allelic recombination operator provides better results on an ad-hoc test problem.

1 Introduction

Given a digraph $G(V, E)$, where $V = \{v_1, \cdots, v_n\}$ is a set of vertices and $E \subseteq V \times V$, a vertex v_j is said to be a descendant of v_i if, and only if, $(v_i, v_j) \in E$, or $(v_k, v_j) \in E$ and v_k is a descendant of v_i. A Directed Acyclic Graph (DAG) is a digraph $DAG(V, E)$, where E verifies that no $v_i, v_j \in V$ exist such that v_i is a descendant of v_j and vice versa.

DAGs are important data structures since they can be used to represent many relevant entities such as *field programmable gate arrays* (FPGAs) [1], Bayesian networks [11, 7], task graphs [2], scene graphs [15], dependence graphs [8, 6] etc. The synthesis of such entities is generally complex, being often approached via evolutionary algorithms (e.g., [5, 9, 12]). For this reason so it is worth analyzing application issues of evolutionary heuristics to this task.

In this work, we focus on DAG recombination from the perspective of an edge-based representation of DAGs. Both this representation and the innards of information transmission during recombination are theoretically analyzed within the framework of Forma Analysis [13]. For this purpose, some background information on Forma Analysis is given (Section 2) prior to the mentioned analysis (Section 3). Subsequently, several variants of recombination are empirically studied in terms of time complexity, edge-transmission ratios and algorithmic performance (Section 4). Finally, some conclusions and future work are outlined (Section 5).

B. Reusch (Ed.): Fuzzy Days 2001, LNCS 2206, pp. 739–748, 2001.

2 Background

Let $\Phi = \{\phi_1, \cdots, \phi_n\}$ be a set of n equivalence relations defined over a discrete search space $\mathcal{S}$. Let Ξ_ϕ be the set of equivalence classes induced by ϕ, and let Ξ_Ψ ($\Psi \subseteq \Phi$) be the set of vectors of equivalence classes induced by equivalence relations in Ψ. Let $[x]_\phi$ be the equivalence class to which x belong under ϕ. If it holds for Φ that for any two different solutions $x, y \in \mathcal{S}$, there exists $\phi \in \Phi$ such that $[x]_\phi \neq [y]_\phi$, then Φ *covers* the search space $\mathcal{S}$. If Φ is independent and covers the search space $\mathcal{S}$, then each solution $x \in \mathcal{S}$ can be represented as a string $\langle [x]_\phi \mid \phi \in \Phi \rangle$, i.e., $x = \langle \eta_1, \eta_2, \cdots, \eta_n \rangle \Leftrightarrow \{x\} = \cap_{i=i}^{n} \eta_i$. Being any ambiguity clear from the context, the same notation is used for equivalence classes and their labels. Each $\phi \in \Phi$ is analogous to a gene, and each $\eta \in \Xi_\phi$ is analogous to an allele for this gene.

The *Dynastic Potential* $\Gamma(\{x,y\})$ of x and y is defined as

$$\Gamma(\{x,y\}) = \bigcap_{\phi \in \Phi} ([x]_\phi \cup [y]_\phi) \ , \tag{1}$$

i.e., the set of solutions that only carry information present in x or y. On the other hand, the *Similarity Set* $\Sigma(\{x,y\})$ is defined as

$$\Sigma(\{x,y\}) = \bigcap_{\phi \in \Phi, [x]_\phi = [y]_\phi} [x]_\phi \ , \tag{2}$$

i.e., the largest forma that contains x and y [14]. Notice that $\Gamma(\{x,y\}) \subseteq \Sigma(\{x,y\})$. If for any alleles η, ζ such that $\eta \cap \zeta \neq \emptyset$, and $x \in \eta$, $y \in \zeta$, it holds that $\eta \cap \zeta \cap \Sigma(\{x,y\}) \neq \emptyset$, then Φ is *separable*. The notation $\eta \triangleright \Psi$ denotes that Ψ is the intersection of several alleles, including η. Similarly, the notation $\Theta \triangleright \Psi$ denotes that Ψ is the intersection of several alleles, including all $\theta \triangleright \Theta$.

3 DAG Recombination

Let $V = \{v_1, \cdots, v_n\}$ be the set of available vertices, and let $\mathcal{S}_V$ be the search space composed of all DAGs defined over V [1]. Given $z \in \mathcal{S}_V$, let $E(z)$ be the set of edges in z. Now let us consider a family of equivalence relations ϕ_{ij} defined over $\mathcal{S}_V$ as

$$\phi_{ij}(x,y) = \texttt{TRUE} \Leftrightarrow [(v_i, v_j) \in E(x) \Leftrightarrow (v_i, v_j) \in E(y)] \ . \tag{3}$$

Each equivalence relation ϕ_{ij} induces two formae ϕ_{ij}^1 and ϕ_{ij}^0, respectively comprising the solutions that include/exclude edge (v_i, v_j). Let $\Phi = \{\phi_{ij} \mid 1 \leq i, j \leq n,\ i \neq j\}$. This set of equivalence relations can be used to induce a representation of solutions in $\mathcal{S}_V$. To show this, it is necessary to prove that (a) Φ is independent, and (b) Φ covers $\mathcal{S}_V$. This is done below.

[1] This search space has an enormous size. There is no closed form for $|\mathcal{S}_V|$, but it grows super-exponentially with $|V|$. For instance, $|\mathcal{S}_V| = 543$ for $|V| = 4$, and $|\mathcal{S}_V| \approx 10^{18}$ for $|V| = 10$ (cf. [10])

Lemma 1. *Let $\Phi' \subset \Phi$, and let $\phi_{ij} \in \Phi - \Phi'$. There always exists an allele $\eta \in \Xi_{\phi_{ij}}$ such that for some $x, y \in \eta$, no allele set $\Omega \in \Xi_{\Phi'}$ can be found for which $\{x, y\} \subseteq \bigcap_{\xi \in \Omega} \xi$.*

Proof. The proof is done by induction on the cardinality of Φ'. Initially, let $\Phi' = \{\phi_{kl}\}$. Let $\eta \equiv \phi^0_{ij}$. Let $x, y \in \eta$ be such that $(k, l) \in E(x)$, and $(k, l) \notin E(y)$. These two solutions exists (for instance, consider $E(x) = \{(k, l)\}$, and $E(y) = \emptyset$). Now, Ω can take two values: $\{\phi^0_{kl}\}$, or $\{\phi^1_{kl}\}$. Clearly, x does not belong to ϕ^0_{kl} in the first case, and the same holds for y with respect to ϕ^1_{kl} in the second case. Hence the base case is established. We now assume that the lemma holds whenever $|\Phi'| = k$. Subsequently, we consider the case $|\Phi'| = k+1$. Let $\Phi' = \Psi \cup \{\phi_{kl}\}$. Then, any $\Omega \in \Xi_{\Phi'}$ can be expressed as $\Theta \cup \{\zeta\}$, where $\Theta \in \Xi_\Psi$ and $\zeta \in \Xi_{\phi_{kl}}$. Assume that the lemma is false in this situation. Thus, for all $\eta \in \Xi_{\phi_{ij}}$ and $x, y \in \eta$, there exists $\Omega \in \Xi_{\Phi'}$ such that $\{x, y\} \subseteq \bigcap_{\xi \in \Omega} \xi$. However, $\{x, y\} \subseteq \bigcap_{\xi \in \Omega} \xi$ implies that $\{x, y\} \subseteq \bigcap_{\xi \in \Theta} \xi$ (recall $\Theta \subset \Omega$). But this is false according to the induction hypothesis. Since we arrive at a contradiction, the lemma must also hold in this case. □

Proposition 1. *The set of equivalence relations Φ is independent and covers the search space $\mathcal{S}_V$.*

Proof. (Independence) It must be shown that given $\phi_{ij} \in \Phi$, no $\Phi' \subseteq (\Phi - \{\phi_{ij}\})$ exists such that $\phi_{ij} \equiv \bigcap_{\psi \in \Phi'} \psi$, i.e., ϕ_{ij} cannot be expressed as the intersection of other equivalence relations in Φ. The proof is done by absurdity. Let $\phi_{ij} \equiv \bigcap_{\psi \in \Phi'} \psi$. This implies that for any $\eta \in \Xi_{\phi_{ij}}$, an allele set $\Omega \in \Xi_{\Phi'}$ exists for which $\eta \equiv \bigcap_{\xi \in \Omega} \xi$. Clearly, this means that for all $x \in \eta$, it holds that $x \in \bigcap_{\xi \in \Omega} \xi$. But this contradicts Lemma 1, so the initial assumption ($\phi_{ij} \equiv \bigcap_{\psi \in \Phi'} \psi$) must be false.

(Coverage) It must be shown that for all $x, y \in \mathcal{S}_V$ ($x \neq y$), there exists $\phi_{ij} \in \Phi$ such that $\phi_{ij}(x, y) = \texttt{FALSE}$. This is easy to prove since –without loss of generality– $x \neq y$ implies that $v_i, v_j \in V$ exist, such that $(v_i, v_j) \in E(x)$ but $(v_i, v_j) \notin E(y)$. Hence, $x \in \phi^1_{ij}$ and $y \in \phi^0_{ij}$, i.e., $\phi_{ij}(x, y) = \texttt{FALSE}$. □

Proposition 1 shows that any DAG $x \in \mathcal{S}_V$ can be univocally and compactly represented as a string $\langle \phi^k \mid \phi \in \Phi, k \in \{0, 1\}\rangle$. As shown below, this representation induced by Φ is not separable.

Proposition 2. *The set of equivalence relations Φ is not separable.*

Proof. The proof is done by example. Let $v_i, v_j, v_k \in V$ be three different vertices. Clearly, it holds that

$$\phi^1_{ij} \cap \phi^1_{jk} \neq \emptyset, \quad \phi^1_{ij} \cap \phi^1_{ki} \neq \emptyset, \quad \phi^1_{jk} \cap \phi^1_{ki} \neq \emptyset, \text{ and} \tag{4}$$

$$\phi^1_{ij} \cap \phi^1_{jk} \cap \phi^1_{ki} = \emptyset \; . \tag{5}$$

Now, let $x \in \phi^1_{ij} \cap \phi^1_{ki}$, and let $y \in \phi^1_{jk} \cap \phi^1_{ki}$. It follows that the similarity set of x and y is a subset of ϕ_{ki}. Then, being ϕ^1_{ij} and ϕ^1_{jk} two compatible formae according to Eq.(4), two solutions $x \in \phi^1_{ij}$ and $y \in \phi^1_{jk}$ exist for which $\Sigma(\{x, y\}) \cap \phi^1_{ij} \cap \phi^1_{jk} = \emptyset$ (recall Eq.(5)). Hence, Φ is not separable. □

The non-separability of Φ implies that a recombination operator processing the induced representation must manipulate macro-units of information in order to preserve feasibility in the descendant being created. These macro-units are termed *transmission sets*, and can be defined as follows:

Definition 1. *The transmission set $T(\Psi,\eta)$ of an allele η is the closure of the following expressions:*

$$\eta \triangleright T(\Psi,\eta) \tag{6}$$

$$[\exists! \eta' \in \Xi_{\varphi'} \ : \ \Psi \cap T(\Psi,\eta) \cap \eta' \neq \emptyset] \Rightarrow \eta' \triangleright T(\Psi,\eta), \tag{7}$$

where Ψ is the partially constructed descendant.

Transmission sets are a more general version of *compatibility sets* [3], in which the descendant is not forced to belong to the dynastic potential of the solutions x and y being recombined. Nevertheless, transmission sets can be shown to preserve feasibility *within* $\Gamma(\{x,y\})$ as long as Ψ is initialized as the similarity set $\Sigma(\{x,y\})$. This is formally established in the following proposition:

Proposition 3. *Let $|\Xi_\phi| = 2$ for all $\phi \in \Psi$. Then, $\Upsilon(\Psi,\eta,x,y) \equiv T(\Psi,\eta)$ for any two solutions x and y, whenever $\Psi \triangleright \Sigma(\{x,y\})$.*

Proof. (Sketch) Essentially, a compatibility set $\Upsilon(\Psi,\eta,x,y)$ is the intersection of all those alleles present in x or y that must be injected in the descendant $z \in \Psi$, so as to ensure feasibility, simultaneously forcing $z \in \Gamma(\{x,y\})$. It can be seen that when $|\Xi_\phi| = 2$, either $x, y \in \eta$ for some $\eta \in \phi$ (and hence $\eta \triangleright \Sigma(\{x,y\})$), or $x \in \eta$ and $y \in \eta'$, $\Xi_\phi = \{\eta,\eta'\}$. This means that any feasible solution $z \in \Sigma(\{x,y\})$ also belongs to $\Gamma(\{x,y\})$, i.e., $\Sigma(\{x,y\}) \equiv \Gamma(\{x,y\})$. Thus, a transmission set will ensure feasibility within $\Gamma(\{x,y\})$ whenever $\Psi \triangleright \Sigma(\{x,y\})$. □

This initial transmission of common alleles has been the approach considered in this work. Within this context, transmission sets must be computed taking into account that if an edge (i,j) is added to the descendant, any edge (j,k) –where k is an ancestor of i in the partially-specified solution– is forbidden. On the contrary, excluding an edge (i,j) from the descendant does not impose any restriction on the addition/excusion of other edges. This can be formalized as follows: let Ψ be a partially specified descendant; let $\Omega(\Psi)$ be a DAG such that $\phi^1_{ij} \triangleright \Psi \Leftrightarrow (i,j) \in E(\Omega(\Psi))$, i.e., $\Omega(\Psi)$ is a graph exclusively composed of the edges already added to the descendant; let C_D be the adjacency matrix of a digraph D, and let C_D^∞ be the transitive closure of C_D; then

$$T(\Psi,\phi^0_{ij}) = \phi^0_{ij}, \quad \text{and} \tag{8}$$

$$T(\Psi,\phi^1_{ij}) = \phi^1_{ij} \cap \bigcap_{C^\oplus_{rs}=1} \phi^0_{sr} \tag{9}$$

where $C^\oplus = C^\infty_{\Omega(\Psi)}$ XOR $C^\infty_{\Omega(\Psi \cap \phi^1_{ij})}$. Now, two main approaches to recombination can be considered on the basis of these units. Both start from $\Sigma(\{x,y\})$; subsequently, an unspecified gene is selected and an allele for that gene (and the corresponding transmission set) is taken from either of the parents. The process is repeated until all genes are specified (gene transmission – GT), or until a certain feasibility criterion is met, being unspecified genes assigned a default value (allele transmission – AT). In the former situation, only ***positive*** alleles (i.e., ϕ^1_{ij}-like alleles) are explicitly transmitted, being the default value a ***negative*** allele (i.e., ϕ^0_{ij}-like). Both approaches are studied below.

4 Experimental Results

The first issue to be tackled is the computational cost of the operators. Calculating the connectivity matrix of a graph can be naïvely done in $O(n^3)$ using Dijkstra's algorithm. Since the number of compatibility sets to be calculated is $O(n^2)$, this would amount to $O(n^5)$, but this is a very overestimated bound (incompatible worst-cases are superimposed).

To obtain a more realistic bound, an empirical assessment has been done. To be precise 200 random DAGs of fixed density and increasing size (5 up to 55 nodes) have been generated and recombined. The results are shown in Fig. 1 for GT and in in Fig. 2 for AT [2]. As it can be seen, a cubic behavior in the number of vertices is found. This cubic pattern is present in both GT and AT, with independence of the graph density of the DAGs recombined. Obviously, the larger the density of the DAGs being recombined, the larger the number of transmission sets that must be computed using Eq. (9). This results in a larger proportionality coefficient. In any case, the grow of this coefficient ceases and stabilizes around $\delta = .5$ (see Fig. 4-left); no significant increase of computational cost is observed in GT beyond that point. Notice also that this cubic behavior in the number of vertices implies a sub-quadratic behaviour ($O(|x|^{3/2})$) in the DAG size since $|x| \in O(n^2)$.

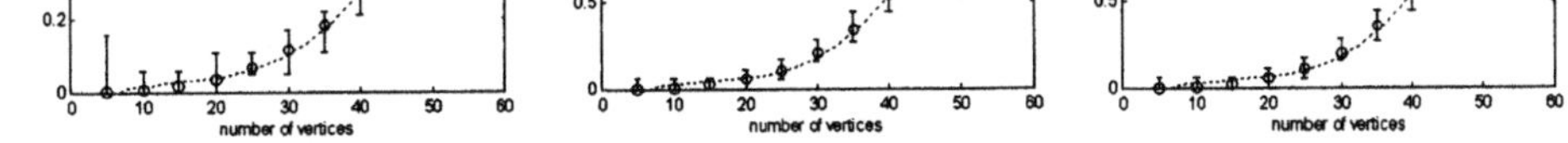

Fig. 1. Time required to recombine two DAGs using GT for increasing graph sizes. The results are shown for different graph densities: $\delta = .1$ (left), $\delta = .5$ (middle), and $\delta = .9$ (right). A fit to a cubic polynomial is shown.

[2] All times have been measured using MatLab® in a Pentium® III 600MHz.

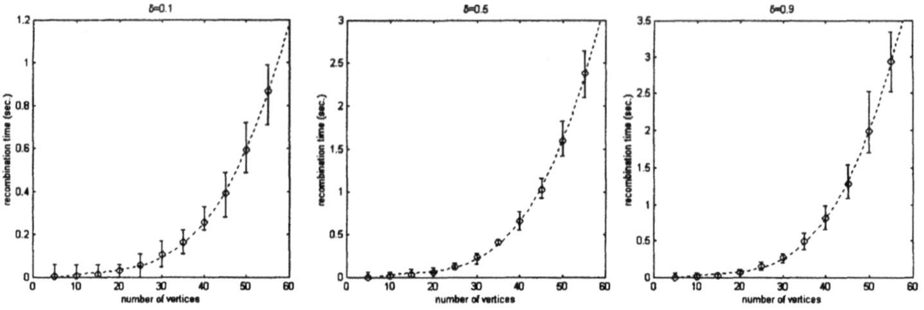

Fig. 2. Time required to recombine two DAGs using AT for increasing graph sizes. The results are shown for different graph densities: $\delta = .1$ (left), $\delta = .5$ (middle), and $\delta = .9$ (right). A fit to a cubic polynomial is shown.

Next, the transmission rates are studied. Due to the problem constraints, the density of the descendant is skewed to be lower that the mean density of the parents when using GT (according to Eq. (9) transmitting an edge of a parent can rule out a number of edges from the other parent). The higher the parents' mean density, the stronger this effect as shown in Fig. 3). The same is true for fixed density and increasing number of vertices. Actually, the descendant density seems to be related to the parents' mean density through an exponential function $\delta_{\text{descendant}} = \delta_{\text{parents}}^{\alpha}$ for some $\alpha < 1$. A fit to such a function reveals that -as in the previous case- the magnitude of α stabilizes around $\delta = .4$ or $\delta = .5$ (see Fig. 4-right). This skew in the generation of descendants can have a detrimental influence in the performance of a GA using GT, were high density DAGs sought. This fact is explored below.

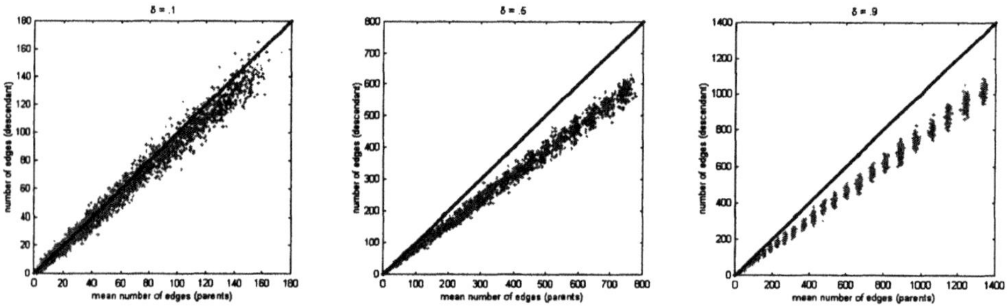

Fig. 3. Progeny density as a function of parental density using GT. The results are shown for different graph densities: $\delta = .1$ (left), $\delta = .5$ (middle), and $\delta = .9$ (right).

Experiments have been done using a steady-state genetic algorithm (binary tournament selection, *popsize*=100, $p_c = .9$, $p_m = 1/n^2$) in a DAG matching

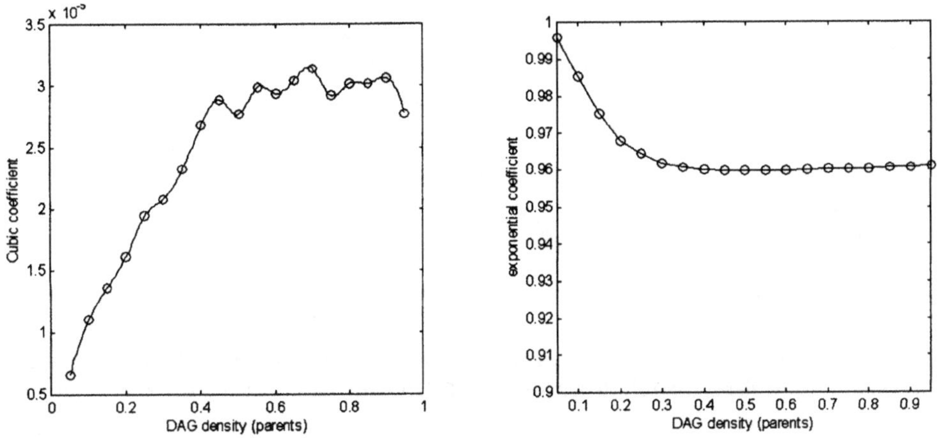

Fig. 4. (Left) Value of the cubic coefficient in the fitted-to-time polynomial. (Right) Value of the exponential coefficient in the fitted-to-density curve.

problem. Three scenarios are considered, consisting of finding a target DAG of a given density ($\delta \in \{.1, .5, .9\}$ in these experiments). The number of vertices is $n = 15$, and the fitness function is the number of different entries in the adjacency matrices of the actual DAG the target DAG. Besides GT and AT, two additional approaches are considered for comparison purposes: UX plus penalty term, and UX plus a repair function. The penalty term is computed in the fist case as the $n^2 \cdot v$, where n is the total number of vertices, and v is the number of vertices connected to an ancestor of itself. As to the repair function, it consists of deleting edges from the graph until no loops remain. In both cases, the initial population is entirely composed of feasible solutions. Finally, the AT operator is adjusted to transmit a number of positive alleles following a distribution centered in the parents' mean number of edges (binomial, $p = 1/2$). This adjusted version of AT produces non-skewed descendants as shown in Fig. 5.

The results averaged for 50 runs are shown in Fig. 6. As it can be seen, AT performs similarly to GT when the density of the target DAG is low (the skew of GT is very small in this case). However, AT is increasingly better when the density of the target DAG is higher, being clearly superior for $\delta = .9$. The same holds with respect to the repairing and penalizing approaches. Again, the difference is stronger when the density of the target DAG is higher, since the chances of getting into the feasible region are lower using random recombination. Notice also that these two latter approaches are straightforwardly defined on this simple problem, but they can be difficult to apply in a more complex situation, e.g., the fitness function might be undefined for infeasible solutions, the repair function might introduce excessive disruption, etc.

Experiments have been also done with GAR (GT starting form a fully unspecified descendant rather than from $\Sigma(\{x, y\})$). The results are virtually identical

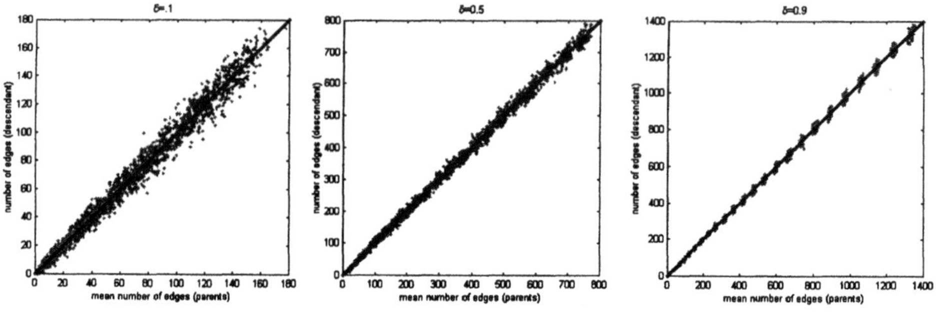

Fig. 5. Progeny density as a function of parental density using adjusted AT. The results are shown for different graph densities: $\delta = .1$ (left), $\delta = .5$ (middle), and $\delta = .9$ (right).

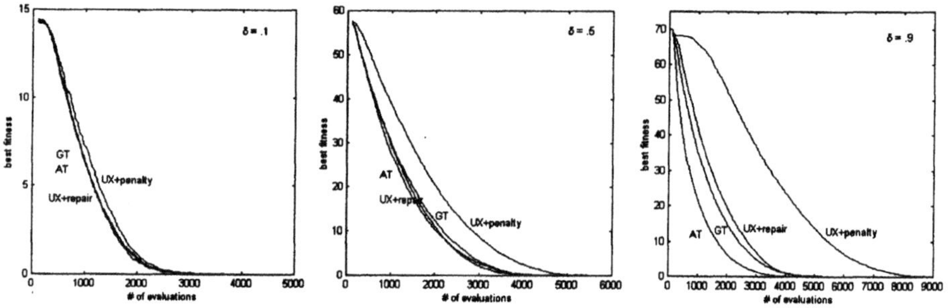

Fig. 6. Results (averaged for 50 runs) of a genetic algorithm using different recombination operators. From left to right: $\delta = .1$, $\delta = .5$, and $\delta = .9$.

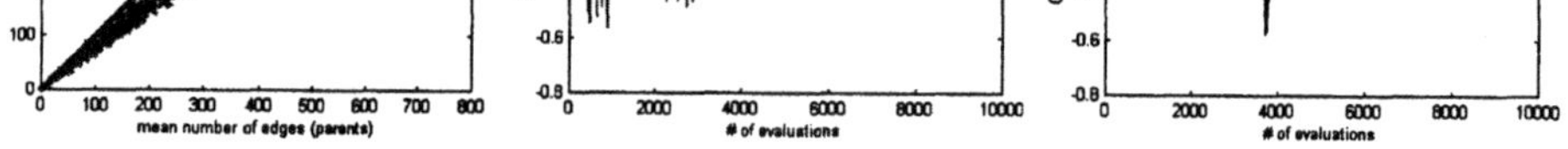

Fig. 7. (Left) Progeny density as a function of parental density using adjusted GAR for $\delta = .5$. (Middle) Best-fitness differential for GT vs. GAR (results averaged for 50 runs). (Right) Mean-fitness differential for GT vs. GAR (results averaged for 50 runs).

to GT (see Fig. 7-middle/right; the fitness differential closely oscillates around zero). This is an expected result, since an off-line assessment shows that the transmission rates of GAR are very similar to GT (Fig. 7-left).

5 Conclusions

This work has studied an edge-based representation of DAGs, and the properties of recombination operators working on it. Regarding this representation, its non-separability has been established, and the structure of the basic units that must be processed in order to maintain feasibility has been described. As to the operators, it has been shown that recombination in the feasible space can be done in sub-quadratic time in the DAG size. It has been also shown that a gene-transmission operator (GT) is skewed to generate DAGs with lower density than the parents average; this can be remedied by using an adjusted version of an allele-transmission operator (AT) that is shown to perform better than both GT and other approaches on a DAG matching problem.

Future work will be directed to confirm these results of other test problems. An especially interesting case is restricting the search to a subset of $\mathcal{S}_V$ (e.g., designing causal networks using low-order statistics to build an undirected skeleton of the network [4]).

Acknowledgement

This work is partially supported by the Spanish *Comisión Interministerial de Ciencia y Tecnología* (CICYT) under grant TIC99-0754-C03-03.

References

1. K. Chen, J. Cong, Y. Ding, A. Kahng, and P. Trajmar. DAG-Map: Graph-based FPGA technology mapping for delay optimization. *IEEE Design and Test of Computers*, pages 7–20, September 1992.
2. M. Cosnard and M. Loi. Automatic task graphs generation techniques. *Parallel Processing Letters*, 5(4):527–538, 1995.
3. C. Cotta and J.M. Troya. On the influence of the representation granularity in heuristic forma recombination. In J. Carroll, E. Damiani, H. Haddad, and D. Oppenheim, editors, *ACM Symposium on Applied Computing 2000*, pages 433–439. ACM Press, 2000.
4. L.M. de Campos and J.F. Huete. A new approach for learning belief networks using independence criteria. *International Journal of Approximate Reasoning*, 24(1):11–37, 2000.
5. H. Ehrenburg. Improved direct acyclic graph handling and the combine operator in genetic programming. In J.R. Koza, D.E. Goldberg, D.B. Fogel, and R.L. Riolo, editors, *Genetic Programming 1996: Proceedings of the First Annual Conference*, pages 285– 291, Stanford University, CA, 1996. MIT Press.

6. S. Handley. On the use of a directed acyclic graph to represent a population of computer programs. In *Proceedings of the 1994 IEEE World Congress on Computational Intelligence*, pages 154–159, Orlando, FL, 1994. IEEE Press.
7. D. Heckerman and M. Wellman. Bayesian networks. *Communications of the ACM*, 38:27–30, 1995.
8. D. Kuck, R. Kuhn, B. Leasure, D. Padua, and M. Wolfe. Dependence graphs and compiler optimizations. In *Proceedings of the Eighth Annual ACM Symposium on Principles of Programming Languages*, pages 207–218, 1981.
9. P. Larrañaga, C. Kuijpers, R. Murga, and Y. Yurramendi. Learning bayesian network structures by searching for the best ordering with genetic algorithms. *IEEE Transactions on System, Man and Cybernetics*, 26(4):487–493, 1996.
10. K. Murphy. An introduction to graphical models. Technical report, Intel Research, May 2001.
11. J. Pearl. *Probabilistic Reasoning in Intelligent Systems.* Morgan Kaufmann, Palo Alto, CA, 1988.
12. R. Poli. Parallel distributed genetic programming. In D. Corne, M. Dorigo, and F. Glover, editors, *New Ideas in Optimization*, pages 403–432. McGraw-Hill, 1999.
13. N.J. Radcliffe. Equivalence class analysis of genetic algorithms. *Complex Systems*, 5:183–205, 1991.
14. N.J. Radcliffe. The algebra of genetic algorithms. *Annals of Mathematics and Artificial Intelligence*, 10:339–384, 1994.
15. R. Turner, Song Li, and E. Gobbetti. Metis - an object-oriented toolkit for constructing virtual reality applications. *Computer Graphics Forum*, 18(2):121–130, 1999.

Rough Set Based Knowledge Discovery of Interface for the Internet Usage among Japanese Elderly Women

Hayato Uchida [1], Yutaka Hata [2], Shinro Matsuura [3], Tadahiro Tsuchikawa [4], Yoshio Morotomi [5] and Hideyasu Aoyama [6]

[1]Department of Health Education, Himeji Institute of Technology
1-1-12 Shinzaike-honcho, Himeji, 670-0092, Japan
Phone: +81 792 92 9367 Fax: +81 792 93 5721,
uchida@hept.himeji-tech.ac.jp
[2]Department of Computer Engineering, Himeji Institute of Technology
[3]Matsuura Clinic, Himeji Medical Association
[4]Department of Architectural Environment, Himeji Institute of Technology
[5]Department of Health Education, Nippon Sport Science University
[6]Department of Hygiene and Preventive Medicine, Okayama University Medical School

Abstract. The world currently faces the new wave of industrial civilization, that is regarded as the changes caused by the Revolution of Information Technology (IT). This study is conducted to clarify how a mobile phone (MP) and the Internet through personal computer (IPC) have been actually used. We also examine what kind of interface (input device) is preferred by the elderly, by using a questionnaire regarding the IT equipment. This study adopts not only the χ^2 test but also the rough set analysis. As a result of the rough set analysis and χ^2 test, those who were interested in a MP preferred, Japanese kana syllabary (JKS) the most. The rough set analysis also discovered the facts that those interested in the keyboard had a preference for the IPC and also that those interested in the JKS and the speech preferred a MP. We can conclude that the complementary use of the rough set analysis and χ^2 test bring more powerful knowledge discovery.

1 Introduction

The world currently faces the new wave of industrial civilization, that is regarded as the changes caused by the Revolution of Information Technology (IT). The Japanese government announced their national scheme called "The IT Strategies" in November 2000, and the official white paper on communication published in 2000 describes that the strategies have had strong influence even on most households.

For example, the IT equipment such as mobile phone (MP) and the Internet have been widely accepted by the consumers. However, the instructions for use are complicated and the actual rate of use is not necessarily high. It is particularly true among the elderly, the number of whom has been increasing. In fact, by the year 2020, the proportion of 65-year-old and over people is projected to comprise 26.9 % of the Japanese population.

B. Reusch (Ed.): Fuzzy Days 2001, LNCS 2206, pp. 749–754, 2001.

This study is conducted to clarify how a MP and the Internet through personal computer (IPC) have been actually used. We also examine what kind of interface (input device) is preferred by the elderly, by using a questionnaire regarding the IT equipment.

We normally use only the method of χ^2 test to analyze the relationship between two. However, this study adopts not only the χ^2 test but also the rough set analysis [1] which employs the concepts of certainty and coverage, so that we can compare the results from these two methods.

2 Subjects and Methods

The survey was done in December 2000 and January 2001. The study sample was forty women, 70-89 years of age (78.5±4.5), who lived in their own houses in Y Town near Himeji City, Hyogo, Japan. This study was limited to women, however, we intend to extend the survey to involve men as well in the future.

We carried out a questionnaire which included the following three categories: the frequency in the current use of a MP, the possible needs in the use of a MP and the IPC in the future, and the most preferred input device including keyboard, touch screen, Japanese kana syllabary (JKS) and speech recognition (speech) for the IPC.

Our analysis method is as follows:

Step (1) employ questionnaire.
Step (2) calculate the frequency in the current use of a MP and the IPC.
Step (3) calculate the frequency in the future needs in a MP and the IPC.
Step (4) calculate the most preferred input device for the IPC
Step (5) classify the subjects into three groups according to the future needs in a MP and the IPC, and the most preferred input device for the IPC
Step (6) use the data of rough set analysis obtained in Step 5.
Step (7) use the data of χ^2 test obtained in Step 5. The magnitude of the relation is evaluated by the odds ratio (OR) [2].

1) Rough set theory

The decision algorithm by the rough set theory is given below:

if (future need, yes) then (input device, +)

For example,
Certainty = If (MP, yes) then (keyboard, +)
Coverage = If (keyboard, +) then (MP, yes)

It was pointed out by Pawlak [1] that the rough set theory offers new insight into Bayes' theorem. The look on Bayes' theorem offered by the rough set theory is completely different to that used in the Baysian data analysis philosophy. It does not refer either to prior or posterior probabilities, inherently associated with Bayesian reasoning, but it reveals some probabilistic structure of the data being analyzed. It states that any data set (decision table) satisfies total probability theorem and Bayes' theorem. This property can be used directly to drawing conclusions from the data without referring to prior knowledge and its revision if new evidence is available [3]. Pawlak [1] proposed a new form of Bayes' theorem. The strength of decision rules is computed from the data or it can be also a subjective assessment. This formula gives new look on Bayesian method of inference and also simplifies essentially computations.

2) χ^2 test

χ^2 value=(a+b+c+d)(ad-bc) ^{2}/(a+b)(c+d)(a+c)(b+d)
OR = ad/bc

Table 1. The Contingency Table.

		keyboard		total
		+	-	
Mobile Phone	yes	a	b	a+b
	no	c	d	c+d
total		a+c	b+d	a+b+c+d

3 Results and Discussion

The frequency in the current use of a MP was 10.0% and nobody used the IPC. However, 60.0% of the subjects recognized their future needs in a MP and 27.5% in the IPC. In terms of the most preferred input device for IPC, the keyboard was selected by 15.0% of the subjects, the touch screen by 40.0%, the speech by 15.0%, and the JKS by 30.0%.

Figure 1 shows the results of the keyboard as an example, which are obtained in Steps 6 and 7.

Certainty = 4/24 = 0.167
Coverage = 4/6 = 0.667
χ^2 value = 10,240/78,336 = 0.13
OR = 1.4

The certainty factor leads to the following conclusion: 16.7% of the subjects who selected a MP preferred the keyboard. On the other hand, the coverage factor leads to the following conclusion: 66.7% of the subjects who preferred the keyboard selected a MP. Table 2 shows that the results of all the input devices, which are obtained in Steps 6 and 7. As a result of the rough set analysis, those who were interested in a MP and the IPC preferred JKS the most. They also preferred the keyboard. χ^2 test, on the other hand, showed a result that there was a significant relation between a MP and JKS (χ^2 = 3.89, OR=5.0: $p<0.01$). The test also indicated that the touch screen had a significant relation with a MP (χ^2 = 9.18, $p<0.01$) and the IPC (χ^2 = 8.84, $p<0.01$). However, ORs were 0.12 respectively, and it means that those interested in a MP and the IPC did not prefer the touch screen. In addition, the rough set analysis discovered the facts that those interested in the keyboard had a preference for the IPC and also that those interested in the JKS and the speech preferred a MP. It is thus obvious that the rough set analysis enabled us to obtain more knowledge.

We found that the complementary use of the rough set analysis and χ^2 test bring more powerful knowledge discovery.

4 Conclusions

This study was conducted to examine what kind of interface (input device) is preferred by the elderly, by using a questionnaire regarding the Information Technology equipment. As a result of the rough set analysis and χ^2 test, those who were interested in a MP preferred JKS the most. The rough set analysis also discovered the facts that those interested in the keyboard had a preference for the IPC and also that those interested in the JKS and the speech preferred a MP. We found that the complementary use of the rough set analysis and χ^2 test bring more powerful knowledge discovery.

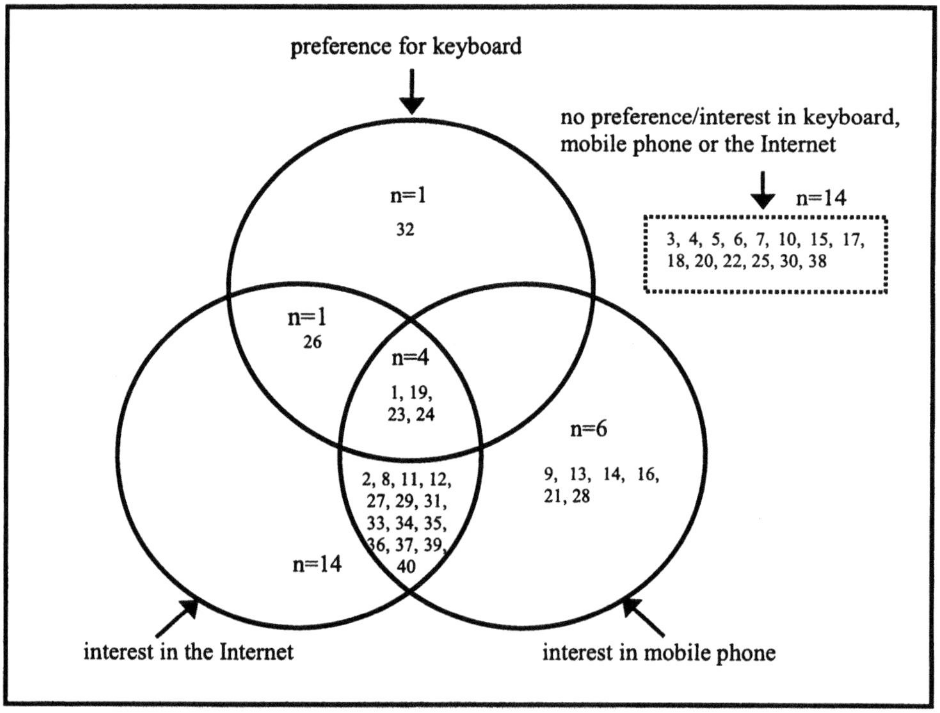

Fig. 1. Relation between IT equipment and input device using the rough set analysis. This represents the results of the keyboard as an example.

Table 2. Relation between IT equipment and input device using the rough set analysis and χ^2 test. This represents the results of all the input devices.

Decision rules	Rough set analysis		χ^2 test	
	Certainty	Coverage	χ^2 value	Odds ratio
1) If (mobile, yes) then (keyboard, +)	0.167	0.667	0.13	1.4
2) If (Internet, yes) then (keyboard, +)	**0.263**	**0.833**	3.64	7.14
3) If (mobile, yes) and (Internet, yes) then (keyboard, +)	**0.222**	0.667	1.34	2.86
4) If (mobile, yes) then (touch screen, +)	**0.208**	0.313	**9.18****	**0.12**
5) If (Internet, yes) then (touch screen, +)	0.158	0.188	**8.84****	**0.12**
6) If (mobile, yes) and (Internet, yes) then (touch screen, +)	0.167	0.188	**7.42****	**0.14**
7) If (mobile, yes) then (JKS, +)	**0.417**	**0.833**	**3.89****	**5.0**
8) If (Internet, yes) then (JKS, +)	**0.421**	0.667	2.53	3.1
9) If (mobile, yes) and (Internet, yes) then (JKS, +)	**0.444**	0.667	3.25	3.6
10) If (mobile, yes) then (speech, +)	**0.208**	**0.833**	1.60	4.0
11) If (Internet, yes) then (speech, +)	0.158	0.500	0.02	1.1
12) If (mobile, yes) and (Internet, yes) then (speech, +)	0.167	0.500	0.07	1.3

JKS = Japanese kana syllabary. speech = speech recognition. **: $p<0.01$

Acknowledgements

This work was supported in part by the grants-in-aid for scientific research (B) from the Japan Ministry of Education, Culture, Sports, Science and Technology (grant number 13490027), and by a grant-in-aid for scientific research from the Japan Ministry of Health, Labor and Welfare.

References

[1] Pawlak, Z., "Decision rules, Bayes' rule and rough sets", New Directions in Rough Sets, Data Mining, and Granular-Soft Computing," Lecture Notes in Artificial Intelligence Vol.1711, pp.1-9 (1999).

[2] Uchida H., Hata Y., Suei K, Nakagawa H. and Aoyama H., "Relation between the IADL and Physical Fitness Tests Focusing on Uncertainty of Answering Questionnaire in Elderly Women," Lecture Notes in Computer Science Vol.1625, Dortmund, Germany, pp.686-688 (1999).
[3] Tsumoto S. and Tanaka H., "Discovery of Functional Components of Protein based on PRIMEROSE and Domain Knowledge Hierarchy," Proceedings of the Workshop on Rough Sets and Soft Computing (RSSC-94), 1994: Lin, T.Y. and Wildberger, A.M. (eds.) Soft Computing, pp. 280-285, SCS (1995).

Fuzzy Limits and Fuzzy Nearness Relation*

Martin Kalina

Dept. Mathematics, Slovak University of Technology, Radlinského 11
813 68 Bratislava, Slovakia

Abstract. The aim of this paper is to introduce the notion of fuzzy limit within the fuzzy analysis developed in the concept of fuzzy nearness relation. The main results concern the corresponding calculus.

1 Introduction

Fuzzy analysis of crisp real-valued functions in the concept of continuity measures has been studied in the papers by Burgin [1], [2], Burgin and Šostak [3] and by Janiš [7], [10]. 'Fuzzy limits' within this concept have been introduced by M. Burgin in [2]. In this paper the notion of 'r-limit' is introduced and it is shown that a function f has the r-limit at a point x if and only if it is bounded in some neighbourhood of x (c.f. Proposition 3.5 in [2]).

The book by Diamond and Kloeden [4] is also devoted to the topic of fuzzy analysis. This book is based mainly on Hausdorff Metric and multifunctions (set-valued functions) – their differentiability and integrability and the results concern fuzzy functions.

The concept of fuzzy analysis of crisp real-valued functions, based on the notion of 'fuzzy nearness relation' has already been studied in the papers by Dobrakovová [5], [6], Janiš [8], [9] and Kalina [11], [12], [13]. The notion of fuzzy nearness relation has been introduced in [11] and further developed by V. Janiš in [8] and by J. Dobrakovová in [5], [6]. In the above mentioned papers by Janiš and by the author of this article, fuzzy nearness based derivatives have been introduced and studied. However, a general notion of fuzzy limit based on fuzzy nearness relations has still been absent. The main aim of this paper is to fill this gap.

2 Fuzzy Limits

First, let us recall the notion of fuzzy nearness relation.

Definition 1 *Let* $\mathcal{N} : \Re \times \Re \to [0;1]$. *We say that* $\mathcal{N}$ *is a relation of fuzzy nearness iff the following holds:*

1. $x\mathcal{N}z = 1 \Leftrightarrow z = x$
2. $x\mathcal{N}z = z\mathcal{N}x$

* Supported by VEGA grants 1/7146/20 and 1/7076/20.

B. Reusch (Ed.): Fuzzy Days 2001, LNCS 2206, pp. 755–761, 2001.

3. *for each* x, z, s, t *if* $x \le s \le t \le z$ *then* $s\mathcal{N}t \ge x\mathcal{N}z$.
4. *for each* $x \in \Re$ *there holds* $\lim_{t\to\pm\infty} x\mathcal{N}t = 0$
5. *for all* $\alpha \in]0;1]$ *and all* $x \in \Re$ *there exist unique* $x_\alpha > x$ *and* $x_{-\alpha} < x$ *such that* $x\mathcal{N}x_\alpha = x\mathcal{N}x_{-\alpha} = \alpha$.

In the following $\mathcal{X}$ and $\mathcal{Y}$ will always denote fuzzy nearness relations connected with the x- and y-axes, respectively and x_α and $x_{-\alpha}$ will denote the uniquely given points from the above definition, which are near to x at the level α w.r.t. $\mathcal{X}$.

Definition 2 *We say that* f *has an* α*-limit at* x *w.r.t.* $(\mathcal{X};\mathcal{Y})$ *iff*

$$f(x_{-\alpha})\ \mathcal{Y}\ f(x_\alpha)\ \ge\ \alpha. \tag{1}$$

Roughly speaking, a function f has an α-limit at x_0, if the values of f at points, which are near to x_0 at the level α, are 'near to each other'.

Definition 3 *We denote*

$$^{\mathcal{X}}\lim_{z\to x} f(z) = \left[\liminf_{\alpha\nearrow 1} \min\{f(x_{-\alpha}); f(x_\alpha)\}; \limsup_{\alpha\nearrow 1} \max\{f(x_{-\alpha}); f(x_\alpha)\}\right]. \tag{2}$$

We say that f *has the fuzzy limit at* x *w.r.t.* $\mathcal{X}$ *iff there exists a* $\mathcal{Y}$ *and an* α_0 *such that for all* $\alpha > \alpha_0$ f *has the* α*-limit at* x *w.r.t.* $(\mathcal{X};\mathcal{Y})$.

I.e. fuzzy limit is an interval (possibly one-point). It can easily be shown that for any fuzzy nearness relations $\mathcal{X}_1$, $\mathcal{X}_2$ and any real-valued function f

$$^{\mathcal{X}_1}\lim_{z\to x} f(z) = {}^{\mathcal{X}_2}\lim_{z\to x} f(z).$$

On the other hand, in general, the existence of the fuzzy limit depends on the fuzzy nearness relation $\mathcal{X}$.

Comment. Let us take a point x_0 and a fuzzy nearness relation $\mathcal{X}$. Then the following holds

$$\frac{df}{d\mathcal{X}}(x_0) = {}^{\mathcal{X}}\lim_{z\to x_0} \frac{f(z)-f(x_0)}{z-x_0},$$

where $\frac{df}{d\mathcal{X}}(x_0)$ is the fuzzy derivative of f at x_0 w.r.t. $\mathcal{X}$, which is defined in [12].

Example 1 *Consider the function* f_1 *given by*

$$f_1(x) = \begin{cases} 1, & \textit{for} \quad x > 0 \\ 0, & \textit{for} \quad x = 0 \\ -1, & \textit{for} \quad x < 0 \end{cases}$$

Then for any fuzzy nearness relation $\mathcal{X}$

$$^{\mathcal{X}}\lim_{z\to 0} f_1(z) = [-1;1],$$

but since $f_1(0_\alpha) = 1$ *and* $f_1(0_{-\alpha}) = -1$ *for any* $\alpha \in]0;1[$, f_1 *has no fuzzy limit at* 0. □

Example 2 *Consider the function f_2 given by*

$$f_2(x) = \begin{cases} 1, & \textit{for} \quad x \quad \textit{being rational} \\ -1, & \textit{for} \quad x \quad \textit{being irrational.} \end{cases}$$

Then for any fuzzy nearness relation $\mathcal{X}$

$$^{\mathcal{X}} \lim_{z \to 0} f_2(z) = [-1; 1].$$

If we take $\mathcal{X}$ such that

$$u \, \mathcal{X} \, v \quad = \quad 1 - \min\{1; |u - v|\} \tag{3}$$

holds, then $f_2(0_\alpha) = f_2(0_{-\alpha})$ for any α. Hence $[-1; 1]$ is the fuzzy limit of f_2 at 0 w.r.t. $\mathcal{X}$.

If we take the fuzzy nearness relation $\mathcal{X}_2$ such that

$$0 \, \mathcal{X}_2 \, u \quad = \quad \begin{cases} 1 - \min\{1; u\} & \textit{if} \quad u \geq 0 \\ 1 - \min\{1; \sqrt{2}|u|\} & \textit{if} \quad u < 0, \end{cases}$$

then for all rational α there holds

$$f_2(0_\alpha) = f_2(1 - \alpha) = 1, \qquad f_2(0_{-\alpha}) = f_2\left(\frac{\alpha - 1}{\sqrt{2}}\right) = -1,$$

hence f_2 has no fuzzy limit w.r.t. $\mathcal{X}_2$. □

Example 3 *Consider the function f_3 given by*

$$f_3(x) = \begin{cases} \sin\left(\frac{1}{x}\right), & \textit{for} \quad x \neq 0 \\ 0, & \textit{for} \quad x = 0 \end{cases}$$

Then for any fuzzy nearness relation $\mathcal{X}$

$$^{\mathcal{X}} \lim_{z \to 0} f_3(z) = [-1; 1].$$

The function f_3 yields the following

$$f_3\left(\frac{1}{x_k}\right) \quad = \quad \begin{cases} 1, & \textit{for} \quad x_k = \frac{2}{k\pi}, & k \quad \textit{being odd positive integers} \\ -1, & \textit{for} \quad x_k = \frac{2}{k\pi}, & k \quad \textit{being odd negative integers} \\ 0, & \textit{for} \quad x_k = \frac{2}{k\pi}, & k \neq 0 \quad \textit{being even integers} \end{cases}$$

If we take $\mathcal{X}$ defined by Formula 3, then for $\beta_k = 1 - \frac{2}{k\pi}$, k being odd naturals, there holds

$$f_3(0_{\beta_k}) = 1, \qquad f_3(0_{-\beta_k}) = -1.$$

Hence f_3 has no fuzzy limit at 0 w.r.t. $\mathcal{X}$.

On the other hand, if we choose $\mathcal{X}_3$ such that

$$0 \, \mathcal{X}_3 \, u \quad = \quad \begin{cases} 1 - \min\{1; 100u\}, & \textit{for} \quad u > 0 \\ 1 - \min\{1; 100v\}, & \textit{for} \quad v > 0 \quad \textit{and} \quad u = \frac{-v}{1 - v\pi} \end{cases}$$

then

$$0_\alpha = \frac{1-\alpha}{100}, \quad \textit{hence} \quad \frac{1}{0_\alpha} = \frac{100}{1-\alpha}$$

$$0_{-\alpha} = \frac{\alpha - 1}{100 - (1-\alpha)\pi}, \quad \textit{hence} \quad \frac{1}{0_{-\alpha}} = \pi - \frac{100}{1-\alpha}$$

I.e.

$$f_3(0_{-\alpha}) = \sin(\pi - 0_\alpha) = \sin(0_\alpha) = f_3(0_\alpha)$$

and the existence of the fuzzy limit of f_3 at 0 w.r.t. $\mathcal{X}_3$ follows. □

Comment to the examples. The function f_1 is continuous at each point except of 0, but the left- and right-hand side limits (crisp) at 0 do not equal to each other. The function f_2 is discontinuous at each point, but for each sequence of points $\{x_i\} \nearrow 0$ such that $\lim_{i\to\infty} f_2(x_i)$ does exist, there exists a sequence $\{\hat{x}_i\} \searrow 0$ such that

$$\lim_{i\to\infty} f_2(x_i) = \lim_{i\to\infty} f_2(\hat{x}_i) \tag{4}$$

The function f_3 is similarly to f_1 continuous at each point except of 0, but there exist sequences of points $\{x_i\} \nearrow 0$ and $\{\hat{x}_i\} \searrow 0$ such that Formula 4 holds for f_3.

Lemma 1 *Let f be any real-valued function and $\mathcal{X}$ any fuzzy nearness relation. Then f has a limit (crisp) at some point x_0 iff f has the fuzzy limit at x_0 w.r.t. $\mathcal{X}$ and the fuzzy limit is a one-point interval.*

Proof. It follows from the fact that

$$\liminf_{x\to x_0} f(x) = \liminf_{\alpha\nearrow 1} \min\{f(x_{-\alpha}); f(x_\alpha)\} =$$
$$\limsup_{\alpha\nearrow 1} \max\{f(x_{-\alpha}); f(x_\alpha)\} = \limsup_{x\to x_0} f(x)$$

and this (c.f. Formula 2) implies the assertion in question. □

Before turning our attention to the calculus we define the multiplication and addition of intervals by

$$[a_1; b_1] * [a_2; b_2] = \tag{5}$$
$$[\min\{a_1 * a_2; a_1 * b_2; b_1 * a_2; b_1 * b_2\}; \max\{a_1 * a_2; a_1 * b_2; b_1 * a_2; b_1 * b_2\}]$$

where $*$ means the multiplication and addition, respectively.

Theorem 1 *Let f and g be any functions having their fuzzy limits at a point x_0 with respect to some fuzzy nearness relation $\mathcal{X}$ such that*

$$\pm\infty \notin {}^{\mathcal{X}}\lim_{z\to x_0} f(z) \quad \& \quad \pm\infty \notin {}^{\mathcal{X}}\lim_{z\to x_0} g(z).$$

Then also $f \cdot g$, $f + g$ and $k \cdot f$ have fuzzy limits at x_0, where $k \in \Re$ and there holds

a. $^{\mathcal{X}}\lim_{z\to x_0}(f(z)\cdot g(z)) \subseteq {}^{\mathcal{X}}\lim_{z\to x_0} f(z) \cdot {}^{\mathcal{X}}\lim_{z\to x_0} g(z)$
b. $^{\mathcal{X}}\lim_{z\to x_0}(f(z)+g(z)) \subseteq {}^{\mathcal{X}}\lim_{z\to x_0} f(z) + {}^{\mathcal{X}}\lim_{z\to x_0} g(z)$
c. *for any $k \in \Re$ there holds*

$$^{\mathcal{X}}\lim_{z\to x_0} k\cdot f(z) = k \cdot {}^{\mathcal{X}}\lim_{z\to x_0} f(z).$$

Proof. Since f and g have their fuzzy limits at x_0, there exist some fuzzy nearness relations $\mathcal{Y}_f$ and $\mathcal{Y}_g$ such that Formula 1 holds for f, g and $\mathcal{Y}_f$, $\mathcal{Y}_g$, respectively. Denote

$$\begin{aligned} {}^{\mathcal{X}}\lim_{z\to x_0} f(z) &= [a_f; b_f] \\ {}^{\mathcal{X}}\lim_{z\to x_0} g(z) &= [a_g; b_g]. \end{aligned}$$

Further denote

$$\begin{aligned} \tau_f(\alpha) &= \inf\{\gamma > 0; (\forall \beta \in [\alpha; 1[)(\gamma > |f(x_\beta) - f(x_{-\beta})|\} \qquad (6) \\ \tau_g(\alpha) &= \inf\{\gamma > 0; (\forall \beta \in [\alpha; 1[)(\gamma > |g(x_\beta) - g(x_{-\beta})|\} \end{aligned}$$

We will show that

$$\lim_{\alpha\nearrow 1} \tau_f(\alpha) = \lim_{\alpha\nearrow 1} \tau_g(\alpha) = 0. \qquad (7)$$

For any α there exists an n such that we can construct a sequence of points $\{z_i\}_{i=0}^{n+1}$ for which

$$z_0 < a_f = z_1 < z_2 < \ldots < z_{n-1} < b_f \le z_n < z_{n+1} \quad \& \quad z_i\ \mathcal{Y}_f\ z_{i+1} = \alpha$$

holds. Put $\varepsilon_f = \max_{0\le i\le n}(z_{i+1} - z_i)$. Then Property 3 of fuzzy nearness relation (Definition 1) implies

$$\tau_f(\alpha) \le 3\varepsilon_f$$

and hence Formula 7 follows for f. Similarly we could treat also the function g.

In order to prove the existence of fuzzy limits for product, addition and multiplication by constants, it is enough to choose fuzzy nearness relations $\mathcal{Y}_a$, $\mathcal{Y}_b$, $\mathcal{Y}_c$, respectively, such that

$$u\ \mathcal{Y}_a\ v = \begin{cases} \alpha, & \text{if } |u-v| = \tau_f(\alpha)\cdot\tau_g(\alpha) \\ 1, & \text{if } u = v \\ 0 & \text{otherwise} \end{cases}$$

$$u\ \mathcal{Y}_b\ v = \begin{cases} \alpha, & \text{if } |u-v| = \tau_f(\alpha)+\tau_g(\alpha) \\ 1, & \text{if } u = v \\ 0 & \text{otherwise} \end{cases}$$

$$u\ \mathcal{Y}_c\ v = \begin{cases} \alpha, & \text{if } |u-v| = k\cdot\tau_f(\alpha) \\ 1, & \text{if } u = v \\ 0 & \text{otherwise} \end{cases}$$

The inequalities **a.**, **b.**, **c.** are implied by Formulas 2 and 5. □

Theorem 2 *Let f has a fuzzy limit at a point x_0 w.r.t. some fuzzy nearness relation $\mathcal{X}$. Denote*

$$[a; b] = {}^{\mathcal{X}}\lim_{z \to x_0} f(z)$$

and assume $-\infty < a \leq b < \infty$. Let $g : \Re \to \Re$ be Lipschitz, i.e. $|g(s) - g(t)| \leq L|s-t|$ for some $L > 0$. Then $f \circ g$ has fuzzy limit at x_0 w.r.t. the fuzzy nearness relation $\mathcal{X}$ and there holds

$${}^{\mathcal{X}}\lim_{z \to x_0} g(f(z)) \subseteq [\inf\{g(s); s \in [a; b]\}; \sup\{g(s); s \in [a; b]\}]$$

Proof. Defining $\tau(\alpha)$ for $\alpha \in]0; 1[$ by Formula 6 we can find a fuzzy nearness relation $\mathcal{Y}_L$ such that

$$u\, \mathcal{Y}_L\, v \quad = \quad \begin{cases} \alpha, & \text{if} \quad |u - v| = L \cdot \tau(\alpha) \\ 1, & \text{if} \quad u = v \\ 0, & \text{otherwise.} \end{cases}$$

Since g is Lipschitz, it is straightforward that

$$g\left(f(x_{-\alpha})\right)\ \mathcal{Y}_L\ g\left(f(x_\alpha)\right) \geq \alpha$$

holds for all $\alpha \in]0; 1[$ and hence $f \circ g$ has its fuzzy limit at x_0 and there holds

$${}^{\mathcal{X}}\lim_{z \to x_0} g(f(z)) \subseteq [\inf\{g(s); s \in [a; b]\}; \sup\{g(s); s \in [a; b]\}]$$

□

Corollary 1 *(Corollary to Theorems 1 and 2) Let f, g have fuzzy limits at a point x_0 w.r.t. $\mathcal{X}$ and let*

$$0 \notin {}^{\mathcal{X}}\lim_{z \to x_0} g(z).$$

Denote

$$[a; b] = {}^{\mathcal{X}}\lim_{z \to x_0} f(z), \qquad [c; d] = {}^{\mathcal{X}}\lim_{z \to x_0} g(z)$$

and assume $-\infty < a \leq b < \infty$ and $-\infty < c \leq d < \infty$. Then $\frac{f}{g}$ has a fuzzy limit at x_0 w.r.t. $\mathcal{X}$. Its value can be estimated by

$${}^{\mathcal{X}}\lim_{z \to x_0} \frac{f(z)}{g(z)} \subseteq \left[\min\left\{\frac{a}{c}; \frac{a}{d}; \frac{b}{c}; \frac{b}{d}\right\}; \max\left\{\frac{a}{c}; \frac{a}{d}; \frac{b}{c}; \frac{b}{d}\right\}\right]$$

Proof. Function $\frac{1}{x}$ is Lipschitz in any interval $[a; b]$ such that $0 \notin [a; b]$. Since g has the fuzzy limit at x_0, by Theorem 2 there exists the fuzzy limit

$${}^{\mathcal{X}}\lim_{z \to x_0} \frac{1}{g(z)} \subseteq \left[\min\left\{\frac{1}{c}, \frac{1}{d}\right\}; \max\left\{\frac{1}{c}, \frac{1}{d}\right\}\right]$$

and by Theorem 1a there exists the fuzzy limit of $\frac{f}{g} = f \cdot \frac{1}{g}$ at x_0 w.r.t. $\mathcal{X}$ and there holds

$${}^{\mathcal{X}}\lim_{z \to x_0} \frac{f(z)}{g(z)} \subseteq \left[\min\left\{\frac{a}{c}; \frac{a}{d}; \frac{b}{c}; \frac{b}{d}\right\}; \max\left\{\frac{a}{c}; \frac{a}{d}; \frac{b}{c}; \frac{b}{d}\right\}\right]$$

□

References

1. Burgin, M.: General approach to continuity measures, Fuzzy sets and Systems **105** (1999), 225–231
2. Burgin, M.: Theory of fuzzy limits, Fuzzy Sets and Systems **115** (2000), 133–443
3. Burgin, M., Šostak, A.: Fuzzification of the theory of continuous functions, Fuzzy Sets and Systems **62** (1994), 71–81
4. Diamond, P., Kloeden, P.: Metric Spaces of Fuzzy Sets, Theory and Applications, World Scientific, Singapore, New Jersey, London, Hong Kong, 1994
5. Dobrakovová, J.: On a fuzzy nearness, In: Proc. Strojné inžinierstvo 1, 1998, 33–37
6. Dobrakovová J.: Nearness, convergence and topology, Busefal **80** (1999), 17–23
7. Janiš, V.: Fixed points of fuzzy functions, Tatra Mountains Math. Publ. **12** (1997), 13–19
8. Janiš, V.: Fuzzy uniformly continuous functions, Tatra Mountains Math. Publ. **14** (1998), 177–180
9. Janiš, V.: Nearness derivatives and fuzzy differentiability, Fuzzy Sets and Systems **108** (1999), 99–102
10. Janiš, V.: Fuzzy mappings and fuzzy methods for crisp mappings, Acta Univ. M. Belii **6** (1998), 31–47
11. Kalina, M.: Derivatives of fuzzy functions and fuzzy derivatives, Tatra Mountains Math. Publ. **12** (1997), 27–34
12. Kalina, M.: On fuzzy smooth functions, Tatra Mountains Math. Publ. **14** (1998), 153–159
13. Kalina, M.: Fuzzy smoothness and sequences of fuzzy smooth functions, Fuzzy Sets and Systems **105** (1999), 233–239

Handwritten Digit Recognition: A Neural Network Demo

Berend-Jan van der Zwaag

EUREGIO Computational Intelligence Center
Dept. of Electrical Engineering
University of Twente
Enschede, the Netherlands
b.j.vanderzwaag@el.utwente.nl

Abstract. A handwritten digit recognition system was used in a demonstration project to visualize artificial neural networks, in particular Kohonen's self-organizing feature map. The purpose of this project was to introduce neural networks through a relatively easy-to-understand application to the general public. This paper describes several techniques used for preprocessing the handwritten digits, as well as a number of ways in which neural networks were used for the recognition task. Whereas the main goal was a purely educational one, a moderate recognition rate of 98% was reached on a test set.

1 Introduction

Handwritten digit recognition is already widely used in the automatic processing of bank cheques, postal addresses, etc. Some of the existing systems include computational intelligence techniques such as artificial neural networks or fuzzy logic, whereas others may just be large lookup tables that contain possible realizations of handwritten digits.

Artificial neural networks have been developed since the 1940s, but only the past fifteen years have they been widely applied in a large variety of disciplines. Originating from the artificial neuron, which is a simple mathematical model of a biological neuron, many varieties of neural networks exist nowadays. Although some are implemented in hardware, the majority are simulated in software.

Artificial neural nets have successfully been applied to handwritten digit recognition numerous times, with very small error margins, see e.g. [2] and [4].

The work described in this paper does not have the intention to compete with existing systems, but merely served to illustrate to the general public how an artificial neural network can be used to recognize handwritten digits. It was part of *NeuroFuzzyRoute in the* EUREGIO, an exposition in the framework of the world exposition EXPO2000 in Hannover.

2 Neural Networks

Artificial neural networks, usually called neural networks (NNs), are systems composed of many simple processing elements (neurons) operating in parallel

B. Reusch (Ed.): Fuzzy Days 2001, LNCS 2206, pp. 762–771, 2001.

whose function is determined by network structure, connection strengths, and the processing performed at computing elements or nodes [1] (other definitions can also be found). NNs exist in many varieties, though they can be categorized into two main groups, where the distinction lies in the learning method:

- supervised learning: the network is trained with examples of input and desired output;
- unsupervised learning: the network tries to organize the input data in a useful way without using external feedback.

In this work, a Kohonen self-organizing map [3] is used, which learns without supervision. The labeling stage is supervised, however, because digit class information is used to label the neurons. The Kohonen network is ideal for demonstrating how information can be stored in a neural network, as the information in the neurons has the same easy-to-interpret format as the input, namely 2-dimensional images of (parts of) numerals.

2.1 Learning

In the learning stage, the network is trained to recognize (parts of) handwritten numerals. This is done by first initializing the network and then repeatedly offering input patterns, determining the winning neuron and adjusting the neurons' feature vectors (see [3] for a more detailed description).

Initialization The self-organizing map consists of $N \times N$ neurons laid out in a rectangular grid. Each neuron contains (basically *is*) a randomly initialized feature vector with as many elements as the number of inputs, i.e., $M \times M$, being the size of the (binary) image that contains (part of) a numeral. The lateral distance between two neurons is the shortest distance between them in the map, usually defined along a straight connecting line or along the gridlines. Depending on the lateral distance between two neurons, they can be either inside or outside each other's neighborhood. The neighborhoods are important for the adjustment of the feature vectors during learning.

Finding the Winning Neuron One by one, training input patterns are offered to the network. For every neuron, the difference (usually Euclidian distance) between its feature vector and the input vector is calculated. The neuron with the smallest difference is then the winning neuron for that particular input example.

Vector Adjustment The winning neuron's feature vector is then adjusted in such a way that the difference with the input example decreases. This is also done for the neurons in the winning neuron's neighborhood, though the larger the lateral distance from the winning neuron, the smaller the adjustment. Often the neurons directly outside the winning neuron's neighborhood are adjusted in the opposite direction. The rate of adjustment as a function of the lateral

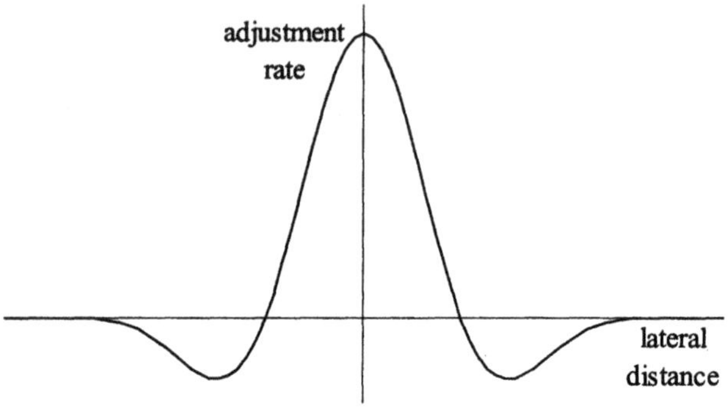

Fig. 1. Illustration of a Mexican hat function

distance to the winning neuron has the shape of a Mexican hat, as can be seen in Fig. 1. Over time, this hat is shrinking both horizontally (the neighborhood is getting smaller) and vertically (the adjustments are getting smaller), until the learning finishes.

2.2 Labeling

Before the network can classify unknown digits, the neurons need to be labeled. This happens after the learning stage has finished. Several ways of doing this are possible, some of which will be explained below. Once the neurons have been labeled, an unknown handwritten digit can be offered to the network. The label of the winning neuron then serves to classify the unknown digit.

Majority Voting Every neuron is assigned a number of counters, one counter for each digit class. Then, from every digit class an equal number of examples is offered to the network. Every time a neuron is the winning neuron, its counter for the corresponding digit class is incremented. After all examples of all classes have been offered, each neuron is labeled according to the class for which its counter was the highest. For example, if a neuron was winning neuron for seven instances of "5", two instances of "6", and never for the remaining digit classes, it will be labeled "5".

Neurons that never were the winning neuron, or for which the above method leads to ambiguities (e.g., as often winner for "5" as for "6"), can be labeled according to the labels of the neighboring neurons using majority voting. This may be done because of the topology preserving property of the self-organizing map: neighboring neurons contain similar information. Alternatively, neurons not labeled in the first round can be labeled "unknown."

Minimum Average Distance Instead of simply incrementing the class counters for winning neurons, more unambiguous labels can be obtained by minimum average distance labeling. In this method, every time an example digit is offered to the network, the corresponding class counters for all neurons are increased by the lateral distance of the neuron to the winning neuron (analogous to the previous method, we use the term "counter" here, even though instead of counting occurrences we now accumulate distances). After all example digits of all classes have been presented, the average distances are calculated by dividing the class counters by the number of example digits for that class. For every neuron, the class with the smallest average lateral distance to the winning neuron forms the label. Neurons that never win but that are close to a winning neuron most of the time, will automatically be labeled accordingly.

Minimum Average Difference Yet another labeling method is labeling by minimum average difference. This method is very similar to the labeling by minimum average distance, but instead of accumulating the lateral distances to the winning neurons, we now accumulate the differences between the presented examples and the neuron's feature vector for each neuron. Again, after all example digits of all classes have been presented, the average differences are calculated by dividing the class counters by the number of example digits for that class. For every neuron, the class with the smallest average difference forms the label.

3 Handwritten Digit Recognition

3.1 Preprocessing

As digitized numerals may originate from a range of sources, it is important to make sure that all digits are preprocessed so as to eliminate extra problems due to non-class-specific differences in size, shear, line thickness, background and digit colors, resolution, etc. A few techniques to minimize these differences are discussed below.

Deskewing To help compensate for different handwriting styles (with different slant), digits that are slanting to the left or to the right are deskewed to put them all upright or to at least make them slant under the same angle. For some examples, see Fig. 2.

The deskewing process works as follows: first, an imaginary line is drawn through the digit so that the total of the horizontal distances from all digit pixels to this line is minimized in a least-squares sense. The imaginary line can be described by $x = ay + b$, where x and y are the horizontal and vertical pixel index, respectively, and a and b are line parameters. Then a side slip is performed: the pixels that form the digit are horizontally shifted to the left or to the right, while keeping their positions relative to the imaginary line constant. This is done in such a way that the imaginary line becomes vertical, with the digit effectively in an upright position. In formula: $x_{new} = x_{old} - ay$ for every pixel, resulting in the new line $x = b$. If the slant is removed by rotation, the recognition problem is likely to become more difficult, due to increased variation

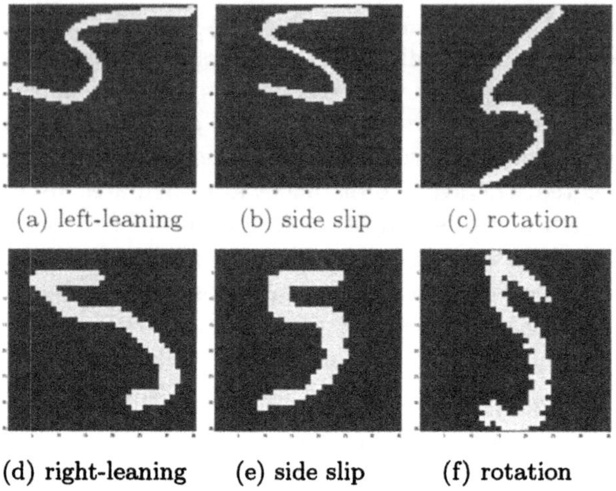

Fig. 2. Two examples of deskewing: (a) and (d): left and right leaning digits, resp.; (b) and (e): deskewed by side slip; (c) and (f): deskewed by rotation

among numerals from the same category. This may become clear by looking at Figs. 2c and 2f.

Thinning To overcome problems related to differences in the line thickness of the written characters, the digit images are thinned to a thickness of one pixel. This can lead to undesired side-effects, such as spurs, as can be seen from Fig. 3b. To correct for this, one can remove spur lines of up to a certain length from the resulting thinned image (Fig. 3c). A negative side-effect of this pruning process is that end-lines that are actually part of the digit are shortened. The length of these digit parts is usually still long enough for correct recognition of the digit, so that the benefits of the thinning process far outweigh the negative sides. Algorithms exist that try to correct for this ; however, the need for pruning can be avoided by smoothing the handwritten digit image before thinning, as seen in Figs. 3d and 3e. Moreover, this method results in an overall smoother thinned image.

Resizing Differences in digit size are due to different image resolutions and handwriting styles. To minimize the effect of these differences, all digits are resampled/rescaled to the same size in pixels, before being fed to the recognizer.

Smoothing Some acquisition methods may cause small artifacts on the digit's edges. These are removed by a smoothing operation. Some preprocessing techniques, including deskewing and resizing, may cause similar artifacts, so smoothing may be useful between several stages of the preprocessing. Effective smoothing can be accomplished by filling tiny gaps in both the foreground and the

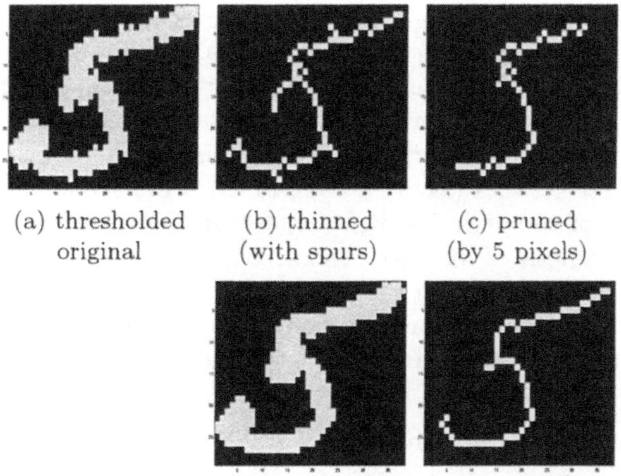

(a) thresholded original (b) thinned (with spurs) (c) pruned (by 5 pixels)

(d) smoothed (e) thinned

Fig. 3. Sequence a-b-c shows the thinning and pruning process; sequence a-d-e shows an alternative: smoothing and thinning without pruning

(a) a gap in the black background (b) not a gap

Fig. 4. Illustration of how 5 can be significantly more than 3 but 6 not significantly more than 2 (see Sect. 3.1, Smoothing)

background. Filling a gap is nothing more than changing the foreground color of a gap pixel to the background color or vice versa. Tiny gaps in the foreground are background pixels with significantly more neighboring foreground pixels than neighboring background pixels. Tiny gaps in the background are defined in a similar way and can in fact also be seen as tiny foreground protrusions. The term "significantly more" cannot be defined as a simple number, but depends on the layout of the pixel neighborhood. For example, in the layout of Fig. 4a, 5 is significantly more than 3 and the central pixel can be safely smoothed away, whereas in the layout of Fig. 4b, 6 is not significantly more than 2 as the central foreground pixel cannot be smoothed away without breaking the foreground connectivity.

4 Network Configurations

4.1 Two NNs in Series

In this setup, one Kohonen map is trained with digit fragments, and another is trained with the ordering information in the trained first map. The idea is that particular combinations of digit fragments are characteristic for a particular digit class; the first network identifies fragments, i.e., not digits, and the second network identifies combinations of fragments, i.e., digits. The second network will need to be labeled in order to be able to classify unknown digits. The first network remains unlabeled. Suppose the digit fragments are $M \times M$ pixels in size, then the first network (say $N \times N$ neurons), will have $M \times M$ inputs, being the pixel values of the digit fragments, meaning that the vectors stored in the neurons have a length of $M \times M$. The second network will then have $N \times N$ inputs, one for every neuron in the first network. Per digit, for every neuron in the first network is recorded for how many digit fragments that neuron was the winning neuron. These totals are input to the second network.

4.2 Template Matching Instead of NNs

Another method for the recognition is to create a template for each digit and to compare unknown digits to these templates to find out which template matches closest. One way of creating such templates is by calculating an average digit from an example set for each digit class.

4.3 Ten NNs in Parallel

In this setup, ten networks are trained for one digit class each. Thus, the networks are pre-labeled. They are trained with digit fragments (as before without location information). For the recognition stage, several methods are available, two of which are discussed below. In the special case that the networks consist of only one neuron each and the digit fragments actually contain the whole digit, this method is similar to the template matching method of Sect. 4.2 with the difference lying therein that creating the templates is now realized by training the networks, rather than calculating the average digits directly.

Minimum Distance Classification For classifying an unknown digit, (fragments of) that digit are offered to all ten networks, and per network the (Euclidian) distances between the winning neurons and the fragments are totaled over all fragments. Then, the label of the network with the smallest total distance gives the class of the unknown digit. In general, all ten networks will have equal numbers of neurons. If this is not the case, the total Euclidian distance is divided by the number of neurons to find the network with the smallest average distance.

Label Classification With label classification, (fragments of) the unknown digit are also offered to all ten networks, but now we are only interested in which network has the winning neuron with the smallest distance to a fragment, measured over all networks. This network is then called the winning network for that fragment. Finally, the label of the most frequently winning network, measured over all fragments, gives the class of the unknown digit.

4.4 Fragmented Digits vs. Whole Digits

Digit Fragments The idea to use fragments of digits rather than the whole digits, originates from the idea that for a specific digit class, digits as a whole show more variation than individual fragments that are characteristic for that class. When using fragments however, a decision needs to be made whether to include any position information or not. For the purpose of demonstrating the working of a Kohonen self-organizing map, the organization of the fragments is expected to be easier to understand if positional information is left out.

Whole Digits When using whole digits as inputs to the networks, what actually happens depends on the training method and the network architecture. With ten small networks, the networks basically store representative examples of the digits during the training stage. These are derived from the example set, although the stored examples are usually not present in the example set themselves, as the network tries to store generalizations of the example digits it receives at its inputs. The networks are pre-labeled and only trained with examples from their respective digit classes.

With one large network, all ten classes of digits are offered to the network during the training stage. The network will then store representative examples of all digits. Depending on the actual size of the network, more complicated numerals or numerals that can appear in several distinctive forms may cause the network to store more examples of that digit class than of another, simpler, class (e.g., consider the class "2", which can be written in very different ways, and the class "0", which most people write in much more similar ways, as can also be seen in Fig. 5).

5 Results

As the goal was a reasonably well working handwritten digit recognizer for demonstration purposes, and due to very limited time constraints, only a small number of experiments have been conducted to enable comparison between the various configurations. An existing set of handwritten digits with 25 digits per class was used. This set, shown in Fig. 5, was split into a training set of 15($\times$10) examples and a test set of 10($\times$10) examples. Previous research ([5] and others) had yielded a maximum 92% correct classification score (percentage correctly recognized digits) on the same test set. Using digit parts, this number now increased to 98% (see Table 1), which was considered sufficient for use in a demonstration. In these experiments, one network was used for each digit class, in total 10 networks. The size of the networks was the same for all 10 networks: 5$\times$5 and 7$\times$7 in two series of experiments. The size of the digits after preprocessing varied between 5$\times$5 and 50$\times$50 pixels and the size of the digit fragments varied between 3$\times$3 and 25$\times$25 pixels. Some results are shown in Table 1. In this table, the best classification results over a range of fragment sizes are shown for a selection of digit sizes, using minimum distance classification (as described in Sect. 4.3).

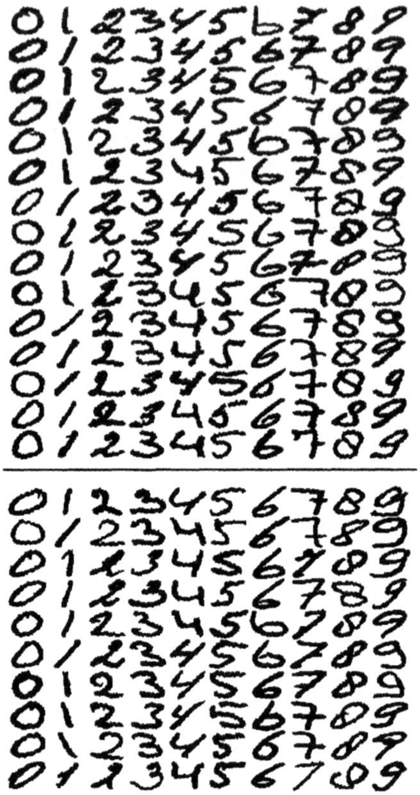

Fig. 5. Set of handwritten digits used for training (*top*) and testing (*bottom*)

A few experiments were conducted using whole digits rather than fragments. Results of these experiments are shown in Table 2. Per digit class one network was used, and classification was done by minimum distance as described in Sect. 4.3. Shown are the performances (on the test set) per digit size for two network sizes. The results are worse than in the experiments with fragmented digits, but strong conclusions cannot be drawn due to the limited number of experiments.

6 Digit Recognizing Demonstration

A continuous presentation has been installed at the Da Vinci Techno Activity Center in Enschede, the Netherlands, where visiting children and adults can learn more about technology. The presentation consists of a slide show explaining the whole Kohonen system of learning and recognizing, and a writing pad with display where visitors can write a digit and see the NN outcome. This NN is the same one that scored best in the experiments: 98% correct recognition rate, see Table 1. In the slide show, a set of unclassified digit fragments is shown

Table 1. Results of experiments with fragmented digits

network	digit size [pixels]				
size	5×5	10×10	15×15	20×20	25×25
5×5	73%	93%	98%	94%	93%
7×7	73%	90%	92%	86%	82%

Table 2. Results of experiments with whole digits

network	digit size [pixels]				
size	5×5	10×10	15×15	20×20	25×25
5×5	63%	82%	91%	92%	93%
7×7	70%	83%	92%	(not used)	

to give people also a chance to determine what digit the fragments come from. The whole presentation was part of "NeuroFuzzyRoute in the EUREGIO," an exposition in the framework of the world exposition EXPO2000 in Hannover.

Unfortunately, the digits written by visitors of the techno activity center have not been recorded, nor is anything known about the recognition rate. However, visitors have reacted enthusiastically to the hands-on demonstration.

7 Conclusions

This paper shows that a Kohonen self-organizing map can be used to recognize digits when given those digits in fragments without positional information. The Kohonen self-organizing map and the handwritten digit recognition have also proven to be a good neural network architecture and application for the purpose of introducing and demonstrating neural networks to the general public.

Acknowledgments

Thanks are due to Andre Beltman who wrote a fast Kohonen NN simulator, *NN Kohinoor*, and to Wubbe Jan Velthuis who designed the slide show.

References

1. DARPA: DARPA Neural Network Study. AFCEA International Press (1988) 60
2. Jain, L.C., Lazzerini, B. (eds.): Knowledge-based Intelligent Techniques in Character Recognition. CRC Press, Boca Raton FL (1999)
3. Kohonen, T.: Self-Organizing Maps. Springer-Verlag, Berlin (1995)
4. LeCun, Y., Jackel, L.D., Bottou, L., Brunot, A., Cortes, C., Denker, J.S., Drucker, H., Guyon, I., Muller, U.A., Sackinger, E., Simard, P., Vapnik, V.: Comparison of Learning Algorithms for Handwritten Digit Recognition. In: Fogelman, F., Gallinari, P. (eds.): Proc. 5th Int. Conf. on Artificial Neural Networks (1995) 53-60
5. Wentink, M.: Master-Slave Kohonen Network Applied to Digit Recognition. Internal Report BSC-021N95, University of Twente (1995)

New Neural Nets

Thomas Kromer

Zentrum für Psychiatrie, Münsterklinik Zwiefalten, 88521 Zwiefalten
Thomas.Kromer@t-online.de
http://home.t-online.de/home/Thomas.Kromer/kromt2.htm

Abstract. Conventional neural networks work by changing the synaptical weights between their neurons. New neural nets (NNN) are presented, using the recording of temporal sequences of activity, generated by various patterns in chains of neurons, to store and reproduce those patterns. In a biological hypothesis, recording of temporal sequences of activity could be done by nucleic acids, each base triplet encoding a certain degree of activity. The specific properties of nucleic acids would enable those systems to perform an associative data processing. Hybridization-processes will select memory-strings with similar sequences, using the high affinity between homologous complementary strings. Using selected homologous memory-strings, the neural chain is able to give an associative „answer" to a presented pattern. Many neural chains, working together, will form a NNN. To demonstrate their potential, the results of a preliminary study are shown, concerning the occurence and treatment of aphasic symptoms in NNN after a network-lesion.

1 Introduction

New neural networks (NNN) work on a new principle (firstly described by Fink[1] for the storing and reproduction of patterns in acoustics). They use the recording of temporal sequences of activity in neural chains to store, compare and reproduce patterns of activity in neural chains. Many neural chains work together, forming a NNN. This paper consists of two main parts: First, the functional principle of NNN is explained. In the second part the results of a little study are shown, the NNN working as pattern-associator, in which aphasic symptoms occur after an extended lesion of the net, the post-lesional performance of the net being improved by new learning under various conditions.

2 Functional Principle

2.1 Basic Units and Procedures

NNN consist of chains of n neurons. Recurrent connections connect the neurons of each chain. The neurons of each neural chain shall be numbered 1-n, they work synchronously. For purpose of learning, all neurons will give their activity centripetally towards neuron 1 to their next-neighboured neuron. Iterating this

B. Reusch (Ed.): Fuzzy Days 2001, LNCS 2206, pp. 772–781, 2001.

procedure, the original pattern of activity will wander step by step along the neural chain in direction to neuron 1. Neuron 1 is working as the memory neuron of the chain. It first records at the beginning its own original state of activity, then the sequence of activity, it will receive from the other neurons. After n iterations of this procedure, neuron 1 will have got a complete record of the original activity-pattern. Recording will be done in computer-simulations quite easily, forming an according string of values.

For pattern reproduction, the connections of the neurons, used for learning, are inhibited, the system now working centrifugally, activity transported towards neuron n. Neuron 1 will select a memory-string, transduce its sequence (beginning at the end of the string; „last in, first out") into the according sequence of activity, which it will send step by step to neuron 2. In perfect reversal of the learning procedure, the original pattern of activity will exactly arise in the neural chain again, without any loss of information.

The memory-neuron will be able to perform associative data- or pattern–processing.

If a pattern A will be presented to the neural chain, an according memory-string A will be produced. Neuron 1 will compare this actual memory-string A to any other memory-strings, formed („learned") earlier. The memory-string X of the long-term memory, showing the least differences to string A will be chosen, the associated pattern X reconstructed in the neural chain. By such a working cycle, the neural chain will be able to give an associative answer to a presented starting pattern.

Figure 1 demonstrates the process of associative completion of a presented pattern by a short neural chain of four neurons.

2.2 The Neural Net

Many neural chains may form a NNN: They cover the surface of the NNN, obeying specific rules of distribution. Such a rule may be a random-distribution as in the study, shown below, it might also be a specific mathematical function, for example a fractal function like the squaring of complex numbers[2].

The NNN consists of many neural columns. Several neurons, belonging to different neural chains, are accorded, due to their localization in the net, to a certain neural column. At the beginning, all neurons of a neural column show the same state of activity, reflecting the activity of the entire column. After the process of producing memory-strings in all neural chains, comparing them to the strings of the long-term memory and chosing those to produce a senseful pattern in each neural chain, the neurons of a neural column show different states of activity, because each neural chain will decide autonomously between the possible patterns and they might come to different decisions. Each neural chain is making thus a proposal, a hypothesis, about the most sensible pattern to the net. These proposals will then be „discussed" within the neural column. The neural columns decide according to certain rules, which state of activity they will represent as a consequence of the activity of their particular neurons. One rule might be the majority decision: A neural column will be set active at the end of a working cycle, if the majority of its neurons will be active. Another rule might set a column active, if any of its neurons will be. The degree of freedom of association between the patterns may be regulated by those rules, depending on the tasks, the neural net is applied to perform.

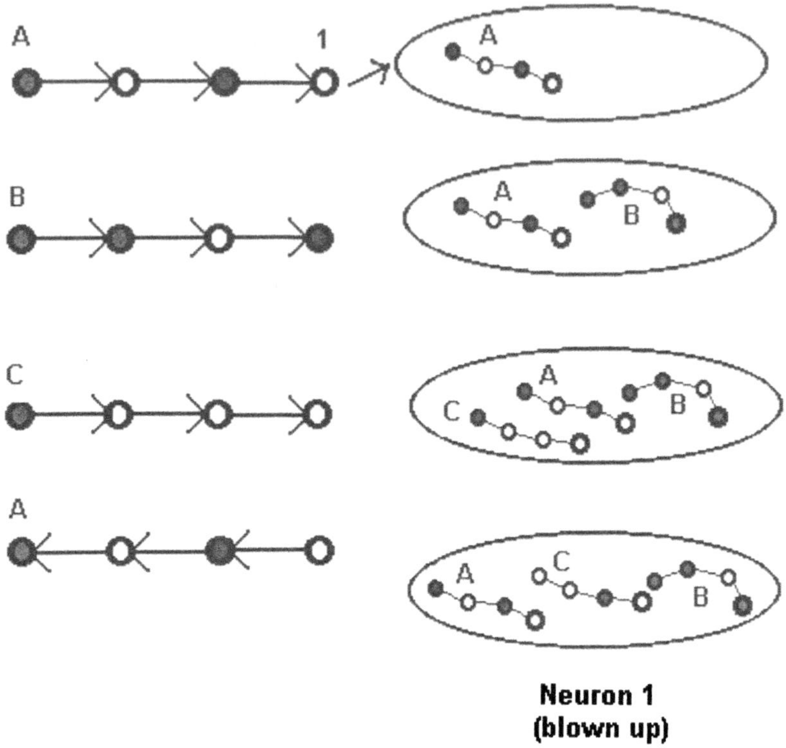

Fig. 1. (From above to below:) Learning of two patterns A and B (Forming according intracellular memory-strings by neuron 1). A new pattern C is presented, the according memory-string C will be compared with string A and B. String A shows the most homologous sequence(differing at only one position), so it will be selected for pattern reconstruction in the neural chain, pattern A representing the associative answer of the neural chain to pattern C.

Exercise of certain patterns could be represented by multiplying the according memory-strings, improving their chance to be selected in the associative process. Memory-strings, which proved to be unuseful could be eliminated. The value of an emotional rating could also be generally attached to a memory-string, supporting the memory- neuron in its task to select the most sensible and useful string.

2.3 The Biological Hypothesis

In biological systems, the recording of temporal sequences of activity in neural chains or neural circuits could be done hypothetically by forming a molecular string, for example a string of a nucleic-acid, in which each base triplet would encode a certain degree of activity. These nucleic-acid-strings, or even better their complementary strings, work as memory-strings, representing the intracellular long-term memory. They could easily be stored in biological neurons and be used to reproduce the according patterns in the neural chain any time later.

Comparison of different memory-strings in the process of developping an associative answer of the neural chain to an actual pattern could be done very elegantly by detecting the memory-string with the most homologous sequence, using the high affinity of homologous complementary strings of nucleic acids. The complementary, best fitting string will be choosen to reproduce its according pattern.

Of course, this whole process is very hypothetical, but all means, neurons need to perform, are not far from being achievable. We know about neural circuits[3] with recurrent connections[4], transformation of extra- into intracellular signals, the involvement of the genetic apparatus of the cells into this signal pathway. Consolidation of long-term memory needs gene-expression, the signal-pathway leading directly from the cell-membrane into the nucleus, mediated by waves of Calcium-cascades wandering along the endoplasmatic reticulum, finally interacting with the Desoxyribo-Nucleic-Acid (DNA) by CREB-Proteins (**c**AMP-**r**esponsive-**e**lement-**b**inding-proteins)[5]. We know, that wide parts of the DNA of the cells do not determine sequences of proteins. Might these parts work as „songbook", providing the possible rhythms of activity of neurons or neural chains? At least in simple organisms we could imagine the limited ways to react to sensory-inputs might be determined by the sequence of the DNA, treasuring well adapted possible rhythms, others having been eliminated during the course of the evolution. This model could explain the observation of inherited behavioural reactions[6,7]. An experimentum crucis might be, whether various sequences of nucleic-acids may cause different rhythms of neural activity.

3 A Preliminary Study

To test the practical applicability of those networks, a short study could be useful. It has been designed to show, how NNN will react to extended lesions and if such tests could work as a model for the occurence and possible treatments of aphasic symptoms in patients, suffering an ischemic stroke or another lesion of speech-relevant areas of the brain.

3.1 Methods and Materials:

The quadratic net is covered by 100x100 neural columns. 32200 neurons form 4600 neural chains, each consisting of 7 neurons. These neurons are randomly distributed over the neural net. After the randomization, 0 to n, in average about 3 neurons belong to each neural column. For better visualization of the data processing, the net is divided into two parts, an input-(consisting of 7000 neural columns) and an output-region (containing 3000 neural columns)(Figure 2).

One working cycle of the net consists of a)presentation of a pattern, b) producing the according memory-string, c) choosing the most homologous string, learnt earlier, d) reproduction of an associative pattern, using the choosen memory-string, e) decision of each neural column, whether it will be set active on base of the state of activity of their neurons.

The final pattern will be taken as starting pattern of the next working cycle, until the fifth working cycle will be finished. Then the reaction of the net will be measured by the number of active neural columns in the different areas of the output-region of the net.

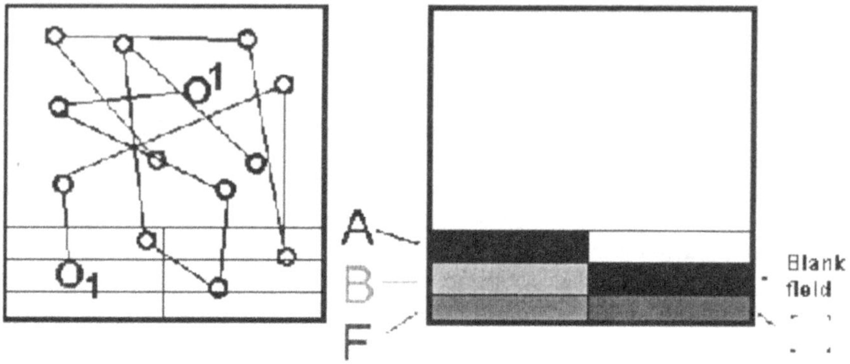

Fig. 2. Structure of the network.Left: 2 neural chains are shown exemplarily (random-distribution of the 7 neurons of each chain). Right: division of the net into two parts: the input-region (upper part) and the output-region, in which 6 signal bars are correlated to the six starting patterns, illustrating the decision of the net between the patterns.

Six different signs are presented to the input region, associated with specific signal-bars in the output-region. (The signal-bars (each consisting of 500 neural columns) allow to measure the „answer" of the net under various conditions, if any pattern is presented to the input region).

Fig. 3. 6 starting patterns, (signs in the input region correlated with the according signal bar in the output-region). (Active neural columns indicated by black colour, scale reduced).

As base of the long-term memory, the according memory-strings are produced. Then several experiments, investigating the reaction of the network under various conditions are performed:

1) The six different signs are presented to the input-region, combined with the associated signal bars in the output-region.
2) The six different signs are presented only to the input-region. The net produces according memory-strings and compares them with the stored strings. After the reproduction of the patterns, according to the selected strings, the activities in the signal bars show, to which degree the presented pattern is correctly recognized by the net.
3) The patterns are presented to the net in the same form like in experiment 2, but incompletely. A large elliptic region(2131 neural columns) in the middle of the input-region will show no activity at the beginning. The neural network is intact.
4) A lesion is produced: The neurons of the elliptic region of experiment 3 are set out of order. The reaction of the net after this lesion is studied, the patterns are presented as in experiment 3.

5) First, the learning procedure of experiment 1 is repeated, now under the condition of the lesioned net. The according memory- strings are available to the long-term memory of the memory-neurons. After this learning procedure, experiment 4 is done again.
6) Like in experiment 5, the six signs are learnt again like in experiment 1 and 5, in this case each sign combined with an additional pattern in the input-region. Afterwards, the input-part of these combined patterns is presented to the network, which is furtherly lesioned as in experiments 4 and 5.

3.2 Results

The activity of the neural columns, belonging to the separated regions of the different signal bars, enables us to measure and visualize the performance of the net in the course of the various experiments very satisfyingly. The six different signs, some of them being quite similar, are very well recognized and reproduced by the intact net, even if they are presented incompletely.

Fig. 4. completion of a pattern, presented incompletely. Correct activation of the neurons of the according signal-region.

A severe loss of performance we observe after the extended lesion in experiment 4-6.

Most neural chains are affected by the lesion, because the lesion concerns about one fifth of the whole net. The neural chains are interrupted at various positions. The memory-strings, produced under these conditions, differ from the strings, associated to the same patterns in the intact net. Therefore wrong patterns are associated to the presented patterns. The result is a paraphasia, as we know from people with aphasia, having suffered an apoplectic insult.

Experiments 5-6 will simulate to some degree the treatment of the aphasia by logopedia. Under the condition of the lesioned net, the original learning is repeated. By this new learning after the lesion, the paraphasia is minimalized. This, because the memory-strings reflect now again the correct association of signs in the input-region with the corresponding signal-bars in the output-region under the conditions of the lesioned net. The performance will not reach the richness of activity of the intact net again, but the recognition of patterns will be improved, being correct in most cases.

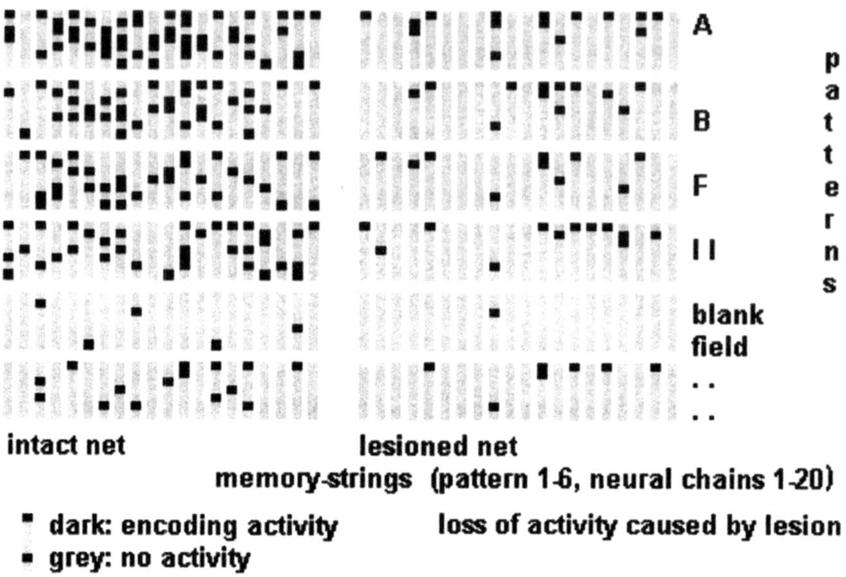

Fig. 5. Memory strings produced by the intact net (left) and by the lesioned net (right). The memory-strings are shown for the six patterns (lines), for the first 20 neural chains(rows). We may see the remarkable loss of activity, encoded after the lesion. The neural chains,(which cannot know,that there had been a lesion),necessarily will accord those strings to wrong patterns, because such strings would encode for quite different input-patterns in the intact net.

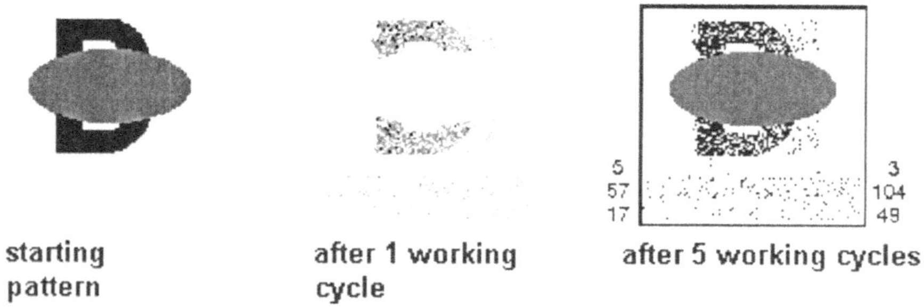

Fig. 6. Occurence of Paraphasia, caused by the lesion (grey elliptical region) of the net. The neural chains decide in their majority, that the input pattern, presented to the input region would be just a blank field (104 neural columns set active in the corresponding signal-region). The memory-strings show so little activity, that they would be produced in an intact net by minimal sensory input. Therefore the decision of the neural chains is reasonable but paraphasic.

An additional remarkable improvement of the performance of the net may be achieved by combining additional specific patterns to the original patterns in the input region. More neurons of the input region will now participate in the process of differentiation. The memory-strings show more structures, which enable the memory-neurons to differentiate to a wider extent between the different strings. In logopedic therapy we know a similar procedure as cued memory, associating words or phonems with additional visual or auditive stimuli.

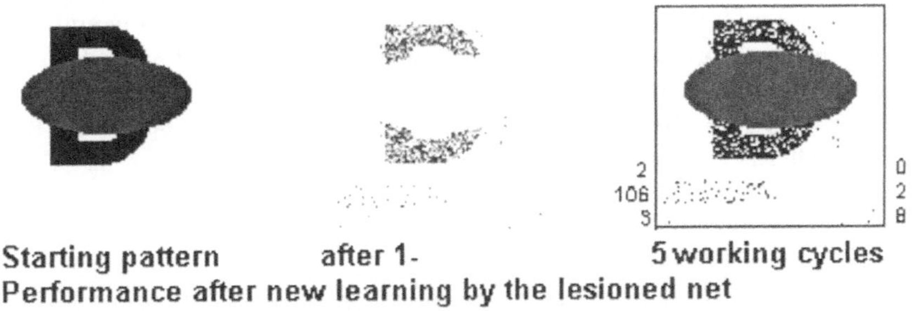

Fig. 7. Repeating the experiment, shown in figure 8 after new learning of the patterns by the lesioned net. There is still little activity(most neural chains severely damaged), but the decision of the net is correct again.

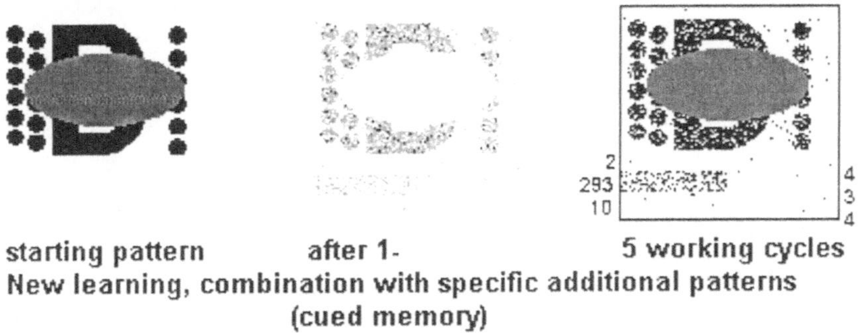

Fig. 8. Example, simulating cued memory: The lesioned net will learn the patterns again, now combined with additional specific patterns in formerly not used parts of the input-region. The performance of the net is furtherly improved by this technique.

4 Conclusions

The presented neural nets work on a new base. They use the temporal sequences of activity in neural chains to produce corresponding memory-strings „intracellularly" in the memory-neurons of the chains. These strings represent the long-term memory. Comparison of the sequences of the memory-strings will enable the neural chains to produce sensible answers to any presented patterns. In the network, the neurons of each neural column will „discuss" these proposals, determining, for instance by rule of majority, the resulting activity of each neural column. Several iterations of this working cycle will lead to the final associative answer of the entire net to a presented pattern. The combined action of all neural chains in a self organizing process will enable these neural nets to perform all tasks, conventional neural nets are able to do. The performance of the net is demonstrating some advantages of this new kind of neural networks: The learning of patterns will be done at only one presentation, the structures of the chains and the whole net remain unchanged by the learning process. At least on the level of the neural chains, several learned patterns will not interfere in pattern reproduction. If a memory-string is selected, the accorded pattern will be reconstructed perfectly in the neural chain, whereas in conventional neural networks, the other patterns are activated to some degree too, with the need to develop sophisticated inhibitory techniques to minimalize this principal effect in neural nets, encoding patterns by changing synaptical weights. These new neural nets learn and reproduce patterns without any changing of synaptical weights.

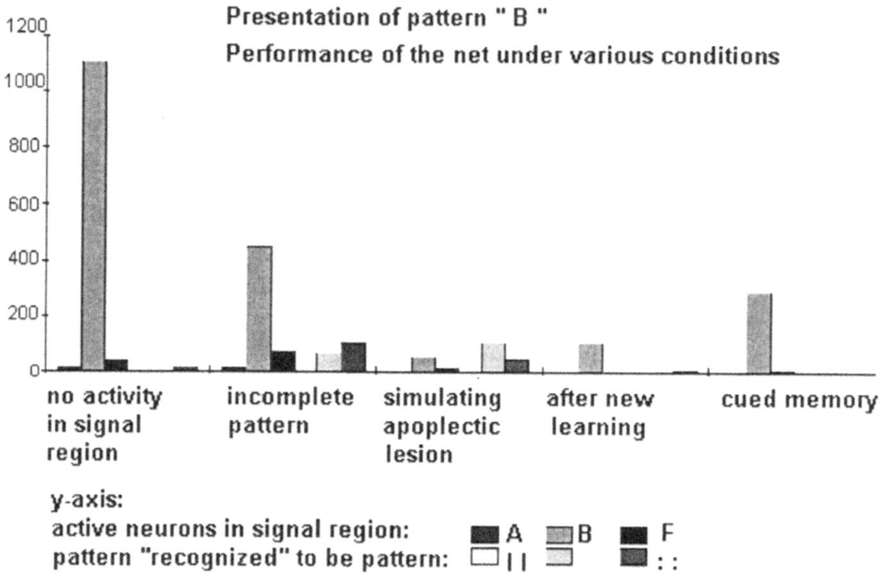

Fig. 9. Diagram, showing the results of different experiments. Each time pattern „B" is presented, showing the correct recognization by the intact net in case of complete or incomplete presentation to the input-region(two sets at the left), occurence of paraphasia caused by the lesion of the net (middle), paraphasia disappearing after new learning under the conditions of the lesioned network, further improvement by the technique of cued memory(right data-set).

They are easy to handle, fast, and offer some concise advantages to conventional neural nets. Hypothetically they may be of biological relevance. Further studies may improve our understanding of the different ways, neural nets may use to perform data-processing.

References

1. Fink M (1996)Time reversal in acoustics , Contemporary Physics 37 , 2 : 95 - 109
2. Kromer T „ Spatial Neural Networks Based On Fractal Algorithms",Computational Intelligence: theory and applications; international conference; proceedings/ 6th Fuzzy Days,

Dortmund, Germany, May 25-27,1999. B. Reusch (ed.). Springer 1999, (lecture notes in computer science; Vol. 1625, pp 603-614)
3. Birbaumer N , Schmidt RF (1996) Biologische Psychologie, p 598 . Springer Berlin
4. Varela,F.J. Resonant cell assemblies: a new approach to cognitive functions and neuronal synchrony.Biol.Res. 28,81-95(1995)
5 Yin, J.C.P., Tully, T. (1996) CREB and the formation of long-term memory. Curr. Opin. Neurobiology 6:264-268
6 Eley, T.C., Plomin, R (1997) Genetic analyses of emotionality. Curr. Opin. Neurobiology 7. 279-284.
7 Greenspan, R.J. (1995) Understanding the genetic construction of behaviour. Sci. Amer. 272(4), 74-79

Inference in Fuzzy Models of Physical Processes

Bohdan S. Butkiewicz[1]

Warsaw University of Technology, Institute of Electronic Systems,
00-665 Warsaw, Poland
"B. Butkiewicz" bb@ise.pw.edu.pl
http://www.ise.pw.edu.pl/~bb/index.html

Abstract. General idea of the paper is comparison of different reasoning methods, which may be used in some types of fuzzy models. Different triangular norms and defuzzification methods were used. It is shown that many reasoning methods give similar results. However, many of them are not very reasonable. Some simple theorems about functions approximated by models are presented. Special attention is applied to modeling of physical processes. Examples of models used in reality are presented. Some of them are build as modifications of Takagi-Sugeno model introduced earlier by author.

1 Introduction

There are many classes of fuzzy models. One of possible classification was given by Pedrycz [10]. He arranged model categories in order of an increased level of structural dependencies. The least structured category appears first on the list

- tabular representations
- fuzzy grammars
- fuzzy relational equations
- fuzzy neural networks
- rule based models
- local regression models
- fuzzy regression models

In the paper, analyze of this categories is performed from point of view of possible applications in modeling of physical processes. Fuzzy grammar [9] models are used for describing time series and signal classifiers. Fuzzy relational equation models were largely studying by Pedrycz, Nola, Hirota, and some others mainly from theoretical point of view. It seems that these models are good for system identification. Tabular, rule based, and local regression models are studying in the paper. Most popular reasoning: Mamdani [8], Larsen [7], Tsukamoto [13], and Takagi-Sugeno [12] are considered. Special attention is applied to inference methods used during approximate reasoning to obtain good results. Generalized Mamdani and generalized Larsen reasoning are used with different triangular norms and some other operations. Comparison of results obtained for some models of physical processes are presented.

B. Reusch (Ed.): Fuzzy Days 2001, LNCS 2206, pp. 782–790, 2001.

2 Description of Models

2.1 Tabular Models

Tabular model has a form of a table, where basic relations between linguistic labels of inputs and outputs are presented. Relations describing dynamics of single input single output (SISO) system of first order may be presented as a single table. Rows and colons denote linguistic values of input and derivative of input. In the table are placed linguistic values of output. The model is very popular in fuzzy modeling and control. It may presents, for example, fuzzy controller of proportional-derivative (PD) type.

Tabular model may be suitable for physical processes, especially when we have no much information about process behavior. Example, steam boiler may be described by quantity of supplied water and fuel, temperature and pressure inside of boiler and quantity of outgoing steam. Using linguistic values small, medium, large, low, medium, high steady-state behavior of boiler may be described. But good description of dynamics requires knowledge about time constant and may be other parameters, so requires analytical description of process. Tabular model is suitable also for simple discrete systems.

Some theoretical and practical results obtained in fuzzy control area [2] [3], and examples presented in the paper show that general Mamdani reasoning, where operations minimum and maximum were replaced by different triangular t-norms and s-norms, is good for this type of model. Moreover, for different triangular norms the results are similar, so it is not very important what pair of norms is used during reasoning. Also the results are identical if rules are used in aggregated form or not. Consider now a simple example of a function $y = f(x)$ with saturation describing by seven rules

R1: if x is PL then y is PL
R2: if x is PM then y is PM
R3: if x is PS then y is PS
R4: if x is ZE then y is ZE
and symmetrically for negative values. If Mamdani, Larsen, Tsukamoto or Takagi-Sugeno model is used then rule weights equals to membership value $\mu_{\{.\}}(x)$ of x for respective set $\{.\}$. Height, areas, and gravity defuzzification methods are weighted means

$$y = \frac{\sum_i y_i w_i}{\sum_i w_i} \tag{1}$$

Also Tsukamoto model use weighted mean. Suppose that membership functions $\mu_{\{.\}}(x)$ have symmetrical triangular shape with trapezes at the end of universum [-10,10]. Thus, membership is linear function $\mu_{\{.\}}(x) = a_i x + b_i$ where a_i, b_i are constant in some regions. If height defuzzification method is used then values y_i are independent on weight w_i, except trapezes. So, $y = \sum y_i(a_i x + b_i)/\sum(a_i x + b_i)$. Similar situation is observed for Larsen model. The results are presented in the Fig. 1. Generally, a theorem may be easy proofed.

Theorem 1

Suppose that membership functions are symmetric and described in different regions by polynomials of n-th order. If Mamdani or Larsen models are used with height defuzzification method then models are described piecewise by rational functions

$$y = \frac{a_n x^n + a_{n-1} x^{n-1} + \ldots + a_0}{b_n x^n + b_{n-1} x^{n-1} + \ldots + b_0} \tag{2}$$

Now, consider Tsukamoto method. Let similarly $\mu_{\{.\}}(x) = a_i x + b_i$. Functions

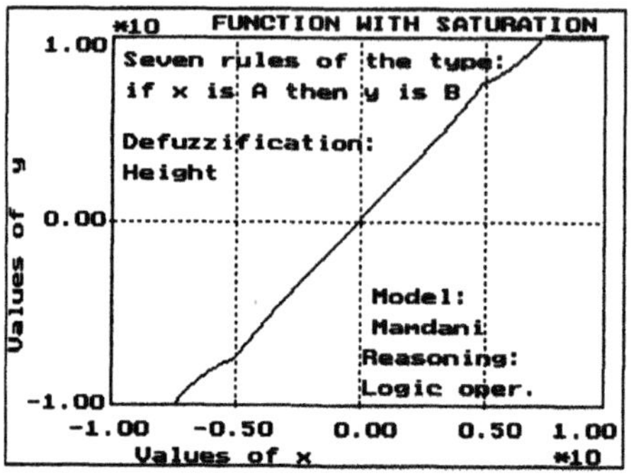

Fig. 1. Mamdani model with height defuzzification, logic (left), Yager (right) operations

$\mu_{\{.\}}(y) = c_i \mid y \mid$ and values $y_i = \mu_{\{.\}}{}^{-1}(w_i) = w_i / c_i$. So, $y = [\sum (a_i x + b_i)^2 / c_i] / \sum (a_i x + b_i)$. Generally, one obtains theorem.

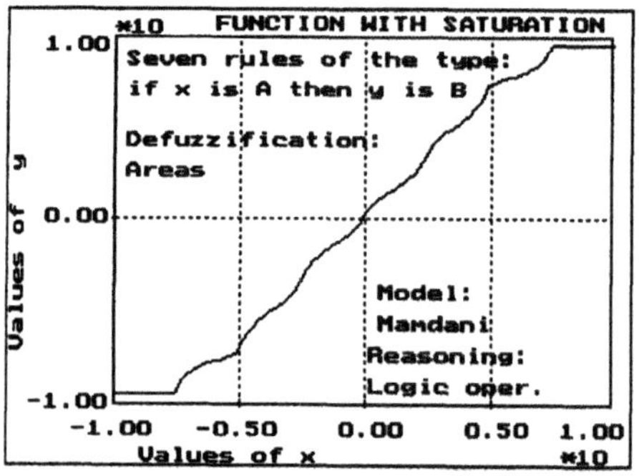

Fig. 2. Mamdani (left) and Larsen (right) models with area defuzzification and logic operations

Theorem 2
If membership functions are symmetric and described in different regions by polynomials of n-th order and Tsukamoto model is used then model is described

piecewise by rational functions

$$y = \frac{a_n x^2 n + a_{2n-1} x^{n-1} + \ldots + a_0}{b_n x^n + b_{n-1} x^{n-1} + \ldots + b_0} \tag{3}$$

Finally, consider Mamdani and Larsen models with areas and gravity methods. For symmetrical triangular shapes values y_i are constant but areas S_i depend on the weights w_i in square $S_i = A(1 - w_i/2)w_i$. Thus, for areas method $y = \sum S_i w_i / \sum S_i = \sum A y_i (a_i x + b_i)[1 - (a_i x + b_i)^2/2] / \sum A(a_i x + bi)[1 - (a_i x + bi)^2/2]$. General dependence is more complicated. An example is presented in the Fig. 2. Two other examples of Mamdani and Tsukamoto models are shown in the Fig. 3. Comparison of models with Mamdani, Larsen and Tsukamoto

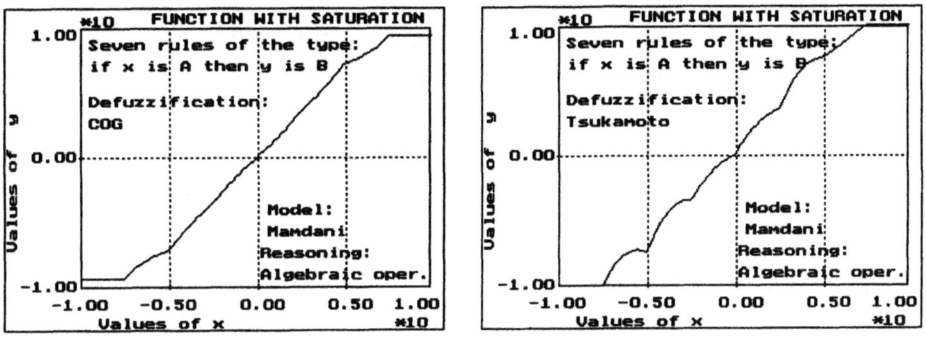

Fig. 3. Mamdani model with gravity defuzzification (left) and Tsukamoto model (right) both with algebraic operations

reasoning with different triangular norms and defuzzification methods shows that choice of triangular norm has no big influence on the result, curves obtained are very similar. Grater influence has defuzzification method, especially when height method is compared with areas and gravity. Mamdani and Larsen models give similar results. Tsukamoto model is different.

Some interesting results concerning equivalence of approximated reasoning using different interpretation of fuzzy if-then rules and aggregation problem are presented in [6].

Sometimes one uses Wang model. It was proofed, (Wang theorem [14]) that any continuous function may be exactly approximated by fuzzy tabular model with gaussian membership functions for input and consequences. Gaussian functions are not very convenient to use and may have nothing common with physical process behavior.

If we have some knowledge about functional relations between process variables it is better to use Takagi Sugeno model.

2.2 Rule-based Models

This type of models is the most popular. This approach to modeling seems more general. The rules may contain heterogeneous form, and different statements Rules can be graduated by word quantifiers. Complex analyze is impossible, because of possible system and rules diversity.

It seems that rule based models are good for situations where non numerical values are expected as decision or model output, example possibility that output take some linguistic value. It may describe some sociological, medical and other decision problems.

An example of model supporting human decision for personnel selection in tourist agency is presented below. Suppose that chief of agency looks for a candidate, which can work as a guide. The candidate ought to:

- known at least two languages among English, French, German, Spain in very high level
- known history and geography of a region in high level
- know-how to use telefax, xserox, computer
- have pleasant sight at least in satisfactory level
- be responsible and patient in high level
- be able to resolve unexpected problems at least in medium level
- ....

Let the candidate fulfill each feature with some level. For example he has a note from an exam or something like this. There are several candidates. Who is the best? Any numerical value of the candidate feature can be treated as fuzzy number with membership $\mu_C(x)$. Any requirement can be described by fuzzy set with membership $\mu_R(x)$. Overlapping membership functions it is possible to find level $l = max\{min[\mu_C(x), \mu_R(x)]\}$ of feature satisfaction. Finding weighted sum $\sum w_i l_i$, where w_i are weight of $i - th$ feature choused before, it is possible to find the best candidate. Other good solution is put $l = max[\mu_C(x)\mu_R(x)]$.

2.3 Local Regression Models

This models are the best if we have some, may be not exact, mathematical description for physical process. However, the model can be used also without this knowledge. Takagi-Sugeno model may be considered under some conditions [15] as universal approximator. Very popular is Takagi-Sugeno-Kang model (TSK) [11] use linear functions as local approximations. Author experience shown that using other than linear functions, example polynomials of second order, without real knowledge of local system behavior, not gives better approximation. Contrary, the results can be considered as bad, i.e. not justified by any reason.
Modification of TSK model
The TSK model may describe sufficiently well any continuouse function $y = f(x_1, x_2, ..., x_n)$. However, the model has one very important inconvenience. Suppose that model describe a function $y = f(x)$ and is composed with two rules
if x is *Small* then $y = a_1 x + b_1$

if x is *Large* then $y = a_2x + b_2$
If conventional triangular or trapezoidal functions are use for membership functions of fuzzy sets of x, here for *Large* and *Small*, see Fig. 4 (left), then non expected effect arise, see Fig. 4 (right) and 5, and [1]. In intermediate region parabolic distortion appears. Thus, the model is worth than conventional crisp model with two linear functions and two separated regions. Of course, it is possible use conventional spline convolution model for better approximation, but it is complicated. Author proposed in [4] some simple modification of TSK model which can avoid inconvenience of TSK model. Suppose that we have some experimental data forming two straight lines with different slopes and an intermediate region. TSK model gives function $g(x)$. Two lines in separate regions may be joined directly in intermediate region by chord line $c(x)$, but this solution is not very good. However, is is possible find better solution. In intermediate region the weighted mean

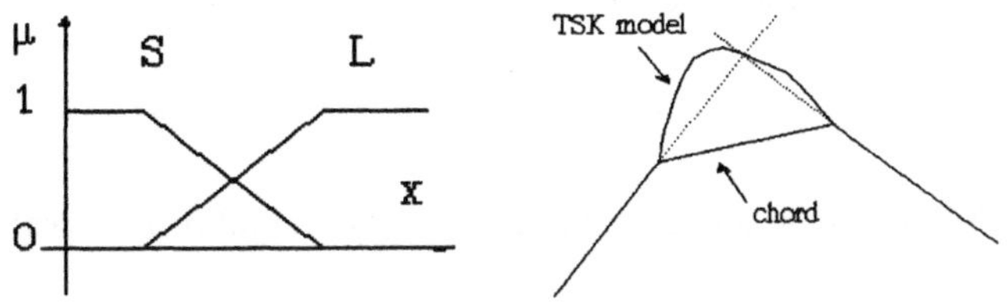

Fig. 4. Membership functions for x (left), TSK model and chord (right)

$$u(x) = \frac{g(x) + \lambda c(x)}{1 + \lambda} \tag{4}$$

can be taken as model value, where λ is a constant choosed experimentally for good approximation of the data. The result is shown in the Fig. 5. The data represent strange effect of optical property relaxation observed in chalkogenide viteouse semiconductor glasses after gamma irradiation [5].

Very interesting property of Takagi-Sugeno model is possibility of knowledge discovering. If mathematical description of some phenomena is known, building this model for unknown process we may verify what phenomena are observed in this process. Presented example shown that two effects are discovered using model built for relaxation process of optical properties in chalcogenides. After gamma irradiation along the time T elapse transparency of semiconductor glass changes in accordance with two different lows. Membership functions of the model give appropriate regions for the lows.

2.4 Fuzzy Regression Models

Sometimes it is not possible to introduce in the model all variables, which have influence on physical process. Simply, these variables are not measured or are not

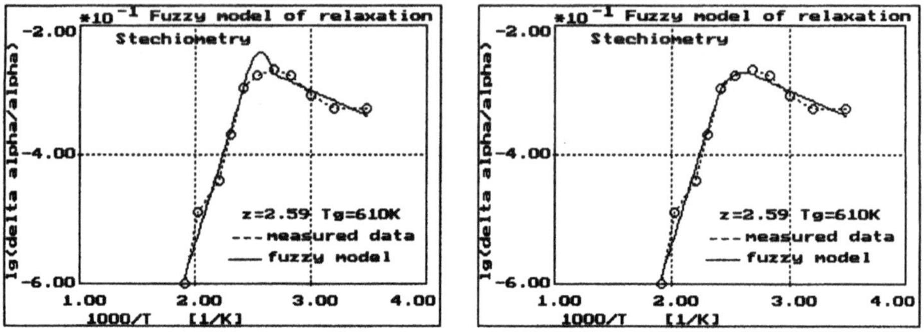

Fig. 5. TSK model without modification (left) and after modification(right)

possible to measure. However, the model must take often in consideration influence of these variables. An example of such situation may be sintering process. It depends on actual total mass of details in the furnace, ambient temperature not possible to preview many days before etc. Reasonable solution is to build fuzzy regression model where conclusion is a fuzzy number or fuzzy function or/and to build fuzzy-probability model where conclusion is random variable or function. In this way additional uncertainty may be introduced.

Numerical considerations concerning inference method are limited to an example of model of sintering process. Different components, as Cu, MnS, C (graphite), StZn are added to iron powder. Exact description of physical and chemical changes during sintering is unknown, because of their complexity. Thus, mathematical model can not be built. Main task of model was preview geometrical changes of detail dimensions after sintering process, taking in consideration proportion of powder components, temperature in sintering zone of the furnace, velocity of tape transporting details in the furnace, and initial density of pressed powder. It was impossible to gathered data in special way. Production process can not be interrupted. However, some specimens with different components were prepared and put in the furnace together with produced details. Thus, some data for the model were very difficult to compare, and results were sometimes discrepant. First, rule-based model was built. Mamdani and Larsen methods are compared. After, modified Takagi Sugeno model is proposed and accepted. The rules have form

if Cu is S and C is S then $\Delta h = f_1(Cu\,,\,C\,,\sigma_1)$
if Cu is S and C is M then $\Delta h = f_2(Cu\,,\,C\,,\sigma_2)$

....

if $StZn$ is S then $\Delta h = f_{10}\,(StZn\,,\sigma_1 0)$

....

if $Temp$ is L and $Time$ is L then $\Delta h = f_{17}\,(Temp\,,\,Time\,,\sigma_{17})$

where Δh describes changes of any parameter, example the height of detail, variables Cu, C , $StZn$, ... contents of the powder components, and variables $Temp$, $Time$ the temperature and time in sintering zone of the furnace. Variables σ_i

have a special task. Functions f_i are fuzzy functions. Its values are fuzzy numbers with trapezoidal shapes of membership functions. Value σ_i desribe width of trapezes, so it describe uncertainty. It is approximate model of the process. Any rule may be considered as multidimensional cloud with different density. Example of this fuzzy surface is presented in the Fig. 6.

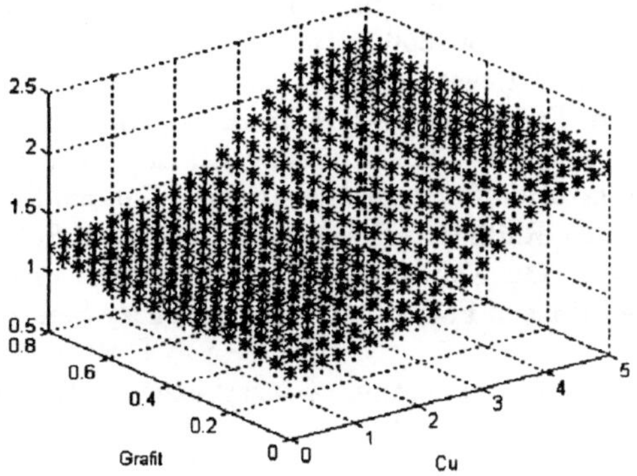

Fig. 6. Example of fuzzy surface for rule 1. Stars represent center value of fuzzy numbers, points minimal and maximal values

3 Conclusion

Physical processes may be modeled in two ways depending on knowledge of mathematical relations between process variables. Without any knowledge, rule-based fuzzy Mamdani and Larsen models or Takagi-Sugeno-Kang fuzzy regression model with linear functions can be build. In many types of models Mamdani reasoning with logic operation (minimum and maximum) is used. Author experience shows that also other triangular norms, ex. algebraic operations, can be used with success. However, many operations give similar results. In the paper only a few examples are presented, but author have tried many other triangular norms. Larsen reasoning is underestimated. It gives also good results and often is easier in numerical applications then Mamdani. If we have some knowledge in the form of functional dependencies for relations between variables, we may use these functions as conclusions in some rules in Takagi-Sugeno fuzzy regression model. Theoretically it is the best solution in this case, but practically it is difficult to find the best shape of membership functions, so estimation of model parameters is complicated. If standard triangular or trapezoidal shape for membership functions is used then strange effect arises in intermediate regions, shown as example presented in the paper. Therefore, author proposed new solution for Takagi-Sugeno model. In intermediate region weighted mean of standard model and chord joining ends of intermediate region is introduced.

References

1. Babuska R.: Fuzzy Modeling for Control, Kluwer Academic Publisher (1998).
2. Butkiewicz B. S.: Steady-State Error of a System with Fuzzy Controller, IEEE Transactions on System, Man, and Cybernetics, Part B: Cybernetics, Vol. 28, No. 6, (1998) 855–860.
3. Butkiewicz B. S.: About Robustness of Fuzzy Logic PD and PID Controller under Changes of Reasoning Methods, European Symp. on Intelligent Techniques, Aachen, Germany (2000) 350–356.
4. Butkiewicz B. S.:Fuzzy Reasoning Methods, its Properties and Applications (in polish), accepted for Prace Naukowe Politechniki Warszawskiej, Elektronika.
5. Butkiewicz B. S., Golovchak R., Kovalskiy A., Shpotyuk O., and Vakiv M.: On the Problem of Relaxation for Radiation-Induced Optical Effects in Some Ternary Chalcogenide Glasses, Radiation Effects and Deffects in Solids, (2000).
6. Czogala E., Leski J., On Equivalence of Approximate Reasoning Results Using Different Interpretations of Fuzzy if-then Rules, Fuzzy Sets and Systems, Vol. 117, (2001) 279–296.
7. Larsen P. M.: Industrial application of fuzzy logic control, Int. J. Man Machine Studies, Vol. 12, No. 1 (1980) 3–10.
8. Mamdani E. H.: Application of fuzzy algorithm for control of simple dynamic plant, Proc. IEE, Vol. 121, No. 12 (1974) 158—1588.
9. Mizumoto M., Toyoda J., Tanaka K.: General formulation of formal grammars, Information Science, Vol. 4 (1972) 87–100.
10. Pedrycz W.: Fuzzy Models: Methodology, Design, Applications and Challenges, in Perdycz W. (ed.) Fuzzy Modelling Paradigms end Practice, Kluwer Acad. Publ. pp 3-22, 1996.
11. Sugeno M., Kang G. T.: Structure Identification of Fuzzy Model, Fuzzy sets and Systems, Vol. 28 (1988) 15–33.
12. Takagi T., Sugeno M.: Fuzzy Identification of Systems and its Application to Modeling and Control, IEEE Trans. on Systems, Man, and Cybernetics, Vol. 15 (1985) 116-132.
13. Tsukamoto Y.: An approach to fuzzy reasoning method, in Fuzzy Set Theory and Applications, Gupta M. M., Ragade R. K., Yager R. R., (eds.), Amsterdam, North-Holland (1979).
14. Wang X. L.: Fuzzy Systems are Universal Approximators, Proc. IEEE Int. Conf. on Fuzzy Systems, San Diego, CA (1992) 1163–1169.
15. Ying H., Ding Y., Li S., Shao S.: Comparison of Necessary Conditions for Typical Takagi-Sugeno and Mamdani Fuzzy System as Universal Approximators, IEEE Trans. on Systems, Man, and Cybernetics-Part A, Vol. 29, No. 5 (1999) 508–514.

About the Combination of Functional Approaches and Fuzzy Reasoning

Hartmut Surmann

GMD - German National Research Center for Information Technology
Schloss Birlinghoven, D-53754 Sankt Augustin, Germany
Phone:+49-2241-142518, Fax: +49-2241-142342
Hartmut.Surmann@gmd.de, http://ais.gmd.de/~surmann

Abstract. Learning fuzzy rule-based systems can lead to very useful descriptions of several problems. Many different alternative descriptions can be generated. In many cases, a simple rule base similar to rule bases designed by humans are preferable since it has a higher possibility of being valid in unforseen cases. Thus, the main idea of this paper is to define a minimal cost function and to generate minimal knowledge bases. Furthermore, this paper shows similarities between the generation of fuzzy systems and the generation of boolean functions on the base of minimal cost functions and it proposes criteria to learn human reasoning fuzzy rules.

1 Introduction

Fuzzy control plays a major role in control theory of the past decade. It is based on fuzzy reasoning which imitates human reasoning procedures by using fuzzy linguistic variables, fuzzy sets, fuzzy modus ponens and e.g. the compositional rule of inference. On one hand fuzzy systems are designed by experts. On the other a lot of scientists try to generate fuzzy systems i.e. fuzzy sets and/or fuzzy rules on the base of I/O data with e.g. neural networks, genetic algorithms or self-made algorithms [1–3]. This is the functional approach to fuzzy systems. The main advantage of these approaches is the improved I/O behavior compared to fuzzy systems designed by experts. The drawback of these functional approaches is the limited usability as an approximation of the human reasoning procedure. To overcome this problem Cheong et. al [4, 5] use very simple systems (triangular membership functions, 50% overlap of the membership functions) to generate interpretable fuzzy rules. This paper presents some remarks on the combination of functional approaches e.g. genetic algorithms together with fuzzy reasoning systems for fuzzy control. Besides the I/O behavior, this paper explains so called structural parameters [6] and the usage of this parameters in functional approaches. One of the parameters is the fuzzy rule based entropy and some operations are "similarity", "universality" and "singularity" of membership functions. Paying attention to these internal parameters helps to close the gap between fuzzy reasoning and functional approaches. Furthermore, the large tuning period, the significant amount of wasted resources and a structure

B. Reusch (Ed.): Fuzzy Days 2001, LNCS 2206, pp. 791–799, 2001.

of unreasonable high complexity could also lead to biasing effects such as the well known over-fitting. It is known [7–9] that a more simple structure leads to a more robust system.

The paper is organized as follows. In section 2 a brief introduction to FRBSs (Fuzzy Rule-Based Systems) together with the definition of optimal FRBS is given . In the next section (3), the criteria for finding minimal systems is presented. The paper finishes with a conclusion in section 5.

2 Basic terms of Fuzzy Rule-Based Systems and affinities to boolean set theory

Fuzzy Systems are based on Fuzzy Set Theory [10, 11]. A number of rules is used to model the way that a system behaves. Fuzzy rules are constructed in the well known IF–THEN form. When crisp values are given as input for a FRBS, they are converted to degrees of membership in the various linguistic values used by the system. This **normalization process** of different universe of discourses to $[0,1]$ allows to compare different input values and to use linguistic concepts like "small", "near", or "approximately". The knowledge representation in terms of understandable linguistic rules differs significantly from the black box approach used by other methods (e.g. Neural Networks).

Boolean set theory on the other hand is a simple and elegant type of algebraic structure. It was found by George Boole (1815 - 1864) and some rules founded by Augustus De Morgan (1806 - 1871). As well as boolean sets $B = \{0,1\}$, fuzzy sets $F = [0,1]$ build an elegant algebraic structure. One aim of the boolean set theory is to build minimal boolean functions on the base of addition or product terms $(x_1x_2 \ldots x_n + \bar{x}_1x_2 \ldots x_n + \ldots)$. The desired function is described by a standard normal form e.g. as product terms and reduced by some algorithms e.g. Quine-McCluskey Tabulation Method (NP-hard).

In FRBS each set of crisp input values activates a number of rules to some degree. The degree to which rules are activated is calculated by combining the degrees of membership (subpremises) with the fuzzy *AND* operator. This leads to the total activation of the rule. Supposing that the k–th rule of the system is:

$$k : \text{IF } x_1 \text{ is } I_{1,k} \text{ AND } x_2 \text{ is } I_{2,k} \text{ AND } \ldots\ x_n \text{ is } I_{n,k} \text{ THEN } y \text{ is } O_k$$

where $x_1, \ldots, x_n$ and y represent input and output variables and $I_{1,k}, \ldots, I_{n,k}$, O_k their respective MFs.

[**Note**: From a structure point of view are the terms $x_1\ is\ I_{1,k} \ldots x_n\ is \ldots$ similar to the product terms in boolean set theory. Only the domain is different.]

Then the extent to which a rule is activated is calculated as:

$$\alpha_k = \mu_{I_{1,k}}(x_1)\ AND\ \mu_{I_{2,k}}(x_2)\ AND\ \ldots\ AND\ \mu_{I_{n,k}}(x_n)$$

where $\mu_{I_{i,k}}(x_i)$ is the membership value of x_i in the $I_{i,k}$ fuzzy set. The result of the *AND* combination is used as a measure for the truth of the rule where the AND operation is a T-Norm e.g minimum. When a rule is activated with a truth value α_k, the inferencing process states that the MF of the output set is:

$$\mu_{O_k}^{inf}(y) = \alpha_k \; AND \; \mu_{O_k}(y),$$

which is the fuzzy result of the rule. Here, the minimum operator *min* is used as an AND operator for both cases.

The total output fuzzy set is calculated by the compositional rule of inference as

$$\mu_O^{total}(y) = \mu_{O_1}^{inf}(y) \; OR \; \mu_{O_2}^{inf}(y) \; OR \; \cdots \; OR \mu_{O_K}^{inf}(y)$$

where it is supposed that the system is described by K rules.

[**Note**: This operation is analog to the unification of product terms of the boolean set theory e.g. $x_1x_2x_3 + \bar{x}_1x_2x_3 + x_1\bar{x}_2x_3 \ldots$]

In order to get a crisp number as output usually a de-fuzzyfication method is used. The calculation of the fuzzy result function $\mu_O^{total}(y)$ and the final crisp value is the bottleneck during computation. Therefore a modified center of gravity algorithm is used to calculate the area A_i and the center of area M_i of each MF before runtime [12]: $A_i = \int O_i(y)dy$, $M_i = \int y \times \mu_{O_i}(y)dy$. The output value is computed as

$$y_{crisp} = \frac{\sum_{i=0}^{K} \alpha_i \times M_i}{\sum_{i=0}^{K} \alpha_i \times A_i} \tag{1}$$

Definition 1 (Fuzzy Rule-Based System). *A fuzzy Rule-Based System* $FRBS = (LV, R, T, I, AGR - DE)$. *It is a 5-Tuple with a set LV of n linguistic input variables and m output variables, a set R of k fuzzy Rules, a T-Norm T, for the combination of rules subpremises, an Implication strategy I verifying* $I(a, 0) = 0 \; if \; a \neq 0$ *(e.g., an R-implication or t-norm), and a aggregation and defuzzification strategy* $AGR - DE$. *This can be a T-CO-Norm* T^*, *for the combination of rules output sets, and a defuzzyfication strategy DE (e.g., center of gravity, maximum or FCOG) or the addition/division of equation 1.*

Theorem 1 (Universal approximator). *Let FRBS be the set of all FS and* $f : U \subseteq I\!R^n \rightarrow I\!R$ *be a continuous function defined on a compact U. For each* $\epsilon > 0$ *there exists a* $FS_\epsilon \in FRBS$ *such that*
$sup\{| \, f(x) - FS_\epsilon(x) \, | \quad | \, x \in U\} \leq \epsilon$.

Wang [13] or Buckley [14] provide the proof. The universal approximator theorem is a very good theoretical result but unfortunately it is not constructive i.e. it can not be used to get a FRBS with a predefined error limit ϵ. Usually, in real world application ϵ is unknown and influenced e.g. by sensor noise so that it does not make sense to fix it.

2.1 Membership function shape

Membership functions will be constructed from probability density functions by $\mu(x) = \lambda p(x)$. The constant λ is calculated using the constraint $\sup \mu = 1$. Here the Gaussian probability distribution $p(x) = \varphi(x; m, \sigma^2)$ is chosen, because this distribution fits in the first approximation a lot of real world problems. Therefore, the membership functions used in the system are defined by

$$\mu(x) = \sigma\sqrt{2\pi}\,\varphi(x; m, \sigma^2) = \exp\left(-\frac{(x-m)^2}{2\sigma^2}\right). \quad (2)$$

The MFs are limited by $\mu(x) = 0\ for\ x \notin [m - 2.9\sigma, m + 2.9\sigma]$. Special margin MFs (smallest and biggest) are defined at the left and right border of a linguistic variable right (Fig. 1):

$$\mu_{\text{left}}(x) = \begin{cases} 1 & , x < 0 \\ e^{-\frac{1}{2}x^2} & , x \geq 0 \end{cases} \qquad \mu_{\text{right}}(x) = \begin{cases} 1 & , x \geq 0 \\ e^{-\frac{1}{2}x^2} & , x < 0 \end{cases} \quad (3)$$

Non–symmetric Gaussian MFs consist of two parts (left and right) with a common mean point but different σ_{left} and σ_{right} for the left and right side. They are used to achieve more flexible systems.

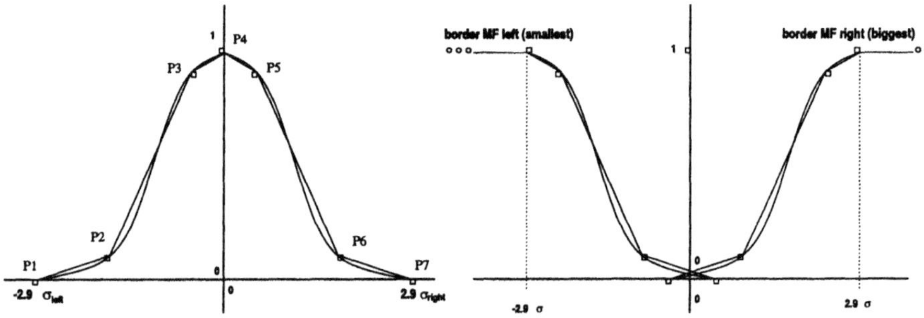

Fig. 1. Linearized model of Gaussian like MFs. Six lines are used to describe them (left). The smallest and biggest MF (ordered by the middle points) are linearized by three lines (right).

2.2 Costs and minimal FRBS

The challenge is to find an automatic learning algorithm which represents a predefined function and which learns and optimizes the structure of the fuzzy rule base, the input and output variables as well as the MFs of a FRBS according to the human reasoning procedure. Therefore, analog to boolean set theory a cost function and minimal system will be defined.

Definition 2 (Complexity Cost Function, *E*). *A FRBS with k fuzzy rules, n linguistic variables, $k * n$ subpremises and $m = m_1 + \ldots + m_n$ MFs is less complex than one with*

1. *$k+1$ fuzzy rules, $k*n$ subpremises and m MFs or*
2. *k fuzzy rules, $(k*n)+1$ subpremises and m MFs or*
3. *k fuzzy rules, $k*n$ subpremises and $m+1$ MFs.*

The weighting of the above three terms depends on the application. Based on the cost function minimal FRBS can be defined.

Definition 3 (Minimal Fuzzy-System). *Let us approximate a real system $f(\boldsymbol{x})$, using a fuzzy system $X(\boldsymbol{x})$ and let $\epsilon > 0$ be a maximal accepted error limit. A fuzzy system X with $sup\{\mid f(\boldsymbol{x}) - X(\boldsymbol{x}) \mid \quad \mid \boldsymbol{x} \in U\} \leq \epsilon$ is called minimal if there is no other fuzzy system Y with $sup\{\mid f(\boldsymbol{x}) - Y(\boldsymbol{x}) \mid \quad \mid \boldsymbol{x} \in U\} \leq \epsilon$ and $E(Y) < E(X)$, where E is the complexity cost function.*

The real system $f(\boldsymbol{x})$ is represented by referential vectors $\boldsymbol{r}$. Alternatively to the absolute metric the Euclidian metric can be used:

$$\sqrt{\sum (f(\boldsymbol{x}) - Y(\boldsymbol{x}))^2} \leq \epsilon. \tag{4}$$

For classification problems the difference between the output of the fuzzy system ($\in [0,1]$) and the optimal classification 0,1 or the number of correct/incorrect classifications can be used. The number of correct/incorrect classifications can be calculated by using a threshold e.g. $= 0.5$ to digitalize the fuzzy output of the fuzzy system.

3 Finding minimal FRBS

Now, finding minimal Fuzzy Systems can be regarded as a search process from a predefined starting point.

3.1 Initial FRBS

For an initial FRBS a predefined number of membership functions is normally distributed over the domain of each variable. The referential vectors are used to select the limits for each variable. Each referential vector is regarded as a potential fuzzy rule. The rule is build from the highest activated MF of each referential vector. Equal fuzzy rules are deleted. Based on this initial rule base the structure of the FRBS is opened i.e. each rule gets its own MFs. Therefore, the number of MFs for each variable and the number of rules are equal after this initialization. Now, minimizing the rule base will close the structure i.e. it will identify similarities between the membership functions and drop out unnecessary rules or variables in a rule.

3.2 Minimizing FRBS

Finding a minimal FRBS is an optimization process of the initial FRBS. Different criteria e.g. the mean square error MSE have to be considered according to definition 3

$$MSE_{pv} = \frac{1}{pv} \sum_{i=1}^{pv} \sum_{j=1}^{n_{out}} (r_{i,j} - o_{i,j})^2, \tag{5}$$

where pv is the number of referential data pairs, n_{out} is the dimension of the output vector, $o_{i,j}$ is the current computed output value of the FRBS and $r_{i,j}$ is the respective referential value ($i = 1 \ldots pv$, $j = 1 \ldots n_{out}$). These criteria are issued for the functional approach.

For the human reasoning approach some criteria for the structure of the FRBS have to be considered. The first criterion (E) for the structure of a FRBS is the *degree of fuzziness* or the *entropy* (reciprocal) of a FRBS and was introduced in [15]. It is defined by the average number of activated rules:

Definition 4 (degree of fuzziness). *The degree of fuzziness of a FRBS is*

$$R_{\phi} = \frac{1}{pv} * \sum_{i=1}^{pv} R_{cur,i}, \tag{6}$$

where $R_{cur,i}$ is the current number of activated rules for the i–th input/output pair and pv the number of those pairs.

The degree of fuzziness is a global criterion for the total rule base. If the number of activated rules is minimal i.e. one, then a FRBS is maximally understandable by humans. FRBS with a high number of activated rules behave like a neural network i.e. a lot of rules (neurons) determine the output values. The degree of fuzziness influences the MFs. If the degree is smaller then the overlap of the MFs is reduced.

Definition 5 (degree of completeness). *The degree of completeness DOC of a rule R_i is defined as*

$$DOC_{R_i} = \frac{NR_i}{pv} \tag{7}$$

where NR_i is the number of examples which activate the rule R_i and pv is the number of training examples (above).

The degree of completeness is a local criterion for each rule and introduced by L. Castillo et al. [16, 5]. For the degree criteria it is important that the Gaussian like membership functions are limited by $\mu(x) = 0 \; for \; x \notin [m - 2.9\sigma, m + 2.9\sigma]$. Otherwise the membership functions and with them the fuzzy rules are often activated with a small degree.

A further criterion (S^{MF}) for the structure of the FRBS is the number of MFs. For that MFs with high degrees of overlap are counted. Such nearly equal MF pairs are desirable, because they are easily unified to only one MF which reduces the cost function (Def. 3(3)). Similar MFs are analogous to implicants

/ primeimplicants in boolean set theory. The criterion for the highest possible number of similar MFs is :

$$S^{MF} = \left(\frac{s_{no}}{(n+m)R_{total}}\right)\gamma \tag{8}$$

where s_{no} is the number of similar MFs i.e. MFs with $m_1/m_2 \approx 1$ and $\sigma_1/\sigma_2 \approx 1$ $n+m$ is the total number of input and output variables, R_{total} is the total number of rules and $\gamma \in \mathbf{R}$ is a predefined weighting factor for the number of similar MFs. Figure 2 shows an example of similar and non similar MFs.

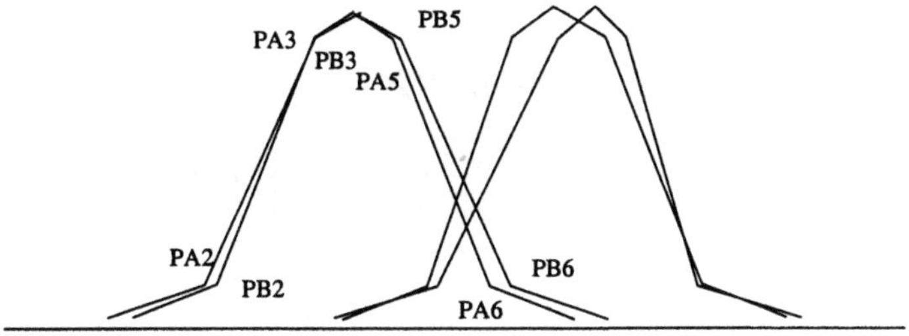

Fig. 2. Example of similar (left) and non similar (right) MFs. The right MFs have non-symmetric widths.

The next criterion (UZ^{MF}) for the structure of the FRBS deals with *never activated* ($\mu(x) = 0, \forall x \in U$) and *always completely activated* ($\mu(x) = 1, \forall x \in U$) MFs. On one hand a subpremise of a rule can be eliminated if the belonging MF is always completely activated for all of the training pairs so that the cost function E decrease (Def. 3(2)). On the other hand MFs which are never activated can be eliminated together with all the rules that they participated to so that the cost function E also decrease (Def. 3(1)). The criterion for the highest possible number of always zero and always one MFs with the predefined weighting factors $\zeta, \eta \in \mathbf{R}$ is:

$$UZ^{MF} = \left(\frac{u_{no}}{(n+m)R_{total}}\right)\zeta + \left(\frac{z_{no}}{(n+m)R_{total}}\right)\eta, \tag{9}$$

where u_{no} is the number of always one MFs, z_{no} is the number of always zero MFs. Always one MFs are replaced by trapeze covering the total variable domain, while zero MFs are replaced by impulses. A collection of MFs with zero, one and similar MFs is shown in figure 3.

An example on a closing step based on the above criteria is presented in Figure 4. On the base of these criteria algorithms can be defined to compute minimal fuzzy systems. The problem is that the computation of minimal fuzzy systems is NP hard. Previous work [15] shows that genetic algorithms are one approximation algorithm to find minimal fuzzy systems but other approaches are desirable.

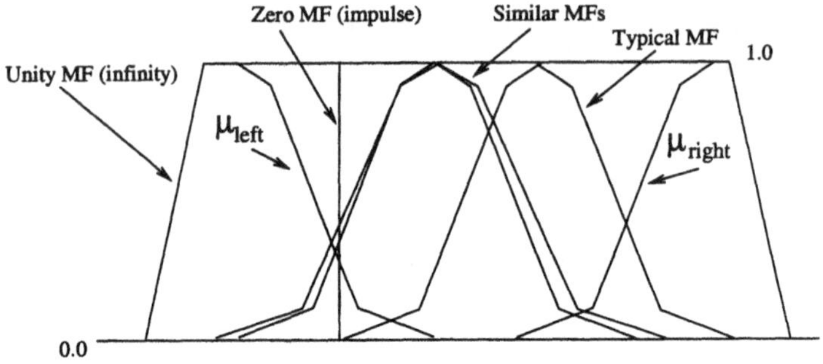

Fig. 3. Different kinds of membership functions.

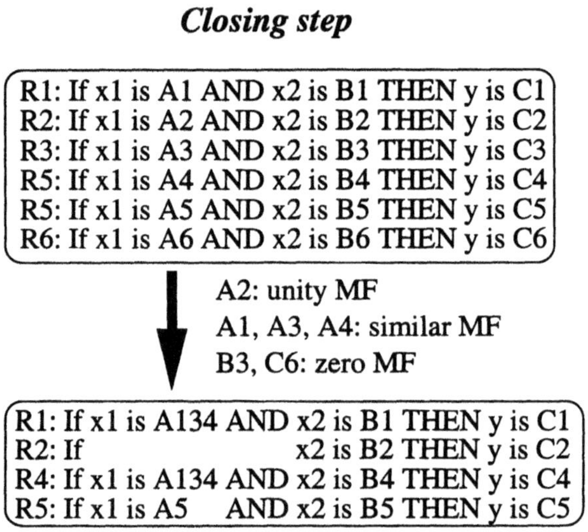

Fig. 4. Example of a closing step. After the closing step two rules and one subpremise are deleted. Three MFs are combined.

4 Concluding remarks

In previous works [6] we have shown that the use of genetic search techniques permits to learn the relevant fuzzy rules, variables and membership functions, This work has presented similarities between the generation of boolean functions and the automatic generation of fuzzy rules and it presents criteria for the generation of simple human like fuzzy rules. Using these criteria as e.g. genetic operators or in the fitness functions generates simpler and more comprehensive knowledge bases. The criterion for the functional approach is a metric (mean square error) between the fuzzy system and the approximated system. The criteria for a minimal fuzzy system i.e. systems similar to systems designed by humans are the rule base entropy (degree of fuzziness) and similarity, universality and pulse membership functions. The last three criteria ensure the simplicity of the variables and the simplicity of the rules. The use of a cost function for fuzzy rules, variables and membership functions allows the learning algorithm to obtain more descriptive and more specific fuzzy rule-based systems.

Acknowledgment Special thanks to Claudia Hirsch for hints and corrections.

References

1. S. K. Halgamuge and M. Glesner, "A fuzzy-neural approach for pattern classification with the generation of rules based on supervised learning," in *Neural Networks & their Applications: Proceedings of Neuro-Nimes '92*, pp. 165 – 173, 1992.
2. F. Hoffmann, "Soft computing techniques for the design of mobile robot behaviors," *Information Sciences*, vol. 122, no. 2-4, pp. 241–258, 2000.
3. M. Krabs and H. Kiendl, "Entwurfskonzept für Fuzzy-Regler mit automatisch generierten Regelsätzen," *VDE-Fachtagung: "'Technische Anwendungen von Fuzzy-Systemen"', Dortmund*, 1992.
4. F. Cheong and R. Lai, "Constraining the optimization of a fuzzy logic controller using an enhanced genetic algorithm," *IEEE Transaction on systems, man and cybernetics – part B: Cybernetics.*, vol. 30, pp. 31–46, February 2000.
5. L. Castillo, A. Gonzalez, and R. Perez, "Includinga simplicity criterion in the selection of the best rule in a genetic fuzzy learning algorithm," *Fuzzy Sets and Systems*, vol. 120, pp. 309 – 321, 2001.
6. H. Surmann and M. Maniadakis, "Learning feed-forward and recurrent fuzzy systems: a genetic approach," *Journal of Systems Architecture, Special issue on evolutionary computing*, vol. 47, no. 7, pp. 535 – 556, 2001.
7. O. Cordon, F. Herrera, and M. Lozano, "On the bidirectional integration of genetic algorithms and fuzzy logic," *http://www.bioele.nuee.nagoya-u.ac.jp/wec2/papers/p003.html*, 1996.
8. T. Takagi and M. Sugeno, "Fuzzy identification of systems and its application to modeling and control," *IEEE Transaction Systen Man and Cyberentic*, pp. 116–132, 1985.
9. S.-T. Song and L.-H. Sheen, "Heuristic fuzzy-neuro network and its application to reactive navigation of a mobile robot," *Fuzzy Sets and Systems*, vol. 110, no. 3, pp. 331 – 340, 2000.
10. L. A. Zadeh, "Fuzzy sets," *Information and Control*, vol. Vol. 8, pp. 338 – 353, 1965.
11. A. Kandel and S. C. Lee, *Fuzzy Switching and Automata: Theory and Applications.* New York: Crane Russak, 1979.
12. H. Surmann and A. P. Ungering, "Fuzzy-rule-based systems on general purpose processors," *IEEE MICRO, Special issue on fuzzy systems*, pp. 40–48, Aug. 1995.
13. L.-X. Wang, "Fuzzy systems are universal approximators," in *First IEEE International Conference on Fuzzy Systems, San Diego*, pp. 1163 –1170, 8. - 12.03.1992.
14. J. Buckley, "Sugeno Type Controllers are Universal Fuzzy Controller," *Fuzzy Sets and Systems*, vol. 53, pp. 299–304, 1993.
15. H. Surmann, *Genetic Optimizing of Fuzzy Rule-Based Systems*, vol. 8 of *Studies in Fuzziness and Soft Computing*, ch. Optimization of Fuzzy Controllers, pp. 389 – 402. Physica-Verlag, Sep. 1996.
16. A. Gonzalez and R. Perez, "Completeness and consistency conditions for learning fuzzy rules," *Fuzzy Sets and Systems*, vol. 96, pp. 37 – 51, 1988.

Expected Utility Networks in Transferable Belief Model

Anna Slobodová*

Institute of Control Theory and Robotics, Slovak Academy of Sciences,
Dúbravská cesta 9, 842 37 Bratislava, Slovakia
utrraslo@nic.savba.sk

Abstract. Traditional Bayesian decision analysis is based on probability theory and utility theory. However, expected utility model can be derived also in the other models proposed to quantify someone's belief. We shall deal only with the transferable belief model. The purpose of this paper is to introduce a new class of graphical representation to simplify the decision process.

1 Introduction

The main objective of decision analysis is to help the decision maker to select an appropriate decision alternative in the face of uncertain environment. Traditional Bayesian decision analysis is based on probability theory and utility theory. Expected utility model can be derived also in the other models proposed to quantify someone's belief. But Bayesians have convincingly showed that if decisions must be "coherent", our belief over the various outcomes must be quantified by probability function. We shall deal only with the transferable belief model ([6]) and use a pignistic probability function to compute the expected utility of each decision alternative.

The purpose of this paper is to introduce a new class of graphical representation to decompose the expected utility function.

2 The Transferable Belief Model

The transferable belief model ([6]) is a mathematical, idealized model that sets out to represent someone's belief. This model claims that beliefs are quantified by a single number (*bel*).

The transferable belief model is based on a two – level model:

– a *credal level* where beliefs are entertained, combined, and updated
– a *pignistic level* where beliefs are used to take decisions (from pignus, a bet in Latin).

* This work has been supported by the grant No. 2/1102/21 of the Scientific Grant Agency of Ministry of Education of Slovak Republik and Slovak Academy of Sciences.

B. Reusch (Ed.): Fuzzy Days 2001, LNCS 2206, pp. 800–806, 2001.

At the credal level beliefs are quantified by belief functions.

In any modeling the uncertainty, it is required that a method should be provided for making decisions. However, decision making requires that we use a probability function to compute the expected utility of each decision alternative. This means that when we make decisions, our uncertainty at the pignistic level has to be quantified by a probability function. Thus, belief held at the credal level must induce a probability measure at the pignistic level. That probability measure is then used to compute the expected utilities.

2.1 Belief Functions

Definition 1. *Let Ω be a nonempty finite set called the frame of discernment. The mapping $bel : 2^{\Omega} \rightarrow [0,1]$ is an (unnormalized) belief function if and only if there exists a basic belief assignment (bba) $m : 2^{\Omega} \rightarrow [0,1]$ such that:*

1. $\sum_{A \subseteq \Omega} m(A) = 1$,
2. $bel(A) = \sum_{B \subseteq A, B \neq \emptyset} m(B)$,
3. $bel(\emptyset) = 0$.

Those subsets such that $m > 0$ are called *the focal elements.* A belief function is normalized if $m(\emptyset) = 0$, in which case $bel(\Omega) = 1$. A *vacuous belief function* is a belief function such that $m(\Omega) = 1$ and $m(A) = 0$, for all $A \neq \Omega$, and it represents total ignorance.

The value *bel(A)* quantifies the strength of the belief that event A occurs. The value *m(A)* , on the other hand, represents the part of belief supporting the fact that A occurs but cannot support any event more specific than A. This part of belief is allocated to A (and A alone) because it cannot support any more specific event (due to a lack of information). Note that m is not the direct counterpart of a probability distribution function p. Both *bel* and P are defined on 2^{Ω}, but m is defined on 2^{Ω} whereas p is defined on Ω.

Consider two distinct pieces of evidence on Ω represented by m_1 and m_2. The belief function $m_{12} = m_1 \oplus m_2$ that quantifies the combined impact of these two pieces of evidence is obtained by the *(unnormalized) Dempster's rule of combination.* Its computation is as follows:

$$\forall A \subseteq \Omega, \;\; m_{12}(A) = (m_1 \oplus m_2)(A) = \sum_{B \cap C = A} m_1(B) m_2(C).$$

2.2 The Pignistic Probability Function Derived from a Belief Function

To map belief functions onto probability functions, many transformations exist. Smets ([5]) requires that the pignistic transformation satisfies the following assumptions.

Proposition 1. *Let F be a finite set and let bel^F be a belief function defined on F. Let $BetP^F$ be a probability function on F. For all $\omega \in F$*

$$BetP^F(\omega) = \Gamma_F(bel^{\Omega})(\omega). \tag{1}$$

The function Γ_F is called the pignistic transformation.

Evaluation of $BetP^F(A)$ for $A \subseteq F$ is obtained by adding the probabilities $BetP^F(\omega)$ for $\omega \in A$.

Proposition 2. *Let bel_1 and bel_2 be two belief functions on the frame of discernment F. Let Γ_F be the pignistic transformation that transforms a belief function over F into a probability function $BetP^F$ over F. Then Γ_F satisfies, for any $\alpha \in [0,1]$,*

$$\Gamma_F(\alpha bel_1 + (1-\alpha)bel_2) = \alpha\Gamma_F(bel_1) + (1-\alpha)\Gamma_F(bel_2) \tag{2}$$

Proposition 3.

$$BetP^F(\Omega) = 1.$$

Proposition 4. *Let R be a permutation function from F to F. The pignistic probability given to the image of $A \subseteq F$ after permutation of the elements of F is the same as the pignistic probability given to A before applying the permutation:*

$$BetP^{F_*}(R(A)) = BetP^F(A),\ \forall A \subseteq F,$$

where $BetP^{F_*}$ is the pignistic probability function on F_* after applying the permutation function.

Proposition 5. *The pignistic probability of an impossible event is zero.*

Proposition 6. *If bel^F happens to be a probability function P defined on F, then $\Gamma_F(P) = P$.*

Under these ssumptions, it is possible to derive uniquely Γ_F. The pignistic transformation is as follows.

Theorem 1. *Let bel^F be a belief function on space F and m^F its related bba. Let $BetP^F = \Gamma_F(bel^F)$. The only solution $BetP^F$ that satisfies the above propositions is:*

$$BetP^F(\omega) = \sum_{A:\omega\in A\subseteq F} \frac{m^F(A)}{|A|(1-m^F(\emptyset))},\ \forall\omega \in F,$$

where $|A|$ is the number of elements of Ω in A, *and*

$$BetP^F(A) = \sum_{\omega\in A} BetP^F(\omega),\ \forall A \subseteq F.$$

2.3 Independence and Non-interactivity in the Transferable Belief Model

The concept of (conditional) independence for variables has not been widely treated in belief-functions theory. ([1]) investigated some ways to define independence relationships between variables when uncertainty is expressed under the form of belief functions. The main problem is stated as follows: let X

and Y be two (logically independent) variables taking their values on the sets $X = \{x_1, x_2, ..., x_n\}$ and $Y = \{y_1, y_2, ..., y_n\}$, respectively. Consider a belief function, representing a global piece of knowledge about the variables X and Y, on the Cartesian product $W = X \times Y$. Then, we want to define the concept of non-interactivity and independence between X and Y.

Definition 2. *Given a bi-dimensional belief functions bel on W associated to its basic belief assignment m, the variables X and Y are said to be non-interactive with respect to m_W if and only if:*

$$m_W(w) = m_X(A)m_Y(B), \quad w = A \times B, A \subseteq X, B \subseteq Y,$$

otherwise $m_W(w) = 0$, where m_X and m_Y are the basic belief assignments obtained by the marginalization of m_W on X *and* Y, *respectively.*

By analogy to stochastic independence for the probabilistic case, we can define belief function independence - that we call "doxastic independence" as follows:

Definition 3. *Given a bi-dimensional belief functions bel_W on $W = X \times Y$ associated to its bba m_W, the variables* X *and* Y *are said to be doxastically independent with respect to m_W if and only if:*

1. *$m_Y(B|A) = m_Y(B), \quad \forall A \subseteq X, \quad \forall B \subseteq Y$, where $m_Y(B|A)$ it the conditional belief function over Y obtained by the Dempster's rule of conditioning, and*
2. *if* X *and* Y *are doxastically independent with respect to both m_1 and m_2, then* X *and* Y *are doxastically independent with respect to $m_1 \oplus m_2$ obtained by applying Dempster's rule of combination on m_1 and m_2.*

3 Expected Utility Networks with Pignistic Probabilities

Modularity in AI allows concise representation of otherwise quite complex concepts.In recent years the fact has been proven that reasoning about probabilities and utilities together requires more than simply gluing together a representation of utility and one of probability. Strategic inference - the reasoning process which underlies rational decision-making - involves the computation of conditional expected utilities, which may not have a modular representation even if probabilities and utilities, taken separately, do.

The purpose of this paper is to introduce a new class of graphical representation, expected utility networks with pignistic probabilities (EUNPP).The basic idea is inspired by [3].

EUNPPs are undirected graphs with two types of arc, representing pignistic probability and utility dependencies, respectively. The representation of utilities is based on a notion of conditional utility independence, and is defined in analogy with its probabilistic counterpart.

What is important about conditionally independent decisions is that they can be effectively decentralized: a single, complicated agent can be replaced by simpler, conditionally independent sub-agents, who can do just as well.

3.1 Conditional Independence of Pignistic Probabilities and Utilities

Strategic inference typically involves computation of the expected utilities for alternative plans of action, and hence what is really important is the ability to decompose the expected utility function. Therefore, we propose a notion of conditional utility independence.

Let $\{X_i\}_{i \in N}$ ($N = \{1, ..., n\}$) be a finite, ordered set of random variables. We assume that they may take only finitely many values. Let $x^0 = (x_1^0, ..., x_n^0)$ be some arbitrary given realization which will act as the reference point. A joint realization $x = (x_1, ..., x_n)$ represents a (global) state or outcome. For any $M \subset N$, we denote by X_M the set $\{X_i\}_{i \in M}$. Let p be a strictly positive pignistic probability measure defined on the Boolean algebra $\mathcal{A}$ generated by X_N, and let u be a (utility) function which assigns to each state x a positive real number. We assume that the decision maker's beliefs and preferences are completely characterized by (p, u), and that p represents the decision maker's prior beliefs, and that for any two pignistic probability measures p' and p'', $p' \succ p''$ (p' is preferred to p'') if and only if $E_{p'}(u) > E_{p''}(u)$, where $E_p(u) = \sum_x u(x)p(x)$. Finally, let

$$q(x_M|x_{N-M}) = \frac{p(x_M, x_{N-M})}{p(x_M^0, x_{N-M})}.$$

The interpretation of q tells us how the pignistic probability changes when the values of X_M are shifted away from the reference point, while the values of X_{N-M} are held fixed at x_{N-M}.

We can define also a corresponding comparison operator for utilities:

$$w(x_M|x_{N-M}) = \frac{u(x_M, x_{N-M})}{u(x_M^0, x_{N-M})}.$$

One way to interpret w is as a measure of the intensity of preference for x_M (with respect to the reference point) conditional on x_{N-M}.

Suppose that $q(x_M|x_{N-M})$ only depends on x_K, where $K \subset N - M$, but not on x_{N-M-K}. This is possible to express by defining new quantities $q(x_M|x_K) = q(x_M|x_{N-M})$, where the conditions x_{N-M-K} are dropped. We can call this notion *p-independence*; it corresponds to conditional probability independence whenever it is defined.

A corresponding notion of conditional utility independence (u - *independence*) is defined accordingly. Suppose that $w(x_M|x_{N-M})$ depends on x_K, but not on x_{N-M-K}. In that case , we again define new quantities $w(x_M|x_K) = w(x_M|x_{N-M})$, and say that X_M is u-independent of X_{N-M-K} given X_K.

We can define an expected utility network as an undirected graph G with two types of arc, representing pignistic probability and utility dependencies, respectively. Each node represents a random variable (say, X_i), and is associated with two positive functions, $q(x_i|x_{P(i)})$ and $w(x_i|x_{U(i)})$, where $P(i)$ denotes the set of nodes directly connected to X_i via probability arcs, and $U(i)$ denotes the corresponding set of nodes directly connected to X_i via utility arcs.

3.2 Conditional Expected Utility

Suppose now that A and B are controllable variables, in the sense that their values can be freely set by the decision maker. A rational decision maker will want to choose values of A and B which maximize expected utility; hence, for each assignment (a, b), the decision maker should compute the corresponding expected utility, and identify an optimal decision. Clearly, the decision process becomes quite cumbersome when there are many decision variables. To reduce its complexity, we seek conditions under which the expected utility calculations can be conveniently decomposed.

Definition 4. *The utility of an event is the expected utility of the event, conditional on the event being true. Formally,*

$$u(E) = \sum_{x \in E} u(x)p(x|E).$$

Corollary 1. *For any nonempty* $E \in \mathcal{A}$ *and for any non-empty, finite partition* $\{E_k\}$ *of* E

$$u(E) = \sum_k u(E_k)p(E_k|E).$$

It is natural to normalize the utility measure around certain values, just as probabilities are normalized to lie between zero and one. Hence, we require that $u(True) = 1$, where *True* denotes the tautological event, or the entire universe.

Bayes' rule does not hold for utilities, but a modified version of it does:

$$u(F|E) = \frac{u(E|F)u(F)}{u(E|F)u(F)p(F|E) + u(E|-F)u(-F)p(-F|E)}$$

Note that conditional utility depends among other things, on conditional pignistic probabilities.

3.3 Conditional Expected Utility Independence

Definition 5. *Two events,* E *and* F, *are said to be* conditionally expected utility independent *given a third event* G *if*

$$u(E \cap F|G) = u(E|G)u(F|G).$$

The very important result is :

Theorem 2. *p- and u- independence jointly imply conditional expected utility independence.*

Hence, the graphical structure of expected utility networks can be exploited to identify conditional expected utility independencies: node separation with respect to both the utility and pignistic probability subgraph implies conditional expected utility independence. The upshot is that, conditional on A, decisions regarding B and C can be effectively decomposed: if both B and C contain variables which are under the control of the decision maker, it is not necessary nor useful to possess information about C in order to decide on B, and vice versa.

References

1. Ben Yaghlane, B., Smets, P., Mellouli, K.: Independence and non-interactivity in the transferable belief model. In: Workshop on conditional independence structures and graphical models. F. Matúš and M. Studený (Eds.). Abstract book. Toronto, Canada, (1999) 4–5
2. Dubois, D. and Prade, H.: Decision Evaluation Methods under Uncertainty and Imprecision. In: Combining Fuzzy Imprecision with Probabilistic Uncertainty in Decision Making. J.Kacprzyk and M. Federizzi (Eds.), Springer, (1987) 48–65
3. Mura, P. and Shoham, Y.: Expected Utility Networks. In: Proceedings of the 15th Conference on Uncertainty in Artificial Intelligence. K.B. Laskey and H. Prade (Eds.), Morgan Kaufmann Publishers, Inc. San Francisco, (1999)
4. Smets, P.: Constructing the Pignistic Probability Function in a Context of Uncertainty. In: Uncertainty in Artificial Intelligence. M.Henrion, R.Shachter, L.N.Kanal and J.Lemmer (Eds.),**5** North- Holland, Amsterdam, (1990) 29–40
5. Smets,P.: Decision Making in a Context where Uncertainty is Represented by Belief Functions. In:Belief Functions in Business Decisions. Srivastava, R.(Eds.) Physica Verlag. Forthcoming. (2001)
6. Smets, P. and Kennes, R.: The Transferable Belief Model. Artif. Intell. **66** (1994) 191 – 234

Data Clustering and Rule Abduction to Facilitate Crime Hot Spot Prediction

Jonathan Corcoran, Ian D. Wilson, Owen M. Lewis, and J. Andrew Ware

Artificial Intelligence Technologies Modelling Group, University of Glamorgan, Pontypridd, Mid Glamorgan, CF37 1DL, UK
{jcorcora, idwilson, jaware}@glam.ac.uk

Abstract. Crime rates differ between types of urban district, and these disparities are best explained by the variation in use of urban sites by differing populations. A database of violent incidents is rich in spatial information and studies have, to date, provided a statistical analysis of the variables within this data. However, a much richer survey can be undertaken by linking this database with other spatial databases, such as the Census of Population, weather and police databases. Coupling Geographical Information Systems (GIS) with Artificial Neural Networks (ANN) offers a means of uncovering hidden relationships and trends within these disparate databases. Therefore, this paper outlines the first stage in the development of such a system, designed to facilitate the prediction of crime hot spots. For this stage, a series of Kohonen Self-Organising Maps (KSOM) will be used to cluster the data in a way that should allow common features to be extracted.

1 Introduction

Techniques for visually displaying crime location have changed a great deal from the time when hand drawn sketches and pin maps were commonplace. The advent of computers and the availability of desktop mapping software have advanced the analytical process, allowing efficient generation of virtual pin maps depicting crime incidents. A logical step beyond visualisation and analysis of trends and patterns is the development of a predictive system capable of forecasting changes to existing hot spots and the evolution of new ones.

Crime prediction in criminological research has been established for a number of decades [1, 2, 3], although its foundation within a geographic and GIS context is still in its infancy.

As predictive crime analysis is a new research area, very little literature currently exists. Olligschlaeger provides an overview of existing forecasting techniques, concluding the time, level of user interaction and the expertise that each demands would be unrealistic for implementation in an operational policing environment. [4] In addition, the inherent inflexibility and inability to dynamically adapt to change would compromise their viability in policing. ANN's are presented as one technique that offers minimal user interaction in addition to dynamic adaptability, and thus a potential operational forecasting solution.

B. Reusch (Ed.): Fuzzy Days 2001, LNCS 2206, pp. 807–821, 2001.

1.1 Potential for Crime Prediction by the Police

A recent survey [5] has highlighted the uptake and use of computer based crime-mapping technologies by British Police Forces. Computerised mapping technologies are rapidly becoming a vital prerequisite for visualisation of incident distributions and assisting in both the identification/allocation of resources and production/evaluation of policing strategies. The ability to efficiently generate simple maps to depict crime location and densities can be used directly to inform police officers and policing strategies, therefore maximising effectiveness and potential. A recent report published by the Home Office [6] has underlined the importance of geographic data for analysis and interpretation of crime at the local level. In addition, the Chief Constable of Kent County Constabulary notes that "*over the last few years, police activity has shifted its centre of balance away from the reactive investigation after events, towards targeting active criminals on the balance of intelligence*". [7]

A natural step beyond visualisation and analysis of past and current incidents is the prediction of future occurrences, thus providing an *early warning system* for the Police. [4] Prediction can help prevent crime in that it facilitates the optimal allocation of limited resources. Prevention might be better than cure, but in the real world, very often, this is under financial constraints. The development of tools for prediction can thus help prevent crime and maintain optimal operational costs.

Prediction requirements for the police have been classified into three main categories according to the period of time involved [8]:

- Short-term (tactical deployment);
- Medium-term (resource allocation);
- Long-term (strategic planning).

The focus for crime forecasting lies with short-term models as police tend to respond to existing and emerging crime patterns on relatively short time-scales for example on the basis of daily, weekly and monthly figures. [9] This paper details a potential prediction framework for short-term, tactical deployment of police resources. The framework will allow identification of risk factors from which probabilities of criminal activity (specifically emergence of hot spots) can be derived and the necessary resources deployed.

1.2 COPDAT

The Crime and Offender Pattern Detection Analysis Technique (COPDAT), outlined in this paper, offers a potential framework that can be applied to geographic prediction. COPDAT entails the implementation of a GIS, integrating various spatial databases to analyse and map the identified trends.

2 Methodology

The volume of crime is insufficient to accurately predict an offence (in terms of location and time) when extrapolating from past offences. Therefore, in the proposed system, the type of crime predicted to take place within a particular time-window is supplemented by a separate prediction of the likely vulnerable areas for the same epoch. In addition, it would seem prudent to have the facility in a finished system to allow information based on police intelligence and experience to be built into the predictive model. The idea is to enhance current police predictive capabilities not replace them.

The spatial framework for the prototype COPDAT conforms to police sector boundaries for Cardiff (the capital city of Wales in the United Kingdom), whereby the volume of crime is sufficient to derive accurate predictions.

2.1 Data Sets

The accurate forecasting of the temporal-geography of crime (where and when a crime is likely to take place) would be of immense benefit, for accurate prediction if acted upon should lead to effective prevention. However, crime prediction frequently relies on the use of data appertaining to past perpetrators and/or past victims. Such data is therefore subject to legal and ethical restriction on its use, resulting in an ethical conundrum [10]. Therefore, the prototype COPDAT involves the use of only two central data sets - one pertaining to crime and the other providing contextual information.

2.2 GIS Techniques

Visual inspection of the various spatio-temporal data sets is a vital pre-requisite in assimilating an understanding of fundamental relationships and trends. The GIS is used as a tool to conduct this basic pattern analysis, including cluster and hot spot techniques.

2.3 Artificial Neural Network Techniques

ANN models provide a mechanism through which the various spatial, non-spatial and temporal data sets can be analysed to identify patterns and trends previously undiscovered. Identification and consolidation of such trends and patterns between the various data sets allows generalisation and subsequent prediction of future events and scenarios based upon that generality. For example, identifying that a specific kind of crime is likely to occur in a certain location type under a given set of ambient conditions allows future incidents to be predicted.

The resultant of this scenario can be to produce a spatial probability matrix encapsulating a risk assessment of the study area. The spatial probability matrix can be directed to the GIS for visualisation in the form of a thematic contour map differentiating different areas with respect to their potential risk. In essence, the ultimate goal of the COPDAT is to learn decision criteria for assigning risk levels to new and future situations. This, for example, may involve identifying and predicting areas most at risk during a hot summer bank holiday football match in the City. Provision of such information is of obvious interest and of operational benefit to the police in terms of both resource allocation and policing strategies.

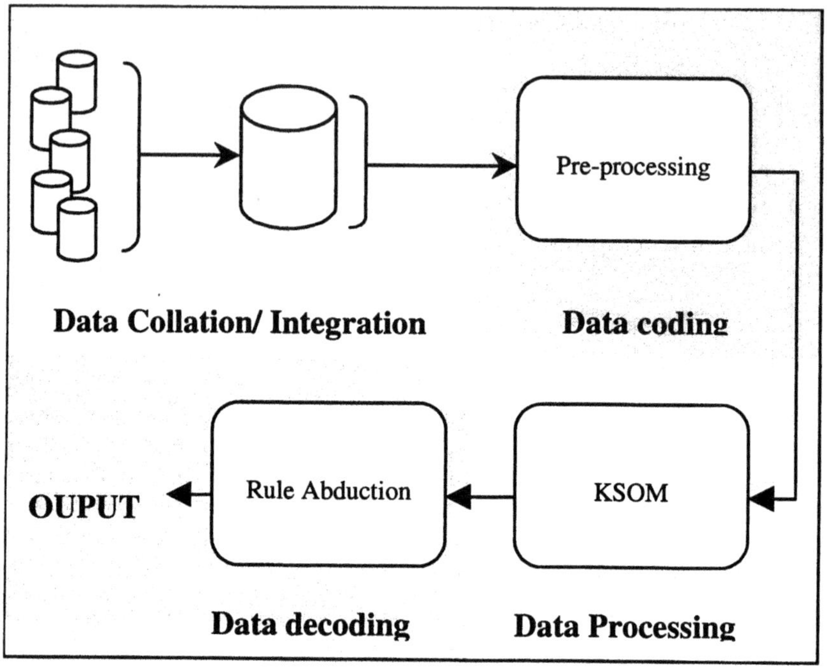

Fig. 1. Data preparation process in relation to ANN processing

2.4 Data Preparation and Processing

Data representation and structuring is of key importance in the production of a robust predictive model. It has been show in previous neural network studies that a certain level of pre-processing of the raw data is advantageous to model accuracy, efficiency and stability. This approach subsequently requires a certain level of post-processing in order to generate the required output values (illustrated in Fig. 1.).

2.5 Data Pre-processing

First, the data will undergo systematic and comprehensive analysis. This is followed, where necessary, by converting the data into an alternative representation more suited for input into an ANN. This sequential process can be broken down to a series of discrete stages:

- Format conversion and integration;
- Error detection and editing;
- Data reduction, transformation and generalisation.

The final stage in the pre-processing is of critical consequence in terms of a successful ANN implementation. The process of feature extraction and encoding of such characteristics impacts upon the ANN's ability to learn and assimilate relationships between salient variables and hence its accuracy in prediction. This process can be further decomposed into three distinct procedures:

1. Transformation and scaling may include:
   - Applying a mathematical function (e.g. logarithm or square) to an input;
   - Scaling the different inputs so that they are all within a fixed range can greatly effect the reliability of an ANN system.
2. Reduction of relevant data includes simple operations such as filtering or taking combinations of inputs to optimise the information content of the data. This is particularly important when the data is noisy or contains irrelevant and potentially erroneous information.
3. Encoding of identified features for input to the ANN. The data types include a range of data categories (discrete, continuous, categorical and symbolic), each to be handled and represented in an explicit manner.

2.6 Clustering Using a Kohonen Self Organising Map

Temporal, spatial and incident data will be clustered using a series of KSOM. These clusters, whose data share the same characteristics, will form the basis for rule abduction. (Note, deduction is where the effect is deduced from the cause - for example, 'the burglary was committed because the house was left unlocked.' Abduction is where the cause is gleaned from analysing the effect – for example, 'sun and alcohol often causes socially unacceptable behaviour'.)

An unsupervised network, such as the Kohonen Self Organising Map (KSOM), organises itself in such a way as to represent classes within a data set. The 2-D KSOM allows classes to be visualised on a feature map, in which similar inputs are spatially clustered. **Fig. 2.** shows a typical 2D KSOM along with an abridged algorithm (Note, the number of nodes are arbitrarily selected for example purposes).

Each output node on the KSOM contains a vector of length 'j', where 'j' is equal to the number of input attributes. Before training begins, the network is placed into an initialised state, i.e. the directions of the vectors in each node are randomised. Training involves passing an input vector into the network through the input nodes. Each node on the KSOM is then compared with the input vector, and the closest node is then changed to be more like the input vector. Neighbouring nodes also become more like the input vector. Iterating this process achieves clustering of similar input vectors in Euclidean space.

2.7 An Overview of the Methodology

The methodology uses a KSOM to find clusters in the input vectors and then the data from each cluster is used to train a separate MLP network. The advantage of using the KSOM for this application is that it can identify clusters within the parent dataset that are difficult to identify using simple sort procedures. Fig. 3. gives an overview of the method. A dataset containing the required elements of the vector v is passed through the KSOM during the training stage and allowed to develop into clusters. After training, the clusters are inspected and the primary clustered features used to describe the sub-datasets. These sub-datasets are then used as the basis for rule abduction.

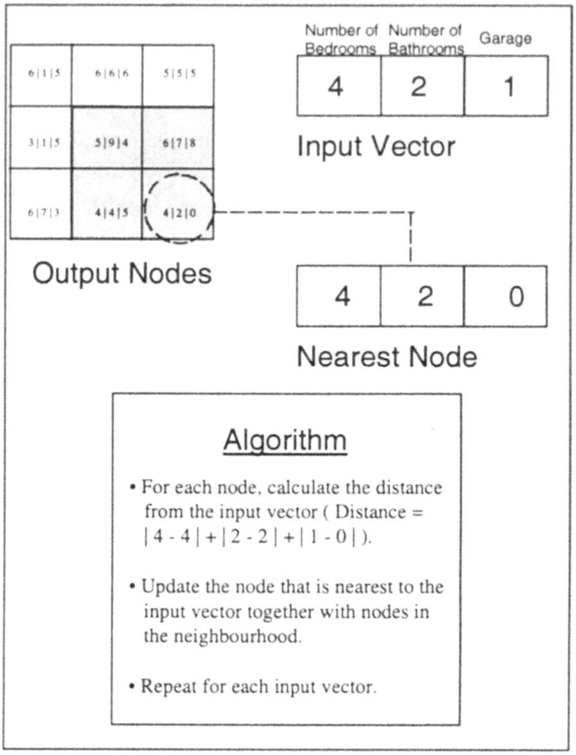

Fig. 2. A Kohonen Self Organising Feature Map

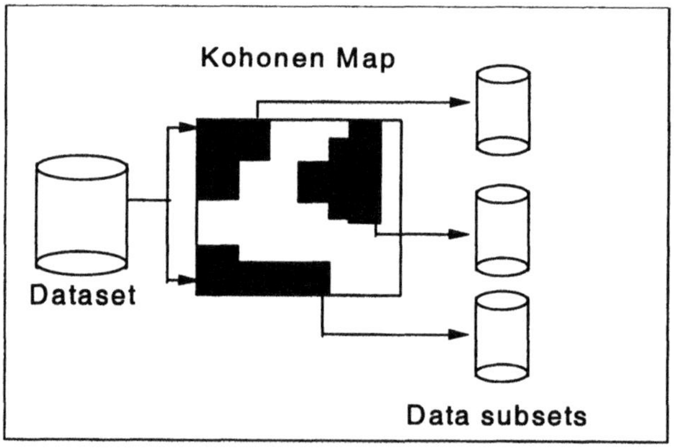

Fig. 3. Using a KSOM to partition a dataset

However, two fundamental problems need to be resolved before this method can be of any use. First, the problem of separating adjacent clusters, and second, the desire to proceed to the abduction phase only using 'good' clusters (see **Fig. 6.**).

The first problem has been recognised in other studies and some guidelines have been provided. [11] In essence, the problem lies in the attribution of boundary nodes to a specific cluster **Fig. 4.** provides an example of a KSOM output with adjacent clusters. There appear to be four classes within the dataset, but there are regions of uncertainty relating to the boundaries of each cluster. (The Digits in each node show the number of vectors mapped to that node.)

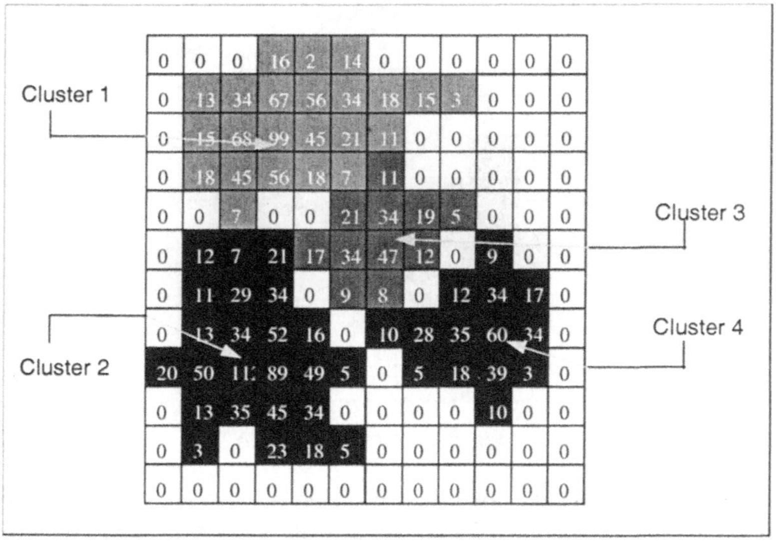

Fig. 4. An example trained kohonen self organising feature map

To overcome this problem a simple method of identifying class boundaries or discriminants can be used, which relies on the fact that a KSOM clusters primarily on binary features. For example, if the type of crime is represented using binary inputs, the KSOM will tend to cluster records according to this attribute. Boundaries between adjacent clusters on a 2D map can then be found by inspecting the records mapped to each node and grouping together nodes that contain the same classification values. However, this level of clustering can be achieved using a multi-level sort procedure. In essence, the binary representation of the data will dictate the make-up of the resulting clusters and more importantly the homogeneity of the data sets.

If the data are represented using continuous inputs, the clusters formed by the KSOM would provide more generalised classes which would be difficult to achieve using a sort procedure. However, the inspection method would no longer identify class boundaries, as the similarities between records would not always be apparent. Clearly, before meaningful training data sets can be formed, the problem of discerning effective class boundaries in a KSOM must be addressed. Ideally, the network adaption rule should cluster similar inputs and clearly distance individual clusters. Zurada [12] *explains "One possible network adaption rule is: A pattern added to the cluster has to be closer to the centre of the cluster than to the centre of any other cluster"*. Using this rule, each node can be examined and the distance from the surrounding centroids can be calculated. (A centroid is taken to be a node, outside any known cluster boundaries, that has the largest number of input vectors mapped to it.) The subject node can then be added to the nearest cluster. **Fig. 5.** illustrates a hypothetical situation where it is unclear where to draw the boundaries around clusters on a KSOM.

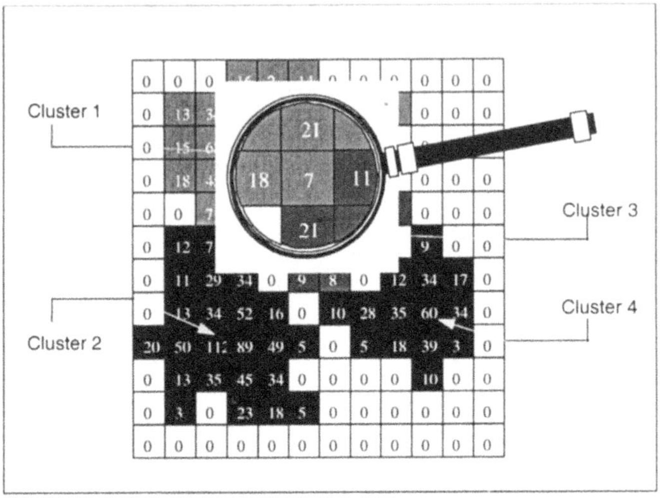

Fig. 5. An Example KSOM

By simply calculating the Euclidean distance of the subject node from the two centroids, the subject node can be assigned to the closest cluster. However, in this application, which aims to generate useful training data sets, the formation of a class boundary for Cluster 1 (including the subject node) may dramatically increase the variance of the training data. This increase will reduce the potential accuracy of the model. In the example, it may have been better to exclude the subject node from either of the clusters, and deem the vectors mapped to the subject node as either being outliers or a separate cluster.

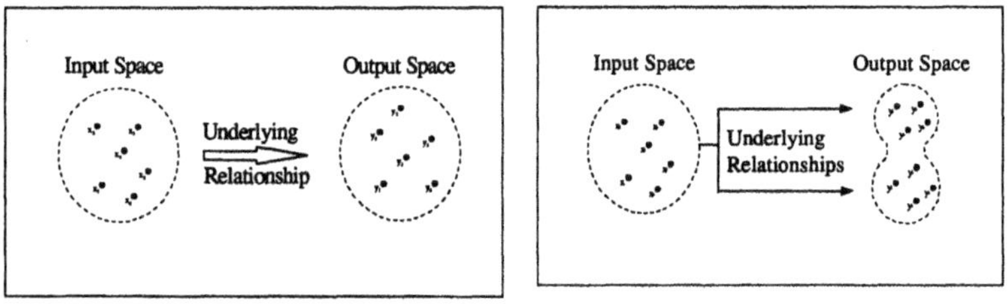

(a)

(a) An Example of a Good Input Cluster. A one-to-one relationship can be established and hence Input Space is homogeneous

(b)

(b) An Example of a Bad Input Cluster. Two or more identical vectors in the Input Space map to different vectors in the Output Space. Hence, the Input Space is not homogeneous.

Fig. 6. Example cluster mappings from input to output space

In addition to identifying boundaries around input clusters, it is also important for this application to match input clusters to appropriate output clusters. In terms of criminal activity, if, for example, the KSOM has clustered crimes from two different locational areas, it is reasonable to expect these crimes to have similar characteristics.

Fig. 6. (a) illustrates a cluster of similar input vectors. When the corresponding data in output space is examined, all the examples describe similar output values. Conversely, **Fig. 6.** (b) shows a situation where the data can only be modelled using two or more functions. The problem now is to estimate the 'usefulness' of a given cluster. There are a number of options available of which the following are the most useful:

- Multi-Layered Perceptron (MLP) Model [13]
- Class Entropy [14]
- R^2 Almy [15]
- Gamma Test [16]

For classification problems, Class Entropy can be used to decide if input clusters are homogenous with respect to output clusters. For example, Quinlan's C4.5 and C5.0 [17] uses Class Entropy to segment the input space until each segment points toward a single class in output space. However, this approach is not applicable for regression problems such as this one and this rules out the use of class entropy.

A quick and easy estimate of the susceptibility of a dataset for function induction can be achieved by executing a multiple regression analysis on the data and use the R^2 value to discern trainable clusters. This technique is useful for data where the function is known to be linear. However, this is not known to be true for crime analysis data.

The Gamma Test. The Gamma test attempts to estimate the best mean square error that can be achieved by any smooth modelling technique using the data. If y is the output of a function then the Gamma test estimates the variance of the part of y that cannot be accounted for by a smooth (differentiable) functional transformation. Thus if $y = f(x) + r$, where the function f is unknown and r is statistical noise, the Gamma test estimates Var(r).

Var(r) provides a lower bound for the mean squared error of the output y, beyond which additional training is of no significant use. Therefore, knowing Var(r) for a data set allows prediction beforehand of what the MSE of the best possible neural network trained on that data would be. The Gamma test provides a method of determining the quality of the data stratification - a good stratification technique will result in a low value of Var(r) for each subset. Interpreting the output from the Gamma test requires considerable care and attention.

The least squares regression line provides two pieces of information. First, the intercept on the Gamma axis is an estimate of the best MSE achievable by any smooth modelling technique. Second, the gradient gives an indication of the complexity of the underlying smooth function running through the data.

The Gamma test may estimate a very low MSE but unfortunately show a high level of complexity that could be difficult to accurately model. It is easier to see this situation when the output from the Gamma test is presented graphically.

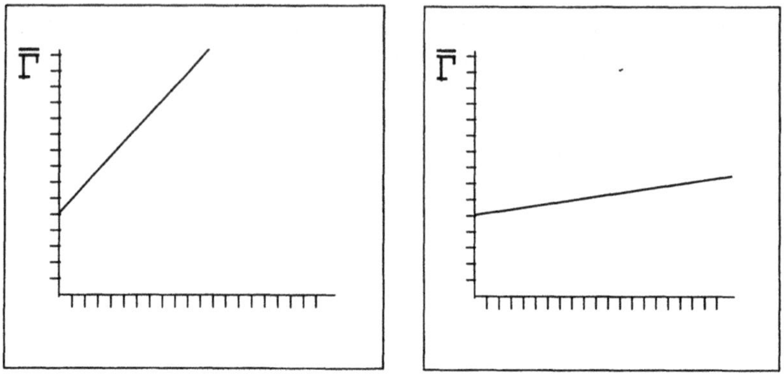

(a) High noise (large Γ value) and high complexity (steep gradient)

(b) High noise (large Γ value) and low complexity (flat gradient)

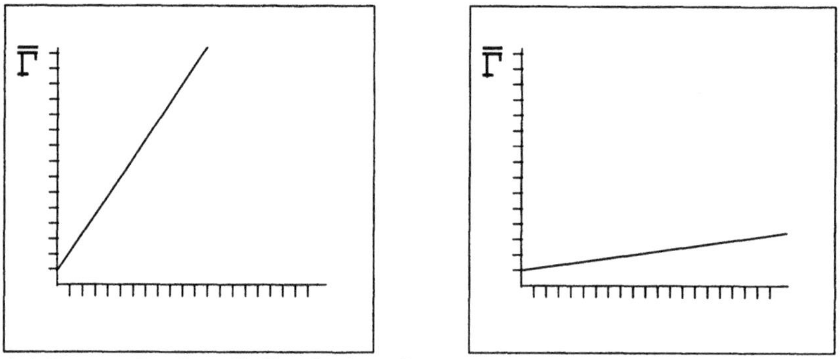

(a) Low noise (small Γ value) and high complexity (steep gradient)

(a) Low noise (large Γ value) and low complexity (steep gradient)

Fig. 7. Interpreting the output from the Gamma Test

A hypothetical example with high noise content and high complexity is shown in **Fig. 7.** (a); high noise and low complexity **Fig. 7.** (b); low noise and high complexity in **Fig. 7.** (c); and, finally, low noise and low complexity (the desired outcome) in **Fig. 7.** (d). In summary, for this methodology to be successful, the following is required:

- class boundaries must be identified around clusters formed by the KSOM over the input space that exclude outliers and nodes from neighbouring clusters, and;
- only 'good' clusters (illustrated in **Fig. 6.**) should go on to form training data sets for subsequent back propagation models.

A Detailed Look at the Methodology. The Gamma test can be used at a number of abstraction levels within the KSOM stratification method:

- Cluster level;
- Node Level;
- Record Level.

Data stratification is achieved at cluster level or at node level, depending on the ease at which cluster boundaries can be determined. The record level gives an indication of outliers.

Cluster Level Analysis. This can be achieved thus:

```
Identify Cluster boundaries in KSOM
For every cluster
 Place records mapped to cluster into a file
 Apply Gamma test to data in the file
 If Var(r) <= some Threshold then
           Use data file as the training set for a MLP
else
     Process at Node Level
 Endif
Next Cluster
```

This level of abstraction is the least computationally intensive as it only requires one pass of the Gamma test for each cluster. The disadvantage with this method is that it is often difficult to identify boundaries manually. In this case the Gamma test should be applied at the Node level.

Node Level Analysis. At this abstraction level, the methodology attempts to identify useful clusters by selecting a centroid and adding neighbouring nodes - where the addition of a node increases the variance significantly it is subsequently removed. This process iterates until the cluster size is maximised within a specified variance threshold. This algorithm identifies useful clusters on a 2D KSOM. This is achieved thus:

```
number_of_clusters:=0
While there are nodes to cluster
    number_of_clusters : = number_of_clusters + 1
    Select the unclustered node with the largest record count
    Apply Gamma test to  the data in the selected node
    If Var(r) <= Threshold then
        nodes_of_interest := None
        (cluster includes only the data from selected node)
        For each unclustered node next to selected node
            Add data from unclustered node to the cluster
            Run gamma test on cluster
            If Var(r) <= Threshold then     Add node number to nodes_of_interest
            else  Remove data from the unclustered node from the cluster
        While nodes_of_interest <> None
            Select c_node from nodes_of_interest
            Remove c_node from nodes_of_interest
            For each unclustered node immediately surrounding c_node
                Add data from unclustered node to the cluster
```

Run gamma test on cluster
If Var(r) <= Threshold then *Add node to nodes_of_interest*
else
Remove data from the node from cluster
Record the boundaries of this cluster
else Process at Record Level

The boundary detection algorithm for a 1D KSOM is very similar except neighbouring nodes are selected progressively further away from the left and the right of the centroid node. This level of analysis is more computationally intensive the cluster level analysis, as it require $m*\sum n_i$ passes of the Gamma test, where 'i' is the number of nodes investigated for cluster 'n' for a KSOM containing 'm' clusters. If using either the cluster level analysis or the node level analysis, useful clusters have been identified, it is then possible to train an independent MLP on each subset. The KSOM is then used to select the appropriate MLP on which to predict the value of a previously unseen example. The resulting system is closely related to a panel judgement system. However, if both methods have still resulted in poor training sets (useless clusters) then the analysis is taken to the most detailed abstraction level, that is the record level.

Record Level Analysis. The record level analysis is the most computationally intensive. The purpose of this level of the methodology is to identify data subsets from examples that have mapped to the same node on the KSOM. This facilitates extraction of outliers from a data set as well as giving some indication as to the examples that require additional features. The algorithm developed for this level of analysis is very similar to that shown for the node level analysis. However now, sets of records are iteratively analysed using the Gamma test. This is achieved thus:

For each node in the KSOM
Apply Gamma test to estimate the variance for the data in node
If Var(r) > Threshold then
For each record at node
Remove record from data set
Apply Gamma test to estimate the variance for the data in node
If New Var(r) < Previous Var(r) then Mark record as outlier
else Add record back into data set
else Proceed at Node Level

This level of analysis will identify the need for additional features and highlight records that may be classed as outliers.

2.8 Post-processing

Natural language rules will be derived from each 'good' cluster found by the KSOM. Each rule will represent a broad and generic definition (that with time can be fine tuned) of a specific sub-model that can be applied to best predict crime, for example:

for Centre of City

if Weather includes Wet
and Day is Friday
and Time is Night
then Problems will include inside pubs (probability 0.9)

for Centre of City
if Weather includes Dry
and Day is Friday
and Time is Night
then Problems will include inside pubs (probability 0.4)

for Centre of City
if Weather includes Dry
and Weather includes Warm
and Day is Friday
and Time is Night
then Problems areas will include outside pubs (probability 0.9)

These rules together with other determining factors (including temporal information such as the day, time and prevailing weather) are then directed to the GIS for production of a thematic crime risk contour map. When used in a predictive manner, the GIS can provide an important visual reference for analysing the relative impact of multiple factors on crime levels in a given area.

3 Implementation Details

A series of KSOMs will provide the mechanism for classifying the heterogeneous nature of crime and criminal activity in a spatial, temporal and contextual framework. The assumption that there is a single simple linear relationship between the various factors is unrealistic and misplaced. In the light of this, a valid approach to assimilate an understanding of crime and criminal activity is to dissemble it into its homogeneous components for independent analysis. This analysis can be used to generate a series of rules and generalisations that can be utilised in a predictive capacity at a broader heterogeneous level.

The KSOM provides an unsupervised technique, through which this stratification process can be automated, resulting in the formulation of homogeneous clusters (illustrated in **Fig. 8.**). Where clusters are not clearly delineated it maybe necessary to calculate a function such as the Euclidean distance or Gamma test [19, 20] in order to assign output to specific clusters.

The user will be able to present the trained suite of KSOMs, in the COPDAT, with structured scenarios from which probabilities and vulnerabilities of criminal activity will be derived and presented to the user via spatial thematic representations. Using standard GIS functionality, the user could subsequently overlay additional information, which may for example involve identification of specific localities, (street names, building etc) for deployment of resources.

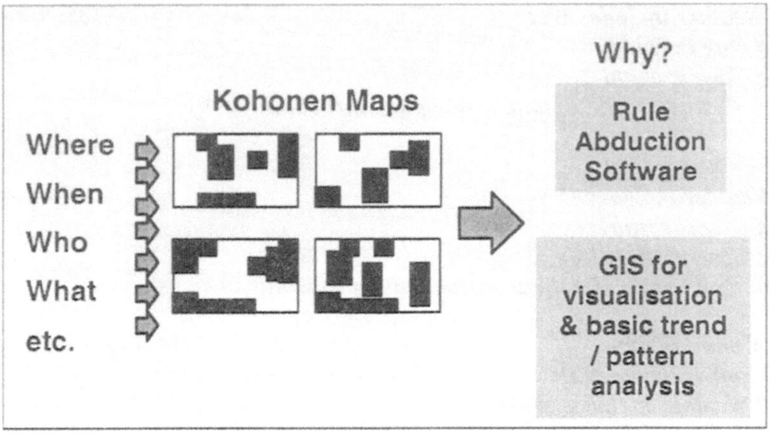

Fig. 8. Proposed COPDAT Hybrid System

4 Summary & Benefits

Coupling the predictive capabilities provided by the proposed COPDAT model with existing GIS techniques would enable law enforcement agencies to more effectively evaluate resource and tactical policies. Crime mapping and analysis using COPDAT will potentially facilitate research, assist with offender management and monitoring, and enable community planners to develop policy and forecast future needs for public safety resources. Provision of such facilities will rely upon use of multiple data sources and be dependent upon the use of pre-defined data processing and encoding techniques that will require new standards and techniques to be integrated into current practices. The Chief Constable of Kent County Constabulary raises this point in respect to police intelligence noting that *"this process is only possible if we can mix and match information across the board. To this end, it is essential that common standards and discipline attach to the intelligence process."* [7]

This paper outlines a potential framework that can be expanded to model and predict a broader variety of crimes based upon a larger mixture of criminal and contextual data sets and a common schema for crime prediction.

References

1. Ohlin, L.E. Duncan, O.D.: The efficiency of prediction in criminology. American Journal of Sociology **5** (1949)
2. Glueck, S.E.: Predicting delinquency and crime. Cambridge, Massachusetts, Harvard University Press (1960)
3. Francis, S.H.: Prediction methods in criminology. Home Office Research Studies (1971)
4. Olligschlaeger, A.M.: Artificial Neural Networks and Crime Mapping. Crime Mapping, Crime Prevention. New York, Criminal Justice Press (1997)
5. Corcoran, J., Ware, J.A.: Application of Computer Crime Mapping Technologies by British Police Forces. (2001) Available from: jcorcora@glam.ac.uk.

6. Home Office: Review of Crime Statistics: A Discussion Document. (2000) http://www.homeoffice.gov.uk/crimprev/crimstco.htm.
7. Phillips, D.: National Intelligence Model, National Criminal Intelligence Service (2000)
8. Gorr, W., Olligschlaeger, A., Thompson, Y.: Assessment of Crime Forecasting Accuracy for Deployment of Police. International Journal of Forecasting (2000)
9. Gorr, W., Olligschlaeger, A.: Crime Hot Spot Forecasting: Modelling and Comparative Evaluation. (1998) http://www.ojp.usdoj.gov/cmrc/grants/welcome.html.
10. Ware, J. A., Corcoran, J.: Forecasting Crime: An Ethical Conundrum. (2001) Available from: jaware@glam.ac.uk.
11. James, H.: An 'Automatic Pilot' for Surveyors. RICS Cutting Edge (1994)
12. Zurada, J.M.: Introduction to Artificial Neural Network Systems. West Publishing Company (1992) 58.
13. Chen, Ke, Xiang, Yu, Huisheng, Chi: Combining Linear Discriminant Functions with Neural Networks for Supervised Learning. Journal of Neural Computing and Applications **6** (1997)
14. Quinlan, J.R.: Induction of Decisions Trees. Machine Learning **1** (1986) 81-106
15. Almy, R., Horbas J., Cusack, M., Gloudemans, R.: The Valuation of Residential Property using Regression Analysis. Computer Assisted Mass Appraisal: An International Review, Ed. McCluskey, WJ and Adair, AS, Ashgate Publishing Company, England. (1998)
16. Stefánsson, A., Kon ar, N., Jones, A.J.: A Note of the Gamma Test. Neural Computing and Applications **5** (1997) 131-133
17. Quinlan, J.R.: C4.5: Programs for Machine Learning. Morgan Kaufmann (1993)
18. Lewis, O.M., Ware, J.A.: A Novel Neural Network Technique for Modelling Data Containing Multiple Functions. Lecture Notes in Computer Science, Vol. 1226, Springer-Verlag (1997) 141-149
19. Lewis, O.M., Ware, J.A., Jenkins, D.H.: A Novel Neural Network Technique for the Valuation of Residential Properties." Journal of Neural Computing and Applications **5** (1997) 224-229.

Computing with Words: Multi-objective GAs for Approximate Reasoning

Marcus Gemeinder

Faculty of Electrical Engineering, Chair of Computer Engineering
Fernuniversität Hagen, Germany
marcus.gemeinder@fernuni-hagen.de

Abstract. Applying the generalised extension principle within the area of Computing with Words (CW) typically leads to complex maximisation problems to be solved. In this paper, a GA based approach to solve such problems is presented. The problems are modelled as multi-objective maximisation problems by employing constraining conditions as additional objectives. Several methods for determining overall fitness according to the different objectives are compared, and a new promising method is introduced.

1 Introduction

A generic problem in CW is the following. From a given set of propositions expressed in a natural language (NL) an answer to a query expressed in NL is to be inferred.

Set of propositions in NL	infer	Answer to query q in NL
Initial Data Set (IDS)	$\Longrightarrow$	*Terminal Data Set* (TDS)

The meanings of the given propositions can be expressed as generalised constraints on a variable X: X *isr* R, with R the constraining relation and *isr* a copula, in which r indicates the type of the constraint (possibilistic, probabilistic etc.). In the remainder of this paper, only constraints of the possibilistic type will be of interest.[1]

One major task in CW is, after translating a set of given propositions into constraints, to derive from these constraints consequent constraints applying the rules of constraint propagation. These rules coincide with the rules of inference in fuzzy logic. Thus, the principal constraint propagation rule is the *generalised extension principle* [5].

[1] In this case, *isd* reduces to *is*, and R is a fuzzy relation that constrains X by playing the role of the possibility distribution of X.

B. Reusch (Ed.): Fuzzy Days 2001, LNCS 2206, pp. 822–828, 2001.

A typical example is the following. The IDS consists of the proposition p:"*most Swedes are tall*", and the query is q:"*what is the average height of Swedes?*" [5]. The proposition p translates to [4]

$$\pi_p(\rho) = \mu_{\text{most}}\left(\int_{h_{\min}}^{h_{\max}} \rho(h)\mu_{\text{tall}}(h)dh \right) , \tag{1}$$

in which $\rho(h)$ is a density function defined on the interval $[h_{\min}, h_{\max}]$, $\rho(h)dh$ is the proportion of Swedes whose height is in the interval $[h, h + dh]$, and π_p is the possibility distribution of ρ induced by p.

The average height of all Swedes following from this density function is

$$\nu = \int_{h_{min}}^{h_{max}} \rho(h)h\, dh \ . \tag{2}$$

The application of the generalised extension principle yields a membership function, which indicates the possible average heights. This membership function is determined by solving the following maximisation problem:

$$\mu_{h_{ave}}(\nu) = \sup_{\rho} \pi_p(\rho) \ , \tag{3}$$

subject to (2).

In general – as one can see easily – the maximisation problem to be solved has the following structure:

$$\mu_q(\nu) = \sup_{\rho} \pi_p(\rho) \tag{4}$$

$$\text{subject to} \qquad \nu = q(\rho) \ . \tag{5}$$

In the sequel, GAs for the solution of this maximisation problem are described.

2 GA Based Problem Solving

In [2] a GA based application to solve this special kind of maximisation problem is presented. A density function $\rho(h)$ is sought maximising (4) and fulfilling the auxiliary condition (5). The algorithm has to be executed several times for different values of ν. The result of each execution is a point $(\nu, \mu_q(\nu))$; together these points form a pointwise representation of the membership function $\mu_q(\nu)$.

2.1 GA Properties

Representation of Solutions. The continuous density function $\rho(u)$[2], defined on the interval $[u_{\min}, u_{\max}]$, is replaced by a discrete function consisting of n_ρ

[2] u denotes an element of U, the universe of discourse.

equidistant Dirac pulses as follows. Firstly, the function is approximated by a step function with n_ρ steps of equal width $\Delta u = \frac{u_{max} - u_{min}}{n_\rho}$ (see Fig. 1(a)). The function value of each step is determined by the value $\rho(u_i)$, with u_i lying in the middle of the interval covered by the respective step. And secondly, a Dirac pulse is positioned at each u_i. Each pulse is weighted by a factor w_i equal to the area of the rectangle under the corresponding part of the step function (see Fig. 1(b)). In combination, the Dirac pulses form a sample ρ_s of the density function ρ (see Fig. 1(c)).[3]

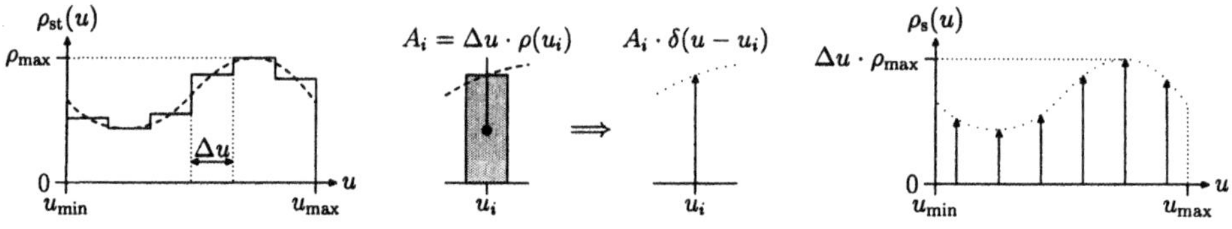

(a) Stepfunction **(b)** Replacement **(c)** Sample Function

Fig. 1. Sampling of the density function $\rho(u)$

Now, each individual is a vector of n_ρ real numbers, the weights w_i.

Genetic Operators. The GA utilises the widely employed standard two-point-crossover for recombination. The operation of mutation is more complicated: to explore new regions of the solution space, it is sometimes useful to have large mutations, but for convergence to an optimum small changes are needed. Consequently, a mutation operator is implemented, which prefers small variations but does not suppress large mutation steps [3]: a value out of an interval $[-r, r]$ is added to the selected variable w_i. Here r is called *mutation range*. This value is determined by a factor $a_i = 2^{-x \cdot k}$, with x being randomly chosen from the interval $[0, 1]$ each time; the *mutation precision* k globally influences the size of mutation changes. The + or − sign is chosen by probability 0.5, each:

$$w_i^{\text{mut}} = w_i \pm r \cdot a_i \quad \text{with} \quad a_i = 2^{-x \cdot k} \quad \text{and} \quad x \in [0, 1] \ . \tag{6}$$

Fitness Evaluation and Selection. For fitness assignment linear ranking is employed. All individuals are ordered with respect to their objective function – the possibility of ρ given p – values. Then, the fitness Φ depends on the rank of an individual within this sorted list, only (see [3], [2]). For selection *stochastic universal sampling* [1] is utilised.

[3] δ is the Dirac distribution.

Handling of Auxiliary Conditions. The considerations made so far are independent from two auxiliary conditions. First, the constraining condition (5), and second, a general condition to be fulfilled by a density function: the integral $\int_{-\infty}^{\infty} \rho(u)du$ has to have the value 1. This second condition is easily fulfilled by normalising the individuals after applying the genetic operators.

Considering the first condition is more complicated. There are two principal methods to choose from: the first one is to punish an offence against such a condition by lowering the corresponding fitness; the second one is to prevent such an offence. In the application described in [2] the second way was chosen. Special operators have been developed, which guarantee that the auxiliary conditions are fulfilled. In case of the example above, an operator has been implemented which shifts the sample functions along the h-axis until the centroid has the right value.

A main advantage of this method is that the resulting performance of the algorithm is very good. But there is obviously a great disadvantage, too: for each class of problems a special operator has to be implemented. The operator mentioned above can be used to solve the example problem, but if the query was, for instance, q:"*How many Swedes are small?*", another operator would be needed.

Owing to this disadvantage, another approach described in the sequel was investigated.

3 Multi-Objective Problem Modelling

In the application presented in this paper the constraining condition (5) is modelled as a further objective function. The optimisation problem (4), (5) is converted into

$$\text{maximise} \quad \mu_q(\nu_{\text{act}}) = \pi_p(\rho) \tag{7}$$
$$\text{minimise} \quad d(\nu_{\text{act}}, \nu) = |\nu_{\text{act}} - \nu| \tag{8}$$
$$\text{subject to} \quad \nu_{\text{act}} = q(\rho) \ . \tag{9}$$

Now, instead of one objective there are two objectives, whereby the second objective – in combination with (9) – makes sure by lowering the fitness values that the constraining condition (5) is regarded. Thereby, the label *act* indicates the actual value of the constraining condition in contrast to the desired value ν.

To have both representations of the maximisation problem absolutely consistent, for the second objective reaching the value 0 has to be demanded. But such an absolute consistence is not necessary. Anyway, as mentioned above, the algorithm has to be executed several times regarding different values of ν. The executions altogether form a pointwise representation of μ_q.

Therefore, in the multi-objective case $\nu_{\text{act}} \approx \nu$ is sufficient, if – and this is the crucial point – for this value ν_{act} the maximum possible value for the first objective is found. If so, the pair consisting of ν_{act} and the corresponding possibility $\pi_p(\rho)$ is part of the pointwise representation of the membership function μ_q – without any loss of accuracy.

In the following section some methods for fitness assignment according to the two objectives are described and rated.

4 Fitness Assignment Strategies

In [6] different strategies for fitness assignment utilised in multi-objective evolutionary algorithms are described. As an example, a weighted-sum approach has been utilised in the application described here. In addition, two other methods have been implemented. The first one combines the different rankings of the individuals according to the different objectives to obtain an overall ranking. The second one is a new grouping-based method, which is very suitable for multi-objective problems like those considered in this paper.

In the sequel, the three assignment strategies are described in more detail and compared with respect to their suitability for the maximisation problem (7), (8), (9) by means of the example described in the Introduction. In any case, linear ranking and stochastic universal sampling are applied after having found the overall ranking order of the individuals. Considering other proposition/query settings yields similar results.

The GA was executed with the following parameters: $N = 12$ individuals with $n_\rho = 15$ entries were used, as probability for recombination the typical value $p_{\text{Cross}} = 0.6$ was chosen, the mutation probability was set to $p_{\text{Mut}} = \frac{1}{15} = \frac{1}{n_\rho}$, the mutation range to $r = 0.05$, the mutation precision to $k = 10$, and finally the selection pressure utilised for linear ranking to $Sp = 1.7$. The interval of interest was set to $[100, 200]$, which means that only heights between 100cm and 200cm were considered. The membership function of the linguistic variable *tall* was set to $\mu_{\text{tall}}(h) = S(h, 165, 175, 185)$ and the quantifier *most* to $\mu_{\textit{most}}(u) = S(u, 0.5, 0.7, 0.9)$, with S being the well-known *S-function* [4].

In the following, the distance $d(\nu_{\text{act}}, \nu)$ is mapped onto the interval $[0, 1]$ to keep the algorithm's parameters independent from the width of the interval of interest, if different settings are considered.

Weighted-Sum. Each objective function value is multiplied by a weight w_i, whereby the second weight is negative according to the fact that this value is to be minimised. The products are added, thus forming an overall objective value.

This method is not very suitable for the given maximisation problem. It is not possible to find a set of weights for which the algorithm works sufficiently in any case.

For instance, for $w_2/w_1 = -1.0$, the algorithm converges to $\nu_{\text{act}} = 173.8$, independent from ν being 160, 165 or 170.[4] For distinctly smaller values of

[4] According to the remark above, the transformed distance – and therefore the second objective value – are equal to 0.038 in the latter case.

ν, no convergence is recognisable. The results are exactly the same if w_2/w_1 is set to -2.0. Increasing w_2/w_1 leads to distances converging to 0.0, but the performance for smaller values ν becomes even worse, the small possibilities (7) become completely covered by the weighted distances. Modifying the shape of the distance function by, for instance, computing the square root before weighting was investigated but abandoned.

Combined Rankings. The individuals are separately sorted according to the two objectives. Different approaches for combining the two obtained rankings have been investigated, for example adding the ranks (here rank 0 is the best one), or adding the ranks after squaring one or both of them.

This approach is not very suitable, either. One obtains optimum individuals with very small distances between ν_{act} and the desired value ν, but there is no convergence to a certain accurate value ν_{act} and – much more important – no absolute convergence to the actual maximum possibility.

Grouping. In the first step, the individuals are assembled into groups depending on the distance between ν_{act} and ν. Thereby, the range of distances is subdivided into intervals of equal width Δd (see Fig. 2(b)), and all individuals whose distances $d(\nu_{\text{act}}, \nu)$ lie in the same interval form a group. After that grouping, the individuals within each group are sorted according to (7).

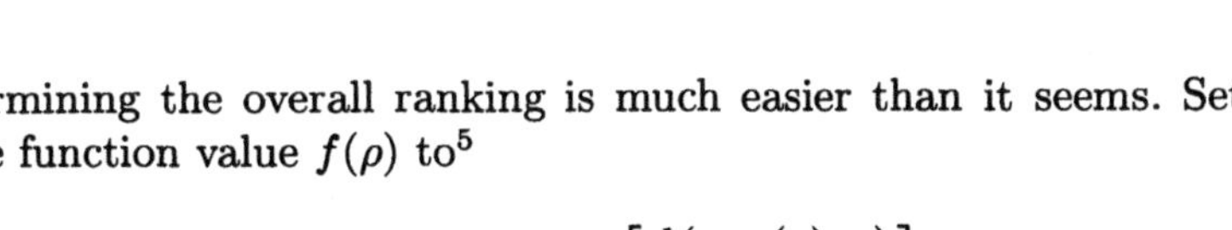

Fig. 2. Grouping of distances

Determining the overall ranking is much easier than it seems. Setting the objective function value $f(\rho)$ to[5]

$$f(\rho) = \pi_p(\rho) - \left\lfloor \frac{d\left(\nu_{\text{act}}(\rho), \nu\right)}{\Delta d} \right\rfloor \tag{10}$$

and sorting according to this value yields the ranking sought.

In contrast to other approaches, this method makes substantial use of the fact that $\nu_{\text{act}} \approx \nu$ is sufficient. If the distance is small enough – where *small enough* is specified by the parameter Δd – only the first objective is decisive. For each combination of ν and Δd the algorithm converges to a certain ν_{act}, and in perfect manner to an accurate value for the membership function $\mu_q(\nu)$.

[5] $\lfloor x \rfloor$ is the largest integer not larger than x.

The performance of the modified GA is – as was expected – worse than the performance of the GA described in [2]. The number of generations required to find the optimum possibility is increased by a factor between 1 and 3, depending on the current ν; but the possibility to use the modified GA in a flexible way totally compensates for this disadvantage.

5 Conclusion

A maximisation problem being very typical within the area of CW was transformed into a representation as a multi-objective maximisation problem. After that, some methods of combining the different objectives were considered and compared with respect to their suitability for the given maximisation problem. Particularly, a new grouping based fitness assignment method was introduced, which produces very promising results.

References

[1] James E. Baker. Reducing bias and inefficiency in the selection algorithm. In John J. Grefenstette, editor, *Genetic Algorithms and their Applications: Proceedings of the 2nd International Conference on Genetic Algorithms.* Lawrence Erlbaum Associoates, Inc., Publishers, 1987.

[2] M. Gemeinder. Fuzzy Constraint Propagation with Evolutionary Algorithms. In D. Reichel, editor, *Proc. East West Fuzzy Colloquium*, Zittau, 2000.

[3] H. Pohlheim. *Evolutionäre Algorithmen: Verfahren, Operatoren und Hinweise für die Praxis.* Springer-Verlag, Berlin Heidelberg New York, 1999.

[4] L. A. Zadeh. A Theory of Approximate Reasoning. In J.Hayes, D. Michie, and L. I. Mikulich, editors, *Machine Intelligence*, volume 9, pages 149–194. Halstead Press, 1979.

[5] L. A. Zadeh. From Computing with Numbers to Computing with Words - From Manipulation of Measurements to Manipulation of Perceptions. *IEEE Transactions on Circuits and Systems*, 45(1), 1999.

[6] E. Zitzler and L. Thiele. Multiobjective Evolutionary Algorithms: A Comparative Case Study and the Strength Pareto Approach. *IEEE Trans. on Evolutionary Computation*, 3(4), 1999.

The Description of All $\vee$-Distributive Triangular Norms of Lengths 2 and 3

Djavvat Khadjiev[1] and Funda Karaçal[2]

[1] Mech. Math. Fac.,
National University of Uzbekistan,
700095 Tashkent – Uzbekistan.
[2] Department of Mathematics,
Faculty of Sciences and Arts,
Karadeniz Tech. Univ., 61080

Abstract. Our main results:

1. The description of all $\vee$-distributive triangular norms of lengths 2 and 3
2. Theorem 3.7 which establishes a connection between direct decompositions of infinite $\vee$- distributive triangular norms on a lattice L and direct decompositions of the biggest element of L.
3. Theorem 4.3 which establishes a connection between direct decompositions of infinite $\vee$-distributive triangular norms on a lattice L and comaximal families in L.

1 Introduction

Triangular norms were introduced by Schweizer and Sklar [1] and have obtained useful applications in Fuzzy Logic Theory and Fuzzy Mathematics [2, 3, 4, 5]. It is an interesting problem – a description of all $\vee$-distributive triangular norms of small lengths. The present paper is devoted this problem. In this paper we also investigate a connection between direct decompositions of infinite $\vee$-distributive t-norms on a lattice L, direct decompositions of the biggest element of L and comaximal families in L. In particularly, we obtain a solution of a problem that was formulated in [5] as an open problem.

2 The Description of All $\vee$-Distributive Triangular Norms of Lengths 2 and 3

Definition 1. [4]*Let* $(L, \leq)$ *be a complete lattice with top and bottom elements* $1, 0$ *correspondingly, then a triangular norm (t–norm, for shortness) is a function satisfying the following properties:*

(1) $T(x, y) = T(y, x), \forall x, y \in L$ (commutativity),
(2) $T(T(x, y), z) = T(x, T(y, z)), \forall x, y, z \in L$ (associativity),
(3) $x \leq y \Rightarrow T(x, z) \leq T(y, z), \forall z \in L$ (monotonicity),

B. Reusch (Ed.): Fuzzy Days 2001, LNCS 2206, pp. 829–833, 2001.

(4) $T(x,1) = x, \forall x \in L$ (neutral element).

Definition 2. *A* $t-$*norm* T *on a lattice* L *is called* $\vee-$*distributive if*

$$T(a \vee b, c) = T(a,c) \vee T(b,c).$$

A $t-$norm T on a lattice L is called infinite $\vee-$distributive if

$$T\left(a, \underset{Q}{\vee} b_\tau\right) = \underset{Q}{\vee} T(a, b_\tau)$$

for all $a, b_\tau \in L, \tau \in Q$.

Let L be a lattice and $d(L)$ be the length of L [6]. Let T be a triangular norm on L. $d(L)$ is called the length of T.

Definition 3. *Triangular norms* T *on a lattice* L *and* P *on a lattice* M *are called isomorphic if there exists a lattice isomorphism* $H : L \longrightarrow M$ *such that* $H(T(a,b)) = P(H(a), H(b)), \forall a, b \in L$.

Let $A = \{0,1\}$, $0 < 1$, be a chain of the length 1 and $A^*(.,.)$ be the $\vee$-distributive $t-$norm on A, defined by $A^*(x,y) = x \wedge y, \forall x, y \in A$. Every $t-$norm of the length 1 is isomorphic to the $t-$norm A^*.

Let $B = \{0, p, 1\}$,$0 < p < 1$, be the chain of the length 2 and, $i = 1, 2$, be the $\vee-$distributive norms, defined by as follows:

(b_1) $B_1^*(1,x) = x, \forall x \in B$,, and otherwise $B_1^*(x,y) = 0$.
(b_2) $B_2^*(x,y) = x \wedge y, \forall x, y \in B$.

Let $A^* \times A^*$ be the direct sum of A^* and A^*. $A^* \times A^*$ is a $\vee-$distributive $t-$norm on a lattice $A \times A$.

Theorem 1. *There exist only three non-isomorphic* $\vee-$*distributive* $t-$*norms of the length* 2*. They are the following:* $A^* \times A^*, B_1^*$ *and* B_2^*.

Let C be a lattice $\{0, a, b_1, b_2, 1\}$,$0 < a < b_1 < 1$, $a < b_2 < 1, a = b_1 \wedge b_2$, $b_1 \vee b_2 = 1$, and $C^*(x,y) = x \wedge y$,$\forall x, y \in C.C^*$ is a $\vee-$distributive $t-$norm on C.

Let D be a chain $\{0, \beta, \alpha, 1\}$,$0 < \beta < \alpha < 1$, and $D_i^*(.,.)$, $i = 1,2,3,4,5,6$, be $\vee$ $-$distributive $t-$norms on D, defined by as follows:

(d_1) $D_1^*(x,y) = x \wedge y, \forall x, y \in D$;
(d_2) $D_2^*(\beta,\beta) = 0$ and otherwise $D_1^*(x,y) = x \wedge y$;
(d_3) $D_3^*(\beta,\alpha) = D_3^*(\beta,\beta) = 0$,and otherwise $D_3^*(x,y) = x \wedge y$;
(d_4) $D_4^*(\alpha,\alpha) = \beta$ and otherwise $D_4^*(x,y) = x \wedge y$;
(d_5) $D_5^*(\alpha,\alpha) = \beta$, $D_5^*(\alpha,\beta) = D_5^*(\beta,\beta) = 0$ and otherwise $D_5^*(x,y) = x \wedge y$.
(d_6) $D_6^*(1,x) = x, \forall x \in D$, and otherwise $D_6^*(x,y) = 0$.

Let E be a lattice $\{0, p_1, p_2, b, 1\}$, $0 < p_1 < b < 1, 0 < p_2 < b, p_1 \wedge p_2 = 0$, $p_1 \vee p_2 = b$, and $E_i^* (.,.)$, $i = 1, 2, 3$, be $\vee-$distributive $t-$norms on E, defined by as follows:

(e_1) $E_1^* (x, y) = x \wedge y, \forall x, y \in E$;

(e_2) $E_2^* (1, x) = x$ and otherwise $E_2^* (x, y) = 0$;

(e_3) $E_3^* (b, b) = p_1, E_3^* (b, p_2) = E_3^* (p_2, p_2) = 0$ and otherwise $E_3^* (x, y) = x \wedge y$.

Let S be a non-empty set,$|S| > 2$ ($|S|$ is a cardinal number of S), and F_S be the lattice $\{0, p_\tau (\tau \in S), b, 1\}$, $0 < p_\tau < b < 1, \forall \tau \in S$,$p_\alpha \wedge p_\beta = 0$ and $p_\alpha \vee p_\beta = b, \forall \alpha \neq \beta$.

Let F_S^* be the $\vee-$distributive $t-$norm on F_S, defined by as follows:

$F_S^* (1, x) = x, \forall \in F_S$, and otherwise$F_S^* (x, y) = 0$.

Proposition 1. *The following $\vee-$distributive $t-$norms of the length 3 are non-isomorphic:*

(1) $A^* \times A^* \times A^*$;
(2) $A^* \times B_1^*$;
(3) $A^* \times B_2^*$;
(4) C^*;
(5) $D_i^*, i = 1, 2, 3, 4, 5, 6$;
(6) $E_i^*, i = 1, 2, 3$;
(7) F_S^* ;F_S^* and F_Q^* are isomorphic iff $|S| = |Q|$.

Theorem 2. *An arbitrary $\vee-$distributive $t-$norm of the length 3 is isomorphic to one of the $\vee-$distributive $t-$norms, given in proposition 2.5.*

Connections between direct decompositions of triangular norms on a lattice, direct decompositions of the neutral 1 of L and comaximal families in L are very important to describe of all $\vee-$distributive $t-$norms of the lengths 2 and 3. Therefore we will give the following subjects.

3 Direct Decompositions of a Triangular Norm on a Lattice

Let $\{a_\tau : \tau \in Q\}$ be a subset of L and $\alpha \in Q$. Put

$$Q(\alpha) = Q / \{\alpha\}, \overline{a_\alpha} = \bigvee_{\tau \in Q(\alpha)} a_\tau.$$

Definition 4. *A subset $\{a_\tau : \tau \in Q\}$ is independent if $a_\alpha \neq 0$ and $\overline{a}_\alpha \wedge a_\alpha = 0$ for all $\alpha \in Q$.*

Definition 5. *A subset $\{a_\tau : \tau \in Q\}$ is a decomposition of the neutral 1 of L if*

$$\bigvee_{\alpha \in Q} a_\alpha = 1.$$

Theorem 3. *Let T be an infinite $\vee-$distributive $t-$norm on a lattice L and $\{a_\tau : \tau \in Q\}$ be a decomposotion of the neutral* 1 *of L. Then the following conditions are equivalent:*

(i) $\{a_\tau : \tau \in Q\}$ is independent;
(ii) $a_\alpha \wedge a_\beta = 0$ for all $\alpha \neq \beta$;
(iii) $T(a_\alpha, a_\beta) = 0$ for all $\alpha \neq \beta$.

Remark 1. But the implication $(iii) \Rightarrow (i)$ is not true if T is not an infinite $\vee-$distributive $t-$norm on a lattice L.

Definition 6. *A subset $\{a_\tau : \tau \in Q\}$ is a direct decomposition of the neutral* 1 *of L if $\bigvee\limits_{\tau \in Q} a_\tau = 1$ and $\{a_\tau : \tau \in Q\}$ is independent.*

Definition 7. *Let $a \in L$. A subset $\{x \in L : x \leq a\}$ is a principal ideal of L.*

Definition 8. *Let T be an infinite $\vee-$distributive $t-$norm on a lattice L. L is a direct product of a family $\{L_\alpha : \alpha \in Q\}$ of principal ideals L_α if:*

(i) For any $x \in L$ there exists the unique family $\{x_\alpha : \alpha \in Q\}$, where $x_\alpha \in L_\alpha$, such that $x = \bigvee\limits_{\alpha \in Q} x_\alpha$. In this case we write $x \approx \{x_\alpha : \alpha \in Q\}$.
(ii) If $x, y \in L$, $x \approx \{x_\alpha : \alpha \in Q\}$, $y \approx \{y_\alpha : \alpha \in Q\}$ then:

$$x \vee y \approx \{x_\alpha \vee y_\alpha : \alpha \in Q\}$$

$$x \wedge y \approx \{x_\alpha \wedge y_\alpha : \alpha \in Q\}$$

$$T(x, y) \approx \{T(x_\alpha, y_\alpha) : \alpha \in Q\}.$$

In this case we write $L = \prod\limits_{Q} L_\alpha$ and call that the $t-$norm T of L is a direct product of $t-$norms T_α of L_α, $\alpha \in Q$, where $T_\alpha = T \downarrow L_\alpha$.

Remark 2. A definition of a direct product of two $t-$norms has given in [5].

Theorem 4. *Let T be an infinite $\vee-$distributive $t-$norm on a lattice L and $\{L_\alpha : \alpha \in Q\}$ be a family of principal ideals of L, where, $L_\alpha = \{x \in L : x \leq a_\alpha\}$, $\alpha \in Q$. Then the following conditions are equivalent:*

(i) T is a direct product of a family $\{T_\alpha : \alpha \in Q\}$ of infinite $\vee-$distributive $t-$norms on a lattice $L_\alpha, \alpha \in Q$, where $T_\alpha = T \downarrow L_\alpha$.

(ii) $\{a_\alpha : \alpha \in Q\}$ is a direct decomposition of the neutral element 1 of L.

Corollary 1. (1)*Let T_1, T_2 be a continuous $\vee-$distributive $t-$norms on $[0,1]$. Then $T_1 \times T_2$ is a continuous $\vee-$distributive $t-$norm on $[0,1]^2$.*

(2)Conversely, let T be an arbitrary continuous $\vee-$distributive $t-$norm on $[0,1]^2$. Then T is a direct product of continuous $t-$norms T_1, T_2 and on $[0,1]$.

Remark 3. This corollary is a solution of the problem that was formulated in [5] as an open problem.

4 Comaximal Elements and Direct Decompositions of the Neutral

Definition 9. *A family* $\{b_\alpha : \alpha \in Q\}$ *is pairwise comaximal if* $b_\alpha \neq 1$ *for all* $\alpha \in Q$ *and* $b_\alpha \vee b_\beta = 1$ *for all* $\alpha \neq \beta$.

Theorem 5. *Let* T *be an infinite* $\vee-$*distributive* $t-$*norm on a lattice* L *and* $\{a_\tau : \tau \in Q\}$ *be a decomposotion of the neutral* 1 *of* L. *Put*

$$b_\alpha = \bigvee_{\beta \in Q(\alpha)} a_\beta.$$

Then $\{b_\alpha : \alpha \in Q\}$ is a pairwise comaximal family in L and

$$\bigwedge_{\alpha \in Q} b_\alpha = 0, a_\alpha = \bigwedge_{\beta \in Q(\alpha)} b_\beta.$$

Theorem 6. *Let* T *be a* $\vee-$*distributive* $t-$*norm on a lattice* L, $\{b_i : i \in \{1, 2, ..., n\}\}$ *be a pairwise comaximal family in* L *and*

$$\bigwedge_{i=1}^{n} b_i = 0.$$

Put $a_j = \bigwedge_{i \in Q(j)} b_i$. Then $\{a_i : i \in \{1, 2, ..., n\}\}$ is a direct decomposotion of the neutral 1 of L and $b_i = \overline{a_i}$ for all $i \in \{1, 2, ..., n\}$.

Theorem 7. *Let* T *be an infinite* $\vee-$*distributive* $t-$*norm on a lattice* L *such that the supremum of atoms of* L *is equal to* 1. *Then* T *is a direct product of a* T_α,$\alpha \in Q$ *such that* T_α *is isomorfic to* A^* *for all* $\alpha \in Q$.

Definition 10. [6]. *An element* $m \in L$ *is maximal if* $m \leq a \leq 1$,$a \in L$ *implies that* $m = a$ *or* $a = 1$.

Theorem 8. *Let* T *be a* $\vee-$*distributive* $t-$*norm on a lattice* L *such that there exists a family* $\{b_1, b_2, ..., b_n\} \subset L$ *of maximal elements* b_i *satisfying the condition* $\bigwedge_{i=1}^{n} b_i = 0$. *Then* T *is a direct product of a family* $\{T_i : i \in \{1, 2, ..., n\}\}$ *of* $\vee-$*distributive* $t-$*norms on a lattice* $L_i, i \in \{1, 2, ..., n\}$ *such that* T_i *is isomorphic to* A^* *for all* $i \in \{1, 2, ..., n\}$.

References

1. B. Schweizer and A. Sklar, Statistical metric spaces, Pacific J. Math., 10 (1960)313-334.
2. S. Ray, Modified TL-subgroups of a group, Fuzzy Sets and Systems 91(1997) 375-387.
3. W.-X.Gu, S.-Y.Li, D.-G.Chen, Y.-H. Lu, The generalized t-norms and TLPF-groups, Fuzzy Sets and Systems, 72 (1995) 357-364.
4. C.A., Drossos, Generalized t-norm structures, Fuzzy Sets and Systems, 104(1999)53-59.
5. De Baets, B., Mesiar, R., Triangular norms on product lattices, Fuzzy Sets and Systems, 104 (1999) 61-75.
6. G. Birkhoff, Lattice Theory, 3 rd edition, Providence, 1967.

Latent Semantic Analysis for German Literature Investigation

Preslav Nakov

Sofia University, Rila Solutions
27, Acad. G. Botchev Str, Sofia, Bulgaria
preslav@rila.bg, preslav@rocketmail.com

Abstract. The paper presents the results of experiments of usage of LSA for analysis of textual data. The method is explained in brief and special attention is pointed on its potential for comparison and investigation of German literature texts. Two hypotheses are tested: 1) the texts by the same author are alike and can be distinguished from the ones by different person; 2) the prose and poetry can be automatically discovered.

1 Introduction

One of the most interesting text features is the text variance. There are always several ways to express the same thought and the authors are forced to choose between different syntactic constructions, synonyms and terminology according to the intended audience and the impact the text must produce. Furnas, Landauer, Gomez and Dumais have shown in [8] that people use the same words to describe the same subject 10-20% of the time.

Authors make their choices according to both the specific text intention and their own subjective preferences. These (denoted as *style*) are consistent (along the text or all the author's oeuvres) and easy to discover for humans but very hard to describe and measure. Researchers in statistical stylistics have concentrated at word-based statistics (word length, word length distribution, long words count, type/token ratios, see [15]), text-based statistics (sentence length, clause complexity, see [12,14]) and statistics based on specific items (pronouns counts, presence/absence of contractions/amplifiers, relative frequency of specific verbs: e.g. *seem*, *appear* etc., see [2,11]). We go different way: Our purpose here is to study the possibilities of using a classic semantic analysis method without additional tuning for automatically discovery of the texts from the same author and to discriminate between prose and poetry.

2 Latent Semantic Analysis

The *Latent Semantic Analysis (LSA)* is a powerful statistical technique for indexing, retrieval and analysis of textual information used in different fields of the human cognition during the last decade. The method is fully automatic and does not use any

B. Reusch (Ed.): Fuzzy Days 2001, LNCS 2206, pp. 834–841, 2001.

preliminary constructed dictionaries, semantic networks, knowledge bases, conceptual hierarchies, grammatical, morphological nor syntactic analysers, etc. The general idea is that there exists a set of latent dependencies between the words and their contexts (phrases, paragraphs and texts). They both are represented in the same semantic space. The identification and proper treatment of the latent dependency permits LSA to deal successfully with the synonymy and partially with the polysemy, which are the major problems with the word-based approaches.

LSA is a two-stage process and includes education and analysis of the indexed data. During the education phase LSA performs an automatic document indexing. The process starts with the construction of a matrix X whose columns are associated with documents, and the rows with terms (words or key-phrases). The cell (i,j) contains the occurrence frequency of term i in document j. The matrix X is then submitted to *singular value decomposition* (*SVD*) which gives as a result three matrices T, D (orthonormal) and S (diagonal), such that $X=TSD^t$. Most of the rows and columns of T, S and D are removed in a way that the matrix $X'=T'S'D'$ is the least squares best-fit approximation of X. This results in the compression of the source space in much smaller one where we have only a limited number of significant factors (generally between 50 and 400). Thus, each term or document is associated a vector (column in the $D'S'$ matrix) of reduced dimensionality, e.g. 100. It is possible to perform a sophisticated SVD, which speeds up the process by directly finding the truncated matrices T', S' and D', see [1].

The second phase is the analysis phase. Most often this includes the study of the proximity between a couple of documents, a couple of words or between a word and a document. A simple mathematical transformation permits to obtain the vector for a non-indexed text. The proximity degree between two documents can be calculated as the dot product between their normalised LSA vectors. The usage of other measures is also possible, e.g.: Euclidean and Manhattan distances, Minkowski measures, Pearson's coefficient etc. [3,13,16,21].

3 Experiments

The experiments were performed on German literature texts we collected on the Web mainly from the following sites [22,23,24]. The file contents were carefully investigated and all index and biographic files were removed. The remaining files were pre-processed and the HTML tags were discarded. The remaining files were pre-processed and the HTML tags were discarded together with the headers, footers and editors' comments leaving just the plain text. We thus obtained the following oeuvres, grouped by author:

- Theodor Fontane: *Effie Briest (181), Der Stechlin (241), Der Schach von Wuthenow (92)*
- Johann Wolfgang von Goethe: *Faust (104), Die Wahlverwandtschaften (152), Die Leiden des jungen Werthers (67), Wilhelm Meisters Lehrjahre (182)*
- Georg Wilhelm Friedrich Hegel: *Phänomenologie des Geistes (90)*
- Heinrich Heine: *Buch der Lieder (245), Neue Gedichte (870)*
- Franz Kafka: *Erzählungen (29), Der Prozeß (126)*
- Karl May: *Vinnetou I, II, III (147)*

Since LSA tries to capture the mutual dependencies between the words and their contexts it is of crucial importance to provide contexts of reasonable size. Usually, when indexing small documents they are passed as they are since it is best to work on the whole document. It is clearly not the case here and we decided to split the documents into chunks of almost equal size. We experimented with different chunk sizes: 1,2,4,8 KB and finally chose 4 KB as most appropriate. The files for each novel were concatenated (if more than one) and split into almost equal parts of approximately 4 KB (In fact the chunk size varies since we do not split the sentences). There are two exceptions: the Goethe and Heine poems were left intact, one poem per file as they originally were.

The words met in just one document were removed since they cannot contribute to the proximity, thus reducing the total different non-stop word forms considered from 72141 to 34191. Someone would argue that these are exactly these words which are characteristic of a text and hence carry important information. This is true in case we want to assign a text a subject category but are of no particular sense when we want to measure similarity between texts using cosine measure.

After the frequency matrix X (2191 × 34191) was built, we divided each row by its entropy and just then performed SVD. [3,5,6,7,13,16] No other words (e.g. stop words) were removed since their frequency of usage could be characteristic for a specific author and thus contribute to the document proximity (Nevertheless the entropy weights them appropriately). No word stemming was performed for the same reasons.

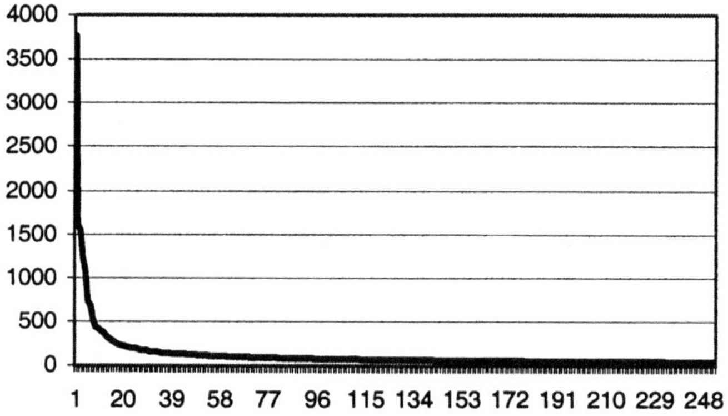

Fig. 1. Choice of similarity measure

A crucial moment when using LSA is the correct choice of dimensionality. Figure 1 shows the top 250 singular values sorted in descending order. The curve goes straight down and then flattens. Intuitively we have to cut the singular values just in the place where the curve behaviour changes. If we cut further we lose important information and if we keep more values we start modelling the noise.

We performed two different space reductions: to space with dimensionality 25 and 100. For each of these cases we calculated the dot product between the normalised vectors for all the document couples. The corresponding correlation matrices (2191×2191) were almost identical. Fig. 2 and 3 show the results in 5 different colours for the five correlation intervals: 87,5-100%, black colour; 75-87,5%, dark grey; 62,5-75%, grey; 50-62,5%, light grey; 0-50%, white.

A well-known feature of LSA is that it maps the semantically related texts next to each other in the vector space. Thus, in our particular case the chunks from the same oeuvre are expected to be more similar to each other than those from different oeuvres.

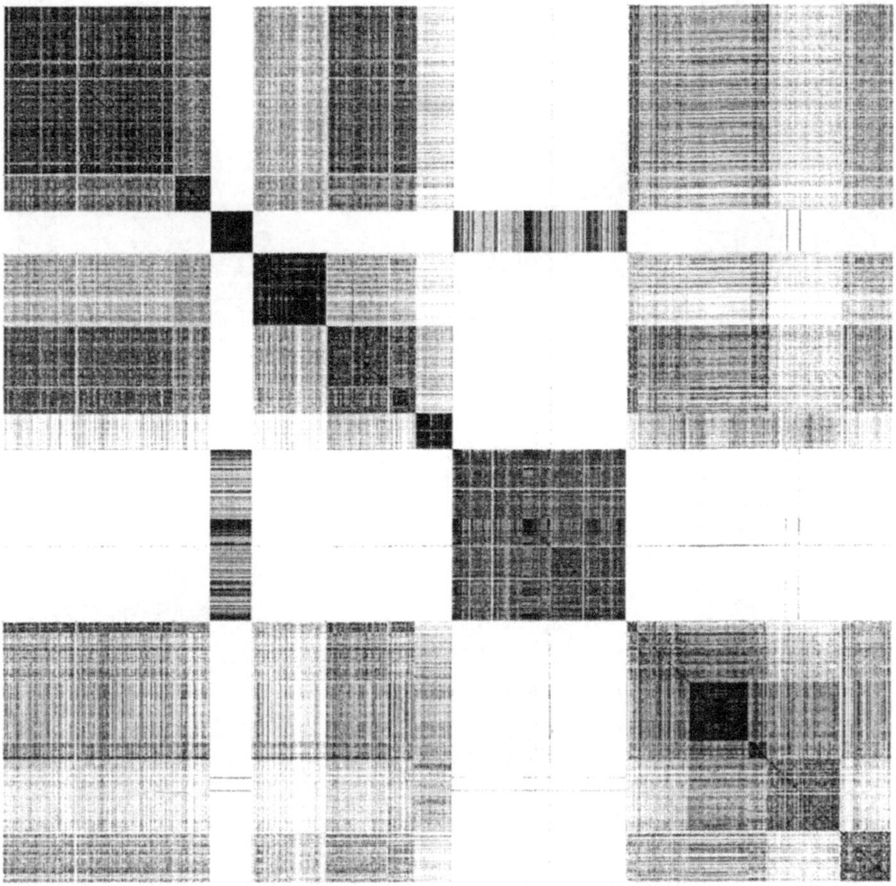

Fig. 2. Correlation matrix map: (vector space dimension 100)

There are several dark squares on the main diagonal, which are clearer on figure 2 than on figure 3 because of the choice of a bit more appropriate vector dimensionality. Let's look at Fig.2 in more details from the upper left towards the down right corner skipping the region in the black square. There are several squares (clusters) and Fontane forms the first one we can see. This is a monolith black block with some artificially added white lines whose function is to separate *Effie Briest* (181), *Der Stechlin* (241), and *Der Schach von Wuthenow* (92), which follow in this order (file counts put in parentheses). The latter is distinguishable and forms a small internal sub-square. The other two oeuvres can be distinguished only thanks to the artificial white line. Thus, the oeuvres by Fontane form a good cluster.

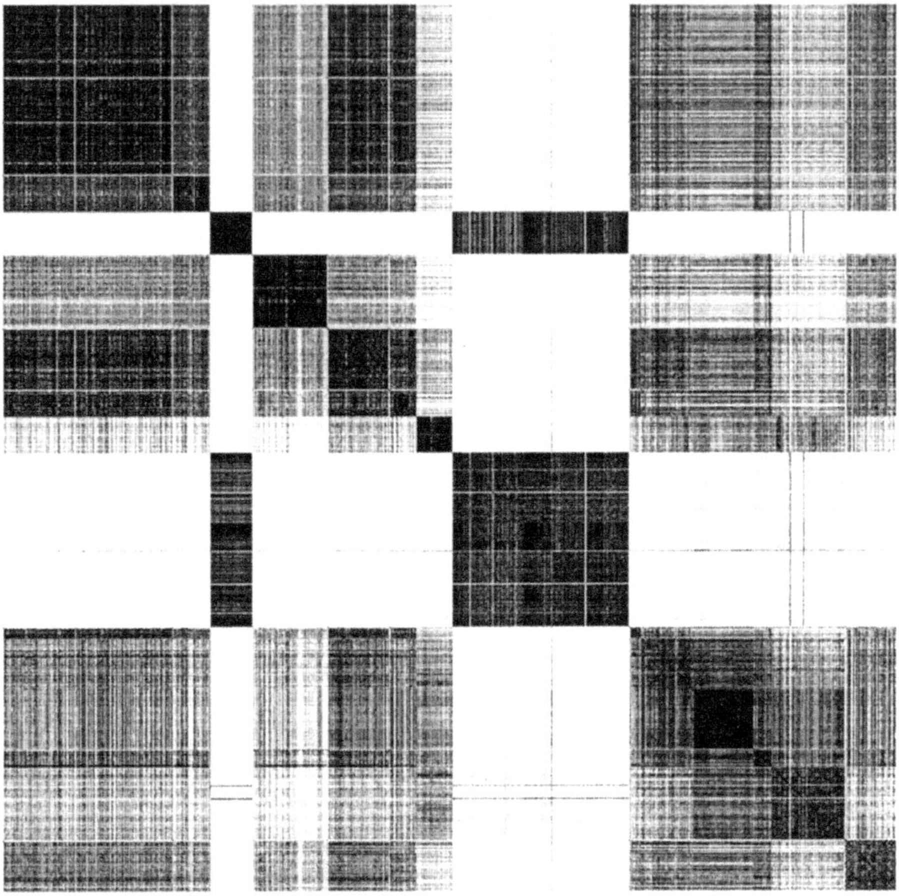

Fig. 3. Correlation matrix map: (vector space dimension 25)

The following small black square represents the Goethe's *Faust* (104). The smooth one just after it has three sub-squares, the first of which corresponds to *Die Wahlverwandtschaften* (152). The second one is split again into two smaller ones corresponding to *Die Leiden des jungen Werthers* (67) and *Wilhelm Meisters Lehrjahre* (152). While each of the last 3 oeuvres forms a well-separated cluster it is normal to expect lower proximity level between texts from different oeuvres. What remains unclear is why *Faust*, although being written by the same author, is so different. The answer is because of the dictionary differences between prose and poetry.

At the end of the previous square there is a small dark cluster not very well separated from the previous ones and representing the Hegel's *Phänomenologie des Geistes* (90). Looking at figure 3 we can see that although it is very similar to the last three Goethe texts their internal proximity is a bit higher. Thus, we can separate the texts by the two authors. We attribute the high proximity level between the Goethe's and Hegel's prose to the fact that they lived in the same historical époque. The same applies in fact to Fontane: looking at the big smooth out-of-diagonal squares we can see the relatively high correlation between the three authors.

The following almost monolith black square contains the Heine's poems: *Buch der Lieder* (245) and *Neue Gedichte* (870). Looking at the out of diagonal elements we can see the high proximity level with the Goethe's poem *Faust.* This must be due mainly to the similar poetic structure. Looking at figure 2 we can see that we are still able to distinguish Goethe's and Heine's poems.

Let us look now at the last big smooth square. It contains the oeuvres by Kafka, May, Keller, Nietzsche and Rilke, in that order. Although it seems difficult for LSA to separate some of them we can distinguish each author from the picture. The darker sub-rectangle in the middle is the Karl May's *Vinnetou* (147). The texts before May belong to Kafka's *Der Prozeß* (126) and *Erzählungen* (29). The *Der Prozeß* (126) can be discovered as a very small (just 29 files) sub-rectangle at the beginning of the smooth one.

Looking carefully one can see a bit bigger small (just 45 files) sub-rectangle just after May: this is the Keller's *Romeo und Julia auf dem Dorfe* (45). Then follow two easier to discover and much bigger sub-rectangles corresponding to Nietzsche's *Also sprach Zarathustra* (179) and Rilke's *Die Aufzeichnungen des Malte Laurids Brigge* (126).

Here we can see again a high semantic proximity between the whole last smooth rectangle and Fontane, Goethe's prose and Hegel. Investigating the texts one can see that all these are prose while the Goethe's *Faust* and Heine's oeuvres are poetry. Thus, we discovered a very good separation between the poetry and the prose among all the texts considered. This gives a strong support for our second hypothesis.

What about the first hypothesis, it seems to hold when comparing the different authors but the judgement whether a particular text belongs to a particular author seems much more difficult. This result is a bit surprising for us since the other researcher experiments on English and Bulgarian (as well as recent experiments on Russian) text collections showed much stronger support for the first hypothesis [19,20]. The problem may be the mixture of prose and poetry we made here: they seem to be very different one from another for this particular text collection, which results in rise of the inner group correlation level harming the support for the first hypothesis. Another problem may be the disproportion of the chunk counts for the different authors.

4 Discussion

As we saw, the different dimensionality reductions reveal different kinds of correlation between the texts. The higher dimensionality matrices show that the texts from the same author are more alike and tend to form separate clusters. When we perform a further reduction we obtain just two classes: the prose and the poetry.

This is consistent with our previous experiments for English, Russian and Bulgarian literature. In general the highest dimensionality matrices show that the texts from the same oeuvre are more alike and tend to form separate clusters. In case the dimensionality is high enough some internal clusters can be discovered inside the same oeuvre. When a further reduction is performed the oeuvres by the same author lose their differences and each author tends to obtain its own cluster (in fact two clusters must be expected if the author is represented by both prose and poetry, as happened above). When we perform a further reduction we obtain just two classes: the prose and the poetry.

5 Conclusion

The experiments performed show that in the general case the selected German authors can be distinguished using LSA but it seems to be hard for some of the authors. On the other hand the selected texts give a strong support for the hypothesis that the prose and poetry can be automatically discovered.

6 Future Work

Additional experiments on new (possibly different language) corpora with new authors have to be performed in order to justify the results obtained and to better study all the factors (e.g. époque, internal oeuvre/author substructure) influencing the text proximity when using LSA.

An interesting possibility is to combine and compare the semantic proximity with the traditional stylistic statistics methods used by [4,10,11,12,14,15].

References

1. Berry M., Do T., O'Brien G., Krishna V., and Sowmini Varadhan, SVDPACKC (Version 1.0) User's Guide. April 1993.
2. Biber D.A typology of English Texts. Linguistics, 27, pp. 3-43. 1989.
3. Deerwester S., Dumais S., Furnas G., Laundauer T., Harshman R. Indexing by Latent Semantic Analysis. Journal of the American Society for Information Sciences, 41 (1990), pp. 391-47.
4. Diab M., Schuster J., Bock P., A Preliminary Statistical Investigation into the impact of an N-Gram Analysis Approach based on Word Syntactic Categories toward Text Author Classification, Proc. Of 6th International Conference on Artificial Intelligence Applications, Cairo, Egypt, 1998.
5. Dumais, S. T. (1993) LSI meets TREC: A status report. In: D. Harman (Ed.), The First Text REtrieval Conference (TREC1). National Institute of Standards and Technology Special Publication 500-207, (pp. 137-152).
6. Dumais, S. T. (1994) Latent Semantic Indexing (LSI) and TREC-2. In: D. Harman (Ed.), The Second Text REtrieval Conference (TREC2), National Institute of Standards and Technology Special Publication 500-215 , (pp. 105-116).
7. Dumais, S. T. (1995) Using LSI for information filtering: TREC-3 experiments. In: D. Harman (Ed.), The Third Text REtrieval Conference (TREC3) National Institute of Standards and Technology Special Publication , in press 1995.

8. Furnas G., Landauer T., Gomez L. and Dumais T. Statistical semantics: Analysis of the Potential Performance of Keyword Information Systems. Bell Syst.Tech.J., 62, Number 6, pp. 1753-1806, 1986.
9. Harman, D. How effective is suffixing? In Journal of The American Society of Information Science. Vol. 42, No 1. 1991.
10. Jiang, J. Using Latent Semantic Indexing for Data Mining. Department of Computer Science, University of Tennessee, December 1997.
11. Karlgen J., Douglas C. Recognizing Text Genres with Simple metrics Using Discriminant Analysis. Proceedings of COLING 94, Kyoto, pp. 1071-1075.
12. Klare G.The Measurement of Readability. Ames: Iowa University Press. 1963.
13. Laudauer T., Foltz P., Laham D. Introduction to Latent Semantic Analysis. Discourse Processes, 25, pp. 259-284.
14. Lorge I. The Lorge Formula for Estimating Difficulty of Reading Materials. New York: Teachers College Press, Columbia University, 1959.
15. Losee R. Text Windows and Phrases Differing by discipline, Location in Document, and Syntactic Structure.Information Processing & Management 32(Nov): 747-67. 1996
16. Nakov P. Getting Better Results with Latent Semantic Indexing. In Proceedings of the Students Presentations at ESSLLI-2000, pp. 156-166, Birmingham, UK, August 2000.
17. Nakov, P. Web-personalisation using extended Boolean operations with Latent Semantic Indexing. Proc. AIMSA-2000, Varna, Bulgaria, Lecture Notes in Artificial Intelligence 1904, Springer 2000, pp. 189-198.
18. Nakov P. Latent Semantic Analysis of Textual Data. In Proceedings of CompSysTech'2000, Sofia, Bulgaria. June 2000.
19. Nakov P. Latent Semantic Analysis for Bulgarian literature. In Proceedings of Spring Conference of Bulgarian Mathematicians Union. Borovetz. Bulgaria. 2001.
20. Nakov P. Latent Semantic Analysis for Russian literature investigation. In Proceedings of the 120 years Bulgarian Naval Academy Conference. Varna. Bulgaria. 2001.
21. http://lsa.colorado.edu
22. http://www.phil-fak.uni-duesseldorf.de/germ5/service/download/index.html
23. http://gutenberg.aol.de
24. http://www.phil.uni-erlangen.de/~p2gerlw/ressourc/archiv.html

Some Possibilities of Improving the CORA Classification Algorithm

Eriks Tipans and Prof. Arkady Borisov

Riga Technical University, Kalku 1, LV-1658, Riga, Latvia
eriks@ibm.cs.ru.lv

Abstract. This paper examines some possibilities of improving the CORA classification algorithm developed by M. Bongard in sixties. The algorithm is based on finding *features* of objects one needs to classify. The theoretical part of this study explains two main shortcomings of the CORA classification algorithm: (1) the algorithm rejects features that at least once appear in the opposite class and thus loses potentially valuable information; and (2) the algorithm has extremely large learning time that can be due to the "combinatorial explosion". The study suggests two methods to overcome these difficulties and to improve the algorithm: (1) *the method of relatively good features* and (2) *the method of sequential covering*. The experimental results demonstrate that both the methods suggested ensure a sufficient algorithm performance's improvement as compared to the classic Bongard algorithm implementation. Best results are achieved when both methods are combined and sufficiently improve the classification precision and reduce learning time.

1 Introduction

The main task of this paper is to study the CORA classification algorithm, in more detail. Besides, it is necessary to try to explain possible shortcomings of the algorithm so as to improve its execution. This algorithm seems very promising as its implementation is rather simple but decision rules (concept generalizations) obtained as a result of its execution are easy to interpret for a human being in contrast to artificial neural networks.

2 Generalization by Features in Bongard's CORA Classification Algorithm

Let us consider the Bongard method, or CORA algorithm. The essence of the algorithm is as follows. The solution of the classification task is based on two procedures: learning and examination which are repeated in the course of generalization as many times as needed. In the course of learning a set of training examples is used that contains examples belonging to a concept to be generalized and also counter examples that do not belong to the concept in question. Learning results in the development of a decision rule, which is able to indicate whether a particular

B. Reusch (Ed.): Fuzzy Days 2001, LNCS 2206, pp. 842–853, 2001.

example belongs to the generalized concept in question, or not. After a decision rule has been developed according to the training set, a stage of examination starts. At this stage, the system is presented a set of examples which certainly contains other examples that differ from those in the training set. If the decision rule demonstrates an acceptable quality of examination set classification in the course of examination, then the procedure of learning is considered to be finished. Otherwise, the decision rule is updated additionally processing a new complemented set of training examples or the same set.

The CORA-3 algorithm in Bongard's realization [1] is as follows.

Learning Algorithm

1. Three columns are taken from a matrix of training examples. It is then checked if the columns contain a combination of zeros and ones that is no less than K times met in vectors of one class and is never met in vectors of the other class. If a combination of this kind has been found, then the combination of zeros and ones available in these three columns will be called a *feature*. The examples that contain this combination of zeros and ones in the corresponding columns will be called examples of which this feature is characteristic. Features A and B are assumed to be equivalent if they are characteristic of the same examples. Feature A is stronger than feature B if it is characteristic of all the examples to which feature B is characteristic and also to some other (or at least to one) examples. It is clear that each feature characterizes objects of only one class.

2. The feature found is compared to all those found before. If this feature turns out to be stronger than any other feature found before, it is written in the place of the weakest feature. If some feature found before is stronger or equivalent to the new one, then the new feature is rejected. If none of these relationships holds, the new feature is written in a new place if there is such.

3. Similarly all column combinations taken by three are checked.

Additional Learning Algorithm

Let us call the number of features characteristic to the given example *the number of votes*, which this example has collected. Those features that in the end of Step 3 have collected no more than n votes will be called lagging features. The number n can be set different for each class depending on the situation.

4. Additional learning is a stage of learning the transition to which occurs not automatically but by a command. When additional learning is performed, K can be changed. Features characterizing no less than K lagging features will be called additional features.

5. Selection of additional features is performed according to Steps 1-3.

6. If the features selected have filled up the memory (64 features for each class in Bongard's realization), then learning is finished. Otherwise, the operator may return to Step 4 (with new K and n) or stop at the result achieved.

7. The result of learning can be stored in the external memory.

Recognition Algorithm

8. Before recognition, either training has to be performed or learning results have to be read from the external memory (see Step 7).

9. When recognition is executed, it is checked how many features voting in favor of one or other class characterize the example being recognized. The example is classified in accordance with the majority of votes.

3 Cross Validation Method to Determine the Accuracy of Classification

The problem is stated as follows. How to test the accuracy of various classification methods so as the testing would be objective, the results would be accurate and would not be based on only two attempts and one could objectively compare different methods with each other? How large a training set and testing set should be? How many times one should perform training and testing to make the result statistically validated? The *cross validation* method seems to be most suitable for this purposes. This method is considered among researchers [5] as sufficiently objective to evaluate the accuracy of classification. The essence of the method is as follows. The whole set of examples is intermixed with the help of random number generator and separated into k nearly equal parts. Each of these k parts is in sequence employed as testing examples provided the classification algorithm learning was performed beforehand using those examples that are not contained in current testing example part.

According to various authors [5], it is most optimal to employ k=10. At this, an optimal relationship between the number of examples of a training set and of a testing set is ensured. In this case the method is called "10-fold cross validation". 90% of examples are used as a training example set, 10% of examples are employed as a testing example set and the cross validation process as a whole consists of 10 iterations. At each iteration, different 10% are taken for a testing set. An average error over those 10 iterations is called classification error.

4 Introduction of the Method of Relatively Good Features

In Bongard' s realization, a conjunction is sought that can no less than K times be met in vectors of one class and is never met in vectors of the other class. Due to this, the algorithm rejects those features that at least once appear in another class and thus loses potentially valuable information. It may be worth softening this strict requirement. For example, Gelfand and Guberman (1976) in the paper [2] dealt with the determination of earthquake epicenters describe an approach similar to CORA in which the conjunction is relatively frequently met in one class and is relatively infrequently met in the other class. An approach of this kind should be effective when classifying new objects, the error on the training set vectors, however, would increase. Let us call it a *method of relatively good features*. The method can be introduced as follows. In addition to an ordinary threshold K, which determines the minimal number of examples that the conjunction found must describe, threshold Ko is set that determines the maximal number of examples of the opposite class, which may be described by the conjunction, found. It is clear that $Ko \ll K$ should hold. A real employable value of Ko has to be found experimentally.

5 Introduction of the Method of Sequential Covering

Practical experiments with the CORA classification algorithm show its essential shortcoming. The algorithm has an extremely large learning time especially if the number of attributes and the length of conjunction is large that can be due to the "combinatorial explosion". The reason for this is the following. The CORA algorithm searches for attribute combinations. If the conjunction length is L but the number of attributes is N, then the algorithm has to examine

$$\frac{N!*2^L}{L!*(N-L)!} \quad (1)$$

possible conjunctions. Moreover, it has also to examine possible conjunctions each of which is searched for over all training set examples. It is possible to reduce the number of operations indirectly, i.e. the above formula is valid, however, the number of training set examples in which new conjunctions has to be sought can be reduced as well as the number of conjunctions found and, hence, the number of conjunctions to be analyzed. This can be achieved by introducing the so-called *sequential covering method* described in [4] where its application to IF-THEN rule generation is presented. The essence of this method is as follows. While analyzing a training example set, the algorithm learns one rule. It then excludes from the training example set all the examples this rule describes. After that, when examining the training set examples remained, the algorithm learns the second rule and again excludes from the training example set all those examples, which are described by this rule. The process continues in iteration until all the examples are excluded from the training example set and the rules that cover all the examples are learned.

This method slightly modified can be adapted to the CORA algorithm. The original sequential covering method excludes examples provided they are described only once. This does not fit the CORA algorithm since it contradicts to the principles of additional learning according to which the examples that have collected less than n votes are considered as lagging and therefore additional learning has to be performed to enable them to collect the votes missing. Due to this the authors of this paper propose to combine these principles and to exclude examples out of the training set when they have collected a specified number of votes, n. This will allow one to reject the stage of additional learning at all by transferring its functions to the sequential covering method thus gaining extra time saving. In this way the sequential covering method is "fused" with the stage of additional learning and is naturally involved in the ordinary learning process of the CORA algorithm thus simplifying this algorithm and at the same time improving its quality and operation speed.

6 Description of Learning and Testing Sets

The experiments have been performed on the basis of example sets taken from the UCI Repository of Machine Learning Databases and Domain Theories available in the Internet [6], [7].

The sets of examples located at this address are well known and cited by many authors. Each of the example sets displays real case studies. The present study employs example sets from two knowledge areas: (1) USA Congress voting results and (2) Tic-tac-toe game. Below one can find a detailed information on each of the sets.

1. USA Congress voting results. The file employed was *house-votes-84.data.* In the congress voting database, each voting case has 16 binary attributes that represent the answers to 16 questions and a class that shows which party has won in voting (0 – democrats or 1 – republicans). 232 examples were used. In 124 cases of them democrats have won (class 0) but in 108 cases republicans have won (class 1). In what follows this database will be called *Votes*.

2. Tic-tac-toe game. The file employed was *tic-tac-toe.data.* The database includes all the 958 positions possible for this game. It is assumed that the crosses go first. If the crosses win, then an example belongs to class 1, if they loose or in the case of drawn play it belongs to class 0. To describe the state of each game square, binary combinations of two attributes were employed as follows: 01 - there is a cross in the square; 11 - there is a naught in the square; 00 - the square is empty. Hence, after the encoding each database example was described by 18 attributes. In what follows this database will be called Tic-tac-toe.

7 Experiments Performed with the CORA Classification Algorithm

In all further experiments, the author's own software program CORA was used to implement the CORA classification algorithm. To check the classification accuracy in all the experiments, the cross-validation method was used and also the learning time spent was measured.

7.1 Determination of Optimal Learning Parameters

The purpose of this experiment is to test the efficiency of the algorithm at different conjunction lengths and different minimal values of threshold K , to determine which conjunctions should be given priority when making selection of conjunctions: those of shorter length or those of larger length as well as to find optimal learning parameters for the given example set.

During the experiment different conjunction lengths were examined, first from 1 through 2 and finally from 1 through 5. The values of threshold K from 10 through 50 with step 10 have been tested. It was also examined which conjunctions should be given priority when making selection of conjunctions: those of shorter length or those of larger length (for example, 1-5 means that priority is given to shorter disjunctions whereas 5-1 means that priority is given to longer disjunctions). In total, 40 experiments have been performed with the example set Votes at different values of the above parameters (see Table 1).

Each row of this table represents the results obtained at 10 training and testing iterations performing cross-validation. The shaded row denotes a variant (No.34) with optimal learning parameters: a low error level is achieved within short learning time.

The same experiments have been performed with the Tic-tac-toe example set (see Table 2).

Table 1. Determination of optimal parameters. Example set: Votes

No.	Conjunction length	Minimal threshold K	Error (%)	Average number of conjunctions	Average time (s)
1.	1-2	10	10.870	15.4	0.0
2.	1-2	20	10.870	12.2	0.1
3.	1-2	30	10.870	10.1	0.0
4.	1-2	40	10.870	8.9	0.1
5.	1-2	50	10.870	5.1	0.0
6.	1-3	10	5.652	45.1	1.9
7.	1-3	20	6.087	26.6	1.1
8.	1-3	30	6.087	19.8	0.7
9.	1-3	40	7.391	14.5	0.5
10.	1-3	50	10.000	7.6	0.4
11.	1-4	10	4.348	82.3	21.7
12.	1-4	20	4.783	41.6	7.9
13.	1-4	30	5.217	31.6	4.2
14.	1-4	40	5.217	20.1	2.8
15.	1-4	50	8.696	8.1	2.1
16.	1-5	10	4.348	103.6	104.2
17.	1-5	20	4.783	52.5	33.8
18.	1-5	30	5.217	39.9	14.8
19.	1-5	40	5.217	24.3	12.4
20.	1-5	50	8.696	8.3	10.6
21.	2-1	10	10.870	16.3	0.1
22.	2-1	20	10.870	13.1	0.1
23.	2-1	30	10.870	10.9	0.0
24.	2-1	40	10.870	9.6	0.0
25.	2-1	50	10.870	5.5	0.0
26.	3-1	10	5.652	54.0	7.8
27.	3-1	20	6.087	33.4	3.7
28.	3-1	30	6.087	24.7	1.7
29.	3-1	40	7.391	18.0	0.8
30.	3-1	50	10.000	10.1	0.4
31.	4-1	10	4.348	90.4	131.9
32.	4-1	20	4.348	47.5	40.1
33.	4-1	30	4.783	35.5	15.7
34.	4-1	40	4.783	24.2	5.5
35.	4-1	50	8.696	11.3	3.4
36.	5-1	10	4.348	110.3	861.8
37.	5-1	20	4.783	57.5	235.5
38.	5-1	30	3.913	43.3	78.8
39.	5-1	40	4.783	27.1	25.9
40.	5-1	50	8.696	12.4	15.7

Table 2. Determination of optimal learning parameters. Example set: Tic-tac-toe

No.	Conjunction length	Minimal threshold K	Error (%)	Average number of conjunctions	Average time (s)
1.	1-2	10	100.000	0.0	0.2
2.	1-2	20	100.000	0.0	0.2
3.	1-2	30	100.000	0.0	0.2
4.	1-2	40	100.000	0.0	0.2
5.	1-2	50	100.000	0.0	0.2
6.	1-3	10	26.316	31.1	2.3
7.	1-3	20	29.579	14.2	2.1
8.	1-3	30	32.211	11.4	2.0
9.	1-3	40	50.421	4.9	2.0
10.	1-3	50	61.263	2.1	2.0
11.	1-4	10	9.474	351.5	133.9
12.	1-4	20	7.053	175.9	37.7
13.	1-4	30	9.789	120.5	25.2
14.	1-4	40	32.105	44.5	17.9
15.	1-4	50	58.211	5.6	16.0
16.	1-5	10	10.947	1150.1	4337.3
17.	1-5	20	4.316	477.5	780.1
18.	1-5	30	6.316	303.7	284.1
19.	1-5	40	27.895	104.1	123.9
20.	1-5	50	57.789	10.0	95.0
21.	2-1	10	100.000	0.0	0.2
22.	2-1	20	100.000	0.0	0.2
23.	2-1	30	100.000	0.0	0.2
24.	2-1	40	100.000	0.0	0.2
25.	2-1	50	100.000	0.0	0.2
26.	3-1	10	26.316	31.1	2.3
27.	3-1	20	29.579	14.2	2.1
28.	3-1	30	32.211	11.4	2.1
29.	3-1	40	50.421	4.9	2.1
30.	3-1	50	61.263	2.1	2.0
31.	4-1	10	9.474	351.5	159.9
32.	4-1	20	7.053	178.6	45.4
33.	4-1	30	9.789	125.4	30.0
34.	4-1	40	32.105	48.5	20.5
35.	4-1	50	58.211	12.4	17.2
36.	5-1	10	10.947	1150.1	5612.4
37.	5-1	20	4.316	477.7	1246.4
38.	5-1	30	6.316	303.9	488.6
39.	5-1	40	27.895	106.6	207.9
40.	5-1	50	57.789	22.6	141.0

For this example set, variant No.13 is considered as optimal.

7.2 Experiments with Additional Learning

After optimal parameters for initial learning have been determined, experiments with additional learning can be performed. In this experiment series a threshold for additional learning is varied so as to find its optimal value for the both example sets (see Table 3 and 4).

Table 3. Experiments with additional learning. Example set: Votes

No.	Conjunction length	Minimal threshold K	Threshold for additional learning	Error (%)	Average number of conjunctions	Average time (s)
1.	4-1	40	35	4.348	30.6	8.2
2.	4-1	40	30	4.348	35.8	8.9
3.	4-1	40	25	4.348	39.0	10.9
4.	4-1	40	20	4.348	46.0	12.7
5.	4-1	40	15	4.348	61.8	17.8
6.	4-1	40	10	3.913	88.0	31.9
7.	4-1	40	5	4.348	126.1	83.5

Table 4. Experiments with additional learning. Example set: Tic-tac-toe

No.	Conjunction length	Minimal threshold K	Threshold for additional learning	Error (%)	Average number of conjunctions	Average time (s)
1.	1-4	30	25	6.421	149.7	47.7
2.	1-4	30	20	7.053	175.9	53.3
3.	1-4	30	15	7.684	246.1	81.6

For the *Votes* set, the additional learning threshold 10 turned out to be optimal but for the Tic-tac-toe set it was 25.

8 Experiments with the Modifications of the CORA Algorithm

The task of this experiment series is to find out whether the introduction of modifications of the standard CORA algorithm was useful.

8.1 Experiments with the Relatively Good Features Method

These experiments were performed to test the efficiency of the relatively good features method. In the experiments the maximal number of examples of the opposite class, Ko, which each conjunction found can describe was varied (see Table 5 and 6).

Table 5. Experiments with the relatively good features method. Example set: Votes

No.	Conjunction length	Minimal threshold K	Maximal Ko	Error (%)	Average number of conjunctions	Average time (s)
1.	4-1	40	1	3.043	29.8	13.6
2.	4-1	40	2	4.783	43.1	21.3
3.	4-1	40	3	6.087	42.4	31.7

Table 6. Experiments with the relatively good features method. Example set: Tic-tac-toe

No.	Conjunction length	Minimal threshold K	Maximal Ko	Error (%)	Average number of conjunctions	Average time (s)
1.	1-4	30	1	11.185	208.4	50.4
2.	1-4	30	2	9.895	227.3	69.7
3.	1-4	30	3	9.158	236.7	86.3
4.	1-4	30	4	12.737	281.3	91.8
5.	1-4	30	5	21.263	365.7	132.7

It can be seen that for the Votes training example set *Ko*=1 turned out to be optimal but for the Tic-tac-toe set it was equal to 3.

8.2 Experiments with the Sequential Covering Method

Let us perform experiments to test the efficiency of the sequential covering method. In these experiments the number of votes, *n*, at which examples are excluded out of a training set will be changed (see Table 7 and 8). The shaded rows represent the three best results in each set.

Table 7. Experiments with the sequential covering method. Example set: Votes

No.	Conjunction length	Minimal threshold K	Number of votes *n*	Errors(%)	Average number of conjunctions	Average time (s)
1.	1-4	20	20	4.783	41.3	7.5
2.	1-4	20	15	4.783	39.9	6.6
3.	1-4	20	10	4.348	33.9	4.2
4.	1-4	20	5	3.913	20.8	1.4
5.	1-4	15	20	4.348	54.9	11.4
6.	1-4	15	15	4.348	48.2	9.3
7.	1-4	15	10	3.913	43.5	4.8
8.	1-4	15	5	5.217	25.2	1.4
9.	1-4	10	20	3.913	73.6	18.4
10.	1-4	10	15	3.478	64.3	14.3
11.	1-4	10	10	3.913	52.2	6.5
12.	1-4	10	5	5.217	29.3	1.6
13.	1-4	5	20	3.913	108.4	37.9
14.	1-4	5	15	3.913	85.1	27.5
15.	1-4	5	10	7.826	55.1	8.1
16.	1-4	5	5	11.739	22.2	2.2

Table 8. Experiments with the sequential covering method. Example set: Tic-tac-toe

No.	Conjunction length	Minimal threshold K	Number of votes *n*	Percentage of errors	Average number of conjunctions	Average time (s)
1.	1-4	20	20	6.737	167.1	34.2
2.	1-4	20	15	6.842	157.6	31.7
3.	1-4	20	10	6.421	131.8	26.1
4.	1-4	20	5	6.842	83.5	18.7
5.	1-4	15	20	7.684	210.8	45.9
6.	1-4	15	15	7.684	197.8	40.8
7.	1-4	15	10	6.737	165.8	31.4
8.	1-4	15	5	6.211	102.9	20.3
9.	1-4	10	20	9.053	291.1	78.0
10.	1-4	10	15	8.842	255.1	61.8
11.	1-4	10	10	7.789	214.8	43.9
12.	1-4	10	5	5.684	131.3	24.6
13.	1-4	5	20	9.684	422.4	154.0
14.	1-4	5	15	9.053	371.3	114.1
15.	1-4	5	10	7.579	273.2	74.8
16.	1-4	5	5	6.526	165.7	35.2

8.3 Experiments with a Combination of Both Methods

The task of these experiments is to test the efficiency of a simultaneous introduction of the method of relatively good features and the sequential covering method. In these experiments the maximal number of examples of the opposite class, Ko, which each conjunction found can describe and the number of votes, *n*, at which examples are excluded out of a training set were varied (see Table 9 and 10).

Table 9. Experiments with a combination of both methods. Example set: Votes

No.	Conjunction length	Minimal threshold K	Number of votes *n*	Maximal Ko	Error (%)	Average number of conjunctions	Average time (s)
1.	1-4	20	5	1	4.783	18.0	0.9
2.	1-4	20	5	2	4.783	16.7	0.8
3.	1-4	20	5	3	3.478	15.7	0.7
4.	1-4	20	5	4	4.783	13.8	0.7
5.	1-4	20	5	5	10.000	11.2	0.7
6.	1-4	15	10	1	3.478	32.6	1.1
7.	1-4	15	10	2	4.348	28.9	0.9
8.	1-4	15	10	3	4.348	26.6	0.8
9.	1-4	15	10	4	3.043	20.6	0.8
10.	1-4	15	10	5	7.391	16.6	0.7
11.	1-4	10	15	1	3.913	53.3	1.7
12.	1-4	10	15	2	21.739	21.9	1.2
13.	1-4	10	15	3	11.304	28.0	0.9
14.	1-4	10	15	4	12.174	23.3	0.9
15.	1-4	10	15	5	17.391	25.5	0.8

Table 10. Experiments with a combination of both methods. Example set: Tic-tac-toe

No.	Conjunction length	Minimal threshold K	Number of votes n	Maximal Ko	Error (%)	Average number of conjunctions	Average time (s)
1.	1-4	20	10	1	6.737	147.5	26.8
2.	1-4	20	10	2	5.263	133.8	25.1
3.	1-4	20	10	3	5.474	147.5	24.5
4.	1-4	20	10	4	6.632	164.9	25.6
5.	1-4	20	10	5	8.632	177.5	24.0
6.	1-4	15	5	1	7.158	110.1	17.7
7.	1-4	15	5	2	5.474	110.8	16.4
8.	1-4	15	5	3	10.211	103.0	15.0
9.	1-4	15	5	4	17.474	91.6	13.3
10.	1-4	15	5	5	21.368	80.5	11.8
11.	1-4	10	5	1	6.842	137.5	19.9
12.	1-4	10	5	2	12.000	110.3	18.7
13.	1-4	10	5	3	23.579	80.8	17.1
14.	1-4	10	5	4	31.789	54.9	14.5
15.	1-4	10	5	5	39.368	36.5	12.9

The experimental results show that better success was achieved when both CORA algorithm modifications suggested were combined. As a result of this combination, both the classification error and learning time were reduced.

Figures 1 and 2 present a summary of the experimental results.

9 Analysis of the Results and Conclusions

The theoretical part of this study explains two main shortcomings of the CORA classification algorithm: (1) the algorithm rejects features that at least once appear in the opposite class and thus loses potentially valuable information; and (2) the algorithm has extremely large learning time that can be due to the "combinatorial explosion". The study suggests two methods to overcome these difficulties and to improve the algorithm: the method of relatively good features and the sequential covering method.

The experimental part of the study presents the experiments performed to examine the usefulness of those methods introduction and also checks their efficiency. The experiments performed made use of example sets from two knowledge areas: (1) USA Congress voting results forecasting (232 examples), and (2) Tic-tac-toe game (958 examples). The experimental results demonstrate that both the methods suggested ensure a sufficient algorithm performance's improvement as compared to the classic Bongard algorithm realization. Best results are achieved when both methods are combined and sufficiently improve the classification precision and reduce learning time.

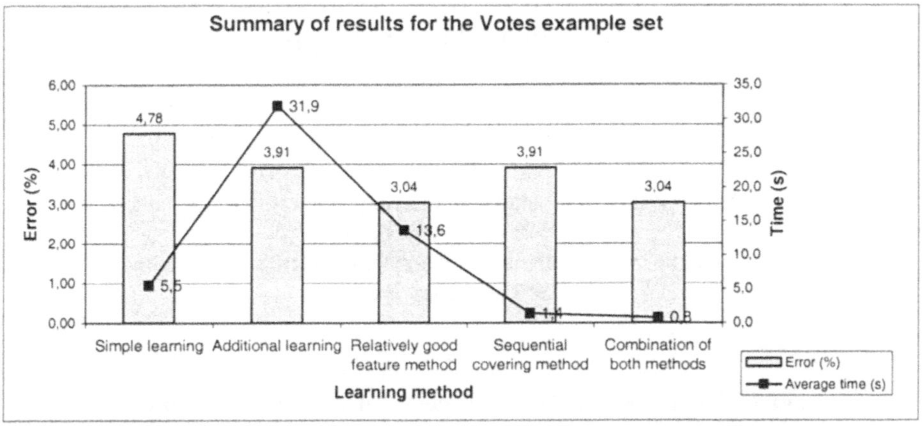

Fig. 1. Summary of experimental results for the Votes example set.

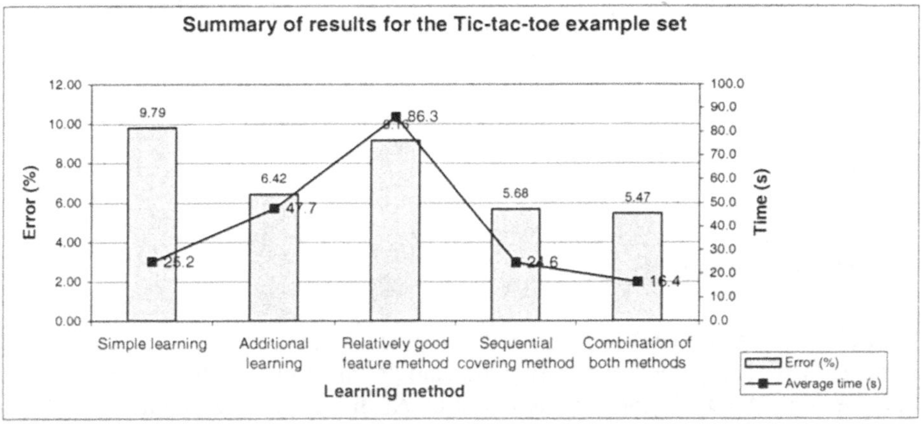

Fig. 2. Summary of experimental results for the Tic-tac-toe example set.

References

1. Bongard, M.: Problem of Recognition. Nauka Publishers, Moscow (1967) (In Russian)
2. Gelfand, M., Guberman, S.: Pattern Recognition Applied to Earthquake Epicenters in California. Physics of the Earth and Planetary Interiors 11 (1976) 227 - 283
3. Gladun, V.: Planning Decisions. Naukova Dumka, Kiev (1987) (In Russian)
4. Mitchell, T. M.: Machine Learning. The Mc Graw - Hill Companies Inc. (1997)
5. Weiss, S., Kapouleas, I.: An Empirical Comparison of Pattern Recognition, Neural Nets, and Machine Learning Classification Methods. Shawlik, I. W., Dietterich, T. G. (editors): Readings in Machine Learning. Morgan Kaufmann Publishers, San Mateo, California (1990) 177 - 183
6. URL: ftp://ftp.ics.uci.edu/pub/machine-learning-databases
7. URL: http://www.ics.uci.edu/~mlearn/MLRepository.html

Computer Aided Selection of Cutting Parameters by Using Fuzzy Logic

Oğuzhan Yılmaz[1], Günseli Görür[1], and Türkay Dereli[2]

[1] Department of Mechanical Engineering, University of Gaziantep
27310 Gaziantep, Turkey
[2] Department of Industrial Engineering, University of Gaziantep
27310 Gaziantep, Turkey

Abstract. This paper introduces a system that has been developed for the selection of cutting parameters of turning operations from the machinability data handbook by using Fuzzy Logic (FL) approach. Main objective of the system is to get rid of the manually input-output oriented handbook, so selection of cutting parameters is made by a user-friendly intelligent system. This system is visual and menu oriented; also usage of the system is very simple.

Key Words: Cutting Parameters, Fuzzy Logic, Computer Aided Selection, Artificial Intelligence.

1 Introduction

Artificial Intelligence (AI), is a new and rapidly developing field of computer science. It deals with computers performing humanlike functions such as reasoning and interpretation. Fuzzy Logic (FL) is a tool of AI that is gaining popularity in recent years. It based on the observation that people make decisions based on imprecise and non-numerical information fuzzy models or sets are mathematical means of representing vagueness and imprecise information, hence the term *fuzzy*. These models have the capability of recognizing, representing, manipulating, interpreting and utilizing data and information that are vague and lack certainty. Linguistic examples are: *a few, almost, all, more, less, very important, good, poor, appropriate, acceptable,* etc [1]. The basic idea behind the FL is quite simple. An exact rule or law cannot be defined for each possible problem. For these cases, humans approximate well solutions to the problems. This approximation is possibly due to the flexibility in the definition of the *words* that constitute the rules. Hence, human logic is very flexible. FL simulates this human logic using a mathematical model.

One of the usage areas of fuzzy logic is the machining science. The machining of metals is a complex physical process, which is influenced by a large number of factors (cutting parameters or conditions). The success of a machining operation depends on the correct choice of these parameters. It has been realized that the proper

B. Reusch (Ed.): Fuzzy Days 2001, LNCS 2206, pp. 854–870, 2001.

selection of machining conditions as the cutting speed, the feed, and the depth of cut in turning has a substantial impact on the overall machining performance which may be assessed in terms of machinability parameters like surface finish, tool-wear rate, dimensional accuracy, cutting power and chip breakability etc.[3]. Also a proper selection not only reduces the machining cost (by means of extended tool life, increased material removal rate, reduced down time and reduced inventory), but also improves the product quality (better surface finish and tighter dimension tolerance). It can be argued that it may be the best investment per capital for improving machining operations [4].

This paper presents a system that has been developed to obtain the cutting conditions (cutting speed, depth of cut and feed rate), for any selected material. With the selection of material type, and the hardness value of the material in terms of Brinell Hardness Number (BHN), selecting the cutting tool type and depth of cut, the program runs based on the Fuzzy Logic model to determine the cutting speed value. The model has used for five fuzzy sets in triangular shapes, five fuzzy experts rules, and centroid defuzzification method. Execution of the system gives the cutting speed value and the feed value that also have been displayed on the screen.

2 Machinability Data Handbook's Approach

The Machining Data Handbook [6] provides starting recommendations for important machining situations in an effort to help personnel with material removal processes. The condensed and integrated source of machining information is useful to manufacturing, process, industrial and method engineers, NC programmers, etc. Data were obtained from the recommendations of metal producers, machine tool manufacturers and cutting tool and cutting fluid manufacturers, as well as from handbooks, machinability reports and miscellaneous literature on machining. The machining data obtained from industry were generally those that are currently being used in production operations [5].

In the Machinability Data Handbook, workpiece material hardness and recommended cutting speeds are given in ranges, as shown in Table 1. Also the handbook approach is presented graphically in Fig. 1 for different types of cutting tool materials. All the cutting speed values and hardness values are given in certain ranges, and approximately for the same material hardness, the handbook gives two different cutting speed values, means that, there are an ambiguity and vague relations. Therefore, determination of a certain cutting speed value corresponding to a given hardness value is very difficult. It is a deficiency in using the handbook especially for an unskilled operator, which is not experienced in his job too much.

In the handbook, material type with its recommended cutting conditions is firstly selected. The cutting speed values are given in ranges and the operator can select any cutting speed value between these cutting speed ranges, according to the specified workpiece material hardness, depth of cut, feed rate, cutting tool type and workpiece material. The selection of the operator will be straight forward or at the middle of the given range. Therefore, to obtain the correct cutting speed value between the specified ranges will be a chance and any wrong selection may lead poor cutting conditions result with poor surface quality, because the cutting speed has an important effect on the cutting conditions.

Table 1. Recommended cutting speed ranges and feed rates for different cutting tools. (Material: carbon steels, wrought, medium carbon, 1 mm depth of cut, hardness range: 125 BHN-425 BHN)

Tool Material	Depth of cut (mm)	Speed range (m/min)	Feed (mm/rev)
High speed steel (ISO S4, S5)	1	18-43	0.18
	4	14-35	0.40
	8	11-27	0.50
	16	15-21	0.75
Carbide tool, uncoated (brazed, ISO P10-P40)	1	87-140	0.18
	4	69-110	0.50
	8	55- 84	0.75
	16	53-66	1.00
Carbide tool, uncoated (indexible, ISO P10-P40)	1	95-180	0.18
	4	76-140	0.50
	8	60-110	0.75
	16	69- 84	1.00
Carbide tool, coated (ISO CP10-CP30)	1	145-260	0.18
	4	90-170	0.40
	8	76-135	0.50
	16	-	-

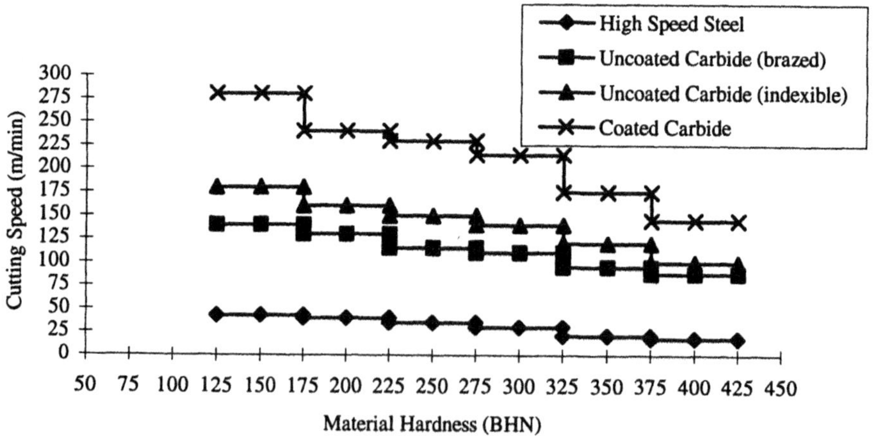

Fig. 1. Recommended cutting speed ranges for different cutting tool materials at 1mm depth of cut machining data handbook

If wrought medium carbon steel with the workpiece material hardness is about 250 BHN, cutting tool type is High Speed Steel (HSS), and the depth of cut is 1 mm, so speed range will be 18-43 m/min. Under the specified cutting conditions, an unskilled operator may select the cutting speed approximately 30 m/min. This type of selection is straight-forward, means that the operator escapes from boundaries for making any mistakes. But, this person can select 43 m/min for fast cutting or 18 m/min for low cutting, but in this case, the user must consider some other machining parameters such as surface roughness, tool wear etc. When this selection is made by an expert operator, the machining parameters such as material hardness, surface roughness, tool wear etc. are considered, and then selection is made more preciously. If the material hardness is *hard*, then an expert's selection will depend upon his experience such that: "*Cutting speed must be low and also feed rate must be small* since surface roughness and tool wear are strictly depends upon mainly these two parameters". If the material hardness is *soft* then the expert's selection will depend on the rule "*cutting speed must be high and feed rate must be high*".

So, the aim of this study is to eliminate the deficiencies of the machining data handbook approach and to get rid of the manually input-output system of the handbook. In the developed system, a definite cutting speed value is determined instead of the range by using fuzzy-expert rules, under the conditions of specified cutting tool, feed rate, depth of cut and hardness range values.

3 Methodology of the Fuzzy Logic to the Selection of Cutting Parameters

3.1 Forming of Fuzzy Sets

In the Handbook, the cutting speed is selected according to workpiece material hardness. For the cutting speed determination, it is essential that all the relevant factors must be considered for any particular application. There are a large number of factors, but to consider all of them is difficult therefore cutting tool type, depth of cut, feed rate are considered as constant. In this study, *material hardness* is the input into the system and *cutting speed* is the output for any specified material type.

Fuzzy sets are used for fuzzifiying processes. The main aim of the fuzzification is to provide a convenient way to create fuzzy relations between input and output variables. In fuzzification operation, more than one fuzzy set are formed. Linguistic variables (or fuzzy expressions) are also used in this operation. Table 2 includes the fuzzy sets of the input (material hardness) and output (cutting speed) of the selection system.

Five fuzzy expressions are extracted and used for both the input and the output. By using these fuzzy sets, input material hardness and output cutting speeds are classified. Number of these sets is dependent on the system behavior; the choice of the number of sets has an essential influence on how fine a control can be obtained [2,5].

Table 2. Fuzzy sets (fuzzy expression or linguistic variables)

Material Hardness	Cutting Speed
Very **H**ard (VH)	**V**ery **L**ow (VL)
HArd (HA)	**L**ow (LO)
Me**D**ium (MD)	**M**e**D**ium (MD)
SOft (SO)	**H**igh (HI)
Very **S**oft (VS)	**V**ery **H**igh (VH)

3.2 Fuzzification

Besides fuzzy sets, the membership functions are absolutely dependent upon the system behavior. There are different types of membership functions (triangular, trapezoidal, z-shaped, s-shaped etc.) but the most used one is the triangular membership function, which is also used in this study, since it is simply formed, and also it is proper for this type of operation [2,5].

Each fuzzy set has members. The members of fuzzy sets are arranged in linearly and equally spaced. The handbook values of medium-carbon wrought steel have a hardness range of 125-425 BHN and cutting speed range of 18-43 m/min. The members of the fuzzy sets for this material are linearly and equally spaced as shown in Table 3, after that the members of fuzzy sets are represented by using the triangular membership functions (trimf). Membership functions regarding to the material hardness and cutting speed membership functions with their ranges are presented in Fig. 2 as an input and output membership functions.

Table 3. Members of fuzzy sets

Fuzzy Sets		Members	Fuzzy Sets		Members
Very Soft	:	125-200	Very Low	:	18-24.25
Soft	:	125-275	Low	:	18-30.50
Medium	:	200-350	Medium	:	24.25-36.75
Hard	:	275-425	High	:	30.50-43
Very Hard	:	350-425	Very High	:	36.75-43

3.3 Extracting the Fuzzy-Expert Rules

Fuzzy rules (IF-THEN) are the interface between input and output relations. The fuzzy rules of this type of cutting conditions selection system must be same as the behavior of an expert operator. Utilization of the IF-THEN rules is very simple in fuzzy logic.

In turning operations, when the material hardness and cutting speed are considered, five general rules are extracted as follows [5]: These rules use fuzzy expression such as *soft, high,* etc., which directly relate to the actions of the lathe operator. The rules are constructed in the form of fuzzy conditional statements.

Rule1: IF Material is VeryHard THEN Speed is VeryLow.
Rule2: IF Material is Hard THEN Speed is Low.
Rule3: IF Material is Medium THEN Speed is Medium.
Rule4: IF Material is Soft THEN Speed is High.
Rule5: IF Material is VerySoft THEN Speed is VeryHigh.

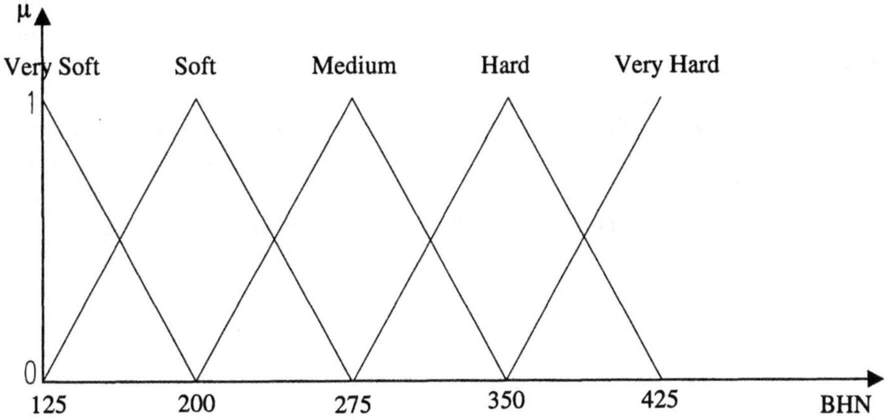

(a) Material Hardness Membership Functions and Their Ranges.

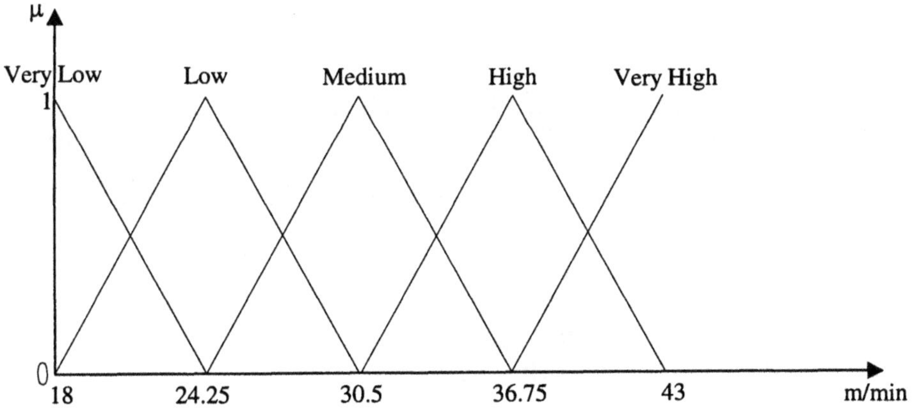

(a) Cutting Speed Membership Functions and Their Ranges.

Fig. 2. Input and output membership functions for medium carbon wrought steel

3.4 Defuzzification

Defuzzification is the reverse of the fuzzification process. The fuzzified subjects are defuzzified for obtaining the precise quantity. Lots of defuzzification methods (Centroid, mean of maxima, weighted average etc.) exists but among these methods, the one which commonly used is the *Centroid Method* [2,5]. First of all, fuzzy sets that contain the input as a member with any membership value are determined then according to these fuzzy sets, the fuzzy rules are employed. The medium carbon steel membership functions that is presented in Fig. 3 for the 125-425 BHN hardness range. If the workpiece material hardness is given as 300 BHN, which is shown in the formed fuzzy sets, the input has one membership value for each fuzzy set that contains the input. One of them is the 0.8 rate by the medium fuzzy set and the other is 0.3 rate of hard fuzzy set between [0,1]. This means that, workpiece material in consideration is hard at a rate of 0.3 and medium at a rate of 0.8. The workpiece material at 300 BHN can be considered both hard and medium hardness rates.

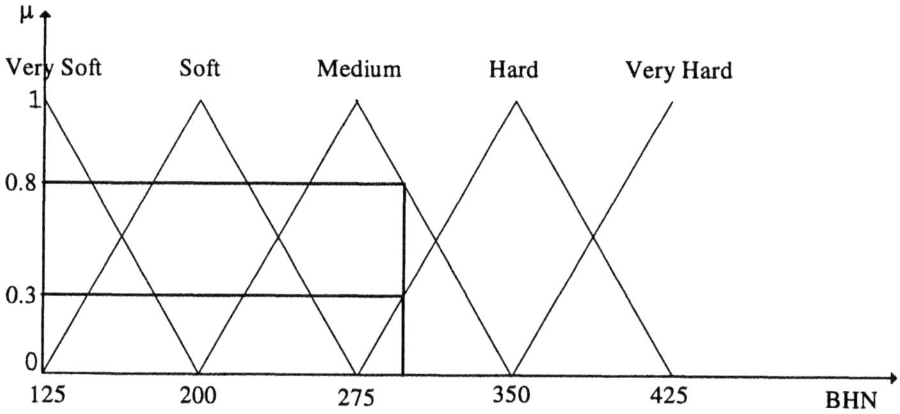

Fig. 3. Membership Values of a 300 BHN Material in Fuzzy Sets.

By using these two rates as 0.8 and 0.3, with the fuzzy expert rules:

Rule2: IF Material is Hard THEN Speed is Low.
Rule3: IF Material is Medium THEN Speed is Medium.

The output precise values are evaluated as shown in Fig. 4. The intersection of the lines at the membership values of material hardness is extended until the lines cut the membership values of cutting speeds by considering the above 2 and Rule 3. The area under the extension lines is obtained for the defuzzification of fuzzy members as shown in Fig. 5. By using the shaded area, cutting speed value will be obtained as a precise quantity for workpiece materials by using the centroid method.

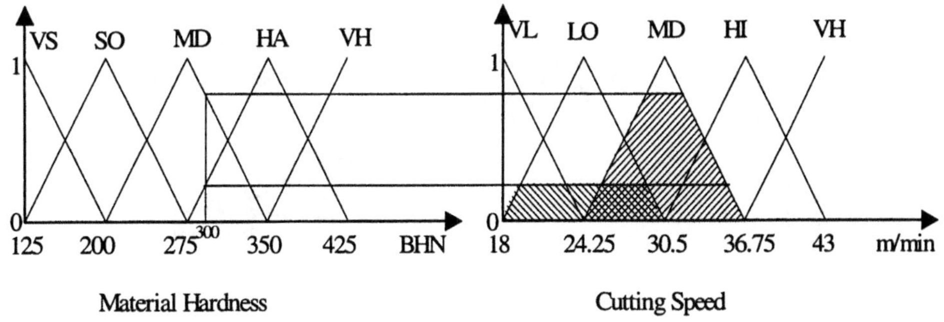

Fig. 4. Evaluation of the Rules for Obtaining the Output Precise Value.

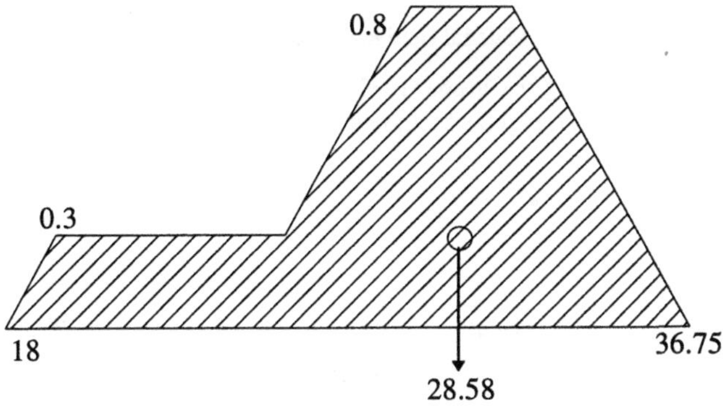

Fig. 5. Precise Quantity Obtained from Output Area.

4 Comparisons of Membership Functions

4.1 Triangular and Trapezoidal Membership Functions

The triangular membership function and trapezoidal membership function are the most usual membership functions. Therefore, in this study, to see the effect of membership functions on the cutting speed results, a wrought, medium carbon steel, with 1 mm depth of cut and High Speed Steel (HSS) tool are selected. The cutting speed results are obtained by using the triangular membership and trapezoidal membership functions.

The cutting speed results obtained by using the triangular and trapezoidal membership functions are presented with the cutting speed results taken from handbook as seen in Fig. 6 and 7 respectively.

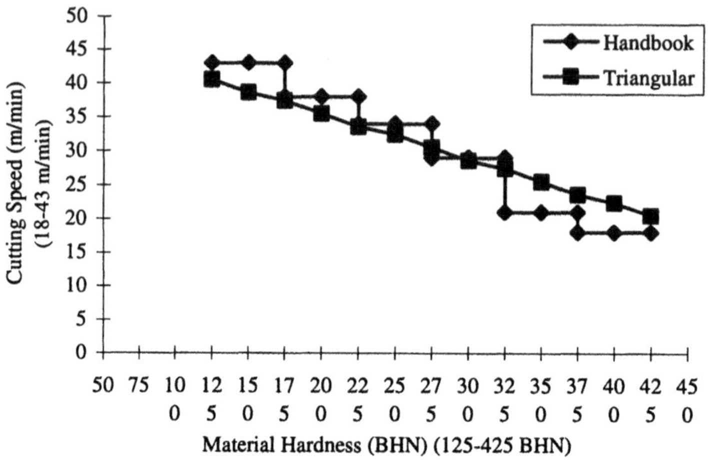

Fig. 6. Fuzzy Cutting Speed for Carbon Steel, wrought, Medium Carbon at 1mm d.o.c using HSS Tool, Handbook and Triangular Membership Function Results

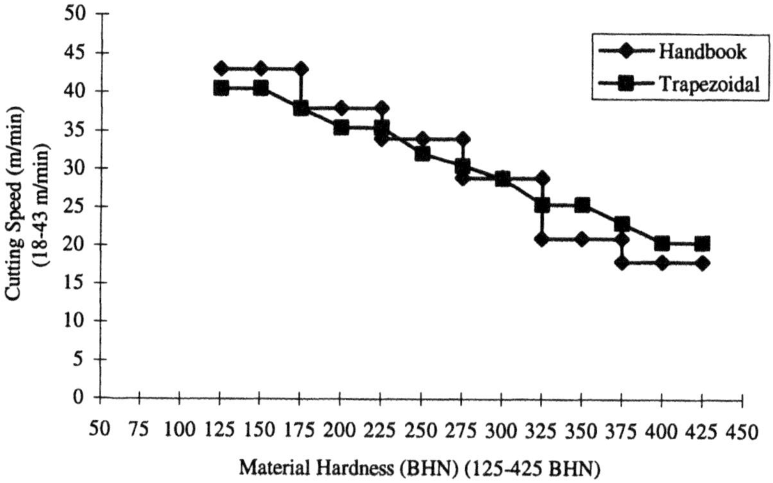

Fig. 7. Fuzzy Cutting Speed for Carbon Steel, wrought, Medium Carbon at 1 mm d.o.c using HSS Tool, Handbook and Trapezoidal Membership Function Results

For the same material and cutting tool type, fuzzy cutting speed results obtained using these two different membership functions with the cutting speed results obtained from handbook are also presented in Fig. 8. As it is seen from these figures, there are no big differences between these cutting speed results. But, triangular membership function's results are more reliable than trapezoidal membership function results, because trapezoidal membership function approach gives same cutting speeds in a range for different hardness values as in the case of handbook cutting speed results. For example, for 125 BHN and 150 BHN values, trapezoidal membership function gives 40.5 m/min cutting speed for both hardness values, but triangular membership function gives 40.5 and 38.58 m/min cutting speeds respectively.

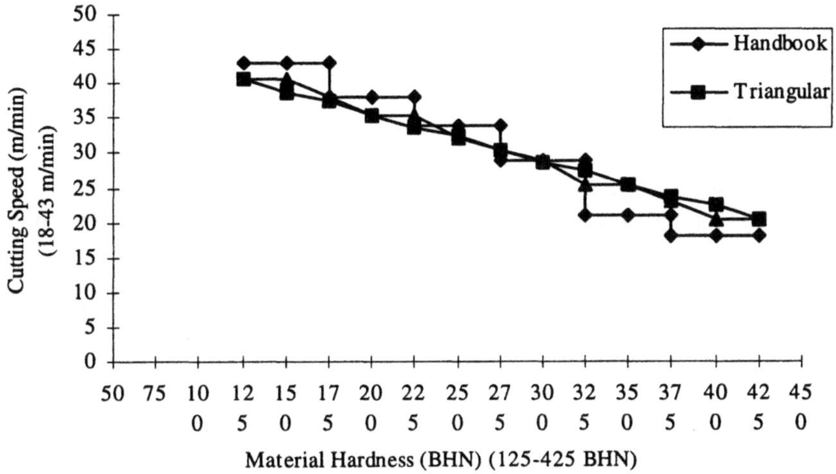

Fig. 8. Fuzzy Cutting Speed for Carbon Steel, wrought, Medium Carbon at 1 mm d.o.c using HSS Tool, Triangular and Trapezoidal Membership Functions Results

Another important point obtained from the Fig. 6-8 is that, at low hardness values, the cutting speed obtained by fuzzy logic is smaller than the handbook speed values (handbook speed value= 43 m/min, fuzzy speed= 40.5 m/min) and at high hardness values the fuzzy cutting speed is greater than the handbook cutting speed values (handbook= 18, fuzzy speed= 20.5). There is a contradiction between the fuzzy speed values and handbook values, because as the hardness increases the speed must be reduced or vice versa. But in the handbook, these cutting speed values are the starting recommended values. Therefore the cutting speed results obtained by using Fuzzy logic can be tolerated at proper conditions.

4.2 Comparison of Gaussian and Triangular Membership Function and the Effect of Number of Fuzzy Sets on Cutting Speed Results

The effect of changing membership functions between the triangular and gaussian has been investigated. Again, a wrought, medium carbon steel with 1 mm d.o.c and HSS tool is selected. The cutting speed results obtained from the handbook and fuzzy cutting speeds are illustrated in Fig. 9-12. There is no much more difference between triangular and gaussian membership functions.

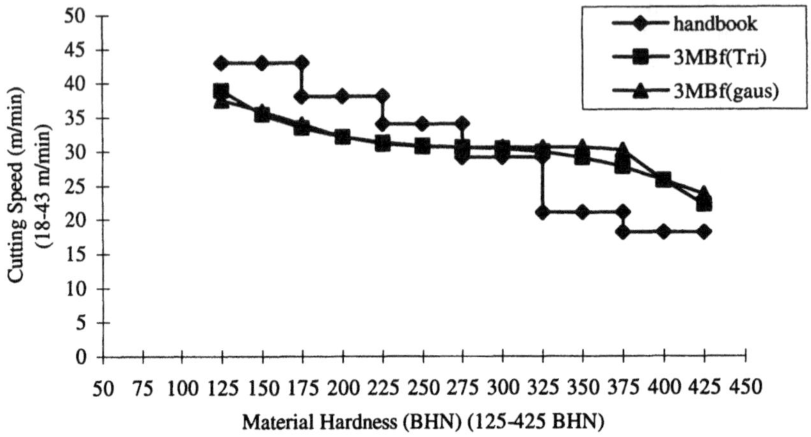

Fig. 9 Fuzzy Cutting Speed for Carbon Steels,wrought medium carbon at 1 mm d.o.c using HSS tool, Handbook, 3Mbf (Trian) and 3Mbf(Gauss) Results

Also the effect of numbers of membership functions on the cutting speed values has been investigated. The fuzzy cutting speed results using HSS tool, 1 mm depth of cut and the fuzzy set numbers for 3, 5, and 7 are obtained for medium carbon steel, and compared with handbook cutting speed results. The results obtained are illustrated in Fig. 9-11. If the numbers of fuzzy sets is 3, the results are much more deviate from the handbook cutting speed values, so they are not recommendable values. If the user examine the fuzzy set results taken with 5, and 7 numbers as shown in Fig. 10-11, that fuzzy cutting speed values are too closely to each other and in the range of handbook results. Therefore, using the 3 numbers of fuzzy set is not suitable, at least it must be 5 or greater than 5. As the number of fuzzy set increased the cutting speed results much more approaches to the handbook cutting values but it brings together too much complexity for the application of fuzzy rules.

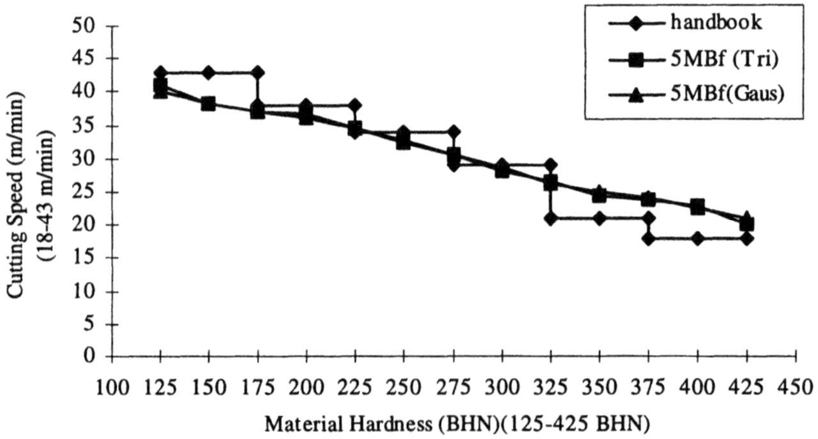

Fig. 10. Fuzzy Cutting Speed for Carbon Steels,wrought medium carbon at 1 mm d.o.c using HSS tool, Handbook, 5Mbf (Trian) and 5Mbf(Gauss) Results

Fuzzy logic based cutting speed values for all the numbers of membership functions as 3, 5, and 7 is presented in Fig. 12. The cutting speed values taken from the handbook are also presented on the same figure. For all other results based on fuzzy logic are taken under the consideration of 5 fuzzy sets, triangular membership function, if-then fuzzy rules and centroid defuzzification method.

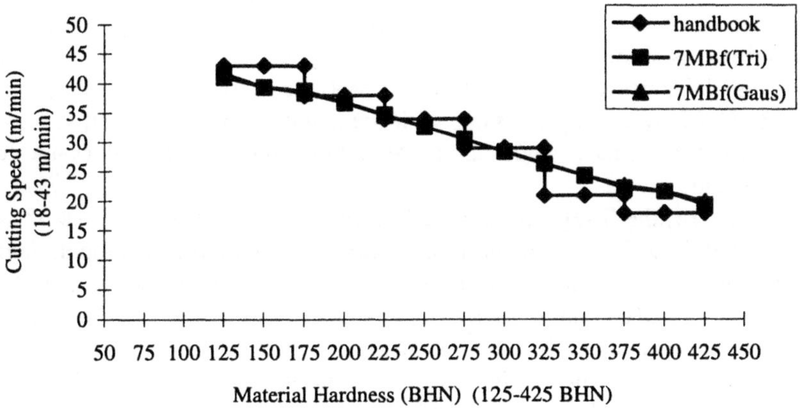

Fig. 11 Fuzzy Cutting Speed for Carbon Steels,wrought medium carbon at 1 mm d.o.c using HSS tool, Handbook, 7Mbf (Trian) and 7Mbf(Gauss) Results

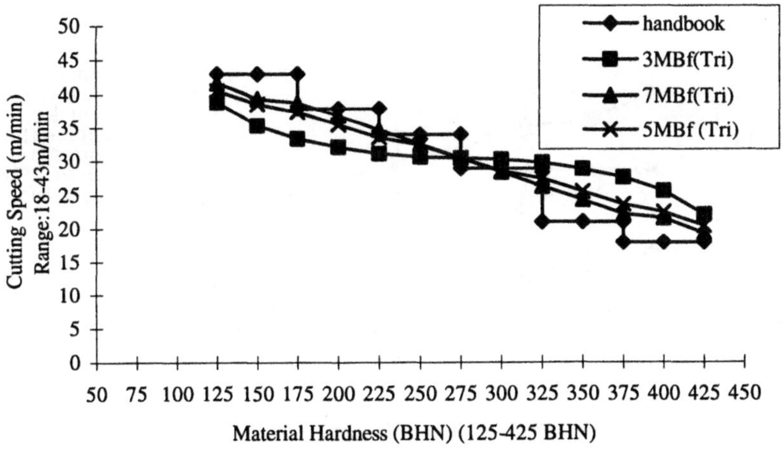

Fig. 12 Fuzzy Cutting Speeds for carbon Steel, wrought medium carbon at 1 mm d.oc using HSS Tool, Handbook, 3,5, and 7 mbf Results

5 Comparisons of Defuzzification Methods

5.1 Comparison of Fuzzy Cutting Results Obtained from Centroid and Mean of Maxima Defuzzification Methods

For the defuzzification of fuzzy results, the centroid defuzzification is used, but to see the effect of defuzzification method on the cutting speed values, the Mean of Maxima defuzzification method is also applied for the same material type and same cutting conditions, as shown in Fig. 13. The cutting speed results obtained by using mean of maxima defuzzification method shows that the starting and ending fuzzy cutting speed values are exactly the same with the handbook cutting speed values. Also as it is seen, as the hardness values are increased the cutting speed values are decreased in between the ranges.

In Fig. 14, the fuzzy cutting speed values obtained by using mean of maxima and centroid defuzzification methods are compared with the handbook cutting speed values. As it is seen from this figure the cutting speed values at the start and end hardness (125 BHN and 325 BHN) points are exactly same with the handbook results and the results obtained by mean of maxima defuzzification method. There is a small difference between the cutting speed results at the start and end hardness points obtained by using the centroid and mean of maxima defuzzification methods. In between the 150 BHN and 300 BHN cutting speed values obtained by handbook, mean of maxima defuzzification method and centroid defuzzification method are not too much differ, they have nearly same cutting speed values.

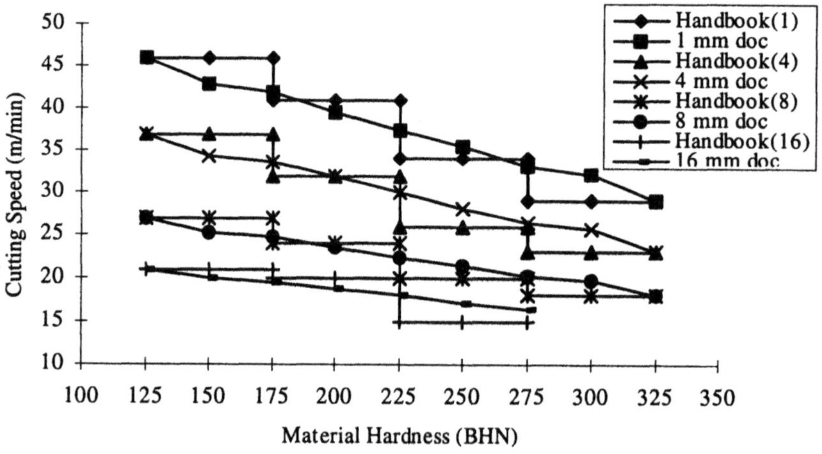

Fig. 13. Fuzzy Cutting Speed for Alloy Steel, Low Carbon at 1 ,4, 8 and 16 mm d.o.c using HSS Tool by using Mean of Maxima Defuzzification Method

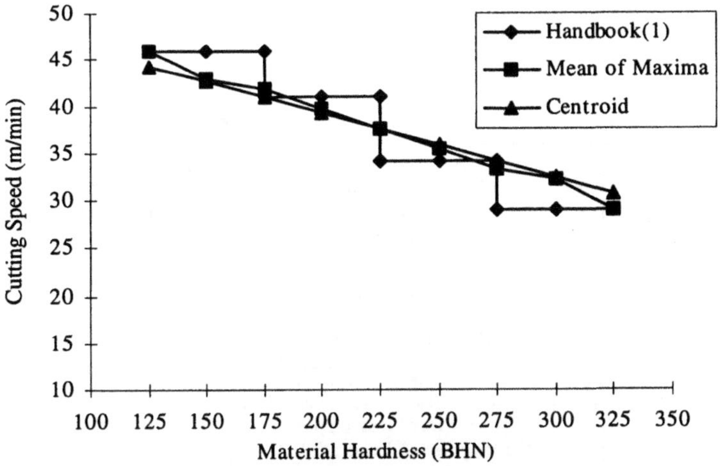

Fig. 14 Fuzzy Cutting Speed for Alloy Steel,Low carbon at 1 mm d.o.c using HSS Tool Comparison between Centroid and Mean of Maxima defuzzification method

6 Fuzzy Logic Based System for Cutting Parameters Selection

The system, called as CAS-CUPFUL (Computer Aided Selection of CUtting Parameters by using FUzzy Logic) has been developed in MATLAB Fuzzy Logic Toolbox and Matlab Programming Language, to facilitate automatic selection of cutting speed for known cutting tool, depth of cut, and hardness of material to be cut. With the entering of material type, the program starts searching the material data from a matrix data file until to obtain the properties of the given material. Next, it takes the material hardness range, cutting speed range and feed rate for the considered material types. From these data, the triangular membership functions for each fuzzy sets, if-then rules, and centroid type defuzzification are evaluated, and the cutting speed is determined.

For this research twelve material types are considered, as Free Machining Steel, Tool Steel, Alloy Steel, High Strength Steel, Malleable Cast Iron etc., for different types of tools as High Speed Steel, Carbide uncoated brazed, Carbide uncoated indexible, Carbide coated tools, and four depths of cut as 1 mm, 4 mm, 8 mm, and 16 mm.

Using the software of Matlab Fuzzy Logic Toolbox, calculations have been carried out to obtain the cutting conditions in terms of cutting speed, feed for different types of cutting tool materials. The system also determines the relationship between the hardness value of the material and the cutting speed.

Simplified flowchart of the developed system is given in Fig. 15. In that system, assigning operations are made from the data file. Also Fuzzy Inference System (FIS) is mentioned in the flowchart, this system is the main module of the Fuzzy Logic Toolbox. All fuzzy operations (fuzzification, fuzzy rules and defuzzification etc.) are performed in that system.

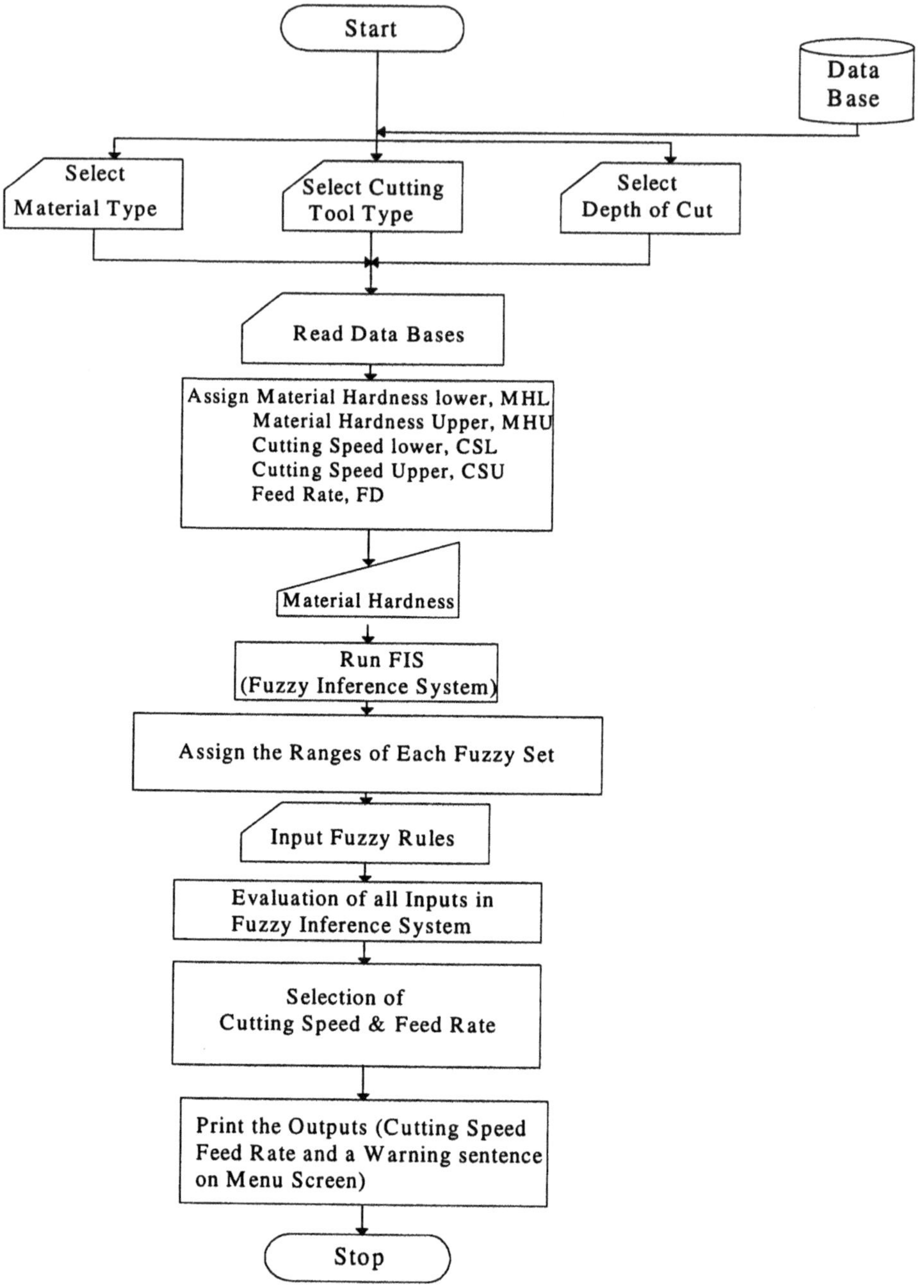

Fig. 15. The simplified flow chart of the developed system.

6.1 Execution of the *CAS-CUPFUL*

CAS-CUPFUL is executed under MATLAB Package Programming Language. When the system is executed, an opening screen including a user-friendly pull-down, listbox, pushbuttons, editable text and static texts menu are immediately displayed on the monitor as shown in Fig. 16.

The opening screen of CAS-CUPFUL has four menus (a listbox, an editable text and two pop-menus), three pushbuttons and four static texts. They can be selected by the use of mouse when the mouse pointer is moved on menu environment (highlighting or pushing on the buttons) and hardness value is given by editing from the keyboard. By pushing first pop-up menu by the mouse, cutting tool types can be easily selected by highlighting, also by pushing another pop-up menu by the mouse; depth of cut value can be easily selected by highlighting. Besides, by using *Your Materials Hardness Range* push-button, hardness range of the selected material can be seen on the static text. *OK* push-button is used for executing the program after all selection is made and entering the hardness value of the selected material from the editable text button. *Close* push-button is used ending of the program and it also close the menu window.

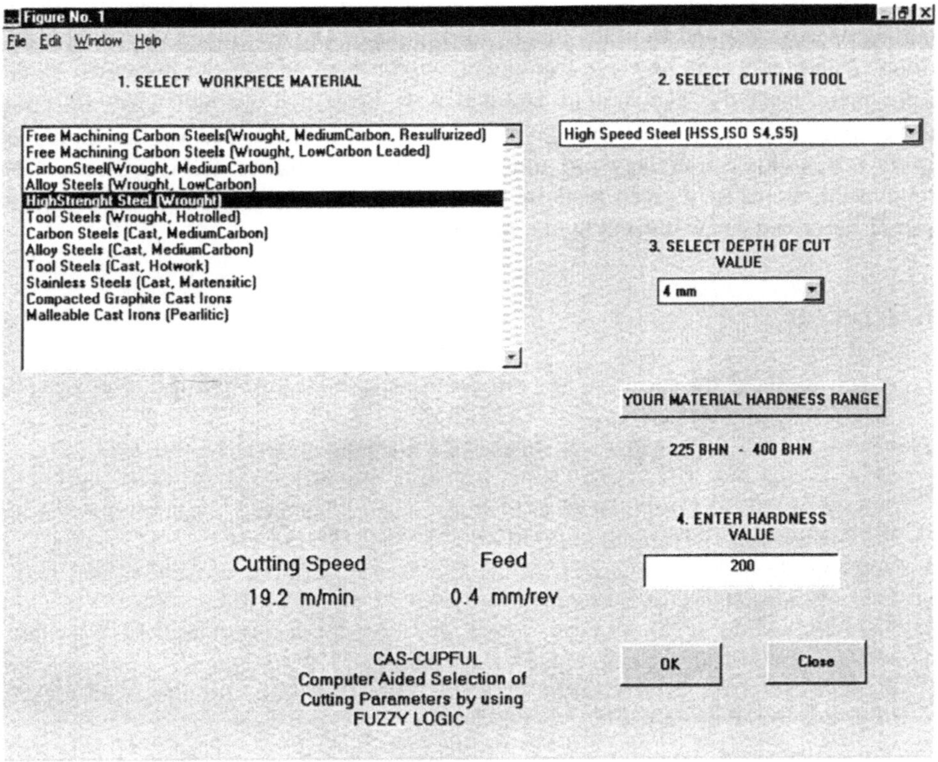

Fig. 16. Menu oriented window of the CAS-CUPFUL on the opening screen

Also, the system menus contain some pull-down menus, which are generated by the MATLAB automatically. These are *File*, *Edit*, *Window* and *Help* menu. But they do not affect the program, since they are independent with the system menu.

7 Conclusion

The scope of this work is the computer-aided selection of cutting parameters by using fuzzy logic. The selection operation has been performed by using fuzzy logic. By means of this system, the handbook's conventional approach is eliminated. Instead of handbook approach, The FL appearance is more flexible, effective and reliable for the solution of complex problems and more tolerant for imprecise data. In application of FL method, one input (material hardness) and one output (cutting speed) were used as in the case of the handbook. Five fuzzy sets for both input and output were used since five fuzzy sets are better or appropriate than less or more than that number. These fuzzy sets were expressed by triangular membership function. Five expert rules were used as an interface between input and output; extracted expert rules imitate the behavior of an expert operator. At the end of these fuzzy operations, centroid defuzzification method is used for obtaining precise value. This method is suggested and its results are more reliable then the others. The results are compared with the cutting speeds obtained from the handbook approach. The developed system is menu oriented and inputs can be given from menu window and outputs are presented on the same menu instantly. The strategy and action of the skilled operator when selecting the cutting speed and feed for a given material can be described by the theory of fuzzy sets, as his/her strategy and action are based on intuition and experience. The relationship between a given material hardness and the recommended cutting speed can be described and evaluated by the theory of fuzzy sets.

References

1. Dereli, T., 1998, "Development of a Process Planning System for Prizmatic Parts", Ph.D. Thesis, University of Gaziantep.
2. Ross, T.J., 1995, Fuzzy Logic with Engineering Applications, McGraw-Hill, USA.
3. Chen, L. and Rao, S.S., 1997, "Manipulation of Uncertainties in the Determination of Optimal Machining Conditions under Multiple Criteria", Journal of Manufacturing Science and Engineering, Transactions of ASME, Vol. 119, pp. 186-192.
4. Yuedong Chen, 1995, "A Fuzzy Expert System for the Design of Machining Operations", Int. J. Machine Tools Manufacturing, Vol. 35, No. 12, pp. 1605-1621.
5. El-Baradie, M.A., 1997, "A Fuzzy Logic Model for Machining Data Selection", Int.J. Machine Tools Manufacturing, Vol. 37, No.9, pp.1353-1372.
6. Machinability Data Center, 1980, Machining Data Handbook, 3rd Edn. Vols. 1& 2, MetCut Research Associates Inc., Cincinati, USA.
7. Von Altrock, C., 1995, Fuzzy Logic & Neuro Fuzzy Applications Explained, Prentice Hall, USA.
8. Zadeh, L.A., 1965, "Fuzzy Sets", Information and Control, Vol. 8, pp. 338-353.

Process Monitoring in Chemical Plants Using Neural Networks

Dr.-Ing. Joachim Neumann, Dr.-Ing. Görge Deerberg, Dr.-Ing. Stefan Schlüter, and Prof. Dr.-Ing. Hans Fahlenkamp

Fraunhofer-Institute for Environmental, Safety and Energy Technology, UMSICHT
Workgroup for dynamic processes
Osterfelder Straße 3, D-46047 Oberhausen, Germany
j.neumann@ct.uni-dortmund.de

Abstract. The suitability of pattern recognition for process monitoring of chemical plants is discussed. Experiments in a miniplant, a pilot plant and simulation studies are carried out. While selecting the required test series of process variables when giving training for neural networks, one tries to use generalized forms of description to illustrate the system in question. It is therefore possible to combine data records originating from various sources. Thus, on the one hand, non-conforming operating conditions have to be simulated in a laboratory or technical system. On the other hand, simulation results might also be used to provide training on neural nets. This combination in the utilized data material permits you to dispense with preparing new physical-chemical models for each data-driven model. The prepared tool is subsequently used as a prototype for hydrogenation in a production system.

Keywords: neural networks, process monitoring, process control system, fault diagnosis, early detection, exothermic reaction, semibatch process

1 Introduction

Intensified international competition forces the chemical and pharmaceutical industry to reorient itself strategically. Lean Management, shorter development times and growing productivity as well as environmental safety factors are decisive for their success. Optimized production and marketing processes and consistent customer orientation are the prerequisites for making one's mark in the marketplace. Controllers and control strategies play a key role in fulfilling marginal conditions for safety and industry. Intelligent components can be used as an addition to an existing process control system here. They are intended to make the assessment of current operating parameters or of those expected in the near future precisely, in order to simplify operator decisions with consequences that can be wide-ranging. Hence it is necessary to develop a process to evaluate current operation conditions, and to link this in the near future with a method for predicting process variables in a second step.

Before the neural based pattern recognition methods can used they require a so called training during which the process behaviour is taught. In certain circumstances operating states and transient phenomenon of non environmentaly acceptable process

B. Reusch (Ed.): Fuzzy Days 2001, LNCS 2206, pp. 871–884, 2001.

courses have to be offered. But it is impossible to bring about this in industrial scaled plants. Therefore the operatability of the prototype of the developed tool must be demonstrated for experimental inaccessible regions via simulation studies, using the in chemical industry widely-used semibatch operational mode. A successful application to a complex hydrogenation process allows the conclusion that the pattern recognition method is transferable to other process types and plants. Fraunhofer UMSICHT develops the described prototype based on neural nets in cooperation with the institute of Safety Research in the Research Center Rossendorf and the Degussa AG in four steps:

1 Feasability study of pattern classification at a lab scale semibatch process
2 Transfer of the procedure from lab scale to pilot scale
3 Transfer to an industrial two-phase process via simulation studies
4 Application of the prototype to pharmaceutical works

By following these steps the pattern recognition methods are used in different plants regarding scale and reaction system. The drafted use of neural nets complement on one another and show a robust way directly to industrial application.

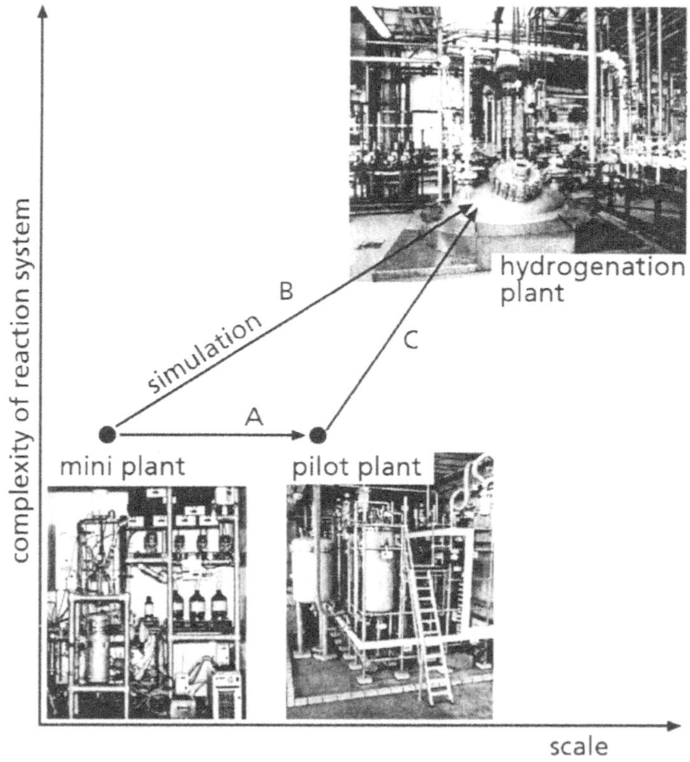

Fig. 1. Strategy to develop an industrial application of the process monitoring system

2 Strategy of Data Management and State Classification

Pattern recognition, for example statistical methods, fuzzy logic, and neural networks, are more and more used for qualitative state classification in technical plants in the form of computer assisted monitoring systems [1]. Fig. 2 shows the general principle of the monitoring system for chemical plants, which is based on neural networks. An artificial neural network consists of a large number of neurons or arithmetic elements that are arranged in a net with an input layer, one or several hidden layers and an output layer. The connection weights determine the response of the network and they must be adjusted for the respective classification task. This is done during a learning phase in which the neural network is trained with process- and plant-data of normal operating states as well as of faulty operating states. Apart from the presentation of the modified data vectors (characteristic feature vectors) to the input layer, the expert knowledge in form of result vectors is presented to the output layer during the training phase. After the training only the inputs are available to the application of the system. On this basis the neural network provides the classification.

In terms of geometry the training corresponds to subdividing the generally high dimensional characteristic space in a way that all feature vectors which represent a condition class exclusively fall in one of these subspaces. The quality of the classification result depends on the selection of the network type, the optimal topology of the network and on using qualified, class specific characteristics. As input values for the neural network measurement results (19 measurement points) are used that are directly extracted from the process. 17 different classes of operation states are distinguished such as *dosing error*, *cooling-breakdown*, *violent release of heat (exothermal)* [2]. This presetting is binary. But the prognosis of the neural net is allowed to be a floating point value for each class.

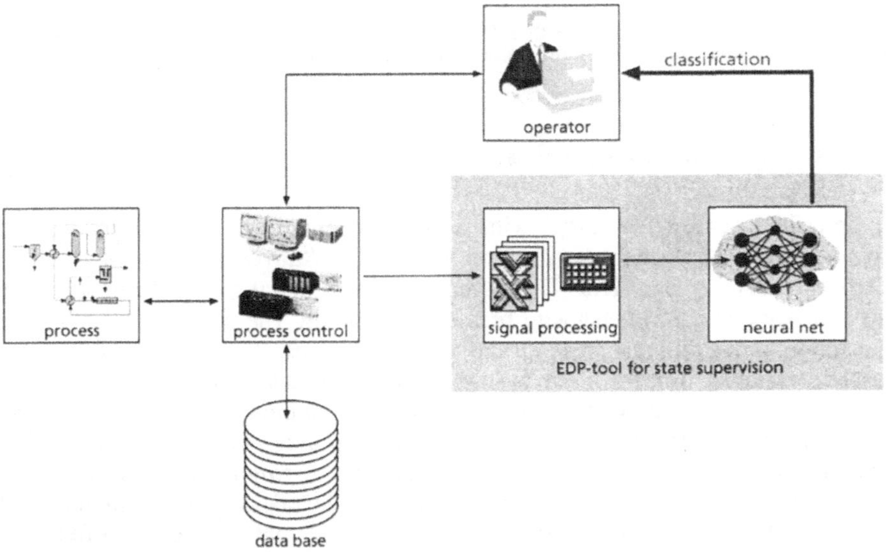

Fig. 2. Application of a neural network trained with an large dataset (off-line). Preparation of classification-results of the network during the operation of the plant by the process control system.

The advantage of the data-driven structural principle is that only less »inside« knowledge about the observed process has to be known. Only observations and experiences are solely trained and recognized and assigned even in complex patterns by the recognition tool. But also it is necessary to offer conditions and transients of undesirable process conditions so that the system can reliably identify these conditions and transients later on. Such conditions might be hard to bring about with a real production process. Therefore process simulation (Fig. 3) is used for the validation of the pattern recognition system before the neural pattern recognition process is used in the real plant for the first time. The simulation allows the intercourse with disorders and therefore without any risk. Apart from that important boundary situations can be tried out and evaluated reliably [3].

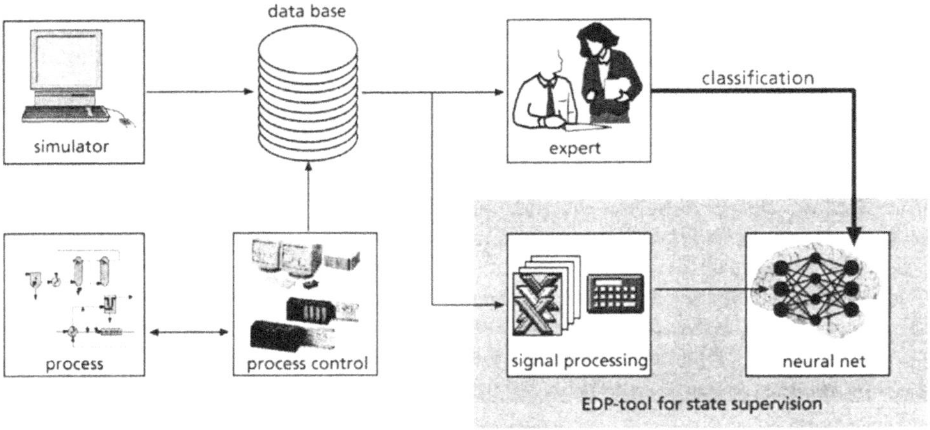

Fig. 3. Teaching phase using *simulated* process data instead of measured data. The results (of classification) can be visualized by the process control system.

3 Investigations at Lab Scale

The described strategy is tested with a reference process. The continuous with Sulphuric Acid catalyzed esterification of Acetic Anhydride with Methanol to Methylacetate and Acetic Acid [7] was chosen. The essential part of the experimental plant is a reaction calorimeter with pressurized reactor. The plant includes a catch-tank, scales and additional heating and control devices. Time-courses of selected measured variables for the used system of medium and reaction by impression disorders are determined by the help of experiments in a miniplant (lab scale). At first the experiment is tested with the process simulator to avoid the damage of the plant. There are scenarios like total *cooling-breakdown*, *stirring-breakdown* or *missing dosage of an educt*. The dynamical simulation takes the engineering components into account as well as all measurement points and the control circuit. Differential-algebraic systems originate from disposing of bland sheets of material and energy for the process components. They are combined in a system of equations and resolved simultaneously by the help of numerical methods. The process simulator is verified by experimental test series to guarantee the accuracy of the prognosis.

The measured variables of the specific perturbed operational mode (semibatch and continuos stirred tank reactor), which result of simulation studies (4) and experimental investigations (1), are combined in a database (6). An allocation of measurements to the different operating states results from the off-line classification (5). Every possible state of the process can be characterized by one or more classes. The aim of the investigations was to demonstrate and to evaluate the capability of different neural nets (2) concerning state assessment (3) based on current process data of the real plant. Thought has been given to various boundary conditions like operational mode and available data for training and test. The best results have been achieved in case of feedforward nets with the RPROP-algorithm [5].

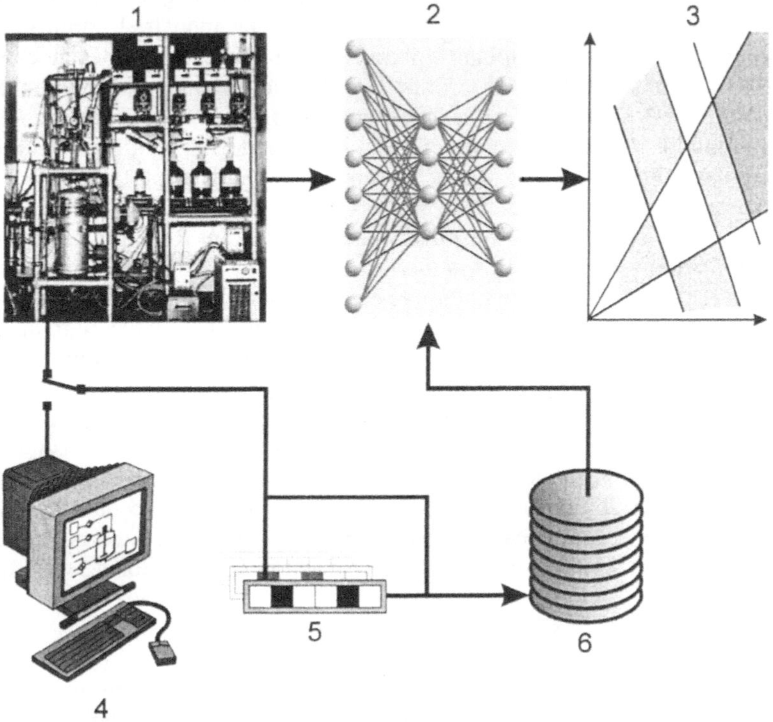

Fig. 4. Field-tested strategy to use experimental and simulation data together in order to train neural nets.

The operability of the methodology has been demonstrated on the basis of the neural network to the identification as directed and non-directed operation of condition. The results with the neural network are almost identical with the experts-presetting. They correctly reproduce the physical-chemical context and thus confirm the potential of neural networks in this domain. A neural network trained with simulations of critical operating modes and with normal operating modes in the real environment can even recognize critical operating modes in the real environment taught only by simulation. The network can be pre-trained with simulations before it is used for diagnosis in the real plant. Critical operating modes do not have to be generated in the real environment to use this technology in a successful way. Based on the simple type of network respectively construction the necessary formulas of calculation can be easily transformed in a program-module. Then they can be integrated in available process control systems. The neural network shows a good behavior of real-time at the forecast by reason of its tight construction. For an acceptable accuracy to the prognosis some hundred training cycles are necessary. This is dependent on the algorithm of training.

4 Application at Pilot Plant

The scale-up from lab scale to pilot plant scale is a major supposition for the acceptance of the pattern recognition methods and the industrial application of this technology. The use of miniplants to train neural nets has to be checked before utilized in a large scale. One sub-ordinate target is to provide a prototype that is easy to apply to the pilot plant. Thus, the substantial awareness of the measurement points in the production plant and the structure of the monitoring tool can be obtained. In this first developement stage the scale-up to the pilot plant takes place without changing the reaction system and the process type.

After the final shutdown of the FIRES plant at the Joint Research Center in Ispra / Italy in 1996 the pilot plant at Fraunhofer UMSICHT is one of the few explosion proofed R&D installations in which investigations with heated, reactive systems can be carried out. The pilot plant (reactor volume 65 litres) is equipped with an industrial process control system to operate and observe the process. The implementation of the operator assistance system can take place very practically [6].

Overall 100 experiments of the disturbed and undisturbed semibatch process in the pilot plant are carried out. The main focus was directed to the isoperibol and isothermal operational mode. Methanol and the catalyst were placed in the vessel. The Acetic Anhydride was dosed into the vessel during a period of up to two hours. Due to the released heat of reaction the temperature rose up to a certain value which was hold up to the end of dosing respectively reaction after the dosing was started.

4.1 Specification with Characteristic Numbers

In order to get a compressed and most possible universal description of the process data, a dimensionless representation was chosen. The used formulation goes back to the off-line safety view of semibatch processes [7]. In the following this method is enhanced to the monitoring of running processes. The generation of characteristic numbers is a kind of preprocessing and is based on an energy balance around the reactor and the cooling circuit. The usage of heat balances is obvious because the process control systems of every plant register temperatures, mass flux an pressure signals. But the measuring equipment to make an on-line analysis (FTIR) is too expensive and is mostly not installed at production plants. After installation of this preprocessing the undisturbed semibatch processes can be traced in a so called phase plot (Fig. 5). Deviations of the expected process course can be identified and located. The results of this strategy are quite good and show the suitability for use in chemical engineering in a pilot plant scale (Fig. 6).

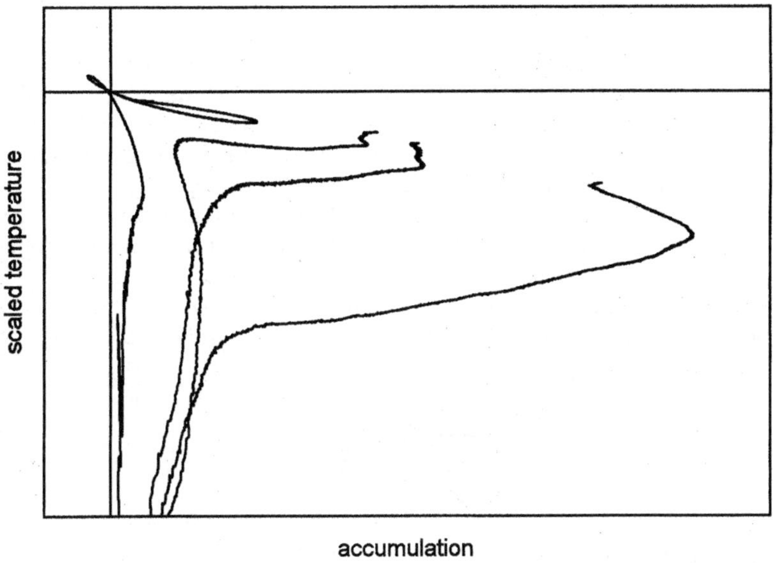

Fig. 5. Family of characteristics of undisturbed semibatch processes. The transformation in the non-dimensional form occurs on-line during the running process.

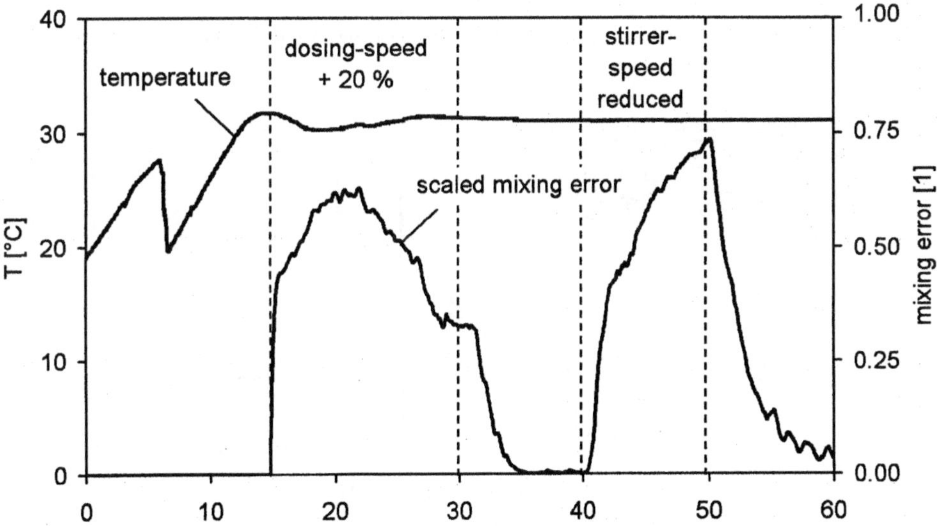

Fig. 6. Inverse transformation into measurement space allows to visualize the measurements and the monitoring result. In this figure the reactor temperature together with a normalized mixing fault is shown. The diagrammed semibatch process includes two faults which were mapped to the normalized mixing error.

4.2 Time-Series Prediction with RBF

Numerous experiments on a technical scale were carried out under modified conditions to create a database for off-line training for neural nets. Neural network training with RBF nets has been carried out and successfully tested in the second step. During the preparation of data material and the combination into the training set a sliding window of five measurement samples come out on top (Fig. 7).

The dimensionless process time, the accumulation and temperature with five time-dependent values are combined to the net input. For classification and prognosis the distance between the sampling points is not equal but rather shorter up to the current value. In order to value the time-series prediction training sets and test sets were created out of real world data with a prediction horizon of 10% of the process time.

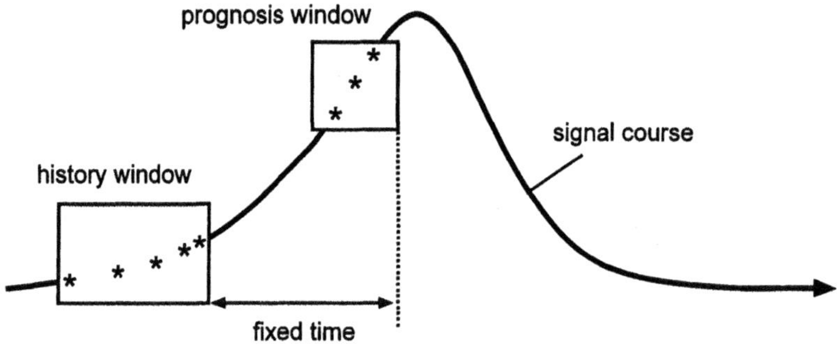

Fig. 7. Sliding window technique for time series prediction.

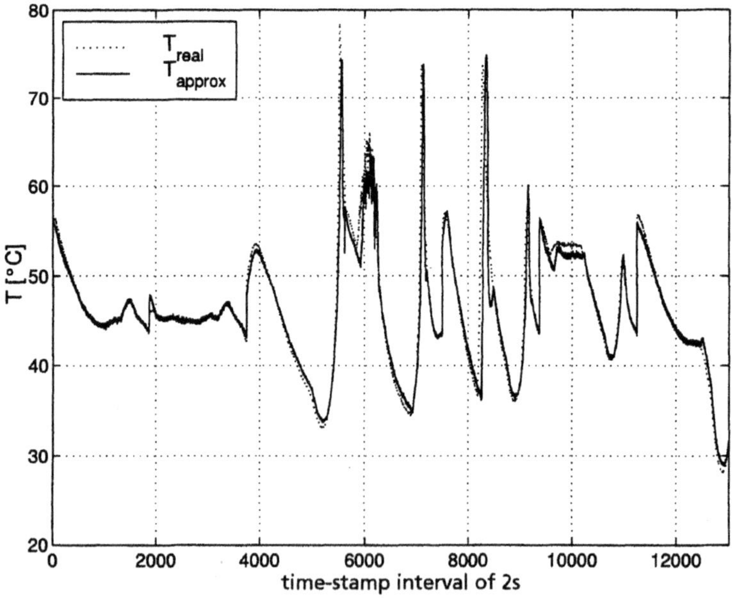

Fig. 8. Temperature prognosis for diverse pertubed process courses of the pilot plant operating. The prognosis horizon of the RBF net with 50 neurons is 3 min.

This demonstrated that extremely accurate predictions for process variables are possible within a technical operation, if the prognosis horizon does not exceed a maximum of 6% of the process time. The neural net based filter for interpreting the actual measurements were successful tested at the running pilot plant. The monitoring tool was integrated to the process control system which was necessary to prepare it for an industrial application.

4.3 Self Organizing Maps

The Kohonen-Maps also well known as Self Organizing Maps (SOM) were primarily developed to visualize high dimensional data spaces. Today SOMs are used for data preparation, modeling and classification [8]. In order to improve the interconnections between measured variables and to detect commanding measurement points the SOM can be of great use. This paper follows the two ideas:

1. The recorded measurement data of numerous succeeded semibatch experiments solely are normalized utilized to apply to a SOM. Since close-by points in the entry will be close-by points at the map, with temporal progression of the process a characteristical track can be kept. Discrepancies in operation are shown as traces leaving the general pathway.
2. In addition to the first method a classification of the operating states is undertaken manual and the data series of deranged process-sequences are called in the SOM-training. According to the issue of the feature map the neurones are compared solitarily with the data-points. A neuron is attributed by class-characteristic, to which the majority of the collated data-points belongs to.

Summarizing the second method a total-units-chart *(U(nit)-matrix)* is sketched, on which is registered the term of the event category assigned to the neurones instead of the potential-palette. In the case that classification upholds topology, at the chart insular regions for every class can be observed. A recently measured data-vector is assigned to that class containing the most contiguous neuron. For preparation of an attribute-chart according to method number one measurement data are standardized linear. Only the data of the reference-runnings are utilized for calculation of SOM. Afterwards each experiment is registered as a trajectory in a diagram. The figures show the unit-matrix (U-Matrix), which mirrors the vicinity-relations as well, containing the neurones, represented as hexagonal arrays. The gradation of the neuron-colour represents a kind of potential-relation: dark sectors graph a culmination of data, bright sectors reflect their boundaries. Trajectories of the reference-experiments do not process across bright regions.

Only at undisturbed experiments the intervals between the marks are well isochronous. Suddenly occuring events, such as the breakdown (complete failure) of the reactor-stirrer, indicate itself firstly vague in an irregular foreward- and backward-moving upon a curve. At the wear off-phase of the disturbance the intervals between the marks differ very strong. Overlappings of the trajectory (cooling- and dosing-breaks) take place. Comparison of the different scaling-methods reveals that the topology-error at logarithmic scaling is most inferior. This depicts the best separation of the classes. The advantage in using of selforganizing feature-charts rests on the simplicity of their use at technical processes. Practically the crude data can be taken

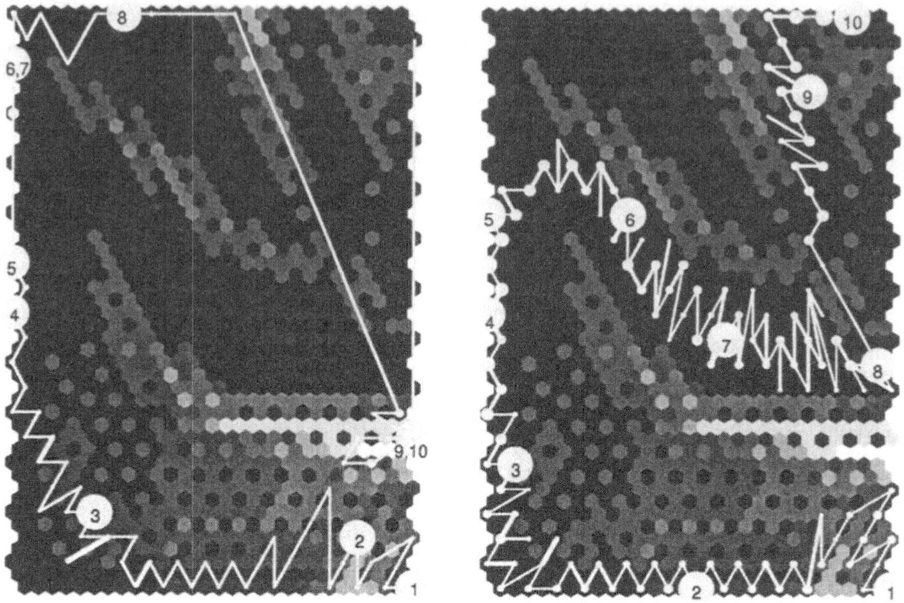

Fig. 9. In this figure the trajectories of two semibatch processes are shown (left experiments on reference (left: reference, right: disturbed process) . In addition to the trajectories there are numbers from 1 to 10 in the feature map to mark the process time equally spaced.

over for calculation of the chart, since necessary standardization can be executed automatically by the software. The lowest topology-error turns the balance towards one of the two forms of standardization. Utilizing the second method at success a correlation between the classes and the incidental data may be made, which can be analysed in both directions: As well as the membership of an unclassified data-vector to a class may be red off the range of values of unique measurands can be indicated depending on the investigated condition-class. However for success the technique's infirmity is the essential condition, that the classes must be separated from each other in fact. They are not allowed to overlap heavily. Hence raised demands are formed at the process-regulation and the formulation of incident-classes. Oscillations around the operating-point excited by regulative interventions, which melt the boundary between the expected and non-expected range. This is a result of the cause-oriented classification in opposition to an effect-oriented formulation.

5 Application at Hydrogenation Plant

In another workstep the application area of the monitoring-tools is extended by dynamic simulation upon a two-phase process in production-scale. Today two-phase driven semibatch-processes, containing gas phase and liquid phase, are mostly used in the chemical industry. For example hydrogenations belong to these processes. Among to the very exothermic reactions proceeding in the liquid phase also the influence of the phase equilibrium and the transport of the substance between gas phase and liquid phase is relevant. A reaction-system, difficult to predict and strongly non-linear, results of the exothermic character of the reaction, the complex transport-processes of the substance, and the effects of the phase equilibrium. By using a model of a production-plant (Fig. 11), validated with real process-data, simulation runs for training and testing of the neural networks may be realized.

Fig. 10. U-matrix of the simulation runs shown as a class membership of the neurons.

As several events may be studied in a detailed way with passable expense, a wide understanding regarding to the consequences of potential failures and the quality of the classification-events is expected of the plant-operator (Fig. 12).

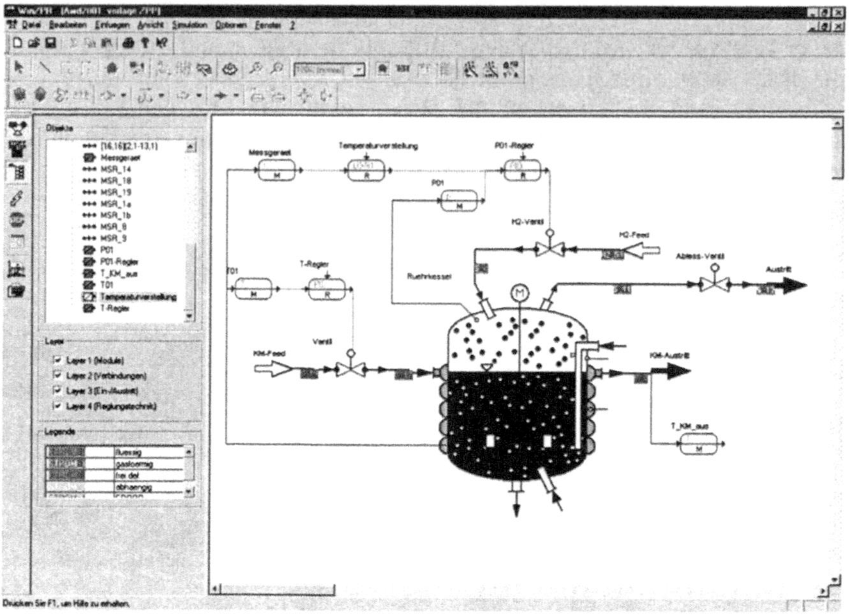

Fig. 11. Simulation of Hydrogenation Process using WinZPR

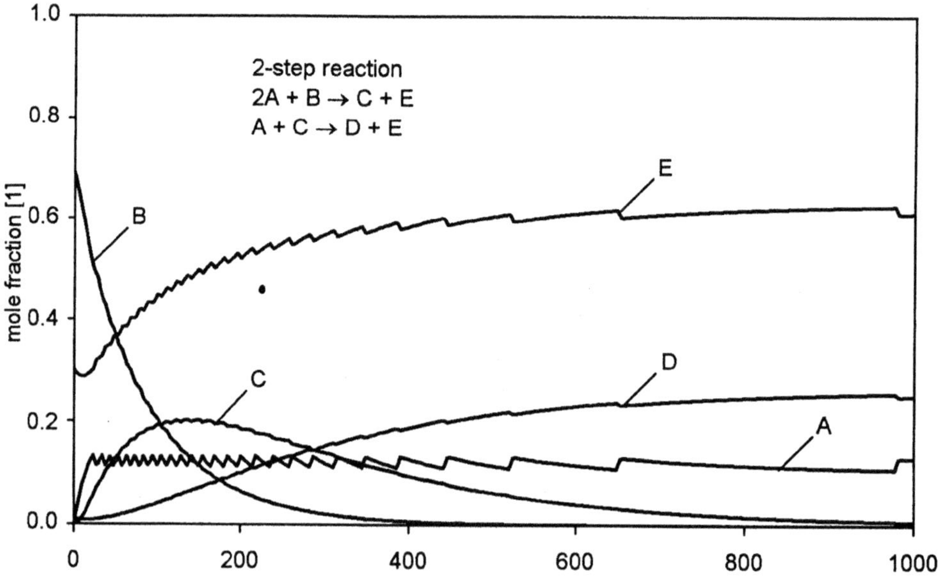

Fig. 12. Exemplarily results of an typical hydrogenation simulation run using WinZPR.

As last workstep the prototype of the process monitoring tool will be utilized in a pharmaceutical plant of the Degussa AG in order to apply it during the running production-process. The declarative-ability and the essential expense for pattern recognition with neural nets for process-monitoring shall be valuated. In addition methods and tool shall be developed which allow an ergonomic and intuitive use of this detection-system. It will be reached by integration of the monitoring-function into the look-and-feel of the process control system (SIMATIC PCS-7). The development of the prototype of equipment condition monitoring system basing on neural networks is the target of this step. This technique, tested in over day operation in a production-plant, shall be upgraded to a tool for operator-support available for the chemical industry.

6 Final Remarks

By the use of the multiple-stage strategy, the monitoring based on neural networks becomes applied to different plants regarding the scale of the plant and the reactive system. Solutions of subtasks at the application of neural networks for monitoring plants complete each other and show a safe way towards industrial use.

During the running process neural networks yield parameters, which are not determinable by measurement or only rudimentary determinable by samples, and thus are available only at off-line mode (model-based measuring). Processes, difficult to observe, can be optimized concerning the conduction-strategy, regarding the quality, and costs by data, made available by neural networks. By further development and transmission of this method to real, multiphase processes in a chemical plant prerequisites become created, for optimal realization of complex chemical processes knowing the parameter limits. The developed tool is applicable for batch- or semibatch processes. Therefore it becomes exemplarily installed and tested at the production process. It shall become utilizeable everywhere for small and medium chemical production.

Acknowledgement

The authors gratefully acknowledge the financial support for this work granted by the Bundesministerium für Bildung und Forschung (bmb+f) under project no. 01RV9801/1. The authors thank all members of the project team for the close collaboration.

References

[1] Fan, J.Y.; Nikolaou, M.; White, R.E.: *An Approach to Fault Diagnosis of Chemical Processes via Neural Networks*, AIChE 39 (1993) 1, pp. 82-88

[2] Neumann, J.; Deerberg, G.; Schlüter, S.: *Früherkennung sicherheitsrelevanter Betriebszustände in Chemieanlagen mit neuronalen Netzen*, TÜ 39 (1998) 10, Springer-VDI-Verlag, pp. 55-58

[3] Neumann, J.: *Zur Früherkennung sicherheitsrelevanter Betriebszustände in Chemieanlagen mit neuronalen Netzen*, PhD thesis University of Dortmund, UMSICHT-Schriftenreihe Band 11, Fraunhofer IRB Verlag, Stuttgart, 1999

[4] Wiss, J.; Stoessel, F.; Killé, G: *Determination of Heats of Reaction under Refluxing Conditions*, Chimia 44, 401-405, 1990

[5] Riedmiller, M.; Braun, H.: *A Direct Adaptive Method for Faster Backpropagation Learning: The RPROP Algorithm*, IEEE 1993, pp. 586-591

[6] Hoff, J.; Neumann, J.; Deerberg, D.; Fahlenkamp, H.; Steiff, A.: *Pilot Plant Tests of Pressure Relief with High Viscosity, Reactive and Non-Reactive Systems*, Chem. Eng. Technol. 24 (2001) in Print

[7] Steinbach, J.: *Safety assessment for chemical processes*, Wiley-VCH, Weinheim, 1999

[8] Kohonen, T.: *Self-organized formation of topologically correct feature maps;* Biological Cybernetics, 43, 59-69, 1982

[6a] Hoff, A.; Neumann, J.; Deerberg, G.; Steiff, A.: *Untersuchungen zur Druckentlastung siedend aufwallender, reaktiver und nichtreaktiver, viskoser Stoffsysteme*, Chem.-Ing.-Tech. 2001, 73, No. 5, 542-546

Fuzzy Probability and Power System Safety

Mariana Dumitrescu and Toader Munteanu

"Dunarea de Jos" Galati University, Electrical Engineering Department
Domneasca Street-111, Galati-6200, Romania, Fax 4.036.460182
mariana.dumitrescu@ugal.ro

Abstract. The paper proposes a fuzzy event-tree analysis of electric power protection system reliability. The elaborated analysis proposes a new algorithm used for fuzzy event-tree calculation. On this purpose a fuzzy logic system (FLS) was elaborated. The proposed SLF has got, as an output element, the electric-power protection system "Safety" instead of the usually "Risk" parameter, used in engineering applications (with, or without fuzzy logic elements). In the paper the FLS output is used for independent analyzing of an electric power protection system. But it is elaborated for a complex analysis including the power system connected with its protection system. An efficient software toll FETA (" Fuzzy Event-Tree Analysis"), for independent analyzing of an electric power protection system "Safety", was built. It is able to compute "General Safety" fuzzy conclusion about the analyzed system and also the General Safety Degree index.

1 Introduction

Fuzzy logic provides an intuitively appealing way of handling the uncertainty, by treating the probability of failure as a fuzzy number. This allows the analyst to specify a range of values with an associated possibility distribution for the failure probabilities. If we associate a triangular membership function with the interval, we assume that the analyst has "more confidence" that the actual parameter lies near the center of the interval than at the edges. Using fuzzy numbers for the verbal statements of an event probability means fuzzy probability event evaluation [5], [8], [2].

A first direction in safety analysis [5] uses fuzzy-set logic to account for imprecision and uncertainly in data while employing this analysis. This means: 1) uncertainty in the probability of failure, and 2) verbal statements for the probabilities and consequences such as low, moderate and high for the impact of certain sequences of events such as normal, alert and abnormal. In this case, for a simple 2-component parallel system, let $\tilde{P}_A$ ("low") and $\tilde{P}_B$ ("high") be the component fuzzy failure probability. Then $\tilde{P}_{sys}$ will be the fuzzy probability $\tilde{P}_A \times \tilde{P}_B$ illustrated in fig.1a.

A second direction in safety analysis [5] uses a linguistic variable to evaluate the fuzzy failure probability. The linguistic values that can be assigned to the variable, called its "term sets", are generally defined as fuzzy sets that act as restrictions on the base values that it represents. Each of these fuzzy sets means a possibility distribution over the domain of the base variable. For example (fig. 1.b) "probability of failure" might be a linguistic variable whose values are from the set "remote" ("r"), "low"

B. Reusch (Ed.): Fuzzy Days 2001, LNCS 2206, pp. 885–889, 2001.

("l"),"high" ("h") and very high ("vh"). Fuzzy failure probability of the simple 2-component parallel system $\tilde{P}_{sys}$ will be the fuzzy probability $\tilde{P}_A \times \tilde{P}_B$ illustrated in fig.1.b [6].

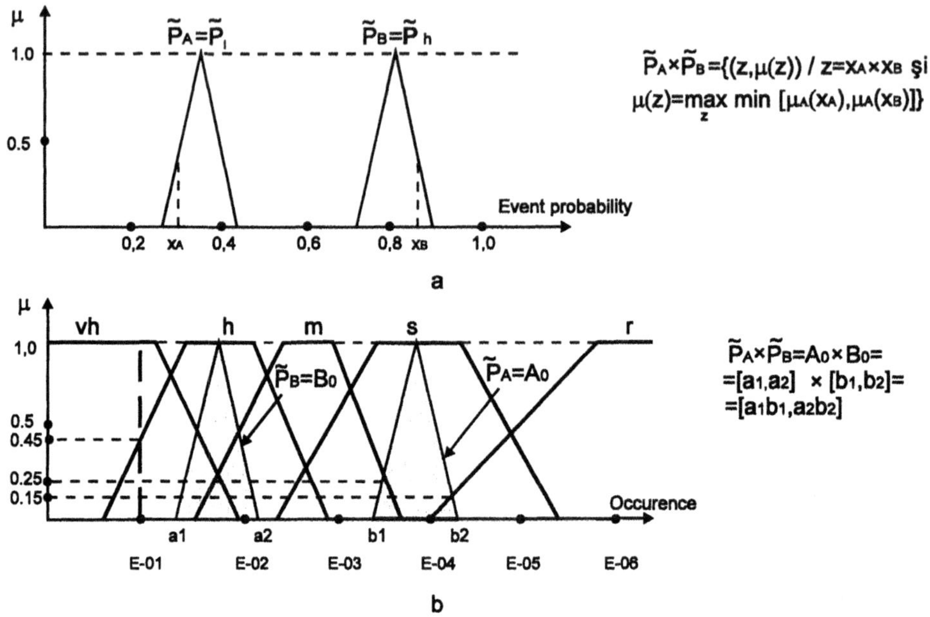

Fig. 1. Fuzzy failure probability evaluation: verbal statement (a), linguistic variable (b).

The paper uses the linguistic variable in fuzzy failure probability evaluation, because this concept is coupled with the possibility theory, which becomes an especially powerful tool for working with uncertainly.

2 A Fuzzy Fault-Tree Analysis for Electric Power Protection System

Event-trees examine sequences of events and their probability of occurrence. They start with some initiating event (say a failure of some kind) and then develop the possible sequences of events into a tree. For example; is the failure detected?, does a safety relay activate? etc. The result at the end of each chain of events is then determined and the probability of each result calculated from the probabilities of the events in sequence leading to it.

The procedure for analyzing a fault tree is precise, but the probabilities on which the methodology is based often are not [2], [1], [7]. So the fuzzy probabilities are used in the analysis of an event- tree, for the electric- power protection system.

Usually the fuzzy event-tree analysis, has the following steps:

- fuzzy "failure" probability and fuzzy "function" probability evaluation, for all reliability block diagram elements;
- fuzzy "occurrence" probability for each path (sequence of events) of the tree;
- fuzzy " consequence" on power system evaluation, after the events sequence achievement;
- fuzzy "risk" on power system for each path of the tree evaluation, depending on the path "occurrence" and path " consequence";
- the tree-paths hierarchy establishing, depending on the path " risk".

The fuzzy probability of an event can be put into following subcategorizes based on the range of probability:

- very high probability: $P_{vh} > 0.9$;
- high probability: $0.8 < P_h < 0.9$;
- moderate probability: $0.4 < P_m < 0.8$;
- low probability: $0.1 < P_l < 0.4$;
- very low probability: $0.0 < P_{vl} < 0.1$.

The fuzzy probability of an event can be evaluated in one of the forms expressed in fig. 1. This paper uses the form from fig. 1b

In order to see which one of the outcome has the highest possibility some applications rank the outcomes on the basis of the maximum probabilities associated with outcomes, or on the basis of the probabilities having maximum grade of membership in the fuzzy probabilities. But both of these approaches may lead to improper ranking of the outcomes. The proper approach of the different outcomes is to consider both the maximum probability, associated with various outcomes and the grad of membership of the rating [5].

Let the fuzzy probability of each outcome be $\tilde{P}_i$. The possible outcome fuzzy set is noted $\tilde{C}_i$. The fuzzy risk of each outcome is;

$$\tilde{R}_i = \tilde{P}_i \times \tilde{C}_i \ , i = \overline{1,n} \qquad (1)$$

The tree - paths ranking evaluation is not enough for the electric - power protection system reliability calculation. A methodology to compute a quantitative index "General Safety Degree" (GSFD), for this kind of system, was developed in [3]. For electric power protection system safety evaluation the next algorithm is proposed (fig.2):

1. linguistic variables for the fuzzy logic system (FLS) parameters construction;
2. FLS inputs "Occurrence" (OC) and "Severity" (SV) evaluation;
3. FLS rule base evaluation and FLS outputs tree - paths " safety" (SF) evaluation;
4. fuzzy general conclusion "General Safety" (GSF), for all tree - paths, evaluation;
5. GSF defuzzification and "General Safety Degree" (GSFD) crisp value calculation.
6. FLS rule base proposal;

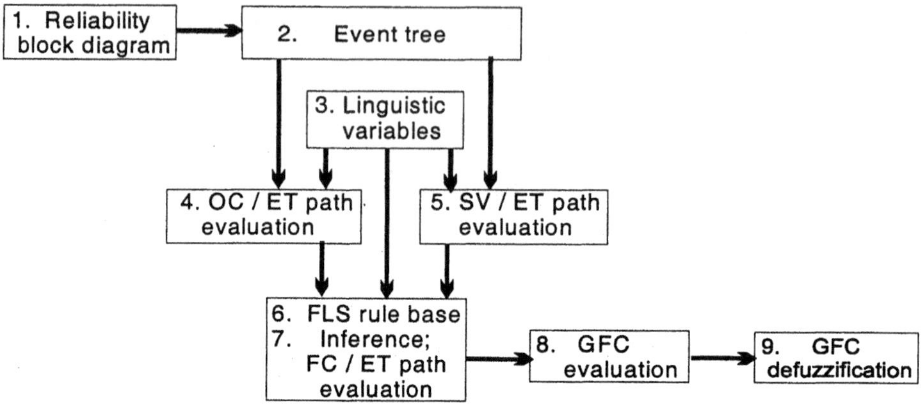

Fig. 2. The steps of the fuzzy event tree analysis.

3 Results

For the fuzzy event- tree analysis of an electric- power protection system, an adequate software tool FETA (Fuzzy Event-Tree Analysis) was created. For an electric generator (EG) protection system and an electric transformer (ET) protection system, the GSFD index presented in fig.3 are computed. The EG-protection system is sensible to the following failure types: low tension level (LT), negative power flow (NPC), external short-circuits (ESC), over-voltage (OV). The ET -protection system is sensible to the following failure types: external short-circuits (ESC), internal short-circuits (ISC) and over-voltage (OV).

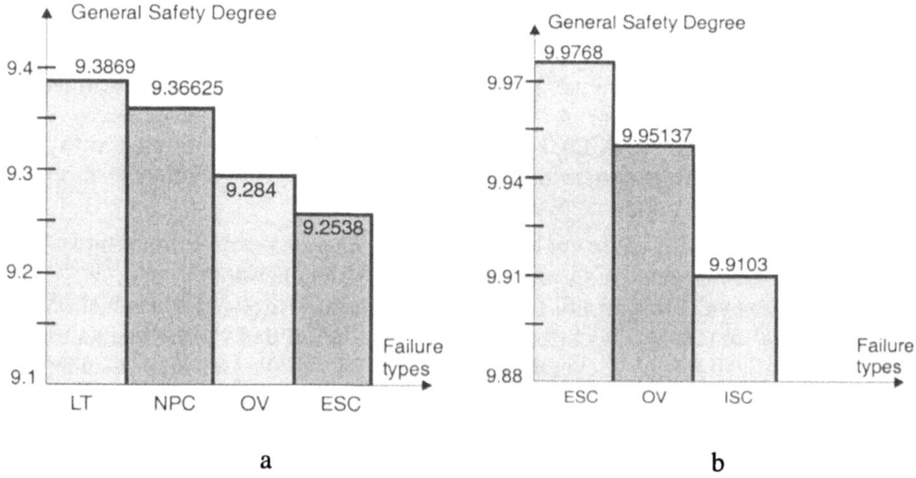

Fig. 3. GSFD index for an electric generator-protection system (a) and for an electric transformer-protection system (b).

4 Conclusion

Using fuzzy probabilities an adequate fuzzy event- tree analysis of electric power protection system reliability, was proposed. An algorithm, using a fuzzy logic system was elaborated. An efficient software toll FETA (" Fuzzy Event-Tree Analysis"), for independent analyzing of an electric power protection system "Safety", was built. It is able to compute " General Safety" fuzzy conclusion about the analyzed system and also the General Safety Degree index.

The proposed SLF uses as an output element the electric- power protection system "Safety" instead of the usually "Risk" parameter, used in engineering applications (with, or without fuzzy logic elements).

References

1. Bastani, F.B. and I.R. Chen: Reliability of Systems with Fuzzy-Failure Criterion. Proc.Ann. Reliability and Maintainability Symp (1994) 265-270.
2. Bowles, J.B. and C.E. Peláez: Application of Fuzzy Logic to Reliability Engineering. Proceedings of the IEEE, 3 (1995) 99-107.
3. Dumitrescu, M. and A.P. Ulmeanu: A fuzzy algorithm for system safety evaluation. Second International conference on Mathematical methods in reliability, MMR'2000, Bordeaux, France (2000) 253-256.
4. Kenarangui, R: Verbal rating of alternative sites using multiple criteria weights. Trans. ANS, 33 (1979) 617-619
5. Kenarangui, R.: Event - tree Analysis by Fuzzy Probability. IEEE Trans. on Reliability, 1 (1991) 45-52
6. Kumar, A. and R.K. Ragade: An Extended Reliability Evaluation Framework for Computer Systems Using Fuzzy Logic. Proc. Ann. Reliability and Maintainability Symp (1994) 317-323.
7. Mendel, J.M. *Fuzzy Logic Systems for Engineering: A Tutorial.* Proceedings of the IEEE, 3 (1995) 390-295
8. Weber, D.P.: Fuzzy Logic for Risk Analysis. Proc. Ann. Reliability and Maintainability Symp, . (1994) 253-259

A Fuzzy Logic System Used in Safety Analysis

Mariana Dumitrescu and Toader Munteanu

"Dunarea de Jos" Galati University, Electrical Engineering Department
Domneasca Street-111, Galati-6200, Romania, Fax 4.036.460182
mariana.dumitrescu@ugal.ro

Abstract. The paper focus on the methodology to elaborate a fuzzy logic system (FLS) used in safety analysis of an electric power protection system. The in FLS parameters Occurrence and Severity and the out FLS parameter Safety are proposed. Also an adequate rule based was elaborated. For safety analysis is used the event-tree method. To elaborate the event tree and to use the proposed FLS on each part of the tree, a software tool "Fuzzy event-tree analysis" (FETA) was created. We can obtain a "Safety" fuzzy set on each path of the tree. All the evaluated "Safety" fuzzy sets are introduced in an algorithm to elaborate the "General Safety" of the system.

1 Introduction

Fuzzy - set logic is used to account for imprecision and uncertainty in data while employing a safety analysis. The fuzzy event-tree allows the use of verbal statement for the probabilities and consequences, such as very high, moderate and low probability [5], [2]. An original fuzzy logic- system enables us to analyse the qualitative evaluation of the event -tree. The technique allows us to develop a fuzzy-event algorithm and to gain quantitative results, as "General Safety Degree" associates to all the paths in the tree. The paper focuses on the proposed fuzzy logic system. The application of fuzzy event-tree algorithm is further demonstrated by using an electric power protection system, to assess the viability of the method in complex situation.

2 A Fuzzy Logic System for the Event-Tree Analysis

The event-tree, associated to the analysed electric power protection system, allows us the system reliability computing. Event-trees generally examine sequences of events and their probability of occurrence.

The probabilities used in an event-tree analysis are not precise [1], [2], [6]. So we use the fuzzy probabilities in the analysis of an event- tree, for the electric power protection system safety evaluation. An adequate fuzzy logic system (FLS), presented in fig.1, is proposed.

The elaboration of the FLS needs the next steps:

- in and out FLS parameters was established;
- in and out membership functions connected to the FLS parameters was developed;
- the FLS rules were proposed.

B. Reusch (Ed.): Fuzzy Days 2001, LNCS 2206, pp. 890–894, 2001.

The FLS in parameters are the OCCURRENCE of the path tree event, the DETECTABILITY of the path tree event and the SEVERITY of the path tree event. The FLS out parameter is SAFETY of the analysed system according to the path tree event.

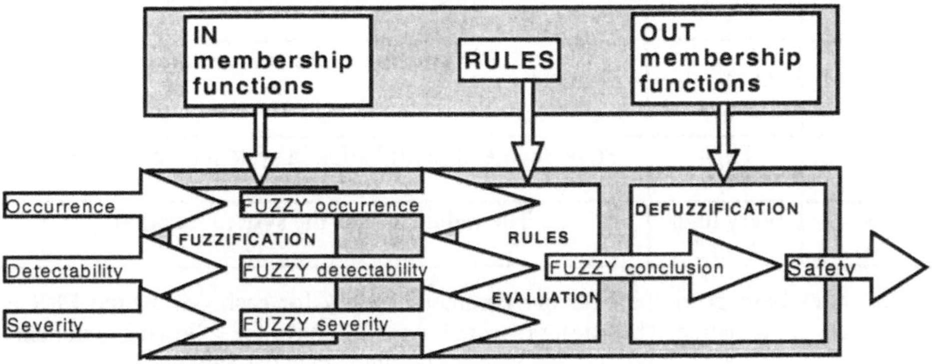

Fig. 1. FLS associated to the fuzzy event tree analysis

The fuzzification process of FLS in parameters uses the linguistic variables. These are elaborated with the proposed FLS in membership functions. The deffuzification process uses a linguistic variable elaborated with the proposed FLS out membership function.

How the Occurrence linguistic variable and the Severity linguistic variable were elaborated is presented in table 1, respectively table 2. The Occurrence uses five fuzzy sets: very low (vl), low (l), moderate (m), high (h) and very high (vh), developed according to the failure probability.

Table 1. Occurrence FLS in parameter elaboration

Rank	Occurrence	Effects	Failure probability
1	Very low " vl "	Failure not possible	< 1 of 10^5
2 3	Low " l "	Low failure rate	1 of10000 1 of 4000
4 5 6	Moderate " m "	Seldom failures	1 of 1000 1 of 80
7 8	High " h "	Usual failures	1 of 40 1 of 20
9 10	Very high " vh "	Very possible failure	1 of 8 1 of 2

Severity FLS in parameters is ranked according to the failure effect associated to the path tree event. It uses five fuzzy sets: very low (vl), low (l), moderate (m), high (h) and very high (vh), developed according to the failure effects.

Table 2. Severity FLS in parameter elaboration

Rank	Severity	Effects
1	Very low " vl "	Minor failure. Not real effects of system performance
2 3	Low " L "	Small deterioration of system performance
4 5 6	Moderate " m "	Perturbation in system functions
7 8	High " h "	High degree of perturbation in system functions, but it still works
9 10	Very high " vh "	The failure affects the system function

The rules base gives the FLS out parameter Safety for each one of the FLS in parameters combination. The rules, presented in table 3, have the following statement: *If* the Occurrence is "low" and the Severity is "high" *then* the Safety is "moderate".

Table 3. Rules base

		Severity " vl"	" l "	" m "	" h "	" vh "
Occurrence	" vl "	-	(1)"vh"	(2)"h"	(3)"im"	(4)"m"
	" l "	-	(5)"h"	(6)"im"	(7)"m"	(8)"l"
	" m "	(9)"im"	(10)"im"	(11)"m"	(12)"l"	-
	" hî "	(13)"hî"	(14)"m"	(15)"im"	(16)"m"	-
	" vh "	(17)"vh"	(18)"vh"	(19)"h"	(20)"im"	-

The defuzzification of the fuzzy conclusion Safety, associated to each path tree event, gives a crisp value of the FLS conclusion. In order to obtain a General fuzzy conclusion for all the paths of the tree, an algorithm was elaborated [3]. Its deffuzifications gives the General Safety Degree (GSFD) of the electric power protection system.

3 Example

An adequate software tool FETA (Fuzzy Event-Tree Analysis) was created for the fuzzy event-tree analysis of an electric power protection system. The electric transformer-protection system from a naval power plant is sensible to the following failures types: external short-circuit (ESC), internal short-circuit (ISC) and over-voltage (OV).

Fig. 2 shows the fuzzy event-tree for the protection system in the external short-circuit case. The system is composed by two fault detector FD_1, FD_2 associated with the relays R_1, R_2 and trip signal TS_1, TS_2 respectively. Given that a fault occurs on the transformer, it is desired to evaluate the probability of successful operation of the protection system. Generally electric power protection system involve the sequential operation of a components and devices set. Event- tree are particular useful because they recognise the sequential operational logic of a system. Starting with initiating fault IF, very low probability is considered for it, because the transformers are rigid systems with no moving parts, thus generally the failure rate for them is very low (the operational rate is very high, noted "Vh" in the event tree). The fault detecting system consists of current transformers, and high impedance relays, where from experience they are reliable, except in case of faults with high currents (the failure rate is moderate, noted "m" in the event tree). The relay/ trip signal device, consists of relay, trip coil, pneumatic systems, along with many moving parts, thus high probability (h) is assumed for its successful operation. Finally, since high technologies have been used in design and manufacturing of the circuit breakers CB, the probability of their successful operation is considered to be very high (the failure rate is low, noted "l" in the event tree).

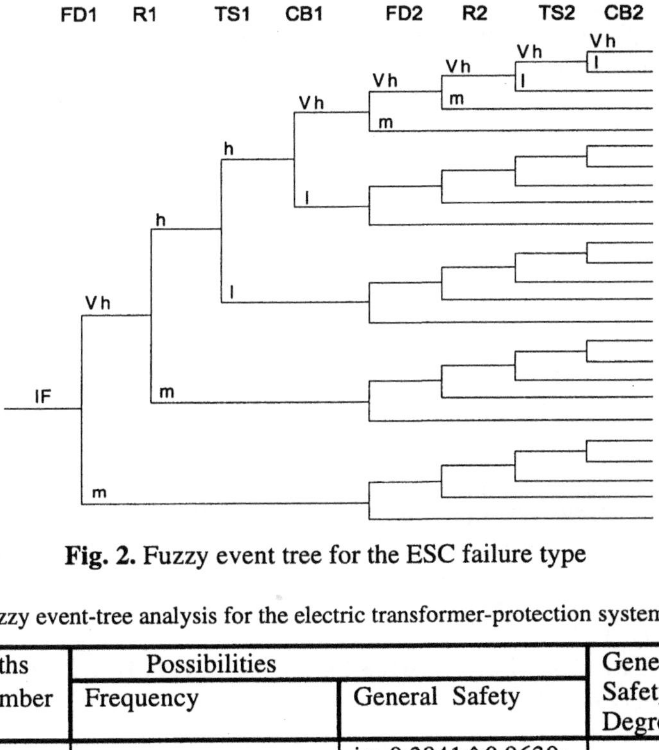

Fig. 2. Fuzzy event tree for the ESC failure type

Table 4. Fuzzy event-tree analysis for the electric transformer-protection system results

Failure Type	Paths number	Possibilities		General Safety Degree
		Frequency	General Safety	
ESC	25	s 1.00000; m 0.50340	im 0.3841;î 0.9630; fî 0.5753	9.07685
OV	25	s 0.61268; m. 0.9862	im 0.3483; î 0.9729 fî 0.5189	9.05137
ISC	139	s 1.00000; m 0.0222	im 0.3655; î 0.9945 fî 0.9468	9.01031
Global Severity Possibilities:		im : 0.05304 ;	î : 0.7365 ; fî : 0.78830 ;	
Global Severity Degree:			9.06945	

The result of the fuzzy event-tree analysis for the electric transformer-protection system are presented in table 4. The path number is the same, 25 paths for ESC and OV failure type, but is very large for ISC failure type. Also the General Safety Degree is the smallest for the ISC failure type. These results confirm the ISC failure greatest severity of all the failure types. The Global Severity for the protection system takes account of the Frequency fuzzy number associated to each of the failure type.

4 Conclusions

The proposed methodology for the safety analysis of an electric power protection is an adequate one. The elaborated fuzzy logic system (FLS) used in safety analysis is able to create the Severity output for each path of the event-tree associated to the protection system of each failure type. To elaborate the event tree and to use the proposed FLS on each part of the tree, a software tool "Fuzzy Event-Tree Analysis" (FETA) was created. All the evaluated "Safety" fuzzy sets are introduced in an algorithm to elaborate the fuzzy "General Safety" of the protection system for each kind of failure type. The fuzzy "Global Safety" for the electric transformer protection system is computed using the FETA software.

References

1. Bastani, F.B. and I.R. Chen: Reliability of Systems with Fuzzy-Failure Criterion. Proc.Ann. Reliability and Maintainability Symp (1994) 265-270.
2. Bowles, J.B. and C.E. Peláez: Application of Fuzzy Logic to Reliability Engineering. Proceedings of the IEEE 3 (1995) 99-107.
3. Dumitrescu, M. and A.P. Ulmeanu: A fuzzy algorithm for system safety evaluation. Second International conference on Mathematical methods in reliability, MMR'200 , Bordeaux, France (2000) 253-256.
4. Kenarangui, R: Verbal rating of alternative sites using multiple critter weights. Trans. ANS, 33 (1979) 617-619.
5. Kenarangui, R.: Event - tree Analysis by Fuzzy Probability. IEEE Trans. on Reliability, 1(1991) 45-52.
6. Mendel, J.M: Fuzzy Logic Systems for Engineering: A Tutorial. Proceedings of the IEEE, 3. (1995) 345-377.

Chaining of Fuzzy Rules in RETE Network

Zenon A. Sosnowski

Department of Computer Science, Technical University of Bialystok
Wiejska 45A, 15-351 Bialystok, Poland
zenon@ii.pb.bialystok.pl

Abstract. A central algorithm in rule-based expert systems is the pattern matching among rule predicates and current data. Systems like OPS5 [1] and its various derivatives use the RETE algorithm for this function. This paper describes and analyses augmentations of the basic RETE algorithm that are incorporated into an experimental fuzzy expert system shell FuzzyCLIPS [2,3] which contains the capabilities of handling fuzzy concepts and reasoning. The new concept of fuzzy RETE network allows all fuzzy rules, with the same linguistic variable, to be activated whenever fuzzy evidence concerning that variable is asserted into to the working memory. The paper presents a mechanism allowing fuzzy evidence to be aggregated when fuzzy inferences are made about the same fuzzy variable by different rules.

1 The RETE Algorithm

The RETE algorithm [4] is an efficient way of performing repeated pattern matches. The basis of this match algorithm is a structure called a RETE network, a data-flow graph with designated input and output nodes. Tokens representing changes to the working memory are sent into the input nodes and from the output nodes emerge tokens describing changes to the conflict sets. A RETE network has saved state that reflects all the tokens that have passed through it. The state of the network represents all the instantiations and partial instantiations according to the current state of working memory. After building such a network, it is possible to use its output to update conflict set incrementally.

A RETE network is made from the LHS's of all the productions. It contains two kind of nodes, alpha and beta. Both node types act as filters, performing tests on incoming tokens, and sending on only those tokens that pass the test. The alpha nodes occur at the beginning of the network and perform tests on individual facts. The final alpha nodes in the tree send their output to beta nodes, which perform the interfact tests. Beta nodes have two inputs, so each represents join which tests for consistent variable bindings between condition elements.

In the RETE network each condition element of the rule is represented by a sequence of tests, each test corresponding to one alpha node. The join beta node holds the test on the variable that is shared by two condition elements of the rule.

B. Reusch (Ed.): Fuzzy Days 2001, LNCS 2206, pp. 895–903, 2001.

2 Fuzzy RETE Network

This section presents an algorithm for dynamically adding new patterns, which represent fuzzy data, to existing RETE network in a way that shares tests and exploits previous match results where feasible [5].

Rule evaluation depends on whether or not antecedents or consequences of the rule contain linguistic variables [6,7]. Three types of rule can be distinguished: CRISP_, FUZZY_CRISP, and FUZZY_FUZZY. If the antecedent of the rule does not contain a linguistic variable, then the type of rule is CRISP_ regardless of whether or not a consequence contains a linguistic variable. If only the antecedent contains a linguistic variable, then the type of rule is FUZZY_CRISP. If both antecedent and consequence contain linguistic variables, then the type of rule is FUZZY_FUZZY.

In this section the rule of the form

$$\textit{if } A \textit{ then } C$$

where A is an antecedent and C is a consequence will be considered. Both antecedent and consequence can have only a single pattern. The method of handling rules, which contain many patterns in the antecedent or consequence, is described in [2, 3].

The inference mechanism realizes a generalized *modus ponens* rule. It may be described as follows.

$$\begin{array}{ll} \textit{if } A \textit{ then } C & CF_r \\ A' & CF_f \\ \hline C' & CF_c \end{array}$$

where:

CF_r is an uncertainty of the rule
CF_f is an uncertainty of the fact
CF_c is an uncertainty of the conclusion

If the type of rule is CRISP_, then A' must be equal to A in order to apply this rule. In that case the conclusion C' is equal to C, and

$$CF_c = CF_r * CF_f \tag{1}$$

where * denotes fuzzy numbers multiplication [8].

The modifications made to standard RETE network affects only rules with fuzzy antecedents. For fuzzy antecedents two alpha nodes are generated. All antecedents with the same fuzzy object share these nodes. First node is a single field node and it represents the name of linguistic variable. The second node is multi-field variable node, which matches all possible values of linguistic variable. The values of fuzzy objects appeared in fuzzy antecedents of LHS are attached to appropriate beta nodes.

To illustrate the organization of fuzzy RETE network, consider the following example where we have shown the condition elements of two rules:

```
(defrule r1
    (speed very fast)
```

```
=>
    ( . . . )
)

(defrule r2
    (speed slow)
=>
    ( . . . )
)
```

When a new working memory element (e.g. (speed *medium*)) is asserted into the working memory, the WME becomes a token that traverses through the fuzzy RETE network (shown in Fig 1) and, subsequently, may be propagated to find out which rules can be activated.

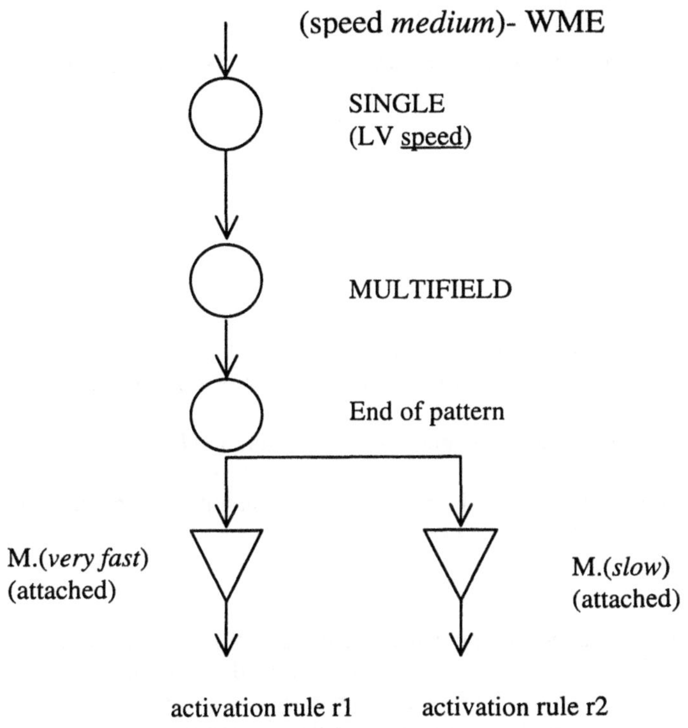

Fig. 1. Fuzzy RETE network for rule r1 and r2

If the type of rule is FUZZY_CRISP, then *A'* must have the same fuzzy object as *A* in order to apply this rule. Values of a fuzzy object in *A* and *A'* represented by fuzzy sets F_a and F_a' do not have to be equal. In that case the conclusion *C'* is equal to *C*, and

$$CF_c = CF_r * CF_f * S \tag{2}$$

where S is a measure of similarity between fuzzy sets F_a and F_a'.

The measure of similarity is based upon the measure of possibility P and the measure of necessity N. It is calculated according to the following formula [9]

$$S = \begin{cases} P(F_a|F_a') & \text{iff} \quad N(F_a|F_a') > 0.5 \\ (N(F_a|F_a') + 0.5) * P(F_a|F_a') & \text{otherwise} \end{cases} \tag{3}$$

where

$$P(Fa|Fa') = \max_{u \in U}(\min(\mu_{F_a}(u), \mu_{F_a'}(u))) \tag{4}$$

and

$$N(F_a|F_a') = 1 - P(\overline{F_a}|F_a') \tag{5}$$

$\overline{F_a}$ is the complement of F_a.

The extended inference engine of FuzzyCLIPS uses the value of fuzzy set attached to appropriate beta node to calculate the measure of similarity between rule antecedent and matching fact in the case of FUZZY_CRISP rule.

If the type of rule is FUZZY_FUZZY, then it was shown in [6] that the fuzzy relation connects the antecedent and consequence of such a rule

$$R = F_a \times F_c \tag{6}$$

where:
F_a is a fuzzy set denoting the value of the fuzzy object of the antecedent,
F_c is a fuzzy set denoting the value of the fuzzy object of the consequence.

The algorithms for forming this relation can be found in [10].

The calculation of conclusion is based upon the compositional rule of inference [6], which can be described as follows:

$$F_c' = F_a' \circ R \tag{7}$$

where F_c' is a fuzzy set denoting the value of the fuzzy object of the conclusion.
The certainty factor of the conclusion is calculated according to (1).
For the FUZZY_FUZZY type rule the attached fuzzy set in beta node is used to make inference based upon the compositional rule of inference.

3 Global Contribution

In a crisp expert system there is never any need to re-assess the facts in the system - once they exist, they exist (unless certainty factors are being used, then certainty factors are modified to account the new evidence and checked if they are larger then a threshold value). But in fuzzy expert systems, refinement of a fuzzy fact may be possible [2,11]. Consequently, the fact concerning fuzzy variable is potentially useful for every rule whose LHS contains this fuzzy variable, and every value of a fuzzy variable may be repeatedly reevaluated. Thus, in the case where the fact concerning a linguistic variable is asserted as a result of performing the RHS action of a rule, this fact is treated as giving contributing evidence about the fuzzy variable (it contributes globally). If information about that fuzzy variable has already been asserted then this new evidence (or information) about the fuzzy variable is combined with the existing information about the fuzzy variable. There are many readily identifiable methods of combining evidence. In FuzzyCLIPS the new value of the fuzzy fact is re-assessed in accordance with the formula

$$\mu_{F_g} = \max(\mu_{F_f}, \mu_{F'_f}) \qquad \textbf{(8)}$$

where μ_{F_g} is the new value of the linguistic variable, μ_{F_f} is the membership function of the existing value of the linguistic variable, and $\mu_{F'_f}$ is the membership function of the linguistic variable to be asserted.

Example. Consider the following three rules:

```
(defrule Rule-1
      (X  A)
 =>
      (Y  B)
)
(defrule Rule-2
      (X  A')
=>
      (Y  B')
)
(defrule Rule-3
      (Y  B)
=>
      (Z  C)
)
```

Assuming that the working memory contains the fact *(X A)* and that the compositional rule of inference is implemented as a sup-min composition it is easy to see that the final conclusions derived for the values of Y and Z do not depend on the order of rules being fired. Let us now examine how the FuzzyCLIPS carries out fuzzy reasoning. The fact *(X A)* enables Rule 1 and Rule 2 to be activated (put on the agenda). Of course we assume that terms A and A' associated with the linguistic variable X have overlapping membership functions. In the next step (does not matter which conflict resolution strategy is applied) Rule1 or Rule 2 fires. After any of this rule fired, the fact *(Y B)* or *(Y B")* would be asserted into the working memory. This new fact would enable Rule 3 (we assume that terms B and B" of the linguistic variable Y have overlapping membership functions). Now we have Rule 1 (or Rule 2) and Rule 3 on the agenda. In the next steps the dictates of fuzzy logic are not violated as the final conclusions derived for the values of Y and Z will not depend on the order of rules being fired. Let us assume that MEA strategy of conflict resolution is applied and the Rule 3 fires first. As a result the fact concerning the value of linguistic variable Z (equal to C if Rule 1 had already fired) would be asserted into the working memory. Now it is time for Rule 2 (or Rule 1) to fire. After this rule fired, the fact concerning the new value of linguistic variable Y refines the existing value of this variable in the working memory according to global contribution. Updating the value of linguistic variable Y enables Rule 3 again. After Rule 3 fired for the second time, a "global contribution" concerning the linguistic variable Z is performed. The final values of linguistic variables Y and Z are equal to those obtained if the order of firing rules would be 1, 2, and 3.

4 An Example

In what follows, we will concentrate on the example and show that the final conclusions do not depend on the order of rules to be fired.

```
Fact-1 (position NW)

(defrule rule-1
    (position NW)
=>
    (distance medium)
    (angle close-to-20)
}

(defrule rule-2
    (position NNW)
=>
    (distance long)
    (angle close-to-30)
}
```

The *position, distance*, and *angle* are linguistic variables. Assume that the depth-first strategy is selected for conflict resolution, meaning that rule matched with the most recent facts will fire first. After Fact-1 have been asserted into the working memory to start the inference engine, rule-1 and rule-2 will be activated and placed in the agenda. Assume that rule-2 is fired first. The firing of this rule cause the fact concerning the value of the linguistic variables *distance* and *angle* to be asserted into the working memory with the appropriate values. Fig. 2 illustrates the inference process resulting from the firing of rule-2.

Current agenda: rule-1.The rule-1 is fired next. As a result the facts concerning the value of the linguistic variables *angle* and *distance* will be updated to those present in the working memory by the global contribution procedure. Fig. 3 illustrates the inference process resulting from the firing of rule-1. Current agenda: empty. The inference process is complete. The results are equal to those obtained if the order of firing rules would be 1 and 2.

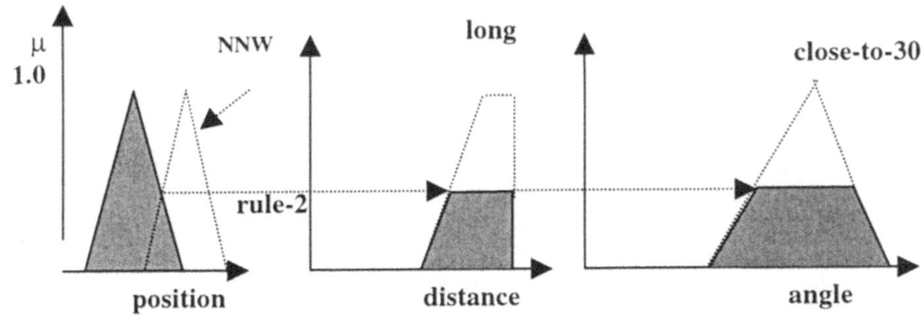

Fig. 2. Illustration of the membership functions calculations for the linguistic variables *angle* and *distance* in the first cycle of the inference engine.

5 Concluding Remarks

The extension of RETE matching algorithm presented here contains the capability of handling fuzzy patterns by RETE network. This allows any mix of fuzzy and normal terms, and uncertainties in the rules and facts of an expert system.

In this paper we described a way the FuzzyCLIPS addresses the issue of allowing fuzzy evidence to be aggregated when fuzzy inferences are made about the same fuzzy variable by different rules when an expert system works in the fuzzy logic controller regime.

The efficiency of pattern matching rapidly decreases as the number of working memory elements increases. This is because the cost of join operations performed in the match process is directly proportional to the square of the number of working memory elements. Moreover inappropriate ordering of conditions generates a large number of intermediate data, which increases the cost of subsequent join operations. To increase the performance of fuzzy pattern matching an optimization algorithm to determine efficient join structures for fuzzy Rete network is being contemplated.

Acknowledgment

Support from the Technical University of Bialystok grant no. W/II/5/99 is greatly appreciated.

References

1. C.L. Forgy. The OPS5 User Manual. *Technical Report*, CMU-CS-79-132, Computer Science Department, Carnegie- Mellon University, 1981.
2. Z.A. Sosnowski. „A fuzzy extension of CLIPS rule-based shell". *Fuzzy Engineering toward Human Friendly Systems*, vol.1, 503-512, 1991.
3. National Research Council, Canada, *FuzzyCLIPS Version 6.04A User's Guide*, 1998.
4. C.L. Forgy. „RETE: A fast algorithm for the many pattern/many object pattern match problem". *Artificial Intelligence*, **19**, 17- 37, 1982.

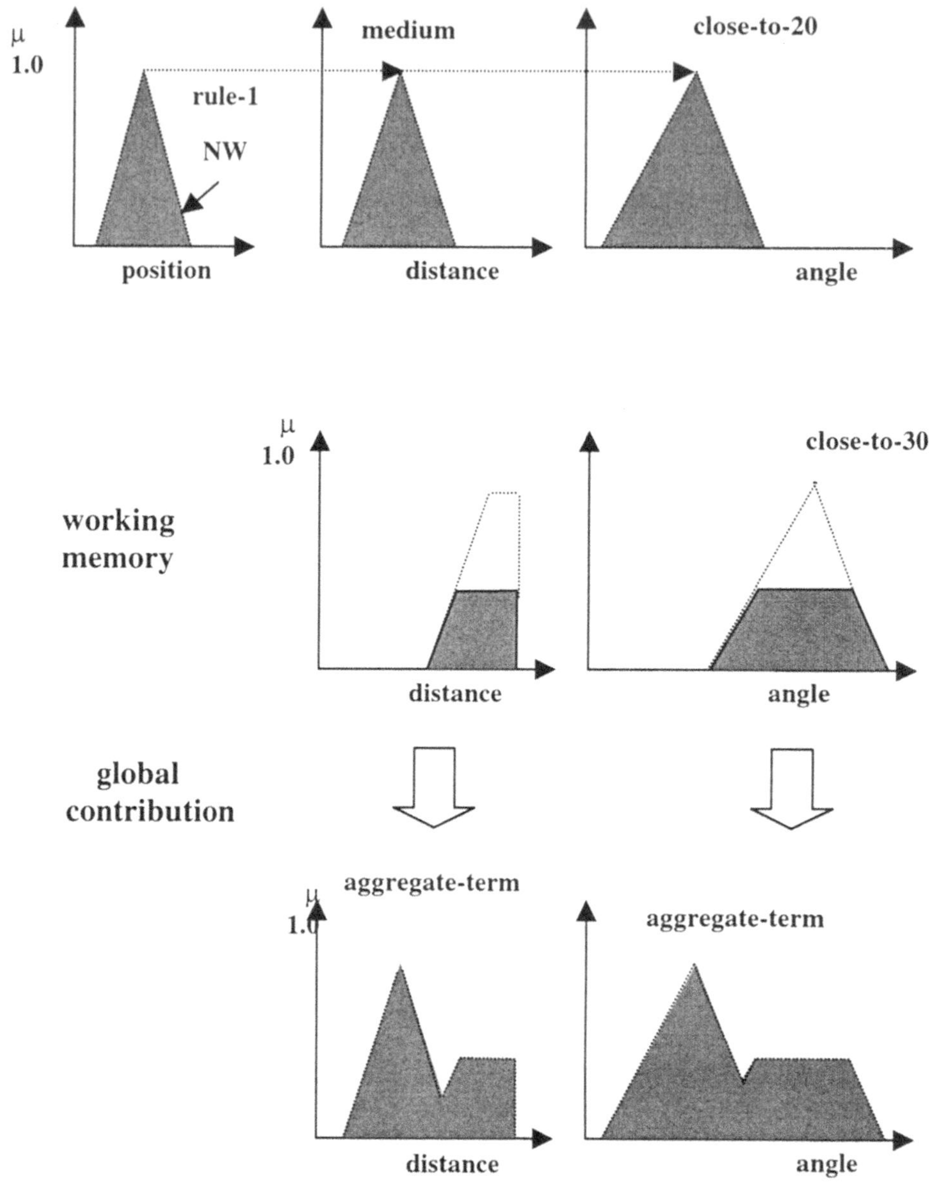

Fig. 3. Illustration of how the membership functions of the linguistic variables *angle* and *distance* are aggregated in the second cycle of the inference engine.

5. Z. A. Sosnowski, "Representation of fuzzy data in RETE network", in Seventh International Fuzzy Systems Association World Congress, Academia, Prague , 1997, vol. II, pp. 504-508.
6. L.A. Zadeh. „ The concept of a linguistic variable and its application to approximate reasoning". *Infor. Sci.,* **8**, 395-460, 1975.

7. Z.A.Sosnowski. „ A linguistic variable in FLIPS programming language". *The Second Joint IFSAEC end EURO-WG Workshop 'Progress in Fuzzy Sets in Europe'*, April 6-8 1988, Vienna, pp. 71-74.
8. A.Kaufman, M.M.Gupta, *Introduction to Fuzzy Arithmetic. Theory and Applications*, Van Nostrand Reinhold, 1985.
9. M.Cayrol, H.Farency and H.Prade. „Fuzzy pattern matching". *Kybernetes*, **11**, 103-106, 1982.
10. M.Mizumoto, S.Fukami, K.Tanaka. „Some methods of fuzzy reasoning". In M.M.Gupota, R.K.Ragade and R.R.Yager, editors, *Advances in Fuzzy Set Theory and Applications*, pp. 117-136. North-Holland, Amsterdam, 1976.
11. J. Pan, G. N. DeSouza, and A. C. Kak, „FuzzyShell: A Large-Scale Expert System Shell Using Fuzzy Logic for Uncertainty Reasoning", *IEEE Trans. Fuzzy Syst.*, **6**, 563-581, 1998.

A Fuzzy Relation for Comparing Intervals

Marius Călin[1] and Dan Gâlea[2]

[1]Department of Informatics,
University of Agricultural Sciences and Veterinary Medicine, Iasi, Romania
Aleea Sadoveanu no. 3, Iasi - 6600, Romania
mcalin@univagro-iasi.ro
[2]Faculty of Automatic Control and Computer Science,
Technical University of Iasi, Romania
Blvd. D. Mangeron nr. 53A, 6600 - Iasi, Romania
dgalea@iit.iit.tuiasi.ro

Abstract. The paper proposes a method for expressing the matching degree of two arbitrary intervals through a *fuzzy relation* having the properties of *reflexivity* and *symmetry*. This method could be useful in situations when one has to compare two entities whose attributes are expressed through intervals and give a measure of their matching degree. Such a real world situation occurs in the mechanical engineering, namely in the steel selection process, when the designer has to determine the steels that are equivalent to a given one. This situation is described, as an application, in the final part of the paper.

1 Introduction

Classic logic operates with assertions that can be only "TRUE" or "FALSE". Usually 1 denotes "TRUE" and 0 denotes "FALSE". However, there are numerous situations when a TRUE/FALSE conclusion can not be drawn. Fuzzy Sets Theory and Fuzzy Logic can be used to represent different linguistic degrees of truth.

For the following discussion, some well-known definitions [3], [4] must be reviewed.

Definition 1. A *fuzzy set* F on U is a mapping $F : U \to [0, 1]$.

Remark. U is the *universe of discourse* and $F(x) \in [0, 1]$ is said to be the *membership degree* of x in the set F.

Definition 2. Let U and V be two arbitrary sets and $U \times V$ their Cartesian product. A *fuzzy relation* (in $U \times V$) is a subset R of $U \times V$.

Remark. When $U = V$ (the relation is defined between elements of the same universe), R is said to be a *fuzzy relation in U*.

Definition 3. Let $R \subseteq X \times Y$ and $S \subseteq Y \times Z$ be two fuzzy relations. The *composition* of the given fuzzy relations, denoted $R \circ S$ is the fuzzy relation

B. Reusch (Ed.): Fuzzy Days 2001, LNCS 2206, pp. 904–916, 2001.

$$R \circ S(x,z) = \sup_{y \in Y} \min(R(x,y), S(y,z)) \quad (1)$$

Definition 4. Let $R \subseteq X \times Y$ be a fuzzy relation and $\alpha \in [0, 1]$. The *α-level* of R is

$$R_\alpha = \{(x,y) \in X \times Y | R(x,y) \geq \alpha\} \quad (2)$$

Definition 5. A *relation of similitude* R in the universe U is a fuzzy relation with the following properties:

1) *reflexivity:* $R(x, x) = 1, x \in U$ (3)

2) *symmetry:* $R(x, y) = R(y, x), x,y \in U$ (4)

3) *transitivity:* $R \circ R \subseteq R$, (5)
that is $R(x,z) \geq \sup_{y \in U} \min(R(x,y), R(y,z)) \quad x,z \in U$.

The *relation of similitude* is an extension to Fuzzy Sets Theory [3] of the classical *relation of equivalence*; it aims to model and evaluate the "resemblance" to each other that could exist between members of a universe of discourse.

If the universe U is finite, $U = \{x_1, ..., x_n\}$, a fuzzy relation R can be described through a $n \times n$ matrix $A_R = (a_{ij})_{i,j=1,...,n}$. The elements a_{ij} express the relation between the corresponding pairs of members of the universe U. They have the property $0 \leq a_{ij} \leq 1$. If R is *reflexive*, then $a_{ii} = 1$, $i = 1, ..., n$. If R is *symmetrical*, then $a_{ij} = a_{ji}$, $i,j = 1,..., n$.

A classical relation of *equivalence* allows the partition of the universe in *classes of equivalence*. In case of (fuzzy) relations of *similitude*, the definition of corresponding *classes of similitude* in possible, but problems rise in handling them, because *classes of similitude* can overlap, unlike *classes of equivalence* that are disjoint. Moreover, when defining a fuzzy relation of similitude in a real world application, it is quite difficult to fulfil the property of transitivity. In [3] was introduced a way to define a (fuzzy) *relation of similitude* as the limit of a sequence of fuzzy relations. Corresponding *classes of similitude* can be then defined. The departing point is a reflexive and symmetrical fuzzy relation. In practice this relation must be defined in a convenient way.

In section 2 a method is proposed to express the matching degree of two arbitrary intervals through a *fuzzy relation* having the properties of *reflexivity* and *symmetry*.

A frequent problem in mechanical engineering design and manufacturing is the replacement of a steel in current use, with an equivalent one, that is another steel having the same behaviour in the given context. Production costs or availability can lead to such a problem. It can not be expected to find another one having identical features, so the search must focus on detecting a replacing steel which is "as close as possible" to the former. This leads to the idea of defining a *degree of equivalence of two steels*. The definition of a degree of equivalence is needed because one can not always give a precise, "yes/no" answer to the question: "are steels X and Y equivalent?" The fuzzy relation defined in section 2 was applied in building a method to determine the chemical equivalence of two steels and the degree of this equivalence. This method is described in section 3 of the paper.

2 Fuzzy Relation for Comparing Intervals

Literature provides numerous methods for comparing intervals. The following approach is based on geometrical remarks regarding the two partial aspects of the topic: the length comparison and the measure of concentricity. Each of them leads to a reflexive and symmetrical fuzzy relation. The overall measure of intervals matching is obtained by combining the two partial ones. Each of these partial measures can also be used separately wherever necessary.

2.1 Matching

Let $[a, b]$ and $[c, d]$ be two intervals that must be compared in order to compute a measure of their matching. Let this matching degree be denoted

$$match([a, b]; [c, d]).$$

The aim is to define *match* as a non-dimensional measure having the following properties:

1) $0 \leq match([a, b]; [c, d]) \leq 1$ where (6)
$match([a, b]; [c, d]) = 0$ means no matching at all (7)
$match([a, b]; [c, d]) = 1$ means a perfect match, iff $a = c$ and $b = d$. (8)

2) $match([a, b]; [a, b]) = 1$ (reflexivity) (9)

3) $match([a, b]; [c, d]) = match([c, d]; [a, b])$ (symmetry) (10)

The *matching degree* is calculated as a product of two partial measures:

$$match([a, b]; [c, d]) = round([a, b]; [c, d]) \cdot conc([a, b]; [c, d]) \tag{11}$$

The component $round([a, b]; [c, d])$ is the measure of the degree in which the lengths of the two intervals match; the name *round* stands for *roundness*. The second component, $conc([a, b]; [c, d])$, expresses the degree of *concentricity* of the intervals. These two are defined further on.

2.2 Roundness

In relation (11), *round* denotes the *roundness* of a rectangle whose sides have the lengths $b - a$ and $d - c$, respectively. The measure *roundness* of some shape was used [2] in literature, within image processing applications. For two dimensions this measure is defined as follows.

Let **R** be a two-dimensional (2D) region. The *roundness* of region **R** can be computed using the following procedure:

- consider a reference circle C_{2D} having
  - centre in the centre of gravity of region **R** ;
  - radius $r = \sqrt{\frac{area(\mathbf{R})}{\pi}}$ (12)
- compute

$$roundness_{2D}(R) = max\left(1 - \frac{area\left(C_{2D} \cap \overline{\mathbf{R}}\right) + area\left(\overline{C_{2D}} \cap \mathbf{R}\right)}{area(\mathbf{R})}, 0\right) \quad (13)$$

- normalise to the interval [0, 1]

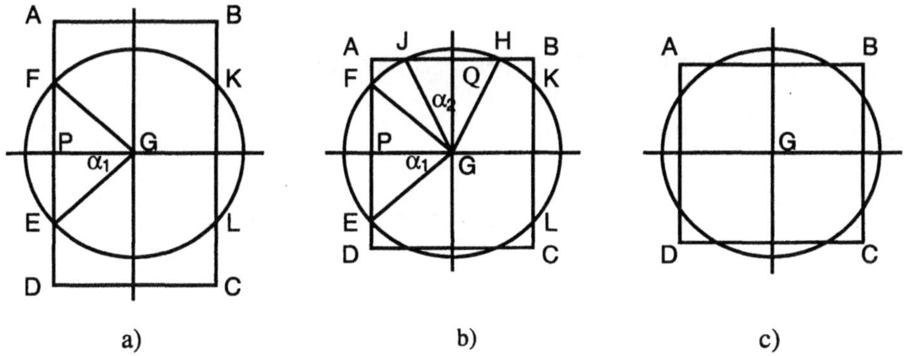

Fig. 1. Computation of the *roundness* of rectangle ABCD

When the region **R** is a rectangle, denoted ABCD, a few special cases appear, which are shown in figure 1 where the angles $\alpha_1 = \angle \mathsf{EGP}$ and $\alpha_2 = \angle \mathsf{JGQ}$ were denoted. There are two more cases that correspond to figures 1a and 1b, in which AB > BC, but the lines of reasoning are the same. In figure 1c is shown the special case when the rectangle ABCD is a square; as it can be expected, this is case of maximum roundness.

Denoting *roundness*(ABCD), the roundness of the rectangle ABCD and also denoting

$$X = \frac{area\left(C_{2D} \cap \overline{\mathsf{ABCD}}\right) + area\left(\overline{C_{2D}} \cap \mathsf{ABCD}\right)}{area(\mathsf{ABCD})} \quad (14)$$

the relation (13) for the roundness becomes:

$$roundness(\mathrm{ABCD}) = max(1 - X, 0) \quad (15)$$

According to figure 1, two distinct cases appear.

Case a (figure 1a)

$$area\left(C_{2D} \cap \overline{\mathsf{ABCD}}\right) = 2 \cdot area(\mathsf{EFP}) \quad (16)$$

$$area\left(C_{2D} \cap \overline{\text{ABCD}}\right) = area(\text{ABCD}) - area(C_{2D}) + 2 \cdot area(\text{EFP}) \tag{17}$$

$$X = \frac{4 \cdot area(\text{EFP})}{area(\text{ABCD})} \tag{18}$$

$$area(\text{EFP}) = \frac{area(\text{ABCD})}{\pi}(\alpha_1 - sin\,\alpha_1\, cos\,\alpha_1) \tag{19}$$

$$X = \frac{4}{\pi}(\alpha_1 - sin\,\alpha_1\, cos\,\alpha_1) \tag{20}$$

Case b (figure 1b). Similar calculations lead to

$$X = \frac{4}{\pi}(\alpha_1 - sin\,\alpha_1\, cos\,\alpha_1 + \alpha_2 - sin\,\alpha_2\, cos\,\alpha_2) \tag{21}$$

Then, α_1 and α_2 must be computed

$$cos\,\alpha_1 = \frac{\text{GP}}{\text{EG}} = \frac{\text{GP}}{radius(C_{2D})} = \text{GP}\sqrt{\frac{\pi}{area(\text{ABCD})}} = \frac{\text{AB}}{2}\sqrt{\frac{\pi}{\text{AB} \cdot \text{BC}}} = \frac{1}{2}\sqrt{\pi p} \tag{22}$$

$$\text{where } p = \frac{\text{AB}}{\text{BC}} \tag{23}$$

Thus,

$$\alpha_1 = arccos\left(\frac{1}{2}\sqrt{\pi p}\right) \tag{24}$$

and, through similar calculations,

$$\alpha_2 = arccos\left(\frac{1}{2}\sqrt{\frac{\pi}{p}}\right) \tag{25}$$

From table 1, where a few computed values of *roundness* are listed, some conclusions can be drawn:

i) the *roundness*(ABCD) becomes zero when its longer side becomes approximately five times bigger than the shorter one;

ii) roundness(ABCD) reaches its maximum, *roundness*(ABCD) ≈ 0.818908, for the square; the normalisation in [0, 1] is shown in the last column of table 1.

Another important conclusion comes from relations (20), (21), (24) and (25):

iii) roundness(ABCD) depends only on the ratio between the height and width of the rectangle.

Table 1. Computed values of the roundness of a rectangle and their normalised values

p = AB/BC	$1/p$	α_1 (rad.)	α_2 (rad.)	*round*	*normalisation*
0.2	5	1.163277	0	0	0
0.210526	4.75	1.152035	0	0.006183	0.00755047
0.666667	1.5	0.761791	0	0.665968	0.81323948
0.785398	1.273239545	0.667457	0	0.769158	0.93924826
0.8	1.25	0.655628	0.135515	0.778422	0.9505613
1	1	0.48166	0.48166	0.818908	1
1.25	0.8	0.135515	0.655628	0.778422	0.9505613
1.27324	0.785398163	0	0.667457	0.769158	0.93924826
1.5	0.666666667	0	0.761791	0.665968	0.81323948
4.75	0.210526316	0	1.152035	0.006183	0.00755047
5	0.2	0	1.163277	0	0

The component *round*([*a*, *b*]; [*c*, *d*]) in relation (11) will be expressed using the roundness of the rectangle that has the two dimensions equal to the lengths of the compared intervals, [a, b] and [c, d] respectively. Conclusion *iii)* drawn before is important because it shows the symmetry of *round*([*a*, *b*]; [*c*, *d*]). The other property of *round*([*a*, *b*]; [*c*, *d*]) must be reflexivity. It comes from *ii)* and the consequent normalisation, which entails *round*([*a*, *b*]; [*c*, *d*]) = 1 iff $a = c$ and $b = d$. This is why a normalisation in [0, 1] of *roundness* was necessary.

2.3 Concentricity

The other element in the calculation of the matching degree (11) is the *concentricity* of the two intervals from each other, *conc*([*a*, *b*]; [*c*, *d*]). The *concentricity* is determined by the centres of the intervals. In figure 2, some cases are shown. The intervals are placed on the real-number axis, in its positive region. This doesn't restrict the generality, and corresponds to the application that will be discussed later. The centres of [*a*, *b*] and [*c*, *d*] have the co-ordinates *m* and *m'*, respectively.

To express the *concentricity*, the points shown in figure 3 were placed in a Cartesian system of co-ordinates. The points A, B and C, D are placed on Oy as if the intervals [*a*, *b*] and [*c*, *d*] were represented on the vertical axis. Again, the two intervals were assumed to be in the positive region of the axis, but this doesn't affect the generality.

Denoting $l_{ab} = b - a$ the length of interval [*a*, *b*] and $l_{cd} = d - c$ the length of [*c*, *d*], for the points in figure 3, the following relations and co-ordinates are defined:

$l_{ab} = b - a = y_B - y_A$ $\quad$ $l_{cd} = d - c = y_D - y_C$

A $(0, y_A)$ B $(0, y_B)$ $\quad$ A' $((l_{ab} + l_{cd})/2, y_A)$ $\quad$ B' $((l_{ab} + l_{cd})/2, y_B)$
C $(0, y_C)$ D $(0, y_D)$ $\quad$ C' $((l_{ab} + l_{cd})/2, y_D)$ $\quad$ D' $((l_{ab} + l_{cd})/2, y_D)$

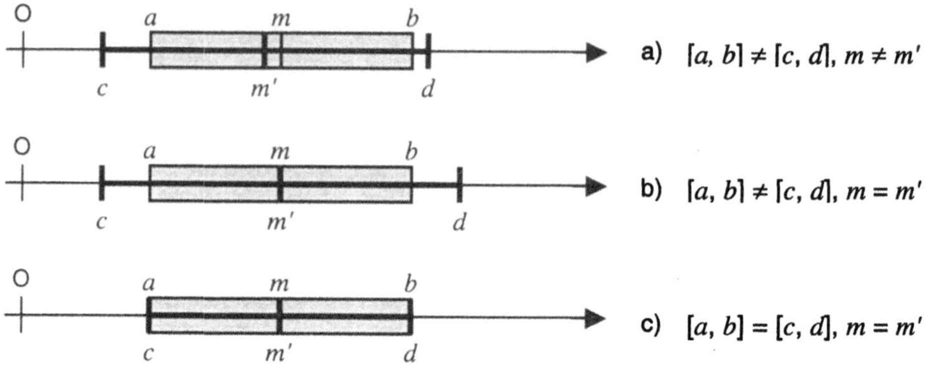

Fig. 2. Different positions of the centres of two intervals

O' $(0, (y_A + y_B)/2)$ M' $((l_{ab} + l_{cd})/4, (y_C + y_D)/2)$

Having defined the points shown in figure 3, the concentricity *conc*, is defined as the cosine of $\delta = \angle MO'M'$.

$$conc([a, b]; [c, d]) = \cos \delta \tag{26}$$

Elementary calculations lead to

$$\cos\delta = \frac{1}{\sqrt{1+tg^2\delta}}, \text{ where } tg\delta = \frac{2}{l_{ab}+l_{cd}}[(y_C + y_D) - (y_A + y_B)] \tag{27}$$

Relation (27) should be used to compute the concentricity of the compared intervals.

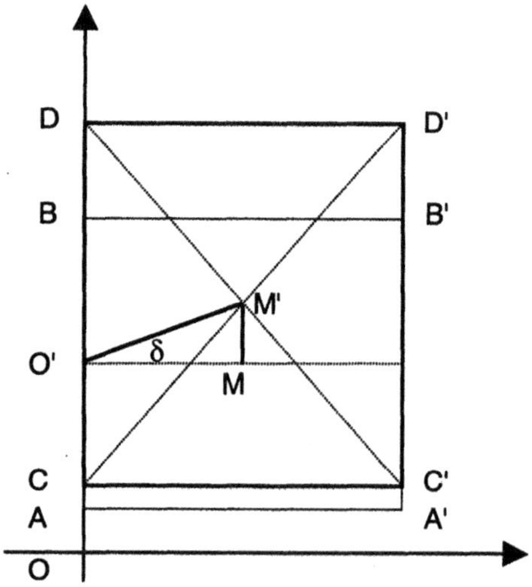

Fig. 3. Construction of the angle δ that expresses the *concentricity* of two intervals

Concentricity $conc([a, b]; [c, d]) = cos\ \delta$ has the following properties:

i) $0 < conc([a, b]; [c, d]) \leq 1$

ii) $conc([a, b]; [a, b]) = 1$

iii) $conc([a, b]; [c, d]) = conc([c, d]; [a, b])$

iv) $conc([a, b]; [c, d]) = conc([a, b]; [c-\alpha, d+\alpha])$

Property *i)* shows the *normalisation* of *conc*, in [0, 1]. Property *ii)* shows the *reflexivity* of *conc*, which means that when the current interval is equal to the reference one, *conc* is equal to 1. Property *iii)* shows the *symmetry* of *conc* that is, when computing the concentricity of two intervals, it doesn't matter which one is the reference.

As well, from *ii)* and *iv)* can be concluded that *concentricity* is equal to 1 for any two intervals that have the same centre ($m = m'$ in figure 2). It means that *conc* is not sufficient to express the whole matching degree between two intervals; it must participate together with *round* in the valuation of the matching degree.

3 The Equivalence Degree of Two Steels

The selection of the most appropriate steel in a certain mechanical design situation is a complex decision task. The technical features of a wide collection of steels must be examined to reach the best solution. Information technology is useful in making such an investigation because it helps processing the great amount of data involved. The usual approach, from this point of view, is the utilisation of a database that stores, for different steel types, the technical features: *chemical composition* and *mechanical properties*. The database is searched using appropriate criteria, which leads to the selection of a set of steels that match, in different degrees, the requirements imposed within the project. This solution saves time and effort, but it has some drawbacks. Most of them come from the lack of exact methods for stating the search criteria, and from the vagueness naturally involved in the linguistic evaluations that any human expert handles. An identification of some of these drawbacks was made in [1] where is suggested a query method based on Fuzzy Sets Theory that should improve the searching process.

As stated in the introductory section, another frequent problem in mechanical engineering design and manufacturing is the replacement of a steel in current use, with an equivalent one, that is another steel that should have the same behaviour in the given context. Such a replacement could be necessary for availability or cost reasons. One the one hand it can not be expected to find a steel with identical features with the original one. One the other hand, in the technical literature of the domain, there is not a precise definition of the notion of *steel equivalence*, even though this concept is often used. Moreover, it's explicitly stated that steel equivalence is rather a vague notion which would be used according to the designer's experience. In this section is proposed a method for computing the *degree of equivalence* between two steels. Such a measure should provide a means to estimate the best decision to make for the replacement of the original steel with another one. This approach is only from the point of view of *chemical composition*. The other main group of characters that define a steel type regards the *mechanical properties*. There are many features in this group that can be treated in a similar way. However, since a part of them require a specific treatment, for the sake of simplicity, the discussion will be restricted on the chemical composition.

The technical features of steels are stored in a database: there is a record for each stored steel. These features are stored in form of values or intervals. Tacking the chemical composition, in the record structure there is a pair of fields that store the minimum and the maximum percentage for each chemical element that might appear: *C-min, C-max, Mn-min, Mn-max, Si-min, Si-max*, etc. The problem of finding, for a given steel, all the other steels in the database that have an *equivalent* chemical composition, means to detect the ones that contain the same chemical elements, in *approximately* the same minimum-maximum percentages, respectively.

Remarks. a) The word "approximately" was highlighted because it expresses a *vague measure that must be quantified*, in order to be automatically processed. *b)* Further on, the given steel will be named "*the reference steel*", and steel in the database, for which the degree of matching with the reference steel must be computed, will be named "*the current steel*".

The elementary step in computing the degree of matching between the reference steel and the current steel is the calculation of degree of matching of two intervals that refer to the percentage of some chemical element: the min-max interval expressing the percentage of the element in the chemical composition of the current steel, and the corresponding min-max interval in the reference steel.

Applying the method that was described in section 2, a matching degree between the corresponding intervals from the reference steel and the current one can be computed. Following this idea, the proposed method performs the following tasks:

1) Through a first filtering, selects those steels from the database, that contain the same chemical elements as the reference steel.

2) For each chemical element in a selected steel, computes a *matching degree* between the respective min-max interval, and the corresponding one from the reference steel.

3) Combines all the matching degrees computed in step 2, giving an *overall matching degree* of the currently examined steel with the reference steel.

Example. In its actual state, the used database includes around 6000 steels from 20 countries. Let the reference steel be the Romanian one AUT 40MN. The chemical composition of this steel type is shown in the first line of table 2. The problem is to find the equivalent ones in the database.

In step 1, from the database were selected all the steels that have the same chemical elements as AUT 40MN. A list of 972 steels resulted.

Since this list is too large, the designer decided to restrict it to the types that contain the given chemical elements in percentages that exceed by 100% the both sides of each min-max interval in the reference steel. These values are shown in the second line of table 2. It has been considered that restriction do not affects the correctness of the process since the matching degrees of the corresponding intervals remain in a reasonable range.

Table 2. Chemical composition of the reference steel and the selection criteria

Country	Type	C-min	C-max	Mn-min	Mn-max	Si-min	Si-max
ROM	AUT 40MN	0.35	0.45	1.2	1.6	0.15	0.35
criteria: ± 100%		0.25	0.55	0.8	2	-0.05	0.55

The resulting list has 43 steels and is shown in table 3. Each of them was given an identification number ID from 0 to 42. The reference steel, the Romanian type AUT 40MN is shown in line 36.

Among the steels selected in step 1, there are those having a satisfactory degree of equivalence with the reference. Their identification is to be performed in the following two steps.

In Step 2 were computed the matching of the intervals [C-min, C-max], [Mn-min, Mn-max], [Si-min, Si-max] from the reference steel to the corresponding ones in the list obtained in step 1. These fuzzy measures were combined in step 3 giving the overall equivalence degree of each selected steel to the reference AUT 40MN. For this combination the multiplication operator was used and the results are shown in table 4.

Sorting table 4 by the column *equivalence* in descending order, the designer would find on the top of the final classification the "nearest" steels to the reference. A brief examination of table 4 shows that steel number 42 (that is ASTM 1541H, produced in USA) has the highest level of equivalence (0.87) with the reference AUT 40MN.

The procedure in step 3 can be further refined and the list resulted in step 2 (containing the matching degrees of all the intervals with the corresponding ones in the reference steel) can be viewed as a Multi-Attribute Decision Model (MADM) problem [6]. Within this framework the designer can consider levels of importance of different chemical elements in the steel composition. Using, for example, the Saaty's Analytic Hierarchy Process (AHP) [5] the designer can assign different levels of (subjective) importance to the chemical components and make a classification with respect to these levels.

Following the concept of relation of similitude that was mentioned in section 1 and applying a procedure that was described in [3], classes of similitude of different levels were determined for the set of 43 steels. However, from engineering point of view the result is debatable and further investigations are necessary.

Table 3. Steels resulted from the database querying process

ID	Country	Type	C-min	C-max	Mn-min	Mn-max	Si-min	Si-max
0	AUS	XK1145	0.42	0.49	1.35	1.65	0.1	0.35
1	AUS	XK1145	0.42	0.49	1.35	1.65	0.1	0.35
2	AUS	XK1335	0.33	0.38	1.4	1.7	0.1	0.35
3	AUS	K1340	0.38	0.43	1.6	1.9	0.1	0.35
4	AUS	XK1340	0.38	0.43	1.4	1.7	0.1	0.35
5	AUS	XK1345	0.43	0.48	1.4	1.7	0.1	0.35
6	DEU	36MN5	0.32	0.4	1.2	1.5	0.15	0.35
7	ESP	36MN6	0.33	0.4	1.3	1.65	0.15	0.4
8	ESP	35MNS6	0.33	0.39	1.3	1.7	0.1	0.4
9	ESP	45MNS6	0.41	0.48	1.3	1.7	0.1	0.4
10	ESP	35MN5DF	0.32	0.39	1.1	1.4	0.15	0.4
11	ESP	46MN4	0.42	0.5	0.9	1.2	0.3	0.5
12	FRA	35M5	0.32	0.38	1.1	1.4	0.15	0.35
13	FRA	40M6	0.37	0.43	1.3	1.7	0.15	0.35
14	FRA	35MF6	0.33	0.39	1.3	1.7	0.1	0.4
15	FRA	45MF6.1	0.41	0.48	1.3	1.7	0.1	0.4
16	FRA	45MF6	0.41	0.48	1.3	1.7	0.1	0.4
17	GBR	120 M 36	0.32	0.4	1	1.4	0.1	0.4
18	GBR	135 M 44	0.4	0.48	1.2	1.5	0.1	0.4
19	GBR	150 M 36	0.32	0.4	1.3	1.7	0.1	0.4
20	GBR	150 M 40	0.36	0.44	1.3	1.7	0.1	0.4
21	GBR	212 M 36	0.32	0.4	1	1.4	0	0.25
22	GBR	212 A 42	0.4	0.45	1	1.3	0	0.25
23	GBR	216 M 44	0.4	0.48	1.2	1.5	0	0.25
24	GBR	216 M 36	0.32	0.4	1.3	1.7	0	0.25
25	GBR	225 M 36	0.32	0.4	1	1.4	0	0.25
26	GBR	226 M 44	0.4	0.48	1.3	1.7	0	0.25
27	HUN	MN2/MN2E	0.33	0.4	1.3	1.65	0	0.4
28	ITA	CF35SMN10	0.32	0.39	1.35	1.65	0	0.3
29	ITA	CF44SMN28	0.4	0.48	1.35	1.65	0	0.3
30	ITA	CF44SMNPB28	0.4	0.48	1.35	1.65	0	0.3
31	ITA	GC 37	0.34	0.4	0.9	1.2	0	0.5
32	JPN	SMN 433(H)	0.29	0.36	1.15	1.65	0.15	0.35
33	JPN	SMN 438(H)	0.34	0.41	1.3	1.7	0.15	0.35
34	JPN	SMN 443(H)	0.39	0.46	1.3	1.7	0.15	0.35
35	RCS	13141	0.27	0.35	1.2	1.6	0.17	0.37
36	ROM	AUT 40MN	0.35	0.45	1.2	1.6	0.15	0.35
37	SWE	1957	0.32	0.39	0.9	1.4	0.15	0.4
38	SWE	2120	0.38	0.45	1.1	1.4	0.1	0.4
39	USA	ASTM 1335	0.33	0.38	1.6	1.9	0.15	0.35
40	USA	ASTM 1340	0.33	0.43	1.6	1.9	0.15	0.35
41	USA	ASTM 1345	0.43	0.48	1.6	1.9	0.15	0.35
42	USA	ASTM 1541H	0.35	0.45	1.25	1.75	0.15	0.35

Table 4. The results of computing the equivalence degrees of the steels in the database with the reference steel AUT 40MN

ID	Equivalence	ID	equivalence	ID	equivalence
0	0.38	15	0.43	30	0.35
1	0.38	16	0.43	31	0.10
2	0.24	17	0.41	32	0.40
3	0.22	18	0.50	33	0.66
4	0.37	19	0.52	34	0.66
5	0.21	20	0.69	35	0.42
6	0.62	21	0.32	36	1.00
7	0.56	22	0.16	37	0.34
8	0.38	23	0.39	38	0.45
9	0.43	24	0.40	39	0.15
10	0.37	25	0.32	40	0.38
11	0.13	26	0.40	41	0.13
12	0.31	27	0.34	42	0.87
13	0.65	28	0.29		
14	0.38	29	0.35		

4 Conclusion

A reflexive and symmetrical fuzzy relation for comparing the matching degree of two intervals was proposed.

The partial components of the proposed relation refer respectively to the lengths and concentricity of the given intervals. They are also symmetrical and reflexive fuzzy relations and can be used separately if needed.

The defined fuzzy relation was applied to define a degree of chemical equivalence of steels. Practical experiments showed this method as a convenient way to evaluate the chemical equivalence of steels.

The method of determining equivalence of steels can be refined by using the Multi-Attribute Decision Model. Existing procedures can be used to assign weights that represent levels of importance of different chemical elements.

References

1. Călin, M., Alexandru, I., Buraga, S.C.: Application of Fuzzy Database Management in Steel Selection. Proc. of Conference on Materials EUROMAT '98, Lisbon, Portugal, July 22-24, (1998)

2. Hiltner, J.: Image Processing with Fuzzy Logic. Lecture notes, Summer School "Intelligent Technologies and Soft Computing", Black Sea University, Romania, Sept. 22-28, (1996)
3. Negoita, C.V., Ralescu, D.A.: Fuzzy Sets and Applications (published in Romanian), Editura Tehnica, Bucharest, (1974)
4. Reusch, B.: Mathematics of Fuzzy Logic. In: Zimmermann, H-J et. al. (eds.) Real World Applications of Intelligent Technologies, Publishing House of the Romanian Academy, (1996).
5. Saaty, T.L.: The Analytic Hierarchy Process: Planning, Priority Setting, Resource Allocation. McGraw - Hill, NY, (1980)
6. Zimmermann, H-J.: Fuzzy Decision Support Systems. In: Zimmermann, H-J., et. al. (eds.). Real World Applications of Intelligent Technologies, Publishing House of the Romanian Academy, (1996)

Tomography in Fractal Neural Nets

Thomas Kromer

Zentrum für Psychiatrie, Münsterklinik Zwiefalten, 88521 Zwiefalten
Thomas.Kromer@t-online.de
http://home.t-online.de/home/Thomas.Kromer/kromt2.htm

Abstract. Tomographic procedures use multiple straight rays, running through or originating from any point P of a structure to many detector elements at the periphery. The measured specific values, are used to reconstruct the structure virtually by tracing the according rays back. In Fractal Neural Nets, whose connectivistic architecture is reflecting fractal functions, especially the squaring of complex numbers, we get the connections ordered in form of binary trees. Activation of any neuron will successively cause a wave of activities spreading over the net, approaching to the periphery. There, neurons may work as detector elements, recording the arriving sequence of activity in form of memory-strings. These memory-strings could be used any time later to activate the net in reversed direction to reconstruct the original pattern by a process, related to tomography, just not being based on straight continous rays, but on saltatory retracable sequences of neural activations.

1 Introduction

Fink[1] has shown a new principle for acoustic media: If a sharp impulse is activating at a certain point P a water-surface or the surface of a silicium waver, it is possible to record at any other point R of the surfaces the sequence of the amplitudes of the interfering and reflected waves. This record may be used to reactivate the surface at point R reversely (Sending the last received amplitudes at first to the surface). The resulting waves will first spread circular over the surface, be reflected at the edge, interfere and quite astonishingly finally will run together at the original point P , causing there a sharp impulse again. Thus the original activation-pattern is reconstructed on the surfaces. This principle may be used to perform a new kind of data-processing in neural nets, based on neural chains[3,4]. Temporal sequences of activations in chains of neurons may be recorded. These records allow to reconstruct the original pattern in the neural chain by reversed activating of the chain. Fractal Neural Nets do not need to wire the neurons of the chains directly, because very elegantly these chains arise functionally, due to the specific structure of connectivity, determined by fractal functions. This shall be shown in this paper.

2 Architecture and Dynamics of Fractal Neural Nets

Fractal algorithms may determine the macroarchitecture of fractal neural networks[2]. The beautiful fractals of Julia sets and the Mandelbrot set are generated by the iteration of function: $f(z)=c+z^2$. In its basic form , c set zero, it describes the squaring

B. Reusch (Ed.): Fuzzy Days 2001, LNCS 2206, pp. 917–923, 2001.

of complex numbers. This function $f(z)=z^2$ is the base for all examples of neural nets, presented in this paper.

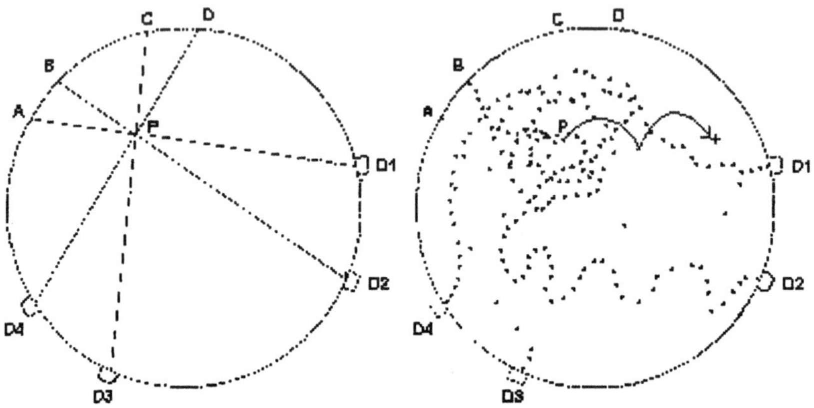

Fig. 1. Tomography uses multiple rays running through or originating from any point P to detectors D at the periphery. Their course has to be reversible, but not necessarily only straight(left), but may also be curved, spiralic or even saltatory (right).The saltatory principle is realized in fractal neural networks.

In a simple model, neurons cover a round plane, representing the region of the complex plane within the unit circle. All neurons z send their axon to a subsequent neuron located at z^2 and to the two neurons at $-z^{1/2}$ and $+z^{1/2}$ (the two square roots of z). All connections should be recurrent, the neurons be working synchronously. The axons of the neurons will follow the course of the trajectories of the function $f(z)=z^2$. Using the opposite direction by the recurrent connections, the course of the neural activity will represent the reversed function $f(z)=+-z^{1/2}$. Based on the geometrical interpretation of the squaring of complex numbers[6] we get a system of segments of logarithmic spirals, shown in figure 2 for the two quadrants with positive imaginary coefficients, leading from the neurons at any point z to the neuron at the according point z^2.

The result will be a macroarchitecture built by many interwoven binary trees[2].

Activation of any neuron z will activate three neurons. First, the neuron at z^2.

Using the reversed direction, $f(z)=+-z^{1/2}$, the two neurons, representing the square roots, will be activated and activate their own square-root-neurons and neuron z in the next step. Thus, step by step, an exponentially growing crowd of neurons, approaching the unit circle, will be activated. All neurons (or only some specialised in the periphery) should be able to record the sequence of activity, they will receive step by step in the sequence of n iterations. The storage-neurons just „listen" to the net, like microphones, recording the incoming „waves" of neural activity digitally on "tapes", the memory-strings. According to the principles shown by Fink[1], the storage-neurons use the „tapes" (the memory-strings). In computer simulations, these memory-strings are just strings of values, in biological systems, they could be molecular strings[4]. They may playing these „tapes" reversely („last in, first out"), to reactivate the net. After n iterations again, the original activation pattern in the net will be reconstructed.

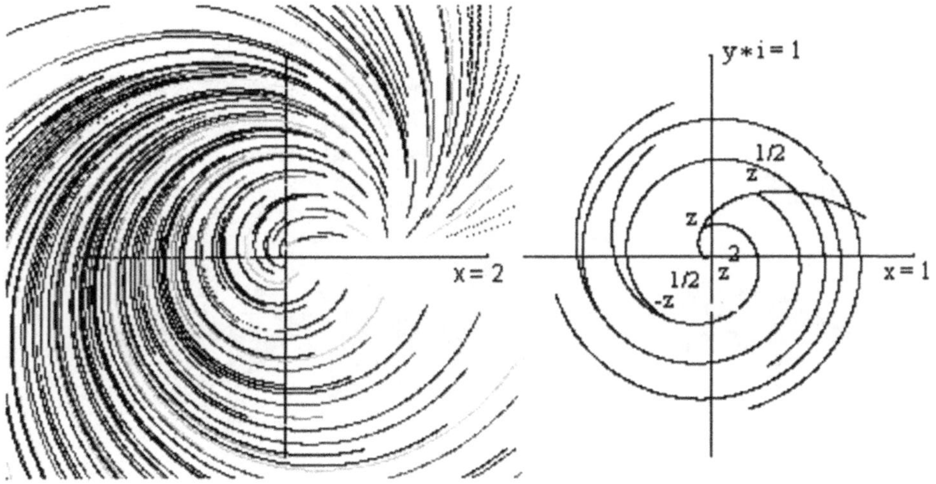

Fig. 2. Left: Trajectorial system of axons leading counterclockwise from any point z to z^2 .Right: Binary tree formed by functions $f(z)=z^2$ and $f(z)=+-z^{1/2}$. (Please notice the different scales)

3 Examples of Fractal Neural Nets

3.1 Methods and Materials

As examples of fractal neural networks two nets are presented:
Net I: consisting of about 31410 neurons, covering a circular layer, representing the unit circle of the complex plane. Each neuron is of quadratic shape, each side measuring 0.01. The temporal sequence of activity will be stored by 704 neurons near the periphery(Fig.3). Learning is done with the function $f(z)=+-z^{1/2}$. The activity spreading out from an active neuron will reach several storage-neurons at different iterations. In the computer program all coordinates, which will be hit in the course of 8 iterations, starting at any neuron, are calculated. After each calculation of these coordinates it is proved, whether a storage-neuron will have been hit. In this case, the storage-neuron will record this hit. Countering the hits at each iteration, they form the memory-strings. Calculating the coordinates, to which activity will be projected, it is assumed, that there will be no errors by rounding. (Assuming, the neurons could do a perfect topolographic representation of intermediate values by means of the overlapping trees of their axonal and dendritic endings, making up a quasi-continuum of synaptical contacts).

Reconstruction of patterns will work with function $f(z)=z^2$. A problem arises: The memory-strings provide no information, at which part of the input region the storage-neuron originally has been hit. Reconstruction starts with the central coordinates of the storage-neuron. To compensate this error to some degree, in the presented illustrations, not only the neuron, directly hit in the course of iterations, will be activated, but also all its directly neighboured neurons.

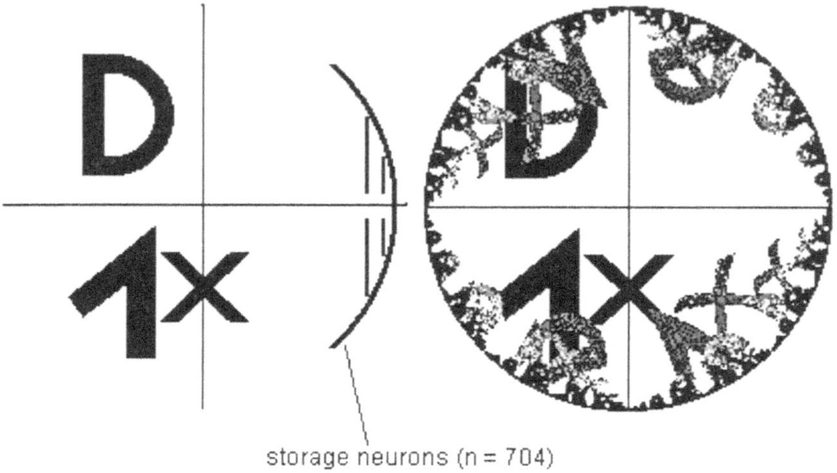

Fig. 3. Learning of a pattern (Net I): Activity is spreading over the net, starting from active neurons (left), approaching the periphery of the unit circle, at last arriving at the storage neurons, which record the according sequences of activity.

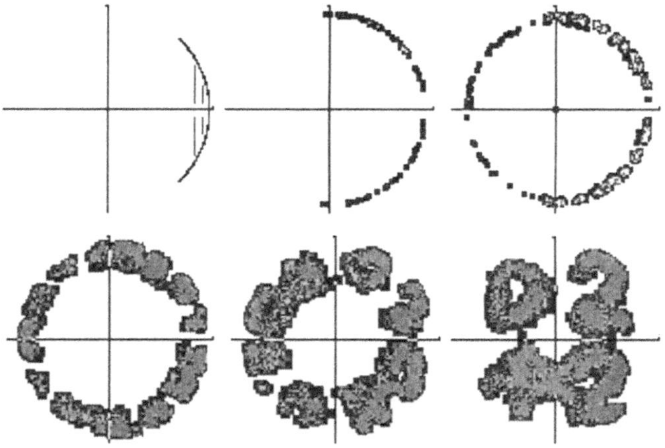

Fig. 4. (Net I) Successive steps of pattern reconstruction. The storage neurons (above, left) use the recorded memory string to activate their subsequent neurons (according function $f(z)=z^2$) reversely. After 6 steps, the original pattern will be reconstructed.

Net II: it is formed by 4970 chains, each consisting of 7 neurons. The neurons are regarded to be fixedly wired within the chains. This makes sure, that each neuron will activate only one neighboured neuron in each direction of the chain. Thus, any projection-errors will be avoided.

In both examples, neurons work synchronously. One working cycle of the nets consists of a) producing a memory-string, b)comparing this to the strings of the long-term memory, c)choosing the most-homologous string, d) taking this string to reconstruct the according pattern, which will work as starting pattern again. To perform pattern completion, two working cycles are completed.

Q-basic has been used as programming language for all programs.

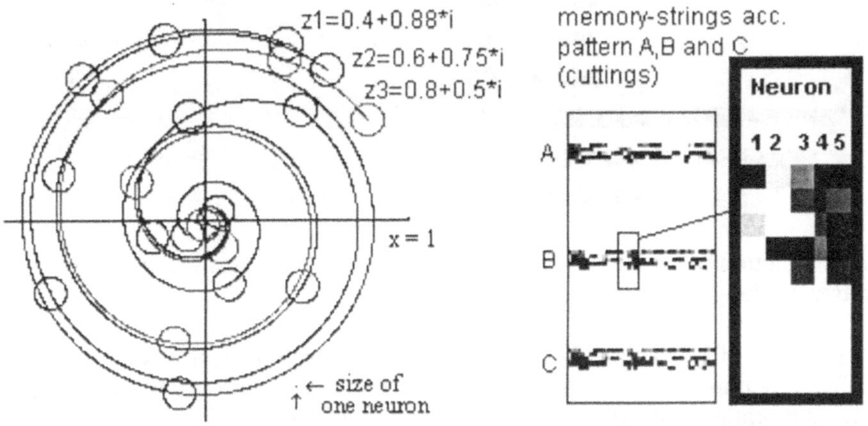

Fig. 5. (Net II) Three of 4970 neural chains, each consisting of 7 neurons (the circles only mark by their centre the position of each neuron in the chains. Coordinates of the storage-neurons are indicated. In the middle of the structure a region of high density of neurons is resulting. Right: Examples of different memory-strings.

3.2 Results

Both presented nets are able to store and to reproduce patterns by the described processes. By comparison and associative selection of similar memory-strings[4], these nets are able to perform associative processing of patterns. As an example, the completion of a pattern by net II is shown. Net II with its directly wired neurons is favourable to perform this task, because its structure is avoiding errors in the reactivation of the net as they occur in Net I. Nevertheless it should be not too difficult, to minimize these projection-errors in nets like Net I, considering the convergence and divergence of structures in the course of iterations. In both examples little use is made of the fact, that in the centre of the nets a region of high density of neurons with small sizes is showing up. In this region, the nets are able to perform high resolution and sharp discrimination between patterns, because nearly all storage neurons will get information about this area and may take part in the associative pattern-processing. This area remembers (occasionally?) to the the Macula densa in the human retina, with its high resolution compared to peripheral regions.

Especially in Net I, the remarkable projection-errors will be minimized near the centre, because of the convergence of the projecting trajectories in this region.

Fig. 6. Completion of a pattern, (one of 5 learnt earlier), presented incompletely (left), within two working cycles of Net II.

The structure of both networks is biomorph, the logarithmic spiral being a pattern, very often resulting in processes of growth. Especially the strictly topographic projections resemble to the somatotopic organization of biological nervous systems.

Recording in biological cells could hypothetically be done by nucleic-acids (RNA or DNA or other molecular strings), each triplet of nucleotides representing a certain degree of activation [3,4].

Because of the disseminated storage of information and the quantity of different ways, which are used for pattern reproduction, these nets are quite well protected against local lesions. The memory-function gets by that point some holographic characteristics. We also may receive a complete pattern, even if we use only a part of the memory-neurons for pattern reproduction. Resolution and contrast will be lowered, but the general stucture of the pattern will be reconstructed (results not shown). Recording the sum of activations, arriving at storage-neurons at any time, even storing and reconstructing of changing patterns should be realizable, in this case, the neurons may even work asynchronously.

Advantageously, each pattern will be learned at one presentation, the structure of the net remains unchanged, because learning does not affect the synaptical connections of the net. The synapses are involved in these nets to guarantee a correct topographic projection by the fine resolution of their projections, made available by the overlapping trees of their axonal and dendritic endings. Synaptical learning might be involved in optimization of often used pathways.

These fractal neural nets differ clearly from neural nets used in fractal image compressing,. For image compression, the structure of the neural net is determined by the structure of the image[5]. Similar areas of the image will be projected onto another in an iterative process. In contrast, in the presented fractal neural nets, the structure may not be changed according to the various patterns, but will be suitable for all patterns.

4 Conclusion

Storing and reconstruction of activation patterns in the presented fractal nets correspond to principles of computer-tomography, just not using straight rays, but saltatory sequences of neural activities, spreading over the neural plane. These sequences may be used to store, compare and reconstruct the patterns of activity.

Fractal neural nets, based on these principles, may perform an effective data-processing.

Simulations show the applicability of the described processes, which may even give us some new ideas about the function of biological neural networks.

References

1 Fink M (1996)Time Reversal in Acoustics , Contemporary Physics 37 , 2 : 95 - 109

2 Kromer T , "Biomorph Neural Networks Based on Fractal Algorithms",Computational Intelligence, proceedings/ 6th Fuzzy Days, Dortmund, Germany, May 25-27,1999. B. Reusch (ed.). Springer 1999, (lecture notes in computer science; Vol. 1625, pp 603-614)

3 Kromer T, "Quechua Quipus, Fraktale Netze und Assoziatives Gedächtnis" ,german abstract, 26. Jahrestagung der Arbeitsgemeinschaft für Aphasieforschung und -behandlung, Konstanz,11/1999

4 Kromer T, New Neural Nets, for publication accepted, proceedings/ 7th Fuzzy Days, Dortmund, Germany, October, 2001. B. Reusch (ed.). Springer 2001, (lecture notes in computer science; same volume)

5 Skarbek W, Fractal Basis Functions for Pattern Recognition, Springer 1997, lecture notes of computer science, 176-189

6 Pieper H (1985) Die komplexen Zahlen . Deutsch , Frankfurt a. Main

Fuzzy Approach in Ultrasound Image Segmentation

Radim Kolář and Jiří Kozumplík

Department of Biomedical Engineering,
Faculty of Electrical Engineering and Computer Science
Brno University of Technology,
Purkyňova 118, Brno, Czech Republic
{kolarr,kozumpli}@dbme.fee.vutbr.cz

Abstract. This contribution present a new approach for ultrasound image segmentation, based on the fuzzy theory. The segmentation is performed on the 2D envelope of radiofrequency (RF) signals instead of the B-mode image. The segmented image is then used to create the sector image.

1 Introduction

The automatic segmentation of ultrasound (US) images is very difficult task, which has not been resolved yet. This is caused by its very poor quality, i.e. low contrast, low signal to noise ratio and presence of speckles. The impact of these properties cause, that the classical method (like 2D convolution operators) cannot be used for the object detection or image segmentation. Therefore some advanced methods have to be utilized for this objective. One of these method is fuzzy based thresholding [1, 2]. The presented method should therefore contribute to solve this difficult task.

2 US Signals

The final 2D images (B-mode images) are the output of standard procedure, which consists of several steps as shown in Fig. 1:

1. A finite number of 1D RF signals are measured from the output of the ultrasound probe. This set of 1D signals (A-scans) is stored in a matrix.
2. The envelope of each RF signal is detected.
3. The data reduction is performed (e.g. downsampling).
4. The magnitude compression is performed (e.g. logarithmic function).
5. The set of these envelopes is converted from polar to rectangular grid (e.g. linear interpolation) and the B-mode image is obtained.

The segmentation is usually performed in the resulted B-mode image in 2D. But it is clear, that the previous processing is nonlinear and also some information loss occurs. Therefore we focused in processing the envelope of the RF signal and the result was then converted to rectangular grid to create the sector segmented image.

B. Reusch (Ed.): Fuzzy Days 2001, LNCS 2206, pp. 924–929, 2001.

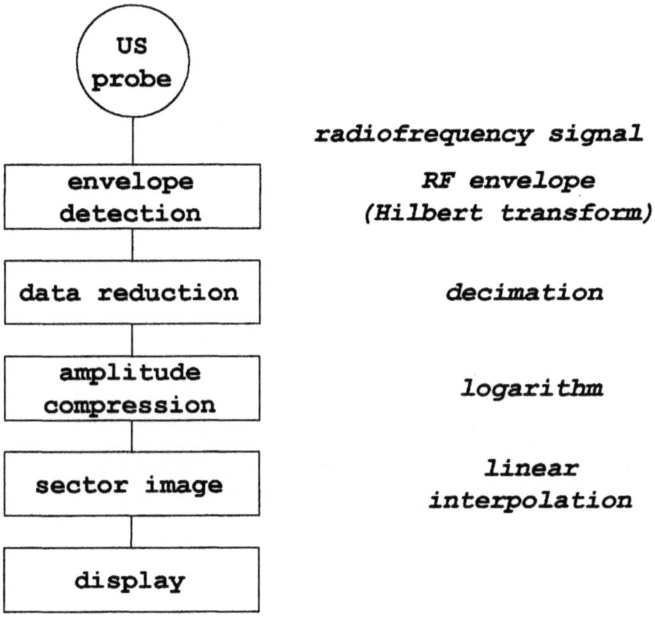

Fig. 1. Processing of the US signals

3 Fuzzy Segmentation

Fuzzy segmentation brings new look on the pixel classification problem. We want to find some optimal pixel representation according to the a priori known membership function, e.g. S-function[1]. This function is defined by these three parameters - the inflection point b and begin and end of the fuzzy region - a, c (see Fig. 2).

When we are looking for these optimal values, we have to determine some criteria, according which, the values are optimal. We have used fuzzy entropy that is defined as [1]

$$H(A, a, b, c) = -\frac{1}{log_2 N} \sum_{i=1}^{N} P(A_i, a, b, c) log_2 P(A_i, a, b, c), \qquad (1)$$

where a, b, c are the S-function parameters, N is a number of sets, to which the pixels are classified and A_i represents particular fuzzy sets. For example if we want to classify the pixels to the three sets, then (A_i,A_2,A_3)=(dark, gray, white). For binary classification only the two sets are used. The value $P()$ represents the probability of fuzzy event based on the membership function. Hence, before this value is evaluated, the histogram of gray values has to be determine. We would like to have maximal fuzzy entropy, when the N and A_i is given.

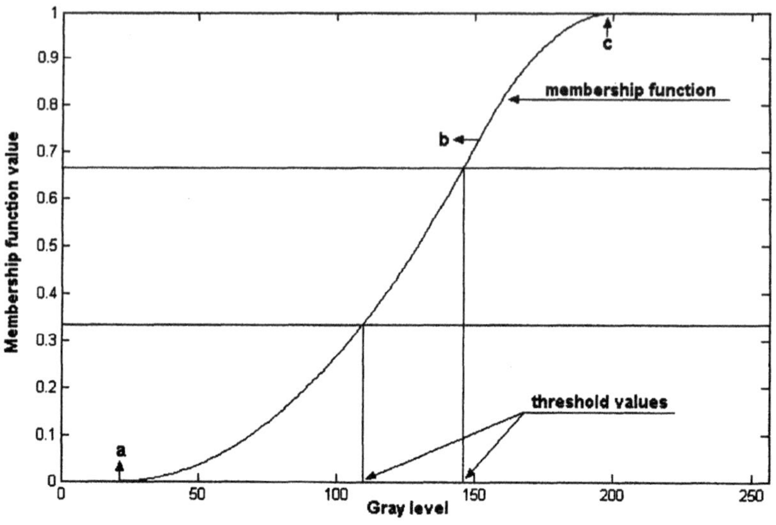

Fig. 2. Determination of the threshold values.

Ones the values a, b, c are assessed, the thresholds for image segmentation can be determine from the boundaries between the particular fuzzy sets (see Fig. 2).

4 Optimization

Because the optimization task has three degrees of freedom, some optimization algorithms have to be used. We have used the genetic algorithm (GA) with binary value representation[1, 3].

The GAs takes an inspiration from nature. Each sought values is a part of chromosomes and a new values (possible solution) are generated by crossovers and mutations from the elit ones. Some parameters need to be defined for GA:

- *object function* is a measure of fitness for each chromosomes.
- *maximal number of generations.*
- *population size* is the number of chromosomes in each generation.
- *probability of crossover* P_c represents probability with which will two chromosomes cross. This operation is performed if some random number is below P_c.
- *probability of mutation* P_m represents probability of changes 0 to 1 or 1 to for each element in chromosome. Mutation is performed if coming random number is below P_m. Next random number then indicate the mutation position.

When the optimization is running, we must ensure, that the new chromosomes always will satisfy the condition $0 \leq a < b < c \leq 255$. The best fitness is shown on the Fig. 3 as a function of generation.

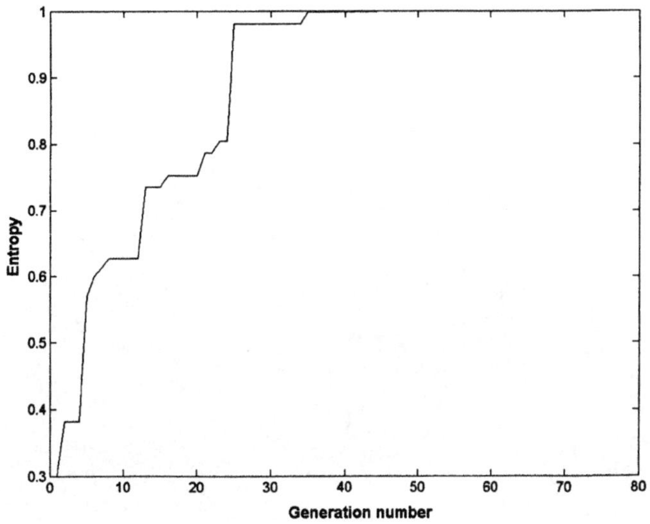

Fig. 3. The best fitness values during the optimalization.

5 Experimental Results

The method described above was used for segmentation of ultrasound images by processing the envelope of the RF signals. The result, i.e. the thresholded image was converted to rectangular grid.

Fig.4 shows a tissue of the pig heart and Fig.5 its binary representation obtained by the optimal thresholding. We can see, that some objects appear more visible than in the original image. Fig.6 shows the 5-level thresholded image and therefore more details are visible. The object borders could be better marked than in the original image.

6 Conclusion

The presented method offers simple and formalized approach to ultrasound image segmentation based on fuzzy entropy maximization. We can choose the number of levels, the type of membership function and also the definition of fuzzy entropy. Also this approach is novel, because we are working with the envelope of the RF signals and the result is converted to the sector image.

Acknowledgment

This work has been supported by the grant No. 102/99/1228 of the Grant Agency of the Czech Republic and states grant No. CEZ J22/98:262200011.

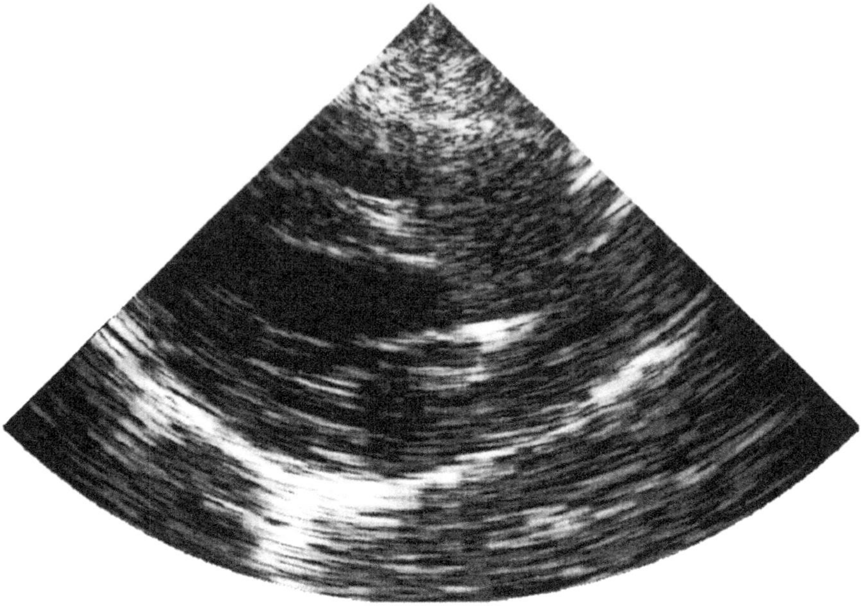

Fig. 4. Original ultrasound image.

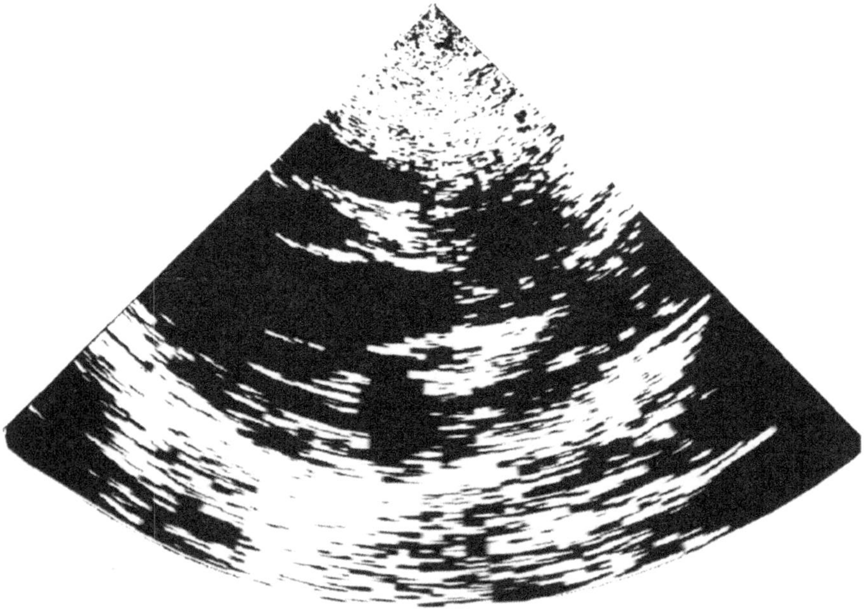

Fig. 5. Binary representation of the ultrasound image.

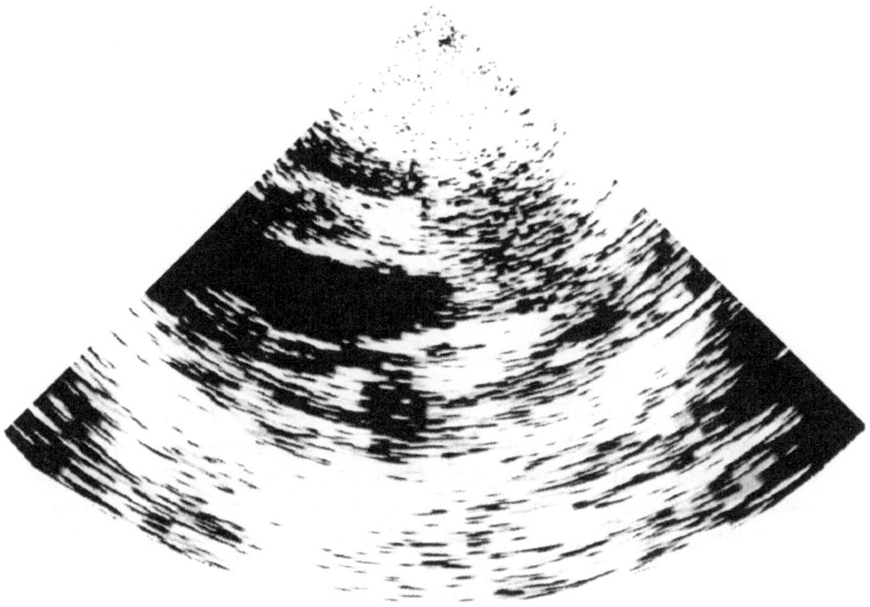

Fig. 6. 5 level representation of the ultrasound image.

References

1. Cheng,H.D., Yen-Hung Chen, Ying Sun,: A novel fuzzy entropyapproach to image enhancement and thresholding. Signal Processing 75, 1999, pp. 277-301
2. Kosko B.: Fuzzy engineering. Prentice-Hall, 1997
3. Houck,Ch.,R., Joines,J.,A., Kay, M., G.: A Genetic algorithm forFunction Optimalization: A Matlab Implementation. http://www.ie.ncsu.edu/mirage/, 1995,

Dynamic Decision Making Based on Partial Probability Information

Dipl.-Inform. Gero Presser

University of Dortmund, Department of Computer Science I,
D-44221 Dortmund, Germany.
Phone: +49 (0)231 755 6309, Fax: +49 (0)231 755 6555
Gero.Presser@QuinScape.de

Abstract. In this paper we introduce Lazy Decision Making which is a framework for dynamic decision making under uncertainty. The key idea is to start with an imprecise description of the agent's knowledge concerning the state of nature (by means of a set of probability distributions) and to successively refine this description (i.e., the corresponding set) until a "good" decision can be made.
The crucial point about this scheme is that the refinement has not to be done in an arbitrary way (which might be extremely inefficient). The algorithm assists the agent by providing him hints on the effect of a refinement of the constraints in question hence guiding him in making the *relevant* things more precise.

1 Motivation

Following the most common approach of decision making under uncertainty (see [1, 8, 6]), the agent's knowledge is represented by means of a probability distribution (defined on the set of states of nature). However, since in many cases it is almost impossible to obtain the necessary probabilities, an approach using *sets* of probability distributions – which are called partial probability information (ppi for short) – is more flexible (see [3]).

In this paper we introduce a framework for *dynamic* decision making based on partial probability information which we denote by *Lazy Decision Making.* The key idea is to start with a quite vague description of the agent's knowledge and to successively make this description more precise until a "good" decision can be made. This refinement is not done in an arbitrary way, the agent is supported by some measures. Hence, only the relevant parts of the description need be refined.

We start with a short introduction to classical decision making under uncertainty (Section 2) and decision making based on partial probability information (Section 3). Next, we introduce Lazy Decision Making (Section 4) and try to illustrate its key elements. However, due to the restricted space we can only address selected topics. The paper closes with a short summary and an outlook, in which we point to topics that are still under research (Section 5).

B. Reusch (Ed.): Fuzzy Days 2001, LNCS 2206, pp. 930–936, 2001.

2 Decision Theory

The aim of prescriptive decision theory is to assist agents in the demanding question how decisions should be made as a matter of rationality (see [1, 8, 6]). An agent, which is supposed to act "rational", has to chose exactly one alternative from a given set of alternatives $\mathcal{A}$. Hence the problem lies in the ranking of the alternatives.

In this paper we exclusively investigate decision making under *uncertainty*, since this branch of decision theory is the origin of Lazy Decision Making.[1]

The most important aspect of decision making under uncertainty is, that the "utility" of an alternative depends on the state of nature which is uncertain. For any possible state $s \in \mathcal{S}$ (where $\mathcal{S}$ denotes the set of possible states of nature) and any alternative $a \in \mathcal{A}$, the utility $u(a, s) \in \mathbb{R}$ of a given s is assumed to be known. If both the set $\mathcal{A}$ and $\mathcal{S}$ are finite, the values $u(a, s)$ can be represented by means of a matrix, the so-called decision matrix.

In decision making under *risk*, the knowledge of the agent concerning the state of nature is described by means of a probability distribution $\pi : \mathcal{S} \to [0, 1]$ on the set of states (i.e., $\sum_{s \in \mathcal{S}} \pi(s) = 1$). Note, that subjective probabilities are permitted (and frequently used). If the agent accepts some well known (and, what is more important, well established) axioms – the so-called axioms of rational choice (see [14, 7]) – and if the utility function u fulfils some specific properties, then it can be proved that the agent should rank the alternatives according to their expected utilities $\sum_{s \in \mathcal{S}} \pi(s)u(a, s)$ (for $a \in \mathcal{A}$). This result is the famous BERNOULLI theorem.

In decision making under *ignorance* it is assumed that the agent exclusively knows $\mathcal{S}$ but has no further knowledge concerning the state of nature. For this somewhat pathological case, many heuristics have been proposed. E.g., following the *Maximin* rule, the agent should chose an alternative $a \in \mathcal{A}$ that maximises the worst-case utility, i.e. $\min\{u(a, s) \mid s \in \mathcal{S}\}$ (or – more general – $\inf\{u(a, s) \mid s \in \mathcal{S}\}$).[2] However, in the case of ignorance there is no such convincing justification as there is in the risk-case for the BERNOULLI principle.

3 Static Decision Making Based on Partial Probability Information

Many important situations seem to be "between" risk and uncertainty: The agent does not know exact probabilities but has "some" knowledge concerning the state of nature. (E.g., he could know that state s_1 is more likely than state

[1] However, there are also applications of Lazy Decision Making in other branches of decision theory like in multiple attribute decision making and in group decision making (see [11]).

[2] Other frequently used heuristics include the *Maximax* rule (the agent maximises the best-case utility $\sup\{u(a, s) \mid s \in \mathcal{S}\}$), the HURWICZ principle (a "linear" combination of the Maximin and Maximax rule) and the LAPLACE rule (where it is assumed that all states are equally probable, i.e. have probability $1/n$). See [15, 2, 6] for details.

s_2.) Very frequently, this case is mapped to a decision making problem under risk by simply estimating the lacking probabilities. However, this does not really meet the situation and is some kind of a temporary solution which can lead to awkward decisions.

A quite versatile solution to this problem is to employ *partial probability information* (ppi for short, see [3]). A ppi is a *set* of probability distributions defined on the set of states $\mathcal{S}$, i.e., a subset of the so called probability simplex

$$\mathbb{V} = \{\pi : \mathcal{S} \to [0,1] \mid \sum_{s \in \mathcal{S}} \pi(s) = 1\}.$$

A ppi should be used to represent the set of all probability distributions that are consistent with the agent's knowledge.

By means of some examples we would like to show how ppi's can be applied in practical settings. In the following we assume that $s_1, s_2 \in \mathcal{S}$.

(1) The ppi $P_1 = \{\pi \in \mathbb{V} \mid \pi(s_1) \geq \pi(s_2)\}$ perfectly meets the situation where the agent is convinced, that s_2 is at most as likely as s_1. Hence it describes a *comparative* probability information.
(2) The ppi $P_2 = \{\pi \in \mathbb{V} \mid \pi(s_1) \in [0.1, 0.3]\}$ can be used when the agent has the opinion that the (maybe subjective) probability of s_1 is between 0.1 and 0.3. Thus, an *interval* probability is described.
(3) Ppi's can easily be combined. E.g., $P_3 = P_1 \cap P_2$ can be used when (1) as well as (2) holds.

Static decision making based on ppi is quite versatile since it covers many important special cases. E.g., decision making under risk and uncertainty as well as comparative probabilities, interval probabilities and DEMPSTER-SHAFER theory of evidence[3] is contained.

KOFLER and MENGES have shown (see [3, 4]), that in a decision making problem with ppi P, the agent should chose an alternative $a \in \mathcal{A}$ that maximises

$$\inf\{\sum_{s \in \mathcal{S}} \pi(s) u(a, s) \mid \pi \in P\}.$$

This rule is called the MaxE_{inf}-principle. However, there are other heuristics that also seem to have some reasonable justification, for example a generalisation of the famous HURWICZ-principle. Most decision rules for decision making based on ppi's are combinations of the BERNOULLI principle and a heuristic for decision making under ignorance.[4]

[3] At least this holds for the probability oriented interpretation of theory of evidence (see [12, 5]).

[4] This point is not surprising since ppi's are somewhat between risk (one probability distribution) and ignorance (the set of all probability distributions, i.e., the probability simplex $\mathbb{V}$).

4 Lazy Decision Making

Using SMETS's terminology (see [13]), a ppi represents *imprecise* knowledge concerning the *uncertain* state of nature. More exactly, probabilities are used to model the uncertainty whereas a set (of probability distributions) is used to deal with imprecision.

Consider the following (dynamic) approach: The agent starts with a decision making problem under ignorance (which corresponds to the completely vacuous ppi $P = \mathbb{V}$). Step-by-step he reduces the size of the ppi and hence gives a more precise description of his knowledge. Eventually, he would obtain a decision making problem under risk (when there is only one probability distribution left in the ppi). This problem under risk can be solved using the BERNOULLI principle.

Lazy Decision Making (see [9, 10]) is based on this idea. The agent starts with an imprecise description of his knowledge and makes this more precise until a "good" alternative can be determined. The algorithm assists the agent by supplying data on which information should be made more precise and on how "good" a proposed alternative really is.

Next, we discuss the approach in more detail. In Lazy Decision Making, only *linear* ppi's are used.[5] We assume that both the set of states $\mathcal{S}$ and the set of alternatives $\mathcal{A}$ are finite and (wlog) $\mathcal{S} = \{1, 2, \ldots n\}$. Hence, we can represent a probability distribution $\pi : \mathcal{S} \to [0, 1]$ by means of a vector $\boldsymbol{\pi} = (\pi(1), \pi(2), \ldots, \pi(n))^T \in \mathbb{R}^n$. A ppi $P \subseteq \mathbb{V}$ is said to be linear if there exist $m \in \mathbb{N}, \boldsymbol{A} \in \mathbb{R}^{n \times m}, \boldsymbol{b} \in \mathbb{R}^m$ such that $P = \{\boldsymbol{\pi} \mid \boldsymbol{A}\boldsymbol{\pi} \le \boldsymbol{b}\}$. We use $\boldsymbol{A}$ and $\boldsymbol{b}$ to represent the linear ppi P. Note, that this representation consists of m contraints (which are somewhat "hidden" using the matrix notation).

With any alternative $a \in \mathcal{A}$ we associate a vector

$$\boldsymbol{a} = (u(a, 1), u(a, 2), \ldots, u(a, n))^T \in \mathbb{R}^n$$

which represents the corresponding (state-dependant) utilities.[6]

During any iteration, the algorithm performs the following steps (which we describe in more detail in a moment):

(1) Eliminate all dominated alternatives.
(2) Compute a suggestion based on a heuristic.
(3) Compute values to assist the user.
(4) If the user makes some refinement: Update ppi and goto (1).

At first, all dominated alternatives are (successively) eliminated.[7] Next, a suggestion $a^* \in \mathcal{A}$ is computed using some heuristic (e.g., the MaxE_{inf}-principle).

[5] Note that this holds for most of the work done in the field of decision making based on ppi's. Linear ppi's are easy to describe (by means of linear constraints) and have a rich expressiveness. In particular, the above mentioned special cases (comparative and interval probabilities, ...) are contained.

[6] Note that, given a probability distribution $\boldsymbol{\pi} \in \mathbb{V}$, $\boldsymbol{a}^T\boldsymbol{\pi}$ denotes the expected utility of alternative a (represented by $\boldsymbol{a}$).

[7] In this context an alternative $a_1 \in \mathcal{A}$ is said to be dominated by $a_2 \in \mathcal{A}$ if $\boldsymbol{a_2}^T\boldsymbol{\pi} \ge \boldsymbol{a_1}^T\boldsymbol{\pi}$ holds for each $\boldsymbol{\pi} \in P$. If a_1 is dominated by a_2, a_1 can be eliminated since a_2 is "at least as good" as a_1.

Probably the most important aspect about the algorithm is, how the agent is being supported in tightening the "right" constraints.[8] For this purpose, different values are computed. Due to the restricted space, we describe the procedure for only one value, the so-called *relative regret value*. Let $\phi : \mathcal{A} \to \mathbb{R}$ be defined by

$$\phi(a) := \sup\{\boldsymbol{a}^T\boldsymbol{\pi} - (\boldsymbol{a}^*)^T\boldsymbol{\pi} \mid \boldsymbol{\pi} \in P\}$$

for all $a \in \mathcal{A}$. The value $\phi(\boldsymbol{a})$ is an upper bound on the lost (expected) utility, that could result from chosing $\boldsymbol{a}^*$ instead of $\boldsymbol{a}$. Let $a^- \in \mathcal{A}$ be such that $\phi(\boldsymbol{a}^-) = \max\{\phi(\boldsymbol{a}) \mid a \in \mathcal{A}\}$. Consequently, $\phi(\boldsymbol{a}^-)$ bounds the potential "damage" that could result from chosing $\boldsymbol{a}^*$. Hence, if $\phi(\boldsymbol{a}^-)$ is "reasonable small", the agent can choose the suggested alternative $\boldsymbol{a}^*$ without having to fear that he "wastes" too much achievable (expected) utility.

The crucial point is that, given $\boldsymbol{a}^-$ and $\boldsymbol{a}^*$, $\phi(\boldsymbol{a}^-)$ can be rewritten as the result of a linear optimisation problem, namely:

$$\max(\boldsymbol{a}^- - \boldsymbol{a}^*)^T\boldsymbol{\pi} \text{ subject to } \boldsymbol{A\pi} \leq \boldsymbol{b}.$$

In theory of (linear) optimisation, the concept of *shadow prices* is well-known. Technically, shadow prices are the solution of the dual problem. For any constraint $\boldsymbol{a}_i^T\boldsymbol{\pi} \leq b_i$ (which is part of $\boldsymbol{Ap} \leq \boldsymbol{b}$) there exists a uniquely determined, non-negative shadow price s_i with the following interpretation: If the constraint is tightened by $\varepsilon > 0$ (i.e., if $\boldsymbol{a}^T\boldsymbol{\pi} \leq b_i$ is replaced by $\boldsymbol{a}_i^T\boldsymbol{\pi} \leq b_i - \varepsilon$), the solution of the optimization problem (its optimal value, i.e. $\phi(\boldsymbol{a}^-)$) will decrease by $s_i\varepsilon$.[9]

The shadow prices can be used for orientation purposes. A (comparatively) high shadow price indicates, that a thightening of the corresponding constraint has a significant effect on $\phi(\boldsymbol{a}^-)$.[10] Since we assume that it is the agent's objective to minimize $\phi(\boldsymbol{a}^-)$, constraints with high shadow prices should be favoured (to be tightened). However, since only the agent knows the real interpretation of the constraints (where they "come from", what they mean), it is up to him to choose reasonable constraints which

(1) are easy to tighten (i.e., even if the constraint is tightened by some $\varepsilon > 0$, it is still consistent with the agent's knowledge),
(2) have a significant impact on $\phi(\boldsymbol{a}^-)$ (i.e., show a high shadow price).

The problem of finding the right constraint to tighten is deliberately left to the agent.[11]

---

[8] The tightening of an existing constraint, which we describe below in more detail, is the most common user action for refining the ppi. However, it is also possible to add new constraints.

[9] There is some additional requirement (concerning the basis of the solution) which we omit here.

[10] Note that in real-world examples with a large amount of constraints, most shadow prices will be zero which significantly reduces the number of constraints to check for further tightening.

[11] Though it would be possible to include measures to numerically grasp point (1), the necessary effort seems to be inappropriate.

Note that $\phi(\boldsymbol{a}^-)$ is just an example of a value that can be used for orientation purposes, there are others that also do well. In our concrete implementation of the framework we actually employ six values and hence the agent has six shadow prices for any constraint that assist him in rating the relevance of the respective constraint.

5 Outlook

In this paper we have introduced Lazy Decision Making and described its most fundamental concepts.[12] Due to the restricted space we had to exclude many aspects that are actually fully developed.[13]

Since Lazy Decision Making is a direct extension to the classical approach, it derives the profound justification of the underlying model. In particular, probabilities are used to represent uncertainty and hence the important BERNOULLI theorem is directly applicable.

The main advantage of our approach is that it significantly helps in reducing the necessary precision needed for making a reasonable decision. The agent does not have to "guess" probabilities (and hence mapping the decision problem to a decision making problem under risk). He can actually start with an imprecise decription of his knowledge (i.e., a "large" ppi) and successively make this description more precise (i.e., reduce the size of the ppi by "moving its borders"). When the agent chooses a suitable objective (in this paper we have only investigated the minimization of $\phi(\boldsymbol{a}^-)$), he can be assisted in finding the relevant constraints (that should be made more precise). Hence, our framework saves the agent from making the wrong constraints more precise (which would have no effect on the quality of the decision measured by means of the chosen objective).

We have implemented Lazy Decision Making as a decision support computer program and tested it for some decision problems.[14] It turned out, that in many problems a good decision can be made after only few iterations (and, what is more important, using a quite imprecise representation of the agent's knowledge).

So far we have found six reasonable objectives that can be applied for obtaining shadow prices and hence ranking the constraints. However, there might be more and – what seems to be even more important – it is by far not obvious if the single values can be combined in a reasonable way which would make it even simpler to find the right constraint to tighten.

---

[12] Note that we have introduced Lazy Decision Making in other papers (cf. [9, 10]) but that the original versions of our framework was based on DEMPSTER-SHAFER theory of evidence (see [12, 5]). It turned out that partial probability information are a much more flexible basis for our approach, hence theory of evidence has been replaced by theory of ppi's.

[13] The results in this paper are part of a PhD Thesis which will be submitted to the University of Dortmund in 2001.

[14] The framework has been implemented using JAVA for the program and XML for representing the data (e.g., the decision problem). By now, it is *not* available on the internet but it is planned to integrate a web interface in the near future.

There are still some other open questions and aspects that might be generalized. For example, it should be possible to use more general ppi's than linear ones. It seems promising to utilize polynomials to bound the ppi's and to compute the shadow prices using LAGRANGE-multiplicators.

The ultimate goal of our framework is to transfer the idea of lazyness to even more components. For example, the granularity of the model could dynamically change during the decision process. It seems promising, when the agent makes the model more precise (e.g., split the set of alternatives or the set of states) not until this enhanced precision it is really needed.

References

1. Jack Hirshleifer and John G. Riley. *The analytics of uncertainty and information.* Cambridge University Press, Cambridge, 1992.
2. Leonid Hurwicz. Optimality criteria for decision making under ignorance. *Cowles Commission discussion paper, Statistics*, No. 370, 1951.
3. Eduard Kofler and Günter Menges. *Entscheiden bei unvollständiger Information.* Springer, Berlin, 1976.
4. Eduard Kofler and Günter Menges. The structuring of uncertainty and the MaxE_{min}-Principle. *Operations Research Verfahren*, 34:223–234, 1979.
5. Jürg Kohlas and Paul-André Monney. Theory of evidence. A survey of its mathematical foundations, applications and computational aspects. *ZOR – Mathematical Methods of Operations Research*, 39:35–68, 1994.
6. Helmut Laux. *Entscheidungstheorie.* Springer-Verlag, Berlin, 1998. 4. Auflage.
7. Robert Duncan Luce and Howard Raiffa. *Games and Decisions: Introduction and critical survey.* Wiley, New York, 1957.
8. John W. Pratt, Howard Raiffa, and Robert Schlaifer. *Introduction to Statistical Decision Theory.* MIT Press, Cambridge, 1996. Second Printing.
9. Gero Presser. Lazy Decision Making. In *Proceedings of the Eight International Conference IPMU*, pages 807–812, 2000.
10. Gero Presser. Lazy Decision Making. In Bernd Reusch, editor, *20. Workshop: Interdisziplinäre Methoden in der Informatik*, 2000. Forschungsberichte der Universität Dortmund.
11. Gero Presser. Lazy Decision Making and multiciteria analysis. In *Proceedings of the Fifth World Multi-Conference SCI*, 2001 (to appear).
12. Glenn Shafer. *A Mathematical Theory of Evidence.* Princetòn University Press, Princeton, NJ, 1976.
13. Philippe Smets. Imperfect information: Imprecision and uncertainty. In Amihai Motro and Philippe Smets, editors, *Uncertainty Management in Information Systems*, pages 225–254, Boston, 1997. Kluwer Academic Publishers.
14. John von Neumann and Oskar Morgenstern. *Theory of Games and Economic Behavior.* Princeton University Press, Princeton, 1943.
15. Abraham Wald. Statistical decision functions which minimize the maximum risk. *Annals of Mathematics*, 46:265–280, 1945.

AMA: An OWA Operator Based on the Majority Process

J.I. Peláez

Dept. of Languages and Computer Sciences
University of Malaga
Malaga 29071. Spain
jignacio@lcc.uma.es

Abstract. The aggregation operators based on the ordered weighted averaging are know as OWA operators. These OWA operators can provide for aggregation lying between the Max, Min operators and the logical *or* and *and* operators. The purpose of this paper is to present a new OWA operator that is based on the majority process. This operator uses the cardinality of the aggregation elements to penalize or reward the weights values.

1 Introduction

The process of information aggregation appears in many applications related to the development of intelligent systems. Ones sees aggregation in neural networks, fuzzy logic controllers, vision systems, expert systems and multi-criteria decision aids. In [3] Yager introduced a new aggregation technique based on the ordered weighted averaging (OWA) operators. In [6] he described their use in modelling natural networks, in [3] their application to multi-criteria decision making was emphasized, [7] their use in data base systems was discussed. O'Hagan [2] has suggested their use in expert systems. Yager and Filev [5] have used these operators to develop flexible models for fuzzy logic controllers. These OWA operators can provide for aggregations lying between the logical *or* and *and.*

A problem that we had encountered in the aggregation process, is how to aggregate the elements that have cardinality greater than one. The most common operators used in the aggregation process [1, 8] produce reasonable results, but, at the same time, when the items to aggregate have cardinality greater than one, they may produce distributed problems. We consider these problems as a variation of the cake cutting problems, where we need to divide a cake in a way that all fellow guests were satisfied with their portions.

For example, if we want to aggregate the following elements: {0.7, 0.7, 0.7, 0.7, 0.5, 0.4, 0.1, 0.1, 0.1}.The classical operators considers nine sets with cardinality one, nevertheless, the majority process considers four sets with cardinality 4, 1, 1 and 3, respectively. If we analyse this example the 55 percent of elements are over 0.4 and the 44 percent are the elements 0.7. Due to, the result should be higher than 0.5 and lower than 0.7, because we obtain a value that represent of majority and the same time the minority. Using the majority process we obtain this type of result (0.551), and we eliminate the distribution problems that other operator may produce.

B. Reusch (Ed.): Fuzzy Days 2001, LNCS 2206, pp. 937–949, 2001.

The paper is organized as follows: In section 2 we show the ordered weighted averaging operators (OWA); in section 3, the dependent weights operators; and in section 4, we present the *additive majority aggregation* (*AMA*).

2 Ordered Weighted Averaging Operators

In this section we review the basic concepts associated with the OWA operators [3].

Definition 1. An OWA operator of dimension n is a mapping

$$F: R^n \to R$$

that has an associated n vector W

$$W = [w_1, w_2, \ldots, w_n]^T$$

Such that:

1. $w_i \in [0,1]$
2. $\sum_{i=1}^{n} w_i = 1$

Furthermore $F(a_1, a_2, \ldots, a_n) = \sum_{j=1}^{n} w_j \cdot b_j$ where b_j is the jth largest of the a_i.

Example 1. Assume

$$W = [0.4, 0.3, 0.2, 0.1]^T$$

then

$$F(0.7, 1, 0.3, 0.6) = (0.4)(1) + (0.3)(0.7) + (0.2)(0.6)$$

$$+(0.1)(0.3) = 0.76$$

A fundamental aspect of this operation is the re-ordering step, in particular an aggregate a_i is not associated with a particular weight w_i, but rather a weight is associated with a particular ordered position of aggregate.

When we view the OWA weights as a column vector we shall find it convenient to refer to the weights with the low indices as weighs at the top and those with the higher indices with weights at the bottom

It is noted that different OWA operators are distinguished by their weighting function. In [3] Yager pointed out three important special cases of OWA aggregations:

1. F^*. In this case

$$W = W^* = [1, 0, \ldots, 0]^T$$

2. F_*. In this case

$$W = W_* = [0, 0, ..., 1]^T$$

3. F_{Ave}. In this case

$$W = W_{ave} = \left[\frac{1}{n}, \frac{1}{n}, \ldots, \frac{1}{n}\right]^T$$

It can easily be seen that

1. $F^*(a_1, a_2, ..., a_n) = \text{Max}(a_i)$
2. $F_*(a_1, a_2, ..., a_n) = \text{Min}(a_i)$
3. $F_{ave}(a_1, a_2, \ldots, a_n) = \frac{1}{n} \cdot \sum_{i=1}^{n} a_i$

A number of important properties can be associated with the OWA operators [3]. We shall now discuss some of these.

Property 1. For any OWA operator F

$$F_*(a_1, a_2, ..., a_n) \leq F(a_1, a_2, ..., a_n) \leq F^*(a_1, a_2, ..., a_n)$$

Thus the upper and lower star OWA operator are boundaries. From the above it became clear that for any F

$$\text{Min}(a_i) \leq F(a_1, a_2, ..., a_n) \leq \text{Max}(a_i)$$

Property 2 (Commutative). The OWA operator can be seen to commutative. Let $< a_1, a_2, ..., a_n >$ be a bag aggregates and let $< d_1, d_2, ..., d_n >$ be a permutation of the a_i. Then for any OWA operator

$$F(a_1, a_2, ..., a_n) = F(d_1, d_2, ..., d_n)$$

Property 3 (Monotonicity). A third characteristic associated with the operators is monotonicity. Assume a_i and c_i are collection of aggregates, i=1,...,n such that for each i, $c_i \leq a_i$. Then

$$F(c_1, c_2, ..., c_n) \leq F(a_1, a_2, ..., a_n)$$

Where F is some fixed weight OWA operator.

Property 4 (Idempotency). Another characteristic associated with these operators is idempotency. If $a_i = a$ for all i=1,..., n, then for any OWA operator

$$F(a, a, ..., a) = a$$

From the above we can see the OWA operators have the basic properties associated with an averaging operator [8].

In [3], Yager defined two important measures associated with an OWA operator. The first measure, called the dispersion (or entropy) of an OWA vector W is defined as

$$Disp(W) = -\sum_{i=1}^{n} w_i \cdot \ln(w_i)$$

We can see when using the OWA operator as an averaging operator Disp(W) measure the degree to which we use all the aggregates equally.

A second measure introduced by Yager [3] which we shall call the orness measure is defined as

$$orness(W) = \frac{1}{n-1} \sum_{i=1}^{n} ((n-i) \cdot w_i)$$

It can easy be shown [3] that

1. orness(W^*) = 1
2. orness(W_*) = 0
3. orness(W_{Ave}) = 0.5

A measure of *andness* can be defined as: andness(W) = 1 – orness(W)

Generally it should be appreciated that an OWA operator with much of the weights near the top will be an "*orlike*" operator, orness(W) ≥ 0.5. At the other extreme when the weights are non-zero near the bottom the OWA operator will be "*andlike*", orness(W) ≤ 0.5.

3 Aggregate Dependent Weights

In this section we shall consider OWA class operators where the weights depend on the aggregates.

In the definition for the OWA operator we indicated that

$$F(a_1, a_2, \ldots, a_n) = \sum_{i=1}^{n} w_i \cdot b_i$$

where b_i is the ith largest of the a_i. The weights were required to satisfy two conditions:

1.- $w_i \in [0,1]$

2.- $\sum_{i=1}^{n} w_i = 1$

In all for the proceeding we have assumed that the weights were fixed given constant values. In this section we shall generalize the concept of OWA aggregation by allowing the weights to be a function of the aggregates of more precisely the ordered aggregates, the b_i. It is still required that the weights satisfy conditions (1) and (2).

$$w_i = f_i(b_1, b_2, \ldots, b_n)$$

In this case

$$F(a_1, a_2, \ldots, a_n) = \sum_{i=1}^{n} f_i(b_1, b_2, \ldots, b_n) \cdot b_i$$

In this case, where the weights depend on the aggregates, many, but not all, of the properties of the OWA operator still hold:

1.- All aggregate dependent operators still lie between F^* and F_*.

2.- The operator is still idempotent

$$F(a, a, \ldots, a) = a$$

3.- The operator is still commutative, the indexing of the a_i's is not important.

One property that is not necessarily retained is that of monotonicity. Assume $A=(a_1, a_2, \ldots, a_n)$ and

$$\hat{A}' = \left(\hat{a}_1, \hat{a}_2, \ldots, \hat{a}_n\right)$$

are two collections aggregates such that $a_i \geq \hat{a}_i$ for all i. If the weights are constant then

$$F(A) \geq \hat{F}\left(\hat{A}\right)$$

We shall subsequently see that this is not the case when the weights are aggregate dependent.

We shall say that an OWA aggregation is near if the aggregated value is independent of the ordering. Let $A = (a_1, a_2, \ldots, a_n)$ be our inputs ordered and $B = (b_1, b_2, \ldots, b_n)$ be our inputs ordered and $C = (c_1, c_2, \ldots, c_n) = \text{Perm}\,(a_1, a_2, \ldots, a_n)$ be an permutation of the input. Formally we shall say that the OWA operator is near if

$$F(a_1, a_2, \ldots, a_n) = \sum_{i=1}^{n} a_i$$

is the same for the any assignment $C = B$.

One simple example of neat OWA operator when $w_i = \frac{1}{n}$. In this case

$$F(a_1, a_2, \ldots, a_n) = \frac{1}{n} \cdot \sum_{i=1}^{n} a_i$$

If the weights w_i's, are fixed, this is the one possible neat OWA operator. However, as we shall see when we allow aggregate dependent weights we can get more neat operators.

One important characteristic of neat OWA aggregators is that they don't need to be ordered. This implies that the formulation for the neat OWA aggregators can be written using the arguments, the a_i's, with introducing the ordered inputs, the b_i's.

3.1 Families of Aggregate Dependent Weight

A first family of aggregate dependent weight that we shall study are the so called BADD OWA operators introduced by Yager y Filev [36].

In the case of these operators we define

$$w_i = \frac{b_i^\alpha}{\sum_{j=1}^{n} b_j^\alpha}$$

It is easy to see that

1.- $w_i \in [0,1]$

2.- $\sum_{i=1}^{n} w_i = 1$

Thus these are an acceptable class of OWA weights. In this case we get

$$F(a_1, a_2 \ldots, a_n) = \frac{\sum_{i=1}^{n} b_i^{\alpha+1}}{\sum_{i=1}^{n} b_i^\alpha}$$

It is easy to see that this is a neat OWA Operators and thus

$$F(a_1, a_2, \ldots, a_n) = \frac{\sum_{i=1}^{n} a_i^{\alpha+1}}{\sum_{i=1}^{n} a_i^{\alpha}}, \quad \alpha \geq 0$$

Hence no ordering is necessary in this process.

We see for the case when $\propto =0$ we get

$$F(a_1, a_2, \ldots, a_n) = \frac{1}{n} \cdot \sum_{i=1}^{n} a_i$$

which is the simple average, F_{ave}. When $\propto = 1$ we get

$$F(a_1, a_2, \ldots, a_n) = \frac{\sum_{i=1}^{n} a_2^2}{\sum_{j=1}^{n} a_j}$$

When $\propto \rightarrow \infty$ we have

$$F(a_1, a_2, \ldots, a_n) = Max[a_i] = F^*(a_1, a_2, \ldots, a_n)$$

It can be seen that for this class of operators $w_i \geq w_j$ for $i > j$, thus follows since $b_i \geq b_j$ for $i < j$. Thus these are buoyancy operators and as such are 'orlike' operators.

Other class of aggregate dependent weights. Here we let

$$w_i = \frac{\left(\frac{1}{b_i}\right)^{\alpha}}{\sum_{j=1}^{n} \left(\frac{1}{b_j}\right)^{\alpha}}$$

These again can be shown to satisfy the conditions of OWA weights. In this case

$$F(a_1, a_2, \ldots, a_n) = \frac{\sum_{i=1}^{n} \left(\frac{1}{b_i}\right)^{\alpha-1}}{\sum_{i=1}^{n} \left(\frac{1}{b_i}\right)^{\alpha}} = \frac{\sum_{i=1}^{n} \left(\frac{1}{a_i}\right)^{\alpha-1}}{\sum_{i=1}^{n} \left(\frac{1}{a_i}\right)^{\alpha}}$$

Thus these are also neat. We see that if $\propto = 0$ we get

$$F(a_1, a_2, \ldots, a_n) = \frac{\sum_{i=1}^{n} \left(\frac{1}{a_i}\right)^{-1}}{\sum_{i=1}^{n} 1} = \frac{\sum_{i=1}^{n} a_i}{n} = \frac{1}{n} \cdot \sum_{i=1}^{n} a_i$$

Thus when $\propto = 0$ we get *Fave*. If $\propto = 1$ we get

$$F(a_1, a_2, \ldots, a_n) = \frac{\sum_{i=1}^{n} 1}{\sum_{i=1}^{n} \frac{1}{a_i}} = \frac{n}{\sum_{i=1}^{n} \frac{1}{a_i}}$$

Which is the Harmonic mean. If we let $\propto \to \infty$ we then get

$$F(a_1, a_2, \ldots, a_n) = a_{min} = F_*$$

Another interesting case of aggregate dependent weights is where

$$w_i = \frac{(1 - b_i)^\alpha}{\sum_{i=1}^{n} (1 - b_i)^\alpha}$$

In this case we get

$$F(a_1, a_2, \ldots, a_n) = \frac{\sum_{i=1}^{n} (1 - a_i)^\alpha \cdot a_i}{\sum_{i=1}^{n} (1 - a_i)^\alpha}$$

Which again is a neat aggregation. When $\propto = 0$ we get *Fave*.

$$F(a_1, a_2, \ldots, a_n) = \frac{1}{n} \cdot \sum_{i=1}^{n} a_i = F_{ave}$$

When $\propto = 1$ we get

$$F(a_1, a_2, \ldots, a_n) = \frac{\sum_{i=1} (a_i - a_i^2)}{n - \sum_{i=1}^{n} a_i}$$

When $\propto \to \infty$ we get

$$F(a_1, a_2, \ldots, a_n) = \mathop{Min}_{i=1,\ldots n} a_i$$

4 Additive Majority Aggregation

The A*MA* operator is a modification of the arithmetic mean, as a arithmetic mean of arithmetic mean, such that, the final result is a weighted arithmetic mean. This operator is a neat OWA operator and theirs weights are calculated on functions of the cardinality of the elements to aggregate.

Definition 1. Let $<a_1, a_2, \ldots, a_n>$ and let $\delta_1, \delta_2, \ldots, \delta_n \in N$, be the cardinality, where $\delta_i \leq \delta_{i+1}$ for all $1 \leq i \leq n\text{-}1$. The AMA operator is calculated as:

$$F_{AMA}(a_1, a_2, \ldots, a_n) = \sum_{i=1}^{n} w_i \cdot b_i = \sum_{i=1}^{n} f_i \cdot (b_1, b_2, \ldots, b_n) \cdot b_i$$

Such that:

1. $w_i \in [0,1]$
2. $\sum_{i=1}^{n} w_i = 1$

Furthermore $F_{MA}(a_1, a_2, \ldots, a_n) = \sum_{j=1}^{n} w_j \cdot b_j$ where w_i*'s* are calculated as:

$$w_i = \begin{cases} \dfrac{1}{d_1} & \text{if } i = 1 \\ \dfrac{1}{d_1} \cdot \dfrac{1 - n^{\delta_2}}{1 - n} & \text{if } i = 2 \\ w_{i-1} + \dfrac{1}{d_{i-1}} \cdot \dfrac{1 - (n - i + 2)^{\delta_i - \delta_{i-1}}}{1 - (n - i + 2)} & \text{if } i > 2 \end{cases}$$

with

$$d_i = \begin{cases} 1 & \text{if } i = 1, n = 1 \\ n^{\delta_2} & \text{if } i = 1, n = 2 \\ n^{\delta_2} \cdot \prod_{j=1}^{n-2} (n-j)^{\delta_{j+2}-\delta_{j+1}} & \text{if } i = 1, n > 2 \\ \prod_{j=i-1}^{n-2} (n-j)^{\delta_{j+2}-\delta_{j+1}} & \text{if } i > 1 \end{cases}$$

Example: Let $< 0.2, 0.2, 0.2, 0.5, 0.6, 0.9, 0.9, 0.9, 0.9 >$ be a bag aggregates then

$$F_{AMA}(0.5, 0.6, 0.2, 0.2, 0.2, 0.9, 0.9, 0.9, 0.9) =$$

$$0.5 \cdot 0.013 + 0.6 \cdot 0.013 + 0.2 \cdot 0.236 + 0.9 \cdot 0.736 = 0.723$$

where:

$$w_1 = \frac{1}{d_1} = \frac{1}{72} = 0.0138$$

$$w_2 = \frac{1}{d_1} \cdot \frac{1 - n^{\delta_2}}{1 - n} = \frac{1}{72} \cdot \frac{1 - 4^1}{1 - 4} = 0.0138$$

$$w_3 = w_2 + \frac{1}{d_2} \cdot \frac{1 - (n - 3 + 2)^{\delta_3 - \delta_2}}{1 - (n - 3 + 2)} =$$

$$0.013 + \frac{1}{18} \cdot \frac{1 - (4 - 3 + 2)^{3-1}}{1 - (4 - 3 + 2)} = 0.236$$

$$w_4 = w_3 + \frac{1}{d_3} \cdot \frac{1 - (n - 4 + 2)^{\delta_4 - \delta_3}}{1 - (n - 4 + 2)} =$$

$$0.236 + \frac{1}{2} \cdot \frac{1 - (4 - 4 + 2)^{4-3}}{1 - (4 - 4 + 2)} = 0.736$$

and

$$d_1 = n^{\delta_2} \cdot \prod_{j=1}^{n-2} (n-j)^{\delta_{j+2}-\delta_{j+1}} = 4^1\left((4-1)^{3-1}(4-2)^{4-3}\right) = 72$$

$$d_2 = \prod_{j=2-1}^{n-2} (n-j)^{\delta_{j+2}-\delta_{j+1}} = (4-1)^{3-1}(4-2)^{4-3} = 18$$

$$d_3 = \prod_{j=3-1}^{n-2} (n-j)^{\delta_{j+2}-\delta_{j+1}} = (4-2)^{4-3} = 2$$

Now we are going to proof as the weights satisfy the conditions 1 and 2.

1. $\sum_{i=1}^{n} w_i = 1.$

Proof.

$$\sum_{i=1}^{n} w_i = w_1 + w_2 + \ldots + w_n =$$

$$\frac{1}{d_1} + \frac{1}{d_1} \cdot \frac{1-n^{\delta_2}}{1-n} + \frac{1}{d_1} \cdot \frac{1-n^{\delta_2}}{1-n} + \frac{1}{d_2} \cdot \frac{1-(n-3+2)^{\delta_3-\delta_2}}{1-(n-3+2)} + \ldots +$$

$$w_{k-1} + \frac{1}{d_{k-1}} \cdot \frac{1-(n-k+2)^{\delta_k-\delta_{k-1}}}{1-(n-k+2)} + \ldots + w_{n-1} + \frac{1}{d_{n-1}} \cdot \frac{1-(n-n+2)^{\delta_n-\delta_{n-1}}}{1-(n-n+2)} =$$

$$\frac{1}{d_1} - \frac{(n-1)}{(n-1)\cdot d_1} + \frac{(n-1)\cdot n^{\delta_2}}{(n-1)\cdot d_1} - \frac{(n-2)}{d_2 \cdot ((n-3+2)-1)} +$$

$$\frac{(n-2)\cdot(n-3+2)^{\delta_3-\delta_2}}{d_2 \cdot ((n-3+2)-1)} - \frac{(n-3)}{d_3 \cdot ((n-4+2)-1)} + \ldots +$$

$$\frac{(n-k-1)\cdot(n-k+2)^{\delta_k-\delta_{k-1}}}{d_{k-1} \cdot ((n-k+2)-1)} - \frac{(n-k+2)}{d_k \cdot ((n-(k+1)+2)-1)} + \ldots +$$

$$\frac{2^{\delta_n-\delta_{n-1}}}{d_{n-1}} = \frac{1}{d_1} - \frac{1}{d_1} + \frac{1}{d_2} - \frac{1}{d_2} + \ldots + \frac{1}{d_k} - \frac{1}{d_k} + \ldots +$$

$$\frac{2^{\delta_n-\delta_{n-1}}}{d_{n-1}} = \frac{2^{\delta_n-\delta_{n-1}}}{d_{n-1}} = \frac{2^{\delta_n-\delta_{n-1}}}{(n-(n-2))^{\delta_n-\delta_{n-1}}} = 1$$

2. $\lambda_i \in [0,1]$

Proof. Trivial with the proof 1.

We point out three important cases for this operator:

1. if $\delta_i = 1$ for all $i=1,..,n$: $F_{MA}(a_1, a_2, ..., a_n) = \frac{1}{n}\sum_i a_i = F_{Ave}$.

2. if $\delta_n \to \infty$: $F_{MA}(a_1, a_2, ..., a_n) = \text{Max}[b_i] = F^*(b_1, b_2, ..., b_n)$.

3. if $\delta_1 \to \infty$: $F_{MA}(a_1, a_2, ..., a_n) = \text{Min}[b_i] = F_*(b_1, b_2, ..., b_n)$.

4.1 Properties

Following, we present the principal properties for *AMA* operator.

Property 1 (Commutative). The *AMA* operator can be seen to commutative. Let $< a_1, a_2, ..., a_n >$ be a bag aggregates and let $< d_1, d_2, ..., d_n >$ be a permutation of the a_i. Then

$$F_{AMA}(a_1, a_2, ..., a_n) = F_{AMA}(d_1, d_2, ..., d_n)$$

Property 2 (Monotonicity). A third characteristic associated with the operators is monotonicity. Assume a_i and c_i are collection of aggregates, i=1,...,n such that for each i, $c_i \leq a_i$. Then

$$F_{AMA}(c_1, c_2, \ldots, c_n) \leq F_{AMA}(a_1, a_2, \ldots, a_n)$$

Property 3 (Idempotency). Another characteristic associated with these operators is idempotency. If $a_i = a$ for all i=1,..., n, then for any OWA operator

$$F_{AMA}(a, a, \ldots, a) = a$$

5 Conclusion

In this paper we show how the classic aggregation operators may produce the distribution problems when we want to aggregate a set of labels where some elements have cardinality greater than one.

We propose a new neat OWA operators (*AMA*) to solve the distribution problems. This operator is based in the majority process. it is very useful as aggregation operator from a computation point of view, since they may be implemented as tables or simple procedures. Therefore it could be applied to Decision-Making or Optimization Problems with different types of information.

Also, we define the procedure for calculating the weights in function of the cardinality of the elements. Further, we present some properties of the operator.

References

1. Delgado M. Verdegay J.L. Vila A. (1993). On aggregation operations of linguistic labels. Int. J. of Intelligent Systems 8. 351-370.
2. O'Hagan. 1988. Aggregating template rule antecedents in real-time expert systems with fuzzy set logic. Proc. 22nd Annual IEEE Asilomar Conf. On Signals, Systems and Computers, Pacific Grove, CA.
3. Yager R. 1988. On ordered weighted averaging aggregation operators in multi-criteria decision making. IEEE Trans. On Systems, Man and Cybernetics 18. 183-190.
4. Yager R. 1991. Fuzzy quotient operators for fuzzy relational data bases. Proc. Int. Fuzzy Engineering Symposium. Yokohama. 289-296.
5. Yager R. Filew D.P. 1992. Fuzzy logic controllers with flexible structures. Proc. Second Int. Cont. on Fuzzy Sets and Neural Networks. Izuka Japan. 317-320.
6. Yager R. 1992. OWA Neurons: A new class of fuzzy neurons. Proc. Int. Joint on Neural Networks. Vol. I. Baltimor. 226-231.
7. Yager R. 1992. Fuzzy quotient operator. Proc. Fourth Int. Conf. On Information Processing a Management of Uncertainty. Palma de Majorca. Spain. 317-322.
8. Yager R. 1993. Families of OWA operators. Fuzzy Sets and Systems. 59. 125-148.
9. Yager R. 2001. On Induced Aggregation Operators. Proceeding of Eurofuse Workshop on Preference Modelling and Applications. Granada.

DSP Implementation of Fuzzy Controlled Magnetic Levitation System

Ahmet Orhan[1], Erhan Akin[2], and Mehmet Karaköse[2]

[1]Firat University
Electric-Electronic Engineering
23119 Elazig/Turkey
aorhan@firat.edu.tr
[2]Firat University
Computer Engineering
23119 Elazig/Turkey
{eakin,mkarakose}@firat.edu.tr

Abstract. This paper presents a fuzzy logic based controller applied to a magnetic levitation (MAGLEV) system. MAGLEV system is nonlinear and complex. It is difficult to obtain exact mathematical model. To improve control performance, an inner current control loop has been used. The proposed controllers for MAGLEV system is simulated by using MATLAB-SIMULINK and implemented on an experimental system using a TMS320C31 digital signal processor.

1 Introduction

Magnetic levitation systems are used in the low noise new railway systems low friction magnetic bearing of large rotating machines and space vehicles. A magnetic levitation system is complex and nonlinear. The mathematical model of MAGLEV system depends on various factors. Using this model, it is difficult to design a classical controller and their results do not yield satisfactory. In this case fuzzy controllers are preferable. Wang has been discussed classic controller design procedure for classroom demonstration and presented analog electronic implementation [1]. Many researchers applied feedback linearizing control to MAGLEV systems [2]. But this control technique does not guarantee robustness in the presence of modeling error and disturbances. Also sliding mode control technique and frequency domain H^{∞} control have been applied to MAGLEV systems [3]. H^{∞} control is an effective method in attenuating the disturbances. Fuzzy controller technique has been studied by various researchers [4,5,6]. This is more convenient for MAGLEV system and easiest to designing the controller.

As known fuzzy controllers are considered as the best method for nonlinear time varying and for mathematical models. A magnetic levitation system is easily handled by human operators and best described quantitatively and makes it possible in many cases to design control systems that are more robust and energy saving.

B. Reusch (Ed.): Fuzzy Days 2001, LNCS 2206, pp. 950–958, 2001.

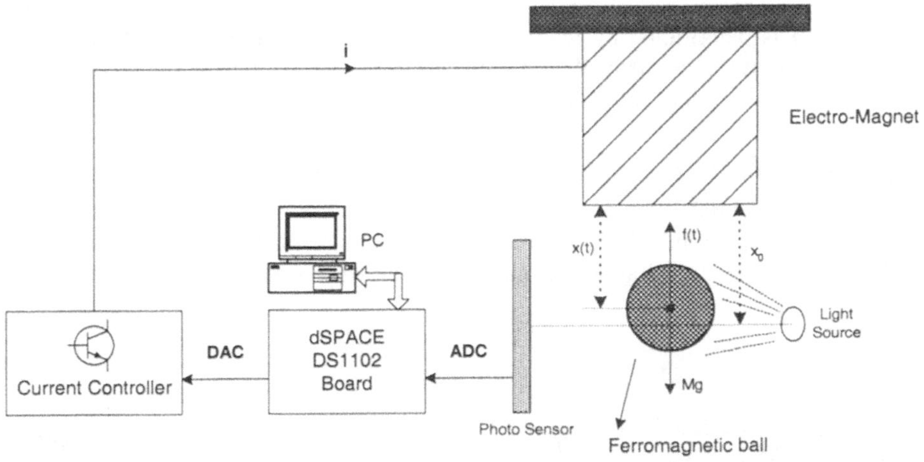

Fig. 1. Block diagram of fuzzy controlled magnetic levitation system

Realized system block diagram is shown in Fig 1. The task of the magnetic levitation system is kept in dynamic balance around its equilibrium position. For position feedback a photo resistor is used.

In this study, the magnetic levitation system firstly is simulated using MATLAB. A fuzzy PD controller has been designed using trial and error methods. In the fuzzy controller five membership functions are used in error and change of error. Development time of the fuzzy controller is shorter than the classical PD controller.

2 Mathematical Model of MAGLEV System

Magnetic repulsion and magnetic attraction are two kinds of magnetic levitation system. In this study, a magnetic attraction system was modeled. Magnetic repulsion system is move complex and requires super conducting material and additional equipment.

In the magnetic levitation system the target is to keep at the ball a reference level ball. For this aim airgap flux density or height of the ball should be observed. For such a system design firstly a mathematical model must be written. An exact mathematical model is very complex. In this case some assumptions are made, magnetic circuit is linear (no saturation), eddy and hysteresis losses are neglected. Magnetic levitation system is expressed with the differential equations below:

$$f = a(\frac{i}{x})^2 \quad \textbf{(1)}$$

$$m\frac{d^2x}{dt^2} = mg - f(x,i) \quad \textbf{(2)}$$

$$e = Ri + \frac{d}{dt}(Li) \quad \textbf{(3)}$$

where m is ball mass as kg and g is gravity acceleration and f is attraction force. The coefficient a is determined by identification procedures, L is the inductance of the system and inductance depends on ball position and L coil inductance. Using linearization and Laplace transform, the system characteristic equation can be written as follows:

$$Q(s) = x_0 LMs^3 + x_0 RMs^2 - Ci_0 Ls - Ci_0 R + G_c(s) Cx_0 B \quad \textbf{(4)}$$

where $G_c(s)$ is classic PD controller. Tuning of the PD controller parameters is difficult. Stability problem have big importance in this magnetic levitation system.

The magnetic levitation system equations can be written as state space form:

$$\frac{dx_1}{dt} = x_2$$

$$\frac{dx_2}{dt} = -\frac{k}{M}\left(\frac{I_0 + x_3}{x_1}\right)^2 + g \quad \textbf{(5)}$$

$$\frac{dx_3}{dt} = -\frac{R}{L}x_3 + \frac{u}{L}$$

In the conventional controller design process the system was linearized at operating point. A state feedback was used to stabilize around equilibrium point. A photo resistor was used as feedback element. The block diagram of MAGLEV system is given in Fig 2.

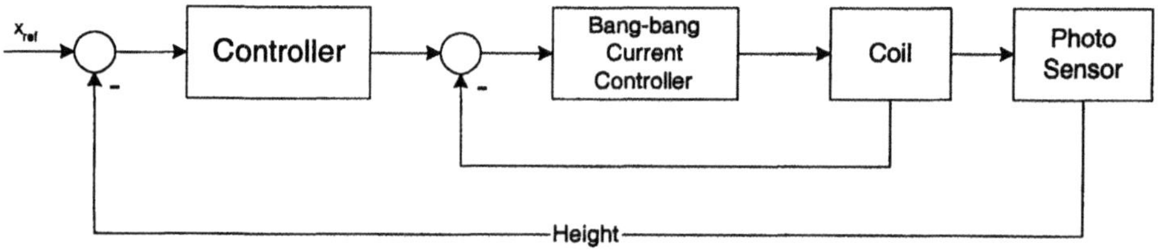

Fig. 2. Block diagram of controlled MAGLEV system

A magnetic levitation system is an unstable system. The ball falls down or track touch. Then the coil gets closer than its designed distance; the current will increase, causing the coil to get even closer until finally the coil touches. When the ball gets further away from the track, the current will decrease and the coil will fall away from the track until the system is no longer operating.

3 Proposed Controllers for MAGLEV System

Fuzzy logic is a technology based on engineering experience and observations. In fuzzy logic, an exact mathematical model is not necessary, because linguistic variables are used in fuzzy logic to define system behavior rapidly. Fuzzy logic is a very recent technology relative to conventional controllers; its areas of application are increasing very quickly. Fuzzy PID, fuzzy PI, fuzzy PD and fuzzy mixed controllers are fuzzy controller design approaches, but unlike conventional controllers the focus is not in the modeling [7]. Some of the problems, such as stability and performance, are encountered both in fuzzy controllers and conventional controllers. Unlike conventional control design, where mathematical models are used to solve these problems, fuzzy controller design involves IF-THEN rules defined by an expert to tackle these problems.

There are two methods that are commonly used to design fuzzy controllers: trial and error method and the theoretical method. In trial and error, IF-THEN rules are defined by using expert knowledge and experience. Then, these rules are applied to the actual system. Unlike the theoretical approach where the parameters are adjusted to guarantee the desired performance, in the fuzzy method the IF-THEN rules are modified until the desired performance is achieved. In practice, both methods can be used to obtain better performance [7].

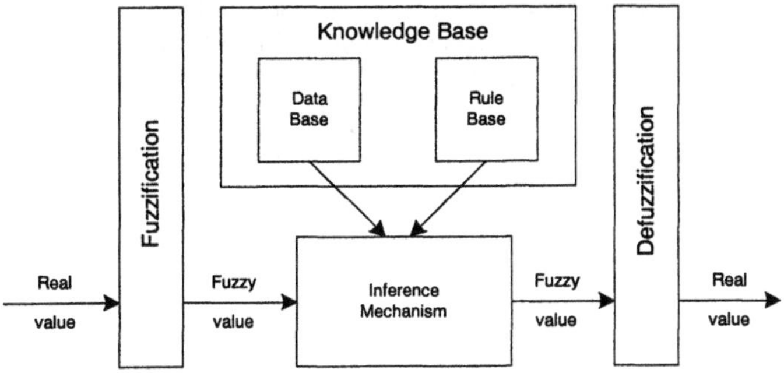

Fig. 3. Block diagram of fuzzy control architecture

The fuzzy controller has four components as shown in Fig. 4. These are:

Fuzzifier: The input values are scaled and grouped into fuzzy sets. In other words, the input values labeled and transformed into linguistic variables.

Inference mechanism: The inference mechanism uses a database and a rule base. The database involves membership functions that are used by the inference mechanism to make fuzzy decisions.

Rule Base: Rule base is a set of IF-THEN rules defined by an expert. The inference mechanism uses these rules.

Defuzzifier: The linguistic variables manipulated by the inference mechanism are converted back to real values.

In a fuzzy controller design, the knowledge and observations of an expert are more important than the underlying mathematical model. This expert knowledge and observation is used while the system is being designed. This kind of approach provides an opportunity to easily embed experience into a controller, which has been gained over a long time. However, it is not possible to obtain automation during controller design.

In this paper, a fuzzy controller was used as Fig. 4. The membership functions of the fuzzy controller used were determined with trial and error method. Unlike conventional controllers, fuzzy controllers are less sensitive to sensor errors and small variations of the parameters.

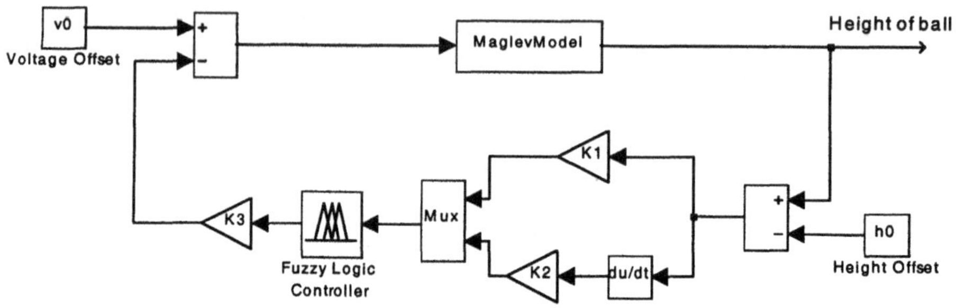

Fig. 4. Block diagram of the fuzzy controller for MAGLEV system

As shown in Fig. 4, first the height of the ball obtained from MaglevModel was compared to reference height. Next, this difference and the derivative of the difference were given as inputs to the fuzzy logic controller. The output of fuzzy controller was compared to a reference voltage. Finally, this voltage was applied to MaglevModel. Each variable of the fuzzy controller was represented by using 5 membership functions at the inputs and output, as shown in Fig. 5a-c. The ranges of the input and output variables were assumed to be [-20,20], [-2, 2] and [-2, 2], respectively. The fuzzy rule base for this fuzzy controller is as shown in Fig. 5d.

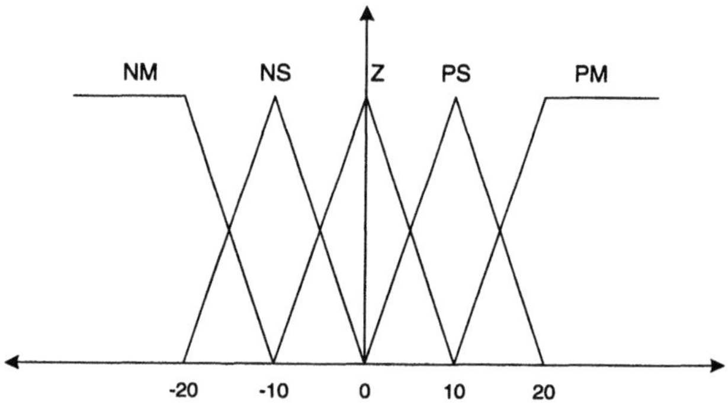

(5a)

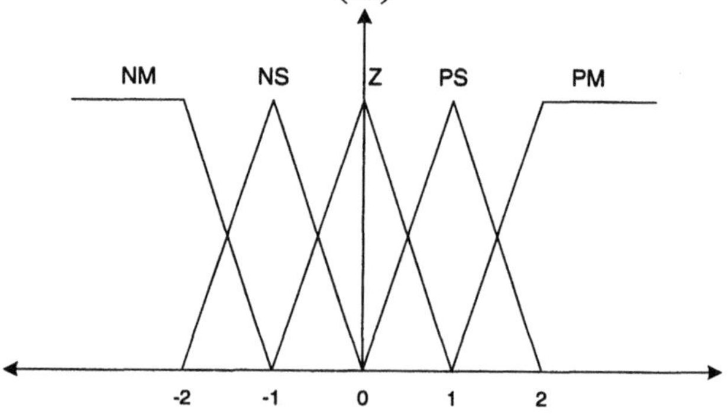

(5b)

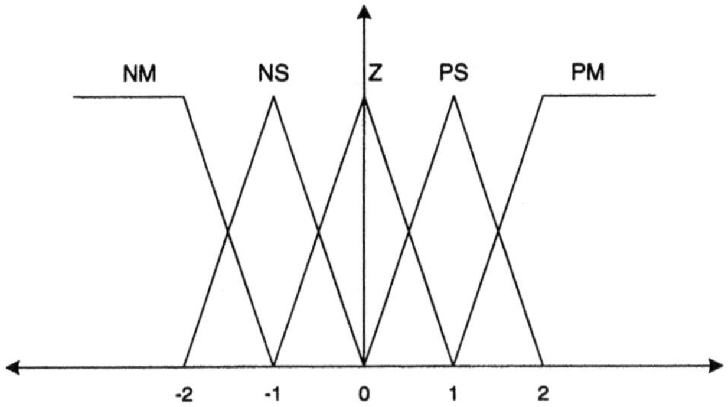

(5c)

error

change of error	NB	NS	Z	PS	PB
NB	PB	PB	PS	PS	Z
NS	PB	PS	PS	Z	Z
Z	PS	PS	Z	NS	NS
PS	PS	Z	Z	NS	NB
PB	Z	NS	NS	NB	NB

(5d)

Fig. 5. Membership functions and rule table of fuzzy controller **(a).** Membership functions of input variable "*error*" **(b).** Membership functions of input variable "*cerror*" **(c).** Membership functions of output variable "*action*" **(d).** Rule table

Thus, after choosing appropriate rules for the process of the system, flexible and robust solutions were obtained. The fuzzy rule base was generated on the basis of the numbers of the sensor positions in the area of the blocked or opened by the ball due to the levitation action.

As known, PID controllers give very good response for a system in MAGLEV system. However, the parameter tuning of a PID controller is very difficult. In this study, a PD controller is used for a MAGLEV system. PD controller produces a voltage according to the height of ball. Used PD controller's parameters were set to optimal values using experimental tuning procedure. In the classical realization a power switch that operate at the active switching region used.

To increase the control performance of MAGLEV system can be added current controller. The bang-bang current controller gives output of the includes a PD controller or a fuzzy controller. Output of PD controller or fuzzy controller mentioned earlier is compare to coil current and a voltage is applied to MaglevModel.

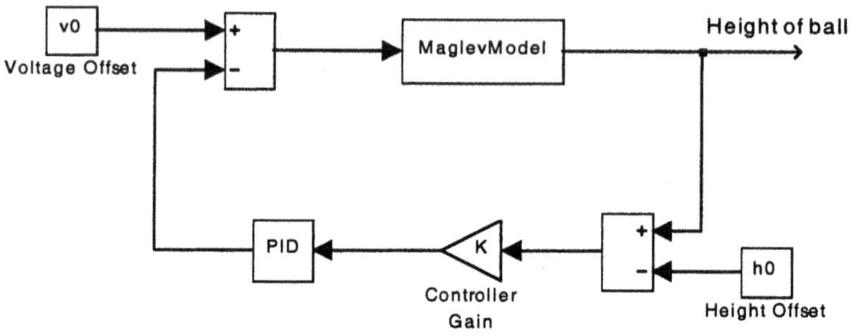

Fig. 6. Block diagram of MAGLEV system with PD controller

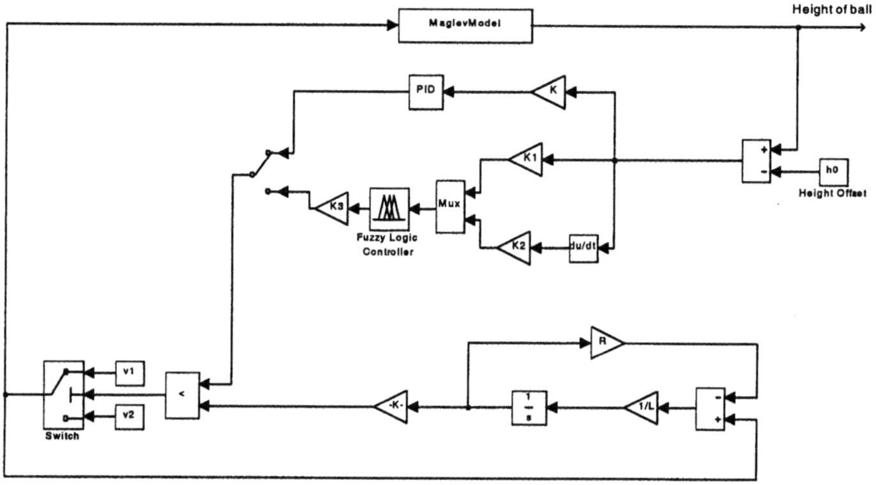

Fig. 7. Block diagram of the MAGLEV system that is used bang-bang current controller

4 Simulation and Experimental Results

Various simulations were carried out by using MATLAB/SIMULINK to assess the performance of proposed controllers for MAGLEV system. Used fuzzy controller is developed by the MATLAB Fuzzy Toolbox [8].

In the simulations, the performance of the fuzzy controller, PD controller and bang-bang controller was compared. The responses of the PD controller are shown in Fig. 8. Simulation results for fuzzy controlled MAGLEV system are given in Fig. 9. As shown in these figures, the performance of the fuzzy-controlled MAGLEV system is better than that of the PD-controlled MAGLEV system. Finally, experimental results for MAGLEV system are given in Fig. 10.

The MAGLEV system was implemented using dSPACE DS1102 board. This board uses TMS320C31 DSP processor. Fuzzy controller execution time is 80 μs. The MAGLEV system with fuzzy control is robust and energy saving.

In Table-1 the parameters of the MAGLEV system used in the simulations and experiments are listed.

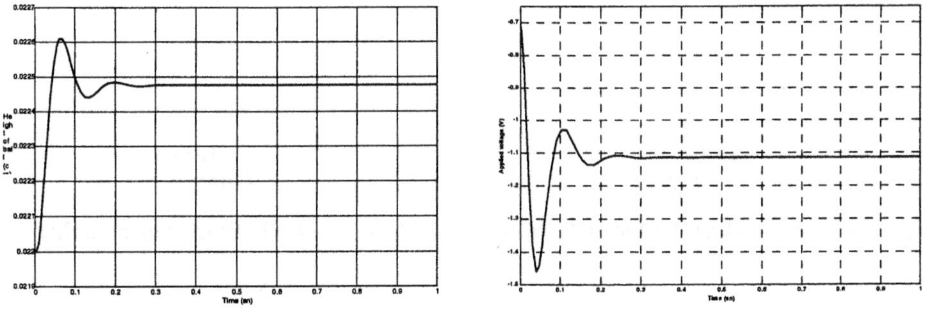

(8a) **(8b)**

Fig. 8. Simulation results of the proposed PD controller: **a)** Height of ball, **b)** Applied voltage to the winding

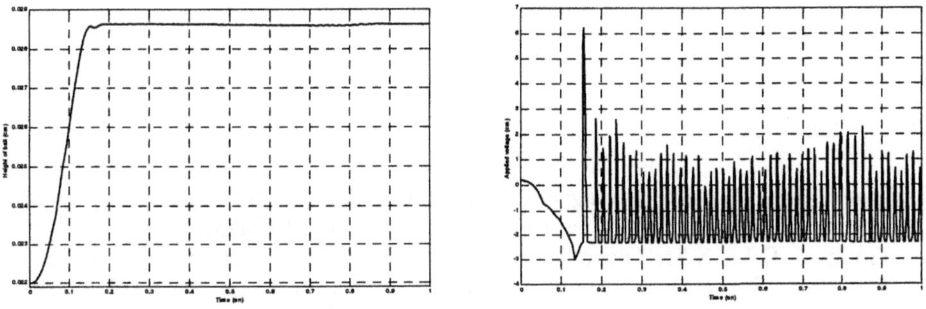

(9a) **(9b)**

Fig. 9. Simulation results of the proposed fuzzy controller: **a)** Height of ball, **b)** Applied voltage to the winding

5 Conclusions

A DSP implementation and simulation of fuzzy controlled magnetic levitation system is presented. Conventional PD controller and fuzzy controller performance are compared. To obtain a robust system in the current control, bang-bang controller has been used. The fuzzy controller has been designed using trial and error method. But it is easier according to PD controller. The PD controller performance depends on the operating parameters of magnetic levitation system. In the stability analysis we observed that in the system under fuzzy controller the error goes to near zero.

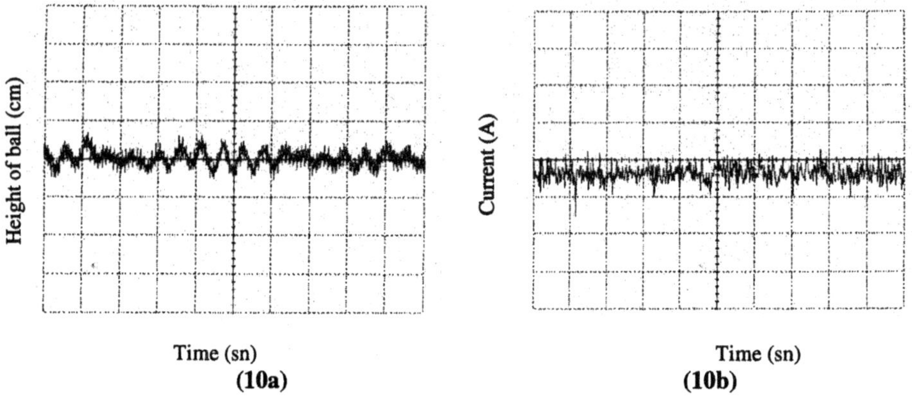

Fig. 10. Experimental result of MAGLEV system: **a)** Height of ball, **b)** Current of controller

Table 1. Parameters of MAGLEV system

Parameters	Value
Equilibrium point x_0	0.03 m
Equilibrium current I_0	0.59 A
Ball mass M	0.0113 g
Coil resistance R	42 Ω
Coil inductance L	0.995 H
G	9.81

References

1. Wang T.H., "Design of magnetic levitation control system", An undergraduate project, IEE Transaction on Education, vol.E-28, no.4, 196-200, 1986.
2. Trumper, D.L., S.M. Olson and P.K. Subrahmanyan, "Linearizing Control of Magnetic Suspension Systems", IEEE Transaction on Control System Technology, vol. 5, no. 4, pp. 427-438, 1997.
3. Shen J., "Nonlinear H^{∞} Control of Magnetic Levitation System", Asian Journal of Control, 2000.
4. Hedjar R., Toumi R., Boucher P., Dumur D., "A finite horizon nonlinear predictive controller for an electromagnetic suspension system", 9th IFAC Symposium Transportation Systems 2000, Braunschweig, Jun 2000.
5. Golob M., "Decomposition of a fuzzy controller based on the inference break-up method", The 3rd On-line World Conference on Soft Computing (WSC3). On the Internet during 21-30 June 1998.
6. Moinuddin, Siddiqui A.S., Sharma A.K., Gupta J.P., "A multi-level fuzzy rule based control strategy for maglev system", EPE Journal, vol.10, no.1, 26-31, 2000.
7. Passino K.M., Yurkovich S., *Fuzzy Control*, Addison-Wesley, 1998.
8. The MathWorks Inc., *The Student Edition of MATLAB Student User Guide*, Prentice Hall, 1992.

Programming Work Simulator of the Ann-Based Integrated Protection System for the Reversible Hydro Generating Sets

A. Halinka and M. Szewczyk

Technical University of Silesia, Gliwice, Poland

Abstract. This paper presents a programming work simulator worked out for verification of the new artificial neural networks – based approach to the decision making systems for the power protection automation purposes.

1 Introduction

High technical requirements of today Electrical Power Systems (EPS) cause the rising importance in the whole control and management strategy of the EPS – especially in the generating nodes area – of objects characterizing high dynamics, large control range, short time periods from stand by to operation and functional flexibility (e.g. possibility of work in generating mode, motor mode, etc.). Such objects include for example hydro power plants equipped with reversible generating sets. But highly valued proprieties of the mentioned above objects achieve at the expense of theirs high structural and functional complexity in comparison to the traditional generating nodes. This provide to the necessity of integrated protection automation system (IPAS) determination dedicated to the complex power plants. In the paper programming work simulator of the ANN – based adaptive protection system for the reversible hydro generating sets is presented.

2 General Conception of the Integrated Protection Automation System for the Reversible Hydro Generating Sets

Possibility of work of the reversible hydro generating sets in some different operation modes and changeable in the wide range frequency of the measurement input signals, which are used by measurement and protection units, provide to the necessity of integrated measurement and protection automation system determination for proper protection in the events of disturbance. Using digital technique and artificial intelligence properties give the possibility of the digital protection systems resolving, dedicated to the complex EPS objects. Such systems have to be characterized by:

- the possibility of proper identification of the current protected object operation mode – *structural criterion* [1],

B. Reusch (Ed.): Fuzzy Days 2001, LNCS 2206, pp. 959–967, 2001.

- the assurance of high accuracy and speed of the measuring and protection algorithms in all possible working states especially during the changing in wide range frequency of input measurement signals – *protection criterion,*
- adaptation of the measurement and protection functions parameters in the protection system executive modules to the changing working conditions (e.g. to the change of operation mode) – *adaptive criterion,*
- fast analyze of information about the disturbances for preventive-restitution automation – *preventive-restitution criterion* [2].

For the taken protected object structure (fig. 1) integrated protection automation system realizing mentioned above criteria was proposed. System is based on seven functional blocks which realizing – in the execution sense – directives taken by ANN – based decision modules.

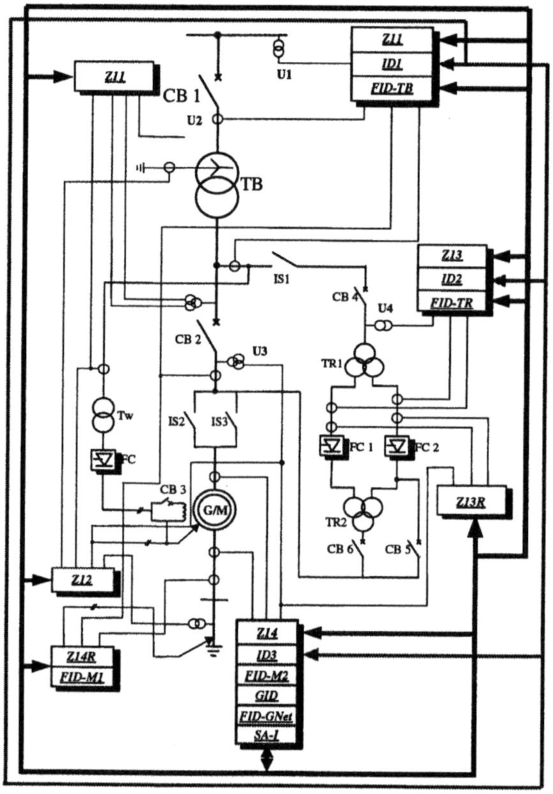

Fig. 1. General integrated protection system structure dedicated to the reversible generating set where:

ID1, ID2, ID3, GID – decision modules realizing identification criterion,
SA-I – decision module; Z11, Z12, Z13, Z13R, Z14, Z14R – executive modules realizing adaptation and protection criterion,
FID-M1, FID-M2, FID-TB, FID-TR, FID-GNet – decision modules realizing preventive-restitution criterion.

Basic decision modules of the identification system [1] realizing *structural criterion* (*ID1*, *ID2*, *ID3*) take the information about the position of switches (from theirs auxiliary contacts), about voltage level (in chosen measurement points) and measurement signals frequency range. To improve the correctness of final decision the global module (*GID*) is basing on provisionally processed information from the basic ones. The main adaptive protection functions parameterization module (*SA-I – protection and adaptation criterion*) is basing on the information from the identification system and some additional signals from the external systems (about the short circuit power level and possibility of generating node operation in the island). The *preventive-restitution criterion* is realizing in modules *FID-...*which are mainly basing on information from the executive protection modules about the picks up and starts of the protection functions.

3 Learning Process of the ANN–Based Decision Modules

As mentioned above all decision modules of the integrated protection automation system for the reversible hydro generating sets are based on ANN structures (fig. 2).

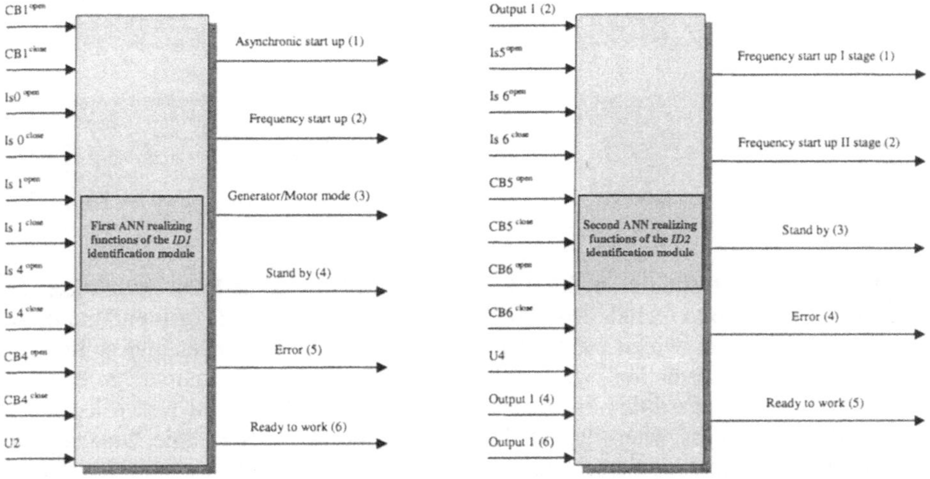

Fig. 2. Functional block scheme of the first (dedicated to the unit transformer) and second (dedicated to the frequency start up system) identification ANNs. Names of the main switches normally opened and closed auxiliary contacts (CBx^{open}, CBx^{close}, Isx^{open}, Isx^{close}) and binary coded information about the voltage level in the particular measuring points (Ux) correspond with elements on the fig 1. Output signal: „Output **n** (**k**)" means signal from output **k** of the identification module **n**.

For the assumed system structure multi layer perceptron artificial networks are proposed to realize functions of the particular decision modules. Levenberg-Marquardt back error propagation algorithm was taken as the method to update weights and biases during the learning process. All networks have three – layer structures. The input and hidden layers have nonlinear transfer functions (hyperbolic tangent sigmoid function) and the output layer has linear transfer function. The number of neurons in each layer depends on number of inputs and outputs of the given module. For all decision modules the knowledge bases were worked out (table 1). They include as many as possible cases of input and output signals combinations which carry correct input – output sequences during all expected protected object working conditions and cases in which some of input signals combinations aren't completely correct (missing, broken or incompatible input signals) in which proper output signals combinations have to be activated according to the current working conditions (i.e. according to the current operating mode).

Table 1. Exemplary fragment of the ANN knowledge base for the module *ID1*

Inputs	1	2	3	4	5	6	7	8	9	10
CB1 open	0	0	0	1	1	1	1	0	0	1
CB1 close	1	1	1	0	0	0	0	1	1	0
Is0 open	1	1	1	1	1	1	1	0	0	1
Is0 close	0	0	0	0	0	0	0	1	1	0
Is1 open	1	1	1	1	1	1	1	0	0	0
Is1 close	0	0	0	0	0	0	0	1	1	1
Is4 open	0	1	0	1	0	1	1	0	0	1
Is4 close	1	0	1	0	1	0	0	1	1	0
CB4 open	0	0	0	1	1	0	1	0	0	1
CB4 close	1	1	1	0	0	1	0	1	1	0
U2 (>0.2Un)	1	1	0	1	1	0	0	0	1	1
Outputs										
Asynchronic start up	1	0	0	0	0	0	0	0	0	0
Frequency start up	0	0	0	1	0	0	1	0	0	0
Generator or Motor	0	0	0	0	0	0	0	0	0	0
Stand by	0	0	0	0	0	0	0	1	0	0
Error	0	1	0	0	1	1	0	0	1	1
Ready to work	0	0	1	0	0	0	0	0	0	0

Next learning and testing bases have to be marked off from the knowledge base. The learning bases have to include from the knowledge bases all cases of input and output signals combinations which carry correct input – output sequences and some cases with "erroneous information" while testing bases are the whole knowledge bases. In such situation the possibility of generalizing events of the neural networks can be achieved. This occurs when during the testing process ANN is able "to generate" correct input – output signal sequence even if such combination was not used during the learning process. The learning process ends when the ANN structure is able "to generate" all the input – output signals sequences combinations from the knowledge base. Exemplary results of the learning process for the two chosen ANN structures realizing functions of the identification module *ID1* are shown on figure 3:

a) first ANN structure: 8 neurons in the input layer, 6 in the hidden and 6 in the output layer; network structure isn't able to recognize all the events from the knowledge base,
b) second ANN structure: 15 neurons in the input layer, 10 in the hidden and 6 in the output layer; network structure is able to recognize all the events from the knowledge base,

SEE (sum squared error) is calculated from the expression:

$$SEE = \sum_{k=1}^{n} (z_k - y_k)^2$$

where:

SSE – sum-squared error
z_k – expected signal level on output k
y_k – level of output k in the given iteration
n – number or ANN outputs

Exemplary results of the testing process are shown on figure 4.

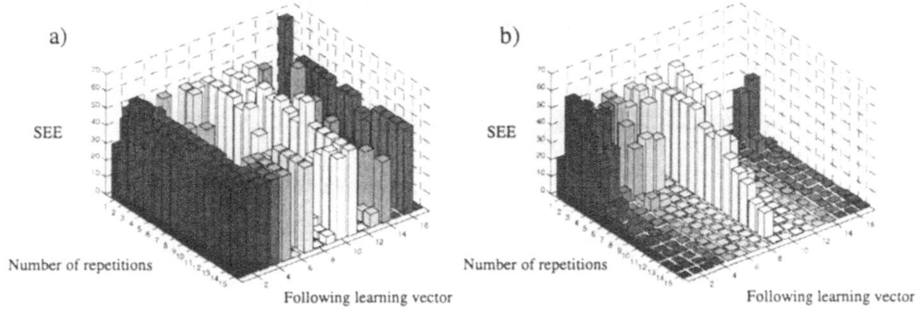

Fig. 3. Exemplary results of the learning process error (SEE) change in dependency of the number of input vectors repetitions

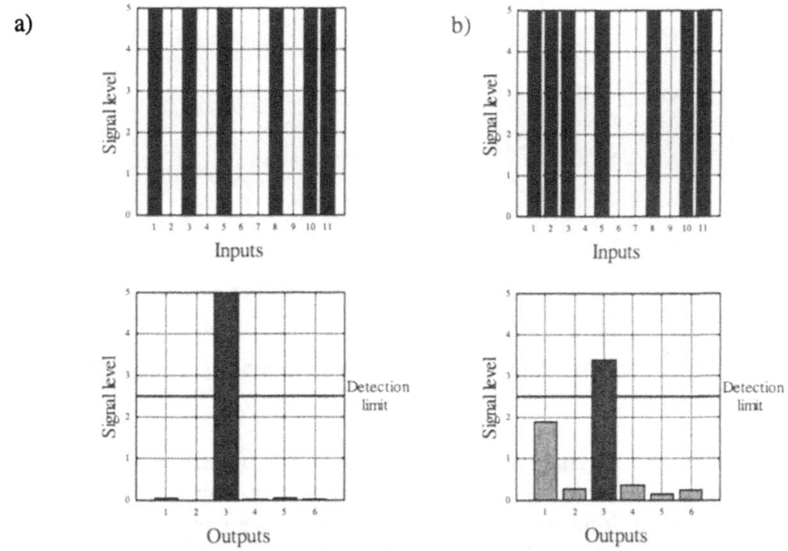

Fig. 4. Module *ID1* response to the correct input signals sequence a) and to the input signals sequence witch include incompatible information about position of circuit breaker 1 b) (inputs 1 and 2 are "high"). Correct identification in both of cases (signal on output 3 above the detection limit)

4 Programming Simulator of the ANN-Based Adaptive Protection System for the Reversible Generating Sets

For the complete verification of proper identification and adaptive protection system functioning the application in the technical language ***Matlab*** was worked out. The application has an ability to put on the individual inputs of the system modules (fig. 5):

- ✓ binary signals from the auxiliary contacts of switches in the closest object environment,
- ✓ binary coded signals about the voltage level in chosen measurement points and the current measurement signals frequency range.

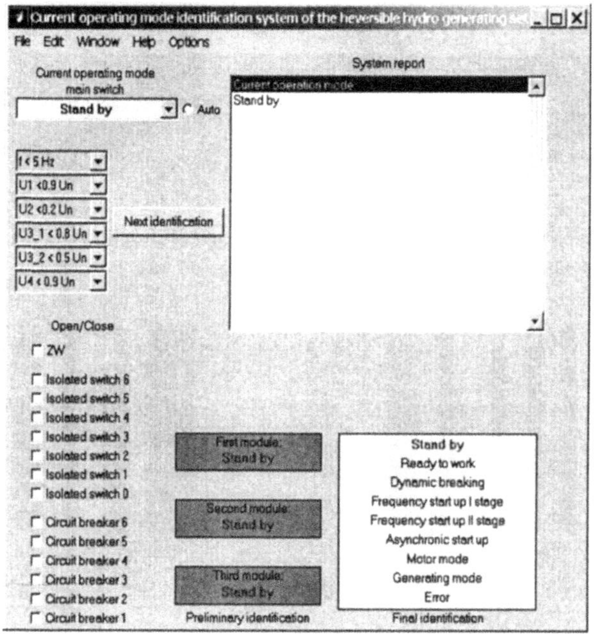

Fig. 5. The main window of the integrated protection automation system programming work simulator

Switching sequence can be manual or automatic. In the manual mode (see fig. 6) opening and closing of switches and selecting the information about voltage level and frequency range can be done by the user. In any time we can see the position of switches on the synoptic scheme (fig. 7). Pushing "Next identification" button cause current operation mode identification according to the switches position, voltage level and frequency range. On the main screen identification results of the particular identification modules and global identification module can be seen. After successful identification proper protection functions are activated in the adaptive protection system executive modules (fig. 8).

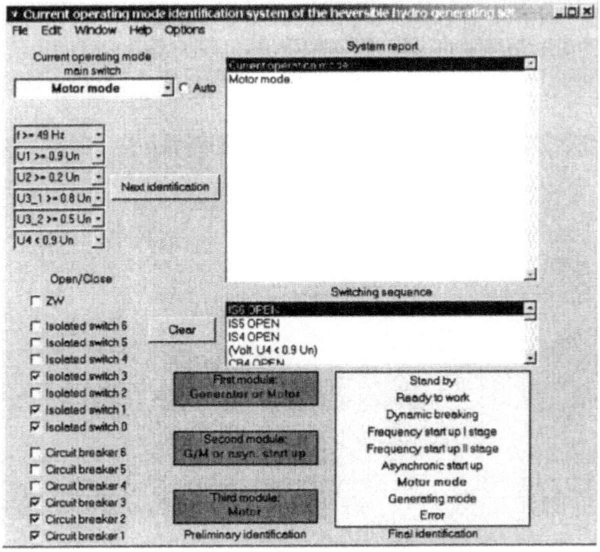

Fig. 6. The main window of the integrated protection automation system programming work simulator in manual mode after pushing "Next identification" button

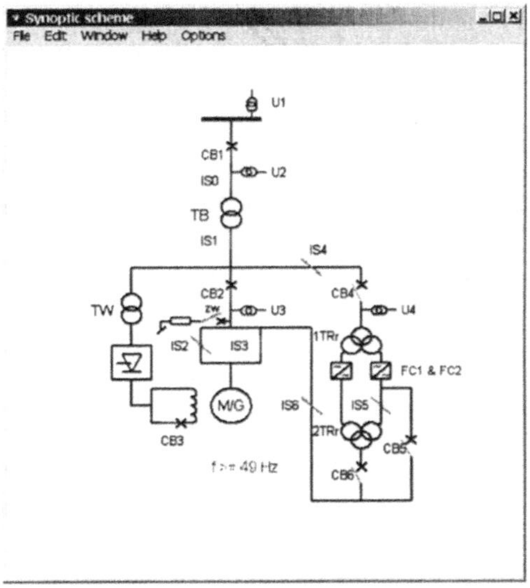

Fig. 7. Protected object synoptic scheme

In the automatic mode (fig. 9) user is able only to "switch" the protected object from one to the another possible operation mode (i.e. from "stand by" to generating mode) by the current operating mode main switch. Switching sequence in this mode is predefined and is the same as it should be in the real object. Also the time of "next identification signal" lunching is fixed. Manual opening and closing of switches and changing the information about voltage level and frequency range is disabled. In any time there is the possibility to stop or pause the automatic switching sequence. After stopping or ending the simulation change to the manual mode is possible.

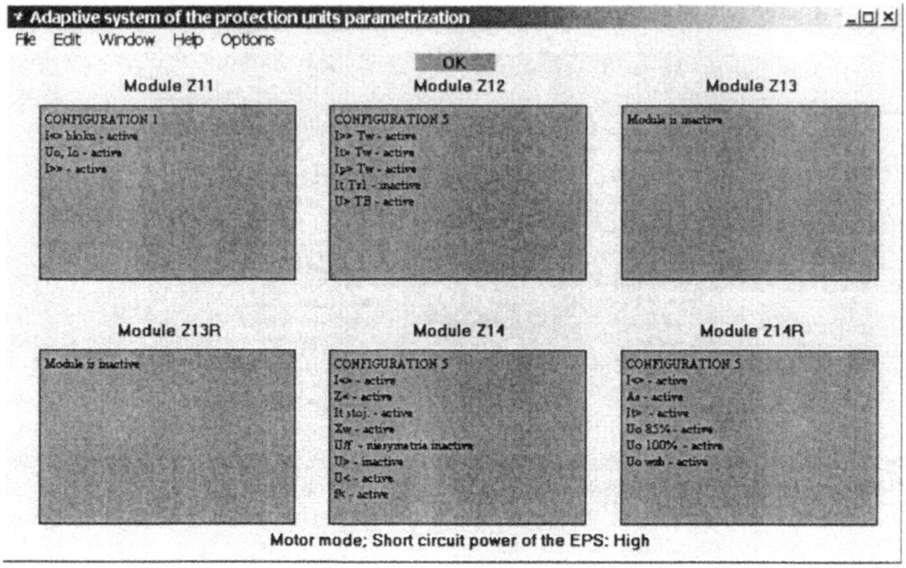

Fig. 8. Active and inactive protection functions in the executive adaptive protection system modules after successful identification process

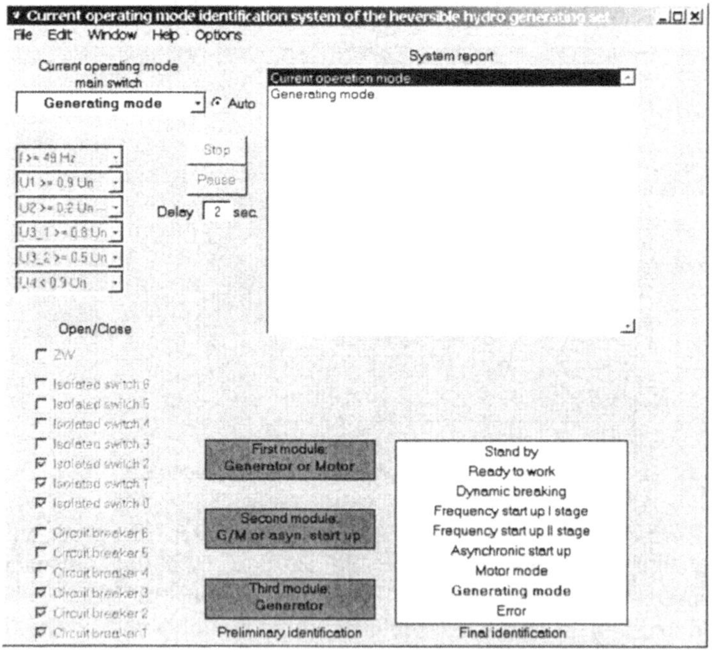

Fig. 9. The main window of the integrated protection automation system programming work simulator in automatic switching sequence mode

The application has also the ability of testing of the integrated adaptive system in the events of disturbance occurring in the input signals such as wrong, missing or incomplete information. So there is a chance to verify the main propriety of the neural networks to the events generalization i.e. to proper identification of the input signals combination to the correct output decision class even in case of missing of such combination during the learning process. This propriety gives the possibility of taking proper decisions in the cases when traditional decision systems (based on typical binary logic or assisted by the expert systems) take the decision "the error occurs" or even take no decision.

5 Conclusions

High structural and functional complexity and changeable in the wide range frequency of the measurement signals (in some operating modes) cause that today protection systems are not capable of protecting the reversible generating sets in all theirs operation modes. The proposition is to apply the integrated protection automation system realizing described in the paper criteria. For the sake of a big number of processing in the system information and required high speed and correctness of taken decisions using the elements of the artificial intelligence in the form of neural networks as the structures realizing decision modules of the mentioned above system is recommended. Complete verification of all realized by the IPAS functions (and not only its individual modules/subsystems) require work out the programming application which enable the possibility of appropriate input signals combination simulation. Only there is the ability of presenting of the output decision signals which are mainly the results of the all subsystems cooperation. Initial simulation results confirm the proper realizing of assumed criteria and admit that choosing neural networks as the decision structures is right.

References

1. A. Halinka, M. Szewczyk, W. Winkler, B. Witek: Identification of complex power system plants operating states for adaptive protection purposes, Proc. of the 11-th Int. Conference on Power System Protection, Bled, Slowenia, 30.09. - 2.10 1998, pp. 133-139
2. A. Halinka, M. Szewczyk, B. Witek : ANN-Based Fault Type and Location Identification System for Autonomus Preventive-Restoration Control of Complex Electrical Power System Plants, Proceedings of the International Conference 6th Fuzzy Days Dortmund, Germany, May 1999, Computational Intelligence Theory and Applications, Springer Verlag, p.682 - 684.

Blending of 2D Objects Using Fuzzy Logic

Ahmet Çınar and Ahmet Arslan

Firat University, Department of Computer Engineering, 23119 Elazığ, Türkiye
acinarfirat.edu.tr

Abstract. In this paper, a method based on fuzzy logic for blending of 2D objects is presented. For this purpose, the curve modeling technique based on control points is used for curve generation where the control points are determined through fuzzy logic technique. As application, 2D object blending operations are implemented.

1 Introduction

Solid modeling is very important subject in computer graphics The one of the most popular problem in the solid modeling is to model objects no have certain mathematical form. To overcome this problem, one of the applied methods is to obtain the new objects by joining smooth surfaces or curves based on given primary objects. This process is called as blending on computer graphics [1,2]. Blending can be defined as joining of objects without sharp edges or corners in the intersection regions of the objects [3,4]. Another definition can be given as follows; intersecting or non-intersecting objects can be joined with curves and surfaces having smooth curves and surfaces [8]. The general blend surfaces was worked by Rockwood [9], constant-radius blending techniques were studied by Rossignac and Requicha [6], Choi and Ju [3], cyclide blends have been studied in Schene [10], Dutta [7].

The first literature on fuzzy logic can be found in Zadeh [12]. Fuzzy logic was used in different areas such as sound detection, pattern recognition, control applications, image processing ... etc. The works about fuzzy curve and surface logic can be found in [5] and fuzzy spline interpolation works was done by Saga [11].

This study presents a method to blend of objects in 2D. At the beginning, the curve is produced afterwards, generated curve is used for blending process.

The structure of the paper is as follows; Section two is concerned about a method, which are developed to create curve in 2-D and fuzzy logic technique. In section three, how to make blending process are mentioned and some sample blended objects were given. Lastly, conclusion was offered.

2 The Curve Modelling Method by Fuzzy Logic Technique

The curve modeling is one of the basic subject in computer graphics. There are generally two curves in computer graphics: approximation and interpolation

B. Reusch (Ed.): Fuzzy Days 2001, LNCS 2206, pp. 968–975, 2001.

curves. While a curve is produced, different methods are implemented. These are; producing the curve via control points, creating the curve to be satisfied C^1 continuous at starting and ending points and generating curve by knowing curvature radius [1]. In this study, the curve producing technique by control points is improved. It is easier than other methods, because first and second order derivation computations don't need.

It is well – known that it is easy to construct the approximation or interpolation curves where starting and ending points are determined. In this study, the approximation curves are preferred for blending. The basis of work is that, in case of knowing the start and end points, least three points, that are computed by means of original points and their auxiliary points, are produced by fuzzy logic technique and approximation curves are generated by using all of points. Afterwards, blend surfaces are produced by means of curves and some blended surfaces are given.

The control points have to be known, while the approximation and interpolation curves are created by means of control points. In this study, where starting and ending points are known, the curve is produced via other points, which are calculated by these points. The basis of study is to compute the interval points by fuzzy logic technique, where starting and ending points are known. While that computation was being done, at least three points except start and end points were used to create curve. The reason is that, local changes on the created curve should be able to satisfy at least three points (Figure 1). Figure 1 depicts this effect. Let $A(u, v)$ and $B(u, v)$ be two points in 2-D. Infinite curves can be created by these two points. If a controlled curve is wanted to be created, then location of auxiliary points (second points) is important.

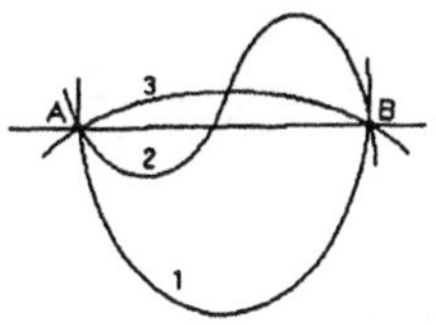

Fig. 1. $A(u, v)$ and $B(u, v)$ two points and infinite curves can be produced by the means of these points.

So, some parameters will be needed to create a controlled curve as well. Therefore, the shape of curve can be satisfied. It is required to use different control points for creating curve 1 on Figure 1. Similarly, curve 2 and 3 require different control points. To compute location of these points by fuzzy logic, we use points called as auxiliary points. The reason of using auxiliary points is that, the control mechanism that will satisfy local changing on the curve, is to obtain. These points will determine shape of curve and set the basis of blending which is given as application.

2.1 Developed Method

Let $A(u,v)$ and $B(u,v)$ be two points in 2-D. Infinite curves can pass from these two points. Let us define two other points to create curve that will satisfy property to be requested and passing from these points. Let $A'(u,v)$ and $B'(u,v)$ be two auxiliary points. Auxiliary points are points at distance k of original $A(u,v)$ and $B(u,v)$ points. Let k be very small such as $k \ll 1$.

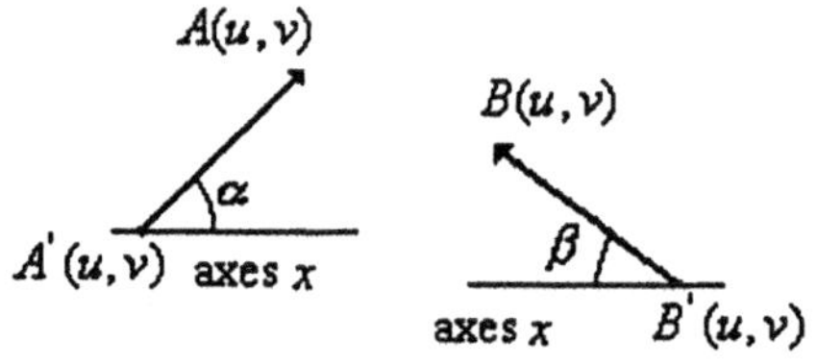

Fig. 2. Case of $A(u,v)$, $B(u,v)$, $A'(u,v)$ and $B'(u,v)$ points.

Having α and β determined, it is not enough to take only these angles to be able to use fuzzy logic technique. The reason is that, if the distance among points is not determined and only angle is evaluated then, undesired cases may occur. So, the curves can not be produced. This case was graphically depicted in figure 3.

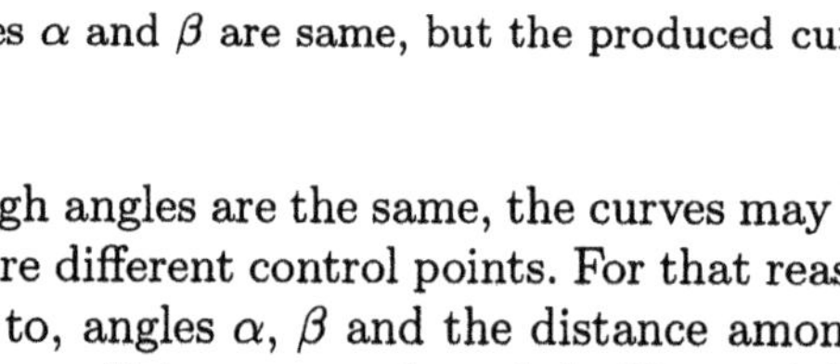

Fig. 3. The angles α and β are same, but the produced curves are different.

If noticed, although angles are the same, the curves may have different shape. Because curves require different control points. For that reason, when fuzzy logic technique is applied to, angles α, β and the distance among points are used as input parameters. These distances are d_1 and d_2. Figure 4 depicts are d_1 and d_2. It is easy to compute distances d_1 and d_2.

If it is noticed to figure 5, the points $M_1(u,v)$, $M_2(u,v)$ and $M_3(u,v)$ are points to be computed. While locations of these points are computed, lines n_1, n_2 and n_3 are used. Figure 5 shows the case of lines n_1, n_2 and n_3. These lines are located as perpendicular to the line connecting the points $A(u,v)$ and $B(u,v)$. The distances m_1, m_2, m_3 and t_1, t_2, and t_3 are important. So, the location of $M_1(u,v)$, $M_2(u,v)$ and $M_3(u,v)$ are very important. Simply, it is easy to

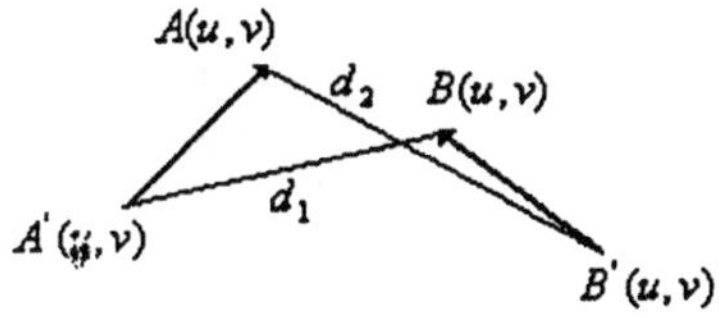

Fig. 4. Distances d_1 and d_2.

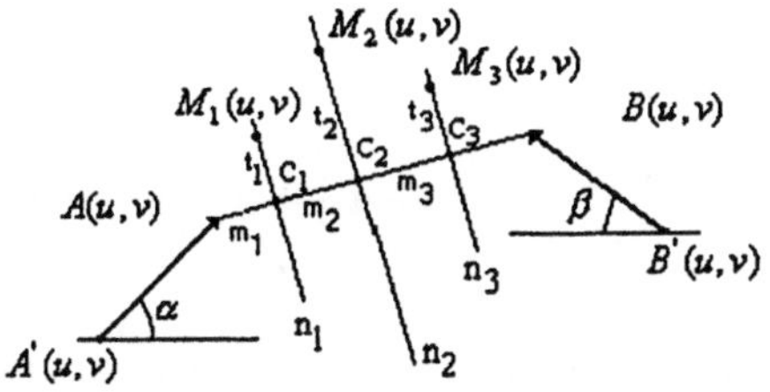

Fig. 5. Case of lines n_1, n_2 and n_3.

compute distances m_1, m_2, m_3 and n_1, n_2, n_3 by fuzzy logic technique. If these distances and points C_1, C_2, C_3 are determined then, computing exact points $M_1(u,v)$, $M_2(u,v)$ and $M_3(u,v)$ can be computed in fuzzy inference machine. Center of sums is used as inference machine. (Equation 1).

$\mu(u_i)$: membership functions.

u_i : The value of membership functions.

$$u^* = \frac{\sum_{i=l}^{l} u_i \cdot \sum_{k=1}^{n} \mu(u_i)}{\sum_{i=1}^{l} \sum_{k=1}^{n} \mu(u_i)} \tag{1}$$

2.2 Rules for Fuzzy Logic and Construction of Membership Functions

In this section, membership functions selected via parameters determined in section 2.1 and produced rule sets by these functions are presented. It is known that, the parameters that will be evaluated for fuzzy logic technique are angle α, β, and distances d_1, d_2. However, there is a problem to select d_1 and d_2. Because distances d_1 and d_2 can change in a very large interval. Instead of d1 and d2 we can use division of d_1 into d_2. Let p be division of d_1 into d_2

$$\begin{aligned} p &= \frac{d_1}{d_2} \qquad d_1 > d_2 \\ p &= \frac{d_2}{d_1} \qquad d_1 < d_2 \end{aligned} \tag{2}$$

So, both mathematical operations and the one of input parameters of fuzzy logic are decreased. Noticed that, there is no case that value p *is* very big. Because

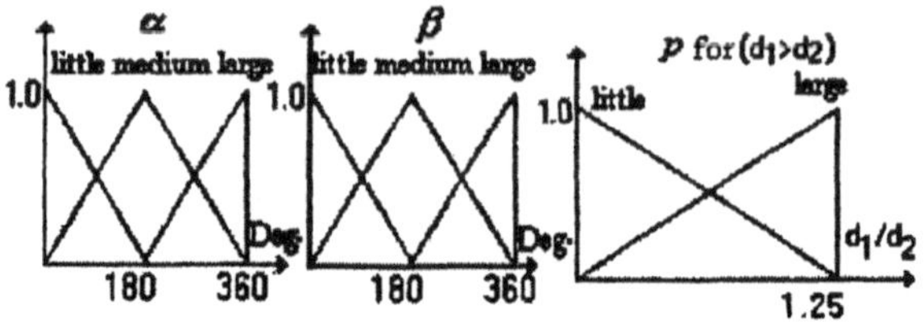

Fig. 6. Membership functions used as input parameters.

Fig. 7. Membership functions used as output parameters.

distance among $A(u,v)$ and $A'(u,v)$ is very small and distance among $B(u,v)$ and $B'(u,v)$ is very small, too. For that reason, distance d_1 is almost equal to distance d_2, and so, it will be selected in interval [0–1.25] of p. Figure 6 shows membership functions used as input parameters and Figure 7 shows membership functions selected as output parameters.

3 Application: Blending in 2D

Applying the method presented in section 2 to blending process require to know exactly some parameters and points $A(u,v)$ and $A'(u,v)$ and $B(u,v)$ and $B'(u,v)$. The method can apply to blending process, after these points and parameters are known.

Let $f(u,v)$ and $g(u,v)$ be two surfaces in 2-D and let $f'(u,v)$ and $g'(u,v)$ be their offset surfaces. It is known that an offset surface is a translated surface at distance d to the normal direction (Figure 9).

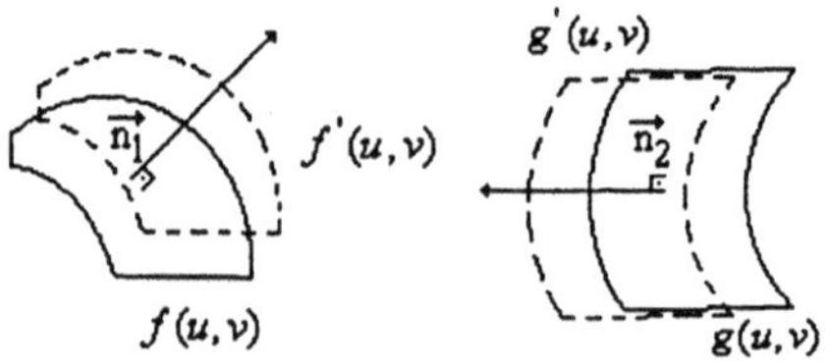

Fig. 9. Original surfaces; $f(u,v)$ and $g(u,v)$ their offset surfaces; $f'(u,v)$ and $g'(u,v)$

Table 1. Fuzzy control rules Inputs Outputs

Inputs			Outputs					
α	β	p	m_1	m_2	m_3	t_1	t_2	t_3
1	1	1	1	2	1	1	2	1
1	1	2	1	1	2	1	2	2
1	2	1	2	3	2	1	2	2
1	2	2	3	1	2	2	1	1
1	3	1	1	2	2	1	3	2
1	3	2	2	3	3	2	3	1
2	1	1	3	2	2	2	1	2
2	1	2	1	1	2	2	1	2
2	2	1	2	3	1	1	1	3
2	2	2	1	2	2	2	2	2
2	3	1	2	2	1	1	2	3
2	3	2	2	1	2	3	1	2
3	1	1	3	2	3	3	1	2
3	1	2	2	1	3	2	3	1
3	2	1	1	2	2	1	2	2
3	2	2	1	2	2	1	3	2
3	3	1	3	1	1	2	1	1
3	3	2	1	3	1	1	2	1

Note 1. For α and β 1, 2 and 3 values show little, medium and large, respectively.
2. For p; the 1, 2 values show little and large, respectively.
3. m_1, m_2, m_3 and t_1, t_2, and t_3: the 1 2, 3 values show little, medium and large, respectively.

Below steps was applied to computing points $A(u,v)$, $B(u,v)$ and $A'(u,v)$ and $B'(u,v)$. At this point, it is not important whether surfaces intersect or not.

$$\begin{aligned} f(u,v) - g'(u,v) &= 0 \\ f'(u,v) - g(u,v) &= 0 \end{aligned} \tag{3}$$

If equation (3) is solved by using Newton-Raphson method, then intersection curves, namely points $A(u,v)$ and $B(u,v)$, can be found. By the help of these points, $A(u,v)$ and $B(u,v)$ can be computed. For this goal, offset surfaces are translated to distance δ. Hence, the points $A'(u,v)$ and $B'(u,v)$, are obtained. However, δ must be very little. Because δ gives only the information on shape of original surface.

In this study, δ is selected in range $0.0001 < \delta < 0.1$ especially for $\delta = 0.02$. Figure 9 shows the computation steps.

After the points $A(u,v)$ and $A'(u,v)$ and $B(u,v)$ and $B'(u,v)$ are computed, three points $M_1(u,v)$, $M_2(u,v)$ and $M_3(u,v)$ can be obtained by the help of developed method in section 2. In order to compute these points, α, β, and p parameters should be given as input parameter of fuzzy logic system. For this purpose, distance d_1 and d_2 is computed. Parameter p and parameters α and β is

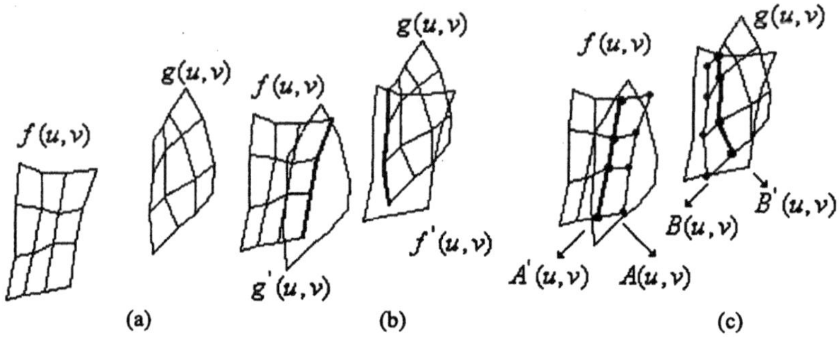

Fig. 9. Computing of points $A(u, v)$, $A'(u, v)$, $B(u, v)$ and $B'(u, v)$.

calculated with respect to distances d_1 and d_2. Therefore, the input parameters of fuzzy logic system are ready. The $M_1(u, v)$, $M_2(u, v)$ and $M_3(u, v)$ points are computed by means of table 1, equation (2), namely inference machine, and output parameters on the figure 7. Finally, 4th degree approximation curve can be computed via equation (4).

$$P(u, v) = (1-u)^4 A(u, v) + 4(1-u)^3 u M_1(u, v) + 6(1-u)^2 u^2 M_2(u, v) \\ +4u^3(1-u) M_3(u, v) + u^4 B(u, v) \tag{4}$$

Figure 10 illustrates the blended shapes in 2D.

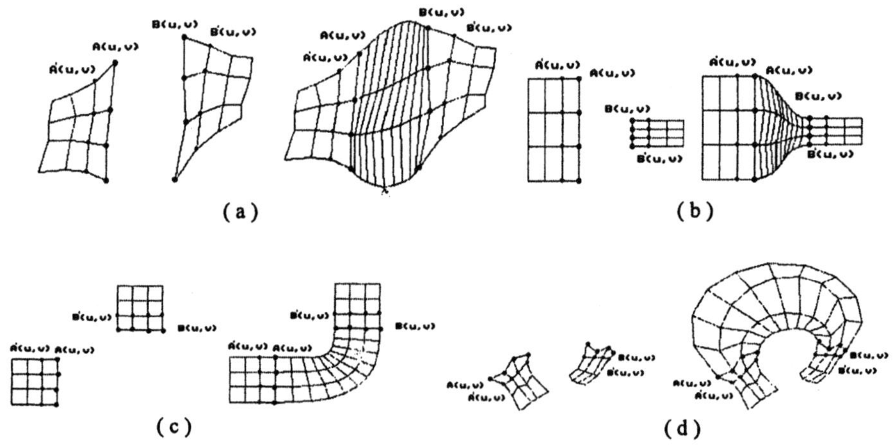

Fig. 10. Some blended objects produced by means of developed method and their blending points.

4 Conclusion

In this paper, a new technique to create an approximation curve was presented. The basic of the developed method is to create an approximation curve by the help of fuzzy logic technique. Produced curves were used for blending in 2-D. As future work, the method will be tried to be applied on blending in 3-D.

References

1. Ahmet Çınar, Ahmet Arslan, "Two Approaches On Blending", ISCIS XI The Eleventh International Symposium On Computer and Information Sciences, November 6–8 Antalya, Turkey, 1996.
2. Farin G., Curves and Surfaces for Computer Aided Geometric Design, Academic Press, Boston, 1990.
3. Choi K.B., Ju Y. S., 'Constant – Radius blending in surface modeling', Computer Aided Design', Vol. 21, No. 4, May, 1989.
4. Filkins P.C., Tuohy S.T., Patrikalakis M. N., 'Computational methods for blending approximations', Engineering With Computer', 9, pp 49–62, 1993.
5. Gao Lun Shan, Kawarada Hideo, 'Applications of fuzzy average to curve and surfaces fitting', 2nd Fuzzy Eng. Sym.IEEE Inter. Con. on Fuzzy systems, Vol. 2, p 971–978, 1995.
6. Rossignac J. R., Requicha A. G. 'Constant radius blending in solid modeling', Comput. Mech. Eng., July, pp 65–73, 1984.
7. Srinivas L. V., Dutta Debasisih, 'Blending and joining using cyclides', ASME Advances In Design Automation, Vol. 2, pp 231–236, 1992.
8. Jung-Hong Chuang, Wei-Chung Wang, Variable-radius blending by constrained spine generation, Visual Computer, Vol: 13, N: 7, 316–329, 1997.
9. Rockwood. A. P., Owen J. C., Blending surfaces in solid modelling, in: Farin G., ed., Geometric Modeling, Algorithms and New Trends, SIAM, Philadelphia, PA, 367–383, 1994.
10. Seth Allen, Debasish Dutta, Cyclides in pure blending I, Computer Aided, Geometric Design, 14, 51–75, 1997.
11. Saga Sato, Makino Hiromi, A Method for Modeling Freehand Curves-the Fuzzy Spline Interpolation, Systems and Computers in Japan, Vol: 26, No:19, pp:1610–1619, 1995.
12. Zadeh L. "Fuzzy sets", Inf. Control, Vol:8, pp:338–353, 1965.

The Application Based on Incremental Knowledge Discovery Algorithm in Color Matching System

HaiYi Zhang[1] and JianDong Bi[2]

[1] Turku Centre for Computer Science TUCS
Department of Information Systems, Abo Akademi University,
Lemminkinengatan 14 B,
Fin-20520 Turku, Finland
zhang.haiyi@abo.fi

[2] Department of computer Science, Harbin Institute of Technology,
HeilongJiang, 150001,China
bjiana@public.hr.hl.cn

Abstract. First, the paper investigates incremental knowledge discovery(IKD) in which the knowledge obtained is reapplied in the discovery process. Then we realize the algorithm application in color matching method (CMM), finding some laws in secondary colors data and their relations with pure color data. Finally, the experiment result that is fairly satisfying is given.

Keyword Knowledge Discovery, Incremental Knowledge Discovery, Color Matching Method, Machine Learning.

1 Introduction

Knowledge discovery in databases is the process of identifying valid, previously unknown and ultimately understandable knowledge hiding in data, by 'mining' in data source. The knowledge found in databases can be used in many areas, such as information management, inquiry response, decision support, process control and etc [2,3]. It has provoked interest of researchers in many fields, including database system, knowledge-base system, artificial intelligence, machine learning, knowledge acquisition, statistics, space databases and data visualization [5,6]. Thanks to its broad prospect, knowledge discovery in databases has become a hot field in the world.

But the theory is not complete and it has few examples of application. Moreover, in all the previous KDD systems, the knowledge acquired from different groups of data is isolated from each other. It is only incorporated into the ultimate performance system, or simply document or reported to interested parties, not fully used in the discovery process. Thus it affects the exactness, systematicness and flexibility of the knowledge systems[7].

B. Reusch (Ed.): Fuzzy Days 2001, LNCS 2206, pp. 976–989, 2001.

In view of this problem, this paper puts forward a new idea of incremental knowledge discovery, in which the knowledge obtained is reapplied in the discovery process. In order to clarify the idea of 'incremental' discovery, this paper constructs a new algorithm based on Bacon algorithm [8,9]by Pat Langley from the points of equation discovery. Then the rule-base is built up and thus the algorithm accomplished.

At last the algorithm is successfully applied to a color matching project, finding some laws in secondary colors data and their relations with pure colors data. The former color matching system is improved according to these laws by replacing the part of manual adjustment with automatic learning and adjustment. The experiment result is fairly satisfying.

This paper has analyzed and evaluated the algorithm, and it has pointed out that in future the research should focus on enlarging its rule-base to enhance its ability of discovering knowledge. In addition, discovering knowledge of more complex description forms should be studied, thus to make the algorithm applicable to broader fields and to give it greater vitality.

2 Incremental Knowledge Discovery Algorithm

2.1 Introduction

In the design of the incremental knowledge discovery algorithm, the data and rule expression forms are same as the forms that BACON algorithm adopted [12,14]. The system employed guiding rule base, and added the basic rules, such as, linear rules, increasing rules and decreasing rules that the BACON algorithm possesses to the guiding rule base, and also plused the discovery polynomial rule. The guiding rule base possesses the uniform formalizing guiding rules. By the combining of the guiding rules in the rule base and introducing the idea of the incremental knowledge discovery, we can strength the ability of knowledge discovery. The rules in the rule base control the execution of the algorithm. The rule base is the core of the whole algorithm.

2.2 The Expression Form of IKD Algorithm

The incremental knowledge discovery algorithm adopted the databases, rule bases and knowledge bases. The database stored a data source, that is, the objects of knowledge discovery. As it is possible to use the previous obtained laws in the incremental knowledge discovery, we must adopt the special form to organize in order to search conveniently.

2.2.1 The Database

The discovery objects are stored in the database. The forms of the discovery objects are data clusters [7]. A data clusters is a set, which is attribute pairs linked samc node. It represents a series of observing values happened at same time. There are two kinds of variables: independent variable and dependent variable. The system controls the attributes of an independent variable. It can change the values of the independent variable and it requires the proper value of the dependent variable. The values of the dependent variable may be numeral and/or name (symbol).

2.2.2 The Rule Base
A rule in the rule base is used for guiding discovery and controlled the execution of the algorithm. The rule is called as a production. The expression forms of the production system are described as following: the production system consists of two parts: One is a condition-action rule set, and other is a dynamic working storage. A production system works in cyclic meaner. In the cycle the conditions of each rule match the current state of working storage. One rule is selected from rules matched successfully to apply. When a rule is applied, Its actions affected the working storage state so that a new rule is matched. This procedure is keep doing until there is no any production matched or input stops.

2.2.3 The Knowledge Base
The knowledge base stores the laws discovered, for instance, the formula in this paper. As the formula discovered is possibly more than one in the same group, and it is possible to discover in more than one group data, we must organize those formula reasonably so that it is convenient to search when we do incremental knowledge discovery. The formula is arranged by the discovered order and coded by a serial number. Besides, we set up a index table, the first item in the table is the name of a independent variable, the second item stores the serial number for the formula that the independent variable is in. We can search the formula accoutring to the independent variable. For example, the knowledge in the knowledge base is
Knowledge 1: $y_1 = a_1x_1 + b_1$ (x_1 is a independent variable, y_2 is a dependent variable)
Knowledge 2: $y_2x_2 = c_1$ (x_2 is a independent variable, y_2 is a dependent variable)
Knowledge 3: $y_1(a_2y_3 + b_2y_3)$ c_2 (x_1 is a independent variable, y_1 and y_3 are dependent variables)
Knowledge 4: $y_2y_1/(x_2 - x_1) = c_3$ (x_1 and x_2 independent variables, y_1 and y_2 is dependent variables)
Here, a_1, a_2, b_1, b_2, c_1, c_2, and c_3 are respectively constants. The indexed table is following form.

The background knowledge is not used in the BACON algorithm. In fact, we can store some background knowledge in the knowledge base. For instance, in the color matching, the principles of absorbing and reflecting of light and equations obtained from energy synthesizing:
$(R_w$ - R$) = (R_w - R_c) + (R_w - R_m) + (R_w - R_y)$

$$(G_w \text{ -G}) = (G_w \text{ - } G_c) + (G_w \text{ - } G_m) + (G_w \text{ - } G_y)$$

$$(B_w \text{ - B}) = (B_w \text{ - } B_c) + (B_w \text{ - } B_m) + (B_w \text{ - } B_y)$$

The equations above are the basic formula in chromatics. Before we do the color matching discovery, they are already existing there. We may put them into the knowledge base in order to use them. With background knowledge to guide the discovery, the ability of system discovery is enhanced greatly.

2.2.4 The Main Steps in the Incremental Knowledge Discovery

In the whole algorithm, discovery with basic rules and incremental discovery are main parts. Here, the incremental discovery is only a step, which is different from the concept of incremental knowledge discovery in the whole algorithm. In fact, the step will also use the basic rules but as it includes the ideas to discovery new knowledge based on the old knowledge, it is the core part of this algorithm.

The procedure of discovery with basic rules

1) Collecting the data

The top goal of IKD algorithm is combining the single independent variable and forming one or more laws. In order to achieve this aim, the system contains a serial of rules which collect systemically data. The collecting method that the system adopts is direct. It starts from keeping one variable unchanged and changing other variables and then trying to discover some laws. In this case the constant values with the variable values are kept. When we have enough the constant values discovered, the system is going to be in the higher layer. Taking them as dependent variable values, the system will try to find the higher layer relationship between the independent variable and the dependent variable which is newly defined. The system adopts the same method to discover the second layer laws. After discovering the second laws, the system will be in further layer repeatedly to do things above until all the independent variables are included in the laws and all the data are summarized. In other word, the procedure of searching laws is embedded in the procedure of searching data spaces

2) Discovery laws

3) Define variables and calculate values

4) Testing laws

The incremental discovery

If discovering some law from a group data, and there exist the relative knowledge in the knowledge, then doing the incremental discovery. Now, we take an example of Color Matching to illustrate this. Suppose that we have discovered the formula:

$R_c = a_1C^2 + b_1\text{C} + c_1$,
$\text{Rm} = a_2M^2 + b_2\text{M} + c_2$,
$\text{Rcm} = a_3(CM)^2 + b_3(\text{CM}) + c_3$.
By searching, we can find the relationship between the formulas. So we can do

Incremental discovery. The first step: Finding some corresponding relationship between the independent variables. For example, Rc, Rm, Rcm are respectively cyan, magenta and the mixed color of cyan and magenta. From this sense, they have some relationship. They will be classified as the same group when we do processing later. One formula might have more one independent variables, it is possibly divided into many group in order to reorder the data. The second step: Calculating the corresponding data according to the values of independent variable and the old formula, and then judging if there is some relationship between independent variables. The relationship is defined by the meanings of those variables in the domain. Normally the background knowledge is given directly. For example, (CM) stands for the mixed color of Cyan and Magenta, that is, Cyan + Magenta, (CM) means C +M. We believe that there is a adding relationship between those three formulas. So we can guess there is an adding relationship between the independent variables, that is, $R_{cm} = R_c + R_m$. If there exist one relationship, we record it down, and reorder the data with this relationship. In our example, we obtain the values of $R_{cm} = (R_c + R_m)$, and form a new group data. The next thing we do is that judging if this group data is all zero or if this group data satisfies the error range allowed. If so, it means that the dependent variables just satisfied the relationship. In our example, $R_{cm} = R_c + R_m$. The formula is put into the knowledge base. If it is not satisfied this condition, the group data is defined as new variables, and then to judge if there exist a relationship between the new variables on the new layer with basic rules.

If there exist no any relationship between the independent variables, we also discover with the basic rules in order to discover some possible knowledge. In some knowledge areas, this step may be omitted because it will spend more time.

Finally, the relationships discovered are put in order. If there is knowledge, it will be put into the knowledge base.

2.2.5 The Error Control

When we do the discovery, most of cases, the data are not accorded with a law completely. Normally we need setting up a allowed error. If any data is in the error range, we believe it is correct. In the procedure of knowledge discovery, the calculating method of an error is determined according to the background of the different domains and the task of discovery.

3 The Rules of The Incremental Knowledge Discovery Algorithm

3.1 The Rules of The Main Algorithm

There are nine rules in main algorithm. They are respectively initialising rule, discovery basic law rule, checking discovery result rule, searching, requesting new data rule, incremental discovery rule, non-incremental discovery rule, ending discovery rule and discovering in the new data rule. These rules are used for controlling the execution of the main algorithm.

At beginning, the initialising rule starts to initialise the database and knowledge base, to prepare the data to discover, and to add the necessary background knowledge into the knowledge base. Then the discovery basic law rule sets up the sub-goal of a law, which is formed by combining the independent variables. The sub-goal will cause the matching with initialising variable that will be given below.

After using the basic rules to discover some knowledge, the algorithm will judge if any knowledge is obtained. If yes, the current status of the working storage is matched with the rule searching the condition, which causes the action adding the knowledge into the knowledge base.

3.2 The Collecting Data Rules

There are five rules for collecting data. They are respectively initialising independent variable rule, modifying the independent variable, recording rule, initialising dependent variable rule and modifying dependent variable rule.

The top goal of the algorithm after discovered the basic law is combining the single independent variable, and forming one or more laws. In order to achieve the goal, the system contains a serial of rules, which can collect data. The first rule applied is the initialising independent variable rule, which will match with the combining goal, and initialise the value of the independent variable and create a goal to peer those values. The modifying the independent variable rule will match with this goal and set up three sub-goals: 1) Combining the other independent variables as a law, 2) Obtaining dependent variable value for the current value of independent variable and 3) peering the other values of the independent variable.

Based on the latest strategy of solving conflict, the first goal will obtain priority. It is satisfied before checking other values of independent variables.

The other two rules will convert the information into a more useful form: the initialising dependent variable rule will initialise a serial of values so that the special dependent variable is stored. The modify dependent variable rule will modify the value of variable.

3.3 Basic Rules

There are eight basic rules. They are respectively discovery rule, constant rule, Checking predicting rule, linea relation rule, defining linea relation rule, modifying linea relation rule, increasing law rule and decreasing law rule.

Those eight rules are used for discovering basic laws, for example, constant law, linea law, increasing e and decreasing law, etc. Here, the basic laws are related to the obtained the knowledge in the procedure of the incremental discovery. As some rules can be called many times, the various formulas may be discovered. In addition to the above eight rules, we also need the rules used to define variables and values, for example, defining ratio and multiplication rule and recording value rule.

Discovering Polynomial Rule

We have 21 rules for discovering the polynomials. They are respectively polynomial rule, calculating error rule, defining error value rule, X constant rule, defining variable rule, Y constant, permuting variable rule, setting value rule, assigning D value rule, etc.

A polynomial is an important formula. Those rules will enhance greatly the ability of the system discovery. Here, discovering polynomial rules adopts a special method, which is efficient for some kinds of polynomials. But it is required for some strict conditions. It needs to be improved further. Conjecturing nature attribute rule

There are five rules about conjecturing nature attribute. They are considering nature attribute rule, defining nature attribute rule, generalizing nature attribute rule, searching nature attribute rule and storing nature attribute rule.

3.4 Incremental Discovery Rules

There are 12 rules doing the incremental discovery. They are respectively corresponding dependent variable, collecting data, independent variable relation, dependent variable relation, processing new data, considering error, recording formula, defining new variable, discovering basic law, independent variable no relation, basic law discover result and ending incremental discovery.

3.5 Setting error rule

There is only one rule about setting error. Normally the discovery task will decide that which error form is used.

4 The Application of The IDK Algorithm in Colour Matching

4.1 The Task of Color Matching

The color matching is a kind of software which efficiently converts one color space to other color space [1,4], for example, from RGB (Red/Green/Blue) color space to CMYK(Cyan/Magenta/Yellow/Black) color space. In this way the computer photograph can be correctly print on the color printer. Normally a color picture on the computer screen is the synthesis of red light, green light and blue light but the picture printed by the color printer Is the mixed color of cyan color, magenta color, yellow color and black color. The RGB color space of the computer screen is different from the CMYK color space of the color printer. So before printing the computer photograph we must do the correct conversion [11,13].

As the different color ink is printed with one level after other level in the real print, there exist osmosed phenomenon and covered phenomenon between them. CMYK is the decreasing color system, the covered phenomenon made the decreasing color function of the covered color weak. The bottom color osmoses to the top-level ink, which makes the decreasing color of the top level ink strength. From this point of the view, the printing result is not the result of CMYK four colors adding up. The error that is caused with such reason must be readjusted by manual, which affects the automatically learning function of the system. So it is necessary to adopt the mixed data and to learn the data and to discover the osmosed laws and covered laws. In this way we can enhance the ability of the automatically learning.

4.2 The Discovery for the Mixed Data

The table 1, 2 and 4 respectively show the data selecting from real domain. Here, CM stands for the mixed color, C and M are Cyan and Magenta color ink, and they are equal quantity.

Table 1. C color data

C	10	20	30	40	50	60	70	80	90	100
Gc	223	218	210	205	194	185	176	169	158	152

Table 2. M color data

M	10	20	30	40	50	60	70	80	90	100
GM	196	162	132	110	82	60	42	30	15	0

Table 3. CN mixed color data

C	10	20	30	40	50	60	70	80	90	100
GCM	195	158	131	107	82	69	61	55	49	46

The independent variables of those data take 10 ad the increasing step. In the real testing CM mixed data takes 5 as the step but C or M single color data take 1 as a step. Besides, we have Y single color data and MY, CY and CMY mixed color data.

Following the IKD algorithm, first we initialise the rule base and knowledge base, and prepare the C, M and CM data in the database. Then we add the background knowledge to the knowledge: (CM) = C + M, the order number is 1, the index table as

Table 4. The knowledge base index after initialising

Variable Name	Knowledge order number
(CM)	1

The formula represents that CM mixed color is the result printing with Cyan and Magenta color. From the quantity of ink they have an adding relationship. After initialising we set up the error, the allowed error is 0.2 times of original data, that is, if the original data is d (d is not 0), then we believe that the result in the range of [d (1-0.2), d(1 + 0.2)] is correct. If d = 0, then we believe that the result in the range of [0, 5] is correct This error is set up according to our experiential results. Normally the quantity of ink is changed in this range, which cannot be seen by human being. Then, the algorithm called the basic rule in the rule base to discover the data in table 1. It discovers the linea relationship between independent variable C and dependent variable Gc, the slope is 0.8 and the displacement is 233.7 So G_c = 0.8 C + 233.7

There is no knowledge related to this in the knowledge base. We add the formula into the knowledge. Now the contents are in the knowledge base:

1. (CM) = C + M

2. G_c = 0.8 C + 233.7

Its index table is

Table 5. The knowledge base index after discovering C data

Independent variable name	Knowledge order number
(CM)	1
C	2

As there is new data, the algorithm will continue to discover the M color data, and then discover that there is a polynomial relationship between the independent variable and dependent variable, that is, $G_M = 0.014M^2$ - 3.66M + 231.7

The independent variable is M in the formula, and dependent variable is GM. There is no any knowledge related to this in the Knowledge base. So, we add the formula into the knowledge base. Now the contents in the knowledge base are:

1. (CM) = C + M

2. G_c = 0.8 C + 233.7

3. $G_M = 0.014M^2$ - 3.66M + 231.7

Its index table is:

Table 6. The knowledge base index after discovering M data

Independent variable name	Knowledge order number
(CM)	1
C	2
M	3

The algorithm still will continue to call the basic rules in the rule base to discover the new data, CM mixed data, and then discover that there is a polynomial relationship between the independent variable (CM) and dependent variable Gcm, that is, $G_{CM} = 0.022\ (CM)^2$ - 4.06 (CM) + 232.3

The independent variable is M in the formula, and dependent variable is G_M. We add the formula into the knowledge base. Now the contents in the knowledge base are:

1 (CM) = C + M

2 G_c = 0.8 C + 233.7

3 $G_M = 0.014M^2$ - 3.66M + 231.7

4 $G_{CM} = 0.022\ (CM)^2$ - 4.06 (CM) + 232.3

Its index table is:

Table 7. The knowledge base index after discovering CM data

Independent variable name	Knowledge order number
(CM)	1
C	2
M	3
(CM)	4

The algorithm seeks the knowledge related to knowledge order number 1 in the knowledge base index. The 1th knowledge is (CM) = C + M. The knowledge is related to C variable and M variable. The formulas corresponding to C variable and M variable are found, that is,

G_c = 0.8 C + 233.7

G_M = 0.014M^2 - 3.66M + 231.7

Then the algorithm does discovering: The first step: To find the corresponding relations between dependent variables. As G_c, G_M and G_CM all stand for the reflection values of green color, the three dependent variables are related.

The second step: To recalculate the values of dependent variables according to the variables. though there exist an error for the knowledge, we believe that it is precise as soon as it is stored in the knowledge. Beside, we recalculate the values but not adopt the original data because some values of independent variables in the original data are no corresponding values of dependent variables. In the color matching, when we measure the mixed color data, in order to avoid more data the value of an independent variable is 5 as the increasing step normally. For instance, when C = 21 and M = 21,there is no values of dependent variables in our mixed color data. So we need to calculate the values.

As there is a relationship, (CM) = C + M, between the independent variables of these three formulas, so the algorithm make a guess: GCM = DC + GM, and record this relation. After arranging, we have a new data: G_{cm} - (G_c + G_m)

All the new data are not in the range of allowed error, we define a new variable T_1 ($T_1 = G_CM$ - (G_c + G_M)). Now the algorithm will do discovering based the new variable T_1 with basic rules. There is a polynomial formula discovered:

T_1 = -0.01 $(CM)^2$ - 0.20 (CM) + 229.10

Finally, we get a formula: G_{CM} - (G_c + G_m) = -0.01 $(CM)^2$ - 0.20 (CM) + 229.10

G_{CM} = (G_c + G_m) -0.01 $(CM)^2$ - 0.20 (CM) + 229.10 (1)

Table 8. Arranging the new data

C, M, (CM)	G_c	G_m	G_{CM}	G_{CM} -(G_c + G_m)
10	225.5	196.5	194.0	228.0
20	217.4	163.9	160.1	221.2
30	209.2	134.2	130.8	212.6
40	201.1	107.1	105.9	202.3
50	192.9	82.8	85.5	190.2
60	184.7	61.2	69.7	176.2
70	176.6	42.4	58.3	160.7
80	168.4	26.2	51.4	143.2
90	160.2	12.8	49.1	123.9
100	152.1	2.2	51.2	103.1

We add the formula into the knowledge base. Now the contents in the knowledge base are:

1 (CM) = C + M

2 G_c = 0.8 C + 233.7

3 G_M = 0.014M^2 - 3.66M + 231.7

4 G_{CM} = 0.022 $(CM)^2$ - 4.06 (CM) + 232.3

5 G_{CM} = (G_c + G_m) -0.01 $(CM)^2$ - 0.20 (CM) + 229.10

Its index table is

Table 9. The knowledge base index after discovering CM data

Independent variable name	Knowledge order number
(CM)	1
C	2
M	3
(CM)	4
(CM)	5

The formula (1) tells us: in the practice, two color inks are printed one layer after other, there exist osmosed phenomenon and covered phenomenon indeed. The reflect value of the green light is not adding relationship between two colors. It must be readjusted. The formula is just the readjusted formula. Following the method above, the algorithm will do discovering for some other variables, which are independent variable C, dependent variables respectively are RC and Bc, independent variable M, dependent variables respectively are RM and BM, independent variable Y, dependent variables respectively are RY and GY, independent variable CM, dependent variables respectively are RCM and GCM, independent variable MY, dependent variables respectively are RMY, GMY and BMY, independent variable CY, dependent variables respectively are RCY, GCY, and BCY, independent variable CMY, dependent variables respectively are RCMY, GCMY and BCMY. We will get the formulas, which are similar to the formulas we obtained above. The details can be found in paper [10].

5 Summaries

Now we analyse the time complexity of the IKD algorithm, here we assume that the attribute of data is 1; the rule number in the rule base is constant m; the data number in the database is a variable, let n stand for the variable. The knowledge number in the knowledge base is F(n), and F(n) is decided by the data number, in other word, it is decided by the cluster number of the data. If the data number in the cluster is at least Lmin, then F(n) = C(n/Lmin) Here C is a constant. The algorithm does discovering with rules, its time complexity is O(m*n), that is, O(n). So, The time complexity of IDK algorithm is O(n). We can see that the time consuming is relying on the size of database.

6 Acknowledgements

The authors would like to thank Ms. Bin Lu for her experimental work and very useful ideas for this project. And the helpful comments received from Prof. Barbro Back, of the Abo Academy University in Finland, are gratefully acknowledged.

References

1. Hong Bozhe. The color adjustment of four-color printer. Japanese picture Committee. 56(2): 112-122, 1993.
2. Ronald J. Brachman and Tej Anand. The Process of Knowledge Discovery in Databases: A Human-Centered Advances in Knowledge Discovery and Data Mining. AAAI/MIT Press. 37-58, 1996.
3. B. Hayes-Roth, K. Pfleger, P. Lalanda, P. Morignot, and M. Balabanovic. A domain-specific software architecture for adaptive intelligent systems. IEEE Transactions on Software Engineering, 21(4):288-301,1995.
4. Mark D. Fairchild. A Novel method for the determination of color matching functions. Color Research and Application. 14(3):122-131, 1989.

5. Usama M. Fayyad, Gregory Piatetsky-Shapiro, and Padhraic Smyth. From Data Mining to Knowledge Discovery: An Overview Advances in Knowledge Discovery and Data Mining. AAAI/MIT Press. 1-36, 1996.
6. K I. Funahashi.On the approximate realization of continuous mapping by neural networks. Neural Networks. (2):183-192, 1989.
7. A. K Jain., R. C Duber. Algorithms for clustering data. Englewood Cliffs. NJ: Prentice-Hall. P42-59, 1988.
8. P. Langley., H. A. Simon., Gray L. Bradshaw. And M. Zytkow. Scientific Discovery 195-213, 1987.
9. P. Langley. back 1: A General Discovery System. In Proceedings of the Second National Conference of the Canadian Society for Computational Studies. 68-89, 1978.
10. B. Lu Research on incremental knowledge discovery algorithm and its application in color matching system Master thesis of Harbin Institute Technology, China, p35-41, 1998.
11. Xiao Sheng, Cun Jing, Xi Duo. High Accyrate Color Transformation based on UCR. Japan Hardcopy 94 papers collection. 177-180, 1994.
12. Jeff Shrager and P. Langley. Computational Models of Scientific Discovery and Theory Formation. Morgan Kaufmann Publishers. 73-97, 1990.
13. G.Spencer, P.Shirley, K. Zimmerman etc. Physically-based Glare Effects for computer Generated Images. Proceedings ACM SIGGRAOH'95. p325-334 (1995).
14. S. Sugiura and Makita T. An improved multilevel error diffusion method. Journal of Imaging Science and Technology. 39(6):495-501, 1995.

Fuzzy Convexity and Connectivity

Antony T. Popov

St. Kliment Ohridski University of Sofia,
Faculty of Mathematics and Informatics,
5, James Bourcher Blvd., BG-1164, Sofia, Bilgaria
atpopov@fmi.uni-sofia.bg,
http://www-it.fmi.uni-sofia.bg/atpopov/

Abstract. In image analysis and pattern recognition fuzzy sets play the role of a good model for segmentation and classifications tasks when the regions and the classes cannot be strictly defined. Shape analysis is of great importance for different applications of image processing, for instance in the recognition of pathological objects on X-ray or microscopic images. Pure mathematical notions like convexity and connectivity play an essential role in shape analysis. In practical image processing, the notions of convexity and connectivity as defined in any textbook in mathematics are rarely encountered, moreover it is not easy to define what exactly a convex discrete arrangement of pixels mean. Also, in real vision systems, whether machine or human, imprecisions are inherent in the spatial and intensity characterisation of the image, there are also effects of noise in sensory transduction and of limits of sampling frequency. Unlike computer systems, human beings are more flexible. In general, they can easily say whether a pattern looks convex or not, or they can specify easily the connected components on a grey-scale or a colour image. Therefore the terms approximate convexity and approximate connectivity based on fuzzy set theory have been introduced.

1 Introduction

In many practical tasks we operate with imprecise or uncertain data, especially when this data comes as an output from some measuring instrument. This problem appears in image processing because of the distortion effect, and often because of non-precise calibration of the camera. Therefore lots of work have been done for the development of algorithms which give reliable results operating with imprecise data.

In image analysis and pattern recognition, fuzzy sets have been found to be a good model for segmentation and classifications tasks when the regions and the classes cannot be strictly defined. Shape analysis is of great importance for different applications of image processing, for instance in the recognition of pathological objects on X-ray or microscopic images. Pure mathematical notions like convexity and connectivity play an essential role in shape analysis. However, images processed by a computer are discrete - they comprise a finite number of elements. Usually these picture elements, called pixels, are arranged regularly

B. Reusch (Ed.): Fuzzy Days 2001, LNCS 2206, pp. 990–1000, 2001.

in a rectangular grid, e.g. an image is represented as a collection of rectangular pixels. Associated with each pixel, there is a non-negative grey level or brightness value - without loss of generality we may consider that this value is between 0 and 1. So, in practical image processing, the notions of convexity and connectivity as defined in any textbook in mathematics are rarely encountered, moreover it is not easy to define what exactly a convex discrete arrangement of pixels mean. Also, in real vision systems, whether machine or human, imprecisions are inherent in the spatial and intensity characterisation of the image. At the low levels of image representation, there are effects of noise in sensory transduction and of limits of sampling frequency. At higher levels of image representation, there are more general imprecisions related to the specifications of image quality. Unlike computer systems, human beings are more flexible. In general, they can easily say whether a pattern looks convex or not, or they can specify easily the connected components on a grey-scale or a colour image. Therefore the terms approximate convexity and approximate connectivity have been introduced. These definitions are strongly based on fuzzy set theory.

The idea of approximate convexity is used in medical image analysis, since as known from the medical experts benign tumors look 'more convex' than the malignant ones. Also, approximate connectivity is used for unsupervised segmentation of the tissues on a tomographic (X-ray, MR, Ultrasound) image.

As we see in this work, fuzzy logic makes possible to join the notions of approximate convexity and connectivity, which is natural from the point of view of the 'crisp' mathematics. As a result, fuzzy implications using variables as 'low convex', medium convex', 'highly convex' and 'low connected', 'medium connected', 'highly connected' are suggested. These results strongly rely on the usage of fuzzy morphological operations, which are found to be a very good alternative of the classical grey-scale morphological operations introduced by Serra.

A new general definion of fuzzy (or approximate) connectivity has been proposed introducing a so-called degree of connectivity of a set, which is a real number within 0 and 1. It works not only with the classical arc-wise crisp connectivity, but with with any connectivity class in the crisp Euclidean space.

2 Mathematical Morphology – Background

Serra [12] and Heijmans [4] have shown that morphological operations can be formulated on any complete lattice. A set $\mathcal{L}$ with a partial ordering "$\leq$" is called a *complete lattice* if every subset $\mathcal{H} \subseteq \mathcal{L}$ has a supremum $\bigvee \mathcal{H} \in \mathcal{L}$ (least upper bound) and infimum (greatest lower bound) $\bigwedge \mathcal{H} \in \mathcal{L}$.

An operator $\varphi : \mathcal{L} \mapsto \mathcal{M}$, where $\mathcal{L}$ and $\mathcal{M}$ are complete lattices, is called dilation if it distributes over arbitrary suprema: $\varphi(\bigvee_{i\in I} X_i) = \bigvee_{i\in I} \varphi(X_i)$, and erosion if it distributes over arbitrary infima. Erosions and dilations are increasing operations [4]. An operator $\psi \in \mathcal{L}^*$ is called a closing if it is increasing, idempotent ($\psi^2 = \psi$) and extensive ($\psi(X) \geq X$). An operator $\psi \in \mathcal{L}^*$ is called an opening if it is increasing, idempotent and anti-extensive ($\psi(X) \leq X$) [4]. A pair of operators (ε, δ), is called an adjunction, if for every two elements $X, Y \in \mathcal{L}$ it follows that $\delta(X) \leq Y \iff X \leq \varepsilon(Y)$. In [4] it is proved that if (ε, δ) is an adjunction then ε is erosion and δ is dilation. If (ε, δ) is an adjunction, then the composition $\varepsilon\delta$ is a closing, and $\delta\varepsilon$ is an opening. As an example, let us consider the lattice $\mathcal{L}$ with elements the subsets of a linear space E, i.e. $\mathcal{L} = \mathcal{P}(E)$. Here and henceforth, by $\mathcal{P}(X)$ we denote the power set of X, i.e. the family of all subsets of the set X . Then every translation-invariant dilation is represented by the standard Minkowski addition: $\delta_A(X) = A \oplus X = X \oplus A$, and its adjoint erosion is given by Minkowski subtraction: $\varepsilon_A(X) = X \ominus A$ [4]. Then closing and opening of A by B are defined as $A \bullet B = (A \oplus B) \ominus B$, $A \circ B = (A \ominus B) \oplus B$. These operations are referred to as *classical* or *binary morphological operations.* Openings and closings are generally used as filters for cleaning binary images.

3 Convexity and Connectivity of Binary Images

It is well known from convex analysis that a crisp compact set X in $\mathbf{R}^d$ is convex if and only if $X = 0.5X \oplus 0.5X$. It means also that X contains the whole segment when the set contains its endpoints. For more information see [7].

In mathematics, the notion of connectivity in $\mathbf{R}^d$ is formalized in the topological framework in two different ways. First, a set is called to be connected when it cannot be partitioned as a union of two open, or two closed sets. In practice, it is more suitable to work with the so - called *arcwise connectivity.* A set X is said to be arcwise-connected when for every two distinct points a and b from X there exists a continuous curve joining a and b and lying entirely in X. Arcwise connectivity is more restrictive than the general one. It is not difficult to show that any arcwise-connected set in $\mathbf{R}^n$ is connected. The opposite is not true in general, but for open sets in $\mathbf{R}^d$ the topological connectedness is equivalent to arcwise connectedness. Following the the works of Serra [11–13] and Haralick and Shapiro [3], an abstract connectivity framework, suited mainly for analysis and processing of binary images has been developed. It is strongly related with the mathematical morphology concepts [11]. The base concept is the *connectivity class*:

Definition 3.1. *Let E be an arbitrary set. A family $\mathcal{C}$ of the subsets of E is called a connectivity class if the following properties hold:*

1. $\emptyset \in \mathcal{C}$ *and* $\{x\} \in \mathcal{C}$ *for every* $x \in E$;
2. *If* $E_i \in \mathcal{C}$ *for* $i \in I$ *and* $\bigcap_{i \in I} E_i \neq \emptyset$, *then* $\bigcup_{i \in I} E_i \in \mathcal{C}$.
3. *If* $E \supseteq A \in \mathcal{C}$, *then for every* $x \in E$ $\tau_x(A) \subset \mathcal{C}$.

The third axiom can be referred to as translation -invariance of the connectivity. It hasn't been imposed in the previous works [13, 3, 9] since it doesn't give any topological impact to the notion of connectivity. However it is essential from morphological point of view since the translation-invariance is one of the main bases of mathematical morphology. This condition can be replaced by more general one, namely *affine invariance of connectivity*, or in the case of arbitrary complete lattice by *T-invariance* with respect to a given abelian group of automorphisms T [4]. In our work it is sufficient to work only with translation-invariant operators.

Given a connectivity class in a universal set E we can define the maximal connected component of a set $A \subseteq E$ containing a particular point x:

$$\gamma_x(A) = \bigcup\{C \in \mathcal{C} \,|x \in C \text{ and } C \subseteq A\}$$

Then it can be proved easily (see [12] or [4]) that :

- For every $x \in E$ γ_x is an algebraic opening in E;
- $\gamma_x(\{x\}) = \{x\}$;
- either $\gamma_x(A) = \gamma_y(A)$ or $\gamma_x(A) \cap \gamma_y(A) = \emptyset$;
- $x \notin A$ implies that $\gamma_x(A) = \emptyset$;
- $\bigcup_{x \in E} \gamma_x(A) = A$.

It is easy to demonstrate that $X \in \mathcal{C}$ if and only if for every two points $x, y \in X$ it follows that $\gamma_x(X) = \gamma_y(X)$.

Theorem 3.2. *If X and A are connected with respect to the connectivity class $\mathcal{C}$, then $X \oplus A$ is connected with respect to $\mathcal{C}$ as well.*

This theorem generalises the result of Theorem 9.59 from [4], where only arcwise connectivity is considered. It follows directly from more general results in [13] and [9].

Let $\mathcal{S}$ be a binary relation between the subsets of a universal set E, i.e. $\mathcal{S} \subseteq \mathcal{P}(E) \times \mathcal{P}(E)$.

Definition 3.3. *We say that $\mathcal{S}$ is a separation is it satisfies the following conditions [3] :*

- *$\mathcal{S}$ is symmetric, which means that $(X,Y) \in \mathcal{S}$ if and only if $(Y,X) \in \mathcal{S}$.*
- *$\mathcal{S}$ is exclusive, which means that $(X,Y) \in \mathcal{S}$ implies $X \cap Y = \emptyset$.*
- *$\mathcal{S}$ is hereditary, which means that $(X,Y) \in \mathcal{S}$ implies $(X',Y') \in \mathcal{S}$ for each $X' \subseteq X$ and $Y' \subseteq Y$.*

We have omitted the last axiom of Haralick and Shapiro as too restrictive for our studies.

Following Ronse [9], we may consider two more axioms characterizing the separation relation:

- If $\bigcap_{i \in I} A_i \neq \emptyset$ and for each $i \in I$ $(A_i, E \setminus A_i) \in \mathcal{S}$, then $\left(\bigcap_{i \in I} A_i, E \setminus \bigcup_{i \in I} A_i\right) \in \mathcal{S}$.
- Given three nonvoid pairwise disjoint sets A, B and C from E, such that $(A,B) \in \mathcal{S}$ and $(A \cup C, B) \in \mathcal{S}$, then we must have $(A, B \cup C) \in \mathcal{S}$.

Every connectivity class defines a separation relation and vice versa. If $\mathcal{C}$ is a connectivity class, then we can define the separation relation $\mathcal{S}$ in the following way: say that $(X,Y) \in \mathcal{S}$ when $\gamma_x(X \cup Y) \cap \gamma_y(X \cup Y) = \emptyset$. Then it is easy to demonstrate that the conditions for $\mathcal{S}$ to be a separation relation are satisfied. Then if $\mathcal{S}$ is a separation defined for the subsets of the universal set E, we can define a connectivity class as $\mathcal{C}$ as any subset of $\mathcal{P}(E)$ with the properties:

1. $\emptyset \in \mathcal{C}$ and $\{x\} \in \mathcal{C}$ for every $x \in E$;

2. A set $\emptyset \neq A \subseteq E$ belongs to $\mathcal{C}$ if for every its two subsets X and Y such that $X \cap Y = \emptyset$, $X \cup Y = A$ it follows that $(X,Y) \notin \mathcal{S}$. Such a set will be called indecomposable.

To show that $\mathcal{C}$ is a connectivity class we must demonstrate that given the indecomposable sets $A_i \subseteq E$ for $i \in I$ and $\bigcap_{i\in I} A_i \neq \emptyset$, then the set $A = \bigcup_{i\in I} A_i$ is indecomposable as well. The last is proved in [9].

If K is a connected structuring element then $X \circ K = \bigcup_i X_i \oplus K$, where X_i are the maximal connected components of X. Then if A is separated from B and the structuring element K is connected it simply follows that $(A \cup B) \circ K = (A \circ K) \cup (B \circ K)$. Then suppose that the random dropout noise N removes points from the ideal image A to produce the observed image B. If the noise N is separated from the complement of A^c, the complement of A, then $A = B \bullet K$. Therefore the connectivity relations give us the possibility of perfect reconstruction of the original image.

For the *epsilon geometry* approach to convexity and connectivity see [8].

4 Fuzzy Morphological Operations

Consider the set E called the universal set. A fuzzy subset A of the universal set E can be considered as a function $\mu_A : E \mapsto [0,1]$, called the membership function of A. $\mu_A(x)$ is called the degree of membership of the point x to the set A. The ordinary subsets of E, sometimes called 'crisp sets', can be considered as a particular case of a fuzzy set with membership function taking only the values 0 and 1. This definition leads to two possible interpretations:

- in image representation the value of the membership function $\mu_A(x)$ at a point x may be interpreted as the grey level value associated with that point of the image plane,
- in pattern recognition, the value $0 \le \mu_A(x) \le 1$ indicates the probability that the point x is in the foreground of an image.

Let $0 < \alpha \le 1$. An α-cut of the set X (denoted by $[X]_\alpha$) is the set of points x, for which $\mu_X(x) \ge \alpha$.

The usual set-theoretical operations can be defined naturally on fuzzy sets: Union and intersection of a collection of fuzzy sets is defined as supremum, resp. infimum of their membership functions. Also, we say that $A \subseteq B$ if $\mu_A(x) \le \mu_B(x)$ for all $x \in E$. The complement of A is the set A^c with membership function $\mu_{A^c}(x) = 1 - \mu_A(x)$ for all $x \in E$.

If the universal set E is linear, like the n-dimensional Euclidean vector space $\mathbf{R}^n$ or the space of integer vectors with length n, then any geometrical transformation, like scaling, translation etc., of a fuzzy set can be defined by transforming its α–cuts. In our work we denote the α–cut of the fuzzy set X by $[X]_\alpha$.

Further we consider a simple way to generate adjoint fuzzy morphological operations.

Say that the function $c(x,y) : [0,1] \times [0,1] \mapsto [0,1]$ is conjunctor if c is increasing in the both arguments, $c(0,1) = c(1,0) = 0$, and $c(1,1) = 1$.

Say that the function $i(x, y) : [0,1] \times [0,1] \mapsto [0,1]$ is implicator if i is increasing in y and decreasing in x, $i(0,0) = i(1,1) = 1$, and $i(1,0) = 0$.

Following the framework of Heijmans, say that the conjucnctor - implication pair is adjoint if $c(b, y) \leq x$ is true if and only if $y \leq i(b, x)$. Then for fixed b function $f(x) = i(b, x)$ is an erosion, and its adjoint dilation is $g(y) = c(b, y)$.

Then having an adjoint conjunctor - implicator pair, as proposed by Heijmans, we can define an adjoint pair of fuzzy erosion and dilation:

$$\mu_{\delta_A(B)}(x) = \sup_y c(\mu_B(x - y), \mu_A(y)), \tag{1}$$

$$\mu_{\varepsilon_A(B)}(x) = \inf_y i(\mu_B(y - x), \mu_A(y)). \tag{2}$$

Heijmans [5] has proposed a number of following conjunctor - implicator pairs to construct morphological operations. Here we give examples of two of them:

$$c(b, y) = \min(b, y),$$

$$i(b, x) = \begin{cases} x & x < b, \\ 1 & x \geq b \end{cases}.$$

These operations are known as operations of Gödel-Brouwer.

$$c(b, y) = \max(0, b + y - 1),$$

$$i(b, x) = \min(1, x - b + 1).$$

These operations are suggested by Lukasiewicz. Sometimes, for simplicity we should write $A(x)$ instead of $\mu_A(x)$. In our work we shall use the first conjunctor - implicator pair. The respective dilation has the form

$$(\delta_B(A))(x) = \sup_b \min(\mu_A(b), \mu_B(x - b)),$$

and is used to study fuzzy convexity [6] . In this case we shall denote $\delta_A(B) = \delta_B(A) = A \oplus B$, because this operation can be obtained also directly from the binary Minkowski addition using the *transfer principle.*

Definition 4.1. *A function $F : \mathbf{R}^n \mapsto \mathbf{R}$ is called upper semicontinuous (u.s.c.) if for any $t \in \mathbf{R}$ such that $t > F(x)$ there exists a neighborhood $B_r(x)$ such that $t > F(y)$ for every point $y \in B_r(x)$.*

It is a well known result from mathematical analysis that if $f_i : \mathbf{R}^d \mapsto [0,1]$ are u.s.c. functions for $i \in I$, then $\inf_i f_i$ is also u.s.c. In image processing it is common to use model based on fuzzy subsets of $\mathbf{R}^d$ which membership function is upper semicontinuous with bounded support From the upper semicontinuity it follows that the α-cuts for $\alpha \in (0,1]$ are compact. The upper semicontinuity condition does not affect the discretization process: Every pixel can be considered as a part of the image, such that all of its points have the same brightness value and therefore the neighbourhood $B_r(x)$ of the pixel centre x can be limited to the pixel itself. Also, given two adjacent pixels, without loss of generality we may suppose that their common boundary segment belongs to the pixel with greater intensity or brightness value. Therefore every real grey - tone image can be regarded as a fuzzy set which membership function is u.s.c. function with bounded support with values in $[0,1]$.

Let $\{A_i\}$ be a given family of fuzzy subsets of $\mathbf{R}^d$ which membership functions are u.s.c. with bounded support. Then from the reasons listed above, it follows that the membership function of their intersection $\inf_i(\mu_{A_i})$ is also u.s.c. with bounded support, and for any $\alpha > 0$, $\bigcap_i [A_i]_\alpha = [\bigcap_i A_i]_\alpha$. It is proved in [6] that if A and B are fuzzy sets which membership functions are upper semicontinuous with bounded support then such is $A \oplus B$, and $[A \oplus B]_\alpha = [A]_\alpha \oplus [B]_\alpha$.

One of the most popular approaches for defining fuzzy morphological operations is based on the so called *inclusion indicators*, specified by a number of axioms in [14]. In the case of fuzzy subsets of a discrete space, which is the most natural situation in image processing tasks, the indicator considered by Kosko can be considered (see also [1]):

$$I^K(A,B) = 1 - \frac{\sum_{x\in\mathcal{S}} \max(0, \mu_A(x) - \mu_B(x))}{\sum_{x\in\mathcal{S}} \mu_A(x)}.$$

It does not satisfy all the axioms proposed by Sinha and Dougherty, thereby making it problematic for the design of morphological operations. But, it has been found to be useful in the construction of measures of approximate convexity because it is easy to compute and is not sensitive to noise [6]. In the case of a continuous universe $\mathcal{U} \subset \mathbf{R}^d$ it takes the form

$$I^K(A,B) = 1 - \frac{\int_{\mathcal{U}} \max(0, \mu_A(x) - \mu_B(x))\, dx}{\int_{\mathcal{U}} \mu_A(x)\, dx}.$$

Therefore, when A and B are crisp compact sets in the Euclidean plane, it takes the form

$$I^K(A,B) = 1 - \frac{\sigma(A \setminus B)}{\sigma(A)},$$

where σ means the area.

5 Fuzzy Convexity

Definition 5.1. *The convexity indicator of the fuzzy set A is defined as*

$$C(A) \quad = \quad I^K(0.5A \oplus 0.5A, A).$$

By definition, a fuzzy subset A of $\mathbf{R}^d$ is called *convex* if

$$\mu_A(\lambda x_1 + (1-\lambda)x_2) \geq \min(\mu_A(x_1), \mu_A(x_2))$$

for all $x_1, x_2 \in \mathbf{R}^d$ and every $\lambda \in [0,1]$. There is a well known result saying that a fuzzy set is convex if and only if its $\alpha-$cuts for $\alpha \in (0,1]$ are convex crisp sets in $\mathbf{R}^d$.

A result proved in [7] says that if A is a fuzzy subset of $\mathbf{R}^d$ whose membership function is upper semicontinuous with bounded support, then $C(A) = 1$. On the contrary, if $C(A) = 1$, then there exists a fuzzy set X whose support has Lebesgue measure 0 such that $A \cup X$ is fuzzy convex. These properties show that the proposed definition of an approximate convexity measure by means of an inclusion indicator and fuzzy dilation is reasonable. In practical image processing we usually deal with objects represented as finite collections of pixels. Even for objects in binary images it is not easy to express formally the meaning of convexity. The task is more difficult in the case of grey-scale images when the objects do not have well defined boundaries. So, if A is a binary or grey-tone image we say that *A is convex of degree α* if $C(A) = \alpha$.

Experiments have shown that the suggested indicator indicator of convexity C is not sensitive to noise. It can be applied also to binary images [7].

6 Fuzzy Connectivity

Definition 6.1. *Following [10], let us say that the points $x, y \in E$ are connected in the fuzzy /set A if there exists a path Γ from x to y such that*

$$\inf_{z \in \Gamma} \mu_A(z) \geq \min[\mu_A(x), \mu_A(y)].$$

Then we may call the fuzzy set A $\alpha-$path-connected if every two points x and y, such that $\mu_A(x) \geq 1-\alpha$ and $\mu_A(y) \geq 1-\alpha$, are connected in A. Let consider further a fuzzy set whose membership function is u.s.c. with bounded support. Consider we are given a connectivity class $\mathcal{C}$ in the universal set E (either $\mathbf{R}^d$ or $\mathbf{Z}^d$). Then we may generalise the upper definition not only for path-connectivity saying that a fuzzy set A is $\alpha-$connected for any $\alpha \in [0, 1)$ if $[A]_\beta \subset \mathcal{C}$ for every positive β less than or equal to $1-\alpha$. For completeness we may call a set A to be 1-connected if its support is in the connectivity class, which means $\bigcup_\alpha [A]_\alpha \subset \mathcal{C}$.

Applying directly Theorem 3.2 we show that if for a given $\alpha \in [0, 1]$ the fuzzy sets X and A are $\alpha-$connected, then $X \oplus A$ is also $\alpha-$connected. So we define the *degree of connectivity* of the fuzzy set A in the following way:

$$dc(A) = 1 - \inf\{\alpha \,|\, A \text{ is } \alpha - \text{connected}\}.$$

It is straightforward to show that $dc(A \oplus B) \leq \min[dc(A), dc(B)]$.

This is a general way to define degree of connectivity, which can be used as a feature in different pattern recognition tasks depending on the basic connectivity class. On Fig.1 one can see an example of X-ray mammogram. The degree of connectivity of the outer darker part is 0.33, the light internal region has degree of connectivity 0.78. The degree of connectivity of the whole image is 0.24 disregarding the background. On the lower part of the image one can see two calcifications (the small light spots at the bottom). Therefore for any small region containing the calcifications, the degree of connectivity will be approximately 0.24. Any of the two calcifications themselves have high degree of connectivity ≈ 0.8. Therefore we can use the ratio between the degree of connectedness and the variance between the maximal and the minimal grey levels of the region as

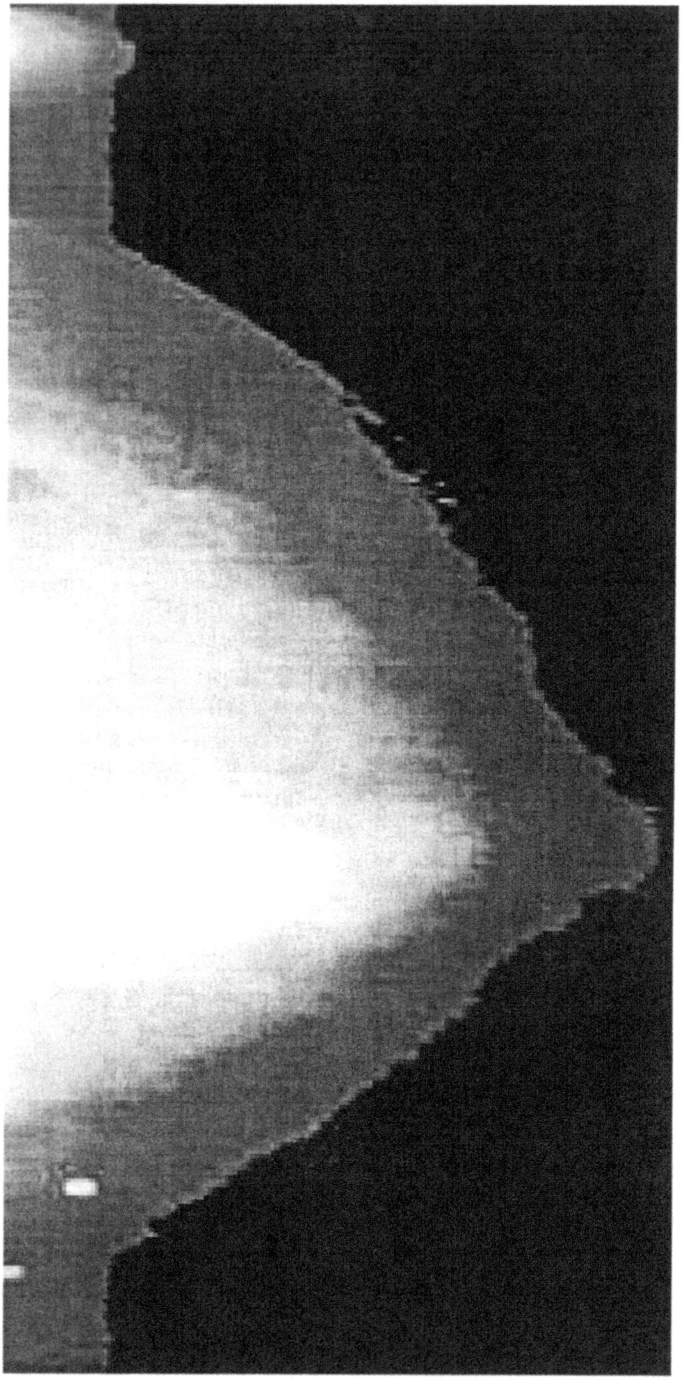

Fig. 1. An example of a mammogram with cancerous calcifications

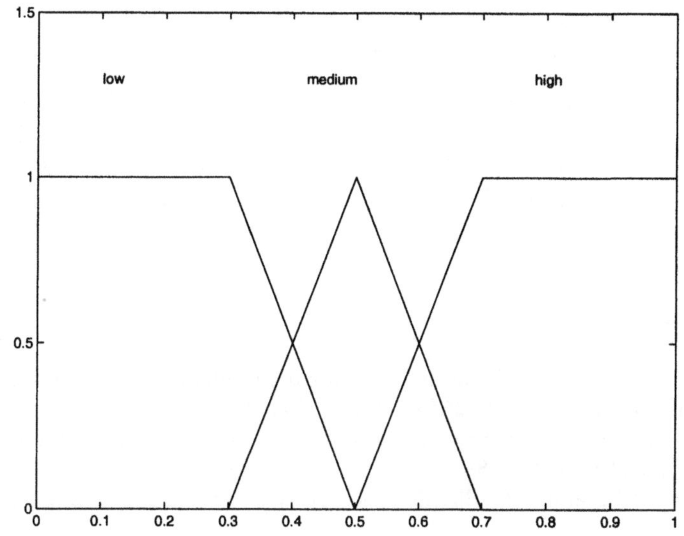

Fig. 2. Graphs of the lingustic variables showing the degrees of convexity or connectivity

a feature of the region. If the value of this feature is small one could suppose the existence of abnormalities in the region.

7 Relations between Convexity and Connectivity, and Further Research

A general definition of convexity and connectivity of fuzzy sets have been established, suited for direct implementation to grey-scale images. Since we work with adjoint fuzzy morphological operations, we have real opening and closing filters (idempotent, increasing, [anti]-extensive), which gives us the opportunity to apply fuzzy closing operation directly to the grey-scale image to remove sparse noise. Such noise may appear with the old CCD cameras when some pixels are dead, but also in the process of photocopying where some ink drops may damage the copy. Also, this noise can be considered as the most severe case of random noise. Therefore, if we demonstrate that a connectivity based filtering works well to such dropout noise, it is most likely that it can apply to any random noise. Since in the crisp case every convex set is connected, it is interesting to find a numerical relation between the values of the degree of convexity and the degree of connectivity of a given fuzzy set. On the second figure one can see graphs of the fuzzy notions 'fuzzy convex' and 'fuzzy connected' (the x - axis repesents the degree of convexity or connectivity) . When using normalized fuzzy sets, i.e. there exists at least one point where the membership function takes vaue 1, then our experiments in the case of arc-wise connectivity, as in the classical discrete analogs (cross- neighbourhood and 3x3 - neighbourhood), show that:

– If the convexity is high then the connectivity is high;
– If the connectivity is low then the convexity is low.

These fuzzy implications are very natural from the 'crisp' mathematical point of view, since we know that if a set is convex it is connected as well. However, our further research should be directed in testing other kind of fuzzy membership functions (not only trapezoidal), quantizing the relations between the geometrical and topological notions as convexity and connectivity.

References

1. Bloch. I, Maître, H.,: Fuzzy mathematical morphologies: A comparative study, Pattern Recognition, **vol. 28(9)** (1995) 1341–1387.
2. Bloch, I.: Fuzzy connectivity and mathematical morphology, Pattern Recognition Letters, **14** (1993) 483–488.
3. Haralick, R.M. and L. G. Shapiro, *Computer and robot vision, volume 1.* Addison-Wesley, 1992.
4. Heijmans, H. J. A. M.: Morphological image operators, Academic Press, Boston 1994.
5. Deng, Ting-Quan, Heijmans, H. J. A. M.: Grey-scale morphology based on fuzzy logic, CWI Report PNA-R0012, Amsterdam, October 2000.
6. Popov, A. T.: On some properties of convexity indicators based on fuzzy morphology, SPIE Proceedings vol. 3168 - Vision Geometry VI, 1996.
7. Popov, A. T.: Convexity indicators based on fuzzy morphology, Pattern Recognition Letters, **18 (3)** (1997) 259 - 267.
8. Popov, A. T.: Aproximate connectivity and mathematical morphology, J. Goutsias, L. Vincent, D. S. Bloomberg (eds), Mathematical Morohology and its Applications to Image and Signal Processing, Kluwer (2000) 149–158.
9. Ronse, C: Set theoretical algebraic approach to connectivity in continuous or digital spaces. Journal of Mathematical Imaging and Vision, **8** (1998) 41 - 58.
10. Rosenfeld, A.: The fuzzy geometry of image subsets. Pattern Recognition Letters, **2** (1984) 311–317.
11. Serra. J.: Image analysis and mathematical morphology, Academic Press, London 1982.
12. Serra, J.: Mathematical morphology for complete lattices. In J. Serra, editor, *Image analysis and mathematical morphology*, vol. 2, Academic Press, London 1988.
13. Serra, J.: Connectivity on complete lattices. Journal of Mathematical Imaging and Vision, **9** (1998) 231 - 251.
14. Sinha, D. , Dougherty, E. R. : Fuzzification of Set Inclusion, Theory and Applications. Fuzzy Sets and Systems, **55** (1993) 15–42.
15. Werman M., Peleg S.: Min - max operators in texture analysis, IEEE Trans. on Pattern Analysis and Machine Intelligence, **7(2)**, (1985) 730–733 .
16. Zadeh, L.: Fuzzy sets and their applications to cognitive processes. Academic Press, London, 1975.

Author Index

Lecture Notes in Computer Science

For information about Vols. 1–2118
please contact your bookseller or Springer-Verlag

Vol. 1962: Y. Frankel (Ed.), Financial Cryptography. Proceedings, 2000. XI, 379 pages. 2001.

Vol. 2119: V. Varadharajan, Y. Mu (Eds.), Information Security and Privacy. Proceedings, 2001. XI, 522 pages. 2001.

Vol. 2120: H.S. Delugach, G. Stumme (Eds.), Conceptual Structures: Broadening the Base. Proceedings, 2001. X, 377 pages. 2001. (Subseries LNAI).

Vol. 2121: C.S. Jensen, M. Schneider, B. Seeger, V.J. Tsotras (Eds.), Advances in Spatial and Temporal Databases. Proceedings, 2001. XI, 543 pages. 2001.

Vol. 2123: P. Perner (Ed.), Machine Learning and Data Mining in Pattern Recognition. Proceedings, 2001. XI, 363 pages. 2001. (Subseries LNAI).

Vol. 2124: W. Skarbek (Ed.), Computer Analysis of Images and Patterns. Proceedings, 2001. XV, 743 pages. 2001.

Vol. 2125: F. Dehne, J.-R. Sack, R. Tamassia (Eds.), Algorithms and Data Structures. Proceedings, 2001. XII, 484 pages. 2001.

Vol. 2126: P. Cousot (Ed.), Static Analysis. Proceedings, 2001. XI, 439 pages. 2001.

Vol. 2127: V. Malyshkin (Ed.), Parallel Computing Technologies. Proceedings, 2001. XII, 516 pages. 2001.

Vol. 2129: M. Goemans, K. Jansen, J.D.P. Rolim, L. Trevisan (Eds.), Approximation, Randomization, and Combinatorial Optimization. Proceedings, 2001. IX, 297 pages. 2001.

Vol. 2130: G. Dorffner, H. Bischof, K. Hornik (Eds.), Artificial Neural Networks – ICANN 2001. Proceedings, 2001. XXII, 1259 pages. 2001.

Vol. 2131: Y. Cotronis, J. Dongarra (Eds.), Recent Advances in Parallel Virtual Machine and Message Passing Interface. Proceedings, 2001. XV, 438 pages. 2001.

Vol. 2132: S.-T. Yuan, M. Yokoo (Eds.), Intelligent Agents. Specification. Modeling, and Application. Proceedings, 2001. X, 237 pages. 2001. (Subseries LNAI).

Vol. 2133: B. Christianson, B. Crispo, J.A. Malcolm, M. Roe (Eds.), Security Protocols. Proceedings, 2001. VIII, 257 pages. 2001.

Vol. 2134: M. Figueiredo, J. Zerubia, A.K. Jain (Eds.), Energy Minimization Methods in Computer Vision and Pattern Recognition. Proceedings, 2001. X, 652 pages. 2001.

Vol. 2136: J. Sgall, A. Pultr, P. Kolman (Eds.), Mathematical Foundations of Computer Science 2001. Proceedings, 2001. XII, 716 pages. 2001.

Vol. 2138: R. Freivalds (Ed.), Fundamentals of Computation Theory. Proceedings, 2001. XIII, 542 pages. 2001.

Vol. 2139: J. Kilian (Ed.), Advances in Cryptology – CRYPTO 2001. Proceedings, 2001. XI, 599 pages. 2001.

Vol. 2140: I. Attali, T. Jensen (Eds.), Java on Smart Cards: Programming and Security. Proceedings, 2001. VIII, 255 pages. 2001.

Vol. 2141: G.S. Brodal, D. Frigioni, A. Marchetti-Spaccamela (Eds.), Algorithm Engineering. Proceedings, 2001. X, 199 pages. 2001.

Vol. 2142: L. Fribourg (Ed.), Computer Science Logic. Proceedings, 2001. XII, 615 pages. 2001.

Vol. 2143: S. Benferhat, P. Besnard (Eds.), Symbolic and Quantitative Approaches to Reasoning with Uncertainty. Proceedings, 2001. XIV, 818 pages. 2001. (Subseries LNAI).

Vol. 2144: T. Margaria, T. Melham (Eds.), Correct Hardware Design and Verification Methods. Proceedings, 2001. XII, 482 pages. 2001.

Vol. 2145: M. Leyton, A Generative Theory of Shape. XVI, 554 pages. 2001.

Vol. 2146: J.H. Silverman (Eds.), Cryptography and Lattices. Proceedings, 2001. VII, 219 pages. 2001.

Vol. 2147: G. Brebner, R. Woods (Eds.), Field-Programmable Logic and Applications. Proceedings, 2001. XV, 665 pages. 2001.

Vol. 2149: O. Gascuel, B.M.E. Moret (Eds.), Algorithms in Bioinformatics. Proceedings, 2001. X, 307 pages. 2001.

Vol. 2150: R. Sakellariou, J. Keane, J. Gurd, L. Freeman (Eds.), Euro-Par 2001 Parallel Processing. Proceedings, 2001. XXX, 943 pages. 2001.

Vol. 2151: A. Caplinskas, J. Eder (Eds.), Advances in Databases and Information Systems. Proceedings, 2001. XIII, 381 pages. 2001.

Vol. 2152: R.J. Boulton, P.B. Jackson (Eds.), Theorem Proving in Higher Order Logics. Proceedings, 2001. X, 395 pages. 2001.

Vol. 2153: A.L. Buchsbaum, J. Snoeyink (Eds.), Algorithm Engineering and Experimentation. Proceedings, 2001. VIII, 231 pages. 2001.

Vol. 2154: K.G. Larsen, M. Nielsen (Eds.), CONCUR 2001 – Concurrency Theory. Proceedings, 2001. XI, 583 pages. 2001.

Vol. 2156: M.I. Smirnov, J. Crowcroft, J. Roberts, F. Boavida (Eds.), Quality of Future Internet Services. Proceedings, 2001. XI, 333 pages. 2001.

Vol. 2157: C. Rouveirol, M. Sebag (Eds.), Inductive Logic Programming. Proceedings, 2001. X, 261 pages. 2001. (Subseries LNAI).

Vol. 2158: D. Shepherd, J. Finney, L. Mathy, N. Race (Eds.), Interactive Distributed Multimedia Systems. Proceedings, 2001. XIII, 258 pages. 2001.

Vol. 2159: J. Kelemen, P. Sosík (Eds.), Advances in Artificial Life. Proceedings, 2001. XIX, 724 pages. 2001. (Subseries LNAI).

Vol. 2161: F. Meyer auf der Heide (Ed.), Algorithms – ESA 2001. Proceedings, 2001. XII, 538 pages. 2001.

Vol. 2162: C. K. Koç, D. Naccache, C. Paar (Eds.), Cryptographic Hardware and Embedded Systems – CHES 2001. Proceedings, 2001. XIV, 411 pages. 2001.

Vol. 2163: P. Constantopoulos, I.T. Sølvberg (Eds.), Research and Advanced Technology for Digital Libraries. Proceedings, 2001. XII, 462 pages. 2001.

Vol. 2164: S. Pierre, R. Glitho (Eds.), Mobile Agents for Telecommunication Applications. Proceedings, 2001. XI, 292 pages. 2001.

Vol. 2165: L. de Alfaro, S. Gilmore (Eds.), Process Algebra and Probabilistic Methods. Proceedings, 2001. XII, 217 pages. 2001.

Vol. 2166: V. Matoušek, P. Mautner, R. Mouček, K. Taušer (Eds.), Text, Speech and Dialogue. Proceedings, 2001. XIII, 452 pages. 2001. (Subseries LNAI).

Vol. 2167: L. De Raedt, P. Flach (Eds.), Machine Learning: ECML 2001. Proceedings, 2001. XVII, 618 pages. 2001. (Subseries LNAI).

Vol. 2168: L. De Raedt, A. Siebes (Eds.), Principles of Data Mining and Knowledge Discovery. Proceedings, 2001. XVII, 510 pages. 2001. (Subseries LNAI).

Vol. 2170: S. Palazzo (Ed.), Evolutionary Trends of the Internet. Proceedings, 2001. XIII, 722 pages. 2001.

Vol. 2172: C. Batini, F. Giunchiglia, P. Giorgini, M. Mecella (Eds.), Cooperative Information Systems. Proceedings, 2001. XI, 450 pages. 2001.

Vol. 2173: T. Eiter, W. Faber, M. Truszczynski (Eds.), Logic Programming and Nonmonotonic Reasoning. Proceedings, 2001. XI, 444 pages. 2001. (Subseries LNAI).

Vol. 2174: F. Baader, G. Brewka, T. Eiter (Eds.), KI 2001: Advances in Artificial Intelligence. Proceedings, 2001. XIII, 471 pages. 2001. (Subseries LNAI).

Vol. 2175: F. Esposito (Ed.), AI*IA 2001: Advances in Artificial Intelligence. Proceedings, 2001. XII, 396 pages. 2001. (Subseries LNAI).

Vol. 2176: K.-D. Althoff, R.L. Feldmann, W. Müller (Eds.), Advances in Learning Software Organizations. Proceedings, 2001. XI, 241 pages. 2001.

Vol. 2177: G. Butler, S. Jarzabek (Eds.), Generative and Component-Based Software Engineering. Proceedings, 2001. X, 203 pages. 2001.

Vol. 2180: J. Welch (Ed.), Distributed Computing. Proceedings, 2001. X, 343 pages. 2001.

Vol. 2181: C. Y. Westort (Ed.), Digital Earth Moving. Proceedings, 2001. XII, 117 pages. 2001.

Vol. 2182: M. Klusch, F. Zambonelli (Eds.), Cooperative Information Agents V. Proceedings, 2001. XII, 288 pages. 2001. (Subseries LNAI).

Vol. 2184: M. Tucci (Ed.), Multimedia Databases and Image Communication. Proceedings, 2001. X, 225 pages. 2001.

Vol. 2185: M. Gogolla, C. Kobryn (Eds.), «UML» 2001 – The Unified Modeling Language. Proceedings, 2001. XIV, 510 pages. 2001.

Vol. 2186: J. Bosch (Ed.), Generative and Component-Based Software Engineering. Proceedings, 2001. VIII, 177 pages. 2001.

Vol. 2187: U. Voges (Ed.), Computer Safety, Reliability and Security. Proceedings, 2001. XVI, 261 pages. 2001.

Vol. 2188: F. Bomarius, S. Komi-Sirviö (Eds.), Product Focused Software Process Improvement. Proceedings, 2001. XI, 382 pages. 2001.

Vol. 2189: F. Hoffmann, D.J. Hand, N. Adams, D. Fisher, G. Guimaraes (Eds.), Advances in Intelligent Data Analysis. Proceedings, 2001. XII, 384 pages. 2001.

Vol. 2190: A. de Antonio, R. Aylett, D. Ballin (Eds.), Intelligent Virtual Agents. Proceedings, 2001. VIII, 245 pages. 2001. (Subseries LNAI).

Vol. 2191: B. Radig, S. Florczyk (Eds.), Pattern Recognition. Proceedings, 2001. XVI, 452 pages. 2001.

Vol. 2192: A. Yonezawa, S. Matsuoka (Eds.), Metalevel Architectures and Separation of Crosscutting Concerns. Proceedings, 2001. XI, 283 pages. 2001.

Vol. 2193: F. Casati, D. Georgakopoulos, M.-C. Shan (Eds.), Technologies for E-Services. Proceedings, 2001. X, 213 pages. 2001.

Vol. 2194: A.K. Datta, T. Herman (Eds.), Self-Stabilizing Systems. Proceedings, 2001. VII, 229 pages. 2001.

Vol. 2195: H.-Y. Shum, M. Liao, S.-F. Chang (Eds.), Advances in Multimedia Information Processing – PCM 2001. Proceedings, 2001. XX, 1149 pages. 2001.

Vol. 2196: W. Taha (Ed.), Semantics, Applications, and Implementation of Program Generation. Proceedings, 2001. X, 219 pages. 2001.

Vol. 2197: O. Balet, G. Subsol, P. Torguet (Eds.), Virtual Storytelling. Proceedings, 2001. XI, 213 pages. 2001.

Vol. 2200: G.I. Davida, Y. Frankel (Eds.), Information Security. Proceedings, 2001. XIII, 554 pages. 2001.

Vol. 2201: G.D. Abowd, B. Brumitt, S. Shafer (Eds.), Ubicomp 2001: Ubiquitous Computing. Proceedings, 2001. XIII, 372 pages. 2001.

Vol. 2202: A. Restivo, S. Ronchi Della Rocca, L. Roversi (Eds.), Theoretical Computer Science. Proceedings, 2001. XI, 440 pages. 2001.

Vol. 2205: D.R. Montello (Ed.), Spatial Information Theory. Proceedings, 2001. XIV, 503 pages. 2001.

Vol. 2206: B. Reusch (Ed.), Computational Intelligence. Proceedings, 2001. XVII, 1003 pages. 2001.

Vol. 2207: I.W. Marshall, S. Nettles, N. Wakamiya (Eds.), Active Networks. Proceedings, 2001. IX, 165 pages. 2001.

Vol. 2208: W.J. Niessen, M.A. Viergever (Eds.), Medical Image Computing and Computer-Assisted Intervention – MICCAI 2001. Proceedings, 2001. XXXV, 1446 pages. 2001.

Vol. 2209: W. Jonker (Ed.), Databases in Telecommunications II. Proceedings, 2001. VII, 179 pages. 2001.

Vol. 2210: Y. Liu, K. Tanaka, M. Iwata, T. Higuchi, M. Yasunaga (Eds.), Evolvable Systems: From Biology to Hardware. Proceedings, 2001. XI, 341 pages. 2001.

Vol. 2211: T.A. Henzinger, C.M. Kirsch (Eds.), Embedded Software. Proceedings, 2001. IX, 504 pages. 2001.

Vol. 2212: W. Lee, L. Mé, A. Wespi (Eds.), Recent Advances in Intrusion Detection. Proceedings, 2001. X, 205 pages. 2001.

Vol. 2213: M.J. van Sinderen, L.J.M. Nieuwenhuis (Eds.), Protocols for Multimedia Systems. Proceedings, 2001. XII, 239 pages. 2001.

GPSR Compliance

The European Union's (EU) General Product Safety Regulation (GPSR) is a set of rules that requires consumer products to be safe and our obligations to ensure this.

If you have any concerns about our products, you can contact us on ProductSafety@springernature.com

In case Publisher is established outside the EU, the EU authorized representative is:

Springer Nature Customer Service Center GmbH
Europaplatz 3
69115 Heidelberg, Germany

Batch number: 10682269

Printed by Printforce, the Netherlands